ADVANCES IN BUSINESS, MANAGEMENT AND ENTREPRENEURSHIP

D1806716

Advances in Business, Management and Entrepreneurship

ISSN : 2639-8249 (Online)
ISSN : 2639-8257

Book series editor:

Ratih Hurriyati
Universitas Pendidikan Indonesia, Bandung, Indonesia

The Global Conference on Business, Management and Entrepreneurship (GCBME) Book Series provides a platform for academicians, educators, researchers, practitioners, managers, graduate students, and entrepreneurs from different cultural backgrounds, presenting and discussing research, developments and innovation in the fields of Business Management and Entrepreneurship. It provides opportunities to exchange new ideas and implement experience, to establish business or research connections and to find global partners for future collaboration.

GCBME Conference Series presents current studies and developments in organizational behavior, leadership and human resources management, innovative operations and supply chain management, and contemporary issues in marketing, accounting, entrepreneurship and financial management.

The series addresses the needs of entrepreneurs, managers, educators, and professional worldwide. The proceedings volumes are published shortly after the conference takes place in electronic form and print.

PROCEEDINGS OF THE 4TH GLOBAL CONFERENCE ON BUSINESS MANAGEMENT & ENTREPRENEURSHIP (GC-BME 4), BANDUNG, INDONESIA, 8 AUGUST 2019

Advances in Business, Management and Entrepreneurship

Editors

Ratih Hurriyati
Universitas Pendidikan Indonesia

Benny Tjahjono
Coventry University

Ade Gafar Abdullah
Universitas Pendidikan Indonesia

Sulastri
Universitas Pendidikan Indonesia

Lisnawati
Universitas Pendidikan Indonesia

CRC Press
Taylor & Francis Group
Boca Raton London New York Leiden

CRC Press is an imprint of the
Taylor & Francis Group, an **informa** business

A BALKEMA BOOK

CRC Press/Balkema is an imprint of the Taylor & Francis Group, an informa business

© 2021 Taylor & Francis Group, London, UK

Typeset by MPS Limited, Chennai, India

Library of Congress Cataloging-in-Publication Data
Applied for

Published by: CRC Press/Balkema
Schipholweg 107C, 2301 XC Leiden, The Netherlands
e-mail: Pub.NL@taylorandfrancis.com
www.routledge.com – www.taylorandfrancis.com

ISBN: 978-0-367-67471-7 (Hbk)
ISBN: 978-1-003-13146-5 (eBook)
DOI: 10.1201/9781003131465
https://doi.org/10.1201/9781003131465

Table of contents

Section 2   Green business

Section 3 Innovation, IT, operations and supply chain management

Section 4 Marketing management

Section 5   Organizational behavior, leadership and human resources management

XI

Section 6   Strategic management, entrepreneurship and contemporary issues

Preface/foreword

The GCBME Book Series aims to promote the quality and methodical reach of GCBME, which is intended as a high-quality scientific contribution to the science of business management and entrepreneurship.

The Contributions are expected to be the main reference articles on the topic of each book. It will be strictly peered reviewed by experts in the fields. This book provides opportunities for the delegates to exchange new ideas and implementation experiences, to establish business or research connection book and to find Global Partners for future collaboration. This book is expected to be held annually and years of 2020 we take the theme of: "Transforming Suistainable Business In The Era Of Society 5.0" GCBME ultimately goals to provide a medium forum for educators, researchers, scholars, managers, graduate students and professional business person from the diverse cultural backgrounds to present and discuss their researches, knowledge and innovation within the fields of business, management and entrepreneurship. I hope the readers can find the sustainability of the usefulness of this book, in developing business analysis and implementation.

The GCBME, covers up major thematic groups yet opens to other relevant topics: Organizational Behavior, Innovation, Marketing Management, Financial Management and Accounting, Strategic Management, Entrepreneurship and Green Business

I hope the readers will find the usefulness of this book in understanding the development of the analysis and application of business management and entrepreneurship for decision support

With warmest regards,
Prof. Dr. Ratih Hurriyati, MP

Advances in Business, Management and Entrepreneurship – Hurriyati et al. (Eds)
© 2021 Taylor & Francis Group, London, ISBN 978-0-367-67471-7

Scientific committee

1) Prof. Toru Matsumoto, M.Eng – The University of Kitakyushu – Japan
2) ASSOC. Prof. Ts Dr. Razali Hassan – Universiti Tun Hussein Onn Malaysia (UTHM) – Malaysia
3) Prof. Dr. Ikuro Yamamoto (Kinjo Gakuin University, Nagoya Japan)
4) Prof. Dr. Taehee Kim, Ph.D (Youngsan University, Busan South Korea)
5) Prof. Dr. Mohamed Dahlan Ibrahim, (Universiti Malaysia Kelantan, Malaysia)
6) Prof. Madya Dr. Lai Chee Sern (UTHM Johor bahru Malaysia)
7) Prof. Madya Dr. Kahirol Bin Mohd Salleh (UTHM Johor bahru Malaysia)
8) Prof. Wann-Yih Wu (Universitas Nanhua Taiwan)
9) Prof. Dr. Nanang Fattah, M.Pd. (Universitas Pendidikan Indonesia, Indonesia)
10) Prof. Dr. Agus Rahayu, M.P. (Universitas Pendidikan Indonesia, Indonesia)
11) Prof. Dr. Tjutju Yuniarsih, S.E., MPd (Universitas Pendidikan Indonesia, Indonesia)
12) Prof. Dr. Disman, M.S. (Universitas Pendidikan Indonesia, Indonesia, Indonesia)
13) Prof. Dr. Suryana, M.S. (UPI, Indonesia)
14) Prof. Dr. Eeng Ahman, M.S. (UPI, Indonesia)
15) Prof. DR. Ratih Hurriyati, M.P. (UPI, Indonesia)
16) Prof. Ina Primiana, S.E., M.T. (UNPAD)
17) Prof Lincoln Arsyad, M.Ec, PhD (UGM)
18) Prof. Gunawan Sumodiningrat, M.Ec, PhD (UGM)
19) Prof Dr Badri Munir Sukoco, Msc, PhD (UNAIR)
20) Prof. Dr. H. Nandan Limakrishna, Ir, MM (Universitas Winaya Mukti)
21) Dr. Phil Dadang Kurnia, M.Sc. (GIZ German)
22) Assoc.Prof. Arry Akhmad Arman, M.T., Dr. (ITB)
23) Assoc.Prof. Dwilarso, MBA, PhD (ITB)
24) Assoc.Prof. Hardianto Iristiadi MSME, PhD (ITB)
25) Assoc.Prof. Rachmawaty Wangsaputra, M.Sc, PhD (ITB)
26) Assoc.Prof. Teungku Ezni Balkiah, M.Sc, PhD (UI)
27) Assoc.Prof. Ruslan Priyadi M.Sc, PhD (UI)
28) Assoc.Prof. Sri Gunawan, MBA, DBA (UNAIR)
29) Assoc.Prof. Yudi Aziz, M.T., PhD (UNPAD)
30) Assoc.Prof. Lili Adiwibowo, M.M., DR. (UPI)
31) Assoc.Prof. Vanessa Gaffar, MBA, DR. (UPI)
32) Assoc.Prof. Chaerul Furqon, M.M., DR. (UPI)
33) Vina Andriany M.Ed, PhD (UPI)
34) Tutin Ariyanti, S.T., M.T., PhD (UPI)

Advances in Business, Management and Entrepreneurship – Hurriyati et al. (Eds)
© 2021 Taylor & Francis Group, London, ISBN 978-0-367-67471-7

Organizing committee

1. Conference Chair
   Prof. Dr. Ratih Hurriyati MP, Universitas Pendidikan Indonesia, INDONESIA
2. Technical Chairperson:
   Assoc. Prof. Lili Adiwibowo Universitas Pendidikan Indonesia, INDONESIA
3. Members:
   Dr. Hari Mulyadi MSi
   Dr. Ade Gafar Abdullah, M.Si
   Dr. Eng Asep Bayu Nandiyanto MSc
   Drs. Rd Dian H Utama, Msi
   Drs. Girang Razati, Ms
   Lisnawati SPd. M
   Sulastri SPd, Mstat., MM
   MasHaryono SPd, MM

Acknowledgements

This book is primary supported by the Universitas Pendidikan Indonesia and Partner supported by the Airlangga University, Padjajaran University, Institut Pendidikan Indonesia, Universitas Garut, Universitas Winayamukti, Sekolah Tinggi Ilmu Ekonomi Indonesia Membangun, Universitas Singaperbangsa Karawang, and Bank BJB. We would like very much to thank CRC Press/Balkema, Taylor & Francis Group for their efforts and cordial cooperation in publishing this book.

Section 1   Financial management and accounting

Advances in Business, Management and Entrepreneurship – Hurriyati et al. (Eds)
© 2021 Taylor & Francis Group, London, ISBN 978-0-367-67471-7

Analysis of macroeconomic variable shocks on the equilibrium of real effective exchange rates in Malaysia

H. Aimon, S.U. Sentosa & M.A. Shahmi
Universitas Negeri Padang, Padang, Indonesia

ABSTRACT: This study investigates the effect and equilibrium of macroeconomic variables on real effective exchange rates (REER) in short and long terms in Malaysia. This study used time series data from 1986–2017, and Johansen-Juselius and error correction model (ECM). There are two main findings in this study. First, economic openness and inflation have a significant effect on real effective exchange rates in Malaysia. Second, in the short term, foreign direct investment (FDI) and inflation disrupt the balance of effective real exchange rates, although in the long terms the inflation will return to its equilibrium. This research is recommended to the government to increase foreign direct investment in Malaysia, because it is a major factor that influences the equilibrium of real effective exchange rates.

1 INTRODUCTION

Real effective exchange rates have an important role in international trade, especially in examining the performance and competitiveness of a country. Appreciation for real effective exchange rates can hinder the pace of the economy due to declining exports and capital inflows (Tsen 2011). For emerging and small-open economy countries, real effective exchange rates have a central role in determining policy designs that aim to maintain balance and shocks in the economy (Yanamandra 2015; Ye, Hutson, & Muckley 2014; Yuan 2011).

Malaysia as one of the emerging markets in Asia has participated in international trade for many years and determined the design of exchange rate policies to promote trade. However, market uncertainty has caused the policy to be unable to effectively maintain exchange rate stability. Experience in the Asian financial crisis of 1997–1998 and the global financial crisis showed that Malaysia was still unable to maintain exchange rate stability and certainly influenced global perceptions of economic performance in this country.

In addition, shocks to macroeconomic variables have an effect on the condition of Malaysia's real effective exchange rate. It was explained that macroeconomic variable shocks such as inflation, economic openness, and foreign investment have an influence on real effective exchange rate appreciation (depreciation) (Bahmani & Motavallizadeh 2017; Blau 2018; Bouraoui & Phisuthtiwatcharavong 2015; Burakov 2018; Chen & Chou 2015). In addition, macroeconomic conditions also determine the exchange rate balance (Comunale 2017).

This study is intended to present a dynamic model in determining the real exchange rate and empirically examine the implications of changes in the determinants of possible short and long term real exchange rate shocks in Malaysia by employing the multivariate Johansen cointegration method and the error correction model (ECM).

2 LITERATURE REVIEW

Based on some relevant theories and research, there are some effects of macroeconomic variable shocks on the real effective exchange rate. First of all, according to the impact of inflation on real effective exchange ratezs, if a country experiences relatively high inflation from other countries, the effective real exchange rate will depreciate (Blau 2018; Daude & Nagengast 2016; Elfaki 2018; Gharaibeh 2017).

Second, Mishkin (2004) explains that high inflation causes pressure on the national currency which leads to depressions in the long run. In addition, the impact of economic openness on real effective exchange rates is that if economic openness increases, then the real effective exchange rate will depreciate (Adusei & Gyapong 2017; Bahmani & Motavallizadeh 2017; Blau 2018; Bouraoui & Phisuthtiwatcharavong 2015). This condition shows that high economic openness implies low trade barriers which are the cause of depreciation of a country's currency (Mishkin & Savastano 2007).

Third, the impact of foreign direct investment on real effective exchange rates is if an increase in investment directly explains the increase in capital flows entering a country (Bahmani & Motavallizadeh 2017;

Blau 2018; Chen & Chou 2015; Comunale 2017), it will cause appreciative pressure on the value of a country's currency (Jongwanich & Kohpaiboon 2013). In addition, foreign direct investment also has an impact on increasing productivity in tradable sectors, which in turn has an impact on strengthening a country's currency values (Buchanan, Le, & Rishi 2012; Jongwanich & Kohpaiboon 2013; Vo 2018) and increased foreign investment directly to export-oriented industries has an impact on the appreciation of real effective exchange rates.

3 METHODS

This study used the multivariate Johansen cointegration method and error correction model (ECM). This research entirely used time series secondary data sources covering years ranging from 1986 to 2017. The model of the long-term relationship between real effective exchange rates is as follows:

$$REER = f(CPI, OP, FDI) \qquad (1)$$

The Real Effective Exchange Rate (REER) is the weighted average of a country's currency with respect to an index or a basket of other major currencies. Weight is determined by comparing the relative trade balance of a country's currency against each country in the index. Economic Openness (OP) is the ratio of the number of exports and imports of GDP. Inflation (INF) uses the consumer price index. Foreign direct investment (FDI) is the ratio of foreign direct investment inflows to GDP. Time series data sources is from world development indicators, world banks.

$$REER_t = \beta_0 + \beta_1 INF_t + \beta_2 OP_t + \beta_3 FDI_t + \varepsilon_t \qquad (2)$$

While ε is a random error. The expected sign of this study is $\beta 1 < 0$ which means the effect of inflation on the real effective exchange rate has a negative coefficient, $\beta 2 < 0$ which means that the effect of economic openness on the real effective exchange rate has a negative coefficient, $\beta 3 > 0$ means that the effect of foreign direct investment on the real effective exchange rate has a positive coefficient.

Equation 3 includes ECt-1, to integrate the short-term dynamics in the long-term function of the real effective exchange rate, so that the error correction model (ECM) is used as follows:

$$\Delta REER_t = \beta_0 + \beta_1 \Delta INF_t + \beta_2 \Delta OP_t + \beta_3 \Delta FDI_t + EC_{t-1+vt} \qquad (3)$$

Estimates of the research model involve three steps, namely: first, the unit test root avoids the wrong regression results. One unit root test used augmented dickey-fuller and Phillips-Perron. Second, variables are integrated in the same order, then cointegration are tested using multivariate Johansen cointegration analysis. Third, cointegrated variables are used to determine error correction models and estimate standard methods.

4 RESULTS AND DISCUSSION

4.1 Unit root test

Unit root test is very important to be done before conducting a cointegration test on the research variable. This test is intended to analyze the possibility of false regression which will show t-statistics and f-statistics that lead to wrong conclusions. So, time series data must be stationary or in the case of non-stationarity, the right methodology must be applied. The study uses the Augmented Dickey-Fuller (ADF) test for unit root tests in Table 1:

Table 1. Unit root test augmented dickey fuller.

Variable	Level	Prob. Values	1st difference	Prob. Values
	Constant		Constant	
REER	−5.358729	0.0001	3.576004	0.0125
INF	1.866038	0.0001	−5.037231	0.0003
OP	−1.864405	0.3437	−3.576004	0.0125
FDI	−2.882535	0.0589	−5.012171	0.0003

4.2 Cointegration test

The Johansen-Juselius cointegration test is applied to find out the equilibrium in the long run. Whether there are any similarities in the movement and stability of the relationship between the variables in this study or not according to the Table 2:

Table 2. Johansen-juselius multivariate cointegration test.

H0	H1	Eigen value	Trace Statistic	0.05 Critical Value	Max-Eigen Statistic	0.05 Critical Value
r=0*	r>0	0.564396	54.55651	47.85613	24.93063	27.58434
r<1	r>1	0.439201	29.62588	47.85613	17.35178	21.13162
r<2	r>2	0.307506	12.27411	29.79707	11.02369	14.26460
r<3	r>3	0.040824	1.250418	15.49471	1.250418	3.841466

Akaike Information Criterion (AIC) and Schwarz information Criterion (SC) from the VAR estimation inform that the optimal lag length used is 2. Table 3 shows that the trace statistic value of 54.55651 is above the critical value of 47.85613 at r = 0, which means that the hypothesis of no cointegration is rejected, and the alternative hypothesis is accepted. The max-eigen statistic value is greater than the value of critical value, meaning the null hypothesis of no cointegration at r = 0 rejected at a 5 percent significant level and supports the alternative hypothesis.

It can be summed up that there is a cointegration relationship on the variables studied. The cointegration relationship states that there is a long-term balance between exogenous and endogenous variables in this study. This is Table 3:

4

Table 3. Estimation of Ordinary least square (OLS).

Variable	Coefficient	Std. Error	t-Statistic	Prob.
C	185.7647*	7.390306	25.13626	0.0000
INF	0.094660*	0.053322	−10.96115	0.0000
OP	−0.188338*	0.032235	−5.842658	0.0000
FDI	0.094660	0.560301	0.168946	0.8671
R-squared	0.857146	F-statistic	56.00128	
Adjusted R-squared	0.841840	Prob (F-statistic)	0.000000	

Estimated Ordinary Least Square (OLS) in Table 3 shows that inflation (INF) and economic openness (OP) significantly affects the long-term real effective exchange rate in Malaysia. Increased inflation has an impact on the appreciation of real effective exchange rates in Malaysia. Every 1 percent increase in inflation, in the long run, will strengthen the real exchange rate by 0.09 percent and every 1 percent increase in economic openness will weaken the real effective exchange rate by 0.18%.

Furthermore, the increase in economic openness contributes significantly to the depreciation of real effective exchange rates in Malaysia. Meanwhile, foreign direct investment in the long term does not significantly influence the appreciation of the real effective exchange rate in Malaysia. This finding supported previous research which found that there was no significant relationship between direct foreign investment and the real effective exchange rate (Jongwanich & Kohpaiboon 2013).

4.3 Estimation of error correction model

The stability of the parameter study model in the long term can be examined with an Error Correction Model (ECM). Thus, it can be identified which variable that affect the real effective exchange rate in the short term and cause a real effective exchange rate shock.

The next discussion is about the short-term relationship between trade openness, the current account balance, foreign direct investment, and inflation in Malaysia. It is known that the equilibrium model of the real effective exchange rate meets market expectations and does not disrupt economic stability. This condition can be seen in the results of the ECM estimation of the real exchange rate in Table 4:

Table 4. Estimation of error correction model.

Variable	Coefficient	Std. Error	t-Statistic	Prob.
C	−1.439020	1.787732	−0.804942	0.4282
D(INF)	−0.031232	0.740175	−0.042195	0.9667
D(OP)	−0.188338*	0.073897	−4.489285	0.0001
D(FDI)	0.409145	0.489550	0.835756	0.4109
RES(-1)	−0.407098	0.153706	−2.648547	0.0136
R-squared	0.528630	F-statistic	56.00128	
Adjusted R-squared	0.456112	(F-statistic) Prob	0.000000	

The estimation results show that error-term lag significantly influences the real effective exchange rate. This explains that there is an imbalance in the short-term relationship between shocks and the flow of foreign direct investment and inflation with the real effective exchange rate. Foreign direct investment and inflation have no significant effect on the real exchange rate. This explains that in the short term, inflation and foreign investment directly disrupt the equilibrium of the real effective exchange rate which support the findings from previous research (Bouraoui & Phisuthtiwatcharavong 2015; Candelon, Kool, & Raabe 2007; Cavaglia & Wolff 1996).

Nevertheless, in the short term, economic openness has a significant effect on the real effective exchange rate in Indonesia. This condition indicates that a policy is needed to control the level of economic openness in Malaysia to maintain a balance on the real effective exchange rate.

The coefficient of determination R2 is 52 percent of the total variation of the real effective exchange rate in Malaysia, which can be explained by the macroeconomic variables investigated. Meanwhile, the term error correction illustrates the proportion of the real exchange rate imbalance that in the long run, can be corrected annually at a significant level of 1 percent. About 48 percent of the imbalance in the real exchange rate shock is being repaired every year in Malaysia.

5 CONCLUSION

The central role of the exchange rate in the evaluation and design of economic policy cannot be separated from the Malaysian economy. Therefore, the government must choose the right policy strategy so that the real effective exchange rate reaches its equilibrium. Based on the research findings and cointegration tests carried out, the equilibrium problem of the real effective exchange rate must be solved. Even though there is a change in balance in the long term that is influenced by inflation, the government must design policies that can affect foreign direct investment to improve the balance of effective real exchange rates.

ACKNOWLEDGMENT

This study is supported by SIMLITABMAS (Sistem Informasi Penelitian dan Pengabdian Kepada Mayarakat) RISTEKDIKTI.

REFERENCES

Adusei, M. & Gyapong, E. Y. 2017. The impact of macroeconomic variables on exchange rate volatility in Ghana: the partial least squares structural equation modelling approach. *Research in international business and finance.* https://doi.org/10.1016/j.ribaf.2017.07.081

5

Bahmani, M., & Motavallizadeh, A. 2017. Exchange rate changes and income distribution in 41 countries: asymmetry analysis. *Quarterly review of economics and finance.* https://doi.org/10.1016/j.qref.2017.11.009

Blau, B. M. 2018. Exchange rate volatility and the stability of stock prices. *International review of economics and finance, 58 October 2015.* 299–311. https://doi.org/10.1016/j.iref.2018.04.002

Bouraoui, T., & Phisuthtiwatcharavong, A. 2015. On the determinants of the THB / USD exchange rate. *Procedia economics and finance*, 30(15):137–145. https://doi.org/10.1016/S2212-5671(15)01277-0

Buchanan, B. G., Le, Q. V., & Rishi, M. 2012. Foreign direct investment and institutional quality: some empirical evidence. *International review of financial analysis*, 21(24), 81–89. https://doi.org/10.1016/j.irfa.2011.10.001

Candelon, B., Kool, C., & Raabe, K. 2007 66. Long-run real exchange rate determinants: evidence from eight new EU member states, *1993–2003*, 35:87–107. https://doi.org/10.1016/j.jce.2006.10.003

Cavaglia, S. M. F. G., & Wolff, C. C. P. 1996. Finance a note on the determinants of unexpected exchange rate movements, 20:179–188.

Chen, S., & Chou, Y. 2015. Revisiting the relationship between exchange rates and fundamentals. *Journal of macroeconomics*, 46:1–22. https://doi.org/10.1016/j.jmacro.2015.07.004

Comunale, M. 2017. Current account and real effective exchange rate misalignments in central eastern EU countries: An update using the macroeconomic balance approach. *Economic systems.* https://doi.org/10.1016/j.ecosys.2017.11.002

Daude, C., & Nagengast, A. 2016. On the effectiveness of exchange rate interventions in emerging markets. *Journal of international money and finance.* https://doi.org/10.1016/j.jimonfin.2016.01.004

Elfaki, K. E. 2018. Determinants of exchange rate stability in Sudan, *April 1991–2016.*

Gharaibeh, A. M. 2017. Fundamental determinants of real effective exchange rate: empirical evidence fundamental determinants of real effective exchange rate: empirical Evidence from Bahrain, *November.*

Jongwanich, J., & Kohpaiboon, A. 2013. Capital flows and real exchange rates in emerging Asian countries. *Journal of asian economics, 24 October 2010*, 138–146. https://doi.org/10.1016/j.asieco.2012.10.006

Mishkin, F. S. 2004. *The economics of money, banking, and financial markets (seventh ed).* New York: Pearson Addison Wesley.

Mishkin, F. S., & Savastano, M. A. 2007. Monetary policy strategies for Latin America. *Monetary Policy Strategy*, 279.

Tsen, W. H. 2011. The real exchange rate determination: an empirical investigation. *International review of economics & finance*, 20(4), 800–811.

Vo, X. V. 2018. Determinants of capital flows to emerging economies – evidence from Vietnam. *Finance research letters.* https://doi.org/10.1016/j.frl.2018.02.031

Yanamandra, V. 2015. Exchange rate changes and inflation in India: what is the extent of. *Economic analysis and policy.* https://doi.org/10.1016/j.eap.2015.07.004

Ye, M., Hutson, E., & Muckley, C. 2014. Exchange rate regimes and foreign exchange exposure: The case of emerging market firms. *Emerging markets review*, 21:156–182. https://doi.org/10.1016/j.ememar.2014.09.001

Yuan, C. 2011. North American journal of economics and finance the exchange rate and macroeconomic determinants: time-varying transitional dynamics. *North American journal of economics and finance*, 22(2):197–220. https://doi.org/10.1016/j.najef.2011.01.005

Advances in Business, Management and Entrepreneurship – Hurriyati et al. (Eds)
© 2021 Taylor & Francis Group, London, ISBN 978-0-367-67471-7

Analysis of bank efficiency in Indonesia

Rahmat
Universitas Pendidikan Indonesia, Bandung, Jawa Barat, Indonesia

ABSTRACT: This research aims to measure the efficiency level of the largest banks in Indonesia according to the categories of ownership, status, and price determination for 2013–2017 using the data envelopment analysis (DEA) method, with the Win4DEAP software. The intermediation approach was used in selecting inputs and outputs variables. Input factors used were assets, funds, and workforce expenses. The output factors used were credit or financing and income. The results showed that (1) the use of input assets, funds, and labor costs on average increased, and in general the use of credit or financing output and income also increased, (2) the use of inputs and outputs of each bank during 2013–2017 by using DEA generally achieved optimum efficiency, and only one bank that has not been fully optimized. The model used was DEA with the assumption of variable return to scale, and (3) based on the efficiency scale of each bank using the DEA method, there were banks that were at the radial movement and slack movement stages. In such circumstances, it can be interpreted that it is necessary to evaluate inputs and outputs for improvement, so that the original value matches the project value to generate optimal efficiency. Compared to the results of the operational efficiency ratio assessment, generally there is no significant difference.

1 INTRODUCTION

Banking as an artery of the economy is demanded to operate at the optimal efficient point in order to bridge the funds owner and those who need funds. Banking efficiency is important in efforts to support economic growth through credit services or healthy financing and low interest rates, so that it can reach all levels of society, especially among micro, small, and medium enterprises. Types of banks can be distinguished in terms of function, ownership, and status. According to Kasmir (2014), types of banks in terms of various aspects, including based on ownership, are as follows:

a. A state-owned bank is a bank whose deed of establishment and bank capital is fully owned by the government, so the profits are owned by the government. The government-owned bank is subsequently called the Persero Bank.

b. National private-owned banks are banks in which all or most of their shares are owned by the national private sector, so that the profits become private property. Private-owned banks are hereinafter referred to as national private commercial banks (BUSN).

c. A cooperative-owned bank is a bank whose shares are owned by a company incorporated as a cooperative.

d. Foreign-owned banks are branches of banks that are abroad, or all of their shares are owned by foreign parties.

e. Mixed-owned banks are banks whose shares are owned by foreign parties and national private

parties, and the majority of shares are held by Indonesian citizens.

Based on data from Indonesian banking statistics on the Financial Services Authority, the types of banks consist of Bank Persero, foreign BUSN, non-foreign BUSN, regional development banks (BPD), mixed banks, and foreign banks. Other categories are Sharia commercial banks, rural credit banks (BPR), and Sharia BPRs.

According to Muliaman, Wimboh, Dhaniel, and Eugenia (2003), banking as a financial institution that has an important role is demanded to have good performance. One of the indicators is its efficiency. The level of efficiency achieved is a reflection of the good quality of performance. The ability to produce maximum output with existing inputs is a measure of expected performance. When efficiency measures are taken, the bank is faced with the conditions of how to obtain an optimal level of output with the existing input level or to get a minimum level of input with a certain level of output.

According to Hasibuan (1984), the notion of efficiency is the best comparison between inputs and outputs (results between profits and sources used), as well as optimal results achieved with the use of limited resources. Inefficiency of banks will lead to a relatively more expensive fund and overhead cost structure, which can result in high lending rates that hinder business development.

Based on the idea that efficiency in banks has an important role in supporting bank business activities and economic systems, it is deemed necessary to conduct bank efficiency studies as practiced by almost every country. In Indonesia, an efficiency assessment has been provided in accordance with Bank Indonesia Circular Number 15/29/ DKBU dated July 31, 2013, which states operational income operational costsor operational efficiency ratio (OER): that in carrying out its operational activities, total operating expenses and total operating income are divided, which is calculated per position (not annualized). According to Kasmir (2014), one of the main activities of banks is channeling funds in the form of loans to those who need them. From this credit, the bank will get a reward in the form of interest. Income from interest is the operational income of the bank because the interest is obtained from its main activities. The OER ratio, called the efficiency ratio, is used to measure the ability of bank management in controlling operational costs to operating income. The smaller this ratio, the more efficient operational costs incurred by the bank concerned, so that the possibility of a bank to be in trouble is smaller too. As a comparison, it also needs to be done with another method in which a data envelopment analysis (DEA) model was used in this study. DEA is an efficiency calculation method that uses a non-parametric approach. DEA measures the efficiency of a decision-making unit (DMU) collection, which is used as a reference in the management of resources (inputs) of the same type that will be able to produce similar outputs, where each relation of the form of the function from input to output is unknown (Coelli 1996).

DEA measures the efficiency of the DMU by maximizing the ratio of weighted outputs to weighted inputs or vice versa. This ratio is normalized based on the best of the units observed, and efficiency is assessed based on intervals between 0 and 1, where 1 represents the efficient unit (Eken & Kale 2011). The analysis is based on an evaluation of the relative efficiency of comparable DMUs. Furthermore, an efficient DMU will form a frontier line. If the DMU is in the frontier line, the DMU can be said to be relatively efficient compared to other DMUs in the sample. The DEA can also indicate a DMU that is a reference for an inefficient DMU (Ascarya, Diana, & Guruh 2008).

There are some considerations regarding the use of DEA, namely (1) assigning weight assessment for each variable determining performance is done objectively, (2) DEA is an analysis of extreme points that are different from the central tendency, so that each observation or unit of economic activity is analyzed individually, (3) DEA forms hypothesis reference (virtual production function) based on observational data available. Various bank efficiency studies in various countries have been carried out. Hasanul Banna, Rubi Ahmad, and Koh (2017) found that bank size, capital adequacy ratio, average return on equity, and the real interest rate significantly influenced bank efficiency in Bangladesh.

Mohamad, Hassan, and Bader (no date) measured and compared cost and profit efficiency in 80 banks in 21 countries of the Islamic Conference Organization

(OIC) consisting of 37 conventional banks and 43 Islamic banks by employing the Stochastic frontier approach. In addition, he assessed the efficiency of these banks based on their size, age, and region.

The findings showed that there was no significant difference between the overall efficiency results of conventional banks versus Islamic banks. However, there is considerable room for increasing cost minimization and profit maximization in both banking systems. Furthermore, this finding did not show a significant difference in the average efficiency score between large banks versus small banks and new banks versus old banks in both banking streams.

In another research, Germanova and Ivanova (2018) used two inputs and two outputs. The inputs were liabilities to banks and customers and operating costs. The outputs were loans and advances to banks and customers and noninterest income. On the output side, capital was also used. The reason for this choice is a small number of banks in the Slovak Republic.

It is clear from this relationship that we can use a maximum of two inputs and two outputs. Inputs are liabilities to banks and customers operating costs. Outputs are loans and advances to banks and customers and noninterest income. In addition, Ayadi (2013) studied the efficiency of Tunisian commercial banks for the period 1996–2010 using the DEA method.

The results showed that the sector's cost efficiency was estimated to reach a score of 41.0%. Market share in terms of savings banks and their involvement was in risky activities, especially in the area of credit that had a negative impact on their efficiency. High bank capitalization positively influenced the latter. In addition, state-owned banks were more efficient than private banks. On the other hand, research in various countries generally using the DEA) model have also been conducted.

Ferrier and Hirschberg (1997) carried out a study with the DEA model involving 94 banks in Italy in 1986 with input variables consisted of the number of employees, capital, consumer deposit accounts, commercial deposit accounts, and industrial deposit accounts, while the output variables were loans (consumer, commercial, and industrial), deposits in other financial institutions, investment, and number of branches.

Puri and Yadav (2013) conducted a study of 17 banks in India in 2010 with labor, fixed assets, and total expenses as input variables and interest income and other income as output variables. Moradi-Motlagh and Saleh (2014) conducted a study of 10 banks in Australia during the period 1997–2005 using expense and noninterest expense as input variables and interest income and noninterest income as output variables.

Hou, Wang, and Zhang (2014) conducted a study involving 44 major banks in China during the period 2007–2011 with deposits, fixed assets, and number of employees as input variables and total net loan and other earning assets as output variables. Řepková (2014) conducted a study of 11 banks in the Czech

Republic during 2003–2012 using labor and deposits as input variable and loans and net interest income as output variables.

Kao and Liu (2014) examined 22 banks in Taiwan during the period 2009–2011 using input labor variables, physical capital, purchased funds and variable output demanded deposits, short-term loans, medium- and long-term loans. Johnes, Izzeldin, and Pappas (2014) conducted research on Islamic banks in 18 countries during the period 2004–2009 using input deposits and short-term funding variables, fixed assets, general and administrative expenses, equity, and variable output for total loans and other earning assets.

2 METHODS

2.1 Research object

The objects of this study were large banks in Indonesia based on the categories of ownership, status, and price determination. The banks in question were Bank Rakyat Indonesia (representing the Persero Bank), Bank Central Asia (representing the Foreign Exchange BUSN Bank), BTPN (representing Non-Foreign Exchange BUSN), DBS Bank (representing mixed Banks), MFUG Bank (representing Foreign Banks), BJB (representing Regional Development Banks), BSM (representing Sharia Commercial Banks), BPR Eka Bumi Arta (representing Conventional BPRs), and Bank Hikmah Insan Karomah (HIK) (representing BPRS).

The type of data used in research was secondary data by using the bank's annual financial statements from Indonesian Banking Statistics Data during the period 2013–2017.

2.2 Data collection techniques

The data collection method was in the form of documentation, obtained from electronic media. The source of the data was from Indonesian Banking Statistics Data, the official website of Bank Indonesia (BI), www.bi.go.id, and the official website of the Financial Services Authority (OJK), www.ojk.go.id.

2.3 Sample determination techniques

Sampling was done by purposive sampling, meaning that the sample selection method is chosen based on certain criteria. The selection of samples is not random; information is obtained with certain considerations.

2.4 Data analysis techniques

DEA method was used to analyze the efficiency of the bank data under study. In calculating the data, WIN4DEAP Version 2 software DEA, which is a linear programming-based technique, was used to measure the efficiency performance of an organizational unit called the DMU.

This technique was used to measure how efficient DMU is in using available resources to produce output. DMU is said to be relatively efficient if the dual value is equal to 1,000 (100% efficiency). Conversely, if the dual value is less than 1,000 (100%), DMU is considered to be inefficient or relatively inefficient. In addition to measuring the relative efficiency level of a DMU against DMU in its group, DEA can also see the source of inefficiency by considering the original value, projected value, radial movement, and slack movement of each input and output.

Each DMU is assumed to be free to determine the weights for each of the existing input and output variables, provided that they are able to fulfill the two conditions required. The efficiency of banking techniques is measured by calculating the ratio between output and input. DEA will count banks that use input n to produce different output m. Through this equation, it can be concluded that banks are said to be efficient if they have a ratio of close to 1,000 or 100%. Conversely, if it is close to 0, it indicates a decrease in a bank's efficiency.

2.5 Operational research variables

2.5.1 Input variable
(1) Assets is the number of assets owned by the bank.
(2) Total funds is the amount of public funds, both individuals and legal entities, that have been successfully collected by banks through fund collection products. The amount of deposits collected from these public funds is divided into several types: current accounts, deposits and savings, and other forms of equivalent.
(3) Labor costs are costs used by the bank for employee costs.

2.5.2 Output variable
(1) Revenue is revenue generated from bank operations. Bank operational activities include a bank's income and other operational income.
(2) Credit/Financing is a product of channeling bank funds to the public, both individuals and legal entities.

3 RESULTS AND DISCUSSION

Efficient values in DEA range from zero to one. An efficient DMU will have a value of 1,000 or 100%, while a value close to zero indicates a lower DMU efficiency. There are two criteria for an efficient DMU: first, if there are no other units or DMU combinations that use the same number of inputs. Second, the amount of output produced is at least equal to the amount of output produced by other DMUs with 1,000 (100%) performance. The measurement results on the

Table 1. The bank's technical efficiency level is based on categories in Indonesia.

No	Bank	Year					Average
		2013	2014	2015	2016	2017	
1	BRI	100,00	100,00	100,00	100,00	100,00	100,00
2	BCA	100,00	100,00	100,00	100,00	100,00	100,00
3	BTPN	100,00	100,00	100,00	100,00	100,00	100,00
4	DBS	100,00	99,70	100,00	100,00	100,00	99,94
5	MFUG	100,00	100,00	100,00	100,00	100,00	100,00
6	BJB	100,00	100,00	100,00	100,00	100,00	100,00
7	BSM	100,00	100,00	100,00	100,00	100,00	100,00
8	BPR EKA	100,00	100,00	100,00	100,00	100,00	100,00
9	BPRS HIKS	100,00	100,00	100,00	100,00	100,00	100,00
	Rata-rata	100,00	99,97	100,00	100,00	100,00	99,99

Table 2. Original value, projected value, radial movement, and slack movement input-output 2014 period (billions of rupees).

Nama Bank DBS	Tingkat Efficiency	Original Value	Projected Value	Radial Movement	Slack Movement
Output Credit	0,997	41.303	41.304	0,000	0,000
Output Income		4.083	5.227	0,000	1.144
Input Asset		65.663	57.149	−225	8.289
Input Dana		44.466	44.314	−152	0,000
Input Biaya		627	625	−2	0,000

efficiency performance of the largest banks in Indonesia based on the categories of ownership, status, and price determination using the DEA method during the 2013–2017 period focused on variable return to scale (VRS) can be seen in Table 1.

3.1 Bank efficiency assessment results

The results of the analysis based on DEA as shown in Table 1 reveal that the efficiency value of the eight biggest banks by category—namely BRI (the biggest bank chosen among the Persero Bank), BCA (the biggest bank chosen among the foreign exchange BUSN), BTPN (the biggest bank chosen among the BUSN non-visa), KCBA MFUG (the largest bank chosen among the Foreign Bank Branch Offices in Indonesia), BJB (the largest bank chosen among the largest BPD in Indonesia), BSM (the largest bank chosen among Syariah Commercial Banks in Indonesia), BPR EKA BUMI ARTA (the largest BPR selected among BPR in Indonesia), and BPRS HIK (the largest BPRS selected among BPRS in Indonesia) during the last five years (2013–2017)—reaches efficiency rate of 1,000 (100%). A bank with an efficiency of 1,000 (100%) in the VRS method is called an efficient DMU, but this DMU is not necessarily efficient if tested with the CRS method. That is, scale efficiency in a DMU is the ratio between efficiency and the CRS assumption of efficiency with the VRS assumption.

One bank—DBS bank—shows inefficiency in 2014 by showing a figure of less than 1,000 (100%) which only reaches 0.997 (99.70%).

Banks with an efficiency value of less than 1,000 (100%) in the VRS method are referred to as inefficient

Table 3. Listing of peers: bank DBS Indonesia.

DMU	Technical Efficiency	Peer Group	Year 2014	%
DBS	0.997	BJB	0.478	47.80
		MFUG	0.174	17.40
		BPRS HIKS	0.348	34.80

DMUs and have not implemented efficiency optimally in Table 2.

In 2014, the performance condition of DBS Bank experienced inefficiencies in output income/revenue, input on assets, funds, and labor costs due to differences between the original value and projected value, according to the results of the DEA analysis in Table 2. Besides that, DBS Bank can optimize efficiency by using a reference bank in accordance with the DEA calculation results. VRS assumption by taking the lambda weight calculation into account results in Table 3.

However, in 2013, 2015, 2016, and 2017, DBS Bank was classified as efficient because the results of the assessment showed 1,000 (100%). In general, large banks in Indonesia are still relatively efficient.

3.2 Comparison of DEA and OER model results

From the results of the efficiency assessment of major banks in Indonesia based on the categories of ownership, status, and price determination using DEA model, assume VRS in a five-year run (2013–2017)

as shown in Table 1, and based on the results of the OER calculation which are based on the letter circular Bank Indonesia number 6/23/DPNP/Year 2004, criteria for evaluating the level of financial soundness of banks is against the results of the ratio for profitability aspects (OER).

The results of the calculation of the major banks efficiency in various categories in Indonesia using the DEA and OER models for the period 2013–2017 can be seen in Table 4 as follows.

Table 4. Calculation results for OER and DEA-VRS average for five years (2013–2017).

Bank	Average-OER	Criteria	Average VRS-DEA	Criteria
BRI	66.36%	Healthy	100.00%	Efficient
BCA	61.22%	Healthy	100.00%	Efficient
BTPN	81.10%	Healthy	100.00%	Efficient
DBS	88.98%	Healthy	99.70%	Inefficient
MFUG	74.96%	Healthy	100.00%	Efficient
BJB	82.36%	Healthy	100.00%	Efficient
BSM	93.59%	Healthy	100.00%	Efficient
BPR EKA	74.54%	Healthy	100.00%	Efficient
BPRS HIKS	72.19%	Healthy	100.00%	Efficient
Rata-rata	77.26%	Healthy	99.97%	Efficient (8) Inefficient (1) DBS (2014)

4 CONCLUSION

Efficiency is a study of how to allocate factors of available resources optimally to produce maximum output. Efficiency studies are also important to measure the potential risk of a central bank/government policy toward changes in banking policy. One way to measure banking performance is efficiency, which can be seen from the use of inputs and outputs used for bank operations. Banking as the pulse of the national economy is demanded to optimize efficiency in an effort to reduce high interest rates, so that it can reach the layers of micro-, small-, and medium entrepreneurs. The results of the analysis and measurement of the efficiency levels of major banks in Indonesia based on the categories of ownership, status, and price determination in the period 2013–2017 consisted of nine banks: BRI, BCA, BTPN, DBS, MFUG, BJB, BSM, BPR EKA, and BPRS HIK. DEA method used with the assumption that VRS is focused on input oriented showed that only one bank, DBS bank, has not reached 100% efficiency level and only occurred once in 2014. Overall, the performance of large-scale banks based on these categories has been efficient.

Banks that have not been optimally efficient can optimize their efficiency by paying attention to input and output and also reference banks from other banks, especially those that have equivalent characteristics. The results of the efficiency assessment use the OER model of large banks in Indonesia based on the

categories of ownership, status, and price determination during the period 2013–2017, with an average value of 77.26% classified as healthy. In general, the principle is directly proportional or there is no difference with the results of the efficient evaluation by using DEA model in the same period.

ACKNOWLEDGMENTS

The author would like to thank the Indonesian University of Education, Global Conference on Business, Management and Entrepreneurship (GC-BME), International Scientific Committee, Organizing Committee, Conference Chair Professor Dr. Ratih Hurriyati MP, Universitas Pendidikan Indonesia, Technical Chairperson Associate Professor Lili Adiwibowo Universitas Pendidikan Indonesia, reviewers, publishers, and those who have helped and cannot be mentioned one by one for the presentation of this paper.

REFERENCES

Ascarya, Diana Y., & Guruh S. R. 2008. *Analysis of conventional banking efficiency and sharia banking in Indonesia with data Development Analysis (DEA)*. Paper in the sharia financial institution current issues book 2009, IAEI team. Jakarta: Kencana prenada media group.

Ayadi, I. 2013, Determinants of Tunisian bank efficiency: a DEA analysis.

Coelli, T. 1996. A guide to DEAP version 2.1: A data envelopment analysis (computer) program. *CEPA Working Paper* 96/08, (1994), 1–49.

Eken & Kale. 2011. Measuring bank branches performance using data envelopment analysis: The case of Turkish bank branches. *African Journal of Business Management*, 5: 889–901.

Ferrier, G. D., & Hirschberg, J. G. 1997. Bootstrapping confidence intervals for linear programming efficiency scores: With an illustration using Italian banking data. *Journal of Productivity Analysis*

Germanova, E., & Ivanova, E. 2018. Efficiency of banks in Slovakia: Measuring by DEA models Eva Grmanová faculty of social economics relationships, *Alexander Dubcek University of Trencin, Slovakia* eva.grmanova@tnuni.sk

Hasanul Banna, Rubi Ahmad, Koh, E. H. Y. 2017. Determinants of commercial banks' efficiency in Bangladesh: Does crisis matter, eric h.y. koh 3 received: april 5, 2017 revised: may 26, 2017 accepted: august 5, 2017.

Hasibuan, M.S. P 1984. *Basic Management, Understanding and Problems*, Jakarta: Publisher Gunung Agung

Hou, X., Wang, Q., & Zhang, Q. 2014. Market structure, risk taking, and the efficiency of Chinese commercial banks. *Emerging Markets Review*, 20, 75–88. https://doi.org/10.1016/j.ememar.2014.06.001

Johnes, J., Izzeldin, M., & Pappas, V. 2014. A comparison of performances of Islamic and conventional banks 2004–2009. *Journal of Economic Behavior & Organization*, 103, 93–107. Https://doi.org/10.1016/j.jebo.2013.07.016

Kao, C., & Liu, S. T. 2014. Multi-period efficiency measurement in data envelopment analysis: The case of Taiwanese commercial banks. Omega, 47, 90–98. https://doi.org/10.1016/j.omega.2013.09.001

Kasmir 2014. Banks & other financial institutions. revised edition, fourteenth printing, PT. Raja Grafindo Persada, Jakarta.

Mohamad, S. Hassan, T., Bader, M. K. I. efficiency of conventional versus Islamic banks: International evidence using the Stochastic Frontier Approach (SFA).

Moradi-Motlagh, A., & Saleh, A. S. 2014. Re-examining the technical efficiency of Australian banks: A bootstrap DEA approach. *Australian Economic Papers*, 53 (1–2), 112–128. https://doi.org/10.1111/1467-8454.12024

Muliaman D. H., Wimboh S., Dhaniel I. & Eugenia M. 2003. Efficiency analysis of the Indonesian banking industry: Use of non-parametric data envelopment analysis (DEA) methods. *Bank Indonesia Research Paper*, Jakarta: Bank Indonesia.

Puri, J., & Yadav, S. P. 2013. A concept of fuzzy mix-efficiency input in fuzzy DEA and its applications in the banking sector. Expert Systems with Applications, 40 (5), 1437–1450. https://doi.org/10.1016/j.eswa.2012.08.047

Řepková, I. (2014). Efficiency of the Czech banking sector employing the DEA window analysis approach. *Procedia Economics and Finance*, 12, 587–596. https://doi.org/10.1016/S2212-5671(14)00383-9

www.bi.go.id. 2018. *Bank Indonesia circular letter number*, 15/29/dkbu. July 31, 2013. Accessed march 2018 from www.bi.go.id

www.ojk.go.id. 2018. Original value, projected value, radial movement, and slack movement input-output. Accessed april 2018.

Form and document process for implementing ISO 37001:2016 anti-bribery management system

R. Aurachman
Telkom University, Bandung, Indonesia

R.A. Zunaidi
Institut Teknologi Telkom, Surabaya, Indonesia

A. Febriani
IT Telkom Purwokerto, Purwokerto, Indoneisa

ABSTRACT: ISO 37001 is an international standard for anti-bribery management system. Some research showed the strengths and links of this standard compared to other standards. Several studies discussed its application in various fields of industry and organizations. This research tries to contribute in designing implementation of standard and generic documents that will later can be applied to any kind of organization. Thus, it is expected to assist organizations in implementing ISO 37001 quickly and precisely.

1 INTRODUCTION

Management System of the Anti-Corruption is a tool that is needed to minimize corruption at the organization. This system could be implemented within the company by Kafel (2016). There is a standard named ISO 37001 Anti-Bribery Management System or known as Anti-Bribery Management System Standards. The efficiency of anti-corruption actions within a company can be improved by anti-corruption management system. Some effective effort is needed to change the thoughts and habits of corruption in organizations, and the success of the system management depends on the government, organization, and society within Kafel (2016).

Some research have discussed ISO 37001, and it is admitted that ISO 37001 is perfectly suitable in a rational administration system and addresses all drivers that lead to corruption behaviors by Valerio (2017). Another advantage of ISO 37001 is because of the global consensus and reach of ISO standards, ISO 37001 seems to be the only anti-bribery management system to be accepted and understood globally by public authorities, enforcement agencies and business partners (Méan & Gehring 2018). ISO 37001, like another anti-corruption standard, forces to implement transparency and bribe-free which has become standard elements of the anti-corruption by Sampson (2015). Another paper discusses the purpose of ISO 37001, how to implement the system, and the benefits that can be gained from implementing it (Khair, Hasnah, Ishak, & Zab 2017). Another research tells

us how to implement ISO 37001 in a specific kind of organization, such as sport organization (Veselovska & Zavadska 2018). Another place was also examined for the implementation of ISO 37001, like in hospitals (Skoczylas, Adam, & Karkowska 2017). Some research describe the involvement of ISO 37001 as a significant step changes or landmarks over recent decades in the fight against bribery of foreign officials and foreign bribery generally by Tiffen (2016).

Based on some of the works above, there is still no research focusing on creating document process that could be applied during implementation of ISO 37001:2016. This research tries to contribute in designing implementation of standard and generic document process that will later could be applied to any kind of organization.

2 METHODS

The method used in this study was conducting semantic study of sentences written in standard documents. The first step taken was examining standard documents. After that, each article and clause was classified into several groups. Based on the classification, then, the work document was designed so that it would help organizations to meet the intended clause. Then for each work document, the contents, table, and guidance was designed in depth based on the clause in detail. At this stage, it was possible for work documents to be mergered, added by new documents, or deleted due to each sentence of clause. The final step was validating and thoroughly checking all work documents and comparing them with the clause, so that the work documents designed will be able to fulfill the clause and ISO 37001: 2016 standard.

3 RESULTS AND DISCUSSION

Based on the review of ISO 37001: 2016 document, a list of documents that need to be prepared by organizations which want to implement these standards was obtained. Documents were designed to be as efficient as possible. It is expected that the organization in implementing ISO 37001 will not be burdened by working on additional business processes. By implementing ISO 37001, organization only has to make few changes in the system and adapt the work method. It is also required to change the way organization record and document each process.

The system which will be applied needs to be as simple as possible so that the organization does not have difficulty in adapting the standard. The prove that adaptation process works well can be seen when organizational members admit the change without resisting. There will be numbers of resistance if the organization needs large number of resources to make changes. Resistance can also occur if changes affect the comfort of organization members. This was experienced before when the new ISO 9001: 2015 system was implemented. Every change needs to be calculated and planned carefully thus the change being made is useful and accurate. The proposed system and documents are designed primarily by adhering to these principles. As long as there are documents that have met the required standards and requirements, or just simply make changes to the format of documents that generally exist, then there is no need to bring up new documents. As the application of new documents and working papers requires socialization, habituation, and training requires funds, adding new thing in the system needs to be minimized.

Table 1 describes that the documents needed to implement the antibribery system based on ISO 37001: 2016. The organization Strategic Plan template, for example, was designed to meet clause 4.1. Understanding the organization. Understanding organization means understanding the strengths and weaknesses of the organization while the organizational context means its threats and opportunities. This is in line with the process of formulating strategies that are generally practiced by an organization. Therefore, it is unnecessary to procure new documents as organization can use and modify existing system documents.

In fulfilling clause 6.2 regarding anti-bribery targets, organization can also use the organization strategic plan document. However, since the anti-bribery target is more dynamic and fluctuative by the time of monitoring, in contrast with the strategy plan which tends to remain steady for a long time, a special document which contains the anti-bribery target plan and its achievement steps is needed.

Clause 4.3, 5.3, 8.1, and 7.2.1 can be fulfilled through Responsible Assignment Matrix document. The document is a table where each row has a business process that should exist in the organization, and each column contains a position in the organization. Furthermore, business processes are mapped and paired to the position using letter symbols: R, A, C, or I; Responsible, Accountable, Consulted, Informed.

In the document, comprehensiveness of anti-bribery management system scope will be mapped to meet the requirement of clause 4.3. Furthermore, in RACI written processes, the decision is made related to what plan to be carried out and what form of control for each process to be established. As stated on the requirement of clause 8.1, each row in RACI matrixes is filled by processes in the organization. There are several processes required to exist based on ISO 3700 clause 5.3.

Risk register documents are needed in the context of organizational risk management. Risk register will consider risk that is related to the application of ISO 37000 and other standards. The risk register is designed to be able to meet the standard requirements of clause 4.4 about description of anti-bribery management system in an organization. The clause stated that anti-bribery management system needs to be designed by reviewing the risks first.

Risk register table needs to contain column with plans for action or follow-up to the existing risks within, in accordance with the requirements of the standard 6.1. These actions are carried out based on the level of risk that is likely to occur and the large impact of these risks.

According to clause 8.2, several processes and systems need to be evaluated and some due diligences need to be done for the control mechanism. Follow-up towards due diligence is designed based on the adequacy of the planned risk mitigation. In the risk register, each line contains a process or activity which is then evaluated for its risk level and written in another specific column. At the end of the risk register table, a follow-up plan for the risk should be planned.

The risk register documents need to be evaluated and updated regularly. Evaluation is based on stake-holder input and evaluation of the planned risk mitigation. As stated in clause 8.8, when there is a quality assurance risk mitigation that is proven to be ineffective, the risk register document needs to be adjusted, adapted to the changes and developments that exist.

Organizations also need internal audit procedure. This internal audit procedure refers to clause 9.2 concerning internal audit. In addition, clause 9.2 regarding internal audit and internal audit procedures are designed to meet clause 10.2 about continuous improvement. Continuous improvement can be done in various situations and processes within the organization. To simplify the process of continuous improvement, the improvement is carried out mainly in the internal audit system. However, organization still need to do continuous improvement using other strategies.

Within an audit, auditor will discover some findings, potential for improvement, and errors. The overall outcomes of the audit are followed up in forms or template about findings and potential of improvements (TLTPP = Tindak lanjut Temuan dan Potensi Perbaikan). This TLTPP form has several columns that must be filled in and implemented so it accommodates clause 10.1 regarding non-conformities and corrective actions. This form also fulfills clause 10.2 which is about continuous improvement.

The TLTPP form regularly records what the findings and potential improvements from the organization are. The form also guides the process which must be followed, as stated in the form fields, as well as the parties who need to be involved in this TLTPP process. Each new finding or new potential improvement is written in a new line on the form and the follow-up continues to be monitored on that line. It is expected that by using this procedure, it will guarantee that every finding and potential improvement is responded quickly and correctly.

In fulfilling clause 8.3 regarding financial control, organizations can modify all sets of procedures relating to financial processes. However, this is not explained in detail because it depends on the context and needs which is different and needs adaption for each organization. This study will not explain in detail how the financial procedures are.

Similarly, the implementation of clause 8.4 concerning non-financial control, is not explained in detail. However, non-financial controls here have detailed implication only on the procurement process in organizations. Although it was said as non-financial controls, not all processes, are regulated in this clause except for finance process. When it is viewed from the clause sentence, it can be seen that the main focus is on the procurement process.

In ensuring organizational compliance with the anti-bribery system, it is not enough to only control the internal organization. Internal organizations are influenced by partners, both partners who supply the needs of organizations and partners who become customers of the organization. Therefore, organizations need to have partner monitoring matrix documents. The document is in the form of a table with time indication columns. The table monitors the progress of partner compliance with anti-bribery practices within the organization. This document is designed in order to meet clauses 8.5.1 and 8.5.2 regarding the application of anti-bribery management system which is managed by administrations and partners, as well as clause 8.6 concerning anti-bribery commitments. This partner monitoring matrix document is updated regularly which records the developments about partner commitment on the anti-bribery standard. The form of monitoring can be done by writing down the results of the audit of the partners and writing the results of the anti-bribery certification that has been independently carried out by the partners.

In implementing ISO 37001: 2016, there are several new functions and processes should be implemented in organizational structure. For instance, the existence of a steering committee board. This steering committee board has a specific function described in clause 5.1.1. So the manifestation of this clause is the job description of the anti-bribery steering committee. The job description has been designed so that it contains the roles mandated by clause 5.1.1.

Slightly different from the steering committee, top management is generally owned by each organization. By implementing ISO 37001, there will be a new job description item that will be delegated to the highest leadership in the organization. That additional item is guided by clause 5.1.2.

Top management is also expected to conduct regular reviews on the organization. In implementing ISO 37000, one of the things that need to be reviewed is the application of ISO 37000. To guarantee that the occurrence of the review process is carried out by the leadership, the Management Review minutes of meeting document need to be standardized. The documents include topics that must be discussed at the management review meeting, guided by clause 9.3.1 about top management review and clause 9.3.2 about the management review of the director. The two clauses are designed to lead to one standardized minutes of meeting which will be carried out together between the steering board and top management, for streamlining business processes and organizational resources. The minutes of meeting document are in the form of a table where each line contains topics that need to be discussed. The columns consist of comments for meeting participants for related topics, plan for follow-up, the person in charge of completing the follow-up plan, deadline, and the status of the discussion whether it has been completed or is still awaiting completion. The document is reviewed and edited at each management review meeting in the hope that the status of each to do list will be done soon.

The anti-bribery system needs to be supported by organizational policies. Organizational policy is a general guideline in designing anti-bribery systems. Procedures and work forms are designed based on the sentence stated in the policy document. Content of the policy, accommodates the clause 5.2 requirements regarding the anti-bribery policy. To ensure compliance, all sentences in clause 5.2 are rewritten in the anti-bribery policy with some adjustments. In addition, the anti-bribery policy was also designed to be able to accommodate clause 8.9 concerning increasing awareness of the anti-bribery system.

In fulfilling clause 8.1, regarding investigations and handling of bribery, a new procedure about enforcement of bribery is needed. For some organizations, this procedure is kind of new thing. However, organizations implementing ISO 37000 are required to pay more attention to violations of bribery. One more form of commitment is forming a standard process to response bribery. In Legal Governance, it is known as the procedural law and order, which regulates the pattern and procedure for prosecuting law violation.

Every individual involved in the organization should realize and understand that anyone, at any time, and anywhere will get the same treatment if they violate anti-bribery commitments. The flow of investigation and trial against violations will be the same for anyone who violates them. The process is made transparent so that no one gets the privilege, or no one gets excessive legal action. By implementing this procedure, it is expected to create certainty and prevent violations of anti-bribery commitments

Based on clause 7.1, resources within the organization must always be monitored and controlled because the lack of resources will cause failure in achieving anti-bribery system target. Guiding the implementation of the process, the implementation of resource matrix documents is proposed. The matrix has rows that list what resources an organization must have and monitor.

While the matrix column contains the status and conditions of the resource over time, there is also a column that explains how to monitor these resources. Each resource that should be monitored can be drawn from the targets of the anti-bribery system so that every provided and maintained resource has a basis and a strong reason. Organization should not provide resources that are not useful or have a low utility level.

Satisfying clause 8.2.1 regarding competency, the organization must have a way to monitor the competencies of organization's members. Monitoring can be done by measuring the competency of each individual, determining the competency target for each position, and measuring the gap between the target and the reality of competencies. To realize this function, a competency gap matrix document is needed. The matrix consists of several tables. The first table contains the competency conditions of each individual. Each row contains individual names and each column contains the measured competency dimension. Likewise, the second table has the same column legend. However, each line contains positions measured by competence while the third table contains a gap value between the first table and the second table. The gap value will be used as a basis for measuring whether the individual is ready to occupy the assigned position or not. Even if there is still a competency gap, it can be seen which competency dimensions it is. Then, the human resource development process will be planned to fulfill that gap.

In the process of hiring and involving new individuals to be part of the organization, an understanding and commitment need to be built so that new individuals are ready to cooperate in maintaining the anti-bribery system according to ISO 37000. Then, according to clause 8.2.2., organizations need to modify the process

of recruiting new human resources. The process that has been running in general within the organization is simply added by the activity of signing a commitment and agreement to be involved in maintaining the anti-bribery system. These include statements of ready to report and cooperate in prosecuting bribery cases.

Document control is a process that needs to be maintained within the organization so that developments and changes in documents can be monitored and maintained. In the document, knowledge and information about the condition and progress of the system are recorded. In accordance with clause 7.5, a document master list template is required. In the template, any document information is controlled by the organization, such as procedures, work instructions, information systems, matrix templates, work forms, guidelines, and other guidelines. The matrix contains information on the condition of each document, when it is made, and what version has been affected. The matrix also explains in the column who are the parties involved in receiving the document and who is authorized on updating document. In addition to guidelines kind of document, document master list may contain a record of evidence of the progress, such as tables that have been filled in or minutes of meetings. The table of documents can be extremely large because it records plenty of information.

Guidelines for communication in the organization are needed, which is designed based on clause 8.4. The guide may appear in the form of simple table where each line shows the type of information. While the column contains detailed explanations of each type of information, such as to whom information is provided, what kind of the media is used to deliver information, etc., so that it can be understood to disseminate information about indications of bribery, what process must be carried out, what media, and whom the person may receive. All plans and designs need to consider the level of risk which has written in risk register.

Each individual who sees a dispute and violation is able to convey the facts they know. The reporting system is designed to be safe and can be applied anonymously to eliminate the risk of reporting. If the reporting process threatens the reporter and not protecting the whistle blower, it will prevent reveling the fact. The violation of law does not get the appropriate action. In fulfilling clause 8.9 about increasing awareness, a reporting form can be applied and can be submitted online and anonymously. This form can use open source technology such as Google Form. On the google form, strict setting can be applied so that those who fill the form do not need to be identified and therefore it eliminates the reluctant of reporting.

4   CONCLUSION

Some organizations have difficulty in implementing the implementation of a standard. An implementation document that can be used as a reference artifact is needed. This research tries to contribute in designing implementation of standard and generic documents that will later could be applied to any kind of organization. It will be able to assist organizations in implementing ISO 37001 quickly and precisely.

However, a work procedure and guidelines are not enough to change the conditions of the organization to perfectly implement ISO 37001. There are aspects of soft systems that need to be considered, reviewed, and designed to be able to accommodate the application of ISO 37001. The implementation of ISO 37001 can make use of digital transformation supported by the implementation of ISO 20000-1 (Siddiqui 2019). Not only ISO 20000-1 is a facility for digital transformation to support the implementation of Anti-Bribery Management System, but also it can be used as the guideline for the implementation of information system security (Tagarev & Polimirova 2019).

REFERENCES

Kafel, P. 2016. Anti-bribery management system tool to increase qualitu of live. *1st international conference on quality of life*, 1–2.

Khair, I. N., Hasnah, H., Ishak, I., & Zab, S. 2017. ISO 37001: anti bribery management system-implementation and benefits. *Journal of governance and integrity (JGI)*, *1*(1), 67–79.

Méan, J. P., & Gehring, H. 2018. Implementing iso 37001 to manage your bribery risks. *Global trade and customs journal*, 13(5), 191–197.

Sampson, S. 2015. The anti-corruption package. *Ephemera: theory and politics in organization*, 15(2), 435–433.

Siddiqui, F. M. 2019. Digital transformation of modern airports by exploiting fog as a service model. *Integrated communications, navigation and surveillance conference* (ICNS), (pp. 1–11).

Skoczylas, P., Adam, T. K., & Karkowska, D. 2017. PN-ISO 37001 management systems for anti-corruption actions and anti-corruption procedure in jan paweł ii hospital in bełchatów. *Agnieszka śliz projekt okładki: marcin szadkowski*, 99.

Tagarev, T., & Polimirova, D. 2019. Main considerations in elaborating organizational information security policies. *Proceedings of the 20th international conference on computer systems and technologies*, (pp. 68–73).

Tiffen, N. 2016. Governance in practice: is the new iso a step change in the foreign bribery journey? *Governance directions*, 68(11), 654.

Valerio, B. (2017). Corruption and iso 37001: a new instrument to prevent it in international entrepreneurship. *World journal of accounting, finance and engineering*, 1–14.

Veselovska, L., & Zavadska, Z. 2018. Implementation of teh ISO 37001: 2016 anti-corruption management system in sport organizations. *Управление экономикой: методы, модели, технологии.*

Advances in Business, Management and Entrepreneurship – Hurriyati et al. (Eds)
© *2021 Taylor & Francis Group, London, ISBN 978-0-367-67471-7*

Interaction of liquidity creation, regulatory capital and risk taking in ASEAN banking industry

S. Fauzie & P. Hidayat
Universitas Sumatera Utara, Medan, North Sumatera, Indonesia

ABSTRACT: The purpose of this study was to find out the interaction among liquidity creation, regulatory capital, and risk taking in the banking industry in ASEAN countries. A panel vector auto regression model was used in this study to see the interaction among those three variables. The research data used was quarterly data from banks listed on the stock exchange in the period 2009–2017. The results of the study indicated that there was a positive reciprocal relationship between regulatory capital and risk taking. Also, the regulatory capital regulations of the previous four quarters reduced liquidity creation. The results also showed that there was no interaction between liquidity creation and risk taking.

1 INTRODUCTION

Several banks in ASEAN countries have implemented capital and liquidity management based on regulations issued by the Basel Committee on Banking Supervision, known as BASEL III, at the end of 2010. This regulation was issued in response to the problems of the global economic crisis that occurred from early 2008 until 2010. The regulations have affected banks to increase the capital adequacy ratio from 8% to 9.5%. This regulation requires banks to have the ability to meet liquidity obligations that are due 30 days ahead in a stress scenario. This is because BASEL III requires the use of two ratios consisting of liquidity coverage ratio (LCR) and net stable funding ratio (NSFR) (DeYoung & Jang 2016). There are several banks in the new ASEAN countries that will implement BASEL III regulations under the policies in their respective countries.

The application of BASEL III has an impact on banks, which must increase the allocation of funds for investment that provides low returns. Therefore, some banks take adverse actions by channeling long-term loans with short-term funding sources. This condition will result in maturity mismatch between long-term investments and short-term obligations as a source of funds called liquidity creation (Tran, Lin, & Nguyen 2016). Furthermore, the placement of funds in long-term investments originating from short-term liabilities will cause liquidity risk (Deriantino, Fauziah, & Surjaningsih 2014). Therefore, if the bank wants to maintain the level of long-term investment, it needs additional capital above the minimum capital regulation regarding the capital adequacy ratio, so that banks can maintain their income from long-term investments such as loans.

The capital increase will have an impact on increasing risk actions by placing funds on low-quality loans. Therefore, increasing the capital buffer shows indications of holding a portfolio of risky assets and reducing aggressive investment (Khan, Scheule, & Wu 2017). This is because an increase in capital above the capital regulation will be able to absorb higher loss reserves so that banks have the power to channel funds (Lee & Hsieh 2013). Similarly, Umar and Sun (2016) suggest a trade-off between high capital and the result of liquidity creation in order to maintain banking stability. The existence of trade-offs between each variable consisting of liquidity creation, capital regulatory, and risk-taking drives this study to examine the interaction of these three variables with each other using the panel vector-autoregressive (PVAR) model.

This research is critical because the application of BASEL III will have an impact on the resilience of banks to be able to maintain their liquidity and capital to create better financial stability. It is expected that this research will contribute to the improvement of regulations and standards set as policies adopted by the central bank in maintaining financial stability.

2 LITERATURE REVIEW

Research on the relationship of liquidity creation and regulatory capital was initiated from the Berger and Bouwman (2009) study, which wished to test a theory called financial fragility-crowding out and risk absorption hypothesis. Financial fragility-crowding out hypothesis stated that increased regulatory capital would reduce liquidity creation, while risk absorption hypothesis developed from Repullo (2004) stated that the higher the capital owned by the bank, the more

Table 1.	Operational definition.	
Variable	Acronym	Definition
Liquidity Creation	LC	Amount of liquidity creation normalized by the total gross asset. Excludes the off-balance sheet activities. from Berger and Bouwman (2009)
Regulatory Capital	CAR	Equity divided by total risk-weighted asset
Risk Taking	RISK	Risk-weighted asset divided by total asset

the ability of banks to create liquidity. The results of the Berger and Bouwman (2009) study showed that the increase in capital will increase liquidity creation in large banks if it enters accounts on the off-balance sheet in measuring liquidity creation.

Distinguin, Roulet, and Tarazi (2013) conducted subsequent research that examined the impact of liquidity creation on regulatory capital buffers on banks that were publicly traded in Europe and the United States by using measures of liquidity creation with on-balance sheet accounts. The results obtained indicated that banks would reduce regulatory capital when banks increased illiquid assets or liquidity creation while the small banks would strengthen their solvency when illiquid assets increased.

Furthermore, Tran et al. (2016) examined the relationship between the liquidity creation and regulatory capital and their impact on profitability. The results of their research showed that the relationship between the liquidity creation and regulatory capital only occurred in small banks and not in the crisis period. Their research also showed a negative relationship between liquidity creation and performance. The results of this research showed that low-capitalized banks would experience an increase in performance if there was an increase in the capital, but an increase in capital would provide a decrease in performance for small capitalized banks.

3 METHODS

3.1 Sample

The sampling from the population of banks listed on the stock exchanges in ASEAN countries used a purposive sampling method that has quarterly financial information data for the period 2009 to 2017. The data sources for this study are published by Bloomberg. The purposive sampling results were 15 banks that had complete data, so the number of observations was 540.

3.2 Operational definition

The operational variables and definitions used in the study are presented in Table 1.

3.3 Analysis

3.3.1 Stationarity test

The first step taken in estimating the PVAR model was to test the stationarity on the data. Stationarity tests can be used using the Levin, Lin, and Chu (2002) test at the same degree (level or different) to obtain a stationary data, where the data variance is not too large and has a tendency to approach the average value.

3.3.2 Determination of optimal lags

Determination of optimal lag was used to determine the optimal number of lags used in stationarity tests by looking at the final prediction error (FPE) correction value or the number of AIC, SIC, and HG, which is the smallest among the proposed lags.

3.3.3 Granger causality

The Granger causality test was used to analyze the causality relationship between variables to be observed. Testing of causality was seen through the F-test and its probability value.

3.3.4 VAR panel analysis

VAR panel analysis was used to see relationships between variables.

$$LC_{it} = \beta 1_{it} + \sum \beta 1 LC_{it} - 1 + \sum \beta 1 CAR_{it} - 1 + 1\varepsilon_{it} \quad (1)$$

$$CAR_{it} = \beta 2_{it} + \sum \beta 2 LC_{it} - 1 + \sum \beta 2 CAR_{it} - 1 + 1\varepsilon_{it} \quad (2)$$

$$CAR_{it} = \beta 1_{it} + \sum \beta 1 CAR_{it} - 1 + \sum \beta 1 RISK_{it} - 1 + 1\varepsilon_{it} \quad (3)$$

$$RISK_{it} = \beta 2_{it} + \sum \beta 2 CAR_{it} - 1 + \sum \beta 2 RISK_{it} - 1 + 1\varepsilon_{it} \quad (4)$$

$$LC_{it} = \beta 1_{it} + \sum \beta 1 LCit_{it} - 1 + \sum \beta 1 RISK_{it} - 1 + 1\varepsilon_{it} \quad (5)$$

$$RISK_{it} = \beta 1_{it} + \sum \beta 2 LC_{t} - 1 + \sum \beta 2 RISK_{it} - 1 + 1\varepsilon_{it} \quad (6)$$

3.3.5 Impulse response function

Impulse response analysis is the effect of innovation on variables that can track the response of endogen variables in the PVAR system.

3.3.6 Variance decomposition

This test was conducted to find out how much is the difference between the variance of a variable caused by innovation (both from self and other variables).

4 RESULT AND DISCUSSION

Sources of financial report data from Bloomberg are in the form of quarterly reports from 2009 to 2017,

Table 2. Descriptive statistics.

| Variable | Descriptive Statistics | | | | |
	Min	Max	Mean	Med	St. Dev
LC	−0.13	0.53	0.18	0.17	0.13
CAR	9.82	51.60	16.83	16.20	3.66
RISK	0.00	1.80	0.66	0.65	0.16

Table 3. Stationarity test.

Levin, Lin, & Chu	Statistic	Prob.**
LC	−5.15487	0.0000
CAR	−6.74722	0.0000
RISK	−4.99791	0.0000

Table 4. Lag optimal.

Lag	7	8
LogL	−281.0280	−249.6140
LR	39.56542	59.08830*
FPE	0.001048	0.000942*
AIC	1.652514	1.545781*
SC	2.287412	2.267255
HQ	1.903455	1.830940*

Table 5. Granger causality.

Null Hypothesis	Prob.
CAR does not Granger Cause LC	0.6150
LC does not Granger Cause CAR	0.7007
RISK does not Granger Cause LC	0.6960
LC does not Granger Cause RISK	0.8629
RISK does not Granger Cause CAR	0.0452
CAR does not Granger Cause RISK	2.E-08

Table 6. Estimate PVAR.

	LC	CAR	RISK
LC(−1)	(0.05066)	(−1.78441)	(−0.07073)
	[11.0984]	[−0.76822]	[−0.05874]
LC(−2)	(0.05955)	(2.09763)	(0.08314)
	[0.84042]	[0.42918]	[0.90139]
LC(−3)	(0.05959)	(2.09908)	(−0.08320)
	[1.41782]	[0.19498]	[−0.76517]
LC(−4)	(0.05966)	(2.10154)	(0.08330)
	[0.74042]	[0.25492]	[0.23873]
LC(−5)	(0.06003)	(−2.11460)	(−0.08381)
	[0.30279]	[−1.51068]	[−0.49193]
LC(−6)	(−0.06279)	(2.21153)	(0.08766)
	[−0.32253]	[1.81267]	[0.03572]
LC(−7)	(0.06771)	(−2.38491)	(0.09453)
	[0.70685]	[−0.49032]	[0.81570]
LC(−8)	(0.05839)	(−2.05673)	(−0.08152)
	[1.61392]	[−0.34931]	[−0.63481]
CAR(−1)	(−0.00136)	(0.04791)	(−0.00190)
	[−0.50095]	[6.67907]	[−1.71431]
CAR(−2)	(0.00143)	(0.05035)	(−0.00200)
	[0.45172]	[1.83258]	[−2.99179]
CAR(−3)	(0.00143)	(−0.05028)	(0.00199)
	[1.36534]	[−1.55896]	[5.97992]
CAR(−4)	(−0.00149)	(0.05257)	(0.00208)
	[−2.05198]	[1.43296]	[1.04136]
CAR(−5)	(−0.00152)	(−0.05338)	(−0.00212)
	[−0.05681]	[−0.02288]	[−1.85414]
CAR(−6)	(−0.00151)	(−0.05318)	(0.00211)
	[0.29144]	[−0.82798]	[0.75494]
CAR(−7)	(0.00144)	(0.05086)	(−0.00202)
	[0.22613]	[1.78517]	[−0.31666]
CAR(−8)	(−0.00136)	(0.04804)	(−0.00190)
	[−0.73323]	[7.08679]	[−0.74951]
RISK(−1)	(0.03606)	(−1.27030)	(0.05035)
	[0.87506]	[−0.46008]	[8.90456]
RISK(−2)	(0.03992)	(1.40601)	(−0.05573)
	[0.28506]	[0.87973]	[−0.44391]
RISK(−3)	(−0.04269)	(1.50353)	(0.05959)
	[−0.03854]	[0.06327]	[1.68624]
RISK(−4)	(−0.04161)	(−1.46558)	(0.05809)
	[−1.04624]	[−1.52022]	[0.67661]
RISK(−5)	(0.03721)	(1.31081)	(0.05196)
	[1.36667]	[1.12718]	[1.91107]
RISK(−6)	(−0.03644)	(1.28351)	(0.05087)
	[−1.53804]	[2.47915]	[2.00452]
RISK(−7)	(0.03584)	(−1.26245)	(0.05004)
	[0.29663]	[−1.43460]	[0.97898]
RISK(−8)	(0.03333)	(−1.17389)	(0.04653)
	[0.69623]	[−1.21988]	[1.86265]

so that data usage was from 2009Q1 to 2017Q4. The sample results showed that only 15 banks were eligible, so that 540 observations were conducted in the study. Descriptive statistics for the variables used are as follows in Table 2.

The following table is the root test output of panel data units using the Levin et al. (2002) method for liquidity creation (LC), regulatory capital (CAR), risk taking (RISK), and the results show that the probability values of all variables are smaller than alpha 0.05 (0.0000 < 0.05). This result shows that all the data is stationary at the level and does not have a long-term or nonintegration relationship (Table 3).

Based on the following testing, it can be concluded that the lag level 8 is the most optimal deadline or good to use in the model. It is known that both are the most

substantial LR value, while FPE, AIC, and HQ, the smallest, are at the lag level 8 in Table 4.

The Granger Causality Test results show that the relationship between CAR and RISK is a two-way causality relationship. RISK affects CAR because the probability value of F-statistic is 0.0452, whereas CAR affects RISK with a probability value F-statistic of 2.E-08 smaller than α of 5% (Table 5).

The following PVAR estimation results show the effect and significance of the lag of a variable on other endogenous variables using a comparison between the absolute value of the t-statistic (the value in the square sign) and the critical value of the rule of thumb of 2.00 in Table 6.

The PVAR estimation results on the relationship between LC and CAR showed that a higher increase in capital decreases flexibility in liquidity creation with an increase in the number of Tier-1 and Tier-2 capital by 1% from the previous four quarters. It encourages banks to invest in liquid assets in the sense of decreasing liquidity creation amounting to -0.00149. These results are more likely to support the financial fragility-crowding out hypothesis compared to the risk absorption hypothesis. These results are in line with the existence of a mutually important two-way relationship between CAR and RISK, which explains that an increase in capital causes an increase in risk taking and vice versa. From Table 6, it can be seen that the capital increase by 1% from the previous three quarters provides flexibility in action, which leads to risks that pushed it to rise by 5.97992, whereas an increase in risk action by 1% from the previous six quarters encourages banks to form reserves that can absorb the risk, so capital increases by 1.28351. These results are more likely to support the risk absorption hypothesis compared to the financial fragility-crowding out hypothesis.

Furthermore, the impulse response function image shows that the shock occurring in the LC has a positive influence on the CAR in the initial period and decreases in the final period, whereas the shock occurring in the CAR gives a negative influence on LC in the initial period and has a positive effect on the final period. This result illustrates that the placement of illiquid investments requires an increase in several reserves in capital to absorb risk, and when reserves have been optimally fulfilled, the bank will increase the placement of liquid assets. This is because the collection of short-term external funds needs to reduce capital costs arising from the addition of capital reserves and leads to the placement of liquid assets.

This result is in accordance with the description of shock that occurs in LC, which has a positive influence on RISK in the initial period and has decreased in the final period. This indicates the existence of capital reserves formed after an increase in LC in the initial period and investing in short-term assets that have a lower risk. Conversely, the shock that occurs in the CAR gives a negative influence on the LC in the initial period and has a positive effect on the final period. In the end, the shock that occurs in the CAR has a positive influence on RISK (Figure 1).

From the results of the variance decomposition test, there is an increase in the RISK explanation of CAR from the beginning to the end of the period. Conversely, there is an increase in CAR's explanation of RISK from period 1 to its peak in period 6, but it decreases in the following period. This result also shows an increase in RISK and CAR against LC from the initial period to the end of the period, in Table 7.

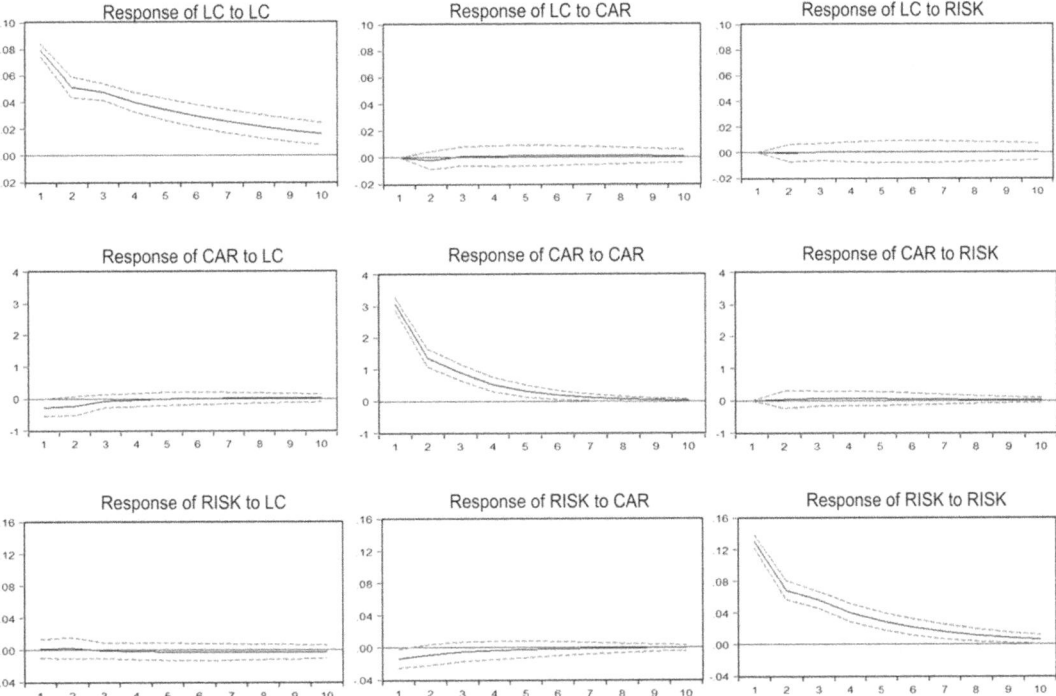

Figure 1.   Impulse response function result.

Table 7. Variance decomposition analysis.

	LC	CAR	RISK
Variance Decomposition of LC:			
Period			
1	100.0000	0.000000	0.000000
2	99.79413	0.060663	0.145211
3	99.55794	0.059328	0.382734
4	99.26246	0.236038	0.501500
5	99.15028	0.352544	0.497181
6	98.89942	0.432243	0.668336
7	98.90977	0.430178	0.660052
8	98.91842	0.422375	0.659203
9	98.78115	0.463867	0.754985
10	98.48748	0.605022	0.907498
Variance Decomposition of CAR:			
Period			
1	0.398520	99.60148	0.000000
2	0.683717	99.26836	0.047919
3	0.696014	99.18256	0.121424
4	0.696807	99.11579	0.187402
5	0.721296	98.69401	0.584695
6	1.215119	98.20194	0.582939
7	1.378926	96.64056	1.980515
8	1.385334	96.52911	2.085560
9	1.310944	96.65760	2.031454
10	1.417135	96.55829	2.024580
Variance Decomposition of RISK_TAKING:			
Period			
1	0.112471	0.982205	98.90532
2	0.117038	2.123404	97.75956
3	0.520963	6.061978	93.41706
4	0.512879	6.995946	92.49117
5	0.497022	8.734580	90.76840
6	0.506363	8.483466	91.01017
7	0.480938	8.070656	91.44841
8	0.821151	7.857428	91.32142
9	0.792826	7.790721	91.41645
10	0.767579	7.750439	91.48198

5 CONCLUSION

This study intends to expand from previous research to know the interaction between liquidity creation and capital regulation by adding risk taking to compare the risk absorption hypothesis, not just the financial fragility-crowding out hypothesis. The research findings showed that the addition of capital regulations supported banks to place funds in risky assets, which caused banks to place liquid assets. In the end, the risk created an increase in capital reserves to absorb the risks that had occurred in strengthening financial stability. These results supported the precision of the risk absorption hypothesis, but these results supported the financial fragility-crowding out hypothesis due to a decrease in liquidity creation caused by additional capital. Therefore, it is necessary to balance off third-party funds by making short-term investments. From these results, it can be concluded that the application of BASEL III regarding the limitation of minimum liquidity and capital can provide resilience to bank stability.

ACKNOWLEDGMENT

This research was funded by Universitas Sumatera Utara through the TALENTA program with contract number 112/UN5.2.3.1/PPM/KP-TALENTA USU/ 2019.

REFERENCES

Berger, A. N., & Bouwman, C. H. S. 2009. Bank liquidity creation. *Review of Financial Studies* 22: 3779–3837.

Deriantino, E., Fauziah, N. R., & Surjaningsih, N. 2014. Interaksi modal bank dan likuiditas. Bank Indonesia: Seminar riset stabilitas sistem keuangan 2015.

DeYoung, R., & Jang, K. Y. 2016. Do banks actively manage their liquidity? *Journal of Banking & Finance* 66: 143–161.

Distinguin, I., Roulet, C., & Tarazi, A. 2013. Bank regulatory capital and liquidity: Evidence from US and European publicly traded banks. *Journal of Banking & Finance* 37(9): 3295–3317.

Khan, M. S., Scheule, H., & Wu, E. 2017. Funding liquidity and bank risk taking. *Journal of Banking & Finance* 82: 203–216.

Lee, C.-C., & Hsieh, M.-F. 2013. The impact of bank capital on profitability and risk in Asian banking. *Journal of International Money and Finance*, 32: 251–281.

Levin, A., Lin, C. F., & Chu, C-S. J. 2002.Unit root tests in panel data: Asymptotic and finite sample properties. *Journal of Econometrics* 108: 1–22.

Repullo, R. 2004. Capital requirements, market power, and risk-taking in banking. *Journal of Financial Intermediation* 13: 156–182.

Tran, V. T., Lin, C.-T., & Nguyen, H. 2016. Liquidity creation, regulatory capital, and bank profitability. *International Review of Financial Analysis* 48: 98–109.

Umar, M., & Sun, G. (2016). Interaction among funding liquidity, liquidity creation and stock liquidity of banks. *Journal of Financial Regulation and Compliance* 24(4): 430–452.

Advances in Business, Management and Entrepreneurship – Hurriyati et al. (Eds)
© 2021 Taylor & Francis Group, London, ISBN 978-0-367-67471-7

An effect of bank efficiency on earnings management at ASEAN banks

F.N. Nasution & S. Fauzie
Universitas Sumatera Utara, Medan, North Sumatera, Indonesia

ABSTRACT: The purpose of this study was to determine the effect of bank efficiency on earnings management behavior at ASEAN banks. Bank efficiency included in this study was cost and profit, where the measurement employed the stochastic frontier analysis model. The data used in this study were banks listed on stock exchanges in ASEAN in the period 2010–2017. Panel data regression analysis was used to test the effect of cost and profit efficiency on earnings management. This study also employed control variables which consisted of bank size, credit risk, liquidity risk, and financial leverage. The results of the study showed that both cost efficiency and profit efficiency had a significant effect on earnings management, and the control variables that had a significant effect on earnings management were credit risk and liquidity risk.

1 INTRODUCTION

A bank can create value for shareholders when their returns exceed the initial investment they have. Returns to shareholders can occur if the bank can provide returns that exceed the expectations of shareholders as reflected in the cost of bank capital. This fulfillment of expectations is reflected in the increase in bank stock prices. Conversely, the management of banks that provides returns below the expectations of shareholders will cause share prices to decline. This condition can cause pressure on the bank's management to take action on earnings management by using discretion in financial statements in misleading bank performance. In general, the discretion carried out by bank managers is on loan accounts consisting of loan loss provisions and loan loss reserves (AB-Hamid, Asid, Sulaiman, Sulaiman, & Bahri 2018; Wu, Ting, Lu, Nourani, & Kweh 2016). This is because most of the funds obtained by banks are channeled to loans. One of the factors that influences earnings management actions is the inability of managers and company staff to maintain an increase in income and profit levels above the increase level in operational costs (Siudek 2008). In other words, earnings management actions occur because of the lack of banks' ability to carry out efficiency by maximizing output using specific inputs or using inputs minimally to produce optimal outputs (Gloria & Irene Rini 2015). This is because decision making requires allocation of inputs, a mix of products that require the collection of deposits, and the distribution of loans or investments that can provide benefits and risks (Guillen, Rengifo, & Ozsoz 2014).

This study involved banks that were listed on stock exchanges in each ASEAN country. ASEAN countries were involved due to the increasing importance of financial integration among ASEAN countries, so that some meetings have been held by finance ministries and central bank governors among ASEAN countries to affirm their commitment to achieve ASEAN financial integration to support economic growth and strengthen financial stability in the ASEAN region. By carrying out financial stability, it is expected that banks in ASEAN can improve their efficiency and reduce earnings management actions that can harm investment liabilities on the stock exchange. This is in line with the Healy and Wahlen (1999) suggestion that market reactions can occur when banks have abnormal and low loan loss provisions. Market reaction assesses that the condition has a relationship to performance and poor cash flow in the future. Therefore, the purpose of this study was to examine the effect of bank efficiency on earnings management actions in ASEAN countries.

This study included two efficiency variables consisting of cost efficiency and profit efficiency using estimation based on stochastic frontier analysis whose approach was taken from the research model of Fu, Lin, and Molyneux (2014) and Naffati, Ben, and Schalck (2011). The model used to look for earnings management indicators was the research model by Cohen, Cornett, Marcus, and Tehranian (2014) in which it seeks discretionary accruals from loan loss provisions of the bank. The results of this study indicated an increase in cost efficiency and profit efficiency resulting in a decrease in earnings management practices.

Some banks in ASEAN countries have already done it, and some of them will implement BASEL III, which requires banks to increase investment in liquid assets, so that they can minimize returns on banking capital and have an impact on earnings management practices.

Therefore, this research is expected to contribute to the determination of the policies and banking standards of each ASEAN country.

2 LITERATURE REVIEW

Several related studies that analyzed the relationship between earnings management and efficiency have been carried out. A research conducted by Shawtari, Saiti, Razak, and Ariff (2015) discussed the efficiency of discretionary loans/allowance for financial losses between conventional and Islamic banks in Yemen. The results of their research showed that both banks used the discretionary provision of loan loss as earnings management, where efficient earnings had a positive effect on the discretionary provision of loan losses.

Research conducted by Hamid, Asid, Sulaiman, Sulaiman, and Bahri (2018) examined the effect of earnings management on cost efficiency, where the results showed that earnings management proxied by loan loss reserves negatively affected cost efficiency while the loan loss provision did not significantly influence cost efficiency. Their results showed that the loan loss reserves had more discretionary management in the preparation of financial statements.

Based on the results of the above research, it shows that efficiency has a positive effect on discretionary loan loss provisions, but not with the opposite effect indicating that the increase in costs results in a decrease in profits, resulting in a low return on equity compared to the cost of equity. Based on this, this study used cost efficiency and profit efficiency to see the consistency of results that both costs and profits provide the same direction of influence on earnings management practices.

3 METHODS

3.1 Sample

The sample used in this study was 58 banks registered in the stock exchanges in each ASEAN country. The purposive sampling method was used, which was selected based on the criteria of banks that publish information in the financial statements used in this study from 2012 to 2017.

3.2 Operational definition

The operational variables and definitions used in the study are presented in the Table 1.

3.3 Analysis

In this study, the search for bank efficiency values consisting of cost efficiency and profit efficiency used

Table 1. Operational definition.

Variable	Acronym	Definition
Earnings Management Disc	LLP	Calculate from equation 2 also 3
Cost Efficiency	Eff_Cot	Calculate from equation 1
Profit Efficiency	Eff Prof	Calculate from equation 1
Bank Size	Size	Natural logarithm of total assets
Financial Leverage	Fin Lev	Total liabilities divided to total equity
Liquidity Risk	Liq Risk	Total loans and total deposits
Credit Risk	Cr Risk	Ratio loans loss reserves to gross loans

the stochastic frontier analysis model adopted from Fu et al. (2014) as follows:

$$
\begin{aligned}
Ln\frac{TC_{it}}{W_{2it}} = {} & \propto_0 + \sum\nolimits_{p=1}^{3} \ln y_{pit} + \sum\nolimits_{m=1}^{2} \delta_m \ln \frac{w_{mit}}{w_{2it}} \\
& + \ln y_{pit} + \frac{1}{2}\sum\nolimits_{p=1}^{3}\sum\nolimits_{p=1}^{3} B_{pq} \ln y_{pit} \ln y_{qit} \\
& + \sum\nolimits_{p=1}^{2}\sum\nolimits_{m=1}^{2} \delta_{mn} \ln \frac{w_{mit}}{w_{2it}} \ln \frac{w_{mit}}{w_{2it}} + n_{1t} \\
& + \frac{1}{2}\eta_{11} + \sum\nolimits_{m=1}^{2} \lambda_{mt} + \ln \frac{w_{mit}}{w_{2it}} \\
& + \sum\nolimits_{p=1}^{3} \zeta_p \ln y_{pit} + \theta_1 \ln E + \mu_{it} + V_{it} \text{ (1)}
\end{aligned}
$$

where TC = total cost; Y1 = total net loans; Y2 = other earning assets; Y3 = noninterest income; W1 = price of purchased funds; W2 = price of physical capital; and W3 = noninterest expense

The difference in finding cost efficiency with profit efficiency is that the efficient dependent variable of profit is TC divided by W2, while profit efficiency is profit divided by W2.

Furthermore, in seeking earnings management, the equations adopted from the research of Cohen et al. (2014) were used as follows:

$$
\begin{aligned}
LLP_{it} = {} & \alpha_{tr} + \beta_1 \mathrm{LnAsset}_{it} + \beta_2 \mathrm{NPL}_{it} + \beta_3 \mathrm{LLR}_{it} \\
& + \beta_4 \mathrm{Loan\,A}_{it} + \beta_5 \mathrm{Loan\,B}_{it} \\
& + \beta_6 \mathrm{Loan\,C}_{it} + \varepsilon_{it} \text{ (2)}
\end{aligned}
$$

where LLP = loan loss provision as a fraction of total loans; $LnAsset$ = the natural log of total asset; NPL = nonperforming loan as a percentage of total loans; LLR = loan loss reserve as a fraction of total loans; $LoanA$ = corporate loan as a fraction of total loans; $LoanB$ = middle commercial loan as a fraction of total loans; and $LoanC$ = consumer loan as a fraction of total loans.

Table 2. Descriptive statistic.

Variable	Descriptive Analysis				
	Min	Max	Mean	Med	St. Dev
Earning	0.06	1.15	0.87	0.90	0.14
E_Biaya	0.48	0.95	0.72	0.72	0.06
E_Laba	0.30	0.66	0.50	0.50	0.05
Cr_Risk	0.05	49.39	2.84	2.17	3.25
Bk_Size	7.84	20.8	12.44	12.35	2.37
Fin_Lev	2.88	18.20	8.22	7.91	2.46
Liq_Risk	0.53	155.7	84.77	86.54	19.26

Table 3. Hausman test.

Variable	Chi-Sq. Statistic	Chi-Sq. d f	Prob.
C	5.1873	6	0.5200

4 RESULTS AND DISCUSSION

This study used data in the form of annual reports sourced from Bloomberg in the period of 2012 to 2017. In seeking cost efficiency and profit, stochastic frontier analysis was used based on the models from Fu et al. (2014), where STATA 15 was used to process the data. After that, the profit management was calculated by using the discretionary loan loss provision variable from Cohen et al. (2014). Eviews 10 was used in processing discretionary loan loss provision data and analyzing regression with panel data models to see the effect of bank efficiency on earnings management. Descriptive statistics from the results of research data processing as many as 58 banks in ASEAN from the period 2012 to 2017 is shown as follows Table 2.

Before regressing the panel data model, a Hausman test was needed, where the test results showed the use of random effect models in the regression analysis performed in Table 3.

The regression results indicate that the two bank efficiencies consisting of cost efficiency and profit efficiency have a negative and significant effect on earnings management, which is proxied by a discretionary loan loss provision. This result is different from the previous results from Shawtari et al. (2015), because in this study discretionary variables from loan loss provisions from conventional banks were used to explain the practice of accrual earnings management in the banking industry (Table 4).

The comparison between these research results with the findings from Hamid, Asid, Sulaiman, Sulaiman, and Bahri (2018) is that loan loss reserve has a negative effect on cost efficiency, but loan loss provisions have no significant effect on cost efficiency. Thus, there is a relationship between the loan loss reserve and loan loss provision through cost efficiency where an increase in

Table 4. The regression test results influence the independent variables on earnings management.

Variable	Coefficient	St. Error	t-Statistic	Prob.
C	5.1607	0.4753	10.8567	0.0000
Eff_Cost	−3.0613	0.3357	−9.1175	0.0000
Eff_Prof	−4.4785	0.4526	−9.8946	0.0000
Size	−0.0069	0.0040	−1.6964	0.0907
Fin_Lev	−0.0004	0.0016	−0.2456	0.8054
Liq_Risk	0.0035	0.0003	9.9489	0.0000
Cr_Risk	−0.0068	0.0011	−6.0532	0.0000
No. of Obs	348			
No. of Banks	58			
Adj. R^2	0,6629			
F Value	114,7280			

loan loss reserve motivates banks to act inadvertently in managing inputs to produce optimal output. This is due to the addition of reserves that increase capital results in risky actions. On the other hand, increasing cost efficiency will reduce earnings management practices because banks have obtained optimal output from minimum inputs, so banks do not need to hide their performance conditions. Seeing the influence of controlling variables, only liquidity risk and credit risk have a significant effect on earnings management, where liquidity risk has a negative effect, while credit risk has a positive effect. These results indicate that an increase in new loans decreases the ratio of non-performing loans, which gives a better performance condition. While credit risk increases due to congestion that provides poor performance conditions and impacts on profits, so the practice of earnings management is used to stabilize earnings movements from one period to another.

5 CONCLUSION

This study showed the cost and profit efficiency in motivating earnings management practices because the output is not achieved optimally by using existing sources (inputs). Poor performance causes bank management to practice earnings management, which has a function as a tool to stabilize income each period. This behavior is reflected in the negative relationship between cost efficiency and profit on earnings management, so that efficiency improvement does not need to be done by earnings management. This is because management does not need to cover up the excellent performance conditions. The findings in this study showed that earnings management behavior served as an intermediary for the reciprocal relationship between loan loss reserve and loan loss provision. It is suggested for further research to use vector autoregression analysis to see a more explicit relationship between earnings management, loan loss provision, and loan loss reserve.

ACKNOWLEDGMENT

This research was funded by Universitas Sumatera Utara through the TALENTA program with contract number: 112/UN5.2.3.1/PPM/KP-TALENTA USU/2019.

REFERENCES

AB-Hamid, M.F., Asid, R., Sulaiman, N.F.C., Sulaiman, W.F.W. & Bahri, E.N.A. 2018. The effect of earnings management on bank efficiency. *Asian Journal of Accounting and Governance*, 10: 73–82.

Cohen, L.J., Cornett, M.M., Marcus, A.J & Tehranian, A. 2014. Bank earnings management and tail risk during the financial crisis. *Journal of Money, Credit and Banking,* 46(1): 171–197.

Fu, X.C., Lin, Y.R., & Molyneux, P. 2014. Bank efficiency and shareholder value in Asia Pacific. *Journal of International Financial Markets, Institutions & Money,* 33: 200–222.

Gloria A. & Irene Rini, D.P. 2015. Faktor-faktor yang mempengaruhi efisiensi bank di *I*ndonesia periode tahun 2008–2012. *Diponegoro Journal of Management*, 6: 1–14.

Guillen, J., Rengifo, E. W., & Ozsoz, E. 2014. Relative power and efficiency as a main determinant of banks' profitability in Latin America. *Borsa Istanbul Review*, 14(2): 119–125.

Healy, P. M., & Wahlen, J. M. 1999. A review of the earnings management literature and its implications for standard setting. *Accounting Horizons,* 13(4): 365–383.

Naffati, A., Ben, F., & Schalck, C. 2011. Earnings management and banking performance: A stochastic-frontier analysis on U.S. bank mergers. *Interdisciplinary Journal of Research in Business*, 1(6): 58–65

Shawtari, F.A.M., Saiti, B., Razak, S.H.A., & Ariff, M. 2015. The impact of efficiency on discretionary loans/finance loss provision: A comparative study of Islamic and conventional banks. *Borsa Istanbul Review*, 15(2): 272–282.

Siudek, T. 2008. Theoretical foundations of bank efficiency and empirical evidence. *Social Research*, 13(3): 150–158.

Wu, Y.-C., Ting, I.W.K., Lu, W. M., Nourani, M., & Kweh, Q. L. 2016. The impact of earnings management on the performance of ASEAN banks. *Economic Modelling*, 53: 156–16

Advances in Business, Management and Entrepreneurship – Hurriyati et al. (Eds)
© *2021 Taylor & Francis Group, London, ISBN 978-0-367-67471-7*

A study of target date fund as an investment instrument for the voluntary pension fund in Indonesia

A. Gunawijaya & A. Suwondo
Universitas Indonesia, Jakarta, Indonesia

ABSTRACT: Pension program participants need an effective pension plan that can give them adequate income replacement to maintain a similar standard of living in retirement. However, the Financial Services Authority report from 2015–2018 reveals that the voluntary pension program participants in Indonesia tend to be risk-averse. They prefer lower risks and lower returns; as shown in the total portfolio, only 3.9% is invested in a riskier instrument with potentially higher returns like stocks while the rest investments are in the money market and fixed income. In countries where the pension industry is more advanced, like the United States, UK, and Canada, pension managers offer target date funds (TDF) for participants who seek to grow assets over a specified period. At the time this study was done, there was no TDF in the Indonesia pension market nor research in Indonesia that focused on them. The objective of this study is to find out whether TDF improves the pension investment returns, TDF response to volatility in the capital market, and TDF cost efficiency. The methodology of this study was a documentary analysis and scenario observation of what the results would be if TDF applies Indonesian capital market historical data. Therefore, the conclusions are rather indicative than definitive. The scenario considered two hypothetical glide paths taken from the U.S. TDF universe, the maximum and the minimum allocation of stocks in the portfolio. The data used Jakarta stock exchange composite index to represent the growth objective in TDF and time deposit to serve the stability objective. The result indicated all of the TDF approaches during the accumulation period had higher average end balances with no worst-case end balances.

1 INTRODUCTION

1.1 *Voluntary pension program in Indonesia*

The social security program in Indonesia, known as *Sistem Jaminan Sosial Nasional* (SJSN), has two mandatory plans that impact the income replacement ratio in retirement: Old Age Security, known as *Jaminan Hari Tua* (JHT), and Pension Plan, known as *Jaminan Pensiun* (JP). The monthly contribution for JHT is 5.7% of wages, while the monthly contribution for JP is 3% of wages. Both in total give income replacement ratio in the range of 35% to 40%, per the official of the BPJS Ketenagakerjaan (BPJS TK) in the 2016 Employee Benefits Forum. Meanwhile, the 2013 HSBC Global Report, based on a survey, said that the desired household income replacement level that people need to feel comfortable in retirement is 78%. To facilitate a higher replacement ratio, employees can voluntarily join a pension program managed by the pension managers. Based on Law number 11 in the year 1992 (UU No. 11/1992), the pension managers that can manage the pension plan for voluntary participants are the Financial Institution Pension Fund, known as Dana Pensiun.

Lembaga Keuangan (DPLK) can only be established by banks or life insurance companies. As of December 2018, there are 24 DPLKs in Indonesia that manage a total of 82.78 trillion rupiahs pension funds from 3,055,617 participants. The Financial Services Authority (*Otoritas Jasa Keuangan*) reports from 2015 to 2018 show that the annual returns on investment of the voluntary pension funds managed by DPLK are within the range of 5.8% to 6.3%. The reports also reveal that most of the participants tend to be risk-averse. They prefer to avoid loss over making a gain—this character of participants usually attaches to investing their pension funds with lower returns. As shown in the December 2018 portfolio, money markets took 61.1%; bonds, notes, and asset-backed securities took 34.25%; riskier risks with potentially higher returns like stocks took only 3.75%; the rests were in other instruments.

1.2 *Voluntary pension program in Indonesia*

The pension funds need to grow, while at the same time, they have to be safe. This statement is the background of the study, with the attachment of risk-averse characteristic, pension program participants in Indonesia need more effective pension plans than those existing available in the pension market. The regulation issued by the Financial Services Authority, POJK No. 3/POJK.05/2015, allows 17 types of investment instrument choices. However, pension managers do not seem to develop many creative and competitive products that can cause pension participants' funds to grow larger. In countries where the pension industry is more advanced, like the United State, UK, and Canada, pension managers offer target date funds (TDF) to participants who seek to grow assets over a specified period. The program has become increasingly popular in recent years. At the time of the crisis of 2008, some of the pension program participants who panicked during the crash disposed of the proportion of stocks in their portfolios, thus losing an opportunity when the market rebounded in 2009. By contrast, after the market weakens, TDF kept buying more shares to maintain the asset allocation as planned. A study by Aon Hewitt and Financial Engines reveals that participants of U.S. pension funds, 401(k), from 2006 to 2012 who used TDF earned higher median annual returns than those who went it alone.

1.3 Problem statement

The preceding narrative raises a question of why there is no DPLK in Indonesia offering TDF to the pension program participants. There are three concerns that can be summarized from casual discussions with relevant stakeholders in the pension industry: does TDF improve the pension investment returns; how is TDF response to volatility in the capital market; and is TDF cost efficient. This study used a literature review and observation with historical data as the basis. Therefore, the conclusions are rather indicative than definitive. It is also expected that the result of this study will trigger the next researches by the academician.

2 LITERATURE REVIEW

2.1 Target date fund

The concept of TDF is that pension program participants, based on their retirement target date, can comfortably select a managed fund with various asset allocations and the pension managers will manage the funds for them throughout a lifetime. The pension funds are grouped into two kinds of funds: the funds which are in a type of investment with growth objective allocation, and the funds which are in a type of investment with stability objective allocation. The adjustment allocation between the two groups within the TDF is automatic over time. The term used for the allocation adjustment over time is the glide path.

One of the basic principles of managing the pension funds investment is that the risk that the pension program participants take when placing the money to work should be appropriately rewarding to them. The idea of TDF is the automatic balance over time from riskier instruments to less risk instrument, so that young pension program participants can enjoy the higher potential return by investing in riskier instruments. The "time," since pension investment takes a long investment period, becomes the protection against risk—and that is what young participants have.

2.2 Glide path

The glide path is the heart of TDF; all TDF has a glide path. The glide path is the path of asset allocation that the TDF follows from a higher risk of investment to become more conservative over time. It represents the balanced mix of the investment funds, including stocks, bonds, and cash equivalents.

In the glide path, the allocation of equities is reflected as a declining percentage in the portfolio as it approaches and can also pass the target retirement date. Different pension fund managers provide different recipes in constructing a glide path. They design their TDF products based on their assumptions and calculations. However, all of those glide paths have the same pattern, which starts with more exposure to equities for younger pension program participants, and over time it balances into more exposure to fixed income and cash for pension program participants near the target retirement date. There are also glide paths that continue to adjust the equity exposure in the portfolio downwards after the participant comes to target retirement date. There are two kinds of pension managers. The first is the pension manager who is passive in managing the portfolio, and the second is the manager who actively manages asset allocations along the glide path within preset limits to respond to volatility in market conditions.

3 METHODS

3.1 Literature study

There are limited researches focusing on TDF. Many resources about TDF are provided by the pension managers as their efforts in educating people on TDF that can be accessed through the Internet. Therefore, besides available papers or research, many Internet materials were used in this study because of its limitless networking of resources. However, since relevance and credibility have become a concern, sources from the Internet were evaluated by selecting materials written only by professionals in TDF and published by the pension managers or professional organizational sites. Whenever there was uncertainty about the credibility of the sources, reference checks were used and advice was asked from experts.

3.2 Observation

An observation with a set of scenarios was run to respond to the concerns mentioned in the problem statement. The observation focused on the glide path was the object of the study. The origin of most literature used in this study was the United States. Thus, the U.S. TDF universe was used as the basis of scenarios reference to build hypothetical glide paths for the

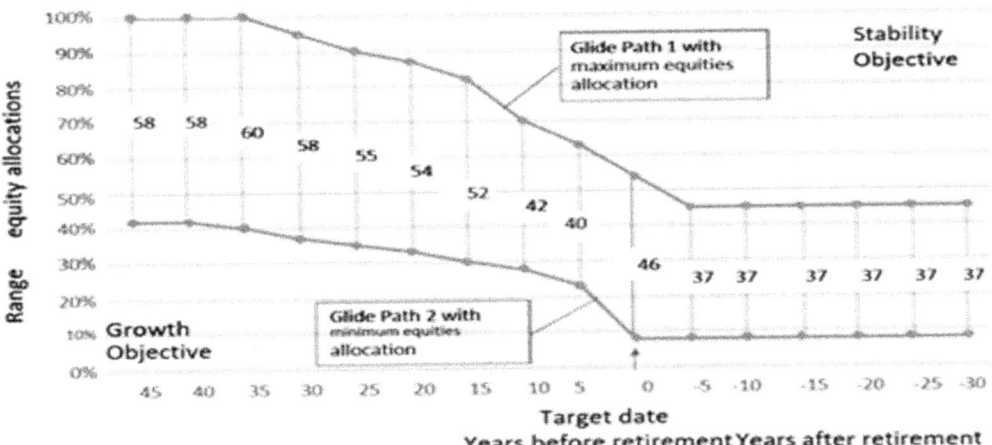

Figure 1. Hypothetical glide path 1 and glide path 2.

observation. The first glide path takes the maximum exposure to equity from the universe, and the second takes the minimum exposure. Figure 1 shows the first and second hypothetical glide paths for the observation.

The third glide path represented participants who were uncomfortable choosing between the riskier instrument and the low-risk instrument and thus decided to maintain a steady course at 50/50, for life. The last glide path represents participants who were cynical of the standard declining risk assets strategies; the path takes the inverse of the first glide path, the one that has a maximum exposure of equity. The logic behind this cynical view is that a pension participant can play more in riskier investment instruments when the assets already grow large. Therefore, when there is a market downturn, the participant will still have assets to invest and expect a market bounce (Figure 1).

4 RESULTS AND DISCUSSIONS

4.1 Scenario 1

Figure 2 shows the result of scenario 1. The result indicates that 100% stocks gives the highest return, and the performance of all glide paths outperforms the return of the 100% time-deposit TDFs.

Figure 3 shows a snapshot of the accumulated funds during the market downturn between 2007 and 2009. The displayed result shows only the portfolio position at the end of the captioned year. In December 2008, the 100% time-deposit outperforms the returns of all of the glide paths with passive manager. There are two glide paths with active manager though that outperform the return of the 100% time-deposit during the downturn period: glide path 2 and glide path 3. The 100% stocks gives the lowest return in 2008. However, after the market bounced back in 2009, the pattern of the returns of TDF formed back to a similar pattern as the returns

in 2007: The 100% stocks gives the highest return, except for glide path 2 with active managers; active managers give better performance than the passive; and the returns of all TDF outperform the 100% time-deposit. An interesting point to highlight is that after the market bounced back in 2009, the 100% stocks had not gotten back to its position in 2007, but TDFs in 2009 showed that they were in better position than their position in 2007, which means not only were they able to recover but also that they continued growing.

4.2 Scenario 2

Figure 4 shows the result of scenario 2. It indicates that both rules of the simple volatility management can give better returns in the case of glide path 3. Market downturn check also shows that the volatility management gives better performance during the downturn. Overall, the 20/80 rule gives higher returns than the 20% rule, but the amount switched is also higher. It means the cost of the 20/80 rule is higher and at the end it impacts the final amount of accumulation.

4.3 Scenario 3

Figure 5 shows the result of scenario 3. This scenario is actually scenario 2 with fund switching fee applied. As seen in the figure, with 2% fee applied to every switching, the returns of glide path 3 with simple volatility management of the 20/80 rule which originally gives the highest return on every run, the net returns are then only close to the return of glide path 3 with active manager. By this case, the volatility management looks meaningless. The return of the 20% rule even gives lower return than the active manager when fund switching fee is applied. This result means volatility management may improve the return of TDF, but if the cost of it is too expensive, the cost may eliminate the benefit resulting from it.

29

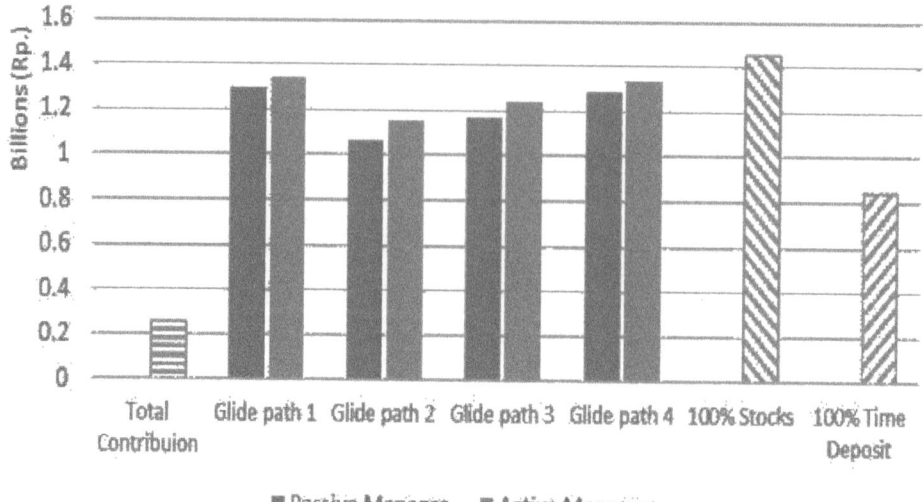

Figure 2.   Result of scenario 1.

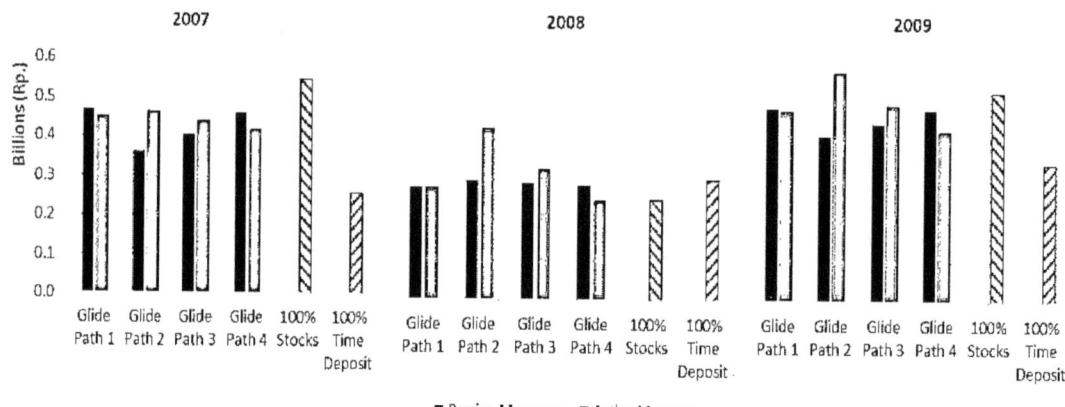

Figure 3.   Snapshot of the accumulated funds 2007–2009.

4.4 *Scenario 4*

Figure 6 shows the result of scenario 4, which is a 10-year decumulation period. The results of the observation show that participants who retired in 1998 have a bigger ending balance after 10-year decumulation than the one who retired in 2008. It is because in the beginning of that period, Indonesia enjoyed the double-digit interest rate. Thus, if the total expenses are lower than the interest rate, it will still let the funds keep growing. That is why with the opening balance of Rp1 billion, after 10-year decumulation of Rp10 million monthlies, the ending balance is bigger than the beginning even without new contributions.

As for the participant who retired in 2008, there was no more double-digit interest rate. Then, it was more challenging for the participant to make sure the funds in the portfolio kept growing while at the same time there was a monthly withdrawal. The analysis of this study then focused on the participant who retired in 2008 because the capital market situation was more relevant to the situation at the time this study was done. In the scenario of decumulation, the results indicate that active managers do not give better return than the passive managers. In the case of glide path 2, the ending balance becomes zero even before completing the 10-year period. It is because the active manager kept the equity portion in the portfolio low following the glide path. So, after being hit in the first half of 2008, the participant did not enjoy the strong bounce back. The same observation with the passive manager gives a different result. The manager did not switch funds, so the participants got a chance to enjoy the strong bounce back after the 2008 downturn. The return is even bigger than the 100% time-deposit.

Glide Path 3

Figure 4. Result of scenario 2.

Glide Path 3

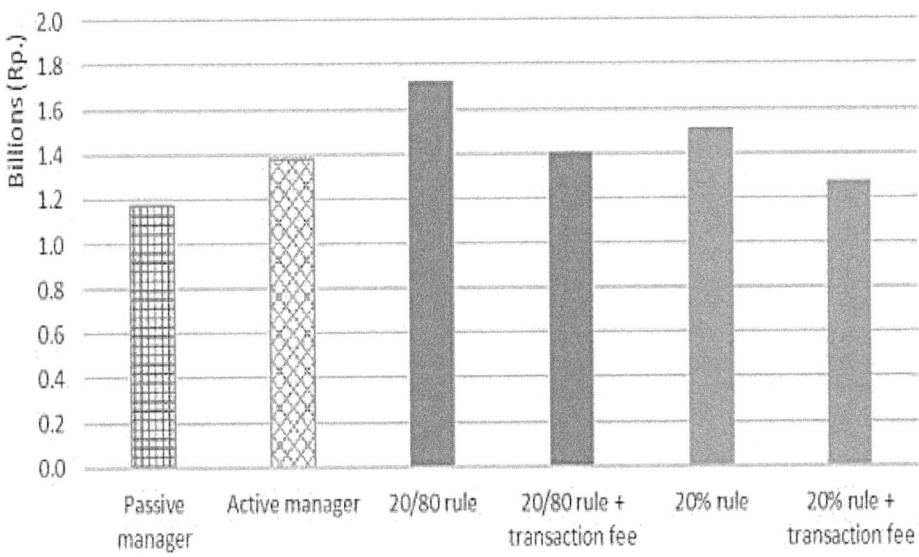

Figure 5. Result of scenario 3.

The result from glide path 1 also gives better returns than the 100% time-deposit. It is because glide path 1 maintains a bigger portion of stocks in the beginning of the decumulation period. However, the return of glide path 1 is below glide path 2 with passive manager. It is because glide path 1 declines its equity portion when it is actually a good time to invest more in stocks.

5 CONCLUSION

5.1 Conclusion

a. The results indicated that TDF gave better pension fund investment returns compared to 100% time deposits in all accumulation scenarios. Since most of the current pension program participants in

31

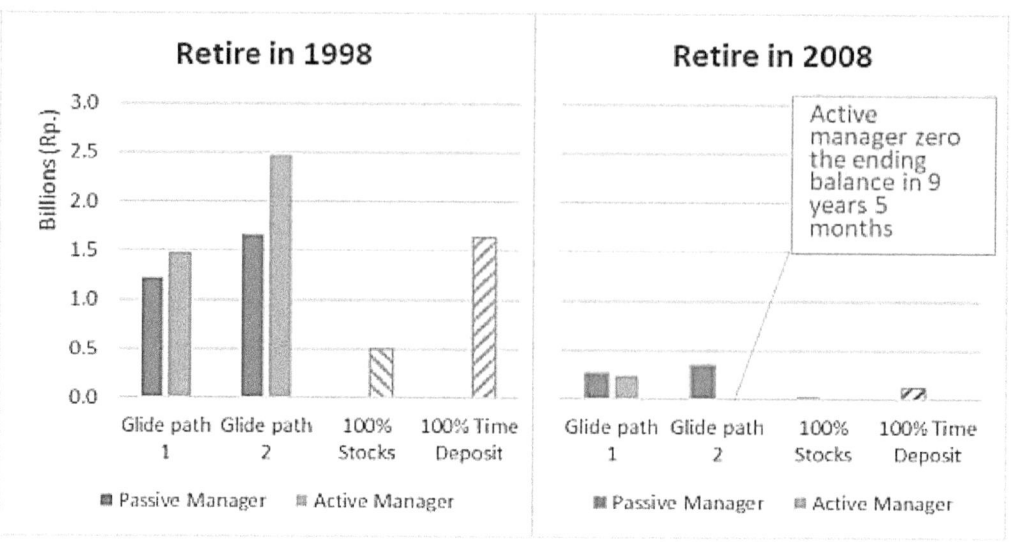

Figure 6.   Results of scenario 4.

Indonesia invest in money markets such as time deposits, TDF would be able to provide better returns in the accumulated funds by gaining more from higher risk investment instruments like stocks during the early stage. There was a special case though, when during the market downturn, all glide paths with passive managers were below the performance of the 100% time-deposit. Except for the glide paths with active managers, there was one glide path that outperformed the 100% time-deposit. This result leads into a suggestion for the next research to study the impact of the TDF returns of the timing of market downturn to the date of retirement. In this study, the market downturn happened in the middle, so we can see that pension program participants by investing in TDF could enjoy the market bounce back and at the end got better results than the 100% time-deposit. The results of the observation also showed that glide paths were not optimized for universally maximizing investment returns. It can be seen from the glide paths that gave different patterns of results in different scenarios. It means pension participants should avoid jumping to conclusions in judging which glide path is good or bad without putting it in a proper context. It is probably because each glide path was designed with an intention of participants segmentation in the United States and based on the historical behavior of funds. This study assumed the U.S. glide paths as universal pension plans by taking only the maximum path and the minimum path from the aggregate of all TDFs. However, for the accumulation period, all of the TDF approaches provided higher average end balances with no worst-case end balances.

b. A volatility management add-on can be installed as a shock absorber to make TDF give a better approach. However, volatility management would add extra costs to the funds. When the active managers already give many funds switching, and consequently higher cost than passive managers, a volatility management add-on could even add more. It is a suggestion for the next research to study if this add-on is worth the benefits.

c. TDF can make investing easier just like mutual funds, and TDF also gives a plus point, the automatic assets allocation decisions. The caveats are that TDF can be expensive; the observation result confirmed it. This study also gives an insight that a passive manager suits the investment instruments that are historically stable. It means the design of the glide path should be able to give the returns that are acceptable and competitive. The pension managers can also use this approach as a target costs control. As for the active managers, it suits the pension participants who have a more significant risk appetite, who believe the opportunity is greatest. Pension program participants and pension managers need to consider, though, the results which are expected to be great should be able to cover costs of funds at the end, especially the switching fees and the administration fees.

5.2   Suggestion

5.2.1   For academician

a. To conduct research about the impact of timing of the market downturn within the TDF life cycle to the investment returns, if the downturn happens in the early stage, middle stage, or near retirement.

b. To conduct research about benchmarking the TDF
c. with the social security system provided by BPJS TK. To conduct research on the ideal income replacement ratio for Indonesian.

d. To conduct research on the actual costs of funds in Indonesia.
e. To conduct research using the Monte Carlo simulations to give some indication of what the future might look like, especially if the correlation of risk persists and the growth, and their particular trends.

5.2.2 For government and regulators
a. To increase pension and investment literacy for Indonesian.
b. To revise the UU No. 11/1992; it is too old and may be obsolete. The regulation needs to be modernized, so the industry can be advancing following the participants' expectations. For example, to allow asset management companies or independent financial company to manage pension funds.

REFERENCES

AON. Media releases. Retrieved from http://aon.mediaroom.com/news-releases?item=136959.

Bank Indonesia: https://www.bi.go.id/id/statis-tik/seki/terkini/moneter/contents/default.aspx.

Blanchett, D. M., 2015. Revisiting the optimal distribution glide path. *Journal of Financial Planning*. February 2015.

BPJS Ketenagakerjaan. (2015, November 24. 07:44:05).

Bruder, B., Culerier, L., & Roncalli, T. 2012. How to design target-date funds? Lyxor Asset Management, Paris.

Cassidy, D. 2014. Reviewing Target Date Funds. Employee Benefit Review; April 2014.

Chan, C.-Y., Chen, H.-C. Chiang, Y. H., & Lai, C. W. 2017. Fund selection in target date funds. *North American Journal of Economics and Finance*, 197–209.

Gottlieb, J. 2008. Benchmarks sought for target-date funds. Pensions & Investments; Chicago Vol. 36, Iss. 2, (Jan 21, 2008): 6,63.

Hoffman, K. (2014, August 27). Target-date funds decoded. Retrieved from.

HSBC. (2013). The future of retirement a new reality. Global report. London, UK: HSBC Insurance Holdings Limited.

Johnson, W. F., Yi, H.-C. 2017. Do target date mutual funds meet their targets? J asset manag (2017) 18:566–579. Macmillan Publishers Ltd 2017.

Kalman, M. B. 2011. The role of the equity risk premium in the shortfall risk of target-date funds. Financial Services Re-View; winter 2011; 20, 4; ProQuest pg. 265.

Kentouris, C. 2010. *Target-date funds target customization.* Securities Industry News; new york Vol. 22, Iss. 6, (Mar 22, 2010): 16.

Martin, I., Wagner, C. 2018. What is the expected return on a stock? *London School of Economics.*

Masters, S. & Fontaine, T. J. 2010. Enhancing target-date funds: Using volatility management for a less volatile ride. *Journal of Pension Planning & Compliance*, 30–50.

Menata persiapan pensiun. Retrieved from: https://www.bpjsketenagakerjaan.go.id/berita/4909/bpjs-ketenagakerjaan-:-menata-persiapan-pensiun.

Oktaviano db hana. (2016, august 31. 21:12 wib). BPJS ketenagakerjaan: replacement ratio income hanya 35%-40%. Retrieved from https://finansial.bisnis.com/read/2016083 1/215/580150/bpjs-ketenagakerjaan-replacement-ratio-income-hanya-35-40.

Otoritas jasa keuangan (OJK). (2018, december). Statistik dana pensiun periode Desember 2018.

Pojk no. 3/pojk.05/2015 concerning pension funds investment.

Pp no. 45/2015 concerning jaminan pensiun.

Radu Constantin Gabudean. 2015. Evaluating target-date portfolios: a practical approach to building family-wide measures. *The Journal of Retirement fall* 15, 80–92.

Saar, H. 2012. Essays on target date funds as a retirement portfolio choice. A dissertation submitted to the graduate division of the University of Hawai'i at Mânoa.

Surz, R. Understanding the hidden risk in target date funds. A publication of paladinregistry.com.

Surz, R., Lohr, J., & Mensack, M., 2016. Fiduciary handbook for understanding and selecting target date funds: It's all about the beneficiaries.

Tang, N. & Lin, Y.-T. 2015. The efficiency of target-date funds. *Journal of Asset Management* 16, 131–148.

Uu no. 11/1992, pp no. 76/1992, and pp no. 77/1992 concerning dana pensiun lembaga keuangan.

Uu no. 40/2004 concerning sistem jaminan sosial nasional pp no. 46/2015 concerning jaminan hari tua.

Vernon, S. (2011, august 2018 2:37pm). Study: 401(k) investors who stayed the course in 2008–09 were big winners. Retrieved from https://www.cbsnews.com/news/study-401k-investors-who-stayed-the-course-in-2008-09-were-big-winners/.

Yahoo finance: https://finance.yahoo.com/quote/%5ejkse/.

Advances in Business, Management and Entrepreneurship – Hurriyati et al. (Eds)
© *2021 Taylor & Francis Group, London, ISBN 978-0-367-67471-7*

Financial performance of waqf institutions in Indonesia

M. Iskandar & N. Nugraha
Universitas Pendidikan Indonesia, Bandung, Indonesia

ABSTRACT: This study analyses the financial performance of waqf institutions in Indonesia related to the effectiveness of its financial models. It also aims to examine the application of good corporate governance (GCG) and intellectual capital (IC) from the manager of waqf institutions. This study used exploratory methods with a qualitative approach. It was conducted in seven waqf institutions in Indonesia. This study found triple-helix concept of the financial performance consisting of productive endowment financing models, the application of GCG, and IC. The concepts found can be used as a basis for the other researchers to carry out empirical research in the field.

1 INTRODUCTION

Waqf (endowments) is an instrument of Islamic Social Finance (ISF) that has become one of the economy pillars since the past. The basic concept of waqf is voluntary, with the aim of getting closer to Allah and giving broad benefits to others (Masyhadi 2019; Sadeq 2002). Waqf can be performed by every Muslim. There is no special requirement for donated asset as long as it is beneficial. The use of donated asset can be adjusted to wakif (people who do waqf) desire. Regarding the process, the donated asset will be managed for the benefit of ummah (people) by nazhir (waqf manager). The net profit is then given to the mauquf 'alaih (person or place that receive waqf).

Waqf has been implemented from 8th century. At that time, Imam Zufar suggested endowment to be invested in businesses with profit sharing mechanisms (mudharabah). The aim is to gain a profit that can be shared to mauquf 'alaih (Cizakca 2002). Unfortunately, there is discussion of the waqh impact to the economy at that time.

Lately, waqf becomes a major concern by many parties. It is a main component of ISF that is inseparable from Islamic Commercial Finance (ICF) in Islamic Financial System (IFS) (Fadilah 2015). It has an important role in the social and economic life (Singer 2006). Through its basic concept, it can accumulate capital in a long period of time. This will improve the financial capabilities of individuals or management organizations that finally give positive contribution to society welfare. In addition, it can also increase holistic financial inclusion and maintain financial system stability.

The economic system of waqf differ from conventional secular economic system. Most of conventional secular economic system is dominated by commercial finance without involving social finance and considering environmental impacts. Conventional commercial finance only focuses on commercial objectives. This is not in line with the triple bottom-line (economic, social and environmental) goals of conventional social finance (Griffin 1994). This is also instable (Rothbard 2008) due to the use of "fractional reserve banking system". In contrast, waqf has social mission of reaching the poor and increasing society welfare. This becomes profitable commercial sustainable missio (Armendáriz et al 2013) in which commercialization (Hamada 2010) makes triple bottom-line in conventional social finance systems impossible to be achieved (Zeller & Meyer 2002).

Islamic commercial finance inherently demonstrates stability (Azis & Osada 2010) This is due to the prohibition of usury, maysir, and risk sharing. The aim is to launch productive investment. Zakat obligation and waqf change the economy and encourage commercial and social investment. In other words, it bridges commercial finance and social finance. Furthermore, Islamic commercial finance invites the public to be more productive in participating in the real sector. It makes partnership-based business activities that are ethical and well-managed. Thus, it can simultaneously achieve the triple bottom-line (Acarya & Guruh 2009).

The practice of combining commercial and social Islamic finance which is supported by the role of zakat institutions, waqf institutions, and Islamic banks has been carried out in various countries. Several countries (such as Egypt, Singapore, Malaysia, and Turkey) have developed waqf. They are able to solve their socio-economic problems. In Indonesia, the importance role of waqf has been proposed in Law No. 41 of 2004. On this law, waqf is viewed not only as ritual worship, but also as the aspect that can give positive contribution to the socio-economic life (Aisyah 2014). Moreover, in

2010, President has launched waqf money movement. The data gained from Indonesian Waqf Agency (BWI) showed that the movement could collect Rp185 billion money waqf and 4.3 billion m2 land waqf until the end of 2015. Regrettably, this number was under the estimated calculation, which was Rp.7.2 trillion per year (Government 2013).

The potential for waqf and zakat fund in Indonesia is basically great. For zakat fund, it is estimated more than Rp.200 trillion. The zakat fund that was distributed to the poor people and to the program was around 4.7 trillion/year (Budiman 2011). Meanwhile, waqf sector has very large assets to manage. In this case, waqf assets in the form of land, schools, mosques, hospitals are estimated to be around Rp 600 trillion (potential waqf money is more than Rp. 60 trillion). This is very promising prospects for further development. In fact, it has not been able to play an effective role in empowering the socio-economic sector (Putra 2009).

There are various problems causing ineffectiveness of waqf. Some of them are: low level of community literacy regarding waqf; public trust in waqf management institutions; habit and lifestyle; the absence of national waqf databases, including data on waqf assets, wakif, etc ; no incentive for wakif; the dualism in authority; the dualism in Indonesian Waqf Agency: as regulators and as operators; position of the Indonesian Waqf Agency as an operator and private Waqf Management Institution; Nazhir has not become a career choice; and weak professionalism of Waqf Management Institutions. Furthermore, there are also problems in financial performance and the model of waqf institutions, the application of good corporate governance, and intellectual capital from HR management of waqf (nazhir). These four problems are interrelated one and another. Good financial performance of waqf institution requires effective planning, implementation and evaluation of financial model. It also needs good corporate governance. Meanwhile, intellectual capital of the waqf manager (Nazhir) has not fully supported the implementation of productive activities in waqf institutions.

Based on the above explanation, this study aims to analyse the financial performance and the financial models used by seven waqf institutions in Indonesia. It also analyses the implementation of good corporate governance principles by using seven criteria such as accountability, responsibility, equitable treatment, transparency, vision to create long-term values, and ethics. Moreover, it also aims to investigate intellectual capital capability of managers in waqf institutions.

Financial performance refers to the act of conducting financial activities. In a broader sense, financial performance refers to the extent to which the financial goals will be, are or have been achieved. This is the process of measuring the results of a company's policies and operations related to monetary or financial aspects (Shafii et al. 2014). As one of the non-profit financial institutions, financial performance of waqf institutions differs from the financial performance of business institutions, in which they have different objectives.

Financial health is crucial to the existence and the operation of any organization. It is even more essential in the case of waqf because productive donated assets are left idle due to insufficient revenue to sustain operational costs (Chowdhury et al. 2011). Financial ratios are very important for evaluating the financial condition of an organization, but those ratios may not be the most important measure of success in non-profit institutions. These institutions must be assessed in a more complex way, which generally deals with aspects of efficiency and effectiveness (Abraham 2003; Keating et al. 2005; Mensah et al. 2008; Mihaiu & Opreana 2010; Sulaiman et al. 2009; Tuckman & Chang 1991). Thus, waqf institutions cannot be evaluated based on financial figures only. For example, high current assets dominated by cash and cash equivalents indicate that the institution may not mobilize its assets.

In nonprofit organizations, there are three main characteristics that are often applied in organizational performance, namely the ability to obtain organizational resources, achieving organizational goals, and organizational efficiency (Hughes 2013). Nonprofits must be able to find proxies to measure efficiency (Adnan et al. 2007). For example, suggests that higher program expenditure ratios show higher cost efficiency. The higher program expenditure ratio to total expenditure indicates that non-profit organizations can run the program with less total expenditure absorbed by management. In short, this is more efficient than one that has a lower program expenditure ratio.

It is (Dewi & Ferdian 2012) who proposed a waqf institution measurement model using six ratios, namely: efficiency ratio program, operating expense efficiency ratio, margin of rental activities ratio, return on investment, fundraising efficiency ratio, and distribution efficiency ratio. As a non-profit institution and institution that might get full support from the government, waqf institutions need to focus on efficiency. In addition, there are four categories of ratios that can represent the efficiency and sustainability of nonprofit institutions through various program implementations, (Epstein et al. 2015) namely: administrative efficiency, program efficiency, fundraising efficiency, and other financial performance measures. Regarding aspects of financial performance, the financial indicators are used to measure efficiency and effectiveness, namely performance efficiency ratio and operating efficiency ratio (Atan et al. 2013).

Waqf institutions in Indonesia strive to improve their financial performance through various productive waqf financing models, ranging from planning, implementation, and evaluation. The purpose of productive endowments here is to improve economic welfare, social welfare, environmental welfare, spiritual welfare, and financial stability. Various criteria for being called productive waqf involve various aspects. First, the waqf management institution can consist of individuals, organizations, social legal entities, cooperative legal entities, and commercial legal entities. Second, waqf assets can be in the form of small business facilities, residential buildings, office building, large commercial centers, and mixed social & community centers. Third, the financing can be done through cash waqf, cash waqf and co-finance, Islamic banks, international institutions, and sukuk. Fourth, management can be done by itself, subsidiary, internal partnership, temporary extern partnership, and permanent extern partnership. Fifth, the expected benefits are related to profitability, increased waqf assets, wakif reach, range of mauqf 'alaih, and increased reputation of Nazhir. Finally, sixth, compliance is related to sharia provisions, legal regulation of waqf management institutions, waqf regulations, land regulations, and regulatory differences. At least, there are two main models of productive waqf financing, namely: (1) simple models and (2) innovative models (Ascarya 2016).

In a modern economy, waqf has taken a new position, both as a product and as a legal entity, especially in the world of Islamic finance. As a result, this scenario automatically demands that institutions be regulated fairly and monitored closely (Abdullah 2015). In this case, corporate governance can be developed for the institution based on current waqf rules and guidelines. Various changes occur in the methods, mechanisms and administration and management of waqf, so that the exploitation of the structure of the waqf institution and its governance is increasingly needed.

Corporate governance, in general, is defined as the principal's rights, roles, and responsibilities of an corporation in line with the provisions underlying the relationship between "a company's management, its board, its shareholders and other stakeholders" (OECD 2004) In this case, a good corporate governance framework can increase the credibility of a particular institution/company and increase the trust of its stakeholders. The appropriate governance framework for waqf institutions will not only contribute to bringing greater accountability and transparency in operations, but will also play an important role in improving the stability of the institution itself (Finnalcial reporting council 2012). In order for waqf institutions to develop consistently and sustainably, there is a need for integration in one system related to the aspects of efficiency, accountability, transparency, monitoring and control mechanisms. For this reason, the need for an appropriate governance framework for waqf institutions cannot be ignored.

The transformation of waqf management into the structure of modern institutions / corporations / companies is important (Marwah & Bolz 2013). This change is in line with the characteristics of waqf related to longer durability aspects. Special characteristics of waqf itself can be strengthened by a strong mechanism of good corporate governance (GCG). It is confirmed that there are similarity between the principles of Western world governance and sharia governance, especially related to fairness, transparency and accountability aspects (Foster & Garduno 2013).

Greater transparency, accountability, and effective management of waqf institutions can be achieved by following the codes and ethical guidelines of corporate governance (Abdullah 2015). In addition, to include the spirit of sharia in the existing structure of governance, certain principles and certain elements related to maqasid al-shari'ah (the highest goal of sharia) can be applied in the waqf institution. This will be realized through the concept of 'adl (fair), sidq (honest), maslahat (benefit), rifq (paying attention to others), trust, accountability, and mu'awanah (cooperation) in the code of ethics (Ascarya 2016). The orientation of the maqasid al-shari'ah must be prioritized from the start.

The implementation of good governance and the application of productive waqf financing models will not work as expected if the waqf institution does not have a waqf manager (nazhir) who has adequate intellectual capital (IC). There is no consensus regarding the exact definition of IC, but basically the researchers agree that IC involves knowledge, experience, and intelligence of employees, systems, processes, procedures, and relationships (Makki 2010). IC is an integration of human capital (HC), structural capital (SC) and capital employed (CE) (Pulic 2004). Corporate governance plays an important role in creating, developing and enhancing ICs that exist in the people, structures and processes of the company. The linkages between good governance, intellectual capital, and performance have been carried out by researchers (Musibah & Alfattani 2014; Nathan 2010; Wahab & Rahman 2015).

Based on some of the literature reviews, the theoretical framework in this study can be presented as a triple-helix financial performance of waqf institutions, as shown in Figure 1.

2 METHODS

This study used exploratory methods with a qualitative approach. This method was used because this study aimed to analyse and describe the data and information according to actual needs. This research was conducted in seven waqf institutions in Indonesia, namely: (1) Daarut Tauhiid Waqf Development Center, (2) Waqf Dompet Waqf (Waqf Tube), (3) Waqf House (RWI), (4) ACT Global Waqf Institution, (5) ESQ 165 Waqf Institution, (6) Waqf Institutions Salman ITB, and (7) Al-Azhar Waqf Institution.

The participant of this study was waqf managers in investigated waqf institutions. This study also used document analysis consisting of official documents both in hard copy, soft copy, and online. The aim was to complete the needs of data and information as well as maintain the accuracy of data and information. To collect the data, three instruments were used. They are: (a) interview and observation records, (b) interview recording equipment, (c) documentation in the form of photographs and other written documents.

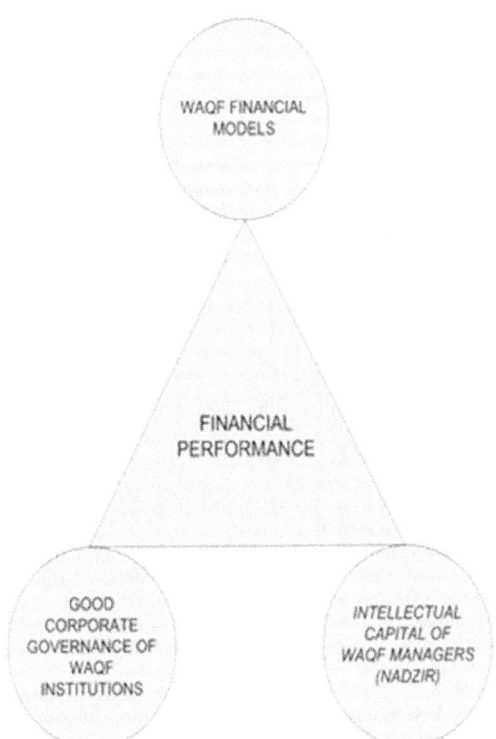

Figure 1. Theoretical framework of waqf institutions financial performance.

Data and information obtained were classified according to research questions and data sources. Thus, it became systematically grouped. Data analysis was done through three stages, namely: (1) data reduction, (2) data presentation, (3) conclusions and verification.

Regarding the focus, this study was limited to answer the following questions:

1. How is the effectiveness and efficiency of waqf institution's financial performance model?

    a. How is the effectiveness and efficiency of waqf acceptance?
    b. How is the effectiveness and efficiency of managing productive assets from waqf?
    c. What are the supporting element and the obstacles in achieving the effectiveness and efficiency of waqf?
    d. What are the supporting element and obstacles in achieving the effectiveness and efficiency of managing productive assets from waqf?

2. How is the effectiveness and efficiency of waqf financial models in waqf institutions?

    a. What are the financial models applied by waqf institution?
    b. How is the effectiveness, efficiency and potential of waqf financial models in waqf institution?

    c. Which parties are involved in planning, implementing, and evaluating waqf financial model?
    d. What are the supporting elements and obstacles in achieving the effectiveness and efficiency of waqf financial models?

3. How is the implementation of good corporate governance principles (accountability, responsibility, equitable treatment, transparency, vision to create long-term values, and ethics) in waqf institutions in Indonesia?

    a. To what extent is the implementation of good corporate governance principles in waqf institutions in Indonesia?
    b. What are the supporting elements and obstacles in the implementation of good corporate governance in waqf institutions in Indonesia?

4. How is the capacity and capability of intellectual capital of waqf manager?

    a. What are supporting elements and obstacles in increasing the capacity and capability of intellectual capital of waqf manager?
    b. To what extent to the capacity and capability of intellectual capital of the waqf manager can support the achievement of financial performance?

3 RESULT AND DISCUSSION

The result of the study answer the formulated research objective. Regarding financial performance, this research supports previous research conducted by Ihsan (2007) and Nahar and Yacoob (2011) which found the financial performance in waqf financial report and manager's ability in managing finances and waqf report. In this case, waqf financial performance need to be seen not only as financial performance, but also as operating performance.

This research also supports the findings of (Atan et al. 2013) which used performance efficiency ratios and operating efficiency ratios to measure performance. These two ratios are considered appropriate to measure the consistency of waqf institution. They provide an overview for the stakeholders related to the mission of agency in distributing the funds. The choice of appropriate ratio can be based on the availability of the fund and financial resource. Thus, there are nine ratios, such as return on assets ratio, viability ratio, primary reserve ratio, net income ratio, operating income ratio, contributed income ratio, debt burden ratio, debt coverage ratio and leverage ratio (Abraham 2006). These ratios can be measured by institutions from various perspectives However, the analysis of financial ratios needs to be supported by qualitative aspects as well as influential factor in financial performance, such as productive endowment models, good corporate governance, and intellectual capital.

Optimal waqf management system cannot be done without cooperation between waqf managers, supervisors, and all related parties. Here, optimal management can be created by developing, implementing, monitoring, and improving supervision tools and policies. Waqf managers and supervisors need to apply strong external controls and risk management to respond to the factors influencing optimal waqf management and supervision. At least, there are three prerequisites for optimal waqf management systems.

First is the framework for formulating waqf management policies. All parties involved and responsible for the overall implementation of the waqf management system must be identified in a framework for formulating waqf policies. This waqf framework is stipulated in endowments, laws, regulations, or other arrangements. The framework reflects the mechanisms for optimal waqf management systems.

Second is well-developed public infrastructure. There are four elements of public infrastructure that support optimal management and supervision of waqf: (i) comprehensive, appropriate, and standardized management and accounting; (ii) external audit systems and independent accountants; (iii) availability of manager who are competent and professional with Islamic ethical standards; and (iv) availability of regional, economic and social statistics.

Third is a clear framework for collecting, investing, managing and distributing waqf activities. A clear framework can optimize the function of waqf as an instrument that can be utilized for society's prosperity and welfare.

This research can be continued with a new focus on the waqf institution's financial performance model. Financial performance of waqf institutions can be said to be good if the activities of collecting, managing and utilizing waqf run well. Management activities, however, requires good governance and intellectual capital of waqf manager.

4 CONCLUSION

This research is a conceptual study which analyses the financial performance of waqf institutions, the effectiveness of waqf financing models, the application of good corporate governance (GCG) in waqf institutions, and intellectual capital of waqf manager. The results of the analysis indicate a linkage in the triple-helix financial performance of waqf institutions. The achievement of financial performance depends on the effectiveness of waqf financing models; implementation of GCG; and intellectual capital of manager. The findings of this research can be used as a basis for conducting empirical research on the same field.

ACKNOWLEDGEMENTS

The authors would like to thank to Management Department of Graduate School, Universitas Pendidikan Indonesia for the opportunity to publish this article.

REFERENCES

Abdullah, M. 2015. A new framework of corporate governance for Waqf: a preliminary proposal. *Islam and Civilisational Renewal* 274(2625):1–18.

Abraham, A. 2003. Financial Sustainability and Accountability: A Model for Non-Profit Organisations. In: AFAANZ 2003. *Conference Proceedings*.

Abraham, A. 2006. Financial Management in the Nonprofit Sectors: A Mission – Based Approach to Ratio Analysis in Membership Organizations. *The Journal of American Academy of Business* 9(2):212–217.

Acarya, D.Y & Guruh, S.R. 2009. Efficiency Analysis of Conventional Banking and Islamic Banking in Indonesia with Data Envelopment Analysis: Current Issues of Islamic Financial Institutions edited by Nurul Huda and Mustofa Edwin Nasution. Jakarta: *Prenada Media Group*.

Adnan, M.A., Maliah, S. & Nor, S.P.M.M.N. 2007 Some Thoughts about Accounting Conceptual Framework and Standards for Awqaf Institutions. *Indonesian Management & Accounting Research* 6(1):43–56.

Aisyah, M. 2014. The role of zakah and binary economics in poverty reduction. Esensi: *Jurnal Bisnis dan Manajemen* 84(2).

Armendáriz, B.D., Espallier, B., Hudon, M & Szafarz, A. 2013. Subsidy uncertainty and microfinance mission drift. *Available at SSRN.*

Ascarya. 2016. Designing Productive Waqf Models. *Ministry of Sharia Economics and Finance Indonesia Bank* 19–22.

Atan, R., Zainon, S., Aliman, S. & Nam, R.Y. 2013. Financial Management in Religious Non-Profit Organizations: A Mission Based Approach to Ratio Analysis. *International Conference on Advanced Computer Science and Electronics Information (ICACSEI).*

Azis, Y & Osada, H. 2010. An empirical study of new value creation in financial service companies using design for Six Sigma approach. International *Journal of Productivity and Quality Management* 27,7(1):104–124.

Budiman, A.A. 2011. Accountability of the waqf management agency. Walisongo: *Journal of Religious Social Research* 7–19 (1):75–102.

Chowdhury, M.S.R., Ghazali, M.F. & Ibrahim, M.F. 2011. Economics of cash waqf management in Malaysia: a proposed cash waqf model for practitioners and future researchers. African *Journal of Business Management* 5(30):12155–12163.

Cizakca, M.A. 2002. History of philanthropic foundations: The Islamic world from the seventh century to the present. Istanbul. *Boðaziçi University Press.*

Dewi, M.K. & Ferdian, I.R. 2012. Evaluating performance of Islamic mutual funds in Indonesia and Malaysia. *Journal of Applied Economics and Business Research* 2(1):11–33.

Epstein, M.J., Buhovac, A.R. & Yuthas, K. 2015. Managing social, environmental and financial performance simultaneously. *Long range planning* 48(1):35–45.

Fadilah, S. 2015. Going Concern: An Implementation in Waqf Institutions (Religious Charitable Endowment). *Procedia-Social and Behavioral Sciences* 25(63):211–356.

Financial Reporting Council. 2012 UK *Corporate Governance Code.*

Foster, S. & Garduno, H. 2013. Groundwater resource governance: Are governments and stakeholders responding to the challenge. *Hydrogeology Journal* 21(2): 317–20.

Government. 2013. Agency Performance Accountability Report: Ministry of Religion in 2013. Jakarta: *Ministry of Religion.*

Griffin, K.M.T.A. 1994.New framework for development cooperation.

Hamada, M. 2010. Commercialization of microfinance in Indonesia: The shortage of funds and the linkage program. *The Developing Economies* 48(1):156–76.

Hughes, J.P. 2013. A new cost efficiency measure for not-for-profit firms: Evidence of a link between inefficiency and large endowments. *Atlantic Economic Journal* 41(3): 279–300.

Keating, E.K., Fische, M., Gordon, T.P. & Greenlee, J.S. 2005. Assessing Financial Vulnerability in the Non-Profit Sector. *Hauser Center for Non-Profit Organisations.*

Makki, M.A. 2010. Impact of corporate governance on intellectual capital efficiency and financial performance (Doctoral dissertation, *National College of Business Administration & Economics Lahore.*

Marwah, H. & Bolz, A.K. 2013. Waqfs and trusts: a comparative study. *Trusts & Trustees* 1–15(10): 811–816.

Masyhadi, A. 2019. The values of sufism teachings of sunan drajat. PhD diss. Surabaya.*UIN Sunan Ampel.*

Mensah, Y.M., Lam, K. & Warner, R. 2008. An Approach To Evaluating Relative Effectiveness In Non-Profit Institutions. *Journal of Public Budgeting, Accounting and Financial Management* 20(3):324–354.

Mihaiu, D.M. & Opreana, A. 2010 Cristescu MP. efficiency, effectiveness and performance of the public sector. *Romanian Journal of Economic Forecasting* 4:132–147.

Musibah, A.S., & Alfattani, W.S. 2014. The mediating effect of financial performance on the relationship between Shariah supervisory board effectiveness, intellectual capital and corporate social responsibility, of Islamic banks in Gulf *Cooperation Council countries. Asian Social Science* 10(17):139.

Nathan, S. 2010. The performance of Shari'ah supervisory boards within Islamic financial institutions in the Gulf Cooperation Council countries. *Corporate Ownership & Control* 247.

OECD (Organisation for Economic Co-operation and Development) (2004) Principles of Corporate Governance, France: *OECD Publications.*

Pulic, A. 2004. Intellectual capital does it create or destroy value Measuring Business Excellence 8(1):62–68.

Putra, S. 2009, Financial Performance of Iron and Steel Industry in India: *An Analytical and Comparative Study of Some Selected Companies During.*

Rothbard, N.M. 2008. The Mystery of Banking (2 ed.). Auburn, Alabama: *Ludwig von Mises Institute.*

Sadeq, M.A. 2002 Waqf, perpetual charity and poverty alleviation. International *Journal of Social Economics* 29(1–2): 135–151.

Shafii, Z., Yunanda, R.A. & Rahman, F.K. 2014. Financial And Operational Measures Of Waqf Performance: The Case Of State Islamic Religion *Council Of Singapore And Malaysia.*

Singer, A. 2006. Charity in Islamic Societies. New York: Cambridge. 2008. Siti Mashitoh M. Waqf in Malaysia: legal and administrative perspectives. Kuala Lumpur: *University of Malaya Press.*

Sulaiman, M., Adnan, M.A. & Nor, S.P.M.M.N. 2009 Trust me! A case study of the International Islamic University Malaysia's waqf fund." *Review of Islamic Economics* 13(1): 69–88.

Tuckman, H. & Chang, C. 1991 A methodology for measuring the financial vulnerability of charitable non-profit organisations. Non-Profit and *Voluntary Sector Quarterly* 20(4): 445–460.

Wahab, N.A. & Rahman, A.R.. 2015 Efficiency of zakat institutions and its determinants. Access to Finance and Human Development Essays on Zakah, *Awqaf and Microfinance* 33.

Zeller, M & Meyer, R.L. 2002. The triangle of microfinance: Financial sustainability, outreach, and impact. *Intl Food Policy Res Inst.*

Advances in Business, Management and Entrepreneurship – Hurriyati et al. (Eds)
© 2021 Taylor & Francis Group, London, ISBN 978-0-367-67471-7

The analysis of financial report by using ratio analysis to assess business performance (case study at PT Tempo Inti Media Tbk)

U. Suherman & I. Solikin
Universitas Pendidikan Indonesia, Bandung, Indonesia

ABSTRACT: Indicators of the success of company performance can be seen from the company's financial statements. Therefore, it is very important to examine the condition of the company through its financial statements. The object in this study was a company that has gone public in Indonesian stock exchange. The data were financial statements in 2017 and 2018. The analytical method used was time series. Meanwhile, the measurement tools used were financial ratios that focused on liquidity ratio, solvability ratio, profitability ratio, and activity ratio. The findings revealed that the financial ratios had changed. However, the company was in a good condition. Thus, the company can continually increase business in the future.

1 INTRODUCTION

Assessing company performance is an important factor in carrying out management functions, particularly controlling functions. The aim is to see how successful the company is in implementing programs that have been created and to assess the level of success and accuracy of the program. Performance evaluation for companies is very necessary if the company wants to survive in an increasingly competitive business competition (Zagloel et al. 2008). For this reason, companies need a method to measure the integrated performance that can describe the condition and development of the companies. The result can be used to determine the companies' policies and strategies in the future.

One way to measure company performance is by knowing the company's financial condition through financial statements. The financial report presents an overview of the financial position of the company's performance in generating profits. It is also conclusions from recording transactions carried out by a company. The financial performance of a company is a primary concern for every stakeholder especially investors. The measurement of the financial health of a company through the reported financial statements gives a qualitative analysis of the company's position as well as an account of how the company has utilised its capital in production (Yusheng 2019).

To understand and interpret financial statements, it is necessary to analyze financial statements. The analysis technique that is often used in analyzing financial statements is ratio analysis. Ratio analysis is an analysis technique to determine the mathematical relationship of certain posts in each element of the financial statements. The results of the calculation of

the ratio will be compared with the previous year, so that changes can be made, whether there is an increase or decrease (Sulasmiyati et al. 2017).

Financial ratios are used to evaluate the financial condition and performance of the company. From the result, it can be seen the health condition of the company (Kasmir 2008). Ratios can also indicate areas that require deeper research and treatment. In relation to the decisions taken by the company, the ratio analysis aims to assess the effectiveness of the decisions that company has taken in order to carry out its company activities (Prastowo & Yulianty 2005).

2 METHODS

This study aimed to explain the description of the financial performance of PT. Tempo Inti Media Tbk, a company that has gone public on the Indonesian stock exchange and is listed on www.idx.co.id. The data was gained by comparing the Financial Report in 2018 with the previous year, 2017. The analysis focused on the company's financial ratio in the form of a balance sheet and profit and loss. It also used secondary data taken from internet.

Based on the description above, this study used descriptive analysis. (Juliansyah 2011) explained that descriptive research is research that seeks to explain a phenomenon, event, event that occurs now. Through descriptive research, researchers tried to describe events and events that became the center of attention without giving special treatment to the event. Descriptive research aimed to explain the existing problem solving based on data

Judging from the type of data, the approach of this research was a qualitative approach, research that intended to understand the phenomenon of what was experienced by the subject of research in a holistic way by means of words and language, in a specific natural context and by utilizing various scientific methods (Meleong 2007).

3 RESULT AND DISCUSSION

Based on the recapitulation, the results of the calculation of financial ratios can be shown in the Table 1.

Table 1. Financial ratio analysis.

Financial Ratio		Year		
		2018	2017	Average
Likuidity Ratio	CR	316,1%	153,7%	234,90%
	QR	165,32%	194,27%	179,80%
Profitability Ratio	GPM	57,77%	58,17%	57,97%
	NPM	8,04%	3,17%	5,61%
	ROI	1,12%	0,46%	0,79%
	ROE	1,95%	1,31%	1,63%
Solvency Ratio	DR	36,15%	64,43%	50,29%
	DER	59,40%	172,87%	116,14%
Activity Ratio	ITO	3,55 Times	3,61 Times	3,58 Times
	TATO	0,15 Times	0,15 Times	0,15 Times

*Source: Survey.

1. Liquidity Ratio Based on the table above, it can be seen that the company's Liquidity Ratio has been maximized. It can be seen in the average Current Ratio of PT. Core media, Tbk. The Current Ratio in 2018 was 316.1%. It exceeded the standard set of 200%, and was greater than in 2017 which only reached 153.7%. This indicates that the company's ability to pay off its smooth debt is maximal. Average Quick Ratio of PT. Tempo Inti Media, Tbk. was 179.80%, greater than the standard set at 100%. This indicates that the company's ability to pay debts is quite good because every 100% current debt is guaranteed by Quick Assets 211.82%.
2. Profitability Ratio The profitability ratio level describe the ability level of PT. Tempo Inti Media, Tbk. in generating profits. The profitability level of PT. Tempo Inti Media, Tbk. can be seen from four types of profitability ratios: Gross Profit Margin (GPM), Net Profit Margin (NPM), Return On Investment (ROI), and Return On Equity (ROE). The calculation results of profitability ratios of PT. Tempo Inti Media, Tbk. will be explained further in the following section. Gross profit margin PT. Tempo Inti Media in 2018 was 57.77%. In 2017, it was 58.17%. This indicates a decrease in the gross profit earned from each sales share. However, Gross profit margin PT. Tempo Inti Media

indicates effectiveness. The greater the Gross profit margin, the better the operating conditions of the company. This means that the cost of sold goods is relatively lower compared to sales. Net Profit Margin PT. Tempo Inti Media in 2018 was 8.04%. On the other hand, in 2017, it was 3.17%. This shows that net income after tax achieved by the company in 2018 has increased further, indicating that the net income of each rupiah obtained from its sales is increasing. Hence, the company's financial performance is getting better. Return on Investmen PT. Tempo Inti Media in 2018 was 1.12%. It is greater than in 2017, which was only 0.46%. The increase of ROI value indicates that PT. Tempo Inti Media can manage the total assets invested in the company to get optimal returns, although the percentage results is still not optimal. The higher the ROI, the better the condition of the company in generating profits. Return on Equity PT. Tempo Inti Media in 2018 was 1.95%. Again, it was greater than in 2017 which was only 1.31%. The increasing ROE value indicates that the level of net income obtained by shareholders on invested capital is increasing.

3. Solvency Ratio On the solvency ratio, the company's Debt Ratio in 2018 was 36.15%. It is lower than in 2017 which was only 64.43%. The Debt Ratio value in 2018 has decreased, below the maximum general standard of 50%. This indicates that the condition of a company has been in the level of a healthy debt ratio. The Debt Equity Ratio value of PT. Tempo Inti Media, Tbk. In 2018 was 59.40. On the contrary, in 2017, it was 172.87%. Even though the value of the Debt Ratio exceeds 50% (general standard), it indicates that the company's operating financing places more emphasis on the use of greater capital rather than the capital owned by the company. However, there is a clear effort from the company to divert financing with owner's equity. Moreover, the financial risks that the company carries are still quite large
4. Activity ratio Activity ratio in 2018 was 3.55 times. Meanwhile, in 2017, it was only 3.61 times. This has increased in the results of Turnover Inventory, indicating that the effectiveness of company management has increased in managing inventory. The value of Total Turnover Turnover is worth one. In 2017, the TATO value was 0.15 times and 2018 was 0.15 times. This indicates that company management have the same behavior in using all assets to create sales. The higher the value of TATO, the more efficient the company in using all assets owned to generate profits. Meanwhile, the percentage result at PT. Tempo remains the same.

4 CONCLUSION

1. The company's ability to pay off its fluency is maximal. The Current Ratio in 2018 was 316.1%. It exceeded the standard set of 200%, and was greater

than in 2017 which only reached 153.7%. The company also have the ability to pay off its smooth debt maximally as the average Quick Ratio of PT. Tempo Inti Media, Tbk is 179.80%, exceeding the standard set at 100%.

2. There is a decrease in gross profit (gross profit margin) obtained from each rupiah sales even though it is only 58.17%. However, the Net Profit Margin achieved by the company in 2018 increased by 8.04% compared to 2017 which was only 3.17%. This indicates that the company's financial performance is getting better. The management of total assets invested in the company to obtain optimal starting profits can be viewed from the increase in Return on Investments value in 2018 (1.12%). The company's ROE value is increasing in 2018. It reached 1.95%. Meanwhile, in 2017, it was 1.31%. This means that the level of net income obtained by shareholders for invested capital is increasing.

3. The condition of the company is classified as a healthy debt ratio, because the Debt Ratio value in 2018 has decreased by 36.15%. It is below the maximum general standard of 50%. The existence of a business in a company to divert financing with own capital is reflected in the decrease of Debt Equity Ratio in 2018, which was 59.40.

4. There is an increase in the results of Inventory Turnover. This indicates the effectiveness of company management in managing inventory. Then, management have the same behavior in using all assets to create sales.

ACKNOWLEDGEMENTS

1. With a decrease in gross profit from sales, even though there are fewer than 58.17%. Become 57.77%. It is hoped that companies in the coming year will continue to increase profits by maximizing existing assets.

2. The company has a Current Ratio of 316.1% in 2018 which can be used as the power to find loan capital to increase sales.

REFERENCES

Juliansyah, N. 2011, Metodologi Penelitian, Jakarta: *Prenada Media Group*

Kasmir. 2008. Analisis Laporan Keuangan. Jakarta. *Rajawali Pers.*

Meleong, L.J. 2007. Metodologi Penelitian Kualitatif, Bandung: *Penerbit PT Remaja Rosdakarya Offset.*

Prastowo, D., & Yulianty, R. 2005. Analisis Laporan Keuangan : Konsep dan Aplikasi. Edisi Kedua. Cetakan Pertama. Yogyakarta: Penerbit *UPP AMP YKPN.*

Sulasmiyati, S., Barus, M.A., & Sudjana, N. 2017. Penggunaan Rasio Keuangan Untuk Mengukur Kinerja Keuangan Perusahaan (Studi Pada Pt. Astra Otoparts, Tbk dan Pt. Goodyer Indonesia, Tbk Yang Go-Publik Di Bursa Efek Indonesia). *Jurnal Administrasi Bisnis* 44(1): 154–163.

Yusheng, M 2019. Financial statement analysis : Principal component analysis (PCA) approach case study on China telecoms industry. Asian *Journal of Accounting* 4 (2): 233–245.

Zagloel, T.Y., Yadrifi., & Laricha, L 2008. Perencanaan Strategi dalam Upaya Menyelaraskan Tujuan Organisasi dan Tujuan Karyawan dengan Pendekatan Total Performance Scorecard. *Jurnal Teknik Industri.* (10):138–150. https://www.idx.co.id.

The effects of growth, managerial ownership, size and efficiency on the firm values and dividend policy

A. Akbar & Kusnendi
Universitas Pendidikan Indonesia, Bandung, Indonesia

ABSTRACT: This study aims to examine the effects of growth, managerial ownership, company size and efficiency on the firm values and dividend policy in the companies listed on Indonesian Stock Exchange in the period 2013–2017. The samples were companies categorized as the members of the LQ45 stock group. The sample was chosen based on three criteria:(1) liquidity and market capitalization, (2) routinely distributing dividends, and (3) originated from various industries. The data obtained from the IDX page (www.idx.co.id). It was then analysed by using path analysis. The results revealed that revenue growth had negative effect on the firm value and dividend policy. On the contrary, managerial ownership had an effect on the firm value. Then, there was no influence of firm size on firm efficiency. Moreover, the size and efficiency of the company had a positive effect on dividend policy. However, managerial ownership and firm value did not change dividend policy.

1 INTRODUCTION

There have been many empirical studies regarding dividend policy, including the influence of dividend policy on firm value (Modigliani & Miller 1961) in which the higher profits expected by investors relating to the reduction of dividend yield (Gordon 1963; Lintner 1962). However, there is a lack of study related to the effects of revenue growth, managerial ownership, and firm size and efficiency on the firm values and dividend policy happening in the companies categorized as the members of the LQ45 stock group.

In responding to the gap, this paper aims to examine the determinant factors of dividend policy in terms of revenue growth, managerial ownership, firm size, company efficiency (total asset turnover), and firm value towards companies that are the members of the LQ45 group in the Indonesian Stock Exchange throughout the period 2013–2017.

1.1 Literature review

Dividend policy is an integral part of the corporate financing decisions (Horne & Wachowicz 2009:501). It will reduce retained earnings (Horne 2002:303) and total internal funding sources. In terms of risk, dividend yield has a lower risk rather than capital gains (Gordon 1963; Lintner 1962).

The high value of the company indicates the prosperity of the company owner. This is reflected in the market price per share of the company issued by PBV (price to book value). From PBV ratios, it can be seen

the relation between company ability to create relative value and the amount of invested capital. Nevertheless, the value of the book may not be in accordance with the value of liquidation (Horne 2002).

Revenue growth reflects the growth of the company's operational activities. The greater the company's income in the future, the more profit obtained. This in turn will result in the greater free cash flow distributed as dividends (Suartawan & Yasa 2016). On the other hand, (Fama & Babiak 1968) stated that profit growth affects dividend policy rather than cash flow. Furthermore, (Amidu & Abor 2006) argued that the higher the growth rate of the company's income, the lower the dividends paid. The same results are also shown by (Nerviana 2016) who noted that there is no relationship between revenue growth (sales) and dividend policy.

For companies that have been long established, the level of managerial ownership can have a positive impact on company performance (Morck et al. 1988) and ultimately increase the value of the company itself (McConaughy et al. 2001). However, there is a negative relationship between managerial ownership of dividend policy (Mehrani 2011). The studies conducted by (Chen et al. 2005; Demsetz & Lehn 1985; Himmelberg et al. 1999; Short et al. 2002) pointed out that there is a positive (even though it is weak) relationship between managerial ownership and dividend policy.

Companies with small market provide higher returns than companies with high capitalization (Horne & Wachowicz 2009). If the size of the company is one of the factors in dividend distribution, the increase of dividends is greater for smaller companies than larger companies (Eddy & Seifert 1988; Michel 1979; Nerviana 2016) provided different opinions, the size of the company apparently does not have an impact on dividend policy.

In short, a company will always strive to achieve efficiency in each of its activities because it is closely related to liquidity, the company's ability to utilize its entire assets (Horne & Wachowicz 2009), and the company's speed in converting various accounts into sales or cash (Megginson & Smart 2009).

1.2 Hypothesis

The hypotheses proposed in the present study are (1)

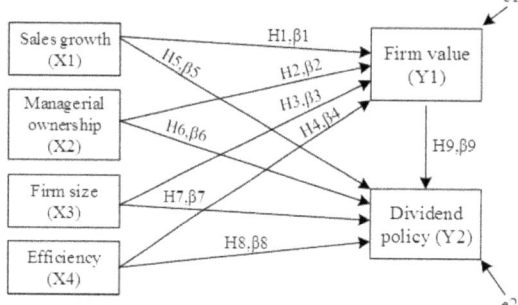

Figure 1.   Research framework model.

H1: Revenue growth has a positive effect on firm value (2) H2: Managerial ownership has a positive effect on firm value, (3) H3: Firm size has a positive effect on firm value, (4) H4: Company efficiency has a positive effect on firm value, (5) H5: Revenue growth has a positive effect on dividend policy, (6) H6: Managerial ownership has a positive effect on dividend policy, (7) H7: Firm size has a positive effect on dividend policy, (8) H8: Company efficiency has a positive effect on dividend policy, (9) H9: Firm value efficiency has a positive effect on dividend policy (Figure 1).

2   METHODS

2.1   Research variables

Firm value (Y1): The fair value of the company was reflected in the price to book ratio. Dividend policy (Y2): Variable dividend policy used dividend payout ratio (DPR). Revenue growth (X1) was calculated by following formula:

$$\text{Revenue Growth} = \frac{(t - (t - 1))}{(t - 1)} \tag{1}$$

t: current year's income
t − 1: previous year's income

Managerial ownership (X2) is the percentage of shareholders from management. Firm size (X3) is calculated by using the application of natural logarithms

of market capitalization values, Size = (Ln) market capitalization. Efficiency (X4) is company's ability to use its assets. The analysis used an activity ratio, namely TATO (total asset turnover = sales/ total assets).

2.2   Population and sampling

The population was 45 companies that became the member of LQ45 group during the period of 2013–2017. The sample was taken by using purposive sampling. There were 36 company chosen as the sample based on criteria such as: (1) the company had active shares and distribu ted dividends for five consecutive years throughout 2013–2017, (2) the company publishes audited financial statements annually during the referred period. To get more accountable and transparent data, the financial statements from companies listed in the LQ-45 Index were accessed through its official website, www.idx.co.id.

2.3   Data analysis techniques

This study used multiple regression with simple mediation as it intended to know the direct and indirect effects of variables and to test theoretical relationships or causal modeling (Hayes 2018; Schumacker & Lomax 2010).

The model of multiple regression equations applied was (Equation 1) $Y1 = \beta1X1 + \beta2X2 + \beta3X3 + \beta4X4 + e1$; (Equation 2) $Y2 = \beta5X1 + \beta6X2 + \beta7X3 + \beta8X4 + \beta9Y1 + e2$. Where X1 = revenue growth, X2 = managerial ownership, X3 = firm size, X4 = efficiency, Y1 = firm value, and Y2 = dividend policy.

3   RESULTS AND DISCUSSION

3.1   Results

Table 1 shows that there is no correlation between the variables. The value was above 0.80. This implies that the research dataset between variables has no multicollinearity problems.

Table 1.   Correlation matrix and descriptive statistics.

	X_1	X_2	X_3	X_4	Y_1	Y_2
X_1	1					
X_2	0.021	1				
X_3	−0.044	0.166	1			
X_4	−0.122	0.294	0.145	1		
Y_1	−0.045	0.349	0.156	0.136	1	
Y_2	−0.082	0.031	0.209	0.315	0.14	1
N	180	180	180	180	180	180
mean	12.25	59.24	31.44	77.01	3.6	37.42
stdev	19.08	13.7	1.32	64.66	21.35	58.86

44

The results of multiple regression analysis model of firm value (Y1) are presented in Table 2.

Table 2. Summary of results of multiple regression analysis of firm value models (Y1).

Model	R	R^2 (R^2 adj)	Beta	t (F)	p
Model 1	0.366	0.134 (0.114)		6.775	0.000
Constant				−2.104	0.037
X1			−0.045	−0.639	0.524
X2			0.327	4.402	0.000
X3			0.097	1.349	0.179
X4			0.020	0.273	0.785
Model 2	0.366	0.134 (0.119)		9.056	0.000
Constant				−2.145	0.033
X1			−0.048	−0.682	0.496
X2			0.333	4.681	0.000
X3			0.099	1.386	0.168
Model 3	0.363	0.131 (0.122)		13.392	0.000
Constant				−2.202	0.029
X2			0.332	4.671	0.000
X3			0.101	1.422	0.157
Model 4	0.349	0.122 (0.117)		24.619	0.000
Constant				−4.294	0.000
X2			0.349	4.962	0.000

Table 2 shows four models. The first model is Y1 = f (X1, X2, X3, X4). The regression analysis result is R2 = 0.134 and p < 0.05. Meanwhile, the result of significant test implies that Ho is rejected. This means that variation Y1 can be explained simultaneously by X1, X2, X3, and X4. The path coefficient test (standardized regression coefficient) reveals that the path coefficient X4 to Y1 is not significant ($\beta4 = 0.020$; t = 0.273; p = 0.785 > 0.05). It means that X4 does not affect Y1. Therefore, X4 is removed from the model.

The second model is Y1 = f (X1, X2, X3). The regression analysis result is R2 = 0.366 and p < 0.05. Meanwhile, the result of significant test implies that Ho is rejected. This means that variation Y1 can be explained together by variations of X1, X2, and X3. The path coefficient test (standardized) shows that path coefficient X1 is not significant ($\beta1 = -0.048$; t = −0.682; p = 0.496 > 0.05). It indicates that X1 does not affect Y1. Hence, X1 is excluded from the model.

The third model is Y1 = f (X2, X3). The regression analysis result is f R2 = 0.131 and p < 0.05. Meanwhile, the result of significant test implies that Ho is rejected. This means that variation Y1 can be explained concurrently by X2, and X3 variations. The path coefficient test (standardized) shows that path coefficient X3 is not significant ($\beta3 = 0.101$; t = 1.422;

p = 0.157 > 0.05). This means that X3 does not affect Y1. Therefore, X3 variable is removed from the model.

The last model is Y1 = f (X2). The regression analysis result is R2 = 0.122 and p < 0.05. After X3 is removed from the model, the R2 value drops from 0.131 to 0.122 (ΔR2 = −0.01). Fortunately, the decrease is not significant (p = 0.157 > 0.05). This means that Model 4 is effective, similar to Models 1, 2, and 3, in explaining the Y1 phenomenon. Model 4 is more parsimony than others.

After X1, X3, and X4 are removed from the model, the adjusted R2 value increases, from 0.114 to 0.117. This means that Model 4, with fewer predictors (only X2), is able to explain Y1 variation more effectively compared to Model 1. Thus, it can be concluded that Model 4 is the best model in explaining Y1 phenomenon

The results of multiple regression analysis models of dividend policy (Y2) are presented in Table 3. There are four models. First, Model 1 is Y2 = f (X1, X2, X3, X4, Y1). The regression analysis result is R2 = 0.147 and p < 0.05. Meanwhile, the result of significant test implies that Ho is rejected. This indicates that Y2 variations can be explained by X1, X2, X3, X4, and Y1. The path coefficient test (standardized) demonstrates that the path coefficient X1 to Y2 is not significant ($\beta5 = -0.029$; t = −0.406; p = 0.685 > 0.05). It means that X1 does not affect Y2. Therefore, X1 is removed from the model.

Second, Model 2 is Y2 = f (X2, X3, X4, Y1). The path coefficient test (standardized regression coefficient) shows that the path coefficient Y1 to Y2 is not significant ($\beta9 = 0.116$; t = 1.546; p = 0.124 > 0.05). Y1 does not affect Y2. Thus, Y1 is removed from the model.

Third, Model 3 is Y2 = f (X2, X3, X4). The results of the path coefficient test (standardized) reveals that the path coefficient X2 is not significant ($\beta6 = -0.092$; t = −1.242; p = 0.216 > 0.05). This means that X2 does not affect Y2, so that X2 variable is removed from the model.

Last, Model 4 is Y2 = f (X3, X4). The regression analysis result is R2 = 0.127 and p < 0.05. After X2 is removed from the model, the R2 value drops from 0.134 to 0.127 (ΔR2 = −0.008). Luckily, the decrease is not significant (p = 0.216 > 0.05). This means that Model 4 is effective, likes Model 1, 2, and 3 in explaining the Y2 phenomenon. However, Model 4 is more parsimony than Model 1, 2, and 3.

The result of the regression analysis shows that after X1, X2, and Y1 are excluded from the model, the adjusted R2 value decreases from 0.122 to 0.117. This means that Model 4 with fewer predictors (containing two predictors, namely X3, X4) is considered more effectively compared to Model 1. Therefore, it can be concluded that Model 4 is the best model in explaining the Y2 phenomenon.

The effect of X3 and X4 on Y2 is 0.167 (4.63%) and 0.291 (8.07%). Moreover, 12.7% of the variations occurred in Y2 can be explained synchronously by

Model	R	R^2 (R^2 adj)	Beta	t (F)	p
Model 1	0,383	0,147 (0,122)		5,992	0,000
Constant				−1,848	0,066
X1			−0,029	−0,406	0,685
X2			−0,127	−1,629	0,105
X3			0,166	2,319	0,022
X4			0,310	4,167	0,000
Y1			0,114	1,521	0,130
Model 2	0,382	0,146 (0,127)		7,484	0,000
Constant				−1,874	0,063
X2			−0,130	−1,669	0,097
X3			0,167	2,337	0,021
X4			0,313	4,265	0,000
Y1			0,116	1,546	0,124
Model 3	0,367	0,134 (0,120)		9,110	0,000
Constant				−2,140	0,034
X2			−0,092	−1,242	0,216
X3			0,179	2,500	0,013
X4			0,317	4,292	0,000
Model 4	0,356	0,127 (0,117)		12,854	0,000
Constant				−2,194	0,030
X3			0,167	2,355	0,020
X4			0,291	4,102	0,000

Table 4. Direct, indirect, and total effects.

The effect between variables	Direct effect (DE)	Indirect effect (IE)	Total effect (TE = DE + IE)
X1 -> Y1	–	–	–
X2 -> Y1	0,327	–	0,327
X3 -> Y1	–	–	–
X4 -> Y1	–	–	–
X1 -> Y2	–	–	–
X2 -> Y2	–	–	–
X3 -> Y2	0,166	–	0,166
X4 -> Y2	0,310	–	0,310
Y1 -> Y2	–	–	–

X3 and X4. A variation Y2 caused by other variables which are not existed in the model is 87.3%. Thus, Y1 does not affect Y2. Meanwhile, the modification of the model with multiple linear regression equations is obtained as follow (i) $Y1 = 0.35X2$; $R2Y1 = 0.12$; (ii) $Y2 = 0.17X3 + 0.29X4$; $R2Y2 = 0.13$.

The direct and indirect effect between variables can be seen in Table 4. There is only X2 variable that has an effect on Y1. In Y2, the variable with strongest effect is X4, followed by variable X3.

The model test reveal that the company value is represented by four determinant factors, such as revenue growth, managerial ownership, company size, and efficiency. Of the four factors, there is only managerial ownership that is able to explain the variations compared to the other factors. This is in line with (McConaughy et al. 2001) who stated that managerial ownership shows that the company has greater value. The managerial ownership here is not those in the management, but the characteristic of group sitting in ownership concentration.

Managerial ownership is determined by the characteristic of a company (Cho 1998). In family company, it has a role to control those who plan to get incentives (Kole 1997). The managers here seek to balance company interests and the welfare of shareholders (Jensen & Meckling 1976). Thus, they strive to control the company. Ownership factors do not have an influence on dividend policy because high managerial ownership will tend to hold profits rather than pay dividends, only if low managerial ownership cannot affect payment of dividends (Mehrani 2011).

Revenue growth (sales) has no impact on company value and dividend policy. These results are in line with (Lau et al. 2002) who performed the studies on the companies in Singapore, but yielded different results compared to the companies in Malaysia. High revenue growth is not necessarily equivalent with high profits. Some companies prefer massive sales with little profit, insufficient profit that has little impact for the company. For this reason, the profitability factor determines company value (Cho & Pucik 2005; Fama & Babiak 1968). The revenue growth does not have an impact on dividend policy because managers will choose profitable investments with the expectation that company performance will be better and the remaining available cash is distributed as dividends (Megginson & Smart 2009). Therefore, high growth company tends to pay lower dividends

Company size tends to reflect company value. Yet, this study found that the size of the company does not affect the value of the company. This is different from the studies conducted by (Barber & Lyon 1997; Fama & French 1992; Short et al. 2002). In their opinion, company size is the total asset owned by the company. Thus, the greater the asset are, the better the company value is.

Regarding dividend policy, this research reveals that there is significant relationship between company size and dividend policy. This is in accordance with the study performed by (Alli et al. 1993) which found the access for large companies to the market, in which there was financial flexibility that prompted them to pay higher dividends. Large market capitalization shows that the company's reputation is maintained by the company and it is characterized by a high dividend pay-out ratio.

Efficiency is needed for the company to manage its assets and to create certain sales volume. In other words, the more efficient the management of a company, the better the company's value in the eyes of investors (Asiri & Hameed 2014; Hall & Brummer 1999). In this study, efficiency does not affect company value. This supports (Gamayuni 2015) research which found that intangible assets (intellectual capital), return on assets, and debt management policies had a significant impact on the firm value. Nevertheless, it is proven that efficiency have an impact on dividend policy. Successful efficiency can reduce the costs and increase the profits. This results in the great advantages. Thus, dividends distributed to investors will also be higher (Suharmanto et al. 2019). As suggested by some studies that dividend policy can affect company value (Baker et al. 2001; Gordon 1963) and not vice versa. On the contrary, this study reveals that there is no relationship between company value and dividend policy. The focus of company is to gain a profit, while dividend policy deals with how much profits distributed to all of the investors.

4 CONCLUSION

From results, it can be inferred that the company value is represented by four determinant factors, such as revenue growth, managerial ownership, company size, and company efficiency. Among this four factors, there is only managerial ownership that has positive and significant impact to the company value. This means that managerial ownership is more effective compared the other factors in determining the company value.

Dividend policy is represented by five variables: revenue growth, managerial ownership, company size, company efficiency, and company value. The first and second variables have negative impact on dividend policy. The company value has an impact, but it is not significant. On the other hand, company size and efficiency have positive and significant impact on dividend police. Among these two variable, the impact of company efficiency is higher.

This study has so many limitation. Further studies are expected to be able to find out more about the other dominant factors influencing company value and dividend policy. Besides, the object of the study can be extended to all companies listed on the Indonesian Stock Exchange without some criteria. It should also be noted that a longer period of time is needed to get a better result.

REFERENCES

Alli, K.L., Khan, A.Q., & Ramirez, G.G. 1993. Determinants of corporate dividend policy: A factorial analysis. *Financial Review* 28(4): 523–547.

Amidu, M., & Abor, J. 2006. Determinants of dividend payout ratios in Ghana. *The journal of risk finance* 7(2): 136–145.

Asiri, B.K., & Hameed, S.A. 2014. Financial ratios and firm's value in the Bahrain Bourse. *Research journal of finance and accounting* 5(7): 1–9.

Baker, H.K., Veit, E.T., & Powell, G.E. 2001. Factors influencing dividend policy decisions of Nasdaq firms. *Financial Review* 36(3): 19–38.

Barber, B.M., & Lyon, J.D. 1997. Detection long run abnormal stock returns: The empirical power and specification of test statistic. *Journal of finance economics* 43(3): 341–372.

Chen, Z., Cheung, Y.L., Stouraitis, A., & Wong, A.W. 2005. Ownership concentration, firm performance, and dividend policy in Hong Kong. *Pacific Basin Finance Journal* 13(4): 431–449.

Cho, H.J., & Pucik, V. 2005. Relationship between innovativeness, quality, growth, profitability, and market value. *Strategic management journal* 26(6): 555–575.

Cho, M.H. 1998. Ownership structure, investment, and the corporate value: an empirical analysis. *Journal of financial economics* 47(1): 103–121.

Demsetz, H., & Lehn, K. 1985. The structure of corporate ownership: Causes and consequences. *Journal of political economy* 93(6): 1155–1177.

Eddy, A., & Seifert, B. 1988. Firm size and dividend announcements. *Journal of Financial Research* 11(4): 295–302.

Fama, E.F., & Babiak, H. 1968. Dividend policy: An empirical analysis. *Journal of the American statistical Association* 63(324): 1132–1161.

Fama E.F., & French, K.H. 1992. The cross section of expected stock returns. *The journal of finance* 47(2).

Gamayuni, R.R. 2015. The effect of intangible asset, financial performance and financial policies on the firm value. International *journal of scientific & technology research* 4(1): 202–212.

Gordon, M.J. 1963. Optimal Investment and Financing Policy. *The Journal of Finance* 18(2): 264–272.

Hall, J.H., & Brummer, L.M. 1999. The relationship between the market value of a company and internal performance measurements. Available at *SSRN*.

Hayes, A.F. (2018). Introduction to Mediation, Moderation,and Conditional Process Analysis: A Regression-Based Approach. New York: *The Guilford Press*.

Himmelberg, C.P., Hubbard, R.G., & Palia, D. 1999. Understanding the determinants of managerial ownership and the link between ownership and performance. *Journal of financial economics* 53(3): 353–384.

Horne, J.C. 2002. Financial Management and Policy (12 ed.). New Jersey. *PrenticeHall,Inc*.

Horne, J.C., & Wachowicz, J.M. 2009. Fundamentals of financial management (13 ed.). Harlow, England. *Pearson Education Limited*.

Jensen, M.C., & Meckling, W.H. 1976. Theory of the firm: Managerial behavior, agency costs and ownership structure. *Journal of financial economics* 3(4): 305–360.

Kole, S.R. 1997. The complexity of compensation contracts. *Journal of Financial Economics* 43(1): 79–104.

Lau, C., Simpso, C., & Schanler, R.J. 2002. Early introduction of oral feeding in pretern infants. *Pediatrics* 110(3): 517–522.

Lintner, J. (1962). Dividends, Earnings, Leverage, Stock Prices and the Supply of Capital to Corporations. *The Review of Economics and Statistics* 44(3): 243–269.

McConaughy, D.L., Matthews, C.H., & Fialko, A.S. (2001). Founding family controlled firms: Performance, risk, and value. *Journal of small business management* 39(1): 31–49.

Megginson, W.L., & Smart, S.B. 2009. Introduction to Corporate Finance. Mason: South-Western Cengage Learning.

Mehrani, S.M. 2011. Ownership structure and dividend policy: Evidence from Iran. African *Journal of Business Management* 5(17): 7516–7525.

Michel, A. (1979). Industry Influence on Dividend Policy. *Financial Management* 8(3): 22–26.

Modigliani, F., & Miller, M.H. 1961, Oct). Dividend Policy, Growth, and the Valuation of Shares. *The Journal of Business* 34(4), 411–433.

Morck, R., Shleifer, A., & Vishny, R.W. 1988. Management ownership and market valuation: An empirical analysis. *Journal of financial economics* 20, 293–315.

Nerviana, R. (2016). The effect of financial ratios and company size on dividend policy. *The Indonesian Accounting Review* 5(1): 23–32.

Schumacker, R.E., & Lomax, R.G. 2010. A beginner's guide to structural equation modeling (3 ed.). New York: *Taylor & Francis Group*.

Short, H., Zhang, H., & Keasey, K. 2002. The link between dividend policy and institutional ownership. *Journal of corporate Finance* 8(2): 105–122.

Suartawan, P.A., & Yasa, G.W. 2016. Pengaruh investment opportunity set dan free cash flow pada kebijakan dividen dan nilai perusahaan. *Jurnal Ilmiah Akuntansi dan Bisnis* 11(2), 63–74.

Suharmanto, A., Widiyanti, M., & Taufik, H. 2019. Analysis of Financial Performance and Opportunity of Investment on Dividend Policy with Profitability As Moderating Variable in LQ45 Company Listed In Indoenesia Stock Exchange. International *Journal of Scientific Research and Engineering Development* 2(1) 183–197.

Advances in Business, Management and Entrepreneurship – Hurriyati et al. (Eds)
© 2021 Taylor & Francis Group, London, ISBN 978-0-367-67471-7

Top management diversity and risk-taking behaviour

Y.S. Ramli & A.A. Hermawati
Universitas Indonesia, Jakarta, Indonesia

ABSTRACT: Risk-taking behaviour is one of the important things in running the company. It is needed especially when the company faces a difficult situation. However, excessive use may negatively affect the company. Risk-taking behaviour is based on risk preferences, which differ among companies as well as individuals. At management level, diversity may be one of the important factors that affect risk-taking behaviour. Previous literature review that studied about diversity in management such as diversity in gender, age, education, experience, etc., shows inconsistent results toward risk-taking behaviour. Based on this inconsistency, the author is interested to analyse the relationship among top management diversity as a decision maker in the company to risk-taking behaviour. This research is conducted to obtain empirical evidence on whether top management diversity is related to risk-taking behaviour. This research used regression on panel data from 333 Indonesian listed non-financial companies for a 3-year period from 2015 to 2017. Dependent variable of this research was risk-taking behaviour, while independent variables were top management diversity, including gender, age, nationality, education background, and experience. Control variables of this research were company size, profitability, and growth. The results showed that top management diversity in gender, age, nationality, education background, and experience were not significantly related to risk-taking behaviour. However, company characteristics such as size, profitability, and growth were significantly related to risk-taking behaviour.

1 INTRODUCTION

According to Indonesia's economic growth data from Central Bureau of Statistics from 2010 to 2017, Indonesia is still recovering from economic downturn which occurred in 2013 to 2015, with the lowest economic growth of 4.79% in 2015. In 2017, Indonesia still had a low level of economic growth around 5.07%, which still needs a lot of work to achieve higher economic growth as has been achieved previously, around 6.19% to 6.81% in 2010 to 2012. There is still much uncertainty on Indonesia's future economic condition. Corporate risk-taking is needed to respond to uncertainty in Indonesia's economic condition.

Risk preference, which differs among companies as well as individuals, directly reflects the operations of a company. Zhang et al. (2016) argued that managerial risk preference differs among management personnel. The main challenge for a corporation is to ensure management configuration which leads to management effectiveness toward corporate risk. According to the Woman in Business research conducted by Grant Thornton International published in early 2019, there is around 20% composition. As the results of more and easier access to better education, lately, there is an increase of participation of the young and higher education in management.

Several management characteristics have received significant attention and are being observed empirically such as size, composition, diversity, and quality (Lenard et al. 2014). Management diversity is one aspect that contributes to management characteristics in relation to risk-taking. Previous research on risk-taking behaviour was conducted by Zhang et al. (2016), which studied risk-taking and its influencing factors (such as gender, age, position, education, industries) on large state-owned enterprises management personnel in China. In Indonesia, there was research from Habinsaran (2016) that studied the association of gender diversity to corporate risk-taking and corporate performance. This research is conducted in response to trend changes in management composition and to extend previous research conducted in Indonesia in terms of diversity variables such as gender, age, nationality, education, and experience.

Risk-taking tendency is associated with management characteristics both from demographic (gender, age, nationality, etc.) and cognitive (education background, experience, etc.). Holmes et al. (2017) argued that males are more likely to engage in risk-taking behaviour than females. Reniers et al. (2016) also found that males tend to take more risks than females. Thus, the increase of females in top management structure may lead to less risk-taking behaviour.

Rolison et al. (2013) showed that higher-age people tend to avoid risk. Pepper and Gore (2014) reported that higher-age management tend to be more risk-averse than lower-age management. This is also in line with Zhang et al. (2016), who found similar results. Thus, the increase of higher-age people in top management composition may lead to less risk-taking behaviour.

Lawal (2018) found positive association between the presence of foreign management and corporate risk-taking. In Indonesia (as a developing country), companies usually have foreign top management from more-developed countries. These foreign top managements tend to be more risk-taking than local top management.

Wang and Zhou (2013) found that education level has negative association with risk-taking. Management with lower education tend to engage with more risk-taking behaviour as they have less ability to reduce and mitigate the risk.

Al-Shammari (2018) found that management with higher experience tend to be more risk-taking. Management with higher experience may use more information to handle risk better. Thus, management with higher experience have better capability in handling risk and are more confident.

Based on the above argumentation, the following hypothesis are formulated:

H1: Female top management is negatively associated with risk-taking behaviour.

H2: Higher-age top management is negatively associated with risk-taking behaviour.

H3: Foreign top management is positively associated with risk-taking behaviour.

H4: Lower-education background top management is positively associated with risk-taking behaviour.

H5: Lower-experience top management is negatively associated with risk-taking behaviour.

2 METHODS

2.1 Research model

$$RISK\_TAKINGit = \alpha + \beta 1 GEND\_DIVit + \beta 2\ AGE\_DIVit + \beta 3 NAT\_DIVit + \beta 4 EDBG\_DIVit + \beta 5\ EXP\_DIVit + \beta 6 COMP\_SIZEit + \beta 7 COMP\_PROFITit + \beta 8 COMP\_GROWTHit + \varepsilon it \tag{1}$$

Note:

RISK_TAKINGit : Risk-taking behaviour is measured by annual mean of daily volatility of stock return.

GEND_DIVit : Gender diversity is measured by calculating the proportion of female top management to the total number of top management.

AGE_DIVit : Age diversity is measured by calculating the proportion of higher-age (more than 40 years old) top management to total number of top management.

NAT_DIVit : Nationality diversity is measured by calculating the proportion of foreign (foreign citizenship) top management to total number of top management.

EDBG_DIVit : Education background diversity is measured by calculating the proportion of higher

education background (master degree or above) top management to total number of top management.

EXP_DIVit : Experience diversity is measured by calculating proportion of higher-experience (more than 10 years working experience in related industry) top management to total number of top management.

COMP_SIZEit : Company size is measured by calculating natural log of a company's total assets.

COMP_PROFITit : Company profitability is measured by return on assets (ROA) of the company.

COMP_GROWTHit : Company growth is measured by sales growth of the company.

ε it : Regression error.

2.2 Variables definition and measurement

Dependent variable in this research was risk-taking behaviour, which is measured by annual mean of daily volatility of stock return. Daily volatility of stock return is measured by calculating the percentage of increment or decrement of daily adjusted stock closing price. Risk-taking behaviour is proxied by annual mean of daily volatility of stock return because stock return represents risk as explained by Belanes and Hachana (2010).

Independent variables used were gender diversity, age diversity, nationality diversity, education background diversity, and experience diversity. Gender diversity is measured by calculating the proportion of female top management to the total number of top management. Age diversity is measured by calculating the proportion of higher-age top management to the total number of top management. Nationality diversity is measured by calculating the proportion of foreign top management to the total number of top management. Educational background diversity is measured by calculating the proportion of higher education background (master degree or above) of top management to the total number of top management. Experience diversity is measured by calculating the proportion of higher experience (more than 10 years in the same industry) top management to total number of top management.

Control variables used were company size, company profitability, and company growth. Firm size is measured by calculating log natural of total assets of a company (Abou-El-Sood 2017; Khaw & Liao 2018). Profitability is measured by using return on assets (ROA) as referred from previous research conducted by Khaw and Liao (2018). Company growth is measured by annual sales growth as referred to research conducted by Sila et al. (2016), which argued that annual company growth can be proxied by annual sales growth.

2.3 Population and sample

Population was all listed non-financial companies on the Indonesia Stock Exchange for the period 2015–2017. Sample selection process was a purposive sampling method, which was excluding several companies with several criteria, such as financial industry companies, delisting or bankrupt companies, and incomplete or unavailable data and information needed within the period 2015–2017.

3 RESULT AND DISCUSSION

After conducting several research tests, the following result is obtained:

Table 1. Regression results.

Variable	Expected Sign	Coefficient	t.	Prob.
CONSTANT		0.0040	2.8012	0.0026
GEND_DIV	−	0.0001	0.3158	0.3762
AGE_DIV	−	−0.0003	−0.3793	0.3523
NAT_DIV	+	−0.0002	−0.4414	0.3296
EDBG_DIV	+	−0.0002	−0.6529	0.2570
EXP_DIV	−	−0.0004	−0.4865	0.3134
COMP_SIZE	−	−0.0001	−2.9163	**0.0018
COMP_PROFIT	−	0.0038	5.4647	**0.0000
COMP_GROWTH	+	0.0001	1.8437	**0.0328
Prob. F		0.0000		
Adj. R squared		0.0295		

Notes:
**$\alpha = 5\%$ (one tailed).
*Source: Output Eviews (Reprocessed by author 2019).

For the first hypothesis, which is female top management is negatively associated with risk-taking behaviour, it is found that female top management do not have negative and significant impact on risk-taking behaviour. For the second hypothesis, which is higher-age top management is negatively associated with risk-taking behaviour, it is found that higher-age top management has negative but insignificant impact on risk-taking behaviour. For the third hypothesis, which is foreign top management is positively associated with risk-taking behaviour, it is found that foreign top management has positive but insignificant impact on risk-taking behaviour. For the fourth hypothesis, which is lower-education background top management is positively associated with risk-taking behaviour, it is found that lower-education background top management don't have positive and significant impact on risk-taking behaviour. For the fifth hypothesis, which is lower-experience top management is negatively associated with risk-taking behaviour, it is found that lower-experience management has negative but insignificant impact on risk-taking behaviour.

The result shows that the presence of females in top management does not provide a significant impact on risk-taking behaviour. The result is in line with several researches which argued that being female has no association with risk-taking. Zhang et al. (2016) found that gender does not affect managerial risk-taking. The result is also consistent with research from Pailing and Reniers (2018), which found that gender has no association with risk-taking. According to Abou (2017), females are not always risk-averse, for example, if the regulation allows them to be more risk-taking. Adams and Funk (2012) argued that female management cannot always avoid risk, especially in a competitive environment. According to the observation on research sample, the percent of females in top management is still low. Most companies' top management is still dominated by males. Therefore, female top management need to be quite competitive to maintain their position or even climb the corporate ladder, which forces them not to be risk-averse. Male domination in top management also prevents female top management from being more risk-averse. Females' top management may be risk-averse but they cannot act due to lack of presence in top management.

The result shows that the presence of higher-age top management does not provide significant impact to risk-taking behaviour. The result is in line with several researches that argued that age has no association with risk-taking. Best and Charness (2015) argued there is no association between age and risk-taking. Shao and Lee (2014) also found that age has no impact to risk-taking. According to the observation on research sample, it is found that most of Indonesian non-financial listed are still dominated by higher-age top management. The presence of lower-age top management may not affect management decisions too much. In addition, management decisions are also taken collectively by a group of top management. Even though there are several researches supporting that higher-age top management are more risk-averse, it may be different when top management takes decision-making collectively. There may be several considerations, such as group company decision, owner preference, company regulation, etc., which strongly weaken the risk-averse character of higher-age top management.

The result shows that the presence of foreign top management does not provide significant impact on risk-taking behaviour. The result is in line with several researches that argued that nationality has no association with risk-taking. Van Slyke (2016) found that nationality difference is not associated with risk-taking behaviour. According to the observation on research sample, it is found that most of Indonesian non-financial listed are still dominated by local top management. The presence of foreign top management may not affect management decision too much. Even though foreign top management come from more-developed countries, the significance level on overall management decision and risk preference is still low. In addition, most of the highest top management such as president director are from local people.

The result shows that the presence of lower education background top management does not provide significant impact on risk-taking behaviour. The result is in line with several researches that argued that education background has no association with risk-taking. Zhang et al. (2016) found there is no association between education background and managerial risk-taking. Van Slyke (2016) also found that education is not associated with risk-taking behavior. According to the observation on research sample, the author found that most of Indonesian non-financial listed are dominated by higher-education background top management. However, the presence of higher-education background top management may not affect companies' decision-making and risk-taking behaviour too much, as companies' decision-making and risk-taking are more practical. Education may provide a conceptual framework and theories that are useful for decision-making, but the decision-making in companies needs further study than conceptual framework and theories. Decision-making in the company is more related to each company's condition, which needs to be assessed and explored specifically.

The result shows that the presence of lower-experience top management does not provide significant impact on risk-taking behaviour. The result is in line with several researches that argued that experience has no association with risk-taking.

Lawal (2018) found that management quality, which is proxied by average years of directors' professional experience including working knowledge, does not affect corporate risk-taking. According to the observation on research sample, it is found that most of Indonesian non-financial listed are dominated by higher-experience top management. Even though there are several researches supporting that higher-experience top management tends to be more risk-taking, there are still several considerations, such as group company decision, owner preference, company regulation, etc., that affect the risk-taking character of higher-experience top management.

4 CONCLUSION

According to the research results, there are several conclusions as follows:

1. There is no evidence that female top management is negatively associated with risk-taking behaviour. In Indonesia, the presence of females in top management is still limited in term of number and frequency. The impact of female top management's presence is also still very limited as the decision-making power is still dominated by male top management, which mostly act as the main top management, such as president director.
2. There is no evidence that higher-age top management is negatively associated with risk-taking behaviour. In Indonesia, the presence of lower age in top management is still limited in terms of number and frequency. The impact of higher-age top management's presence is also still biased as most of the company's top management is still dominated by higher-age top management.

3. There is no evidence that foreign top management is positively associated with risk-taking behaviour. In Indonesia, the presence of foreigners in top management is still limited in term of power by local top management who have more important role in company's decision making.
4. There is no evidence that lower-education background top management is positively associated with risk-taking behaviour. In Indonesia, education background does not significantly affect risk-taking behaviour since higher-education background top management still needs to learn about the company and industry.
5. There is no evidence that lower-experience top management is negatively associated with risk-taking behaviour. Thus, it can be concluded that in Indonesia, top management experience does not significantly affect decision-making including risk-taking. Company characteristics have more significant impact on risk-taking behaviour than management characteristics such as experience.

This research has several limitations that need to be improved in the next researches that have similar topic and discussion. The limitations of this research are as follow:

1. Measurement of higher age (40 years old) management still needs stronger evidence and argumentation to support the research.
2. Measurement of higher experience (more than 10 years) still needs stronger evidence and argumentation to support the research.
3. This research is only conducted using panel data from a 3-year period, 2015 to 2017. Panel data from a longer comparative period may provide stronger research evidence.

Based on the above limitations, there are several recommendation for further research:

1. Future research can use age classification to elaborate more on the evidence and argumentation for the classification of higher age such as median of the data.
2. Future research that uses experience classification may elaborate more on the evidence and argumentation for the classification of higher experience such as median of the data.
3. Future research that studies management characteristics need to use panel data from a longer period such as 5 years to provide stronger research evidence.

REFERENCES

Abou, E.SH. (2017). Corporate governance and risk taking: the role of board gender diversity. *Pacific Accounting Review*.

Adams, R. & Funk, P. (2012). Beyond the glass ceiling: does gender matter. *Management Science* 58(2): 219–235.

Al-Shammari, H.(2018). CEO incentive compensation and risk-taking behavior: The moderating role of CEO characteristics. *Academy of Strategic Management Journal.* 17: 1–16.

Belanes, A., & Hachana, R. (2010). Corporate governance and managerial risktaking in Tunisia: An agency perspective. *Journal Of Global Business Administration.* 2(1): 1–18.

Best, R., & Charness, N. (2015). Age differences in the effect of framing on risky choice: A meta-analysis. *Psychology and Aging.* 30: 688–698.

Habinsaran, M.K. (2016). Pengaruh Gender Diversity terhadap Corporate Risk Taking dan Corporate Performance. *Thesis, Universitas Indonesia.* Jakarta.

Holmes, J., Rawsthorne, M., Paxton, K., Luscombe, G., Hawke, C., Ivers, R., Skinner, R., & Steinbeck, K. (2017). Risk-taking behaviours among younger adolescents in rural and regional *New South Wales.*

Khaw, K.L.H., & Liao, J. (2018). Board gender diversity and its risk monitoring role: Is it significant? *Asian Academy of Management Journal of Accounting and Finance.* 14(1): 83–106.

Lawal, B.(2018). Board rudiments and the executive attitude towards corporate risk-taking. International *Journal of Financial Research.* 9 (2): 1–16.

Lenard, M. J., Yu, B., & York, E. A. (2014). Impact of board gender diversity on firm risk. *Managerial Finance.* 40(8): 787–803.

Pailing, A. N., & Reniers, R. L. E. P. (2018). Depressive and socially anxious symptoms, psychosocial maturity, and risk perception: Associations with risk-taking behaviour. *PLoS ONE* 13(8).

Pepper, A., & Gore, J. (2014). The economic psychology of incentives: An international study of top managers. *Journal of World Business.* 49: 350–361.

Reniers, R.L.E.P., Murphy, L., Lin, A., Bartolomé, S.P., & Wood, S.J. (2016). Risk perception and risk-taking behaviour during adolescence: The influence of personality and gender. *PLoS ONE* 11(4).

Rolison, J. J., Hanoch, Y., Wood, S., & Liu, P.J. (2013). Risk-taking differences across the adult life span: A question of age and domain. *The Journals of Gerontology Series B: Psychological Sciences and Social Sciences.* 69 :870–880.

Shao, R., & Lee, T.M.C. (2014). Aging and risk taking: Toward an integration of cognitive, emotional, and neurobiological perspectives. *Neuroscience and Neuroeconomics.* 3: 47–62.

Sila, V., Gonzalez, A., & Hagendorff, J. (2016). Women on board: Does boardroom gender diversity affect firm risk? *Journal of Corporate Finance.* 36: 26–53.

van Slyke, J.K. (2016). Examining the roles of nationality, socioeconomic status, and setting in gambling expectancies and behavior among men. *Dissertation* 1–127.

Wang, Y.H., & Zhou, W. (2013). An empirical study on the relationship between the educational background of top decision makers and their corporate risk tolerance in China. *Education and Economy.* 6: 30–35.

Zhang, Y., Luan, H., Shao, W., & Xu, Y. (2016). Managerial risk preference and its influencing factors: *Analysis of large state-owned enterprises management personnel in China. Risk Management Article,* 18(2–3): 135–158.

Evidence from Indonesia of spillover effects among macroeconomic variables in emerging markets

E. Mahpudin, S. Suhono & N. Nursito
Universitas Singaperbangsa Karawang , Indonesia

ABSTRACT: This study aims to conduct forecasting using the graph method and multiple regression modeling, assisted by statistical software. This study uses data from 5 of 7 developing market economies, except for Russia and Mexico. The author considers that each of the 5 emerging markets is closely integrated with other countries, especially with other emerging market countries. By capitalizing on market size and integration between countries, the authors empirically estimate spillover effects through this study using the forecasting graph model to provide additional input and information relating to monetary macroeconomics in emerging countries. The variables used in this study include the money supply, the inflation rate, the composite stock price index, and the interest rate and expected return in the 5 emerging market countries.

1 INTRODUCTION

In today's world economy, there are several terms related to the state and the market. In the 70s and the 80s, several countries in the world used the terms developed countries, developing countries, and less-developed countries to refer to the state and the market. When it comes to the market, there will be something that gives potential benefits or risks, leading to all kinds of possible violations. Some economists have described the market as something that is developing. This understanding is considered by some as possibly misleading, but there is no guarantee if a country is in a developing or an underdeveloped position or if it will enter a more developed position. A country may move from a more developed position into a less-developed one and even go further into undeveloped one altogether. All of the aforementioned is related to the economic process that occurs in a country, be it a successful investment process or a successful process of controlling some of the country's economic variables.

Researchers and writers such as Paul Samuelson, Ben Bernanke, Gregory Mankiw, and Christopher Sims have conducted many studies on emerging markets. The professors with Princeton University and Harvard Business School conducted scientific studies in various countries in Europe, Asia, and the Arabian Peninsula. Recently, Vercueil (2012) proposed a pragmatic economic theory for developing countries based on the distinction between developing economies and markets. This research is conducted with an introduction and proposal of pragmatism through addressing the influence of financial characteristics.

The proposal of the applied, pragmatic theory has introduced various terms in modern economics, such as a middle-income model, income based on PPP theory for each population. Over the past few years, there has been very rapid economic growth and development everywhere, making the income gap between developed and developing countries smaller. The process of reincarnation in a financial institution owned by a country becomes visible because of the direction of transformation. The occurrence of such economic reincarnation spreads evenly, making contributions to developing and developed countries in the global economy (Kvint 2009).

Rapid rates of growth in several countries make developments in the foreign investment market (known as FDI) an interesting topic to talk about. Many emerging giant companies have branches and representatives in several countries in the world, signifying the rapid growth of the global economy, including markets in developing countries that have a high rate of return on investment and risk. Various efforts are carried out by, among others, prioritizing development, strengthening the empirical research side of emerging markets, and forming a better, stronger basis. With this, the process of understanding and discussion of each organization in the economy can be integrated. The identification of the main players in the global market and something that is considered prominent makes the market drivers more comprehensive. The entry of new players in the market and the dominance of old players have solidified the landscape for business economics in the last few years. Furthermore, the focus of economic problems becomes clear, even though it was initially difficult to find out an integrated relationship between markets in developing countries and those in developed countries. Small examples can be seen from work and safety standards that are increasingly developing and progressing. This can indirectly have an impact on the environmental system in a country if this small thing is implemented properly and correctly (Kong et al. 2016).

Some economic studies (Zongfang 2011) review and highlight developing markets or developing countries as objects of study. Various transaction analyses see such issues as credit, sales, and risk management as an important part of modern business activities. These studies explain that several problems arise when the internal management mechanism of organizations in large companies become poorly organized. Inadequacy of system and poor risk management are likely to occur, impairing business activities and preventing them from running properly. Marketing theories on how to make products quickly penetrate the market and save time in terms of distribution can create very good effects for market participants, especially those in developing countries. Given that, the studies review various risk analysis models and propose some economic models for emerging markets. One of the reviewed analysis models is one for assessing risks and analyzing the costs incurred, starting from credit submission until credit repayment. The occurrence of a theoretical relationship between credit consumers and credit companies from developing markets results in ownership of accounts receivable and costs incurred when granting credit to financial markets. Next, the emergence of expressions of high receivables ownership and deduction of small credit costs lead to asymmetry and spillover in financial transactions. The application of the calculation methods of adjusted earnings and the concept of risk management when credit is proposed to raise the name of risk capital for applying for credit. This is applied to prevent companies from suffering losses, which are likely to occur because the process of applying and selling credit is not done as planned.

Other researchers who also review developing markets/developing countries are Zongfang (2011 2015). The research focuses on the opening of substantial improvements in economic policy over the years and the recognition of fundamental values in developing markets. Under certain conditions, the above-mentioned are better-valued and able to make developing countries position themselves as equals with developed countries. To move towards this better direction, some developing countries are expected to be able to withstand the economic crisis that occurs, either now or shortly. Almost most developing countries have developed models, policies, and proposals to withstand such an economic crisis. With a variety of economic framework models, those countries have created a good and careful governance system to survive a turmoil that is likely to occur and to set a condition where disarray in developing countries is not allowed to last long. Various equilibrium models for supply, flow, and policy framework is implemented

to strengthen local capital markets and make financial relationships more solid. Thus, the process of product diversification and export value from developing countries to global markets is directed.

Other studies that also review emerging markets include Jebran (20170, Bae & Zhnag (2015), and Srivastava, et al (2015). Using samples from several developing countries, their research still concerns the spillover effects of a developing country's market during the crisis periods before and after the Asian crisis.

2 METHODS

The author used data on monetary economic variables such as the amount of money in circulation, inflation rates, interest rates, expected returns, and stock prices in developing countries. This study used a regression analysis model to forecast the economy. This analysis model is also widely used in understanding an independent variable about a determinant dependent variable and in seeing the exploration of these two variables in a significant relationship. In unfavorable conditions, this analysis model can also be used to make a forecast visible through graphics as is the case in the present study (Armstrong 2012; Cook &Weisberg 1982; Freedman 2005).

3 RESULTS AND DISCUSSION

In the current market economy, a newly-born theory explains that a developing country will have an advanced economy if it can run and grow like an industrial process or a company expanding its business. Developing countries that experience such phases can run if the world economy and political elements that limit it can be integrated with its economic development. Because developing countries have lower income per capita, the social-political element is very crucial. The unemployment rate, inflation rate, economic growth, and investment can work. Developed countries on average have a very high growth rate, and the income per capita for each population is also very high, whereas developing countries in Africa, Asia, Eastern Europe, and Latin America are heading toward an advanced economy. Current studies are usually based on the importance and elements of the economy, especially those relating to monetary elements in developing countries. Using data analysis techniques and the proposed data analysis model, the author tries to forecast matters related to monetary variables in several developing countries.

The regression analysis model proposed generated results showing differences due to the easing of the model proposed concerning the significant value of probability, adjusted r-square, and value of Durbin Watson. Hence, the forecasting can be generated by the graphic method as shown in Figures 1 and 2.

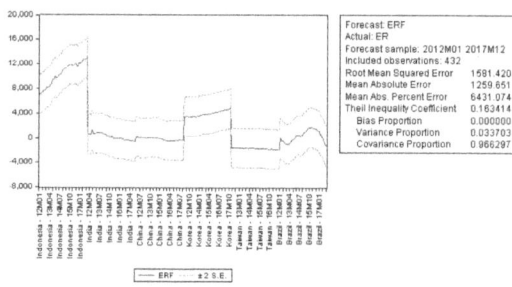

Forecast: ERF	
Actual: ER	
Forecast sample: 2012M01 2017M12	
Included observations: 432	
Root Mean Squared Error	1591.420
Mean Absolute Error	1259.651
Mean Abs. Percent Error	6431.074
Theil Inequality Coefficient	0.163414
Bias Proportion	0.000000
Variance Proportion	0.033703
Covariance Proportion	0.966297

Figure 1. Forecast from regression model 1.

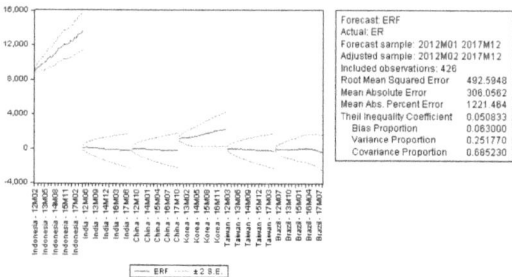

Forecast: ERF	
Actual: ER	
Forecast sample: 2012M01 2017M12	
Adjusted sample: 2012M02 2017M12	
Included observations: 426	
Root Mean Squared Error	492.5948
Mean Absolute Error	306.0562
Mean Abs. Percent Error	1221.464
Theil Inequality Coefficient	0.050833
Bias Proportion	0.083000
Variance Proportion	0.251770
Covariance Proportion	0.685230

Figure 2. Forecast from regression model 2.

4 CONCLUSION

This study seeks to investigate the forecasting function of several monetary variables in developing countries, especially the amount of money, inflation rates, interest rates, stock prices, and expected returns. Spillover effects can be seen from the generated volatility by using the proposed regression model, the results of which can be seen clearly in Figures 1 and 2 in this study. This study highlights some key findings that are considered to be interesting inputs. The discovery of two-way volatility spillover between the upper and lower limits for the forecasting model using the graph method can be used as a measure and comparison rate in predicting the monetary variables in developing countries in the future. This research is only limited to proposals and not to implications for monetary policy, especially in developing countries.

REFERENCES

Armstrong, J. Scott. 2012. "Illusions in Regression Analysis". *International Journal of Forecasting (forthcoming). 28* (3): 689. doi:10.1016/j.ijforecast.2012.02.001.

Bae, K. H., & Zhang, X. 2015. The cost of stock market integration in emerging markets. *Asia-Pacific Journal of Financial Studies, 44*(1), 1–23.

Cerutti, E., Claessens, S., & Laeven, L. 2017. The use and effectiveness of macroprudential policies: New evidence. *Journal of Financial Stability, 28*, 203–224.

Cook, D.S. and Weisberg, S. 1982, Criticism and Influence Analysis in Regression, *Sociological Methodology*, 13. pp. 313–361.

Freedman, D.A. 2005. *Statistical Models: Theory and Practice*, Cambridge University Press.

Jebran, K., Chen, S., Ullah, I. and Mirza, S.S., 2017. Does volatility spillover among stock markets varies from normal to turbulent periods? Evidence from emerging markets of Asia. *The Journal of Finance and Data Science*, 3(1–4), pp. 20–30.

Kenc, T., Erdem, F.P. and Ünalmıþ, Ý., 2016. Resilience of emerging market economies to global financial conditions. *Central Bank Review, 16*(1), pp. 1–6.

Kong, J., Zhou, Y., Lai, H., Zhang, F. and Zhou, Z., 2016. Analysis of Credit Sale Risk of Emerging Market Product. *Procedia Computer Science, 91*, pp. 362–371.

Kvint, V., 2010. *The global emerging market: Strategic management and economics*. Routledge.

Srivastava, A., Bhatia, S. and Gupta, P., 2015. Financial crisis and stock market integration: An analysis of select economies. *Global Business Review, 16*(6), pp. 1127–1142.

Vercueil, J. 2012. *Les pays émergents. Brésil–Russie–Inde–Chine... Mutations économiques et nouveaux défis "(Emerging Countries. Brazil - Russia - India - China. Economic change and new challenges*. Bréal,

Zongfang, Z. 2011. *Enterprise credit sale risk evaluation and management*. Beijing: Economic Science Press.

Zongfang, Z. 2015. *Emerging market projects risk evaluation and comprehensive management*. Beijing: Economic Management Press.

The model of financial report quality (an empirical study of financial report statement in West Java Province)

W. Roswinna & D.K. Priatna
Universitas Winaya Mukti, Sumedang, Indonesia

ABSTRACT: The purpose of this research is to know and analyze the influence of effectivity of internal control of the commitment of the organization as the application of an accounting system and accountability on the quality of government financial report in West Java. Methods used in this research was descriptive method and explanatory survey. The unit of analysis in this research was the government officials in West Java Province with a sample size of 370 people. The analysis method used was descriptive analysis and analysis of structural equation model. The research found that to improve the quality of a financial statement that dominant financial report formed by dimensions can be compared to the reports on the previous period financial report financial or other entities reporting in general (Y7). By implementing effective internal control, organization commitment, and the application of an accounting system, it can promote accountability for government financial report in West Java Province.

1 INTRODUCTION

Good governance is the issue raised in the management of public administration today. In the second semester of 2013, the Audit Board of the Republic of Indonesia (BPK) has examined some financial reports in which the district government/city in West Java still have to revise over the assessment of their financial report in 2008 until 2013. Both district and cities in West Java financial reports generally were judged to be reasonable with exceptions (WDP) at 52% of 33 entities.

Many factors that could cause financial reports produced to be less or not qualified at a local government such as low accountability financial report. This indicated that there are still many government financial statements that are still lack of the data, commitment, and the changes application accounting system of the central and regional governments.

Based on the background and the identification of the problems above, research problems can be formulated as follows:

1) Is there influence of the effectiveness of internal control over the accountability government financial statements in West Java?
2) Is there commitment of organization over the accountability of government financial statements in West Java?
3) Is there application of an accounting system over the accountability of government financial statements in West Java?
4) Is there influence of the effectiveness of internal control, commitment of organization, and application accounting system together over the accountability of government financial statements in West Java?
5) Is there influence of the effectiveness of internal control on the quality of government financial statements in West Java?
6) Is there commitment of organization on the quality of government financial statements in West Java?
7) Is there application of accounting system on the quality of government financial statements in West Java?
8) Is there accountability on the quality of financial report on the government in West Java?
9) Is there influence of the effectiveness of internal control, commitment of organization, application of accounting system, and accountability together on the quality of government financial statements in West Java?

2 LITERATURE REVIEW

According to Mulyadi (2008), the internal control was a process integral to the action of and activities undertaken by the leader and all of the employees to give confidence over the achievement of the aims of the organization through activities effectively and efficiently. Steers (1985) stated that commitment of an organization was an identification (a sense of

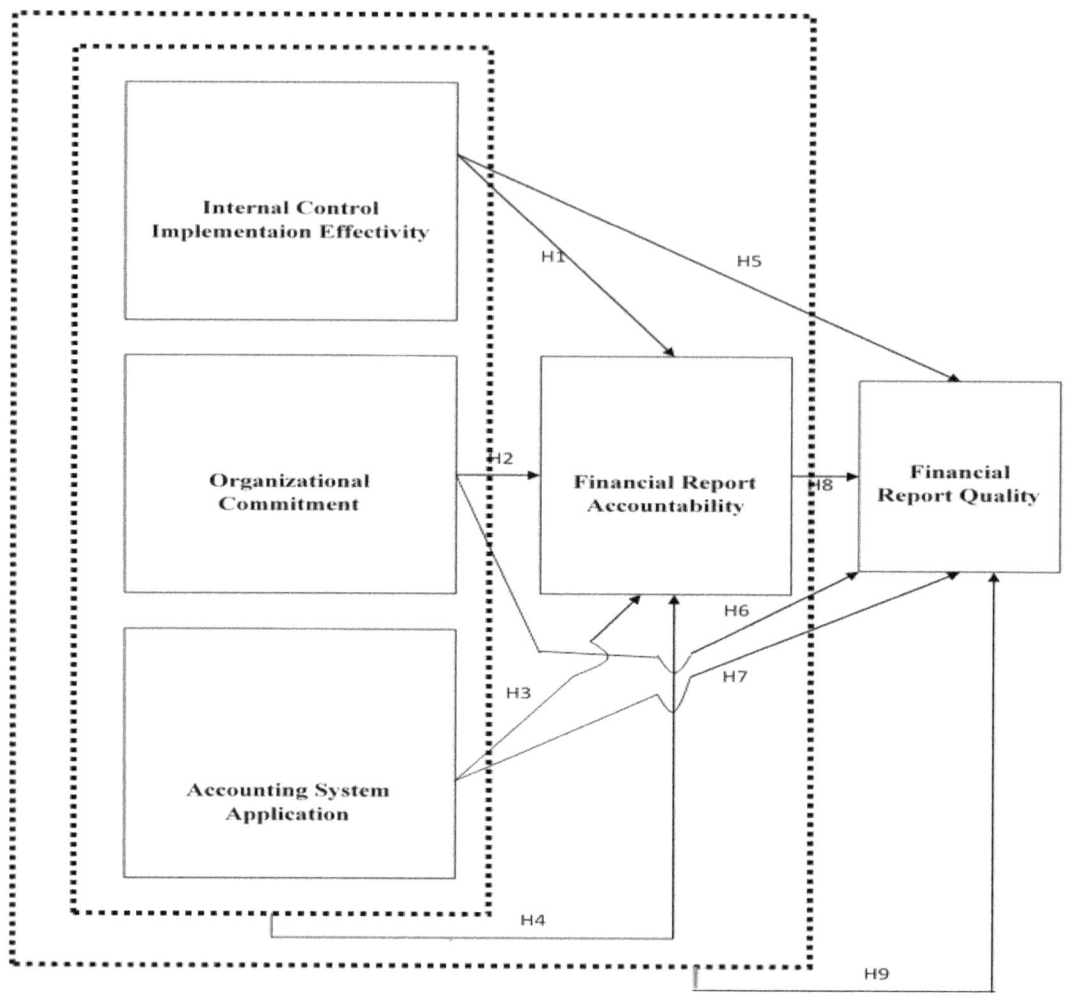

Figure 1. Conceptual framework.

confidence in the values of organization), engagements (a willingness to do their best for the sake of the interests of the organization),and loyalty (desire to become a member of an organization).

The relationship of research variables can be seen in Figure 1.

3 METHODS

This research was descriptive and verifiable. The research method used was a descriptive survey method and the method was an explanatory survey. The type of investigation in this research was the causality. The unit of analysis in this study is the individual, that is, the employees of the Regional Financial Department of West Java Province. Time horizon in this research is cross-sectional, namely, information from most of the population (a sample of respondents) was collected directly from the location of the empirical basis, with the purpose to find out the opinions of most of the population toward the object being studied.

The population in this research was SKPD officials from 33 districts in West Java Province, in the amount of 900 officials.

4 RESULTS AND DISCUSSION

The full model of equation structural equation modeling on the LISREL 8.70 software obtained two diagram models, the standardized model and t-values model; each model is shown in Figures 2 and 3.

Based on Figures 2 and 3, these calculations raised testing of γ and β parameters (coefficient factor) structural on the exogenous and endogenous models. The testing is aimed to know the influence of one variable

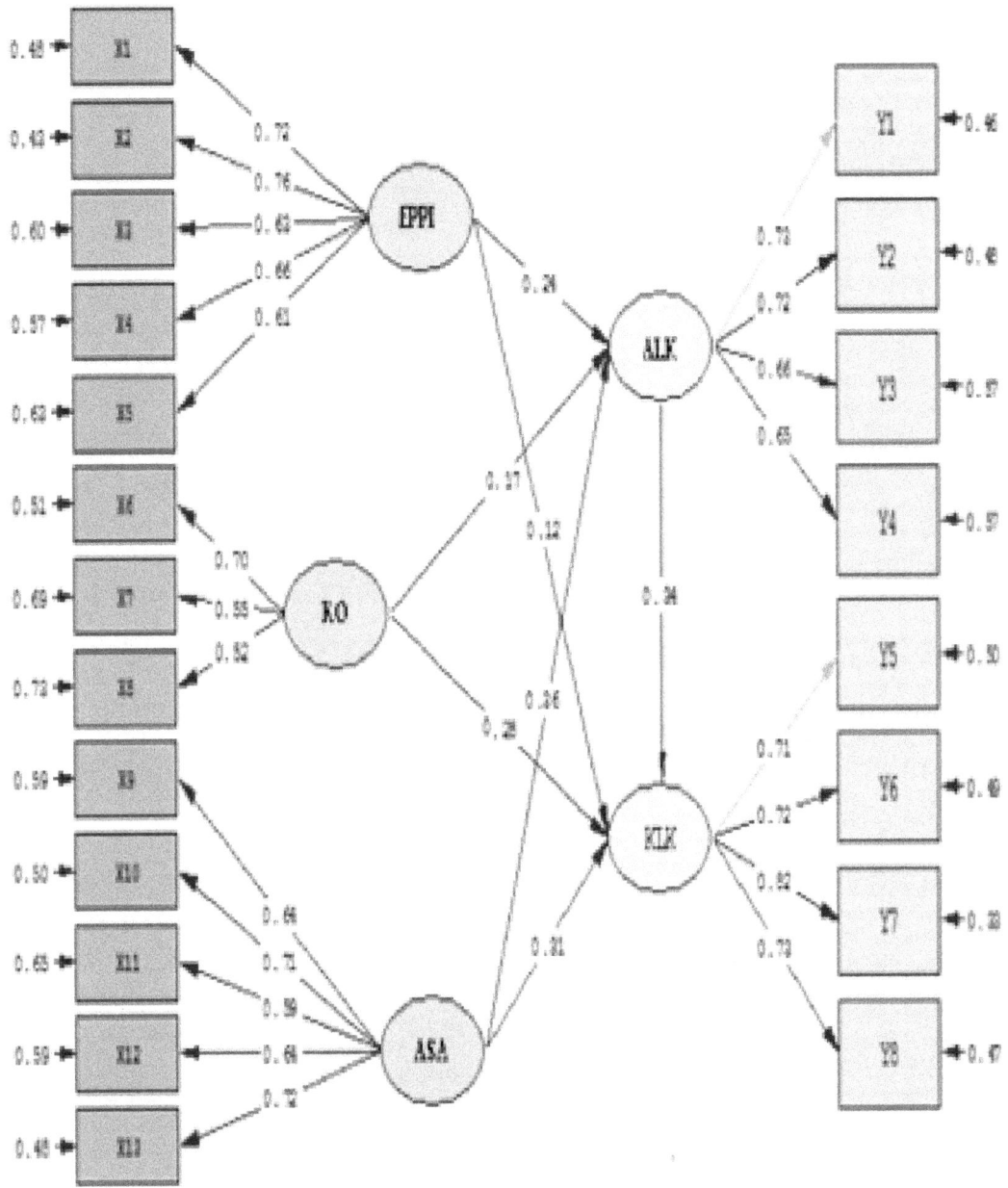

Chi-Square=634.55, df=179, P-value=0.00000, RMSEA=0.068

Figure 2. Standardized model second order.
Keterangan:
EPPI = Internal Control Implementation Effectivity
 KO = Organizational Commitment
ASA = Accounting System Application
ALK = Financial Report Accountability
KLK = Financial Report Quality

latent on other variables latent. Testing parameters γ and β are the regression coefficient standardized test (standardized regression weight) to exogen variable and endogenous, as can be seen in Table 1.

Based on the results obtained, the conclusions of all hypothesis, hypothesis one (H1) through hypothesis nine (H9), were accepted. The results of testing each hypothesis can be seen in Table 2.

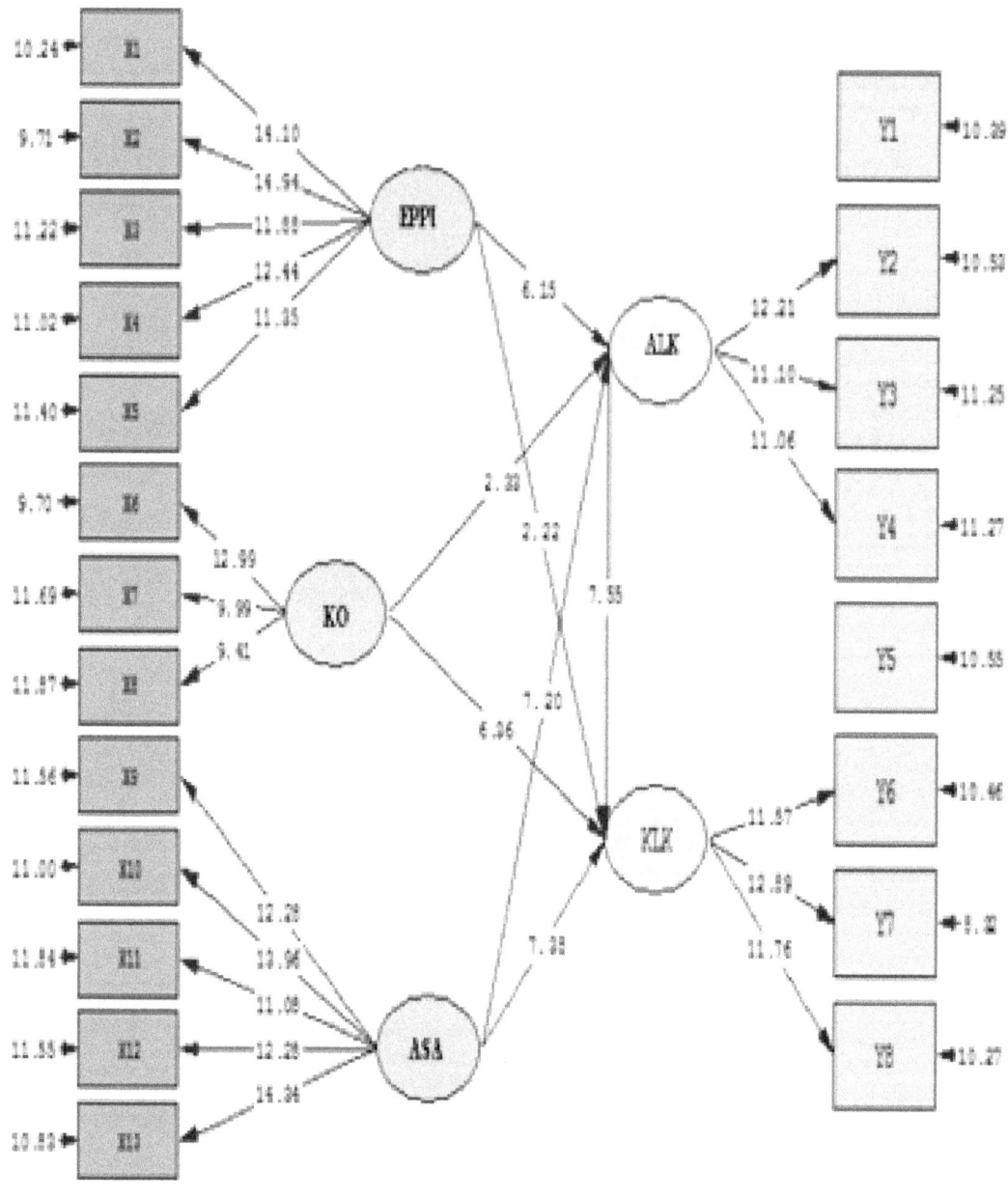

Chi-Square=634.50, df=176, P-value=0.00000, RMSEA=0.001

Figure 3. t-value model.

4.1 *Direct and indirect effect*

The accountability of financial statements (ALK) mediates the effectiveness of internal control (EPPI) on the quality of financial statements (KLK) as much as $0.24 \times 0.34 = 0.0816$ or 8.16%, while the direct effect of the effectiveness of internal control (EPPI) on the quality of financial statements (KLK) is as much as 0.122 or 0.0144 or 1.44%, so that indirect effect through ALK becomes larger, so that ALK can

be considered as variable mediating for the influence of EPPI to KLK.

Similarly, the accountability financial statements (ALK) mediate the influence of the application of accounting system (ASA) on the quality of financial statements (KLK) as much as $0.36 \times 0.34 = 0.1224$ or 12.24%, while the direct effect of the application accounting system on the quality of financial statements (KLK) as much as 0.312 or 0.0961 or 9.61%,

Table 1. Structural model statistical test.

Structure Model

Endogent Variables	Exogent/Endogent Variables	Coefficient Factor (Standardized)/ R^2	t_{hitung}/F_{hitung}	Result
ALK	EPPI	0.24	6.15	Significant (Parcial)
	KO	0.17	2.33	Significant (Parcial)
	ASA	0.36	7.20	Significant (Parcial)
KLK	EPPI, KO, ASA	0.45	99.82	Significant (Simultaneous)
	EPPI	0.12	2.22	Significant (Parcial)
	KO	0.28	6.36	Significant (Parcial)
	ASA	0.31	7.38	Significant (Parcial)
	ALK	0.34	7.55	Significant (Parcial)
	EPPI, KA, ASA, ALK	0.75	273.75	Significant (Simultaneous)

Table 2. Hypothetical test result.

Hyp	Description	Result
H1	Is the effectiveness of internal control over the accountability financial report on the government in West Java.	H_0 rejected; H_a accepted
H2	Is the commitment of organization over the accountability financial report on the government in West Java.	H_0 rejected; H_a accepted
H3	Is the application accounting system over the accountability financial report on the government in West Java.	H_0 rejected; H_a accepted
H4	Is the internal the effectiveness of control, commitment of organization and application accounting system together over the accountability financial report on the government in West Java.	H_0 rejected; H_a accepted
H5	Is the effectiveness of internal control on the quality of financial report on the government in West Java.	H_0 rejected; H_a accepted
H6	Is the commitment of organization on the quality of financial report on the government in West Java.	H_0 rejected; H_a accepted
H7	Is the application accounting system on the quality of financial report on the government in West Java.	H_0 rejected; H_a accepted
H8	Is the accountability financial report on the quality of financial report on the government in West Java.	H_0 rejected; H_a accepted
H9	Is the internal the effectiveness of control, commitment of organization, application accounting system, and accountability financial report together on the quality of financial report on the government in West Java.	H_0 rejected; H_a accepted

so that indirect effect through ALK becomes larger. Thus, ALK is considered a mediating variable for the influence of hope in KLK.

5 CONCLUSION

1) There is an effectiveness of internal control over the accountability financial report on the government in West Java.
2) There is commitment of organization over the accountability financial report on the government in West Java.
3) There is an application of accounting system over the accountability financial report on the government in West Java.
4) There is an effectiveness of control, commitment of organization, and application accounting system together over the accountability financial report on the government in West Java.

5) There is effectiveness of internal control on the quality of financial report on the government in West Java.
6) There is commitment of organization on the quality of financial report on the government in West Java.
7) There is an application of accounting system on the quality of financial report on the government in West Java.
8) There is accountability of financial report on the quality of financial report on the government in West Java.
9) There is effectiveness of internal control, commitment of organization, application accounting system, and accountability financial report together on the quality of financial report on the government in West Java.

REFERENCE

Mulyadi. 2008. Sistem akuntansi. Jakarta: Salemba Empat.

What factors influence the accounting information systems

A.F. Anggraeni
Politeknik LP3I, Bandung, Indonesia

M.L. Yulianti
Universitas Winaya Mukti, Sumedang, Indonesia

ABSTRACT: A good decision-making process is supported by accounting information quality produced by accounting information systems quality. Information technology and organizational structure are considered as factors to be considered whether the accounting information systems is in good quality or not. This study sought to examine empirically the quality of accounting information systems. Furthermore, this study examined the relationship of information technology in the quality of accounting information systems and organizational structure in accounting information systems quality. We used PLS structural equation modelling analysis to examine 78 responses from accounting and finance units on 28 universities in Bandung City, West Java Province, Indonesia. Results indicated a significant relationship between the overall influences of accounting information systems quality. Accounting information systems quality was significantly affected by information technology and organizational structure. Therefore, many universities in Bandung City, West Java Province, Indonesia must repair information technology and organizational structure to improve accounting information systems quality.

1 INTRODUCTION

In globalization, the world has become very complex (Hoffer et al. 2011), where the existence of information is important for individuals (Azhar 2013). As revealed by Davis and Olson (1985), the processing of data into information is the main activity carried out by each individual. The processing of data into information provides benefits not only for individuals but also for organizations (Hoffer et al. 2011). Hoffer, Wilkinson et al. (2000) state that information is crucial for individuals and organizations. As Azhar (2013) argues, information is very important in all aspects of life, both for individuals and organizations.

Organization members use that information in making decisions (Wilkinson et al. 2000). A similar statement by Romney and Steinbart (2015) says that when making decisions, managers and non-managers as organization members need useful information. Useful information depends on the quality inherent in the information (Wilkinson et al. 2000).

In fact, there are problems about unqualified and not useful information, as stated by Thomas Suyatno (2015) as chairperson (ABPPTSI), that there are problems with private universities in Indonesia, about financial management delaying while preparing financial reports. In line with Harry Azhar Azis's Opinion (2015) which explained that there was a debt problem to third parties in three ministries of 1.21 trillion rupiah that could not be traced and not supported by sufficient evidence.

Nonquality accounting information makes an organization not synergized, so that organizations make decisions that deviate from it (Azhar 2013). Quality accounting information comes from a good data procession, which is a representation of a quality accounting information system (Wilkinson et al. 2000). Without quality accounting information systems, data will be difficult to transform into quality accounting information (Valacich et al. 2016).

But nonintegrated accounting information systems are explained by Bambang (2015) as an e-Government Directorate from the Indonesia Ministry of Communication and Information that, so far, e-Government implemented by the government is considered to run individually. Then, the inaccessible accounting information system, explained by Hermawan (2014) as the head of the Bandung Regency Fisheries and Agriculture Office, that the banking system is difficult for farmers to access, especially small farmers, so it has a negative impact on the sustainability of its business. Accounting information systems cannot be relied upon, as explained by Harry Azhar Azis (2015) that recording of asset transfers of the Ministry of Energy and Mineral Resources cannot be explained because it is not supported yet by an adequate system to guarantee data accuracy. Muhammad Nasir (2015), as minister of Research, Technology, stated the phenomenon of inflexible accounting information systems and universities (MENRISTEKDIKTI) that the financial system of legal entity state universities (PTN-BH) has been less flexible and is feared to hamper their information systems. A similar statement was made by Salis S. Aprilian (2016), as the president director of the NGL Agency, that both systems (FLNG and OLNG) do not have high flexibility, are not reliable, and do not have broad double effects.

Problematic accounting information systems are caused by several factors, including information technology (Obrien & Marakas 2011). Laudon and Laudon (2016) suggest that important dimensions in accounting information systems are overall organizations, management, and information technology. As also stated by Effy (2009), the accounting information system runs well with the support of information technology.

Other factors that influence accounting information systems are the organizational environment, as explained by Wilkinson et al. (2000), which states that the business organization environment and organizational characteristics are precisely designed by the accounting information system in accordance with the needs of the organization. Accounting information systems in each organization reflect the main dominant resources and specific needs of the organization (Wilkinson et al. 2000). Thus, the design of information systems in organizations including accounting information systems must consider the nature and characteristics of several other subsystems expressed through organizational structure (Wilkinson et al. 2000).

2 METHODS

Adopting causal study theoretical basis, this study proposes to investigate the research model shown in Figure 1. The information technology and organizational structures are considered factors that influence the quality of accounting information systems.

2.1 Information technology and accounting information systems quality

Information technology influences the quality of accounting information systems, where Hart and Gregor (2011) argue that the use of information technology has an impact on efforts to improve the quality of information systems. Agreeing with Hart and Gregor, Wilkinson et al. (2000) suggest that information technology has a significant influence on accounting information systems. Then Thompson and Baril (2002) state that information technology, when used as part of an accounting information system, can monitor changes in an organization for sudden customer preferences, so organizations can respond quickly and increase the level of flexibility and responsiveness of the organization. With the opinion of Reynolds, Rapina (2015), Sacer and Oulic (2013), and Nelsi (2013) explained in research that information technology influences the quality of accounting information systems in an organization

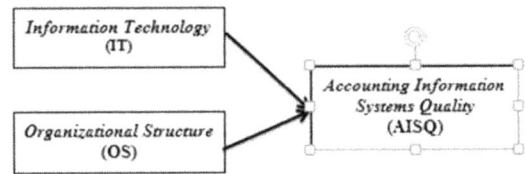

Figure 1. Research model.

Based on these arguments, therefore this study proposes that:

H1: Information technology will positively influence accounting information systems quality

2.2 Organizational structure and accounting information system quality

Organizational structure has a significant influence on accounting information systems and their components (Wilkinson et al. 2000). In line with his opinion, Wilkinson, Laudon, and Laudon (2016) reveal that accounting information systems in business companies usually reflect the type of organizational structure. Then, as supported by Stair and Reynolds (2010), organizational structure depends on the objectives and management approach and has an impact on the use of accounting information systems. The opinions of these experts are supported by Rapina's (2015) research, Sri Dewi Anggadini (2015), and Inta Budi Setya Nusa (2015), which explain that organizational structure has an influence on the quality of accounting information systems.

Based on these arguments, therefore, this study proposes that:

H2: Organizational structure will positively influence accounting information systems quality

This study used employees of the Financial and Ac-counting Unit on Higher Education Institutions in Bandung City, West Java Province, Indonesia as data sources to test its hypotheses. In collaboration with each of the employees of the financing and accounting units, they were sent to participate on a paper-based survey. In the survey instrument, the participants were requested to answer the questions based on their own organization. The identity of the organization was not released. Participants must select one perspective to answer the questions.

There are two main reasons why this study used the employees of financing and accounting units. First, individuals who are using the accounting information systems in organizations are the employees of financing and accounting units. Second, the employees of financing and accounting units will choose correct answers while researchers ask about the quality of accounting information systems. Third, researchers found many phenomena about accounting information systems quality on higher education institutions in Indonesia, particularly in Bandung City. For these reasons, the employees of financing and accounting units were regarded as appropriate subjects for this study.

Table 1. Average variance extract.

Dimension	AVE	Explanation
Functionality	0.820	Valid
Compatibility	0.894	Valid
Maintainability	0.667	Valid
Chain of Command	0.857	Valid
Span of Control	0.793	Valid
Accessibility	0.763	Valid
Flexibility	0.710	Valid
Integration	0.942	Valid
Reliability	0.644	Valid

Table 2. Composite reliability.

Dimension	CR	Explanation
Functionality	0.901	Reliable
Compatibility	0.944	Reliable
Maintainability	0.800	Reliable
Chain of Command	0.923	Reliable
Span of Control	0.884	Reliable
Accessibility	0.866	Reliable
Flexibility	0.830	Reliable
Integration	0.891	Reliable
Reliability	0.783	Reliable

2.3 Operational measures of the study variable

The following variable was adopted from prior studies and prior theory in textbooks. All the variables (dependent and independent variables) were measured using five- to seven-point Likert scales.

Dependent Variable

Accounting information system quality (AISQ) was measured using eight questions based on four characteristics: integration, accessibility, flexibility, and reliability (Azhar 2013; Barrier 2002; Bocij et al. 2015; Heidmann 2008; Khosrow 2007; Stair & Reynolds 2012).

Independent Variable

Information technology was measured using six questions based on their indicators: capacity, speed, conformance to standards, interoperability, modularity, scalability (Alter 2002; Khosrow 2007; Thompson & Baril 2003).

Organizational structure was measured using six questions based on their indicators such as ability to communicate, degree of specialization, unity of command, and authority (Gibson et al. 2012; Griffin & Moorhead 2014; Ivancevich et al. 2014; Mullins 2014; Robbin & Judge 2012; Robbins & Judge 2013).

2.4 Reliability of the measures

Based on the data of the survey instrument from PLS-SEM statistical analysis, Tables 1 and 2 show that the reliability estimates (composite reliability [CR]) and the validity of loading factors estimates (average variance extracted [AVE]) of the measures of the planned variable are well above acceptable.

3 RESULTS AND DISCUSSION

3.1 Response rate

The invitations of paper-based survey were sent out to 150 members of financing and accounting units of higher education institutions in Bandung City. The total of completed and useable responses was 78, thus the response rate for this survey is 52%.

3.2 Sample characteristics

Regarding the financing and accounting background of respondents, the study revealed that there were 25 men (32.10%), 49 women (62.80%), and 4 unknowns (5.10%). The data indicates that employees with positions as manager were 21 people (26.90%), as staff were 54 people (69.20%), and unknowns were 3 people (3.80%). The education background of the financing and accounting units showed that 7 people (9.0%) had diplomas, 15 people (19.20%) had bachelor of financing/accounting degrees, 16 people (20.50%) had masters of financing/accounting, and 3 people (3.80%) were unknown. The experience of financing and accounting units reported in this study results were approximately 26 people with 1 to 5 years work experience (33.30%) and 19 people with more than 20 years work experience (24.40%).

3.3 Data examination

The structural equation modeling technique (SEM) was used to test hypotheses in this model. Hair et al. (2014) state there are two reason that a model must be using SEM: first, the variable was unobservable and second, the research approach was causal study. In this case, the sample in this study is limited, as researchers were using the PLS-SEM approach to test hypotheses in this model.

3.4 Model assessment

3.4.1 Measurement Model Fit

In this study, the entire first-order measurement model is the relationship between dimension and the indicators, and it is reflective. In measuring, the quality of an indicator is seen from the results of validity and reliability of indicators as well as the results of validity and reliability of one indicator and others to form dimensions.

**First-Order Measurement
Indicators Validity**

The validity of each reflective indicator can be seen from the significance of the loading factor. If testing is significant, then the indicator is said to be valid in measuring the dimensions or constructs mentioned. Based on the results of testing validity using a significance level of 0.05, it can be concluded that all

Table 3. Sample characteristics.

Sex	Frequencies	Percentage
Man	25	32.10%
Woman	49	62.80%
Unknown	4	5.10%

Positions	Frequencies	Percentage
Manager	21	26.90%
Staff	54	69.20%
Unknown	3	3.80%

Education	Frequencies	Percentage
High School	7	9.00%
Diploma Finance/Accounting	15	19.20%
Bachelor Finance/Accounting	37	47.40%
Master Finance/Accounting	16	20.50%
Doctoral Finance/Accounting	0	0
Unknown	3	3.80%

Experience	Frequencies	Percentage
1–5 years	26	33.30%
6–10 years	12	15.40%
11–15 years	8	10.30%
16–20 years	7	9.00%
>20 years	19	24.40%
Unknown	6	7.70%

Table 4. Dimensions of information technology coefficients compatibility.

	Chain of Command	
Coefficients	DCOC1	DCOC2
Estimation	0.934	0.918
STDEV	0.013	0.018
t-Statistics	69.430	51.290
P-Value	0.00	0.00
Sig.	Sig	Sig.
R2	0.872	0.843

Table 5. Dimensions of information technology.

	Functionality	
Coefficients	DF1	DF2
Estimation	0.947	0.944
STDEV	0.011	0.013
t-Statistics	88.622	71.018
P-Value	0.00	0.00
Sig.	Sig	Sig.
R2	0.897	0.891

Table 6. Dimensions of information technology.

	Maintainability	
Coefficients	DM1	DM2
Estimation	0.796	0.837
STDEV	0.05	0.038
t-Statistics	15.840	22.286
P-Value	0.00	0.00
Sig.	Sig	Sig.
R2	0.634	0.701

Table 7. Dimensions of organizational structure.

	Span of Control	
Coefficients	DSPOC1	DSPOC2
Estimation	0.876	0.905
STDEV	0.034	0.021
t-Statistics	25.517	43.911
P-Value	0.00	0.00
Sig.	Sig	Sig.
R2	0.737	0.819

Table 8. Dimensions of organizational structure.

	Compatibility	
Coefficients	DK1	DK2
Estimation	0.914	0.896
STDEV	0.015	0.027
t-Statistics	61.459	33.037
P-Value	0.00	0.00
Sig.	Sig	Sig.
R2	0.835	0.803

loading factors differ from zero (significant), so it can be stated that all indicators are valid for measur-ing their dimensions. Based on Tables 4–11, the value of R2 is more than zero with a significance level of 0.05. So the indicators used in this study are significant.

Indicators Reliability
Measurement of the reliability of indicators can be seen from the value of R2. An indicator is said to be reliable if the R2 value is not less than 0.50. Based on the results of testing the reliability of indicators presented in Table 12–20, it can be concluded that all indicators have R2 values of more than 0.50. Therefore, all indicators have a good level of reliability.

Convergence Validity of Indicators in Each Di-mension
Based on the results of data processing in Tables 12–20, it can be seen that all dimensions have AVE values more than 0.50, which means that all dimensions have indicators with good convergence validity levels.

Table 9. Dimensions of accounting information systems quality.

| Coefficients | Accessibility | |
	DA1	DA2
Estimation	0.878	0.869
STDEV	0.021	0.029
t-Statistics	42.001	29.629
P-Value	0.00	0.00
Sig.	Sig	Sig.
R2	0.771	0.755

Table 10. Dimensions of accounting information systems quality.

| Coefficients | Flexibility | |
	DF1	DF2
Estimation	0.859	0.826
STDEV	0.035	0.042
t-Statistics	24.684	19.533
P-Value	0.00	0.00
Sig.	Sig	Sig.
R2	0.738	0.683

Table 11. Dimensions of accounting information systems quality.

| Coefficients | Span of Control | | | |
	DSPOC1	DSPOC2	Explanation	
R2	0.737	0.819	Valid	Reliable
CR	0.884		Valid	Reliable
AVE	0.793		Valid	Reliable

Table 12. Dimensions of information technology.

| Coefficients | Functionality | | | |
	DF1	DF2	Explanation	
R2	0.897	0.891	Valid	Reliable
CR		0.901	Valid	Reliable
AVE		0.820	Valid	Reliable

Table 13. Dimensions of information technology.

| Coefficients | Compatibility | | | |
	DK1	DK2	Explanation	
R2	0.835	0.803	Valid	Reliable
CR		0.944	Valid	Reliable
AVE		0.894	Valid	Reliable

Internal Consistency Reliability of Indicators in Each Dimension

Internal consistency reliability in each dimension is measured through composite reliability with minimum value of 0.70. The results are shown in

Table 14. Dimensions of information technology.

| Coefficients | Maintainability | | | |
	DK1	DK2	Explanation	
R2	0.634	0.701	Valid	Reliable
CR		0.800	Valid	Reliable
AVE		0.667	Valid	Reliable

Table 15. Dimensions of organizational structure.

| Coefficients | Chain of Command | | | |
	DCOC1	DCOC2	Explanation	
R2	0.872	0.843	Valid	Reliable
CR		0.923	Valid	Reliable
AVE		0.857	Valid	Reliable

Table 16. Dimensions of organizational structure.

| Coefficients | Reliability | |
	DR1	DR2
Estimation	0.760	0.843
STDEV	0.06	0.054
t-Statistics	12.744	15.733
P-Value	0.00	0.00
Sig.	Sig	Sig.
R2	0.578	0.711

Table 17. Dimensions of accounting information systems quality.

| Coefficients | Accessibility | | | |
	DA1	DA2	Explanation	
R2	0.771	0.755	Valid	Reliable
CR		0.866	Valid	Reliable
AVE		0.763	Valid	Reliable

Tables 12–20, where the CR value is greater than 0.70, which means that the indicators of each dimension have good internal consistency.

Table 18. Dimensions of accounting information systems quality.

| Coefficients | Flexibility | | | |
	DF1	DF2	Explanation	
R2	0.738	0.683	Valid	Reliable
CR		0.830	Valid	Reliable
AVE		0.710	Valid	Reliable

Discriminant Validity

Evaluation of discriminant validity in each indicator on dimensions can be seen based on Dell-Larcker Criteria in Table 21. The criteria value in a dimension must be larger to the side itself compared to other dimensions (Hair et al. 2014). Table 21 shows that all the criteria values for each dimension (which are in the main diagonal) are greater than the criteria for the other dimensions (outside the main diagonal). Therefore, indicators on these dimensions have good discriminant validity.

Second-Order Measurement Assessment

In evaluating the second stage of the measurement model (i.e., the relationship between dimensions and constructs), in this study there is only one type of measurement model to be discussed, namely the reflective measurement model.

Dimensions Validity

The validity of each dimension can be seen from the significance of the loading factor. If the significance testing is fish, then the dimension is said to be valid in measuring the construct. Based on the results of testing validity using a significance level of 0.05, it can be concluded that all loading factors differ from zero (significant) so that it can be stated that all dimensions are valid for measuring the construct. Based on Tables 22–27, the value of R2 is more than zero with a significance level of 0.05. So the dimensions used in this study are significant.

Dimensions Reliability

Measurement of the reliability of a dimension can be seen from the value of R2. An indicator is said to be reliable if the R2 value is not less than 0.50. Based on the results of testing the reliability of indicators presented in Tables 22–27, it can be concluded that all of them are assessed as having the value R2 (column d) of more than 0.50. Therefore, all dimensions have a good level of reliability

Table 19. Dimensions of accounting information systems quality.

	Integration			
Coefficients	DI1	DI2	Explanation	
R2	0.869	0.903	Valid	Reliable
CR	0.891		Valid	Reliable
AVE	0.942		Valid	Reliable

Table 20. Dimensions of accounting information systems quality.

	Reliability			
Coefficients	DR1	DR2	Explanation	
R2	0.578	0.711	Valid	Reliable
CR	0.783		Valid	Reliable
AVE	0.644		Valid	Reliable

Table 22. Information technology.

	Information Technology		
Coefficients	Funct	Comp	Maint
Estimated	0.925	0.872	0.863
STDEV	0.012	0.038	0.056
t-Statistics	74.335	22.829	15.550
P-Value	0.00	0.00	0.00
Sig.	Sig	Sig.	Sig.
R2	0.836	0.776	0.775

Table 21. Discriminant validity.

Fornell Larcker Criterion Untuk Validitas Diskriminasi Model Pengukuran Tahap Pertama

	Funct	Comp	Maint	Chain OC	Span OC	Access	Flexi	Integ	Relin	Acc	Comp	Relev	Timel
Funct	0.945												
Comp	0.716	0.905											
Maint	0.717	0.618	0.817										
Chain OC	0.321	0.414	0.469	0.926									
Span OC	0.567	0.502	0.227	0.606	0.89								
Acc-Ess	0.584	0.420	0.452	0.413	0.413	0.873							
Flexi	0.310	0.324	0.209	0.701	0.643	0.653	0.842						
Integ	0.449	0.301	0.421	0.582	0.564	0.628	0.673	0.944					
Relin	0.513	0.441	0.363	0.519	0.708	0.647	0.642	0.553	0.803				
Acc	0.575	0.460	0.532	0.338	0.520	0.635	0.461	0.461	0.635	0.890			
Comp	0.495	0.345	0.316	0.314	0.567	0.620	0.414	0.414	0.699	0.77	0.899		
Relev	0.539	0.366	0.489	0.421	0.534	0.704	0.588	0.588	0.699	0.784	0.786	0.840	
Timel	0.493	0.298	0.310	0.421	0.644	0.507	0.507	0.507	0.685	0.690	0.789	0.750	0.932

Convergence Validity of Dimensions in Each Construct

Based on the results of data processing in Tables 22–27, it can be seen that all dimensions have AVE

Table 23. Organizational structure.

	Organizational Structure	
Coefficients	Chain of Command	Span of Control
Estimated	0.869	0.898
STDEV	0.042	0.022
t-Statistics	20.550	40.333
P-Value	0.00	0.00
Sig.	Sig	Sig.
R2	0.790	0.805

Table 24. Accounting information system quality.

	Accounting Information Systems Quality			
Coefficients	Access	Flexi	Integ	Reliable
Estimated	0.857	0.865	0.818	0.844
STDEV	0.043	0.032	0.047	0.035
t-Statistics	19.869	26.944	17.428	25.858
P-Value	0.00	0.00	0.00	0.00
Sig.	Sig.	Sig.	Sig.	Sig.
R2	0.745	0.770	0.682	0.693

Table 25. Information technology.

	Information Technology				
Coefficients	Funct	Comp	Maint	Explanation	
R2	0.856	0.760	0.745	Valid	Reliable
CR		0.918		Valid	Reliable
AVE		0.788		Valid	Reliable

Table 26. Organizational structure.

	Organizational Structure			
Coefficients	Chain of Command	Span of Control	Explanation	
R2	0.755	0.806	Valid	Reliable
CR		0.877	Valid	Reliable
AVE		0.781	Valid	Reliable

Table 27. Accounting information system quality.

	Accounting Information Systems Quality					
Coefficients	Access	Flexi	Integ	Reliable	Explanation	
R2	0.734	0.748	0.669	0.712	Valid	Reliable
CR			0.910		Valid	Reliable
AVE			0.717		Valid	Reliable

values more than 0.50, which means that all dimensions have indicators with good convergence validity levels.

Convergence Reliability of Dimensions in Each Construct

Internal consistency reliability of dimensions in each construct is measured through composite reliability with a minimum value of 0.70. Tables 22–27, column g, show the results of data processing, where the CR value is greater than 0.70, which means that the dimensions of each construct have good internal consistency.

3.4.2 *Structural Model Fit*

Evaluation of structural models is the result of analyzing the relationship of each construct with the objectives set in this study, namely, to determine the effect of information technology and organizational structure on the quality of accounting information systems and their impact on the quality of accounting information. To find out the effect of information technology and organizational structure on the quality of accounting information systems and their impact on accounting information quality, a predictive analysis tool, Partial Least Structural Modeling (PLS-SEM), was used.

Before testing the structural model in this study, an evaluation was carried out first to find out whether there was a correlation between the independent variables, namely, information technology and organizational structure.

Based on the results of testing the hypothesis in Table 28, it can be seen that the value of P-Value is 0.042. This shows that the P-Value is smaller than 0.05, which means that information technology influences the quality of accounting information systems at

Table 28. Structural model.

	Causal Effect	
Coefficients	IT-SIAQ	SO-SIAQ
Estimation	0.173	0.528
STDEV	0.083	0.065
Statistics-t	2.071	8.119
P-Value	0.042	0.000
Sig.	Sig.	Sig
VIP	1.411	1.411
Collinearity	Non-Kol	Non-Kol

a significance level of 5%. The effect is positive with a low influence of 0.173. This condition means that for every increase in 1 standard deviation of information technology scores on the quality of accounting information systems then causes an increase in the accounting information system quality score on average by 0.069, assuming other variables are in constant condition. That is, the better the application of information technology, the more quality the accounting information system is applied.

Based on the results of testing the hypothesis in Table 28, it can be seen that the P-Value value is 0.00. This shows that the P-Value is smaller than 0.05, which means that the Organizational Structure affects the Quality of Accounting Information Systems at a significance level of 5%. The effect is positive with moderate influence of 0.528. This condition means that for every increase in 1 standard deviation score of organizational structure in the quality of accounting information systems then causes an increase in the score of accounting information system quality on average by 0.065 assuming other variables are in constant condition. That is, the better the implementation of the organizational structure, the more quality the accounting information system is applied.

Influence Information Technology to Accounting Information Systems Quality
Based on the results of descriptive statistical analysis, it shows that the average score of the answers to the majority of higher education analysis units that make up the sample of the study on the quality of accounting information systems are worth 3.59 in the good category. This condition is because there are several analysis units of universities that have not accommodated information technology in the accounting information system that they use for 1.5%.

Following is the description of each dimension represented by indicators in reflecting the influence of information technology on the quality of accounting information systems:

(a) The functionality dimension, measured using the capacity and speed indicators, which are reflections of information technology variables, has an average value of answers to the universities' unit of analysis, which is a sample of 3.65 and 3.45 in the good category. This can be interpreted that the accounting information system used by the majority of universities, which are the research samples, has adopted a good information technology device, reflected in the capacity and speed of the information technology devices that support the accounting information system to be of high quality.

(b) The compatibility dimension measured using the conformance to standard and interoperability indicators has an average value of answers in the unit of analysis of universities, which is a sample of 3.65 and 3.62 in the good category. This can be interpreted that the accounting information system used by the majority of universities that are the research samples have adopted a good information technology device, reflected in the suitability of standards and interoperability of the information technology devices that support the accounting information system to be of high quality.

(c) The maintainability dimension measured using the modularity and scalability indicators has an average value of answers to the universities' unit of analysis, which is a sample of 3.54 and 3.63 in the good category. This can be interpreted that the accounting information systems used by the majority of universities and institutions, which are the research samples, have adopted a good information technology device, reflected in the modularity of space and scalability of information technology devices that support the accounting information system to be of high quality.

Based on the results of descriptive statistical analysis on the dimensions above, in general the unit of analysis of universities, which is the sample of research, has adopted a good information technology device and can improve the quality of accounting information systems. However, from the results of the calculation of the relative frequency there are about 1.5% of the analysis units of universities that have not adopted information technology properly, so that the accounting information system used is not yet qualified.

According to Hart & Gregor (2011), it is said that the use of information technology has an impact on efforts to improve the quality of accountant information systems. Agreeing with Hart and Gregor, Wilkinson et al. (2000) suggests that information technology has a significant influence on accounting information systems. Thompson and Baril (2002) state that information technology, when used as part of an accounting information system, can monitor changes in an organization for sudden customer preferences, so organizations can respond quickly and increase the level of flexibility and responsiveness of the organization. These underlying theories are reinforced and in line with the research conducted by Rapina (2015), Sacer & Oulic (2013), and Nelsi (2013), which requires that information technology influences the quality of accounting information systems in an organization.

Theories and empirical evidence from previous studies do not adequately meet to answer the problem formulation regarding the unit of analysis of higher education, which is the research sample. The results of hypothesis testing to explain that the path correlation coefficient show a value of 0.173 with low influence criteria, which means that information technology contributes relatively low in improving the quality of accounting information systems.

To follow up on the results of hypothesis testing, the author conducted a brief interview for the manager/head of the Accounting and Finance and Staff Section. From the results of the interviews conducted, the causes of information technology did not affect the quality of accounting information systems or interpreted information technology to contribute low in/on accounting information system quality is the number of units of analysis of universities, which is the research samples still using information technology that is not up-to-date, even in some universities that are research samples that have not used accounting information systems or manual systems. These results, financial statements, or accounting information produced are not qualified.

Influence Organizational Structure to Account-ing Information Systems Quality

Based on the results of descriptive statistical analysis, it shows that the average score of the answers to the majority of higher education analysis units that are the sample of the study of accounting information system quality are worth 3.75 in the good category. The convention is because there are several university analysis units that have not accommodated the organizational structure in the accounting information system that they use for 2.25%.

Following is the description of each dimension represented by indicators in reflecting the influence of organizational structure on the quality of accounting information systems:

(a) The dimension of chain of command measured using the indicators of unity of command and authority, which is a reflection of the organizational structure variable, has an average value of answers to the unit of analysis of universities, which is a sample of 3.88 and 3.68 with good categories. This can be interpreted that the accounting information system used by the majority of universities that are the sample of research have adopted the organizational structure well, reflected in the command unit of each unit and authority of the organizational structure adopted can support the accounting information system to be of high quality.

(b) The dimensions of span of control, which are measured using the ability to communicate and degree of specialization indicators, have an average value of answers in the unit of analysis of universities, which is a sample of 3.65 and 3.83 in the good category. This can be interpreted that the accounting information system used by the majority of universities that are the sample of research have adopted a good organizational structure, reflected in the ability of each unit to communicate, and the level of specialization of employees in the organizational structure that can support the accounting information system to be qualified.

Based on the results of descriptive statistical analysis on the above dimensions, in general the unit of analysis of higher education, which is the sample of the study, has adopted the organizational structure well and can improve the quality of accounting information

systems. However, from the results of the calculation of the relative frequency there is around 2.25% of the unit of analysis of universities that have not adopted the organizational structure into the accounting information system used, so that the accounting information system used is not yet qualified.

According to Wilkinson et al. (2000), the organizational structure has a significant influence on the accounting information system and its components. In line with his opinion, Wilkinson et al (2016) reveal that accounting information systems in business companies usually reflect the type of organizational structure. Then, supported by Stair and Reynolds (2010), organizational structure depends on the objectives and management approach and has an impact on the use of accounting information systems. These underlying theories are supported by Rapina's (2015) research, Sri Dewi Anggadini (2015), and Inta Budi Setya Nusa (2015), which explains that organizational structure influences the quality of accounting information systems.

The theory and empirical evidence from research previously did not meet enough to answer the problems related to the unit of analysis of universities that became the research sample. The results of hypothesis testing explain that the path correlation coefficient shows a value of 0.528 with quite high influence criteria, which means that the organizational structure contributes relatively low in/on the improvement in the quality of accounting information system.

To follow up on the results of testing the hypothesis, the author conducted a brief interview for the manager/head of the Accounting and Finance and Staff Section. From the results of the interviews conducted, the cause of the organizational structure that contributed low in/on the quality of accounting information systems was the number of units of analysis of universities that became research samples have an organizational structure that is "fat," even in some universities that are the sample of the study still have employees who have more than one job description and employees do not fully carry out the functions of an organizational structure because they are constrained by inappropriate educational backgrounds. These results, financial statements, or accounting information produced are not qualified.

4 CONCLUSION

4.1 *Summary*

1) Information technology influences the quality of accounting information systems. The quality of accounting information systems caused by information technology has not yet been fully implemented in the accounting information system.

2) Organizational structure influences the quality of accounting information systems. Organizational structures that have not been fully adopted by the

accounting information system cause the lack of quality accounting information systems.

4.2 *Contribution*

Based on the results of the research and discussion and conclusions, the authors can provide the following suggestions:

1) Information technology influences the quality of accounting information systems so that it has an impact on improving the quality of accounting information. Therefore, there are several considerations for developing information technology, as follows:

   (a) Concerning aspects of functionality are reflected in capacity and speed. This capacity and speed will support employees on carrying out their duties, which intersect with accounting information systems. By improving the capacity and speed of an information technology device, the accounting information system becomes a quality of its application.
   (b) Supports compatibility aspects that are reflected by adjusting to standards and interoperability. Adjustments to standards and interoperability support employees in carrying out their functions in the organization. This is because the organization is able to provide an accounting information system in synergy with the information technology applied, so that the accounting information system becomes qualified for producing accounting information.
   (c) Implementing information technology devices are easily maintained, which is reflected by the creation of each subsystem space and is easily upgraded. These result in employees more easily adjusting themselves to changes in the organization related to the preparation of accounting information in the organization. The more quality accounting information systems supported by adequate information technology devices, the more quality accounting information is produced.

2) Organizational structure affects the quality of accounting information systems so that it has an impact on improving the quality of accounting information. Therefore, there are several considerations for developing an organizational structure, as follows:

   (a) Responsibility for the work of each manager/section head and staff must be clearly demonstrated to the application of the accounting information system. Individual duties and authorities of the organization are contained in documents that have been signed by the authorities. The management range in the organization shows the position of the boss and the caretaker that must be considered that every job carried out should be clearly accountable to whom, so that the use of an accounting information system is directed and able to produce quality accounting information
   (b) Accounting information systems must be able to help a leader in paying attention to effectiveness and efficiency when overseeing a number of people that he leads. The use of this accounting information system can help optimally the supervision carried out by the leader.

4.3 *Limitation and recommendation future study*

The results of this study are expected to be useful for the academic world and for other researchers in developing the knowledge described as follows:

1) By striving to fulfill scientific research characteristics, namely generalizability and replicability. The researchers suggest to other researchers to do the research again with the same research method, but on different units of analysis and samples, with the aim of producing the same results (replicability). Research carried out on an ongoing basis can increase the confidence in research that has been carried out, and the usefulness of the research can be accepted with a broad scope of applicability in various types of organizations (generalizability).

2) This research is deemed not to reveal fully the variables relating to and affecting the quality of the accounting information system. Therefore, other researchers are recommended to examining other variables outside of this study, to enrich scientific knowledge

REFERENCES

Alter, S. 2002. Information systems: the foundation of e-business. Fourth Edition. USA: Prentice Hall.

Azhar, S. 2013. Accounting information systems: development risk management structures. Bandung: Lingga Jaya.

Bambang, D.A. 2015. Smart government accounting information systems: structure development risk control. [Online]. Retrieved from http://www.babelprov.go.id/content/smart-government-system- integrated.

Barrier, T. 2002. Human computer interaction development and management. USA: IRM Press.

Bocij, P., Greasley, A., & Hickie, S. 2015. Business information systems. Fifth edition. USA: Pearson Education Limited.

Effy, O. 2009. Management information systems. Sixth edition. USA: Cengage Learning

Gibson, J.L., Ivancevich, J.M., Donnelly Jr, J.H., Konopaske, R. 2012. Organizations: behavior, structure, and processes. New York: McGraw Hill.

Griffin, R.W., & Moorhead, G. 2014. Organizational behavior: managing people, and organizations. 11th edition. USA: Cengage Learning.

Hair, J.F, Hult, T.M., Ringle, C.M., & Sarstedt, M. 2014. A primer on Partial Least Square Structural Equation Mod-Ellung (PLS-SEM). USA: SAGEPUB.

Hart, D.N & Gregor, S.D. 2011. Information systems foundation: constructing and critising. Australia: ANUE PRESS.

Heidmann, M. 2008. The role of management accounting information system in strategic sensemaking. Germany: GMWV Fachverlage GmbH.

Hoffer, J.A., Ramesh, V., & Topi, H. 2011. Modern database management. 10th edition. New York: Pearson Learning.

Inta, B.S.N. 2015. Influence of organizational culture and structure on quality of accounting information system. *International Journal of Scientic and Technology Research* 4(5).

Khosrow, M. 2007. Dictionary of information science and technology. 1. USA: Idea Group Reference.

Laudon, K.C., & Laudon, J.P. 2016. Management information system: managing the digital film. 14th edition. USA: Pearson Learning.

Mullins, L.J. 2014. Management and organizational behavior. Pearson Spain: Education Limited.

Nasir, M. 2015. Menristekdikti: ptn-bh finance is less flexible. [Online]. Retrived from http://www.antaranews.com/berita/517386/menristekdikti-keuangan-ptn-bh-kurang-fleksibel.

Nelsi, W. 2013. The effect of information technology on the quality of information system and its impact on the quality of accounting information. *Research Journal of Finance and Accounting* 4(5). [Online]. Retrieved from www.iiste.org.

Obrien, J.A., & Marakas, G.M. 2011. Management information systems. 10th edition. New York: McGraw Hill.

Rapina. 2015. Factors that effect accounting systems and accounting information (survey on local bank in Bandung-Indonesia). *Australian Journal of Basic and Applied Sciences*: 78–86.

Micro credit agreement between individual customers and business entities with banking institutions in the legal perspective of business contract design

D. Anggraenia, A.E. Suyono & H. Suyanto
Universitas Pamulang, Tangerang Selatan, Indonesia

N.L. Krisna
Univeristas Persada Indonesia YAI, Jakarta, Indonesia

ABSTRACT: As a business entity, banking institutions that include state owned banks, national private banks, and foreign private banks, play a vital role in both maintaining funding sustainability and raising funds from and by the community. The existence of the banking world is an absolute prerequisite for the growth of the economy of a country. There should be various regulations and legal standards to update the development of the banking world. In addition, the bank should be selective in giving micro credit approval to either individual customers or micro business entities to avoid the risk of bad debt. Furthermore, the guarantee of immovable assets is crucial in dismissing all adverse effects to occur when the default takes place.

1 INTRODUCTION

Banking is an institution that relies on public trust. This institution is an entity that collects and distributes funds from and to the community. The existence of banks both state-owned banks (BUMN), national private banks and foreign private banks that have established and obtained operational licenses in Indonesia have become part of the driving system of the Indonesian economy. With a system that has been integrated with all elements and joints of payments for all interconnected economic system transactions between one institution and another, it is clear that the role of banking is very crucial and very vital. Without it there will be stall in the payment system which will result in the stagnation of the system and function of the economy on a very massive scale and fatal and detrimental to the joints of the economy of a country. The experience of the 1998 monetary crisis which affected the fall of the economy and the value of our currency on foreign currencies, especially the US dollar, is evidence of the importance of maintaining the integrity of domestic banking in the global economy.

Returning to the function of banking as an institution channeling funds to the public in the form of granting credit in the form of working capital loans (micro) certainly cannot be separated from the regulations stipulated in Law No. 10 of 1998 changes to Law No. 7 of 1992 concerning Banking. In Article 1 point 11 of Law No. 10 of 1998 regarding amendments to Law No. 7 of 1992 concerning banking, it is assumed that credit is the provision of money or equivalent claims based on agreements or agreements between banks and other parties that require the borrowing party to pay off its debt after a certain period of time accompanied by interest.

Based on the above understanding shows that the achievement that must be carried out by the debtor for the credit given to him is not merely to pay off the debt but also accompanied by interest in accordance with the agreement previously agreed upon. And besides that, there is also a guarantee of assets by the debtor as a preventive measure when the debtor is unable to pay off his debts (bankruptcy) and defaults of course with the fundamentals and auction process that are transparent so as not to harm one party. According to Hasanudin Rahman, the above is caused by: "Bank position in accordance with financial institutions operational activities are within the scope of business raise funds from the community and manage these funds by replanting them to community (in the form of credit) arrived the funds returned to the bank."

With the application of prudent principles and creditable risk management, banking institutions are able to capture business partners both individuals and MSME entrepreneurs. And grow to be the leading institution that has a vision that is in line with the booming and rapidly growing world of entrepreneurship. And the potential for bad debt can be minimized as early as possible.

2 RESULTS AND DISCUSSION

History and Regulation in the Micro Credit Agreement Between Individual Customers/Micro Business

Entities and Banking Institutions in Indonesia are as follows:

2.1 History of microcredit

Microcredit was first introduced by a Professor named Muhammad Yunus in the early 70s. This man born June 28, 1940 is a professor in economics at one of the universities in Bangladesh. One time Muhammad Yunus got a drastic inspiration or change, he was moved to see residents around his campus in poor condition, even people who starved to death, but he could not do anything to be able to help local residents. Muhammad Yunus made a change movement on the basis of conscience and knowledge in the economic field. Finally he tried to go to the slum and make an initial move. The professor thought why this poor person did not have the same opportunity as himself or anyone else in terms of access to banking, the professor invited one of the people and brought it to the bank to apply for credit with a personal guarantee (the debt of the poor was guaranteed by Professor Muhammad Jonah). The person is fostered and guided in entrepreneurship and miraculously the person is able to pay the principal debt and interest. This is what solidifies the thinking of the Professor. The movement to borrow funds and then carry out business assistance was carried out continuously and finally in 1983, Prof. Muhammad Yunus founded a bank called Grameen Bank.

2.2 Regulation on microcredit in Indonesia

While the concept of microcredit has now undergone a shift in functions and receiving segments. Micro credit developed in the Indonesian banking world clearly differ substantially from what has been developed by Professor Yunus in Bangladesh. The cultural conditions and also the level of the Indonesian economy allow micro-credit to be given to the MSME business sector, which is already rapidly following the 1998 monetary crisis. Microcredit is a product of the government's attention and has potential in a banking institution.

2.3 Micro credit agreement

Microcredit agreement is a consensual agreement between micro business actors (individuals or business entities) in this case called Debtors with Creditors (in this case the Bank) which give birth to a legal event in the form of accounts payable debt, where the Debtor is obliged to repay the loan provided by the creditor, based on the terms and conditions agreed upon by the parties.

2.4 Principles in the micro credit agreement

Refer to Bank Indonesia Regulation Number 7/12/pbi/2015 concerning Amendments to Bank Indonesia Regulation Number 14/22/pbi/2012 concerning Provision of Credit or Financing by Commercial Banks and Technical Assistance in the Context of the Development of Micro, Small and Medium Enterprises – general contract/agreement (article 1338 Civil Code).

In fact and practice in the field, that the form and material of the Credit Agreement are not always the same, and are always conditioned by the products and facilities provided. Guarantee clause.

To guarantee payment from a given loan, the Debtor is asked to submit a guarantee to the Bank where the guarantee will be bound as stipulated in the laws and regulations. For the Customer who obtains several facilities (the loan is not in one agreement) where each facility is guaranteed different should also include provisions regarding Cross Collateral. The use of cross collateral clauses provides additional benefits for existing guarantees.

2.5 Compensation clause

This Article concerning Compensation is regulated in connection with the existence of articles 1425 to 1429 of the Civil Code concerning debt compensation. This Compensation Clause contains the agreement of the Debtor to release his rights stipulated in the article, so that the Debtor cannot compensate the accounts receivable he has to the Bank (if any) with his debt to the Bank.

2.6 Transfer of rights

The purpose of the inclusion of this rights transfer clause is that the Debtor has given approval to the Bank to transfer the loan to a third party without changing the conditions previously agreed upon. While the Debtor cannot transfer the loan to another party without approval from the Bank.

2.7 Negligence clause

This clause lists some conditions that could cause the debtor to be in a default state or in a default state so that all Debtor obligations become due and must be repaid instantly and simultaneously, without the need for a bailiff's reprimand or other letter similar to that if one of the following events occurs: Payment Default/negligent repayment of obligations, violation of the terms of the Agreement, giving incorrect information, financial condition, bonafide and solvency of the debtor in such a way that can result in the debtor not being able to pay the debt again, the debtor is declared bankrupt or asks for the postponement of debt repayment, the debtor is dissolved or decides to disband, the debtor assets are totally or partially seized by the competent agency and is deemed to be reduced so that it can endanger Credit Returns, the guarantee is confiscated by the authorized agency, or damaged or destroyed for any reason, the debtor or the Guarantor is negligent of other agreements, especially agreements that can If the debtor is obliged to pay a certain amount.

In the event that the value of the assets/assets owned by the Debtor according to the Bank's assessment decreases, the actions that can be taken by the Bank if the Debtor carries out negligence is to stop giving credit facilities if it has not been disbursed. execution of the Guarantee if the Debtor is unable to repay the loan in full.

2.8 *Additional and closing provisions clause*

In the last part of the credit agreement it is regulated regarding the provisions that have not been specifically accommodated in the standard clauses in the credit agreement. This clause is intended to regulate the terms and conditions that deviate from other terms and conditions that have been printed in the credit agreement. This clause includes:

1) Choice of Law (Choice of Law)
   In this clause the parties determine certain laws to be applied if there are differences in interpretation or if there is a dispute between the parties regarding the agreement.
2) Choice of Dispute Resolution Forum (Choice Of Forum)
   This clause is intended if a dispute occurs, the Parties have agreed to resolve the issue through a mutually agreed institution. The choice of the institution (forum) for resolving disputes is usually Court or Arbitration, specifically for Arbitration, where the Arbitration is intended. In addition to the Court and Arbitration, there has also been a discourse on the use of the Alternative Dispute Resolution (ADR) mechanism, but this institution is not well known in Indonesia and its decision does not yet have definite legal force.

2.9 *Other provisions that must be considered in the micro credit agreement*

With the enactment of Law No. 8 of 1999 dated April 20, 1999 concerning Consumer Protection ("UUPK"), the contents of the credit agreement must also fulfill the provisions in the UUPK, such as regarding the inclusion of a standard clause. Where in Article 18 paragraph (1) UUPK states that in a credit agreement it is prohibited to include a standard clause, among others, stating the consumer's submission to regulations in the form of new rules, additions, and/or further amendments made unilaterally by business actors in the time consumers utilize services he bought. Stating that the consumer gives power to the business actor imposition of mortgage rights, liens, or collateral rights on goods purchased by consumers in installments.

3 CONCLUSION

Business transactions also include the fulfillment and provision of guarantees for loans made by banking institutions with related business partners. The provision of Micro Loans to MSME players is an implementation of the mandate of the Law, especially regarding Micro and Medium Enterprises (MSMEs) with banking institutions as credit channeling institutions and as well as community fund raising institutions. Some regulations concerning the provision of micro credit guarantees and MSME businesses are rolled out, given their significant role in supporting the country's economy. With 20 million MSME actors and extraordinary economic potential both in absorbing labor and contributing to national GDP that cannot be underestimated, micro-entrepreneurs both individuals and microenterprises need to get support from the government.

REFERENCES

Peraturan Bank Indonesia Nomor 7/12/pbi/2015 Tentang Perubahan Atas Peraturan Bank Indonesia Nomor 14/22/pbi/2012 Tentang Pemberian Kredit Atau Pembiayaan Oleh Bank Umum Dan Bantuan Teknis Dalam Rangka Pengembangan Usaha Mikro, Kecil, dan Menengah.

Rahman, H. No Year. Aspek-aspek hukum pemberian kredit perbankan di Indonesia. Bandung: Citra Aditya Bakti.

Undang Undang Nomor 20 Tahun 2008 tentang Usaha Mikro, Kecil, dan Menengah (UMKM).

Bank income structure in Indonesia: An analysis of theory of structure-conduct-performance

D.A. Suryanto
Universitas Pendidikan Indonesia, Bandung, Indonesia

ABSTRACT: This study aims to analyze empirically the phenomenon of bank income in the period 2013–2017 using the structure-conduct-performance theory. The method used is the panel regression model. The results of the study found that the banks' interest income in Indonesia can be explained by the variable of loan market share, bank efficiency supported by economic growth in a positive direction. Likewise, the banks' non-interest income in Indonesia can be explained by ownership of ATMs, bank efficiency supported by economic growth in a positive direction. The linearity of the influence of the independent variables included in the structural and behavioral aspects of the performance aspects shows that the explanation of changes in interest income and non-interest income can be explained by the structure-conduct-performance theory.

1 INTRODUCTION

Indonesian banking is one of the countries that have the highest NIM level in the world. Based on the data from the World Bank, the ratio of the net interest margin of banks in Indonesia is the highest in the range of 5.6 to 5.8%, far above neighboring countries such as Malaysia which is only 2%, Thailand with 2.5%, Singapore in the range of 1.5–2%, and the Philippines with a range of 4%. Korea is under 2%, and even Japan is under 1%. In the annual report issued by nine banks listed on the stock exchange, which are state-owned private companies, most banks have high NIMs in the last five years, which is above five percent. The high condition of NIM reflects that the banking industry in Indonesia is faced with a relatively high level of efficiency and risk. Therefore, banks need to look for other sources of income in the form of fee-based income.

Empirically, the non-interest income of banks in Indonesia in the last five years has increased. Apart from being caused by issues of efficiency and the level of risk in interest income, the increase in non-interest income was due to the increasingly widespread electronic-based banking services. The following Figure 1 shows the growth of the bank's non-interest income in Indonesia.

The empirical description of interest and non-interest income in banks in Indonesia shows that the structure of bank income has undergone a shift. Definitely, this is an interesting situation to be studied further regarding the factors that cause a shift in the structure of bank income in Indonesia. The results of research on the factors of interest income and non-interest income viewed from the standpoint

Figure 1. Growth of non-interest income.

of a structural-conduct-performance theory are still debatable.

Moreover, the current trend in banks in various countries is oriented towards non-interest income as an effort to maintain business continuity (DeYoung 2004). Therefore, this study aims to conduct an empirical examination of market share, bank efficiency, and the rate of economic growth in Indonesia in relation to interest income and non-interest income in commercial banks in Indonesia.

Theory of Structure – Conduct – Performance (SCP).

The market structure parameters used in this study are market share based on the logarithmic variant of bank size. In this study, the size of the bank is proxied by the number of loans channeled per year (Beck 2008; Rodriguez et al. 2009). Another proxy used to measure market structure is the provision of automated teller machines (ATMs) (Abedifar et al. 2018; Alubisia 2018; DeYoung 2004; Sathye & Sathye 2017).

In this study, the conduct of the banks studied based on aspects of price competition and service quality competition observed based on bank efficiency (Sahoo & Mishra 2012; Samad 2008). The bank efficiency proxy uses a proxy for operational costs per operating income (BOPO) as a proxy used by the Financial Services Authority (OJK).

The final component in SCP's theory is bank performance. Bank performance in this study is measured by interest income and non-interest income. Interest income is income earned by the bank for lending, while non-interest income is income earned by the bank in providing various financial facilities such as various payments for utilities, transfers, and other services (Abedifar et al. 2018; DeYoung 2004).

1.1 Theory of bank intermediation

The banking intermediation function has changed due to changes in the economic environment and the development of financial markets, especially in developed countries such as countries in the European Union (Bikker & Wesseling 2003). These factors tend to reduce transaction costs and asymmetric information between savers and investors, and this is contrary to the classical financial intermediation function.

1.2 Market share and revenues

Market share can be used in assessing the level of company competition. Companies with a broad market share are often identified as companies that will lead the market. Neuberger (1998) also found market share to be a parameter in assessing bank conduct and performance. Parameters in assessing bank performance can be seen from the side of bank income divided into interest income and non-interest income.

Growth in interest income as the main source of banking according to Maudos & Solís (2009) is determined by market structure variables. The level of banking competition is reflected in the alpha and beta coefficients for the deposit and loan markets. Studies on market capability in Indonesia is based on ownership of technology that enables transactions in banking services without limited service time, including ownership of ATMs.

1.3 Bank efficiency and bank revenues

The current measurement of bank efficiency in Indonesia prefers to use BOPO size as an implementation of an allocative efficiency model, which compares operational costs as input factors to operating income as an output factor (Wijaya 2009). Theoretically, banks must maintain a positive margin to cover their operational costs. The higher the operational costs, the higher the level of NIM that must be determined by

the bank. Efficiency is also expected to affect non-interest income. This is in line with the findings of Engle et al. (2014) who explained that bank efficiency is a factor that accelerates banks to obtain non-interest income.

The bank's efforts to make efficiency and increase non-interest income must also be supported by economic factors. Gross Domestic Product (GDP) as a parameter in assessing economic growth is expected to be a stimulus to increase people's income.

2 METHOD

This study aims to empirically examine the variables of market share, the level of efficiency of interest income, and non-interest income by placing gross domestic product as a moderating variable. The number of public banks analyzed was 82 national commercial banks and foreign banks operating in Indonesia with a five-year observation period in 2013–2017. The variables and their measurement in this study are explained as follow:

Table 1. Variables.

Variable	Measurement
Dependent Variable:	
Interest Income (NIM)	Net interest income/Average Earning Assets
Non-Interest Income (NOII)	Operating Income Excluding Interest/Operating Income
Independent Variable:	
Market Share (MS)	Ln Total Loan per Bank Number of Bank ATM outlets
Bank Efficiency (BOPO)	Operating Cost/Operating
Gross Domestic Product (PDB)	Income Household consumption (C) + Government consumption (G) + Investment (I) + (Export–Import)

Panel data regression was used as a data analysis technique, with input in the form of a pool data between bank sample units with the observed period. The development of panel data regression models is as follows.

(1) Model 1

$$NIM_{i,t} = \beta_0 + \beta_1 MS\_LOAN_{i,t} + \beta_2 BOPO_{i,t} + \beta_3 PDB_{i,t} + \beta_4 MS\_LOAN_{i,t}^* PDB_t + \beta_5 BOPO_{i,t}^* PDB_t + \varepsilon_{i,t}$$

Where:

$NIM_{i,t}$ = Interest income of bank i in period t,

$MS\_LOAN_{i,t}$ = Loan market share of bank i in period t,

$BOPO_{i,t}$ = Operating cost per operating income of bank i in period t,

PDB_t = Indonesia gross domestic product in period t,

$\varepsilon_{i,t}$ = error term of bank i in period t.

Table 2. Panel regression model of banks NIM in Indonesia.

Variable	Coefficient	Std. Error	t-Statistic	Prob.
C	48.51019	4.437874	10.93095	0.0000
MS-LOAN?	1.594423	0.280948	5.675154	0.0000
BOPO?	−0.042285	0.012335	−3.428023	0.0007
PDB?	4.099482	0.733292	5.590520	0.0000
LOAN?*PDB?	0.165790	0.042210	3.927784	0.0001
BOPO?*PDB?	0.008554	0.004328	1.976502	0.0489

R-squared	0.935859	Mean dependent var	29.35543
Adjusted R-squared	0.918782	S.D. dependent var	22.85511
S.E. of regression	6.659381	Sum squared resid	14324.20
F-statistic	54.80010	Durbin-Watson stat	1.836198
Prob (F-statistic)	0.000000		

(2) Model 2

$$NOII_{i,t} = \beta_0 + \beta_1 MS\_ATM_{i,t} + \beta_2 BOPO_{i,t} + \beta_3 PDB_{i,t} + \beta_4 MS^*_{i,t} PDBi,t + \beta_5 BOPO^*_{i,t} PDB + \varepsilon_{i,t}$$

Where:

$NOII_{i,t}$ = Non-interest income of bank i in period t,

$MS\_ATM_{i,t}$ = Market share of ATM services of bank i in period t,

3 RESULTS

The following table presents the results of panel data regression calculation of the bank interest income model in Indonesia

Specifically, Table 1 informs the loan market share has a positive effect on interest income. The wider the market share of loans held by banks, the higher the opportunity for banks to earn interest income. This finding supports the structure-conduct-performance theory which explains the linear relationship between the structure of the banking market which is proxied by the loan market share towards the achievement of banking performance proxied by NIM.

In addition to the loan market share, the probability value indicates lower than 5% alpha means BOPO is a variable that affects the NIM in a positive direction. The more efficient the bank's operations, the greater the expected rate of interest profit will be obtained. Macroeconomic conditions proxied by GDP show that they are also variables that determine the NIM variable.

Table 1 explains the positive relationship between GDP and NIM. In the empirical context in Indonesia, by observing the development of high NIMs, maybe that the macroeconomy is not conducive to encourage bank efficiency. Table 1 shows GDP as a variable that can moderate bank efficiency, meaning that both of these factors must go in harmony. The acquisition of

Table 3. Panel regression model of NOOI of Banks in Indonesia.

Variable	Coefficient	Std. Error	t-Statistic	Prob.
C	5.552938	0.469771	11.82053	0.0000
MS-LOAN?	0.067238	0.006430	10.45699	0.0000
BOPO?	0.022802	0.009528	2.393061	0.0172
PDB?	0.565138	0.068209	8.285399	0.0000
LOAN?*PDB?	0.009620	0.000931	10.33388	0.0000
BOPO?*PDB?	0.003725	0.001396	2.667407	0.0080

R-squared	0.704224	Mean dependent Var	8.795231
Adjusted R-squared	0.700564	S.D. dependent var	32.41192
S.E. of regression	0.368160	Akaike info Criterion	−0.454736
Sum squared resid	54.75902	Schwarz criterion	−0.395963
Log likelihood	99.22092	Hannan-Quinn criter.	−0.431484
F-statistic	192.3799	Durbin-Watson Stat	0.231529
Prob(F-statistic)	0.000000		

the value of determination can be categorized that the relationship of the independent variable studied has a strong influence on the dependent variable. Observed variables become important parameters in explaining the NOII changes in banks in Indonesia.

In other words, the more efficient the management of the bank, the greater the non-interest income. Economic conditions proxied by GDP also contribute to determining NOII variables (see Table 1, probability value of GDP to NOII). Even GDP is also able to moderate the influence of ATM on NOII and BOPO against NOII. With these findings, it can be explained that GDP is another important requirement for banks to increase non-interest income. Banks with adequate macroeconomic environments have a higher chance of increasing their income (Demirguc-Kunt & Huizinga 1999).

The findings in model 1 and model 2 appear to have a positive effect of independent variables on the dependent variable. This finding indicates that the explanation of the structure of bank income in Indonesia in the form of interest income and non-interest income can be explained using structure-conduct-performance theory.

In term of intermediation, the role of banks as intermediary institutions in Indonesia seems to be moving towards a change in classical intermediation towards modern intermediation with the nuances of information technology (Bikker & Wesseling 2003).

4 CONCLUSION

The results of the study found that interest income in banks in Indonesia can be explained by the variable of loan market share, bank efficiency, supported by economic growth in a positive direction. Likewise, the non-interest income in banks in Indonesia can be explained by ownership of ATMs, bank efficiency, supported by economic growth in a positive direction. The linearity of the influence of the independent variables included in the structural and behavioral aspects of the performance aspects shows that the explanation of changes in interest income and non-interest income can be explained by the structure-conduct-performance theory.

The implications of this research are certainly expected to be a consideration for banks to implement strategies for efficiency, especially in terms of creating interest income which currently exceeds the provisions of the OJK. However, efficiency is also important to be considered by banks to avoid the risk of using ATM as a means of banking information technology to efforts to increase non-interest income.

However, Table 2 informs that the existence of an ATM is considered capable of being the driving force for banks to obtain non-interest income. As with interest income, efficiency as measured by BOPO also contributes to positive non-interest income (see Table 3, the probability value of BOPO to NOII).

REFERENCES

Abedifar, P., Molyneux, P., & Tarazi, A. (2018). Non-Interest Income and Bank Lending. *Journal of Banking and Finance*, 87, 411–426. https://doi.org/10.1016/j.jbankfin. 2017.11.003

Alubisia, L. B. (2018). Effect of Technology Based Financial Innovations on Non-Interest Income of Commercial Banks in Kenya. *European Scientific Journal*, 14(7), 337–349. https://doi.org/10.19044/esj.2018.v14n7p337

Beck, T. (2008). *Bank Competition and Financial Stability: Friends or Foes*? (Policy Reserach Working Paper No. WPS4656). Washington DC, United States of America.

Dabla-Norris, E., & Floerkemeier, H. (2007). *Bank Efficiency and Market Structure: What Determines Banking Spreads in Armenia*? (No. 07). New York, United State of America.

Demirguc-Kunt, A., & Huizinga, H (1999), Determinants of Commercial BAnk Interest Margins and Profitability: Some International Evidence, *The World Bank Economic Review*, 13 ((2), 379–389.

DeYoung, R. (2004). Non Interest Income and Financial Performance at U.S. Commercial Banks. *Financial Review*, 39(1), 101–127.

Engle, R., Moshirian, F., Sahgal, S., & Zhang, B. (2014). *Banks Non-Interest Income and Global Financial Stability*.

Gordo, G. M. (2013). Estimating Philippines Bank Efficiencies Using Frontier Analysis. *Philippines Management Review*, 20, 17–36.

Hahm, J. (2008). Determinants and Consequences of Non-Interest Income Diversification of Commercial Banks in OECD Countries. *Journal of International Economic Studies*, 12(1), 3–32.

Islam, S., & Nishiyama, S. (2016). The Determinant of Bank Net Interest Margins: A Panel Evidence from South Asian Countries. *Research in International Business and Finance*, 16, 1–38. https://doi.org/10.1016/j.ribaf.2016. 01.024

Maudos, J., & Solís, L. (2009). The Determinants of Net Interest Income in the Mexican Banking System: An Integrated Model. *Journal of Banking and Finance*, 33(10), 1920–1931.

Meslier, C., Tacneng, R., & Tarazi, A. (2014). Is Bank Income Diversification Beneficial? Evidence from an Emerging Economy. *Journal of International Financial Markets, Institutions & Money*, 31, 97–126. https://doi.org/10.1016/j.intfin.2014.03.007.

Neuberger, D. (1998). Industrial Organization of Banking: A Review. *International Journal of The Economics of Business*, 5(1), 97–118.

Rodriguez, E. J. J., Dominguez, J. M. F., & Marin, J. L. M. (2009). Comparative Analysis of Operational Risk Approaches Within Basel Regulatory Framework: Case Study of Spanish Saving Bank. *Journal of Financial Management & Analysis*, 22(1), 1–15.

Sahoo, D., & Mishra, P. (2012). Structure, Conduct and Performance of Indian Banking Sector. *Review of Economic Perspectives*, 12(4), 235–264. https://doi.org/10.2478/v10135-012-0011-9

Sathye, S., & Sathye, M. (2017). Do ATMs Increase Technical Efficiency of Banks in a Developing Country? Evidence from Indian Banks. *Australian Accounting Review*, 27(80), 101–111. https://doi.org/10.1111/auar.1211

The effect of macroeconomic factors and stock index using against Jakarta stock price index

T.L. Situngkir & I. Mubarokah
Universitas Singaperbangsa Karawang, Karawang, Indonesia

ABSTRACT: Stock Market can be considered as the financial institution to get fund so stability of stock market become important to be researched including intern macro variable such as interest rate, exchange rate, consumer price index, and extern macro variable such as DJI, STI and HSE. The objective of this research is to analyze whether interest rate, exchange rate, consumer price index, Dow Jones Index, Strait Times Index and Hang Seng Index, each has a significant effect on Composite Stock Index. The methodology of analysis of this research is Error Correction Model. The result of research found that in short term interest rate has positive and significant on Composite Stock Market while in long term it has negative and no significant. Exchange rate has negative and significant in short term and long term on Composite Stock Index. In short term consumer price index has negative and no significant while in long term it has positive and significant on Composite Stock Index. In short term Dow Jones positive and no significant while in long term it has positive and significant on Composite Stock Index. In short and long term Strait Time Index has positive and significant on Composite Stock Index. In short and long term Hang Seng Index negative and no significant on Composite Stock Index. This research only covers period 2006–2016.

1 INTRODUCTION

The financial industry sector is very closely related to economic growth and development that means closely related to the capital market. Capital Market is one indicator of economic growth and the economic milestone of a country. The capital market has an important role of investment for development. The value of stock prices is one of the important thing to consideration. The increase of global world situation has an impact on economic value, where stock prices are not only influenced by economic conditions and domestic phenomena, economic turmoil and extraordinary events that occur abroad but also influenced by the value of shares. Therefore, efforts to create a stable investment climate by the government are important to be realized immediately, because it is closely related to the improvement of the domestic macroeconomic conditions. The more stable macroeconomic conditions, the more secure and comfortable about the funds invested.

The capital market is one of the driving forces of the economy in a country, because the capital market is a means of forming capital and public participation in mobilizing funds to support development financing in a country, because almost all industries in a country are represented by the capital market. Referring to the current literature, there are causal relationships which includes: (1) the monetary system is a stimulation of economic growth, (2) economic growth enhances

financial development, (3) the relationship between economic growth and financial development influences other factors. Arbitrage Pricing Model (APT) by Tandelilin (2010) found number of indicators that were taken into consideration by investors to do investment in the capital market, fundamental matters that were also should be considered in technical naturally, namely macroeconomic variables: inflation, interest rates, and foreign exchange rates, otherwise external factors also have great potential for changes in stock prices in the capital market. Likewise, foreign stock indices sometimes participate in influencing volatility in the domestic stock price index. This situation is evidenced by previous research between macroeconomic relations and foreign stock price indices to influence the stock price index in a country. Information from research results becomes important for capital owners (investors) in determining their investments so that the results can be maximized and minimize the risks that may occur. Therefore, researchers conducted research related to the influence of macroeconomic factors that affect the joint stock index.

Macroeconomic Factors Affecting Stock Prices.

1.1 Exchange rate (exchange rate)

According to Sukirno (2012) the exchange rate is a value of domestic currency needed to get a unit of foreign currency. Being influenced by changes in demand and supply of goods traded among various countries, foreign exchange rates are also influenced by long-term and short-term capital flows.

1.2 Inflation

An index calculates the average price change of goods and services consumed by households in a certain period of time. CPI is an indicator to measure inflation. Changes in CPI over time describe the rate of increase (inflation) or the rate of decline (deflation) of goods and services. In simple inflation is a phenomenon of continuous price increases.

1.3 Interest rate (BI Rate)

The BI Rate is a policy interest rate to reflect the movement of monetary policy by Indonesian banks and be announced SBI interest rates used by Bank Indonesia as an instrument to control inflation. If the interest rate rises, it will directly increase the interest expense (Sunariyah 2004).

1.4 Composite stock price index

The joint stock price index is a series of historical information regarding the movement of a joint stock price, up to a certain date. The composite stock price index reflects a value of measurement of the performance of stock on the stock exchange (Sunariyah 2004). The foreign share price index used are The Straits Times Index (STI), the Dow Jones Industrial Average (DJIA), and the Hang Seng Index (HSI).

2 METHODS

2.1 Population and sample

The method of data collection is done by particulary sampling which is the selection of sample members based on certain criteria or characteristics possessed by the sample so that they can obtain a representative sample. The sample chosen is the time series data for monthly periods 2006–2016. The data used are secondary data from the Indonesian Financial Economic Statistics (SEKI) issued by Bank Indonesia for IHGS data, inflation, interest rates, exchange rates and exchange rates.

Data analysis method. The analytical tool in this study is a multiple regression dynamic model, namely Error Correction Model. The model is as follows:

$$\Delta \text{IHSG} = bo + b1\Delta\text{BIRATE} + b2\Delta\text{IHK} + b3\Delta\text{NT}$$
$$+ b4\Delta\text{DJI} + b5\Delta\text{STI} + b6\Delta\text{HSE3}$$
$$+ b7\text{BINFt} + b8\text{BSBI} + b9\text{BNT}$$
$$+ b10\text{DJI} + b11\text{STI}$$
$$+ b12\text{BHSEt} - 1 + E \qquad (1)$$

Where JCI = Jakarta Composite Index; CPI = Consumer price index; SBI = BI rate; NT = Exchange rates; DJI = Down Jones Index; STI = Straits Time Index; HSE = Hang Seng Index; b1 to b5 are short-term coefficients; b6 to b12 is the long term coefficient.

3 RESULTS AND DISCUSSION

3.1 Descriptive statistics

The results of processing for descriptive statistics of research variables can be seen in Table 1. For the IHSG variables obtained an average value of 3471.98 and a standard deviation of 1337.39 indicates that there are considerable fluctuations in the JCI movement during the period 2006.1 to 2016.12. This can be seen from the minimum value of 1216.14 and the maximum value of the JCI of 5518.68. If viewed according to the JCI development during the period of January 2006 to December 2016, the performance of the capital market in Indonesia as a whole experienced an increase in performance as indicated by the JCI movement which increased from year to year. For more details, see Table 1.

3.2 Hypothesis results

The processing of the ECM model is carried out in 3 stages, namely unit root testing and degree of integration, co integration testing and ECM model hypothesis testing for both short and long term. The results of the stock price index equation can be displayed in Table 2 below:

The analysis of the short term equations performed significant positive effect only on interest rates on the first lag and two lags, while the Dow Jones price index, Hang Seng price index, consumer price index, strait time index, and exchange rate have no significant effect on price index movements joint stock With a coefficient on lag 1st (first) worth 0.370490, if the interest rate rises 1 (one) will increase of 0.370490 on the joint stock price index, also on the second lag with a coefficient of 0.274399 interest rates will positively affect the stock price index 0.274399.

The analyzing of the long term equations only the Dow Jones price index is not significant, while the Hang Seng price index, consumer price index, strait

Table 1. Statistik desktiptif variabel penelitian.

Variable	Mean	Median	Max	Min	St.Dev
IHSG	3471.98	3805.24	5518.68	1216.14	1337.59
BIRATE	7.51	7.38	12.75	4.75	1.75
IHK	96.61	95.18	126.71	68.18	17.48
KURS	10447.50	9552.50	14657.00	8508.00	1716.39

Table 2. Results ECM estimation in equations of the composite stock price index short term result.

Short term result

Variable	Coefficient	T-Static	Remark
D (LBIRATE(-1))	0.370490	3.80739	SIGN
D (LBIRATE(-2))	0.274399	2.75226	SIGN
D (LBIRATE(-3))	−0.137101	1.39188	NO SIGN
D (LDJI(-1))	0.066189	0.72491	NO SIGN
D (LDJI (-2))	0.075611	0.83901	NO SIGN
D (LDJI(-3))	−0.102877	1.12045	NO SIGN
D (LHSE(-1))	−0.031392	0.39147	NO SIGN
D (LHSE (-2))	−0.035006	0.42887	NO SIGN
D (LHSE (-3))	−0.087777	1.14289	NO SIGN
D (LIHK(-1))	0.998182	1.90779	NO SIGN
D (LIHK (-2))	−0.917496	1.58025	NO SIGN
D (LIHK (-3))	0.823377	1.51139	NO SIGN
D (LKURS(-1))	0.013754	0.12760	NO SIGN
D (LKURS (-2))	0.103377	0.97912	NO SIGN
D (LKURS (-3))	−0.105238	1.00419	NO SIGN
D (LSTI(-1))	−0.045211	-0.43947	NO SIGN
D (LSTI (-2))	0.052820	0.50337	NO SIGN
D (LSTI (-3))	0.106818	1.03790	NO SIGN
C	−0.007981	1.73985	NO SIGN
CointEq1	0.012200	1.06522	NO SIGN

Longterm result

Variable	Coefficient	T-Static	Remark
LBIRATE(-1)	1.000000		
LDJI(-1)	−1.054840	1.80982	NO SIGN
LHSE(-1)	5.123615	7.21209	SIGN
LIHK1(-1)	−12.48696	3.55894	SIGN
LKURS(-1)	−1.282139	2.25137	SIGN
LSTI(-1)	0.068519	3.91139	SIGN
C	74.40290		

time price index, and exchange rate significantly influence the movement of the composite stock price index. With the coefficient on lag 1st (first) worth 0.370490, if the interest rate rises 1 (one) will increase 0.370490

on the joint stock price index, also in the second lag with a coefficient value of 0.274399 interest rates will affect the movement of the stock price index of 0.274399 positively.

4 CONCLUSION

Simultaneously in the short and long term proven that the overall variables used have significants impact on the Jakarta Composite Index. The results of the study was found that the exchange rate is an independent variable that has significant negative effect on the index of the joint stock price in the long and short term. This shows that the stability of the foreign exchange market is very influential on the capital market in Indonesia. Monetary authorities as interested parties in making macro policies must pay attention to the stability of the rupiah exchange rate against other foreign currencies.

The Straits Time Index proved to have a significant effect on the short-term and long-term to composite stock price index, this finding indicates that the STI is a benchmark for the Indonesian capital market. It is evident that Singapore is Indonesia's main trading partner otherwise activities that occur in Singapore have a significant impact on capital market activities.

REFERENCES

Sukirno, S. 2012. *Makro ekonomi teori pengantar*. Jakarta: Raja Grafindo Persada.
Sunariyah, S. 2004. *Pengantar pengetahuan pasar modal*. Yogyakarta: UPP AMP YKPN.
Tandelilin, E. 2010. *Portofolio dan investasi: teori dan aplikasi*. Yogyakarta: Kanisius.

Trickledown effect with corruption disclosure model through forensic auditing to the procurement of goods and services project

D. Hamdani & I. Alamsyah
Sekolah Tinggi Ilmu Ekonomi (STIE) INABA, Bandung, Indonesia

ABSTRACT: Fraud occurs because of the opportunity and encouragement so that the authority misused, it looks as if true and natural, the action is a deliberate error and fraud. The purpose of the research is to know whether corruption can affect potential PDRB. This type of research is a quantitative study (traditional method), with descriptive methods and verification. The nature of the study "hypothesis testing", which analyzes the influence or relationship between variables based on empirical data through hypothesis testing. The results of the study with the Two-State DEA regression Model showed that the higher the corruption would have an impact on the decrease in efficiency, with the estimation of the corruption Model; Corruption has scraped the potential for the gross regional domestic product (PDRB). The research concludes that corruption can have a bad impact on the excavation of economic potential, so corruption can cause a trickle-down effect is proven.

1 INTRODUCTION

Indonesia is a democratic country with dynamic political changes, including freedom of speech and freedom of opinion. The current condition is relatively better than the previous era, where state officials did not show the grandeur and be more supportive of the community. The state's financial system is currently governed by a strict procedure and supervised by the Corruption Eradication Commission (KPK). However, along with the improvement of the state's financial system, the quantity and quality of corruption in Indonesia remain quite dominant especially in the procurement of goods and services in each province in Indonesia.

Data Transparency International recorded the Indonesian Corruption Perception Index (CPI) score of 2017 is 37 and ranks 96 from 180 countries.

In Indonesia, corruption includes certain sectors, such as the economy, educational, transportation, social, and health.

Table 1. Indonesia corruption perception index and ranking table.

Description	2017	2016	2015	2014	2013	2012
Score	37	37	36	34	32	32
World Rank	96	96	96	103	103	103

Note: *) From 180 countries.
Source: Transparancy International. 2017

The highest form of corruption in the procurement of goods and services since it is simple and vulnerable to be corrupted through an office interference mode in competing for tender procurement of goods and services, or even by marking up the price.

The data in Table 2 is relevant with the fact of large numbers of poor people in West Java from 2013 to 2017. The data do not show significant improvement. It presents that West Java has a higher position compared to other provinces in term of absolute, comparative, and competitive aspects.

According to Rimawan (2011), corruption in Indonesia is done with sophisticated strategies along

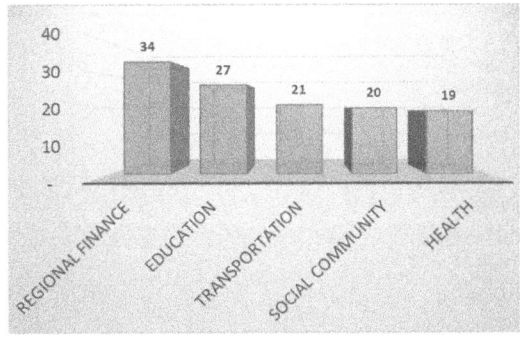

Figure 1. Total of corruption cases in various sectors in Indonesia.

with the advanced information system, so the prevention should be done extraordinarily since it cannot be handled in a traditional manner.

Generally, fraud in Indonesia is the procurement of goods and services and affiliated directly with the third party (suppliers). Even though some districts and cities have implemented the e-procurement system to minimize face to face interaction, the fraud cases are found in various forms.

Fraud can be done by entering the supplier's name in a match with low credentials. It aims to forfeit the chance from the engagement holder. Consequently, the prospective winners who already know the rules can win the game easily.

The concept of prevention and eradication of corruption in criminal acts using conventional means will not produce maximum results, but rather it needs more sophisticated handling and comprehensive. The concept should be designed through renewal innovation (novelty innovation) to prepare the large forensic database auditorial system.

Therefore, this study aims to answer some research questions, such as (1) How to budget income and expenditure areas outside the procurement budget of goods and Jas, budget procurement of goods and services, forensic audit, e-Procurement, economic growth has an effect on financial efficiency Area, (2) How

Table 2. Corruption facts 2004–2018 in Indonesia.

Corruption case explanation	Number of cases
Procurement of goods and services	180
Licensing	22
Bribery	507
Charges	21
Budget abuse	46
Money laundering	29
Hinder	9
Total:	814

Table 3. Corruption facts 2004–2018 in Indonesia by position.

Position	Number
Board members	229
Head of Board/Ministry	26
Ambassador	4
Commissioner	7
Governor	20
Mayor/deputy Mayor	91
Eselon I / II / III	192
Judge	19
Prosecutor	2
Lawyer	10
Private	214
Other	90
Corporation	4
Total:	915

Source: KPK, 2018

to lose due to the Multiplier effect on investments and consumption of regencies and cities in West Java province, (3) The extent of the effectiveness of forensic Audit can prevent corruption in regencies and cities in West Java province, (4) How much economic loss potential because of corruption for the community in the district and city in West Java province.

2 METHOD

The study belongs to the quantitative research category or is often referred to as a traditional non-experimental method that is included in the category of associative research.

According to Sugiyono (2014), quantitative research methods can be interpreted as a research method based on the philosophy of positivism and quantitative data analysis/statistics. It aims to test a hypothesis.

Descriptive and verification were employed as the research method in this study aiming at providing an overview and finding verification or researching the existence of empirical theories with relationship variables formulated in hypotheses. Since this study attempts to test hypothesis, the researcher analyses the influence or relationship between empirical data-based variables through hypothesis testing to test factors that affect efficiency, corruption, and other potential economic losses.

The regression analysis technique involved data panel by combining cross-sectoral data with Time Series data. To figure out the efficiency of two regional government, the model of Hodrick and Prescot (HP) filters employed with multivariate Total workforce productivity.

Moreover, this study also employed primary and secondary data with documentation, interviews, and observations as research instruments. The main data purposed to answer and prove hypotheses about the factors encouraged the corruption cases by following The Fraud Diamond Model.

To analyse the data, two stage data analysis paired with the data panel was conducted to identify the Trickle-Down Effect accurately and measurably.

3 RESULTS AND DISCUSSION

The regional financial efficiency in this study refers to input-output, using a deterministic model. By knowing the comparison of expenditure and realization of regional acceptance, the level of efficiency of regional financial management is given the assessment using assessment criteria. The criteria of research assessment refer to the decree of the Minister of Home Affairs No. 690,900,327 the year 1994 on the guidelines of valuation and financial performance compiled as in Table 4.

Analysis of the technical efficiency level of local Government (PEMDA) in the West Java Province covers two phases. In the first phase, this study measures the level of technical efficiency with a non-parametric approach to data-wrapped analysis by using the Data Envelope Analysis/DEA software. The second phase is determining the analysis of the level of technical efficiency to investigate the relationship and the nature of the regression panel data by using the EViews 6 Enterprise Edition software.

The real level value or significance level (α) in this study is 5% or 0.05, and the Ftable value is determined based on the magnitude of α (degree of freedom), the Ftable or F value is 0.05, 9,474 and calculated with the formula F. INV. RT resulting in The value of 1.8996, this test uses one of the Eviews functions to obtain calculations and probabilities.

For the value of n the normality test, and the real level value (α) in this study is 5% or 0.05. The value is based on the α value with the degree of freedom as many independent K variables. In Multicollinearity, perform a regression of DF1 for the same duplicates as the number of independent variables and intercept (k) minus two or $DF1 = k - 2 = (5 + 1) - 2 = 4$. Whereas DF2 for Denumerators equals the sample observation ($n = i \times t$) minus the number of independent variables and intercept (k) minus one or $DF2 = (10 \times 48) - (5 + 1) - 1 = 473$.

The F-table or F value (0.05.4, 473) is calculated with the formula F. INV. RT yields a value of 2.3908. The autocorrelation, dU and dL values are determined by the n number of samples as much as 480 and k the number of independent variables and constants (6). So obtained the value dU = 1.8712 and dL = 1.8291.

The value of the APBD variable, APBJ variable, e-Procurement, economic growth or T (0.05, 474) is calculated by the-T formula.

The data gained from INV function in Microsoft Excel software 2016 reveal the value of 1.6481. For corruption, the value of the table based on the magnitude of α with the freedom level equals the number of observations ($n = ixt$) or equal to $DF = n - k = (10x48) - (5 + 1) = 474$.

A t table or T value (0.05, 474) is calculated with the T formula, and by using the INV function in Microsoft Excel 2016, a value of $-1,6481$ is obtained.

Economic growth variables (eg), goods and services budget procurement (APBJ), corruption (cast),

E-Procurement (ePro), budget income and expenditure (APBD) simultaneously have a positive influence unless variable corruption (cast) has an impact Significant influence on LGs efficiency in West Java province.

4 CONCLUSION

The results explained that the regional budget of income and expenditure outside the budget of goods and services Procurement (APBJ without APBD), budget procurement of goods and services (APBJ), forensic Audit (AF), E-Procurement (ePro), economic growth (EG) has a significant effect on the regional financial efficiency ((EKD). Losses caused by corruption in the effect of the investment Multiplier (INV) and consumption (Kon) of the district communities and cities in West Java province resulted in the weakening purchasing power of the people, thereby lowering the economic growth. The potential economic losses due to corruption in district communities and cities in West Java province resulted in the expenditure of the local government in which the actual < Budget resulting in the growth of Regional infrastructure to be hampered.

REFERENCES

Abidin, Zaenal & Endri. 2009. Kinerja Efisiensi Teknis Bank Pembangunan Daerah: Pendekatan Data Envelopment Analysis (DEA). Jurnal Akuntansi dan Keuangan, Vol. 11, No. 1,2009.

Anner & Sarah, E. (1990). Experience Effects in Auditing: The Role of Task-Specific Knowledge. Journal The Accounting Review. Vol. 65, No. 65.

Chen, Elder, Liu. (2005). Auditor Independence, Audit Quality and Auditor-Client Negotiation Outcome: Some Evidence From Taiwan. Asia-Pacific. Journal of Accounting and Economics Symposium.

Cooper, Donald R., Emory, C. William. 1996. Business Research Methods. Jakarta: Erlangga

COSO. (2011). COSO Enterprise Risk Management: Establishing Effective Governance Risk, and Compliance processes. Second Edition. John Wiley & Soms. Canada.

Dara & Saidin. (2013). The Relationship Between Audit Experience and Internal Audit Efectiveness in the public Sector Organizations. International Journal of Academic Research in Accounting, Finance and Management Sciences. Vol. 3.

Dickins & O'Reilly. (2009). The Qualifications And Independence Of Internal Auditors. Journal Business And Economics – Accounting Vol. 24.

Elder, Beasly, Arens. (2010). Auditing and Assurance Services: An Integrated Appproach. Thirteenth Edition. Pearson Prentice Hall. USA.

Elliott & Jacobson. (1998). Audit Independence Concepts. The CPA Journal. Vol 68, No. 12.

Iskandar, Rahmat & Ismail. (2010). The Relationship Between Audit Clientt Satifaction and Audit Quality Attributes: Case of Malaysian Listed Comforties. Journal of Economics and Management. Vol4, No. 1.

Table 4. Criteria of regional financial efficiency.

% Financial Effiency Region	Criteria
Up–100%	Not Efficient
90%–100%	Less Efficient
80%–90%	Enough
60%–80%	Efficient
Under 60%	Very Efficient

Source: The Minister of Home Affairs, 1994.

Jeaning, Pany & Reckers. (2006). *Strong Corporate Gover-mance and Audit Firm Rotation: Effect on Judge Inde-pendence Perception and Litigation Judgments. Journal* Accounting Horizons. Vol. 20, No. 3.

Kym Boon, Jill Mc Kinnon, Philip Ross. (2008).*Audit Service Quality in Compulsory Audit Tendering: Preparer Per-ceptions and Satifaction.* Journal Accounting Research. Vol. 21.

Mardiasmo. (2004). Akuntani Sektor Publik. Andi. Yogyakarta.

Mc Daniel. (1990). *The Effects of Time Pressure and audit program structure on audit performance.* Journal of accounting Research. Vol.28, No.2. USA.

Rimawan. (2011). Penegakan Hukum dan Pencegahan Tin-dak Kejahatan dalam Tinjauan Ilmu Ekonomi. Jurnal EBNEWS, Edisi 9 Tahun 2011.

Saleh, K. Wantjik, (1983). Tindak Pidana Korupsi dan Suap. Jakarta.Ghalia Indonesia.

Sugiyono. (2014). Metode Penelitian Pendidikan Pendekatan Kuantitatif, Kualitatif, dan R&D. Bandung. Alfabeta.

The Committee of Sponsoring Organizations of the Tread-way Commission (COSO). (2013). *Internal Control – Integrated Framework.* New York. AIGPA. Publicaton Division

Innovation of the pad of the original income (PAD) through the giving of incentive motor vehicle taxes (PKB) paid for advanced and special rates on the choice of police number as mass customization

Kasir, A. Kusumardani & D. Syarief
STIE Indonesia Membangun, Bandug, Indonesia

ABSTRACT: This study aims to examine the factors that influence economic growth and local revenue originating from the potential of motor vehicle tax both Two-Wheel (R-2) and Four-Wheel (R-4) with Analog Bond systems and special rates on police numbers selection in order to assess the potential impact of fiscal and economic growth on the welfare of the people in West Java, especially the southern region. This study takes a population of all cities/districts in the West Java region. The research method used is descriptive and verification method, and the data taken is secondary data. Meanwhile the analytical tool used uses Data Panel Regression, as well as to calculate tax potential using the Hodrick and Prescott Filter (HP Filter). The results of his research show that the provision of tax incentives for motorized vehicles that are paid for in advance and special rates for police numbers have a significant effect on local revenue and economic growth.

1 INTRODUCTION

West Java is one of the most populous provinces in Indonesia, which currently reaches 47.379 thousand people, with an average population growth rate of 2016 of 1.43%. (West Java in Figures - BPS, 2016). Regional Original Revenue (PAD) obtained by West Java province for 2016 for several regions, namely the northern region obtained PAD of 10.745 billion, the southern region amounted to 2.342 billion and the western region amounted to 4.119 billion. (West Java in Figures - BPS, 2016). Thus the revenue of PAD for the southern region is the smallest compared to the area of North Sumatra and the western region. In extracting the potential of local taxes (motor vehicle tax) which is one source of regional income in the form of regional taxes, of course, it can be used to increase revenue from the taxation sector. Especially in the period of 5 (five) years the number of motorized vehicles in the area of West Java increases every year. In 2016 the number of motorized vehicles of various types of vehicles, both commercial and personal, as well as both two-wheeled and four-wheeled vehicles, were 11.025.188 with an increase of 12.55% annually (West Java in Figures - BPS, 2016). So that by looking at the potential that exists both the large population and the number of motorized vehicles which are increasing from year to year by the West Java Government can be used to increase regional income which can later be used to develop still underdeveloped regions (southern West Java) so that welfare local people are increasing too.

Furthermore Mardiasmo (2002) describes that the potential of local tax revenue is the strength that exists in an area to produce a certain amount of revenue.

To see the potential sources of regional revenue, knowledge of the development of several variables that can be controlled and variables that cannot be controlled is needed which can influence the strength of local revenue sources. Another opinion that is similar about regional tax is stated by Haryadi (2002) outlining the improvement of regional tax management and regional retribution in accordance with their potential to provide additional Regional Original Income, but on the contrary if there is no potential it will cause losses because of its potential not being utilized optimally. Furthermore Mulyanto (2002) describes the potential dimensions of regional taxes and regional retributions referring to the level of looking for trends from various types of regional taxes and regional levies that apply in each region in the region. In exploring the potential for tax revenue through motor vehicle taxation, so far it has been carried out by several cities/regencies and provinces in Indonesia and abroad in a way that has been commonly done, namely the elimination of motor vehicle tax (more than one vehicle ownership (progressive tax), fines/sanctions for late payments, and transfer fees). Even though there are other ways that can be done by the local government in exploring the potential of taxes from motorized vehicles, namely by way of payments for motorized vehicle tax, both two-wheeled and four-wheeled vehicles, which are only done once a year. Payment can be made upfront, where the rates will be applied later, analogous to local government bonds to the community (motor vehicle owner taxpayers), for example, the payment discount is adjusted to the applicable deposit rates in the banking system. Payment of up-front motor vehicle tax, for example, payments for a 2-year tax period are paid at once. Likewise, for tax periods 3, 4 and 5 years also be treated at once. Besides through upfront payments, there are also other potentials of motor vehicle taxation, namely in terms of giving/registering selected numbers. Because all this time making or re-registering selected numbers (specifically) was not entered into the official tax receipt. The rates are specifically applied. Thus the results can be used as revenue revenues for regional governments in addition to the receipt of other income that has been carried out.

Some references and research conducted by previous researchers related to extracting the potential of Regional Original Revenue (PAD) through Motor Vehicle Tax for the development of underdeveloped areas so that the community becomes prosperous are carried out by Kristine (2017), writing about the frequency of vehicle tax payments in the country ASEAN, where payment of motor vehicle tax in each ASEAN country is different. Like the country of Vietnam in payments can be one month to 30 months. Research related to the contribution of PKB that has an impact on PAD for regional development is carried out by Herliene & Tarmizi (2013), whose result is that the per capita income of a region is determined by PKB. Whereas previous studies related to tax revenues through the improvement of the local tax system which results in the development of areas that are still underdeveloped were carried out by Feiyang (2017), whose results were improvements in the local taxation system carried out by regional governments, the results of which will be used to increase development to the regions left behind, so that the infrastructure built can narrow the gap with other regions. From the description above it is very clear that some phenomena in the province of West Java are quite unique so that creative and innovative solutions are needed.

2 METHODS

The data used in this study is secondary data, where this data is data that has been collected by other parties. The secondary data that will be used in this study include the Regency/City GRDP data, Regency/City Population, Regency/City Regional Original Income, Regency/City Motor Vehicle Tax in West Java Province obtained through various agencies such as the Central Agency Central and West Java Statistics (BPS), Bappeda, Regional Government Work Units (SKPD) and the Regional Investment Promotion and Investment Coordinating Board (BKPPMD) which were the subjects in this study As for the objects of this study are Economic Growth, GDP, Population, Regional Original Income (PAD), Two-wheeled Vehicle Tax (PKB), Four-wheeled or more Motorized Vehicle Tax (PKB), two-wheeled motor vehicle tax (PKB) prepaid, motorized vehicle tax (PKB) for four wheels or more upfront, two-wheeled motorized vehicle tax with a choice number, four-wheeled or more motorized vehicle tax with a choice number, in the West Java Province.

Whereas to test the factors that influence Economic Growth, and Regional Original Revenue (PAD), a regression analysis technique with Data Panel is used that combines cross-sectoral data (Cross Section) with time series data, while estimating Potential PAD sourced from PKB receivables, multivariate Hodrick and Prescot (HP) Filters will be used.

3 RESULTS AND DISCUSSION

The data used in this study is secondary data and is done by doing calculations by connecting between variables that are calculated by estimating the regional economic growth model and estimating income estimation models for the regions. Also by using Hodrick Presscot (HP). Where the result is that the receipt of tax on 2-Wheeled and 4-Wheeled Motorized vehicles affects regional original income and economic growth. And PKB Wheels 2 and 4 which are prepaid affect regional income and regional economic growth significantly, as well as Wheel 2 and 4 PKB with community choice numbers also significantly influence regional income and regional economic growth.

4 CONCLUSION

The conclusion of this study is that all local taxes obtained through the existing tax potential, namely by providing tax incentives for motorized vehicles both two-wheeled and four-wheeled vehicles which are paid in advance and special tariffs on selected numbers greatly influence regional income and regional economic growth. So that the incoming funds can be used for public welfare.

REFERENCES

Badan Pusat Statistik Provinsi Jawa Barat. 2016. *Jawa Barat dalam angka tahun 2016*. Bandung: BPS Provinsi Jawa Barat.

Feiyang, C. 2017. Local tax scale and its economic effects in China. *Modern Economy* 8(3): 445–457.

Haryadi, B. 2002. Analisis pengaruh fiscal stress terhadap kinerja keuangan pemerintah kabupaten/kota dalam menghadapi pelaksanaan otonomi daerah. *Simposium Nasional Akuntansi V Semarang*.

Herliene, Y.A. & Tarmizi, H.B. 2013. Kontribusi pajak kendaraan bermotor terhadap PAD dan dampaknya bagi pengembangan wilayah di Sumatera Utara. *Jurnal Ekonomi* 16(3).

Kristine, V.M. 2017. The road tax or motor vehicle user's charge in selected ASEAN member-countries. *NTRC Tax Research Journal XXIX*.

Mardiasmo, M. 2002. *Ekonomi dan manajemen keuangan daerah*. Yogyakarta: Penerbit ANDI.

Mulyanto, M. 2002. Potensi pajak daerah dan retribusi daerah di kawasan Subosuka Wonosaren Provinsi Jawa tengah (the potential of local government revenue: a case study in Subosuka Wonosaren District, Central Java Provincies). Surakarta: Economics Faculty SebelasMaret University.

Advances in Business, Management and Entrepreneurship – Hurriyati et al. (Eds)
© 2021 Taylor & Francis Group, London, ISBN 978-0-367-67471-7

The effect of macroeconomics on the performance of commercial banks in Indonesia

A. Fuad & D. Disman
Universitas Pendidikan Indonesia, Bandung, Indonesia

ABSTRACT: This paper analyzes the effect of macroeconomic indicators, including gross domestic product (GDP), inflation, Bank Indonesia interest rates (BI rates), Jakarta's composite stock index, exchange rates, and crude oil prices on the performance of commercial banks in Indonesia. We use the Vector Error Correction Model (VECM) on monthly banking data from 2012–2018 and obtain several research findings. First, the impulse response function shows the greatest response to the bank's efficiency performance, which is proxied by the BOPO ratio due to the influence of shocks in the macroeconomy. The author believes the volatility of this bank efficiency indicator represents the inefficiency of commercial banks in Indonesia. Second, the author believes the application of Bank Indonesia interest rates as an effective monetary instrument according to the results of the study that Bank Indonesia's interest rate shocks generally provide the biggest response from most bank performance indicators.

1 INTRODUCTION

In the past 25 years, there have been a number of banking vulnerabilities in various countries in the world. Caprio & Klingebiel (2003) recorded 117 cases of systemic banking crises and 51 cases of non-systemic banking crises in developed and emerging market countries since 1970. Systemic is defined as a situation where all or a large portion of capital in the banking system has been eroded (Haldane et al. 2005). This banking vulnerability has a large cost for the government in the form of a fiscal resolution cost. These costs are incurred to improve the banking system including the cost of bank recapitulation and payments to depositors through the guarantee agency scheme (Apriadi et al. 2017).

In this case, the currency crisis is defined as a nominal depreciation in the domestic currency (against the USD) of 25% combined with a 10% increase in the rate of depreciation in the year of the banking crisis (Apriadi et al. 2017). The latter condition is designed to exclude the influence of countries with high inflation that have a trend of high depreciation rates (Frankel & Rose 1996). For example, the cost of cumulative resolution of the crisis in Indonesia that began in 1997 was around 50% of gross domestic product (GDP), while the crisis in Turkey was 30% of GDP (Haldane et al. 2005).

Started by Aharony & Swary (1983) and Swary (1986), a series of papers have applied a study methodology to the effects of certain bank failures or bad news for certain banks on the stock prices of other banks (see, for example, Slovin et al. 1999; Wall & Petersen 1990). In another series of papers, various regression approaches are used to link abnormal bank stock returns with asset side risks, including those related to aggregate shock (for example, Kho et al. 2000).

The relationship of banking activity with macroeconomic variables is further discussed by Albertazzi and Gambacorta (2009) and most recently by Kok et al. (2015). The first two find bank return on assets (ROA), which is positively related to GDP growth, identifying increases in net interest income (NII), and reducing provisions as the main channel in influencing bank returns on assets. Borio et al. (2015) found evidence that there was no effect of GDP growth on interest rates, yield slope, and bank profitability. Several different results of research were conducted in various countries, like the results of the study of Alessandri & Nelson (2015) with a sample of British banks for the period between 1992Q1 and 2009Q3 that found that interest rates had a negative effect on NII in the short term but were positive in the long term. Meanwhile, Albertazzi & Gambacorta (2009) consider the lagging effect of macroeconomic variables, which may have different effects in the short and medium term.

Frequently used indicators of bank performance are ROA and return on equity (ROE) (Abiodun 2012; Ali et al. 2011; Alper & Anbar 2011; Athanasoglou et al. 2005; Bonin et al. 2003; Ghazali 2008; Gizycki 2001; Mirzaei et al. 2011; Rumler & Waschiczek 2010; Sastrosuwito & Suzuki 2011). Some previous research using ROA (El-Moussawi 2012; Hamadi & Awdeh 2012; Naceur 2003) added a net interest margin (NIM) variable as a proxy for performance proxies. Other researchers Schinasi (2005), Kool (2006), and Festic and Beco (2008) use the non-performing loan variable (NPL) as an indicator of bank performance.

In various other studies, macroeconomic variables that are often used as determinants of banking performance are national income or economic growth, inflation, and interest rates. Naceur (2003) uses GDP per capita growth and inflation as macro variables that affect the performance of the banking system. Ali et al. (2011) and Mirzaei et al. (2011) used economic growth and inflation variables, while researchers such as Gizycki (2001) and Hamadi and Awdeh (2012) used other macro variables such as interest rates. Festic and Beco (2008) and De Bock and Demyanets (2012) use the exchange rate variable.

The purpose of this paper is to examine the effect of macroeconomic shock conditions on the performance of commercial banks in Indonesia. This is to see how sensitive bank performance indicators are to external shock variables, particularly from macroeconomic factors. This is the main focus of the study, although the authors are aware that the possibility of a reverse effect, which is an indicator of bank performance, through credit channels, can also affect macroeconomic variables, such as the Clower constraint theory as proposed by Robert Clower in 1967 (Blancard & Fischer 1998). Macroeconomics variables used include inflation, policy interest rates (BI rate), and exchange rates. The Composite Stock Price Index (CSPI) will be used as a proxy for external stock market shocks and energy prices as a proxy for world crude oil prices. Proxies for bank performance are deposits of third-party funds (DPK), credit, and financial ratios such as loan to deposit ratio (LDR), return on assets (ROA), net interest margin (NIM), non-performing loans (NPL), and ratios liquid assets.

Based on its main purpose, the definition of bank performance is how the bank operates to obtain the highest profit potential (Mishkin 2001). Guerrieri and Welch (2012) use four proxies for banking performance, namely total net off offs, prestipulated net income, NIM, and tier-1 capital ratio. Meanwhile, Awojobi and Amel (2011) use the capital adequacy ratio (CAR) variable as an indicator of bank performance because they are more focused on the efficiency of risk management. In this case, the author uses the following banking performance variables in this study.

Return on Assets (ROA). In some of the previous literature, overall bank performance is measured using a proxy for profitability indicators. The two main indicators of performance are ROA and ROE (Ali et al. 2011). ROA was chosen over ROE because it ignored the risks associated with high leverage and financial leverage (Athanasoglou et al. 2008; Bashir 2003; Davydenko 2010; Demirgüç-Kunt & Huizinga 1998; Flamini et al. 2009; Sastrosuwito & Suzuki 2011). In addition, the multiplier of high-equity securities does not distort ROA and, therefore, seems to be a better measure to represent the ability of companies to produce returns on their portfolio assets (Rivard & Thomas 1997).

Net Interest Margin (NIM). Naceur (2003) uses the NIM variable as a proxy for banking performance (e.g., the case in Tunisia). NIM is more focused on the

benefits of activities that generate interest. Meanwhile, Gerlach, Peng, and Shu (2005) use the NIM and NPL variables as factors that are considered to represent profitability for cases in Hong Kong SAR using panel data. Likewise, Hamadi and Awdeh (2012) and Saad and El-Moussawi (2012) use NIM as an indicator of banking performance in Lebanon.

Capital Adequacy Ratio (CAR). A strong capital structure is one of the main variables that provide additional strength to drive profitability and withstand financial instability. In this regard, a number of studies, including those for transition economies, found a positive impact on bank capitalization on profitability (Athanasoglou et al. 2008; Berger 1995; Capraru & Ihnatov 2014; Demirgüç-Kunt & Huizinga 1998; Djalilov & Piesse 2016; Fries & Taci 2002; Nuriyeva 2014; Petria et al. 2015).

Non-Performing Loans (NPL), Loan-to-Deposit Ratio (LDR), and Asset Liquidity Ratio. Schinasi (2005), Kool (2006), and Festic and Beco (2008) use the NPL variable as one indicator of bank performance with the confirmation that the NPL is able to measure the quality of the balance sheet. Several influential studies, including those in the transition economy, have found a negative relationship between credit risk, liquidity, and profitability, and concluded that increased exposure to credit risk is generally associated with reduced bank income (Athanasoglou et al. 2006, 2008; Capraru & Ihnatov 2014; Davydenko 2010; Djalilov & Piesse 2016; Petria et al. 2015; Roman & Sargu 2015). Another study found that appreciation of the effective exchange rate did not reduce the NPL ratio. In addition, rising unemployment and inflation worsen the NPL ratio, while faster GDP growth slows the NPL ratio (Baboucek & Jancar 2005).

Credit and Deposits. A higher loan-to-asset ratio is expected to affect profitability positively unless the bank takes an unacceptable level of risk. A positive impact is expected because interest on loans is one of the main sources of bank profits. Athanasoglou et al. (2008), Davydenko (2010), and Bashir (2003) found a positive effect of loans on profitability. The effect of deposits on bank profitability is ambiguous and highly depends on the characteristics of the bank and the country's macroeconomics. For example, Davydenko (2010) found a negative effect of deposits on the profitability of banks in Ukraine, one of the transition economies. The same thing was found by Demirgüç-Kunt and Huizinga (1998) and Alper and Anbar (2011), while Sun et al. (2017) found a positive relationship.

Gross Domestic Product (GDP). Economic activity, measured by growth rates or components of the GDP cycle, is believed to influence bank performance positively (Athanasoglou et al. 2006, 2008; Davydenko 2010; Dietrich & Wanzenried 2011; Flamini et al. 2009; Goddard et al. 2004; Trujillo-Ponce 2013). Demand for loans and other bank services increases during the cycle of economic activity rises and decreases during the decline. The author believes that the cycle component is better than the growth rate

Table 1. Variables and data source.

Variables	Formula	Data Source
Dependent Variables		
CAR (%)	Modal/ATMR	SPI BI/OJK
ROA (%)	Earning before tax/Average asset	SPI BI/OJK
BOPO (%)	Operating costs /Operating income	SPI BI/OJK
NIM (%)	Net interest income (annualized)/Average earning assets	SPI BI/OJK
LDR (%)	Credit/Third-Party Funds (DPK)	SPI BI/OJK
Liquid Asset (%)	Liquid assets/Total assets	SPI BI/OJK
NPL (%)	Non-performing loans/ Total loans	SPI BI/OJK
Credit (Billion Rp)	Total loans to non-banks	SPI BI/OJK
Deposit (Billion Rp)	Total third-party fund	SPI BI/OJK
Independent Variables		
IHSG	The Composite Stock Price Index (CSPI)	IDX
Inflation (%)	Inflation rate	SEKI-BI
GDP (%)	Level of GDP	BPS
Interest Rate (%)	Bank Indonesia interest rates	SEKI-BI
Exchange rate (Rp/USD)	The exchange rate of Rp to USD	SEKI-BI
Crude Oil Price ($/bbl)	Petroleum prices	World Bank

because the former reflects the ups and downs. Therefore, the measure of economic activity in this study is the GDP cycle because it is the country-specific feature discussed.

Inflation Rate. Demirgüç-Kunt and Huizinga (1998) and Molyneux and Thornton (1992) found that inflation has a significant positive impact on profitability because of the additional benefits extracted from asymmetric information. Also, Djalilov and Piesse (2016), Petria et al. (2015), Capraru and Ihnatov (2014), Athanasoglou et al. (2006), and Munteanu (2012) found a positive impact of inflation on bank profitability in European countries, where a number of transition economies were included. In contrast, Seferli (2010), Fries and Taci (2002), Munteanu (2012), and Davydenko (2010) concluded that, with increasing inflation, bank profitability declined.

Changes in Exchange Rates and Changes in Oil Prices. Ibrahimov (2016) and Bayramov (2014) found that currency devaluation has a significant negative effect on the profitability of Azerbaijan's banks. In contrast, Davydenko (2010) concludes that Hryvna depreciation increases the profitability of Ukrainian banks. Whether a country is an oil exporter or not, the increase in oil prices has positive implications for the entire economy, including the banking sector. Ibrahimov (2016) also found a statistically significant positive effect of oil prices on the profitability of Azerbaijan's banks.

Some previous literature reviews the relationship of variables with macroeconomics, such as Greenwald and Stiglitz (1988) in Blancard and Fischer (1998), the role of credit in the business cycle, especially in the transmission of monetary policy affecting the economy. Macro indicators such as GDP growth per capita and inflation do not have a significant impact on NIM, but inflation is found to have a significant negative effect on ROA (Naceur 2003). Research on banking in Australia shows that the internal variables of each bank cause variation in credit risk and profitability (Gizycki 2001). This is different from Naceur (2003), where macro indicators are used as explanatory variables to have a strong influence on credit risk and profitability. The interest rate variable does not affect ROA but inflation has a positive effect on ROA property prices.

GDP has a positive effect on bank CARs in Nigeria, but the inflation rate has no significant effect on CAR (Awojobi & Amel 2011). In line with that, Abiodun Research (2012) found that there are no macro variables (economic growth, inflation, interest rates, and exchange rates) that significantly influence ROA. Other research states that economic growth and exchange rate growth have a significant negative effect on NPL (De Bock and Demyanets 2012). In line with Gerlach et al. (2005), using the panel data method states that increasing economic growth and inflation rates reduce NPLs, short-term interest rates have a positive effect on NPLs and economic growth, inflation, and short-term interest rates have a positive impact on NIM.

This paper continues with the following sections, in which the second section explains the methodology; the third section presents results and discussion, the fourth section presents conclusions; and the fifth section presents acknowledgment of this study.

2 METHODOLOGY

The data used in this study are time series data with monthly periods from January 2012 to December 2018, in the form of macroeconomic and financial variable indicator data and banking performance indicators. Banking performance indicators used in this

Table 2. Analysis of variance decomposition (average of 25 months, in percent).

	OWN	IHSG	Inflation	GDP	BI Rate	Exchange rate	Oil Price
CAR	71.36	4.77	1.62	1.68	12.47	5.69	2.41
ROA	81.59	3.05	4.34	1.08	0.21	1.66	8.07
BOPO	92.01	0.81	1.29	2.62	2.49	0.27	0.51
NIM	61.88	2.21	2.17	6.35	2.07	0.49	24.83
LDR	86.77	1.04	0.71	1.36	0.21	6.19	3.71
NPL	17.35	5.11	10.64	4.49	59.58	2.72	0.09
Credit	65.96	7.88	6.69	8.98	3.00	3.95	3.55
Deposit	78.94	4.26	4.64	4.58	4.12	2.99	0.47

study include financial ratios such as ROA, NIM, NPL, and LDR as well as non-ratio data, such as the number of loans, and the number of deposits at commercial banks in Indonesia. The macro variables used are from national income (GDP), inflation, interest rate policies, exchange rates, stock indices, and world oil prices.

The econometric method used is the Vector Error Correction Model (VECM) to observe the macro effects of the performance of commercial banks. The use of the VECM model is because the variables used have the potential to have stationary data at the first derivative level (first difference). The macroeconomic variables that are used also have simultaneous facts, therefore allowing variable endogeneity problems to be solved through VECM modeling.

The VECM model can generally be represented as follows (Enders 2004):

$$\Delta y_t = a_0 + a_{1t} + \Pi y_{t-1} + \sum_{i=1}^{p-1} \Gamma_i \Delta y_{t-1} + u_t \quad (1)$$

In which is an intercept vector column with size (nx1), is the coefficient vector for time trend (t), $\Pi = \alpha\beta'$ where α is the adjustment matrix and β' contains the long-term cointegration equation, is a regression coefficient matrix measured by (nxn), is an error matrix.

Furthermore, by using the VECM method, the proportion of the impact of changes in a variable can be seen if there is a shock or change in the variable in a period. So, by analyzing the results of variance decomposition, we can measure the estimated error variance of a variable, namely how high the difference between before and after a shock occurs, both originating from the variable itself or from other variables. The author uses the eViews version 10 application in processing data to match the expectations in the methodology.

3 RESULTS

3.1 Impulse response function (IRF) analysis of the effects of macroeconomic variables on bank performance indicators

IRF analysis shows the response of a variable in the system is the result of shocks from other variables. All tests were carried out in this study, starting with non-stationary data testing, VAR stability testing, optimal lag testing, and cointegration testing. Based on IRF

analysis, the authors found that BOPO responded to the highest shock to macroeconomic variables. BOPO responded positively to shocks from the BI rate. The biggest response of BOPO to shocks from the BI rate is compared to other macroeconomic variables. BI rate is the basis for determining the interest rate for loans and deposits, so if the BI rate rises, the cost of funds will rise and the BOPO will also rise.

Other findings of this study indicate that inflation and GDP have a relatively low effect on bank performance, which was confirmed by several previous studies, including Naceur (2003), Sastrosuwito and Suzuki (2011), and Abiodun (2012). Meanwhile, according to Mirzaei et al. (2011), inflation and economic growth only affect the performance of banks in developed countries and do not affect developing countries. However, GDP and exchange rates have a positive effect on the NPL. This is different from what was found by De Bock and Demyanets (2012), that shocks to economic growth negatively affect NPL, where it is concluded that NPL is countercyclical.

NIM responds to large and negative shocks to inflation but with positive responses to GDP shocks. In line with the results of the research, Saad and El-Moussawi (2012) found the same thing with the effect of using GDP, but the results are contrary to inflation. While in Lebanon, GDP has a positive and significant effect on NIM, but the opposite effect is not significant for inflation.

Deposits and credit rates respond negatively to shocks in the capital market, which is proxied by CSPI. This is in line with Wachtel (2003), that bank financing still dominates financing for both individuals and companies. Funding new investments can be obtained from several sources such as banks or through the capital market to obtain new shares. On the other hand, companies that also act as bank creditors can also be involved in the stock market. When the stock market rises, there is potential for the diversion of company funds in banks to become equity investment instruments, which will harm bank performance.

3.2 Variance decomposition (VD) analysis of the influence of macroeconomic variables on bank performance indicators

The VD proportion analysis describes sequential movements due to the shock of the variable itself when compared to the shock of other variables. Based on VD analysis, the finding is that the BI rate has an influence on NPL. An increase in the BI rate will raise credit interest rates. The higher the loan interest rate, the potential for an increase in NPLs. JCI has a strong influence on credit and BOPO in line with the findings of Wachtel (2003). Likewise, the performance of banks proxied by NIM and deposit rates confirms the findings of Saad and El-Moussawi (2012) that GDP has a positive effect on NIM.

Macroeconomic indicators that give shocks to the performance of commercial banks in Indonesia in the top 5% are as follows: (1) CSPI against BOPO, LDR, and credit; (2) Inflation of BOPO, CAR, ROA, LDR, liquid assets, and DPK; (3) GDP for NIM, LDR, liquid assets, and deposits (DPK); (4) BI rate for CAR, liquid assets, and NPL; (5) exchange rates against BOPO and liquid assets; and (6) oil prices against credit.

4 CONCLUSION

In general, among all macro shocks, the BI rate is an indicator that has the largest variable response to bank performance indicators. The BI rate is the most powerful instrument to maintain financial sector stability, especially banking. In other words, the use of the BI rate policy as a monetary instrument can be maintained. The banking performance indicator that is the most unstable against macroeconomic variable shocks is BOPO, which is an indicator of efficiency. BOPO fluctuation shows that commercial banks in Indonesia are still relatively inefficient. This result is supported by Alfin, Siregar, and Hasanah (2015), which shows that commercial banks in Indonesia as a whole have not operated efficiently.

REFERENCES

Abiodun, B.Y. 2012. The Determinants of Bank's Profitability in Nigeria. *Journal of Money, Investment and Banking*, No. 24, pp. 6–16.

Aharony, Joseph and Itzhak Swary. 1983. Contagion E§ects of Bank Failures: Evidence from Capital Markets. *Journal of Business*, 56(3), 305–317

Albertazzi, U. and Gambacorta, L. 2009. Bank profitability and the business cycle. *Journal of Financial Stability*. 393–409

Alfin, A., Siregar, H., and Hasanah, H. 2015. Faktor-faktor yang Mempengaruhi Efisiensi BiayaPerbankan di Kawasan ASEAN-5. *Monograf*, Departemen Ilmu Ekonomi, Fakultas Ekonomi dan Manajemen, IPB. Bogor.

Ali, K., Akhtar, M.F., and Ahmed, H.Z. 2011. Bank-Specific and Macroeconomic Indicatorsof Profiability – Empirical Evidence from The Commercial Banks of Pakistan. *International Journal of Business and Social Science*. Vol.2, No. 6, pp. 235–242.

Alper, D. and Anbar, A. 2011. Bank Specific and Macroeconomic Determinants ofCommercial Bank Profitability: Empirical Evidence fromTurkey. *Business and Economics Research Journal*, Vol. 2, pp. 139–152.

Apriadi, Sembel, Sentosa & Firdaus; 2017.Kompetisi dan Stabilitas Perbankan Di Indonesia suatu pendekatan

analisis panel vector Autoregression. *Jurnal Manajemen Untar*.Vol 21 No.1

Athanasoglou, P.P., Brissimis, S.N., and Delis, M.D. 2005. Bank-specific, Industry-Specific and Macroeconomic Determinants of Bank Profitability. *Bank of Greece Working Paper No.25*. pp.4–35.

Awojobi, O. and Amel. 2011. Analyzing Risk Management in Banks: Evidence of BankEfficiency and Macroeconomic Impact.*Journal of Money, Investment and Banking*, Vol.22, pp. 147–162.

Baboucek, I., & Jancar, M. 2005. Effects of macroeconomic shock to the quality of the aggregate loan portfolio. *Czech National Bank, Working Paper Series, 1*, 1–62

Bank Indonesia. 2011. Peraturan Bank Indonesia Nomor 13/1/PBI/2011 tentang Penilaian Tingkat Kesehatan Bank Umum. Bank Indonesia, Jakarta.

Bashir, Abdel-Hameed M. 2003. Determinants of Profitability in Islamic Banks: Some Evidence from the Middle East. Islamic Economic Studies 11: 32–57.

Blancard, O.J. and Fischer, S. 1998. *Lectures on Macroeconomics*. Cambridge, MA and London. The MIT Press.

Bonin, J.P., Hasan, I., and Wachtel, P. 2003. Bank Performance, Efficiency, and Ownership inTransition Countries. Makalah yang diseminarkan pada *The Ninth Dubrovnik EconomicConference*, Dubrovnik, 26–28 June.

Borio, C., Gambacorta, L. and Hofman, B., 2015. The influence of monetary policy on bank profitability. *BIS Working Papers*, pp. 1–35.

Calomiris, C., Orphanides, A., and Sharpe, S. 1997. Leverage as a State Variable for Employment,Inventory Accumulation and Fixed Investment dalam F Capie and G Woods (editor), *Asset Prices and The Real Economy*, Macmillan Press, London.

Capraru, Bogdan, and Iulian Ihnatov. 2014. Banks' Profitability in Selected Central and Eastern European Countries. *Procedia Economics and Finance* 16: 587–91.

Caprio, G. and Klingebiel, D. 2003. Episodes of Systemic and Borderline Financial Crises. World Bank, January 22. http://go.worldbank.org/5DYGICS7B0

Clair, R.S.T. 2004. Macroeconomic Determinants of Banking Financial Performance and Resilience in Singapore. *MAS Staff Paper*, No. 38. Monetary Authority of Singapore.

Davydenko, Antonina. 2010. Determinants of Bank Profitability in Ukraine. Undergraduate Economic Review 7: 2.

De Bock, R. and Demyanets, A. 2012. Bank Assets Quality in Emerging Markets: Determinants and Spillovers. *IMF Working Paper*, No. 71, pp.1–26. **412**

Dedi Rosadi. 2012. Ekonometrika dan Analisis Runtun Waktu Terapan dengan Eviews. Yogyakarta : Andi Offset

Demirgüç-Kunt, Asli and Harry Huizinga 1999. "Determinants of commercial bank interest margins and profitability: some international evidence", *World Bank Economic Review*, 13(2), pp 379–408

Dietrich, Andreas, and Gabrielle Wanzenried. 2011. Determinants of bank profitability before and during the crisis: Evidence from Switzerland. *Journal of International Financial Markets, Institutions and Money* 21: 307–27.

Djalilov, Khurshid, and Jenifer Piesse. 2016. Determinants of bank profitability in transition countries: What matters most? *Research in International Business and Finance* 38: 69–82.

Enders W. 2004. *Applied Econometric Time Series*. Alabama: University of Alabama.[FEM]. Fakultas Ekonomi dan Manajemen. 2012. *Perkembangan Makroekonomi and Keuanganserta Implikasinya terhadap Bisnis BRI*. Institut Pertanian Bogor.

Festic, M. and Beco, J. 2008. The Banking Sector and Macroeconomic Performance in Central European Economies. *Czech Jornal of Economics and Finance*. No. 58, Vol. 3–4, pp. 131–151.

Flamini, Valentina, Calvin McDonald, and Liliana Schumacher. 2009. The Determinants of Commercial Bank Profitability in Sub-Saharan Africa. *IMF Working Paper* No 09/15. Washington: International Monetary Fund (IMF).

Fries, Steven, and Anita Taci. 2002. Banking Reform and Development in Transition Economies. London: *European Bank for Reconstruction and Development*

Gerlach, S., Peng, W., and Shu, C. 2005. Macroeconomic Conditions and Banking Performance in Hong Kong SAR: A Panel Data Study. *BIS paper*, No.22, Monetary and Economic Department, Bank for International Settlements, Swiss.

Ghazali, M.B. 2008. The Bank-Specific and Macroeconomic Determinants of Islamic Bank Profitability: Some International Evidence. *Thesis*. Faculty of Business and Accountancy, University of Malaya.

Gizycki, M. 2001. The Effect of Macroeconomic Condition on Banks Risk and Profitability. *Research Discussion Paper,* No. 06, System Stability Department, Reserve Bank of Australia.

Goddard, John, Phil Molyneux, and JohnWilson. 2004. The Profitability of European Banks: A Cross-Sectional and Dynamic Panel Analysis. The Manchester School 72: 363–381.

Guerrieri, L. and Welch, M. 2012. Can Macro Variables Used in Stress Testing Forecast the Performance of Banks?. *Finance and Economics Discussion Series*, No 49, Divisions of Research & Statistics and Monetary Affairs, Federal Reserve Board.

Haldane, Hoggarth, Saporta, Sinclair. 2005. Financial Stability and Bank Solvency. University of Birmingham. Bank of England

Hamadi, H. and Awdeh, A. 2012. The Determinants of Bank Net Interest Margin:Evidence from the Lebanese Banking Sector. *Journal of Money, Investment and Banking*, Vol.23, pp. 85–98.

Kaufman. G. 1998. Central Banks, Asset Bubbles, and Financial Stability. *Federal Reserve Bank of Chicago Working Paper*, WP98/12.

Kho, Lee and Stulz, 2000. US Bank Crises, and Bailouts: From Mexico to LTCM. National Bereau of Economic Research. Working Paper 7529 http://www.nber.org/papers/w7529

Kok, C., Móré, C. and Pancaro, C., 2015. Bank profitability challenges in euro area banks: The role of cyclical and structural factors. *Financial Stability Review*, pp. 148-158.

Kool, C., 2006. An analysis of financial stability indicators in European banking: The role of common factors. *Tjalling C. Koopmans Research Institute,* Utrecht, Utrect School of Economics, Discussion paper no. 12.

Lowe, P. and Rohling, T. 1993. Agency Costs, Balance Sheets, and The Business Cycle. *Reserve Bank of Australia Discussion Paper* No. 9331.

Mishkin FS. 2001. *The Economics of Money, Banking, and Financial Markets*. New York : Columbia University.

Mirzaei, A., Liu, G. and Moore, T. 2011. Does Market Structure Matter on Banks' Profitability and Stability? Emerging versus Advanced Economies. *Economics and Finance Working* Paper, No. 11–12. Pp. 1–40.

Molyneux, Philip, and John Thornton. 1992. Determinants of European bank profitability: A note. Journal of Banking and Finance 16: 1173–178.

Munteanu, Ionica. 2012. Bank liquidity and its determinants in Romania. Procedia Economics and Finance 3: 993–998.

Ibrahimov, Anar. 2016. The Impact of Devaluation and Oil Price on the Banking Sector of Azerbaijan. Master's dissertation, Porto University, Porto, Portugal.

Naceur, S.B. 2003. *The Determinants of The Tunisian Banking Industry Profitability: Panel Evidence*. Department of Finance, UniversitéLibre de Tunis.

Nuriyeva, Zülfiyye. 2014. Factors Affecting the Profitability of Azerbaijan Banking System. Master's dissertation, Eastern Mediterranean University, Gazimağusa, North Cyprus, Turkey, July.

Pasiouras, F. and Kosmidou, K. 2007. Factors influencing the profitability of domestic andforeign commercial banks in the European Union, *International Business and Finance*, No. 21, 222–237.

Petria, Nicolae, Bogdan Capraru, and Iulian Ihnatov. 2015. Determinants of banks' profitability: Evidence from EU 27 banking systems. Procedia Economics and Finance 20: 518–524.

Philipp Hartmann, Stefan Straetmans, and Casper de Vries, 2005. Banking System Stability: A Cross-Atlantic Perspective, National Bureau Of Economic Research

Rivard, Richard J., and Christopher R. Thomas. 1997. The effect of interstate banking on large bank holding company profitability and risk. Journal of Economics and Business 49: 61–76.

Roman, Angela, and Alina Camelia Sargu. 2015. The impact of bank-specific factors on the commercial banks liquidity: Empirical evidence from CEE countries. *Procedia Economics and Finance* 20: 571–579

Rumler, F. and Waschiczek, W. 2010. The Impact of Economic Factors on Bank Profits. *Monetary Policy and The Economy Q4*, pp. 49–67.

Saad, W. and El-Moussawi, C. 2012. The Determinants of Net Interest Margins ofCommercial Banks in Lebanon.*Journal of Money, Investment and Banking*, Vol.23, pp. 118–132.

Sastrosuwito, Suminto, and Yasushi Suzuki. 2011. Post Crisis Indonesian Banking System Profitability: Bank-Specific, Industry-Specific, and Macroeconomic Determinants. Paper presented at the 2nd International Research Symposium in Service Management, Yogyakarta, Indonesia, July 26–30.

Schinasi, J. G., 2005. Preserving financial stability. Washington, *International Monetary Fund*, economic issues, no. 36.

Shaher, T.A., Kasawneh, O. and Salem, R. 2011. The Major Factors that Affect Banks' Performance in Middle Eastern Countries.*Journal of Money, Investment and Banking*, Vol. 20, pp. 101–109.

Slovin, Myron, Marie Sushka and John Polonchek. 1999. ìAn Analysis of Contagion and Competitive Eﬀects at Commercial Banks.îJournal of Financial Economics, 54, 197–225.

Sukarman, W. 2014. *Liberalisasi Perbankan Indonesia – Suatu Telaah Ekonomi Politik*. Jakarta (ID): KPG.

Swary, Itzhak. 1986. ìStock Market Reaction to Regulatory Action in the Continental Illinois Crisis.îJournal of Business, 59(3), 451–473.

Trujillo-Ponce, Antonio. 2013. What determines the profitability of banks? Evidence from Spain. Accounting and Finance 53: 561–86.

Wall, L. and D. Peterson, 1990, The effect of continental Illinois failure on the financial performance of other banks, *Journal of Monetary Economics (August)*, 77–99.

Advances in Business, Management and Entrepreneurship – Hurriyati et al. (Eds)
© 2021 Taylor & Francis Group, London, ISBN 978-0-367-67471-7

The passive and active optimal portfolios

D.F. Salim, D. Disman & I. Waspada
Universitas Pendidikan Indonesia, Bandung, Indonesia

ABSTRACT: The portfolio was first introduced by Markowitz in 1952, where the journals that are used as a reference for almost all researchers who research about a portfolio, the Portfolio Selection. Then the mixing from various types of stocks can be done to eliminate the similar homogeneity. If the mixing practice is done, then the risk that will arise from the investment will be accumulated. This study uses economic value added (EVA) and market value added (MVA) ratios to form active and passive portfolios to obtain an optimal portfolio. The samples used were 24 stocks, which entered consistently in the LQ 45 for the 2014–2018 period. As a conclusion, this study supports the previous research, which states that the active strategy gives a higher return than the passive strategy. High EVA portfolio gets the highest return on the passive portfolio strategy, then low EVA gets the highest return on the active portfolio.

1 INTRODUCTION

The portfolio is interesting to do for discussing investing. The portfolio is useful for diversifying the level of risk which arises from investment activities. The portfolio was introduced for the first time by Markowitz in 1952, the journals that are used as a reference for almost all researchers who research about portfolio, Portfolio Selection. Then the mixing from various types of stocks can be done to eliminate the similar homogeneity. If the mixing practice is done, then the risk that will arise from the investment will be accumulated (Sharpe 1964). There is no efficient portfolio of each individual because of the efficient nature formed due to a combination of several stocks, which is formed in the Roll portfolio (1977).

Testing the performance of a portfolio cannot be done with traditional capital asset pricing model (CAPM) methods because in a portfolio it is designed with beta efficient variance Roll (1978). The return rate of a portfolio can be designed in such a way by looking at the returns that the market gets, if the market returns, which is obtained by when the market is higher when compared to portfolio returns. Then the investors can rearrange the composition of stocks in the portfolio (Bollerslev et al. 1988). In Bai et al. (2009), the portfolio optimization can be done by mixing the common stock, preferred stock, bonds, and other types of investments.

The measurement of portfolio risk level can be done by using the VROL value, the greater the VROL value, the better the performance of its portfolio will be (Fitriaty et al. 2012). Portfolios' insecurities have more than two compositions aiming to reduce risk and increase the returns for investors (Uyar 2012). Portfolios that

have undervalued stock prices provide higher expected returns (Hidayat & Hendrawan 2017). An active portfolio strategy provides a higher expected return than a passive portfolio (Hendrawan & Salim 2017). The optimal portfolio is also designed with EVA, return on assets (ROA), return on equity (ROE) ratio, and the result is low ROA, high EVA. ROE can provide a higher expected return than the Indeks Harga Saham Gabungan (IHSG) market (Salim 2019). EVA and market value added (MVA) were chosen as calculation tools because a lot of research discusses EVA and MVA as measures of company performance, both internal and external, such as research from Romplo (2009). EVA is used to measure the added value of universities in Thailand with nonprofit institutions, and Romplo uses the academic value-added ratio (AVAR) to measure the performance of universities in Thailand. EVA calculation also cannot be released from the help of investment managers, who always use several funds to be reinvested to get profits in the future (Young 1997).

EVA and PDB significantly affect MVA. The company managers are advised in EVA testing so it will be better to calculate the MVA together (Zaima 2008). Positive EVA value is obtained by the company after the company has made the merger process (Sa'diyah et al. 2015). EVA is an information guide for the company to invest. There are many types of calculations to calculate economic value-added for companies such as CVA, MVA, SVA, and Rona (Berzakova et al. 2015).

Since some researchers discuss it consistently, this study intends to examine some points, namely the results of expected return and portfolio risk using the EVA and MVA ratio and the results of passive or active strategies done in investing with the formation of EVA and MVA ratio.

1.1 Portfolio theory

The optimal portfolio for the first time was introduced by Markowitz (1952) in the famous journal Portfolio Selection. In this optimal portfolio theory, Markowitz suggests two things, where the first to analyze the performance of a company in the future will be entered into the portfolio, then the second is to determine the decision based on expected return and variance of securities.

1.2 Return and risk

Risk and return cannot be released because the return is the result of an investment, then every investment made by an investor will certainly have a risk. Risks arise when the unexpected does not occur or is not achieved, and then the risk will disappear or diminish by diversifying all the investments that are owned with those by portfolio theory. Several risks arise when investing. The first is a systemic risk where the systematic risk cannot be eliminated because it originates from the market or from outside. Then the risk is not systematic where the risks arising from internal companies can be reduced by diversification (Husnan 2015).

1.3 Economic value added

Economic value added (EVA) is one model that can calculate how much economic value is added from a company in an operating period. The value of EVA itself depends on the company's decision in adding investment to add value in the future to fulfill the interests of investors, management, and company owners. EVA itself was first introduced in the journal of Stephen et al. (1996).

1.4 Market value added

Market value added (MVA) approach is a calculation that comes from outside the company because it measures the company's value from the stock market, how much the increased value is from a company in the stock market. The value of MVA itself can be a benchmark by investors to invest in certain stocks because the greater the value of MVA, the company is declared good and has added value in the stock market.

1.5 Framework

Picture 1 is a framework in which investment holds back current consumption. It aims to be used in the future, and its value will increase from the initial investment. This research offers an investment where it can be done by making optimal portfolio formation in stocks that are included in LQ 45. The portfolio is formed by calculating EVA and MVA. This research offers portfolio composition that consists of passive portfolios where the portfolio is formed based on

MVA and EVA value. The portfolio will not change its composition from 2014–2019. Then the active portfolio changes annually according to the amount of EVA and MVA each year. Each portfolio will be tested for its performance using Sharpe; after conducting the Sharpe test, it will form an optimal portfolio of this research that is supported by previous studies such as Hidayat and Hendrawan (2017), Hendrawan & Salim (2019), and Salim (2019).

2 METHOD

This research is a quantitative study that follows the EVA and MVA ratios as the secondary data of annual reports from each sample. All shares included in the LQ 45 index were involved as the research participants because they were categorized into LQ45 from 2013–2018 with 24 shares.

The formation of a portfolio will be divided based on the results of EVA and MVA calculations by using MS Excel. The results then will be divided based on the value of High EVA, Medium EVA, Low EVA, High MVA, Medium MVA, and Low MVA.

3 RESULTS

From Table 1, it is shown that the High MVA portfolio has the highest return, which is 43.75%. This portfolio return can beat the expected IHSG market return where IHSG received an expected return of 25.34%, then Low EVA has a negative return rate of −5.26%. This is due to the composition of that portfolio which has a calculated value of $E(R_i)$ negative averages from 2014–2018. The average return obtained by the passive portfolio is 25.39%.

Table 1 shows that the High MVA portfolio has the highest return which is 43.75%. This portfolio return can beat the expected IHSG market return where IHSG received an expected return of 25.34%, then Low EVA has a negative return rate of −5.26% This is due to the composition of that portfolio which has a calculated value of $E(R_i)$ negative averages from 2014–2018 (Hendrawan & Salim 2017) which states that the active strategy gives a higher return than the passive strategy.

Table 1. Passive portfolio return.

Portfolio	Return	Risk	Sharpe
HIGH MVA	43,75%	2,06%	1,17%
HIGH EVA	40,74%	2,30%	1,05%
MEDIUM EVA	35,05%	2,32%	0,71%
LOW MVA	31,69%	1,49%	0,80%
IHSG	25,34%	8,63%	0,71%
MEDIUM MVA	6,43%	2,97%	−0,48%
LOW EVA	−5,26%	1,81%	−0,72%
AVERAGE	25,40%	2,16%	0,42%

The average return obtained by the passive portfolio is 25.39%.

High EVA portfolio gets the highest return in the passive portfolio strategy, then Low EVA gets the highest return on the active portfolio. The average return obtained by the passive portfolio is 25.39%. The calculation of passive portfolio risk can be seen in Table 1, which is in line with Table 1 where the sequence gained from the calculated value of E (Rp) portfolio. The highest risk results are obtained by the IHSG market by 8.63%, and the lowest risk is achieved by a Low MVA portfolio. The average risk obtained by the passive portfolio is 2.16%. These results prove that the portfolio can be diversified/reduce the level of risk that will arise in an investment.

In addition, the performance test calculations using Sharpe revealed that High MVA has the highest performance results of 1.17% This performance is in line with the return results obtained by a high MVA portfolio The return results can be seen in Table 1, as well as the Low EVA performance that also has a negative result of -0.72%. Then the average performance gained by the passive portfolio is 0.42%.

The calculation results of active portfolio returns which can be seen in Table 2, the Low EVA portfolio gets the highest results by 45.94% active portfolio the composition changes every year based on the calculation of the value of each EVA and MVA ratio each year during the 2014–2018 period. The average active portfolio return is 25.49%. Based on the results of active portfolio risk calculations in Table 2, IHSG got the highest results by 8.62%; this is in line with the passive portfolio. Then the risk of Medium EVA has the lowest risk level by 4.5%.

In Table 2, the calculation of High Sharpe MVA performance gets the greatest results by 3.22%, the results of the High MVA portfolio performance could exceed the 2.3% IHSG market performance, the average Sharpe's active portfolio performance is 0.92%. You see the results of Sharpe EVA Low-performance calculation −0.06% These results were obtained because in 2015, the performance of all portfolios received negative results, and all of the Low EVA portfolios had the greatest negative results when compared to other portfolios. Low EVA portfolio conditions began to improve in the following years, evidenced by the recomposition that is done every year that makes the Low EVA portfolio have the highest return compared to other portfolios that can be seen in Table 2.

The optimal portfolio is also designed with EVA, ROA, ROE ratio, and the result is low ROA, high EVA. ROE may provide a higher expected return than the IHSG market (Salim 2019)

Table 3 present a comparison between active and passive portfolios, return quantities, and risk. It also implies that Sharpe active portfolio is superior than the passive one. It is supported by Hendrawan and Salim (2017), which suggest that active portfolio strategies produce better expected returns than passive portfolios.

4 CONCLUSIONS

Portfolios are theories that can reduce the level of risk by getting certain returns that can be obtained by investment. This research offers to investors to choose the portfolio composition that can be a reference for stock investments, especially in the Indonesia Stock Exchange (IDX). Following the research objectives written above, this study found that the High MVA ratio has the highest rate of return on the passive strategy that is equal to 43.75% with a risk level of 2.06%, then Sharpe's performance of 1.17% based on Sharpe's performance calculation results in line with the calculation of returns obtained by High MVA, then the active portfolio strategy formed by EVA Low has the highest return of 45.94% with a risk level of 7.15% according to the theory of high risk, high return.

In the comparison between which is the best, active or passive portfolio strategy, this study supports research conducted by Hendrawan & Salim (2019), which states that active strategy gives a higher rate of return when compared to passive strategy. With this statement, this study obtained the same results that can be seen from the comparison results in Table 3.

For further research, you can use other ratios such as PER, PBV, EPS, liquidity ratio, gross profit margin, and choose an index such as comparing between the two industrial sectors created by its portfolio. Investors can use the guidelines as decision making in whether using active or passive strategies in investing in the capital market or other investments. The thing that needs to be distinguished is that the investigation, which is done from the beginning of buying to the end of the investment, has not modified its portfolio composition. However, active strategy portfolio composition is changed regularly with the aim of getting

Table 2. Active portfolio returns.

Portfolio	Return	Risk	Sharpe
LOW EVA	45,94%	7,15%	−0,06%
HIGH EVA	42,17%	5,37%	2,40%
HIGH MVA	37,66%	5,64%	3,22%
IHSG	25,24%	8,62%	2,39%
LOW MVA	18,13%	5,50%	0,68%
MEDIUM EVA	5,77%	4,50%	−0,55%
MEDIUM MVA	3,28%	6,42%	−0,15%
Average	25,49%	5,76%	0,92%

Table 3. Portfolio comparison.

	Passive	Active
Return	25,40%	25,49%
Risk	2,16%	5,76%
Sharpe	0,42%	0,92%

short-term profits so that the risk for large losses can be prevented quickly.

REFERENCES

Bai, Z., Liu, H., & Wong, W. 2009. On The Markowitz Mean-Variance Analysis Of Self-Financing Portfolios. *Risk and Decision Analysis*. 1 (1), 35–42. https://doi.org/10.3233/RDA-2008-0004

Berzakova, V., V. Bartosova & E. Kicova. (2015). Modification of EVA in value based management. *Procedia Economics and Finance*, 26 pp. 317–324.

Bollerslev tim, Robert F Engle, dan Jeffey M. Wooldridge. (1988). A Capital Asset Pricing Model with Time-Varying Covariances. *Jurnal of Political Economy*. vol.96 no 1.

Fitriaty, Tona Aurora Lubis., Pungki Rekno Asih. 2012. Analisis Kinerja Portofolio Optimal pada Saham-Saham Jakarta Islamic Index (JII) Periode 2010–2012.*Mankeu*, Vol 3 No.1 hlm. 374–463. Universitas Jambi

Hendrawan, Riko dan Dwi Fitrizal Salim. 2017. Optimizing Active and Passive Stocks Portfolio Formed Tobin's Q and Price Earning Ratio Model Stocks on Kompas Index-100 Period 2012–2017. *Internasional Jurnal of Applied Business and Economic Reseach*. (ISSN: 0972-7302)

Hidayat, Firman dan Riko Hendrawan. 2017. Performance Comparison Simulation of the Stock Portfolio Active and Passive Strategy Formed with Price Earnings Ratio, Price Book Value, and Price Earning Growth Ratio Stocks on LQ-45 Index Period 2011–2016. *Jurnal Internasional Journal of Economi Perspectives*. (ISSN: 1307-1637)

Husnan, Suad. 2015. *Dasar-dasar Teori Portofolio & Analisis Sekuritas*. Edisi ke 5. Yogyakarta. UPP STIM YKPN.

Markowitz, H., 1952. Portofolio Selection, *Journal of Finance*.7, 77–91.

Rompho, Nopadol. 2009. Application of the Economic Value Added (EVA) Protocol in a University Setting as a Capital Budgeting Tool. *Journal of Financial Reporting and Accounting* Vol. 7 Iss 2 pp. 1 – 17.

Roll, R. 1977. A Critique Of The Asset Pricing Theory's Tests; Part 1: On Past And Potential Testability Of The Theory. *Journal of Financial Economics*. 4, 129–176.

Roll, R. 1978. Ambiguity When Performance Is Measured By The Securities Market Line. *Journal of Finance*. 33, 1051–1069.

Sa'diyah, Halimatus., Raden Rustam Hidayat., Achmad Husaini. 2015. Analisis Dampak Merger Terhadap Economic Value Added (Eva) Dan Market Value Added (MVA) (Studi Pada Perusahaan Di Bursa Efek Indonesia Yang Melakukan Merger Tahun 2011). *Jurnal Administrasi Bisnis* (JAB)|Vol. 24 No. 1 Juli 2015

Stephen F. O'Byrne, Stern Stewart & Co. 1996. EVA®AND MARKET VALUE. *Journal Of Applied Corporate Finance* Volume 9 Number 1 SPRING 1996

Sharpe, William F. 1964. Capital Asset Prices: A Theory Of Market Equilibrium Under Conditions Of Risk. *The journal of FINANCE*. Vol. XIX no.3

Salim, Dwi Fitrizal. 2019. Perancangan Portofolio Optimal Dengan Menggunakan Return On Assets, Return On Equity dan Economic Value Added Pada Jakarta Ismaic Index Periode 2014–2018. *JURNAL RISET AKUNTANSI DAN KEUANGAN* 7 (1), 2019, 43–54

Uyar, Umut., Sinem Guler Kangalli. Markowitz Modeline Dayali Optimal Portfoy Seciminde Islem Hacmi Kisiti. *Ege Academic Review*. 2012, vol. 12, issue 2, 183–192

Young, D. (1997). Economic value added: a primer for European managers. *European Management Journal* 15(4), 335–343.

Zaima, Janis K. 2008. Portfolio Investing with EVA. *The Journal Of Portfolio Management* SPRING 2008 34.3:34–40.

Advances in Business, Management and Entrepreneurship – Hurriyati et al. (Eds)
© *2021 Taylor & Francis Group, London, ISBN 978-0-367-67471-7*

How do firm's specific factors affect capital structure? Empirical study on 50 biggest market capitalization of the Indonesia stock exchange (2013–2018)

S. Heliola, D. Disman & I. Waspada
Universitas Pendidikan Indonesia, Bandung, Indonesia

ABSTRACT: Capital structure decisions are much influenced by firm-specific factors, whereby these factors are supported by the existing Capital Structure Theory. This study aimed to look at how firm-specific factors supporting the Pecking Order Theory from the 50 Biggest Market Capitalization of the Indonesia Stock Exchange for the period 2013–2018. These specific factors chosen were Tangibility Asset, Market to Book Ratio, Size, Profitability, and Liquidity. The results suggest that tangibility asset, profitability, and liquidity were having a negative relationship with the leverage, whereas the market to book ratio was positively related to the leverage. The size in this study did not appear to be significantly related to leverage. The results of this study supported The Pecking Order Theory.

1 INTRODUCTION

According to statistical data released by The Financial Services Authority (OJK), from 2013 to 2018 growth of productive credit in Indonesia has declined dramatically. The productive credit growth rate in 2013 has reached 25% and fell to its lowest point in the last 5 years to 7% (2017) and in 2018 started to increase at 12%. Similarly, the average growth of listed company assets in Indonesia, have shown a drastic decline from 23.7% in 2013 to 13% in 2014. The trend in the decline of productive credit growth and the company's assets in the period of 2013–2018 has shown a relatively similar and their average growth of 13%. Such a phenomenon questions the relationship between debt and equity.

The combination financing sources through debt or capital is known as the capital structure. Capital Structure Theory firstly appeared in 1958 by Franco Modigliani and Merton Miller (MM) that capital structure does not affect the value of the company, known as The Capital Structure Irrelevance Theory. The trade-off theory is the development of the MM Theory by taking into consideration the effects of taxes. Then Jensen and Meckling (1976) developed another theory on capital structure, titled The Agency Theory. The Agency Theory is based on agency problems. Myers & Majluf (1984) proposed The Pecking Order Theory that companies will prioritize internal financing before external financing.

There are various Capital structure studies that have shown the presence of firm's specific factors in influencing capital structure, thus including researches carried out by Rajan and Zingales (1995); Booth et al.

(2001); Hall et al. (2004); Fan et al. (2012); Çekrezi (2013); Onofreia et al. (2013); Singh (2016) and Li & Islam (2019). The firm's specific factors that were identified as determinants of capital structure by previous researchers were factors that largely supported Trade-Off Theory, Pecking Order Theory, and Agency Theory.

Before determining the specific factors to be tested for their influence on capital structure, it is necessary to determine the leverage ratio first. Various leverage ratio measurement tools have been used by researchers. Measurements were done based on book values, it always refers back by measuring what has happened, whereas measurement focusing on market values are generally looking forward in which is considered on seeing the future opportunities (Frank and Goyal 2009). Booth et al. (2001); Fan et al. (2012); Li & Islam (2019), using book values and market values. And many other studies use book value for financial stability (Kumar 2005). In the research of Rajan and Zingales (1995); Hall et al. (2004); Kumar (2005); Onofreia et al. (2013); Çekrezi (2013); Thippayana (2014); Singh (2016) calculates the company's financial leverage with book value, as well as in this study.

So many combinations of firm's specific factors that influence leverage have been tested by researchers. Tangibility Asset, Size, Profitability, and Market to Book Ratio are used by Rajan and Zingales (1995) when looking at how specific factors affect corporate leverage in G7 countries. Likewise, Fan et al. (2012) choose the same firm's specific factors in 39 developed and developing countries, and Li & Islam (2019) in Australia. Hall et al. (2004) added factor of company age. Psillaki & Daskalakis (2009) complements risk factors and liquidity factors by Onofreia et al. (2013). Referring to this study, Tangibility Asset, Market to Book Ratio, Size, Profitability, and Liquidity were used as company-specific factors in examining the effect on leverage. Also in this study, the risk factor has been ignored because from the previous studies only in the study of Psillaki & Daskalakis (2009) one country has shown its influence, whereas the age factor of the company was not used because in the study of Hall et al., (2004), it looked at the age of the company related to the strength of internal funding factor, whereas in this study the opinion can be represented by the company's liquidity factor.

The purpose of this study was to look at a firm's specific factors that influence the choice of corporate financing in Indonesia so that companies can determine whether the choice of asset financing through debt is the right choice. By using panel data regression, this study discussed how the firm's specific factors influence the leverage of companies in Indonesia. 50 Biggest Market Capitalization companies which were non-financial and were listed on the Indonesia Stock Exchange (IDX) between 2013 and 2018, were chosen as the sample.

1.1 Capital structure theory

The first theory of capital structure is MM Theory that coined by Modigliani & Miller (1958) that capital structure was irrelevant or did not affect the value of the company. Then trade-off theory is a development of the MM theorem by taking the effects of taxes and bankruptcy costs. The Trade-off Theory states that debt will increase the value of the company but only to a certain point, thus the companies need to balance the benefits and costs of their financing choices. The company prefers debt financing rather than issuing equity by utilizing the profits from the debt tax shield (Myers 2001). Through research by Baxter (1967); Kraus & Litzenberger (1973), Trade-off Theory shows that companies chose their capital structure by balancing the benefits of borrowing, especially tax savings, with costs associated with loans including bankruptcy costs.

The Agency Theory was presented by Jensen & Meckling (1976), which had an idea that the interests of corporate managers and shareholders were not aligned. Managers tend to maximize their utility rather than the value of the company in contrast to shareholders, in which managers developed a tendency to retain resources so that they have control over it. Debt can be considered as a way to reduce agency conflict with free cash flow. If the company uses debt, then the manager will be forced to issue cash from the company to pay interest. The next capital structure theory is The Pecking Order Theory which was presented by Myers & Majluf (1984) and Myers (1984). This theory cannot determine the optimal point of capital structure. The Pecking Order Theory ranks the financing sources,

where the company prioritizes financing from internal sources then external financing.

1.2 Leverage and firm's specific factors

The leverage ratio illustrates how the composition of a company's capital structure. Several alternative definitions of leverage have been used in the literature. This ratio can be shown from the calculation of total debt to assets (Çekrezi 2013; Delcoure 2007; Hall et al. 2004; Onofreia et al. 2013; Psillaki & Daskalakis 2009; Rajan & Zingales 1995; Singh 2016), or towards company equity (Fan et al. 2012; Kester 1986; Thippayana 2014). Debt can also be separated between short-term debt and long-term debt (Çekrezi 2013; Delcoure 2007). For consideration of values used, some studies use book values (Çekrezi 2013; Hall et al. 2004; Psillaki & Daskalakis 2009; Rajan and Zingales 1995) and some use market values (Fan et al. 2010; Krishnan & Moyer 1997). The choice of ratio used by the researcher depends on the main focus that will be reviewed in his research. This research used the Debt to Asset Ratio to show the company's leverage.

Existing capital structure literature has raised many firms' specific factors such as liquidity, asset tangibility, profitability, company size, growth opportunities, market to book ratio, risk probability, company age, and various other factors. In this study, Tangibility Asset, Market to Book Ratio, Size, Profitability, and Liquidity was determined as the firm's specific factors. These specific factors have been used in the research of Harris & Raviv (1991); Rajan & Zingales (1995); Antoniou et al. (2002); Deesomsak et al. (2004); De Jong et al. (2008); Fan et al. (2012); Çekrezi (2013); Onofreia et al. (2015); Singh (2016), Li & Islam (2019).

Tangibility Asset or some researchers call it an asset structure, describes how much tangibility asset influence on corporate financing through debt and equity. The TradeOff Theory shows a positive relationship between leverage and asset reliability (Çekrezi 2013; Fan et al. 2012; Rajan & Zingales 1995) because tangible assets can generally be used as collateral to reduce credit risk, in contrast to Agency Theory and Pecking Order Theory predicts negative relationships (Onofreia et al. 2015; Psillaki & Daskalakis 2009; Singh 2016). According to The Pecking Order Theory, companies with low tangibility asset will be more sensitive to information asymmetry, and companies will prioritize internal financing, but choose debt financing over equity when external finance is needed.

Market to Book Ratio (MBR) is one of the ratios that can describe the growth of a company's assets. Rajan & Zingales (1995) have explained two reasons for the negative relationship between MBR and leverage. First, when MBR increases, the cost of financial difficulties also increases. Second, companies prefer issuing equity when shares are valued higher (over-valued). Previous researchers have shown the relationship of Market to Book Ratio with leverage because it helped to explain the market expectations of the value of investment opportunities and company growth (Antoniou et al. 2002). MBR has a positive impact on leverage (Getzmann 2010; Ozkan 2001), but negative impact shown by Rajan and Zingales (1995), Frank and Goyal (2003), Li & Islam (2019).

Firm size most commonly has been used as determinants of the capital structure, either by using the logarithm of assets or logarithm of sales. The size of the company has been used by researchers to see the tendency of companies in debt, as in The Trade-off Theory, large companies tend to have higher debt because generally they have a better credit rating compared to smaller companies. Positive affect shows by Harris Raviv 1991; Rajan & Zingales 1995; Booth et al. 2001; Antoniou et al. 2002; Fan et al. 2012; Onofreia et al. 2015; Singh 2016; Li & Islam 2019. Unlike the case of the Pecking Order Theory, it has been interpreted as predicting an inverse relationship between leverage and firm size (Frank & Goyal 2009).

Profitability is one factor that widely used. Pecking Oder Theory states that companies that have high profits will have a negative relationship to leverage and negative relationship mostly reported by the empirical literature (Deesomsak et al. 2004). On another hand, The Trade-off Theory shows the relationship between the two variables is positive because the expected profitability is higher than the debt benefits and lower costs of financial difficulties. Static trade-off theory predicts that profitable firms should have more debt (Frank & Goyal 2008:175).

Many researchers choose Liquidity as one of the firm-specific factors because it is one of the factors to prove the Pecking Order Theory. Liquidity ratios reflect the speed of current assets covering the current liabilities, thus companies with high liquidity prefer to go with low debt, which is a negative relationship and it has been elaborated by most researchers (Deesomsak et al. 2004). In Agency theory, which considers conflicts of interest between shareholders and managers in the concept of free cash flow, managers tend to hold back resources by having control over them. Debt can be considered as a way to reduce agency conflict with free cash flow (Jensen & Meckling 1976).

1.3 Hypothesis

Referring to the literature on capital structure debate, the hypothesis that we built in this study is as follows

HO 1: Asset Tangibility is negatively related to capital structure.
HO 2: Market to Book Ratio is positively related to capital structure.
HO 3: Size is negatively related to capital structure.
HO 4: Profitability is negatively related to capital structure.

HO 5: Liquidity is negatively related to capital structure.

2 METHOD

2.1 Sample

The sample of this study covers 24 companies from The 50 Biggest Market Capitalization during the period 2013–2018 on the Indonesia Stock Exchange. The study excluded companies in the category of financial companies, because of their different and tight policies from regulators.

Table 1. Samples.

Criteria	
Population 50 Biggest Market Capitalization 50	50
Financial companies	(6)
Non Financial companies	44
Not consistent as 50 Biggest Market Capitalization 2013–2018	(20)
Sample	24

2.2 Variables

The dependent variable in this study is the capital structure which is the debt ratio (TDR), the ratio between total debt and total assets. For independent variables are firm's specific factors in this study were selected based on previous research are Tangibility Asset, Market to Book Ratio, Size Profitability, and Liquidity. The measurement of the dependent and independent variables can be explained in the table below:

Table 2. Calculation of variable values.

Variabel		Indicator
Y	Debt to Asset Ratio (DAR)	Total Debt/ Total Asset
X1	Tangibility Asset (TANG)	Fixed Asset/ Total Asset
X2	Market to Book Ratio (MTB)	Market value of equity/ book value of equity
X3	Corporate Size (SIZE)	Ln(Total Asset)
X4	Profitability (PROF)	Net Income/ Total Asset
X5	Liquidity (LIQ)	Current Asset/ Current Liabilities

2.3 Methodology

To measure the strength of the relationship between the variable Y with the variable X in this study, panel data regression was used. The econometric functional models used to determine the firm's specific factors influencing leverage is as follows:

$$DAR = a + b1TANG + b2MBR + b3SIZE + b4PROF + b5LIQ + e \quad (1)$$

3 RESULTS

3.1 Descriptive analysis

Descriptive statistics highlight several important indicators to help explain the general picture of research results.

Table 3. Descriptive statistics.

	DAR	TANG	MBR	SIZE	PROF	LIQ
Mean	0.430	0.567	7.358	31.115	0.125	2.243
Min	0.121	0.115	0.356	28.209	−0.134	0.380
Max	0.785	0.901	82.444	33.474	0.657	7.930

(source IDX 2013–2018: data processed)

Table 3. Reports summary statistics for the variables used in the study. The table shows that the average DAR 0.43, it means the sample uses 0.43 times debt of its assets, thus means the sample company only uses 43% of the value of its assets to guarantee its debt. The amount of fixed assets owned is more than 50% of the total assets of the sample company, this shows that the company does not invest too much in the form of fixed assets. When compared to the large percentage of fixed assets with DAR, it illustrates that 86% of fixed assets have been utilized to guarantee the debt. The sample company also looks very liquid, reflected by current assets more than 2 times greater than current debt as indicated by an average liquidity ratio of 2.2427.

The value of the company's equity at market value is better than the book value, on average, shown at 7.3584, this shows the market's interest to invest in sample companies. Based on the average natural logarithm of total assets is 31.1153, the minimum number 28.2088 is not far from the average. The sample can be categorized as large companies, so the sample companies might get ease in debt. There are still companies in a loss that show a negative at minimum ROA (−0.1343), but in general, the ROA of the sample company can be categorized as good at the level of 12.45%.

3.2 Regression results

The impact of firm's specific factors on leverage such as the regression analysis of panel data are presented in the table below:

Table 4. Regression summary.

Model	t	Sig.	Collinearity Statistics	
			Tolerance	VIF
(Constant)	3.973	0		
TANG	−3.184	0.002	0.742	1.347
MBR	5.500	0	0.397	2.518
SIZE	−1.687	0.094	0.665	1.503
PROF	−5.524	0	0.352	2.84
LIQ	−14.100	0	0.726	1.378

a. Dependent Variable: DAR

From Table 4 can be explained that the four variables that represent the firm's specific factors have a significant influence on leverage. Asset tangibility, profitability, and liquidity have a negative influence on a capital structure which can be interpreted in the opposite direction. For each increase in assets, tangibility, profitability, and liquidity, the company will reduce the amount of debt. Following study hypothesis, the effect of negative profitability (Çekrezi 2013; Fan et al. 2012; Hall et al. 2004; Li & Islam 2019; Onofreia et al. 2015; Psillaki & Daskalakis 2009; Rajan & Zinggales 1995; Singh 2016) and liquidity have a negative effect which is supported by the results of the research by Çekrezi (2013), Onofreia et al. (2015), Singh (2016). The effect of Tangibility Asset also same as with the hypothesis used in this study, empirical finding shows negative relationship supported the findings of Psillaki & Daskalakis (2009), Onofreia et al. (2015), Singh (2016).

The significant positive effect is shown by the market to book ratio. Market to Book Ratio (MBR) which as we predict in our hypothesis. Booth's (2001) study, it showed a positive MBR for the most part. While company size is the only variable that does not have a significant influence on leverage, this result does not fit predicted by the hypothesis. These results supporting by Onofreia et al. (2015).

Table 5. R square.

Model	R	R Square	Adjusted R Square	Std. Error of the Estimate	Durbin-Watson
1	.830a	.689	.677	.1061674	2.451

a. Predictors: (Constant), MBR, TANG, SIZE, LIQ, ROA
b. Dependent Variable: DAR

From the Summary model Table 5. shows that the independent variables influence 69.9% of leverage, this can be interpreted that company-specific factors Tangibility Asset, Market to Book Ratio, Size, Profitability, and Liquidity are influential factors in determining capital structure.

4 CONCLUSION

The purpose of this study is to see how the firm's specific factors influence the capital structure of the 50 Biggest Market Capitalizations in the Indonesia Stock Exchange. Our regression analysis has shown that almost all factors have a significant influence on capital structure.

1. Tangibility Assets are negatively related to capital structure, the same as our hypothesis.
2. There is a positive relationship between leverage and Market to Book Ratio and the hypothesis that estimates a positive relationship.
3. Firm size does not have a significant effect on capital structure, the study hypothesis have predicted a negative relationship
4. Profitability and Liquidity are negatively related to leverage, these result consistent with the hypothesis that researchers predicted.
5. R square of 0.689 indicates that the variable Tangibility Assets, Market to Book Ratio, Size, Profitability, and Liquidity, have 68.9% its influence on the capital structure.

Some of the variables in the results of this study consistent from the initial hypothesis of the researcher, namely Asset Tangibility, Market to Book Ratio, Profitability, and Liquidity. In general, these results support the Pecking Order Theory in the decision of the company's capital structure at The 50 Biggest Market Capitalization on the Indonesia Stock Exchange for the period 2013–2018. It supports previous research on the Pecking Order Theory that companies tend to reduce their leverage if asset tangibility, profitability, and liquidity increase.

Unexpectedly, the negative direction for the size of the company, this specific factor does not have a significant influence on leverage. We can explain that the average of a firm's size can be categorized as large companies, so the sample companies might get ease in debt, thus the size factor does not influence capital structure decisions. Although the Market to book ratio shows market expectations of the value of investment opportunities.

The results of this study indicate that the findings are consistent with studies of capital structure in other developed and developing countries that have been done by previous researchers. The results of this study can have implications for important policies for corporate financial managers in Indonesia in making capital structure decisions.

REFERENCES

Antoniou, A., Guney,Y., Paudyal, K., 2002. Determinants of Corporate Capital Structure: Evidence from European Countries. *Working Paper, University of Durham.*

Baxter, N.D. 1967. Leverage, Risk of Ruin and The Cost of Capital. *The Journal of Finance, 22, 395–403.*

Booth, L., Aivazian, V., Demirguc-Kunt, A., & Maksimovic, V. 2001. Capital Structure in Developing Countries. *The Journal of Finance, 56, 87–130.*

Bursa Efek Indonesia. 2019. [Online]. http://www.idx.co.id

Çekrezi, Anila. 2013. Impact Of Firm Specific Factors On Capital Structure Decision: An Empirical Study Of Albanian Firms. *European Journal of Sustainable Development (2013), 2, 4, 135–148.*

Deesomsak, R., Paudyal, K., Pescetto, G. 2004. The Determinants of Capital Structure: Evidence from The Asia Pacific Region. *Journal of Multinational Financial Management, 14, 387–405.*

De Jong, A., Kabir, R., & Nguyen, T. T. 2008. Capital Structure Around The World: The Roles of Firm and Country-Specific Determinants. *Journal of Banking & Finance, 32, 1954–1969.*

Delcoure, N. 2007. The determinants of capital structure in transitional economies. *International Review of Economics & Finance, 16(3), 400–415.*

Fan, J., Titman, S., & Twite, G. 2012. An international comparison of capital structure and debt maturity choices. *Journal of Financial and Quantitative Analysis, 47(1), 23–56.*

Frank, M. Z., & Goyal, V. K. 2009. Capital Structure Decisions: Which Factors are Reliably Important? *Financial Management, 38(1), 1–37.*

Getzmann, Lang, dan Spemann. 2010. *Determinants of The Target Capital Structure and Adjustment Speed – Evidence from Asian Capital Markets (versi elektronik).* Asian Finance Symposium.

Hall, G., Hutchinson, P., & Michaelas, N. 2004. Determinants of the capital structures of european SMEs. *Journal of Business Finance & Accounting, 31, 711–728.*

Harris, M., & Raviv, A. 1991. The Theory of Capital Structure. *The Journal of Finance, 46, 297–355.*

Jensen, M., & Meckling, W. 1976. Theory of The Firm: Managerial Behaviour, Agency Costs and Ownership Structure. *Journal of Financial Economics, 3, 305–360.*

Kester, W.C. (1986). Capital and Ownership Structure: A Comparison of United States and Japanese Manufacturing Corporations. *Financial Management, 25, Spring, 5–16.*

Kraus, A., dan Litzenberger, R.H. 1973. A State-Preference Model of Optimal Financial Leverage. *Journal of Finance. September: 911–922.*

Krishnan, V. S., dan R. C. Moyers. 1997. Performance, Capital Structure and Home Country: An Analysis Of Asian Corporations. *Global Finance Journal 8(1): 129–143.*

Kumar, J. 2005. Capital Structure and Corporate Governance. Xavier Institute of Management, India, unpublished paper.

Li, Larry & Islam, Silvia Z. 2019. Firm and Industry Specific Determinants of Capital Structure: Evidence from The Australian Market. *International Review of Economics and Finance 59 (2019) 425–437.*

Modigliani, F., & Miller, M. 1958. The Cost of Capital, Corporation Finance and The Theory of Investment. *The American Economic Review, 48, 261–297.*

Myers, S. 1984. The capital structure puzzle. *The Journal of Finance, 39, 575–592.*

Myers, S.C., Majluf, N.S. 1984, Corporate Financing and Investment Decisions When Firms Have Information That Investors Do Not Have. *Journal of Financial Economics, 13(2), 187–221.*

Myers, S. C. 2001. Capital Structure. *The Journal of Economic Perspectives, 15(2), 81–102*

Onofreia, M., Tudoseb, M.B., Durdureanub, C., Anton, S.G., S. 2013. Determinant Factors of Firm Leverage:

An Empirical Analysis at Iasi County Level. *Procedia Economics and Finance 20 (2015) 460–466*

Otoritas Jasa Keuangan. http://www.ojk.go.id

Ozkan, Aydin. 2001. Determinants of Capital Structure and Adjusment To Long Run Target: Evidence from UK Company Panel Data. *Journal Business Finance & Accounting, 28 (1) & (2), January/ March 175–196.*

Psillaki, M., & Daskalakis, D. 2009. Are The Determinants of Capital Structure Country or Firm Specific? *Small Business Economics, 33(3), 319–333.*

Rajan, R., & Zingales, L. 1995. What Do We Know About Capital Structure: Some Evidence from International Data. *The Journal of Finance, 51, 1421–1460.*

Singh, D. 2016: A Panel Data Analysis of Capital Structure Determinants: An Empirical Study of Non-Financial Firms in Oman. *International Journal of Economics and Financial Issues, 2016, 6(4), 1650–1656.*

Thippayana, P. 2014. Determinants of Capital Structure in Thailand. *Procedia – Social and Behavioral Sciences 143 (2014) 1074–1077.*

Advances in Business, Management and Entrepreneurship – Hurriyati et al. (Eds)
© 2021 Taylor & Francis Group, London, ISBN 978-0-367-67471-7

Capital structure policy of manufacturing companies in Indonesia

Maya Sari & S. Sulastri
Universitas Pendidikan Indonesia, Bandung, Indonesia

ABSTRACT: Capital structure policy has a strategic role to create corporate value. The policy is intended to select and determine the best composition of the use of funding for the company's operational and investment activities. This study aims to analyze the company's capital structure policy through an analysis of determinants of capital structure and the moderating effect of firm size on the relationship of capital structure and its determinant factors. This research was conducted at manufacturing companies in Indonesia that are listed on the IDX. Based on the purposive sampling approach 884 units of analysis were obtained from 68 manufacturing companies in Indonesia. The research variables used consisted of liquidity, profitability, institutional ownership, tangibility, and efficiency as independent variables, company size as moderation variables, and capital structure as the dependent variable. The research method uses panel data regression. Chow test and Hausman tests were conducted to test the panel data model that will be used. Based on the random effect model, the results of changes in capital structure can be explained by changes in liquidity, profitability, institutional ownership, tangibility, and efficiency which are moderated by company size. Research findings indicate a stronger Pecking Order pattern in manufacturing companies. The use of Pecking Order patterns in manufacturing companies is intended to maintain financial flexibility in anticipation of debt restrictions by creditors when investing in technology. To improve capital structure policy, manufacturing companies in Indonesia not only focus on liquidity performance, profitability, institutional ownership, tangibility, and efficiency but must also consider the size of the company when making capital structure policies.

1 INTRODUCTION

The development of stock performance in Indonesia showed an increase over the 2014–2017 period marked by an average growth in the performance of the Composite Stock Price Index of 5.62% annually. The manufacturing industry is one industry that contributed to the increase in the performance of these shares with an average sectoral stock performance growth of 5.14%, even in 2016 the growth of the performance of these industrial stocks far exceeds the performance of the Indonesian capital market as a whole (modified from https://finance.yahoo.com). The high growth market shares show an increase in the value of the company which has an impact on improving the welfare of the owner. (Setiadharma & Machali 2017). Improved stock performance in the manufacturing industry describes the company's success in implementing the company's strategic policies. Capital structure policy is one of the policies that play a role in creating corporate value. Capital structure policy is a policy that aims to select and determine the best composition of the use of internal and external funds for the company's operational and investment activities (Bandyopadhyay & Barua 2016). Choosing the right funding source will result in a low cost of

capital which can ultimately optimize the value of the company (Chadha & Sharma 2015).

Studies in various countries including Indonesia found two widely used capital structure policy patterns namely Pecking Order (POT) and Trade-Off Pattern (TOT). The TOT pattern will seek an optimal capital structure by balancing the benefits and sacrifices arising from the use of debt. The use of debt will increase the value of the company to a certain point, after which the addition of debt will have the opposite effect on value of the company (Abel 2018). The POT pattern shows the sequence of funding sources used by the company. Internal funds will be used first followed by the use of debt and the last is the issuance of capital stock. Internal funding sources were chosen as priority funding sources because they have the lowest cost of capital (Adair & Adaskou 2015). The asymmetric information assumption underlying the POT pattern causes the use of debt is a positive signal related to company performance in the future, to increase the value of the company (Bukit et al. 2018) and issuance of shares can be a negative signal related to the company's current and future performance (Almeida et al. 2016).

Various studies have found the manufacturing industry has a higher capital structure compared to other industrial sectors (Acaravci 2015). This high capital structure is generally owned by companies that use the TOT pattern (Muritala 2018). Other studies have also found POT patterns in companies in the manufacturing industry as found in the research (Chen 2004). The inconsistencies found in the various studies have led to the question of what capital structure policy is used in manufacturing companies in Indonesia. In this study, capital structure policy is analyzed by looking at the influence of determinants of capital structure. The determinant factors used in research are consistently used in various studies, namely liquidity, profitability, institutional ownership, tangibility, and efficiency. In the TOT approach, companies that have high liquidity, profitability, tangibility, and efficiency, can generate more cash flow, low risk of financial difficulties and high debt guarantees, so companies can use additional debt in larger amounts (Basu & Rajeev 2013; Huang & Song 2006; Tamulyte 2012). In the TOT approach, high institutional ownership can reduce agency conflict between shareholders and creditors so that companies can increase the use of debt with a low cost of debt (Arslan et al. 2014).

In the POT pattern, an increase in liquidity, profitability, efficiency, tangibility, and institutional ownership will have an impact on reducing the proportion of debt. Increased liquidity, profitability, and efficiency will increase cash flow and reduce the risk of financial difficulties so that funding needs can be obtained from retained earnings (Liu & Ren 2009; Suko 2004; Zheng 2013). High tangibility is generally owned by large companies that have a large source of internal funding, so the need for debt use is also low (Liu & Ren 2009). The use of low debt will also be carried out on companies that have high institutional ownership. Institutional ownership can replace the use of debt as a control to reduce agency costs (Hasan et al. 2014).

The existence of an empirical gap related to capital structure policy as explained in the previous section, also raises the question, in what situations each of these capital structure policy patterns will be used. Capital structure theory is known as conditional theory, so to find out the situational factors that are relevant to the capital structure policy patterns that are used, then in capital structure policy research, it is important to analyze moderation factors (Qamar et al. 2016). Several studies have shown a moderating effect of firm size on the relationship capital structure and its determinant factors. Based on this background, this study will analyze the capital structure policy of manufacturing companies in Indonesia through the analysis of determinants of a capital structure consisting of liquidity, profitability, institutional ownership, tangibility, and efficiency, and examine the moderating effect of firm size on the relationship of capital structure with factors the determinant (Hussain & Matlay 2007). Based on the results of tests conducted, it can be found whether the capital structure policy of manufacturing companies in Indonesia is in line with the Pecking Order or Trade-off pattern.

2 METHOD

This research was conducted at manufacturing companies in Indonesia that are listed on the IDX. The data used are secondary data during the period of 2005–2017. Based on the purposive sampling approach 884 units of analysis were obtained from 68 manufacturing companies in Indonesia. The research variables used to consist of liquidity (CR), profitability ROA), institutional ownership (IO), Tangibility (TANG), and efficiency (ATO) as independent variables. The dependent variable of this study is the capital structure (DAR). This study also tested the moderating effect of company size (SIZE) on the relationship of capital structure with liquidity, profitability, tangibility, institutional ownership, tangibility, and efficiency. The analytical method used is a panel data regression analysis model. The panel data model developed in this study is as follows:

$$
\begin{aligned}
DAR_{i,t} = {} & \beta 0 + \beta 1 CR_{i,t} + \beta 2 ROA_{i,t} + \beta 3 IO_{i,t} \\
& + \beta 4 TANG_{i,t} + \beta 5 ATO_{i,t} + \beta 6 SIZE_{i,t} \\
& + \beta 7 CR_{i,t}.SIZE_{i,t} + \beta 8 ROA_{i,t}.SIZE_{i,t} \\
& + \beta 9 IO_{i,t}.SIZE_{i,t} + \beta 10 TANG_{i,t}.SIZE_{i,t} \\
& + \beta 11 ATO_{i,t}.SIZE_{i,t} + \varepsilon_{i,t}
\end{aligned}
$$

The following formulas are used to calculate variables:

DAR = Total Debt/Total Asset
CR = Current asset/current liabilities
ROA = Earning before taxed/ total asset
TANG = Fixed Asset/Total Asset
ATO = Sales/Asset
IO = Shares Owned By Institutional Investors/ Shares Outstanding
SIZE = Ln Asset

Testing panel data models using the Chow test and Hausman test. Based on the test results, the best model will be selected whether the common effect, fixed effect model or random effect model. After the best model is found, the next step will be to test the hypothesis by conducting Simultaneous Test (F Test), determination of the Coefficient of Determination (R2) and t-test.

3 RESULTS AND DISCUSSION

Testing the capital structure policy of manufacturing companies in Indonesia will be analyzed by testing the effect of liquidity, profitability, institutional ownership, tangibility, and efficiency on capital structure and testing the effect of company size moderation on capital structure and the factors that influence it. The initial stage of the analysis carried out the feasibility test panel data model using the Chow test. The test results obtained the probability of cross-section F and cross-sectional chi-square $0.0000 < \alpha$ (0.05), which means the panel data model that is more used is the fixed effect. The suitability of the next model was tested by using the Hausman test method to determine the appropriate model, both fixed and random effects. The Hausman test results show a random cross-section probability of $0.2903 > \alpha$ (0.05), it can be concluded that the Random Effect Model is more feasible to use than the Fixed Effect Model. The last test performed the Lagrange multiplier (LM) test and obtained a Breusch-pagan Cross-section probability value of $0.0000 < \alpha$ (0.05) which meant that the Random Effect was more feasible to use.

Based on the results of the feasibility testing model, then Table 1 shows the results of the panel data regression analysis using random effects:

Table 1. Regression Analysis Panel Data Determinants of Capital Structure and Moderation Effects.

Variable	Coefficient	Std. Error	t-Statistic	Prob.
C	2.292349	1.177716	1.946437	0.0519
CR	0.231764	0.097349	2.380759	0.0175
ROA	−3.361874	0.865227	−3.885539	0.0001
IO	−2.309767	1.396999	−1.653378	0.0986
TANG	−2.387727	1.305752	−1.828623	0.0678
ATO	0.111690	0.264459	0.422333	0.6729
SIZE	−0.123528	0.083047	−1.487448	0.1373
CR_SIZE	−0.019698	0.007354	−2.678683	0.0075
ROA_SIZE	0.203681	0.055308	3.682693	0.0002
IO_SIZE	0.174427	0.098644	1.768245	0.0774
TANG_SIZE	0.179632	0.090995	1.974095	0.0487
ATO_SIZE	−0.008496	0.020448	−0.415512	0.6779

R-squared	0.545831	Mean dependent var	0.578664
Adjusted R-squared	0.501824	S.D. dependent var	0.679371
S.E. of regression	0.479511	Akaike info criterion	1.453017
Sum squared resid	185.0940	Schwarz criterion	1.880587
Log likelihood	−563.2334	Hannan-Quinn criter.	1.616492
F-statistic	12.40341	Durbin-Watson stat	1.116547
Prob (F-statistic)	0.000000		

Table 1 shows an F-statistic with a sig F of 0,000, which means that the research model developed meets the model suitability (goodness of fit) at the level of sig $0.0000 < \alpha$ (0.05) which means that changes in capital structure can be explained by changes in liquidity, profitability, institutional ownership, tangibility, and efficiency are moderated by company size. Partial testing shows that liquidity and efficiency have a positive effect on capital structure, while profitability, institutional ownership, and tangibility have a negative effect on capital structure. The findings of the study indicate that an increase in liquidity and efficiency will have an impact on increasing capital structure while increasing profitability performance, institutional ownership and efficiency will have an impact on decreasing capital

structure. The results of testing the size of the company as an independent variable shows the size of the company has a negative effect on capital structure.

As for the results of testing the company size as a moderator variable shows the size of the company provides a moderating effect that reinforces the negative influence of profitability, institutional ownership, and tangibility, on the contrary, the size of the company weakens the positive influence of liquidity and efficiency on capital structure. The study shows the negative effect of profitability, institutional ownership, and tangibility of the capital structure is in line with the Pecking Order pattern. The results of this study are in line with (Hasan et al. 2014; Huang & Song 2006). In manufacturing companies, an increase in profitability will increase the compan's cash inflows that act as a source of corporate internal funding that will be used first for operational and investment activities, so the debt needs will be reduced. Manufacturing companies with high institutional ownership can be used to replace the debt function as controllers of opportunistic behavior by managers that can harm the interests of the owner so that companies with high institutional ownership can reduce the use of debt so that the cost of capital used by the company will also decrease. The negative influence of tangibility on capital structure in manufacturing companies in Indonesia shows that the manufacturing companies that become the research sample are large companies, which have sufficient internal funding sources, so the need for the use of debt or other external funds decreases.

Testing the effect of company size moderation on the relationship of capital structure with its determinant factors indicates a strong Pecking Order pattern in large companies in the manufacturing industry. Although the capital structure of manufacturing companies is relatively high, the use of debt will decrease in line with increasing company size. The findings of the research findings show that manufacturing companies with larger business scale, have the policy to use debt that is relatively lower compared to companies with company size, although both companies have the same level of profitability, institutional ownership, and tangibility. The results of this study are in line with Byoun (2007) dan Hossain & Ayub (2012). Companies with different sizes have different financial flexibility. The manufacturing industry is one of the industries with a level of competition that is characterized by rapid changes in markets and technology that is faster than other industries. For companies to adapt to these changes, manufacturing companies are required to be able to maintain their financial flexibility optimally. Large companies have a much greater debt capacity when the need for technology investment must be done, to maintain financial flexibility but optimal, large companies will use free cash flow to foster working capital and investment, so that when large funding needs are needed, then the debt capacity is still sufficient to be able to obtain funding sources at a low cost, and the existence of debt restrictions by creditors can be avoided.

4 CONCLUSION

This study aims to analyze capital structure policies in manufacturing companies in Indonesia. The results of testing with a random effect model show that changes in capital structure can be explained by changes in liquidity, profitability, institutional ownership, tangibility, and efficiency which are moderated by company size. Research findings indicate that the Pecking Order pattern is stronger in manufacturing companies. The use of the Pecking Order pattern in manufacturing companies is intended to maintain financial flexibility in anticipation of debt restrictions by creditors when investing in technology. Manufacturing companies must also consider the size of the company when making capital structure policies.

REFERENCES

Abel, A. B. (2018). Optimal debt and profitability in the trade-off theory. *The Journal of Finance*, 73(1), 95–143.

Acaravci, S. K. (2015). The determinants of capital structure: Evidence from the Turkish manufacturing sector. *International Journal of Economics and Financial Issues*, 5(1), 158–171.

Almeida, H., Fos, V., & Kronlund, M. (2016). The real effects of share repurchases. *Journal of Financial Economics*, 119(1), 168–185.

Arslan, M., Phil, M., & Zaman, R. (2014). Relationship between Capital Structure and Ownership Structure: A Comparative Study of Textile and Non Textile Manufacturing Firms. *Public Policy and Administration Research*, 4(11), 53–63

Bandyopadhyay, A., & Barua, N. M. (2016). Factors determining capital structure and corporate performance in India: Studying the business cycle effects. *The Quarterly Review of Economics and Finance*, 61, 160–172.

Basu & Rajeev (2013) *"Determinants of capital structure of Indian corporate sector evidence of regulatory impact"*, working paper The Institute for Social and Economic Change, Bangalore

Bukit, R. B., Nasution, F. N., Ginting, P., Sambath, P., & Nurzaimah, M. (2018, January). The Influence of Firm Performance, Firm Size and Debt Monitoring on Firm Value: The Moderating Role of Earnings Management. In *1st Economics and Business International Conference 2017 (EBIC 2017)*. Atlantis Press.

Byoun, S. (2007). Financial Flexibility, Firm Size and Capital Structure. *Journal of Economic Literature*

Chadha, S., & Sharma, A. K. (2015). Capital structure and firm performance: Empirical evidence from India. *Vision*, 19(4), 295–302.

Chen, J. (2004). "Determinants of capital structure of Chinese-listed companies". *Journal of Business Research*, 57, 1341–51.

Hasan, M. B., Ahsan, A. M., Rahaman, M. A., & Alam, M. N. (2014). Influence of capital structure on firm performance: Evidence from Bangladesh. *International Journal of Business and Management*, 9(5), 184.

Hossain, Faruk & Ayub Ali, (2012),"Impact of Firm Specific Factors on CapitalStructure Decision: An Empirical Study of Bangladeshi Companies, International", *Journal of Business Research and Management (IJBRM)*

Huang, Samuel G.H., and Frank M. Song, (2006), "The Determinants of Capital Structure:Evidence from China", *China EconomicReview*, 17,14–35.

Hussain, J., & Matlay, H. (2007). Financing preferences of ethnic minority owner/managers in the UK. *Journal of Small Business and Enterprise Development*, 14(3), 487–500.

Li-Ju Chen, & Chang Jung Christian, (2012), "How the Pecking-Order *Theory* Explain Capital Structure", *The Journal Of International Management Studies ISSN 1993*–1034

Liu, Yuanxin and Ren, Jing (2009), "An Empirical Analysis on the Capital Structure of Chinese Listed IT Companies", *International Journal of Business and Management*, 4, (8), 46–51.

Muritala, T. A. (2018). An empirical analysis of capital structure on firms' performance in Nigeria. *IJAME*.

Qamar, M. A. J., Farooq, U., Afzal, H., & Akhtar, W. (2016). Determinants of Debt Financing and Their Moderating Role to Leverage-Performance Relation: An Emerging Market Review. *International Journal of Economics and Finance*, 8(5), 300.

Setiadharma, S., & Machali, M. (2017). The effect of asset structure and firm size on firm value with capital structure as intervening variable. *Journal of Business & Financial Affairs*, 6(4), 1–5.

Suko, A. Nugroho, (2004). *Analisis Faktor-Faktor Yang Mempengaruhi Struktur Modal* (Perusahaan Properti Yang GO-PUBLIC Di Bursa Efek Jakarta Untuk Periode Tahun 1994 – 2004) Thesis Universitas Diponegoro Semarang: tidak diterbitkan

Tamulyte, J. (2012). The determinants of capital structure in the Baltic States and Russia. *Electronic Plublications of Pan-European Institute*.

Zheng, M. (2013). Empirical research of the impact of capital structure on agency cost of Chinese listed companies. *International Journal of Economics and Finance*, 5(10), 118.

Website: https://finance.yahoo.com

Dividend policy and company value analysis based on debt and profitability policies (empirical study of registered manufacturing companies)

Suparno & E. Mahpudin
Universitas Singaperbangsa, Karawang, Indonesia

ABSTRACT: This study aims to examine the effect of debt policy proxied with debt to equity ratio (DER) and profitability which proxied with return on equity (ROE) to firm value projected with price book value (PBV) with dividend policy as a moderating variable. Empirical studies on research use sampling technique using purposive sampling of 175 samples of manufacture companies listed in the Indonesia Stock Exchange years 2011–2017. The method of analysis of this study uses moderated regression analysis (MRA) using an analysis tool that is E-Views. The results of this study are that debt policy has a significant positive effect on corporate value, and profitability has a significant positive effect on corporate value. Dividend policy strengthens the influence of debt policy on corporate value, and dividend policy strengthens the influence of profitability on corporate value.

1 INTRODUCTION

Normatively, the main purpose of establishing a company is generally for the benefit of shareholders through increasing the value of the company. The value of the company has a very important role because the high value of the company will be followed by the high prosperity of shareholders (Brigham & Houston 2010). Company value in a broad sense is a price they are willingly paid by the candidate buyer if the company is sold (Husnan & Pudjiastuti 2008). The value of the company in the narrow sense associated with public companies is the value of the company identical to the stock price, and thus increasing the value of the company means increasing stock prices (Brigham & Houston 2006). This study focused on the value of the company reflected in the stock price. The company's share price individually or in combination from various sectors in the capital market are meters that are very important to assess the market response to companies listing on the IDX, and in macroeconomic context this can be a parameter in assessing the growth of an economy. Empirically, the market response to issuers in the Indonesian capital market up to 2017 showed better growth with an average development of 33% after several years of decline, both caused by national, regional, and global conditions, which the dilator was behind by political, economic, and social nuances and security throughout 2010 to 2017. However, in contrast to developments in the BEI in the manufacturing sector in 2011 to June 2015 when the IDX composite manufacturing sector experienced a fluctuating development, stock prices from 2011 to 2017 increased by an average of 24.10%. However, the average JCI increase in manufacturing is still below the average JCI of all issuers at 33.00%.

The decline in the BEI caused investors to sell shares, causing some stock prices to decline, which ultimately affected the value of the company. Several other factors that can affect the value of the company include debt policy and profitability. In debt policy, to expand its business, the company carries out various ways to meet capital requirements. One of the policies taken by the company is by using external funds (debt). Fulfillment of capital originating from external funding sources will determine the company's ability to carry out its activities while increasing the company's financial risk. In certain compositions, debt will increase the productivity of the company to increase the company's ability to generate profits for shareholders. However, if the composition of debt becomes excessive, then what happens is an increase in financial risk that can increase the risk of bankruptcy. Therefore, management must be careful in determining its debt policy to increase the value of the company (Martini & Rihardjo 2014).

Profitability is the ability of a company to obtain profits related to sales, total assets, and own capital. One important indicator for investors in assessing company prospects in the future is to look at the growth in profitability from previous years, whether it has increased or declined. The increase in profitability from the previous year indicates that the company's prospects are good so investors will respond positively and the company's value will increase. If profitability decreases from the previous year, then this can lead to a negative response and the company's value will decrease. Profitability also determines how much the return will be received by the investor for the investment made (Pratiwi & Mertha 2017).

Another factor that influences company value is dividend policy, namely, the decision whether the profits obtained by the company will be distributed to the shareholders as dividends or held in the form of retained earnings for future investment financing that "dividend policy is a decision about how much current profit will be paid as dividends rather than being held for reinvestment in the company." (Brigham & Houston 2006). If the company increases dividend payments, it might be interpreted by investors as a signal of management's expectation that the company's performance will improve in the future, so that dividend policy has an influence on company value (Myers & Majluf 2002).

2 METHODS

The object in this study is a company listed on the IDX manufacturing sector during the period 2011–2017. The type of data used is quantitative data in the form of secondary data with data sources originating from IDX. The sampling technique uses purposive sampling method, taking into account the following criteria:

1. Manufacturing companies that are listed on the IDX continuously during the 2011–2017 period;
2. Manufacturing companies that publish continuous financial reports during the 2011–2017 period;
3. Manufacturing companies that earn profits continuously during the 2011–2017 period;
4. Companies that pay dividends continuously during the 2011–2017 period.

2.1 Operational variable

2.1.1 Independent variable (X)
1. Debt Policy (X1)
Debt policy is another way for agency intermediate problems to increase debt (Jensen & Meckling 1976). That is, in large companies, the smaller the unemployed funds that can be used by companies to reduce the costs required or vice versa. According to Jansen (1986), changes to reduce free cash flow from bonds, which are used by managers to prove that they will not waste company funds, and they struggle to take risks if they cannot manage the company seriously. Stocks, policies to increase spending, can reduce supervision because third parties who spend funds will lend management so that loans are not misused. Debt policies are all types of loans made by companies, both short term and long term (Indahningrum & Handayani 2009; Nasser & Firlano 2006) from corporate debt policies (Nuringsih 2005). Formula DER: Total Debt: Equity.

2. Profitability (X2)
Profitability is revenue and loss of income during the reporting period. Profitability analysis is very important for creditors and investors. For creditors, profit is

the source of interest and principal payments. As for investor equity, according to Saidi (2004), profitability is the ability of a company to make a profit. Investors invest in a company to get a decision. The higher the company generates profits, the greater is expected by investors, thus making the company value better. The more profits earned, the greater the company's ability to pay dividends, and this increases the increase in firm value. With a high profit ratio owned by the company, it will attract investors to be interested in the company. With the following formula: ROE: After-Tax Profit: Owner's equity.

2.1.2 Dependent variable (Y)
The dependent variable is the main variable, which is the factor that applies in the investigation (Uma Sekaran 2006). The dependent variable in this study is firm value, according to Nurainun and Sinta (2007); Andinata (2010) company value is the price that investors are willing to pay to own a company. The value of the company is reflected in the stock price. If the value of the company is good, that is, having good performance and prospects, then investors are willing to pay more to buy the shares. The value of the company in this study is measured by the price to book value (PBV) ratio because it is related to the growth of its capital compared to market value with the value of the book. Price to book value ratio is a ratio that is often used to determine the value of a company and make investment decisions by comparing market prices' year end shares with company book value. In this study the price to book value (PBV): Share Price: Share Book Value.

2.1.3 Moderate variable (Z)
The moderate variable (Z) is a fully approved variable that can attach or license the relationship between the independent variable X and the dependent variable Y (Sugiyono 2010). In this study, the moderating variable is dividend policy (Brigham & Houston 2006). Dividend policies are "Decisions about how much profit today will be approved as dividends from those agreed to be reinvested in the company" (Nurainun & Sinta 2007). The author uses dividend policy as a moderating variable based on several reasons, such as according to Meythi (2012); Myron Gordon & John Lintner (2007); Brigham & Gapensi (1996); and I Made Sudana (2011). This ratio is seen in the portion received as dividends to investors. Other parts that are not distributed will be reinvested into the company (Hanafi 2004) Companies that have high growth rates will have a low dividend payout ratio or vice versa. Dividend payments are also a company dividend policy (Alwi 2003; Ciaran Walsh 2003), Formula: DPR: Dividend Per Share: Earnings Per Share.

2.2 Analysis method

To process data and draw conclusions, the researcher uses a multiple linear regression analysis (MRA) with the E-Views program. This analysis is used to determine the effect of debt policy and profitability on firm value with dividend policy as a moderating variable in manufacturing companies listed on the Stock Exchange for the period 2011–2017 with the regression equation as follows:

$$PBV = a + b1DER + b2ROE + e \text{ (Model1)}$$
$$PBV = a + b1DER + b2DPR + b3DER.DPR + e \text{ (Model2)}$$
$$PBV = a + b1ROE + b2DPR + b4ROE.DPR + e \text{ (Model3)} \quad (1)$$

Descriptive Statistics

Descriptive statistical tests are a description of the study consisting of the number of samples, minimum values, average values (average), and standard deviations. Chi-square probability value test results $0.2190 > 0.05$ means there is no heteroscedasticity. The Automatic Correlation test results obtained from the Daubin Weston of 1.8368 DU values obtained from the Durbin Weston table of 1.5789, DL values obtained from Durbin Weston 1.6632. Then obtained $1.6632 < 18368 < 2.2411$. This shows that there is no autocorrelation. Based on test results that the Middle DER VIF value is 1.023169, ROE is 1.078922, DPR is 1.071330, it can be concluded that the data of this study passed the multicollinearity test because of each variable < 10, and the DPR.

3 RESULTS AND DISCUSSION

Results

3.1 *Model 1*

Table 1. F Test result.

Model	F-Statistic	Prob(F-Statistic)
1	36,71850	0,000000

Based on F Test, results show that the F value is 36.72 with a probability level of 0.00, where the number is less than 0.05, and it was concluded that there was an influence between DER and ROE on PBV so that regression model 1 was feasible to use.

Table 2. Coefficient of determination result.

Model	R-Square	Adjusted R-Square
1	0,478494	0,465775

Based on the above, it shows that the coefficient of determination (Adjusted R-square) is 0.465775 or 46.58%. This means that the value variable of the

company can be explained by the variables of debt policy and profitability; 46.58% and the rest 53.42% are explained by other variables outside the variables in the research model.

Table 3. T test result.

Variable	Coefficient	t-Statistic	Prob.
C	−3,645949	−3,692353	0,0004
DER	4,763882	2,233777	0,0032
ROE	31,49308	7,564821	0,0000
DPR	6,295301	5,422944	0,0000

Based on the results of the T-test above, the DER variable shows a significant value of $0.0032 < \alpha\ 0.05$ with the direction of the positive regression coefficient, which indicates that the DER variable has a significant positive effect on the firm value variable, so hypothesis 1 is accepted. ROE variable shows a significant value of $0.0000 < \alpha\ 0.05$ with the direction of the positive regression coefficient, which indicates that the ROE variable has a significant positive effect on the firm value variable, so hypothesis 1 is accepted.

3.2 *Model 2*

Table 4. F Test result.

Model	F-Statistik	Prob(F-Statistic)
2	33,1014805	0,000000

Based on Table 4, result shows the F value of 33.10 with a probability level of 0.00, where the number is less than 0.05, so it is concluded that the regression model is feasible to use.

Table 5. Coefficient of determination result.

Model	R-Square	Adjusted R-Square
2	49.583178	0,000000

Based on Table 5, it can be seen that the coefficient of determination (Adjusted R-squared) is 49.583178 or 49.58%%. This means that the value variable of the company can be explained by the variable debt policy and profitability, which is moderated by dividend policy of 49.58%, while the remaining 50.42% is explained by the other variable outside the variables in the research model. The coefficient of determination in equation 2 is greater than the value of the coefficient of determination in equation 1, which only has a determination coefficient of 46.58%, meaning that the presence of a dividend policy moderation variable can strengthen the influence of the DER variable on the PBV variable.

111

Table 6.   T-test result.

Variable	Coefficient	t-Statistic	Prob.
C	−1,74566	−1,34269	0,24706
DPR	1,234308	0,042519	1,06612
DER*DPR	17,22754	2,306712	0,04191

From the results of the T-test above, the DPR variable shows a significant value of $1.06612 > \alpha$ 0.05 with a regression coefficient 0.042519. This indicates that the DER variable does not affect company value, so hypothesis 2 is not accepted. DER variables that are moderated by the DPR show a value of $0.042 < \alpha$ 0.05, which is significant with the regression coefficient 2.306712. It can be concluded that hypothesis 2 is accepted. Thus, it can be interpreted that the DPR can moderate (strengthen) the influence of DER on the value of the company.

3.3   Model 3

Table 7.   F test result.

Model	F-Statistic	Prob(F-Statistic)
3	37,98339	0,000000

Based on the table above, F test result showing the calculated F value is 37.98 with a probability level of 0.00 so that the value is less than 0.05. It was concluded that there was influence of Debt Policy (DER) and Profitability (ROE) toward Corporate Value (PBV) so model 3 is feasible to use.

Table 8.   Coefficient of determination result.

Model	R-Square	Adjusted R-Square
3	0,584087	0,567172

The table above shows that the coefficient of determination (Adjusted R-square) is 0.567171 or 56.71%. This means that the variable (Y) value of the company can be explained by the variable debt policy and profitability 56.71% while 43.29% is explained by variables outside the variables in the research model. The terminated coefficient value in model 3 is greater than the terminated coefficient value in model 1, which is equal to 46.58%, which means that the variable dividend policy (DPR) is able to moderate the Effect of Profitability (ROE) against company value (PBV).

Table 9.   T-test result.

Variable	Coefficient	t-Statistic	Prob.
C	−1,313517	−1,216326	0,2262
DPR	1,868097	1,145987	0,2540
ROE*DPR	34,56912	3,695242	0,0003

The result of the T-test above shows that the DPR variable shows a significant value of $0.0027 < \alpha$ 0.05 with the direction of the positive regression coefficient, which indicates that the DER variable has a significant positive effect on the firm value variable, so hypothesis 1 is accepted. ROE variable shows a significant value of $0.0000 < \alpha$ 0.05 with the direction of the positive regression coefficient, which indicates that the ROE variable has a significant positive effect on the firm value variable, so hypothesis 1 is accepted.

Discussion

3.4   Effect of debt policy on company values

Based on the signaling theory, the increase in debt made by the company is a signal given by managers to investors. Managers have confidence that the company has good prospects in the future because they feel they can pay their obligations. The addition of debt is also used by companies to carry out their operational activities or if the company is expanding so that it will increase the company's capacity. Thus, investors can capture the signal given by the manager and will respond positively so that the stock price will rise and will be followed by the value of the company going up as well.

3.5   Profitability has a positive impact on company values

The company's ability to generate profits is an indicator of assessing the company in fulfilling obligations for shareholders and at the same time describes the company's prospects in the future. The increasing profitability will attract investors to invest in the company so that the demand for shares will increase. With the increase in investment, the higher will be the share price of the company. If the stock price of a company increases then it is a sign that the value of the company also increases.

3.6   Dividend policy moderates the effects of debt policy on company values

Based on the signaling theory, the addition of debt made by the company is a signal given by managers to investors. Managers have confidence that the company will have good prospects in the future because they feel they can pay their obligations. The addition of debt is also used by companies to carry out their operational activities, or the company is expanding so that it will increase the company's capacity. Thus, investors capture the signal given by the manager and will respond positively so that the stock price will rise and will be followed by the value of the company rising as well. This explanation is reinforced by the existence of a dividend policy, based on signaling theory investors, which will mean that the company has good prospects in the future as the company distributes dividends. The prospect improvement in the future coupled with dividend distribution will increasingly attract investors; this shows that the company can manage debt well so that it can pay the principal and interest and still be able to pay dividends. Thus, many investors are interested in these shares so that the stock price will rise and will be followed by the value of the company rising as well.

3.7 Dividend policy moderates the effects of profitability on company values

The company's ability to generate profits is an indicator of the company's ability to fulfill obligations for shareholders and can show the company's prospects in the future. Increasing profitability will attract investors to invest in the company so that demand for shares will increase. With the increase in investment, the higher will be the share price of the company. If the stock price of a company rises then it confirms that the value of the company also rises. This influence is further strengthened by the existence of dividend policy. Based on the signaling theory, the market will interpret dividend payments as a signal about the company's bright future prospects. With the increase in dividend payments to shareholders, investors predict that the profits that have been earned by the company will continue or even get better. The better the prospects of the company, the company will be considered profitable by investors, and as a result investors are interested in buying company shares and increasing demand for company shares. This will increase stock prices and company value.

4 CONCLUSION

This study seeks to answer the research objectives: 1) to provide empirical evidence that the variable-variable debt policy and corporate value profitability affect the manufacturing sector in the Stock Exchange in 2014–2017; 2) provide empirical evidence that the dividend policy variable moderates the effect of debt policy and profitability on the corporate value of the manufacturing sector on the IDX in 2014–2017. Based on data analysis and findings and discussion, the following conclusions can be drawn.

1. Equation Model 1: The two independent variables used, debt policy and profitability, partially affect the value of the company: a) Dividend payments motivate manufacturing company investors on the IDX to buy shares of companies that pay dividends; b) Large companies tend to distribute high dividends to maintain a reputation among investors, while small companies tend to allocate profits to retain earnings to increase company assets so that companies tend to distribute low dividends. Investors are more interested in companies that distribute high dividends; c) Profitability can signal positive information for its investment. Therefore, when the company achieves high return (ROE), it will be captured by investors on the IDX as a positive sign of future dividend yields; d) Conditions of rising and falling values of manufacturing companies on the IDX can be influenced jointly by debt policy and profitability.

2. Equation Model 2: Both independent variables were used, debt policy and profitability, with variable moderation dividend policy otherwise significantly affecting the value of the company: a) Second independent variable debt policy and profitability together by the variable dividend policy is immediate in explaining the impact on company value; b) Investors tend to pay more attention to the debt policy. When large manufacturing companies on the IDX tend to be stable, the higher is the level of investor confidence in the company's ability to provide dividends; c) Profitability is important for investors because the greater dividends (dividend payout) will further save capital costs; on the other hand, managers can increase ownership due to dividend receipts as a result of high profits. So, dividend policy can moderate the relationship between profitability and company value.

REFERENCES

Alwi, I.Z. 2003. Pasar modal teori dan aplikasi, cetakan pertama. Jakarta: Yayasan Pancur Siwah.

Andinata. 2010. Analisis pengaruh profitabilitas dan kebijakan dividen terhadap nilai perusahaan manufaktur. *Jurnal Ekonomi*.

Brigham, & Houston. 2010. Dasar – dasar manajemen keuangan Edisi 10 Jilid 1. Jakarta: Salemba Empat.

Brigham, E.F., & Houston, J.F. 2006. Dasar-dasar manajemen keuangan Jilid 1 Edisi Kesepuluh. Jakarta: Salemba Empat.

Brigham, E.F., & Gapensi, L.C. 1996. Intermediate financial management 5th edision. Sea Harbor Drive: The Dryden Press.

Gordon, M., & Lintner, J. 2007. Security price, risk, and maximal Gains from diversification. *The Journal of Finance* 20(4): 587–615. [Online]. Retrieved from http//doi.wiley.com/101111/j.1540-6261.1965.tb02930.x.

Husnan, & Pudjiastuti. 2008. Manajemen Keuangan Teori dan Penerapan (Keputusan Jangka Panjang). Yogyakarta : BPFE.

Indahningrum, & Handayani. 2009. Pengaruh kepemilikan manajerial, kepemilikan institusional, deviden, pertumbuhan perusahaan, free cash flow, dan profitabilitas terhadap kebijakan hutang perusahaan. *Jurnal Bisnis dan Akuntansi* 11 (3): 189–207.

Jensen, & Meckling. 1976. Theory of the firm: managerial behavior, agency, and ownership structure. *Journal of Financial Economics* 4(4): 305–360.

Martini, P.D., & Riharjo, I.B. 2014. Pengaruh kebijakan hutang dan profitabilitas terhadap nilai perusahaan: kebijakan dividen sebagai variabel moderasi. *Jurnal Ilmu & Riset Akuntansi* 3(2).

Meythi. 2012. Dampak interaksi antara kebijakan utang dan kebijakan dividen dalam menilai perusahaan. *Jurnal Keuangan dan Perbankan* 16(3): 407–414.

Myers, & Majluf. 2002. Corporate financing and investment decision when firm have information that investor do not have. *Journal of Financial Economic* 13: 419–453.

Nasser, & Firlano. 2006. Pengaruh struktur kepemilikan dan dewan komisaris indepenen terhadap nilai perusahaan dengan manajemen laba dan kebijakan hutang sebagai variabel intervening, media riset akuntansi, auditing dan informasi 8(1): 1–27.

Nurainun Bangun, & Sinta Wati. 2007. Analisis pengaruh profitabilitas dan kebijakan dividen terhadap nilai perusahaan perdagangan, jasa, dan investasi yang terdaftar di bursa efek Jakarta. *Jurnal Akuntansi* 11(2).

Nuringsih, K. 2005. Analisis pengaruh kepemilikan manajerial, Kebijakan utang, roa dan ukuran perusahaan Terhadap kebijakan dividen: studi 1995–1996. Jurnal Akuntansi dan Keuangan Indonesia Juli-Desember 2005, Vol. 2, No. 2, pp. 103–123

Pratiwi, N.P., & Mertha, M. 2017. Pengaruh kebijakan hutang dan profitabilitas pada nilai perusahaan dengan kebijakan dividen sebagai variabel moderasi. *E- Jurnal Akuntansi Universitas Udayana* 20(2).

Saidi. 2004. Faktor-faktor yang mempengaruhi struktur modal pada perusahaan manufaktur go publik di BEJ tahun 1997–2002. *Jurnal Bisnis dan Ekonomi* 11(1).

Sekaran, U. 2006. Recearch methods for business (motodologi penelitian bisnis. Jakarta: Salemba Empat.

Sugiyono. 2010. Metode penelitian kuantitatif, kualitatif dan R&D. Bandung.

Walsh, C. 2003. On the efficiency of internal andexternal corporate control mechanisms. *Academy of Management Review* 15: 421–458.

Advances in Business, Management and Entrepreneurship – Hurriyati et al. (Eds)
© 2021 Taylor & Francis Group, London, ISBN 978-0-367-67471-7

The effect of the role-playing learning method on financial literation based on the level of parents' education

A. Fauziyah & S. Sulastri
Universitas Pendidikan Indonesia, Bandung, Indonesia

ABSTRACT: This study aims to improve students' financial literacy through role-playing learning methods in early childhood education. Aside from the learning method, this study also notes students' external factors, namely the level of parents' education. The research method that will be used is a quasi-experimental study using factorial between-subject design using two-way ANOVA. The results of the study show that 1) there are differences in financial literacy in the class that uses the method of role-playing learning with classes that use conventional learning methods; 2) there are differences in the increase in financial literacy with the level of parents' education of high, medium, and low; and 3) there are interactions between the role-playing distribution method, the level of parents' education, and financial literacy.

1 INTRODUCTION

Financial literacy is an effective tool that can enhance economic development. Lately, financial literacy is considered to be more important because there is a change in the economic environment due to the global recession. As a result, consumers are not able to make sound financial decisions, which leads to making errors in financial decision making that affects a country's financial instability (Assad 2015; Ijevleva & Arefjevs 2014; INFE 2008).

Financial literacy is a combination of awareness, knowledge, skills, attitudes, and behaviors that a person needs to have to make sound financial decisions and ultimately achieve individual financial well-being (Bhabha et al. 2014; Opletalova 2015). Someone with high financial literacy will know how to manage their financial resources, will tend to behave frugally, and will have more responsibility for financial planning. The Minister of Finance of the Czech Republic (MFCR) as conveyed by Opletalova (2014) added that a person's financial literacy not only makes them able to manage their finances but also able to manage their family's finances, including managing assets and liabilities that will change their financial life situation for the better.

Today, the level of financial literacy in Indonesia is only 29.66% (OJK 2018). Planting financial literacy values as early as possible in children will greatly affect their understanding and knowledge of financial literacy and also the level of welfare in the future. Cognitive traits in children who are still concrete thinkers and still in development stages are still very effective in instilling financial literacy values. Even though the family is the first community in instilling financial literacy values, the role of the school is also very important in providing knowledge of financial education for children.

In education, there is a process of learning interaction between students and teachers. As the process of interaction takes place, there is management of the learning process where learning becomes more meaningful and able to improve learning outcomes that are not only seen from how students can have extensive knowledge and understanding, but how students can have attitudes, behaviors, and special skills. After the learning process is completed, it will be able to direct students in problem-solving and better communication (Calderon 2013; Nasrallah 2014; Shupe 2007).

In previous research, financial education was introduced through inquiry type cooperative learning models, case studies based on daily life, didactic games, dramatic education, training, instruction, and visits to financial institutions (Maurer 2014; Opletalova 2015; Walstad et al. 2010). In Indonesia, the Financial Services Authority Institute provides training to several schools and even launched a book that is expected to increase student financial literacy. But still, these efforts have not shown the results because financial education does not necessarily produce responsible behavior (Braunstein & Welch 2002; Robb & Woodyard 2011).

Role-playing method is a method that involves interaction between two or more children about a topic or situation, in which the child performs each role according to the character he plays as they interact with each other in an open role. This role-playing method is based on the assumption that it is possible to create authentic analogies in real-life problem situations (Mulia 2017). The use of financial education learning can be used as the role-playing learning method with basic competencies such as recognizing needs, desires, and self-interest, which is considered very appropriate because it can produce financial attitudes and behaviors from an early age.

To produce optimal results from the implementation of the role-playing learning method, the parents' participation in instilling financial literacy values is needed. Haditono (2006) and Hurlock (1974) stated that the environment closest to children is family, and a background factor in the level of parental education is something that has a big influence on the child's development. This background level of parental education is positively correlated with the way they care for children. This means that the higher the education of the parents, the better the way the child is nurtured and consequently the child's development is affected positively. On the other hand, the lower the level of parental education, the less the child is nurtured and consequently the child's development is affected negatively.

2 METHODS

A quasi-experiment was employed as a research design to find out the treatment effect. This study used a factorial design 3x2. The research variable, X1: Role Playing Learning Method (independent variable) as treatment, X2: Education Level of Parents (High, Medium, and Low) is an independent variable as a factor, Y: Financial Literacy as the dependent variable.

This research was conducted in Wanita PUI Kindergarten by using one experimental class that is class B2 and one control class that is class B1. The selection of Class B itself is based on the Basic Competencies that will be used in connection with Financial Literacy.

The experimental class was treated using the role-playing learning method, while the experimental class was treated using the conventional method. In addition to using learning methods, this study also looked at aspects of parental education levels as one of the factors in efforts to improve student financial literacy. Data regarding the level of education of parents of students is obtained from data of students registered in Wanita PUI Kindergarten and obtained with the approval of the principal and parents of the students themselves.

More details of how factorial designs can be seen are shown in Table 1.

Table 1. Factorial experimental design.

	Method(A)	
	Role Playing (Experiment Class) (A1)	Conventional (Control Class) (A2)
Factor (B)		
Education High (B_1)	A1B1	A2B1
Level of Medium (B_2)	A1B2	A2B2
Parents Low (B_3)	A1B3	A2B3

3 RESULTS AND DISCUSSION

This research was conducted at the Wanita PUI Kindergarten by using one experimental class, namely class B2, and one control class, namely class B1. The experimental class is given treatment by using the role-playing learning method, while the experimental class is given treatment using conventional methods. In addition to using the learning method, this study also looks at aspects of the level of parental education as one factor in efforts to improve student financial literacy. At the level of education of parents, Class B1 consists of 14 students where 5 parents have high education level (bachelor degree, master degree, and postgraduate), 8 parents have a moderate level of education (Junior and Senior High School), and 1 parent has a low level of education (elementary school); whereas the B2 class consisted of 14 students, which consisted of 8 parents of students having a higher education level, 5 parents having a moderate level of education, and 1 parent of having a low level of education. The following are presented in Figure 1.

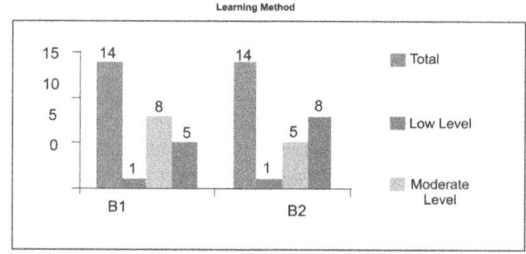

Figure 1. Parents' educational level.

The hypothesis testing uses analysis of variance between subject design. ANOVA in this experimental study is used to test the main and interaction effects of one or more nonmetric or categorical independent variables whose categories are more than two against one independent variable metric (interval, ratio). The independent variable is called a factor, and this study involves two factors: method and parents' education level. The following is the result of the calculation of ANOVA between subject design.

Based on the results of the ANOVA between subject design test, the significance value is <0.05 so that the three proposed hypotheses are accepted, and the interaction between learning methods and parents' education level about student financial literacy is seen in the following profile picture plot of financial literacy.

The influence of parents' education level as one of the individual factors that influence financial literacy aside from the learning method is a factor that can optimize the increase in financial literacy.

Based on the results of hypothesis testing regarding the interaction of learning methods and parents' education level on financial literacy, the value of F = 6.866

Table 2. Partial results of testing ANOVA between subject design.

Source	F	Sig.
Corrected Model	1442,806	0,000
Intercept	330030,960	0,000
Method	3269,136	0,000
Parents' Level Education	66,429	0,000
Method * Parents' Level Education	6,886	0,005
Error		
Total		
Corrected Total		

a. R Squared = ,997 (Adjusted R Squared = ,996)

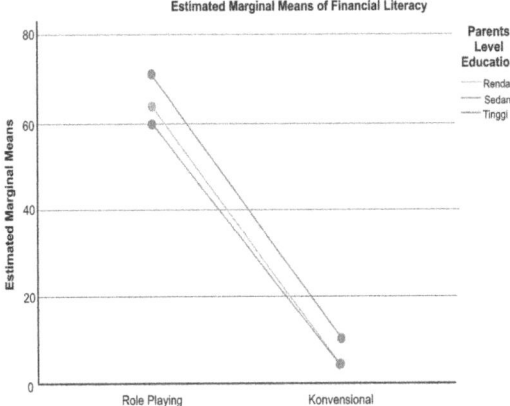

Figure 2. Profile plots interaction learning method and parents' level of education.

and $p = 0.005 < 0.05$ means that the hypothesis is accepted, that there is the interaction of role-playing method with parents' education level on students' financial literacy in subjects recognizing the needs, desires, and self-interests.

Financial literacy is not only related to financial knowledge and behavior. Through financial knowledge and behavior, students are expected to be able to make more responsible financial decisions. O'Connell (2008) says that knowing financial matters is indeed important, but knowledge alone cannot guarantee the best financial decisions. To improve students' abilities in terms of financial decision making, embed financial literacy values as early as possible. Responsible financial behavior cannot be built if it is not supported by the knowledge, skills, and experience provided by the family as the first community. Through the use of role-playing learning methods, the learning process not only results in increased knowledge but is accompanied by financial behavior when compared to the learning process that uses conventional learning methods. The existence of financial education delivered with appropriate learning methods can enhance more responsible financial behavior.

Investing financial literacy values as early as possible in children will greatly affect the understanding and knowledge of financial literacy and also the level of welfare in the future. The family, which is the first community, is a very effective place to instill financial literacy values (Rapih 2016). The higher the level of the parents' education will affect the awareness of parents to provide direct experience with money earlier to children when financial risk is still very low. This is under what was stated by Lusardi & Mitchell (2009), that a highly educated family environment will give effect to the intelligence, attitude, and behavior of a child in managing his finances.

For this reason, the influence of the use of role-playing learning methods based on the level of education of students' parents on financial literacy in basic competencies recognizing the needs, desires, and self-interest is one important factor to improve student financial behavior from an early age.

4 CONCLUSION

Based on the results of data processing and discussion of results, the conclusions in this study are as follows: that for the results of financial literacy to be more optimal, it should be prepared with an adequate allocation of time in the use of role-playing learning methods. Facilities and infrastructure during the learning activities must also be prepared for each class so that they have the same level of financial literacy between one class and another.

Although the role-playing learning method is student-centered and provides opportunities for students to play a role in learning activities, the participation of teachers, principals, and parents of students is still needed so that the achievement of financial literacy can create better and responsible entrepreneurial behavior.

For students with low financial literacy results, it is necessary to develop and apply other learning methods that can help improve student financial literacy, while for students with high financial literacy levels, good and responsible financial behavior must be maintained even though the student has finished completing formal education.

REFERENCES

Bhabha, J.I et al. 2014. Impact of financial literacy on saving-investment behavior of working women in the developing countries. *Research Journal of Finance and Accounting* 13(5): 118–122.

Braunstein, S., & Welch, C. 2002. Financial literacy: An overview of practice, research, and policy. *Federal Reserve Bulletin*: 445–457.

Calderon, O. 2013. Direct and indirect measures of learning outcomes in an msw program : what do we actually measure?. *Journal of Social Work Education*.

Haditono. 2006. Psikologi erkembangan. Yogyakarta: Gadjah Mada University Press.

Hurlock. 1974. Psikologi perkembangan suatu pendekatan sepanjang rentang kehidupan. Jakarta: Erlangga.

Ijevleva, K., & Arefjevs, I. 2014. Analysis of the aggregate financial behavior of costumers using the transtheoretical model of change. *Social and Behavioral Sciences* 156: 435–438.

International Network on Financial Education (INFE). 2008. Financial Education in Schools. OECD.

Lusardi, A., & Mitchel, O.S. 2009. Financial literacy among the young: evidence and implications or consumer policy. in pendison research working paper. University of Penyslavia, Pension Research Council.

Maurer, T.W. 2014. Process oriented guided-inquiry learning in financial literacy education. NC State University, Departement of 4-H Youth Development and Family & Consumer Sciences.

Mulia, A. 2017. Pendidikan karakter anak usia dini melalui metode role playing. *Prosiding Seminar Nasional Tahunan Fakultas Ilmu Sosial Universitas Negeri Medan Tahun 2017* 1(1): 379–383.

Nasrallah, R. 2014. Learning outcomes'rolein higher education teaching. *Education, Business and Society: Contemporary Middle Eastern Issues* 7(4): 257–276 ©. [Online]. Retrived from doi 10.1108/EBS-03-2014-0016.

O'Connell, A. 2008. Evaluating the effectiveness of financial education programmes. *OECD Journal: General Papers* 3: 9–51

Opletalova, A. 2015. Financial education and financial literacy in the Czech education system. *Procedia Social and Behavioral Sciences* 171: 1176–1184.

Rapih, S. 2016. Pendidikan literasi keuangan pada anak :mengapa dan bagaimana?. *Scholaria* 6:14–28.

Robb, & Woddyard. 2011. Financial knowledge and best practice behavior. *Journal of Financial Counseling and Planning*. 22(1).

Shupe, D. 2007. Significantly better: The benefits for an academic institution focused on student learning outcomes. *On the Horizon* 15(2): 48–57.

Walstad et al. 2010. The effects of financial education on the financial knowledge of high school students. *The Journal of Consumer Affairs* 44(2). ISSN 0022–0078.

Advances in Business, Management and Entrepreneurship – Hurriyati et al. (Eds)
© 2021 Taylor & Francis Group, London, ISBN 978-0-367-67471-7

Does size matter? Indonesian banking efficiency measurements using two-stage network DEA (2013–2018)

P. Darmanto & B.C. Siahaan
Universitas Indonesia, Depok, Indonesia

ABSTRACT: This study compares the efficiency of different groups of ownership and bank size and investigates the productivity change during the period, pursuant to fulfilment of regulations issued in 2012 concerning "Business Activities and Office Networks Based on Bank Core Capital" and "Minimum Capital Adequacy Requirement for Commercial Banks" that ignite the change of banks' strategy, capitals, and ownership, as well as attracting merger and acquisitions with more foreign investments. The measurement method of the bank efficiency adopts the two-stage network data envelopment analysis (DEA) model developed by Liang et al. (2008) to obtain intermediation and operational efficiencies to establish the overall bank efficiency. The boot-strapped truncated regression algorithm as proposed by Simar and Wilson (2007) was employed to examine the exogenous factors to the efficiencies. The study employs 105 conventional banks operating in Indonesia since 2013, which suggests Indonesia banking efficiencies have been improving, as evidenced with improving overall efficiency scores and the gap efficiencies between intermediary and operating functions narrowed during the observed period (2013–2018), and the study concluded that bank size affects those efficiencies.

1   INTRODUCTION

Efficiency has become an important word in banking. Studying the efficiency of banks is very important because it enables us to determine how banks should react to different problems, and how probable they are to survive them. Banking plays a key position in the economy and also carries an increasingly important role in the development of financial systems.

The Indonesia Banking Sector has become the backbone of the financial system of the Indonesian economy. Over the past 30 years, various structural and organizational reforms have been carried out, such as the establishment of OJK, the Indonesia's Financial Services Authority, and the independence of Bank Indonesia as the Central Bank from the government has improved the efficiency and stability of the Indonesian banking sector. Pangestu and Habir (2002) stated the liberalization of the banking sector after 1988 had increased the presence of foreign banks. It was hoped that foreign competition would transfer technology through technical assistance or foreign staff movements to local banks. Foreign competition took major and increased risks and costs of competing domestic and international banks concentrated primarily on the corporate industry, particularly in the company of multinational corporations operating in Indonesia. There was intense competition for business in this market. Foreign investment banks and commercial banks help the top Indonesian corporations by tapping foreign and domestic capital markets through the issuance

of equity or debt instruments either long-term bonds and short-term commercial paper.

Banking competition in Indonesia has been increasingly apparent since the openness of Indonesian banking that was initiated with the issuance of a policy package on 1 June 1983 (PAKJUN) with the aim of modernizing the banking, which was then continued by the October Package (PAKTO) on 27 October 1988. Indonesia Banking competition remains, as the number of banks has been significantly decreasing from 145 banks in 2001 to 115 banks in 2018 owned by the government, private sector, region, mixed sector/joint venture and foreign sector. It demonstrated Indonesian banking consolidation is on-going and competition remains.

Indonesia learned from the regional and global financial crisis in 1998 and 2008. Bank Indonesia in 2012 issued regulation number14/26/PBI/2012 concerning Business Activities and Office Networks Based on Bank Core Capital and number 14/18/PBI/2012 concerning Minimum Capital Adequacy Requirement for Commercial Banks that was amended by regulation issued by OJK no. 34/POJK.03/2016 in order to establish a stronger banking platform in Indonesia by strengthen its capital structure. These regulations define banking activities based on its capital size, namely BUKU (Bank Umum Berdasarkan Kegiatan Usaha or Classification of Commercial Banks based on the Business Activities). The first regulation permitted business activities of a Conventional Bank determined by reference to

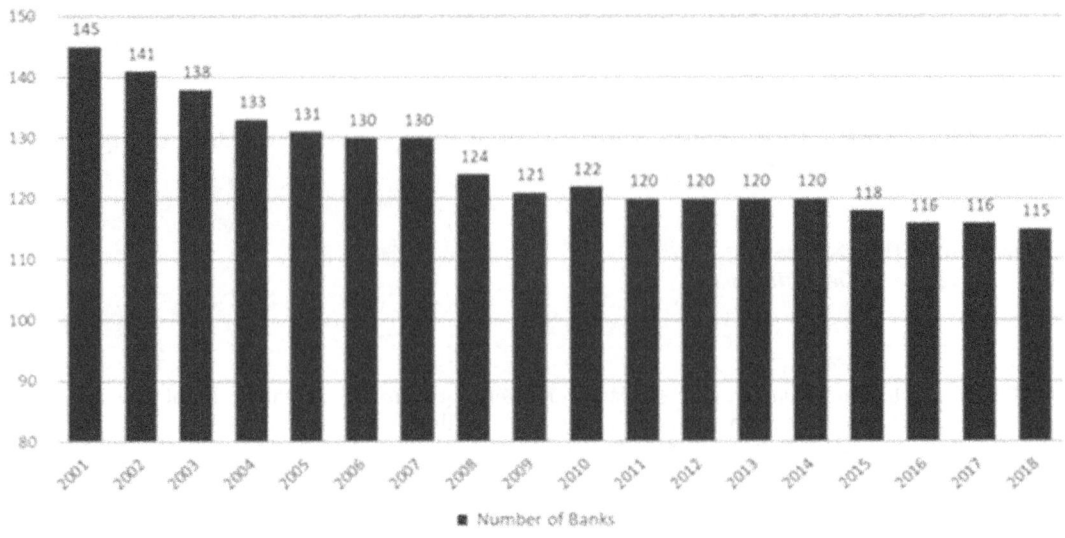

■ Number of Banks

Figure 1. Number of commercial banks in Indonesia, 2018. * Source: OJK (2018).

the BUKU classification, that it may attract mergers and acquisitions (M&A) and foreign investments since the issuance. The latter regulation incorporates the requirement to comply with the implementation of Basel III. It is a revision of Basel II, which contains preventive measures to avoid the banking crisis.

1.1 Literature review

1.1.1 Measuring bank efficiency
What drives the efficiency of a bank? Understanding factors of efficiency of banking is important because the efficiency of banks shapes the health of banks and is a very important driving factor in a country's economic growth. Literature in this area has been growing rapidly, capturing positive and negative empirical relationships between bank efficiency, the market, and various bank characteristics. With financial deregulation and market integration, the scope of bank activities has been completely reshaped from traditional intermediation products to new forms of business (Allen & Gale 2000; Levine 2005).

Despite the extensive literature on bank efficiency, there has not been a comprehensive study of whether diversification and ownership of bank income increases or impedes efficiency. Most research focuses on bank technical efficiency (Drake et al. 2009). Whether banks use fewer inputs to produce a certain amount of output or maximizes the output quantity given a certain number of inputs indicates efficiency. If data of the price for input and/or output is available, it can measure the costs and/or profit efficiency. Efficiency of cost is a product of technical efficiency. The allocative efficiency refers to the ability of banks to use the optimal mix of inputs given the price of each.

As a result, cost efficiency shows the ability of banks to provide services without wasting resources as a result of technical or allocative inefficiencies.

1.1.2 Methods of measuring banking efficiency
It has become an interesting discussion on the exact definition of input and output (Fethi & Pasiouras 2010). Assumptions of the input and output are almost as many as data envelopment analysis (DEA) applications. Berger and Humphrey (1992) identified two significant approaches in selecting inputs and outputs: production approaches and intermediary approaches. The first approach assumes that banks are producing loans and providing deposit services using labor and capital as inputs, and that the number and type of transactions or documents processed are used as measuring outputs. The latter approach views banks as financial intermediaries between savers and investors. Neither one of the two approaches is perfect because they cannot fully capture the dual role of financial institutions as providers of transaction/document processing services as well as financial intermediaries (Berger & Humphrey, 1992). The production approach might be somewhat better at evaluating bank branch efficiency, and the intermediation approach might be a better choice when evaluating financial institutions holistically. In addition, there are several obstacles in gathering information about the detailed flow of transactions needed in the production approach. Considering the above, the intermediation approach is preferred in the literature.

Cook et al. (2010) say that efficiency measurement is divided into two orientations, namely efficiency that has input orientation and output orientation. Chen et al. (2014) said that the approach with input orientation requires a combination of inputs to be used so that the output produced is maximal. While the output orientation approach is explained without changing the number of inputs used, the output can be increased. Managing both parameters will provide the most efficient system.

Berger and Humphrey (1992) emphasis that efficiency measurement is a useful tool for reallocating available resources in the company. This process minimizes input and maximizes the output produced and therefore the company remains competitive in the market for a long time. Efficient companies are more profitable and therefore efficiency measurement is very important for business executives who want to achieve set targets. Increasing efficiency means increasing company profitability. Management can use efficiency to compare business performance with the industry's highest results.

Farrell (1957) suggested the idea of a nonparametric efficiency approach that was enhanced by Charnes et al. (1978), who presented DEA to assess the performance of a group of homogeneous entities, namely DMUs, that could translate many inputs into several outputs.

2 METHODS

2.1 *Two-stage network DEA*

The DEA has been extended in later years to explore the efficiency of two-stage procedures or network structures. In contrast to traditional DEA research, which treats manufacturing as a one-step method, Liang et al. (2008) proposed that by analyzing this two-stage process or network structure, the internal structure of the DMU could be examined by explicitly modeling two stages. As seen in Figure 2, in the simplest two-stage structure, all outputs from the first stage are seen as intermediate steps, which are inputs to the second stage.

The two-stage network system employs a standard DEA model applied separately at each stage. Such approaches treat stages in a two-stage process as operating independently of each other (Kao & Hwang 2008).

The two-stage production structure for banks has been discussed by Seiford and Zhu (1999), Ho and Zhu (2004), and Gulati and Kumar (2010); however, they implemented the model independently at each stage.

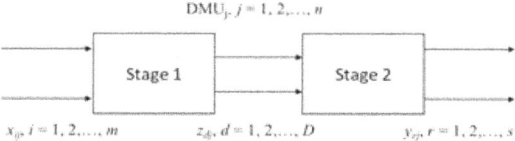

DMU$_j$, $j = 1, 2, ..., n$

Figure 2.   The two-stage NDEA model.
* Source: Liang et al. 20018.

The potential for conflict that arises because of intermediary steps was not addressed in these studies. Stage 1 managers may seek to maximize intermediate output production, and stage 2 managers may seek to minimize their use, which raises a conflict (Fukuyama & Weber 2012). Research efforts have been carried out in recent years to measure the efficiency of DMU utilizing the two-stage data envelopment analysis (NDEA) model. Kao and Hwang (2008) and Liang et al. (2008) may be the first to consider a two-stage production process in which the output of the first substage is used as input in the following substage. A survey article by Cook et al. (2010) and Halkos et al. (2014) referred to various two-stage NDEA models and their applications.

2.2 *Data-making units and the DEA model*

Assume that the production method consists of a two-stage series as shown in Figure 1. Using the notation in Liang et al. (2008), we assume that each DMUj (j = 1, 2,..., n) has the input m xij, (i = 1, 2,..., m) to stage 1, and D produces zdj, (d = 1, 2,..., D) from stage 1 which is known as an intermediate step. This D output is then input to stage 2. The output of stage 2 is yrj, (r = 1, 2, ..., s). For DMUj, we declare the efficiency scores for the first and second stages. On the basis of the input-oriented DEA model from Charnes et al. (1978), we define:

$$\theta_j^1 = \frac{\sum_{d=1}^{D} \omega_d z_{dj}}{\sum_{i=1}^{m} v_i x_{ij}} \quad (1)$$

$$\theta_j^2 = \frac{\sum_{r=1}^{s} u_r y_{rj}}{\sum_{d=1}^{D} \tilde{\omega}_d Z_{dj}} \quad (2)$$

where v_i, u_r, w_d, and $\tilde{\omega}_d$ are not known to be nonnegative weights. To connect two subprocesses with the entire production process, Liang et al. (2008) set $w_d = \tilde{\omega}_d$ in (3.1) and states the efficiency of the entire production process (θ_j) as a product of subprocess efficiency, e.g., $\theta_j = \theta_j^1 \times \theta^1$. Because of the assumption $w_d = \tilde{\omega}_d$ on (3.1), for a specific DMUo, $\theta_j^1 \times \theta_j^2$ becomes $\sum_{r=1}^{s} u_r y_{ro} / \sum_{i=1}^{m} v_i x_{io}$ which is the overall efficiency specified in the model of Kao and Hwang (2008).

By adding the convexity constraint ($\sum \lambda j = 1$) in the standard DEA approach, we got a variable returns to scale (VRS) envelope model. Under the standard DEA approach, by introducing independent variables, the VRS multiplier model can be obtained.

2.3 *Bootstrapped truncated regression*

An inherent issue in DEA is when DMUs are in a dominance set: when no combination of other DMUs exist with lower inputs for the same, outputs are assigned efficiencies and other DMUs are expressed in terms of this dominant set. These DMUs are not necessarily efficient when they are merely dominant, which means that no other DMUs were found to be more efficient. If the DMUs of the dominant set are in reality less than efficient, DEA overestimates their efficiency. Likewise, this applies for the other nondominated DMUs. It means DEA efficiency scores overestimate efficiency and are biased (Alirezaee et al. 1998).

As an alternative, this study employs a bootstrap truncated regression (BTR) as proposed by Simar and Wilson (2007). The method consists of Algorithm #1 and Algorithm #2, where the bias corrected procedures are incorporated therein.

2.4 Data making units

This study employs samples of conventional banks in Indonesia. There are 108 samples in 2013 to 100 samples by 2018 based as per OJK's Statistik Perbankan Indonesia portal. DEA requires non-zero and negative values, hence samples used are as detailed in Table 1.

Table 1. Number of DMUs employed.

Conventional Banks	2013	2014	2015	2016	2017	2018
Number of Samples	108	105	105	103	102	100
– with zero value	–	–	–	–	1	1
– with negative value	–	–	–	–	–	–
Total DMUs	108	105	105	103	101	99

*Source: Research results.

Table 2. Number of DMUs employed.

Variable Name	Description
Inputs:	
– Core Capital	Disclosed reserves and ordinary shares
– Employee	Total employee expenses
– Deposits	Total third-party funds
Intermediary:	
– Advances	Total outstanding credits & loans
– Investments	Total investments
Outputs:	
– NII	Net interest income
– Non-NII	Other income

* Source: Research results.

2.5 Research variables

Table 2 describes the input, intermediation, and output variables used in previous DEA applications. We adopt variables used by Gulati and Kumar (2017), employing three inputs, two intermediary, and two output variables as follows.

3 RESULTS AND DISCUSSION

3.1 Descriptive statistics and DEA results

Table 3 shows that Indonesian banking heterogenous with very large variance data, for example core capital employed between two banks, could be IDR164 trillion difference, or a bank can generate NII of IDR10 billion, while others can generate IDR106.3 trillion.

Figure 3 shows that overall efficiency scores of banking in Indonesia are improving and the gap between intermediation and operating efficiencies is narrowed down during the observation period.

Overall efficiency scores of banking in Indonesia in 2013, presented in Figure 4, demonstrated that the average operating efficiency score was low at 0.376 and the average intermediary efficiency score was at 0.806. Although on average, banks in Indonesia were well delivering their intermediation function in the year, their operations were far from efficient. However, we can see that the number of banks grouped in the Ace quadrant have been well performing above their peers.

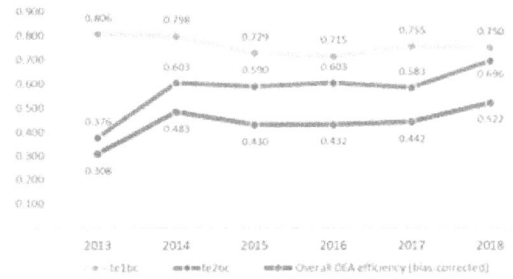

Figure 3. Banking efficiencies bias-corrected, 2013–2018.
* Source: Research results.

Table 3. Descriptive statistics.

IDR'b	Core Capital	Emply.	Depos.	Advc.	Investment	NII	Non-NII
Mean	8.431	839	42.380	39.658	9.157	4.720	1.679
Median	1.740	200	8.708	8.815	1.947	1.117	126
S/D	21.821	2.158	113.014	99.017	23.527	11.890	4.331
Min.	104	6	28	34	16	10	1
Max.	164.925	20.753	898.033	804.357	185.114	106.337	38.463

* Source: Research results.

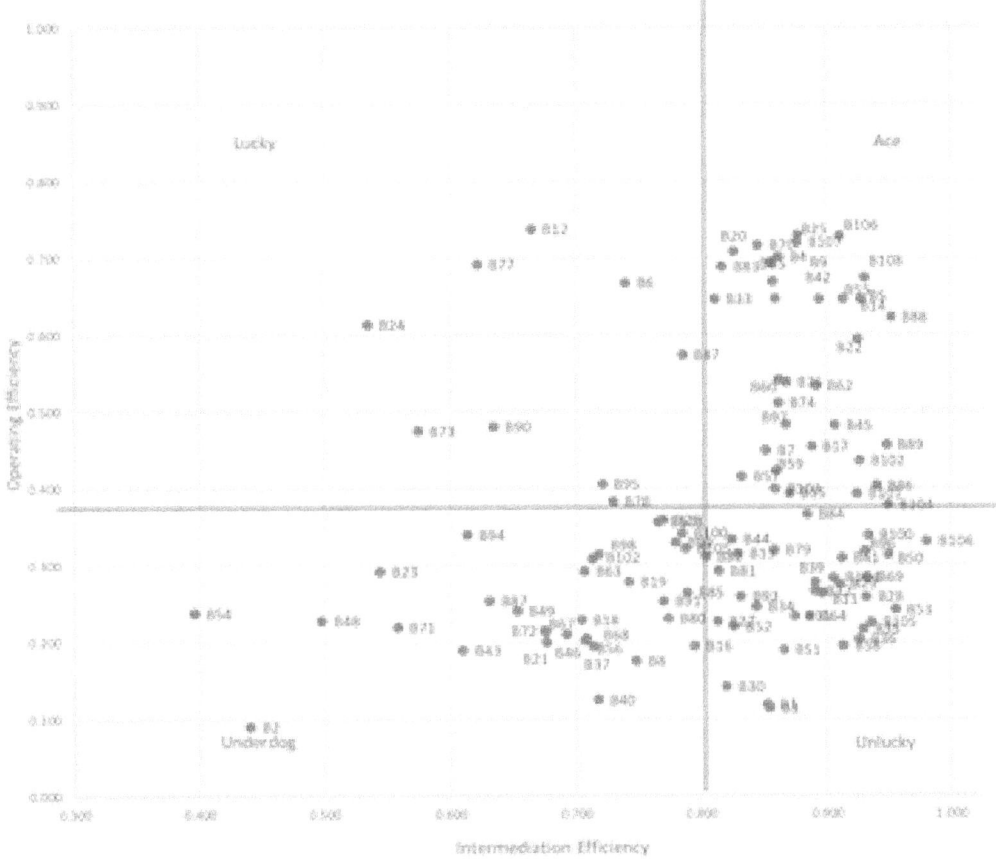

Figure 4. Bank positions in the 4-Qs (2013). *Source: Research results.

Figure 5 projects scatter plots of overall efficiency scores of banking in Indonesia in 2018, which is pointing out that in general, Indonesian banking efficiency has been much improving with average intermediation efficiency score at 0.75 and the average operating efficiency score significantly improved to 0.696, much better than earlier years. Despite that the average intermediation scores is relatively lower than 2013's, the average operating scores have improved since 2015.

Overall, banks in Indonesia have improved their operating efficiency scores, from 0.376 in 2013 to 0.696 in 2018. Numbers of banks that fall in quadrant Underdog and Unlucky shall take necessary actions and set right their strategy for better positions in coming years.

3.2 Bootstrapped truncated regression results

θ_1 and θ_2 are the reciprocal of intermediation (θ_1) and operating (θ_2) efficiency scores. Blank and Valdmanis (2010) suggested that in the bootstrapped truncated regression, the use of reciprocal of efficiency scores as a dependent variable can be considered.

To examine key exogenous factors of the operating scores as the θ_j^1 we utilize seven variables as per the following model:

$$\theta_j^2 = \beta_0 + \beta_1 \text{SIZE}_j + \beta_2 \text{ROA}_j + \beta_3 \text{RISK}_j$$
$$+ \beta_4 \text{DIVERSE}_j + \beta_5 \text{FOREIGN}_j + \varepsilon_j^2 \quad (3)$$

To examine key exogenous factors of the operating scores as the θ_j^0 this study utilizes five variables as per the following model:

$$\theta_j^0 = \gamma_0 + \gamma_1 \text{SIZE}_j + \gamma_2 \text{LIQUID}_j + \gamma_3 \text{ROA}_j$$
$$+ \gamma_4 \text{RISK}_j + \gamma_5 \text{DIVERSE}_j + \gamma_6 \text{PRIORITY}_j$$
$$+ \gamma_7 \text{IC}_j + \gamma_8 \text{FOREIGN}_j + \varepsilon_j^0 \quad (4)$$

Table 4 shows that banking efficiency, as financial intermediary, in Indonesia was statistically affected by its size, in particular from 2014 to 2018.

Table 5 shows that operating efficiency of banking in Indonesia is influenced by size (from 2013 to 2017) and their ability to diversify their revenue stream, while foreign ownership was negatively affecting operating efficiency, in particular during 2014 to 2018.

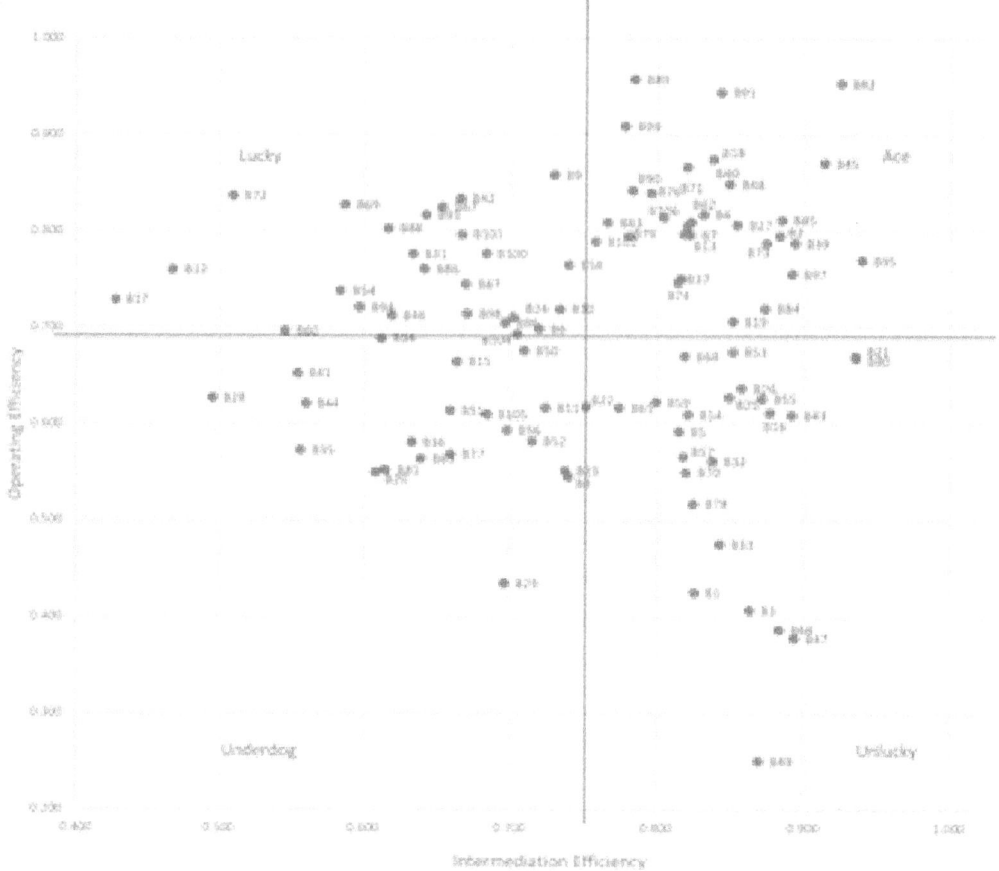

Figure 5. Bank positions in the 4-Qs (2018) *Source: Research results.

Table 4. Result of BTR of the intermediary efficiency scores.

Variables	2013	2014	2015	2016	2017	2018
SIZE	4.6E-07	3.4E-07	3.5E-07	2.3E-07	1.1E-07	1.9E-07
	(0.055)	(0.025)	(0.001)	(0.009)	(0.038)	(0.023)
LIQUID	0.0008	0.0010	0.0006	0.0003	0.0005	0.00028
	(0.077)	(0.012)	(0.052)	(0.197)	(0.103)	(0.207)
ROA	−0.0110	−0.0109	−0.0341	−0.0060	0.0006	0.00679
	(0.354)	(0.317)	(0.000)	(0.243)	(0.939)	(0.484)
DIVERSE	0.3591	0.8675	0.1652	0.2825	0.2823	0.11436
	(0.493)	(0.097)	(0.471)	(0.206)	(0.270)	(0.589)
PRIORITY	0.0005	0.0002	−0.0004	0.0001	0.0005	−0.00005
	(0.534)	(0.774)	(0.427)	(0.812)	(0.432)	(0.950)
IC	1.2578	1.5863	0.5650	0.3840	0.7695	1.67161
	(0.195)	(0.075)	(0.430)	(0.634)	(0.409)	(0.213)
FOREIGN	−0.0247	−0.0345	0.0124	0.0418	0.0427	0.06809
	(0.628)	(0.389)	(0.742)	(0.198)	(0.223)	(0.058)
Constant	0.7084	0.6440	0.6770	0.6420	0.6365	0.62070
	(0.000)	(0.000)	(0.000)	(0.000)	(0.000)	(0.000)

* Source: Research results.

Table 5. Result of BTR of the operating efficiency scores.

Variables	2013	2014	2015	2016	2017	2018
SIZE	8.0E-07	3.4E-07	3.0E-07	2.1E-07	1.3E-07	7.2E-08
	(0.000)	(0.005)	(0.008)	(0.014)	(0.070)	(0.330)
ROA	0.0033	0.0004	0.0111	0.0024	0.0166	0.0080
	(0.692)	(0.965)	(0.127)	(0.679)	(0.024)	(0.432)
RISK	0.0192	0.0061	0.0172	0.0123	0.0188	0.0071
	(0.008)	(0.334)	(0.000)	(0.024)	(0.002)	(0.358)
DIVERSE	2.4450	2.7676	0.8844	0.9850	0.5746	0.7113
	(0.000)	(0.017)	(0.000)	(0.000)	(0.007)	(0.002)
FOREIGN	−0.0350	−0.1637	−0.1912	−0.1605	−0.1406	−0.1626
	(0.310)	(0.000)	(0.000)	(0.000)	(0.000)	(0.000)
Constant	0.2550	0.5649	0.5418	0.5673	0.5205	0.6839
	(0.000)	(0.000)	(0.000)	(0.000)	(0.000)	(0.000)

* Source: Research results.

4 CONCLUSION

Conventional banks in Indonesia were not fully efficient during the earlier observation period of 2013 to 2018. However, this study suggests that Indonesia Bank efficiency significantly improved in 2018. Implementation of BAON regulation and Basel III has provided positive contribution to the banking efficiency, as efficiency gap between intermediary and operating efficiencies narrowed down from 0.430 in 2013 to 0.05 in 2018, both functions can be interpreted to be more correlated. Those regulations have driven banks to improve their capital, which increases their size, both in self-equity increase or synergize through M&A.

This study concurred that size does matter: bank size statistically affected banking efficiency in Indonesia from 2013 to 2018. Thus, banks shall consider improved their capital to have better assets (bank size) for better efficiency in delivering both roles, as (i) financial intermediary, and (ii) to generate income from its operation. In addition, the study found that diverse and intermediary cost variables are also statistically influential to the overall efficiency of conventional banks operating, while there is no statistical evidence that foreign banks have positive influence on banking efficiencies.

REFERENCES

Alirezaee, M.R., Howland, M., & Van de Panne, C. 1998. Sampling size and efficiency bias in data envelopment analysis. *Advances in Decision Sciences* 2(1): 51–64.

Allen, A. & Gale, G. 2000. Competition and financial stability. *Journal Money, Credit, and Banking* 36: 453–480.

Berger, B. & Humphrey, H. 1992. Efficiency of financial institutions: international survey and directions for future research. *European Journal of Operational Research* 98(2): 175–212.

Blank, J.L. & Valdmanis, V.G. 2010. Environmental factors and productivity on Dutch hospitals: a semi-parametric approach. *Health care management science* 13(1): 27–34.

Charnes. A., Cooper, W.W., & Rhodes, E. 1978. Measuring the efficiency of decision making units. *European Journal of Operational Research* 2: 429–444.

Cook, W.D., Liang, L., & Zhu, J. 2010. Measuring performance of two-stage network structures by DEA: a review and future perspective. *Omega* 38(6): 423–430.

Chen, Y., Cook, W.D., Kao, C., & Zhu, J. 2014. Network DEA pitfalls: divisional efficiency and frontier projection under general network structures. *European Journal of Operational Research* 226(3): 507–515.

Drake, L., Hall, M.J., & Simper, R. 2009. Bank modelling methodologies: A comparative non-parametric analysis of efficiency in the Japanese banking sector. *Journal of International Financial Markets, Institutions and Money* 19(1): 1–15.

Farrell. M.J. 1957. Assessing bank efficiency and performance with operational research and artificial intelligence techniques: a survey. *European Journal of Operational Research* 204(2): 189–198.

Fethi, M.D. & Pasiouras, F. 2010. Assessing bank efficiency and performance with operational research and artificial intelligence techniques: A survey. *European journal of operational research* 204(2): 189–198.

Fukuyama, H. & Weber, W.L. 2012. Estimating two-stage network technology inefficiency: an application to cooperative Shinkin banks in Japan. *International Journal of Operations Research and Information Systems (IJORIS)* 3(2): 1–23.

Gulati, R. & Kumar, S. 2017. Analysing banks' intermediation and operating efficiencies using the two-stage network DEA model. *International Journal of Productivity and Performance Management* 4: 500–516.

Halkos, G.E., Tzeremes, N.G., & Kourtzidis, S.A. 2014. A unified classification of two-stage DEA models. *Surveys in operations research and management science* 19(1): 1–16.

Ho, C.T. & Zhu, D.S. 2004. Performance measurement of Taiwan's commercial banks. *International Journal of Productivity and Performance Management* 53(5): 425–433.

Kao, C. & Hwang, S.N. 2008. Efficiency decomposition in two-stage data envelopment analysis: An application to non-life insurance companies in Taiwan. *European journal of operational research* 185(1): 418–429.

Levine, R. 2005. *Finance and growth: theory and evidence*. Amsterdam: North-Holland.

Liang et al. 2008. *Geomorphic evolution of the Sahara and the Nile in M.A.J. Williams & H. Faure (eds), The Sahara and the Nile: 21–35*. Rotterdam: Balkema.

Pangestu, M. M. & Habir, M.M. 2002. The boom, bust and restructuring of Indonesian banks (No. 2–66). *International Monetary Fund*.

Seiford, L. M. & Zhu, J. 1999. Profitability and marketability of the top 55 US commercial banks. *Management science* 45(9): 1270–1288.

Simar, L. & Wilson, P.W. 2007. Estimation and inference in two-stage, semi-parametric models of production processes. *Journal of econometrics* 136(1): 31–64.

Advances in Business, Management and Entrepreneurship – Hurriyati et al. (Eds)
© 2021 Taylor & Francis Group, London, ISBN 978-0-367-67471-7

The impact of financial derivatives markets on economic growth

P. Fariska, N. Nugraha & I. Solikin
Universitas Pendidikan Indonesia, Bandung, Indonesia

M.A. Rohandi
Universitas Islam Bandung, Bandung, Indonesia

ABSTRACT: Over the past years, derivatives markets have played a vital role in financial systems and greatly contributed to various aspects of economic growth. A positive contribution of derivatives markets on economic growth in market economy in developing countries is less evident with more recent data. This paper investigates the dynamic relationship between financial derivatives market and economic growth in Indonesia. We used a Granger-Causality test in the framework of vector autoregression (VAR) and impulse response function (IRF) through vector error correction model (VECM) to examine this casual and dynamic relation for the period of 2014Q1 to 2018Q4. Derivatives market had a significant negative effect in the long run on economic growth in Indonesia but had a positive effect in the short term. We also found that response received by economic growth due to derivatives market shock was convergence. It tended to be negatively affected and then changed into positive. The response due to shocks given will eventually disappear so that the shock does not leave a permanent effect.

1 INTRODUCTION

Over the past years, derivatives markets have played a vital role in financial systems and greatly contributed to various aspects of economic growth. Derivatives is one of the successful innovations in capital market, but the question is whether derivatives markets can play an important role in economic growth in developing countries.

Many studies have examined and supported the relationship between financial intermediation and economic growth (Goldsmith 1969; McKinnon 1973; Schumpeter 1911/1959) or questioned if financial development enhances growth or vice versa (Greenwood & Jovanovic 1990; Pagano 1993). Research up to the mid-1990s showed a positive relationship between financial sector and economic growth. Finance does not only follow growth; finance seems importantly to lead economic growth (King & Levine 1993). Furthermore, research conducted by Rousseau and Wachtel (2009) shows a weakening relationship between financial sector and economic growth; the impact depends on economic crisis and non-crisis observations.

Moreover, many experts (Atje & Jovanovic 1993; Arestis et al. 2001; Baier et al. 2004; Beck & Levine 2004; Demirgüç-Kunt & Levine 1996; Harris 1997; Levine & Zervos 1998; Rousseau & Wachtel 1998) have examined the relationship between stock market development and economic growth.

There are empirical evidences to examine the relationship between economic growth and financial derivatives markets (Acemoglu & Zilibotti 1997; Beck & Levine 2004; Coskun et al. 2017; Haiss & Sammer 2010; Kolb 2003; Menyah et al. 2014; Rousseau & Wachtel 2009; Sendeniz-Yüncü et al. 2018). But only a limited body of empirical evidence exists to examine derivatives market, economic growth, and macroeconomic indicator (Vo et al. 2019).

The purpose of this paper is to investigate the dynamic relationship between financial derivatives markets on economic growth in developing countries. Indonesia is considered to be one of the developing countries which has experienced a decline in the last five years according to IMF in regional economic outlook, in 2018.

Therefore, two questions were the focus of the study. First, is there a relationship among derivatives market, economic growth, and macroeconomic indicator? Second, how do shocks caused by the derivatives market have an impact on economic growth and macroeconomic indicator? The first question involves derivatives market, economic growth, and macroeconomic indicator in Indonesia. To address the first question, financial derivatives market volume, interest rates, and inflation as macroeconomic indicator and gross domestic product (GDP) were examined. GDP is one indicator that can describe economic growth in a country. Indonesian GDP has a slow increase. It can be seen in Table 1.

Table 1. Indonesian GDP over the last five years.

(Billion Rp)				
2018	2017	2016	2015	2014
Q4 3.798.675	3.489.915	3.193.904	2.939.559	2.697.695
Q3 3.841.755	3.503.439	3.205.019	2.990.645	2.746.762
Q2 3.685.273	3.366.096	3.073.537	2.867.948	2.618.947
Q1 3.511.654	3.227.762	2.929.269	2.728.181	2.506.300

Source: Badan Pusat Statistik (BPS 2018).

Varieties of macroeconomic factors as control variables in the investigation of the effects of financial development on economic growth are used by Thumrongvit et al. (2013) and Ruiz (2018). Moreover, macroeconomic determinants Granger cause economic growth (Pradhan et al. 2014). Development of interest rates and inflation as macroeconomic indicators in Indonesia can be seen in the following graph (Figure 1).

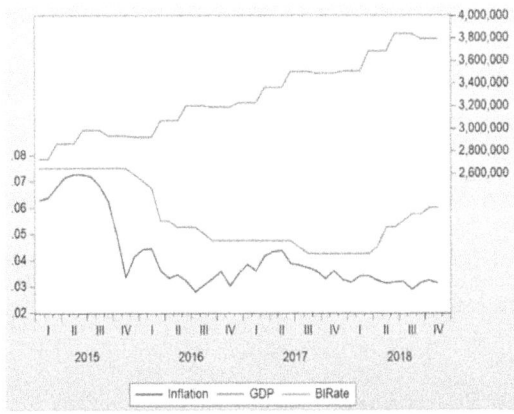

Figure 1. Development interest rate, inflation and GDP.

The second question involves the identification of shocks caused by derivatives market in the short term and long term. A panel vector autoregressive (VAR) and Impulse Response Function (IRF) through Vector Error Correction Model (VECM) were employed. These formulations are rarely used in the finance-growth literature.

2 METHODS

The impact of financial derivatives market on economic growth and the role of macroeconomic indicator in the relationship between derivatives market and economic growth using time-series analysis in short term and long term was shown. To find out the impact on the short term, the impulse response function (IRF) is used

through the Vector Error Correction Model (VECM), Impulse response analysis is used to see the impact of shocks from one variable on another variable, and the VAR model is used to see long-term impact.

There are several important analyses in the VAR model, four of which are forecasting, impulse response, forecast decomposition variance, and causality test (Juanda & Junaidi 2012). The most important thing in estimating the VAR model is determining the lag, and in determining the optimal lag using several criteria, namely LR (Sequential Modified Likelihood Ratio Test) with the greatest value or in the AIC (Akaike Information Criterion), SC (Schwarz Information Criterion), FPE (Final Prediction Criterion), and HQ (Hannan-Quinn Information Criterion) which have the smallest value.

To access causal relationships among derivative markets, macroeconomic indicators, and economic growth, the causality test was applied. The causality test is a test to determine the causal relationship between variables from the VAR system that are tested using the Granger causality test.

2.1 Research question and purpose hypotheses

VAR was used in order to detect the direction of causality among the variables. Evidently, among other things, this study melds several strands of the literature. Thus, the following three hypotheses were proposed:

H1: Derivatives market Granger causes economic growth and vice versa.

H2: A macroeconomic variable Granger causes economic growth and vice versa.

H3: Derivatives market Granger causes a macroeconomic variable and vice versa.

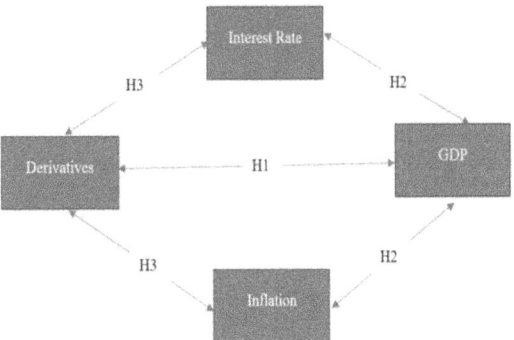

Figure 2. The conceptual framework of causality between variables.

2.2 Data, variables, and model specification

The analysis utilized monthly time series data over the period 2015Q1 to 2018Q4. This study considered Indonesia as a sample country.

The variables used in this study are that derivatives market (derivative) is the real value of derivatives in

127

Table 2. Definition of variables.

Variable	Definition
Derivative	Financial derivatives market volume
GDP	Value of final goods and services produced by various production units in the territory of a country within a year.
Interest Rate	Rates that reflect the attitude of monetary policy set by bank Indonesia.
Inflation	Indicates level of change and is considered to occur if the process of price increases takes place continuously and influences each other.

Indonesia stock exchange, GDP, and two other macro-economic variables, inflation rate (inflation) and BI rate (interest rate).

Based on literature review, the following empirical models describe the relationship among economic growth, financial derivatives market, and two other macroeconomic variables; the regression specification is as follow:

$$GDP = f\{derivatives, interest\ rate, inflation\} \quad (1)$$

$$GDP = a0 + a1\ derivative + 2\ interest\ rate$$
$$+ a3\ inflation + \varepsilon \quad (2)$$

GDP is not always a dependent variable; causal relationship as shown in Figure 2 describes the possibility that the direction of causation between variables may proceed simultaneously in one direction or in both directions.

3 RESULTS AND DISCUSSION

First, for the long run, this study found negative impact of the financial derivatives market on economic growth in Indonesia. This implies that an excessive increase in the development of the derivatives market can cause adverse effects in the long run. This result supports previous theories carried out by Baluch and Ariff (2007) and Vo et al. (2019) concerning the beneficial effects of the derivatives market on economic growth. From the equation model, it is known that inflation and BI rates have a significant positive impact on economic growth.

Based on the summary of statistic results, R-squared for the regression model indicates that 0,549982 effected economic growth. It explains the variability of the response data around its means. Derivatives variable periods t-1 and t-2 have a different impact on economic growth, adverse effects from t-2 are higher than t-1. The inflation variable periods t-1 and t-2 also have a different impact on economic growth, an higher increase occurred in t-2, but the interest rate variable period t-1 has a higher impact on economic growth than t-2.

Second, this study examined the short-term relationship among variables and presents the result from cumulative orthogonalized impulse response function. It carried out IRFs of the dynamic relationship. The response GDP is positively affected by its own shock as well as GDP to derivatives, interest rate, and inflation. Economic growth tended to be positively influenced by financial derivative markets and macroeconomic indicators over the past 10 months and is convergence, which means that economic growth tends to be negatively affected then turn positive and in convergence.

GDP in a first period was negatively affected by its own shock, positively in second period, and in the third period the shock will be convergence, as well as GDP to derivatives. But response GDP to interest rate shock are positively affected in the first and second periods and then in convergence after that. Response GDP to inflation is positive in a first period then negatively affected in the second period and then will be in convergence in third period.

Last, this study estimated causal relations among variables using the Granger causality test. From the causality test, it was found that there was a causal link between economic growth and the financial derivatives market, with Prob. value 0,0318 significance at 5% level but not vice versa. Second, it can be concluded that there is a causal link between economic growth and interest rates, with Prob. Value of 0,0146 significance at 5% level but not vice versa. But there is no causal link between economic growth and inflation and vice versa. Finally, there is no causal link between the financial derivatives market and interest rates and vice versa, there is no causal link between inflation and the financial derivatives market and not vice versa, and all not significance at 1%, 5%, and 10% levels.

4 CONCLUSION

This study finds that the financial derivatives market has a significant negative effect in the long run on economic growth in Indonesia, but there is a positive influence when viewed from short-term effects. Importantly, it also finds that the response received by GDP due to derivative market shocks over the past 10 months is convergence, which means that economic growth tends to be negatively affected and then turns positive and in convergence; this also applies to inflation and interest rates.

Moreover, the Granger causality test between the variables showed that economic growth has a significant effect on derivative markets but not vice versa, as well as interest rates. It can be concluded that in Indonesia, economic growth has a direct impact on the derivative markets and interest rates of Bank Indonesia. Finally, there is no causal link between financial derivative markets and interest rates and vice versa but there is no causal link between inflation and financial derivatives market and not vice versa.

The results of this study contribute to previous studies conducted by Baluch and Ariff (2007), Haiss and Sammer (2010), Menyah et al. (2014), Coskun et al. (2017), and Vo et al. (2019), who argue that derivative markets contribute to economic development, and strengthen research conducted by the IMF Regional Economic Outlook in 2018 which states that long-term GDP has far less impact due to the existence of expiration periods that affect the behavior of companies and households.

REFERENCES

Acemoglu, D., & Zilibotti, F. 1997. Was prometheus unbound by chance? risk, diversification, and growth. *Journal of Political Economy* 105: 709–51.

Arestis, P., Demetriades, P., & Luintel, K. 2001. Financial development and economic growth: the role of stock markets. *Journal of Money, Credit and Banking* 33(1): 16–41.

Atje, R. & Jovanovic, B. 1993. Stock markets and development. *European Economic Review* 37: 632–640.

Badan Pusat Statistik. 2018. Indonesian GDP. [Online]. Retrived from www.bps.go.id.

Baier, S.L., G.P. Dwyer Jr., & Tamura. 2004. Does opening a stock exchange increase economic growth?. *Journal of International Money and Finance* 23: 311–331.

Baluch, A., & Ariff. M. 2007. Derivative markets and economic growth: is there a relationship?. Bond University Globalisation & Development Centre Working Paper Series No. 13.

Beck, T., & Levine, R. 2004. Stock markets, banks, and growth: panel evidence. *Journal of Banking and Finance* 28(3): 423–442.

Coskun, Yener, Ünal, Murat & Ulussever, T. 2017. Capital market and economic growth nexus: evidence from Turkey. *Central Bank Review* 17: 19–29.

Demirgüç-Kunt, A., & Levine, R. 1996. Stock markets corporate finance and economic growth. *World Bank Economic Review* 10(2): 223–240.

Goldsmith, R.W. 1969. Financial structure and development. CT: Yale University Press, New Haven.

Greenwood, J., & Jovanovic, B. 1990. Financial development, growth, and the distribution of income. *Journal of Political Economy* 98(5): 1076–1107.

Haiss, P., & Sammer B. 2010. The impact of derivatives markets on financial integration, risk, and economic growth. [Online]. Retrived from http://dx.doi.org/10.2139/ssrn.1720586.

Harris, R.D.F. 1997. Stock markets and development: a reassessment. *European Economic Review* 41: 139–146.

International Monetary Fund. 2018. IMF Regional Economic Outlook in 2018. [Online]. Retrived from www.imf.org.

Juanda, B., & Junaidi. 2012. Ekonometrika deret waktu. IPB Press: IPB Bogor.

King, R., & Levine, R. 1993. Finance and growth: schumpeter might be right. *The Quarterly Journal of Economics* 108(3): 717–737.

Kolb, W. 2003. Futures, options, and swaps 4th edition. Oxford: Blackwell Publishing.

Levine, R. 1991. Stock markets, growth, and tax policy. *Journal of Finance* 46(4): 1445–1465.

Levine, R., & S. Zervos. 1998. Stock markets, banks, and economic growth, *The American Economic Review* 88(3): 537–558.

McKinnon, R. 1973. Money and capital in economic development. Washington DC: Brookings Institution.

Menyah, K., Nazlioglu, S., & Wolde-Rufael, Y. 2014. Financial development, trade openness and economic growth in African countries: new insights from a panel causality approach. *Economic Modelling* 37(2): 386–394.

Pagano, M. 1993. Financial markets and growth: an overview. *European Economic Review* 37(2-3): 613–622.

Pradhan, R.P., Mak B,A., John H.H, & Bahmani, S. 2014. Causal nexus between economic growth, banking sector development, stock market development, and other macroeconomic variables: the case of Asean countries. *Review of Financial Economics* 23: 155–73.

Rousseau, P.L., & Wachtel, P. 1998. Financial intermediation and economic performance: historical evidence from five industrialized countries. *Journal of Money, Credit and Banking* 34(4): 657–678.

Rousseau, P., & Wachtel, P. 2009. What is happening to the impact of financial deepening on economic growth?. *Economic Inquiry* 48. [Online]. Retrieved from Doi:10.1111/j.1465-7295.2009.0019z.x.

Ruiz, J.L. 2018. Financial development, institutional investors, and economic growth. *International Review of Economics and Finance* 54: 218–24.

Schumpeter, J. 1911. Theorie der wirtschaftlichen entwicklung. Leipzig: Duncker and Humblot.

Schumpeter, J. 1959. The theory of economic development. Harvard University Press.

Sendeniz-Yüncü, İlkay, L.A., & Kürşat Aydoğan. 2018. Do stock index futures affect economic growth? evidence from 32 countries. *Emerging Markets Finance and Trade* 54: 410–29.

Thumrongvit, P., Yoonbai, K. & Pyun, C.S. 2013. Linking the missing market: the effect of bond markets on economic growth. *International Review of Economics and Finance* 27: 529–41.

Vo, D.H., Huynh, S.V., Vo, A.T., & Ha, D.T. 2019. The importance of the financial derivatives markets to economic development in the world's four major economies. *Journal of Risk and Financial Management* 12(35). [Online]. Retrieved from: doi:10.3390/jrfm12010035.

How does working capital and sales volume increase net income?

G.S. Manda, D.J. Suyaman & R.L. Batu
Universitas Singaperbangsa, Karawang, Indonesia

ABSTRACT: PT Tri Jaya Teknik Karawang have experienced unstable profit that impacted the selling price and caused debt. This condition affected working volume, sales volume, and net income. The purpose of this study is to determine whether the working capital and sales volume affect the net income either partially or simultaneously. There were two independent variables and one dependent variable. Independent variables were working capital (X1) and sales volume (X2), and the dependent variable was net profit (Y). The study used a quantitative method, in which normality test and classical assumption test were used. The data were analyzed using descriptive and verification analysis. The finding showed that there was partial influence of working capital on net income; and sales volume on net income. Meanwhile, the result of an F-test revealed $0.010 < 0.05$. This meant simultaneously influence of working capital and sales volume on net profit. The coefficient of determination test showed that working capital and sales volume had a strong influence on net profit that was equal to 73,1%. The other percentage (26,9%) was influenced by the other variable, which was not examined.

1 INTRODUCTION

1.1 *Background*

Global economic development drives Indonesia to become a developing country in the industrial sector (Teruel & Salono 2012). This can be shown by the emergence of various competition in the industrial sector that produce products and implement various strategies to maintain and be in competition. The aim is to gain profits, or to roll out the profits in the business cycle for a long term.

A company requires profits to survive. The size of profits gained determines the success of management. Management can use information from company financial statements to predict the size of profits in the future (Asriyanti & Syafruddin 2017).

One of the prominent subsectors in Indonesia, including in Java, is the processing sector. Natural resources in Indonesia can produce various basic needs. Therefore, many companies in Indonesia carry out activities to change a basic item mechanically, chemically, or by hand so that it becomes finished goods or semifinished goods, and/or make less-value goods to be higher value goods which are closer to the end user (Edi & Saad 2014; Teruel & Salono 2012).

Central Statistics Agency explained that the number of large processing industries in Java continually increase. This is due to several factors. One is that entrepreneurs prefer to establish their companies on Java Island rather than on the other islands. Java has many large cities with industrial area, such as Karawang Regency. This eases the process of exporting goods as the facilities and infrastructure are more complete, ranging from transportation to managing costs. This condition finally can minimize company budgets and add company profits.

Working capital is one of the important components in carrying out the company's business activities, as it is contained in financial statements. The used working capital is expected to be able to return to the company in a short time through sales. It will spin continuously in every period and can be reallocated to finance the company's operations. Companies need working capital to pay their day-to-day operations, such as providing raw materials, purchasing raw materials, paying labor wages, paying employee salaries, and so forth.

Central Statistics Agency also stated that there is an increase of mediumsized large industry in Karawang every year. This includes the establishment of PT Tri Jaya Teknik Karawang. Similar to the other companies, the company has working capital and sales volume contained in financial statements that can affect net income in each period. Working capital and sales volume are related to net income. Net income is the excess of income for costs in a certain period after deducting income tax which is presented in the form of income statement, a part of financial statements.

The following table provides working capital, sales volume, and net income of PT Tri Jaya Teknik Karawang for the period 2012–2016.

The table shows that working capital, sales volume, and net income of PT Tri Jaya Teknik Karawang are unstable. The working capital decreased in the first semester of 2013, first and second semester of 2014, and first semester of 2016. The sales volume also decreased in the first semester of 2013, 2014, and

Table 1. PT. Tri Jaya Teknik report.

Year	Sem	Working Capital (Rp)	Sales Volume (Rp)	Net Income (Rp.)
2012	I	194.778.723	1.889.784.900	226.773.866
	II	238.062.883	3.366.743.385	232.723.615
2013	I	216.414.518	2.520.809.620	230.118.109
	II	377.609.679	3.281.214.430	333.962.973
2014	I	213.151.855	1.446.508.619	321.817.647
	II	210.139.159	2.334.621.646	217.850.802
2015	I	407.008.032	709.729.393	335.245.070
	II	497.454.262	4.534.113.703	223.496.713
2016	I	333.536.563	1.919.816.914	444.916.938
	II	519.425.046	3.422.517.125	524.186.043

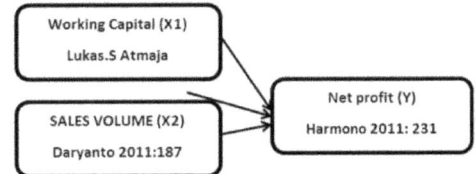

Figure 1. Theoretical framework.

2016. A decrease also occurred on net income in the first semester of 2013, first and second semester of 2014, and second semester of 2015.

This indicates a problem in the company. A decrease of working capital, sales volume, and net income are caused by the same selling price of the products in each year, whereas the material price is continually increasing. This is basically done to keep competitiveness with the other companies, so that the customers can be maintained. Unfortunately, this condition forces the debt. The working capital in the next period should be used to cover company debt first. Then, the working capital that should be used to finance operational activities in the following period also decreases. This impacts the number of sales volume and net income gained.

Rizal (2016) stated that the variables of business capital and sales have a significant influence on operating income. This is in accordance with Flower (2017), who argued that simultaneously working capital and sales have a significant effect on net profit; and partially working capital and sales have a significant positive effect on the net profit of the food and beverage subsector that is available on the Indonesia Stock Exchange (Neag 2014).

1.2 Research problem

Based on the described background, the formulation of the research problems are as follows:

1. How is working capital at PT Tri Jaya Teknik Karawang?
2. How is sales volume at PT Tri Jaya Teknik Karawang?
3. How is net income at PT Tri Jaya Teknik Karawang?
4. To what extent does working capital and sales volume increase net income?

The conceptual framework or frame of mind is a synthesis or extrapolation of a theoretical review that reflects the interrelationship between the investigated variables and is a demand for solving research problems.

Based on the data of PT Tri Jaya Teknik Karawang, there is an assumption that there is an influence between working capital and sales volume on net profit. Working capital and sales volume, however, describe a relationship between certain and the other factors.

Regarding this, Riyanto explained that every company needs working capital to finance its daily operations, such as providing raw materials, purchasing raw materials, paying laborers' or employee salaries, and so forth. The spent money is expected to be able to re-enter the company through the volume of product sales in a short time. This money is reused to finance further operations, and it continually spins in every period (Riyanto 2013). Working capital management is essential in any kind of enterprise, regardless of its size or sector (Motlicek & Martinovicova 2014).

Sales volume is the peak activity in an effort to achieve the desired target. Sales volume is a measure that indicates the amount of goods or services sold (Tjiptono 2012).

Net income is a profit that has been reduced by costs that are the burden of the company in a certain period, including taxes (Marpaung 2019). To be able to achieve maximum net profit in planning and realization, management can take several steps, one of which is by optimally increasing sales volume (Hidayabti et al. 2016). Skousen stated that net income is the difference between the amount of revenue earned by a business unit during a certain period and the number of costs that can be applied to the financial statement (Marpaung 2019).

2 METHODS

This study used the financial statement of PT Tri Jaya Teknik Karawang for the period 2012–2016. The working capital, sales volume, and net profit were used as the primary data. As it was considered a small sample, t distribution was used. Meanwhile, the secondary data was obtained through related documents, books, and records that were either published or nonpublished.

a. The normality test, especially Komogorov-Smirnov, was used. This test was aimed to test whether in the regression model, the residual confounding variable had a normal distribution; if this assumption was violated then the statistical test becomes invalid for a small sample number (Ghozali 2013).

b. SPSS software was conducted to process the data statistically. The data was considered to be normally distributed if the Asymp Sig (2-tailed) value of the

Kolmogorov-Smirnov calculation results was more than $1/2\alpha$.

c. Multicollinearity test was performed to test whether the regression model found a correlation between independent variables. Regarding this, this study used the value of VIF (variance inflation factor) and tolerance values.

3 RESULT AND DISCUSSION

3.1 *The effect of working capital on net income*

The analysis result shows that t-count is 4.349. When it is compared to the value of table, it is equal to 2,262. In other words, the value of t-count is greater than the value of t-table. Then, the significance value of 0.003 is smaller than 0.05; the positive effect is 1.032. This indicates that if there is an increase in working capital it will be followed by an increase in net income.

Based on the above result, H0 is rejected and H1 is accepted. Therefore, it can be concluded that there is the influence of working capital on net income at PT Tri Jaya Teknik Karawang for 2012–2016.

The result also shows that the coefficient of determination or R Square (R2) is 0.318 or equal to 31.8%. This implies that the influence of working capital on net income is 31.8%. The remaining percentage (68.2%) is influenced by the other variables or interfering variables (error variables).

3.2 *The effect of sales volume on net income*

The value of t-count is −3,276. When it is compared to the value of table, it is equal to 2,262. In other words, the value of t-count is greater than the value of t-table. The significance value of 0.014 is smaller than 0.05; then the negative effect is −0.105. This means that if there is an increase in the sales volume, it will be followed by a decrease in net income.

The above result implies that H0 is rejected and H2 is accepted. It can be concluded that there is an influence of sales volume on net income at Tri Jaya Teknik Karawang in the period 2012–2016.

The result also shows that the coefficient of determination or R Square (R2) is 0.003 or 0.3%. This implies that the effect of sales volume on net income is 0.3%. The remaining percentage (99.7%) is influenced by the other variables or interfering variables (error variables).

3.3 *The effect of working capital and sales volume on net income*

The calculated F value is 9,493. The F-table value at the 5% significance level and the degrees of freedom df1 = 2 and df2 = 7 was 4.737 (F (2; 7)). The value of F-count with F-table is then compared, so that the value of 9.493 > 4.737 is obtained. The value of F-count is greater than the value of F-table. The significance value of 0.010 is smaller than 0.05. Thus, Ho is rejected, and H3 is accepted. It can be concluded

that there is a simultaneous influence of working capital and sales volume on net income at the Tri Jaya Teknik Karawang PT for the period 2012–2016.

The result shows that the coefficient of determination or R Square (R2) is 0.731 or equal to 73.1%. This indicates that the simultaneous influence of working capital and sales volume on net income is 73.1%. The remaining percentage (26.9%) is influenced by the other uninvestigated variables or interfering variables (error variables).

The independent variables in this study are working capital and sales volume or variables X1 and X2. The value of the correlation between the two variables is 0.855. This value is in the category (0.800–1,000) of very strong correlation relationship. The correlation is also positive. It means that the correlation between the working capital variable and sales volume is in the same direction. In other words, the greater value of the working capital variable will be followed by the increase of sales volume variable, and vice versa.

The value of R2 is 0.731 or 73.1%. This means that the independent variables which consist of working capital and sales volume has 73.1% simultaneous effect on the dependent variable, namely net profit. The remaining percentage (26.9%) is influenced by the other factor outside two investigated independent variables.

4 CONCLUSION

There are three important points from this study:

1. There is partial influence of working capital towards net income at PT Tri Jaya Teknik Karawang for the period 2012–2016.
2. There is partial influence of sales volume on net income at PT Tri Jaya Teknik Karawang for the period 2012–2016.
3. There is simultaneous influence of working capital and sales volume on net income at PT Tri Jaya Teknik Karawang for the period 2012–2016.

REFERENCES

Asriyanti, E. & Syafruddin. 2017. "Pengaruh Harga Jual, Volume Penjualan Dan Biaya Operasional Terhadap Profitabilitas Perusahaan Pada PT. Prisma Danta Abadi (Tahun 2014–2016) Effect Of Sell Prices, Sales Volume And Operational Costs On Company Profitability In. Prisma Danta Abadi. *Measurement Journal* 11(1): 33–50.

Edi, N. & Saad, N.M. 2014. "Working Capital Management: The Effect of Market Valuation and Working Capital Management: The Effect of Market Valuation and Profitability in Malaysia." *International Journal of Business and Management* 5: 140–47.

Ghozali, I. 2013. Aplikasi analisis Multivariate dengan progam IBM SPSS, Edisi Ketujuh. Semarang: *Univeritas Diponegoro*.

Hidayanti, F., Yahdi, M. Biaya Operasional, and Laba Bersih. E-ISSN:2622-304X, P-ISSN: 2622-3031 Available [Online]. Retrieved from Http://Proceedings.Stiewid

yaga malumajang.Ac.Id/Index.Php/Progress" 1, no. 1 (2016): 399–406.

Marpaung, N. 2019. pengaruh modal kerja dan volume penjualan properti yang terdaftar di bursa efek indonesia program studi keuangan perbankan. *Universitas Komputer Indonesia.* 8(2).

Neag, R. 2014. "The Effects of IFRS on Net Income and Equity: Evidence from Romanian Listed Companies." *Procedia Economics and Finance* 15(14): 1787–90.

Nia, Nahid Maleki, Hossein Asgari Alouj, and Azam Ghezelbash. "An Analytical Review of the Effect of Working Capital Development on Financial Performance Measures," no. April 2018 (2012). https://doi.org/10.2139/ssrn.2007821.

Motlicek, Z & Martinovicova, D. 2014. Impact of working capital management on sales of enterprises focusing on the manufacture of machinery and equipment in the czech republic. Acta Universitas Agriculturae et silviculturae mendelianae brunensis. 62(4): 677–84.

Riyanto, B. 2013. Dasar-dasar pembelanjaan Perusahaan Edisi 4. Yogyakarta: *BPFE. 4th ed. Yogyakarta.*

Teruel, P.J.G. & Salono, P.M. 2012. "Effects of Working Capital Management on SME Profitability. *Internasional journal of managerial finance.*

Tjiptono, Fandy. 2012. Pemasaran Strategik. Yogyaka: *Andi Offset.*

Lecturer: rational or irrational

D. Andriani
Universitas Komputer Indonesia, Bandung, Indonesia

N. Nugraha
Universitas Pendidikan Indonesia, Bandung, Indonesia

ABSTRACT: This study aims to examine the relationship of cognitive bias with investment decisions on lecturers. Investment activities can sometimes cause errors or bias in the process of perceiving information relating to investment. Cognitive bias is an error in the thinking of someone who tends to be illogical in understanding something. Lecturers, as educators, must think logically and rationally in determining investment decisions. This will be tested whether lecturers are generally more rational or irrational in determining investment decisions. The sample of respondents was 70 lecturers. Data collection comes from the results of questionnaires. Multiple regression analysis was used to identify 7 cognitive biases that influence investment decisions, namely Overconfidence, Conservatism, Anchoring, Herding, Availability, Representativeness, and Mental Accounting. In addition, this study also identified which bias is the most dominant effect on investment decisions. The results showed that lecturers experience cognitive bias so that it can be said, they act irrationally in determining investment decisions. The insignificant biases in investment decisions are Overconfidence and Representativeness, because these biases have a negative effect, and the dominant cognitive bias is Mental Accounting.

1 INTRODUCTION

The investment decision for each person will be different. Rational thinking investors are principal investors who expect maximum profits with certain risks or certain benefits with minimal risk. In addition, to think rationally, Schwarz & Clore (1996) said that psychological factors also influence investment decisions. The things such as emotions and moods can affect a person's behaviour in making investment decisions.

According to Chandra (2014), decision making for each investor will be different, but in the reality of the decision-making process, investors experience a perception of bias. Biased perception is a psychological tendency of a person to lose the objectivity of perception. In a state of bias, a person will believe in his own ability to evaluate events accurately, as well as an assessment of the situation, even though what actually happens will be affected by bias (there is an emotional element that influences or deviates perceptions) in making decisions to place funds that will be invested.

A lecturer, as an educator, should be more logical and rational in determining investment decisions. In this study, we will examine whether bias perceptions can influence their decisions in investing. The perception bias that will be discussed in this study is overconfidence, conservatism, anchoring, herding, availability, representativeness, and mental accounting.

2 METHODS

The sampling technique used the cluster method and the samples taken were 70 lecturers from several departments at Universitas Komputer Indonesia.

2.1 Concept measurement

The concepts that were measured in this study were overconfidence, conservatism, anchoring, herding, availability, representativeness, and mental accounting and investment decisions. The variable scale used was a Likert scale with 5 scales namely, scale 1 (strongly disagree), scale 2 (disagree), scale 3 (sufficient), scale 4 (agree) and scale 5 (strongly agree). According to Sugiyono (2010), the interval to find out the average yield of each variable can be known by using the formula:

$$I = \frac{H - L}{K} \tag{1}$$

Where: I = Interval, H = High Score, L = Low Score, K = Classification that will be made.

Based on the formula, the interval is:

$$I = \frac{5 - 1}{5} = 0.8 \tag{2}$$

So that the classification can be determined as follows:

Table 1. Range interval.

Range	Category
1,00–1,80	Strongly Disagree
1,81–2,60	Disagree
2,61–3,40	Sufficient
3,41–4,20	Agree
4,21–5,00	Strongly Agree

This Likert Scale, respondents to express approval or disagreement from each of a series of statements regarding the object of stimulation. The interval range is also used to group respondents' answer categories.

2.2 Variables

The questionnaire given to respondents was based on 7 (seven) perceptions of bias, namely overconfidence, conservatism, anchoring, herding, availability, representativeness, and mental accounting as X variables (independent) and investment decisions as Y (dependent) variables.

2.3 The analysis technique

The analysis technique in this study used quantitative analysis techniques. According to Sugiyono, quantitative analysis was carried out using descriptive statistics that presented data in the form of tables and frequency distributions. The analytical tool used in this study was the multiple regression analysis.

Multiple regression was used to identify the influence of each variable that affects investment decisions, namely excessive trust, conservatism, restraint, grazing, availability, representation, and mental accounting. This research model is illustrated in Figure 1.

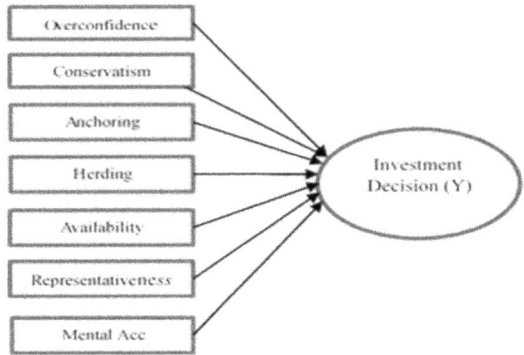

Figure 1. Research model.

Regression equations modeled in this study are as follows:

$$KI = a + b1Oc + b2Cv + b3Ac + b4Hd + b5Av + b6Rp + b7Ma + e \qquad (3)$$

Where:
KI = Investment Decision (Y)
a = Constant
b = Koefisien regresi
Oc = Variable of overconfidence (X1)
Cv = Variable of conservatism (X2)
Ac = Variable of anchoring (X3)
Hd = Variable of herding (X4)
Av = Variable of availability (X5)
Rp = Variable of representativeness (X6)
Ma = Variable of mental accounting (X7)
e = Error term

The analytical tool used to test this regression was the test of validity and reliability, classic assumption test and regression test. According to Singarimbun & Effendi (1989), validity tests are used so that measurements are completely free of systematic errors and random errors, while the reliability test is an index that shows the extent to which a measuring instrument can be trusted or reliable. According to Arikunto (2010) the classic assumption test is used to test the data to meet the Best Linear Unbiased Estimator (BLUE) criteria, so that it can produce valid estimator parameters.

3 RESULTS AND DISCUSSION

3.1 Validity and reliability test results (pilot test) and classical test

Questionnaires for 70 lecturers showed the following results; 7 questions about perceptual bias and 3 questions about investment decisions. Validity test showed valid for all questions > 0.235 and reliability test shows valid with Cronbach's Alpha > 0.700. Classic tests include normality, linearity, multicollinearity, autocorrelation, and heteroscedasticity were carried out and have met the requirements.

3.2 Headings multiple regression analysis

The amount of R square was 0.549, this means that variations in lecturer decision making can be explained by the 7 biases, while the rest (100%–54.9%) were explained by other factors outside the model. From the ANOVA test or F test, the calculated value F and significance (sig <0.05), the regression model was used to predict investment decision making, or it can be said to be biased perceptions together influencing investment decision making. The equation of the multiple regression model is as follow:

$$KI = 4.278 + 0.410 \, Ov - 0.226 \, Cv + 0.446 \, Ac$$
$$-0.230 \, Hd + 0.327 \, Av - 0.06 \, Rp$$
$$+0.476 \, Ma + s \qquad (4)$$

The equation stated that the lecturers were irrational because it was positively influenced by the bias of overconfidence, anchoring, availability and mental accounting in investment decision making.

3.3 Discussion

Lusardi & Mitchell (2006) show that knowledge encourages someone to make financial planning and minimize mistakes in decision making. Lecturers have a high level of education, according to Heath & Tversky (1991) the higher the level of education will increase the level of confidence because the higher education makes an investor feel more competent in almost all fields including finance. That is, lecturers have a tendency to experience overconfidence bias in making investment decisions because they are confident in their thinking. Similarly, anchoring bias in the research of Bhandari & Deaves (2006) shows that it can increase if it has a higher education and this is experienced by investors who graduate.

According to Tversky & Kahneman (1974), individuals rely on comfort based on experience and the amount of information available to make decisions. Availability bias is cognitive bias where people tend to estimate the likelihood of results based on how close or familiar these results occur in our lives. This means that a lecturer will make decisions based on experience or gather available information, then estimate the likelihood that will occur without discussion with an expert person. Dewanti (2018) said that the demographic variables of respondents (age, education, income per year, total wealth, and financial wealth) do not affect mental accounting, but in this study greatly affect investment decisions because the lecturers have more than one account in a bank that is distinguished for routine expenses or investment plans.

4 CONCLUSION

Lecturers experience perceptions of bias in making investment decisions so that it can be said the lecturer acts irrationally in making investment decisions. Variables that are not significant to investment decisions are, conservatism, herding and representativeness variables because these variables negatively influence the decision, meaning that these variables have no effect on investment decisions. Based on the multiple linear regression equation, the lecturers could be said to be irrational because it was positively influenced by the bias of overconfidence, anchoring, availability and mental accounting in investment decision making. The dominant perception was Mental Accounting.

REFERENCES

Arikunto, S. 2010. Prosedur penelitian suatu pendekatan praktik. Jakarta: PT. Rineka Cipta.

Chandra, C. 2014. Persepsi bias investor dalam keputusan investasi pada masyarakat yang berusia produktif di Surabaya. *Jurnal FINESTA* 2: 109–113.

Dewanti, P.W. 2018. Mental accounting dalam proses pengambilan keputusan investasi. *Jurnal Akuntansi* 6(1). Universitas Negeri Yogyakarta.

Heath, C., & Tversky. 1991. Preference and belief ambiguity and competence choice under uncertainty. *Journal of Risk and Uncertainty* 4: 5–28.

Lusardi, A., & Mitchell, O.S. 2006. financial literacy and retairment prepairedness: evidence and implications for financial education. *Business Economics*: 35–44.

Schwarz, N., & Clore, G.L. 1996. Feelings and phenomenal experiences. Dalam E.T. Higgins & A. Kruglanski (Eds.), Social psychology, Handbook of basic principles: 433–465. New York: Guilford.

Singarimbun, M., & Effendi, S. 1989. Metode penelitian survai. Jakarta.

Sugiyono, 2010. Metode penelitian pendidikan pendekatan kuantitatif, kualitatif, dan R&D. 93. Bandung: Alfabeta.

Tversky, A. & Kahneman, D. 1974. Judgment under Uncertainty: Heuristics and Biases. *Science, New Series* 185 (4157): 1124–1131.

Tversky, A., & Kahneman, D. 1981. The framing of decisions and the rationality of choice. *Science* 211: 453–458.

Section 2   Green business

Advances in Business, Management and Entrepreneurship – Hurriyati et al. (Eds)
© 2021 Taylor & Francis Group, London, ISBN 978-0-367-67471-7

Drivers of green product adoption: Green perceived quality, green satisfaction and green trust

A. Widodo & R. Yusiana
Telkom University, Bandung, Indonesia

ABSTRACT: This study discusses green which is perceived to be quality in green trust and the role of mediation through green satisfaction. Research carried out on gunny sack consumers who are creative industries in fashion. Furthermore, this study tries to publish literature on green marketing and marketing upwards. This study uses Smart-PLS 3.2.7 modeling to conduct empirical studies. The results show that the perception of green quality affects the negative green trust. In addition, this study proves that the perceived quality of green is influenced by green satisfaction and green satisfaction that positively influences green trust. Therefore, the perceived quality of green and green satisfaction in increasing green trust greatly helps consumers in buying green products, with support from other consumers that can be a stimulus in the use of green products.

1 INTRODUCTION

Today, in Indonesia the growth of business where business increases every year that is more than just a creative economic sector. According to Statistical Data and Results of the Creative Economy Survey (2017), the Creative Economy Agency (Bekraf) in collaboration with the Central Statistics Agency (BPS) shows that in 2016 there was an increase in the GDP of the creative economy sector. The number of creative economy industries spurs for every industry that exists to be able to compete with competitors in the same industry, one of which is fashion or fashion industry.

According to the results of the 2017 creative economy special survey, West Java contributed 33.56% or US $6.499 billion from Indonesia's creative economy. Fashion manufacturing industries are the biggest contributors to creative economy exports. The number of fashion industry players, making textile waste is currently the second largest waste in the world. This condition creates the concept of green fashion. Green fashion or eco-friendly fashion is often called eco-fashion is not always related to using organic cotton (a material produced without pesticides, herbicides or chemical fertilizers but good for health and the environment) or recycled materials (O'Malley 2018).

Green fashion is not just a fashion trend but more than that. The fashion industry players want to give a different view of fashion that is still fashionable but still pay attention to the existing environmental conditions (Roloff 2018). Therefore, there are currently many campaigns on green marketing to introduce and provide education to the public regarding green fashion. Where the product produced must meet the criteria as an environmentally friendly product. Furthermore, companies now have to function a wastewater treatment plant which is useful to restore the function of a polluted environment where textile companies are still less concerned about the environment.

At present, eco-fashion business aims to reduce fashion waste, protect the environment and the health of users of fashion products. Rumah Karung Goni is an eco-fashion business that maintains a commitment to the environment. This business is included in slow fashion, where the basic ingredients of its products are on Rumah Karung Goni, which are made from plant fibers that are environmentally friendly.

Today's green marketing is the main topic of society in anticipating the destruction of nature where, a marketing of products that are assumed to be safe for the environment (Jain & Kaur 2003, Rahbar & Wahid 2011). Whereas according to Zubair (2014) said that people who care about the environment and then receive green marketing are increasing, this is also supported by scientific changes and lifestyle changes of people who are increasingly concerned with the environment, they consider this will create credibility and attitude positive for the environment.

According to Tjiptono (2014), perceived quality is a consumer perception of the overall quality or superiority of a product. Usually because of the lack of knowledge of the buyer about the attributes or characteristics of the product to be purchased, the buyer estimates the quality from the aspect of price, advertising, company reputation, and the country of manufacture. Perceived quality is "Consumer's opinion of a product's (or a brand's) ability to fulfill his or her expectations. It is the current public image, and is

Figure 1. Framework.

based on the current (or brands) current public image, a consumer experience with the firm's other products, and the opinion leaders, consumer's peer group, and others" (Unknown 2018). Chen & Chang (2013) mention that green perceived quality are consumer ratings on product quality related to environmental aspects. So that it can be interpreted that green perceived quality is a view or perception of consumers to see the quality of a product related to the environment.

According to Kotler & Keller (2016) in general customer satisfaction is a person feeling of pleasure or disappointment resulting from comparing a product received performance (or outcome) in relations to the person's expectation. The existence of green satisfaction on consumers will strengthen consumer relations with existing green companies so as to bring satisfaction when consuming or using a product that is environmentally friendly.

Green Trust, according to Chen in Chen (2013), is the desire of consumers to depend on a product or service based on the beliefs or results of expectations of the product's credibility, strengths, and product capabilities for the environment. When consumer confidence in a product increases, anxiety and uncertainty will decrease, and the integrity of a brand and company will be stronger (Chen & Quester 2015). According to Kim et al. In Chen & Quester (2015), there are several dimensions in building customer trust: (1) Affect-based, related to the impact of out-side third parties influencing consumer perceptions of a business; (2) Experienced-based, overall consumer evaluation of the cumulative interactions of consumer with companies; (3) Cognition-based or observation based, the impact of consumer interaction directly with the seller in consumer perceptions; (4) Personality-oriented, personal characters and consumer shopping habits. Customer satisfaction with environmentally friendly products will lead to the next level, namely customer trust in green products.

2 METHODS

This research uses descriptive causality type with quantitative methods. According to Sugiyono (2014) descriptive research is a study used to determine the value of independent variables, either a variable or more (independent) without making comparisons, or connecting between one with another variable. While the causal relationship according to Sugiyono (2014) is a causal relationship. So here there are independent variables (variables that influence) and dependent (influenced).

Quantitative research methods can be interpreted as research methods that are based on positivist philosophy, used to examine certain populations or samples, sampling techniques are generally done randomly, data collection uses research instruments, data analysis is quantitative/statistical in order to test the hypothesis has been determined (Sugiyono 2014).

In this study, the authors took the consumer population of on Rumah Karung Goni that had purchased and also used a green product from the Goni Sack House. Where the number of consumers of the on Rumah Karung Goni is known through consumer data owned by the on Rumah Karung Goni.

Partial Least Square (PLS), which is one of the variant based SEM (Structural Equation Modeling) statistical methods designed to complete multiple regression when specific data problems occur, such as small research sample sizes, missing values and multicollinearity. Variant-based PLS which can simultaneously test the measurement model while testing structural measurement models (Abdillah & Hartono 2015).

The purpose of PLS is to predict the effect of X on Y and explain the theoretical relationship between two variables. The measurement model is used to test the validity and reliability, while the structural model is used for the causality test (hypothesis testing with prediction models). The formal model defines latent variables as linear aggregates of the indicators. The weight estimate to create a score component for latent variables is based on how the inner model (the structural model that connects between latent variables) and the outer model (the measurement model that is the relationship between indicators and constructs) is specified as the residual variance of the dependent variable.

3 RESULTS AND DISCUSSION

In this study, primary data was collected to find out the Green Perceived Quality of Green Trust through Green Satisfaction on Goni Sack Consumers through distributing questionnaires to 250 respondents. To find out the respondent's personal data, a questionnaire was asked based on the socio-demographic.

Based on the results obtained from questionnaires and screening obtained respondent data as much as 234. Where the results consist of the presence of data in incomplete filling as many as 8 respondents. Furthermore, there are statements from respondents who have a number of replenishments that does not meet criteria such as incomplete respondent data, and incomplete statement of content of 8 respondents.

Researchers used primary data to determine the effect of Green Perceived Quality mediated by Green Satisfaction on Green Trust for consumers of on Rumah Karung Goni products. The following describes the characteristics of respondents based on gender, occupation, and income or income per month, and the last education used.

Table 1.   Profile of the respondents, n = 234.

Category	Description	No. of Respondents	%
Gender	Male	147	63
	Female	87	37
Level of Income (Monthly Income)	<1.000.000 IDR	65	28
	1.000.001–2.500.000 IDR	103	44
	2.000.001–5.000.000 IDR	35	15
	5.000.001–7.000.000 IDR	19	8
	Above 7.000.000 IDR	12	5
Highest Educational Completed Today	Senior High School	47	20
	Diploma	23	10
	Bachelor Degree	161	69
	Masters Degree	3	1
Occupation	Student	155	66
	Civil servant	21	9
	Private employees	33	14
	Entrepreneur	14	6
	Other	11	5

Respondents' data by sex showed that female respondents' sex dominated more than 63%. Furthermore, respondents' characteristics based on occupation were the highest number of respondents with 66% of respondents as students. Then for characteristics based on income or monthly income is more dominated by respondents' income of 1.000.001–2.500.000, IDR as much as 44%.

Furthermore, the characteristics based on recent education were dominated by respondents with final S1 education with 69% of respondents.

This is reinforced by Septifani et al. (2014) who stated that respondents taken in the study were customers who had consumed environmentally friendly products that were young. Based on the results of the study, respondents who consumed more green-products were female customers, with an age group of 21–25 years, and had income < 1.000.000.00, IDR. Women tend to have qualities that are gentle, caring, caring, caring for the environment and easily sympathetic. Students with a 21–25 years age range are classified as young and productive. At that age, many respondents work and socialize with peers and interact with various media, so it is quite easy to receive and understand environmental issues.

Based on the results of the respondents' characteristics in this study, it can be concluded that the majority of respondents who fill are students in terms of knowledge and awareness have begun to grow and care about the types of products consumed are no exception consumers of gunny sacks, where the product is one product environmentally friendly and has a variety of advantages, so that in this study has been explicitly drawn from the respondents and has begun

Table 2.   Measurement model assessment, N = 234.

Latent Variable	Question Item	Indicator Reliability	Convergent Validity	Internal Consistency Reliability	
		Loading Factor (>0.7)	AVE (>0.5)	Composite Reliability (>0.7)	Cronbach's Alpha (>0.7)
Green Perceived Quality	GPQ-1	0.765			
	GPQ-2	0.852			
	GPQ-3	0.736	0.640	0.898	0.859
	GPQ-4	0.798			
	GPQ-5	0.841			
Green Satisfaction	GS-1	0.733			
	GS-2	0.822			
	GS-3	0.755			
	GS-4	0.783	0.619	0.936	0.923
	GS-5	0.772			
	GS-6	0.781			
	GS-7	0.819			
	GS-8	0.802			
Green Trust	GT-1	0.836			
	GT-2	0.716			
	GT-3	0.775			
	GT-4	0.790			
	GT-5	0.793			
	GT-6	0.837			
	GT-7	0.823			
	GT-8	0.817			

to be aware of the environment where, the awareness that arises comes from the environment and social.

In the green trust variable, the R-Square value is 0.967. This means that the green trust variable has an influence of 96.7% and the remaining 3.3% is influenced by other variables outside the research. Indicators of green trust are dominated by product trust, quality and brand. Then the indicators used in this study can be said to only explain 97.7% as a green trust factor and 3.3% explained by other factors out-side of this study.

Based on the value of t-statistic (to) the test results on each hypothesis indicate that the Green Perceived Quality of the Green Trust, shows a value of 1.832> tα value of 1.96; thus, H0 is accepted and H1 is rejected. This means that the Green Perceived Quality variable does not have a positive and significant effect on the Green Trust. Furthermore, the Green Perceived Quality of Green Satisfaction shows a value of 42.008> tα value of 1.96; thus, H0 is rejected and H1 is accepted. This means that Green Perceived Quality Variables have a positive and significant effect on Green Satisfaction. Green Satisfaction with Green Trust shows a value of 42.664> tα value of 1.96; thus, H0 is rejected and H1 is accepted. This means that the Green Satisfaction variable has a positive and significant effect on the Green Trust. Green Perceived Quality of Green Trust mediated by Green Satisfaction shows a value of 30.226> tα value of 1.96; thus, H0 is rejected and H1 is accepted. This means that Green Perceived Quality Variables have a positive and significant effect on the Green Trust variable, which is mediated by Green Satisfaction.

REFERENCES

Abdillah, W. & Hartono, J. 2015. *Partial Least Square (PLS): alternatif Structural Equation Modeling (SEM) dalam penelitian bisnis*. Yogyakarta: Penerbit Andi.

Badan Pusat Statistik Jawa Barat. 2017. *Proyeksi Jumlah Penduduk Provinsi Jawa Barat, 2010–2016*. [Online] Retrieved from https://jabar.bps.go.id/Subjek/view/id/12#subjekViewTab3|accordion-daftar-subjek1.

Chen, S.C. & Quester, P.G. 2015. The relative contribution of love and trust towards customer loyalty. *Australasian Marketing Journal (AMJ)* 23(1): 13–18.

Chen, Y.S. 2013. Towards green loyalty:driving from green perceived value, green satisfaction, and green trust. *Sustainable Development* 21(5): 294–308.

Chen, Y.S. & Chang, C.H. 2013. Greenwash and green trust: The mediation effect of green consumer confusion and green perceived risk. *Journal of Business Ethics* 114(3): 489–500.

Jain, S.K. & Kaur, G. 2003. Strategic green marketing: How should business firms go about adopting it? *The Indian Journal of Commerce* 55(4): 1–16.

Kotler, P. & Keller, K.L. 2016. *Marketing Management*. New Jersey: Pearson Pretice Hall.

O'Malley, M. 2018. *Maker monday featuring thread ethic*. [Online]. Retrieved from implififabric.com/blogs/blog/maker-monday-featuring-threadethic?_pos=1&_sid=cb430c305&_ss=r.

Rahbar, E. & Wahid, N.A. 2011. Investigation of green marketing tools' effect on consumers' purchase behavior. *Business Strategy Series* 12(2): 73–83.

Roloff, L.Y. 2018. *How to fashion a better world*. [Online]. Retrieved from https://www.greenpeace.org/international/story/17710/how-to-fashion-a-better-world/.

Septifani, R., Achmadi, F. & Santoso, I. 2014. Pegaruh green marketing, pengetahuan dan minat membeli terhadap keputusan pembelian. *Jurnal Manajemen Teknologi* 13(2): 201–218.

Sugiyono, P.D. 2014. *Metode Penelitian Kuantitatif kualitatif dan R&D*. Bandung: Alfabeta.

Tjiptono, F. 2014. *Pemasaran Jasa*. Jakarta: Gramedia.

Unknown, n.d. 2018. *Perceived quality*. [Online]. Retrieved from http://www.businessdictionary.com/definition/perceived-quality.html.

Zubair, T.M. 2014. Impact of green advertisement and green brand awareness on green satisfaction with mediating effect of buying behavior. *Journal of Managerial Sciences* 8(2).

Advances in Business, Management and Entrepreneurship – Hurriyati et al. (Eds)
© 2021 Taylor & Francis Group, London, ISBN 978-0-367-67471-7

Drivers of growing green purchase behavior: Role of green perceived value and green trust

R. Yusiana & A. Widodo
Telkom University, Bandung, Indonesia

ABSTRACT: The restaurant business is growing rapidly supported by numerous people who like to shop and eat outside their homes. A large community of nature lovers is very enthusiastic about the emergence of restaurants that provide environmentally friendly products. Public awareness to help protect the environment is increasing because of environmental damage that is currently happening, such as global warming and the greenhouse gas effect. Kehidupan Tidak Pernah Berakhir (KTPB) is a vegan restaurant in Bandung, which provides education about green products and healthy lifestyles. This study summarized the literature on green marketing and marketing relations into the research framework at the green purchase behavior, using two constructs: green perceived value and green trust. This study used a quantitative method with descriptive research on 100 consumers of KTPB and was with Partial Least Square (PLS), Smart PLS 3.2.7. Empirical results showed that green perceived value positively affected green purchase behavior and green trust. This study indicated that the relationship between the green purchase behavior and antecedent (green perceived value) was mediated by green trust. Therefore, investing resources to increase green perceived value and green trust is very helpful to increase the green purchase behavior.

1 INTRODUCTION

In this globalization era, competition among companies is getting tighter. They compete in planning and organizing marketing activities by displaying superiority that differentiates the products offered by their company with competitors. Companies also understand consumer behavior by trying to meet the needs and desires of consumers for survival and the company's main goal is to get profits.

The current global warming is becoming a serious conversation. It is changing the climate on life on earth. The occurrence of melting glaciers, rising sea levels, deforested forests, and endangered wildlife are the result of global warming. This happens because of the lack of human awareness to protect the environment, deforestation, excessive fuel consumption, waste incineration, and many other causes.

The existence of these environmental issues sparked the concept of green marketing, which deals with environmentally friendly concepts that are not harmful to the environment. Green marketing refers to eco products, phosphate free, recyclable, rechargeable, safe for the ozone layer, and environmentally friendly. One of the lifestyles in the world is vegan. Vegetarian lifestyle is relevant to these environmental issues. This is because activities in the agricultural sector contribute 9 percent to carbon dioxide, 65 percent nitrous oxide and 37 percent methane gas. Nitrous oxide gas

is sent by manure, 296 times more likely to cause greenhouse gases than CO2. This description does not include its contribution to soil and water pollution. Research by Gidon Eshel and Pamela Martin of the University of Chicago concluded changing animal diet to vegetarianism was 50 percent more effective in preventing global warming than replacing SUVs with hybrid cars.

Today a vegan restaurant is developing and there are around 400 vegetarian restaurants in Indonesia. Vegan restaurants usually use local, organic, and / or plants grown in their own garden. One of them is Restoran Kehidupan Tidak Pernah Berakhir (KTPB). This restaurant only provides foods that come from various kinds of plants such as vegetables, rice, tubers, and beans.

A vegan or similar restaurant that is developing today is an alternative healthy life choice for all people who care about the environment. It aims to not add to the effects of global warming by consuming green products without animal preparations. This restaurant also seeks to encourage consumers to protect the environment by promoting information on vegan benefits and the importance of consuming green products.

Chen & Chang (2012) research showed that Green Perceived Value will positively influence Green Trust and Green Purchase Intentions, while Green Perceived Risk had a negative impact on both. Chen & Chang research also showed the relationship between Green

Purchase Intentions and their antecedents: Green Perceived Value and Green Perceived Risk, partly mediated by Green Trusts.

The purpose of this study was to determine the effect of Green Perceived Value and Green Trust on Green Purchase Behavior of KTPB consumers. This research framework can help companies to improve Green Purchase Behavior by their determinants: Green Perceived Value and Green Trust.

2 LITERATURE REVIEW

According to Dahlstrom (2011), Green marketing is the process of planning and implementing a marketing mix to facilitate consumption, production, distribution, promotion, packaging, and reclamation of products in a way that is sensitive or responsive to ecological interests. The green product is divided into 3 dimensions:

1. Products that can be recycled
2. Durable products
3. Products that are guaranteed to be safe for the environment.

Green perceived value is an overall assessment by consumers of all benefits received and what is sacrificed is based on the desire of the environment with the expectation of sustainability for all the needs of the green (Chen & Chang 2012).

According to Chen & Chang (2012), Green trust is the willingness to use some reliable specific services or brands that are believed to have a positive impact. The following are the indicators:

1. Organic claim,
   Organic claim is the consumer's belief in an organic product based on its environmentally friendly recognition. Organic products are products that do not contain harmful substances, such as poisons or chemicals and provide benefits that are good for human health. Besides being good for human health, organic products also have a positive impact on the environment.
2. Reputation
   Reputation is the consumer's belief in the reputation of organic products. Someone will consume products in a place that has a good reputation for something that has been known to people, to satisfy their wants and needs.
3. Environmental performance
   Environmental performance is the consumer's belief in the performance of organic products on the environment. Such materials used can be recycled or use recycled materials, so they can reduce waste that is not useful and will help maintain the environment.
4. Environmental commitments
   Environmental commitments are consumer beliefs about the commitment of organic products to environmental protection. Someone has faith that by consuming organic products on a regular basis can generate a positive impact on preserving the environment.

According to Xu (2013) Green purchase behavior can be interpreted into an act of consuming the products conservable, beneficial for the environment and responding to environmental concerns.

According to Lee (2008) factors that contribute to influencing the green purchasing behavior are:

1. Social influence
2. Environmental attitude
3. Environmental concern
4. Perceived environmental responsibility
5. Perceived Seriousness of Environmental Problems
6. Perceived Effectiveness of Environmental behavior
7. Concern of Self-image in Environmental Protection

3 METHODS

This research used quantitative methods. Based on the philosophy of positivism, quantitative research used to examine a particular population or sample, data collection using research instruments, quantitative/statistical data analysis, with the aim to test the hypotheses that have been determined (Sugiyono 2015).

This research used Green Perceived Value as independent variable and Green Purchase Behavior as dependent variable and Green Trust as mediation variable. The sample used was 100 which consisted of the consumers of Kehidupan Tidak Pernah Berakhir restaurant in West Java.

The data used in this study were primary data gained from distributing questionnaires and secondary data were books, literature, journals, scientific papers or previous research. The analysis used in this study was Partial Least Square (PLS) with Smart PLS 3.2.7. Partial Least Square (PLS), which is one of the variant based SEM (Structural Equation Modeling) statistical methods designed to complete multiple regression when specific data problems occur, such as small research sample sizes, missing values and multicollinearity. Variant-based PLS which can simultaneously test the measurement model while testing structural measurement models (Abdillah & Hartono 2015).

The research study was conducted by convenience sampling method where questionnaires were distributed using google forms and offline technology. The questionnaire used was developed using previous research and designed according to the related research variables for this study. As an assessment, respondents was asked questions to indicate the level of suitability, with each statement using 5 Likert Scale points.

4 RESULTS AND DISCUSSION

Based on the distribution of questionnaires, the characteristics of respondents as shown in Table 1.

The R-square PLS model can be evaluated by looking at predictive Q-square relevance for the variable model. The Q-square measures how well the observations produced by the model and also the parameter estimates. Q-square value was greater than 0 (zero), meaning that the model had a predictive value of relevance, while the Q-square value was less than 0 (zero) meaning that the model lacked predictive relevance. The data of total effects can be seen in Table 2, and data of R-square can be seen in Table 3.

The effect of green perceived value on the green purchase behavior mediated by green trust was not significant because the p value was above 0.5, meaning that consumers ignored green trust, because consumers already had a perceived value that caused positive behavior in purchasing. This assumption was proven by the amount of the total effects of 0.777. This result was reinforced by research from Lobo & Greenland (2017), which stated that the greater value on the use of environmentally friendly products tend to resulted in green purchasing behavior, this occurred because of positive environmental attitudes and the existence of a relationship between the determinants of purchasing decisions (such as perceived benefits).

The test results using PLS Algorithm in the Smart-PLS 3.2.7 Program in this study as shown in Figure 1.

Based on the bootstrapping in Figure 1 it can be seen that T-count of each relationship between variables was greater than the T-Table at 1.96, suggesting that the relationship or the influence of each variable

was significant at 5%. The data of total indirect effects can be seen in Table 4, and data of Cronbach's alpha and composite reliability can be seen in Table 5.

From the results of the PLS analysis, it can be seen that there were positive influences on the hypotheses H3. This was supported by the value of t analysis which was greater than the value of t table (1.96). In addition, it can also be seen that the value of R Square of green purchase behavior was 60%. The R square value shows how much the dependent variable explains the independent variable, a high R square value also indicates that overall, almost all variables can be explained in this theory or construct so that the possibility of the influence of other variables is quite small. Green trust value was 2%, which meant consumers did not have confidence in the green products when making in making purchases, but these consumers already had a perceived value for green products so that they directly

Table 3. R-Square.

Variabel	R Square	R Square Adjusted
Green Purchase Behavior	0.608	0.600
Green Trust	0.025	0.015

Table 1. Profile of the respondents, n = 100.

Characteristics of Respondents	Information	%
Gender	Female	60
Age	24–34 Years	38
Current Education	Bachelor	51
Occupation	Private Employees	43
Earnings/month	5.000.000–10.500.000 IDR	44
Purchasing intensity	More than 1 time	60
Domicile	Bandung	30

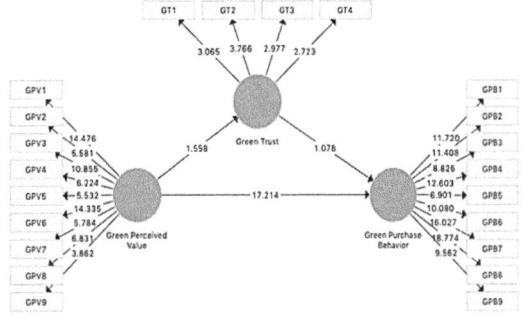

Figure 1. Output Bootstrapping.

Table 2. Total effects.

Mean, STDEV, T-Values, P-Values

	Original Sample	Sample Mean	Standard Deviation (STDEV)	T Statistics	P Values
Green Perceived Value → Green Purchase Behavior	0.777	0.789	0.041	19.029	0.000
Green Purchase Behavior → Green Trust	0.157	0.205	0.101	1.558	0.120
Green Trust → Green Purchase Behavior	0.065	0.067	0.060	1.078	0.281

Table 4. Total indirect effects.

Mean, STDEV, T-Values, P-Values

	Original Sample	Sample Mean	Standard Deviation (STDEV)	T Statistics	P Values
Green Perceived Value Green Purchase Behavior Green Trust	0.010	0.014	0.014	0.738	0.461

Table 5. Cronbach's alpha and composite reliability.

Construct	Cronbach's alpha	Composite Reliability
Green Perceived Value	0.812	0.855
Green Purchase Behavior	0.876	0.900
Green Trust	0.623	0.779

lead to the purchase behavior in vegan or organic restaurants.

This is in line with research from Chen & Chang (2012), companies must increase green perceived value and green trust to increase buying interest in green products. Marketers can develop marketing strategies by increasing green perceived value and helping potential customers develop green trust to build long-term relationships in the environmental era.

5 CONCLUSION

Based on the results of research that has been done, can be drawn conclusion as follows:

1. Effect of Green Perceived Value on Green Purchase Behavior based on the results of data processing and correlation between variables showed t of $17.214 > t\alpha$ value of 1.96 which meant that the Green Perceived Value variable had a positive and significant effect on Green Purchase Behavior
2. Effect of Green Perceived Value on Green Trust based on the results of data processing and correlation between variables showed t equal to $1.558 > t\alpha$ value of 1.96 which meant that the Green Perceived Value variable had a positive effect, but not significant to the Green Trust.

3. Effect of Green Trust on Green Purchase Behavior based on the results of data processing and correlation between variables showed t of $1.078 > t\alpha$ value of 1.96 which meant Green Trust Variables had a positive effect, but not significant to Green Purchase Behavior.
4. Effect of Green Perceived Value on Green Purchase Behavior mediated by Green Trust based on the results of data processing and correlation between variables showed t equal to $19,029 > t\alpha$ value of 1,96 which meant that the Green Perceived Value variable had a positive and significant effect on the Green Purchase Behavior variable mediated by Green Trust.

REFERENCES

Abdillah, W. & Hartono, J. 2015. Partial Least Square (PLS): alternatif structural equation modeling (SEM) dalam penelitian bisnis. Yogyakarta: Penerbit Andi, 22, 103–150.
Chen, Y.S. & Chang, C.H. 2012. Enhance green purchase intentions: the roles of green perceived value, green perceived risk, and green trust. *Management Decision*. 50 (3), pp. 502–520.
Dahlstrom, R. 2011. Green Marketing Theory. *Practice and Strategies, Cengage Learning*. New Delhi, 163–164.
Lee, K. 2008. Opportunities for green marketing: young consumers. *Marketing Intelligence & Planning*, 26 (6), pp. 573–586.
Lobo, A. & Greenland, S. 2017. The influence of cultural values on green purchase behavior. *Marketing Intelligence & Planning*. Vol. 35 Issue. 3, pp. 377–396, https://doi.org/10.1108/MIP-08-2016-0131.
Sugiyono. 2015. Metode Penelitian Bisnis. Bandung: Alfabeta.
Xu, Y. 2013. The research analysis of the green label's impact on the consumer purchase behavior. *International Business Management*.

Ecotourism of Sukorejo Liberica Coffee Village as a CSR strategy of petroChina Jabung

P.M. Agustini, E.J. Mihardja & T. Widiastuti
Universitas Bakrie, Jakarta, Indonesia

ABSTRACT: SKK Migas and PetroChina Jabung have supported the development of Village Liberica Coffee Sukorejo Ecotourism through CSR funds. Sukorejo Ecotourism is a pilot project based on the coffee farmer community in Betara, Tanjung Jabung Barat. The main based of this research in SWOT Analysis. The method used in this study was in-depth interviews and FGD conducted in local society and other stakeholder groups. The results of this study indicated that (1) ecotourism was formed because of the tradition of coffee gardening and commodities experiencing marketing problems (2) the main factors driving the development of ecotourism was the strong desire of the coffee farmers group, young people who wanted economic improvement and increased the fame of the coffee region (3) the main obstacle in forming ecotourism was human resources capabilities that still needed improvement especially the skills to provide superior quality tourism services for tourists and activity skill of social media. This study provides theoretical and practical implications in the field of CSR, specifically regarding CSR strategies for agricultural economy on peat lands.

1 INTRODUCTION

Currently, CSR has meaning associated with strategy in many aspects and became a national issue. In its implementation, CSR requires a variety of strategies to suit the expectations of many parties, which cover its scope, form of activities, and the methods it uses. The diversity of CSR beneficiary groups is also one of the factors that determine CSR programs.

CSR affects the image of the company. Based on the results of research conducted by Bin Cha Jae & Mi Na Jo (2019) economic, policy and legal responsibilities significantly influenced the company's image. Economic responsibility must be met before other social responsibilities.

In the oil and gas industry operating in Indonesia, CSR is an obligation that must be fulfilled. Oil and gas companies in India also carry out CSR activities (Rueth 2019). The activities and impacts of CSR activities have been reported on company folders.

CSR in Indonesia is not only in the form of philanthropy but also as a value creation. CSR as a value creation is based on the odds that corporate success and social welfare are independent (Fontaine 2013). Like Fontaine's research, the approach to increase interest in CSR is created share value or CSV.

The landscape approach is carried out by company that carry out Corporate Social Responsibility activities in the Tanjung Jabung Barat area as the main principle in contributing to land use and empowerment. This principle is an approach that

ensures the synergy between various land uses by various parties or stakeholders that complement and do not overlap and pay attention to various ecological, environmental, socio-cultural factors and also ensure that the economic goals of each stakeholder can be achieved.

The people of West Tanjung Jabung Regency have been cultivating coffee for generations around 70 years ago. Coffee is grown on peat land with high acidic soil conditions. At present, the development of liberica coffee is purely carried out by 6 groups of coffee farmers in the West Tanjung Jabung Regency. Farmers began to work together in an effort to improve the quality of coffee and think about ways to protect public Liberica coffee products. Current community constraints, marketing for the type of Liberica coffee is still difficult.

SKK MIGAS supports the development of the liberica coffee ecotourism pilot based on the coffee farmer community in Mekar Jaya Village, Betara District, West Tanjung Jabung, Jambi, which is in the upstream miga industrial work area of Jabung PetroChina International Ltd. Community-based liberica coffee ecotourism pilot development is funded through the Corporate Social Responsibility (CSR) fund PetroChina Jabung, as a company that engaged in the upstream oil and gas industry, located in the block of Jabung.

In the framework of developing the Liberica coffee ecotourism pilot, the PetroChina International Jabung Ltd first synchronized and committed with the local government and the community of coffee farmers in the village.

Based on the description of the background of the problem, this paper explains: (1) how does Sukorejo Ecotourism develop? (2) what are the factors that drive the development of Ecotourism? (3) what are the obstacles to develop Ecotourism?.

2 LITERATURE REVIEW

2.1 Strategy of Corporate Social Responsibility (CSR)

CSR is the company's commitment to act ethically, operate legally, and contribute to improve the economy coupled with increasing quality of life, both employees and their families, the local community, and society at large. The World Business Council for Sustainable Development (1999) defines CSR as the continuing commitment by business to behave in an ethical and contextual way to economic development while improving the quality of life of the workforce and at large. The priorities of CSR activities according to this institution include: human rights, employee rights, environmental protection, supplier relations, community involvement, stakeholder rights, CSR performance monitoring and assessment. CSR means that a corporation should be held accountable for any of its actions that affect people, their communities, and their environment (Lawrence & Weber 2005).

CSR through the community development and environmental preservation program has strategic significance, both viewed in macro terms in relation to regional development and development in Indonesia, as well as in the micro effort in realizing togetherness and social cultural legitimacy, especially in the working area of the Cooperation Contract Contractor Oil and gas. That is, the seriousness of the company to play a role in community empowerment, environmental preservation and regional development in Indonesia is an important part in linking the development chain in Indonesia.

In implementing CSR, the right strategy is needed, so that it is in accordance with the company's vision and mission in the CSR field. Strategy is an overall approach that is related to planning and executing an activity carried out in a certain period of time. In order for the implementation of CSR to run according to ideas and its basic concepts, it requires extra ideas and strategies. The extra strategy includes four main agendas, namely: guidelines and codes of conduct, corporate management systems and policies, leadership strategies in CSR, and commitment and partnerships among stakeholders. Strategy is the ability to see the direction to be addressed, and to do the things needed to stay on track and achieve the intended goals (Watson 2013: 18) CSR strategies are realized in various CSR programs. In order to achieve optimum results, it requires synergies with all stakeholders, including local governments, in implementing CSR. It is very important to eliminate the assumption as if the company gets an additional "burden" on economic social and environmental issues that should be a shared responsibility. Ideal conditions are that each stakeholder has a different role, and in turn the company will carry out its business activities smoothly in accordance with existing rules while taking into account the conditions of its partners. While the government and the community also provide support and benefit more widely both social, economic and environmental aspects.

2.2 Ecotourism

Ecotourism can be broadly defined as nature-based tourism that does not result in the negative environmental, economic, and social impacts that are associated with mass tourism (Duffy 2002) Ecotourism is promoted as a form of travel that brings only benefits to the host societies, because ecotourism are thought to be culturally aware of "ethical travelers" who are keen to reduce negative impacts on environment.

Ecotourism is a form of travel to natural areas that is carried out with the aim of conserving the environment and preserving the lives and welfare of the local population. At present, ecotourism has developed. This tour is not just for observing birds, riding horses, tracing the tracks in the wilderness, but has been related to the concept of forest preservation and local residents. Ecotourism is then a mix of growing interests from environmental, economic, and social concerns. Ecotourism cannot be separated from conservation. Therefore, ecotourism is referred to as a form of responsible travel. This is in line with what was stated in The International Ecotourism Society (2015), ecotourism is a journey that is responsible for natural areas that spread the environment, sustain the welfare of the local community, involving interpretation and environmental education.

Understanding of ecotourism has evolved over time. However, in essence, the notion of ecotourism is a form of tourism that is responsible for the preservation of natural arenas, benefits economically, and maintains cultural integrity for the local community. On the basis of this understanding, the form of ecotourism is basically a form of conservation movement carried out by the world's population.

Ecotourism developments also imply smaller and more specialist forms of accommodation. Effect of mass tourism that the build of accommodation for tourists has on the environment. Eco tourist accommodation is intended to be sited in wilderness areas, and they tend to be smaller and locally owned, providing local employment and being spread over a wide area rather than clusters in one resort by Moscardo (Duffy 2002). The difficulties associated with conventional tourism definitely apply to ecotourism because it form one part of a global tourism business and also one of the asset growing sector of the tourism industry from Cater (Duffy 2002). However, ecotourism is a strategy of development.

2.3 SWOT analysis

The initial formulation of the strategy is done by analyzing the situation. Situation analysis requires company leaders to think strategically to find compatibility between external opportunities and internal forces, besides paying attention to external threats and internal weakness. SWOT is an acronym for company's Strengths, Weakness, Opportunities, and Threats (Hunger & Wheelen 2009). So, the SWOT analysis is used to identify the company's distinctive competence, namely the specific expertise and sources of excellence that the company has to take on various opportunities, with a level of risk that can be overcome.

3 METHODS

This research was conducted using a case study method. Case study (Yin 2014) is used as a comprehensive explanation relating to various aspects of a person, a group, an organization, a program, or a community situation that is examined, pursued, and explored as deeply as possible. Data collection techniques were carried out through in-depth interviews and focus group discussion, conducted in local society and other stakeholder groups such as community leaders, communities, coffee farmer group, local government, CSR officer, and management consultant ecotourism. The respondents were chosen because they can provide information and relating to CSR strategies and perceptions of program beneficiaries. Data were analyzed with interactive models from Miles and Huberman.

4 RESULTS AND DISCUSSION

4.1 Pilot program of sukorejo liberica coffee ecotourism

One program that has been carried out by the company in West Tanjung Jabung is the Coffee Development Program Liberica Tungkal Komposit.

The selection of the Liberica Coffee Village development program was based on a simple SWOT analysis (Wheelen & Hunger 2012) include:

1. Strength: (a) The company's management commitment is related to the contribution of the institution in the form of a CSR program as stated in existing policies, (b) The pillars of CSR have been formulated as a reference in the allocation and focus of CSR programs, (c) oil and gas companies have a high awareness of the social sphere to the public's, (d) availability of resources and funding related to CSR programs, and (e) the existence of liberica coffee farmer empowerment programs and other CSR programs that are running well.
2. Weakness: (a) Limited human resources both in terms of number and skills in managing CSR programs, (b) The communication strategy and program dissemination have not been optimal, (c) Communication with partners is not continuous,

and (d) The unavailability of instruments for measuring the success of CSR activities.
3. Opportunity: (a) The existence of program's partners has a good reputation so far, (b) Economic, educational, and environmental issues that are the main problems in the community are opportunities for companies to work on , (c) The rapid progress of digital media can be utilized in managing the program and its communication aspects, (d) Government policies and programs that support people's economy and environmental preservation are opportunities for CSR programs that can be initiated by the company, and (e) Open partnership with universities consisting of younger generation.
4. Threat: (a) Public perception that has not fully enlightened and understood the existence of the company along with its roles and functions, (b) There are high expectations for the existence of a CSR program, both related to the nature of the program that is sustainable, its fields and aspects of its management, and (c) There is a perception of programs that have got charity and grant. This is a challenge in designing sustainable programs.

Liberica Coffee Development in West Tanjung Jabung is carried out by conducting training and coaching, as well as bringing coffee farmers together with downstream industries such as coffee shop owners and coffee lover communities. In this way, communication between the Liberica Coffee farmers and the owners of outlets and the coffee lover community is more intimate. In addition, the Liberica Coffee farmers gain insight into the quality requirements that must be met and the technical aspects of marketing to be marketed in modern coffee outlets, such as by introducing Liberica Coffee to the coffee lover community directly with the liberica coffee taste test session. Thus, coffee farmers get advice and input for the development of Liberica Coffee in its production aspects.

Liberica coffee development carried out in Tanjung Jabung Barat requires an adoption process to be developed with appropriate standards in business model. This is in line with research conducted by Ingenbleek & Reinder (2012) that the creation of a sustainable coffee market was determined by retailers and coffee roasters.

The development of Liberica coffee in West Tanjung Jabung, in addition to aspects of cultivation and marketing, also continued with the eco-development program. Ecotourism development was carried out in Mekar Jaya Village, Betara District, West Tanjung Jabung, Jambi under the name Liberica Sukorejo Coffee Village Ecotourism (see Picture 1).

Village Liberica Coffee Sukorejo that was developed in Betara, Tanjung Jabung Barat (Figure 1), included the following spots: MSME outlets, Bonsai, Coffee Garden, Liberica Coffee Nursery, Paristo Liberica Coffee, Tourist Attractions, Natural Twin Civet Coffee, Poktan Sri Utomo III, Sido Poktan Appears, UKM Bolu and Brownies, UKM Sale, UKM Taro Chips, UKM Sweet Potato Chips, UKM Tempe

Chips, Mama Banana Chips, Painting, and Tempe Chips.

4.2 The factors that drive the development of sukorejo village coffee liberica.

Sukorejo Ecotourism was formed because of the tradition of coffee gardening and commodities experiencing marketing problem. The tradition of coffee gardening began before 1980. It began with conventional gardening to meet household needs, to finally be traded. However, Liberica coffee has marketing problems now.

4.2.1 The tradition of liberica coffee gardening
Liberica coffee plants can grow well in land such as in Kuala Tungkal which is a lowland and a type of peat soil with high acid levels. Based on the research of Musdalifah (2017) Unja said that Liberica coffee plantations experienced several development phases, namely:

a. Phase before 1980, when Liberica coffee was first planted, where plantations were still on a small scale.
b. Phase of the 1980s, where this year was the beginning of the cultivation of Liberica coffee which was cultivated on its own with a garden system and began to become a commodity traded.
c. At present, Liberica coffee plantations have brought prosperity to the people of Kuala Tungkal, especially the central areas for coffee cultivation, namely Betara, Bram Itam, Pengabun, and Senyerang. There are at least 2700 ha to the Liberica coffee bun in Betara District which is managed by 16 farmer groups, with production of coppers reaching 1277 tons per year (Plantation Office Tanjung jabung Barat 2016).

Coffee plants belonging to the community are endangered, when it was found about 40% of the plants contracted white fungus attacks. This disease causes the leaves of the coffee plants to dry out and fall out, the plants also no longer produce fruit even though they are already in their productive age, i.e 4-6 years. This PetroChina Jabung CSR gives attention by organizing field schools with the development of biological agency laboratories in the farmer group in the form of trichoderma spp. Trichoderma spp is an antagonistic fungus that has the ability to suppress the development and spread of white root fungus (rigidoporus lignosus).

Nowadays it has also been developed in Indonesia and in the world which is made from coffee called cascara tea. This tea can be developed in areas that use liberica coffee peels.

4.2.2 Liberica coffee commodities experiencing marketing problem.
High expectations from the existence of Liberica coffee are not in line with the expected economic benefits.

The price of Liberica coffee is relatively low, so the motivation of farmers to maintain this commodity is low. The length of the trading system chain is considered to be a contributing factor. The price of freshly picked coffee beans ranges from Rp. 13,000 to Rp. 16,000/ kg. While those who have already set up a coffee bean, range from Rp. 30.000 to Rp. 34,000/ kg. This low price causes the coffee plant to be converted into areca palm and palm oil, because it is considered to have higher economic value.

Based on the condition of Liberica's coffee commodities and its lack of marketing, PetroChina Jabung, through its CSR program, is committed to develop sustainable economic -based local economic agriculture, including Liberica coffee. Liberica coffee not only needs to be preserved but must provide added value to the welfare of farmers.

To overcome the problem of Liberica coffee marketing, PetroChina Jabung CSR conducts cultivation programs and group strengthening programs. The Liberica coffee cultivation program is carried out by building field schools and biological agency laboratories, while group strengthening programs are carried out by forming Sido Muncul farmer groups, in the village of Mekar Jaya, Betara. This farmer group consists of 46 people, each of which has a partner of 2 ha of land planted with Liberika coffee which is aged 6-8 years. The institutional arrangement is done by assisting Sido Muncul Cooperative. The hope is that with this program, Liberica coffee can be produced with premium quality, so that it can enter the export market.

4.3 Factors supporting development of sukorejo liberica coffee village

The main factors driving the development of ecotourism is the strong desire of the coffee farmers group, young people who want economic improvement and increase the fame of the coffee region. Generally, Liberica coffee farmers have about 2 hectares of land. The younger generation tends to be active in organization or forum and they want to make Liberica Coffee more widely known to the public.

4.3.1 The strong desire of the liberica coffee farmers group.
Liberica Ecotourism is expected to provide benefits as a place of education for the community regarding the nursery process to the extent of Liberica coffee marketing. It is expected to improve the welfare of the community and also maintain the conservation value of the ecosystem of the coffee area.

In Jambi, liberica coffee producers are concentrated in the Tanjung Jabung area, where more than 50% of this area is peat land with a height of 0-100 above sea level (mdpl).

The liberica ecotourism pilot development program is in line with the conditions of the community, especially in the Mekar Jaya Village, where the community is generally living, namely 78% as garden farmers,

64% of households have Liberica coffee farms with an average area of 2 ha. In the Mekar Jaya region, generally the types of Liberica coffee planted are around 8-10 years old, so community-based Liberica ecotourism programs are considered very potential.

These coffee farmers groups are willing to become beneficiaries of the cultivation program and institutional strengthening carried out by PetroChina Jabung. The farmers were very enthusiastic in following the entire range of training and mentoring program processes. The support of the farmers' wives is very strong, where they were involved in the shared learning process created in this program. The farmers were also eager to take part in the program to increase coffee ore processing, so the coffee produced can be categorized in the premium class.

4.3.2 Role of the young people

Liberica coffee is not as famous as Arabica and Robusta. The results of the cupping test for the coffee taste that was assisted by PetroChina Jabung showed encouraging results, namely fulfilling grade 8 with the title of excellence. This provides a trigger for farmers to be able to produce Liberica premium quality coffee, and make the younger generation want to help market and introduce Liberica coffee. The young generation wants Liberica coffee and the Betara region to become famous. One of them took the initiative to build a store to help market products in Jabung area, especially Liberica coffee. Development is not only related to liberica coffee commodities, but also other aspects that can synergize with the tourism industry, such as Jambi Creative Program. In the Jambi Creative program, 15 MSMEs have been able to support Liberica coffee ecotourism.

PetroChina Jabung has also helped to open outlets run MSME in Betara younger generation, in which one of the activities is to manage coffee shop. In addition, this MSME outlet becomes the entry point for entering the Liberica Sukorejo Coffee Tourism Area. The young generation has a role in the spots of Ecotourism, whose purpose is to give a positive impression to tourists visiting Liberica Coffee Village Ecotourism.

4.4 Factors that constrain development of village liberica coffee Sukerejo

4.4.1 Need of improvement to reach superior quality tourism services

The program for developing ecotourism Libericacoffee faces various obstacles. The main obstacle is human resources capabilities that still need superior quality tourism services for tourists. The main challenge of ecotourism is to provide a memorable experience for tourists. Therefore, every spot visited by tourists must provide superior service. The lack of tourism management skills can cause tourists to be reluctant to come back. Positive tourism experiences are an important aspect of developing the next Sukorejo ecotourism.

4.4.2 The low use of social media to disseminate sukorejo ecotourism and liberica coffee information.

Liberica Sukorejo ecotourism coffee is still rarely found on social media or in conventional media. The familiarity of a new tourist destination is influenced by the skills of the surrounding community to disseminate information about tourist destinations. The lack of instagramable spots is an obstacle to being able to produce interesting spots that can increase tourists desire to visit Sukorejo. Therefore, training on the use of social media for managers and communities is needed.

5 CONCLUSION

(1) Liberica Coffee Sukorejo Ecotourism Village is a CSR Strategy of PetroCnina Jabung. The ecotourism development strategy is based on the results of the SWOT analysis and other studies. The spirit of environmental preservation and improvement of economy of the Tanjung Jabung Barat community were important.
(3) Ecotourism was formed because of the tradition of coffee gardening and commodities experiencing marketing problems.
(3) The main factor driving the development of ecotourism was the strong desire of the coffee farmers group, young people who want economic improvement and increase the fame of the coffee region
(4) The main obstacle in forming ecotourism was human resources capabilities that still need improvement especially the skills to provide superior quality tourism services for tourists and activity skill of social media.

ACKNOWLEDGEMENT

The study was funded by Ministry of Research and Universities through the Higher Education Primary Research Scheme in 2019.

REFERENCES

Bin Cha Jae & Mi Na Jo.2019. The effect of the corporate social responsibility of franchise coffee shops on corporate image and behavioral intention. *Journal sustainability*, page 2-16.DPI:10.3390/sl 1236849.

Duffy, Roseleen. 2002. A trip too far: ecoturism, politics and exploitation. *Earthscan publications ltd.*

Fontaine, Michael.2013. Corporate social responsibility and sustainability: the new bottom line? *International journal of business and social science*, Vol 4, No.4, April 2013, page. 110.119.

Hunger, J. David & Thomas L. Wheelen.2009. *Strategic management and business policy*, 12 th edition. Prentice hall.

Ingenbleek, Paul T.M-Machiel J. Reinders . 2012. The development a market for sustainable coffee in the netherlands; rethinking the contribution of fair trade. *Journal business ethics* (2013) 113:461-474.DOI 10.1007/s1 0551-012-1316-4.

Lawrence, A.T & J. Weber. 2005. *Business and society: stakeholder, ethics, public policy*. Mc graw hill.

Musdalifah. 2017. Perkebunan kopi liberika rakyat kuala tungkal 1980-2015. Universitas Jambi.

Rueth, Rene.2019. Business and management: csr, sustainability, ethics & governance-csr in india between tradition, cultural, influence, social structure, and economic growth: a status quo analysis on csr engagement in india and a critical evaluation on the new csr law.DOI.10.1007/978-3-319-41781-3. Springer, Cham.

Wheelen, Thomas L. & J. David Hunger.2012. *Strategic management and business policy-toward global sustainability*. Pearson.

Yin, Robert.K. 2014. *Case study research design and methods* (5th edition). Thousand oaks, ca: sage.

Knowledge and lifestyle: A study of green consumer behavior

Kurniawati & Susanti
Universitas Pendidikan Indonesia, Bandung, Indonesia

ABSTRACT: This research is motivated by the complexity of human activity and lifestyle. It is also driven by the development of consumption concepts which lead to consumerist behavior, natural resources exploitation, environmental degradation, and natural resources scarcity. Thus, it aims to analyze the society's understanding of green product and green behavior as well as consumption lifestyle. It also attempts to investigate the influence of green product knowledge on green consumer behavior. The research was based on surveys. Questionnaires were distributed to 225 people in Bandung. The results showed that most respondents had a good knowledge of green product and green behavior. Unfortunately, it was found that there was low influence of green product knowledge on people's lifestyles

1 INTRODUCTION

1.1 *The new consumption paradigm*

The development of civilization in the era of communication technology has an impact on the fast-paced, practical and integrated social life. It also influences the consumption purpose. Basically, the purpose of consumption is to meet the needs (Barnetti 2003; Katzner 2014; Ramya & Kaliyamurthy 2018). To date, it has shifted into communicating identity, such as lifestyle, and delivering the message. This is the focus of this research.

Modernity changes the life pattern and rhythm. Life tends to be fast-paced, practical, materialistic, and hedonic. This causes changes in consumption decision making (Ngafifi 2017; Safuwan 2007). People consume goods that are not environment-friendly and are unhealthy. Both consciously and unconsciously, they consume goods that affects the quality of the environment and human life. Junk food and fast food are unhealthy products, so they are categorized as nonenvironment-friendly products. Junk food and fast food are unhealthy foods because they contain low nutrition, high fat, low fiber, high salt, high sugar, high additives, high calories, and low vitamins and minerals (Oetoro 2013). This food can cause obesity (Damapoli et al. 2013).

In response to this condition, the concept of green consumer behavior emerges. Green consumer behavior concept has two manifestations. One of them is environment-friendly behavior. This behavior communicates consumers' values and concerns the social environment and natural environment. Second, green consumption refers to the consumption of products that have "green" labels; has environment-friendly

inputs, production technique, and consumption manner.

1.2 *Determine of green consumer behavior*

There are two factors that influence green customers behavior. The first factor is related to internal factor. It is knowledge. Knowledge is an important factor in building green consumer behavior (Aertsean et al. 2011; Bockman & Sirotnik 2009; Chan & Lau 2000; Fotopoulos & Krystallis 2002; Kaufman 2014). It will form judgments or evaluations of action regarding right and wrong, which affect consumer's environmental behavior (Astors & Legendare 2009; Brecard, et al. 2009; Ellen 1994; Mustafa et al. 2007). The next factor is external factor, which is social influences (Chan & Lau 2004; Follows, Jobber 2000; Pelsmaker et al. 2005). Social influence is a person's lifestyle that can be influenced by knowledge. (Mowen & Minor 2002) mentioned that lifestyle influences a person's consumption behavior patterns. Lifestyle, according to (Kotler & Keller 2007), is a pattern of life which is expressed in activities, interests, and opinions. Lifestyle describes the entire pattern of how someone acts and interacts in the world. According to (Assael 1984), it is recognizable by how people spend their time (activity), what people consider as important in the environment (interest), and what people think about themselves and the world around them (opinion).

This research aims to find out the influence of life-style and knowledge on consumer environment-friendly behavior. The formulated hypotheses are: (1) hypothesis 1: Knowledge influences green consumer behavior; (2) hypothesis 2: Lifestyle influences Green

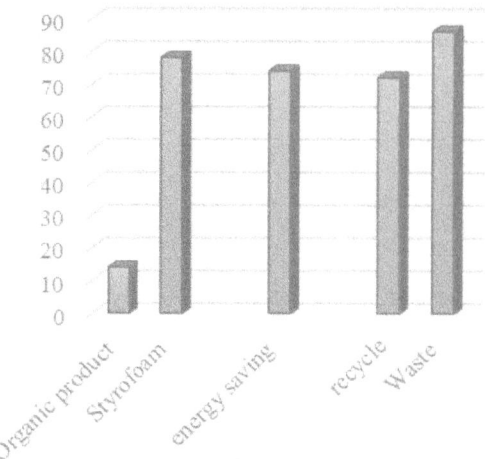

Figure 1. Percentage of Green Consumer Behavior Level.

Consumer Behavior; (3) hypothesis 3: Knowledge influences green consumer behavior through lifestyle. The result is expected to be an input to encourage smart consumers.

2 METHODS

This research was based on surveys. The respondents were 225 consumers in Bandung. They were between the ages of 20–35 years old. They had minimum undergraduate education. To collect the data needed, interviews and questionnaires were used. The questionnaire consisted of questions related to knowledge and lifestyle. In the case of knowledge, it investigated respondents' knowledge of the product ingredients, product production process, and green packaging products. In terms of lifestyle, it asked the respondent's activity intensity, opinion, and interest of green and nongreen products. The data were then analyzed descriptively and inferentially. Descriptive analysis was done by using centralized tendency statistics, while inferentially analysis was done through structural equation modeling partial least squares.

3 RESULTS AND DISCUSSION

The results include the analysis of consumer behavior, knowledge, and lifestyle. It also explains the result of hypothesis testing.

3.1 Analysis of consumer green behavior

Green consumer behavior is consumer behavior in purchasing, using, and managing waste. It includes the intensity of using green products and managing wasted goods (garbage or refuse). The results are presented in Figure 1.

The figure shows that respondents have low green behavior. The intensity of purchasing organic goods,

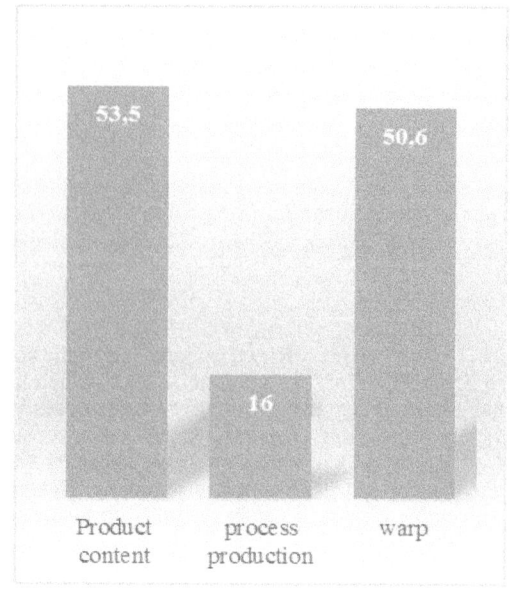

Figure 2. Percentages of Consumer Knowledge Levels.

household appliances, electronics, environment-friendly gadgets, and recyclable items is 14%. The intensity of using Styrofoam packaging is 78%. The intensity of using environment-friendly tools and materials is 74%. The intensity of waste management by applying recycling methods is 72%. Meanwhile, the intensity of disposing waste is relatively high at 86%.

3.2 Analysis of consumer knowledge

In this case, knowledge is consumer knowledge of the product ingredients, product production process, and green packaging product. The percentages of consumer knowledge levels can be seen as follows:

Figure 2 reveals percentages of consumer knowledge of product ingredients. The percentage of consumers who have good knowledge about the dangers of soft drinks and the benefits of organic food for health is 75%. However, 78% of consumers have limited to low knowledge about the effects of junk food, cotton, and chocolate. In terms of knowledge, the product production process was relatively low. There are 84% of consumers who do not know that mineral water, cotton, and chocolate are not produced in environment-friendly ways. In terms of packaging, most consumers (78%) agreed that some efforts needed to be made to reduce packaging. Besides, there are 77.7% of consumers who believe that the product packaging that is labeled green product is a truly environment-friendly.

3.3 Analysis of green lifestyle

Green lifestyle is measured by the intensity of environment-friendly activities, consumer opinion

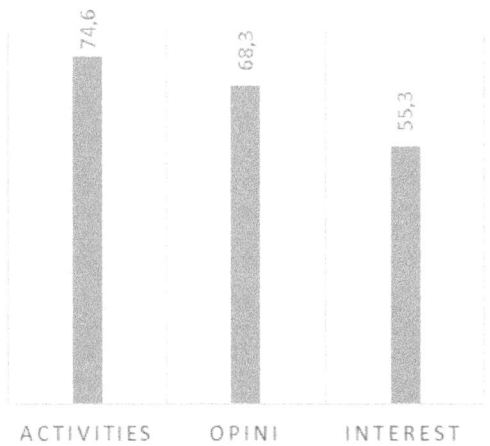

Figure 3. Percentage of Consumer Green Life Style.

about environment-friendly behavior, and consumer interest in green behavior.

The study revealed that activity that is not environment-friendly is relatively high. The intensity of purchasing fast food is 73%. The intensity of using body care products is 68.5% and the intensity of using of nonrecycled material is high. In terms of opinion, 78% of consumers believe that consuming fast food is a practical action and in accordance with the fast-paced rhythm of life. There are 82% of consumers who believe in the relation of vehicles to prestige. There are 45% of consumers prioritizing the consumption of environment-friendly products. Lifestyle is shown by an interest in environment-friendly behavior. The data show that 83% of consumers want to simplify their practical consumption pattern, 36% want to consume organic products, and 47% want to separate organic goods and inorganic materials.

3.4 Hypothesis testing

Hypothesis testing was performed on the bootstrapping sample on the overall model. The value used to test the hypothesis is the path coefficient and the value of R2 (coefficient of determination). Endogenous constructs are used to evaluate models with a significance value of p. Based on data analysis using SEM PLS, it is known that 37.7% of green consumer behavior is influenced by knowledge and lifestyle. In the model of the relationship between variables, product knowledge and lifestyle affect consumer behavior. In this case, knowledge and lifestyle have influences of 10.4% on green consumer behavior. Knowledge level affects 84.5% of consumer green behavior. Knowledge influences 57% of lifestyle, and lifestyle affects 31.7% of consumer green behavior. The results show that consumers have relatively low knowledge about the impact of junk food, cotton, mineral water, and chocolate on the environment. This indicates that consumers do not know the ingredient of those products as well as their production processes.

3.5 Discussion

The low consumption of organic goods based on interviews is due to lack of knowledge, high prices, and the difficulties to gain the goods. Fast food is food that is not friendly to environmental health because it has low nutritional content, is high in sugar, and low in fiber. Consumers tend to simplify consumption more practically. They need food that is easy and fast, such as unhealthy food.

Knowledge of product content is also related to consumer behavior in consuming organic ingredients. In this case, the higher the product knowledge, the more selective the community in food selection. Based on the results of the study, 84% consumers know the information of environment-friendly products based on packaging labels. For products that do not have a green label, consumers (75%) know that organic products are environment-friendly and healthy. However, many people do not know that products made from cotton (clothing, diapers, cosmetic cotton, etc.), chocolate, and junk food are not environment-friendly. This reflects that the concept of green products in the community is still limited to the results of its products that have an impact on the environment. Whereas the concept of a green product is a product that has good inputs, processes, and outputs that do not have a bad effect on the environment or economic and social conditions.

Knowledge of the production process influences consumer behavior in purchasing goods. Based on the results of the study, the low level of knowledge about the production process impacts to the low intensity of purchasing environment-friendly goods, low intensity of purchasing green products, low intensity of saving behavior toward the use of fuel, and poor waste management behavior. Mineral water is produced from the exploitation of the environment in protected forest areas. Consumers also do not understand that cotton (which is used for shirts, disposable diapers, beauty products, and tissue) is an environment-unfriendly product. To produce cotton requires a large amount of fertilizers and insecticides that cause water, soil, and air pollution.

There is low knowledge of environment-friendly product packaging. Most of the consumers purchase nonenvironment-friendly packaging foods such as Styrofoam and plastic. The high activity is using electricity equipment, fuel-based product, and not managing waste properly.

There is diversity of consumers in meeting their needs. It is mostly influenced by a lifestyle, someone's interest, and opinion or view of the product to be purchased. All of these influence consumers' purchasing decisions. Therefore, each consumer has different abilities in making decisions. The environment-friendly lifestyle, based on research results, is in a low category. The fast-paced rhythm of life demand consumers to meet their needs instantly and ignore health aspects. The concept of consumption has shifted from meeting the needs to communicating self-identity that often leads to consuming products that are not environment-friendly. Chocolate consumption, for example, is enjoyed by some people as delicious product. However, its production is accompanied by slavery, low wages, and other adverse socio-economic impacts on some segments of society (Supriatna 2018).

Consumers often think that T-shirts made from cotton or 100% cotton are environment-friendly products as there is information of 100% cotton. This green label refracts the understanding that products made from cotton are environment-friendly products

Lifestyle is also indicated by an interest in environment-friendly behavior. The results of the study revealed that consumers do not wish to consume organic goods, have not tried to reduce waste by separating organic and nonorganic waste, or intend to change how they consume goods.

4 CONCLUSION

Based on the results of the research, it can be concluded that: (1) consumers have good knowledge in terms of product ingredients, but limited knowledge in terms of environment-friendly production processes. (2) In the case of lifestyle, most of the consumers do not have environment-friendly lifestyles. It is seen from the activities, opinions, and interests. (3) Knowledge and lifestyle affect green consumer behavior. In this case, knowledge has more influence on consumer green behavior than lifestyle.

REFERENCES

Aertsean, W., Muys, B.B., Kint. V. &. Orshove, J.V. 2012. Effects of Scale and Scaling in Predictive Modelling of Forest Site Productivity. *Environmental Modelling and Software* 31:19–27.

Assael, H. 1984.Consumer Behavior and Marketing Action. (Second Editions). Boston: Kent Publishing Company.

Barnetti, W. 2003. The Modern Theory of Consumer Behavior: Ordinal Orcardinal The Quarterly *Journal Of Austrian Economics* 6(1): 41–65.

Bockman, S. & Sirotnik, B 2009. Going green from left to center stage: An empirical perspective. *Journal of American Academy of Business* 14(2):8–17.

Brecard, D., Boubaker H.S.L., Yves, P. & Frederic, S 2009. Determinants Of Demand For Green Products. *An Application To Eco-Label Demand For Fish In Europe Ecological Economics* 69(1): 115–125.

Chan, R.Y. & Lau.2000. Anteceden Of Green Purchase Survey In China. *Journal Of Consumer Marketing* 17(4): 338–357.

Chen J. & Lobo. A.2012. Marketing of organic food in urban China: An analysis of Consumers' lifestyle segments. *Journal of International Marketing and Exporting* 17(1).

Damopoli, W., Nelly, M. & Masi, G. 2013. Hubungan Konsumsi Fastfood Dengan Kejadian Obesitas Pada Anak SD Di Kota Manado *ejournal Keperawatan (E-Kp)* 1(1).

Ellen, P. 1994. Do we know what we need to know? objective and subjective knowledge effects on proctological behaviors. journal of business research, 30(1):43–52.

Follows, S.B. & Jobber, D 2000 Environmentally Responsible Purchase Behavior: A Test Of A Consumer Model. *European Journal Of Marketing* 34:723–746.

Fotopoulos, C. & Krystallis, A. 2002 Purchasing Motives And Profile Of The Greek Organic Consumer: A Countrywide Survey. *British Food Journal* 104(9):730–765.

Katzner. 2014. Ordinal Utility and the Traditional Theory of Consumer Demand. *Real-World Economics Review* (67).

Kaufman, N. 2012. Factors Affecting Consumers Green Purchasing Behavior. *An Integrated Conceptual Framework. Amifiteatru Economic* 14(31).

Kotler, P. & Keller, K.L. 2007. Manajemen Pemasaran jild 1, edisi kedua belas. Jakarta: *PT. Indeks*.

Mowen, J.C. & Minor, M. 2002. Perilaku Konsumen. Jakatra. *Penerbit Erlangga* 282.

Mustafa, A.F., Zhao, X. & Zhang, R.H 2007. Effect of feeding oilseeds on nutrient utilization by lacting ewes. *Small ruminat research* 67(2–3):307–311

Ngafifi, M. 2017. Kemajuan Teknologi Dan Pola Hidup Manusia Dalam Perspektif Sosial Budaya.

Oetoro, S. 2013. Parengkuan, E., Parengkuan, J. 2012. Smart Eating. Jakarta *Gramedia Pustaka Utama*.

Ramya, N. & Kaliyamurthy. 2018. Approaches To Consumer Behaviour-A Review *Journal Of Management (Jom)* 5(4):125–137.

Safuwan, a. 2007. Gaya hidup, konsumerisme dan modernitas. *Jurnal suwa*.

Supriatna, J. 2018. Konservasi biodiversitas: Teori dan praktik di indonesia. *Yayasan pustaka obor Indonesia*.

Model of business development strategy for cassava farming technology

N. Sembiring
Universitas Tama Jagakarsa, Jakarta, Indonesia

N.L. Krisna
Universitas Persada Indonesia Y.A.I, Jakarta, Indonesia

ABSTRACT: Agricultural development which includes the food crops and horticulture sub-sectors is one of the strategic sub-sectors which economically, ecologically, and socio-culturally plays an important role in national development. Cassava is the third-most important food crop after rice and corn. Cassava is beneficial and has a fairly low planting and maintenance costs. The purpose of this research is to know the role of a farmers' group (Gapoktan) in increasing cassava farming income. The research method used in this study was a quantitative method with a descriptive correlational survey research approach. The dynamics of the farmers' group are proven to play a positive and significant role in increasing the income of Medan Cassava. The dynamics of the farmers' group continues to change according to the economic environment and the conditions of the farmers themselves. Thus, the development of the farmers in this case becomes very important, because the dynamics of the farmer groups play a big role in increasing income.

1 INTRODUCTION

Agricultural development which includes the food crops and horticulture sub-sectors is one of the strategic sub-sectors which economically, ecologically, and socio-culturally plays an important role in national development. Cassava is the third-most important food crop after rice and corn. Cassava is beneficial and has a fairly low planting and maintenance costs (Yue et al. 2010).

Food is a basic human need so that the availability of food, especially rice for Indonesian communities, must always be guaranteed. With the fulfilment of the community's needs of food, the community will get a calmer life and will better be able to play a role in development. Food crops are an important sector because it is a group of crops that produce food as a source of energy to sustain human life. Indonesia currently prioritizes four types of food crops: rice, corn, soybeans, and cassava (Kapinga et al. 2007).

However, some problems in marketing agricultural commodities are found in developing countries in general and in Indonesia in particular, such as: (a) the unavailability of agricultural commodities in sufficient quantities continuously; (b) price fluctuations; (c) the implementation of inefficient marketing; (d) inadequate marketing facilities; (e) dispersed location of producers and consumers; (f) incomplete market information; (g) lack of knowledge of marketing; and (h) lack of response from producers to market demand (MacNeal 2013).

The increasing number of populations, increasing living standards of many people, and the benefit of public welfare are some reasons for people to consume cassava so that the increase in demand for this commodity is very large. The strength of the cassava market can also be seen from the growth and development of companies that process cassava into various types of food products, whether in the form of snacks or other food products. It is clear that cassava is a very promising multifunctional product (Londhe et al. 2013).

Cassava farmers sell their agricultural products to traders; traders who want to buy can come directly to farmers. There is price bargaining and after the price agreement is reached, the cassava will be handed over to the trader. Farmers' motivation to increase the production and quality of cassava produced is largely determined by the high and low prices received. The high and low prices received by farmers are closely related to the condition of the market structure and the amount of marketing margins, so that the price increase in marketing cassava can be achieved if the market structure and causes of high marketing margins are known (Orinda 2013).

Increased cassava production can be done through expansion of planting areas. Besides that, cassava production can still be improved by upgrading the production technology at the farm level. The production of Medan cassava tends to increase and decrease. This is because the land used for the crops has been widely used by farmers to plant timber plants. This affects the productivity and production of cassava in the Toba Subdistrict. This matter must be the concern of the relevant agencies so that Toba can increase the production of cassava without the need for expansion of farming land.

2 METHODS

The research method used in this study was a quantitative method with a descriptive correlational survey research approach, which is to see the role of Gapoktan and the dynamics of farmer groups in increasing cassava farming income. The research units in this paper activity were members of the farmer group who carried out the cassava farming activities in Medan. The object of this research is the role of Gapoktan and the dynamics of the farmer group in increasing the revenue of Medan cassava farming.

The data collected consists of primary data and secondary data. Primary data collection was done by field observation, direct interviews, and through questionnaires. Secondary data was obtained from related agencies to this research and literature study.

3 RESULTS AND DISCUSSION

3.1 *The role of Gapoktan in increasing farming income*

Based on the results of multiple regression analysis, the hypothesis summary model shows that the contribution of Gapoktan (X1) variable value in increasing farm income (Y) is 21.4%, while the remaining 78% is caused by other factors. Probability value 0.05>0.008 (i.e. significant), this shows that Gapoktan has a significant role in increasing farm income.

The results of the coefficient table on Gapoktan variables explained that the constant values (a) = 58.183 and b1 = 0.583 and the t-count value and significance level of 0.008, the equation can be obtained from the table: $\approx \partial = 58.183 + 0.583$ X1. So, every increase in the role of Gapoktan in one unit, then farm income will increase by 0.583, with (a) constant = 58,183. The magnitude of the role of Gapoktan variable (X1) in increasing the farming income variable (Y) shown in the standardized coefficients is 0.463.

The regression coefficient of 0.583 states that with each addition (because of the + sign) the role of the Gapoktan will increase farming income by 0.583. On the contrary, if the role of Gapoktan decreases by 1 unit, then farm income is also predicted to decrease by 0.583, so the sign (+) indicates the direction of the unidirectional relationship, where the increase or decrease in Gapoktan role variable (X1) will result in an increase or decrease in farming income variable (Y). T test is done to test the significance of constants and variables of farm income. Regression equation (Y= 58.183 + 0 583 X1) which can be obtained later will be tested whether it is indeed valid to predict the variables of

farm income which, in other words, will test whether Gapoktan can really play a role in increasing farm income (Blankenhorn 2007).

3.2 *The role of farmer group dynamics in increasing farm income*

The results of the multiple regression analysis hypothesis summary model shows that the contribution of the value of the farmer group dynamics (X2) in increasing farm income (Y) is affected by 48.6%, while the remaining 51.4% is caused by other factors. The probability value is (Œ±) 0.05>0.001 (significant) and indicates that the dynamics of the farmer group have a significant role in increasing farm income. In the table of farmers' group dynamics coefficients explained that the constant values, (a) = 65,379 and b2 = 1,166 and the farmer group dynamics t-value and the significance level 0.001, the equation above can be obtained: $\approx \partial = 65,379 + 1,166$ X2. So, if every increase in farmer group dynamics is a unit, then farm income will increase by 1,166, delivering a constant of 65,379. The magnitude of the role of farmer group dynamics (X2) in increasing farm income (Y) in the standardized coefficients is 0.697.

The regression coefficient of 1.166 states that each addition (due to the + sign) of the farmer group dynamics of 1.166 will increase farming income by 1,166. Conversely, if the farmer group dynamics fall by 1 unit, then farm income is also predicted to experience a decline of 1,166, so a sign (+) indicates the direction of a unidirectional relationship, where the increase or decrease in farmer group dynamics (X2) will result in an increase or decrease in farm income (Y). T-test was done to test the significance of constants and variables of farm income.

Regression equation (UY = 65,379 + 1,166 X2) which can then be tested whether it is indeed valid to predict the variables of farm income, in other words, it will be tested whether the dynamics of farmer groups can really play a role in increasing farm income. This is in line with the theory put forward by Leilani & Hasan (2020) Farmer group dynamics is a dynamic that makes it possible to become farmers' business in increasing farm income (Guttormsson 2011; Orinda 2013).

3.3 *Gapoktan hypothesis analysis, farmer group dynamics together on farming revenue*

Based on the summary model, Gapoktan and farmer group dynamics together play a role in increasing farm income. The contribution of the value of Gapoktan (X1) and farmer group dynamics (X2) in increasing farm income (Y) by 22.7%, while the remaining 77.3% is caused by other factors. At Gapoktan coefficients value and farmer group dynamics together explained that the constant values, (a) = 65.909 and b1 = 0.897 and b2 = 2.165 as well as t-count Gapoktan 3.013, significance level of 0.009 and farmer group dynamics t-count of 7.449, significance level of 0.002, from the table can be obtained the regression equation: $\approx \partial = 65.909 + 0.897\ X1 + 2.165\ X2$. So, for every increase in the role of Gapoktan by unit, then farm income will increase by 0.897, with the assumption of constant farmer group dynamics, while each increase in farmer group dynamics by unit, then farm income will increase by 2.165.

Gapoktan regression coefficient of 0.897 states that each addition (because of the + sign) the role of Gapoktan is 0.897 will increase farming income by 0.897. Conversely, if Gapoktan falls by 1 unit, then farm income is also predicted to have a decrease of 0.897, so a (+) sign indicates the direction of a unidirectional relationship, where the increase in Gapoktan variable (X1) will result in an increase in farming income variable (Y).

Farmer group dynamics regression coefficient of 2.165 states that each addition (due to the + sign) of farmer group dynamics of 2.165 will increase farm income by 2.165. Conversely, if the farmer group dynamics fall by 1 unit, then farm income is also predicted to experience a decrease of 2,165, so the sign (+) indicates the direction of the unidirectional relationship, where the increase in farmer group dynamics variables (X2) will lead to an increase in farming income variable (Y).

The results of the standardized coefficients regression analysis showed that the magnitude of the role of Gapoktan (X1) and farmer group dynamics (X2) together in increasing farm income (Y) was 0.325 Gapoktan (X1), 0.594 farmer group dynamics (X2). The dynamics of the farmer group are dominant in increasing farm income rather than the role of Gapoktan. However, the effectiveness of farmer group dynamics will increase if supported by the active role of Gapoktan (He et al. 2013).

4 CONCLUSION

Gapoktan is proven to have a positive and significant role in increasing the farming income of Medan cassava. This shows that the existence of Gapoktan is still needed by the farmers of cassava for increasing their farming sales. The dynamics of the farmer group proved to play a positive and significant role in increasing the farming income of Medan cassava. The dynamics of the farmer group continues to change according to the economic environment and the conditions of the farmers themselves, so that the development of the farmers in this case becomes very important.

The potential for cassava farming is promising with a large size of land in the urban area so it is natural still allows for the planting of large amounts of cassava, but there are several obstacles, such as: capital, marketing, and training for cassava farmers.

REFERENCES

Blankenhorn, S.U. 2007. Seaweed farming and artisanal fisheries in an Indonesian seagrass bed: Complementary or competitive usages?

Guttormsson, L. 2011. Population, households and fisheries in the Parish of Hvalsnes, Southwestern Iceland 1750–1850. *Acta Borealia*. [Online]. Retrieved from https://doi.org/10.1080/08003831.2011.626935.

Kapinga, R., Kingamkono, R., Msabaha, M., Ndunguru, J., Lemaga, B., & Tusiime, G. 2007. Tropical root and tuber crops as human staple wood. *Plant Breeding*. [Online]. Retrieved from https://doi.org/10.1017/S0014479709007832.

Leilani, A., & Hasan, O. S. (2020). Analisis Dinamika Kelompok Pada Kelompok Tani Mekar Sari Desa Purwasari Kecamatan Dramaga Kabupaten Bogor. Jurnal Penyuluhan Pertanian, 1(1), 18–27.

Londhe, D., Nalawade, S., Pawar, G., Atkari, V., & Wandkar, S. 2013. Grader: A review of different methods of grading for fruits and vegetables. *Agricultural Engineering International: CIGR Journal*.

MacNeal, T. 2013. Growth prospects brightening for global MIM market. *Metal Powder Report*. [Online]. Retrieved from https://doi.org/10.1016/S0026-0657(13)70017-1.

Orinda, M.A. 2013. Analysis of factors influencing sweet potato value addition amongst smallholders Farmers in Rachuonyo South District , Kenya.

Yue, C., Grebitus, C., Bruhn, M., & Jensen, H. H. 2010. Marketing organic and conventional potatoes in Germany. *Journal of International Food and Agribusiness Marketing*. [Online]. Retrieved from https://doi.org/10.1080/08974430903373060.

Advances in Business, Management and Entrepreneurship – Hurriyati et al. (Eds)
© 2021 Taylor & Francis Group, London, ISBN 978-0-367-67471-7

Green entrepreneurship ecosystem factors in making ecopreneur roadmap for recycling business

K. Khoirunnisa & R.L. Nugroho
Telkom University, Bandung, Indonesia

ABSTRACT: The purpose of this study is to identify the dimensions of the green entrepreneurship ecosystem consist of markets, infrastructure, innovation, governance-regulations, geographic location, visibility, networks, and financing, to create ecopreneur roadmap in developing recycling business. This research is a qualitative research where data obtained from interviews conducted to five respondents related to entrepreneur and ecopreneur activities. Based on data processing obtained from the interviews, it was found that the dimensions of the green entrepreneurship ecosystem are identified within the ecopreneur and could be used as a direction or roadmap for ecopreneurs who want to develop the business of recycled products.

1 INTRODUCTION

The amount of waste generated in Indonesia will continue to increase if the handling has not been carried out maximally. It is predicted that in 2019, the amount of waste generated in Indonesia will be 67.10 million tons of waste per year (Geotimes 2015). In 2015, Indonesia produced 38.50 million tons/year of waste (KLHK 2015). Java Island became the biggest contributor by producing 21.20 million tons/year because Java Island is the island with the largest population in Indonesia, which is 145 million people or 56% of the total population in Indonesia. West Java is the province with the highest population of 46 million (Badan Pusat Statistik 2017).

Bandung is the city with the largest population in West Java, which is as much as 2.40 million (Badan Pusat Statistik 2017). The population in the city of Bandung is expected to increase by 2.60% per year so that the volume of waste is expected to increase significantly with the increase in population. The garbage problem is not only the duty of the government but is also a shared responsibility, including the citizens of Bandung (Lubis 2015). The amount of waste in Bandung currently reaches 1,568 tons/day (PD Kebersihan Kota Bandung 2016) with the largest contribution from residential waste. From that total amount of waste, only 1,200 tons which can be transported to the Sarimukti Final Disposal Site. The rest, as many as 150–250 tons were processed by residents, 150–250 tons of other waste were not transported and disposed of in wild landfills (TPS) (National Geographic Indonesia 2014).

With the increasing volume of waste, PD Kebersihan Kota Bandung has a Long-Term Development Plan (RPJP) for 2005–2025 related to the waste management target, namely: waste management with 3R (40%), waste final processing with environmentally friendly technology (30%), and the sanitary landfill (20%).

The target of waste management with Reduce, Reuse, and Recycle (3R) is only 30% in the Regional Medium-Term Development Plan (RPJMD) of Bandung for 2014–2018. But in 2005–2025 of Long-Term Development Plan (RPJP), waste management with 3R increased to 40%, hence this method has the largest portion of waste management in the city of Bandung. From the description of the problem of a waste generation that faced the city of Bandung, there was a concern for the community to help manage waste by doing a recycling business. People or entities that provide goods and services that are environmentally friendly, use recycle technology, green concepts, green foods are called ecopreneur (Schaper 2002). One of the entities that do business in recycling waste into handicrafts in the city of Bandung is Chilaz Recycled Newspaper Craft (Chilaz).

Researchers have conducted pre-research at the home of the Chilaz owner. Based on informal interviews between researchers and research objects, the researchers found that in running a business to recycle newspaper waste into handicrafts, Chilaz experienced problems including limited human resources, limitations in making training models for people interested in making handicrafts from newspaper waste, and limitations in terms of selling their products to a wider market. Based on these problems, a roadmap that contains guidance for ecopreneur is needed in developing a recycling business. The concept of the green entrepreneurship ecosystem is deemed appropriate to be used as a research variable following the phenomenon revealed by Chilaz owners in the pre-research interview.

1.1 Ecopreneur

Ecopreneur is an individual/unit that runs a business that is not only to generate profits but also integrated into the scope of concern for the environment (Schuyler 1998). Another definition of ecopreneur is a person/entity that provides goods and services that are environmentally friendly, uses recycle technology, the concept of green, green food (Schaper 2002). Ecopreneur has the potential to solve environmental problems and gradually develop the earth's ecosystem (Cohen & Winn 2007). Based on Liz & David (2002) ecopreneurship means that entrepreneurship is focused on understanding the concept of green, where it moves the environment or sustainable ecology into their products and processes. According to Isaak (1998, p.113), "to become an ecopreneur is an existentialist commitment in which the entrepreneur knows he or she will never reach the ideal, but that very idea of sustainability gives meaning to everything the ecopreneur does upon the earth". Ecopreneurs are crucial change agents that strive to achieve a more sustainable future by transforming their businesses or creating new green startups (Walley et al. 2010).

1.2 Green entrepreneurship

Sometimes the term "ecopreneurship" refers to "green entrepreneurship" (Schaper 2002) or "entrepreneurship through environmental lenses". While entrepreneurs might have ideas for change, they cannot drive the change alone. The green entrepreneurship ecosystem needs to be favorable. Non-entrepreneur-specific general context includes available infrastructure, governance and regulations, markets, innovations, and geographical location. Direct factors influencing entrepreneurs include financing, entrepreneurial training and capacity building, culture, networks, technical support, and exposure of entrepreneurs i.e. visibility. In each country and for each economic sector, the entrepreneurship ecosystem can be unique. (Switchmed 2015).

The framework of this study departs from the phenomenon of solid waste in the city of Bandung and Long-Term Development Plan (RPJP) of PD Kebersihan Kota Bandung in 2005–2025 related to the waste management target. Waste management with 3R has a larger percentage target, which is 40%. This condition triggered the ecopreneur to carry out the waste recycling business. But in carrying out this business, the ecopreneur still has several obstacles and does not yet have a roadmap as a guide. In this study, the authors adopted a framework adapted from Switchmed (2015) that focuses on aspects of markets, infrastructure, innovation, governance-regulations, geographic

Table 1. Green Entrepreneurship Ecosystems (Switchmed 2015).

Markets	In the entrepreneurial process, it is not only environmental and social challenges that require action, but also potential customer needs that are not met. They represent market opportunities that may be found and handled by entrepreneurs. Entrepreneurs need to find and meet market needs to satisfy them and do business.
Infrastructure	Physical structures or facilities of organizations such as shops, workshops and physical assets such as manufacturing facilities, buildings, vehicles, machinery, systems, point-of- sale systems, and distribution networks
Innovation	Innovation gives entrepreneurs a competitive advantage by distinguishing it through a unique value proposition. If sustainability is the ultimate goal of green entrepreneurs, innovation is thus their means.
Governance-Regulations	The business operates under the laws that applied in every country, which means that it must comply with certain rules and legal requirements. This involves when you want to start a business, when you have to decide on a legal form that is suitable for the business you are living in (different in each country) such as: sole proprietorship, partnership, limited liability company, corporation, cooperative or social company.
Geographic locations	Driven by the sustainability motto 'global thinking, acting locally', impact of green business in society is intrinsically linked to the local community, which refers to the geographical area where green entrepreneurs operate. In this way, thelegacy of social responsibility is formed by the green entrepreneurs involved with it: the creation of local jobs, contributions to increasing awareness and education of local residents, partnerships with local businesses and public administration.
Visibility	Add new potential stakeholders, collaborate, and help others to give more visibility (conditions that can be seen)
Networks	Consists of all activities needed to create a collaborative environment (which can be physical or digital) where values are exchanged. It's about maintaining an operational network that is profitable for reaching consumers, promoting sales and communicating with stakeholders. Some examples are the market for goods and services or the web community for environmental issues or social issues.
Financing	Some business models rely heavily on financial resources and/or guarantees: such as cash, credit lines, or stock options for hiring employees. At present, a variety of non-conventional funding mechanisms have arrived for the benefit of green business and economic democratization, including: 1. Crowd-funding 2. Financial cooperatives 3. Micro credits 4. Ethical bank

location, visibility, networks and financing. The author will use these aspects to create an ecopreneurship roadmap in the development of recycling businesses. The aspects will be explained in Table 1.

2 METHODOLOGY

2.1 *Research approach*

This study uses a qualitative method. Qualitative methods in research on entrepreneurship are defined as research that studies certain phenomena from the circumstances, the daily life of a person (Neergaard & Ulhøi 2007). A qualitative approach will allow a more flexible way to retrieve information from informants following the actual event.

2.2 *Research method*

This research uses a case study as the research method. Case studies are considered appropriate for research in the field of entrepreneurship because they can accommodate holistic thinking in one or several cases simultaneously by providing local theory. These include involvement, participation, and intervention that provide experience and some heterogeneous realities (Neergaard & Ulhøi 2007:288).

The data are collected through in-depth interviews, participant observation, and document studies. The informants are ecopreneur (Chilaz), APWI representative (Association of Indonesian Entrepreneurs), and institutions that support businesses in the field of creative industries (BEKRAF).

2.3 *Data validity*

The author will use data triangulation techniques for the validity and reliability of this study. Based on Sekaran and Bougie (2010: 384) triangulation is a technique that is also often associated with validity and reliability in qualitative research.

3 RESULTS

Based on the results of interviews and observations with respondents, the authors made an ecopreneur roadmap in developing a recycling business using the five stages of growth in a small business approach or five stages of small business growth (Lewis & Churchill 1983).

In the first stage, the main problem of business is getting customers and providing products or services (Lewis & Churchill 1983). If it is associated with the green entrepreneurship ecosystem variable, then at this stage ecopreneur focuses on financing and markets. Financing, funding to buy raw materials that support the process of producing recycled works until the marketing process. Markets, ecopreneur must know well who is the target market for recycled products. Based on the results of interviews, the best target markets are those who understand or are happy with the art, tourists, and also someone aware of environmental issues and supports recycled works. After knowing the target market, then it is to determine where the product will be marketed. The best way of marketing is through exhibitions and selling online (based on the results of interviews).

In the second stage, businesses have shown that their business is a business entity that can be applied. The business has quite a several customers and how to maintain is to satisfies them with the products or services provided (Lewis & Churchill 1983). If it is associated with the green entrepreneurship ecosystem variable, then at this stage the ecopreneur focuses on visibility and networks. After getting capital for the production process and knowing where to market its recycled products, ecopreneur must also actively seek and join communities of other entrepreneurs (entrepreneurs or ecopreneur), so that more relations and information related to the development of their recycling business. Also, ecopreneur can begin to work with potential parties who can support their recycling efforts.

In the third stage, the decision faced by the owner at this stage is whether to exploit the company's achievements and expand them, or simply keep the company stable but still profitable (Lewis & Churchill 1983). If it is associated with the green entrepreneurship ecosystem variable, then at this stage ecopreneur focuses on infrastructure and innovation. Ecopreneur must begin to innovate on products created then ecopreneur must start looking for a place to become a workshop for the recycling business. At the workshop not only as a place to showcase or sell their products, but also can be a place to do training in making recycled products.

In the fourth stage, the main problem is how businesses can grow rapidly and how to finance that growth (Lewis & Churchill 1983). If it is associated with the green entrepreneurship ecosystem variable, then at this stage ecopreneur focuses on governance regulation and geographic location. After the recycling business goes well, ecopreneur can start looking for support from the government in terms of marketing to tourists. Besides, in the environment where the business stands, ecopreneur can start training neighbors who want to learn how to make recycled products so that in the future when more and more demand for products, ecopreneur can use the workforce in their business environment.

In the fifth stage, the biggest concern of companies entering this stage is to consolidate and control financial benefits caused by rapid growth, and to maintain profits from small size, including flexibility of response and entrepreneurial spirit (Lewis & Churchill 1983). If it is associated with the green entrepreneurship ecosystem variable, then at this stage ecopreneur focuses back on financing variables and networks. Financing, ecopreneur may receive requests to become investors in other companies. Likewise, ecopreneur might look for investors or want to invest their money in stocks or return to business. Ecopreneur already knows the amount of financial strength he has, so he can assess the extent to which his business can finance and make better business planning. For example, planning to increase the number of employees or buy a new tool for operating the business. For networks, when businesses have succeeded in developing, every entrepreneur must also hope that his business can go international and be known throughout the world, as well as ecopreneur. Relations are important in the business world, with good relationships, the recycling business will run well and get several benefits such as additional capital, promotional and advertising media, cooperation in the safekeeping of goods, or other cooperation.

4 DISCUSSION AND CONCLUSIONS

Road mapping is a process that takes many forms and concerns various layers from whole industries and technologies to a single business unit inside a corporation. It has been seen as a method of (future) research itself (Gordon 1994), also as a method of processing and organizing data obtained by various means of study. The roadmap in this paper refers to the order of priorities that should be taken by ecopreneurs based on factors that exist in the green entrepreneurship ecosystem. It can be concluded that the results of the study show green entrepreneurship ecosystem factors, namely markets, infrastructure, innovation, governance regulations, geographic location, visibility, networks, and financing are identified within ecopreneur and can be used as a roadmap for the development of recycling businesses.

Apart from using the green entrepreneurship ecosystem as a roadmap reference, creating a roadmap for ecopreneurs can also be done by understanding a sustainable business model. But there is little research conducted in the field of scalable business models for sustainability (Täuscher & Abdelkafi 2018). More research is done on scaling strategies for social and sustainable enterprises, however not on how the business model should be designed (Heinecke & Mayer 2012). This made it difficult to find success factors on how to design a scalable business model. This research used literature mostly in the field of traditional business models and scalability (Lund & Nielsen 2018; Nielsen & Lund 2018; Stampfl et al. 2013). Therefore, future research should deal more intensively with designing scalable business models for sustainability (Van der Horst 2019).

REFERENCES

Badan Pusat Statistik Kota Bandung. 2017. Kota Bandung Dalam Angka 2016.

Cohen, B., & Winn, M. I. 2007. Market imperfections, opportunity and sustainable entrepreneurship. *Journal of business venturing, 22*(1), 29–49.

Geotimes. 2015. "2019, Produksi Sampah di Indonesia 67,1 Juta Ton sampah Per Tahun". Retrieved from http://geotimes.co.id/2019-produksi-sampah-di-indonesia-671-juta-ton- sampah-per-tahun/.

Gordon, T. J. 1994. The delphi method. *Futures research methodology, 2*(3), 1–30.

Heinecke, A., & Mayer, J. 2012. Social Entrepreneurship and Social Business. *In Social Entrepreneurship and Social Business* (pp. 191–209).

Isaak, R. 1998. *Green Logic: Ecopreneurship, Theory and Ethics*, Greenleaf Publishing, Sheffield, UK.

Lewis, V. L., & Churchill, N. C. 1983. The five stages of small business growth. *Harvard business review, 61*(3), 30–50.

Liz & David. 2002. *Opportunists, Champions, Mavericks. A Typology of Green Entrepreneurs.* Greenleaf Publishing.

Lubis, R. L. 2015. The Triple Drivers of Ecopreneurial Action for Taking The Recycling Habits to The Next Level: A Case of Bandung City, Indonesia. *International Journal of Multidisciplinary Thought, 5*(2), 17–48.

Lund, M., & Nielsen, C. 2018. The Concept of Business Model Scalability. *Journal of Business Models, 6*(1), 1–18.

National Geographic Indonesia. 2014. "Setiap Hari 400 Ton Sampah di Kota Bandung Tak Terangkut". Retrieved from http://nationalgeographic.co.id/berita/2014/09/setiap-hari-400-ton- sampah-di-kota-bandung-tak-terangkut/.

Neergaard, H. & Ulhøi, J. P. 2007. *Handbook of Qualitative Research Methods in Entrepreneurship.* Cheltenham, UK: Edward Elgar.

PD Kebersihan Kota Bandung. 2016. Laporan PD Kebersihan Kota Bandung 2016.

Schaper, M. 2002. The Essence of Ecopreneurship. *Greener Management International, 38*, 26–30.

Sekaran, Uma & Bougie, Roger. (2010). Research Method for Business: a skill – building approach - 6th ed. United Kingdom: John Wiley & Sons Ltd.

Switchmed. (2015). Building Green entrepreneurship ecosystem in the Mediterranean. Retrieved from https://www.Switchmed.eu/en/media/Leaflets%2C%20 factsheets%2C%20brochures/Switchme d_ge_al_en.pdf.

Täuscher, K., & Abdelkafi, N. 2018. Scalability and robustness of business models for sustainability: A simulation experiment. *Journal of Cleaner Production, 170*, 654–664.

Walley, L., Taylor, D., & Greig, K. (2010). Beyond the visionary champion: Testing a typology of green entrepreneurs. *Making ecopreneurs: developing sustainable entrepreneurship, 2*, 59–74.

Van der Horst, Y.N. (2019). A roadmap to scalability: a tool for new ventures to design scalable business models for sustainability.

Section 3   Innovation, IT, operations and supply chain management

Study on determining priority of Base Transceiver Station (BTS) construction

W. Nugroho

Sepuluh November Institute of Technology, Surabaya, Indonesia

ABSTRACT: For sustainable development, the selection of Base Transceiver Station construction should consider points of view from all stakeholders. Mobile operators consider the best location from a maximum profit standpoint for the long term, considering the variables of government interest, the suitability of land use and environmental allocations. It is necessary to analyze the decision-making in the appropriate location selection, from the view-point of government, operators, or other relevant stakeholders. This study used an empirical approach based on statistical methods in the multi-criteria decision-making system. The data arranged in the form of variables then elaborated to be the factor analysis of these variables to obtain which variables that play a role in the construction of BTS. Based on the value of eigenvalue of both factors, the cumulative percentage of both factors is 80.61 percent, it can be concluded that the formed factors are enough to represent the diversity of variables – the origin variable.

1 INTRODUCTION

1.1 Background

The purpose of optimizing cellular networks is to achieve a certain quality of service (QoS) objectives. For mobile operators, an optimized design will in-crease overall network reliability. Researchers in recent years have explored some of the most difficult optimization problems that have arisen in cellular network design, namely, facility placement and frequency determination (Yu & Kim 2013). Base Transceiver Station (BTS) is one of the most important facilities in cellular telecommunication. Construction of this facility can cause negative impacts and conflicts in the future if there is a discrepancy in the selection of locations both from the perspective of government, community, and cellular operators (Shahbazi-Gahrouei et al. 2014). Therefore, there is a need for research to formulate location criteria ac-cording to stakeholders.

1.2 Literature review

The terminology of Base Transceiver Station (BTS) is becoming popular in the era of the current mobile expansion. BTS serves to bridge the communication device with the network to another network. A range of BTS emissions can be called cells. Mobile communication is a modern communication that supports high mobility. From several BTS then controlled by one Base Station Controller (BSC) connected with microwave or fiber-optic connection. BTS serves as an interconnection between mobile system infrastruture with Out Station. BTS must always monitor the Out Station that enters or out of the BTS cell. The wide range of BTS is strongly influenced by the environment, including topography and high buildings. BTS is instrumental in maintaining GSM quality, especially in terms of hopping frequencies and diversity antennas (Asif & Khanzada 2015).

BTS can be seen from the base of the inside of the Base Station Service (BSS) network as the connection equipment between Base Station Controller and Mobile Station. BTS function in BSS is BTS interacts directly with Mobile Subscriber through the radio interface. BTS must be able to communicate with the mobile device in a coverage area and able to meet the traffic channel for communication including voice and text data exchange. To have a continuous connection between BTS and mobile devices, the maximum distance between BTS and mobile de-vices must be set. This determination should be based on minimum signal-to-noise-ratio at the receiving end and maximum transmission power at the sender side. In the BTS construction plan, the cover-age area is usually approx-imated by the hexagonal model and ideal coverage approach, but in reality, the coverage area is irregular, depending on the environment, site topography and land morphology (Bai & Heath 2015).

Wireless network implementation should consider the cost of equipment procurement and supporting infrastructure (Dasare et al. 2014). The determinants of BTS tower location consist of land price, tower type, tower height, high population density, spatial pattern, service area coverage, and safety (Sulistyarso & Dynastya 2013). Gacovski & Cvetanoski (2006) also mentioned five criteria for determining base station location consist of investment cost, coverage area, population coverage, location affordability, and signal interference (Gacovski & Cvetanoski 2006). BTS construction cost is driven by different factors depending on the characteristics of the base stations deployed. When site density increases, operational and transmission costs tend to dominate rather than radio equipment and site costs. The results also show how, for different capacity requirements, the costs can be minimized by a proper selection of for example macro, micro and pico base stations. In many scenarios the macro base stations yield the lowest cost, indicating that coverage (cell range) is an important parameter when designing wireless systems. (Johansson et al. 2004).

Some researchers consider the problem of the availability of electric power in their research. Gonzales-Brevis et al. (2011) examined the combined problems of BTS locations mainly related to optimal power allocation, to optimize the energy efficiency of cellular wireless networks. A simple method for estimating the costs of building and operating a cellular mobile network is proposed. Using empirical data from a third-generation mobile system. The research shows that the transfer of tissue from a small number of high-power cells to a large number of small cells can increase the energy efficiency of the network (González-Brevis et al. 2011). As more locations gain access to telecommunication, there is a growing demand to provide energy in a reliable, efficient and environmentally friendly manner while effectively addressing growing energy needs. Erratic power supply and rising operation costs (OPEX) have in-creased the need to harness local renewable energy sources. Thus, identifying the right generator schedule with the renewable system to reduce OPEX is a priority for operators and vendors (Oviroh & Jen 2018).

1.3 *Aim of the study*

Based on the background above, from previous studies, it is found that the variables affecting the BTS construction are as follows:

a. cost of delivery
b. time of delivery
c. customer density
d. social support
e. government support
f. power availability

There has never been a combination study of various variables in the analysis of decision-making and this study raises the following questions:

1. What are the main factors that formed by all the variables above?
2. What factors play an important role in decision making for the best priority of BTS construction?

2 METHODS

2.1 *Exploratory factor analysis*

This study used an empirical approach based on statistical and mathematical methods in the decision-making system as a step in solving the problem. Types of data were qualitative data and quantitative data. Qualitative data was used to determine the weight criteria that affect the determination of BTS tower location. Obtained from the respondent's perception in determining the comparison between criteria and sub-criteria (in pairwise comparison process). Meanwhile, quantitative data consisted of procurement costs, procurement duration, and customer density.

This study used questionnaires that were filled by stakeholders from the parties involved in the construction of BTS towers in Central Java and Yogyakarta. The questionnaire was expected to explore the score or weight of each location to these variables. Then, factor analysis was done on these variables to obtain which variable role in the construction of BTS tower. So that the next process not included the whole factor, but only influential factors.

Factor analysis is one of the statistical techniques that can be used to provide a relatively simple description through the reduction of the number of variables called factors. Factor analysis is a procedure for identifying items or variables based on their similarities. The similarity is shown by a high correlation value. Items that have a high correlation will form a crowd of factors (Härdle & Simar 2013).

The basic principle in factor analysis was to simplify the description of data by reducing the number of variables/dimensions. Factor analysis in this study was an exploratory factor analysis and performed with SPSS software tool. The steps in performing factor analysis were as follows (Hair et al. 2006):

1. Conducting a correlation test between variables of origin with the aim that depreciation of variable factor analysis becomes more simple and useful, without losing much information before.
2. The data feasibility test (using factor basis) is suitable for factor analysis.
3. Finding the root of the characteristics and matrix Σ or R.
4. Sort the root traits that are formed from the largest to the smallest.
5. Finding the proportion of diversity that is useful to know how many factors will be formed.
6. Allocate each variable of origin into the factor according to the loading value.
7. If there is a loading value that is identical or almost the same then do the rotation either by orthogonal or non-orthogonal.
8. Once convinced by the factors formed, then given the identity by way of giving the number/name on the factor by looking at the constituent variables factor.

These variables are then elaborated for factor analysis of these variables to determine which variables play a role in the construction of BTS towers. In addition to identifying variables that play a significant role, the process of factor analysis can also explain the commonalities between variables, then the commonly related variables are grouped into one group of factors, while the variables that do not play a significant role in the activities will be eliminated. This insignificant variable does not mean that the literature that states the variable is in the wrong role, but the variable is not appropriate to apply to the condition of this study so that its influence is not significant. Thus, the factor analysis process plays a role in the preparation of hierarchical structures that will be used in decision analysis using the Analytic Hierarchy Process (AHP) method.

2.2 Multi-Criteria Decision-Making (MCDM)

Decision Support System is defined as a computer-based system that helps in the decision-making process. This system is adaptive, interactive, flexible, specifically developed to support solutions of unstructured management problems in improving the quality of decision making (Kusumadewi 2004). Analytic Hierarchy Process (AHP) is a decision-making system that uses measurement theory through pairwise comparisons and the results depend on experts who decrease the AHP priority scale is al-so a multi-criteria decision-making approach in several factors based on hierarchical structure (Saaty 1990, 2002).

AHP is used as a multi-criteria decision-making method to determine best locations in health, construction, and various purposes. In the AHP method the comparison, weight of each of the criteria analyzed and Consistency Ratio (CR) of the criteria pair comparison (Dehe & Bamford 2015).

The procedure in determining the comparison of each criterion in AHP is as follows:

a. Decomposition of decision-making issues into a hierarchy.
b. Make pairwise comparisons and assign priorities among elements in the hierarchy.
c. Judgments to get the weight of each criterion.
d. Checks and analysis of judgment consistency.

3 RESULTS AND DISCUSSION

This study found two variables that have the lowest MSA value are cost of delivery variable and government support variable. Since there is a variable whose MSA value is <0.5, there must be a variable that is omitted and for subsequent retesting of the rest of the variables. The elimination of these two variables does not mean that in the implementation will eliminate the consideration of the cost and government support, but the two are used as criteria in the initial screening

Table 1. Commonalities.

Variables	Initial	Extraction
Power	1.000	.744
Time	1.000	.858
Customer	1.000	.750
Social	1.000	.873

Table 2. Component matrix.

	Component	
Variables	1	2
Power	−.620	−.599
Time	.843	−.385
Customer	.576	.647
Social	−.744	.565

prior to priority analysis. Thus, it should be the location of the BTS to be selected, qualified in terms of cost-efficient and complete permissions.

After the process of elimination of two variables, and reanalysis with only four variables, the KMO value increases to = 0.539 (eligible above 0.5). Similarly, the value of significance of 0.045, which is be-low the value of alfa = 0.05. Thus, the elimination of these two variables increases the KMO value. This may indicate that these four variables are 'more than enough' feasible for factor analysis. The four variables have MSA value > 0,5, then the variable can be analyzed further.

From the total values in the commonalities table, it is found that the four variables have large commonali-ties (> 0.5). This can be interpreted that the entire variable used has a strong relationship with the fac-tors that are formed. In other words, the greater the value of commonalities the better the factor analysis, because of the greater the characteristic of the origi-nal variable that can be represented by the factors formed.

The amount of diversity that can be explained by Factor 1 of 49.497 percent, while the diversity that can be explained the cumulation Factor 1 and 2 by 80.610 percent. Based on the reason that the value of eigenvalue of both factors is more than 1 and the cumulative percentage of both factors is 80.610 percent, it can be concluded that both factors are enough to represent the diversity of the original.

The table 2 shows the magnitude of the correlation of each variable in the factor formed. The values of the correlation coefficient between the variables with the factors that are formed (loading factor) can be seen in Component matrix table. Both factors pro-duce a factor loading matrix whose values are the correlation coefficients between the variables and those factors. When viewed the variables that corre-late to each factor, the loading factor generated has not been able to give meaning as expected. This is evident from the power variables in which the corre-lation of this variable with factor 1 is 0.620, while the factor 2 is -0.599 (negative sign only indicates the direction of correlation), so it is difficult to de-cide whether the power variable is inserted into fac-tor 1 or factor 2. Each factor cannot be interpreted clearly so that need to be rotation by a varimax method. Varimax rotation is orthogonal rotation which makes the number of variants of the loading factor in each factor will be maximum, whereas the original variables will only have a high correlation and strong with certain factors (the correlation is close to 1) and certainly have a weak correlation with the factor other (the correlation is close to 0). Such a thing has not been achieved in the matrix component (table 2).

Table 3. Rotated component matrix.

| Variables | Component | |
	1	2
Power	.127	−.853
Time	−.902	.209
Customer	−.062	.864
Social	.934	−.006

It is seen that the rotated factor loading has given meaning as expected and every factor can be clearly interpreted. It is also seen that each variable is only strongly correlated with one factor alone (no varia-bles are correlated <0.5 in both factors). Thus, it is more appropriate to use the rotated factor loading because each factor is able to explain the initial vari-ability of all variables appropriately and the result is as follows:

1. Factor 1 have a strong correlation with time and social variables. Factor 1 is hereafter named as The Environmental Factor.
2. Factor 2have a strong correlation with power and customer variable. Factor 2 hereinafter named as The Support Factor.

Table 4. Component transformation matrix.

Component	1	2
1	−.793	.610
2	.610	.793

From the analysis of pairwise comparison with AHP method on the factor and variable as the analysis process for the questionnaire data, the following re-sults are obtained:

1. There is an assessment of the same level of impor-tance between environmental factor and support factor.
2. Social support and customer density are the most important variable for the respondents. This is evidenced by the results of pairwise comparison

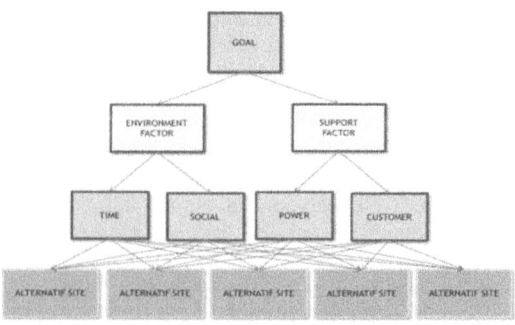

Figure 1. The hierarchy of pairwise comparison.

which social and customer each has a ratio of 0.333, while time and power criteria each have a ratio of 0.167.

4 CONCLUSION

The conclusions obtained in this study are as follows:

1. Factors influencing BTS site selection consist of two factors:
   a. Environmental factor, that has a strong corre-lation with time of delivery and social support variable, and
   b. Support factor, that has a strong correlation with power and customer density variable.
2. Based on the value of eigenvalue of both factors, the cumulative percentage of both factors is 80.61 percent, it can be concluded that the two factors are enough to represent the diversity of variables – the origin variable.
3. There is an assessment of the same level of impor-tance between environmental factor and support factor.
4. Social and customer are the most important variable in BTS construction.

ACKNOWLEDGMENTS

We hereby thank you to the Civil Engineering De-partment, Sepuluh November Institute of Technology, for supporting the publication of this.

REFERENCES

Asif, R. & Khanzada, F. 2015. Cellular base station pow-ered by hybrid energy options. *International Journal of Computer Applications*.
Bai, T. & Heath, R.W. 2015. Coverage and rate analysis for millimeter-wave cellular networks. *IEEE Transactions on Wireless Communications*.
Dasare, H.D., Murthy, R.V., Shanker, R., & Nagadevara, V. 2014. Antennae location methodology for a telecom operator in India. *SSRN Electronic Journal*.

Dehe, B. & Bamford, D. 2015. Development, test and comparison of two Multiple Criteria Decision Analysis (MCDA) models: A case of healthcare infrastructure location. Expert *Systems with Applications*.

Gacovski, Z. & Cvetanoski, H. 2006. Fuzzy decision-making for selection of mobile base station location. *Proceedings of the International Conference on Information Technology Interfaces, ITI*.

González-Brevis, P., Gondzio, J., Fan, Y., Poor, H.V., Thompson, J., Krikidis, I., & Chung, P.J. 2011. Base station location optimization for minimal energy consumption in wireless networks. *IEEE Vehicular Technology Conference*.

Hair, J.F., Black, W.C., Babin, B.J., & Anderson, R.E. 2006. Multivariate data analysis (6th ed.).

Härdle, W.K. & Simar, L. 2013. Applied multivariate statistical analysis. *In Applied Multivariate Statistical Analysis*.

Johansson, K., Furuskär, A., Karlsson, P., & Zander, J. 2004. Relation between base station characteristics and cost structure in cellular systems. *IEEE International Symposium on Personal, Indoor and Mobile Radio Communications, PIMRC*.

Kusumadewi, S. 2004. Fuzzy quantification theory I untuk analisis hubungan antara penilaian kinerja dosen oleh mahasiswa, kehadiran dosen, dan nilai kelulusan mahasiswa. *Media Informatika* 2(1).

Oviroh, P.O. & Jen, T.C. 2018. The energy cost analysis of hybrid systems and diesel generators in powering selected base transceiver station locations in Nigeria. *Energies*.

Saaty, T.L. 1990. How to make a decision: The analytic hierarchy process. *European Journal of Operational Research*.

Saaty, T.L. 2002. Decision making with the Analytic Hierarchy Process. *Scientia Iranica*.

Shahbazi-Gahrouei, D., Karbalae, M., Moradi, H.A., & Baradaran-Ghahfarokhi, M. 2014. Retracted article: Health effects of living near mobile phone Base Transceiver Station (BTS) antennae: a report from Isfahan, Iran. *Electromagnetic biology and medicine* 33(3): 206–210.

Sulistyarso, H. & Dynastya. 2013. Model lokasi menara BTS ditinjau dari faktor-faktor penentu lokasi menara BTS di Surabaya. *Jurnal Teknik Pomits*.

Advances in Business, Management and Entrepreneurship – Hurriyati et al. (Eds)
© 2021 Taylor & Francis Group, London, ISBN 978-0-367-67471-7

Outpatient waiting time and length of stay in hospitals at West Java

D.A. Farmaciawaty, M.H. Basri, I.N. Rachmania & F.B. Widjaja
Institut Teknologi Bandung, Bandung, Indonesia

ABSTRACT: Patient queues have become a problem in healthcare delivery systems. Long queues will result in a delay of a patient receiving services as well as an economic loss since patients lose their productivity time. Patient waiting time has been one among the other indicators to evaluate the healthcare providers' performance and quality. This research aims to investigate waiting time or length of stay spent by outpatient customers of several hospitals in West Java. Data were collected through the electronic record from the hospital's information system and analyzed using the Business Process Model and Descriptive Statistics. From the re-search, it was found that hospitals still were not able to fulfill the outpatient waiting time and length of stay minimum service standard set by the government and the hospital itself, which should be less or equal with 60 minutes for outpatient waiting time and 1,5 hours for outpatient length of stay.

1 INTRODUCTION

The healthcare industry is an industry that is ex-pected by the community to provide excellent quali-ty services. The standard of quality service is deter-mined by the government, one of which is the patient's queue. Long patient queues will result in longer patient wait-ing times. For some healthcare providers, the speed of response to serving patients is essential and is one of the focuses of the im-provement of these health care facilities. The patient queue needs to be improved since long queues will result in the delay of a patient receiving services as well as an economic loss since patients lose their productivity time and affect the whole satisfaction of patients toward health care ser-vices (Bleustein et al. 2014). Moreover, in general, waiting experiences are typically negative and affect the overall satisfaction of consumers with the service encounter (Bielen & Demoulin 2007).

To improve healthcare service, patient waiting time has become one of among the other indicators that have been the target of improvement in health ser-vices. It is in accordance with Kepmenkes No. 129 of 2008 con-cerning Hospital Minimum Service Standards (SPM). Based on the Minister of Health Decree, the opera-tional definition of waiting time is the time required by the outpatient from the registra-tion until served by a physician. Meanwhile, the standard of patient maxi-mum waiting time is less than equal to 60 minutes (≤ 60 minutes). Sometimes, some hospitals use the length of stay of outpatient as their waiting time indicator. The length of stay of outpatient is measured from when the patient gets the queue number until they go home.

Outpatient waiting time has become a concern to hospitals since it measures organizational efficiency because the number of outpatients is usually higher than the number of inpatient in a hospital (Kujala et al. 2006). In outpatient care, the patient does not need a medical service that requires a prolonged stay at a hospital or clinic. However, despite the importance of improving the patient queue, waiting time is still a challenge for the hospital's management. It is diffi-cult to predict the exact number and time of arrival of patients since patients arrived randomly.

Waiting time in hospitals has become a challenge for the hospital's management. Despite the technologi-cal advancement in medical care, the patient frequently has to wait during the process of acquiring service in the hospital. According to (Barlow 2002), these wait-ing experiences are typically negative and have been shown to affect the customers' overall satisfac-tion and their evaluation of the service encounter. Moreover, (Ward 2017) stated that longer waiting times had been associated with longer length of stay and risk of patient safety.

Outpatient service is one of the essential activities of a hospital that provide many services such as diag-nostic, curative, preventive, and rehabilitative services. Outpatient service has many processes in it and also consists of several specialties with a comprehensive range of support (Mital 2010). In addition, (Pillay et al. 2011) argued that the hospital's management has become more concerned with the outpatient waiting time because it measures organizational efficiency. He stated that the critical aspect of outpatient service is the excessive waiting time, which is a major complaint of patients. Extra waiting time is also non-value adding time because during this period, resources are not used to improve patients' condition. This excessive waiting time is a lose-lose strategy, while patients lose valuable time, and hospitals lose their patients and reputation. In line with (Meng et al. 2015), long waiting time occurs due to various reasons such as limited capacity, variable demand, and inefficient operations management. Therefore, hospital managers are challenged to reduce waiting times by introducing interventions at various time points. (Meng et al. 2015) also stated before introducing the interventions; it is essential to explore significant factors, such as mean waiting time, its variance, and time points, associated with the maximum waiting time policy.

Queueing theory is one of the advanced mathematical technique that can estimate waiting times. In general, a queueing system has two main components; customers and servers. Since customers arrive randomly, and there is variability in the system, the delays are highly variable and depend on the number of servers (Bittencourt et al. 2018). Queueing theory has its origins in the early 1900. Since then, queueing theory has been well developed and applied in many areas, as well as healthcare settings (Palvannan & Teow 2012).

According to (Liu & D'Aunno 2012), a queueing model can be used to translate the arrival patterns and processing times to estimate important system performance measures, such as average customer waiting times and the likelihood of a random customer encountering zero delays, for any number of servers. (Palvannan & Teow 2012) explained that queueing analysis can help to quantify the appropriate service capacity so that the patient's waiting time is acceptable and help to analyze the impact of change in demand or service factors. Moreover, he sumed up that 'queuing theory,' is the equation that defines the relationship between demand, capacity, and queues/wait time when there is significant variability. Thus, this established theory helps us to quantify the appropriate service capacity to meet the patient demand, balancing system utilization, and the pa-tient's wait time.

On the other hand, (Barlow 2002) argued that traditional queueing theory and productivity measurement techniques only look at the speed of service, the actual waiting times, the effectiveness of both services and waiting time. He stated that more appropriate and increasingly recognized approach aims to manage the customers' experience and levels of satisfaction.

To improve healthcare service, patient waiting time has become one of among the other indicators that have been the target of improvement in health services. It is in accordance with Kepmenkes No. 129 of 2008 concerning Hospital Minimum Service Standards (SPM). Based on the Minister of Health Decree, the operational definition of waiting time is the time required by the outpatient from the registration until served by a physician. Meanwhile, the standard of patient maximum waiting time is less than equal to 60 minutes ($\leq$60 minutes). Sometimes, some hospitals use the length of stay of outpatient as their waiting time indicator. The length of stay of outpatient is measured from when the patient gets the queue number until they go home.

Table 1. Methodology.

Hospitals	Status	Size	Point of Recorded Data	Data	Analysis
Hospital A	Private	General hospital, district level	The first point of service is registration, then service at the cashier and the last point is service at the pharmacy.	Data were collected from 1st January 2019 until 30th April 2019. The data contain 106.392 rows.	Length of Stay
Hospital B	Public	General Hospital, national level, hospital category A	The first point is getting queue number, the second point is getting called at registration service, and the third point is at finishing registration service	Data were collected from January to June 2018 at 6 a.m.−1 p.m. The data contain 17,376 rows.	Outpatient waiting time
Hospital C	Public	Special Hospital, national level, hospital category A	The first point is when Medical Patient Record is ordered, the second point is when Medical Patient Record is distributed, and the third point is when Medical Patient Record is received at the clinic.	Data were collected from 1st September 2014-30th September 2014. The data contain 10,476 rows	Outpatient waiting time

Outpatient waiting time has become a concern to hospitals since it measures organizational efficiency because the number of outpatients is usually higher than the number of inpatient in a hospital (Kujala et al. 2006). In outpatient care, the patient does not need a medical service that requires a prolonged stay at a hospital or clinic. However, despite the importance of improving the patient queue, waiting time is still a challenge for the hospital's management. It is difficult to predict the exact number and time of arrival of patients since patients arrived randomly.

Waiting time in hospitals has become a challenge for the hospital's management. Despite the technological advancement in medical care, the patient frequently has to wait during the process of acquiring service in the hospital. According to (Barlow 2002), these waiting experiences are typically negative and have been shown to affect the customers' overall satisfaction and their evaluation of the service en-counter. Moreover, (Ward 2017) stated that longer waiting times had been associated with longer length of stay and risk of patient safety.

Outpatient service is one of the essential activities of a hospital that provide many services such as diagnostic, curative, preventive, and rehabilitative services. Outpatient service has many processes in it and also consists of several specialties with a comprehensive range of support (Mital 2010). In addition, (Pillay et al. 2011) argued that the hospital's management has become more concerned with the outpatient waiting time because it measures organizational efficiency. He stated that the critical aspect of outpatient service is the excessive waiting time, which is a major complaint of patients. Extra waiting time is also non-value adding time because during this period, resources are not used to improve patients' condition. This excessive waiting time is a lose-lose strategy, while patients lose valuable time, and hospitals lose their patients and reputation. In line with (Meng et al. 2015), long waiting time occurs due to various reasons such as limited capacity, variable demand, and inefficient operations management. Therefore, hospital managers are challenged to reduce waiting times by introducing interventions at various time points. (Meng et al. 2015) also stated before introducing the interventions; it is essential to explore significant factors, such as mean waiting time, its variance, and time points, associated with the maximum waiting time policy.

Queueing theory is one of the advanced mathematical technique that can estimate waiting times. In general, a queueing system has two main components; customers and servers. Since customers arrive randomly, and there is variability in the system, the delays are highly variable and depend on the number of servers (Bittencourt et al. 2018). Queueing theory has its origins in the early 1900s. Since then, queueing theory has been well developed and applied in many areas, as well as healthcare settings (Palvannan & Teow 2012).

According to (Liu & Aunno 2012), a queueing model can be used to translate the arrival patterns and processing times to estimate important system performance measures, such as average customer waiting times and the likelihood of a random customer encountering zero delays, for any number of servers. (Palvannan & Teow 2012) explained that queueing analysis can help to quantify the appropriate service capacity so that the patient's waiting time is acceptable and help to analyze the impact of change in demand or service factors. Moreover, he summed up that 'queuing theory,' is the equation that defines the relationship between demand, capacity, and queues/wait time when there is significant variability. Thus, this established theory helps us to quantify the appropriate service capacity to meet the patient demand, balancing system utilization, and the pa-tient's wait time.

On the other hand, (Barlow 2002) argued that traditional queueing theory and productivity measurement techniques only look at the speed of service, the actual waiting times, the effectiveness of both services and waiting time. He stated that more appropriate and increasingly recognized approach aims to manage the customers' experience and levels of satisfaction.

To improve waiting time, first (Meng et al. 2015) suggestion), we need to define the waiting time's significant factors at the hospital. Therefore, this research aims to investigate and explore waiting time or length of stay spends by outpatient customers of several hospitals in West Java. To carry out this research, data were collected through the electronic record from the hospitals' information system and analyzed by using the Business Process Model and descriptive statistic.

This paper is organized as follows. Section 1 describes the problem, objective, and the importance to conduct this research. Section 2 describes the literature review of outpatient service waiting time, as well as literature in queuing theory. Section 3 illustrates the overview of the methodology. Section 4 presents the results and findings. Then section 5 presents the conclusion, and further research is also detailed.

2 METHODS

Healthcare services in Indonesia are organized into three levels, which are national-level (Level 3), state-level (Level 2), and district-level (Level 1). National level hospitals provide a comprehensive range of tertiary care services, which serves as the National Referral Centre. While state-level hospitals provide a comprehensive range of secondary services. On the other hand, district-level healthcare services only provide basic patient care services.

This study was carried out in one private hospital at the district level and two public hospitals at the na-tional level in West Java. We obtain both primary and secondary data to support this study. First, we conducted an interview with the healthcare services management officers of the three hospitals to identify the process of outpatient service. Then, the waiting time and length of stay data were collected from the electronic record from the hospitals' information system. To obtain these data, we interviewed the information system staff from the three hospitals. The methodology used to collect secondary data is described in Table 1.

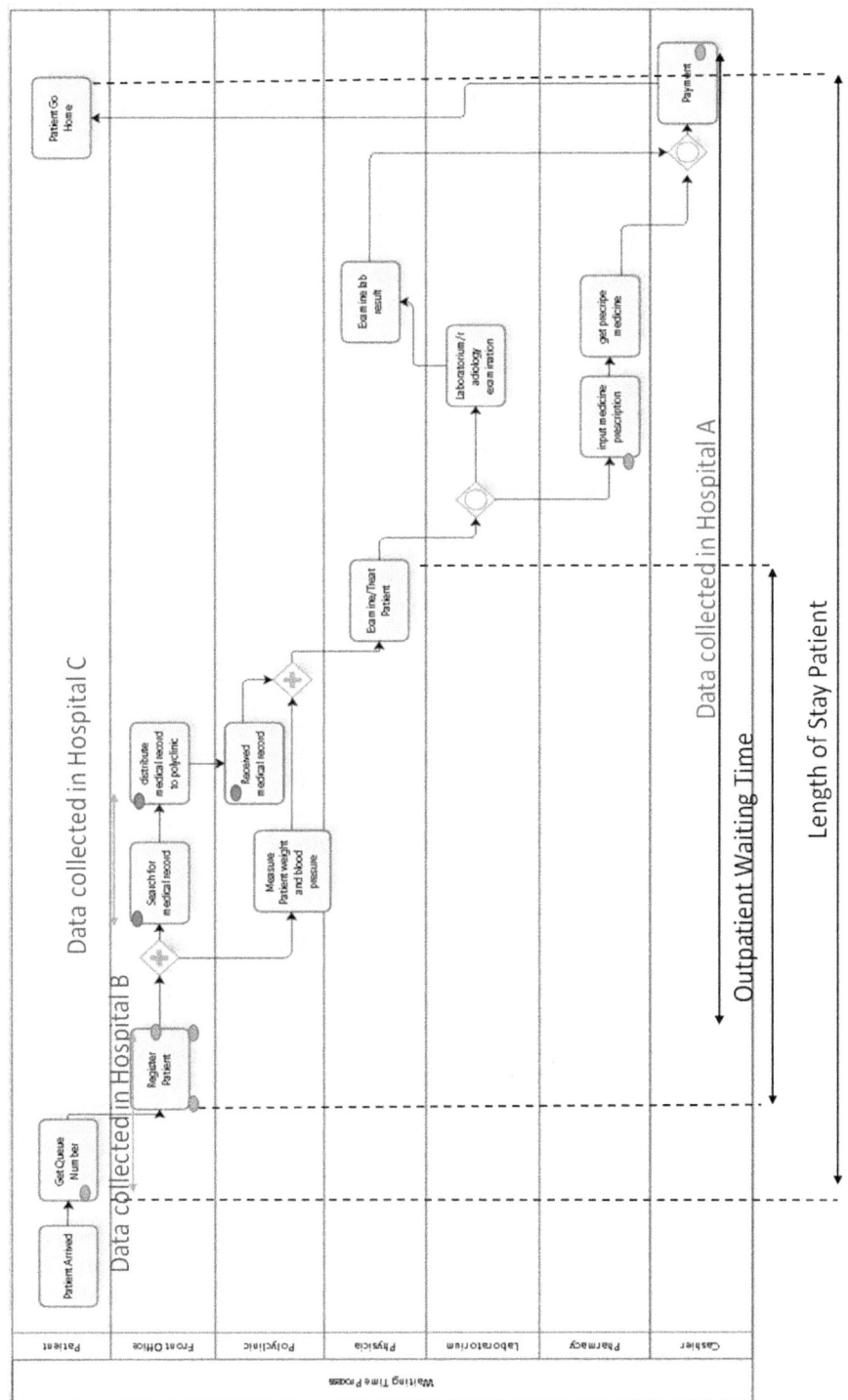

Figure 1. Waiting time business process of outpatient.

Table 2. Outpatient waiting time and LOS.

Hospital	Average Outpatient Waiting Time/LOS (hours)	Standard Deviation Outpatient Waiting time/LOS (hours)	Maximum Outpatient Waiting Time/LOS (hours)	Minimum Outpatient Waiting time/LOS (hours/minutes)
Hospital A	3,50 hr	3,33 hr	23,54 hr	6,00 minutes
Hospital B	5,10 hr	2,05 hr	10,60 hr	1,67 minutes
Hospital C	0,74 hr	0,73 hr (44,4 minutes)	7,05 hr (44,28 minutes)	4,23 minutes

3 RESULTS AND DISCUSSION

3.1 Waiting time business process

From Figure 1, we can see that outpatient waiting time for the hospitals is started from the register point until the examination of the patient by the physician. The waiting time in accordance with the regulation stated by the ministry of health. However, the outpatient waiting time indicator measured from the two public hospitals is not the full process. From the recorded electronic data, in Hospital B, outpatient waiting time is measured started from the patient gets the queue number until the patient receive service in register patient point. In Hospital C, outpatient waiting time is measured from searching medical records until the nurse at the physician clinic receives the medical record.

Meanwhile, the private hospital has its own indicator, which is the length of stay of outpatient. The length of stay process should be started when the patient gets the queue number until the patient goes home. However, in this case, we cannot measure the full process of LOS as well since the data that can be retrieved is only from when the patient finished the register service, inputted medicine prescription, and did the payment.

3.2 Outpatient waiting time and length of stay

From Table 2, we can see that the average length of stay of the outpatient at hospital A is 3,5 hours. It is higher than the LOS hospital's policy, which is 1,5 hours. Not to mention that the data taken is not the actual LOS, therefore average LOS data could be much higher than 3,5 hours. In addition, the Standard deviation of LOS is also high (3,3 hrs), which means that LOS time data is fluctuating highly. The maximum number of LOS is 23,54 hours, and the minimum is 6 minutes.

The average waiting time for hospital B is 5,1 hours (it is higher than government policy for outpatient waiting time, which is less than 1 hour). Data taken

from hospital B is also not the entire process of outpatient waiting time, only from getting the queue number until finishing the registration process. Thus, the actual waiting time for hospital B is much higher than 5,1 hours. The maximum waiting time for the patient to wait in this system is 10, 6 hours.

The average waiting time for hospital C is 44,4 minutes. However, data taken from hospital C is also not from the entire process. Therefore, the average waiting time for Hospital C can be higher than one hour. Not to mention that the data are taken only from the medical record distribution process, not the registration process.

4 CONCLUSION

To conclude, the patient queue is an important indicator to measure healthcare service performance by the hospital. This indicator needs to be improved since long queues will result in a delay of a patient receiving services as well as an economic loss since patients lose their productivity time. From this research, it can be concluded that the average of outpatient waiting time and length of stay in the three hospitals is above the standard of outpatient maximum waiting time set by the government and length of stay indicator set by the hospital itself.

For further research, the investigation of the root cause of the long waiting time should be conducted to recommend a new process to improve this situation.

REFERENCES

Barlow, G.L. 2002. Auditing hospital queuing. *Managerial Auditing Journal* (17)7: 397–403.

Bielen, F. & Demoulin, N. 2007. Waiting time influence on the satisfaction loyalty relationship in services. *Managing service quality An Internasional Journal:* (17)2.

Bittencourt, O., Verter, V., Yalovsky, M.J.I.J.O.P. & Management, P. 2018. Hospital capacity management based on the queueing theor (67)2:224–238.

Bleustein, C., Rothschild, D.B., Valen, A., Valatis, E., Schweitzer, L. & Jones, R.J.T.A.J.O.M.C 2014. Wait times, patient satisfaction scores, and the perception of care (20): 393–400.

Kujala, J., Lillrank, P., Kronström, V.. & Peltokorpi, A.J.J.O.H.O. 2006. Time-based management of patient processes (20)6: 512–524.

Liu, N. & Aunno, T.D. 2012. The productivity and cost efficiency of models for involving nurse practitioners in primary care: a parspective from queueing analysis. *Helth services research* 47(20: 595–613.

Meng, F., Teow, K. L., Ooi, C. K., Heng, B. H. & Tay, S. Y. 2015. Analysis of patient waiting time governed by a generic maximum waiting time policy with general phase-type approximations. *Heath care management science* (18)3 : 267–278.

Mital, K..M. 2010. Queuing analysis for outpatient and inpatient services. *Management decision* (48)3: 419–439.

Palvannan, R.K. & Teow, K. L. 2012. Queueing for health-care. *Journal of medical systems* (36)2: 541–547.

Pillay, D.I.M.S., Johari, D.M.G.R., Hazilah, A.M N., Hassan A.A.A., Abu , B.A., Salikin, F. & Ismefariana, W.I W. J. I. I O.H.C.Q.A. 2011. Hospital waiting time: the forgotten premise of healthcare service delivery? (24)7: 506–522.

Ward, P.R. 2017. 'Waiting for'and 'waiting in'public and private hospitals: *a qualitative study of patient trust in South Australia* (17)1 :333.

Advances in Business, Management and Entrepreneurship – Hurriyati et al. (Eds)
© 2021 Taylor & Francis Group, London, ISBN 978-0-367-67471-7

Stakeholder analysis of national medicine supply system

D.A. Farmaciawaty, M.H. Basri, I.N. Rachmania & F.B. Widjaja
Institut Teknologi Bandung, Bandung, Indonesia

ABSTRACT: Efficient national medicine supply system is needed to avoid shortages and stock-outs of medicines. According to WHO, one factor contributing to the availability and accessibility of medicines is the co-ordination between healthcare partners and institutions. Therefore, this research aims to understand and evaluate the medicine supply system stakeholder's behavior, intentions, interrelations, and influence. Data were collected through focus group discussion and in-depth interviews with healthcare providers, public health officials, and medicine suppliers in West Java. To carry out this research, all key (primary and secondary) stakeholders were identified. Then, stakeholder's power, interest, and influence were analyzed by using stakeholder analysis matrix. The results can be used to understand stakeholder's power and interest, the relationship among them, and the issues they care most.

1 INTRODUCTION

According to Health Law No. 36, 2009 and the Decree of Ministry of Health about national medicine policy, the government has to ensure the availability of medicine, especially essential medicines. Since the enactment of the law on regional government in 2004, the provision or supply of essential medicine as well as the budget have become the responsibility of the regional government. On the other side, the central government must provide the medicines for health program and buffer stock. The government have to maintain the medicine supply since its shortage affects patients' health and life (Uthayakumar & Priyan 2013).

The national medicine supply plays an essential role in the overall healthcare system. Inadequacy of national medicine stock can cause some problems that are often associated with the risk of patient safety (Uthayakumar & Priyan 2013). Donelle et al. (2018) highlighted the importance of keeping a stable medicine supply to keep healthcare costs down, avoid expensive price for sudden emergencies, and maintain access to medications for the entire population. Thus, designing a supply chain that remains resilient to different types of disruptions becomes a significant challenge for all stakeholders.

According to (Yang et al. 2016), there are some factors contributing to disruptions in the availability of medicines, such as 1) disruptions in the supply of raw or bulk materials; 2) manufacturing difficulties and regulatory issues (such as a capacity crunch); 3) voluntary recalls; 4) change in product formulation or manufacturer; 5) manufacturers' production decisions and economics (such as the low drug price or demand); 6) industry consolidations; 7) restricted drug production distribution and allocation; 8) inventory management (which may appear at any node of the supply chain, such as manufacturers, wholesalers, or hospitals/retailers); 9) unexpected increases in demand and shifts in clinical practice; 10) gray market or alternative distributors; and 11) natural disasters.

The availability of medicine stock becomes a complex global problem. The World Health Organization (WHO) addresses that medicine shortages has been reported for several years, affecting more than 20 countries worldwide (Bogaert et al. 2015). (Yang et al. 2016) stated that both developing and developed countries were affected by medicine shortage problems, which seemed to be worse in recent years. Medicine shortages are defined as a situation in which the current or projected demand for medicine at the user level is inadequately met.

(Walker et al. 2017) pointed out the medicine supply problem relating to its stakeholders. In this case, they found that stakeholders suggested to evaluate two areas in solving the inadequate quantification and communication of stock levels at various facilities. These areas were human resources and accurate reporting or forecasting. In stakeholders' opinion, the information on stock status was often unclear and too late to employ conservative behaviors. Regarding this, central authorities have concerns for incorrect report in peripheral healthcare. On the other hand, (Bogaert et al. 2015) explained the problems from the side of the manufacturer, such as raw material or APIs, inadequately sized production facilities, global sourcing, production or quality issues, manufacturing capacity issues, or non-compliance with the applicable regulatory standard. This is in line with (Yang et al. 2016). He stated that in terms of the manufacturing process, only a few companies are qualified to produce some materials, which means that they can monopolize the product quickly. Furthermore, he also explained the complicated and lengthy approval process of imported drugs. C. (Yang et al. 2016) also explained some problems from the hospital pharmacist side, such as severe delays in payment by hospitals, national formulary and selective drug use by hospitals, and also selective drug distribution by wholesalers. It aggravated by insufficient information sharing among different parties of the supply chain.

From the side of government, (Yang et al. 2012) stated that government intervention in network relations is widespread. It affects the whole industry and network structures. In turn, policy-makers are challenged to reduce health care expenses without compromising the quality and availability of the products (Bogaert et al. 2015).

The operations of medicine supply involve numerous stakeholders. According to (Yang et al. 2012), this has made the healthcare system relatively more complicated than general corporate network systems. This situation also happens in Indonesia. Currently, the national medicine supply system in Indonesia follows the regulations of the National Health Insurance system (JKN) as part of the government's attempt to fulfill the universal health coverage. The implementation of this new policy affects the interests, behavior, and attitude of almost all stakeholders in the healthcare system. Numerous problems related to the medicine supply system arise. In addition, the demand for medicines has dramatically increased, especially for generic medicines.

Even though the government must ensure the availability of medicine, the primary condition of medicine supply is still in the understock stage. According to WHO (2015), the median availability of selected generic medicines in Indonesia in 2007–2013 is 57.8% for private healthcare and 65.5% for the public healthcare provider. In addition, the performance report issued by the Ministry of Health (2018) also mentioned that the availability of selected essential medicine and vaccine at the primary healthcare center (puskesmas) is still around 80%.

According to WHO (2015), to avoid shortages and stock out of medicines, there should be efficient national medicine supply. It will improve access (availability, affordability (lower price), accessibility (able to distribute to a remote area), and acceptability) to medicines. One factor contributing to ensure the availability and accessibility of medicines is the coordination of healthcare partners and institutions. Transparency, communication, coordination is essential among stakeholders to resolve shortages problem. However, to do this, identification of stakeholders involved in the national medicine supply as well as their characteristics should be done. Therefore, the objective of this research is to do a stakeholder analysis in order to understand and evaluate the medicine

supply system stakeholder's behavior, intentions, interrelations, and influence.

2 METHODS

To identify the stakeholders of the national medicine supply and their characteristics, stakeholder analysis was conducted. Stakeholder analysis is a method used to gain knowledge and information related to relevant actors (Brugha & Varvasovszky 2000).The objective of this method is to understand the actor's behaviour, intentions, interrelations, agendas, interests, and influences (Brugha & Varvasovszky 2000).

Stakeholder analysis is usually conducted to analyze policy, predict policy development, implement a specific policy or project, or obtain an organizational advantage in one's dealing with other stakeholders (Varvasovszky & Brugha 2000). To do this analysis, the identification of the stakeholders' characteristics, such as their power/influence, interest, position, involvement, and impact on the issue should be done (Varvasovszky & Brugha 2000).

A qualitative methodological approach was conducted. A focus group discussion (FGD) and semi-structured interviews were conducted to gain information from different stakeholders. The FGD was conducted with management officer from one public hospital and three private hospitals in Bandung, and two representatives from Indonesia Association Hospitals (Perhimpunan Rumah Sakit Seluruh Indonesia, PERSI) in West Java. Meanwhile, a semistructured interview was done to one pharmaceutical wholesaler in Bandung. The analysis of secondary data was employed to support the primary data Table 1.

3 RESULTS AND DISCUSSION

3.1 *Stakeholder identification*

categories, which are the government and private sectors. From the government sector, there were 15 stakeholders of medicine supply system. Meanwhile, from the private sector, there were 7 stakeholders. The roles of the stakeholders are described based on information obtained from the literature studies and interview/FGD. The stakeholders' information and their role can be seen in Table 1.

3.2 *Grid analysis of stakeholder's power and interest*

Based on the above table, it can be seen stakeholders have power/influence and interest in the national drug supply system. The data is gained through the questions: which stakeholders can influence policy?; how strong is their influence?; which stakeholders who support, block, and neutral about the efficient

Table 1. Stakeholder's identification.

Stakeholder	Role
Public/Government Sector	
Ministry of Health	Ensure provision/supply, equity, affordability of health program medicines, and provide buffer stock. Issue policies related to drug supply
Pharmacy unit at Health Office	Conduct selection, planning, and procurement of government drugs
Government Pharmacy Installation	Manage government-owned drug supplies (storing and distributing drugs to government hospitals/puskesmas)
Public hospital/primary healthcare center	Manage medicine supplies at government healthcare facilities (Selecting, planning RKOs, holding, storing, distributing)
Government Health Office (Provincial, District/City)	Guarantee the provision/supply, equity, affordability, and budget management for the procurement of essential medicines, Delivering RKO to Ministry of Health
Pharmacy higher education institution	The institution that provides pharmaceutical personnel
Pharmacy staff (Pharmacists and pharmaceutical technical personnel)	Do pharmaceutical work
Government Goods and *Services Procurement Policy (LKKP)*	Procurement of e-catalog medicine Conduct government drug procurement auction
National Agency of Drug and Food Control (BPOM)	Drug control/monitoring (issuing marketing licenses for medicinal products, conducting intelligence and investigations in the field of drug control)
Health Security Administrative Body (BPJS)	Issue a policy to pay for drug claims from healthcare facilities Pay for drug claims from healthcare facilities.
Customs and Excise	Check the distribution of drugs or medicinal materials imported from abroad at the airport
Formulary national team	Create and update a list of essential medicines that are included in the list of formulary national drugs
Indonesia Pharmacist Association	Association of human resource who do pharmaceutical work
Importer	Import drug or drug's raw material
Physician	Prescribed drugs to the patient

continued.

Table 1. Continued

Stakeholder	Role
Private Sector	
Public/Government Sector	
Raw material pharmaceutical manufacturer	Providing/produce medicinal raw materials
Public/Government Sector	
Association of Pharmacy Entrepreneur	Association of pharmacy's company
Drug Distributor	Procurement, storing, distributing medicines to healthcare facilities
Private hospital/ clinic pharmacies	Manage medicine supplies in private hospitals/clinics (Selecting, plan RKO, holding, storing, distributing)
Pharmacies	Manage drug supplies at the pharmacy store (holding includes compiling RKO, storing, distributing/selling drugs to the community)
Drug Store	Store and sell drugs in the category of over-the-counter drugs

supply of medicines?; which stakeholders have an interest in the efficient supply of medicines?; and which stakeholders are in favor of an efficient drug supply system?. From this information, a map showing their influence and interest on the national drug inventory system is made (Figure 1)

The "Interest" axis shows the extent to which stakeholders are affected by efficient medicine supply and how much they have interests in policy changes that will affect the availability of medicines. There are three categories of interest: "low" interest if they experience losses because of the efficient medicine supply system; "medium" if they do not get any benefits or losses; and "high" if they are benefited from the efficient medicine supply system. Mean-while, the Power/Influence" axis measures the influence that stakeholders have on the national medicine supply system, or how far they can support or inhibit the achievement of an efficient national medicine supply system. We identify stakeholders as having "low" power/influence if they cannot impose influence/policies on other parties; "medium" if they can impose some aspects of policies on other parties; and "high" if they can impose influence/policies on other parties.

Many stakeholders are involved in the process of the national medicine supply system. (Yang et al. 2012) argued that different stakeholders in the health care system have different concerns and interests. The finding in this research support this statement.

From the map, it can be seen that the ministry of health is at the very top of the list of powers and interests in the realization of efficient national medicine supply. This fact indicates that the ministry of health will be the decision-maker that determines the state of the national medicine supply. Stakeholders that have high power/influence but moderate interest in the realization of an efficient national medicine supply is BPOM. This stakeholder must always be maintained to remain satisfied and should be appointed as patrons or supporters for policy changes or the implementation of planned strategies.

Stakeholders that have moderate power/influence and have moderate to high interests in the realization of efficient national medicine supply are pharmacy unit at Health Office, LKPP, Health Security Administrative Body, importer, physician, and formulary national team. These stakeholders are those whose opinions must be taken into account and whose alignments must always be maintained in an effort to realize an efficient national drug supply. The assistance from these stakeholders can influence the success of the planned research/policy changes/strategies.

Stakeholders that have moderate power/influence but low interest are drug distributors, the raw material pharmaceutical manufacturer, pharmacy/drug manufacturer, and Association of Pharmacy Entrepreneur. These stakeholders need to be maintained its partisanship to the efficient national drug supply policies. They can support or even inhibit the policies since the efficient drug supply can affect their business profit. Therefore the government needs to collaborate with these stakeholders to achieve the success of the policies or strategies.

Stakeholders that have low power/influence but have a high interest are drug stores, pharmacies stores, pharmacies of public hospital/primary healthcare centre, government pharmacy installation, and private hospitals/clinic pharmacies. These stake-holders can support or even inhibit the research/changes or planned policies/strategies. If they are moved, they can form a coalition that can influence the change. There are also stakeholders who are neutral as they have their own agenda or interests. Stakeholders whose power/influence is low but have moderate interests are higher education institutions, pharmacy personnel, IAI, NCC, customs and excise. These dstake-holders can maintain alignments with the planned activities/policies.

4 CONCLUSION

To conclude, an efficient national medicine supply system is needed to avoid shortages and stock-outs of medicines. Ministry of Health is the key to achieve this condition because they have power and significant influence in making and establishing the regulation regarding medicine supply system. To realize this achievement, The Ministry of Health needs a support from other stakeholders, such as: BPOM, pharmacy unit at Health Office, LKPP, Health Security Administrative Body, importer, physician, pharmacies, formulary national team, industry or drug manufacturers, raw material pharmaceutical manufacturers, drug distributor, and Association of Pharmacy Entrepreneur.

REFERENCES

Bogaert, P., Bochenek, T., Prokop, A.. & Pilc, A. 2015. A qualitative approach to a better understanding of the problems underlying drug shortages, as viewed from Belgian, French, and the European Union's Perspectives. *PloS one* 10(5).

Brugha, R. & Varvasovszky, Z. 2000. Stakeholder analysis: a review. *Health policy and planning* 15(3):239–246.

Donelle, J., Duffin, J., Pipitone, Jonathan, P. & White-Guay, B. 2018. Assessing Canada's Drug Shortage Problem. *CD howe institue commentary*.

Organization, W.H. 2015. Technical consultation on preventing and managing global stock-outs of medicines. Geneva: *World Health Organization*.

Uthayakumar, R. & Priyan, S. 2013. Pharmaceutical supply chain and inventory management strategies: optimization for a pharmaceutical company and a hospital. *Operations Research for Health Care* 2(3):52–64.

Varvasovszky, Z. & Brugha, R. 2000. A stakeholder analysis. Health policy and planning, 15(3):338–345.

Walker, J., Chaar, B.B., Vera, N., Pillai, A.S., Lim, J.S., Bero, L. & Moles, R.J. 2017. Medicine shortages in Fiji: A qualitative exploration of stakeholders views *PloS one*, 12(6).

WHO. 2015. Median Availability of Selected Generic Medicines Data by country. [Online]. Retrieved from http://apps.who.int/gho/data/node.main.488?lang=en

Yang, C., Wu, L., Cai, W., Zhu, W., Shen, Q., Li, Z. & Fang, Y. 2016. Current situation, determinants, and solutions to drug shortages in Shaanxi Province, China: a qualitative study. *PloS one* 11(10).

Yang, W.H., Hu, J.-S. & Chou, Y.-Y. 2012. Analysis of network type exchange in the health care system: a stakeholder approach. *Journal of medical systems* 36(3):1569–1581.

Advances in Business, Management and Entrepreneurship – Hurriyati et al. (Eds)
© 2021 Taylor & Francis Group, London, ISBN 978-0-367-67471-7

The customer satisfaction model on online delivery service

N. Nurjanah & A.M. Rezza
Politeknik Pos Indonesia, Bandung, Indonesia

ABSTRACT: The purpose of this research is to develop a model which describes the factors that affect the customer satisfaction using online delivery services, case study on Go-Send Services. The Factors that influence customer satisfaction are identified through understanding the influence of company image and service quality. This research also indicates that customer retention is influenced by customer satisfaction simultaneously and service quality directly. Research model testing was conducted by using the approach of Structural Equation Model with sample data obtained based on online survey of 150 respondents who have been using Go-send delivery Services. The result shows that service quality has a positive impact on company image and customer satisfaction. It also shows that there is a positive relationship between company image and customer satisfaction. Moreover, it also explores a negative relationship between customer satisfaction and customer retention.

1 INTRODUCTION

Many researches were conducted in examining service quality, company image, customer satisfaction, and customer retention (Al-Tit 2015; Ilmaniati & Wiratmadja 2016; Ladhari et al. 2013; Stan et al. 2013; Zeithaml et al. 1996). Zeithaml et al. (1996) defined service quality as a customers' judgement regarding products overall excellence. Meanwhile, Gronroos (1984) categorized service quality as a combination of technical quality (the outcome of the service performance) and functional quality (the manner in which the service is delivered).

Company Image is defined as the net result of the interaction of all experience, impressions, beliefs, feeling and knowledge that people have toward the company (Ladhari et al. 2013; Worcester 1997). Company Image as the overall perception and impression appears when the company name is mentioned (Ilmaniati & Wiratmadja 2016). The company's image is related to physical and behavioral attributes of the company such as business name, building, product or service variation, and interaction with customer (Stan et al. 2013).

Customer satisfaction is an overall attitude formed based on the experience after the purchase of a product or use a service (Fornell 1992). Customer satisfaction has a positive influence on customer retention (Al-Tit 2015). Customer retention indicates consumers intention to repurchase a service from the service provider (Morgan & Hunt No Year).

Gronroos (1984) argues that both technical quality (what customers receive from the service experience) and functional quality (the manner in which the service is delivered) contribute to build the company image.

Ladhari et al. (2013) believe that the perceived quality of a service determines the image.

The services quality are closely related to customer satisfaction which affects actual behavior from customers to stay (customer retention) or move. Al-Tit (2015) reported that customer satisfaction mediates the relationship between service quality and consumer retention. Service quality is related to customer retention. On individual level behavioral intentions, service quality perceptions can be viewed as signals of retention or defection and are desirable to monitor (Zeithaml et al. 1996).

Go-Send is a famous online delivery services. Customers use go-send to do shipment within the city. The service provided by Go-Send are the same day delivery service dan instant delivery service where customers can choose the type of service based on their need. Thus, this research would discover the relationship between company image, service quality, customer satisfaction, and customer retention in online transport delivery service.

2 METHODS

This research conducted an online survey to collected data. The target population of this study consists of all customers who have used Go-Send delivery service. The survey was conducted in June 2019. According to Gojek Indonesia, Go-Send delivery service can be classified into two types: same-day delivery service and instant delivery service. The number of samples or the respondents were 150 respondents. The data analysis technique employed in this research was structural equation model.

Table 1.	Model fit indicators.		
Measure	Estimate	Threshold	Interpretation
CFI	0.919	>0.95	Acceptable
SRMR	0.064	<0.08	Excellent

Table 2.	Regression weight.			
	Est.	S.E	C.R	P Value
SQ → CI	.783	.111	7.070	***
SQ → CS	.653	.089	7.334	***
SQ → CR	.674	.105	6.439	***
CI → CS	.648	.233	2.781	0.005
CS → CR	−.048	.327	−.147	0.883

Table 3.	Standardized direct effect.			
	SQ	CI	CS	CR
CI	.688	.000	.000	.000
CS	.484	.533	.000	.000
CR	−.026	.000	.480	.000

Table 4.	Standardized indirect effect.			
	SQ	CI	CS	CR
CI	.000	.000	.000	.000
CS	.367	.000	.000	.000
CR	.408	.256	.000	.000

3 RESULTS AND DISCUSSION

The examination of the model fit was conducted using AMOS Version 22. The Cut off criteria for fit indexes in covariance structure analysis used fit index combination of Comparative Fit Index (CFI) and the Standardized Root Mean Square Residual (SRMR). The expected value CFI is 0.96 or higher and the expected value for SRMR is 0.09 or lower. It shows that value is less than 0.1. Table 1 shows that the values are satisfactory and suggest an adequate fit of the model.

Hair et al. (2017) states that the path coefficients values range from −1 to +1. If the values is closer to +1, it means that the relationship between the two constructs is strong. Meanwhile if the value is closer to −1, it indicates that the relationship is negative. The result of the path analysis (Table 3) shows that service quality has a positive and significant impact on company image (b = 0.738; p < 0.05) which supports H1.

The study presumed that both service quality (H2) and company image (H4) would have a positive influence on customer satisfaction. This result shows that service quality has a positive impact on customer satisfaction (b= 0.653, p < 0.005), this result (Table 3) is in agreement with Al-Tit (2015) and Zeithamal & Bitner (1996). Company image has a positive and significant impact on customer satisfaction (b= 0.648, p < 0.005). Ladhari et al. (2013) stated that corporate image becomes an important tool for companies to strengthen, retain customers and maximize profits. Good image will help to develop and maintain relationships with consumers, which can affect customer satisfaction (Stan et al. 2013).

In this study, customer retention acts as a measure of customer's intention to repurchase with the service on the future. The result (Table 2) shows that service quality has a positive impact on customer retention

(b= 0,674, p<0.005) thus, it supports H3. The same result was found by Al-Tit (2015).

In terms of the relationship between customer satisfaction and consumer retention, the findings (Table 3) demonstrates that customer satisfaction has a negative and not significant impact on customer retention (b= −0,048, p>0.05) thus, it ignores H5. This result indicates that even though customers are satisfied with the service, the customers does not always repurchase the same service. This can be influenced by various factors, such as the emergence of competitors that offer cheaper price for the same services. However in Tables 3 and 4, customer satisfaction has mediation effect on customer retention. This result shows that other factors influence the relationship and requires further.

4 CONCLUSION

This study develops and empirically tests a conceptual model of customer retention, while considering three antecedents: service quality, customer satisfaction, and company image. The result shows that service quality has a positive and significant impact on company image, customer satisfaction and customer retention. Company image has a positive and significant impact on customer satisfaction. Customer satisfaction has a negative and not significant impact on customer retention. This result is different with Al-Tit (2015) whose findings demonstated that customer satisfaction has a positive influence on customer retention.

ACKNOWLEDGEMENT

Special thanks is given to DIKTI, and Ali Mohamad Rezza as the team of this article for their support so that this article can be completed and submitted in this conference.

REFERENCES

Al-Tit, A. A. 2015. The effect on service and food quality on customer satisfaction and hence customer retention. *Asian Social Science* 11: 129.

Fornell, C. 1992. A national customer satisfaction barometer: The Swedish experience. *Journal of Marketing*: 6–21.

Gronroos, C. 1984. A service quality model and its marketing implication. *European Journal od Marketing*: 36–44.

Hair, J., Hult, G., Ringle, C., & M, S. 2017). A primer on partial leas squares structural equation modeling *second edition*. Sage.

Ilmaniati, A., & Wiratmadja, I. I. 2016. Pengembangan model loyalitas nasabah dan perbankan syariah (Bank Syariah Mandiri). *Journal of Engineering and Management in Industrial System*: 94–101.

Ladhari, R., Souiden, N., & Ladhari, I. 2013. Determinants of loyalty and recomendation: the role of perceived service quality, emotion satisfaction, and image. *Journal of Financial Service Marketing*: 145–159.

Morgan, R. M., & Hunt, S. D. No Year. The commitment trust theory of Relationship Marketing. *Journal of Marketing*: 20–38.

Stan, V., Caemmerer, B., & Cattan-Jallet, R. 2013. Customer loyalty developmnet: the role of switching costs. *The Journal of Applied Business Research*: 1541–1554.

Worcester, R. M. 1997. Managing the image of your bank: the glue that binds. *International Journal of Bank Marketing*: 146–152.

Zeithamal, V. A., & Bitner, M. J. 1996. *Services marketing*. New York: McGraw-Hill.

Zeithaml, V. A., Berry, L., & Parasuraman. 1996. The behavioral consequences of service quality. *Journal of Marketing* 60: 31–46.

Inventory planning policy for spare parts using classification scheme model toward stochastic demand and lead time in aircraft industry

Z. Suryaputri, A.Y. Ridwan & B. Sentosa
Universitas Telkom, Bandung, Indonesia

ABSTRACT: Spare parts are important components in maintenance, repairing, and operating activities in manufacturing companies. There are some cases in the aircraft industry in which the lead time and demand of spare parts is stochastic and lead to stockouts. Due to this condition, research is needed. The research is aimed to obtain an optimal inventory policy under stochastic demand and lead time. Inventory policy was done by classifying the spare parts using classification scheme model for stochastic demand and lead time with continuous review (s, Q) policy to address the problem. A distribution approach was conducted to describe the pattern of demand of spare parts and to define the fit equation so the optimal inventory system was capable of generating order quantities and reorder points to minimize stockouts that were caused by stochastic demand and lead time. The results of the inventory policy calculation provided to increase 100% from actual fill rate.

1 INTRODUCTION

In the production process, the performance of machines needs to be maintained by carrying out maintenance and repairing activities so that the production process can run smoothly. Spare parts are important components in those activities, therefore, the availability of spare parts must be maintained. The downtime of spare parts can cause loss of income, customer dissatisfaction, and be very expensive (Cavalieri et al. 2008; Driessen et al. 2014). The supply chain of aircraft companies faces significant challenges in providing rapid repair and maintenance services while minimizing costs (Liu et al. 2013). Therefore, a strategy of stocking spare parts has an important impact on the productivity and efficiency of industrial plants and an important role in inventory management.

The demand of spare parts has specific characteristics and is different from other materials such as work-in-process. A large proportion of the inventory is described as having an intermittent or a slow-moving demand pattern (Eaves & Kingsman 2004). Spare parts can be classified into intermittent, irregular, slow moving, or other demand patterns. An intermittent demand pattern has variable demand sizes, and the demand appears randomly and sometimes there are certain periods with no transactions. An intermittent demand occurs when the frequency of customer requests varies. A slow-moving demand pattern always has low demand sizes. It occurs when it does not have a large variation between demand and quantity of demand. Materials that have intermittent demand patterns can be classified as having intermittent demand, erratic/irregular demand, lumpy demand, and slow-moving demand (Ghobbar & Friend 2002).

Also, the lead time is a random variable (Silver et al. 2017). Eaves & Kingsman (2004) propose an analytical method for classifying demand by elaborating the variance components of lead time demand (LTD) into transaction variability, and the demand size variability and the lead time variability into smooth, irregular, slow moving, or intermittent patterns. In addition, Silver et al. (2017) propose LTD components to establish the boundaries of slow-moving and fast-moving items.

The common model that is used for inventory is EOQ. The EOQ model was presented originally by Ford Whitman Harris in 1913. Along with the evolution of the industry, the EOQ model had been developed by many researchers. Chuang & Chiang (2015) investigated the EOQ model that was applied for stochastic demands in General Motors' dealerships. In their paper, they assumed that according to the fit statistics, involving the EOQ model into inventory management could lead to better strategic decision making. Widyadana et al. (2017) solve the inventory policy for stochastic and intermittent demand at a bicycle shop using Monte Carlo simulation and an evolutionary algorithm. They compare three continuous review polices, (s, Q), (s, S), and (S, T) policies, and the result shows that the performance of the (s, Q) policy is the best for all products that have stochastic and intermittent demand. Meanwhile Nugraha & Wijaya (2015) implement the (s, Q) policy and EOQ model to minimize total inventory cost and also to determine order quantity and the time for effective ordering for maintenance support parts in a manufacture company.

The main objective of this research was to propose an inventory policy for spare parts. This research proposed an inventory policy for spare parts in an aircraft industry in Indonesia by considering stochastic demand and lead time. Specifically, this paper aimed to solve inventory problems in real situations where the demand and lead time of spare parts is stochastic. This paper considered a statistical approach to determine the pattern of demand and a modelling LTD analytical method to classify the spare parts, while using (s, Q) policy for slow moving and intermittent items.

2 METHOD

The following was the methodology of the study along with a list of the data collected:

a) Data of production machinery spare parts demand.
b) Lead time data for each spare part.
c) Data of storage costs, the costs incurred due to storing spare parts in the warehouse.
d) Data on shortage costs, costs that must be incurred by the company when unable to meet the needs of the production machinery spare part products needed.
e) Data message costs, costs that must be incurred by the company to bring in spare part products.
f) Data on spare part/unit prices determined by each vendor/supplier.

There were five steps in this study, those are:

1) Classifying the spare part demand patterns using coefficient variance (CV) classification. Inputs from this classification were lead time data, spare part demand data, and time data between spare part demand transactions obtained from demand data.
2) Testing the distribution of spare part product demand data, which has a slow-moving and intermittent pattern. Inputs from this distribution test were classified demands that have intermittent and slow-moving demand patterns. The results of the data distribution test showed the data demand distribution pattern of each spare part product.
3) Calculating the total inventory costs and actual condition fill rate, which included calculating inventory costs and the percentage of timely fulfillment of demand on actual conditions for comparative analysis with proposed conditions.
4) Calculating the order quantity, reorder point, and proposed safety stock. In this study, the variables that had been obtained from the calculation results were distributed into the equation carried out iteratively until the values obtained were convergent (Nahmias & Olsen, 2015). The safety stock was obtained from the reorder point, which was reduced by the average demand during the lead time (the result of multiplying between the average demand and the average lead time).
5) Calculating the total inventory cost and fill rate of the proposed condition, after getting a large reorder point, order quantity, and safety stock. Then, calculating the total cost of inventory and the fill rate at the proposed condition. The results of this step were used in a comparison at the stage of data analysis of proposals for existing conditions.

3 RESULTS AND DISCUSSION

3.1 *Result*

The continuous review inventory (s, Q) policy was applied using data from the aircraft industry. Spare part products were used, where stockouts happened and caused downtime.

3.1.1 *Demand classification*
The data were processed by looking for the value of the coefficient variation. Also, for the distribution test, this study used SPSS software and the Kolmogorov–Smirnov test. Furthermore, products were classified based on the value of each coefficient variation. For example is spare part number A36E031, where:

Coefficient Variation Demand $= 13.47$
Coefficient Variation Transaction $= 0.35$
Coefficient Variation Lead Time $= 0$
Class $=$ Irregular
Distribution $=$ Poisson

Materials that have intermittent demand patterns can be classified under intermittent demand, erratic/irregular demand, lumpy demand, and slow-moving demand (Ghobbar & Friend 2002).

3.1.2 *Lead time analysis*
Lead time is a fundamental component in any inventory control system. If demand and lead time are known, inventory management can be planned with good accuracy. However, this was not a common situation in inventory management. For most spare parts, demand and lead time are stochastic (Conceicao et al. 2015). Since the lead time of spare parts was uncertain, we must know the demand that will occur, so that stockouts due to high demand during lead time can be minimized.

In this paper, we use spare part number A36E031 as an example to show the calculation. For this spare part, the mean DL is 1 unit. This means that the demand that occurs during lead time is 1 unit.

3.1.3 *Parameter inventory*
Here is the example of spare part A36E031 to determination the order quantity (Q) and reorder point (s) spare part for the proposed system by (s, Q) policy.

The steps taken in the determination of Q* and s* are as follows:
Iteration 1
Step 1: Calculate order quantity with EOQ.

$$Q = \sqrt{\frac{2KD}{h}}$$

$$Q = \frac{2(\text{Rp } 5,675,000)(2)}{(\text{Rp } 2,639,536.4)}$$

$$Q = 2.9 \approx 3\text{pcs} \tag{1}$$

Step 2: Calculate reorder point, with substituting the EOQ for Q.

$$1 - F(s) = \frac{Qh}{pD}$$

$$1 - F(s) = \frac{3(\text{Rp } 2,639,536.4)}{(\text{Rp } 18,370,705)(2)} = 0.2107$$

$$F(s) = 0.7893 \tag{2}$$

Because the lead time is uncertain, the demand during lead time is $\mu = 0.5$ in the cumulative Poisson table, the value 0.7893 is between F (0) = 0.6065 and F (1) = 0.9098. According to Nahmias & Olsen (2015), rounding to a greater value is done. Then the value of F (1) with s = 1 is obtained from the cumulative Poisson table for reorder points with the value F (1) = 0.9098.

Step 3: Calculate expected shortage per cycle.

For demand that has a Poisson distribution, with the value F (s) obtained from the previous step, namely F (1) = 0.9098, then 1−F (s) = 0.0902. And for F (s + 1) =
F (2), where s is 1 as obtained from the cumulative Poisson table, the formula is as follows:

$$n(s) = \mu[1 - F(s)] - s[1 - F(s + 1)]$$

$$n(s) = 0.5 \times (0.0902) - 1(1 - 0.9865)$$

$$n(s) = 0.307 \approx 0 \tag{3}$$

Step 4: Calculate order quantity.

$$Q = \sqrt{\frac{2D[K + p\,n(s)]}{h}}$$

$$Q = \sqrt{\frac{2(2)[(\text{Rp } 5,675,000) + (\text{Rp } 18,370,705)(0)]}{(\text{Rp } 2,639,536.4}})$$

$$Q = 2.9 \approx 3 \text{ pcs} \tag{4}$$

Iteration 2

Step 1: With Q = 3 from iteration 1.

$$1 - F(s) = \frac{Qh}{pD}$$

$$1 - F(s) = \frac{3(\text{Rp}2,639,536.4)}{(\text{Rp}18,370,705)(2)} = 0.2107$$

$$F(s) = 0.7893 \tag{5}$$

Because the lead time is uncertain, the demand during lead time is $\mu = 0.5$ in the cumulative Poisson table, the value 0.7893 is between F(0) = 0.6065 and F (1) = 0.9098. Then the value of F (1) with s = 1 is obtained from the cumulative Poisson table for reorder points with the value F (1) = 0.9098.

Step 2: Calculate expected shortage per cycle.

With the value F (s) obtained from the previous step, namely F (1) = 0.9098, then 1-F (s) = 0.0902. And for

F (s + 1) = F (2) = 0.9865 where s is 1 as obtained from the cumulative Poisson table, the formula is as follows:

$$n(s) = \mu[1 - F(s)] - s[1 - F(s + 1)]$$

$$n(s) = 0.5 \times (0.0902) - 1(1 - 0.9865)$$

$$n(s) = 0.307 \approx 0 \tag{6}$$

Step 3: Calculate order quantity.

$$Q = \sqrt{\frac{2D[K + p\,n(s)]}{h}}$$

$$Q = \sqrt{\frac{2(2)\,[(\text{Rp } 5,675,000) + \text{Rp } 18,370.705)(0)]}{(\text{Rp } 2,639,536.4)}}$$

$$Q = 2.9 \approx 3 \text{ pcs} \tag{7}$$

The values of Q and s are convergent with an order quantity of 3 unit and reorder point *s* of 1 unit for spare part no. A36E031.

Next is to determine the safety stock (SS) for the proposed system as follows:

$$SS = s - \mu$$
$$SS = 1 - 1$$
$$SS = 0 \text{ unit}$$

The results of the calculation of the parameters above, obtained:

$$\text{Order Quantity (Q)} = 3\text{unit}$$
$$\text{Reorder point (s)} = 1\text{unit}$$
$$\text{Safety Stock} = 0\text{unit}$$
$$\text{Fill Rate} = 100\%$$

3.2 Discussion

In Table 1, it can be seen the comparison between the total inventory cost of the actual inventory system with the proposed inventory system of all spare parts. The total inventory cost was reduced and the proposed system saved 70% and the fill rate of all spare parts increased 100% from the actual fill rate.

Sensitivity Analysis of Component of Inventory Cost

It can be concluded that the parameters of holding cost, ordering costs, and shortage costs were sensitive parameters and have a significant effect on total inventory costs.

In the process of calculation, the efficiency of total inventory costs and increase in fill rate were achieved. This paper used a statistical approach to determine the pattern of demand, modelling LTD analytical method to classify the spare parts, and used (s, Q) policy for slow-moving and intermittent items to calculate the proposed inventory costs. The explanation of these are as follows:

3.2.1 Demand classification

The spare parts used in the aircraft industry were classified by LTD. The classification was done by classifying the transaction variability, demand size variability, and lead time variability (Eaves & Kingsman 2004) as shown in Table 1.

Table 1. Cost comparison between actual and pro-posed inventory system.

Component of Inventory System	Actual Inventory System	Proposed Inventory System
Holding cost	Rp 23,180,890	Rp 24,113,349
Ordering cost	Rp 36,303,739,43	Rp 36,421,989
Shortage cost	Rp 145,386,905	Rp –
Total inventory cost	Rp 204,871,534	Rp 60,535,338
Save	Rp 144,336,196.30	
% Save	70%	

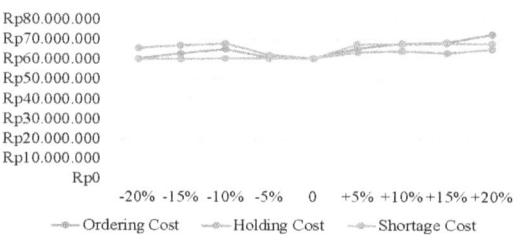

Figure 1. Sensitivity analysis.

Table 2. Coefficient variance classification.

Transaction Variability	Demand size Variability	Lead Time Variability	Demand Classes
≤0.74	≤0.10	–	Smooth
≤0.74	>0.10	–	Irregular
>0.74	≤ 0.10	–	Slow moving
>0.74	>0.10	≤0.53	Mildly intermittent
>0.74	>0.10	>0.53	Highly intermittent

The equation for the coefficient variance is as follows:

$$CV = \sigma/\bar{x} \qquad (8)$$

Where, σ is the standard deviation of the parameter and $\bar{x}$ is the mean of the parameter. Inventory problems with uncertainty lead time and the demand classification as intermittent/slow moving can be solved by method (s, Q) policy with a distribution approach to calculate safety stock, reorder point, and order quantity so that optimal inventory costs are obtained (Conceicao et al. 2015; Nugraha & Wijaya 2015).

3.2.2 Demand lead time

Demand and lead time are independent random variables (Silver et al. 2017). In calculating the mean, variance, and standard deviation of the request during the interval, the lead time was as follows:

$$\mu = \lambda \times \mu_\tau \qquad (9)$$

$$Var = \sigma^2 = (\mu_\tau \times v^2) + (\lambda^2 \times \sigma_\tau^2) \qquad (10)$$

$$\sigma L = \sqrt{Var\,(DL)} \qquad (11)$$

Where, μ and σ^2 were the mean and variance of the demand during the lead time interval, λ and v^2 were the mean and variance of demand, and μ_τ and σ_τ^2 were the mean and variance of lead time.

3.2.3 Continuous review (s, Q) policy

In this policy, Q was the amount of quantity that was ordered when the stock level reached s value. The policy combined with the distribution approach, used the parameter of each distribution that fit to the pattern of demand to define the appropriate formula. Figure 1 shows a graphic image of this policy (Silver et al. 2017):

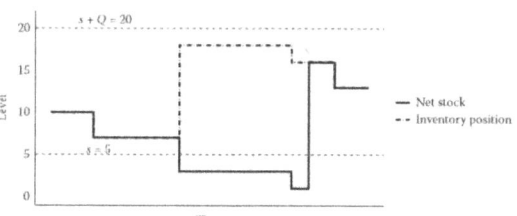

Figure 2. Continuous review (s, Q).

The order quantity Q and reorder point R were determined by optimizing total inventory cost G (s, Q) (Nahmias & Olsen 2015) as shown in the following model:

$$G(s,Q) = h\left(\frac{Q}{2} + s - DL\right) + \frac{KD}{Q} + \frac{p\lambda n(s)}{Q} \qquad (12)$$

To get optimal value, the determination of the Q and s values was done iteratively until the values obtained were convergent (Nahmias & Olsen 2015). Here is the step:

Step 1: Calculate the initial Q value with the EOQ method:

$$Q = \frac{\sqrt{2KD}}{h} \qquad (13)$$

Step 2: Calculate the s value by substituting the Q value for the equation and $F(s)$ is replaced with cumulative distribution function of demand:

$$1 - F(s) = \frac{Qn}{pD} \tag{14}$$

Step 3: Substitute the s value and replace $F(s)$ with probability density function of demand to calculate the value of expected shortage per cycle $n(s)$ for Poisson distribution:

$$n(s) = \mu[1 - F(s)] - s[1 - F(s+1)] \tag{15}$$

Step 4: Define Q with this equation:

$$Q = \sqrt{\frac{2D[K + p\,n(s)]}{h}} \tag{16}$$

Step 5: Return to Step 2 and continue until the convergent Q and s results are obtained.

Safety Stock: Safety stock as additional inventory is provided to protect against the possibility of being out of stock. Following is the equation for safety stock:

$$SS = s - DL \tag{17}$$

Where s is the reorder point, which is the point at which the order will be made and DL is the result of multiplication between the mean of demand and the mean of lead time.

3.2.4 *Inventory cost*

The objective of a spare parts inventory system is to minimize inventory costs and increase service level. There are relevant costs associated with an inventory system (Silver et al. 2017):

a) Order Cost (OC) with K as the order cost/order, λ as the demand per unit of time, and Q as the order quantity as follows:

$$OC = \frac{K\lambda}{Q} \tag{18}$$

b) Carrying Cost/Holding Cost (HC) with h as the holding cost, SS as the safety stock, and Q as the order quantity as follows:

$$HC = h\left(SS + \frac{Q}{2}\right) \tag{19}$$

c) Shortage Cost (SC) with P as the shortage cost per unit per period, with λ as the demand per unit of time, $N(s)$ as the expected shortage cost per cycle, and Q as the order quantity as follows:

$$SC = \left(\frac{\lambda n(s)}{Q}\right)P \tag{20}$$

4   CONCLUSION

Based on the research objectives that have been formulated, an inventory policy can be obtained with the result of parameters in the form of order quantity, reorder point, the amount of safety stock, and the results of increase fill rate and savings made on the total cost of the proposed policy inventory. This policy was based on the results of a continuous review (s, Q) policy for cases of stockouts and the lead time was uncertain. Inventory policy was exemplified in stock no. A36E031, and the results of the calculation showed that the value of order quantity was 3 units, reorder point was 1 unit, and safety stock was 0 units. An order was made if the inventory position had reached the reorder point. Furthermore, fill rate and total inventory cost in one year using the proposed system for all spare parts increased the fill rate by 100% and saved 70% in actual total inventory cost.

ACKNOWLEDGEMENT

We are grateful to Indonesia Aircraft Industry that provided the data for this study. We also acknowledge the financial support provided by the Faculty of Industrial Engineering of Telkom University, Bandung. Finally, we thank the anonymous reviewers for their constructive comments, which helped us to improve the manuscript.

REFERENCES

Cavalieri, S. M., Garetti, M., Macchi, & Pinto, R. 2008. A decision making framework for managing maintenance spare parts. *Production Planning & Control*: 379–399.

Chuang, C. H., & Chiang, C. Y. 2015. Dynamic and stochastic behavior of coefficient of demand uncertainty incorporated EOQ variables: An application in finished goods inventory from general motors' dealerships. *International Journal of Production Economics*: 95–109.

Conceicao, S. V., Silva, C. D., & Nunes, R. 2015. A demand classification scheme for spare part inventory model subject to stochastic demand and lead time. *Production Planning & Control*.

Driessen, M., Arts, J., Houtum, G. J., & Rustenburg, J. W. 2014. Maintenance spare part planning and control: A framework for control and agenda for future research. *Production Planning & Control*: 1–20.

Eaves, A., & Kingsman, B. 2004. Forecasting for the ordering and stock holding of spare parts. *Journal of the Operational Research Society* 55: 431–437.

Ghobbar, A., & Friend, C. 2002. Sources of intermittent demand for aircraft spare parts within airline operations. *Journal of Air Transport Management, 8*(4): 221–231.

Liu, P., Huang, S. H., Mokasdar, A., Zhou, H., & Hou, L. 2013. The Impact of additive manufacturing in the aircraft spare parts supply chain: Supply chain operation reference (Scor) model based analysis. *Production Planning & Control*: 1169–1181.

Nahmias, S., & Olsen, T. L. 2015. Production and operations analysis (7th edn). United States of America: Waveland Press Inc.

Nugraha, S. W., & Wijaya, A. R. 2015. Penentuan safety stock, reorder point dan order quantity suku cadang mesin produksi berdasarkan ketidakpastian demand dan lead time pada perusahaan manufaktur (studi kasus di PT Wijaya Karya Beton PPB Boyolali). *Seminar Nasional Teknik Industri Universitas Gadjah Mada 2015.*

Silver, E. A., Pyke, D. F., & Thomas, D. J. 2017. Inventory and production management in supply chains (4th edn). New York: CRC Press.

Widyadana, I. G., Tanudireja, A. D., & Teng, H. M. 2017. Optimal inventory policy for stochastic demand using Monte Carlo simulation and Evolutionary Algorithm. *JIRAE*: 8–11.

Innovation of bank performance index model based on the peer group model of banks operating in Indonesia

Sugiarto & F. Budhijono
Universitas Prasetiya Mulya, Tangerang, Indonesia

A. Nursiana
Sekolah Tinggi Ilmu Ekonomi Wiyatamandala, Jakarta, Indonesia

ABSTRACT: For the purposes of evaluating and monitoring the bank's performance comprehensively, a Bank Performance Index is needed. The authors developed a model as a bank performance index based on the bank's peer group model. This model will contribute significantly to national interests as a source of reference for other banks in assessing their performance and can also be used by Otoritas Jasa Keuangan (OJK, Financial Services Authority of Indonesia) as a guideline in assessing the comprehensive performance of banks operating in Indonesia. The establishment of the Bank Performance Index model was based on 15 banks in the peer group of Bank Jasa Jakarta, a private bank operating in Indonesia. Performance data from December 2015 to June 2018 were taken from Otoritas Jasa Keuangan website. The bank's performance index model developed by the authors is proven to be able to represent the prevailing reality.

1 INTRODUCTION

A bank's performance should be evaluated by considering benchmarking its performance to the performance of the peer group banks (Peraturan Bank Indonesia-Bank Indonesia Regulation, Nomor 13/1/PBI/2011). Bank Indonesia (BI) as the Indonesian bank authority has published an official listing of peer group banks based on bank's core capital. Bank Indonesia categorizes banks operating in Indonesia in 4 groups under the name BUKU (Bank Umum Kelompok Usaha – Commercial Bank Business Group) I, II, III and IV. Banks that are classified under BUKU I are banks with core capital below IDR 1 trillion. Banks that are classified under BUKU II have a minimum core capital of IDR 1 trillion to under IDR 5 trillion. Banks that are classified under BUKU III are banks with core capital of IDR 5 trillion to under IDR 30 trillion and banks that are classified under BUKU IV are banks with minimum core capital of IDR 30 trillion.

The determination of the bank's peer group as issued by Bank Indonesia is doubted to match the bank's needs due to lack in specifics in scope for certain banks. With the wide range encompassed it will be difficult for banks operating in Indonesia to evaluate their performance based on the banks' peer groups. However the majority of banks operating in Indonesia do use the peer group banks as determined by Bank Indonesia as a reference in evaluating and making performance comparisons since there has been no standard provision imposed by Bank Indonesia in

setting the bank peer groups yet. Many banks operating in Indonesia have also subjectively set their own peer groups which are not yet proven as scientifically accountable. Taking into consideration, the important role of bank's peer group as the benchmark for objective performance monitoring, the determination of bank's peer group should be done objectively. The author formulated an innovation by developing a peer group bank mod-el based on probabilistic approaches (Sugiarto 2017a). The bank's peer group model developed by the author had been tested using cross section (Sugiarto et al. 2018) and time series data (Sugiarto 2018) and proven to be a reliable and accountable model (Sugiarto et al. 2018). By using this bank's peer group model, it is possible to find out the performance ratings of each bank compared to the performance of other banks in the peer group related to the variables such as Non Performing Loan (NPL), Capital Adequacy Ratio (CAR), Return on Assets (ROA) and Return on Equity (ROE) (Sugiarto 2018) from thereof stated as concerned variables. However for the purpose of evaluating and monitoring the overall performance of all concerned variables, a model that can reflect the bank's performance comprehensively will be needed. The question is whether it is possible to develop a model which can be used widely or in the peer group. For this purpose, the authors are innovating to develop a model for a bank performance index based on the peer group model already developed by the first author, which will be the novelty of this research. This research aims to develop an innovative model as a bank

191

performance appraisal index based on equalizing the amount of core capital. The model formed is intended to be a model that can be used generally by banks in assessing their performance. If the model developed by authors proves to be a reliable model, this model will contribute significantly to national interests since it can be a source of reference for other banks operating in Indonesia in assessing their performance and can also be used by OJK as a guideline in assessing the comprehensive performance of banks operating in Indonesia.

However its reliability will only be known after the testing is done. A study was conducted using Bank Jasa Jakarta (BJJ), a private bank operating in Indonesia, as the bank that will be used for testing this model. Two reasons for this choice of bank were first, in the previous research BJJ's peer group bank was already formed on the basis of a peer group bank model formulated by the authors. The second reason was because BJJ is a bank with excellent and consistent performance. Throughout its history, BJJ has succeeded in maintaining its position as a healthy and solid bank, and consistently implementing prudent policies aimed at maintaining credit quality and strong capital. and managing a healthy liquidity position. In the achievement of good performance, BJJ has received many awards from several banking institutions. If the test results obtained from using this bank's performance index model show that BJJ's performance is in line with the awards given for BJJ's performance, this indicates that the bank's performance index model used is a reliable model.

2 METHODS

This research was conducted to produce a Bank Performance Index model on the basis of a bank peer group model formulated by Sugiarto (2017b). The establishment of the Bank Performance Index Model will be based on the 15 banks in the peer group of BJJ. Performance data from December 2015 to June 2018 were taken from bank publication reports from the Otoritas Jasa Keuangan (OJK Financial Services Authority of Indonesia).

Performance monitoring was carried out on the variables of NPL, CAR, ROA and ROE. Bank's equalization performance was carried out on the basis of the core capital of each bank towards the core capital of BJJ. In the following stage, the data that had undergone equalization was converted into performance values. When doing the conversion, the range value for each variable that was of concern was obtained from the data of December 2015 until June 2018. The lowest conversion value was zero and the highest value was 100. Further, ratings were carried out. Rating 1 was given to bank that had the highest value in the bank's performance index and rating 15 was given to bank with the lowest bank's performance index. To reinforce the argument why BJJ deserve to win so many awards,

Table 1. NPL performance rating for the period of December 2015 to December 2016.

Bank Name	Dec-15	Mar-16	Jun-16	Sep-16	Dec-16
BJJ	4	3	2	2	2
Bnparahyangan	12	13	13	13	13
BOII	14	14	15	15	15
Bumi arta	7	6	8	9	7
Capital	8	4	4	4	10
Ganesha	15	15	9	7	5
Index	9	7	11	11	9
Mas	3	2	5	3	8
Maspion	6	5	6	5	4
Mayora	5	10	10	8	11
Metro express	2	9	3	10	3
Nationalnobu	1	1	1	1	1
Sahabat sampoerna	11	12	12	12	12
SBI Indonesia	13	11	14	14	14
Windu (CCBI)	10	8	7	6	6

the formation of the bank's performance index will use December 2017 performance data. If the results obtained from bank's performance index showed that BJJ's performance was in line with the awards given regarding BJJ's performance, it meant that the bank's performance index model was a representative model.

3 RESULTS AND DISCUSSION

Tables 1 and 2 showed the quarterly perfor-mance ratings of banks that were members of the BJJ peer group with respect to the NPL variables for the period of December 2015 to June 2018.

Tables 3 and 4 showed the quarterly performance ratings of banks that are members of the BJJ peer group with respect to the CAR variable for the period December 2015 to June 2018.

Throughout the period of observation on BJJ's performance in relation to CAR, this variable showed improvement from 5th place in December 2015 to 4th place in June 2018. BJJ's CAR perfor-mance rating has never been lower than 5th.

Tables 5 and 6 showed the quarterly perfor-mance ratings of the banks that were members of the BJJ peer group with respect to the ROA variable for the period December 2015 to June 2018.

Throughout the period of observation, BJJ's performance in relation to ROA was always at a first rank rating to 3rd. Although BJJ's performance rating due to ROA's performance had dropped to the 3rd place in March 2017, it had undergone improve-ments after that and had never been lower than 2nd.

Table 2. NPL performance rating for the period of December 2016 to June 2018.

Bank Name	Mar-17	Jun-17	Sep-17	Dec-17	Mar-18	Jun-18
BJJ	2	3	3	3	3	2
Bnparahyangan	14	15	15	14	15	15
BOII	15	13	14	15	14	14
Bumi arta	5	8	10	6	7	6
Capital	9	9	8	9	9	8
Ganesha	4	5	5	4	4	13
Index	10	10	9	10	10	9
Mas	7	6	6	5	5	4
Maspion	3	4	4	8	6	5
Mayora	12	11	11	12	13	10
Metro express	8	2	2	2	2	1
Nationalnobu	1	1	1	1	1	3
Sahabat sampoerna	11	12	12	11	11	12
SBI Indonesia	13	14	13	13	12	11
Windu (CCBI)	6	7	7	7	8	7

Table 4. CAR performance rating for the period of December 2016 to June 2018.

Bank Name	Mar-17	Jun-17	Sep-17	Dec-17	Mar-18	Jun-18
BJJ	5	4	4	4	4	4
Bnparahyangan	12	12	12	13	10	10
BOII	2	2	2	1	1	1
Bumi arta	10	9	8	8	8	8
Capital	14	14	14	14	14	14
Ganesha	4	3	3	3	3	3
Index	9	8	7	6	9	9
Mas	8	10	11	11	12	12
Maspion	7	7	6	9	6	6
Mayora	6	6	5	5	5	5
Metro express	3	5	10	10	13	13
Nationalnobu	11	11	9	7	7	7
Sahabat sampoerna	13	13	13	12	11	11
SBI Indonesia	1	1	1	2	2	2
Windu (CCBI)	15	15	15	15	15	15

Table 3. CAR performance rating for the period of December 2015 to December 2016.

Bank Name	Dec-15	Mar-16	Jun-16	Sep-16	Dec-16
BJJ	5	4	4	5	5
Bnparahyangan	13	13	13	13	12
BOII	9	6	5	4	4
Bumi arta	10	11	11	11	11
Capital	14	14	14	14	14
Ganesha	2	2	2	2	3
Index	8	8	9	9	9
Mas	4	5	6	7	7
Maspion	7	9	8	8	8
Mayora	12	7	7	6	6
Metro express	1	1	1	1	2
Nationalnobu	6	10	10	10	10
Sahabat sampoerna	11	12	12	12	13
SBI Indonesia	3	3	3	3	1
Windu (CCBI)	15	15	15	15	15

Table 5. ROA performance rating for the period of December 2015 to December 2016.

Bank Name	Dec-15	Mar-16	Jun-16	Sep-16	Dec-16
BJJ	1	2	1	1	1
Bnparahyangan	10	15	11	11	14
BOII	14	8	15	15	15
Bumi arta	7	7	8	8	7
Capital	9	13	9	9	9
Ganesha	3	1	4	4	3
Index	2	6	6	5	2
Mas	5	5	5	6	5
Maspion	6	4	3	3	4
Mayora	8	9	7	7	6
Metro express	11	3	2	2	11
Nationalnobu	13	14	14	12	10
Sahabat sampoerna	4	10	10	10	8
SBI Indonesia	15	11	13	14	13
Windu (CCBI)	12	12	12	13	12

Tables 7 and 8 showed the quarterly perfor-mance ratings of banks that were members of the BJJ peer group with respect to the ROE variable for the period December 2015 to June 2018. Throughout the period of observation, BJJ's performance in rela-tion to ROE was always at the first and the 2nd rank and in the last three quarters of the period, the per-formance of the BJJ's ROE had always been in the first rank.

Tables 1–8, showed the performance ratings of each bank compared to the performance of banks in the BJJ peer group, with respect to the determined vari-ables of concern. By using this information, it could be seen in which variables a particular bank may need serious improvement compared to other banks in the peer group. Overall, BJJ's performance with respect to the four variables of concern showed performance stability.

Table 6. ROA performance rating for the period of December 2016 to June 2018.

Bank Name	Mar-17	Jun-17	Sep-17	Dec-17	Mar-18	Jun-18
BJJ	3	2	1	1	2	2
Bnparahyangan	15	15	14	14	13	15
BOII	1	7	15	15	5	6
Bumi arta	8	8	7	6	4	3
Capital	10	10	10	11	11	11
Ganesha	2	3	3	3	3	5
Index	4	4	4	5	7	8
Mas	7	6	6	7	9	7
Maspion	6	5	5	4	6	4
Mayora	9	11	8	8	8	10
Metro express	14	14	11	9	14	12
Nationalnobu	13	13	13	12	12	13
Sahabat sampoerna	11	9	9	10	10	9
SBI Indonesia	5	1	2	2	1	1
Windu (CCBI)	12	12	12	13	15	14

Table 8. ROE performance rating for the period of December 2016 to June 2018.

Bank Name	Mar-17	Jun-17	Sep-17	Dec-17	Mar-18	Jun-18
BJJ	2	2	2	1	1	1
Bnparahyangan	15	15	14	14	13	15
BOII	3	8	15	15	8	10
Bumi arta	9	10	8	4	6	6
Capital	1	4	7	8	5	4
Ganesha	5	7	6	7	3	9
Index	4	5	4	5	9	7
Mas	8	3	3	2	4	2
Maspion	7	6	5	3	7	5
Mayora	11	11	10	9	11	11
Metro express	14	14	13	13	15	14
Nationalnobu	13	13	12	11	12	12
Sahabat sampoerna	10	9	9	10	10	8
SBI Indonesia	6	1	1	6	2	3
Windu (CCBI)	12	12	11	12	14	13

Table 7. ROE performance rating for the period of December 2015 to December 2016.

Bank Name	Dec-15	Mar-16	Jun-16	Sep-16	Dec-16
BJJ	2	2	1	1	1
Bnparahyangan	9	15	9	11	12
BOII	14	4	15	15	15
Bumi arta	7	5	8	8	4
Capital	5	9	5	4	8
Ganesha	1	1	4	5	7
Index	3	6	3	3	2
Mas	11	8	6	6	6
Maspion	6	3	2	2	3
Mayora	8	10	7	7	5
Metro express	12	12	12	10	11
Nationalnobu	13	14	14	12	10
Sahabat sampoerna	4	7	10	9	9
SBI Indonesia	15	13	13	14	14
Windu (CCBI)	10	11	11	13	13

Table 9. Bank performance index (NPL & CAR).

Bank Name	NPL conversion	NPL Score	CAR conversion	CAR Score
Maspion	3.209	97.800	0.744	11.200
Sahabat Sampoerna	5.561	96.100	0.624	8.500
Metro Express	0.655	99.500	0.630	8.800
Ganesha	1.777	98.800	1.078	18.800
Mayora	6.330	95.500	0.865	14.100
BOII	34.615	75.900	1.523	28.900
Bank Jasa Jakarta (BJJ)	1.000	99.300	1.000	18.000
SBI Indonesia	7.022	95.100	1.328	24.500
National NOBU	0.088	99.900	0.772	11.900
BN Parahyangan	14.166	90.100	0.616	8.100
Bumi Arta	3.060	97.900	0.754	11.500
Index	4.479	96.900	0.798	12.500
MAS	2.746	98.100	0.624	8.500
Windu (CCBI)	3.082	97.900	0.258	0.200
Capital	3.249	97.300	0.432	4.200

In order to monitor the reliability of the bank per-formance index model, during the period of observa-tion, an analysis was conducted for December 2017. Results showed that Awards of appreciation with regard to excellence in performance of BJJ in 2017 on review were all supported by empirical data rein-forced by the bank performance index model.

As shown in Table 9 up to Table 11, for December 2017 on the basis of the bank performance index model, it was found that the aggregate performance of BJJ in relation to the observed variables was in the first rating. Thus the bank performance index model formed on the basis of the peer group bank model was able to represent the prevailing reality and reinforced the argument why BJJ had received so many awards so far. The bank performance index model developed

Table 10. Bank performance index (ROA & ROE).

Bank Name	ROA conversion	ROA Score	ROE conversion	ROE Score
Maspion	0.686	93.600	0.798	91.100
Sahabat Sampoerna	0.253	90.200	0.368	89.400
Metro Express	0.253	90.250	0.103	88.500
Ganesha	0.708	93.600	0.632	90.400
Mayora	0.349	91.000	0.403	89.500
BOII	−4.956	50.200	-8.417	57.500
Bank Jasa Jakarta (BJJ)	1.000	96.200	1.000	92.000
SBI Indonesia	0.988	95.800	0.697	90.600
National NOBU	0.172	89.600	0.283	89.100
BN Parahyangan	−0.394	85.800	-0.681	85.700
Bumi Arta	0.633	93.200	0.751	90.900
Index	0.653	93.460	0.727	90.700
MAS	0.583	92.650	0.890	91.400
Windu (CCBI)	0.110	89.100	0.148	88.600
Capital	0.188	89.700	0.504	90.000

Table 11. Bank performance index (comprehensive).

Bank Name	Performance score	Performace Index	Rating
Maspion	75.884	0.96343	4
Sahabat Sampoerna	73.563	0.93396	11
Metro Express	74.818	0.94990	10
Ganesha	77.910	0.989157	3
Mayora	74.819	0.94991	9
BOII	56.345	0.715364	15
Bank Jasa Jakarta (BJJ)	78.764	1	1
SBI Indonesia	78.330	0.99448	2
National NOBU	75.628	0.96018	7
BN Parahyangan	69.573	0.88331	14
Bumi Arta	75.869	0.96324	5
Index	75.742	0.96163	6
MAS	75.207	0.95483	8
Windu (CCBI)	71.973	0.91378	13
Capital	73.015	0.92700	12

confirmed that BJJ deserved the awards because its performance was indeed excel-lent.

Although the core capital of Bank Jasa Jakarta (BJJ) was not the largest amongst the existing banks in the peer group, BJJ was able to be in the first rat-ing according to the management of NPL, CAR, ROA and ROE. BJJ's performance had been rein-forced by numerous awards of achievement received in recog-nition for the BJJ's performance. In 2017 Bank Jasa Jakarta consistently maintained its excel-lent perfor-mance and received many Indonesia Banking Awards in 2017 for the following categories, The Most Reliable Bank, The Most Effi-cient Bank, Best Bank, Indonesia and Best nonpub-lic Bank in BUKU 2. BJJ was also recognised and awarded in the Bisnis Indonesia Bank-ing Award 2017 for Most Efficient Bank in BUKU 2. It also won the Diamond Trophy in Infobank Awards 2017 for the category of Bank Performance Award for successfully maintaining "Excellent" performance for 20 consecutive years.

4 CONCLUSION

The findings of this study indicate that it is possible to develop a model for bank performance index based on the peer group model. A good model has the abil-ity to represent what happens in reality. The bank's performance index model proposed by the authors in this study has proven to be able to represent the pre-vailing reality. This model confirms that BJJ indeed deserves various awards for its achievements. With regard to this model, the achievement of BJJ's per-formance was rated excellent resulting from the bank's performance index compared to other banks in the peer group. Even though BJJ's core capital was in 7th rank, BJJ's performance proved to be able to outperform other banks with larger core capital. On the basis of its achievements, BJJ indeed deserves various awards for excellence in achievement. The novelty of this study is the establishment of a bank performance index model formulated on the basis of the bank's peer group. Com-pared to ordinary performance comparisons that only use financial ratio data, the results achieved by this model are more realistic as benchmarking was carried out after equalization of core capital, so that the per-formance of one bank, and other banks, is measured by equalizing financial capabilities.

ACKNOWLEDGEMENT

The authors sincerely expresses appreciation and thank you to Bank Jasa Jakarta for being a role mod-el in proving the model developed by the authors.

REFERENCES

Bank Indonesia. 2011. Peraturan Bank Indonesia Nomor 13/1/PBI/2011 tentang Penilaian Tingkat Kesehatan Bank Umum.
OJK Financial Services Authority of Indonesia. Bank pub-lication reports. [Online]. Retrieved from www.ojk.go.id/ id/kanal/perbankan/data-dan-statistik/laporan-keuangan-perbankan/default.aspx)Sugiarto. 2016. Performance eval-uation of Indonesian banks and foreign banks operating

in Indonesia related to classification of capital. *Advances in Economics, Business and Management Research 15. 1st Global Conference on Business, Management and Entrepreneurship* (GCBME-16).

Sugiarto. 2017a. Formulasi model peer group bank.Workshop model peer group bank, Bank Jasa Jakarta. Jakarta.

Sugiarto. 2017b. Core capital performance evaluation of banks operating in Indonesia. *International Journal of Applied Business and Economic Research* 15 (6). Scopus Index Journal, 2017 **ISSN:** 0972-7302.

Sugiarto. 2018. Time series analysis of performance consistency of bank peer group model. *The 2nd International Conference on E-Business and Internet* (ICEBI 2018). National Taipei University of Business, Taiwan.

Sugiarto, & Suryadi, K. 2018. Peer group model as a reference of banks performance assessment. *The 3rd Global Conference on Business, Management, and Entrepreneurship*. Grand Mercure Hotel, Bandung

Sugiarto, Adinoto, N., & Fongnawati, B. 2018. Model peer group sebagai patok duga penilaian kinerja bank. Laporan Akhir Penelitian Berbasis Kompetensi. STIE Wiyatamandala.

Advances in Business, Management and Entrepreneurship – Hurriyati et al. (Eds)
© 2021 Taylor & Francis Group, London, ISBN 978-0-367-67471-7

Supply chain performance evaluation for tea commodities through the unified modeling language

M.A. Sultan, C. Furqon, A. Ciptagustia & M.T. Hidayat
Universitas Pendidikan Indonesia, Bandung, Indonesia

ABSTRACT: This study aims to obtain an overview and evaluate the supply chain system used by PT X, and analyze the optimization of tea production with a supply chain management approach. The concept of supply chain management used is the Unified Modeling Language (UML) method. PT X, as one of the plantation companies, is used as the unit of analysis. The research methodology used is the qualitative method approach with plantation side confirmation. The results showed that there were ten business people involved in the system. And the results also show that the actors are still not optimal so that communication between these actors needs to be improved.

1 INTRODUCTION

Competition in the distribution of products in the global market is increasing. The tendency of the trend of increasingly shorter product innovation has an impact on product life cycles that are getting shorter, while continued customer expectations have an impact on changing the company's focus, not only on internal business factors, but the role of supply chain management is also considered important (Adianto 2015). Integration between internal company actors and supply chain members will certainly have an impact of good supply chain performance. If the performance of the supply chain is not maintained properly it will certainly have an impact of inefficiency and could potentially be detrimental. Therefore, controlling business processes from upstream to downstream becomes an important priority. SCM is a series of planning, coordinating, and controlling all business processes and activities in the supply chain to create the best consumer value with efficient costs while still meeting all the needs of other stakeholders in the supply chain (Bozarth et al. 2008). Value, or better known as added value (Lambert et al. 1998), is something to be obtained for consumers and is reflected in the revenue generated by the company. To produce optimal value, three decisions must be determined by executives, namely how the network structure, business processes, and management components of the supply chain work (Dharwiyanti & Wahono 2003; Furqon 2014).

This study aims to determine the optimization of supply chain management in a plantation company and to find a picture of the distribution path of the commodity in the plantation company. Therefore, the right method for obtaining the optimization description is by using the UML method (Dharwiyanti & Wahono 2003).

UML is used to photograph the extent to which supply chain activities are proceeding accordingly. There are two types of diagrams in UML, the first is a diagram that models the structure of the system and the second is a diagram that models the behavior of the system. The use of prototyping methodology makes the modeling process unnecessary detailed.

To give a model of the service-based SCM system discussed in this paper, two types of diagrams are used to model each characteristic. Modeling the structure of the system uses class diagrams, which are representations of object-oriented based modeling, and to use behavior models using case diagrams. There is no data modeling in the form of an entity-relationship diagram because of the stability of the structure of data storage (Dharwiyanti & Wahono 2003).

The tea industry is one of the industries affected by the changes in business processes, therefore, some improvement is needed.

Tea growth still depends on the size of the plantation area with natural conditions, constant temperature, and humidity, which are ideal conditions for tea plant growth. However, the area of tea plantations is decreasing and the weather and water supply levels do not always remain stable, while the demand for tea is increasing day by day.

2 METHOD

This research was conducted at a plantation company in West Java. It focuses on one superior commodity (black tea products) because this research uses the UML approach. To obtain an overview of supply chain management flow, observation and taking photographs were employed as research instruments because they exposed directly the site. Interviews with the plantation actors were also utilized to support the data from the observations and documentation.

The research method involved observation and photographing directly the field and using interview techniques involving the plantation actors, especially in West Java plantations. As a research design, this study involved a mixed-method because it provides comprehensive, valid, reliable, and objective data.

3 RESULTS AND DISCUSSION

Supply chain observation at PT X with the UML approach involved a series of mapping activities in planning, coordinating, and controlling all business processes and activities in the supply chain to create the best consumer value with cost-efficiency while still meeting all the needs of other stakeholders in the supply chain (Meindl & Chopra 2005). Value or added value (Porter 1998) is the main element for consumers and is reflected in the revenue generated by the company. To produce optimal value, it is necessary to have a good mapping of each element of the supply chain actors involved from the supply chain (Lambert & Cooper 2000).

The following is an illustration of the business people involved in the supply chain management development system at PT X at Jawa Barat. There are ten business people involved in the system, namely Admin, Reception, Withering, Milling, Oxidation, Drying, Separation, Packing, Storage, and Sales.

Definition of the Actor

1. Admin has the authority to manage system user data or employees of PT. Perkebunan Nusantara.
2. Reception Section has access rights to input material acceptance data.
3. Withering Section has the authority to input material data in the withering process.
4. Milling Section has access rights to manage the grinding process data.
5. Oxidation Section has access rights to manage the oxidation process data.
6. Drying Section has access rights to manage the data drying process.
7. Separation Section has access rights to manage the data sorting process.
8. Packaging Section has access rights to manage the data packaging process.
9. Storage Section has access rights to manage the data storage processes.
10. Sales Department has access rights to manage the sales process data.

Use case Diagram

A use case diagram illustrates how actors interact with the system. It is made according to business processes gained from the analysis of the ongoing system. This diagram is described by the actor and use case.

From the observation, there are ten actors in the system where the interrelationships between sub-systems are as follows:

From the definition explained above, it can be seen the interaction between Actors and the System, Actors in Reception with Acceptance Data Systems, then forwarded to Actors with Impaction. Interaction with Withering Data Systems, forwarded to Milling Actors interacting with Milling Data Systems, forwarded to Oxidation Actors interacting with Oxidation Data System, forwarded to the Drying Section Actor interacts with the Drying Data System, forwarded to the Separation Section Actor interacts with the Separation Data System, and finally passed on to the Packaging Section Actor that interacts with the Packaging Data System.

This connection illustrates the need for strong integration in creating optimal supply chains. This diagram presents the principles of object oriented with the UML, which helps the modeling process in the analysis stage (Hilman et al. 2012).

Activity Diagram

The activity diagram illustrates the various activity streams in the system, from the initial flow, the decisions that might occur, and how they end.

Supply chain activities at PT X's plantations are described from the plantation to the production activities at the factory, from withering tea to packaging tea, while storing and selling activities are at the warehouse.

In addition, PT X has 3 distributors for distribution, where each distributor has their own distribution goals. Team 1 has to distribute to the local area, Team 2 distributes in the regional area, and Team 3 is responsible for the international market.

It can be concluded the flow and supply chain management of PT X, especially in the commodity of black tea. PT X manages 26 black tea plantations that are still in operation today. Of the 26 gardens, the tea produced in one harvest period will be processed in the factory.

Tea processing in the factory consists of seven stages, namely (1) the tea is received by the factory from the garden, (2) the tea undergoes a process of withering, (3) the tea undergoes a grinding process, (4) the tea undergoes an oxidation process, (5) the tea undergoes a drying process, (6) the tea undergoes a separation process, and finally (7) the tea is packed in a sack.

After being processed in the factory, the tea is then delivered to PT X to undergo the process of checking and lab testing. After being tested, the tea that passes the lab test will be moved to the warehouse and the marketing department will start marketing the product to various sales alternatives, such as the tea auction in Jakarta or via direct sales to consumers, both domestic and abroad.

4 CONCLUSION

From the observations, it is illustrated that the activities of each actor are crucial in creating an optimal supply chain. Therefore, it is necessary to design a technology-based integration of each activity in relation to one another to improve the optimization of the supply chain. However, to ensure which activities need to be considered for improvement, further research is required.

REFERENCES

Adianto, Joko. 2015. Analisa Sistem Berorientasi Objek. Dokumen Tips, 11 Juli 2015, accessed in January 5, 2019 <https://dokumen.tips/documents/modul-uml-55a2309be0d89.html>

Bozarth, Cecil, and Robert Handfield. 2008. Introduction to Operations and Supply Chain Management 2nd edition. New Jersey: Pearson Education Inc.

Chopra, S., and Meindl, P. (2001). Supply chain management: Strategy, Planning, and Operations. New Jersey – Prentice Hall.

Dharwiyanti, Sri., dan Wahono, Romi Satria. (2003). Pengantar Unified Modeling Language. Dilihat pada December 4, 2018 <www.IlmuKomputer.com>

Furqon, Chairul. (2014). "Analisis Manajemen dan Kinerja Rantai Pasokan Agribisnis Buah Stroberi Di Kabupaten Bandung". *IMAGE*, *3* Nomor 2.

Heizer, Jay, dan Render, Barry. (2014). Operations Management: Sustainability and Supply Chain Management, 11th Edition, *Pearson Education, Inc.*

M. Hilman, F. Setiadi, I. Sarika, J. Budiasto, dan R. Alfian (2012). Supply chain management berbasis layanan: desain dan implementasi prototipe sistem. *Journal of Information Systems*, Volume 8, Issue 2

Lambert, D., dan Cooper, M. (2000). Issues in Supply chain management. *Journal of Industrial Marketing Management*, Volume 29, pp. 65–83.

D. M. Lambert., J. R. Stock., dan L. M. Ellram. (1998) Fundamentals of Logistic Management. McGraw-Hill.

Meindl, P., dan Chopra, S. (2005). Supply chain management: Strategy, Planning, and Operation. *Pearson Education International/Prentice Hall.*

Porter, M. (1998). Competitive Advantage: Creating and Sustaining Superior Performance: With a New Introduction. Free Pr.

Advances in Business, Management and Entrepreneurship – Hurriyati et al. (Eds)
© *2021 Taylor & Francis Group, London, ISBN 978-0-367-67471-7*

Digital marketing strategy based on user experience to increase user growth and engagement start-up in Tasikmalaya

A. Hermawan, S.S. Maesaroh & B.M. Purwaamijaya
Universitas Pendidikan Indonesia, Bandung, Indonesia

ABSTRACT: The era of the Industrial Revolution 4.0 has almost been felt at this time. The era was marked by the use of digital technology that encouraged automation and data exchange in manufacturing technology. This era often raises concerns for many parties about human work that will be replaced by the latest technology. Even so, when more and more new digital-based businesses are emerging that are opening up new jobs, this new digital-based business is often referred to as Startup. A startup is a term for entrepreneurs who are starting a business in the digital field. In starting a digital business, you must have the ability in the IT field. There are many services offered, such as website creation, software, mobile applications, and others. Startup prioritizes interesting and creative content so that people are interested. To get started, start-ups need to think about strategies. The strategy itself is in the form of an action plan to achieve the desired goals, and it is suggested that these objectives must be measured properly. For example: through a digital marketing campaign that runs for 2 months, start-ups want to produce a trial increase of products by 15%. But the strategy will also depend on the scale of the business. At the startup level, having focused and simple goals is highly recommended. Even in the implementation of the pioneering Start-up, now it must be more responsive to users or consumers. This is referred to as User Experience. User experience is the attitude, behavior, and emotions of users when using a product, system, or service. This experience involves individual perceptions related to perceived benefits, the convenience obtained. User experience is very dynamic, over time the user-perceived perceptions can change with changing environments, habits, and values.

1 INTRODUCTION

In this globalization era, the rapid advancement of technology, communication, and the internet makes the system of exchanging information in the world like no space and boundaries. As we know, now all things can be accessed online, so that indirectly changing people's behavior becomes more independent. The development of technological advances, communication, and the internet is often referred to as the development of the digital world. According to Glossary Digital Marketing, the digital world is a general description that relates to modernization as well as the devices within it, the container for modern humans doing all their activities. From this fact it can be concluded that the digital world is a condition where humans have the behavior to adopt technological developments, communications, and the internet that exists today to meet their daily needs.

The rate of adoption of digital technology in Indonesia is quite high. Minister of Communication and Information Rudiantara, at the World Economic Forum 2019, said that Indonesia's population is more than 264 million people, 2/3 of them have received good communication technology penetration (aged 15–64 years), then 105 millions of them active internet users, 173 million are among active mobile users, 96 million are active social media users, and 28 million are among active e-commerce consumers. With this data, it is no wonder that the development of businesses that are oriented towards digital technology is increasingly in demand by the people of Indonesia. This is evidenced by the emergence of new professions such as Digital Entrepreneurs. Digital Entrepreneurs are entrepreneurs who have optimized the digital world to develop traditional technologies that already exist and have been used before, as a basis for developing a business that is being run. This business development is commonly known as Digital Business. Digital business can be an alternative to overcoming unemployment since it can easily be applied to novice business people because, with the technology that is connected online without any limits on space, time, and place, it can make business activities more effective and efficient and have reach broad.

But in running a Digital Business, in reality, it is not as easy as imagined. This is because start-up entrepreneurs do not understand what concepts should be used in running their business. According to PT Exa-Bytes Network Indonesia, 2017, the inability to run a business is caused by businesses not doing a good market analysis. As you know, market analysis before starting a business is very necessary to find out what is needed by consumers and what we can prepare to meet the needs of these consumers. To analyze whether the products offered by business actors have met the needs of consumers or not, then business actors can engage prospective consumers directly to experience the benefits of the products offered through User Experience. According to ISO 9241-210 (2009), User Experience (UX) is the perception and response of users as a reaction from the use of a product, system, or service.

Another obstacle found in business failures is when the results of the market analysis using User Experience do not show good results, such as the reaction of dissatisfied potential customers to a product, system, or service offered because it does not match the expectations of potential customers. This affects the sale of a product that will not be too promising. Therefore, to help realize the current market trends, strategies need to be made that can wrap a product so that it is ready to be marketed to the wider community. Based on the background that has been described, researchers need to research "Digital Marketing Strategy Based on User Experience to Increase User Growth and Engagement Start-Up in Tasikmalaya."

2 METHODS

User Experience Research is one of the investments in product development, including what is important in the product design and design process. Data collection and analysis is a way to improve usability, using a series of research techniques

The purpose of UX research is to assist the product development process based on user understanding in using the product, the results can be reused to make decisions in making the product itself. By interacting with users, it is hoped that they can get a better sense of product design and design for potential users so that they can leave viewpoints, assumptions, and personal expectations based on this research.

Researchers use this UX research to identify as well as develop capital for digital marketing strategies that will be used at a startup. This UX research involves users who are also practitioners in the fields of start-up, technology, and marketing.

This research uses a qualitative descriptive approach. The method used is a survey through the distribution of questionnaires and the involvement of industry players in a discussion group (FGD) and in-depth interviews with users who are also practitioners. The purpose of this study is for researchers to develop potential start-ups in the City of Tasikmalaya and also get an overview of perceptions of digital marketing strategies carried out by start-up in Indonesia.

3 RESULTS AND DISCUSSION

Indonesia currently has a lot of start-ups from various sectors, from trade, transportation, and even education.

The development of start-up in Indonesia is so rapid, of course, it becomes a question why the newly pioneered company is very fast growing and even appears in many regions. This has encouraged researchers to examine the phenomenon. How to start-up can compete quickly and compete with companies that have long operated.

Researchers conducted a study of marketing tools used by 3 Unicorn start-ups from Indonesia, namely Gojek, Traveloka, and Bukalapak. The reason why the researchers chose the 3 start-ups is that they are original Indonesian start-up, so the corporate behavior towards consumers is certainly well tested and 3 start-ups have also been proven to have success and growth. which is very high up to now.

3.1 Digital marketing channel

Digital Marketing Channel is a technologybased marketing channel that can be accessed by anyone, anytime and anywhere. In choosing a digital marketing channel, a marketer must know where the target market for the product is located, wherein each digital marketing channel will be different how to use and marketing strategies that must be done. In this study, researchers have conducted digital marketing channel research on well-known start-ups in Indonesia, namely Gojek, Bukalapak, and Traveloka. This research resulted in the average engagement achieved by the three start-ups, as a result the three start-ups received around 73% engagement resulting from their digital marketing activities.

As for the results of this study, based on the user that has been studied shows that the user gets the first time about the start-up from the following channels.

Based on a survey conducted to users, the highest number comes from social media as the most dominant channel compared to Digital Ads and Digital Conversations. Although the three percentages are not very significant.

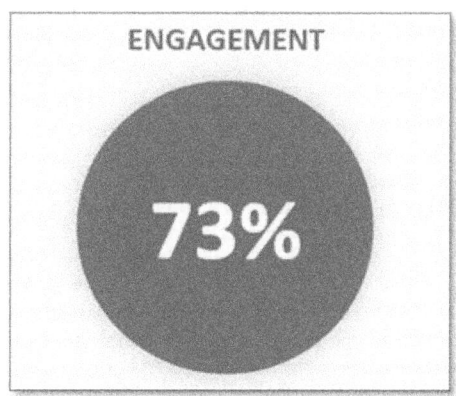

Figure 1. The results of research on startup regarding engagement.

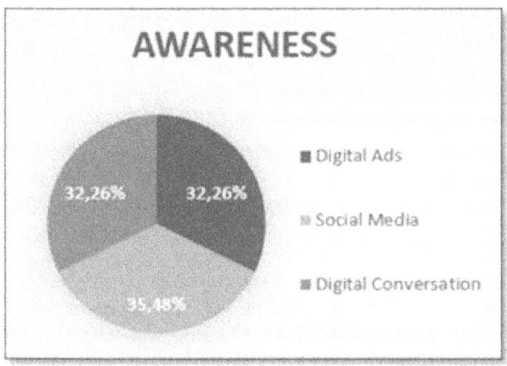

Figure 2.   Research results on startup awareness based on channel.

3.1.1   *Digital ads*

Digital Advertising or Digital Ads is an advertising system that uses digital media as its main tools. Digital Ads come along with the development of increasingly rapid technology, advertisements that are usually present only offline through banners, magazines or newspapers, now come with digital packages that are launched online. For example: Websites, television and advertisements on applications (online promotion).

3.1.2   *Social media*

As an online human-to-human portal, social media is the best channel to use. Social media can bring customers directly with the desired product, even prospective customers can get testimonials from other customers. For product owners, social media is very important because the information provided can come and arrive quickly to the customer. Example: Instagram, Facebook, Twitter and Linkedin.

Table 1.   List of social media audiences wearesocial 2019.

No	Social media advertising audiences	
1	Facebook	130 Million
2	Instagram	62 Million
3	Twitter	6,43 Million
4	Linked in	3,80 Million

3.1.3   *Digital conversation (WOM)*

WOM or Word of Mouth is a communication process in the form of providing recommendations both individually and in groups of a product or service that aims to provide information personally (Kotler & Keller 2009). As part of marketing communication, word of mouth is one of the most influential strategies in consumers' decisions to use products or services. The development of the technology world certainly has a big hand in terms of Digital Conversation, nowadays humans can carry out digital conversations with many tools. So that humans can interact with each other in

one portal, be it a website or an application. Example: Whatsapp, Line, Blog, Website and other portals that support interaction with one another.

3.2   *Content strategy*

In digital marketing, content is a very important thing that must be considered by companies. Content in the digital world is a tool to convey the value and differentiation of products from companies that will create company positioning in the minds of the public. Content in digital marketing must pay attention to several aspects such as the following aspects:

Table 2.   List of content marketing aspect.

No	Content aspect
1	Content that matches the character and brand
2	Content preferred by consumers
3	Content that becomes the top of mind

From these aspects, researchers divide the type of content marketing into 4 parts of content commonly used for start-up namely Education, Inspiring, Promotion and Entertain, the results of the study show that the content created by Gojek, Bukalapak, and Traveloka are as follows.

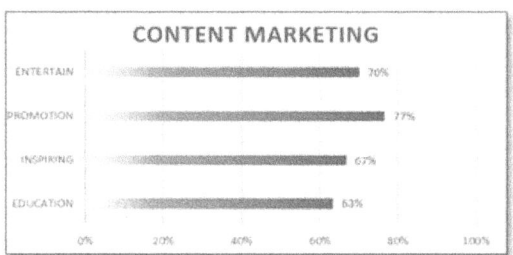

Figure 3.   Content marketing that attracts attention based on user experience.

From the results that have been studied, the most interesting content is promotional content with 77%, Entertain content with 70%, Inspiring content 67% and educational content with 63%. Promotional content has more influence for engagement with users, which can directly impact startup growth.

3.3   *Content strategy*

In carrying out content strategy, the company must monitor so that the planned strategy can run accordingly. To monitor the planned content strategy, the company needs to conduct regular monitoring. Stages of monitoring content on social media can be done starting from the stage of making content strategies as well, are as follows.

By using content marketing monitoring, start-up can evaluate their digital marketing periodically and

Figure 4.   Step by step content marketing strategy.

neatly, so that later use can help to decide on the next marketing strategy. Monitoring is very important to be done to understand the customer as a tool for developing products and future marketing strategies.

4   CONCLUSION

Based on research conducted on digital marketing strategies based on user experience to increase user growth and engagement start-up in the city of Tasik-malaya, it can be concluded that digital marketing strategies that can be applied by start-up in Tasik-malaya City are based on studies from well-known start-up in Indonesia such as Gojek, Bukalapak and Traveloka have several stages, which are as follows:

1. Choose the right marketing channel
    Start-up must choose the right marketing channel for the product being sold. How to choose the right marketing channel (digital marketing channel) for a product is by analyzing the segmentation and target of the product, then searching for channels through the user characteristics of each channel that is sim-ilar to the target market that has been selected. By choosing the right marketing channel, then the start up will get the right target market too.
2. Creating interesting content
    After selecting the right digital marketing channel, start-up needs to create content that can attract con-sumers. The results of this research indicate that at the moment, the most popular content is content that contains elements of sales or promotion. But to strengthen the promotional content, start-up must also deliver content that shows product value and differentiation so that it will create a strong posi-tioning in the minds of consumers regarding the products being sold.
3. Periodic monitoring
    So that all digital marketing channel planning and selected content can run as expected, the start-up needs to monitor the strategy for all digital marketing activities undertaken. The function of monitoring this strategy is to find out whether the public is still interested in the content offered by the company and whether the digital marketing channel used is still relevant to the chosen target market.

REFERENCES

Kemp, S. 2019. Report state internet 2019. [Online]. Retrieved from https://wearesocial.com/special-reports/state-internet-q2-2019.
Kotler, P., & Keller, K.L. 2009. Manajemen pemasaran (Thir-teenth ed.). London United Kingdom: Pearson Education.

Advances in Business, Management and Entrepreneurship – Hurriyati et al. (Eds)
© 2021 Taylor & Francis Group, London, ISBN 978-0-367-67471-7

Fuzzy AHP multi criteria decision making approach for supplier selection of blanket engineering service tender in PT. ABC

R. Fitriyadi
Universitas Indonesia, Depok, Indonesia

M. Hamsal
Universitas Bina Nusantara, Jakarta, Indonesia

ABSTRACT: PT. ABC operates in Indonesia under a Production Sharing Contract with the Indonesian oil and gas regulatory body (SKK Migas); therefore, all supplier selection processes shall follow government regulation PTK007. The success of plant performance and efficiency of the company is highly dependent on the selection of suppliers. The current supplier selection process caused several unperformed contract executions. Supplier selection is one of the fields to which the Fuzzy Analytic Hierarchy Process (AHP) approach has been applied to select the best supplier and providing the most satisfaction for the criteria determined. Fuzzy AHP is an extension of the classical AHP method where the fuzziness of the decision makers is considered. This paper gives recommendation on supplier selection improvement to help the decision makers to evaluate a set of alternatives using a Fuzzy AHP approach and Price-Quality Method without violation of PTK007.

1 INTRODUCTION

PT. ABC is a global energy company that operates in Indonesia under a Production Sharing Contract (PSC) with the Indonesian oil and gas regulatory body (SKK Migas 2017). There were several supplier's unperformed cases that gave operational impact to PT. ABC, including project delays, late payment to subcontractors, incompetence of supplier's project management and personnel, and inefficiency of manpower and equipment utilization, which caused additional costs to be incurred. Given the magnitude of the risk of such a supplier selection system, PT. ABC needs to improve their supplier selection processes to get the best supplier providing the most satisfaction for the company. It is taken into consideration to be one of the most valuable processes of the purchasing function, which gives direct impact to the configuration and operational performance of a supply chain (Barbosa et al. 2017). In fact, in the literature, the two most important aspects of the supplier selection process are (a) the selected criteria and (b) the choice of a method to rank the available bidders (Allouche & Jouli 2017). Therefore, this paper aims at applying Fuzzy AHP to calculate the correct weighing for technical criteria and to combine technical and commercial evaluation score (Price-Quality method) without violation to the Government of Indonesia's guidelines in supply chain management (PTK007) to

make sure the cost is recoverable by the Government of Indonesia.

PTK007 is intended to provide an integrated and clear legal basis for governance, technical and administrative implementation guidelines of upstream oil and gas business activities in the territory of the Republic of Indonesia in the procurement of goods/services, and to realize the principle basis of supply chain management (PTK007 page 1). PTK007 segregates between technical and commercial evaluation. Bid evaluation is carried out starting from the technical bid evaluation and continuing with commercial bid evaluation. Consequently, supplier selection processes at PT. ABC are more focused on price. The evaluation is based on the lowest bid price through evaluation of the bidder whose technical score is equal to or exceeds the minimum threshold. The lowest bid price evaluation system encourages competition amongst bidders and maintains transparency, but project delays and material late delivery, project cost increment, tendency to compromise quality, and adversarial relationship among contracting parties are major drawbacks (Shrestha 2014). Wan (2008) said that the lowest bid price evaluation system is not an efficient sourcing system anymore; the supplier selection process is a multiple criteria decision-making process.

Supplier selection is not a simple clerical process, but a product of Multi Criteria Decision Making

(MCDM). According to Mukherjee (2017), pairwise comparison is an intrinsic characteristic of human beings, and most of MCDM approaches favor the use of a pairwise comparison matrix to get a consensus through different methods of normalization and aggregation. The AHP method was proposed and developed by Thomas L. Saaty and is an effective problem-solving methodology that needs to be evaluated by linguistic variables. It uses a 1–9 scale for comparing two factors. In supplier selection, the factors compared are the criteria to be evaluated during the tender process.

Fuzzy AHP is an advanced analytical method developed from the traditional AHP method, since traditional AHP does not include the vagueness of personal judgments. It has been improved by the fuzzy logic approach sets theory developed by Zadeh. Basically, the Fuzzy AHP method elaborates a traditional AHP method into the fuzzy domain by using fuzzy numbers for calculating instead of real numbers (Özdağoğlu & Güzin 2007). The Fuzzy AHP approach smooths the way for defining criteria weights by aggregation of judgments from decision makers. The approach determined criteria weights for supplier selection. The results showed that the proposed fuzzy AHP method is superior to the traditional AHP in terms of improved quality of criteria prioritization (Ozyurek & Erdal 2018). In AHP, these numerical values are exact numbers, whereas in the Fuzzy AHP method, they are intervals between two numbers with the most likely value. In Fuzzy AHP, the pairwise comparisons are implemented through the linguistic variables, which are represented by Triangular Fuzzy Numbers (Wang & Tsai 2018).

The Blanket Engineering Services, one of PT. ABC's tenders, was used as the case study, which had been executed during the research period, and already decided on before the research:

a. There were six qualified bidders who passed the qualification process;
b. There were four technical evaluation criteria to be evaluated (Experiences and technical capability, Personnel qualification, Support facilities, and Health-Safety-Security-Environment (HSSE);
c. The weighing of the Price-Quality method has a 60% technical score and a 40% commercial score.

2 METHOD

There were five steps in conducting the research:

1. Getting pairwise comparison data for each technical criterion through questionnaires;
2. Defining weighing for each technical criterion through the MCDM technique using the Fuzzy AHP process;
3. Conducting a technical evaluation process and getting the Technical Evaluation Score. Only the

bidders who passed the minimum technical passing grade continued for commercial evaluation;
4. Conducting a commercial evaluation process and getting the Commercial Evaluation Score. The bid prices from bidders were normalized by local content and the company status of each bidder as part of the PTK007 process;
5. Executing the Price-Quality method to combine the Technical Evaluation Score and Commercial Evaluation Score. The proposed tender winner was the bidder who had the highest score.

3 RESULTS AND DISCUSSION

3.1 Questionnaires

In order to define the weighing factor for each technical criterion, questionnaires were distributed to the experts to determine the comparison and correlation between two criteria based on their Saaty score. The experts closely managed the supplier selection and/or contract execution to make sure they had a deep understanding of the scopes; what PT. ABC needed to achieve with this contract and what was expected from the supplier. The experts also had valid certification in one reputation of international project management or PTK007 competency from the National Professional Certification Body (Badan Nasional Sertifikasi Profesi).

From the geometric mean of the questionnaire results, the pairwise comparison matrix can be delineated as per Table 1.

Table 1. AHP pairwise comparison matrix.

	Experiences and technical capability	Personal qualification	Support facilities	Health, safety, and environment
Experiences and technical capability	1.00	2.29	4.40	4.71
Personal qualification	0.44	1.00	4.23	4.31
Support facilities	0.23	0.24	1.00	3.43
Health, safety, and environment	0.21	0.23	0.29	1.00

3.2 Consistency check

The maximum eigenvalue (λmax) was used to check the consistency of judgment (Mukherjee 2017):

– Normalize the matrix from Table 1:

$$u_i = \frac{\left[\sum_{j=1}^{n} b_{ij}\right]^{1/n}}{\sum_{i=1}^{n} \left[\sum_{j=1}^{n} b_{ij}\right]^{1/n}} \quad \forall i,j = 1,2,3\ldots,n \quad (1)$$

$Qb1 = (1.00 \times 2.29 \times 4.40 \times 4.71)^{1/4} = 2.6259$
$Qb2 = (0.44 \times 1.00 \times 4.23 \times 4.31)^{1/4} = 1.6797$
$Qb3 = (0.23 \times 0.24 \times 1.00 \times 3.43)^{1/4} = 0.6550$
$Qb4 = (0.21 \times 0.23 \times 0.29 \times 1.00)^{1/4} = 0.3461$
$\Sigma Qb = Qb1 + Qb2 + Qb3 + Qb4 = 5.3067$
$u1 = Qb1/\Sigma Qb = 2.6259/5.3067 = 0.4948$
$u2 = Qb2/\Sigma Qb = 1.6797/5.3067 = 0.3165$
$u3 = Qb3/\Sigma Qb = 0.6550/5.3067 = 0.1234$
$u4 = Qb4/\Sigma Qb = 0.3461/5.3067 = 0.0652$

– R denotes the n-dimensional column vector describing the sum of:

$$R = [R_i]_{n \times 1} = Bu^T = i = 1,2,\ldots n$$

$$Bu^T = \begin{bmatrix} 1 & b_{12} & \cdots & b_{1n} \\ b_{21} & 1 & \cdots & b_{2n} \\ b_{n1} & b_{n2} & \cdots & 1 \end{bmatrix} \begin{bmatrix} u_1 & u_2 & \cdots & u_n \end{bmatrix} = \begin{bmatrix} R_1 \\ R_2 \\ \vdots \\ R_n \end{bmatrix} \quad (2)$$

$$\begin{bmatrix} 1.00 & 2.29 & 4.40 & 4.71 \\ 0.44 & 1.00 & 4.23 & 4.31 \\ 0.23 & 0.24 & 1.00 & 3.43 \\ 0.21 & 0.23 & 0.29 & 1.00 \end{bmatrix} \times \begin{bmatrix} 0.4948 \\ 0.3165 \\ 0.1234 \\ 0.0652 \end{bmatrix} = \begin{bmatrix} 2.0714 \\ 1.3361 \\ 0.5344 \\ 0.2797 \end{bmatrix}$$

$$\begin{bmatrix} 2.0714 \\ 1.3361 \\ 0.5344 \\ 0.2797 \end{bmatrix} \div \begin{bmatrix} 0.4948 \\ 0.3165 \\ 0.1234 \\ 0.0652 \end{bmatrix} = \begin{bmatrix} 4.1862 \\ 4.2213 \\ 4.3298 \\ 4.2884 \end{bmatrix}$$

– Calculate maximum eigenvalue λmax:

$$\lambda_{\max} = \frac{\sum_{i=1}^{n} Ru_i}{n} \quad i = 1,2,3,\ldots n \quad (3)$$

$\lambda_{\max} = (4.1862 + 4.2213 + 4.3298 + 4.2884)/4 = 4.26$

– Calculate Consistency Index (CI):

$$CI = \frac{\lambda_{\max} - n}{n - 1} \quad (4)$$

$CI = (4.26 - 4)/(4 - 1) = 0.08$

– Calculate consistency ratio (CR):

$$CR = \frac{CI}{R1} \quad (5)$$

$CR = 0.08/0.9 = 0.09$

Table 2. Average index for randomly generated weights.

Matrix (N)	2	3	4	5	6	7	8	9	10
Random Index (RI)	0.00	0.58	0.90	1.12	1.24	1.35	1.41	1.45	1.49

Source: Jain, Rajeev., Singh, A. R., Mishra, P. K., 2013.

Table 3. Linguistic terms and the corresponding TFN.

Saaty Scale	Definition	TFN Scale
1	quite important	(1, 1, 1)
3	Weekly important	(2, 3, 4)
5	Fairly important	(4, 5, 6)
7	Strongly important	(6, 7, 8)
9	Absolutely important	(9, 9, 9)
2	The intermittent	(1, 2, 3)
4	value between	(3, 4, 5)
6	two adjacent	(5, 6, 7)
8	scales	(7, 8, 9)

Source: Wang & Tsai, 2018.

Since the consistency ratio was below 0.10, it means that the matrix is reasonably consistent, so the researchers continued to the next tender process.

3.3 Fuzzified pairwise comparison matrix

The pairwise comparison matrix based on Triangular Fuzzy Number (TFN) can be converted from the questionnaire results using linguistic terms and the corresponding TFN shown in Table 3.

– From the Geometric Mean of the pairwise comparison matrix based on the TFN for each criterion, the TFN pairwise comparison matrix can be created (Table 4):
– Create Fuzzy Geometric Mean as a pairwise comparison and relative scores:

$$\widetilde{Q}_b = (l_b, m_b, u_b)$$

$$l_b = (l_{b1} \otimes l_{b2} \cdots \otimes l_{bi})^{\frac{1}{i}}, b = 1,2,\ldots i$$

$$m_b = (m_{b1} \otimes m_{b2} \cdots \otimes m_{bi})^{\frac{1}{i}}, b = 1,2,\ldots i$$

$$u_b = (u_{b1} \otimes u_{b2} \cdots \otimes u_{bi})^{\frac{1}{i}}, b = 1,2,\ldots I \quad (6)$$

Fuzzy Geometric Mean for row criterion of Experiences and Technical Capability:

Column-l: $(1.00 \times 2.20 \times 3.83 \times 4.20)1/4 = 2.44$
Column-m: $(1.00 \times 2.29 \times 4.40 \times 4.71)1/4 = 2.63$
Column-u: $(1.00 \times 2.38 \times 4.96 \times 5.16)1/4 = 2.79$

– The values in each column "Fuzzy Geometric Mean" in Table 5 are summed using Equation 7:

$$\tilde{Q}_y = \left(\sum_{b=1}^{i} l_b, \sum_{b=1}^{i} m_b, \sum_{b=1}^{i} u_b \right) \quad (7)$$

– Calculate Fuzzy Weight:

$$\tilde{R} = \frac{\tilde{Q}_b}{\tilde{Q}_y} = \frac{(l_b, m_b, u_b)}{\sum_{b=1}^{i} l_b \sum_{b=1}^{i} m_b \sum_{b=1}^{i} u_b}$$

$$= \left[\frac{l_b}{\sum_{b=1}^{i} u_b}, \frac{l_b}{\sum_{b=1}^{i} m_b}, \frac{l_b}{\sum_{b=1}^{i} l_b} \right] \quad (8)$$

For "Experiences and Technical Capability," Column-l shows $2.44/5.75 = 0.42$, Column-m shows $2.63/5.31 = 0.49$, and Column-u shows $2.79/4.85 = 0.58$. Using the same calculation for

Table 4. TFN Pairwise comparison matrix.

	Experience and technical capability			Personnel qualifications			Support facilities			Health, safety, and environment		
	I	III	II	I	III	II	I	III	II	I	III	II
Experience and technical capability	1.00	1.00	1.00	2.20	2.29	2.38	3.82	4.40	4.96	4.20	4.71	5.16
Personnel qualifications	0.42	0.44	0.45	1.00	1.00	1.00	3.66	4.23	4.76	3.42	4.31	5.10
Support facilities	0.20	0.23	0.26	0.21	0.24	0.27	1.00	1.00	1.00	2.74	3.43	4.03
Health, safety, and environment	0.19	0.21	0.24	0.20	0.23	0.29	0.25	0.29	0.36	1.00	1.00	1.00

Table 5. Normalized weighing of fuzzy AHP.

	Fuzzy geometric mean			Fuzzy weight			Centre of area	Normalize fuzzy weight
	I	III	II	I	III	II		
Experience and technical capability	2.44	2.63	2.79	0.42	0.49	0.58	0.50	0.49
Personnel qualifications	1.51	1.68	1.82	0.26	0.32	0.38	0.32	0.32
Support facilities	0.58	0.65	0.73	0.10	0.12	0.15	0.13	0.12
Health, safety, and environment	0.31	0.35	0.40	0.05	0.07	0.08	0.07	0.07

Table 6. Final technical evaluation results.

Bidders	Experience and technical capability (49%)	Personnel qualifications (32%)	Support facilities (12%)	Health, safety, and environment (7%)	Technical evaluation score	Pass/fail
PT. SM	95.20	85.10	100.00	70.60	90.97	Pass
PT. TE	100.00	95.40	100.00	78.70	97.13	Pass
PT. DE	92.70	68.30	100.00	72.90	84.59	Fail due to personnel qualification not meeting minimum score of 85
PT. RPE	0.00	0.00	0.00	0.00	0.00	Did not submit bid
PT. BE	97.60	88.80	100.00	75.20	93.63	Pass
PT. SE	100.00	91.40	100.00	74.10	90.62	Pass

other criteria, the results can be seen in Table 5 column "Fuzzy Weight."

- Defuzzification technique by the center of area method using Equation 9 (Diouf & Choonjong 2018):

$$Q_i = \frac{(w_{i1} + w_{i2} + w_{i3})}{3}, i = 1, 2, 3, \ldots, n \quad (9)$$

Experiences and Technical Capability $= (0.42 + 0.49 + 0.58)/3 = 0.50$. Using the same calculation for other criteria, and the results can be seen in Table 5 column "Centre of Area".

- To find the Normalize Fuzzy Weight, each element of the center of area was divided by the sum of the Center Area column:
Experiences and Technical Capability $= 0.50/(0.50 + 0.32 + 0.13 + 0.07) = 0.49$. Using the same calculation for other criteria, and the results can be seen in Table 5 column "Normalize Fuzzy Weight."

3.4 Detail tender process

3.4.1 Technical evaluation
There were six qualified bidders who passed the qualification process, but only five bidders submitted a bid and only four bidders passed the technical evaluation as mentioned in Table 6.

Since all bidders had local content commitment for more than 30%, all bidders got local content and company status price preferences as mentioned in Table 7.

3.4.2 Price-quality method
Weighted technical score is calculated based on the multiplication between the technical evaluation score of each bidder (as mentioned in Table 6) and the weighing percentage (0.6).

The calculation of the weighted commercial score follows two steps:

Table 7. Commercial evaluation result.

Component		Bidders			
		PT. BE	PT. SE	PT. TE	PT. SM
Goods cost component (IDR)	a	0.00	0.00	0.00	0.00
Service cost component (IDR)	b	33,969,190,050,00	35,997,153,750,00	40,274,161,996,00	47,691,646,635,00
Non-cost component (IDR)	c	602,432,450,00	3,999,683,750,00	56,241,254,00	1,907,665,865,00
Total bid price (IDR)	d	34,571,622,500,00	39,996,837,500,00	40,330,403,250,00	49,599,312,500,00
Local content	e	96.68%	88.88%	93.48%	97.90%
Local content preference – goods	f	15.00%	15.00%	15.00%	15.00%
Local content preference – services	g	7.50%	7.50%	7.50%	7.50%
Local content target from government	h	55.00%	55.00%	55.00%	55.00%
Company status (local, national, foreign)	i	Local	Local	Local	Local
Company classification in APDN (Preferred, maximize, empowered). APDN – local product appreciation	j	Preferred	Preferred	Preferred	Preferred
Commitment execution by local company	k	50.00%	100.00%	100.00%	98.00%
Commitment execution within Indonesian territory	l	50.00%	100.00%	100.00%	98.00%
Company status preference	m	7.50%	7.50%	7.50%	7.50%
Bid price evaluation good cost component	n	0.00	0.00	0.00	0.00
Bid price evaluation service cost component	o	31,599,246,558,14	33,485,724,418,60	37,464,336,740,47	44,364,322,451,16
Bid price evaluation after company status preference	p	29,394,647,961,06	31,149,511,087,07	34,850,545,805,08	41,269,137,163,87
Commercial score	q	29,997,080,411,06	35,149,194,837,07	34,906,787,059,08	43,176,803,028,87

Step 1: Normalize all commercial scores in Table 8 based on the lowest commercial score, which is the score from PT. BE (IDR 29,997,080,411.06).

Step 2: The weighted commercial score is calculated based on the multiplication between the normalized commercial score of each bidder in Step 1 and the weighing percentage (0.4).

The final scores of each bidder are the sum of the weighted technical score and the weighted commercial score, which can be found in Table 8.

The owner's estimation (OE) is IDR 45,000,000,000. The OE is used as a tool to assess the fairness of bid offers, and as a reference in determining the prospective tender winner. Although the bid of PT. BE, as the 1st Rank (IDR 34,571,622,500.00), is lower than the OE, PT. ABC conducted price negotiations, but

Table 8. Technical–commercial calculation result.

Component	Bidders			
	PT. BE	PT. SE	PT. TE	PT. SM
Technical evaluation (weight 60%)				
Technical evaluation score	93.63	90.62	97.13	90.97
Technical score (weighted)	56.18	54.37	58.28	54.58
Commercial evaluation (weight 40%)				
Commercial score	29,997,080,411.06	35,149,194,837.07	34,906,787,059.08	43,176,803,028.87
Normalized commercial score	100.00	85.34	85.93	69.47
Commercial score (weighted)	40.00	34.14	34.37	27.79
Price-quality				
Technical and commercial combination	96.18	88.15	92.65	82.37
Rank	I	III	II	IV

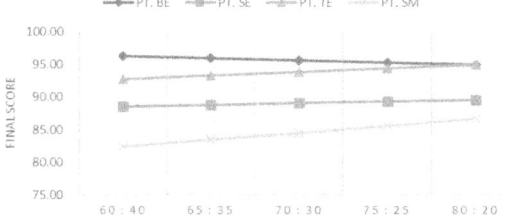

Figure 1. Sensitive analysis.

there was no reduction in cost. Therefore, PT. BE was proposed as the tender winner.

3.4.3 Sensitivity analysis
Referring to PTK007, the weight of the technical score was between 0.60 and 0.80, while the commercial score was between 0.20 and 0.40. There are several alternative weighing combinations of technical and commercial scores. To review the sensitivity analysis of the combinations to the tender result, five alternative combinations were used, namely 60:40, 65:35, 70:30, 75:25, and 80:20. The results can be found in Figure 1.

Although the highest score for all alternative combinations was still PT. BE, there was a tendency for score differences to be less. Even the difference in scores between PT. BE and PT. TE for the combination of 80:20 is only 0.01%.

4 CONCLUSION

Fuzzy AHP can be utilized to define the weighing of technical criterion without violation to PTK007 and can be applied to all tender processes in PT. ABC, involving the stakeholders as experts. The stakeholders can come from among users, procurement, tender committee, decision makers, and other related functions. It is required to make sure that the technical criteria and weighing factor are suitable for the scope of works and acquire the best supplier to achieve the contract execution that meets company expectation.

Collaboration between all stakeholders is the key to the success of implementation of the recommendation to solve the business challenge. Based on PTK007, the Price-Quality Method is only for consultancy service types of contracts, therefore, further research is expected to include the application of Fuzzy AHP for technical criteria selection combining with Total Cost Ownership evaluation for tenders in companies under PSC contracts. Fuzzy AHP can also be exercised during prequalification processes to assess the suitability of administration requirements, bidder experiences, and financial performance.

ACKNOWLEDGEMENTS

This research was supported by the Procurement and Supply Chain Management and Project and Modification teams of PT. ABC. We thank our colleagues who provided insight and expertise that greatly assisted the research.

We would like to also show our gratitude to Rofiqoh Rokhim, SE, SIP, DEA, Ph.D., and Rizqiah Insanita, S.T., M.M., from the Faculty of Economics and Business, Universitas Indonesia, for comments that greatly improved the manuscript.

REFERENCES

Allouche, A. M., & Jouili, T. 2017. Supplier selection problem: A fuzzy multicriteria approach. *Southern African Business Review* 21. ISSN: 1998-8125.
Barbosa de Santis, Golliat, & Pestana de Aguiar, E. 2017. Multicriteria supplier selection using Fuzzy Analytic Hierarchy Process: Case study from a Brazilian railway operator. *Brazilian Journal of Operations & Production Management* 14: 428–437.
Diouf, M., & Choonjong, K. 2018. Fuzzy AHP, DEA, and managerial analysis for supplier selection and development from the perspective of open innovation. *MDPI Journal Sustainability*. ISSN 2071-1050.
Jain, Rajeev., Singh, A. R., & Mishra, P. K. 2013. Prioritization of supplier selection criteria: A Fuzzy-AHP approach. *MIT International Journal of Mechanical Engineering* 3(1): 34–42.

Mukherjee, K. 2017. Supplier selection, an MCDA-based approach, springer, studies in systems, decision and control. 88.

Özdağoğlu, A., & Güzin. 2007. Comparison of AHP and Fuzzy AHP for the multicriteria decision making processes with linguistic evaluations. *İstanbul Ticaret Üniversitesi Fen Bilimleri Dergisi Yıl* 6: 65–85

Ozyurek, I., & Erdal, M. 2018. Assessment of qualification criteria described in public procurement law code 4734 in construction works by Analytic Hierarchy Process (AHP). *Journal of Science* 31(2): 437–454.

Shrestha, S. K. 2014. Average bid method – an alternative to low bid method in public sector construction procurement in Nepal. *Journal of the Institute of Engineering* 10(1): 125–129.

SKK Migas. 2017. Pedoman Tata Kerja Nomor PTK-007/ SKKMA0000/2017/S0 (Revisi 04) Buku Kedua dan Petunjuk Pelaksanaan Tender Nomor EDR-0167/ SKKMH0000/2017/S7 (PTK007).

Wan Lung Ng. 2008. An efficient and simple model for multiple criteria supplier selection problem. *European Journal of Operational Research* 186: 1059–1067

Wang, T. C., & Tsai, S. Y. 2018. Solar panel supplier selection for the photovoltaic system design by using Fuzzy Multi-Criteria Decision Making (MCDM) approaches. Energies 11: 1989. [Online]. Retrived from doi:10.3390/ en11081989.

Section 4   Marketing management

The challenge of visitors experience: A case study at Museum Nasional, Jakarta, Indonesia

Mudji Sabar
Universitas Mercu Buana, Jakarta, Indonesia

ABSTRACT: This research is aimed to explore the visitors' experience at Museum Nasional (MN) in Jakarta, Indonesia. This is a case study employing descriptive analysis. The sample was 251 respondents. This study applied a convenient sampling technique as a method to collect the primary data. This study revealed the visitors' satisfaction in terms of museum's cleanliness, service delivery, collection display, and ambience facilities. It also showed the visitors' intention to recommend others to visit MN. The majority of the visitors (83%) felt that the high density of the museum caused them feel less comfortable. The research recommends MN management to take managerial efforts to improve the level of the visitors' experiences. They should apply digital reservation system with the principle "first come, first serve". In addition, the management should innovatively create new attractions or novelties based on the visitors' preferences in order to maintain the attractiveness of the museum.

1 INTRODUCTION

1.1 *Museum's name and exhibition*

Museum Nasional (MN) was launched on April 24th, 1778. Formerly, this museum was named Museum Gajah (Elephant Museum) as there was an elephant statue, an honourable gift from King Chulalongkong of Thailand, placed in front of the building. The government has declared Museum Nasional as an official name of this museum since May 28th, 1979.

The storyline of the exhibition was designed based on Koentjaraningrat's classification in term of seven substances of cultural elements, i.e. religious system and religious ceremonies; societal systems and organizations; knowledge system; language; arts; livelihood systems; as well as systems of technology and tools or equipment. Besides, there is a special category of collections, namely treasures and ceramics collections, which were placed on the 4th floor of the building. MN building consists of seven levels. Four of which are for permanent exhibitions. The layout of the four levels are as follows:

a. Level-1: Man and Environment
b. Level-2: Knowledge, Technology and Economy
c. Level-3: Social Organizations and Settlement Patterns
d. Level-4: Treasures and Ceramics

1.2 *Visitors'experie*

This term refers to the feeling after visiting the museum. The Expectation Disconfirmation Theory mentions three possibilities of the visitors' experience. First is positive disconfirmation, where visitors feel satisfied after visiting the museum. Second is neutral disconfirmation, where visitors feel neutral. Third is negative disconfirmation, where visitors feel dissatisfied with the museum services (Ujakpa et al. 2017).

Kotler & Keller (2009) said that customers' satisfaction refers to the feeling of happiness. In this case, when the visitors experience higher values from the museum they visit, they tend to have higher satisfactory. On the contrary, they will feel unhappy when they get less benefit than what they really expect (Kotler & Keller 2009).

The practices of quality management system determine the level of visitors' satisfaction. This relates to the museum's artefact exhibition, the ambiance facilities, the cleanliness, the professionalism and hospitality of the museum's officers, as well as physical facilities. There are two functions of the museums, i.e. as public education institution and as a recreation place for family members or colleagues (Boylan 2004).

In addition, the implementation of museum internal regulation to maintain the museum carrying capacity (MCC) is also important to increase the level of visitors' satisfaction. The density of visitors at certain time may influence the level of the visitors' comfort. The study of Gua Pindul in 2016 showed the evidence that over carrying capacity lessen the visitors' feeling of comfort (Sabar 2016). Accordingly, the museum management must anticipate such condition just to maintain the visitors' feeling of satisfaction and comfort.

2 METHODS

The population of this research was the visitors of MN. The average number of visitors was 314,200 people per year (calculated from MN Visitors, 2014–2016, Berita Jakarta, July 31st, 2017). Using Slovin formula (Tejada & Punzalan 2012) with this number of population and sampling error of 5%, the sample size was 313 people; however, the respondents who completely filled.

3 RESULT AND DISCUSSION

3.1 Visitors' profile

Table 1 indicates the following findings:

a. The main visitors of MN were young women coming from Jakarta and surrounding cities.
b. They visited the museum with small groups of family members or friends.
c. The visitors' main motive was recreation (Boylan 2004).
d. The length of visit was from 1 to 2 hours.
e. The visitors were not frequent visitors.

Table 1 shows that respondents only spent 1 to 2 hours in the museum. This length is too short to fully comprehend the displayed artefacts. Here, the visitor might only see the artefact they like without completely learn it. Thus, it can be inferred that MN is predominantly functioned as public recreation spot instead of public education. This is in line with respondents' main motive to visit museum, which is for recreations. Questionnaire were only 251 people. This research applied accidental sampling method to collect primary data using a list of guided questionnaire. The criteria of the respondents are as follows (Packer & Ballantyne 2002):

a. Adult (mature enough to answer the questionnaire).
b. Independent (the informants must have free time to answer the questionnaire).

Table 1. Visitors' Profile.

Criteria	Description	Frequency	%
Gender	Women	154	61.35
Age	18–30 years	160	63.75
Place of Origin	DKI Jakarta	113	45.02
	BoDeTaBek	94	37.45
Acquaintance	Friends	108	43.03
	Family Members	77	30.68
Visit	First Visit	79	31.47
	Second Visit	79	31.47
Motive	Recreation or Holiday	153	60.56
Length of Visit (LoV)	2 Hours	94	37.45
	1 Hour	79	31.47

Source: Primary Data derived from the questionnaires. August 2018.

c. Alone (more preferable) or with small group of visitors of maximum 3 people.
d. Not occupied by certain duty or job, such as the tour guides, the researchers, the curators, etc.
e. Willing to response the questionnaire.

These criteria should be applied in order to find the right respondents to provide information needed by the research. The respondents should be questioned in advance before answering the questionnaires to ensure they comply the criteria. Therefore, only those who suited the criteria were taken as the respondents of this research. The others who did not fulfil the criteria were not taken as respondents and excluded from the research.

3.2 Visitors' experience

3.2.1 The Visitors' satisfaction

MN has a good access. It is located in very strategic location, which is near to the palace state and National Monument (Monas). Either private vehicles or tourist buses can use parking lot facility at the area of Monas. Besides, there are a lot of public transportation passing through this site.

The research discovered that the respondents (99%) felt satisfied after visiting the museum This is in line with their perception towards the cleanliness of the museum (Table 2). All the respondents (99%–100%) perceived that the museum's facilities, especially toilets and prayer rooms, were clean and having a good aroma. It means that MN management has a high concern on its facilities.

The respondents (100%) also had a good perception towards the museum officers. They argued that the officers dressed very well. They were grooming, and their costumes were neat. The officers also served them with good manners. They were very kind and friendly. The visitors also thought that the officers had a good knowledge of the artefacts in the museum. They knew how to serve their visitors.

The respondents mentioned that MN's collections were good. The artefacts were well maintained and displayed. They seemed to be satisfied with the ambience of the museum. The galleries were well lighted, and the temperature was well managed. They also perceived that MN was able to design attractive display of museum's artefacts.

The research also showed that most of the respondents (98%) would like to recommend their colleagues to visit MN. This statement indicates the tendency of respondents' loyalty. However, there is a hypothetical question whether the respondents actually have a great intention to revisit the museum or not. Table 1 confirms that the probability of MN's respondents to revisit the museum was too low. Most of respondents (69,94%) declared that they had visited MN once or twice. Unfortunately, they looked unsure whether they would like to revisit the destination.

Regarding this, (Som et al. 2011), on their study of 'Tourist Satisfaction and Repeat Visitation', discovered the evidence indicating that the visitors may have intention to recommend other people to come because they want to share their positive experience to the other. However, it is no guarantee that

Table 2. Visitors' Experience.

Questionnaires	SD	D	A	SA	Total
The Cleanliness					
Toilet facilities are clean	0%	1%	59%	40%	100%
Toilet facilities are free from bad odor	0%	2%	61%	37%	100%
Prayer rooms are clean	0%	0%	61%	39%	100%
The Service Delivery					
Museum officers are dressed well	0%	0%	56%	44%	100%
Museum officers serve visitors with good attitude	0%	0%	49%	51%	100%
Museum officers are professional	0%	0%	52%	48%	100%
MN Collections					
The collections are well maintained	0%	1%	49%	50%	100%
The display of the collections is attractive	0%	0%	53%	47%	100%
The lighting of the galleries is good	0%	0%	51%	49%	100%
The temperature of the galleries is good	0%	0%	55%	45%	100%
The Visitors' Satisfaction					
I get the value I need	0%	1%	55%	44%	100%
Questionnaires	SD	D	A	SA	Total
I would like to recommend my colleaques to visit NM	0%	2%	54%	44%	100%
The Visitors' Comfort					
I feel the museum is overcrowded	2%	15%	75%	8%	100%
I feel the number visitors disturb my comfort	2%	36%	47%	11%	100%

Source: Primary Data derived from the questionnaires. August 2018.
Notes:
● SD=Strongly Disagree; D = Disagree; A = Agree; SA = Strongly Agree
● Total respondents = 251 respondents

they would like to revisit the destination. People may revisit the destination when they assume that there are new things they never see before. When every attraction remains the same, the visitors would be reluctant to come again to the destination. This is quite logical. New things or novelties will trigger people to come again. Novelties are effective instruments to attract people to come. Novelties within the museum may relate to renewed facilities, renewed attractions, renewed display, new ways to serve the visitors, or the introduction of modern technology application in museum services.

In response to the above result, the museum management must dynamically absorb what people need and want. In doing so, the management must conduct a research to understand the people preferences. By understanding the customers' needs, their preferences, and their psychographics characteristics, it will make the management easier to create novelties (Dragicevic et al. 2012). In this era, improving the museum's artefact exhibition is important. However, the improvement must be based on the visitors' orientation. Otherwise, the improvement will be useless.

Furthermore, most of the visitors (70%) were willing to pay more money to help MN in funding its operational budget (Table 2). This indicates the visitors' awareness of public donation. This awareness is good and should be enhanced.

3.3 The visitors' comfort

Related to the density, most of the respondents (83%) mentioned that the museum was over crowded. More than a half of them felt less comfortable with this situation (58%). This result is contradicted with the previous findings about respondents' satisfaction. Logically, when they are satisfied, they should also be comfortable with the service provided by the museum. Thus, the feeling of comfort may function as one of visitors' satisfaction indicators.

According to Sabar (2016), most of the visitors feel less comfortable because of the destination's density. Thus, the feeling of comfort may strengthen or even lessen the level of the visitors' satisfaction.

One of the problems faced by MN is a large group of visitors that come nearly at the same time in the weekends and holidays. The peak hours usually occurs from 11.00 to 16.00, while the museum opens from 09.00 to 17.00. This condition often causes visitor density.

Besides, the average number of visitors per day was around 873 people (calculated from MN visitors from 2014–2016, Berita Jakarta, July 31st 2017). In the weekends and holidays, the number may increase. Ironically, the museum officers, who are mostly Civil Servants, are off in the weekends and holidays. Even though there are some freelance guides who come to assist the museum in the weekends and holidays, the number is not adequate enough to serve the large number of visitors.

To tackle the situation, MN has to arrange the maximum number and the flow of the visitors. The museum's guides, in this case, must manage the visitors' stay at every gallery in order to ensure that they have adequate time and equal opportunity to enjoy the museum's collections. They have to make sure that there is no visitor density in a gallery. By doing this, the management can provide greater opportunity for the visitors to have higher level of experience.

One way that can be done by management is by grouping the number of visitors and hiring freelance guides. The management, here, can divide the visitors in a group. One group may consists of 25 people. By assuming the total number of visitor 837 people, there will be 35 groups per a day. Most of the visitor stay at the museum for 1–2 hours. If the average time of trip in museum is 1,5 hours and the entrance time of each group is 15 minutes, it can cover 25 group of visitors from 09.00 to 17.00 o'clock. This calculation only cover 71% of 873 visitors. If the museum want to accommodate all visitors, the number of visitors in each group should be 25 to 35 people. Consequently, the density may occur and visitors' experience will decrease. Thus, the manager should have 10–11 guides to handles 35 group of visitors. This is not easy for manager to find such number of voluntary guides.

The above findings indicate two problems that should be solved by MN management. First problem is related to the comfort. To solve this issue, the management should be able to improve the visitors' experience by deploying new ways to increase the level of visitors' comfort. Concerning this, the management should conduct the following actions:

a. Control the museum's density.
   It is necessary that the management should limit the total number of visitors. This is to ensure that the total number of visitors does not exceed the Museum Carrying Capacity (Sabar 2016). In this context, it is advisable to introduce electronic registration system. Group of tourists can register their entrances to the museum via this system, and the system will treat the registration by using the principle "first come first serve". This system has the capability to manage the registration much better than the manual system. This system can also determine the maximum number of visitors who are allowed to visit the museum. When the number of registrars is already beyond the capacity of the museum, the system will automatically advise the registrars to choose another available dates.
b. Smoothen the route of visitors.
   The management has to arrange the route of visitors during their journey at the museum. The visitors may be grouped into batches. The entrance time of every batch should also be regulated. The length of stay at every gallery must be limited too.
c. Prevent the bottleneck at the entrance gate.
   The electronic registration system can prevent the museum from having high density of visitors as well as the bottleneck condition at the museum. In

addition, this is quite necessary for the management to perform some cultural attractions. This is meant to disperse the visitor to prevent bottleneck condition. While waiting for the entrance call, the visitors may enjoy the cultural performances. The performance spots should also be distributed into some points with adequate distance from one spot to another. This approach can also be used to prevent visitors from being disheartened because they have to queue for a long time at the entrance gate.

The second problem concern to the visitors' intention to revisit the museum. Most of the visitors confirm that they are happy with the museum services and they intend to recommend others to come to the museum. However, there is no guarantee that the visitors would like to come again to the museum. Hence, to attract visitors' intention to be back again to the museum, the management should initiate novelties within the museum. The research discovers that the main visitors of the museum are young ladies coming from DKI Jakarta and surrounding cities (Table 1). It is imaginable that the preferences that meet their characteristics are special. Speculatively, metropolitan young ladies may be familiar with the usage of social media. To satisfy their expectations, the management should redesign the museum services by providing some photo spots where visitors can take photographs and upload them through their social media. In addition, there must also be meeting facilities for their togetherness with family members or colleagues along their stay at the museum (Boylan 2004).

The management should be high-tech oriented in order to renew the museum services. The management may renew the museum facilities, the display of the artefacts, the lighting system, the artefact catalogues, and all other things related to museum services by utilizing digital technology. In this era, high-tech orientation is prerequisite condition to attract people to visit and revisit the museums (Dragicevic et al. 2012).

4 CONCLUSION

4.1 conclusion

a. The research concludes that visitors are satisfied with the museum services due to their perceptions on the cleanliness, the service delivery, the collection display, the ambience facilities, and the willingness to recommend others.
b. The visitors intend to recommend their friends to visit MN. However, this does not guarantee that they would like to revisit MN.
c. The challenge faced by MN in the future is the visitors' comfort due to the density in museum.

4.2 Recommendations

The management is advised to renew the MN services by introducing these following novelties:

a. Digital Registration System, which is designed to prevent the museum from having over crowded visitors.
b. Cultural Performances, which are meant to provide visitors with alternative performances while waiting for entrance call of the museum.
c. Novelty Facilities, such as photo galleries, meeting spots for family and colleague gathering within the area of museum, as well as more digital facilities of the museum.
d. Some managerial efforts to improve MN services should be conducted and primarily addressed to young women as the main market of MN. These efforts may cover promotional efforts as well as educational efforts related to museum services.

REFERENCES

Berita Jakarta. 2017. Data Kunjungan Museum Nasional 2014–2016 (2014 = 245,848 visitors; 2015 = 297,134 visitors; 2016 = 399,618 visitors).

Boylan, P.J. 2004. Running a Museum: *A Practical Handbook*. *ICOM*, France.

Dragicevic, M., Letunic, S. & Pisarovic, A. 2012. Tourists' Experiences and Expectations towards Museums and Art Galleries Empirical Research Carried Out in Dubrovnic. *Recent Advances in Business Management and Marketing* 225–232.

Kotler, P & Keller, K.L. 2009. Marketing Management. *Pearson Prentice Hall 13rd Edition New Jersey*, USA.

Packer, J. & Ballantyne, R. 2002. Motivational Factors and Visitors Experience: *A Comparison of Three Sites. Curator* 45(3): 183–189.

Sabar, M. 2016. Carrying Capacity and Visitors' Satisfaction of the Ecotourism Object of Cave Pindul Tubing, Gunung Kidul, Yogyakarta, Indonesia. *Academy of Strategic Management Journal* 15(3).

Som, M., Puad, A.B. Bader, M. 2011. Tourist Satisfaction and Repeat Visitation; Toward a New Comprehensive Model. *World Academy of Science, Engineering and Technology International Journal of Economics and Management Engineering* 5(2).

Tejada, J.J. & Punzalan, J.R.B. 2012. On the Issue of Slovin Formula. *The Philippine Statistician* 61(1): 120–136.

Ujakpa, Martin, M., Uushini & Naemi, M. 2017. Customer Satisfaction Assessment Using the *SERVEQUAL Model*. *Science Journal of Business and Management* 5(5): 194–198.

Advances in Business, Management and Entrepreneurship – Hurriyati et al. (Eds)
© 2021 Taylor & Francis Group, London, ISBN 978-0-367-67471-7

The influence of customer needs, service distribution, and corporate governance on superior service: A study at Husein Sastranegara Airport, Bandung

A. Raras & E. Herlinawati
STIE Inaba, Bandung, Indonesia

ABSTRACT: This study aims to analyze the influence of customer needs, service distribution, and corporate governance on superior service. This research was conducted to 100 customers at Husein Sastranegara Airport, Bandung. Data processing from questionnaires, interviews, and research documents processed was done by using the SmartPLS 3 program. In this study, it was proven that customer needs did not have a significant effect on superior service. Moreover, service distribution did not have a significant effect on superior service, and corporate governance had a significant effect on superior service.

1 INTRODUCTION

1.1 *Airport operators in Indonesia*

Commercial airports in Indonesia can be managed by state-owned enterprises and private business entities (Law of the Republic of Indonesia, number 1, 2009, about aviation). Requirements that must be met by a business entity that will manage the airport include an Airport Business Entity (Badan Usaha Bandar Udara) certificate, KM 47 years (2002), issued by the Ministry of Transportation. In Indonesia, there are two commercial airport operators, PT Angkasa Pura I and PT Angkasa Pura II (AP II) as shown in Table 1.

1.2 *Airport operators in Indonesia*

Husein Sastranegara Airport is a civil enclave airport, that hitches a ride at the military airport (Law of the Republic of Indonesia, number 1, 2009). The physical airport is still an old legacy, based on SKEP/347/XII/99 concerning the design standards and/or airport engineering facilities and equipment SNI 03-7046-2004 about the Airport Passenger Terminals (Correia & Wirasinghe 2013). This makes disruption of customer needs, service distribution, and corporate governance towards superior service. Furthermore, AP II invests in improvements, terminal expansion, and the provision of new physical facilities in order to improve superior service (Horonjeff et al. 2010; Senguttuvan 2006).

This study used four latent variables, namely Customer needs (X1) with indicators of physical needs (X1.1), operational needs (X1.2), and additional needs (X1.3) (Griffin et al. 2010; Anggono Raras 2007); Service distribution (X2) with indicators of air side

Table 1. Airport listing by Angkasa Pura II.

Airports	cities
Husein sastranegara	Bandung
Soekarno-hatta	Tangerang
Halim perdanakusuma	Jakarta
Sultan thaha	Jambi
Kualanamu	Medan
Sultan mahmud badaruddin ii	Palembang
Sultan syarif kasim ii	Pekanbaru
Minangkabau	Padang
Supadio	Pontianak
Raja haji fisabilillah	Tanjung pinang
Depati amir	Bangka
Sultan iskandar muda	Banda aceh
Sisimangaraja xii	Siborong-borong
Kertajati	Majalengka
Banyuwangi	Banyuwangi
Tjilik riwut	Palangka raya
Raden inten ii	Bandar lampung
Fatmawati	Bengkulu
Hanandjoeddin	Tanjung pandan

indicator (X2.1), terminal (X2.2) (Andersson et al. 2000), and air side (X2.3) (Ashford et al. 2011); Corporate governance (X3) with indicators of transparency (X3.1), accountability (X3.2), responsibility (X3.3), independence (X3.4), and fairness (X3.5) (Mas 2005; Ministry of Transportation 2009). And this study used independent variables of Service superior (Y) with indicators of quality (Y1), being on time (Y2), bomfortable (Y3), ethic (Y4), and friendly (Y5), (Tjiptono 2005; Yen & Teng 2003; Zeithami & Bitner 2004).

Table 2. Variables and indicator study.	
Variables	Indicators
Customer needs (X1)	Physical needs (X1.1) Operational needs (X1.2), Additional needs (X1.3)
Service distribution (X2)	Land side (X2.1), Terminal (X2.2) Air side (X2.3),
Corporate governance (X3)	Transparence (X3.1), Accountability (X3.2) Responsibility (X3.3), Independence (X3.4) Fairness (X3.5)
Service superior (Y)	Quality (Y1), On time (Y2), Comfortable (Y3), Ethic (Y4), Friendly (Y5),

Table 3. Characteristics of respondents.

Name	Percentages
Gender	
Male	57
Female	43
Education	
High School-D3	65
S1	30
S2–S3	35
Occupation	
Private	33
Student	22
Government	19
Others	26
Ages	
0–17	10
18–55	60
56–80	30

2 METHOD

2.1 Model of study

This research was a joint research, which employed descriptive and verification research. Descriptive research is used to measure the outer model, which measures the value of the indicators of each latent variable. Verification was used to examine the influence between the variables studied. The research model used was the Structure Equation Model (SEM) with Smart Partial Least Square (SmartPLS) 3, as shown in Table 2.

2.2 Hypotheses

There are three hypotheses proposed, namely the influence of Customer needs on superior service, the effect of service distribution on superior service, and the influence of corporate governance on superior service (Airport Cooperative Research Program Report 76; ACRP 76, 2012).

3 RESULTS AND DISCUSSION

3.1 Hypothesis

The characteristics of respondents in this study were divided into four, namely gender, education, occupation, and age. Gender consisted of 43% female and 57% male. Education consisted of 65% in high school-D3, 30% in undergraduate study, and 35% in postgraduate study. Occupations consisted of 33% private company, 22% student, 19% governmental, and 26% others. Ages consisted of three age groups, namely 10% at 0–17 years old, 60% at 18–55 years old, and 30% at 56–80 years old as shown in Table 3.

Table 4. Validity test outer loadings.

Variable/ Indicators	Result	Cut off value	Remarks
X1.1	0.687	>(0.6–0.7)	Valid
X1.2	0.887	>(0.6–0.7)	Valid
X1.3	0.811	>(0.6–0.7)	Valid

Table 5. Validity test outer loadings.

Variable/ Indicators	Result	Cut off value	Remarks
X2.1	0.505	>(0.6–0.7)	dibaikan
X2.2	0.887	>(0.6–0.7)	Valid
X2.3	0.811	>(0.6–0.7)	Valid

3.2 Descriptive test

Validity test outer loading in the customer needs variable (X1) for X1.1 is 0.687, X1.2 is 0.887, and X1.3 is 0.811, in which the values of X1, X2, and X3 are greater than the cut off value of 0.6–0.7. Hence, these three indicators are valid. The Validity Test Outer Loadings are shown in Table 4.

Validity test outer loading in the service distribution variable (X2) for X2.1 is 0.505, X1.2 is 0.865, and X1.3 is 0.855. The value of X1 is below the cut off value, so the X2.1 indicator can be ignored. The values of X2.2 and X2.3 are greater than the cut off value of 0.6–0.7. Hence these two indicators are valid. The Validity Test Outer Loadings can be seen in the following Table 5.

Validity test outer loading in the Corporate governance (X3) variable gets a 0.865 score for X3.1, 0.828 for X3.2, 0.839 for X3.3, 0.805 for X3.4, and 0.784 for X3.5; the value of X3.1 to X3.5 is greater than the cut

Table 6. Validity test outer loadings.

Variable/ Indicators	Result	Cut off value	Remarks
X3.1	0.865	>(0.6–0.7)	Valid
X3.2	0.828	>(0.6–0.7)	Valid
X3.3	0.839	>(0.6–0.7)	Valid
X3.4	0.805	>(0.6–0.7)	Valid
X3.5	0.784	>(0.6–0.7)	Valid

Table 7. Validity test outer loadings.

Variable/ Indicators	Result	Cut off value	Remarks
Y.1	0.865	>(0.6–0.7)	Valid
Y.2	0.828	>(0.6–0.7)	Valid
Y.3	0.839	>(0.6–0.7)	Valid
Y.4	0.805	>(0.6–0.7)	Valid
Y.5	0.784	>(0.6–0.7)	Valid

Table 8. R square test.

Latent Variable	Result	Cut off value		Remarks
Service superior	0.737	0.67	0.75	Strong
		0.33	0.50	Moderate
		0.19	0.25	Weak

Table 9. Path coefficients test.

Latent Variable	P Value	Cut Off Value	Remarks
X1 → Y	0.650	<0.05	Insignificant
X2 → Y	0.516	<0.05	Insignificant
X3 → Y	0.000	<0.05	Significant

Table 10. Path coefficients.

Latent Variable	Result	Remarks
X1 → Y	0.051	Insignificant
X2 → Y	−0.083	Insignificant
X3 → Y	0.840	Significant

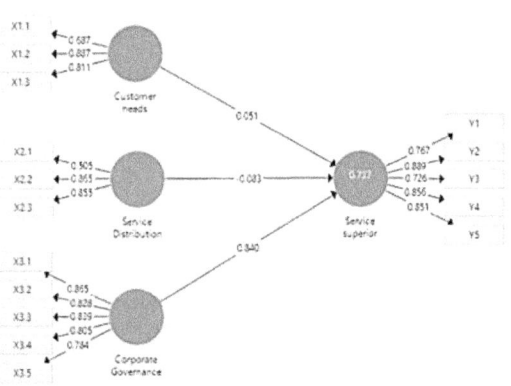

Figure 1. Result of the model study.

off value of 0.6–0.7, so that these five indicators are valid. See Table 6 for the Validity Test Outer Loadings:

Validity test outer loading on the variable Corporate governance (Y3) obtains 0.767 for the Y.1 results, 0.889 for Y.2, 0.726 for Y.3, 0.8565 for Y.4, and 0.851 for Y.5; the value of Y.1 to Y.5 is greater than the cut off value of 0.6–0.7; so these five indicators are valid. See Table 7 for Validity Test Outer Loadings:

3.3 *Hypothesis*

An R Square Test is a test is to find out how much the superior service latent variable can be influenced by the latent variables of customer needs, Service distribution, and corporate governance. The results show that it reaches 73.7%, which is considered strong, as seen in Table 8.

The Path Coefficients Test shows the magnitude of the influence of the latent variables of Customer needs (X1), Service distribution (X2), and Corporate governance (X3) on the latent variable of Service superior (Y), respectively by 0.650; 0.516; and 0,000, meaning that the latent variables X1 and X2 have no significant effect, while the variable X3 has a significant effect as shown in Table 9.

In calculating using the PLS Algorithm on Path Coefficients, it shows the magnitude of the influence of the latent variables of Customer needs (X1), Service distribution (X2), and Corporate governance (X3) to the latent variable of superior Service (Y) respectively by 0.051, −0.83, and 0.840. This means that the latent variables X1 and X2 have no significant effect, while the X3 variable has a significant effect as shown in Table 10.

The new relationships that could be examined include the relationship between corporate governance and service distribution and the relationship between corporate governance and customer needs as shown in Figure 1.

4 CONCLUSION

The results of the analysis on the research model proved that airport terminal development construction supported customer needs. There was a high value level on physical needs, operation needs, and additional needs. This also occurred for service distribution, corporate governance, and superior service.

The high value levels of superior service were in quality, on-time service, comfort, ethics, and pleasure. The influence of the latent variable of customer needs on superior service was insignificant, the effect of the latent variable of service distribution on superior service was not significant, and the influence of the latent variable of corporate governance on superior service was significant. This study is still not perfect, but it can be an opportunity for further research, namely by increasing the relationship between existing latent variables.

REFERENCES

Airport Cooperative Research Program Report 76 (ACRP 76), 2012. Addressing uncertainty about future airport activity levels in airport decision making, transportation research board.

Andersson, Karl, Carr, Francis, Feron, Eric, Hall William D. 2000. Analysis and modeling of ground operations at hub airports, 3rd USA/Europe air traffic management R&D seminar.

Anggono Raras T. S., 2007. Effect of customer needs and service marketing mix on customer value and its impact on customer loyalty, dissertation, Padjadjaran University.

Angkasa Pura II, 2010. Peraturan operasional bandar udara.

Angkasa Pura II, 2020. Daftar bandar udara yang dikelola.

Ashford, Norman J, Mumayiz, Saleh, Wright, Paul H. 2011. Airport engineering planning, design, and development of 21st century airports, John Wiley & Sons, inc.

Correia, Anderson, Ribeiro, Wirasinghe, S. C. 2013. Modeling airport landside performance, in modelling and managing airport performance, eds.

Fandy Tjiptono, Prinsip-prinsip Total Quality Service, Yogyakarta: CV Andi Offset. 2005.

Griffin, Katy J, Yu,Peter, Rappaport, David B. 2010. Evaluating Surface Optimization Techniques Using a Fast-time Airport Surface Simulation, 10th AIAA Aviation Technology, Integration, and Operations (ATIO) Conference.

Horonjeff, Robert, MCKelvey, Francis X, Sproule, William J, Young, Seth B. 2010. Planning and design of airport, the Mcgraw-Hill companies, inc.

Kementerian Perhubungan, 2009. Peraturan tata kelola bandar udara.

KM 47 tahun 2002 tentang sertifikasi operasi bandar udara. senguttuvan, p.s., 2006, Economic of the airport capacity system in the growing demand of air traffic: A global view.

SKEP/347/XII/99 Standar rancang bangun dan atau rekayasa fasilitas dan peralatan bandara SNI 03-7046-2004 tentang terminal penumpang bandar udara.

SKEP/77/VI/2005 Tentang persyaratan teknis pen-goperasian fasilitas teknik bandar udara

Solak, Senay, Clarke, John-Paul B, Johnson, Ellis L. 2009. Airport terminal capacity planning transportation research part B 43, pp. 659–676.

Undang-undang RI Nomor: 1 Tahun 2009 tentang Penerbangan.

Yen, Jin-Ru, Teng, Chung-Hsiang. 2003. Effect of spatial congestion in the level of service at airport passenger terminals.

Zeithami, Valarei A., Mary jo Bitner. 2004. *Service marketing: Integrating customer focus across the fim*, 3rd edn, mc graw hill, New York.

How can service quality create customer loyalty?

L. Lindayani, N.A. Hamdani, A.O. Herlianti, G.A.F. Maulani & T.M.S. Mubarok
Universitas Garut, Garut, Indonesia

ABSTRACT: An effective way to exploit a business potential is to provide quality services so that customers will be loyal in using the offered products or services. This quantitative study is aimed at examining how service quality increases customer loyalty. The samples are 166 pet shop customers. Data were analyzed using path analysis by means of SPSS. It is revealed that customer loyalty is indicated by how properly pet shop keepers treat pet animals and how they communicate with the customers. Customer loyalty to a pet shop can also be facilitated by several factors such as provision of free consultancy, provision of animal playground, competitive prices, and provision of animal care services.

1 INTRODUCTION

Pet ownership has become more popular. However, taking care of a pet can be burdensome especially for busy people. Therefore, pet care service has been a mushrooming business lately. The services may include grooming, daily care, and pet clinic. Nevertheless, pet care service business has always been restricted by limited qualified human resources and facilities. This leads to customer dissatisfaction, which in turn affects customer loyalty (Namin 2017).

Marketing experts argue that customer loyalty is the real goal of a set of marketing activities, not customer satisfaction (Wai Lai 2019) because satisfaction does not guarantee repeat purchase. Loyal customers are those who make repeat purchases (Nastasoiu & Vandenbosch 2019). However, in the long run, it is necessary for businesses to improve customer satisfaction to maintain customer loyalty (Borishade et al. 2018).

In business, customer loyalty means sustainability (El-Adly 2018). Therefore, business entities will always strive to improve it (García-Fernández et al. 2018). The higher the level of customer loyalty, the more sustainable a business will be (Marakanon & Panjakajornsak 2017). In brief, customer loyalty is a strong commitment from customers to come back consistently in the future (Srivastava & Rai 2018) despite competition (Wolter et al. 2017).

In business, service is everything (Rasheed & Abadi 2015). Service is what makes customers satisfied. The same applies to pet shop businesses. Service quality plays a very important role in their customer loyalty and repeat purchase (Xie 2011). A previous study shows that a gap between customer expectation and perception leads to low customer satisfaction (Uzunboylu 2016). One possible method for gathering information about customer perceptions and expectations is SERVQUAL (Yousapronpaiboon 2014). SERVQUAL identifies five dimensions of a service quality including tangibles, reliability, responsiveness, assurance, and empathy. Tangibles can be indicated by physical facilities, equipment's, materials, and personnel, reliability refers to reliable and accurate service performance, responsiveness can be measured by personnel willingness to help customers and to provide prompt service, assurance means knowledge and courtesy of personnel and their ability to convey trust and confidence, and empathy refers to the ability of personnel to recognize customer needs (Al-Fawzan 2005). This study is conducted to examine the effect of service quality on customer loyalty in pet shop businesses.

2 METHODS

To achieve the purpose, this study was carried out using an explanatory survey. Questionnaires were addressed to 166 pet shop customers in Garut, Indonesia, selected as respondents using accidental simple random sampling. Data were analyzed using path analysis by means of SPSS. The data collection method in this study was using a questionnaire method. In this study, the technique used is Probability Sampling, which is a sampling technique to provide equal opportunities for each member of the population to be selected as sample members with Accidental Simple Random Sampling, a sampling method by selecting who coincidentally exists or is encountered.

3 RESULTS AND DISCUSSION

The influence of service quality on customer loyalty was measured using an F-test. The result is presented in Table 1.

Table 1. Result of F-test.

ANOVA[a]

Model	Sum of Squares	Df	Mean Square	F	Sig.
Regression	9574.846	1	9574.846	162.302	.000[b]
Residual	7551.224	128	58.994		
Total	17126.069	129			

a. Dependent variable: Customer loyalty
b. Predictors: (Constant), Service quality

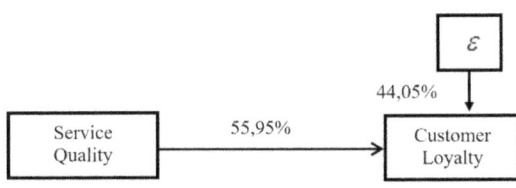

Figure 1. Effect of service quality on customer loyalty.

Table 1 shows that, using the probability (Sig) of 0.000 and significance level of 0.05, service quality had positive influence on customer loyalty in pet shop businesses.

The observed F value was 165.257, higher than the critical F of 3.915. Therefore, it justifies the previous conclusion that service quality had positive influence on customer loyalty in pet shop businesses.

The coefficient of determination (D) was calculated using the following formula.

$$D = r^2 \times 100\% \qquad (1)$$

Note:
r^2 = correlation coefficient (R square)
   The result is as follows:

$D = r^2 \times 100\%$
$D = (0.748)^2 \times 100\%$
$D = 0.5595 \times 100\%$
$D = 55.95\%$

In other words, customer loyalty is influenced by service quality as much as 55.95%, the other 44.05% is influenced by other factors as illustrated in Figure 1.

The questionnaires revealed that, among five service quality dimensions, tangibles received the highest score by 1,760 or 64.47%, followed by responsiveness by 1,740 or 63.74%, assurance by 1,769 or 64.780%, and empathy by 1,701 or 62.31%. Overall, generally pet shops in Garut offered sufficiently good services to the customers with the service quality score of 6,970 or 63.83%.

In order to improve customer satisfaction, good communication should come along with quality customer service. And quality service my improve purchase intention. Service quality captures customer expectations and perceptions of a service.

Customer loyalty was measured using 18 questionnaire items which included the following dimensions: repeat purchase, immunity, refers to other dan purchase across product and service line. The result of questionnaire revealed that the dimension repeat purchase scored the highest by 1,773 or 64.95%, followed by immunity by 1,767 or 64.54%, refers to other by 1,751 or 64.14%, and purchase across product and service line by 1,773 or 64.45%. Overall, the customers were sufficiently loyal with the customer loyalty score of 6,459 or 64,52%. This shows that the customers were satisfied so that they made repeat purchases (Chen 2015). Repeat purchase is closely linked with brand loyalty factors, which most companies strive for (Gamboa & Gonçalves 2014).

Unlike trial purchase, where customers buy products in small quantities and without any commitment (Kreis & Mafael 2014), repeat purchase indicates that the products are agreeable to the customers and that they are willing to repurchase in a greater quantity (Toufaily et al. 2013).

Our finding that service quality had significant influence on customer loyalty is in line with some previous studies (e.g. Uzunboylu 2016). In other words, service quality can create customer loyalty. Service quality can help companies win the competition or at least build relationship with the customers (Rasheed & Abadi 2015), and interpersonal communication can help improve customer satisfaction and loyalty (Pan et al. 2012).

4 CONCLUSION

The finding of this study shows that service quality contributes to customer loyalty. In order to improve their service quality, pet shop businesses need to take into account the following factors: good physical appearance of their shops to elicit customer purchase interests, quality equipment, competent personnel, dependable and accurate service, committed and prompt service, knowledgeable and trusted employees, attentive employees, and so on. This study has also found that pet shop businesses in Garut provides sufficiently good service, leading to enough customer satisfaction. In its turn, this customer satisfaction creates customer loyalty, indicated by repeat purchase. To conclude, customers will be loyal provided that the offered products or services meet their expectations.

REFERENCES

Al-Fawzan, M.A. 2005. Assessing service quality in a Saudi bank. *Journal of King Saud University-Engineering Sciences*, 18(1), 101–114.
Borishade, T.T. et al. 2018. A dataset of customer loyalty and variation in perception of customers across demographic characteristics in healthcare sector of Nigeria. *Data in Brief* 20: 353–357. https://doi.org/10.1016/j.dib.2018.08.014.

Chen, S.C. 2015. Customer value and customer loyalty: is competition a missing link? *Journal of Retailing and Consumer Services* 22: 107–116. http://dx.doi.org/10.1016/j.jretconser.2014.10.007.

El-Adly, M.I. 2018. Modelling the relationship between hotel perceived value, customer satisfaction, and customer loyalty. *Journal of Retailing and Consumer Services* (xxxx): 0–1. https://doi.org/10.1016/j.jretconser.2018.07.007.

Gamboa, A.M & Gonçalves, H.M. 2014. Customer loyalty through social networks: lessons from Zara on Facebook. *Business Horizons* 57(6): 709–717.

García-Fernández, J., Gálvez-Ruíz, P., Vélez-Colón, L. & Bernal-García, A. 2018. Antecedents of Customer Loyalty. *Contemporary Sport Marketing* 109(2002): 139–155.

Kreis, H. & Mafael, A. 2014. The influence of customer loyalty program design on the relationship between customer motives and value perception. *Journal of Retailing and Consumer Services* 21(4): 590–600. http://dx.doi.org/10.1016/j.jretconser.2014.04.006.

Marakanon, L. & Panjakajornsak, V. 2017. Perceived quality, perceived risk and customer trust affecting customer loyalty of environmentally friendly electronics products. *Kasetsart Journal of Social Sciences* 38(1): 24–30. http://dx.doi.org/10.1016/j.kjss.2016.08.012.

Namin, A. 2017. Revisiting customers' perception of service quality in fast food restaurants. *Journal of Retailing and Consumer Services* 34(September 2016): 70–81. http://dx.doi.org/10.1016/j.jretconser.2016.09.008.

Nastasoiu, A. & Vandenbosch, M. 2019. Competing with loyalty: How to design successful customer loyalty reward programs. *Business Horizons* 62(2): 207–214. https://doi.org/10.1016/j.bushor.2018.11.002.

Pan, Y., Sheng, S. & Xie, F.T. 2012. Antecedents of customer loyalty: an empirical synthesis and reexamination. *Journal of Retailing and Consumer Services* 19(1): 150–158. http://dx.doi.org/10.1016/j.jretconser.2011.11.004.

Rasheed, F.A. & Abadi, M.F. 2015. Impact of service quality, trust and perceived value on customer loyalty in Malaysia Services Industries. *Procedia – Social and Behavioral Sciences* 164(August): 298–304. http://dx.doi.org/10.1016/j.sbspro.2014.11.080.

Srivastava, M. & Rai, A.K. 2018. Mechanics of Engendering Customer Loyalty: A Conceptual Framework. *IIMB Management Review* 30(3): 207–218. https://doi.org/10.1016/j.iimb.2018.05.002.

Toufaily, E., Ricard, L. & Perrien, J. 2013. Customer loyalty to a commercial website: Descriptive meta-analysis of the empirical literature and proposal of an integrative Model. *Journal of Business Research* 66(9): 1436–1437. http://dx.doi.org/10.1016/j.jbusres.2012.05.011.

Uzunboylu, N. 2016. Service Quality in International Conference Industry; A Case Study of WCES 2015. *Procedia Economics and Finance* 39(November 2015): 44–56.

Wai Lai, I.K. 2019. Hotel image and reputation on building customer loyalty: An Empirical Study in Macau. *Journal of Hospitality and Tourism Management* 38(December 2018): 111–121. https://doi.org/10.1016/j.jhtm.2019.01.003.

Wolter, J.S., Bock, D., Smith, J.S. & Cronin, J.J. 2017. Creating ultimate customer loyalty through loyalty conviction and customer-company Identification. *Journal of Retailing* 93(4): 458–476. https://doi.org/10.1016/j.jretai.2017.08.004.

Xie, X. 2011. Service Quality Measurement from Customer Perception Based on Services Science , Management and Engineering. 1: 337–343.

Yousapronpaiboon, K. 2014. SERVQUAL: Measuring Higher Education Service Quality in Thailand. *Procedia – Social and Behavioral Sciences* 116: 1088–1095. http://dx.doi.org/10.1016/j.sbspro.2014.01.350.

The effect of attitude toward Instagram ads on brand attitude and engagement behavior

A.W. Saragih & Y. Alversia
Universitas Indonesia, Depok, Indonesia

ABSTRACT: The rapid growth of social media users in Indonesia has created a better chance for advertising a brand that was previously on a general Internet platform into social media platforms. Social network is basically a two-way communication medium, however it is not suitable if advertisement on social media only used a one-way promotion tool. In this study, the researcher used Instagram as the social media network as the advertising feature on this platform offers more features for users to engage with compared to other social media. The findings indicate that users' attitude toward the advertisement, which influenced by Informativeness, Irritation and Entertainment influenced positively on users' attitude toward the brand and users' engagement behavior. A total of 195 respondents who were actively using Instagram participated to fill out an online questionnaire and the data were analyzed using SEM (Structural Equation Modeling).

1 INTRODUCTION

1.1 Background

Technology has changed the old ways used in advertising. Marketers and advertisers get convenience in advertising, especially in terms of accessibility and the amount of media that can be used. Many companies nowadays are using the Internet as one of their advertising media, and enjoy the benefits of online technology. The Internet has become a popular platform in the advertising world because of the flexibility in terms of content, and ease in controlling the material of the advertisement (Lim et al. 2010).

Facebook, Instagram, Twitter, etc. recently got a lot of attention as they offer a feature to advertise. Those platforms are one of the latest forms of Internet advertising using social media or networks. Social networking has become one of the most popular and phenomenal online activities in the world, 60% of Internet users are actively spending their time on the internet (Nielsen 2011). The ease of accessing social networking has become one of the causes of the widespread use of this medium in the advertising world. Nowadays, marketers integrate social media networking with their marketing strategy, and plan to increase the social media budget. Marketers know that the use of social networking media can create opportunities for marketers to increase revenue and brand value, and deepen relationships between companies and customers. Marketers who recognize the importance of social media will be better able to anticipate future market and technological changes.

Advertising through social networking media is quite new, so customer acceptance of this kind of advertising is still important to note, both by advertisers and providers of the platform. Beyond all the ease and benefits of approaching customers through social networks, it is important to know whether those social network users can accept advertising activities as part of the social networking site. What the marketer fears is that advertising through social networking can interfere with the user's personal space, and their ads are placed in unwanted media by the customer (Kelly et al. 2010). The attitude of the customer toward an object on its base will affect his or her attitude towards other objects, which are still related (Wahid & Ahmed 2011). So, if the customer likes and has an emotional closeness with an advert, basically it will produce a positive attitude toward the product of the brand.

1.2 Previous research

Display advertising has been studied in several journals in various perspectives, such as how customer attitude to display advertising (Roberts 2010) and how the effectiveness of customer response (Ghajarzadeh & Sahebjamnia 2010). Both studies are structured based on display advertising on Facebook and traditional websites. How customers react to advertising in social media is the key to utilizing social media as part of integrating marketing communication (Taylor et al. 2011). Getting more consumer engagement with the brand advertisement will increase brand awareness or, even more, increase the chance for a brand product to be purchased.

Some of the studied respondents had a negative attitude toward advertising through social media because they were considered excessive (Halalau & Kornias

2012). In this study, the researcher chose to study Instagram as a platform. There are more features for a brand or company to put an advertisement on Instagram, which has more engageable features compared to Facebook or Twitter.

2 METHOD

2.1 Data collection

The type of data used in this study are primary data and secondary data. Primary data was be used in testing the proposed hypothesis, while secondary data will strengthen the formation of the hypothesis. Primary data was obtained from questionnaires which are distributed to Instagram users. The questionnaire contained questions related to behavior in using Instagram social media as part of their lifestyle. In addition, there was various questions raised about the user's attitude toward marketing communications carried out via Instagram, in this case limited to the fashion industry. Secondary data was obtained from various journals regarding IMC and research regarding marketing communication through other social media that have been done by previous researchers. Population in this study were all Instagram users in Indonesia and it used 201 questionnaire respondents as the sample.

2.2 Data analysis

The analysis of data in this research was conducted in some steps: validity test and reliability test using SPSS tool and also analyze the measurement and structural model using structural equation modeling (SEM).

2.3 Research model

According to Mackenzie & Lutz (1989), Ad Credibility, Ad Perceptions, Attitude toward Advertiser, Attitude toward Advertising and Mood are 5 independent variables that become antecedents of Attitude towards the advert. Ducoffe (1996) shows that factors such as Entertainment, Informativeness, and Irritation are the starting points to explain how consumers get value from advertising that affects Attitude toward Advertising. This study put 3 factors that affect Attitude toward Advertising from Ducoffe (1996) as independent variables to determine Attitude toward the Ad.

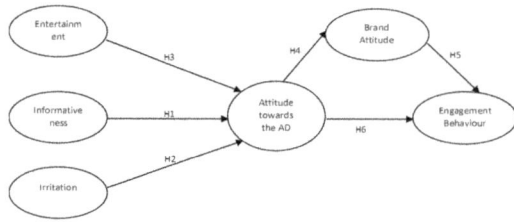

Figure 1. Research model.

Table 1. Variables in research model.

Variables
Informativeness (IF)
Irritation (IR)
Entertainment (EN)
Attitude toward the Ad (AAd)
Brand Attitude (BA)
Engagement Behavior (EB)

2.4 Research hypotheses

The hypotheses of this research include:

- H1: Informativeness has a positive influence on Attitude towards the Ad.
- H2: Irritation negatively influences Attitude towards the Ad.
- H3: Entertainment has a positive influence on Attitude towards the Ad.
- H4: Attitude towards the Ad in the Instagram has a positive influence on Brand Attitude.
- H5: Brand Attitude has a positive influence on Engagement Behavior.
- H6: Attitude towards the Ad in the Instagram has a positive influence on Engagement Behavior.

2.5 Questionnaire design

This research questionnaire design that was used for this research was divided into:

1. Screening: in this segment there are 3 screening questions. If the respondents fail to answer the correct screening questions, they will be taken out. The screening questions are:
   a. Do you have an Instagram Account?
   b. Have you ever found an Advertisement on your Instagram Timeline?
   c. Have you actively used Instagram in the previous month?
2. Questions: in this segment respondents were asked questions related to the advertisement on Instagram. There were 23 parameters distributed into 6 variables from the variable operationalization. The response form was used to test of the hypotheses and research questions. The researcher used 6 points Likert scale for the strict questions in the online survey questionnaire, eliminating the option for respondents to answer the middle neutral answer.

3 RESULTS & DISCUSSION

3.1 Questionnaire result

From the 201 distributed questionnaire, 195 respondents passed the screening and finished the survey (97%).

Table 2. Descriptive result.

Indicator	Mean
IF1	4.03
IF2	3.89
IF3	3.70
IR1	2.30
IR2	3.16
IR3	2.96
EN1	3.60
EN2	3.40
EN3	3.45
AAd1	3.50
AAd2	3.54
AAd3	3.73
AAd4	3.77
BA1	4.15
BA2	4.03
BA3	4.16
BA4	4.01
EB1	3.01
EB2	2.82
EB3	3.26
EB4	3.73
EB5	2.66
EB6	2.59

3.2 Hypotheses test

For a hypothesis to be accepted, the C.R (T-Value) score need to be above 1.645 while p-value must be smaller than 0.05. From the 6 hypotheses, all hypotheses are accepted.

Table 3. Hypotheses test.

Code	Hypotheses	Estimate	C.R	P	Remarks
H1	AAD <— IF	0.197	3.262	0.001	Accept
H2	AAD <— IR	−0.124	−3.505	***	Accept
H3	AAD <— EN	0.758	11.647	***	Accept
H4	BA <— Aad	0.589	9.231	***	Accept
H5	EB <— BA	0.262	3.02	0.003	Accept
H5	EB <— AAD	0.322	3.738	***	Accept

3.3 Antecedents of attitude toward the ad

Previous studies indicate that customers believe that newspapers have the most informative, reliable and credible ads, while television and radio rates are less (Ducoffe 1996). On the other hand, television advertising is the most entertaining (Larkin 1979). Instagram advertisements basically have what television ads have, appealing visuals and sounds. From all the hypotheses which relationship has proven, the researcher summarized all the antecedents that positively influence attitude toward the ad and compared each antecedents/variable influence on them. From 3 variables that influence brand attitude, Entertainment has the highest influence score (76.9%), followed

Table 4. Attitude toward the ad standardized regression weight summary.

Regression	Estimate
AAd <— IF	0.193
AAd <— IR	−0.124
AAd <— EN	0.769
AAd4 <— AAd	0.887
AAd3 <— AAd	0.895
AAd2 <— AAd	0.907
AAd1 <— AAd	0.915

by Informativeness (19.3%) and negatively, Irritation (−12.4%). This result implies that when an advertiser wants to adapt a similar integration of method in a marketing campaign, the advertisement has to be entertaining and not to induce negative evaluation from consumer. From the 3 indicators that influence Entertainment, EN1 (Instagram Advertisement is enjoyable) influence 84.2%, EN2 (Instagram Advertisement is entertaining) influence 93.4%, EN3 (Instagram Advertisement is pleasing) influence 90.5%. Based on this result, the researcher concluded that the entertainment of the Instagram Advertisement plays an important role in shaping the users' perception of how entertaining the advertisement.

3.4 Attitude toward the ad influence on brand attitude

Advertising messages also influenced the connection between Ad attitude and brand behavior, particularly if customers are unfamiliar with the advertised brand due to the absence of previous understanding on which to base their attitude toward brand evaluation. They are more likely to rely on Attitude toward the Ad in forming their Attitude toward the brand (Campbell & Keller 2003; Machleit & Wilson 1988). The results of the study show that Attitude toward the Ad (AAd) has a positive effect on Brand Attitude (BA), with a total of 58.9%, which is a considerable influence when compared with the magnitude of the influence rate among other variables in the model. Even when compared with the effect of AAd on Engagement Behavior (EB) which is only 32.2%, the figure is fairly large. This discovery is interesting, because it turns out that the effect of AAd on BA is greater than that of the EB. That is, this can be a suggestion for brands to improve users' attitudes toward their brand by increasing the attitude of users to the advertisement they made. This finding is something considerable for all brand owners,

3.5 Brand attitude influence on engagement behavior

The influence of BA on EB was 26.2%. When compared with the effect of AAd on EB by 32.2%, the effect of BA on EB is quite similar. However, be-cause

227

in this study the BA value was obtained through the influence of AAd, so that the influence of BA on the EB was an indirect effect of AAd on the EB and to compare its influence on the EB must also go through indirect calculations.

The approach calculates the indirect effect by multiplying two regression coefficients (Sobel 1982). The two coefficients are obtained from two regression. Using the Sobel approach, a product is formed by multiplying two coefficients together, the partial regression effect for BA predicting EB and the simple coefficient for AAd predicting BA:

$$(0.589)(0.262) = 0.154 \text{ or } 15.4\% \qquad (1)$$

Compared to direct effect of AAd on EB, indirect effect is lower. It can be concluded that for users to engage with the advertisement, their attitude toward the advertisement influence is higher than user attitude toward the brand. Even so, attitude toward the advertisement influence on brand attitude is something considerable for a brand, especially a newcomer, that they can get a good brand attitude by maximizing their effort on making an advertisement that could be perceived positively by Instagram users.

4 CONCLUSION

Based on the data analysis of the results of the discussions that have been carried out previously, it can be stated that:

1. Findings from the hypotheses result show that Informativeness has a positive influence on Attitude toward the Ad. This relationship is proven by the 0.001 p-value and with t-value 3.262 C.R. Informativeness has a 19.7% influence on Attitude toward the Ad. Looking at the indicators that influence Informativeness, an indicator shows the biggest influence than others; IF1 (Instagram Advertisement is a good source of information) influence 84.9%. This indicator shows the importance of the information that provided on the advertisement, whether it could convince users to perceive that the advertisement is a good source or not.
2. Findings from the hypotheses result show that Irritation has a negative influence on Attitude toward the Ad. This relationship is proven by the <0.05 p-value and with t-value −3.505. Irritation has a −12.4% influence on Attitude toward the Ad. Looking at the indicators that influence Irritation, an indicator shows the biggest influence than others; IR3 (Instagram Advertisement is irritating) influence 92.1%. This indicator shows that some Instagram users find advertisement on Instagram is irritating and this finding can be a consideration for Instagram.
3. Findings from the hypotheses result show that Entertainment has a positive influence on Attitude toward the Ad. This relationship is proven with

<0.05 p-value and with t-value 11.647. Entertainment has a 75.8% influence on attitude toward the Ad. While this finding is expected, this relationship is aligned with a lot of previous research that has explained how brand attitude will directly influence purchase intention. Looking at the indicators that influence Entertainment, EN2 (Instagram Advertisement is entertaining) shows the biggest influence than others with 92.1%, compared to EN1 (Instagram advertisement is enjoyable) and EN3 (Instagram advertisement is pleasing).

4. Findings from the hypotheses result show that Attitude toward the Ad has a positive influence on Brand Attitude. This relationship is proven with <0.05 p-value and with t-value 9.231 Attitude to-ward the Ad has a 58.9% influence on Brand Attitude. This is an interesting finding as the influence is big, compared to another inter-variables influences in this research. Looking at the indicators that influence Attitude toward the Ad, AAd1 (Instagram users like the Instagram Advertisement) shows the biggest influence than others, with influence 91.3%. It can be concluded that users' attitude toward brands can be shaped when they like the advertisement.
5. Findings from the hypotheses result show that Brand Attitude has a positive influence on Engagement Behavior. This relationship is proven with 0.003 p-value and with t-value 3.02. Brand Attitude has a 26.2% influence on Engagement Behavior. It shows that users' willingness to engage with the advertisement influenced by their perception of brand that being advertised.
6. Findings from the hypotheses result show that Attitude toward the Ad has a positive influence on Engagement Behavior. This relationship is proven with <0.05 p-value and with t-value 3.738. Attitude toward the Ad has a 32.2% influence on Engagement Behavior. It shows that how users perceived the advertisement affect their willingness to engage.

REFERENCES

Campbell, M.C. & Keller, K.L. 2003. Brand familiarity and advertising repetition effects. *Journal of Consumer Research* 30(2): 292–304.

Ducoffe, R.H. 1996. How Consumers Assess the Value of Advertising. *Journal of Current Issues and Research in Advertising* 17: 1–18.

Ghajarzadeh, A. & Sahebjamnia, N. 2010. A new model of online advertising effectiveness on customer responsiveness: a case of laptop companies in Malaysia. *Journal of Business and Policy Research* 5(2): 237–261

Halalau, R. & Kornias, G. 2012. Factors influencing users' attitude towards display advertising on Facebook. *Internationella Handelshögskolan Högskolan i Jönköping*.

Kelly, K., Kerr, G., & Drennan, J. 2010. Avoidance of advertising in social networking sites: the teenage perspective. *Journal of Interactive Advertising* 10(2): 16–27.

Larkin, E.F. 1979. Consumer perceptions of the media and their advertising content. *Journal of Advertising* 8(2): 5–7.

Lim, Y.M., Seng Yap, C., & Chai Lau, T. 2010. Response to Internet advertising among Malaysian young consumers. *Cross Cultural Communication* 6(2): 93–99.

Mackenzie, S.B. & Lutz, R.J. 1989. An empirical examination of the structural antecedents of attitude toward the ad in an advertising pretesting context. *Journal of Marketing* 53: 48–65.

Machleit, K.A. & Wilson, R.D. 1988. Emotional feelings and attitude toward the advertisement: The roles of brand familiarity and repetition. *Journal of advertising* 17(3): 27–35.

Nielsen, N. 2011. *State of the media: the social media report: Q3 2011.* [Online]. Retrieved from http://www.nielsen.com/nielsenwire/social/.

Roberts, K.K. 2010. Privacy and perception: How Facebook advertising affects its users. *The Elon Journal of Undergraduate Research in Communications* 1(1).

Sobel, M.E. 1982. Asymptotic confidence intervals for indirect effects in structural equation models. *Sociological methodology* 13: 290–312.

Taylor, D., Lewin, E., & Strutton, D. 2011. Friends, Fans, and Followers: do ads work on social networks? How gender and age shape receptivity. *Journal of Advertising Research* 51(1): 258–275.

Wahid, N.A. & Ahmed, M. 2011. The effect of attitude toward advertisement on Yemeni female consumes' attitude toward brand and purchase intention. *Global Business and Management Research: An International Journal.*

Major online purchase: A study on the effect of brand and positive review on iGeneration

I. Permana, R. Setiawan & A. Solihat
Universitas Garut, Garut, Indonesia

ABSTRACT: iGeneration refers to those who experience early exposure to sophisticated technology and gadgets. They do multitask using a smartphone and explore using a personal computer. This paper explains iGenertion's behavior in making online purchases of major products for clothing. The sample is 200 people. The results of data analysis show that brand and positive reviews have an influence on online purchases of major products by iGeneration at a 95% confidence level. Therefore, it is necessary for companies to build brand and attract positive reviews to improve sales performance.

1 INTRODUCTION

Online purchases are now growing rapidly along with technological advancements. In the process of buying online, consumers get more offers than better products and services. The Ministry of Communication and Information of the Republic of Indonesia in (Kominfo 2018), citing that during 2018 Indonesia claimed the 6th rank in the world with a total of 123 million internet users per month. Survey report of the Indonesian Internet Service Providers Association shows that internet users in Indonesia during 2018 are dominated by residents aged 15–19 years (Apjii 2017). It can be interpreted as internet users mostly born between 1999–2003. Based on the year of birth on the theory of generation (Rahman & Mannan 2018), generations can be divided into three categories namely Why Generation (born 1982–1985), Millennium (born 1985–1999) and iGeneration (born 1999–2002).

iGeneration has priority for self, socializing and education with the main purchases of appearance, clothing, cars, recreation, hobbies and traveling (Ruane & Wallace 2013). In Indonesia during 2018, clothing became the largest commodity purchased online about 14.6% (Apjii 2017). The most commercial internet content used by internet users in Indonesia is Shopee about 11.2% (Apjii 2017).

1.1 *Online purchase*

The stage that is passed by customers who make online purchases is the need recognition, information search, alternative evaluation, buying decisions and post-purchase behavior (Donald 2014). The internet allows consumers to shop whenever and wherever they want (Yoganes 2011).

Table 1. Generations and related term.

Author(s)	Term	Statement
(Dumeresque 2012)	iGeneration	Generally characterized by Net Generatin/Digital Natives are use and familiarity with digital technologies, communications and social media.
(Kilian 2012)	Millenials	Millenials generation adopt social media both passively and actively. They do not always use social media, but they do so quite often.
(Ruane & Wallace 2013)	Generation Y	Generation Y love fashion and shopping. Further, generation Y's woman use of the internet for purchasing fashion brands.
(Reis et al. 2015)	Generation Y	Exhibit relatively low trust in brands and advertising. They trust the opinions of other consumers more than what brands say about themselves on social media.

According to Schiffman & Kanuk in (Yohanes 2011), three levels of purchasing decisions are extensive problem solving, limited problem solving and routinized response behavior. Extensive problem solving is that customers do not yet have criteria for evaluating product categories or brands that meet so that they require more information. Limited problem solving is that customers have evaluated product categories and various brands so that the information received is only in addition. Routinized response behavior is customers who already have some experience with product and brand categories so that they don't need a lot of information. For online purchases, the intensity of information search will be seen from the intensity of the use of commercial internet content.

1.2 Brand

Every business has a different purpose in the commercial internet context. Trust in the products and brands offered causes customers to accept claims for products and services (Rowley 2009). Online reviews create a brand impression in the minds of customers (Chakraborty 2019). According to Stokes, well-known brands have the potential to provide information, make advertisements, word of mouth communication, prior purchases or use more brands (Park & Lennon 2009). According to Chaudri and Holbrook, e-trust can reduce customer uncertainty by relying on a trusted brand (Ha 2004). So that in the activity of buying online brand trust and brand familiarity it becomes very important. Brand trust is the willingness to trust and rely on brands based on the belief that brands can be trusted, safe and honest even though there is risk or uncertainty (Becerra & Badrinarayanan 2013).

Tam said that brand familiarity is the accumulation of customer experience of a brand that can increase confidence in making purchases (Lin 2013). Brand familiarity connects online brand experience with online purchases because customers are more interested in well-known brands when opening commercial internet content (Pappas 2018). The element of brand familiarity is that it can direct the customer's attention and is more often seen in advertisements (Mason-Jones & T 1999).

1.3 e-Trust

Information quality plays an important role in purchasing as consumer behavior. Mayer et.al said, states that the willingness of a party to provide information in the hope that the other party will take the same action regardless of the ability to monitor or control it is called trust (Chen 2017). Online buyers for fashion products consider information based on volume comments on commercial internet content (Pappas 2018). Cognitive compatibility can occur when individuals use a five-star rating system to help them make purchasing decisions (Chen 2017). Receiving information from other consumers influences online purchasing decisions on local brand fashion (Rahman & Mannan 2018). Cezar and Ogut, stated that the high number of recommendations and ratings in business internet content had a positive impact on online purchases (Mason-Jones & T 1999). Indicators of customer trust in commercial internet content are trustworthy, honest, keep promises, keep obligations and have a reliable infrastructure (Rahman & Mannan 2018).

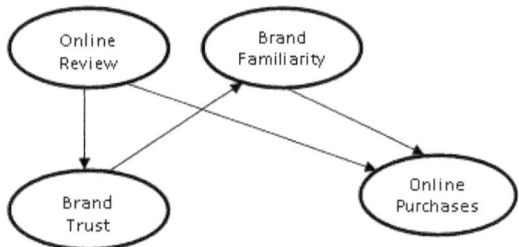

Figure 1. Research framework.

2 METHODS

The research design used in this study was a causal design to collect and structure data that allowed researchers to understand the causal relationship of several variables being studied. Determination of sample size in this study was 200 online buyers who had a representation of the population at an error rate of 5%. The sample was determined by using purpose sampling techniques with consideration of the age of 15–19 years who bought clothes online and had a Shopee account.

Operationalization of variables in this study consists of three variables, namely brand, positive review and major online purchases. Meanwhile the technique of collecting data using an online questionnaire. The measurement scale used is the Likert scale because consumer behavior is an attitude. The decision of testing the validity and reliability of the instrument is declared significant if the tvalue > 1.701 and rvalue > 0.361. Data analysis in this study is a Structural Equation Modeling based on a variant in the form of Partial Least Square.

3 RESULTS AND DISCUSSION

This study investigates how brand and positive review influences consumer's online buying behavior.

Based on the data in Table 2, it can be stated that:

1) Brand familiarity do not affect to online clothing purchase decisions in the iGeneration group. In this group, 35% chose to buy unbranded clothes, 39% did not always buy branded clothes, 12% only bought one branded clothes and 10% chose to buy branded clothes in the Shopee application. Previous research have reported finding brand familiarity was positively related to online phurchase intention (Yoo & Kim 2014; Yu et al. 2017). Brand familiarity indirectly influences online purchase intentions, but brand familiarity is first processed into image elaboration as a stimulus.

2) Brand trust affects brand familiarity for the iGeneration group. In this group, 25% really want a brand that can be trusted, 58% want a brand that can be trusted and 10% simply want a brand that

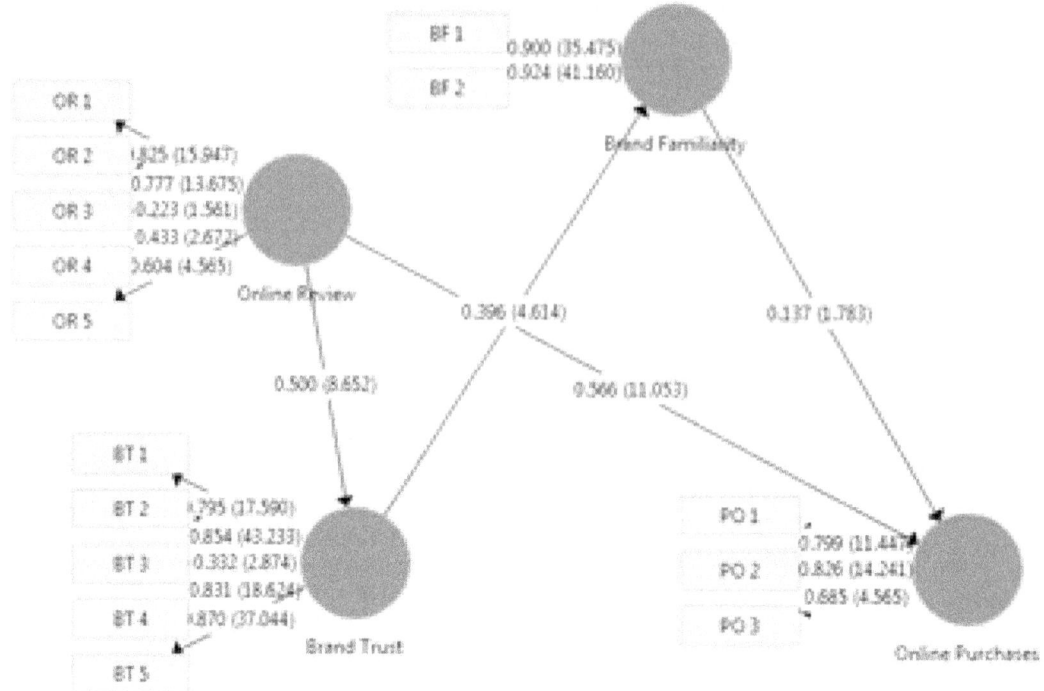

Figure 2. Measurement result.

Table 2. Hypotheses result.

Hypotheses Path	β	tvalue	Label
BF -> OP	0.136	1.783	Rejected
BT -> BF	0.394	4.614	Accepted
OR -> BT	0.531	8.652	Accepted
OR -> OP	0.587	11.053	Accepted

can be trusted for clothing online purchasing at the Shopee app. Previous research have reported finding brand trust was positvely related to brand familiarity (Bapat 2017; Chinomona & Maziriri 2017).

3) Online review affects brand trust in the iGeneration group. In this group, 33% consider highly reviews from previous buyers, 38% consider reviews from previous buyers and 23% consider enough reviews from online clothing buyers in the shopee app. Strengthened by data about 53% which states that the use of a five-star rating to grow brand trust of themself. Previous research have reported finding online review was positively related to brand trust (Bhandari & Rodgers 2017).

Online review of online purchases in the iGeneration group. In this group, the percentage of trust in online reviews is the same as the effect of online reviews on brand trust but has a higher analytical value. Because the information needed by iGenerasi in this

case is a photographic image of clothing purchased by another customer in the review column. Buying adds to their trust in major online purchases.

4 CONCLUSION

Overall, it can be concluded that the level of major online purchasing decisions for iGeneration group is limited problem solving. iGeneration is a customer who has evaluated various brand fashion product categories so that information received in online reviews is only in addition. iGeneration expressed pleasure with the Shopee application because it is reliable and provides a pleasant shopping experience for its age.

ACKNOWLEDGEMENTS

Research on major online purchases on iGeneration by involving brand variables and positive reviews, as the author's contribution to developing marketing knowledge. Besides that, research with this title has never been made before.

REFERENCES

APJII 2017. *Infografis penetrasi & perilaku pengguna internet Indonesia*. [Online]. Retrieved from www.apjii.or.id.

Bapat, D. 2017. Impact of brand familiarity on brands experience dimensions for financial services brands. *Int J Bank Mark* 35(4): 637–48.

Becerra, E.P. & Badrinarayanan, V. 2013. The influence of brand trust and brand identification on brand evangelism. *J Prod Brand Manag* 22(5): 371–83.

Bhandari, M. & Rodgers, S. 2017. What does the brand say? Effects of brand feedback to negative eWOM on brand trust and purchase intentions. *Int J Advert* 487:0–17.

Chakraborty, U. 2019. The impact of source credible online reviews on purchase intention the mediating roles of brand equity dimensions. *J Res Interact Mark* 13(2):142–61.

Chen, C.W. 2017. Five-star or thumbs-up? The influence of rating system types on users' perceptions of information quality, cognitive effort, enjoyment and continuance intention. *Internet Res* 27(3): 478–94.

Chinomona, E. & Maziriri, E.T. 2017. The influence of brand trust, brand familiarity and brand experience on brand attachment: A case of consumers in the Gauteng Province of South Africa. *Journal of Economics and Behavioral Studies* 9(1): 69–81.

Donald, F.K. 2014. *Entrepreneurship: Theory, process, practise*. South-Western: Cengage Learning.

Dumeresque, D. 2012. The net generation: Its impact on the business landscape. *Strateg Dir* 28(9): 3–5.

Ha, H.Y. 2004. Factors influencing consumer perceptions of brand trust online. *J Prod Brand Manag* 13(5): 329–42.

Kilian, T., Hennigs, N., & Langner, S. 2012. Do millennials read books or blogs? Introducing a media usage typology of the internet generation. *J Consum Mark* 29(2): 114–24.

Kominfo 2018. *Komunikasi dan informatika republik indonesia. pengguna internet Indonesia nomor enam dunia*. [Online]. Retrieved from www.kominfo.go.id.

Lin, Y.C. 2013. Evaluation of co-branded hotels in the Taiwanese market: The role of brand familiarity and brand fit. *Int J Contemp Hosp Manag* 25(3): 346–64.

Mason-Jones, D.R.R. & T. 1999. Analytic perspectives on online purchasing in hotels. *Int J Contemp Hosp Manag*.

Pappas, I.O. 2018. User experience in personalized online shopping: a fuzzy-set analysis. *Eur J Mark* 52(7–8): 1679–703.

Park, M. & Lennon, S.J. 2009. Brand name and promotion in online shopping contexts. *J Fash Mark Manag an Int J* 13(2): 149–60.

Rahman, M.S. & Mannan M. 2018. Consumer online purchase behavior of local fashion clothing brands. *J Fash Mark Manag an Int J* 22(3): 404–19.

Reis, S.R., Ting, Z.T., Proença, J., & Jay, K. 2015. Why are Generation Y consumers the most likely to complain and repurchase. *J Serv Manag*.

Rowley, J. 2009. Online branding strategies of UK fashion retailers. *Internet Res* 19(3): 348–69.

Ruane, L. & Wallace E. 2013. Generation Y females online: Insights from brand narratives. *Qual Mark Res* 16(3): 315–35.

Yohanes, S 2008 *Keputusan membeli secara online dan faktor-faktor yang mempengaruhinya*. [Online]. Retrieved from http://download.portalgaruda.org/article.php?article =7425&val=544&title=Keputusan Membeli Secara Online dan Faktor-Faktor yang Mempengaruhinya.

Yoo, J. & Kim, M 2014. *The effects of online product presentationon consumer responses: A mental imagery perspective*. [Online]. Retrieved from http://dx.doi.org/ 10.1016/j.jbusres.2014.03.006.

Yu, U., Cho, E., & Johnson, K.K.P. 2017. Effects of brand familiarity and brand loyalty on imagery elaboration in online apparel shopping. *J Glob Fash Mark* 2685:1–14.

Could brand image affect companies facing competitiveness?

A. Yusuf & R.L. Batu
Universitas Singaperbangsa Karawang, Karawang, Indonesia

ABSTRACT: Indonesia is one of the countries targeted by foreign companies to open fast-food restaurant branches in their effort for business expansion. Consumerism and lifestyle in the community create a high competition for fast food restaurants in Indonesia. This study examines foreign branded fast-food restaurants in Purwasuka (Purwakarta, Subang, Karawang). The purpose of this study was to determine the effect of brand image on fast-food restaurants competitiveness. This research was quantitative research conducted using descriptive and verification analysis. The type of data used in this study was primary and secondary data. Data were processed using external model analysis and structural model analysis, which was analyzed by PLS software. The results of this study indicated that there is a positive influence of brand image on competitiveness.

1 INTRODUCTION

1.1 Background

The fast-food industry is growing fast all over the world. Gofton (1995) in Shamoon et al. (2012) states that although there are many criticisms of fast food concerning nutrition, it has no negative points.

Fast food restaurants continue to collaborate and improve marketing strategies. Fast-food franchises have easily become popular food places. Their standard menu, recognizable signs, and integrated advertising strategies make brand names from fast-food restaurants like McD, A&W, and KFC famous (Lifestyle.kompas.com 2018).

In the last few years, fast food in America has turned into a staple food for consumers who want instant fun and also become a popular industry because of their advertisements and promotional materials, promote salads, and calorie calculation. McDonald's, the biggest burger brand that have survived for the past four years, have seen a decline in consumers at several outlets in the United States and is trying to increase burger meat (Vice.com 2018).

Fast food restaurant companies are changing business strategies in the United States and some developed countries. In contrast, fast food entrepreneurs find new customers with old strategies after opening branches in other parts of the world to make them easier to penetrate. Very low consumer awareness is a major problem with the risk of the products they sell. (Vice.com 2018). Indonesia is one of the destination countries to open fast-food restaurants, which can be seen from the number of fast-food restaurant brands in Indonesia. This is supported by the majority of Indonesians who are more likely to visit fast food outlets than other types of restaurant when they want to eat outside

Table 1. Consumer purchasing priorities in 2016.

Restaurant	Percentage
Fast-Food Outlet	80%
Food Court	61%
Cafe	22%
Fine Dining	1%

of their homes. (Tribunnews.com 2018). Table 1 shows the consumer purchasing priorities in 2016.

The fast-food restaurant business in Indonesia is increasingly in line with the expansion of each company's business interest. One of the companies that issued the new package was PT. Pioneerindo Gourmet Tbk (PTSP) as the holder of the CFC brand. This company relies on a cheap package program called Commander during Ramadan and Eid. The new menu strategy is also an option taken by PT. Holders of the KFC Fast Food Indonesia Tbk (FAST) brand. (Ekonomi.kompas.com 2018).

Eating out is part of the modern lifestyle or a necessity of recent times. Fast food is more comfortable and more satisfying (Shamoon et al. 2012). Short ordering time, excellent service, taste, and quality offered are some reasons why choosing fast food. This lifestyle change is increasingly being used as an opportunity by companies offering fast food products as their opportunity to open and establish new outlets to areas throughout Indonesia (Amran 2017).

Business strategies represent the general direction of the company, and various types of business strategies influence the way in which companies enter and articulate information from the environment (Matsuno & Mentzer 2000; Murray et al. 2011).

1.2 *Problem solving*

The formulation of the problem is based on the identification of research problems described above, therefore the formulation of the research problem is:

1. How the brand image of fast food restaurants is described
2. How the competitiveness of fast-food restaurants is defined
3. How strong brand image influences competitiveness of fast-food restaurants in Karawang

2 LITERATURE REVIEW

Marketing Strategy according to Wijayanti (2014) in Untari et al. (2017) is a guideline or basis for making marketing plans for a product and marketing tactics.

Brand image is "The set of beliefs held about a particular brand" (Kotler & Amstrong 2014). Ehsan et al. (2012) explain that brand image has three dimensions, namely attributes, benefits, and evaluation.

Market orientation is a concept that is commonly used by all business organizations. Various theories, concepts, definitions, and dimensions of market orientation are widely expressed in a variety of literature, both textbooks and scientific journals. However, in practice, not all theories and concepts can be applied to every type and size of business units. Adjustments need to be made in applying market orientation measurement methods. The concepts that have been modified according to the characteristics, needs, situations, and conditions of the microenvironment, both internal and external or external environmental conditions, are called constructs. The measurement of business units market orientation is course based on constructs that have been made with certain indicators (Hidayat 2015).

Brand competitiveness is ownership of a brand's competitive advantage. Competition is a market condition, while competitiveness is about the ability to create competitive advantage (Winzar et al. 2018).

One of the things to surpass competitors is to provide better value through a combination of price and product/service quality. Competitiveness is a relative construct. Competitive advantage is only meaningful when comparing competing brands (Winzar et al. 2018).

Competitiveness, being a favorable comparison with other providers, is a potential driver of customer loyalty (Baumann et al. 2015; Jo et al. 2015; Kumar 2002). According to Porter (1990), "Competitive strategy is the search for favorable competitive positions in an industry, the fundamental arena where competition occurs. The competitive strategy aims to build a profitable and sustainable position against predictions that determine industrial competition". There are two generic competitive strategies, i.e. cost leadership and differentiation. There are several arguments against using generic strategies such as "companies try to outperform their rivals to achieve a greater share of existing demand (Aura 2019; Kim & Mauborgne 2005; Nithisathian et al. 2012). According to Muhardi (2007) in Rohmanudin (2017) operational competitiveness is an operating function that is not only internally oriented but also proactively responds to target markets.

Factors in brand image greatly affect consumer perceptions of a product brand. Factors in brand image can create a positive perception therefore brand image can be improved so that a product can gain market share (Pajrin 2016).

When consumers think of a brand that has a good image, it affects the competitiveness of a brand in the market. The higher the brand image of a product, the higher the competitiveness of the product in the market. This opinion is confirmed by the results of previous studies.

A research by Pradini (2012) stated that there is an effect of service quality and brand image on KFC consumer repurchase interest in Salatiga. This is because consumer confidence in the KFC brand was very inherent and they truly believed that KFC brand was halal and good, thus influencing repurchase interests.

Another research by Putri (2015) found that the variables of product quality, brand image and company image simultaneously had a significant influence on customer loyalty.

Rohmanudin (2017) stated that the application of marketing strategies that had been carried out by the furniture industry in Wayhalim Regency, Bandar Lampung, namely: using direct sales, word of mouth promotions, promotions using business cards, catalogs, and opening performance space.

3 METHODS

The independent variable (X) of this research is brand image, and the dependent variable (Y) is competitiveness. The respondent of this research was a fast-food restaurant in the development area of West Java.

This research used descriptive and verificative research approach in which the variables were examined for their relationship. The aim was to present a structured, factual, picture of the facts of the relationship between the variables examined. The type of data used in this study was primary and secondary data.

The population in this study was fast-food restaurants in one development area of West Java, PURWA-SUKA (Purwakarta, Subang, and Karawang). Sample requirements for fulfilling PLS (Partial Least Square) calculations is 30–100 respondents. (Batu et al. 2016).

Based on the definition of the proposed sample and the minimum requirement for the number of samples to be analyzed, the sample used in this study was 30 fast-food restaurants in the development area. Data collection techniques in this study

Step 1

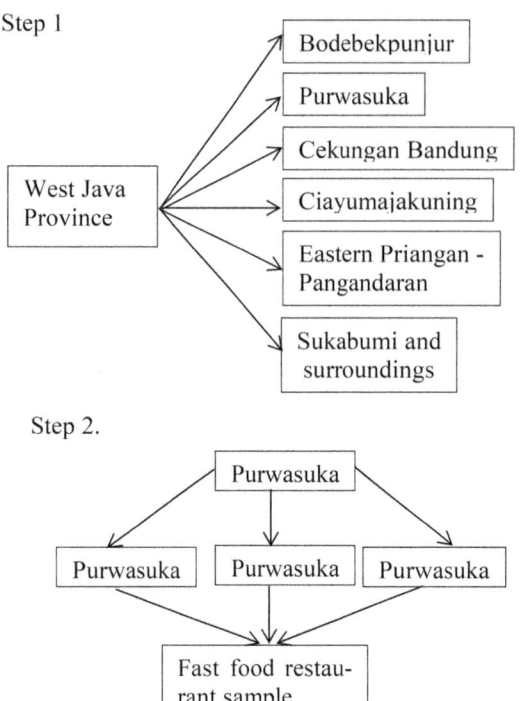

Step 2.

Figure 1. Cluster sampling technique.

Table 2. Outer loading.

	Brand Image	Competitiveness
X1	0,895	
X2	0,897	
X3	0,920	
Y1		0,884
Y2		0,820
Y3		0,773
Y4		0,999

Source: Data processing results, 2018

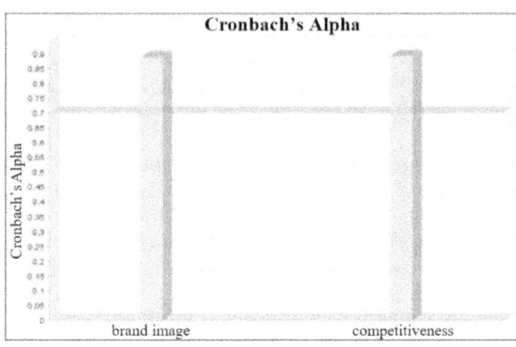

Figure 2. Cronbach's alpha. Source: Data processing results, 2018.

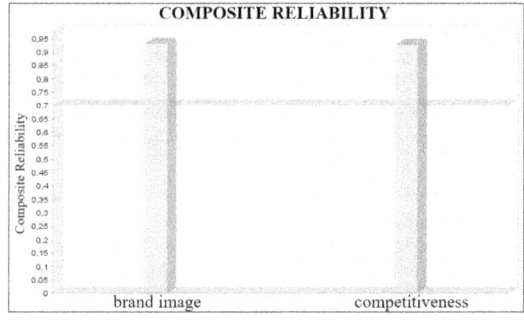

Figure 3. Composite reliability. Source: Data processing results, 2018.

were questionnaires, literature studies, and literature studies.

This study used a probability sampling technique with cluster sampling drawn in advance to determine which regions had the most fast-food restaurants. The steps taken in the cluster sampling technique explained in Figure 1.

4 RESULTS AND DISCUSSION

4.1 Validity and reliability

The test of convergent validity in PLS with reflective indicators is assessed based on loading factors (correlation between item scores/component scores with construct scores) indicators that measure the construct.

Based on Table 2, it can be seen that all dimensions have an outer loading value > 0.7, then the measurements of a high-correlated construct.

Reliability testing in PLS can use two methods, namely Cronbach's alpha and composite reliability. Rule of thumb alpha value or composite reliability must be more than 0.7. However, the value of 0.6 is still acceptable (Hair et al. 2014).

Based on Figure 2 and Figure 3 it can be seen that cronbach's alpha and composite reliability have values > 0.7, so the variables show accuracy, consistency and determination in making measurements.

4.2 Descriptive analysis

Overall brand image assessment which consisted of strengths, uniqueness and superiority can be explained based on the scores obtained from the data recapitulation, where the values are compared with the standard score criteria. The standard score is obtained through the calculation of the ideal score with the smallest score. The dimension that obtains the highest average score is the strength with a percentage of 33.5%, while the dimension with the lowest average score is the advantage with a percentage of 32.4%.

The brand image continuum value is 1180 with a percentage of 87.4% which, according to the research data, is considered very high. It means that the dimensions of research strength, uniqueness, and excellence have been well studied.

Overall competitiveness variables consisting of condition factors, demand factors, supporting factors & related industries, corporate and competing strategic factors is also assessed based on the scores obtained. The dimension that obtains the highest average score is the demand factor with a percentage of 26.2% while the dimension with the lowest average score is supporting industry with the percentage of 22.9%.

The competitiveness continuum value is 1544 with a percentage of 85.8% which is considered very high category. It means that competitiveness, demand, support and related factors, company strategy and competitor research have been running well in this study.

4.3 Hypothesis result

The results of this study are supported by previous research. A research by Pradini (2012) stated that there was an effect of service quality and brand image on KFC consumer repurchase interest in Salatiga. In her study, consumers' confidence in KFC brand was very inherent and they truly believed in the quality and halalness of the brand, thus influenced repurchase interests (Pradini 2012). In line with Pradini, Putri (2015) proved that the variables of product quality, brand image and company image simultaneously had a significant influence on customer loyalty. In addition, the latest research by Rohmanudin (2017) studied the implementation of marketing strategies that have been carried out by the furniture industry in Wayhalim Regency, Bandar Lampung using direct sales, word of mouth promotions, promotions using business cards, catalogs, and opening performance spaces.

Previous research has explained that the company's brand image and marketing strategy influences repurchase interests or consumer loyalty which indirectly increases the company's competitiveness. Fast food restaurants are made the object of this study because the number of fast food restaurants today is increasingly high competitiveness of fast food restaurant companies. After conducting research, the researchers obtained a result which proves that there is a direct effect of brand image on the company's competitiveness as much as 58.4% as shown in Figure 4.

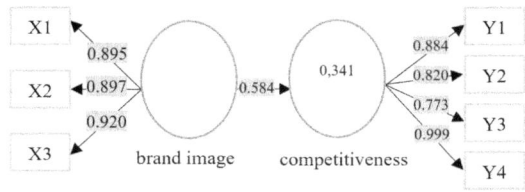

Figure 4. Processing hypothesis testing. Source: Data processing results, 2018.

5 CONCLUSION

Based on the results of research, it can be concluded as follows:

1. Based on the results of data processing that has been done, it can be seen that the brand image variable understudy has run well, and the dimension that has the highest average score is strength. The dimension that gets the lowest average score is excellence.
2. Competitiveness variables was measured through the dimensions of conditions factors, demand factors, supporting factors & related industries, corporate and competitor's strategic factors. Based on the results of data processing, it can be seen that the competitiveness variable understudy has run well, and the dimension that has the highest average score is the demand factor. The dimensions that get the lowest average score are supporting factors and related industries.
3. There is a positive influence of brand image on the competitiveness of fast-food restaurants in Purwasuka in 2019. If the brand image is positive, then competitiveness increases, conversely if brand image is negative, competitiveness decreases.

REFERENCES

Amran, F.R. 2017. Analisis strategi bisnis McDonald's di Labuan Ratu.

Aura, A. 2019. Pengaruh bauran pemasaran jasa terhadap tingkat daya saing pada orang gym di Kota Bekasi. http://repository.unpas.ac.id/45253/.

Baumann, C., Hoadley, S., Hamin, H. & Nugraha, A. 2015. Competitiveness vis-à-vis service quality as drivers of customer loyalty mediated by perceptions of regulation and stability in steady and volatile markets. *Journal of Retailing and Consumer Services* 36, no. September 2015 : 62–74. https://doi.org/10.1016/j.jretconser.2016.12.005.

Ehsan, M.M., Ghafoor, M.M. & Iqbal, H.K. 2012. Impact of brand image, service quality and price on customer satisfaction in Pakistan Telecommunication Sector" 3, no. 23: 123–29.

Ekonomi.kompas.com. Accessed on 24 Oktober 2018 at 10:09 WIB.

Hair, J.F., William, C.B., Barry, J.B. & Anderson, R.E. 2014. Multivariate Data Analysis. 7 Edition. United States of America: Pearson Education Limited.

Hidayat, C. 2015. Pengukuran orientasi pasar pada jenis usaha mikro, kecil, dan menengah. *Business and Management* 6, no. 9: 250–57.

Jo, M., Booms, B.H., Tetreault, M.S. & Bitner, M.J. 2015. Service encounter: diagnosing favorable and unfavorable incidents. 54, no. 1: 71–84.

Kim, W.C, & Mauborgne, R. 2005. Blue ocean strategy: how to create uncontested market space and make the competition irrelevant. Boston: Harvard Business School.

Kotler, P. & Amstrong, G. 2014. Principles of Marketing. 15th ed. United States of America: Pearson Education.

Kumar, P. 2002. On the Relationship Between. https://doi.org/10.1177/1094670502005001006.

Lifestyle.kompas.com. Accessed on 23 Oktober 2018 at 21:22 WIB

Matsuno, K. & Mentzer, J.T. 2000. The effects of strategy type on the market orientation – Performance. 64, no. October: 1–16.

Murray, J.Y., Gao, G.Y. & Kotabe, M. 2011. Market orientation and performance of export ventures: The process through marketing capabilities and competitive advantages. 252–69. https://doi.org/10.1007/s11747-010-0195-4.

Nithisathian, K., Takala, J., Rattanakomut, S., Walsh, J., Wu, Q. & Liu, Y. 2012. Operational competitiveness development in turbulent business environment: a case study in thailand fine gold jewelery export industry. No. 3: 53–62. https://doi.org/10.2478/v10270-012-0024-y.

Pajrin, N. 2016. Pengaruh citra merek (brand image), kualitas pelayanan, lokasi, dan faktor pribadi terhadap keputusan pembelian konsumen pada KFC Gelael Bandar Lampung tahun 2015.

Porter, M.E. 1990. The Competitive Advantage of Nations. New York: Free Press.

Pradini, A.L.W. 2012. Analisis pengaruh kualitas layanan dan brand image terhadap minat beli ulang pada restoran Kentucky Fried Chicken (KFC) Salatiga.

Putri, A.P. 2015. Pengaruh kualitas produk, citra merek dan citra perusahaan terhadap loyalitas pelanggan Amanda Brownies Surabaya. http://eprints.perbanas.ac.id/696/.

Rohmanudin. 2017. Analisis strategi pemasaran dalam meningkatkan daya saing.

Shamoon, S., Tehseen, S. & Nousheen, A. 2012. Bugs and buds in fast food industry of Pakistan: Effect of SWOT on the performance of fast food industry in Pakistan. *School of Doctorial Studies (European Union) Journal*, no. 4.

Tribunnews.com. Accessed on 24 Oktober 2018 at 09:55 WIB.

Untari, Shinta Nurafni, Sutrisno Djaja, and Joko Widodo. "Strategi Pemasaran Mobil Merek Daihatsu Pada Dealer" 11 (2017): 82–88. https://doi.org/10.19184/jpe.v11i2.6451.

Vice.com. Accessed on 24 Oktober 2018 at 9:35 WIB.

Vice.com. Accessed on 24 Oktober 2018 at 21:52 WIB.

Winzar, H., Baumann, C. & Chu, W. 2018. Brand competitiveness introducing the customer-based brand value. https://doi.org/10.1108/IJCHM-11-2016-0619.

Marketing strategy based on consumer behavior in facing competition between coffee shops

A. Solihat & N.A. Hamdani
Universitas Garut, Garut, Indonesia

ABSTRACT: The coffee shop has become a promising business. This is indicated by an increase in the amount of consumption coffee of Indonesian people up to 1.79 kg/capita/year in 2019. The increase in the amount of coffee consumption and the establishment of coffee shop businesses has encouraged a shift in people's lifestyles to make coffee as a necessity, social recognition and new trends in community. Coffee shop business has opened new opportunities in developing and establishing coffee shops in various regions in Indonesia, including in Garut, West Java. Due to this phenomenon, the coffee shop is very tempting as a research subject in coffee sales marketing strategies. The purpose of this study was to determine marketing strategies based on consumer behaviour from various circles in several coffee shops in Garut district. The two coffee shops included were Collega Coffe Shop and Kopilogi Coffee Shop. The survey was conducted by interviewing 25 visitors at the Collega Coffe Shop and Kopilogi Coffee Shop. The results of the study showed that a few visitors understood the different taste and variant of coffee offered, but visitors chose a coffee shop because of the facilities offered that provide convenience for visitors/ consumers. The marketing strategy applied by coffee shop to maintain business and to increase its turnover business should cover strategic product, price, place and promotion.

1 INTRODUCTION

Indonesia has an attractive tourism sector for domestic and foreign tourists. The increasing number of tourists visiting Indonesia has provided a widely business opportunity for Indonesian people. Business development in the 21st Century has developed very rapidly. One of the business sectors that has a great opportunity is business in the culinary field. One of the most sought businesses is to open a cafe/coffee shop. People in business try to make an interesting concept to make consumers come to their place. Place-based marketing is a widespread and acknowledged strategy for value-adding at points of consumption (Wright & Aklimawati 2018).

Coffee plants in Indonesia has been cultivated since ages. Coffee plants become a source of people's main income and has been used to increase foreign exchange through exports of raw seeds and coffee beans processed. Arabica and robusta coffee are two types of coffee with the most highly economic value and widely traded. Coffee plants are usually processed into drinks and its flavor make them different from any other beverages. Besides, coffee is a natural source of caffeine that can reduce fatigue. Coffee consumers consider drinking coffee not only to quench thirst, but also to supplement daily life activities such as meetings, dating, doing assignments, business meetings and others.

Based on data from the Central Bureau of Statistics, the level of coffee consumption of Indonesian people from 2015 to 2019 is projected to increase by 1.79 kg/capita/month. An increase of coffee consumption and a shift in people's lifestyles makes coffee drinking as a part of their daily needs. This certainly makes an opportunity for business people to develop a coffee shop business.

Based on the Agriculture Data and Information System Center (Figure 1), coffee consumption in 2016 reached around 250 thousand tons and grew by 10.54 percent to 276 thousand tons. Indonesia's coffee consumption in the period of 2016–2021 is predicted to grow with an average of 8.22%/year. In 2021, coffee supply is predicted to reach 795 thousand tons with consumption of 370 thousand tons, resulting in a surplus of 425 thousand tons. About 94.5 coffee production in Indonesia is supplied by smallholder coffee producers.

The increasing number of coffee shops occurs in various cities, both in large cities and small cities, including Garut. Garut is a region surrounded by mountains so that many hot springs spa are visited by tourists. Due to location advantages, coffee shop business in Garut has become increased sector. Among many number of coffee shops in Garut, there are two coffee shops which have the most visitors. Those are the Coffee Collega located on Pahlawan Street No. 55 and Kopilogi located on Cikuray Street No. 42.

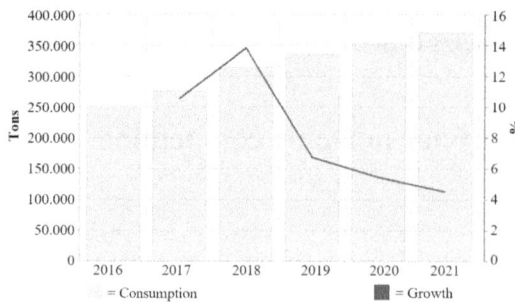

Figure 1. National coffee consumption projection 2016–2021.

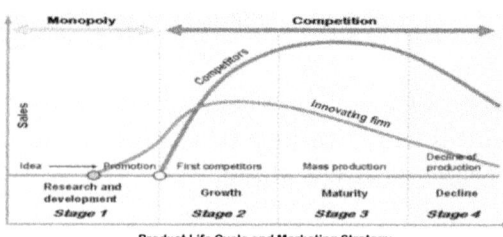

Figure 2. Product life cycle.

The location of the two mentioned coffee shops are very competitive. This can be seen from the large number of coffee shops around the location that offer various concepts to capture and maintain the market. Coffee shop owners are required to pay attention to marketing strategies, because marketing strategies are the key to be able to maintain and develop the business. Therefore the authors conducted research on Marketing Strategy and Consumer Behavior in Coffee Shops Business.

2 METHODS

The object investigated in this study was the best marketing strategy based on consumer behavior in Colega cafe and Kopilogi coffee shop. This research was a descriptive type using a qualitative approach, aimed to find out a problem and to obtain information and data available at the research location. Data sources used in this study consisted of primary and secondary data. The sampling technique used in this study was purposive with snowball sampling technique. The sample selection at the time of the research setting was based on the key informant requirement, thus they were the consumers who visited the two mentioned coffee shops. The data was collected by documentation, interview, and observation.

3 RESULTS AND DISCUSSION

3.1 *Coffee shop*

Millennial lifestyle trends make business in the culinary field, especially the Coffee Shop to grow faster in each region. Based on Figure 2 the product life cycle consists of several stages, i.e. introduction, growth, maturity and decline.

In the maturity stage of, many competitors open a coffee shop business. Therefore, to avoid the decline of the business, the companies must be able to innovate both in menu variants and consumer convenience.

3.2 *Consumer behavior*

The variables discussed in this study was consumer purchasing behavior divided into several stages; (1) information seeking, (2) buying decisions, (3) post-purchase behavior. The author asked several questions related to each stage in the process of consumer purchasing decisions.

3.3 *Marketing strategy*

The marketing strategies used in colega coffe shop and kopilogi coffee shop were product strategy, pricing strategy, location/location strategy and promotion strategy.

4 CONCLUSION

Based on the results of the discussion on consumer characteristics, consumer behavior and marketing strategies in the coffee colleague and copilogy, this research has concluded that:

a) The characteristics of consumers in the colega coffe shop and kopilogi coffee shop, there were more male visitors than female and most of them are teenagers (mostly students who live in downtown).
b) 75% of visitors to the coffee shop did not understand coffee. The main reason they came to the coffee shop was because of comfortable place, wifi connection and the place to hang out with friends.
c) The strategies required to increase sales need to focus on marketing strategies; product strategy, pricing strategy, place strategy and promotion strategy.

REFERENCE

Wright, J. & Aklimawati, L. 2018. Geographical indications and value capture in the Indonesia Cofee Sector. 59 (January): 35–48.

Can electrocardiographic response define the effectiveness of an advertisement video?

A.W. Arifin & A.D. Hanani
Universitas Indonesia, Depok, Indonesia

ABSTRACT: There are some mass media that we can use to advertise our organization, product, service, or idea, such as television, radio, magazine, etc. That media transmits the advertising message to large group of population and some are expected to be potential customers. The company must conduct some research to determine which media can generate the most effective results. Integrated media mix is one factor that is also involved in media decisions. The company can do pretesting of finished ads, such as theater test, on-air test, and physiological measures. The less common method of pretesting finished ads is physiological measurement. This research aims to examine and assess the advertising communication effectiveness toward physiological response using single channel electrocardiographic measurements that measure the most effective scene of a video; and influences advertisement effectiveness that consists of some variables like attitude toward the advertisement and purchase intention. This study was experimental research using a deep interview process for the electrocardiograph response. It was conducted with 35 respondents using single channel (lead I) portable electrocardiograph. The aim was to know the most effective scene in 4 Indonesian instant noodle advertisement videos. After knowing the most effective scene and conducting video editing process, only 1 video was considered as the most appropriate by experts and was used as the stimulus for the next process. Questionnaires were also conducted for 2 stimuli (original and edited advertisement video) to measure some variables. It was distributed to 125 respondents to find the difference for both videos. The respondents' result was then analyzed by using structural equation modeling. The result of this research showed that the majority of the hypotheses in stimulus I and stimulus II about the significant influence for all variables was accepted. This research also showed that the most effective scene from electrocardiograph response significantly influenced the variables relation in the structural model and the mean scores of the variables measured in this research.

1 INTRODUCTION

1.1 *Background*

Belch & Belch (2015) explained that advertising is such a critical part in integrated marketing communications programs because it can be the most cost-effective program to reach a large number of customers and the most valuable tool for building company or brand equity. It also creates differentiation of advertising purpose for each company or product. After the company advertised the product, one of the important parts of advertising is how to determine the effectiveness of the communications program. To measure the effectiveness of communications, there are some elements that the company should know. The source factor is the first element. The company must know the effectiveness of the spokesperson that is being used and how the market responds to him or her. Negative publicity can easily change the value of source. The next element is message variables. The company must know how strong the message to at-tract the public is or how clear the advertising content delivers the message. Media strategies are also one of the important elements that

the company must know when measuring communication effectiveness. The company must conduct some research to determine which media can generate the most effective results. Integrated media mix is another factor involved in media decisions.

The company can do pretesting of finished ads, such as theater test, on-air test, and physiological measures. The less common method of pretesting finished ads are physiological measurement (that also involve some research of psychological responses measurement). Some common responses to be measured are pupil dilatation (Pupillometrics), galvanic skin response (GSR) or electro-dermal response (EDR), eye tracking, and brain waves (Electroencephalographic or EEG).

Venkatraman et al. (2015) explained that Neuroscientific techniques such as functional magnetic resonance imaging techniques (fMRI) are also increasingly being used to assess physiological responses. FMRI is possibly the most promising and reliable method of neuromarketing but most of the current measurements still rely on ECG and GSR tests that measure a single dimension. The speed of heart rate (pulse) is measured with EKG by using external skin electrodes to measure the electrical activity of the heart. When the body responds to external stimulus, there are two antagonistic systems that control heart rate. The first one is the parasympathetic nervous system (PNS) and the other is the sympathetic nervous system (SNS). The effects of these two antagonistic systems are called heart rate acceleration.

Slower heart rate or deceleration of heart rate is referred to the PNS that make calm and relaxed state. When the heart rate has a deceleration in response to advertising, it implies that there is an increase in ability to focus and will provide an independent measure of attention. On the other hand, the activation of the SNS will increase the heart rate because of the respiration that specifically called respiratory sinus arrhythmia. The increase of heartbeat is also used as a measure of the affective processes.

1.2 Previous research

There are numerous researches about advertising effectiveness toward cognitive response and about neuromarketing. One of the previous researches used as reference is the study conducted by Sebastian (2013). The study was about neuromarketing and evaluation of cognitive and emotional responses of consumers to market stimuli. The result becomes the basis to understand decision-making mechanisms and also to assess the knowledge level on consumer's behavior. Neuromarketing also can help marketers to understand how unconscious mind processing influence the decision to purchase, providing a better understanding of consumer's thoughts, emotions, feelings, needs, and motivation in relation to the marketing products.

Venkatraman et al. (2015) explained that there are various methods that have been developed to assess advertising effectiveness, ranging from traditional self-reported measures to eye the neurophysiological tools. The research found the potential of neurophysiological measures to complement traditional measures in improving the predictive power of advertising success models. The study also demonstrated the potential of neuroscience applied to marketing research and practice by extending existing measures, helping enrich marketing theories, and improving models of marketing success. Neurophysiological methods are typically more expensive and less accessible to traditional methods. It is explained that some methods could be conducted on neurophysiological variables like eye tracking, heart rate or pulse, skin conductance response, EEG, and fMRI.

The effect of advertisement sensuality level toward cognitive responses found that there was average significant difference toward the cognitive responses (attitude toward brand, attitude toward ad, and purchase intention). The lower the sensuality level is, the higher the average value to the cognitive responses (Anggraini 2013).

2 METHOD

2.1 Data collection

Data collection was conducted in several steps:

a. Physiological Response: The Physiological Response process was conducted for minimum 30 people using single channel (lead I) electrocardiograph responses in order to get information about the most effective scene in the advertising video. Respondents' heart rates were recorded before and while they watched the advertising video using video player on notebook computer.

b. Pretest questionnaire: A pretest requires a minimum of 30 respondents. The respondents for the pretest and the actual survey were taken from the same population. Pretest was needed in a quantitative research involving questionnaires to identify and eliminate the parameter or indicator in the questionnaire that showed low validity and reliability scores.

c. Video editing: The original advertising video was edited by deleting the most effective durations from the electrocardiograph result.

d. Post-test questionnaire. This questionnaire was used as the tool to measures the variables that were analyzed to generate the result of this research. This questionnaire was distributed into 2 groups (125 respondents in each group). The first group viewed the original advertising video, while the second group viewed the edited advertising video.

2.2 Data analysis

The analysis of this research was conducted in some steps: coding and analyzing strategy decision, using descriptive analysis, using electrocardiographic analysis, and also comparing mean analysis by using structural equation modeling (SEM). SEM was used to analyze the data gathered from the research questionnaire.

2.3 Research conceptual model

Source-oriented thoughts and advertisement execution thoughts were the variables that the researcher used in the conceptual model to measure the attitude toward the advertisement and its impact on purchase intention. This model was adapted from Mashwama & Vuyelwa (2016) and Belch & Belch (2015). However, the model did not measure the product/message thoughts and brand attitude. Mashwama & Vuyelwa (2016) explained that attitude toward advertisement will influence the purchase intention. Belch & Belch (2015) also explained that attitude toward the advertisement was influenced by source-oriented thoughts and advertisement execution thoughts. The research model is shown in Figure 1.

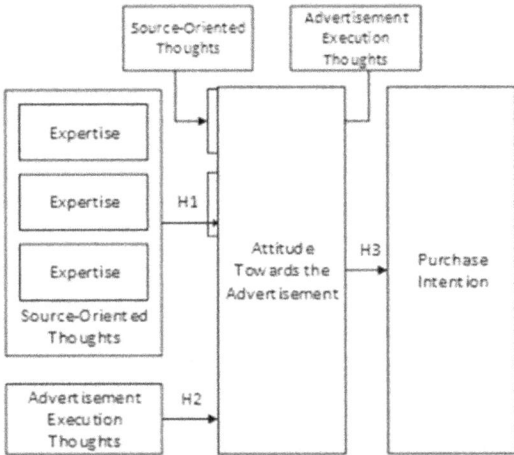

Figure 1. Research model.

2.4 Research hypotheses

The hypotheses of this research are:

H1: Source-oriented thoughts have a significant influence on attitude toward the advertisement.

H2: Advertisement execution thoughts have a significant influence on attitude toward the advertisement.

H3: Attitude toward the advertisement has a significant influence on purchase intention.

The research question is adapted from Venkatraman et al. (2015) research that the electrocardiograph response can define the respondent affection toward the advertisement. Thus the researcher proposes the research question: What is the most effective scene giving significant influence on advertisement effectiveness?

2.5 Questionnaire design

The design that was used for this research is divided into:

1. Screening: In this segment, the researcher gave three screening questions. If the respondents failed to answer the correct screening questions, they were redirected to the end of the questionnaire. The screening questions are:
   a. Do you know the brand of the product?
   b. Do you know the name of the actor?
2. Questions: In this segment, respondents were asked some questions related to some variables of advertisement effectiveness. There were 19 parameters distributed into 4 variables from the variable operationalization. The responses were used to test the hypotheses and to answer the research question. The researcher used 6 points Likert scale for the strict questions in the online survey questionnaire, eliminating the option for respondents to answer neutrally.

2.6 Advertisement video

The advertisement video that was used to measure the physiological response is Indonesian instant noodle, "Mie Sedaap Ayam Bawang (Actor: Chicco Jerikho)".

3 RESULTS AND DISCUSSION

3.1 Electrocardiographic response result

After measuring the respondents' heartbeats using portable ECG, the result showed fully interesting response on the duration 00:12:00 until 00:15:00 (4 respondents showed the same results) and potentially interesting response on duration 00:11:00 and 00:16:00 until 00:18:00 (3 respondents showed the same results). Based on the results, the researcher conducted video editing to delete the most effective scene from the advertisement video. The deleted durations are 00:11:86 until 00:16:00.

3.2 Questionnaire result

In questionnaire I, from 151 distributed questionnaires, 136 passed the screening and finished the surveys (90.066%). In questionnaire II, from 157 distributed questionnaires, 139 passed the screening and finished the surveys (88.535%). The descriptive result of the questionnaire is provided in Table 1.

The comparison showed that the video with deleted interesting edits (Stimulus II) mostly have lower mean score than the original video (Stimulus I). There is only 1 Indicator (5.3%) which has same score, while the other 18 Indicators (94.7%) have lower mean scores.

3.3 Hypotheses test results

3.3.1 Hypothesis 1
The hypothesis 1 of this research is that source-oriented thoughts (Tso) have a significant influence on attitude toward the advertisement (Aad). Stimulus I shows that the p-value of this relationship is 0.010 (below 0.05) with 2.573 C. R (above 1.645), so this hypothesis is accepted with 24.7% influence of the attitude on the advertisement. Meanwhile, Stimulus II shows that the p-value of this relationship is 0.011 (below 0.05) with 2.537 C. R (above 1.645), so this hypothesis is also accepted with 19.1% influence of the attitude on the advertisement as 19.1%.

3.3.2 Hypothesis 2
The hypotheses 2 of this research is that advertisement execution thoughts (Tae) have a significant influence on attitude toward the advertisement (Aad). Stimulus I shows that the p-value of this relationship is less than 0.001 (below 0.05) with 7.172 C. R (above 1.645), so this hypothesis is accepted with 72.8% influence of the attitude toward the advertisement. Whereas Stimulus II shows that the p-value of this relationship is less than 0.001 (below 0.05) with 8.675 C. R (above 1.645), so

Table 1. Descriptive result.

Indicator	Stimulus I	Stimulus II	Comparison
TSOA1	4.82	4.59	The video with deleted interesting duration has lower score
TSOA2	4.35	4.18	The video with deleted interesting duration has lower score
TSOA3	4.82	4.46	The video with deleted interesting duration has lower score
TSOT1	3.87	3.76	The video with deleted interesting duration has lower score
TSOT2	4.37	4.12	The video with deleted interesting duration has lower score
TSOT3	4.28	4.07	The video with deleted interesting duration has lower score
TSOE1	4.76	4.43	The video with deleted interesting duration has lower score
TSOE2	4.40	4.19	The video with deleted interesting duration has lower score
TSOE3	4.34	4.16	The video with deleted interesting duration has lower score
TAE1	4.52	4.41	The video with deleted interesting duration has lower score
TAE2	3.99	3.99	The video with deleted interesting duration has lower score
TAE3	4.37	4.24	The video with deleted interesting duration has lower score
AAD1	4.38	4.27	The video with deleted interesting duration has lower score
AAD2	4.20	4.11	The video with deleted interesting duration has lower score
AAD3	4.33	4.13	The video with deleted interesting duration has lower score
PI1	4.24	4.05	The video with deleted interesting duration has lower score
PI2	4.18	4.13	The video with deleted interesting duration has lower score
PI3	4.17	4.06	The video with deleted interesting duration has lower score
PI4	3.85	3.73	The video with deleted interesting duration has lower score

this hypothesis is also accepted with 73.3% influence of attitude toward the advertisement.

3.3.3 Hypothesis 3

The hypotheses 3 of this research is that attitude toward the advertisement (Aad) has a significant influence on purchase intention (PI). Stimulus I shows that the p-value of this relationship is less than 0.001 (below 0.05) with 10.175 C. R (above 1.645), so this hypothesis is accepted with75.2% influence to the purchase intention. Meanwhile, stimulus II shows that the p-value of this relationship is less than 0.001 (below 0.05) with 8.550 C. R (above 1.645), so this hypothesis is also accepted 67.0% influence to the purchase intention. It can be concluded that attitude toward the advertisement in the advertising video has a significant influence to the purchase intention. The indicator

with the highest influence on attitude towards the advertisement is from indicator AAD2 "I like this advertisement very much" (94.3% on stimulus I and 97.6% on stimulus II). The audience response for this advertisement is that they have high appreciation for the advertisement. The relationship of attitude toward the advertisement with purchase intention has been already proved in many previous research studies. This result is as expected.

3.3.4 Research finding

This research attempted to find out the most effective scene that gives significant influence on advertisement effectiveness. The first stimulus is the most effective scene that significantly influence the variables relation in the structural model. This research reveals that the influence of variables relation decreases on stimulus II. This is because the most effective scene deleted are Tso that influence the Aad. Moreover, it is also the Aad that influences PI.

The decrease of Tso influence on Aad occurs because the interesting act done by the actor is deleted on the stimulus II. This decreases the relation be-tween the actor and audience. Meanwhile, the de-crease of Aad influence on PI is caused by there being no effective scene that is considered as an interesting segment. It is based on EEG measurement. This result indicates that stimulus II has less interesting scene than stimulus I, so that audience prefer stimulus I rather than stimulus II.

3.4 Discussion

3.4.1 Source-oriented thoughts

Even though the advertisement video has a significant influence in forming Tso, trustworthiness (Tsot) basically plays the biggest role. For the Indonesian instant noodle target market, using an actor that is honest, dependable, and trustworthy is more effective than using a more attractive or more expert actor to create higher Tso.

3.4.2 Advertisement execution thoughts

Even though the advertisement video has a significant influence in forming advertisement execution thoughts (Tae), interesting advertisements (Tae3) have the biggest influence. For the Indonesian instant noodle target market, making an interesting advertisement video is more effective than making a recommended or good advertisement to create higher Tae. To fulfil this need, the company/brand should create a good story, graphic, video effect, cinematic angle, and other interesting factors for the advertisement video.

3.4.3 Attitude toward the advertisement

Even though the advertisement video has a significant influence in forming Aad, the customer's appreciation (Aad2) plays the key role. For the Indonesian instant noodle tar-et market, making an advertisement video liked by the customers is more effective than making an advertisement that generates positive and favorable attitudes from the audience to create higher Aad. In response to this, the company/brand can conduct some research to know what kind of advertisement video most audiences currently like before making a new advertisement video. This is because the audience appreciation and positive attitude plays a crucial role in shaping their attitude toward the advertisement.

3.4.4 *Purchase intention*

Even though the advertisement video has a significant influence in forming PI, the customer's intention to try (PI2) and buying interest (PI3) have greater influence. For the Indonesian instant noodle target market, making an advertisement video that can generate customer's intention to try and buy a product is more effective than making an advertisement that generates customers' brand consideration and willingness to recommend the brand to create higher PI. Regarding this, the company/brand can conduct research to know what kind of advertisement video that mostly succeeds to create audience intention to try and buy from other FMCG product's advertisement video result.

4 CONCLUSION

From the hypotheses test results, the researcher aims to answer the research questions of this study with additional insights.

a. The first research question is "How to analyze the influence of source-oriented thoughts on attitude towards the advertisement?" The result shows that source-oriented thoughts have significant influence on attitude toward the advertisement.
b. The second research question is "How to analyze the influence of advertisement execution thoughts on attitude toward the advertisement?" The result reveals that advertisement execution thoughts have significant influence on attitude toward the advertisement.
c. The third research question is "How to analyze the influence of attitude toward the advertisement on purchase intention?" The result finds out that attitude toward the advertisement has significant influence on purchase intention.

d. The fourth research question is "How to analyze the influence of most effective durations influence on advertisement effectiveness?" After measuring the EEG response by capturing the duration when the SNS triggered, it can be discovered what is the most effective scene/duration is for the audience. This effective duration has significant influence on advertisement effectiveness.

ACKNOWLEDGEMENTS

I would like to thank Allah SWT to give me willingness and power to be able to finish this paper. I also would like to thank my family, I'm very grateful for the support and prayers that you gave. I also would like to acknowledge the all the people who have contributed directly or indirectly to this paper especially Mr. Alberto Daniel Hanani as my supervisor, to guide me through this process in unconventional setting and context.

REFERENCES

Anggraini, A.G. 2013. Analysis of The Effect of Advertisement Sensuality Level and Model Utilization Type Toward Cognitive Responses, Attitudes, and Purchase Intention. *Thesis. University of Indonesia.*
Belch, G.E. & Belch, M.A. 2015. Advertising and Promotion: An Integrated Marketing Communications Perspective 10th Global Edition. *Boston McGraw-Hill.*
Mashwama, V. 2016. Brand Endorsements: A Study into the Opportunity of Using Spokes-Characters as Brand Endorsers. *University of the Witwatersrand.*
Sebastian, V. 2013. Neuromarketing and Evaluation of Cognitive and Emotional Responses of Consumers to Marketing Stimuli. *Journal of Social and Behavioral Sciences. Elsevier, Ltd.*
Venkatraman, V., Dimoka, A., Pavlou, P.A., Vo, K., Hamptom, W., Bollinger, B., Ishihara, M. &. Winer, R.S. 2015. Predicting Advertising Success Beyond Traditional Measures: New Insights from Neurophysiological Methods and Market Response Modeling. *Journal of Marketing Research. American Marketing Association* 52(4):436–452

Advances in Business, Management and Entrepreneurship – Hurriyati et al. (Eds)
© 2021 Taylor & Francis Group, London, ISBN 978-0-367-67471-7

Brand image and service quality on customer loyality through customer satisfaction

A. Sudarso, L. Suryati & I.P.N. Sitepu
Sekolah Tinggi Ilmu Ekonomi IBBI, Medan, Indonesia

ABSTRACT: This study aims to determine the direct influence of Brand Image and Service Quality toward Customer Satisfaction and Customer Loyalty. This study also examines the influence of Brand Image and Service Quality indirectly to customer loyalty through customer satisfaction. The research method used is quantitative descriptive research method. This research data is obtained from the questionnaires. The population was customers of Pijer Podi Kekelengen Rural Bank as many as 342 people and 77 people were taken as sample by using the Slovin formula. The data analysis used WarpPLS approach. WarpPLS is one of the variance-based SEM statistical methods designed to solve multiple regressions when specific data problems occur in a very small sample size, missing values and multicollinearity. PLS is an alternative approach that shifts from a Covarian-based SEM approach to a variance-based. The results showed that there was a significant direct impact between brand image and service quality on customer satisfaction as well as on customer loyalty. In addition, Customer Satisfaction influenced Customer Loyalty. Brand Image had positive and significant indirect impact on customer loyalty through customer satisfaction. Service quality also had a positive and significant indirect impact on customer loyalty through customer satisfaction.

1 INTRODUCTION

Banking is a service industry that is very important in supporting the development of financing programs, either as a fund collector, as an investment and working capital financing institution or as an institution that facilitates the flow of money from the community and towards society. In this case, the bank is a public financial intermediary and a development tool in operational activities, one of the bank's funding sources comes from the community which is collected in the form of savings.

The object of the research is Pijer Podi Kekelengen Rural Bank where the company is engaged in banking activity. The pre survey results found that the efforts made by the company to increase customer loyalty do not get the expected results. This was the results of the interim responses from 30 respondents using the services of Pijer Podi Kekelengen Rural Bank regarding customer loyalty.

1.1 Brand image

According to Kotler in (Sulistian 2011) brand image is as a set of beliefs, ideas, and impressions held by a person towards a brand, therefore the attitudes and actions of consumers towards a brand are largely determined by the brand's image.

1.2 Service quality

Service Quality According to Kotler (2002) is any action or activity that can be offered by a party to another party, which is basically intangible and does not result in any ownership.

1.3 Customer satisfaction

Customer Satisfaction According to Philip Kotler (2002) is a feeling of pleasure or disappointment of someone who comes from a comparison between his impression of the performance (or results) of a product and its expectations whereas Customer satisfaction according to Zulian Yamit (2005) is a post-purchase evaluation or evaluation results after comparing what is felt with his expectations".

1.4 Customer loyalty

Loyalty according to Kotler and Keller (2009) is "a commitment that is held in depth to buy or support a product or service that is preferred in the future even though the influence of the situation and marketing efforts have the potential to cause customers to switch".

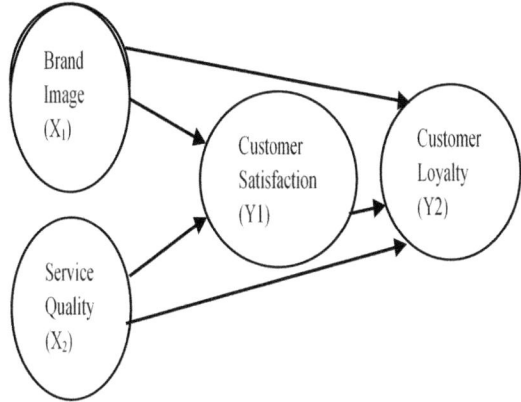

Figure 1. Framework.

1.5 *Framework*

The framework to explain the relationship between the dependent and independent variables is explained in the following Figure 1.

2 METHODS

2.1 *Method used*

In this study, researchers used quantitative descriptive research methods. The population used in this study is 342 customers of PT BPR Pijer Podi Kekelengen Simpang Selayang Branch. The sample in this study was 77 customers.

In this study the data analysis method used is structural equation modeling least squares (SEM-PLS) using WarpPLS software.

3 RESULTS AND DISCUSSION

Based on the results of the direct effect in Figures 2 and Figure 3, it is known:

The path coefficient of the Brand Image (X1) against Customer Satisfaction(Y1) is 0,279, which is positive, with a P-Values value of 0.040 < 0.05, then Brand Image(X1) has a significant effect on Customer Satisfaction (Y1).

The path coefficient from Service Quality(X2) to Customer Satisfaction(Y1) is 0,486, which is positive, with P-Values value <0.001 which means <0.05, then Service Quality(X1) has an effect significant to Customer Satisfaction (Y1).

The path coefficient of the Brand Image(X1) against Customer Loyalty(Y2) is 0,406, which is positive, with the value of P-Values 0.004 < 0.05, then Brand Image(X1) has a significant effect on Customer Loyalty(Y2).

The path coefficient of Service Quality(X2) to Customer Loyalty(Y2) is 0,264, which is positive, with a P-Values value of 0.046 < 0.05, so Service

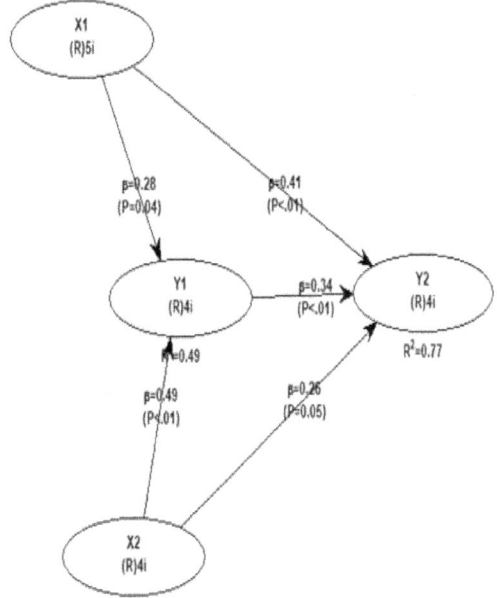

Figure 2. Test results for direct effect.

WarpPLS 3.0 - Path coefficients and P values

Close Help

Path coefficients

	X1	X2	Y1	Y2
X1				
X2				
Y1	0.279	0.486		
Y2	0.406	0.264	0.337	

P values

	X1	X2	Y1	Y2
X1				
X2				
Y1	0.040	<0.001		
Y2	0.004	0.046	0.009	

Figure 3. Results of direct effect testing.

Quality(X2) has a significant effect on Loyalty Customer (Y2).

The path coefficient of Customer Satisfaction (Y1) to Customer Loyalty (Y2) is 0,337, which is positive, with a P-Values value of 0.009 < 0.05, so Customer Satisfaction(Y1) has a significant effect on Customer Loyalty(Y2).

According the results in Figure 2, note:

The Value of R-Square on Customer Satisfaction (Y1) is 0.49 which means that the variable Brand (X1) and Quality of Service(X2) is able to influence the Customer Satisfaction (Y1) by 49%.

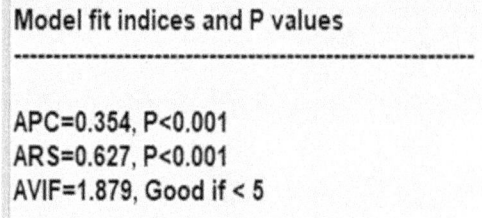

Model fit indices and P values

APC=0.354, P<0.001
ARS=0.627, P<0.001
AVIF=1.879, Good if < 5

Figure 4. Goodness of fit test.

The value of R-Square on Customer Loyalty (Y2) is 0.77 which means that the Brand Image(X1), Service Quality (X2) and Customer Satisfaction (Y1) can influence Customer Loyalty (Y2) by 70 %.

3.1 Compatibility testing model (goodness of fit)

Figure 4 presents the results of the WarpPLS for the goodness of fit test.

Based on Figure 4, it is known that the probability (p-values) of APC and ARS are significant, namely $P < 0.001$ which means <0.05 and $P < 0.001$ <0.05, and the AVIF value $= 1.879$ is less than 5. This means the model submitted has been fit (Sholihin & Ratmono 2013).

3.2 Testing of indirect effects

Significance Test of Customer Satisfaction (Y1) in Mediating the Effect of Brand Image (X1), Service Quality (X2) on Customer Loyalty (Y2)

Next, the test of the significance of indirect effects, namely: Customer Satisfaction (Y1) significantly mediates the relationship of Brand Image (X1) to Customer Loyalty (Y2). Customer Satisfaction (Y1) significantly mediates the relationship of Service Quality (X2) to Customer Loyalty (Y2).

Figure 5 presents the results of testing the significance of mediation.

Based on the results of testing the indirect effect in Figure 5, the following results are obtained.

It is known the indirect effect of Brand Image (X1) on Customer Loyalty (Y2) through Customer Satisfaction (Y1) of 0.210, with P-Values $0.047 < 0.05$ significance level, then Customer Satisfaction (Y1) is significant in mediating the effect of Brand Image (X1) to Customer Loyalty (Y2).

It is known the indirect effect of Service Quality (X2) on Customer Loyalty (Y2) through Customer Satisfaction (Y1) of 0.366, with P-Values $0.002 < 0.05$ significance level, then Customer Satisfaction (Y1) is significant in mediating the effect of Service Quality (X2) towards Customer Loyalty (Y2).

3.3 Discussion

Testing of Hypothesis 1 The path coefficient of the Brand Image(X1) of Customer Satisfaction(Y1) is 0, 279, which is positive value, with a P-Values value of $0.040 < 0.05$, then Brand Image (X1) significant

```
* Indirect and total effects *
==============================

Indirect effects for paths with 2 segments
------------------------------------------

              X1       X2       Y1       Y2
X1
X2
Y1
Y2         0.210    0.366

Number of paths with 2 segments
-------------------------------

              X1       X2       Y1       Y2
X1
X2
Y1
Y2            1        1

P values of indirect effects for paths with 2 segments
------------------------------------------------------

              X1       X2       Y1       Y2
X1
X2
Y1
Y2         0.047    0.002
```

Figure 5. Testing of indirect effects (mediating).

effect on Customer Satisfaction (Y1). Regarding the diversity of products offered can be utilized properly, and products offered by following the development of information technology, customer satisfaction can be achieved. These results are in line with previous studies namely Suciati & Nur Hidayah (2013), Satriyanti (2012), Dennisa & Santoso (2016), Gedalia & Hartono (2015), Tyas & Kenny (2016) and Sulibhavi & Shivashankar (2017) which stated that Brand image influences customer satisfaction.

Testing of Hypothesis 2 Path coefficient from Service Quality (X2) to Customer Satisfaction (Y1) is 0, 486, which is positive value, with P-Values value of 0.001 which means <0.05, then Service Quality (X1) has an effect significant to Customer Satisfaction (Y1). In connection with the facilities of the bank that are adequate and provide clear information to the customers, customer satisfaction can be achieved. These results are in line with the research of Sandriana (2016), Sarivastava (2016), Gedalia & Hartono (2015), Tyas & Kenny (2016), Satriyati (2012), Dennisa & Santoso (2016), and Iqbal et al. (2018) proving that quality service affects customer satisfaction.

Testing of Hypothesis 3 Path coefficient of Brand Image(X1) to Customer Loyalty(Y2) is 0, 406, which is positive value, with P-Values $0.004 < 0.05$, Brand (X1) significant effect on Customer Loyalty (Y2). Regarding the diversity of products offered can be used properly, and products offered by following the development of information technology, customer loyalty can be achieved. This result is in line with the research of Gedalia & Hartono (2015), Satriyanti (2012), Dennisa & Santoso (2016), Tyas & Kenny (2016) proving that Brand Image influences Customer Loyalty.

Testing of Hypothesis 4 Path coefficient of Service Quality (X2) to Customer Loyalty (Y2) is 0, 264, which is positive value, with P-Values value $0.046 < 0.05$, then Service Quality (X2) significant effect on Customer Loyalty (Y2). In connection with the facilities of the bank that are adequate and provide clear information to the customer, customer loyalty can be achieved. This result is in line with the research of Gedalia & Hartono (2015), Tyas & Kenny (2016), Satriyanti (2012), Dennisa & Santoso (2016) proving that Service Quality influences Customer Loyalty.

Testing of Hypothesis 5 Path coefficient of Customer Satisfaction(Y1) to Customer Loyalty (Y2) is 0, 337, which is positive value, with a P-Values value of $0.009 < 0.05$, Customer Satisfaction (Y1) significant effect on Customer Loyalty (Y2). Regarding customers who have a positive impression of the quality of products and services, feel the benefits of using products and services, Customer Loyalty can be achieved. This result is in line with the research of Mohammad Shoki (2013), Satriyanti (2012), Dennisa & Santoso (2016), Tyas & Kenny (2016), Gedalia & Hartono (2015) proving that service quality and customer satisfaction affect customer loyalty.

Testing of Hypothesis 6 shows the indirect influence of Brand Image (X1) on Customer Loyalty (Y2) through Customer Satisfaction (Y1) of 0.210, with P-Values $0.047 < 0.05$ significance level, then Customer Satisfaction (Y1) is significant in mediating the influence of Brand Image (X1) against Customer Loyalty (Y2). Regarding the diversity of products offered can be used properly, and products offered by following the development of information technology and customers who have a positive impression on the quality of products and services, feel the benefits of using products and services, Customer Loyalty can be achieved. This result is in line with the research of Dennisa & Santoso (2016), Gedalia & Hartono (2015), Tyas & Kenny (2016) and Sulibhavi & Shivashankar (2017) proving that Customer Satisfaction mediates the Effect of Brand Image on Customer Loyalty.

Testing of Hypothesis 7 shows the indirect effect of Service Quality (X2) on Customer Loyalty (Y2) through Customer Satisfaction (Y1) of 0.366, with P-Values $0.002 < 0.05$ significance level, then Customer Satisfaction (Y1) is significant in mediating the influence of Service Quality (X2) against Customer Loyalty (Y2). In connection with the facilities of the bank that are adequate and provide clear information to customers and customers who have a positive impression of the quality of products and services, feel the benefits of using products and services, Customer Loyalty can be achieved. This result is in line with the research of Gedalia & Hartono (2015), Tyas & Kenny (2016), Dennisa & Santoso (2016) prove that Customer Satisfaction mediates the Effect of Service Quality on Customer Loyalty.

4 CONCLUSION

Based on the results of the analysis and discussion that has been conducted, it can be concluded that:

1. Brand Image has a direct and significant effect on Customer Satisfaction;
2. Service Quality directly and significantly influences Customer Satisfaction;
3. Brand Image has a direct and significant effect on Customer Loyalty;
4. Service Quality has a direct and significant effect on Customer Loyalty;
5. Customer Satisfaction directly and significantly influences Customer Loyalty;
6. Brand image has an indirect and significant effect on Customer Loyalty through Customer Satisfaction;
7. Service Quality has an indirect and significant effect on Customer Loyalty through Customer Satisfaction.

REFERENCES

Dennisa, E.A., & Santoso, S.B. 2016. Analisis pengaruh kualitas produk, kualitas layanan dan citra merek terhadap loyalitas pelanggan melalui kepuasan pelanggan sebagai variable intervening (Studi pada Klinik Kecantikan Cosmedic Semarang). *Journal of Management Diponegoro* 5(3): 1–13.

Gedalia, C.C., & Hartono, S. 2015. Pengaruh kualitas layanan dan brand image terhadap loyalitas konsumen dengan kepuasan konsumen sebagai variable intervening (Studi kasus monopole coffee lab Surabaya). 3(3): 1–10.

Iqbal, Ali, M., Murni, Yanti, & Sulistyowati. 2018. Analysis of the influence of brand image and customer value on customer satisfaction and its impact on customer loyalty. *International Journal of Economics, Business and Management Research*, 2(4): 343–355.

Kotler, P. 2002. Manajemen Pemasaran, Analisa Perencanaan, Implementasi dan Control Edisi ke IX Jilid 2. Jakarta.

Kotler, P. 2002. Marketing management Milenium Edition North Western University New Jersey: Prentice Hall Inc.

Kotler, & Keller. 2009. Mangement pemasaran, Jilid I Edisi ke 13. Jakarta.

Sandriana, M. 2016. Customer loyalty as the implications of price fairness determined by relationship marketing and service quality of airline service. *South East Asia Journal of Contemporary Business, Economics and Law* 11(2): 43–51.

Satriyanti, E.O. 2012. Pengaruh kualitas layanan, kepuasan nasabah dan citra bank terhadap loyalitas nasabah bank muamalat di Surabaya. *Journal of Business and Banking* 2(2):171–184.

Suciati, & Hidayah, N. 2013. Analisis jalur brand image sebagai anteseden loyalitas (Studi pada Program Pascasarjana Universitas Terbuka). *Journal of Education and Culture* 19(1): 1–17.

Sulibhavi, B., & K Shivashankar. 2017. The impact of brand image on customer's loyalty towards private label brands: the mediatong effect of satisfaction. Hubli-Dharwad Conglomerate City of Karnataka 5(8): 43–50.

Sulistian. 2011. Pengaruh brand image terhadap loyalitas pelanggan Rokok Gudang Garam Filter. Kuningan: Fakultas Ekonomi dan Bisnis Universitas Kuningan.

Tyas, AAWP., & Kenny, A. 2016. Pengaruh *service quality dan brand image* terhadap loyalitas melalui kepuasan konsumen pada Inul Vizta Karaoke di Jakarta. *Jurnal Ekonomi* 7(2): 82–91.

Yamit, Z. 2005. Manajemen kualitas produk dan jasa. Jakarta: Ekonisia.

Factors affecting customer's purchase decision to apply household loan credit

I.S. Nusannas, R. Hurriyati & M.A. Sultan
Universitas Pendidikan Indonesia, Bandung, Indonesia

ABSTRACT: This Study is to find out the influence of service quality and promotion price in making Household Loan Credit (KPR) decisions. Although profitable, research on the decision to apply for a KPR in West Java which is published in leading journals is still not widely available. This study intends to examine the effect of service quality and promotion prices on the decision to apply for a mortgage at a private bank in the Kiaracondong area of West Java. The research model has a dependent variable namely the decision to submit a mortgage, and two independent variables namely service quality and promotion price. Statistics used were multiple linear regression analysis. The model estimation used SPSS version 24. The data were taken by distributing questionnaires to 46 customers whose KPR applications were approved during the promotion program. The results of the study stated that the service quality and promotion price had a positive and significant effect on the decision to submit a KPR

1 INTRODUCTION

The primary human needs which consist of the needs for clothing, food and housing is a need that must be met in order to create prosperity for people's lives in various economic level. One of the most crucial needs is the need for housing that can be reached even by those in middle and lower income level of economy. This is because housing is needed for survival both individually and with family. Everyone's motivation to fulfill what is needed and desired tends to change with the number of alternatives. Therefore to make a purchase decision process on a particular product or service, it is necessary to match with their needs. Parasuraman et al. (in Freddy Rangkuti 1997) explained that quality is the standardization of the total value of a good service to harmonize good relations between customers and companies. This is a benchmark for companies to understand customer expectations and needs. Service quality is obtained by comparing customer perceptions. Service quality is the main thing that is taken seriously by the company, which involves all resources owned by the company. There are five dimensions that need to be considered in the measurement of service quality, those are: tangibility, reliability, assurance, responsiveness, and empathy. The service sector has a competitive advantage that cannot be ignored. One of the most needed service sectors is services in the housing sector. It becomes a government policy to meet housing needs by cooperating with other parties. In this case, the government acts as a builder and "climate" creators that encourages national housing development efforts and raise public awareness of the nature of healthy settlements.

Therefore, many banks offer household loan credit (KPR) services. Developers also compete with each other offering quality service, to always pay attention to the needs and desires of customers and to try to meet expectations by providing more satisfying services than those carried out by competitors, especially if many competitors are located closely.

The decision to buy at the time of the promotion has been widely researched by experts such as Shiv & Fedorikhin (1999) finding differences in the decision to buy consumer goods. Prelec & Loewenstein (1998) research concludes that customers prefer credit purchases over cash. Mette et al. (2018) obtained the results of research on the decision to take credit influenced by discounts or because there is a certain period of time to pay. Promotion interest offered by banks, welcomed by customers. In fact, customers are waiting for cheap interest times that usually occur in February to April each year. Although time is limited, customers sometimes wait for the next promotion period to buy. The influence of financial knowledge, impulsivity, and materiality on credit buying decisions was investigated by (Mette et al. 2018). Research still leaves ample room for enriching the treasury of knowledge. For this reason, this research will try to look at the quality of service and promotional pricing on the decision to buy a mortgage.

1.1 Service quality

Service characteristic that can only be felt (Parasuraman et al. 1985), cannot be measured by comparison between hope and perception (Parasuraman et al.

1990). Service must be something appear from heart and will arrive at the heart of customer. Service quality is a method to fill in the hope of customer and the need of customers. The customer compared between the result and the hope in which the customer receive a qualified service or not (Mulyono 2008). Service quality start from a need of customer and the end at customer perception (Kotler 1997). Service quality is a level of measurement for service quality that assumed related to price development (Roderick & Gregory 2008).

According to the opinion above, service quality can be concluded as a level of advantage or measurement that is expected to fill a want of customer that related to price development. Kotler & Keller (2009) divided the dimension of service quality into five, they are: Tangible; the appearance of physical facilities equipment, staff, and communication. Reability; the ability to perform the promised service dependably and accurately, responsiveness; the willingness to help customers and provide prompt service, assurance; the knowledge and courtesy of employee and their ability to convey trust and confidence, emphaty; the provision of caring, individualized attention to customer.

Allred & Addams (2000) doing survey to the customers of Credit Union and Bank to measure service quality. The result is service quality at credit union better for accesibilty and easyness, Allred & Addams (2000) sugested credit union and bank to focus at customer expectation, evaluating service progress, and customer retain strategy. Service quality at bank also wrote by Spathis et al. (2004) according to customer perception at Greece bank. The result shown that gender effected service process. Culiberg & Rojšek (2010) survey of service quality in Slovenian bank, found that assurance and empathy, reliability and responsiveness, access, tangibles, and service range significantly affected customer satisfaction. In Bahrain commercial banl, a study by Ramez (2011) found that reliability and promise are important factors for customer. In Bangladesh, Akhtar (2011) focus to relation between service quality satisfaction, and loyalty. The result there is the relation between quality, satisfaction and loyalty positif and significant.

1.2 *Promotional price*

According to Swastha & Irawan (2005) promotion is a form of marketing communication that aims to encourage demand. Marketing communication is a marketing activity that seeks to disseminate information, influence and or remind the target market of the company and its products to be willing to accept, buy, and be loyal to the products or services offered by the company.

Promotions are used to increase sales of a product or brand usually in the short term (Wierenga & Soethoudt 2010). A promotion is a customer knowledge that is beneficial when promotion prices are launched and it can create customer behavior (Yusuf 2010).

A study conducted by (Wierenga & Soethoudt 2010) found that in 1997–2004 with the subject of consumable consumer goods, 75% of consumers purchased during promotions while the other 25% purchased in activities other than promotions. The research of Preston, Dwyer, and Rudelius in 1978 (Santini et al. 2015) also found there was an increase in the number of account when 50% discount were made. Boscetti (2012), in Santini et al. (2015) also confirms there is a positive link between promotions and the submission of loans to financial services. If discount and promotion are managed properly, they can be treated as effective ways to increase sales.

1.3 *Purchasing decision*

Purchasing decision is individual activity in making decision to directly buy products or services offered by the sellers. According to Kotler & Armstrong (2001) purchase decision is the stage in the buyer decision-making process where the consumer actually buys. Consumer decision is being influenced by consumer characteristics and also products, prices, places and promotions. Thus, producing purchasing decisions can be based on product choices, brand choices, channel choices, time of purchase and number of purchases (Kotler 1997). Service quality is a benchmark in determining whether a person purchases or not, because through the quality of service they will be able to assess their performance and feel satisfied with the services provided by the company (Nizar & Sugiono 2011). In terms of consumer behavior, there are many influences that underlie a person in making a purchase decision on a product or service. Usually, consumer buying behavior begins and is influenced by the number of stimuli from external factor, both in the form of marketing stimuli and environments. These stimuli are then processed within themselves, according to their personal characteristics, before the purchase decision is finally made.

According to Schiffman and Kanuk in Suwandi (2007), the purchasing decision process is influenced by psychological elements that determine the type of purchase they make including perception, personality and attitude. The need is the gap between factual circumstances and the desired condition of consumers. This need can be felt both through stimuli from outside and inside the consumer. Information seeking before purchasing information is needed as a tool for consideration of various alternatives. This information, collected in more than one number that can have similarities, complements and even differs in its existence. Information equations support the power of trust meanwhile differences is needed for evaluating conformity with the needs and desires of consumers. Evaluating alternative comparisons of the various available alternatives can be done so that the best choice can be obtained. According to Setiadi (2003), the stages of a specific purchasing process consist of three activities that take place in the process of purchasing decisions made by consumers, namely: 1) Consumer routines in making purchases 2) Quality of purchasing decision 3) Consumer's commitment or loyalty (Hahn, 2002 in Lembang & Sugiono 2010). Generally, there are five kinds of roles that a person can do. The five roles include (Amstrong & Philip 1996) 1) Initiator, which is the person who is first aware of a desire or need that has not been fulfilled and proposes an idea to buy an item or certain services. 2) Giving influence, namely people whose views, advice or opinions influence purchasing decisions. 3) Decision maker, namely the person who determines the purchasing decision. 4) Buyer, namely the person who made the actual purchase. 5) Users, namely people who use or consume goods or services purchased.

Meanwhile, the indicators of purchasing decisions used in this study are 1) the need and desire for a product, 2) the desire to try, 3) the stability of the quality of a product repeated purchase decision (Yusuf 2011).

2 METHODS

This study used qualitative and quantitative data. The types of data used in this study were qualitative data that was data consisting of responses/perceptions from customers about the effect of service quality, promotion prices, on household loan credit purchase decisions. Meanwhile, quantitative data was data consisting of numbers based on questionnaires regarding the identity of respondents and qualitative data that was converted into numbers.

The sources of data were 1) primary data based on the results of direct interviews from filling out the questionnaire and 2) secondary data that obtained from company data in the form of general condition of BCA Kiaracondong Branch. To add, the sample in this study were 46 people.

3 RESULTS AND DISCUSSION

From the results of the calculation, it can be explained that the magnitude of the correlation or relationship (R) value was 0.820. Based on the percentage of the effect of the X1 variable on the Y variable which was called the coefficient of determination was about 0.673 which meant that the effect of variable X1 on Y variable was 67.3%, while 32.7% was influenced by constructs outside of the research. From the results of these calculations, it can be found that the effect of Service Quality has the same positive result as revealed by (Kotler 1997).

Service quality is the fulfillment of consumer expectations or consumer needs that compares the results with expectations and determine (Mulyono 2008). Quality of service must begin with customer needs and end with customer perception (Kotler 1997). Service Quality is a measure of the level of service quality that is assumed to be related to price development (Roderick & Gregory 2008). Based on the opinion above, it can be concluded that Service Quality is the level of excellence/size expected to meet customer desires related to price developments.

From the results of statistical calculations, it can be obtained that the results of the relationship (R) is equal to 0.359. Based on the presentation of the effect of variable X2 on the Y variable called the coefficient of determination which is the result of the reduction in the value of R. The value of R2 or KD (coefficient of determination) was 0.429 which meant that the influence of variable X2 on variable Y was 42.9%, while 57.1% was influenced by constructs outside the research. This means the variable Promotion Price has a significant effect on Consumer's Purchase Decisions.

Similar to the research revealed by Boscetti 2012, in Santini et al. (2015) also confirmed the positive relationship between price promotions and proposals for loans to financial services. If discount and promotion were managed properly, they are effective ways to increase sales. In general, it can be interpreted that the more the value of promotion price, the higher the influence on the value of customer purchasing decisions.

Model Summary results inform the effect of all independent variables on the dependent variable. The influence is symbolized by R (correlation). As it can be seen in the model summary table, the value in column R is 0.824, it means that the effect of the Service Quality, Promotion Price variable on Purchasing Decisions is 82.4% (0.824 × 100%) but that value can be said to be "contaminated" by various disturbances that are might cause measurement error, for that researchers provide an alternative R Square value as a comparison of the accuracy of the effect. It can be seen that the value of R Square is 0.679 which means 67.9%. This value is greater than the R value due to the adjustment. For more accurate prediction where the influence of researchers is also based on the Adjusted R Square value, the R Square value has been more adjusted and is usually the most accurate. It can be seen that the Adjusted R Square value of 0.664 or 66.4% influences the variable Service Quality and Promotion Price on the Consumer's Purchase Decision variable. Purchasing decisions are individual activities in making direct decisions to buy products or services offered by sellers. According to Kotller & Armstrong (2001) it is the stage in the decision making process of buyers where consumers actually buy. Consumer decisions in buying beside being influenced by consumer characteristics, it can also be influenced by things raised by companies that include products, prices, places and promotions. The above variables mutually influence the purchasing decision process in order to produce a purchase decision based on product choices, brand choices, supplier choices, time of purchase, number of purchases (Kotler 1997). In general, it can be interpreted that the more the value of Service Quality and Promotion Price increases, the higher the influence on the value of customer purchasing decisions.

4 CONCLUSION

Considering the above findings, this study contributes to the academic body of knowledge by giving more knowledge into the selection criteria, which influence consumers to decide to apply for a mortgage. At the same time, this research has several implications for marketers in this case the bank, because it reveals which service attributes are most important for customers and determines their decision to decide on a mortgage. An accurate appreciation of the selection criteria can help banks in their efforts to offer, which will motivate customer choices. According to research findings, service quality, referring to consumers' judgments about the overall superiority or superiority of the entity, continues to be the main competitor of profits in the bank market, because it is the first choice criterion for customers in building long-term relationships with banks through their mortgage loans. In fact, service quality and price promotion remain the key to gaining a competitive advantage. This research provides strong evidence that maintaining high quality customer service and providing price promotions are an inseparable component for building long-term customer relationships that benefit the bank. From the results of the study it can be concluded that the Service Quality Variable has a positive and significant effect on Consumer Decisions Submitting Household Loan, the promotion price variable has a positive and significant effect on consumers' decision to submit a mortgage, the service quality variable and the promotion price have a positive effect on the consumer's decision to submit a household loans at BCA Kiaracondong Branch.

REFERENCES

Akhtar, J. 2011. Determinants of service quality and their relationship with behavioural outcomes: empirical study of the private commercial banks in Bangladesh. *International Journal of Business and Management* 6(11): 146.

Allred, A.T. & Addams, H.L. 2000. Service quality at banks and credit unions: what do their customers say?. Managing Service Quality: An International Journal.

Amstrong, G. & Philip, K. 1996. *Dasar-dasar pemasaran.*

Culiberg, B. & Rojšek, I. 2010. Identifying service quality dimensions as antecedents to customer satisfaction in retail banking. *Economic and business review* 12(3): 151–166.

Kotler, P. 1997. *Manajemen pemasaran: analisis, perencanaan, implementasi, dan kontrol.* Jakarta: Prenhallindo.

Kotler, P. & Armstrong, G. 2001) *Principles of marketings.* USA: Prentice Hall

Kotler, P. & Keller, K.L. 2009. *Marketing management.* Praha: Grada.

Lembang, R.D. & Sugiono, S. 2010. Analisis pengaruh kualitas produk, harga, promotionsi, dan cuaca terhadap keputusan pembelian teh siap minum dalam kemasan merek teh botol sosro (studi kasus pada mahasiswa fakultas ekonomi s1 reguler ii Universitas Diponegoro) (Doctoral dissertation, Perpustakaan Fakultas Ekonomi UNDIP).

Mette, F.M.B., de Matos, C.A., Rohden, S.F., & Ponchio, M.C. 2018. Explanatory merchanisms of the decision to buy on credit: the role of materialism, impulsivity and financial kowledge. *Journal of Behavioral and Experimental Finance* 21(2019): 15–21.

Mulyono, B.H. 2008. Analisis pengaruh kualitas produk dan kualitas layanan terhadap kepuasan konsumen (studi kasus pada perumahan Puri Mediterania Semarang) (Doctoral dissertation, program Pascasarjana Universitas Diponegoro).

Nizar, R.K. & Sugiono, S. 2011. Analisis pengaruh harga, kualitas layanan, dan lokasi terhadap keputusan pembelian minyak tanah non subsidi (studi kasus pada Pangkalan Minyak Tanah Di Jalan Gor No 129 Kudus) (Doctoral dissertation, Universitas Diponegoro).

Parasuraman, A., Zeithaml, V.A., & Berry, L.L. 1985. A conceptual model of service quality and its implications for future research. *Journal of marketing* 49(4): 41–50.

Parasuraman, A., Zeithaml, V., & Berry, L.L. 1990. Delivering quality service: Balancing customer perceptions and expectations.

Perin, M.G. & Vieira, V.A. 2015. An analysis of the influence of discount sales promotion in consumer buying intent and the moderating effects of attractiveness. *Revista de Administração (São Paulo)* 50(4): 416–431.

Prelec, D. & Loewenstein, G. 1998. The red and the black: Mental accounting of savings and debt. *Marketing Science* 17(1): 4–28.

Ramez, W.S. 2011. Customers' socio-economic characteristics and the perception of service quality of Bahraini commercial banks. *International Journal of Business and Management* 6(10): 113.

Rangkuti, F. 1997. *Riset pemasaran.* Bandung: Gramedia Pustaka Utama.

Roderick, B. & Gregory, B.J. 2008. Investigating the service: a customer value perspective. *Journal of Business Research* 62(7): 345–355.

Setiadi, N.J. 2003. *Perilaku konsumen: Konsep dan implikasi untuk strategi dan penelitian pemasaran.* Jakarta: Prenada Media.

Shiv, B. & Fedorikhin, A. 1999. Heart and mind in conflict: The interplay of affect and cognition in consumer decision making. *Journal of consumer Research* 26(3): 278–292.

Spathis, C., Petridou, E. & Glaveli, N. 2004. Managing service quality in banks: customers' gender effects. *Managing Service Quality: An International Journal.*

Suwandi, I.M.D. 2007. *Citra perusahaan, seri manajemen pemasaran.*

Swastha, B.D.H. & Irawan, M.B.A. 2005. *Manajemen pemasaran modern.* Yogyakarta: Liberty.

Wierenga, B. & Soethoudt, H. 2010. Sales promotions and channel coordination. *Journal of the Academy of Marketing Science* 38(3): 383–397.

Yusuf, J.B. 2010. Ethical implications of sales promotion in Ghana: Islamic perspective. *Journal of Islamic marketing.*

Yusuf, R.P. 2011. Consumer willingness to pay for poultry products from biosecure farms in Bali. *55th National Conference of the Australian Agricultural and Resource Economics Society.*

Advances in Business, Management and Entrepreneurship – Hurriyati et al. (Eds)
© 2021 Taylor & Francis Group, London, ISBN 978-0-367-67471-7

Consumer behavior analysis of Tokopedia application online marketplace using Unified Theory of Acceptance and Use of Technology 2 (UTAUT2)

I. Indrawati & E.P. Sumendap
Universitas Telkom, Bandung, Indonesia

ABSTRACT: The use of smartphones in Indonesia is currently higher compared to the use of personal computer. The company began to transfer its business from the site into an application, which also aims for the ease and mobility of consumers in online transactions. Tokopedia is one of the companies providing online trading services through websites and applications. This research was conducted to analyze the consumer behavior in using Tokopedia application. The population in this research is people in Indonesia who are included into new user category or are still learning in using Tokopedia application. The sample used in this study were 400 respondents. The results reveal that there are five factors in the UTAUT 2 Model that significantly affect the intent of the behavior of the use of Tokopedia application

1 INTRODUCTION

Mobile device in Indonesia currently has a higher duration of the use and frequency compared to the use of computers and laptops (Asosiasi Penyelenggara Jasa Internet Indonesia/APJII 2014). The company began to transfer its business from the site to the application, which also aims for the ease and mobility of consumers in buy and sell transactions. Currently Tokopedia also focuses on improving the features of the service on the mobile application. The rank of Tokopedia app in Indonesia is one of the highest, with 10 million downloads and 4.4 ratings. Consumer preferences and behaviour in using the Tokopedia app are considered important to know.

This research was conducted to analyse consumer behaviour in using Tokopedia application, especially the late majority segment. According to Teddy Arifianto in mix.co.id, nowadays e-commerce users are more than just an early adopter. The definition of E-Commerce according to Laudon & Laudon (2015) is to use internet media and websites to conduct business transactions between organizations and individuals. Users often use more than one online marketplace. Users can easily compare prices and web by using mobile apps, where they are free to choose more options and easy access. Meanwhile, according to Bayu in mix.co.id, although the number of users in online marketplace has been very much, there are still users who have not been touched well like early and late majority segment. This research is done as a reference for the company in order to create the right marketing strategy for the late majority segment. E-commerce penetration is potentially able to enter the early and late majority segments, as well as the penetration of social media applications in Indonesia.

This research was created to analyze consumer behaviour on Tokopedia applications using UTAUT2 theory. The theory of UTAUT2 is used to analyze system acceptance in technology of users in an organization.

2 METHODS

In this study, the authors modified the theory of Model UTAUT 2 by Venkatesh et al. (2012) based on research needs. This study used the UTAUT 2 Model because this model was the latest theory in technology acceptance and is based on eight theories of technology acceptance; Theory of Reasoned Action (TRA), Technology Acceptance Model (TAM), Motivational Model (MM), Theory of Planned Behavior (TPB), Combined TAM-TPB (C-TAM-TPB), Personal Computer Utilization (MPCU), Theory Diffusion of Innovation (IDT) and Social Cognitive Theory (SCT).

The framework of this study modified the UTAUT2 model based on research needs.

In this study, researchers selected Performance expectancy (PE) variables, Effort Expectancy (EE), Social influence (SI), Price value (PV), and Perceived of Security and Risk to measure the use behavior through Continuance intention on Tokopedia users. This study adapted Continuance Intention with Use Behavior because this study wants to analyze the factors that influence Tokopedia application users. In addition, respondents from this study were users in the

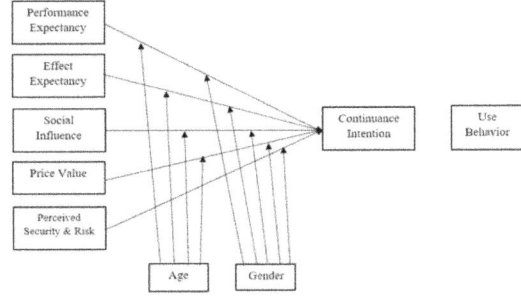

Figure 1. UTAUT2 model based on research.

Table 1. Path coefficients and t-value.

Hypothesis	Path Coefficients	T Statistic	T Table	Hypothesis Test Result
CI–UB	0.873	72.464	1.645	Significant
PSR–CI	0.390	6.141	1.645	Significant
PE–CI	0.183	3.925	1.645	Significant
PV–CI	0.187	3.813	1.645	Significant
SI–CI	0.129	2.280	1.645	Significant
EE–CI	0.103	1.740	1.645	Significant

late majority group, which are users who had just used the Tokopedia application while most people had used it. This study also aimed to find out whether existing customers wanted to continue using Tokopedia.

The UTAUT 2 model has three moderate variables, namely age, gender and experience. However, the authors only applied two moderating variables, age and gender. This was because this research was a cross sectional study whereas if using moderating experience variables, the writer must take the method of data collection periodically which the author did not run in this study. Experience variables require longitudinal studies, which were research methods in which data is collected for the same subject repeatedly over a period of time. Therefore, the author did not include experience as a moderating variable. The modified UTAUT 2 model from this study was consisted of 5 independent variables, 2 moderate variables, and 2 dependent variables.

3 RESULTS AND DISCUSSION

In this study, there are 5 independent variables and 2 dependent variables. From processing descriptive analysis, the results show that the variables had a high category. Descriptive analysis can be described as the respondent's assessment of the intention of continuation in the adoption of Tokopedia applications in Indonesia. The percentage details of the variables are Performance Expectancy with 81.4%, Effort Expectancy with 78.9%, Social Influence with 69.6%, Price Value with 77.5%, Perceived Security & Risk with 70.6%, Continuance Intention with 73.0%, and Use Behavior with 70.6%. The highest percentage was Effort Expectancy, which indicated that respondents considered the ease of use of the Tokopedia application to reach their expectations. The lowest percentage is Social Influence even though it was still in the high category.

Furthermore, to test the research hypothesis, this study collected data from 400 valid respondents who were Tokopedia users. The respondents were selected using a purposive sampling technique throughout Indonesia through online surveys using a questionnaire. To analyze the data collected, the author uses Smart PLS 3.0 software. In PLS, there are two different testing models that are carried out, namely the inner and outer models. In the analysis phase of the measurement model (outer model), there were two things that was analyzed, namely the analysis of validity (Convergent Validity, Discriminant Validity) and reliability analysis (Cronbach's Alpha, and Composite Reliability). Whereas in the analysis of the structural model (inner model), there were two things that became testing tools, namely R-square analysis (R2), Goodness of Fit (GoF) and statistical t-test both for direct influence, indirect influence and moderating influence from gender and age using multigroup analysis, which was obtained using Bootstraping calculations in the Smart-PLS application. The path coefficient must have a t-value of at least 1,645, each considered a significant level at the 95% level. Table 1 shows the path coefficients and t-values of the model as a result of bootstrap.

All variables had a positive and significant influence on Continuance Intention and Use Behavior. The value variable Continuance Intention (CI) has an R2 value of 0.663, where the value indicated that the variable Continuance Intention can be explained by the Variable of Performance Expectancy (PE), Effort Expectancy (EE), Social Influence (SI), Price Value (PV) and Perceived Security & Risk (PSR) which were amounted to 66.3% and the remaining 33.7% was explained by other factors not examined. The Use Behavior (UB) variable has a value of 0.763 which showed that the Use Behavior Variable can be explained by Continuance Intention Variables of 76.3%, the remaining 23.7% is explained by other factors that was not examined. To measure the effect of age and sex as a moderating variable, this study used a group comparison approach. The study carried out the following steps to test the differences between sub-groups: 1) Divide the sample into groups (for example, men and women; young and adult, low and high income) 2) Calculate each group in a separate model on SmartPLS. 3) Compare line differences using the method proposed by (Chin 2000).

$$t = \frac{\text{Path sample 1–Path sample 2}}{\sqrt{s.e^2 \text{sample 1} - s.e^2 \text{sample 2}}} \quad (1)$$

Table 2. Comparison of age and gender.

Compared paths	t-value from compared paths	Result	Compared Paths	t-value Compared Paths	Result
PE–CI	14.46	Significant	PSR–CI	23.78	Significant
PV–CI	11.41	Significant	SI–CI	14.8	Significant
EE–CI	5.45	Significant	PE–CI	10.59	Significant
PSR–CI	0.92	Significant	EE–CI	4.71	Significant
SI–CI	−7.87	Significant	PV–CI	−28.25	Significant

4 CONCLUSION

Based on the results of data processing and analysis and hypothesis testing in this study, according to the formulation of the problem and the research objectives, the researcher concluded the following:

1. The variables that significantly influenced Continuance Intention were Performance Expectancy, Effort Expectancy, Social Influence, Price Value and Perceived Security & Risk. Perceived Security & Risk was the most influential variable with the highest path coefficient value of 0.39, then the magnitude of the influence was followed by Price Value (0.18), Performance Expectancy (0.18), Social Influence (0.12), while Effort Expectancy (0.10) had the smallest effect. The variable that significantly influenced Use Behavior was Continuance Intention with an influence level of 0.87. The variables had positive influence on the tendency of users to continue to use the Tokopedia application in the future, while Continuance Intention has a positive influence on actual use of Tokopedia application according to (Indrawati & Marhaeni 2015).
2. Based on the results of hypothesis testing, Age was proven to moderate Performance Expectancy, Effort Expectancy, Social Influence and Price Value on Continuance Intention on the Tokopedia application. While Gender was proven to moderate Performance Expectancy, Effort Expectancy, Social Influence, Price Value and Perceived of Security and Risk towards Continuance Intention on the Tokopedia application.

The first factor that must be considered by Tokopedia is Social Influence. Tokopedia must increase social influence variables. There is a limit where members of social networks, such as family, friends, coworkers influence each other's behavior when using the Tokopedia application. Tokopedia management will be better to have more interaction with several communities or companies and the first interaction to be carried out is to people who have an important role in the community or company, so that they will eventually influence each other. In Addition, for further research, it is suggested to analyze e-commerce with a balanced composition of respondents, both in terms of age and sex, so that the results shown can be more representative and the suggestions given can be more specific.

ACKNOWLEDGEMENTS

This research was supported by Telkom University through Master of Management Telkom University.

REFERENCES

Asosiasi Penyelenggara Jasa Internet Indonesia/APJII. 2014. *Penggunaan internet sektor bisnis* 2013. [Online]. Retrieved from https://apjii.or.id/downfile/file/Survei PenggunaanInternetSektorBisnis2013.pdf.

Chin, W. 2000. *Frequently asked questions – Partial Least Squares & PLS- Graph.* [Online]. Retrieved from http://disc-nt.cba.uh.edu/chin/plsfac/plsfac.htm.

Indrawati, I. & Marhaeni, G.A.M.M. 2015.Measurement for analyzing instant messenger application adoption using a unified theory of acceptance and use of technology 2. *International Business Management* 96.

Laudon, K.C. & Laudon, J.P. 2015. *Management information systems (13th. Ed)*. Jakarta: Salemba empat.

Venkatesh, V., Thong, J.Y.L., & Xu, X. 2012. Consumer acceptance and use of information technology: Extending the unified theory of acceptance and use of technology. *MIS Quarterly* 36(1): 157–178.

School fees as a determinant in choosing a University: A study at Politeknik Pos in Indonesia

A.D. Anggraeni, R. Hurriyati & H. Hendrayati
Universitas Pendidikan Indonesia, Bandung, Indonesia

ABSTRACT: The growth of universities in Indonesia, especially in Bandung, has affected high level of competition both in state and private universities, particularly in attracting prospective students. Some colleges that do not have adequate competitiveness will experience the impact of this competition in the form of a lack of number of registering students. This issue occurs at the Politeknik Pos Indonesia. The purpose of this study is to describe and measure the influence of school fees in choosing university. It employed descriptive and verification method in which respondents were determined by using Slovin formula resulted in 266 students, and data analysis was done by using regression analysis to test the hypothesis. Based on the results of the partial regression analysis, there was a positive and significant effect of: (1) school fees to the decision to continue the study, which showed strong influence; and (2) simultaneous school fees to the decision to continue the study, which showed strong influence

1 INTRODUCTION

Higher education institutions will utilize all of their potentials and advantages as positive selling value. However, institutions lacking capability and competitiveness will get affected by the competition, which resulted in low number of registering students. The problem of decreasing interest also occurred in private higher education institutions, especially poly-technic in Bandung. Politeknik Pajajaran Insan Cinta Bangsa Bandung experienced a decrease of number of students in 2014. In 2013 the number of students was 452, and it declined to 415 students. Mean-while, Politeknik Pos Indonesia Bandung experienced a fairly serious problem. There was a significant decline in the number of students in three years. In 2012, there were 967 registered students in D-3 and D-4 Study Programs. There was an increase in 2013 to as many as 975. However, in 2014, the number of students dropped to 795. Similar issue was also experienced by higher institutions in other countries.

The Hossler and Gallagher Model, as one of the stage in Rasch Model to assess factors that influence students' choice in determining universities in Kenya (Zamri & Nordin 2013). It was found that prospective students take prices into consideration. It can be said that this stage is one of the most influential stages, as states that price is the value of an item expressed in money (Alma 2014).

In a research conducted by Cokgezen (2012), it was found that the factors determining the selection of universities in Turkey were costs, distance from home, and the quality. Of these three factors, costs were considered as an influential factor. In choosing private universities, prices were considered as the most sensitive one.

The influence of pricing (reasonable price, scholarship, financial aid and reasonable accommodation fees) that is offered by the university is one of the issues found by researchers as the most influencing factors towards college choice decision (Connie et al. 2018). According to Chapman (1981), when the costs become the obstacle that stops students to further studying in a college, then financial aid shall to be provided in order to reduce the burden in terms of costs. Whereas, if the financial support (like scholarship) is well managed, then the students can benefit by having more options to study. In fact, state universities seem to offer cheaper education fees compared to private universities, which put private education institution to face challenges, seen from price advantage. However, if the program or facilities offered by the particular private universities are able to help the consumer to gain more in return, then the pricing offered is supposed to be more convenient (Das et al. 2009).

It was reviewed by Joseph & Joseph (2000) that cost-related issues seem to have more importance. For instance, Houston (1979) found that they were at the bottom of the scale, while in Webb (1993), Joseph & Joseph (1998) they are one of the most important elements.

Politeknik Pos Indonesia implemented the School Fees Strategy so that prospective students feel they have made a good decision in choosing a college. It prioritized the strategy in 2015 to draw the interest of prospective students to choose universities. In terms of school fees, Politeknik Pos Indonesia competed with other polytechnics in Bandung. School Fees at Politeknik Pos Indonesia was regarded as affordable compared to other Polytechnics in Bandung.

1.1 *Decision to continue study*

According to Djaali (2008), interest is basically the acceptance of a relationship between oneself and something outside oneself. Interest is a very influential in having achievements in a job, position, or career. People who are not interested in a job will probably be unable to complete the job properly.

To fulfill this purpose, consumers can make decisions based on the following aspects:

1. Choice of product
   Students can make decisions to use a product or their money for other purposes. Therefore, companies must be able to focus their attention on people who are interested in making study decisions and other alternatives they consider.
2. Choice of brand
   Students can make decisions about which brands to buy. Each brand has differences. In this case, companies must know how consumers choose a brand.
3. Choice of distributor
   Students must make a decision on where a product will be consumed.
4. Quantity
   Students can make decisions about how much product they will consume.
5. Time
   Students can make decisions about when to decide. This aspect is related to involve the availability of money to buy a product.
6. Payment method
   Students must make decisions about the payment method for products consumed, whether full payment or in installments. This decision will influence decisions on sales and amounts. In this case, companies must know the students' desire in the method of payment.

1.2 *Tuition fees*

Price is one element of the marketing mix that produces revenue. Prices are perhaps the easiest element of the marketing program that can be adjusted with its: product features, distribution channels, and also communications. In addition, price also shows functional and emotional values of product. A well-designed and marketed product can command a price and reap big profits. However, new economic realities have caused many consumers to pinch pennies, and many have had to carefully review their pricing strategies as a result (Kotler & Keller 2016).

Zeithaml in Lovelock (2010) elaborates four expression values in general:

1. Value is a low price
2. Value us anything I want in a product
3. Value is the quality I get from the price I paid
4. Value is anything I get from what I give.

In this study, the value concept for customer was used as a dimension of tuition variable, which is the value obtained by customers from the fee being paid.

2 METHODS

The object of this study is the effect of school fees on the decision to continue study at Politeknik Pos Indonesia. The independent variable in this study is school fees. The dependent variable is the decision to continue study that includes: choice of product, choice of brand, and payment method at Politeknik Pos Indonesia.

The research method used is the Descriptive and Verification Method. The data analysis methods are Successive Interval Method, Simple Regression Analysis, Correlation, T-test, F-test, and Coefficient of Determination. In addition, the sampling technique used is probability sampling.

The number of respondents, based on the calculation using Slovin formula, was 266 students. The data analysis used regression to test the hypothesis.

3 RESULTS AND DISCUSSION

3.1 *Respondents' assessment of school fees at Politeknik Pos Indonesia*

The result shows that the majority of respondents stated that School Fees at Politeknik Pos Indonesia were very good. The result indicated that prospective students considered that the price suited the program, the quality of education, and the environment offered by Politeknik Pos Indonesia. This is due to School Fees offered by Politeknik Pos Indonesia was considered quite competitive with other private universities. Additionally, a very supportive learning environment and the distance from the city center made the respondents comfortable in carrying out learning. This is in accordance with Lovelock (2010:163) that the value obtained by consumers from the money being paid. In meaning that the consumers assessed the suitability of money being paid and the things they obtain.

3.2 *Respondent's assessment of the decision to continue study*

The result above shows that the majority of respondents stated that the choice of products at Politeknik Pos Indonesia was good. Kotler and Keller (2012:184) state that consumer can make decision to use certain products or use the money for other purposes. This result indicated that a consideration in choosing a polytechnic education program is due to its program that prepares the graduates to work, has collaboration with many industries, and provides opportunities for internships in national and international companies. This was because Politeknik Pos Indonesia is a vocational education program that prepares students to be ready for work, and has collaboration with several leading companies.

Furthermore, the majority of respondents stated that the choice of brand in Politeknik Pos Indonesia was very good. Kotler & Keller (2012) state that consumer make decision about the products of which brand to buy, since each brand has its own uniqueness. The result indicated that one of the considerations in choosing Politeknik Pos Indonesia was because of its uniqueness and its focus on the logistics and supply chain management fields. Another consideration was that it has a collaboration with Pos Indonesia, Ltd., which was rated very well by prospective students. The reason would be because Politeknik Pos Indonesia has the one and only Business Logistic Study Program in Indonesia and many logistics companies throughout Indonesia and inter-national companies are looking for its prospective graduates. Moreover, every year, 60 best graduates are admitted to work at Pos Indonesia, Ltd.

The result shows that the majority of respondents stated that the payment method at Politeknik Pos Indonesia was very good. Kotler & Keller (2012) talk about the decision in the method payment of the products. The result found that the payment of School Fees at Politeknik Pos Indonesia can be done in full payment or in installments. The vast alternatives of tuition payment method were valued very well by the prospective students of Politeknik Pos Indonesia. This was because the payment can be made in all branches of post offices and BNI banks throughout Indonesia.

In addition, the calculation results showed that School Fees affected the decision to continue study and the effect was positive and strong. This means that if the stipulated school fees are appropriate, the number of prospective students to continue their studies will increase, and vice versa.

School Fees affected the decision to continue the study, both partially and simultaneously. The relationship was positive, which means that the relation-ship of School Fees to the Decision to Continue Study was very strong.

4 CONCLUSION

The calculation results showed that School Fees affected the Decision to Continue Study and the effect is positive and strong. This means that if the stipulated School Fees are appropriate, the number of prospective students to continue their studies will increase, and vice versa.

School Fees affected the decision to continue the study, both partially and simultaneously. The relationship was positive, which means that the relation-ship of School Fees to the Decision to Continue Study was very strong.

ACKNOWLEDGMENTS

This research was supported by the Pos Indonesia Polytechnic, Business Management Study Program and the Indonesia University of Education.

REFERENCES

Alma, B. 2014. *Manajemen pemasaran dan pemasaran jasa [marketing management and marketing services]*. Bandung: Alfabeta.

Chapman, D. 1981. A model of student college choice. *The Journal of Higher Education* 52(5): 490.

Cokgezen, M. 2012. Determinant of university choice: a study on Economic Department in Turkey. *Research Gate*

Connie, G., Subramanian, P., Ranom, R. (2018). Exploring key factors influencing university choice: an empirical study on Malaysia Students. *IDCMBMSS: Malaysia.*

Das, C., Joseph, S., Rani, N., & Abdullah, S. 2009. Higher learning institutions of choice: a conceptual review of consideration sets.

Djaali. 2008. *Psikologi pendidikan [education psychology]*. Jakarta: Bumi Aksara.

Houston, M. 1979. Cognitive structure and information search patterns of prospective graduate business students. *Advances in Consumer Research* 7: 552–557.

Joseph, M. & Joseph B. 1998. Identifying need of potential students in tertiary education for strategy development. *Quality Assurance in Education* 6(2): 90–96.

Joseph, M. & Joseph, B. 2000. Indonesian students' perceptions of choice criteria in the selection of a tertiary institution: Strategic implications. *International Journal of Educational Management* 14(1): 40–44.

Kotler, P. & Keller, K.L. 2012. *Marketing Management*. England: Pearson.

Kotler, P. & Keller, K.L. 2016. *Marketing management (15e Edition)*. England: Pearson.

Lovelock, C. 2010. *Service marketing: people, technology, strategy (eighth edition)*.USA: Prentice Hall.

Webb, M. 1993. Variables influencing graduate business students' college selections. *College and University* 68(1): 38–46.

Zamri, A. & Nordin. 2013. Assessing factors influencing students' choice of Malaysian public university: a rasch model analysis. *International Journal of Applied Psychology.*

Advances in Business, Management and Entrepreneurship – Hurriyati et al. (Eds)
© 2021 Taylor & Francis Group, London, ISBN 978-0-367-67471-7

Analysis and design of Integrated Marketing Communication (IMC) for Small and Medium Enterprise (SME) Logistics Service Provider (LSP) form web-based application perspectives

S.H. Suarsa & R. Hurriyati
Universitas Pendidikan Indonesia, Bandung, Indonesia

S. Nirwan
Politeknis Pos Indonesia, Bandung, Indonesia

M.M. Munawar
STIE Dr KHEZ Muttaqien, Purwakarta, Indonesia

ABSTRACT: This research is amed to design and determine consumer responses on IMC activities carried out by SME LSP and obtain the concept of a campaign by using IMC through creative, persuasive, and communicative media in providing information in order to attract consumers to buy and increase sales. Small industries generally do not have market information because of their limited ability, thus it caused difficulty for small enterprises to expand the market access. The use of information technology has an urgent need for SMEs, especially SME LSP, so they can compete and increase their business capacity. To see opportunities and information from various sources used in decision making by SME LSP, information technology is needed. The method used in the design of the IMC model is the Unified Process and DFD that aim to describe business processes and design. This research used 10 SMEs LSP in Bandung as sample. The results show that the IMC model that can be used by SMEs, as well as identifying the need to build a web site design. The IMC model is implemented in the form of a prototype application with IMC-based software development methods.

1 INTRODUCTION

Integrated Marketing Communication (IMC) activities are essential so that people know the ability of the products and information offered by the company. The more often marketing communication made by the company, the higher the achievement of corporate objectives in communicating their products to the public and it will increase the number of buyers. Increasing the number of buyers will ultimately increase the purchasing decision (Densa & Relawan 2016; Prabela et al. 2016; Suarsa 2012).

Not only the big companies that require IMC, small and medium companies all also needs to perform activities of IMC to be better-known products. Small and Medium or Small Medium Enterprise (SME) in the business sector that became one of the foundations of the real sector of the Indonesian economy. This effort contributes based on data from the Central Statistics Agency (BPS) of almost 60%, with employment practically 85% of Gross Domestic Product (GDP) of Indonesia in BPS 2015. SME is currently experiencing a significant challenge both internally in terms of the quality and continuity of products as well as external challenges, including the entry into force of the ASEAN-China Free Trade Agreement (ACFTA) that

began in 2010. The ASEAN Economic Community (AEC) already came into effect in Indonesia at the end of 2015, so the the challenge of products and services would be higher.

SMEs in the middle of the competition is open and competitive. Website Business competitiveness has been defined by Cuevas-Vargas et al. (2016) as the ability of a company to perform a well adapted or the ability of a company to compete with one another. The utilization of information technology has an urgent requirement for SME, particularly SME Logistic Service Provider (LSP), so that it can compete and increase their business capacity. Technology and information systems are needed, especially to see the opportunities and information from various sources to be used for decision-making by SME of LSP.

According to Hsu (2012) competitive strategy consists of leadership cost and differentiation. In Sari & Hanoum (2012) competitiveness can be measured through organizational performance such as 1) Increasing number of business partners/partners owned by SMEs after using the internet 2) Increasing the number of orders after using the internet 3) Increasing the amount of production after using internet 4) Increasing revenue (earnings) after using the internet 5) Increasing the profit earned by the SMEs after using the internet 6) Increasing the scale of business experienced by SMEs after using the internet.

Sparkes & Thomas (2001) also found that the use of the internet as a Critical Success Factor (CSF) in Agrifood SMEs in Welsh can respond to challenges such as enabling small Agri-food companies to gain access to the internet, developing "user-friendly" Web sites, links to overseas markets via the internet, and build long-term relationships of sustainable customers through e-commerce and e-communication. Maharani et al. (2012) explained that some of the following indicators can measure the implementation of social networking on an SME, they are: 1) As a means of communication between business actors and their customers 2) Media promotion 3) As a means to obtain market research 4) Knowing the development or strategy of competitors 5) The company's brand is better known, strong and kind. Meanwhile, Sari & Hanoum (2012) stated that internet usage is measured by Internal factor dimension and external factor dimension, such as: 1) How often use the internet in business activity 2) How satisfied internet usage in business 3) How convenient internet usage in trade.

However, the phenomenon that exists today is many SME of LSPs have not and do not use the help of technology and information systems in promoting their products and services. Technology and information systems that can be utilized by SME of LSP, including the website. Website is chosen as one of the media to carry out the sale, because the site has some advantages compared with other promotional media such as having extensive coverage, having advantages in terms of audio-visual, interactive, communicative where promotional costs are cheaper than the promotional tools of traditional (brochures, leaflets, banners, and so on) (Clow et al. 2012). To add, the IMC program will be more easily integrated with web-based application programs. This research aims to design how the IMC strategy is supported by the application of information technology in the form of a web-based information system that can increase sales of SMEs, especially SME of LSP.

2 METHODS

IMC design methodology is divided into several stages set out as a systematic process in the making of this design, as see in Figure 1.

1) Initiation stage: this stage is a general introduction to the topic discussed. This stage is also the stage to coordinate with the team to perform the duties and finalize the plan. This stage includes conducting an inventory of needs and drawing up a more detailed schedule of activities.

2) Phase defining Issues and data collection: this step was aimed to define the problem objectively. The problem was defined through the introduction of the field in addition to the problems derived from the study of literature and see the existing

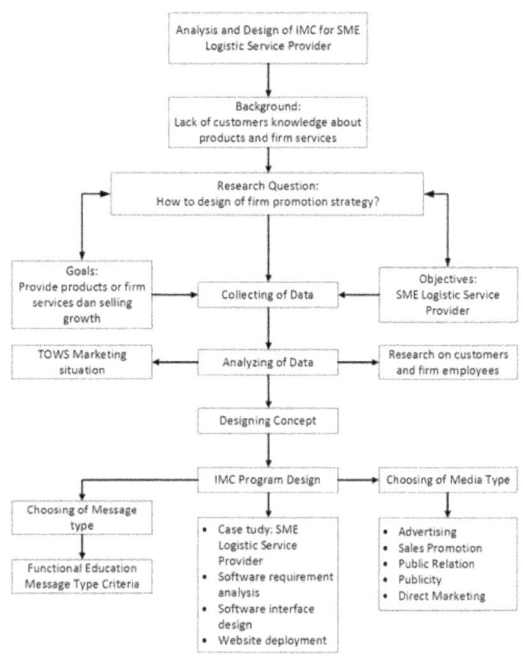

Figure 1. Research methodology.
*Source: Data analysis, 2016.

condition of the technologies and available resources. It aimed to determine the problem and create restrictions. The data were collected by the study of literature and surveys to potray a picture of the existing problems in the field. Brief report was then prepared to see the progress. To add, the field survey were conducted to 10 SME of LSP in Bandung.

3) Determining the stage of specification: It based on the design of IMC has been formulated, it can be determined the specification of a prototype web IMC that will be designed and manufactured. This specification included the specification of workflow processes, hardware, and software web application.

4) Phase design software and hardware infrastructure: It based on the technical requirements. Infrastructure design method was adapted to the needs of field implementation.

In the process of software development required specific methods. The software development approach to method development model adopted the Unified Process (UP). UP that was appeared as the software development process is called iterative (repetitive). UP was widely used for building object-oriented systems and component-based systems (Kroll 2003). Characteristic of the UP and the approach of the modern software development process to another was a software development iterative (repetitive) and incremental (gradual). With this approach, the event of software was organized into smaller projects within a specified time frame referred to as iteration. The results of each iteration were tested, integrated, and was part of the overall software deliverable. Each iteration had activity requirements analysis, design, implementation, and testing of its own. The system gradually increased until all system development needs were met.

3 RESULTS AND DISCUSSION

SME LSP is carrying out various activities to retain customers with promotional activities via the IMC (Integrated Marketing Communication). However, based on previous studies (Suarsa & Asrofah 2014), promotional activities through advertising media are education conducted by SME LSP is usually still minimal. Some of these promotional activities, as see in Table 1.

Table 1. Promotional activity.

No	Activities	Goals
1	Magazine (Advertising)	Brand awareness
2	Direct Promotion (by Sales)	Selling
3	Sales Promotion (by sales)	Selling
4	Website Marketing	Selling

*Source: Data analysis 2016.

Table 1 showed a standard promotional activity carried out by SME of LSP. There has been no promotion of innovation by the company in communicating the services to customers. However, SME of LSP has fairly competitive products and has several functions that have a good bargaining power. For instance, providing big sales, facilities for consumers who want to give fast service, systems to track the shipment through the website, smartphone, and internet, although limited, or some product for purchase very quickly.

In the various activities, many customers do not know about some of the services that are owned by SME of LSP. It is caused due to the lack of marketing communications carried out by SME of LSP that aimed to provide value in the form of information about all the services provided to customers.

From the observations, the design of IMC can be seen in Table 2.

A quality product or service the maximum would be useless if it is not promoted or communicated to customers. IMC is the right strategy to meet the needs of information products, as well as to improve the brand image, which has the ultimate goal is to increase sales (Morissan 2010). The design of IMC is applied in the application form of the company's website.

From the results of the design on the next IMC defined functional requirements of the software into the website, as see in Table 3.

Software website was divided into several modules. The modules that were stated in Table 3 above are advertising module, sales promotions module, public relations module, personal selling module, and direct marketing module. Each module was managed by each

Table 2. Designing IMC SME LSP.

Activities	Design	Medium	Goals
Advertising	Preparation and dissemination of information regarding some of the services that are offered SME LSP. Spread either directly by employees/mailing.	Brochure, Poster.	AIDA*), Brand Image, Relationship, Knowledge.
	Create and deploy TVC broadcast content of company profile, product & services, instructions and system usability for costumers.	Video Presentation.	AIDA, Brand Image, Relationship, Knowledge.
	Using a few souvenirs as a medium to deliver information about products/ services SME LSP.	Calendar Gift, Mouse Pad, Stationary.	AIDA, Brand Image, Relationship, Knowledge.
Sales Promotions	Rebate program for costumers.	Brochure, Poster, Mailing.	AIDA, Brand Image, Relationship.
Sales Promotions	Provide rebates for costumers who make regular deliveries.	Brochure, Poster, Mailing.	AIDA, Brand Image, Relationship.
Public Relations	The creation and dissemination of newsletters/monthly magazine with the provision of information about products, services, systems, excellence, explanations promotion and travel schedule delivery.	Magazine, Bulletin, Calendar Gift, Poster, Mailing.	Information, AIDA, Brand Image.
	Dissemination of information using email media made any change of important information for costumers, related to price, product, promotion, service, and more.	Mailing Internet.	Relationship, Loyalty, Satisfaction.
Personal Selling/ Direct Promotion	Provision of information by doing direct communication by sales withhold a meeting or Customers visit.	Telephone Meeting Room.	Relationship, Satisfaction.
	To follow up on a regular basis by providing useful information to costumers and Q&A service has been established.	Exhibition	Relationship, Satisfaction, AIDA.

*AIDA: Attention, Interest, Desire, and Action.

Table 3. Software functional requirement.

No	Main Function of IMC Web	Description
1	Modul Advertising	Dissemination of information in the form of the company's products and services for consumers, and dissemination of information related to the price, terms of products, services, and other promotions.
2	Modul Sales Promotions	Managing sales programs and tracking systems, for instance, a specific rebate program for consumers who first use of air freight, sea, and land as well as for consumers who have done several transactions.
3	Modul Public Relations	Dissemination of information programs associated with the company's CSR (Corporate Social Responsibility), company news, awarding, event, providing information about products and services company, excellence, promotion, and explanation itinerary by air and sea transport.
4	Modul Personal Selling	Provision of information by performing direct communication by sales through messenger programs, interactive forums, and chats.
5	Modul Direct Marketing	To follow up regularly to provide useful information to customers and ask for a response to the service has been obtained (online questionnaire).

*Source: Data analysis 2016.

of the functions in the company with features that aimed to facilitate the communication activity with customers.

The design of website were:

1) The design of promotional media such as websites that provide information about products, services, news, deals, promotions, transaction history, and company profile.
2) Providing multimedia features as well as the dissemination of information promos and offers to customers through the web page.
3) The design features chat and messenger programs to communicate with customers and visitors to the company's web page.
4) Planning the provision of an interactive forum for FAQ (Frequently Ask Question) and the various problems faced by customers.

The analysis is the first step to the construction site for the design, and even implementation website development will not be appropriately realized without any analysis of the systems that are being used today.

In general, websites that will be built are designed to facilitate communication between the marketing company with customers, in general, to obtain information about the company's products and services online.

For users of this website, the user is granted access to view the latest information, especially information about the company's products along with services. Users who register will be given access to more that can communicate directly with sales and can ask questions through the discussion forum provided on the website.

Promotion with advertising and personal selling modules via the Internet includes the development of marketing strategies using better technology than the media information manually, like sticking a brochure in a public place. These web-based services can be fully integrated into media campaigns. Thus, customers or potential customers can more easily gain access to information about products and services SME LSP.

In the early stages of system design has been carried out various activities in order fulfillment system to be built. Activities may include determining the background, problem identification, limitation of the scope, the analysis system is currently running, and analysis of current procedures. Before moving to the next stage, it will be evaluated the results of the investigation and the planning of activities for the next step.

The analysis describes what activities are carried out to build a system. This process produced a list of requirements, data flow diagrams, process descriptions, and data modeling. For more details can be seen in Table 4.

Table 4. Analysis activities and process specification.

No	Analysis Activities	Output
–	Analysis of System Requirement	List of Requirement
DFD-1	Website Login Authentication Process	Data Flow Diagram Process Specification
DFD-2.1	Manage Promotion	Data Flow Diagram Process Spesification
DFD-2.2	Distribution of Promo Information	Data Flow Diagram Process Specification
DFD-2.3	Manage Offering to Customer	Data Flow Diagram Process Specification
DFD-2.4	Distribution of Offering Information to Customer	Data Flow Diagram Process Specification
DFD-2.5	Display Promo and Offering	Data Flow Diagram Process Specification
DFD-3.1	Manage Customer Registration	Data Flow Diagram Process Spesification
DFD-3.2	Manage Chat Facilities	Data Flow Diagram Process Specification
DFD-3.3	Manage Discussion Forum	Data Flow Diagram Process Spesification
DFD-3.4	Display Transaction History	Data Flow Diagram Process Specification
DFD-4.1	Manage Service Profile	Data Flow Diagram Process Specification
DFD-4.2	Manage Product Catalog	Data Flow Diagram Process Specification
DFD-4.3	Manage Sales Profile	Data Flow Diagram Process Spesification
DFD-5	Manage Company News	Data Flow Diagram Process Specification
–	Data Model	Conceptual Data Model

*Source: Data analysis 2016.

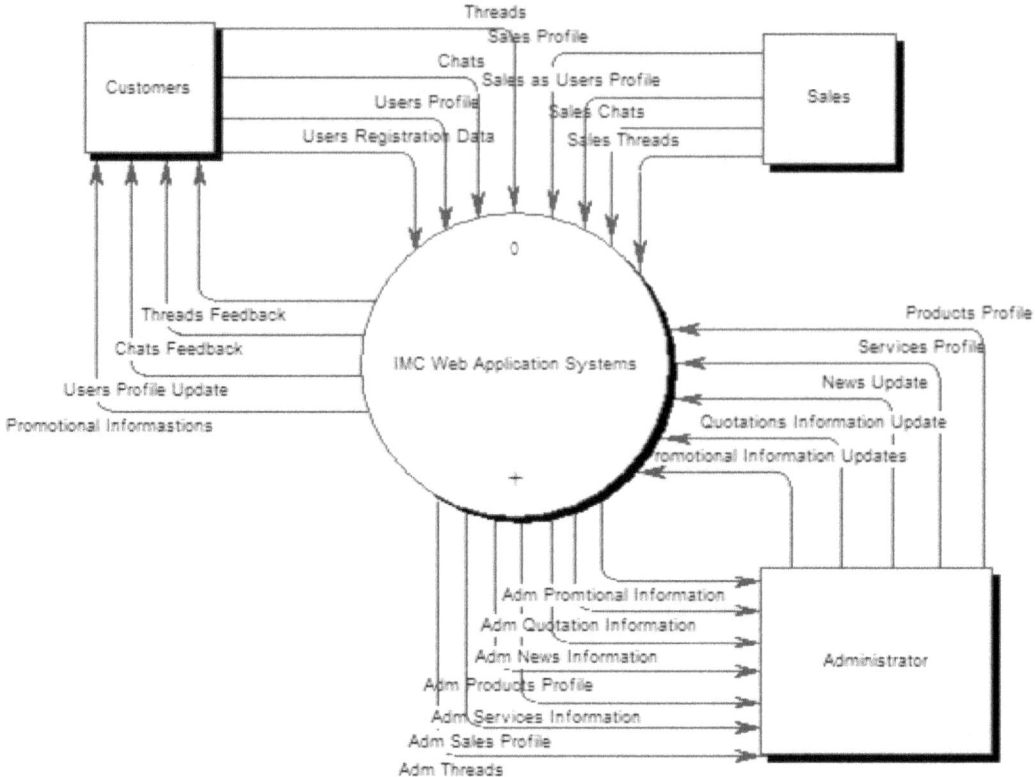

Figure 2. Contextual diagram for web IMC. *Source: Data analysis 2016.

Data Flow Diagrams (DFD) at the web application level contextual diagram IMC could be seen in Figure 2.

Application has three entities, namely:

1) Web administration is a person who has access rights to manage sales data, product data, data services, data news, bidding data, and the data also promo.
2) Sales are people who have access rights to receive information and promo deals from admin to then be forwarded to the user. Sales can view information forum and leave a response or responses to any questions put by the user either through the forum or chat. Besides, sales can also provide the latest information through a discussion that has been submitted.
3) Customers are people who have access rights to receive information from the sales offers and promos. Customers can view and purchase order transaction history, discuss topics like the forums, do chat (direct communication) with sales.

Practical application is further described by the next level Data Flow Diagram, as shown in the following Figure 3.

After designing a new system procedure, the next step is to develop a database, which will make it easier to create databases and programs to be designed. Database design in IMC web application system is intended to allow the operation and implementation can be obtained with accurate information and comfortable in data manipulation processes.

Conceptual Data Model (CDM) is a data model that is based on the facts and data requirements described in the form of essential objects that are related to each other in which the objects named entity (entity). CDM is used to describe in detail the structure of the database in the form of logic. In Figure 4, the CDM of the web application system. There are 12 entities that each relate to one another. Each object has attributes and has unique characteristics that are used as the primary key. In related entities, the primary key to each of these entities into a foreign key to another entity.

The final results of the analysis of the design are the design of web applications that meet the needs of users and displays the features that are required in IMC activities. These features are designed can be seen in Table 5.

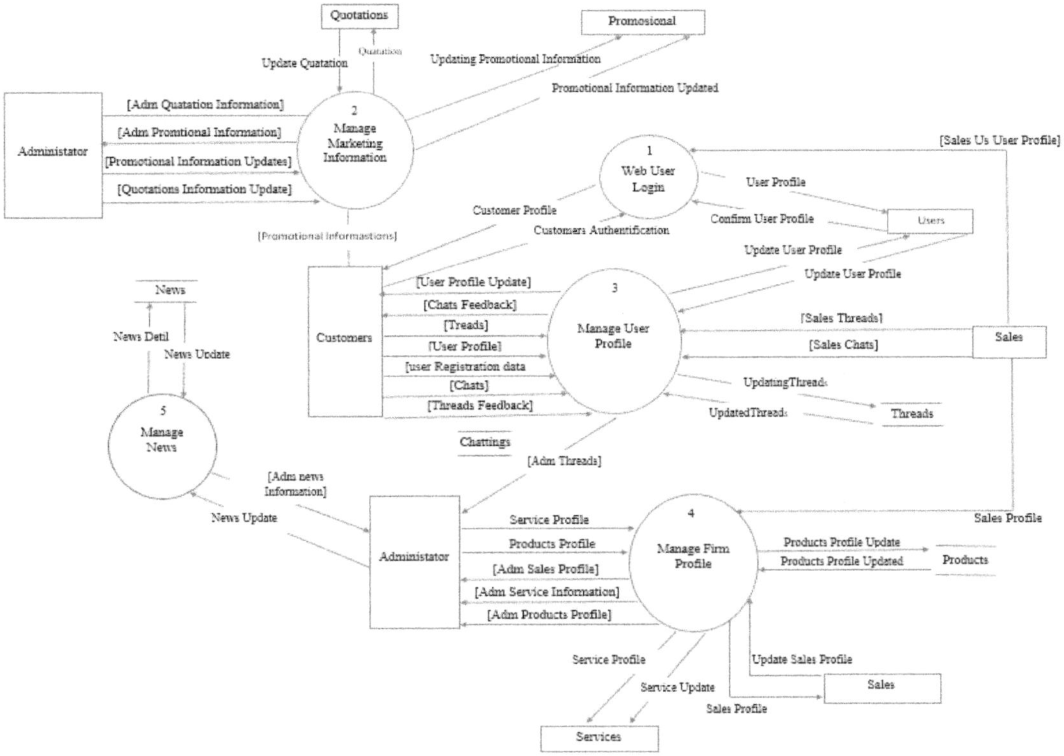

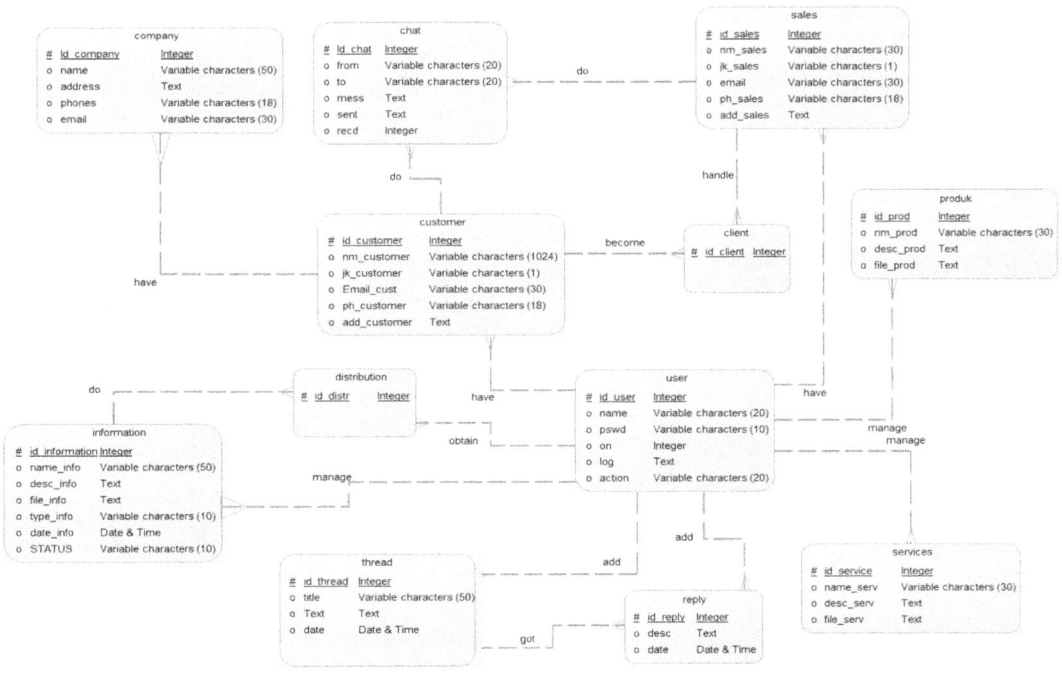

Figure 3.　Contextual diagram for web IMC. *Source: Data analysis 2016.

Figure 4.　Contextual data model for web IMC. *Source: Data analysis 2016.

Table 5. Design implementation of web IMC.

No	Design of Web IMC
1	Registration, which is to enter a new customer profile in the system. In this process displays customer data input form, with personalized customer profiles or corporate customer.
2	Login: process input username and password as part of the login process. In this process, the display form data input the user a username and password that will do the validation process. The validation process is done after the username and password are entered; if valid, the user gets the appropriate level of user access rights.
3	Managing News is managing news profile.
4	Manage Offering Product or Services is managing data by displaying the deals related features available deals on the application form the add and change bidding data. Also, there are features broadcast/share bid data, the publication of data on deals, drafting bidding data, and preview bidding data.
5	Managing Promotion: displays the features associated promo is provided on the application form the add and change the data promos. Also, there are features of the data broadcast promo, promo publication of data, drafting promo of data, and the data preview promo.
6	Display Promotions and Offers: displaying information promos and deals that have been spread by admin will appear on the homepage and customer sales.
7	Distributing of offering product and services: the distribution of the data captor process information dissemination bid by admin to user sales and customer. Data transmitted by pressing the button on the page broadcast deals. Data on sales spread to the porch, customer homepage, and the customer email.
8	Distributing Promotion: process data distribution process information dissemination promo undertaken by admin to user sales and customer. Data transmitted by pressing the broadcast on the promo page. Data transmitted to the porch sales, customer homepage, as well as customer emails.
9	Managing the Discussion Forum: displays the features in the discussion forum. Sales and customer user can add topics to the new discussion forums and answering the replies to the topic being discussed.
10	Manage Chat (online conversations with customers in service): displays chat conducted by customers with sales and sales by admin. In this process, there is a feature to display the message history chat activity ever undertaken by admin, sales and customer.
11	Transaction history displays transaction history obtained from the orders placed by the customer. Transaction history data is displayed on the home sales and the customer.
12	Manage Catalog Products: displaying products related features such as product catalogs and modify profiles added products. Also, there are features product publications, drafting product, and product preview
13	Manage various service: displays the features added services company in the form and change the service profile. Also, there are features of the service publications, drafting services, and service preview.
14	Managing the Sales profile: displays the features related sales. Admin can display sales data and also a list of client data that is handled by a salesperson.
15	System integration is the process of integrating the various features on the IMC Website so that it can be easily managed.

*Source: Data Analysis 2016.

4 CONCLUSION

SME of LSP is carrying out various activities to retain customers with IMC, one of which is the implementation of the promotion. Many of SME of LSP still ignore this process, because they focus on the company's operational processes, so we need a guide for designing integrated IMC activities. These studies suggest an optimal IMC design for SME LSP.

An effective implementation are also needed an exposure and a detail interface design specification, so that it becomes an easy reference in the implementation, integration, testing, and deployment of Web applications for SME of LSP IMC. Analysis and design of IMC in this research is intended for the SMEs of LSP to have an overview of the activities in an integrated campaign that contained in the application of the company's website.

ACKNOWLEDGEMENTS

Special thanks are given to the team of this article; Politeknik Pos Indonesia, Prof. Dr. Hj. Ratih Hurriyati, M.P., Saepudin Nirwan, S.Kom., M.Kom., Muh. Meki Munawar, S.Pd., MM., for their support in this article, so that it can be completed and submitted in this conference.

REFERENCES

Clow, C., Kenneth, E. & Donald, B. 2012. Integrated advertising, promotion, and marketing communications. England: Pearson Education Limited.

Cuevas-Vargas, H., Estrada, S. & Larios-Gómez, E. 2016. The effects of ICTs as innovation facilitators for a greater business performance. *Procedia Computer Science* 91: 47–56.

Densa, E.P.B. & Relawan, I.N. 2016. Pengaruh komunikasi pemasaran terhadap keputusan pembelian CD Jkt48 di Bandung. *E-Proceeding of Management* 3(3).

Hsu, S.H. 2012. Factors influencing on online shopping attitude and intention of mongolian consumers. *The Journal of International Management Studies* 7(2): 167–176.

Kroll, K. 2003. The rational unified process made easy: A practitioner's guide tothe RUP. USA: Addison Wesley.

Maharani, M., Ali, A.H.N., & Astuti, H.M. 2012. Faktor-faktor pengaruh media sosial terhadap keunggulan bersaing: studi kasus Coffee Toffee Indonesia. *Jurnal Teknik ITS* 1(1): A567–A572.

Morissan, M. 2010. *Periklanan: Komunikasi pemasaran terpadu*. Jakarta: Kencana.

Prabela, C.V.E., Srikandi, K. & Mawardi, M.K. 2016. Pengaruh integrated marketing communication (IMC) dan public relations terhadap citra merek dan keputusan pembelian (survei pada pengunjung Harris Hotel & Conventions Malang). *Jurnal Administrasi Bisnis (JAB)* 35(2).

Sari, R.M. & Hanoum, S. 2012. Analisis faktor-faktor yang mempengaruhi penggunaan internet terhadap peningkatan kinerja UKM menggunakan metode structural equation modelling. *Jurnal Teknik ITS* 1: 488–493.

Sparkes, A. & Thomas, B. 2001. The use of the Internet as a critical success factor for the marketing of Welsh Agri-food SMEs in the twenty-first century. *British Food Journal* 103(5): 331–347.

Suarsa, S.H. 2012. *Komunikasi pemasaran terpadu: Teori dan praktek.* Bandung: Divisi Buku Manajemen Bisnis & Pemasaran.

Suarsa, S.H. & Asrofah, A. 2014. Perancangan strategi promosi dengan model integrated marketing communications di PT KN Sigma Trans. *Prosiding Seminar Nasional Penelitian dan Pengabdian Kepada Masyarakat LPPM UNISBA*: 265–272.

The consumer's shopping intention during omnichannel shopping journey: Case study Sephora Ritel

N. Nurjanah, R. Hurriyati & M.A. Sultan
Universitas Pendidikan Indonesia, Bandung, Indonesia

ABSTRACT: The purpose of this research is to develop a model which describes the factors that affect the consumer's shopping behavior on Sephora omnichannel. The factors influencing consumers behavior are identified through understanding about the influence of performance expectancy, effort expectancy, social influence, hedonic motivation, and habit on shopping intention in the Sephora retail omnichannel distribution channel. Research model testing was conducted using the approach of Partial least Square (PLS) with sample data obtained based through online survey of 100 respondents who had been shopping on Sephora's online channel. The results of the research indicate a positive relationship of effort expectancy by shopping intention, a positive relationship between hedonic motivation with shopping intention, a positive relationship between habit with shopping intention. Negative relationship occurred between performance expectancy and shopping intention.

1   INTRODUCTION

Recent research has suggested that intention drivers vary considerably over time in various fields of study (Ajzen 2005; Sultan et al. 2012; Venkatesh et al. 2003; Wiraatmadja et al. 2017). Ajzen (2005) suggests that intention is the main causal mechanism behind the enactment of behavior. The intention to shopping on one or all of the channels provided by the company is very influential on consumer decisions. Consumers uses several channels during buying process, both online and offline channels such as website, social media, store mobile apps, catalogue and many more. In addition, consumers switch easily and continuously between these channels and they experience all channels together as one complete channel (Van 2013). The consumers decides which channel to use and search for information or buy product based on their wants and needs at a moment (Kollmann et al. 2012). For examples, the consumers can found information online to take the advantage that it saves both time and effort but they don't buy through online but offlines.

Recently, new media have emerged resulting in a closer connection between retailer and consumers, decreased retail's use of social media (Van 2013). Social media can be defined as media for social interaction, accessible by personal computer, tablet, and smartphone (Markova & Petkovska-Mircevska 2013; Van 2013). New media and mobile devices have created a new type of shopper. They have strong emotional bond with his/her smartphone and social media. They always online, has fast access to information for compare and evaluate price, uses several channel to shop and buy, shares his product experience through social

media, and so on, anytime and anywhere (Van 2013). In the old day, gathering and investigating product information can be done in-store, but today's consumers make most use of the wealth of information that is available online (rai et al. 2019).

The consumers start "a shopping journey" in social media (Kaczorowska 2017). Availability of online channels and social media has changed consumers decision making process, starting from recognition products to the end of the purchase. In this research, the consumers use online channels (social media) to recognition product anytime and the intention to buy the product and the channels they use for gathering information. Whether consumers choose for social media depend on their social media performance expectancy, Social media's Effort Expectancy, social media's Hedonic motivation, and Habit on using social media. This study wants to know the relationship between intention and performance expectancy on using social media, Effort Expectancy on using social media, Hedonic motivation on using social media, and Habit.

1.1   *Performance expectancy*

Performance expectancy is defined as the degree to which using a technology will provide benefits to consumers in performing certain activities (Venkatesh et al. 2003). The term performance expectancy in (Venkatesh et al. 2003) emerges from the combination of five factors that helped in the formation of perceived ease of use (technology acceptance model), external motivation (motivational model), job fit (personal computer utilization model), relative advantage (innovation diffusion theory) as well as outcome expectancy (social cognition theory).

The performance expectancy construct is the strongest predictor of intention and remain significant at all points of measurement in both voluntary and mandatory settings (Venkatesh et al. 2003). The performance expectancy has positive effect on intention (Sair & Danish 2018) in this research performance expectancy has implication for the use of social media among shopping intention. This is simply because the way consumers use social media (they can rely on social media to access information about product before going shopping). These is reason to expect that the positive relationship between performance expectancy and shopping intention.

H1. The positif relationship between performance expectancy on using social media and shopping intention.

1.2 *Effort expectancy*

Effort expectancy is the degree of ease associated with consumers' use of technology (Venkatesh et al. 2003). Effort expectancy is based on the idea that there are relationship between the effort put forth at work, the performance achieved from the effort, and the reward received from the effort (Ghalanadari 2012).

(Venkatesh et al. 2003) predict that the effort expectancy constructs significant to intention in both voluntary and mandatory usage contexts, however each one is significant only during the first time period, and becoming nonsignificant over periods of extended and sustained usage. These is a reason to expect that the positive relationship between effort expectancy and shopping intention.

H2. The positif influence between effort expectancy on using social media and shopping intention at lates stage of experince.

1.3 *Hedonic motivation*

Hedonic motivation is defined as the fun or pleasure derived from using a technology (Brown & Venkatesh 2005). Kaczmarek (2017) defined hedonic motivation as the willingness to initiate behaviors that enhance positive experince (pleasant or good) and behaviors that decrese negative experince. Kahnemen (1999) proposed that the experience of good versus bad is the essence of hedonism. Good experinece defined by its utility reflected in the willingness to continue action and sustain current experience.

Syafita (2018) indicate hedonic motivation as a sense of pleasure that is felt in the process of choosing a product. Costumer's perceived hedonic motivation is the most important driving factor in carrying out the shoping process (Yang 2010). Venkatesh et al. (2012) measured the effect of hedonic motivation on behavioral intention to use technology. Yang (2010) found that hedonic motivation influenced consumers intention to uses online shopping. These is reason to expect that the positive relationship between hedonic motivation and shopping intention.

H3. The positif influence between hedonic motivation on using social media and shopping intention.

1.4 *Habit*

Habit has been defined as the extent to which people tend to perform behaviours automatically because of learning (Limayem et al. 2007). Habit is created by consistently performing the same behaviour in a regular cotext (Henderson et al. 2014). The issue of whether the effect of habit operates directly on intention has been extensively discusses in prior research (Ajzen 2002; Kim et al. 2005; Limayem et al. 2007; Venkatesh et al. 2012). Habit will have stronger effect on intention for more experienced consumers (Venkatesh et al. 2012). Tobias (2009) resulted that habitual customers automatically perform behaviours without actively forming intentions or fully considering competing alternatives. These researches predict relationship between habit and shopping intention.

H4. The positif relationship between habit on using social media and shopping intention.

2 METHODS

This research was conducted in Sephoras Omnichannel through an online survey to collected data. The population in this study are all the consumers who have shopped on Sephoras. Online survey was conducted in the period between April–May 2019. The method used in the study was questionnaire used semantic differential scale "1–7". The number of samples or the respondents is 100 respondents. Data Analysis technique used in this research is partial least square using Smart PLS.

3 RESULTS AND DISCUSSION

The demographic profile of the respondents presented in Table 1.

The measurement model examined using SPSS software to asses reliability and validity before testing the various structural models. Table 2 present information about reliability and validity measurement model, 30 respondents used to test reliability and validity studies. Items declared valid if correlation between items with overall items in an indicator are significant. Significant indicate that Spearman Correlation koefisien > rtabel (0.361, 30 samples with $\alpha = 0.5$). Reliability indicate that Cronbach Alpha was greater than 0.6 (30 Samples). Based on the result, measurement model was valid and reliable. This result indicating that if the research repeated with different times and dimensions will produce the same conclusion (Ghozali 2002).

Table 1. Demographic details of responden.

Indicators	Code	Percentage (%)
Age	≤19	4
	20–29	77
	30–39	16
	40–49	2
	≥50	1
Gender	Male	36
	Female	64
Registered	Bandung	48
Address	Jabodetabek	35
	Priangan Timur	7
	Other	10
Media Social	Instagram @Sephora	44
Used	Sephora Mobile Apss	42
	Facebook @Sephora	14

*Result of data collection 2011.

Table 2. The validity and reliability.

Indicators	Items	Spearman Koefisien	Cronbach's Alpha
Performance	PEE1	0.453	0.635
Expectancy	PEE2	0.569	
Effort Expectancy	EFR1	0.379	0.679
	EFR2	0.381	
Hedonic Motivation	MOT1	0.374	0.786
	MOT2	0.365	
	MOT3	0.565	
Habit	HAB1	0.447	0.640
	HAB2	0.364	
	HAB3	0.773	
Shopping Intention	INT2	1.000	

*Result of SPSS test 2019.

Table 3. Path coefficient.

	Path Coefficient	t-satistic	P Values
PEE->INT	−0.023	0.493	>0.01
EFR->INT	0.405	8.273	<0.01
MOT->INT	0.133	1.969	<0.05
HAB->INT	0.712	11.886	<0.01

*Result of SmarPLS test 2019.

This research used partial least square (PLS) to test our model because this research have quite a number of interaction terms and PLS capable of testing these effects (Chin et al. 2003; Venkatesh et al. 2012). Test hypothesized relationship using Partial Least Square (PLS) is presented in Table 3.

The Value of R Square are 0.810 indicated reliability of this result. The value of the effect size for each path can be seen in Table 4.

H1. Performance expectancy on using social media and shopping intention had a negative impact. The

Table 4. Effect size.

No	Jalur	Effect Size	
1	Performance Expectancy → Shopping Intention	0.001	Small
2	Effort Expectancy → Shopping Intention	0.501	Substantial
3	Hedonic Motivation → Shopping Intention	0.066	Small
4	Habit → Shopping Intention	1.623	Substantial

*Result of SmarPLS test 2019.

result shows on Tabel 3 and Tabel 4 that there was non-significant negative relationship between performance expectancy and shopping intention. (Path coeficien −0.023, P-Value >0.05), The relationship between performance expectancy and shopping intention had small effect size (f2 = 0.001).

H2. Effort expectancy on using social media and shopping intention at lates stage of experince had a positive impact.The result show in Table 3 and Table 4, that there was significant positive relationship between effort expectancy and shopping intention (Path coeficien 0.405, P-Value <0.001). The relationship between effort Expectancy and shopping intention had substantial effect size (f2 = 0.501).

H3. Hedonic motivation on using social media and shopping intention had a positive impact. The result shows in Table 3 and Table 4, that there was significant positive relationship between hedonic motivation and shopping intention (Path coefficient 0.133, P-Value <0.05). The relationship between hedonic motivation and shopping intention had small effect size (f2 = 1.623).

H4. Habit on using social media and shopping intention had a positive impact. The result shows in Table 3 and Table 4, that there was significant positive relationship between habit and shopping intention (Path coeficien 0.712, P-Value <0.001), The relationship between habit and shopping intention has substantial effect size (f2 = 0.066).

4 CONCLUSION

This research adopted combination of UTAUT and UTAUT 2 model to identify factors influencing shopping intention. The variables are performance expectancy, effort expectancy, hedonic motivation, and habit. Result suggested the non-significant negative effect of performance expectancy using social media on consumers shopping intention. It can be interpreted that benefits on using social media does not relate to consumers in performing shopping activities. It also indicates that respondents' characteristic influenced this relationship. Effort expectancy can be distinguished which one that had a positive and significant effect on shopping intention. Thus if users feel comfortable using social media, they will be willing to go shopping in these services. This result is consistent with that of (Ghalandari 2012; Sair & Danish 2018; Venkatesh et al. 2003; Venkatesh et al. 2012).

Hedonic motivation had a positive significant effect on shopping intention. This result indicate that using social media make good experience for consumers. Habit had a positive significant effect on shopping intention. The strength of the resulting habit on shopping intention. Habits provide a strong memory advantage over compering behaviors, intention accessibility decays as habits develop (Tobias 2009). It means once the consumers' using social media to gathering information become habit, that can develop the shopping intention.

ACKNOWLEDGEMENTS

Special thanks is given to: Prof. Dr. Hj. Ratih Hurriyati, Dr. Mokh Adib Sultan, and DIM UPI 2019 for their support in this article so that it can be completed and submitted in this conference.

REFERENCES

Ajzen, I. 2002. Residual effects of past on later behavior: Habituation and reasoned action perspectives. *Personality and Social Psyxhology Review* 6(2): 107–122.

Ajzen, I. 2005. *Attitudes, Personality, and Behavior (2nd Ed.)*. Berkshire: Open University Press.

Brown, S.A. & Venkatesh, V. 2005. Model of adoption of technology in households: A baseline model test and extension incorporating household life cycle. *MIS Quarterly*: 399–426.

Chin, W.W., Marcolin, B.L., & Newsted, P.R. 2003. A partial least squares latent variable modeling approach for measring interaction effect: Result from a monte carlo simulation study and an electroic mall emotion/adoption study. *Imformation System Research* 14(2): 189–217.

Ghalanadari, K. 2012. The effect of performance expectancy, effort expectancy, social influence and facilitating conditions on acceptance of e-banking services in Iran: The moderating role of age and gender. *Middle East Journal of Scientific Research*: 801–807.

Ghozali, I. 2002. *Aplikasi analisis multivariate program SPSS*. Semarang: Badan Penerbit Universitas Diponegoro.

Henderson, C., Steinhoff, L., & Palmatier, P. 2014. Consequances of customer engagement: how customer engagement alters the effect of habit, dependence, and relationship-based intrinsic loyalty. *Marketing Science Institute Working Paper Series*.

Kaczmarek, L.D. 2017. Hedonic motivation: Encyclopedia of personality and individual difference. *Encyclopedia Of Personality and Individual Difference*.

Kaczorowska, S.D. 2017. Consumer perspective of omnichannel commerce. *Management*.

Kahnemen, D. 1999. *Objective happiness*. New York: Russell Sage Foundation.

Kim, T.G., Bock, G.W., Zmud, R.W., & Lee, J.N. 2005. Behavioral intention formation in knowledge sharing: Examining the roles of extrinsic motivators, social psychological forces and organizational cliate. *MIS Quarterly*: 87–111.

Kollmann, T., Kuckertz, A., & Kayser, I. 2012. Cannibalization or synergy? consumers' channel selection in online–offline multichannel systems. *Journal of Retailing and Consumer Services* 19(2): 186–194.

Limayem, M., Hirt, S., & Cheung, C. 2007. How habit limits the predictive power of intention: The case of information systems continuance. *MIS Quarterly*: 705–737.

Markova, S. & Petkovska-Mircevska, T. 2013. Social media and supply chain. *Amfiteatru Economic Journal* 15(33): 89–102.

Rai, H., Mommens, K., Verlinde, S., & Macharis, C. 2019. How does consumers's omnichannel shopping behavior translate into travel and transport impacts? Case-study of a footwear retailer in Belgium. *Sustainability*: 25–34.

Sair, S.A. & Danish, R.G. 2018. Effect of performance expectancy and effort expectancy on the obile commerce adption intention through personal innovativeness among Pakistani consumers. *Pakistan Journal of Commerce and social sciences (PJCSS)* 12(2):501–520.

Sultan, M.A., Haryanto, B., Haryono, T., & Riani, A. 2012. Proses pembentukan perilaku niat wisatawan berkunjung kembali. *Jurnal Siasat Bisnis*: 107–118.

Syafita, J.D. 2018. Utilitarian and hedonic values that influence customer satisfaction and their impact on the repurchase intention: Online survey towards Berrybenka fashion e-commerce's buyer. *Russian Journal of Agricultural and Socio-Economic Sciences* 73(1).

Tobias, R. 2009. Changing behaviour by memory aids: A social psycological model of prospective memory and habit development tested with dynamic field data. *Psycological Riview*: 408–438.

Van, D.W. 2013. Omni channel shopping behaviour during the customer journey: An empirical study into the contribution of omni channel shopping characteristics during the customer journey by consumers segment. *Eindhoven University of Technology*.

Venkatesh, V., Morris, M., Davis, G., & Davis, F. 2003. User cceptance of information technology: Toward a unified view. *Management Information System Quarterly*: 425–478.

Venkatesh, V., Thong, J., & Xu, X. 2012. Consumer acceptance and use of information technology: Extending the unified theory of acceptance and use of technology. *Management Information System Quarterly*.

Wiraatmadja, I.I., Nurjanah, N., & Kurniawati, A. 2017. Model penerimaan petani terhadap teknologi sistem pertanian organik di kabupaten tasikmalaya. *Jurnal Manajemen Teknologi*: 81–91.

Yang, K. 2010. Determinants of US consumer mobile shopping services adoption: Implications for designing mobile shopping service. *Journal of Consumer Marketing*.

Advances in Business, Management and Entrepreneurship – Hurriyati et al. (Eds)
© 2021 Taylor & Francis Group, London, ISBN 978-0-367-67471-7

Purchasing decision analysis of Wuling Confero (Survey at Wuling Suci Branch Office Bandung)

M. Rahayu, R. Hurriyati & M.A. Sultan
Universitas Pendidikan Indonesia, Bandung, Indonesia

ABSTRACT: This research tests the impact of product quality and brand image on purchasing decision through purchasing intention as a mediating variable. The population were visitors of Kantor Wuling Suci Bandung Branch Office. The sample were 152 visitors who were chosen by Slovin equation. The data was collected by survey with questionnaire method. This research used regression and path analysis method to analysis the impact of product quality and brand image on purchasing intention simultaneously and partially, and figure out the impact of purchasing intention on purchasing decision. The result of this research are product quality effects purchasing intention partially, brand image effects purchasing intention partially, product quality and brand image effect purchasing intention simultaneously, purchasing intention effects purchasing decision.

1 INTRODUCTION

There is a diversity of research results regarding factors that can influence purchasing decisions. The main factor which determines consumer purchasing decisions is psychological factors, named purchase intention (Sangadji et al. 2013). This opinion is supported by Prasad (2018) that purchase intention influences car purchase decisions. Some factors that can influence purchase intention include product quality (Aryadhe & Rastini 2016; Mashahadi & Mohayidin 2015; Sultan et al. 2013), and brand image (Wang & Tsai 2014). However, the results of study from Eze et al. (2012) found that product quality and brand image did not have a strong influence on purchasing decisions.

Based on the results of the study, it opens up space in this study to analyze how prospective buyers do purchase decisions.

This study uses the opinions of several experts regarding product quality, brand image, purchase intention and purchase decision.

1.1 Product quality

Product quality is something that can be measured and the number of its size includes performance, features, reliability, conformity with specifications, durability, service ability, aesthetics, perceive quality. Tjiptono & Chandra (2015) stated that in research for product quality, it can be used 5 (five) measures, they are: principal performance, additional features, reliability, serviceability and aesthetics.

Some previous studies found that product quality has an effect on purchase intention (Aryadhe & Rastini 2016; Mashahadi & Mohayidin 2015; Seng & Husin 2015; Sultan et al. 2012). This proves that the higher quality of the product, the higher purchase intention will become.

1.2 Brand image

According to Aaker in Sangadji et al. (2013) brand image is a set of unique associations that marketing officers want to create and maintain. Brand image is the consumer's assessment of the brand. If consumers' ratings are bad, the product image will be bad. If the co-assessment is good, the product image will be good. From this opinion, it can be explained that brand image is a consumer's bad or good picture of products or services they've enjoyed (Sangadji et al. 2013).

The brand image dimensions consist of brand associations, brand association supports, brand association strengths, and uniqueness of brand associations (Sangadji et al. 2013). Previous research stated that brand image influences purchase intention (Joghee & Dube 2016; Khan & Supinit 2015; Malik et al. 2013; Shehzad et al. 2014). This shows that the higher the brand image, the higher the purchase intention will become.

1.3 Purchase intention

Purchase intention is the possibility that the customer will buy a product or service. The intention of a large purchase leads to a purchase decision (Wang & Tsai 2014). According to Wang & Tsai (2014) the dimensions of purchase intention consist of 9 (nine) dimensions but purchase intention in this study uses 4 (four) dimensions, they are considerations of prospective buyers, willingness to buy Wuling Confero, prices in accordance with the expectations of the buyer, prospective buyers will buy at the show room. The results of the Prasda study 2019 show that purchase intentions influence purchase decisions. It can be explained that the higher the purchase intention, the higher the purchase decision will become.

1.4 Purchase decisions

Purchase decisions are consumer activities in the form of introduction to needs, search for product/service information, purchase decisions, and post-purchase information. Purchase decisions consist of several stages of activity: Introduction of needs, information seeking, evaluation of alternatives, purchase decisions and post-purchase behavior (Kotler & Armstrong 2014). In needs of recognition activities, prospective buyers will recognize and determine their needs about car, so that prospective buyers will find the information from mass media and social media. In alternative evaluation activities, the prospective buyer will evaluate the car before making a choice, named buying or not buying or making a purchase decision. In the end, after the purchase, the buyer can get satisfaction or not.

Based on the results of previous studies, the hypotheses' in this study are:

H1: There is a positive relationship between Product Quality and Brand Image.

H2: Brand Product and Image Quality has a positive effect on the Purchase Intention simultaneously.

H3: Purchase intention has a positive effect on the purchase decision.

1.5 Research framework

Based on the Hypotheses compiled, the model formed is as illustrated.

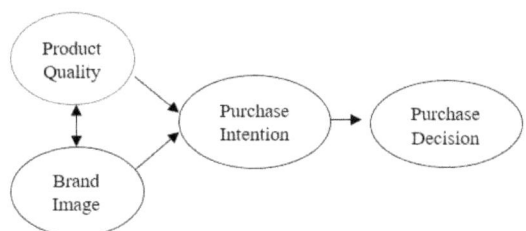

Figure 1. Research framework.

Research methods: This study uses quantitative, descriptive and verification methods. Data are obtained by distributing questionnaires. A number of samples are based on Slovin formula. Respondents are determined based on random sampling. The results are presented in quantitative data (numbers) and then interpreted the meanings. The study used survey and data processing methods using the Excel application program, SPSS version 23 and Amos version 23.

2 RESULTS AND DISCUSSION

2.1 Respondents data

The number of respondents were 152 with the details of men as many as 102 people (67.11%) are the highest number of respondents. Female respondents were 50 people (32, 89%), the ages of respondents are the most respondents aged 21 years to 30 years as many as 79 people (51.97%), respondents aged 31 years to 40 years are 45 people (29.61%) and those who aged 41 years to 50 years are 28 people (18.42%). The number of respondents who work as entrepreneurs are the highest number of respondents with 33 people (33.33%), followed by respondents who work as members of the National Police 32 people (32.32%), Public Servants 31 people (31,31%), 30 members of TNI (Indonesian National Army) (30.30%) and the last 29 private employees (29.29%). The highest education degree of respondents is Doctoral degree with a number of respondents are 2 people (2.02%), Respondents with Master diplomas are 15 people (15.15%), the diploma that the respondents have the most is Bachelor degree with the number of respondents 41 people (41.41%), D-III respondents are as many as 39 people (39.39%), Senior High School respondents are as many as 38 people (38.38%) and DI respondents are as many as 17 people (17.17%).

2.2 Test of research instruments

Table 1. Validity test.

Variable	Score	Result
Product Quality	0.627–0.742	Valid
Brand Image	0.567–0.682	Valid
Purchase Intention	0.505–0.696	Valid
Purchase Decision	0.649–0.735	Valid

Table 1 data shows a score above 0.3. So the data is declared valid.

Table 2. Reliability test.

Variable	Score Cronbach Alpha	Result
Product Quality	0.862	Reliable
Brand Image	0.796	Reliable
Purchase Intention	0.782	Reliable
Purchase Decision	0.857	Reliable

The data in Table 2 shows that the Cronbach Alpha value is higher than 0.7. It means that all data is reliable.

2.3 Normality test

Table 3. Normality test.

Variable	Asymp	Result
Product Quality	0.200	Normal
Brand Image	0.200	Normal
Purchase Intention	0.200	Normal
Purchase Decision	0.200	Normal

The data in Table 3 shows the asymptotic value as large as 0.200 > 0.05. So, the data has been normally distributed.

2.4 Correlation value

Table 4. Correlation test.

		Brand Image
	Pearson Correlation	.840**
Product Quality	Sig. (2-tailed)	.000
	N	152

Based on Table 4 correlation value of Product Quality with Brand Image is 0.840. This means that product quality has a very strong relationship with the brand image.

2.5 Path analysis

This study has 2 (two) sub-structures. Sub-structure 1 discusses the path coefficient of product quality and brand image of purchase intentions. Sub-structure 2 discusses the path coefficients of purchase intentions on purchase decision.

Sub Structure 1.

Table 5 presents data about the value of standardized beta column coefficients which shows the magnitude of the path coefficient of the product quality variable and brand image.

Table 5. Standardized coefficients.

Model	Beta	T	Sig
Constant			
Product Quality	.465	6.209	.000
Brand Image	.440	5.874	.000

The Standardized Coefficient value in the beta column for product quality is 0.465 and for product image is 0.440. This means that the product quality coefficient value is 0.465 and the brand coefficient of the brand is 0.440. This shows that the quality of Wuling

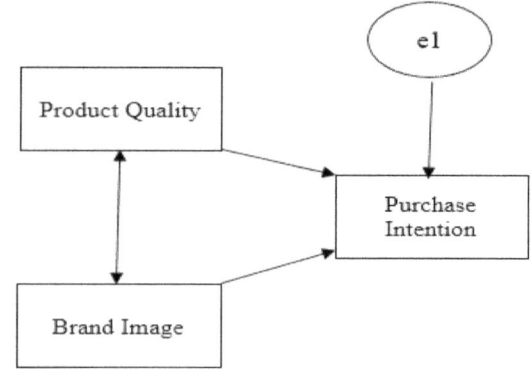

Figure 2. Effect of product quality and brand image on purchase intention.

Confero products has a greater influence on purchase intentions compared to brand images.

Based on Table 5, a sub-structure research model 1 can be built that influences product quality and brand image on purchase intention.

Based on Figure 2, structural equations can be arranged as follows:

$$X3 = \rho x3x1X1 + \rho x3x2X2 + \varepsilon 1 \tag{1}$$

$$X3 = 0.465X1 + 0.440X2 + \varepsilon 1 \tag{2}$$

Information: X1 = Product Quality; X2 = Brand Image; X3 = Purchase Intention.

Based on Table 5 and Figure 2, there are direct effects and indirect effects of product quality. The results of brand image influnced the purchase intention are presented in Table 6.

Table 6. The path analysis.

Variable	Direct Effects	Indirect Effects			
		X1	X2	Skore	Total
Product Quality	21.62		17.18	17.18	38.80
Brand Image	19.32	17.18		17.18	36.50
Total	40.94	17.18	17.18	34.36	75.30

Table 6 explains that the quality of the product gives a direct influence on purchase intentions of 21.62% and provide indirect influence through a brand image of 17.18%. So the direct and indirect influence of product quality on purchase intentions are 21.62% + 17.18% = 38.80%.

Brand image has a direct influence on purchase intention of 19.32%, and gives indirect influence through product quality variables of 17.18%. So, the influence of brand image on purchase intentions directly and indirectly is 36.50%. Thus it can be explained if you want to increase the intention to buy visitors, then what must be improved first is the quality of the product because the percentage effect on purchase intention is greater than the percentage of brand image.

Respondents' results showed that the lowest score of reliability indicators was 3.99 (dimensions of product quality), 3.88 low prices with a score of 3.88 scores (brand image dimensions).

The number of direct and indirect effects of product quality and brand image on purchase intention is $38.80\% + 36.50\% = 75.30\%$. This is the same as the determination coefficient value of 75.30%. But if you want to improve product quality and brand image together, then the indicator of the brand image that must be repaired to an expensive price. Then you should need a cheaper pricing strategy from competing brands. Or the price is almost the same as the competitive price, only the purchase of Wuling is equipped with a 1-year maintenance guarantee, providing a spare tire and Umrah gifts, which are more attractive to prospective buyers.

Other factors that influence intention purchases of 24.70% include factors in product promotion, service networks, location of show rooms, easiness of obtaining spare parts.

Sub Structure 2.

Table 7. The path purchase intention to purchase decision.

| Model | Standardized Coefficients | | |
	Beta	T	Sig.
1 (Constant)		11.597	.000
Purchase Intention	.938	33.153	.000

Table 7 explains that the value of the path coefficient of purchase intention towards purchasing decisions is 0.938 or 93.8%.

Table 8. R Square.

Model	R	R Square
1	.938[a]	.880

Table 8 explains that the value of $R = 0.938$ means that the relationship between purchase intention and purchasing decision is very strong. R2 (R square) value is 0.880 or 88%. This means that the coefficient of determination is 88%. The meaning of the coefficient of determination 88% is the purchase intention contributes to the purchase decision of 88%. The remaining 12% is caused by other variables not discussed in this study, for example the price of cars or car models. In Tables 7 and 8, we can obtain a picture of the model of the influence of purchase intention on purchasing decisions.

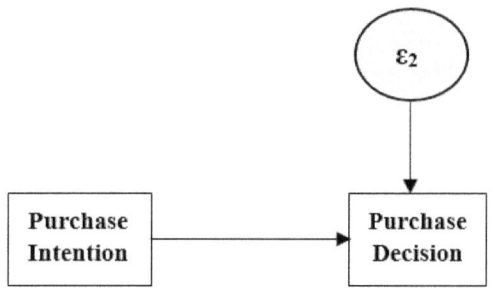

Figure 3. Effect purchase intention on purchase decision.

Using the data in Table 8 as a sample to make structural equations as follows:

$$Y = \rho X3 + \varepsilon 2 \tag{3}$$

$$Y = 0.938\,(X3) + \varepsilon 2 \tag{4}$$

Information $Y = $ Purchase Decision; $X3 = $ Purchase Intention.

Overall Model Image.

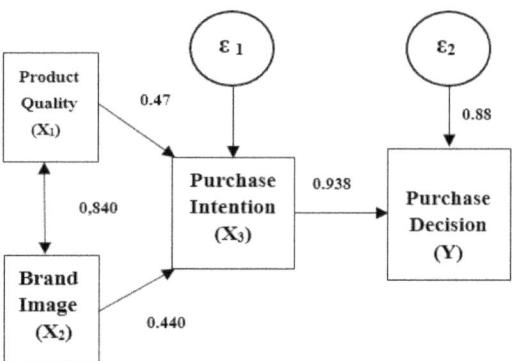

Figure 4. Overall model.

Figure 4 explains that product quality has a correlation with the brand image as much as 0.840 grouped as a very strong relationship. Contributions to the influence of product quality and brand image are 75.30%. The contribution of the purchase intention to purchase decisions is as much as 88%.

2.6 Hypotheses test

Correlation between product's quality with brand image:

Ho: ρ X1X2 $= 0$ There is no relationship between product Quality and brand image.

Ha: ρ x1x2 $\neq 0$ There is a relationship between product quality and brand image.

Effect of Product Quality (X1), Brand Image (X2) on Purchase Intention (X3).

The effect of product quality (X1) and brand image (X2) on purchase intention (X3) simultaneously, the hypotheses are as follows:

Ho: ρ x3x1 $= \rho$ x3x2 $= 0$ There is no influence of product quality (X1), brand image (X2), on purchase intention (X3).

Ha: There is at least one ρ x3xi $\neq 0$; where i $= \rho$ x3x1, ρ x3x2 There is an influence of product quality (X1), brand image (X2), on purchase intention (X3).

The test using the F test is done by comparing the value of F count with the value of F table. If the value of F count> value of F table then Ho is rejected, meaning that there is an influence of product quality and brand image on purchase intention.

Table 9. Simultaneity test.

Model	Mean Square	F	Sig.
Regression	1940.401	227.525	.000[b]
Residual	8.528		

Value F Calculate $= 227.525$. The table F value for 152 respondents is reduced by two by 3.90. So, F count> F table. Thus, it can be proven that there is an influence of product quality and brand image on purchase intentions simultaneously, and positively. A significance value of $0.000 < 0.005$ indicates that the influence of product quality and brand image on purchase intention is a significant influence.

Table 10. Coefficients test.

Model		Standardized Coefficients		
		Beta	T	Sig.
1	(Constant)		.890	.375
	Product Quality	.465	6.209	.000
	Brand Image	.440	5.874	.000

Structural Hypotheses Test 2, Effect of Purchase Intention on Purchase Decisions.

Ho: ρyx3 $= 0$ There is no effect of purchase intention variable (X3) on purchase decision variables (Y).

Ha: ρyx3 $\neq 0$ There is the influence of purchase intention variable (X3) on the purchase decision variable (Y).

The value of t count is 33.153. It's from 152 people minus 2 by 0.676. t count > t table. So it can be proven that there is a positive influence of purchase intentions on purchasing decisions. Significance value $0.000 < 0.005$ here is a significant influence of purchase intention on purchasing decisions.

This study proves that purchase intentions influence purchasing decisions. Therefore, to increase the purchasing decision to buy a Wuling car, it is necessary

to increase the purchase intention of visitors to the Wuling Office in the Bandung branch. One indicator of purchase intention that gets the smallest score is the consideration of visitors. That is, visitors are still doubtful of whether buying a Wuling Confero car will benefit greatly or otherwise will cause a loss. To eliminate visitors 'doubts, we recommend that PT SGMW provide assurance that the Wuling Confero car is a reliable car in the class, sturdy, and still cheaper compared to its competitors' prices.

The statement that gets the lowest score from the purchase decision variable is an evaluation of the information on the advantages of Wuling Confero. This shows that visitors do not evaluate the information received regarding Wuling Confero. Therefore, PT SGMW needs to convey information about the advantages of Wuling Confero through mass media and social media.

3 CONCLUSION

This research has proven that there .is a positive relationship between product quality and brand image. Likewise, product quality and brand image have a positive influence on the purchase intention of Wuling Confero simultaneously and the purchase intention has a positive effect on the purchase decision of Wuling Confero. Thus, the results of the study support the results of research conducted by (Aryadhe & Rastini 2016; Joghee & Dube 2016; Khan & Supinit 2015; Malik et al. 2013; Mashahadi & Mohayidin 2015; Seng & Husin 2015; Shehzad et al. 2014; Sultan et al. 2012; Tariz & Woodman 2013; Wang & Tsai 2014).

REFERENCES

Aryadhe, P. & Rastini, N.M. 2016. Kualitas pelayanan, kualitas produk, citra merk terhadap niat beli ulang di PT Agung Toyota Denpasar. *Fakultas Ekonomi Bisnis Universitas Negeri Udayana*: 32–41.

Eze, U.C., Tan, C.B., & Yeo, A.L.Y. 2012. Purchasing cosmetic product: A preliminary perspective of Gen-Y. *Contemporary Management Research* 8(1).

Joghee, S. & Dube, A.R. 2016. Brand image & reflections: An empirical study in UAE with car buyers of UAE nationals. *International journal of Economics, Commerce and Management* 4(3): 401–414.

Khan, F.H. & Supinit, V. 2015. What effects the purchase decision of the car in Thailand. *International Journal Marketing*: 17–25.

Kotler, P. & Armstrong, G. 2014. *Prinsip-prinsip pemasaran*. Jakarta: Erlangga

Malik, M.E., Ghofoor, M.M., Hafiz, K.L., Riaz, U., Hassan, N.U., Mustafa, M., & Shahbaz, S. 2013. Importance of brand awareness and brand loyalty in assessing purchasing intention of consumer. *International Journal of Business and Social Science*: 43–51.

Mashahadi, F. & Mohayidin, M.G. 2015. Consumers buying behavior toward local and imported cars: An outlook after the implementation of ASEAN free trade agreement

(Afta) in Malaysia. OUM International Journal of Business and Management 1(1): 1–11.

Prasad, P. 2018. *Eliminating friction in automobile path to purchasing*. New York: Nielsen Publisher.

Sangadji, S., Etta, M., & Sopiah, S. 2013. *Perilaku kosumen*. Yogyakarta: Andi.

Seng, L.C. & Husin, Z. 2015. Product and price influence on car purchasing intention in Malaysia. *International Research Journal of Interdisciplinary & Multydisciplinary Studies (IRJIMS)* 1(7):108–119.

Shehzad, U., Ahmad, S., Iqbal, K., Nawaz, M., & Usman, S. 2014. Influence of brand name on consumer choice & decision. *Journal of Business and Management* 16(6): 72–76.

Sultan, M.A., Haryono, T., Haryanto, B., & Riani, A.L. 2012. Proses pembentukan perilaku niat wisatawan berkunjung kembali. *Siasat Bisnis* 16(1).

Tariz, S. & Woodman, J. 2013. Using mixed method in health research. *JRSM Short Report* 4(6).

Tjiptono, F. & Chandra, G. 2015. *Service management*. Yogyakarta: Andi.

Wang, Y.H. & Tsai, C.F. 2014. The Relationship between brand image and purchase intention: Evidence from award winning mutual funds. *The International Journal of Business and Finance Research* 8(2): 27–40.

Advances in Business, Management and Entrepreneurship – Hurriyati et al. (Eds)
© 2021 Taylor & Francis Group, London, ISBN 978-0-367-67471-7

Emotional preferences towards e-learning based on analytic hierarchy process and Kansei for decision making

Y. Sudaryo & N.A. Sofiati
STIE INABA Business School, Bandung, Indonesia

A. Hadiana
Indonesian Institute of Sciences, Jakarta, Indonesia

ABSTRACT: User Interface Design (UID) could induce critical emotional experience and impression in users the first time they run a software system. This paper aimed at discovering relationship between UID and users' emotional experience in an e-Learning system. Kansei Engineering was adopted as methodology to analyze users' emotional experience towards the UID. This research implemented a combination approach of Kansei Engineering and Analytic Hierarchy Process to analyze students' emotional experiences as e-Learning users in a higher learning institution and determine an open source e-Learning system design that suits their positive emotional experiences to catch their attention. The research found that there were two critical students' emotional factors, i.e., the factors of "clear" and "pleasant". These two factors had big impacts in the selection of e-Learning system with factor "clear" having the largest impact of all. The results suggested preferred e-Learning system for students based on those evoking students' positive emotional experiences.

1 INTRODUCTION

E-Learning system in educational institution plays an important role in providing better learning environment over internet/intranet, because of its potential to support students' learning activities and ultimately enhance their learning performance (Donkin & Askew 2017; Liang et al. 2017). Today an abundance of open source and proprietary e-Learning systems exist. In general, most of them are similar to each other and have standard functions required to perform learning activities via internet/intranet. It has become harder to determine which ones are appealing and bring positive experience to students, as positive experience appears to be the first phase to success of any electronic or online systems. This has forced educational institution to find the best mechanism to assess, analyze, and finally select the most suitable one/s to be implemented in the institution. It has become a critical matter for educational institution, as the implementation of e-Learning system needs to be tailored to students' implicit requirements.

Software requirements consist of functional aspect and emotional aspect. However, emotional aspect lacks developer's attention, and e-Learning design and development mostly focuses on functional requirements. Considering about implementation of e-Learning system based on students' emotional preferences is important to ensure what they need implicitly (Hadiana & Lokman 2016; Redzuan et al. 2011;

Sihombing et al. 2013), and it could encourage the students to enjoy the system and gather more knowledge during learning.

Nowadays, there have been many kinds of e-Learning system, and all of them functionally have the same facilities for learning such as material learning management, discussion, task management, and so forth. It needs a decision method to determine the suitable e-Learning system for students based on their emotional feeling toward the e-Learning system through its interface.

Kansei Engineering basically is a kind of approach that has been widely used in the research of product development including software due to psychological aspects or emotional factors (Nagamachi & Lokman 2011; Nagamachi & Lokman 2015). Kansei Engineering has an ability to describe users' emotional feeling into an emotional concept of software including e-Learning system (Norman 2004). Especially when the users of e-Learning system are students who need it in order to support learning activity.

The goal of designing product including software is to make an object can be understandable easily and communicate effectively (Norman 2004; Tharangie et al. 2008). Software designers have to consider users' emotions and communicate with them to observe what they need implicitly, because based on feedback from users, it helps to establish the directions of future products (Crilly et al. 2009). The image of product is subjective, related with users' experience, and emotional factors (Dehaene 2001). Kansei words in Kansei Engineering are used widely to represent users' perception for the products. Emotional perception using Kansei words provides good efficiency and satisfaction (Chen et al. 2006).

A lot of research adopted Kansei Engineering in the field of e-Learning system in order to support and enhance learning performance. Kansei Engineering basically is targeted to observe what students completely desire about e-Learning based on their learning experiences and tries to translate it into element designs of e-Learning system. Most of evaluations in Kansei Engineering use multivariate statistical method such as principal component analysis and factor analysis (Chaminda et al. 2009; Chen 2013; Sandanayake & Madurapperuma 2009; Redzuan et al. 2014). Multivariate method explores psychological aspects presented by Kansei Words towards specimens in order to find kind of element designs related to e-Learning system. However, multivariate method has difficulty to find a decision for selecting a specimen of e-Learning system. Before implementing an open source e-Learning system, an education institution needs to take a proper decision by involving students as learning actors.

Research Sato et al. (2007) applied Kansei Engineering into decision support system, but it still used multivariate analysis to select an object, so we considered to use Analytic Hierarchy Process (AHP) as the method to select an object. AHP generally consider non-emotional factors as attributes such as price, popularity, etc (Saaty 2003).

This research attempted to apply Kansei Engineering into a decision support method of AHP in order to provide an education institution such as university to consider students' emotional factors based on their learning experiences. Combination of AHP and Kansei Engineering provides a suitable decision about what kind of e-Learning system is better to be implemented to support learning activity. By concerning students' emotional factors as the fundamental factors on decision of e-Learning system, the recommended e-Learning system would be used for long time.

2 METHODS

This research introduced AHP instead of statistical multivariate in evaluation of collected data questionnaire. Figure 1 shows the systematical processes of evaluation in this research to process data questionnaire step by step from raw data questionnaire to final decision.

Five alternative popular open source e-Learning systems such as Moodle (www.moodle.org), ATutor (www.atutor.ca), ILIAS (www.atutor.ca), Dokeos (www.dokeos.com), and Opigno (www.opigno.org) are selected as Kansei specimens based on their popularity in around academic institutions' environment. The specimens are selected based on their design characteristics such as colour balance, page layout,

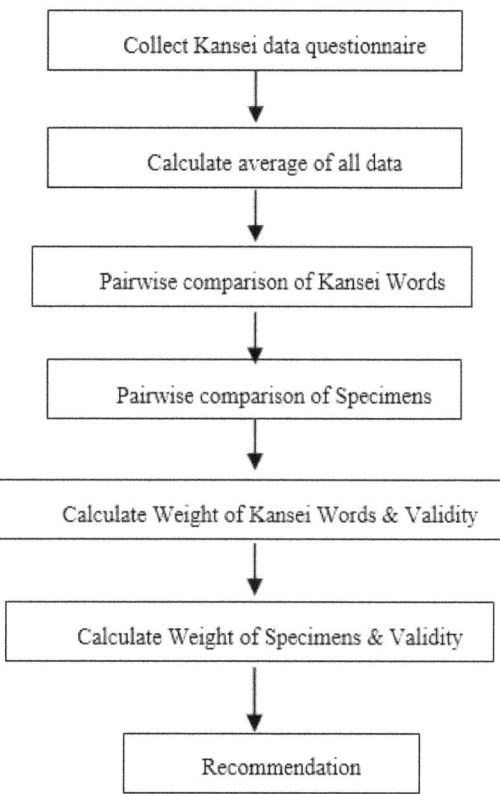

Figure 1. Research methodology.

text, etc. Eight Kansei Words are selected to represent students' emotional factors toward to e-Learning systems. The research constructs an instrument that consists of each Kansei Words with five points Semantic Differential Scale to measure students' feelings after they look and explore each specimen for a while. One hundred students participated for collecting data questionnaire.

The evaluation begins by calculating the average data of each Kansei Words. First of all, we have to prepare pairwise comparison of Kansei Words and pairwise comparison of specimens based on each Kansei Words to create matrix comparison. Using matrix comparison of Kansei Words we calculated the weight of each Kansei Words, and also calculate the weight of each specimens based on each Kansei Words. Finally, we combine the weight calculation of Kansei Words and specimens in order to find an e-Learning system as specimen to be selected as recommendation. In AHP, it is important to check the validity of Kansei Words and specimens in order to ensure that data consistency and the results are reliable.

3 RESULTS AND DISCUSSION

The first step is to gather adjective words that have closed relationship to e-Learning system according to

Table 1. Dataset of all respondents' average.

E-Learning

Kansei Words	Moodle	ATutor	ILIAS	Dokeos	Opigno	average
1. Dynamic	3.75	3.67	3.77	2.28	2.36	3.17
2. Pleasant	4.02	3.94	2.96	2.86	4.18	3.59
3. Simple	3.47	2.31	2.25	3.75	3.62	2.98
4. Clear	3.76	3.68	3.92	3.86	3.79	3.80
5. Harmony	3.81	3.73	3.80	2.11	2.24	3.14
6. Unique	2.36	3.82	3.70	2.17	2.25	2.86
7. Formal	2.45	3.68	2.23	3.72	2.16	2.85
8. Excellent	2.80	2.87	3.16	2.23	2.20	2.65

Table 2. Kansei words pairwise comparison step 1.

Kansei Words	1	2	3	4	5	6	7	8
1	0.00	−0.42	0.19	−0.63	0.03	0.31	0.32	0.52
2		0.00	0.61	−0.21	0.45	0.73	0.74	0.94
3			0.00	−0.82	−0.16	0.12	0.13	0.33
4				0.00	0.66	0.94	0.95	1.15
5					0.00	0.28	0.29	0.49
6						0.00	0.01	0.21
7							0.00	0.20
8								0.00

Table 3. Kansei words pairwise comparison step 2.

Kansei Words	1	2	3	4	5	6	7	8
1	1	−1	1	−2	1	1	1	2
2		1	2	−1	1	2	2	2
3			1	−2	−1	1	1	1
4				1	2	2	2	2
5					1	1	1	1
6						1	1	1
7							1	1
8								1

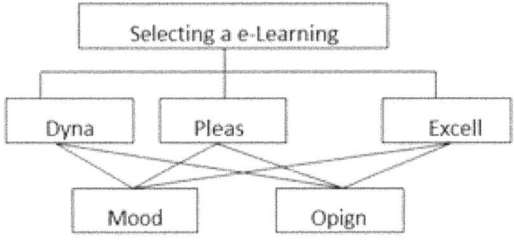

Figure 2. Kansei words based AHP.

Table 4. Specimens pairwise comparison step 1.

Dynamic	Moodle	ATutor	ILIAS	Dokeos	Opigno
Moodle	0.00	0.08	−0.02	1.67	1.39
ATutor		0.00	−0.10	1.39	1.31
ILIAS			0.00	1.49	1.41
Dokeos				0.00	−0.08
Opigno					0.00

Table 5. Specimens pairwise comparison step 2.

Dynamic	Moodle	ATutor	ILIAS	Dokeos	Opigno
Moodle	1	1	−1	3	2
ATutor		1	−1	2	2
ILIAS			1	2	2
Dokeos				1	−1
Opigno					1

discussion with educational experts. We restricted only using eight adjective words as Kansei Words to express students' emotional feeling toward e-Learning system.

We collected all data questionnaire from one hundred undergraduate students as respondents, and then we calculated the average values as shown in Table 1. Each Kansei Words have average's value between 2 and 4 because we used five points semantic scale that commonly used in Kansei research.

For further analysis using AHP, we needed to calculate the average of Kansei Words from all specimens as shown at column average in Table 1. This average's values are used to calculate pairwise comparison between Kansei Words. The average of Kansei Words at each column of specimens are used to calculate the pairwise comparison between specimens per each Kansei Words. These two types of pairwise comparison are fundamental components to calculate and find the final decision in AHP as described in Figure 2.

3.1 Kansei words' weight calculation

Using data at column average in Table 1 then the pair comparison between Kansei Words are performed. Pair comparisons are conducted by substitution operator between two Kansei Words. For example, calculation between dynamic and informative is $3.29 - 3.85 = -0.56$, calculation between dynamic and beautiful is $3.29 - 2.65 = 0.64$. All substitution results are shown in Table 2.

3.2 Specimens' weight calculation

The next preparation before starting AHP calculation, it has to calculate pairwise comparison between two specimens based on each Kansei Words, for example Table 4 shows the pairwise comparison due to Kansei Word Dynamic using data in Table 1, and the complete pairwise comparison are shown in Table 5 and Table 6.

Using the same method, we also have to calculate pairwise comparison between specimens based on

Table 6. Specimens pairwise comparison step 3.

Dynamic	Moodle	ATutor	ILIAS	Dokeos	Opigno
Moodle	1	1	1	3	2
ATutor		1	1	2	2
ILIAS			1	2	2
Dokeos				1	1
Opigno					1

Table 7. Specimens pairwise comparison step 3.

Pleasant	Moodle	ATutor	ILIAS	Dokeos	Opigno
Moodle	1	1	3	3	1
ATutor		1	2	2	1
ILIAS			1	1	1/2
Dokeos				1	1/2
Opigno					1

Table 8. Specimens pairwise comparison step 3.

Simple	Moodle	ATutor	ILIAS	Dokeos	Opigno
Moodle	1	2	2	1	1
ATutor		1	1	1/2	1/2
ILIAS			1	1/3	1/2
Dokeos				1	1
Opigno					1

Table 9. Specimens pairwise comparison.

Clear	Moodle	ATutor	ILIAS	Dokeos	Opigno
Moodle	1	1	1	1	1
ATutor		1	1	1	1
ILIAS			1	1	1
Dokeos				1	1
Opigno					1

Table 10. Specimens pairwise comparison step 3.

Harmony	Moodle	ATutor	ILIAS	Dokeos	Opigno
Moodle	1	1	1	3	3
ATutor		1	1	3	2
ILIAS			1	3	3
Dokeos				1	1
Opigno					1

Table 11. Specimens pairwise comparison step 3.

Unique	Moodle	ATutor	ILIAS	Dokeos	Opigno
Moodle	1	1/2	1/2	1	1
ATutor		1	1	3	3
ILIAS			1	3	2
Dokeos				1	1
Opigno					1

Table 12. Specimens pairwise comparison step 3.

Formal	Moodle	ATutor	ILIAS	Dokeos	Opigno
Moodle	1	1/2	1	1/2	1
ATutor		1	2	1	3
ILIAS			1	1/2	1
Dokeos				1	3
Opigno					1

Table 13. Specimens pairwise comparison step 3.

Excellent	Moodle	ATutor	ILIAS	Dokeos	Opigno
Moodle	1	1	1	2	2
ATutor		1	1	2	2
ILIAS			1	2	2
Dokeos				1	1
Opigno					1

Table 14. Data validity.

Kansei Words	CI	CR	Alpha
Dynamic	0.0049	0.0044	5.0194
Pleasant	0.0064	0.0057	5.0255
Simple	0.0049	0.0044	5.0194
Clear	0.0000	0.0000	5.0000
Harmony	0.0053	0.0048	5.0212
Unique	0.0082	0.0074	5.0329
Formal	0.0069	0.0062	5.0277
Excellent	0.0000	0.0000	5.0000

remaining Kansei Words, and the results are shown in Tables 7–13.

3.3 *AHP calculation*

This research conducts AHP calculation based on data in Tables 4 and 6 (including remaining Kansei Words comparison based on specimen). It is important in AHP to check data validity by calculating Consistency Index (CI) and Consistency Ratio (CR).

CI, CR, and Alpha calculation of Table 4 has values of 0.0111, 0.0079, and 8.0775 respectively. Table 7 shows the values of CI, CR, and Alpha for all Kansei Words. According to these results, all values of CR are less than 0.1, it means that pairwise comparison between Kansei Words are reliable and can be validated to continue further analysis. Table 14 shows CI, CR, and Alpha of specimens based on each Kansei Words. The results show that all values of CR are less than 0.1, then all data in this research have good validity.

$$A = \begin{pmatrix} 0.0 & 0.0 & 0.0 & 0.0 & 0.0 & 0.0 & 0.0 & 0.0 \\ 331 & 533 & 247 & 4 & 276 & 137 & 139 & 231 \\ 0.0 & 0.0 & 0.0 & 0.0 & 0.0 & 0.0 & 0.0 & 0.0 \\ 303 & 450 & 123 & 4 & 256 & 323 & 302 & 231 \\ 0.0 & 0.0 & 0.0 & 0.0 & 0.0 & 0.0 & 0.0 & 0.0 \\ 303 & 207 & 114 & 4 & 276 & 297 & 139 & 231 \\ 0.0 & 0.0 & 0.0 & 0.0 & 0.0 & 0.0 & 0.0 & 0.0 \\ 140 & 207 & 269 & 4 & 092 & 117 & 302 & 116 \\ 0.0 & 0.0 & 0.0 & 0.0 & 0.0 & 0.0 & 0.0 & 0.0 \\ 152 & 450 & 247 & 4 & 101 & 126 & 119 & 116 \end{pmatrix}$$

$$B = \begin{pmatrix} 0.1229 \\ 0.1848 \\ 0.1000 \\ 0.2000 \\ 0.1000 \\ 0.1000 \\ 0.1000 \\ 0.0924 \end{pmatrix} \qquad AxB = \begin{pmatrix} 0.2293 \\ 0.2387 \\ 0.1968 \\ 0.1643 \\ 0.1710 \end{pmatrix}$$

Figure 3.   Kansei words' ranking.

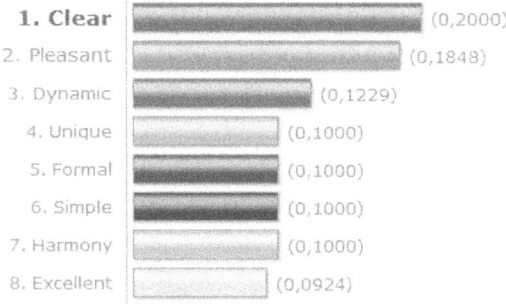

Figure 4.   Kansei words' ranking.

Figure 2 shows the final calculation of AHP. Weight of specimens per Kansei Words are in matrix A, and weight of Kansei Words are in matrix B. The last calculation to find ranks of e-Learning systems as specimens is a multiplied to B.

Final calculations of Kansei Words using AHP are described in Figure 3 to show Kansei Words' impact, and Figure 4 to show rank of e-Learning system.

Figure 4 shows the weight of each Kansei Words. It explains which of students' emotional factors has great impact on selection of e-Learning system. According to Figure 4, we found that there are two students' emotion (presented by pleasant and clear). The two Kansei Words have big impacts on selecting e-Learning system. However, the students' emotion of clear has the biggest impact to prefer a learning system.

Figure 5 shows the final result of this research to show which e-Learning system to be selected based on what students desire for. According Figure 5 the alternative e-Learning system to be considered based on students' emotion are ATutor and Moodle. However, students trended to select ATutor as e-Learning system in order to support their learning activities via

Figure 5.   Specimens' ranking.

internet. Moodle can be considered as alternative system because the difference between the both systems was only 0.0094.

According to these results we found that in order to enhance the design of user interface of e-Learning system it is important to consider the students' emotional perspectives, in this case presented by Kansei Words of clear and pleasant. In this research, another emotion that also should be considered alternatively in designing e-Learning system in the future is dynamic. This research also urges e-Learning developers to pay more attention to students' implicit requirements such as emotional factors.

4   CONCLUSION

This research attempts to incorporate Kansei Engineering into AHP in order to find the most suitable e-Learning system based on students' emotional factors. Kansei Words related to open source e-Learning system were used to describe students' emotions. Eight adjective words were used as Kansei Words, and five open source e-Learning system were used as specimens. Kansei Words were involved as attributes of AHP in considering a final decision. Kansei Word of clear has the biggest impact on selecting e-Learning system. ATutor e-Learning system is a recommended kind to be implemented as a learning tool at the institution.

Further research could also highlight the importance to assess e-Learning systems and Kansei at larger scale to generalize findings. Furthermore, comparison of AHP with other decision method could also be of importance, as the findings would be useful to find insights whether more decision method could be combined to find clearer decisions.

ACKNOWLEDGEMENTS

The author would like to thank to students of STIE INABA Bandung and STMIK LIKMI Bandung for collecting data, and LIPI Bandung for their great support in discussions and their cooperation during this research. The research also acknowledges the support given by Malaysia Association of Kansei Engineering (MAKE), and Research Initiative Group of Kansei/Affective Engineering (RIG KAE), UiTM.

REFERENCES

Chaminda, H.T., Basnayake, A., Madurapperuma, A., & Osano, M. 2009. An interactive E-Learning system using Kansei engineering. *International Conference on Biometrics and Kansei Engineering*: 157–162.

Chen, C.H., Khoo, L.P., & Yan, W. 2006. An investigation into affective design using sorting technique and Kohonen self-organising map. *Advances in Engineering Software* 37(5): 334–349.

Chen, Y. 2013. Research on optimized design of Kansei engineering-based web interface. *International Conference on Computational and Information Sciences*: 1709–1712.

Crilly, N., Moultrie, J., & Clarkson, P.J. 2009. Shaping things: intended consumer response and the other determinants of product form. *Design studies* 30(3): 224–254.

Dehaene, S. 2001. *The cognitive neuroscience of consciousness*. Cambridge: MIT Press.

Donkin, R. & Askew, E. 2017. An evaluation of formative "in-class" versus "E-learning" activities to benefit student learning outcomes in biomedical sciences. *Journal of Biomedical Education.*

Hadiana, A. & Lokman, A.M. 2016. Kansei Evaluation in Open Source E-Learning System. *Jurnal Teknologi* 78(12–3).

Liang, K., Zhang, Y., He, Y., Zhou, Y., Tan, W., & Li, X. 2017. Online behavior analysis-based student profile for intelligent E-learning. *Journal of Electrical and Computer Engineering.*

Nagamachi, M. & Lokman, A.M. 2011. *"Innovations of Kansei Engineering," Industrial Innovation Series, Adedeji B.* Florida: Taylor & Francis Group.

Nagamachi, M. & Lokman, A.M. 2015. *Kansei innovation: Practical design applications for product and service development (Vol. 32)*. USA: CRC Press.

Norman, D. A. 2004. *Emotional design: Why we love (or hate) everyday things*. USA: Basic Civitas Books.

Redzuan, F., Lokman, A.M., & Othman, Z.A. 2014. Kansei semantic space for emotion in online learning. *3rd International Conference on User Science and Engineering (i-USEr)*: 168–173.

Redzuan, F., Lokman, A.M., Othman, Z.A., & Abdullah, S. 2011. Kansei design model for e-learning: A preliminary finding. *In Proceeding of the 10th European Conference on e-Learning (ECEL-2011)*: 685–696.

Sandanayake, T.C. & Madurapperuma, A.P. 2009. Conceptual model for e-learning systems using kansei engineering techniques: A Preliminary Study. *International Conference on Biometrics and Kansei Engineering*: 148–152.

Sato, N., Anse, M., & Tabe, T. 2007. A Method for Constructing a Movie-Selection Support System Based on Kansei Engineering. *Symposium on Human Interface and the Management of Information*: 526–534.

Saaty, T.L. 2003. Decision-making with the AHP: Why is the principal eigenvector Necessary. *European journal of operational research* 145(1): 85–91.

Sihombing, H., Yuhazri, M.Y., Yahaya, S.H., & Syaifoelida, F. 2013. The kansei design characteristics towards learning style. *Journal of Engineering.*

Tharangie, K.G.D., Irfan, C.M.A., Marasinghe, C.A., & Yamada, K. 2008. Kansei engineering assessing system to enhance the usability in e-learning web interfaces: Colour basis. *In 16th International Conference on Computers in Education* 1(1): 145–150.

Advances in Business, Management and Entrepreneurship – Hurriyati et al. (Eds)
© 2021 Taylor & Francis Group, London, ISBN 978-0-367-67471-7

Satisfaction and revisit intention of tourists in Bandung

M.M. Munawar, R. Hurriyati & M.A. Sultan
Universitas Pendidikan Indonesia, Bandung, Indonesia

S.H. Suarsa
Politeknik Pos Indonesia, Bandung, Indonesia

ABSTRACT: The main objective of this study is to determine the tourists' intention to revisit Bandung and subsequently examine the various predictors including food image, destination image, towards their visit satisfaction. The focus of this research is tourism in the Bandung city, with a sample of 200 tourists who have the intention revisiting Bandung. The sample technique used Convenient Sampling. Data collection is done through the questionnaire method and statistical analysis method used in this research is using Structural Equation Modeling (SEM) to analyze and elaborate all observed variables. The results of this research showed that the food image and destination image have a significant positive influence on tourist satisfaction and revisit intention. The destination image also showed a positive significant effect on tourist satisfaction but showed have not significant effect on revisit intention. Tourist satisfaction has a positive significant influence on revisit intention.

1 INTRODUCTION

There is a variety of theoretical review on tourist behavior, especially behavior of revisit intention of tourists in Bandung. Various methods, models, and predicator variables in shaping tourist behavior create debates among researchers (Chen & Tsai 2007; Sultan et al. 2012). This shows that previous research cannot be implemented in all phenomena that might happen as the result of the various methods, models, and predicator variables. Therefore, this condition provides an opportunity for other research to give alternative models that can explain various conditions and phenomena happened not only in recent context but also in the future.

Bandung tourism is one of destinations for vacation in West Java that has culinary variety and it becomes one of excellences in the city. Recently Bandung is one of tourism destinations that have the highest index in Indonesia, that is 95.30, and respectively followed by Bali with 87.65 and Yogyakarta 85.68 (BPS Jawa Barat 2018). Competition to become the best tourist destination city, required a strategic that must be offered by each city so that it will be able to attract tourists visiting the city. Because of the importance of this strategic, the aim of this research is to get the right strategy to attract tourists visiting the city of Bandung. The choice of Bandung as research object with various culinary tourisms, various habits of tourists in assessing food image and destination image and tourist satisfaction in having vacation aims to avoid bias model in this research.

A number of previous research shows that expectation, quality perception, value perception, destination image, budget and risk are good predictors on tourist satisfaction (Aliman et al. 2016). Other research mentions that image of food; tourist satisfaction is positive predictor to create loyal tourists on their choice (Astuti et al. 2015). Other research unveils that quality perception; value perception and tourist satisfaction are positive predictors on tourist behavior of revisit to the destinations (Sultan et al. 2012).

Causality relation from a number of previous researches display positive relation pattern as this happens because food image and destination image from tourism impacts on tourist satisfaction. Therefore, the better the food image and destination image, the higher tourism satisfaction. Thus, tourism behavior of visit intention increases as well.

In previous research, not all researchers unveil variables of food image, destination image and tourist satisfaction as good predictors in predicting tourism aspiration behavior to revisit. Some previous theoretical reviews show that food image variable has positive impact on tourist satisfaction and destination image has positive impact on tourist satisfaction, but both predictors do not become predictors on behavior of revisit intention (Aliman et al. 2016; Astuti et al. 2015).

Based on the above phenomena, this research deploys four variables that are predicted to be used as model construct. Food image is the first variable that is considered. This variable is one that contributes to the forming of satisfaction and in the end will influence tourist intention behavior (Astuti et al. 2015). Good or bad food image will influence tourist satisfaction and behavior of revisit intention.

The second variable in this research is destination image of the tourism objects. This variable is also important as food image in the tourism object because the better the destination image, the higher the tourist satisfaction (Aliman et al. 2016). With better increase of tourist satisfaction, it will impact the revisit intention.

Beside those two variables, tourist satisfaction becomes the third variable considered in this research. This variable is used as variable that mediates food image and destination image on the behavior of revisit intention, because tourist satisfaction is a good predictor of revisit intention (Sultan et al. 2012). The higher the tourist satisfaction, the higher the tourists revisit intention to the destinations.

The last variable from this research is behavior of revisit intention. This variable is used as purpose variable to predict true tourist behavior to be accurately measured. By analyzing causality relation formed, it is expected that the model developed can predict tourist intention behavior in the future. (De Rojas & Camarero 2008; Sultan et al. 2012). Therefore the better food image, destination image and tourist satisfaction will have great influence in the high tourist revisit intention.

1.1 Food image

Image is a total perception of an object that is formed by information processing for various sources (Setiadi 2013). Image represents simplification of many associations and information chunks that are connected in a certain place. It is a product of mind that attempts to process and frame a big number of data about a certain place (Kotler 2002). Image of a place is formed by combination of belief, idea, knowledge, and feeling about. For this study, image is specifically related to tourists' evaluation perception on food, therefore food image is used. Food image is a holistic assessment of a trip based on experiences of customers or tourists on what they spend in terms of money or time (Chi et al. 2013).

Tourists develop food image in their mind based on their perception on food attributes. Factors of environment and personal as the start for forming the brand image are important, because the factors influence somebody's perception (Walter 1974). Environment factors that influences food image are not only technical attribute of a product, of which the factor can be controlled by the producers, but also social and culture. Personal factors are the mental readiness of customers in processing perception, experience, feeling, need and motivations from consumers. Image is the final product from initial attitude and knowledge that is formed from repetition of dynamic process because of experiences (Arnould et al. 2004).

Many researchers have studied food, and suggested that food image can be measured through regular price of item, richness of taste, high nutrition, hospitality of food provider and herbs and spices (Chi et al. 2013). But some other research states that food image can be measured through four indicators, they are food quality, attraction, price and uniqueness (Chi et al. 2013).

1.2 Tourist satisfaction

Previous research shows that tourist satisfaction is the result of correlation between perception of service quality in the location and value perception taken by the tourists in the area (Sultan et al. 2012). The review of previous research shows that tourist satisfaction is an important variable for the success of tourism object marketing, because it highly influences destination choice, product consumption and revisit decision (Kozak & Rimmington 2000).

Satisfaction is feeling level of costumers after consuming a product or a service by comparing performance of the product/service accepted and expected (Kotler & Armstrong 2010). Customer satisfaction is customer's response on disconfirmation felt between expectation and actual performance from a product after using it, if a tourist has a positive experience and trust on a food or culinary product, he or she will be motivated to try the product again in a certain city (Ryu & Jang 2006).

Positive image will influence tourist satisfaction and in the end satisfaction will influence tourist loyalty (Chi et al. 2013). Therefore, food providers in this case culinary producers in Bandung City should be able to create positive impression to tourists so they will expect to try the food again in the city. And when they are satisfied, they will become loyal customers and will come to visit again and buy the culinary products and in the end they may recommend the product to their colleagues. Thus, the hypothesis is formulated as follows:

H1: Food image positively influences tourist satisfaction.

1.3 Destination image

Several previous theoretical reviews focus on destination image that has important role on customer satisfaction (Aliman et al. 2016). Destination image is all mental representation from tourists or feeling perception and knowledge on a specific destination (Assaker et al. 2011; Fakeye & Crompton 1991). Based on previous research, destination image development can be different between tourists who come for the first time and those who come regularly and even tourism mostly contributes to destination image as a whole (Liu et al. 2017). Tourists can be considered making decision of their journey based on destination image and it has important role in their decision making because it can be said that destination image directs emotional bonding from tourists that then influence tourist satisfaction (Silva et al. 2013). Destination image can be assessed from cognitive and affective perspectives (Fakeye & Crompton 1991) and previous research shows that affective image component has a bigger influence than cognitive component (Stylidis et al. 2017). In the context of influencing revisit intention, destination image has indeed significant influence (Prayag & Ryan 2012).

Indicator from destination image used in this study is cognitive and affective perspectives (Fakeye & Crompton 1991). Previous research shows that destination image positively influences whole tourists and directs to tourist satisfaction (Bigne et al. 2001; Pratminingsih & Boediprasetya 2014). Destination image is regarded the most important variable in creating tourist satisfaction. Therefore, the hypothesis formulated is:

H2: Destination image has positive influence on tourist satisfaction.

1.4 Revisit intention

Revisit intention behavior is defined as visitor assessment on the correlation to revisit the same destination or their willingness to recommend a certain destination to other people (Chen & Tsai 2007). A previous theoretical study defines intention behavior as individual who anticipates and plan future behavior. The intention behavior is recommended to be a central factor that is strongly related to observed behavior (Swan & Trawick 1981). Another theoretical study considers tourist satisfaction variable to be used as important variable that influence intention behavior, especially revisit intention behavior. There are two variables that are used to predict tourist intention behavior; they are tourist satisfaction and food image (Astuti et al. 2015).

Besides investigating the influence of food image on tourist satisfaction, theoretical reviews also indicate that there is a correlation between food image and intention behavior and in this case is tourist revisit intention. Therefore, the hypothesis formulated as follows:

H3: Food image has a positive influence on tourist revisit intention behavior.

Several theoretical reviews show that concept of future behavior from tourists does not only influence by food image. Destination image is an important variable that decides tourist revisit intention (Aliman et al. 2016). Destination image is regarded to influence not only tourist satisfaction but also tourist revisit intention behavior. Therefore the hypothesis is formulated as follows:

H4: Destination image has a positive influence on tourist revisit intention behavior.

Tourist future behavior can happen in their decision to revisit or intention to recommend or not to other people. Some literature reviews prove the correlation between satisfaction and repurchase intention (Baker & Crompton 2000; Lee et al. 2004; Petrick 2004; Petrick & Backman 2002; Sultan et al. 2012; Tam 2000).

In marketing theory, satisfaction is an important variable for the success of a business. This shows that there is tendency of literature review that studies measurement of satisfaction that focus on the concept of loyalty, because loyalty is the key in business success (Chi & Qu 2008). This happens because when tourists feel satisfied, what will they do is to come again to the place and recommend the place to others.

If tourists do not feel satisfied, it is a big possibility that they will not come again and will not recommend to others. This phenomenon will make tourist satisfaction and good predictor variable on tourist intention behavior. Therefore the hypothesis is formulated as follows:

H5: Tourist satisfaction has a positive influence on revisit intention behavior.

1.5 Research model

Based on several theoretical reviews and casual connection that is hypothesized, the following is the model designed to show phenomena hypothesized (Figure 1).

This model shows that there is a correlation from a variable with others. The variable correlation starts from food image that has a positive influence on tourist satisfaction and revisit intention. Then destination image has a positive influence on tourist satisfaction and revisit intention. And the last part is there is a positive influence of tourist satisfaction on revisit intention.

2 METHODS

Focus of this research is phenomenon in Bandung City with 200 respondents as sample, comprising tourists who have been to Bandung City for vacation. Questionnaires were distributed through electronic media from 27 until 30 April 2019 and suitable data of 200 samples were collected. To conduct validity test, confirmatory factor analysis was used and for reliability test was construct reliability (CR). It can be reliable if the score is >0.5. Statistical analysis used in this research is Structural Equation Modeling (SEM) with hypotheses testing was done with AMOS version 23 to analyze causal relation in the proposed structured model.

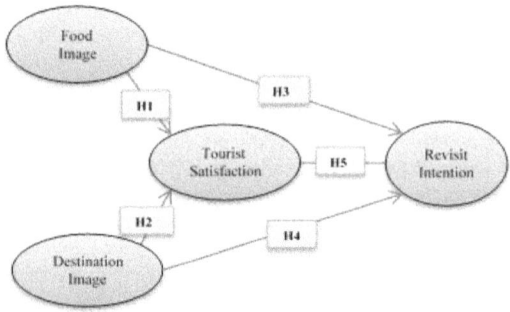

Figure 1. Research model: satisfaction and revisit intention of tourists in Bandung.

3 RESULTS AND DISCUSSION

3.1 *Respondent identification*

Table 1 is the complete result to explain description of respondents in this research. The research respondents are tourists who have visited Bandung. Samples are 200 people and have fulfilled minimum criteria for data analysis using SEM. Descriptive statistical analysis aims to discover respondent characteristics, which is identification related to some aspects including sex, age, education, occupation, visit frequency and domicile of respondents. Table 1 shows that female becomes majority from the sample of this research with total female respondent is 102 people (51%). In terms of age majority of respondents are in between 16–40 years old about 96 people (48%) and then in between 26–40 with 72 people (36%) in the second place and the last one is over 40 years old with 32 people (16%). In terms of education level the majority is undergraduate degree with 107 people (53.3%) and then high school graduates 44 people (22%), besides diploma, master and doctoral degrees in total 49 people (23.5%).

In terms of occupation the majority of respondents is private employees comprising 86 people (43%),

student/university students 53 people (26.5%), Civil Servants/State Owned enterprise employee 34 people (17%) and others are retired/ not working and entrepreneurs are 27 people (13.5%). Majority of responded in terms of visit frequency to Bandung City is 192 people (96%) visiting Bandung for more than once meanwhile the rest which is 8 people stating that they have just come to Bandung once. Respondents' domicile is mostly from Purwakarta 85 people (42.5%), Jakarta 17 people (8.5%) and Cianjur 9 people (4.5%) and the rest is 89 people (41.8%) from different parts of cities in Java or outside of Java.

3.2 *Validity and reliability tests*

Table 2 shows the result of validity and reliability of this research. Validity test result shows that all question items have loading factor value above minimum benchmark, which is 0.5 showing that all questions in this research are valid. Reliability test result also shows that average of Cronbach's alpha for every variable showing the value above 0.5, this means that all statements in the research can be categorized reliable and shows that all question items used are consistent.

Table 1. Respondent identification.

Descriptive Statistics		N	%
Sex	Male	98	49
	Female	102	51
		200	100
Age	18–25 years old	72	36
	26–40 years old	96	48
	Above 40 years old	32	16
		200	100
Education	High School	44	22
	Diploma (D1–D4)	16	8
	Undergraduate Degree	107	53.5
	Master Degree	31	15.5
	Doctoral Degree	2	1
		200	100
Occupation	Students/University Students	53	26.5
	Civil Servants/State Owned enterprise employee	34	17
	Private Employee	86	43
	Entrepreneurs	16	8
	Retired/not working	11	5.5
		200	100
Visit Frequency	First Visit	8	4
	More than 1 visit	192	96
		200	100
Domicile	Purwakarta	85	45.2
	Jakarta	17	8.5
	Cianjur	9	4.5
	Other cities	89	41.8
		200	100

* Source: Analyzed primary data 2019.

Table 2. Validity and reliability test result.

Variable			Validity Loading Factor (A)	Desc.	Reliability CR	Ave	Desc.
FI1	←	FI	0,801	Valid	0,919	0,590	Reliable
FI2	←	FI	0,833	Valid			
FI3	←	FI	0,796	Valid			
FI4	←	FI	0,773	Valid			
FI5	←	FI	0,81	Valid			
FI6	←	FI	0,567	Valid			
FI7	←	FI	0,811	Valid			
FI8	←	FI	0,724	Valid			
Total			6,115				
Total Square			37,39				
DI4	←	DI	0,787	Valid	0,908	0,586	Reliable
DI3	←	DI	0,769	Valid			
DI2	←	DI	0,814	Valid			
DI1	←	DI	0,757	Valid			
DI5	←	DI	0,775	Valid			
DI6	←	DI	0,7	Valid			
DI7	←	DI	0,752	Valid			
Total			5,354				
Total Square			28,66				
TS1	←	TS	0,886	Valid	0,928	0,812	Reliable
TS2	←	TS	0,907	Valid			
TS3	←	TS	0,911	Valid			
Total			2,704				
Total Square			7,311				
RI1	←	RI	0,872	Valid	0,938	0,791	Reliable
RI2	←	RI	0,902	Valid			
RI3	←	RI	0,891	Valid			
RI4	←	RI	0,893	Valid			
Total			3,558				
Total Sqaure			12,65				

*Source: Analyzed primary data 2019.

Table 3. Fit model test result.

Criteria	Result	Recommendation
Goodness of Fit Index (GFI)	0.808	<0.90
Parsimonious Goodness of Fit Index (PGFI)	0.648	0.50–1.00
Adjusted Goodness of Fit Index (AGFI)	0.86	>0.80
Comparative Fit Index (CFI)	0.912	>0.90
Tucker Lewis Index (TLI)	0.9	>0.90
Root Mean Square of Approximation (RMSEA)	0.061	<0.08

* Source: Analyzed primary data 2019.

Table 4. Hypotheses test result.

Hypotheses	Estimation	S.E.	C.R.	P	Conclusion
H1 FI → TS	0.422	0.089	4.741	***	Significant
H2 DI → TS	0.522	0.088	5.92	***	Significant
H3 FI → RI	0.376	0.073	5.187	***	Significant
H4 DI → RI	−0.13	0.074	−1.764	0,078	Not Significant
H5 TS → RI	0.697	0.081	8.621	***	Significant

*Source: Analyzed primary data 2019.

3.3 Model fit test

In fit test of the model used in this research is shown in Table 3. In general based on model fit test with goodness of fit index (GFI), parsimonius goodness of fit index (PGFI), adjusted goodness of fit index (AGFI), comparative fit index (CFI), tucker lewis index (TLI) and root mean square of approximation (RMSEA), all model fit tests shows value with range that can be recommended from each test. Therefore it can be concluded that all reseach models are fit.

3.4 Hypotheses test

This research attempts to propose five hypotheses based on observations and also theoritical approach about tourist revisit intention. Table 4 shows complete results from hypotheses analysis that was conducted.

The first hypotheses test result, is food image has a positive influence to tourist satisfaction. Based on the claculation in Table 4 it is shown that C.R. value for food image to tourist satisfaction is 4.741 with significance level p < 0.05, therefore it shows that hypotheses 1 is proved to have an influence. This possibly happens because food image aspect is the aspect that has the strongest influence on tourist satisfaction. Thus, this aspect enhances food image that will significantly impacts on the high level of tourist satisfaction.

The research result from the first hypotheses shows quality pattern consistency that elicits the ability of this research to support significant causal pattern phenomenon as conducted in previous research that food

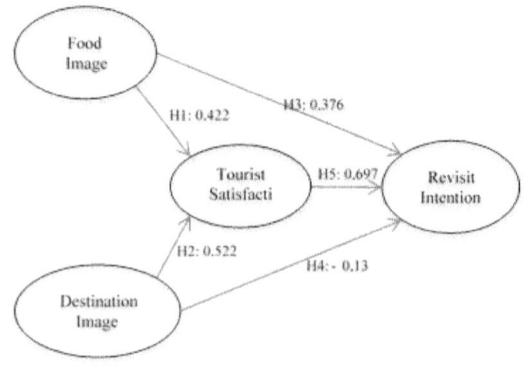

Figure 2. Result of research model.

image is the most positive influental aspect on tourist satisfaction (Astuti et al. 2015).

Result of the second hypothesis that is destination image has a positive influence on tourist satisfaction. The calculation on Table 4 shows that the value of C.R. destination image on tourist satisfaction is 5.92 with significance level is p < 0.05. Therefore, this shows that hypotheses 2 is proved to have influence. This possibly happens because destination image aspect is the aspect that has strong influence on tourist satisfaction. Thus, high the destination image can significantly impact on the high level of tourist satisfaction.

Research result of the test that was conducted to see the causal pattern consistency that shows the research ability in supporting causal pattern phenomenon as stated in a previous research stating that destination image has positive influence on tourist satisfaction (Aliman et al. 2016).

Result of the third hypothesis test, that is food image, has a positive influence on revisit intention. The calculation on Table 4 shows that the value of C.R. food image one revisit intention is 5.187 with significance level is p < 0.05. Therefore this shows that hypothesis 3 is proven to be influential. This can happen because the food image aspect has a strong influence on the revisit intention. Thus this aspect will enhance food image that will impact on the high level of tourist revisit intention.

Research result on the test done has show causal pattern consistency that indicates the research ability to support causal pattern phenomenon is significant as stated by a previous research that food image has a positive influence on the revisit intention (Hwang et al. 2018).

Result of the forth hypothesis test, that is destination image, has a positive influence on revisit intention. The calculation on Table 4 shows that the value C.R. of destination image on revisit intention is −.764 with level of significance p > 0.05. Therefore this shows that hypotheses 4 is proven to be not influential. This can happen because destination image aspect is the aspect that might not have influence on tourist revisit intention, other aspects are may still needed to ensure tourists to have revisit intention, for example tourist satisfaction aspect to create revisit intention.

Research result of the test done to indentify the inconsistency of causal pattern that shows the inability of this research in supporting causal pattern phenomena is significant as stated by a previous research (Wibowo et al. 2016). However, another research elicits consistency of relation pattern that is not significant on both aspects (Herstanti et al. 2014). This indicates that further study is needed to scrutiny the influence of destination image aspect on tourist revisit intention. Thus it is expected that future research can result in the model that is able to explain this phenomenon.

Result of the fifth hypotheses, that is tourist satisfaction, has positive influence on revisit intention. The calculation in the Table 4 shows that the value of C.R. of tourist satisfaction on revisit intention is 8.621 with level of significance $p < 0.05$. Therefore this shows that the hypotheses 5 is proven to be inflluential. This can happen because tourist satisfaction aspect has a strong influence on revisit intention. Thus this aspect can enhance tourist satisfaction that can have significant impact on the high tourist revisit intention.

Research result on the test conducted to see causal pattern consistency shows that the ability of this research to support causal pattern phenomenon is significant as stated by a previous research that tourist satisfaction has a positive influence on revisit intention (Oh 2000; Sultan et al. 2012). But different from other research that causal patter is not significant to both aspects (Lee et al. 2008; Mittal & Walfried 1998). This indicates the need of further study to observe the existance of tourist satisfaction on revisit intention, so that it is expected that future research can create model that is able to explain this tested phenomenon.

4 CONCLUSION

From the research it can be concluded that the model construct that is tested in development of revisit intention that is influenced by the interaction of three variables, they are food image, destination image and tourist satisfaction. The result shows that food image and tourist satisfaction can create tourist revisit intention. Food image and destination image can create tourist satisfaction. Meanwhile destination image has negative influence in developing tourist revisit intention.

This research has some limitation. First is that purpose variable in this research, that is tourist revisit intention. Therefore this research only focuses on the variable with sampling technique of convenience sampling respondents who have visited Bandung. The second is the scope of tourism object is Bandung city so that the characteristics of culture, social norm and perception from tourist as respondents on the tourism objets in Bandung has influence on general result. This research limitation is in line with aspects of culture, geographical feature and demography that are things that cannot be avoided in this research.

ACKNOWLEDGMENTS

Special thanks is given to the team of this article: Prof. Dr. Hj. Ratih Hurriyati, M.P, Dr. M. Adib Sultan, MT. and Senny Handayani Suarsa, SE., MM. for their support in this article so that it can be completed and submitted in this conference.

REFERENCES

Aliman, N.K., Hashim, S.M., Wahid, S.D. M., & Harudin, S. 2016. Tourists satisfaction with a destination: An investigation on visitors to Langkawi Island. *International Journal of Marketing Studies* 8(3): 173–188.

Arnould, E., Price, L., & Zinkhan, G. 2004. *Consumers*. New York: McGraw-Hill.

Assaker, G., Vinzi, V.E., & O'Connor, P. 2011. Examining the effect of novelty seeking, satisfaction, and destination image on tourists' return pattern: A two factor, non-linear latent growth model. *Tourism management* 32(4): 890–901.

Baker, D.A. & Crompton, J.L. 2000. Quality, satisfaction and behavioral intentions. *Annals of Tourism Research* 27(3): 785–804.

Bigne, J.E., Sanchez, M.I., & Sanchez, J. 2001. Tourism image, evaluation variables and after purchase behaviour: Inter-relationship. *Tourism management* 22(6): 607–616.

BPS Provinsi Jawa Barat 2018. *Jumlah kunjungan wisatawan ke obyek wisata*. [Online]. Retrieved from https://jabar.bps.go.id/statictable/2018/03/23/475/jumlah-kunjungan-wisatawan-ke-obyek-wisata-menurut.html.

Chen, C.F. & Tsai, D. 2007. How destination image and evaluative factors affect behavioral intentions. *Tourism Management* 28: 1115–1122.

Chi, C.G.Q., Chua, B.L., Othman, M., & Karim, S.A. 2013. Investigating the structural relationships between food image, food satisfaction, culinary quality, and behavioral intentions: The case of Malaysia. *International Journal of Hospitality & Tourism Administration* 14(2): 99–120.

Chi, G.Q., Chua, B.L., Othman, M., & Karim, S.A. 2013. Investigating the structural relationships between food image, food satisfaction, culinary quality and behavioral intentions: The case of Malaysia. *International Journal of Hospitality & Tourism* 14(99–120).

Chi, C.G. & Qu, H. 2008. Examining the structural relationships of destination image, tourist satisfaction and destination loyalty: An integrated approach. *Tourism Management* 29: 624–636.

De Rojas, C. & Camarero, C. 2008. Visitors' experience, mood and satisfaction in a heritage context: Evidence from an interpretation center. *Tourism Management* 29: 525–537.

Fakeye, P.C. & Crompton, J.L. 1991. Image differences between prospective, first-time, and repeat visitors to the Lower Rio Grande Valley. Journal of Travel Research, 30 (2), 10–16.

Herstanti, Ghassani et al, 2014. Three Modified Models to Predict Intention of Indonesia to Revisit Sydney. European Journal of Bussiness and Mnagement. Vol 6, No. 25

Hwang, B. Y., Jun, H. J., Chang, M. H., & Kim, D. C. (2018). Efficiency Analysis of the Royalty System from the Perspective of Open Innovation. Journal of Open Innovation: Technology, Market, and Complexity 4(3): 22.

Kotler, P. 2002. *Marketing places*. New York: Simon and Schuster.

Kotler, P. & Armstrong, G. 2010. *Principles of marketing*. England: Pearson education.

Lee, J., Graefe, A.R., & Burns, R.C. 2004. Service quality, satisfaction, and behavioral intention among forest visitors. *Journal of Travel and Tourism Marketing* 17(1): 73–82.

Lee, S.Y., Huh, J., & Sung-Kwon, H. 2008. Determining behavioral intention to visit a festival among first-time and repeat visitors. *International Journal of Tourism Sciences* 8(1): 39–55.

Liu, X., Li, J., & Kim, W.G. 2017. The role of travel experience in the structural relationships among tourists' perceived image, satisfaction, and behavioral intentions. *Tourism and Hospitality Research* 17(2): 135–146

Mittal, B. & Walfried M.L. 1998. Why do customers switch? The dynamics of Satisfaction versus Loyalty. *Journal of Services Marketing* 12 (3): 177–194.

Oh, H. 2000. The effect of brand class, brand awareness, and price on customer value and behavioral intentions. *Journal of Hospitality and Tourism Research* 24(2): 136–162.

Pratminingsih, S.A., & Puspitasari, D.M. 2015. Developing Bandung as culinary destination. *The Fourth International Conference on Entrepreneurship and Business Management (ICEBM 2015), Universitas Tarumanagara, Universiti Sains Malaysia, Dusit Thani College, Kun Shan University, Universitas Ciputra*.

Pratminingsih, S.A. & Boediprasetya, A. 2014. Predicting consumer behavioral intention in Bandung ethnic restaurant. *Annual Symposium on Management and Social Sciences (ASMSS), Asia-Pacific Conference on Management on Business (APCMB), Korea E-trade Research Institute (KETRI)*.

bibitem Prayag, G. & Ryan, C. 2012. Antecedents of tourists' loyalty to Mauritius: The role and influence of destination image, place attachment, personal involvement, and satisfaction. *Journal of travel research* 51(3): 342–356.

Petrick, J.F. 2004. The roles of quality, value and satisfaction in predicting cruise passengers' behavioral intentions. *Journal of Travel Research* 42(4): 397–407.

Petrick, J.F. & Backman, B. 2002. An exaniation of the construct of perceived value or the prediction of golf travelers' intentions to revisit. *Journal of TravelResearch* 41(1): 38–45.

Ryu, K., & Jang, S. (2006). Intention to experience local cuisine in a travel destination: The modified theory of reasoned action. Journal of Hospitality & Tourism Research, 30(4), 507–516.

Setiadi, N. 2013. *Perilaku konsumen : Perspektif kontemporer pada motif, tujuan, dan keinginan konsumen*, Jakarta: Kencana Prenada Media.

Silva, C., Kastenholz, E., & Abrantes, J. L. 2013. Place attachment, destination image and impacts of tourism in mountain destinations. *Anatolia* 24(1): 17–29.

Stylidis, D., Shani, A., & Belhassen, Y. 2017. Testing an integrated destination image model across residents and tourists. *Tourism Management* 58: 184–195.

Sultan, M.A., Haryono, T., Haryanto, B., & Riani, A.L. 2012. Proses Pembentukan Perilaku Niat Wisatawan Berkunjung Kembali. *Jurnal Siasat Bisnis*.107–118

Swan, J.E. & Trawick, I.F. 1981. Disconfirmation of expectations and satisfaction with a retail service. *Journal of Retail* 57(3): 49–67.

Tam, J.L.M. 2000. The effects of service quality, perceived value and customer satisfaction on behavioral intentions. *Journal of Hospitality and Leisure Marketing* 6 (4): 31–43.

Walters, G. 1974. *Consumer behaviour theory & practice*. Ilinois: Richard D.Irwin Inc.

Wibowo, S.F., Sazali, A., & RP, A.K. 2016. The influence of destination image and tourist satisfaction toward revisit intention of Setu Babakan Betawi Cultural Village. *JRMSI-Jurnal Riset Manajemen Sains Indonesia* 7(1), 136–156.

Advances in Business, Management and Entrepreneurship – Hurriyati et al. (Eds)
© 2021 Taylor & Francis Group, London, ISBN 978-0-367-67471-7

Online shopping usage behavior analysis

M. Amin, R. Hurriyati & M.A. Sultan
Universitas Pendidikan Indonesia, Bandung, Indonesia

ABSTRACT: This study aims to analyze behavioral variables, Behavior intention to use online shopping (X1), Perceive behavioral control using online shopping (X2), Attitudes toward behavior online shopping (X3), perceived ease of using online shopping applications (X4) and Perceive usefulness online shopping (X5) as a predictor of the use of online shopping and correlation between variables. Based on the testing of multiple linear regression, the results showed that the dependent variable behavior intention to use online shopping, perceive behavior control using online shopping behavior, attitudes toward the use of online shopping, perceived ease of using online shopping applications and the perception of the benefits of using online shopping could be predictors of actual system use online shopping. Based on the results of the validity test with the Pearson correlation method, the results show that there is a relationship between each variable.

1 INTRODUCTION

In Technology Acceptance Model, Davis (1989) found that actual system use and behavior intention have an influence on the use of a technology. Attitude toward behavior using technology and perceive usefulness variable have a positive influence on the intention to use technology. Perceive ease of use in using technology affect the behavior and perceive usefulness of using technology.

In this study, perceptions of behavioral control variables will be added and the extent to which they can be predictors of behavior using online shopping applications.

1.1 Theory Acceptance Model (TAM)

Theory Acceptance Model by Davis (1989) is used to find out why consumers / individuals accept or reject an information technology. Theory Acceptance Model assesses whether there is a causal relationship between design system features, perceived usefulness, perceived ease of use, attitudes toward behavior, behavior intention and actual system use. TAM model is as shown in Figure 1.

1.2 Bank Indonesia's initiative on fintech

1. Facilitator: Bank Indonesia is a facilitator in terms of providing land for payment traffic Intelligent business analyst. In collaboration with the authorities and international agencies, Bank Indonesia is an analyst for FinTech related businesses to provide views and direction on how to create a safe and orderly payment system.

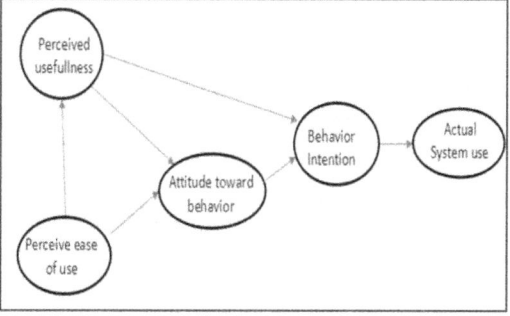

Figure 1. Theory acceptance model (Davis, 1989).

2. Assessment: Bank Indonesia conducts monitoring and assessment of each business activity involving FinTech and its payment system using technology.
3. Coordination and Communication: Bank Indonesia maintains relations with relevant authorities to support the existence of FinTech payment systems in Indonesia. Bank Indonesia is also committed to supporting business people in Indonesia by providing regular guidance on FinTech (Bank Indonesia website).

2 METHODS

2.1 General description of respondents

Questionnaires were distributed to consumers in various cities in Indonesia online through the google form application from 4 until 12 April 2019 with 131 questionnaires filled out by respondents. The general description of the respondent's data is as follows:

Table 1. Profil responden.

Gender:	Amount
Male	76
Female	53
Age:	
<59 year (baby boomers)	2
39–58 yers (generation Z)	79
25–38 year (generation Y)	31
9–24 year (generation X)	8
Education:	
Elementary School	–
Middle School	1
High School	13
University	113
Domicile:	
Sumatera	11
Jawa	108
Bali	1
Kalimantan	5
Papua	4
Overseas (Jeddah)	1

* Source: questionnaire data.

2.2 Operational variables

Actual system used variables online shopping (Y), which were the factors that caused consumers used online shopping applications with available financial technology facilities.

Behavior intention online shopping (X1) was a process when the consumers considered the price, product completeness, time, effectiveness and other factors before decided to use online shopping

Perceive behavior control (X2) was a social factor that encouraged consumers to use online shopping applications such as social influences from closest person, family, and other factors that influenced consumer behavior control.

Attitudes toward behavior (X3) is how consumers acted based on their beliefs that became the background factors such as age, gender, religion, education, experience, information from electronic media, price of goods and other factors.

Perceive usefulness online shopping application (X4) is benefits obtained by consumers in online shopping, in terms of more efficient time, cheaper prices, the opportunity to choose more and varied products, system online payment (fintech).

Perceive ease of use online shopping (X5) is about convenience perceived by consumers in using online shopping applications such as the ease of getting online applications, flexibility, understanding in using online shopping applications. Perceive ease of use online shopping was an individual belief that the use of the application was not trouble-some or did not require a large effort.

2.3 Test validity and reliability of the questionnaire

Validity test using the Pearson correlation method for each variable obtained on Pearson's correlation level where the result showed above 0.254, so that it was declared valid. The value of r statistical table with two-sided test; $N = 131$ or $df = 131 - 2 = 129$ with a significant level of 0.01 was 0.254. The value of r calculated correlation between the smallest variables was the correlation between Intention (X1) and Perceive ease of use (X4) which was equal to 0.506. So that, all variables were declared valid because the correlation value was greater than 0.254.

Reliability test of the questionnaire with Cronbach Alpha method was used by generating a cronbach alpha value of 0.901, which meant the level of reliability of the variable was good because it was above 0.8.

2.4 Residual normality test, multicollinearity test and autocorrelation test

Residual normality test to find out whether the Actual system using online shopping applications (Y) had a residual value that was normally distributed, e.g. data spread around the diagonal line and followed the direction of the normal P-P Plot of Regression Standardized Residual.

Multicollinearity test between the independent variables and SPSS, with the tolerance values of variables X1, X2, X3, X4 and X5 were above 0.1 and VIF less than 10, which meant that the online shopping model did not have a multicollinearity problem.

Autocorrelation test was to find out whether there was a correlation of the residuals for observations between variables by performing the Durbin-Watson test (DW test). The Dublin Watson value calculated at 1.936 was between the values of $dU = 1.693$ and $4\,dU = 6.722$. The dL and dU values in the Durbin Watson table at the significance of 0.05 with the number of data ($n = 131$) and the number of independent variables ($k = 5$) obtained results namely $dL = 1.557$, $dU = 1.693$, $4dU = 6.772$ and $4dL = 6.228$. This meant that there was no problem of autocorrelation in the use of online shopping behavior model.

2.5 Analysis of multiple linear regression

Multiple linear regression equations of the independent variables and dependent variables were as follows:

$$Y = b0 + b1X1 + b2X2 + b3X3 + b4X4 + b5X5$$
(1)

Where Y = Actual System Use (Y); b0 = constanta; b1, b2, b3, b4, b5 = regression coefficient; X1 = Behavior

Intention; X2 = Perceive behavior control; X3 = Attitude toward behavior; X4 = Perceive ease of use; X5 = Perceive usefullness.

Multiple regression coefficient equation results were as follows:

$$Y = 0{,}747 + 0{,}347X1 - 0{,}032X2 + 0{,}024X3 \\ + 0{,}45X4 + 0{,}247X5 \qquad (2)$$

2.6 *Analysis of the coefficient determination and F table r square test*

Analysis of R Square or the coefficient of determination was used to find out how much the percentage of the influence of independent variables together on the independent variable. From the regression calculation using SPSS the adjusted R Square value was 0.731 which meant that the contribution of the influence of the independent variable was 73.1% on the dependent variable while the rest was influenced by other factors not examined.

The F test was used to test the effect of independent variables together on the dependent variable. From the calculation obtained F count 71.581, while t table at the significance level of 0.05 with df1 $= k - 1$, where k was the number of independent variables so that df1 $= 5 - 1 = 4$. Then df2 $= n - k - 1$ where n was the number data from respondents is 131, so that df2 $= 131 - 5 - 1 = 125$ and F table obtained a value of 2.444. The criterion was that if F count > from F table then H was rejected which meant that the independent variables together had the same effect on the dependent.

3 RESULTS & DISCUSSION

There was a correlation between variables with a correlation value above 0.25 so that this study was valid and all independent variables including perceived behavior control online shopping had an influence on the dependent variable.

Based on the testing of multiple linear regression, the F test results obtained, independent variable was behavior intention to use online shopping (X1), perceive behavior control online shopping (X2), attitude toward behavior online shopping (X3), Perceive ease of use online shopping (X4) and Perceive usefulness online shopping (X5) together affected the actual system use online shopping (Y) as shown in Figure 2.

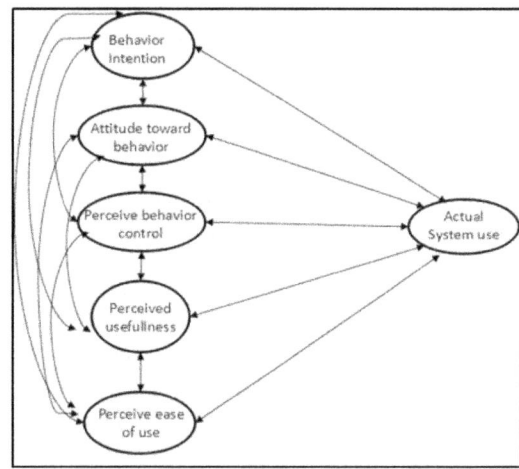

Figure 2. Online shopping usage analysis model.

4 CONCLUSION

This research is conducted to analyze the use of online shopping applications and the correlation between independents variable and dependent variable. All independent variables have an influence on the dependent variable. Perceive behavioral control variables that adds in the analysis model together with other variables has a positive influence on Actual system use online shopping applications.

ACKNOWLEDGEMENTS

Thank you to Ratih Hurriyati and M. Adib Sultan as a lecturer in Management Science at the Bandung University of Indonesia, who has provided guidance in research.

REFERENCE

Davis, F. 1989. Perceive usefulness Perceive ease of use and user acceptance technology model. *MIS Quarterly*: 319–340.

Advances in Business, Management and Entrepreneurship – Hurriyati et al. (Eds)
© 2021 Taylor & Francis Group, London, ISBN 978-0-367-67471-7

The influence of perceived service quality and perceived price on students' satisfaction and recommendation

N.A. Sugiharto & R. Hurriyati
Universitas Pendidikan Indonesia, Bandung, Indonesia

ABSTRACT: This study examines the influence of perceived service quality and perceived price on students' satisfaction and recommendation in higher education institution. The sample consisted of 155 students who majority had learnt in Poltekpos and STIMLOG of Bandung. The sample was taken by convenience sampling. The data was collected by survey with questioner method. Path analysis with Amos technique was the statistical method chosen to measure influence of perceived service quality and perceived price on students' satisfaction for recommendation to others. The results indicate that perceived service quality has a significantly positive influence on students' satisfaction, but perceived price has negative influence on students' satisfaction. The result also indicate that students' satisfaction has significantly positive influence on recommendation to others.

1 INTRODUCTION

Student satisfaction is important to be discussed, considering that there is a good effect if the students are satisfied, and vice versa (Sumaedi et al. 2011).

"Psychologists have found that student satisfaction helps to build self-confidence, and that self-confidence helps students develop useful skills, acquire knowledge". On the other hand, student's dissatisfaction can lead to negative activities, such as a bad grade, an unpleasant relationship between the student and the staff, faculty, and friends (Athiyaman 1997; Letcher & Neves 2010;. Based on the previous explanation, it is important to discuss about factors which determine students' satisfaction.

Students' perceived service quality has more effect on student's satisfaction compared to students' perceived price fairness (Sumaedi et al. 2011). Students who had satisfactory on educational experiences are expected to recommend the university to others, and this is also an important predictor of attending the same institution in the future (Boulding et al. 1993).

Study on influence of perceived service quality and perceived price on students' satisfaction has been done a lot, including whether after students satisfied, will the students give a recommendations to others.

The purpose of this study is to examine how the influence of perceived service quality and price perception on student satisfaction and on student recommendations to others in higher education institutions.

1.1 Students' perceived service quality

Perceived quality is a general overall appraisal of service (Ismail et al. 2009). Perceived quality should be conceptualized as "similar to an attitude" approach (Cronin & Taylor 1992). Perceived quality is defined as the difference between customer expectation and customer perception towards service performance. If customer perception is higher than customer expectation, the customer will have higher perceived quality and vice versa (Parasuraman et al. 1988, 1994). While the perceived service quality is defined as "a global judgment, or attitude, relating to the superiority of the service" (Parasuraman et al. 1988). Kang & James (2004) assert that perceived service quality is the core issue of service quality in the services marketing literature.

1.2 Students' perceived price

Price is the amount of money or goods needed to acquire some combination of other goods and its companying services (Hanif et al. 2010). Furthermore, according to Kotler & Armstrong (2010), Hanif et al. (2010) price is the amount of money charged for a product or service, or the sum of the values that customer exchange for the benefits of having or using the product or service. In the other hand, perceived price defines as customer perception about what is sacrificed to obtain a product or service (Aga & Safakli 2007; Lien & Yu 2001; Zeithaml 1988). According Lien & Yu (2001), perceived price can be measured by the fairness of price to be paid. Thus, the more reasonable or the cheaper the price paid, the more satisfied the customer on the price of a product or service (Clemes et al. 2008).

1.3 Students' satisfaction

Satisfaction as the emotional evaluation shows how far consumers believe that the use of the services can generate positive feelings. This means customer satisfaction is related to customer's emotional evaluation. Furthermore, some experts, such as (Oliver 1980; Tse et al. 1988; Yi 1990) believe that customer's satisfaction lies in the "disconfirmation of consumer expectations" paradigm while a positive disconfirmation leads to customer satisfaction and negative disconfirmation will lead to customer dissatisfaction (Ismail et al. 2009; Jamali 2005). This means satisfaction is a function of customer experience and expectations of various services outcomes.

1.4 Recommendation

Study involved university students and identified strong links between service quality and favourable future behavioural intentions and their strategic importance to the university. The favourable future behavioural intentions included praising the university, planning to pledge to contribute money to the class upon graduation, and planning to recommend to an employer as a good place from which to recruit (Boulding et al. 1993).

Students who had satisfactory educational experiences are expected to recommend the university to others, and this is also an important predictor of attending the same institution in the future (Boulding et al. 1993).

The Intention to Attend the same university in the future is positively influenced by students' level of Satisfaction. Similarly, the Intention to Recommend the university to other people is also positively influenced by students' level of satisfaction (Kao 2007).

Bone (1992) has contended that high levels of service quality leads to perceived value as well, consequent increased satisfaction and stimulates positive word of mouth, whilst researchers such as Soares & Costa (2008) have empirically demonstrated perceived value to be consistent antecedent of word-of-mouth activity. Much of the literature has focused on positive behavioural outcomes resulting from positive assessments of value.

2 METHODS

Lien & Yu (2001) reported, "In most service industry marketing literature, perceived service quality captures the spot light, while perceived product quality is absent. For most service industries providing intangible service and tangible goods, these two forms of products both play important roles in consumer satisfaction and loyalty". As educational institution can be categorized as pure services (Oldfield & Baron 2000; Solomon et al. 1985).

The exploratory and confirmatory factor analyses have shown that service quality is made up of six latent dimensions (teaching, administrative services, academic facilities, campus infrastructure, support services and internationalization). These results could help leaders of institution to better recognize

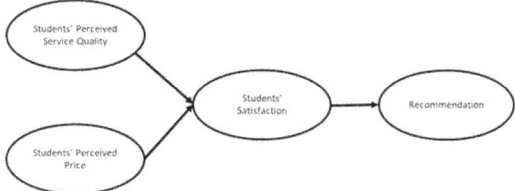

Figure 1. Research model.

the factors contributing to service quality, so that they are able to discreetly provide better services that enhance student satisfaction, motivation and loyalty (Annamdevula & Bellamkonda 2014).

Student perceived price was measured by using statement about the total price/cost should be borne by the students, including the tuition fee, cost of books and teaching materials that modified from the work of (Kao 2007).

Parasuraman et al. (1994) satisfaction model, whereas, customer satisfaction was influenced by three abstract concepts, i.e. perceived product quality, perceived service quality and perceived price. On certain service industries, the characteristic of their product made it possible to combine two variables (perceived product quality, perceived service quality) into a single variable, namely perceived service quality (Clemes et al. 2008; Natalisa & Subroto 2003). Moreover, Lien & Yu 2001).

The Intention to recommend (future attendance) construct was measured using two item, they were intention to choose some university and intention to continue for further studies at the same university (Kao 2007).

The research model showing relationship between all variables which have been discussed above is presented in figure 1.

The following hypotheses are:

H1. Students perceived service quality has a significantly positive influence on students' satisfaction.
H2. Students perceived price has a negative influence on students' satisfaction.
H3. Students perceived service quality and perceived price together have significantly positive influence on students' satisfaction.
H4. Students' satisfaction has a significantly positive influence on recommendation.

2.1 Sample and data collection

The sample for this study was 155 students who majority had learnt in Politeknik Pos Indonesia and Sekolah Tinggi Manajemen Logistik Bandung. Data were collected through online questionnaire. In measuring the items representing the constructs shown in the research model, the study used the multi-scaling method, namely likert scale (7-point).

Table 1. Regression weights.

			Estimate	P
Student' Satisfaction	←	Perceived service quality	1.059	***
Student' Satisfaction	←	Perceived price	−.036	-407
Recommendation	←	Student' Satisfaction	.897	***

*Result of statistical test 2019.

2.2 Data analysis method

This study used path analysis with AMOS and SPSS version 23 test. First of all, validity and reliability test were conducted. Validity test was conducted by item analysis (item to total analysis) approach; meanwhile reliability test was conducted by alpha Cronbach approach. Second, some data assumption tests, i.e. normality test, multicolinearity test, autocorrelation test, and linearity test, were conducted. Lastly, a multiple regression analysis was conducted to test the hypothesis. All data analysis was conducted with 95% significant level.

3 RESULTS AND DISCUSSION

3.1 Validity, reliability, and normality test result

From the validity test, it showed that each the question items had significant levels (p-value) less than 0.05. This means that each question items is valid. From the reliability test, the alpha value for each constructs ranged from 0.763 until 0.914 which was greater than 0.7. This means that the measurement scales are stable and consistent in measuring the construct. The normality test used in this study was Kolmogorov-Smirnov method and from the test conducted, the result obtained asymp.sig. (Alpha) values of 200, which was greater than 0.05. This means that the curve of standardized residual value shapes a normal spread.

3.2 Hypotheses testing

The results are shown in table 1.

1. Positive and significant influence between students perceived service quality and students' satisfaction (support the hypotheses H1).
2. Negative influence between students perceived price and students' satisfaction (support the hypotheses H2).
3. Positive and significant influence between students' satisfaction and recommendation (support the hypotheses H4).

Furthermore, the results for hypotheses H3 with SPSS test can be seen on table 2.

From table 2 appears that students perceived service quality and perceived price together have positive

Table 2. Anova results for regression analysis.

ANOVA[a]

Model	Sum of Squares	df	Mean Square	F	Sig.
Regression	168.413	2	84.207	151.063	.000[b]
Residual	84.729	152	.557		
Total	253143	154			

a. Dependent variable: Student' satisfaction.
b. Predictors: (Constant), students' perceived price, students' perceived service quality.

significantly influence on students' satisfaction (support the hypotheses H3).

3.3 Discussion

This research shows student perceived quality has a positive significant influence toward student satisfaction. Thus, any increase in student perceived quality will result in an increase of student satisfaction. Previous literatures showed that student perceived quality had a positive significant effect on student satisfaction (Hanief et al. 2010; Letcher & Neves 2010; Yunus et al. 2010).

Another finding of this research is the student perceived price has a negative influence on student satisfaction. It means that any increase in student perceived price will result in the decline of student satisfaction.

This finding is in line with the findings of some service marketing researcher like (Malik et al. 2012) who said that brand image, service quality and price are correlated to customer satisfaction. Increasement in price has shown to have a negative impact on customer satisfaction. Kao (2007) shows that student perceived price has no impact on student satisfaction by several reasons. First, this research is conducted in a developing country. Meanwhile, Kao's research is conducted on developed country. In developed countries, education cost is not a big deal for them compared to those on developing countries. This makes the student in developing countries more concern about education cost. Second, according to Kao (2007), the students who become the object of his study didn't pay the tuition fee by themselves or just paid in some portion, so they lack of sensitiveness towards this matter.

The influence of Students' satisfaction on recommendation in several findin also described as word-of-mouth intention, was an issue of considerable interest to researchers and marketing practitioners. Whilst early research in this area tended to focus on the negative aspects like customer complaining behaviour (Gronhaug & Kvitastein 1991), the focus had swiftly progressed toward investigating the factors that influence customers to make positive recommendations.

In addition, According to Lam et al. (2004), customer loyalty was evaluated by two dimensions:

recommend and patronage. Customer loyalty (recommend) can be understand as follow: after customers satisfy about the service of this firm, they will recommend or will tell other people about this service and the other people can start using this service through their recommend in the other hand Customer loyalty (patronage) state that after customers satisfy with this service, they may continue using this service or using more services of this firm.

4 CONCLUSSION

The research result shows that students' perceived service quality has a positive significantly influence on student's satisfaction, and students' satisfaction has positive significantly influence on recommendation to others. Besides that, students' perceived price has a negative influence on student's satisfaction, so service quality has more effect on student's satisfaction. Hence, in order to generate student satisfaction, the managers of higher education institution should always consider and improve their students' perceived quality with fairness price.

REFERENCES

Aga, M. & Safakli, O.V. 2007. An empirical investigation of service quality and customer satisfaction in profesional accounting firms: Evidence from North Cyprus. *Problems and Perspectives in Management* 5(3).

Annamdevula, S. & Bellamkonda, R.S. 2014. Effect of student perceived service quality on student satisfaction, loyalty and motivation in Indian universities Development of Hieduqual. *Journal of Modelling in Management*.

Athiyaman, A. 1997. Linking student satisfaction and service quality perceptions: The case of University Education. *European Journal of Marketing*.

Bone, P. F. 1992. Determinants of WOM communication during product consumption. *In Sherry, J.F., Sternthal, B. (Eds), Advances in Consumer Research*: 579-83).

Boulding, W., Kalra, A., Staelin, R., & Zeithaml, V.A. 1993. A dynamic process model of service quality: from expectations to behavioral intensions. *Journal of Marketing Research* 30 (1).

Clemes, M.D., Gan, C., Kao, T.H., & Choong, M. 2008. An empirical analysis of customer satisfaction in International Air Travel. *Innovative Marketing* 4(2).

Cronin, J.J. & Taylor, S.A. 1992. Measuring service quality: a reexamination and extension. *Journal of Marketing* 56(3).

Gronhaug, K. & Kvitastein, O. 1991. Purchases and complaints: A logitmodel analysis. *Psychology & Marketing* 8(1).

Hanif, M., Hafeez, S., & Riaz, A. 2010. Factors affecting customer satisfaction. *International Research Journal of Finance and Economics* 60(1), 44-52.

Ismail, A., Madi, M., & Francis, S.K. 2009. Exploring the relationships among service quality features, perceived value and customer satisfaction. *JIEM* 2(1).

Jamali, A. 2005. Study of customer satisfaction in the context of a public private partnership. *International Journal of Quality & Reliability Management* 24 (4).

Kang, G.D. & James, J. 2004. Service quality dimensions: An examination of Gronroos's service quality model. *Managing Service Quality* 14(4).

Kao, T.H. 2007. University students' satisfaction: An empirical analysis. *Master of Commerce and Management, Thesis Lincoln University*.

Kotler, P. & Armstrong, G. 2010. Principles of marketing. Pearson education.

Lam, S.Y., Shankar, V.M., Erramilli, K., & Murthy, B. 2004. Customer value, satisfaction, loyalty, and switching costs: an illustration from a business-to-business service context. *Journal of the Academy of Marketing Science* 32(3).

Letcher, D.W. & Neves J.S. 2010. Determinant of undergraduate business student satisfaction. *Research in Higher Education Journal*.

Lien, T.B. & Yu, C.C. 2001. An integrated model for the effects of perceived product, perceived service quality, and perceived price fairness on customer satisfaction and loyalty. *Journal of Consumer Satisfaction, Dissatisfaction, and Complaining Behaviour* 14.

Malik M.E., Ghafoor, M.M., & Iqbal, H.K. 2012. Impact of brand image, service quality, and price on customer satisfaction in Pakistan telecommunication sector. *International Journal of Business and Social Science*.

Natalisa, D. & Subroto, B. 2003. Effects of management commitment on service quality to increase customer satisfaction of domestic airlines in Indonesia. *Singapore Management Review* 25 (1).

Oldfield, B.M. & Baron, S. 2000. Student perceptions of service quality in a UK university business and management faculty. *Quality Assurance in Education* 8(2).

Oliver. 1980. A cognitive model of the antecedents and consequences of satisfaction decision. *Journal of Marketing Research* 17.

Parasuraman, A., Zeithaml, V.A., & Berry, L.L. 1988. Servqual: a multiple-item scale for measuring consumer perceptions of service quality. *Journal of Retailing* 64(1).

Parasuraman, A., Zeithaml, V.A., & Berry, L.L. 1994. Reassessment of expectations as a comparison standard in measuring service quality: implications for further research. *Journal of Marketing* 58.

Soares, A.A.C. & Costa, F.J. 2008. The influence of perceived value and customer satisfaction on the word of mouth behaviour: An analysis in academies of gymnastics. *Review of Business Management* 10(28).

Solomon, M.R., Surprenant, C., Czepiel, J.A., & Gutman, E.G. 1985. A role theory perspective on dyadic interactions: The service encounter. *Journal of marketing* 49(1): 99-111.

Sumaedi, S., Yuda Bakti, I.G.M., & Metasari, N. 2011. The effect of student's perceived service quality and perceived price on student satisfaction. *Management Science and Engineering*.

Tse, D.K., David, K., & William, P.C. 1988. Model of consumer satisfaction formation: An extension. *Journal of Marketing Research* 25.

Yi, Y. 1990. Cognitive and affective priming effects of the context for print advertisements. *Journal of Advertising* 19(2).

Yunus, N.K.Y., Ishak, S., & Abdul Razak, A.Z.A. 2010. Motivation, empowerment, service quality and polytechnic students' level of satisfaction in Malaysia. *International Journal of Business and Social Science* 1(1)

Zeithaml, V.A. 1988. Consumer perceptions of price, quality and value: a means-end model and synthesis of evidence. *Journal of Marketing* 52.

Effect of viral marketing on changes in consumer behavior and decision in using smartphone

S. Santoso, R. Hurriyati & M.A. Sultan
Universitas Pendidikan Indonesia, Bandung, Indonesia

ABSTRACT: This research evaluates the effect of Viral Marketing on changes in community behavior and the decision to use smartphones based on the Technology Acceptance Model. There were 100 questionnaires distributed to respondents who were active social media users seeking for information about smartphones. This research was conducted due to the phenomenon in the smartphone user community in Indonesia that considers smartphones as the fourth basic necessity in addition to clothing, food, and housing. This research employed a survey approach by applying questionnaires and statistical analysis, especially quantitative descriptive analysis using SmartPLS. The results showed that Trust had a strong influence on Perceived Usefulness and Behavioral Intention to Use. Social Norm had an effect on Perceived Usefulness and Perceived Ease to Use. Behavioral Control had an effect on Perceived Ease to Use and Perceived Usefulness. It is also seen that Behavioral Intention to Use was strongly influenced by Perceived Usefulness.

1 INTRODUCTION

The phenomenon of the smartphone existence draws the interest of marketing researchers. The most popular research topics focus on the application of new technologies in various fields and the adoption of new technologies. Smartphones are very useful in the field of medicine (Parker et al. 2010), agriculture (Molina-Martíneza & Ruiz-Canalesb 2009), and the food industry (Mattolia et al. 2009).

In mid-2000, special devices to access cellular internet appeared. The first smartphone model was available in 2002, but smartphones were extensively used since 2008, especially after the release of Apple's first iPhone model (LaRue et al. 2010). Smartphones have changed the way of communicating between individuals, especially young people. The many features provided by the producers have pampered the users. Smartphones that facilitate all media and communication activities in one hand have resulted in the teenagers get increasingly carried away by the technology in their smartphones, in contrast to the earlier generations. Especially for teenagers, smartphones are now a part of their lives. In fact, a large number of teenagers depend on their smartphones. There are many benefits that the smartphone users can enjoy. However, the negative impacts they carry could be detrimental. All this is due to globalization.

The Technology Acceptance Model (TAM) was developed from a theory in psychology that explains the behavior of computer users based on the beliefs, attitudes, intentions, and relationships of user behavior. The purpose of the model is to explain the main factors of Information Technology user behavior in the actual use of Information Technology in more detail. The acceptance level of information technology users is determined by six constructs, namely external, perceived ease of use, perceived usefulness, Attitude toward using, behavioral intention, and actual usage variables (Davis 1989). TAM states that perceived usefulness is influenced by social norms and behavioral control whereas perceived enjoyment is not influenced by them. Teenagers who use smartphones from utilitarian needs are not hedonic. (Alt et al. 2012).

Aghdaie et al. (2012) evaluated the effect of consumer trust on the acceptance of viral marketing. By applying TAM, this goal has been followed by examining the relationship of Trust as external variables and Perceived Usefulness, Perceived Ease of Use, Attitude toward Using, Intention, and Actual Usage as the original variable of the model. Given that consumer involvement in viral marketing campaigns is only referred to as users, Perceived Ease to Use and Trust are only the independent variables, and Actual Usage is only the dependent variable, the remaining research variables include Perceived Usefulness, Attitude toward Using, and Intention to Use play both roles. From the results of the hypothesis, it shows that Trust does not affect Perceived Usefulness, however Trust influences Attitude and Intention. The purpose of this study was to determine the effect of attitudes and intentions in determining the choice to use Smartphone.

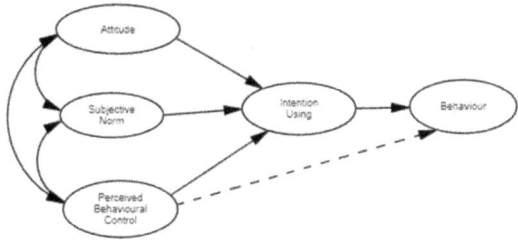

Figure 1.   The theory of planned behavior (Ajzen 1991).

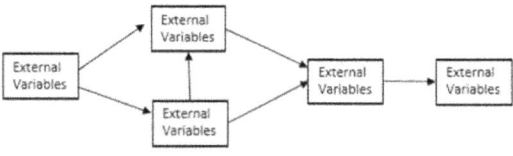

Figure 2.   Final version of Technology Acceptance Model (TAM) (Venkatesh & Davis 1996).

1.1   Literature review

1.1.1   Planned behavior theory

Planned Behavior Theory (Ajzen 1985, 1987) is the development of the Theory of Reasonable Action (Ajzen & Fishbein 1980) stating that the most important factor that will produce behavior is individual intention. On the other hand, the intention factor is influenced by three interconnected things, namely attitude, subjective norms, and perceived behavioral control. Perceive behavioral control refers the consumer or individual perceptions about ease or difficulty in performing a behavior.

Ajzen (1991) developed the Theory of Planned Behavior that is about three factors that determine the behavioral intention of a person's attitude towards that behavior. The first two factors are the same with the Theory of Reasonable Action (Fishbein & Ajzen 1975). The third factor, known as perceived behavioral control, is the control felt by the users that can limit their behavior.

1.1.2   Technology acceptance model

Technology Acceptance Model (TAM) was introduced by Davis (1986) for his doctoral proposal. As an adaptation of the Theory of Reasonable Action, TAM is specifically designed to model user's acceptance of systems or information technology. The final version of TAM was formulated by Venkatesh & Davis (1996), as depicted in Figure 2, after their main findings that showed that both Perceived Usefulness and Perceived Ease of Use had direct influence on Behavior Intention, thus eliminating the need for constructive attitudes.

1.1.3   Viral marketing

Viral marketing is "marketing techniques to exploit pre-existing social networks to increase brand awareness, through a process similar to the spread of epidemics" (Datta et al. 2005). Viral marketing can be

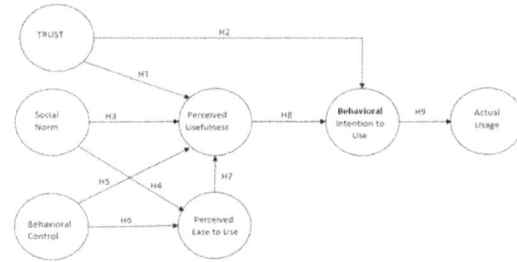

Figure 3.   Conceptual model.

interpreted as "increasing the network using words" of the mouth (Datta et al. 2005).

Viral marketing essentially means the rebirth of word of the mouth, making breakthroughs, and communication strategies. "Viral marketing operates by drawing good friends into other people's marketing targets in the database" (Bulkeley 2002). Richardson & Bachman (2004) wrote that the term viral marketing was popularized for the first time by Steve Juvertson. Juvertson and his colleagues were the owners of Hotmail and the line "Get your personal email for free on Hotmail" was their idea. Hotmail easily created a tribute process, Juvertson's process of viral marketing in 1997 in the space bulletin explained the phenomenon of Hotmail's success.

2   METHODS

2.1   Hypothesis

The purpose of this research is to evaluate the effect of Viral Marketing on changes in community behavior and decision on using smartphone. TAM was applied to find out the relationship between Trust, as an external variable, and Social Norms and Behavior Control. Perceived Usefulness, Perceived Ease of Use, Intention to Use, and Actual Use are the original variables. The model of the research considers the consumer involvement in the promotions through viral marketing which will then lead the consumers to use smartphones, as depicted in Figure 3.

Trust, Social Norms, and Behavior Control are the independent variables. Actual Use is the dependent variable. Other variables, namely Perceived Ease of Use and Behavior Intention to Use have strong roles in the Actual Use.

2.2   Research methodology

This research employed quantitative, descriptive, and verification methods. Data were obtained by distributing questionnaires. A total of 100 respondents were active Instagram users who lived in various regions in Indonesia. The data obtained were processed using the SmartPLS application.

This research can be regarded an as applied research from a goal perspective and descriptive survey with respect to the nature and method. As the research instrument, a self-managed questionnaire consisted of a series of questions, which were designed and used to collect data to analyze the hypotheses quantitatively.

3 RESULTS AND DISCUSSION

3.1 *Respondents*

The demographic profile of the 100 respondents is as follows: male (53%) and female (47%); age 13–20 years old (24%), 21–35 (25%), 36–45 (21%), 46–55 (21%), and 55 and older (9%); senior high school education level (14%), diploma-3 (21%), bachelor degree (36%), and master degree (47%); experience using smartphone 1–5 years (23%), 6–10 years (47%), 11–15 years (32%), 16–20 years (14%), and 20 years and more (2%). All of the questions used 7-point Likert scale (1 = "Low" and 7 = "High").

The respondents were 100 people between 13 to 55 years old. The questionnaires were distributed to

Table 1. Demographic profile of respondents.

Variable	Frequency	Percent
Gender		
Male	53	53%
Female	47	47%
Age (years)		
13–20	24	24%
21–35	25	25%
36–45	21	21%
46–55	21	21%
More than 55	9	9%
Education		
Senior High School	14	14%
3-year diploma	21	21%
Bachelor degree	36	36%
Master	47	47%
Doctoral	0	
Experience Using a Smartphone		
1–5	23	23%
6–10	47	47%
11–15	32	32%
16–20	14	14%
20 year or more	2	2%

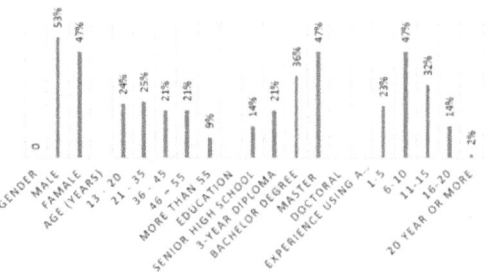

Figure 4. Demographic profile of respondents.

Table 2. Outer loading.

Variable	Outer Loading	Variable	Outer Loading
TR1	0.819	PU3	0.883
TR2	0.806	PU4	0.713
TR3	0.804	PEU1	0.835
TR4	0.716	PEU2	0.906
SN1	0.743	PEU3	0.841
SN2	0.874	BIU1	0.752
BC1	0.854	BIU2	0.910
BC2	0.758	BIU3	0.802
BC3	0.782	AU1	0.939
PU1	0.888	AU2	0.920
PU2	0.912		

the respondents spread across various locations in Indonesia using Google Forms.

3.2 *Measurement*

3.2.1 *Outer model test*

In the initial stage, the model was tested using PLS to determine the presence of inter construction collinearity and model prediction capabilities (Marko et al. 2017). Evaluation of the model is presented in the following indicators:

3.2.1.1 Reliability indicator

The purpose of Reliability Measurement Indicators is to assess whether the indicator of the latent variable is reliable or not. Whether the indicators are reliable or not can be seen from the loading value out of each indicator. A loading value above 0.7 indicates that the construct can explain more than 50% of the indicator variant (Wong 2013).

Table 2 presents that the external loading value for all variables is above 0.7, which means that the construct is able to explain more than 50% of the indicator's variance.

3.2.1.2 Internal consistency reliability

Internal Consistency Reliability measures how indicators can measure latent constructs (Marko et al. 2017). The value used is composite reliability and Cronbach alpha. As for the value of composite reliability, Marko et al. (2017) stated that values between 0.6–0.7 were considered to have good reliability, and for Cronbach's alpha values are expected to be larger than 0.7 (Ghozali & Latan 2015).

Table 3. Cronbach's Alpha, rho_A, Composite Reliability, and Average Variance Extracted (AVE) Value.

Table 3 shows that the Cronbach's alpha values obtained for all variables are greater than 0.7 and for composite values for all variables are above 0.7. Thus, it can be assumed that the model has good reliability.

3.2.1.3 Convergent validity

Convergent validity is determined based on the principle that measurement constructs must be highly correlated (Ghozali & Latan 2015). Based on the value

Table 3. Cronbach's alpha, rho_A.

	Cronbach's Alpha	Rho_A	Composite Reliability	Average Variance Extracted (AVE)
Actual Usage	0.843	0.854	0.927	0.864
Behavioral Control	0.719	0.732	0.841	0.638
Behavioral Intention to Use	0.745	0.768	0.855	0.665
Perceived Ease To Use	0.826	0.837	0.896	0.742
Perceived Usefulness	0.874	0.905	0.913	0.727
Social Norm	0.489	0.520	0.793	0.658
Trust	0.797	0.799	0.867	0.620

Table 4. Coefficient of determination.

	R square	R square adjusted
Actual Usage	0.497	0.491
Behavioral Intention to Use	0.185	0.177
Perceived Ease to Use	0.467	0.451
Perceived Usefulness	0.540	0.526

of Average Variance Extracted (AVE), the expected AVE value is equal to 0.5 or more, which means that the construct can explain 50% or more of the item's variance. From Table 1, it can be seen that the AVE value obtained for each variable is above 0.5.

3.2.1.4 Discriminat validity

Discriminant validity has a purpose to determine reflective indicators which is a good construct measure based on the principle that each indicator must have a high correlation with the construct. Different construct measures cannot be highly correlated (Ghozali & Latan 2015). In the SmartPLS 3.2.8 Application, the discriminant validity test was carried out using the value of cross-loadings, and Fornell-Larcker Criteria (Henseler et al. 2015).

3.2.2 *Inner model test*

After testing the external model, the next step is to test the inner model or the predictive ability of the model. Model prediction capabilities can be assessed using two criteria, namely the coefficient of determination (R2), and cross-validated redundancy (Q2).

3.2.2.1 Coefficient of determination

This value is the way to assess how much endogenous constructs can be explained by exogenous construct. The expected value is between 0 and 1.

From Table 4, the coefficient of determination (R2) of this model is 0.399. Chin (1998) in Ghozali & Latan (2015) categorizes its value as moderate.

3.2.2.2 Cross-validated redundancy

This value is used to determine predictive relevance. The expected Q2 value is greater than 0, indicating that the model has accurate predictive relevance for certain constructions (Marko et al. 2017).

Table 5. Cross-validated redudancy.

	SSO	SSE	Q^2
Actual Usage	200.000	118.746	0.406
Behavioral Control	300.000	300.000	
Behavioral Intention to Use	300.000	267.972	0.107
Perceived Ease to Use	300.000	208.375	0.305
Perceived Usefulness	400.000	258.081	0.355
Social Norm	200.000	200.00	
Trust	400.000	400.000	

3.2.3 *Model fit*

To measure the fit of the model in SmartPLS, the value of the Standardized Root Mean Square Residual (SRMR) is used, which is the difference between the observed correlation and the model that states the correlation matrix. Thus, it is possible to assess the magnitude of the average difference between the observed and expected correlations as the absolute size of the fit criteria (model). The expected value is less than 0.1 or 0.08 is the fit criteria (Trial 2017).

4 CONCLUSION

This research has proven that the quality of Perception, Behavioral Control affected the Perceived Ease to Use and Perceived Usefulness. Actual usage was influenced by Behavioral Intention to Use and Trust, while Trust had a strong influence on Perceived Ease to Use, Behavioral Intention to Use and Perceived Usefulness. Cronbach's alpha value obtained by the variables was greater than 0.7 and for composite values almost all variables are above 0.7. Therefore, it can be assumed that the model was considered to have good reliability, except the Social Norm variable with the value of Cronbach Alpha = 0.489, Rho_A = 0.520. This is due to the difference in the respondents' location which means that there was different understanding of social norms between each locations. As a result, this research can be improved.

REFERENCES

Aghdaie, S.F.A., Sanayei, A., & Etebari, M. 2012. Evaluation of the consumers' trust effect on viral marketing acceptance based on the technology acceptance model. *International Journal of Marketing Studies* 4(6): 79.

Ajzen I. & Fishbein. 1980. *Understanding attitude and predicting social behavior*. New Jersey: Prentice Hall.

Ajzen, I. 1985. *From intentions to action: a theory of planned behavior. In J. Huhl, & J. Beckman (Eds.), Will; performance; control (psychology); motivation (psychology)*.

Ajzen, I. 1987. *Attitudes, traits, and actions: Dispositional prediction of behavior in personality and social psychology*.

Ajzen, I. 1991. The theory of planned behavior", Organizational Behavior and Human Decision Processes, Vol. 50.

Alt, M.A., Pal, Z., & Seer, L. 2012. Using the theory of technology acceptance model to explain teen-agers adoption of smartphone in Transylvania.

Bulkeley, W. 2002. E-commerce: Advertisers find a friend in viral marketing.

Datta, P.R., Chowdhury, D.N., & Chakaraborty, B.R. 2005. Viral marketing: new form of word of mouth through internet. *The Business Review, Cambridge*.

Davis, F. 1989. Perceived usefullness, perceived ease of use, and user acceptance technology model. *MIS Quarterly*.

Fishbein, M. & Ajzen,I 1975. Belief, attitude, intention, and behavior: an introduction to teory and reaserch.

Ghozali, I. & Latan, H. 2015. *Partial Least Squares, konsep, teknik, dan aplikasi menggunakan program smartpls 2.0 m3 untuk penelitian empiris*. Semarang: Badan Penerbit Universitas Diponegoro Semarang.

Henseler, J., Ringle, C.M., & Sarstedt, M. 2015. A new criterion for assessing discriminant validity in variance-based structural equation modeling. *Journal of the academy of marketing science* 43(1): 115–135.

LaRue, M.E., Mitchell, M.A., Terhorst, L., & Karimi, A.H. 2010. Assessing mobile phone communication utility preferences in a social support network. *Telematics & Informatics* 27(4).

Marko, S., Christian, M.R., & Joseph, F.H. 2017. Partial least squares structural equation modeling.

Mattolia, V., Mazzolai, B., Mondini, A., Zampolli, S. & Dario, P. 2009. Flexible tag datalogger for food logistics.

Molina-Martíneza, J.M. & Ruiz-Canalesb, A. 2009. Pocket PC software to evaluate drip irrigation lateral diameters with online emitters. *Computers and Electronics in Agriculture* 69(1).

Parker, A., Rubinfeld, I., Azuh, O., Blyden, D., Falvo, A., Horst, M., & Patton, P. 2010. What ring tone should be used for patient safety? Early results with a Blackberry-based telementoring safety solution. *The American journal of surgery* 199(3): 336–341.

Richardson, M.P. & Bachman, E. 2004. *Viral marketing dalam seminggu. Alih bahasa: Rekha Trimaryoan*. Jakarta: Prestasi Pustaka Publisher.

Trial, D. 2017. Model Fit. [Online]. Retrieved from https://www.smartpls.com/documentation/algorithms-and-techniques/model- fit.

Venkatesh, V. & Davis, F.D. 1996. A model of the antecedents of perceived ease of use: Development and test.

Wong, K.K. 2013. Partial Least Squares Structural Equation Modeling (PLS-SEM) Techniques Using SmartPLS.

Advances in Business, Management and Entrepreneurship – Hurriyati et al. (Eds)
© 2021 Taylor & Francis Group, London, ISBN 978-0-367-67471-7

Customer-based brand equity in Indonesian logistics courier company

D.L. Sumarna, R. Hurryati & M.A. Sultan
Universitas Pendidikan Indonesia, Bandung, Indonesia

ABSTRACT: The purpose of this study is to determine the correlation of Brand Factor to Customer-Based Brand Equity from logistics Courier Company in Indonesia. This research used Partial Least Square with sample data from an online survey of 100 respondents who had experience with online shopping in Indonesia. This study indicates positive impact and significant between perceived value to customer-based brand equity; positive impact but not significant between brand loyalty with customer-based brand equity; and positive impact and significant between brand awareness/association with customer-based brand equity.

1 INTRODUCTION

There is a diversity of research results regarding Customer-Based Brand Equity (CBBE). Aaker (1991) uses brand awareness, brand association, perceived quality, and brand loyalty to measure customer-based brand equity, which is also used by Lee et al. 2011; Haryanto 2015; Setiawan & Rachmawati 2017; and Nguyen & Luu 2018. Yoo et al. (2000) combines two variables, Brand Awareness and Brand Associations into one variable, namely Brand Awareness/Association to measure Brand Equity. Haryanto (2015) in their research, found a relationship between brand associations to brand equity and brand loyalty to brand equity, but they didn't find an influence between brand awareness to brand equity.

This study investigated the relationship between perceived quality, brand loyalty, and brand association/awareness with brand equity in Indonesian logistics courier companies.

1.1 Brand equity

The definition of Brand Equity according to Aaker (1991) is *"a set of assets and liabilities that add to or subtract from the product or service to a company and/or its customer."* At present, Brand Equity is based on two main perspectives, namely the financial and consumer perspectives (Nguyen & Luu 2018). In this study, what was discussed was Brand Equity based on the customer's perspective. Customer-Based Brand Equity is defined as the different impact of brand knowledge on consumer responses to brand marketing (Keller 1993).

Factors that influence Brand Equity are Brand Awareness, Brand Associations, Perceived Quality, and Brand loyalty (Aaker 1991; Haryanto 2015; Lee et al. 2011; Nguyen & Luu 2018; Setiawan & Rachmawati 2017). Yoo et al. (2000) combine two variables,

Brand Awareness and Brand Associations into one variable, namely Brand Awareness/Association.

In this study, five variables were used to measure CBBE as used by Davis et al. (2009) and Nguyen & Luu (2018), namely the level of willingness to pay more, the level of perceived differences, the level of perceived benefits, the tendency to switch to products that have the same features, and the tendency to switch to products that are equally good.

1.2 Perceived quality

Keller (2013) defines perceived quality as customer recognition of the overall quality or superiority of a service or product compared to alternatives. Nguyen & Luu (2018) use 4 variables to explain this factor, namely buying X was a really good decision, X is a well-known brand, X's products are of high quality, and price for X is reasonable.

H1: Perceived Quality has a positive impact on Brand Equity.

1.3 Brand loyalty

Nguyen & Luu (2018) define Brand Loyalty as a promise that is upheld to buy back a product/service in the future. Yoo et al. (2000) use three variables to explain this factor, namely I consider myself to be loyal to X; X would be my first choice; and I will not buy other brands if X is available at the store.

H2: Brand loyalty has a positive impact on Brand Equity.

1.4 Brand association and awareness

Aaker (1991) defines the Brand Association as everything that is connected to the customer's memory to a brand. As for brand awareness, it is defined as the ability of potential customers to recognize or remember that a brand is part of a particular product category.

To explain these two factors, Yoo et al. (2000) use 6 variables, namely I know what X looks like, I can recognize X among other competing brands, I am aware of X, some characteristics of X come to my mind quickly, I can quickly recall the symbol or logo of X, I have difficulty in imagining X in my mind.

H3: Brand Association/Awareness has a positive influence on Brand Equity.

2 METHODS

Based on the hypotheses compiled, the model formed is as illustrated in Figure 1.

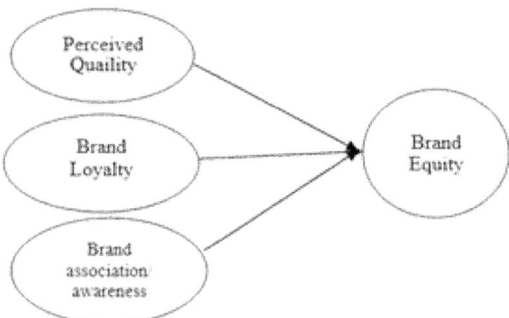

Figure 1. Conceptual framework.

In this research three hypotheses will be tested:

- H1: Perceived Quality has a positive impact on Brand Equity.
- H2: Brand Loyalty has a positive impact on Brand Equity
- H3: Brand Association/Awareness has a positive influence on Brand Equity

2.1 Research method

This study uses quantitative, descriptive, and verification methods. The data were obtained by distributing questionnaires. As many as 100 respondents of the online purchasing actors were collected. The data obtained were processed using the SmartPLS application.

3 RESULTS AND DISCUSSION

3.1 Respondents (descriptive statistic)

The number of respondents was 100 people with the details of men who were the most respondents as many as 60 people (60%), and female respondents as many as 40 people (40%). The ages of the respondents were under the age of 23 as many as 27 people (27%), between 24 years and 40 years as many as 14 people (14%), between 41 years and 55 years who were

the most respondents as many as 58 people (58%), and those above 59 years old as many as 1 person (1%).

Monthly income from respondents of less than Rp.3,200,000 as much as 25, between Rp.3,200,000 and Rp.5,000,000 as many as 8 people, between Rp.5,000,000 and 10,000,000 as many as 46 people, which constitute the highest number, and above Rp.10,000,000 as many as 21 people. The number of online shopping per month was 1 to 3 times as many as 88 people (88%), which was the highest number, between 4 and 8 times as many as 6 people (6%), and more than 8 times per month as many as 6 people (6%).

3.2 Measurement model

3.2.1 Outer model test

In the initial stage, testing the PLS model aimed to determine the existence of inter-construct collinearity and predictive ability of the model (Sarstedt et al. 2017). Evaluation of this model is seen from the following indicators:

- Reliability Indicator
  Reliability indicators aimed to assess whether the measurement indicators for latent variables were reliable or not. Whether determining indicators were reliable or not can be seen from the value of outer loading of each indicator. A loading value above 0.7 indicates that the construct is able to explain more than 50% of the indicator variance. (Wong K. K. 2013).
  Table 1 shows that the outer loading values for all variables are above 0.7, which means the construct is able to explain more than 50% of the indicator variance.

Table 1. Outer loading.

Variable	Outer Loading
PQ 1	0.911
PQ 2	0.704
PQ 3	0.909
PQ 4	0.827
BL 1	0.721
BL 2	0.880
BL 3	0.855
BAA1	0.887
BAA2	0.814
BAA3	0.873
BAA4	0.894
BAA5	0.831
BAA6	0.880
BE1	0.758
BE2	0.752
BE3	0.798
BE4	0.752
BE5	0.754

- Internal Consistency Reliability
  Internal consistency reliability measured how capable indicators measure their latent constructs (Sarstedt et al. 2017). The value used was composite

Table 2. Cronbach's alpha, rho_a, composite reliability, and average variance extracted value.

	Cronbach's Alpha	Rho_A	Composite Reliability	Average Variance Extracted
BAA	0.932	0.934	0.946	0.746
BE	0.823	0.828	0.874	0.582
BL	0.763	0.804	0.861	0.675
PQ	0.863	0.907	0.906	0.709

reliability and Cronbach's alpha. Composite reliability values (Sarstedt et al. 2017) reveal that values between 0.6 and 0.7 are considered to have good reliability, and for the expected Cronbach's alpha value above 0.7 (Ghozali et al. 2015).

Table 2 shows the Cronbach's alpha values obtained for all variables are greater than 0.7 and for the composite reliability values the value for all variables is above 0.7, so we can assume that the models are considered to have good reliability.

- Internal Consistency Reliability
  Convergent validity was determined based on the principle that the measurements of a construct should be highly correlated (Ghozali et al. 2015). Judging from the value of average variance extracted (AVE), the expected AVE value was equal to 0.5 or more, which means that the construct can explain 50% or more of the item variance. From Table 1 it can be seen that the AVE value obtained for each variable is above 0.5.

- Discriminant Validity
  Discriminant validity aimed to determine a reflective indicator if it was a true good measure of the construct based on the principle. Each indicator must have a high correlation to the construct. Different construct gauges should not be highly correlated (Ghozali et al. 2015). In the SmartPLS 3.2.8 application, the discriminant validity test was carried out using the value of cross loadings, and Fornell–Larcker criterion (Henseler et al. 2015).

The expected value of cross loadings was greater than 0.7 and construct correlation with the measurement item is greater than the other constructs. Table 3 shows that the value of cross loadings was greater than 0.7 and construct correlation with the measurement item was greater than the other constructs, which meant that each indicator correlated highly with the construct.

Another method for assessing discriminant validity in PLS was to look at the Fornell–Larcker criterion value. If the AVE square root value of each construct was greater than the correlation value between constructs and other constructs in the model, it can be said that the model had good discriminant validity values of Fornell–Larcker (Wong 2013). Table 4 shows this, so it can be concluded that the model has good discriminant validity.

Table 3. Cross loadings.

	BAA	BE	BL	PQ
BAA1	0.887	0.424	0.260	0.508
BAA2	0.814	0.520	0.269	0.533
BAA3	0.873	0.462	0.264	0.496
BAA4	0.894	0.403	0.298	0.442
BAA5	0.831	0.420	0.295	0.413
BAA6	0.880	0.448	0.271	0.416
BE1	0.400	0.758	0.284	0.484
BE2	0.452	0.752	0.228	0.507
BE3	0.482	0.798	0.241	0.456
BE4	0.318	0.752	0.247	0.368
BE5	0.289	0.754	0.315	0.326
BL1	0.346	0.196	0.721	0.415
BL2	0.276	0.289	0.880	0.350
BL3	0.209	0.330	0.885	0.246
PQ1	0.504	0.591	0.443	0.911
PQ2	0.308	0.292	0.198	0.704
PQ3	0.484	0.515	0.249	0.909
PQ4	0.506	0.464	0.373	0.827

Table 4. Fornell–Larcker criterion.

	BAA	BE	BL	PQ
BAA	0.864			
BE	0.522	0.763		
BL	0.320	0.341	0.822	
PQ	0.546	0.573	0.389	0.842

Table 5. Coefficient of determination.

	R square	R square adj
BE	0.399	0.381

3.2.2 Inner model test

After testing the outer model, the next step is to test the inner model or predictive ability of the model. Model prediction capabilities can be assessed from two criteria, namely the coefficient of determination (R2), and cross-validated redundancy (Q2).

- Coefficient of Determination
  This value is a way to assess how large an endogenous construct can be explained by exogenous constructs. The expected value is between 0 and 1.
  From Table 5, the coefficient of determination (R2) of this model was 0.399. Ghozali et al. (2015) categorize the value as moderate.

- Cross-validated Redundancy
  This value was used to know predictive relevance. Expected Q2 value was greater than 0, indicating that the model had an accurate predictive relevance to certain constructs (Sarstedt et al. 2017).

Table 6. Cross-validated redundancy.

	SSO	SSE	Q^2
BAA	600.000	600.000	
BE	500.000	397.156	0.206
BL	300.000	300.000	
PQ	400.000	400.000	

Table 7. Model fit.

	Saturated Model	Estimated Model
SRMR	0.088	0.088

Table 8. Path coefficient.

	O	M	STDEV	T Stat	P Values
BAA->BE	0.281	0.277	0.088	3.203	0.001
BL->BE	0.103	0.121	0..092	1.125	0.261
PQ->BE	0.380	0.392	0.097	3.908	0.000

It can be seen in Table 6 that the model has a Q2 value greater than 0, which means that the model has accurate predictive relevance to the construct.

3.2.3 Model fit

To measure the model fit in SmartPLS, the value of the standardized root mean square residual (SRMR) was used, which is the difference between the observed correlation and the model that states the correlation matrix. Thus, it is possible to assess the magnitude of the average difference between the observed and expected correlations as the absolute size of the (model) match criteria. The expected value is a value less than 0.1, or 0.08 is a fit criterion (Smartpls 2014).

3.3 Path coefficient

After the model has been tested, measurements of path coefficients are carried out between constructs to see the significance and strength of the relationship, as well as to test the hypothesis.

Hair et al. (2017) state that the path coefficient values range from −1 to +1, where if closer to +1 the relationship between the two constructs is stronger, and the closer the relationship to −1 indicates that the relationship is negative. The significance is seen from P values whose value is smaller than 0.005.

H1: Perceived Quality has a positive impact on Brand Equity

For the first hypothesis, from Table 8, it can be seen that the path coefficient obtained for Perceived Quality on Brand Equity is 0.392 and P value is 0.0; which means that H1 can be accepted or Perceived Quality has a positive impact on Brand Equity and the effect is significant.

H2: Brand Loyalty has a positive impact on Brand Equity.

For the second hypothesis, also from Table 8, it can be seen that the path coefficient obtained for Brand Loyalty toward Brand Equity is 0.121 and P value is 0.261; which means that there is not enough evidence to accept H2 or Brand Loyalty has a positive impact on Brand Equity and the effect is not significant.

H3: Brand Association/Awareness has a positive impact on Brand Equity

For the third hypothesis, from Figure 1 it can be seen that the path coefficient obtained for Brand Association/Awareness of Brand Equity is 0.277 and P value is 0.001; which means that there is enough evidence to accept H3 or Brand Association/Awareness has a positive influence on Brand Equity and the effect is significant.

3.4 Discussion

In this study, five variables were used to measure CBBE as used by Davis et al. (2009) and Nguyen & Luu (2018), namely the level of willingness to pay more, the level of perceived differences, the level of perceived benefits, the tendency to switch to products that have the same features, and the tendency to switch to products that are equally good. Results show that the variables used to measure CBBE are valid and reliable, the same as results obtained by Davis et al. 2009; Nguyen & Luu 2018.

Uses four variables to explain Perceived Quality, namely buying X was a really good decision, X is a well-known brand, X's products are of high quality, and price for X is reasonable. Results show that the variables used to measure Perceived Quality are valid and reliable, the same as result obtained by Nguyen & Luu 2018.

Use three variables to explain Brand Loyalty, namely I consider myself to be loyal to X, X would be my first choice, and I will not buy other brands if X is available at the store. Results show that the variables used to measure Brand Loyalty are valid and reliable, the same as results obtained by Yoo et al. 2000.

To explain Brand Association/Awareness, six variables are used, namely I know what X looks like, I can recognize X among other competing brands, I am aware of X, some characteristics of X come to my mind quickly, I can quickly recall the symbol or logo of X, I have difficulty in imagining X in my mind. Results show that the variables used to measure Brand Association/Awareness are valid and reliable, the same as results obtained by Yoo et al. 2000.

Hypothesis testing results show that Perceive Quality, Brand Loyalty, and Brand Association/Awareness have a positive effect on Brand Equity, but it is not significantly for the relationship between Brand Loyalty and Brand Equity. Different with Haryanto (2015), who found a relationship between Brand Association to Brand Equity and Brand Loyalty to Brand Equity, but there was no influence between Brand Awareness to Brand Equity. The characteristics of the respondents taken are thought to influence differences in the results obtained.

4 CONCLUSION

This research has proven that Perceived Quality, Brand Loyalty, and Brand Association/Awareness have a positive effect on Brand Equity, but it is not significantly for the relationship between Brand Loyalty and Brand Equity. This is thought to be influenced by the characteristics of respondents taken in relation to the frequency of online shopping. 88% of respondents have a monthly shopping frequency of up to 3 times per month.

REFERENCES

Aaker, D. 1991. Managing brand equity: Capitalizing on the value of a brand name. *New York: The Free Press*.

Davis, D., Donna, F., Golicic, G., Susan, L., Marquardt, M., & Adam, A. 2009. Measuring brand equity for logistic services. *The International Journal of Logistics Management* 20(2).

Ghozali, G., Imam, I., Latan, L., & Hengky, H. 2015. *Partial least squares, konsep, teknik, dan aplikasi menggunakan program Smartpls 2.0 m3 untuk penelitian empiris*. Semarang: Badan Penerbit Universitas Diponegoro Semarang

Hair, J. F., Hult, G. T. M., Ringle C. M., & Sarstedt, M. 2017. *A Primer on partial least squares structural equation modeling. 2nd ed.* Thousand Oaks: Sage.

Haryanto, I. 2015. Pengaruh country of origin image terhadap brand equity melalui mediasi elemen brand associations, brand loyalty dan brand awareness pada air conditioner (ac) merek LG di Surabaya. *Jurnal Ilmiah Mahasiswa Universitas Surabaya* 4(2).

Henseler J., Ringle C. M., & Sarstedt M. 2015. A new criterion for assessing discriminant validity in variance-based structural equation modeling. *Journal of the Academy of Marketing Science* 43.

Keller, K. L. 1993. Conceptualizing, measuring, and managing Customer-Based Brand Equity. *Journal of Marketing* 57(1).

Lee, C. G., Yew, L., & Chieng, F. 2011. Dimension of customer-based brand equity: A study on Malaysian brands. *IBIMA Publishing; Journal of Marketing Research and Case Studies*.

Lee, C. G., Yew, L., & Chieng, F. 2011. Customer-based brand equity: A literature review. *Brands researchers' world; Journal of Art Science & Commerce*.

Nguyen, C. Q. T. & Luu, T. D. 2018. Factors affecting consumer-based brand equity of Vietnamese pharmaceutical companies Dalat University. *Journal of Science* 8(1): 145–147.

Ringle, C. M., Sarstedt, M., Mitchell, R., & Gudergan, S. P. 2018. Partial least squares structural equation modeling in HRM research. *The International Journal of Human Resource Management*: 1–27.

Setiawan, N. & Rachmawati, I. 2017. Pengaruh customer-based brand equity terhadap customer satisfaction pada operator seluler di Indonesia. *EProceedings of Management* 4(1).

Smartpls. 2014. *Model fit*. [Online]. Retrieved from https://www.smartpls.com/documentation/algorithms-and-techniques/model-fit.

Wong, K. K. K. (2013). Partial least squares structural equation modeling (PLS-SEM) techniques using SmartPLS. *Marketing Bulletin*, 24(1), 1–32.

Yoo, B., Donthu, N., & Sungho, L. 2000. An examination of selected marketing mix elements and brand equity. *Journal of the Academy of Marketing Science* 28(195).

Advances in Business, Management and Entrepreneurship – Hurriyati et al. (Eds)
© 2021 Taylor & Francis Group, London, ISBN 978-0-367-67471-7

Do brand image and brand awareness influence brand loyalty?

R. Setiawan, R. Hurriyati & A. Rahayu
Universitas Pendidikan Indonesia, Bandung, Indonesia

ABSTRACT: Brand image and brand awareness play a central role in customers' purchase-decision making. The purpose of this study is to examine the influence of brand image and brand awareness on brand loyalty. Data were collected through a survey on 133 sport motorcycle enthusiasts in Indonesia. The results revealed that brand image and brand awareness had positive influence on brand loyalty. This may imply that sport motorcycle companies should strive to establish brand image and brand awareness to promote brand loyalty.

1 INTRODUCTION

One of the fast-growing motorcycle industries in Indonesia is Kawasaki. PT Kawasaki Indonesia engages in the automotive sector which mostly releases automotive products in the form of sport motorcycles. Motorcycles manufactured by Kawasaki Indonesia are limited in number compared to those of produced by other motorcycle companies that have long been established in Indonesia such as Honda or Yamaha. Kawasaki Indonesia is indeed aware of the fact that creating a motorbike requires innovation. Without continuous effort for improvement, products created will not grow and develop which may decrease the appeal of the product to the consumers. In other words, if the company only produces the same product, the consumers may change their decision to buy the product. Such knowledge is important to marketers because it can aid in managerial actions meant to shape consumers' image of the brand, and thereby enable managers to reap the benefits associated with a strong brand image (Krishnamurthy & Kumar 2018).

The following are the data on Sport Bike sales in Indonesia from 2016 to 2018.

Table 1. Sport bike sales from 2016 to 2018.

Years	Yamaha YZF-R25	Honda CBR250R	Kawasaki Ninja 250SL
2016	4.002	1540	2.254
2017	5.150	2.698	259
2018	777	5.419	1.691

* Source AISI

Although Kawasaki Ninja 250SL gained improvement in 2018, it experienced a very high decline in 2017 making it inferior to its competitors. The decline clearly becomes a problem for PT. Kawasaki. Brand loyalty is one of the marketing concepts that have received much attention from researchers in the marketing discipline (Adam et al. 2018). Branding has been around for centuries as a means to distinguish the goods of one producer from those of another (Keller 2013). Aaker imagined brand equity as a set of assets and indicated these assets as brand loyalty, brand awareness, brand associations, perceived quality and other proprietary brand assets. On the other hand, Keller stated that brand knowledge plays a critical role in the formation of brand equity because it is the prerequisite of brand equity, which is supposed to create a distinctive effect, and brand knowledge creates brand awareness and brand image (Dedeoğlu et al. 2019).In the automotive industry, the level of brand loyalty has mostly been revealed by attitudinal studies; that is, statements from customers on intended repurchase behavior (Jørgensen et al. 2016). In order to survive and thrive in business, companies should be able to maintain brand loyalty, a behavior of prioritizing a brand by making repeated purchases. Therefore, companies should compete in maintaining loyal customers, one of which is by paying attention to brand competition to provide an image for their consumers. Brand will distinguish one product from another. If a brand is able to meet consumers' expectations and provide good quality, consumers will likely be more confident with their.

Keller states that brand image is considered among the top concepts in the field of marketing and is referred to as the perceptions of the brand by the consumers as represented by the brand associations in the consumers' memory (Mohammed & Rashid 2018). Brand image is a widely recognized and a salient component of marketing (Dirsehan & Kurtulus 2018). Dhillon states that Image is an important element of a hotel, a brand acts as the most influential element in services because of its natural uniqueness like perishability, inseparability, tangibility and heterogeneity (Lahap et al. 2016). The rationale for brand images in extant branding literature suggests that strong brand images create trust, stability, and differentiation (Rindell & Iglesias 2014). One of the key consequences of a good brand image is reputation (Swati et al. 2018). Brand awareness is related to the strength of the brand node or trace in memory, which we can measure as the consumer's ability to identify the brand under different conditions (Keller 2013). To build the dream, the difference between the levels of brand awareness and brand penetration is of paramount importance (Kapferer & Valette-Florence 2018). Awareness, on the other hand, means the influence that the brand creates in the costumers mind (Yildirim & Aydın 2012). Brand awareness is strongly related to performance in business markets (Homburg et al. 2010). Creating brand awareness is usually the first step in building a brand. Brand awareness is defined as "the ability of the decision-makers in [an] organizational buying center to recognize or recall a brand (Wang et al. 2016)

Brand loyalty, the third brand asset category, is excluded from many conceptualizations of brand equity. There are at least two reasons; they are appropriate and useful to include. Brand loyalty is a key consideration when placing a value on a brand that is bought or sold, because a highly loyal customer can be expected to generate a very predictable sales and profit stream (Aaker 1996). Anderson and Srinivasan state that Brand loyalty is a consumers' favorable attitude toward a brand that results intentions to repurchase and recommend (Yeh et al. 2016). Brand loyalty is usually measured from two perspectives: behavioral loyalty and attitudinal loyalty (Lu & Xu 2015). There are two different aspects of brand loyalty – behavioral and attitudinal. Behavioral, or purchase, loyalty consists of repeated purchases of the brand, whereas attitudinal loyalty refers to the psychological commitment that a consumer makes in the purchase act, such as intentions to purchase and intentions to recommend without necessarily taking the actual repeat purchase behavior into account (Su & Chang 2017).

2 METHODS

The present study utilized quantitative approach and was designed as a survey research in which it took sampling and used a questionnaire for data collection. The population of this study was a member of Ninja Motor Club in Garut Regency. In determining the sample size, the study utilized Slovin formula through which 133 people were selected. In selecting the respondents, this study used accidental sampling.

3 RESULTS AND DISCUSSION

3.1 *Multiple regression test*

It can be seen from Table 2 that the result of multiple linear regression is as follows:

$$Y = 31.41 + 0.40X1 + 0.26X2 \qquad (1)$$

Table 2. Multiple regression test.

Model	Unstandardized Coefficients		Standardized Coefficients		
	B	Std. Error	Beta	t	Sig.
(Constant)	31.419	3.406		9.224	.000
Brand I mage	.409	.153	.238	2.670	.009
Brand Awareness	.260	.084	.277	3.108	.002

a. Dependent Variable: Brand Loyalty
* Source: Primary data processed in 2018.

The result indicates that the intercept or the constant without the influence of brand image and brand awareness is 31.41. The value 0.40 for brand image indicates that there is a relationship between brand image and brand loyalty. Every 1% increase in brand image will increase brand loyalty by 40%. Meanwhile, the value 0.26 for brand awareness means that there is a relationship between brand awareness and brand loyalty. Every 1% increase in brand awareness will increase brand loyalty by 26%.

Hypothesis testing t (t test).

Table 3. T-test (coefficients).

Model	Unstandardized Coefficients		Standardized Coefficients		
	B	Std. Error	Beta	t	Sig.
(Constant)	31.419	3.406		9.224	.000
Brand Image	.409	.153	.238	2.670	.009
Brand Awareness	.260	.084	.277	3.108	.002

a. Dependent Variable: Brand Loyalty.

The results of the t-test shown in Table 3 indicates that there are 2 independent variables (X) that have significant influence on the dependent variable (Y). The results are explained as follows:

a. Brand Image (X1)
  Brand image variable (X1) has significance (sig.) value 0.009. This t sig. value is lower than 5% (0.009 <0.005) indicating that brand image (X1) has a significant influence on brand loyalty. The amount of influence is 0.409.
b. Brand Awareness (X2)
  Brand awareness variable (X2) has significance (sig.) value 0.002. This sig. value is lower than 5% (0.002 <0.005) which means that brand awareness (X2) has a significant influence on brand loyalty. The amount of influence is 0.260.

Table 4. Results of the determination coefficient (model summary).

Model Summary

Model	R	R-Square	Adjusted R Square	Std. Error of the Estimate
1	.612[a]	.451	.327	3.478

a. Predictors: (Constant), Brand Awareness, Brand Image
b. Dependent Variable: Brand Loyalty
*Source: Primary data analyzed in 2018

3.2 Hypothesis testing (F test)

F-test was conducted to determine the effect of all independent variables on the dependent variable. In the formulated hypothesis, it is predicted that all variables: brand image and brand awareness influence brand loyalty. The results of hypothesis testing using F-test. The results of calculation presented in Table 4 shows that the F value is 0.001 which is lower than 5% $(0.001 < 0.05)$. Therefore, it can be concluded that H1 is accepted and H0 is rejected. In other words, all variables: brand image and brand awareness influence Y variable.

3.3 Coefficient of determination

In the present study, the coefficient of correlation (R) and determination (R2) were determined by the help of SPSS. The correlation coefficient (R) can give an idea of how strong the relationship between these variables is. The results of the coefficient of determination can be seen in the following table:

Table 4 shows that the coefficient of determination (R2) or R square is 0.45. This means that 45% variation in brand loyalty is influenced by brand image and brand awareness, while the remaining 55% is caused by other factors beyond the model such as brand trust, customer perception of quality and etc. In the table, it can also be seen that the R value is 0.61 which means that the correlation between brand image and brand awareness on brand loyalty is strong. The determination coefficient of 0.327 shows that the magnitude of influence of brand image and brand awareness on brand loyalty is 32% and the remaining 68% is influenced by other factors not examined in the study. This suggests that favorable evaluation of all brand benefits positively affects brand loyalty irrespective of the effect it has on the development of consumerbrand relationships (Giovanis & Athanasopoulou 2017). However, it is possible that a measure considering the actual behaviour of individuals would render different results. In this regard, it would be of particular interest to consider longitudinal measures assessing the brand loyalty of consumers (Ferreira & Coelho 2015).

4 CONCLUSION

Based on the discussion on the influence of brand image and brand awareness on brand loyalty on Kawasaki Ninja motorbike (a study on the ninja motor club in Garut), there are some conclusions generated. Consumers were satisfied with the design that Kawasaki has in the current type of the sport motorbike. However, in regard to the brand association, Kawasaki is still considered to have lack of a good image in the eyes of consumers. The plausible explanation for this finding is the fact that Kawasaki motorcycles are well known as sport bikes which are mostly preferred by young people making the adults or women feel unfit to choose a motorbike from Kawasaki. Moreover, with respect to brand recall dimension, the respondents mostly disagree with the statement that Kawasaki motorcycles are ones to be used every day. The response may stem from the fact that the respondents do not use the sport bike as an everyday transportation. Finally, in regard to habitual buyer dimension, the respondents disagree if in the future they must buy a motorbike with the Kawasaki brand. The respondents stated that there are still good motorbikes with better specifications and more affordable prices than those of from the Kawasaki brand. There is an influence of brand image on brand loyalty for Kawasaki Ninja in Ninja motorbike club in Garut. There is an influence of brand awareness on brand loyalty for Kawasaki Ninja in Ninja motorbike club in Garut. There is an influence of brand image and brand awareness for brand loyalty on Kawasaki Ninja in Ninja motorbike club in Garut.

ACKNOWLEDGEMENTS

This research work is supported by Faculty of Entrepreneur Garut University.

REFERENCES

Aaker, D.A. 1996. Building strong brands. New York: The Free Press.

Adam, D.R., Ofori, K.S., Okoe, A.F., & Boateng, H. 2018. Effects of structural and bonding-based attachment on brand loyalty. African Journal of Economic and Management Studies 9(3): 305–318.

Dedeoğlu, B.B., Van Niekerk, M., Weinland, J., & Celuch, K. 2019. Reconceptualizing customer-based destination brand equity. Journal of Destination Marketing and Management 11(April): 211–230.

Dirsehan, T. & Kurtuluş, S. 2018. Measuring brand image using a cognitive approach: Representing brands as a network in the Turkish airline industry. Journal of Air Transport Management 67: 85–93.

Ferreira, A.G. & Coelho, F.J. 2015. Product involvement, price perceptions, and brand loyalty. Journal of Product and Brand Management 24(4): 349–364.

Giovanis, A. & Athanasopoulou, P. 2017. Gen Y-ers' brand loyalty drivers in emerging devices. Marketing Intelligence and Planning 35(6): 805–821.

Homburg, C., Klarmann, M., & Schmitt, J. 2010. Brand awareness in business markets: When is it related to firm performance? *International Journal of Research in Marketing* 27(3): 201–212.

Jørgensen, F., Mathisen, T.A., & Pedersen, H. 2016. Brand loyalty among Norwegian car owners. *Journal of Retailing and Consumer Services* 31: 256–264.

Kapferer, J.N. & Valette-Florence, P. 2018. The impact of brand penetration and awareness on luxury brand desirabilit: A cross country analysis of the relevance of the rarity principle. *Journal of Business Research* 83: 38–50.

Keller, K.L. 2013. *Strategic brand management 4Th. Pearson education limited (4th ed., Vol. 58)*. London.

Krishnamurthy, A. & Kumar, S.R. 2018. Electronic word-of-mouth and the brand image: Exploring the moderating role of involvement through a consumer expectations lens. *Journal of Retailing and Consumer Services*, 43: 149–156.

Lahap, J., Ramli, N.S., Said, N.M., Radzi, S.M., & Zain, R.A. 2016. A Study of brand image towards customer's satisfaction in the Malaysian Hotel Industry. *Procedia – Social and Behavioral Sciences* 224: 149–157.

Lu, J. & Xu, Y. (2015. Chinese young consumers' brand loyalty toward sportswear products: A perspective of self-congruity. *Journal of Product and Brand Management* 24(4): 365–376.

Mohammed, A. & Rashid, B. 2018. A conceptual model of corporate social responsibility dimensions, brand image, and customer satisfaction in Malaysian hotel industry. *Kasetsart Journal of Social Sciences* 39(2): 358–364.

Rindell, A. & Iglesias, O. 2014. Context and time in brand image constructions. *Journal of Organizational Change Management* 27(5): 756–768.

Su, J. & Chang, A. 2017. Factors affecting college students' brand loyalty toward fast fashion. *International Journal of Retail & Distribution Management* 46(1): 90–107.

Swati, P., Satyendra, C.P., Andrea, B.X.T. 2018. University brand image as competitive advantage: a two country study. *International Journal of Educational Management* 22(4):288–299.

Wang, Y., Hsiao, S.H., Yang, Z., & Hajli, N. 2016. The impact of sellers' social influence on the co-creation of innovation with customers and brand awareness in online communities. *Industrial Marketing Management* 54: 56–70.

Yeh, C.H., Wang, Y.S., & Yieh, K. 2016. Predicting smartphone brand loyalty: Consumer value and consumer-brand identification perspectives. *International Journal of Information Management* 36(3): 245–257.

Yildirim, Y. & Aydın, K. 2012. The Role of Popular TV Series and TV Series Characters in Creating Brand Awareness. *Procedia – Social and Behavioral Sciences* 62: 695–705.

Advances in Business, Management and Entrepreneurship – Hurriyati et al. (Eds)
© 2021 Taylor & Francis Group, London, ISBN 978-0-367-67471-7

Halal tourism: Service differentiation challenges

D. Mutmainnah, R. Hurriyati & M.A. Sultan
Universitas Pendidikan Indonesia, Bandung, Indonesia

ABSTRACT: This study is aimed to test the effect of service quality and competitive advantage on business performance and also explains the relationship between competitive advantage and business performance due to service differentiation as a moderator. The number of samples in this study were 100 people who had visited provinces that were declared as halal tourist destinations in Indonesia. To test the empirical model, partial least square structural equation modeling (PLS-SEM) analysis was used. The results indicate that service quality has an effect on competitive advantage. The essence of halal tourism emphasizes Moslem principles in managing tourism through polite and friendly services and also service differentiation for all tourists and the surrounding environment. Therefore, to maintain Indonesia as a first-world's halal tourist destination, which has good business performance, the development strategy is directed at fulfilling the tourism competitiveness index as its main indicators, especially improving the service quality of tourism business.

1 INTRODUCTION

The importance of the analyzing of business performance is reflected to examine how businesses are offering service quality to gain competitive advantage and ensure business performance (Wijetunge 2016). Many studies related to corporate performance are conducted in the service industry (Sihite 2018), thus, it becomes the important attention on the further research. Furthermore, the study which explained about competitive advantage also becomes the important attention on the creation of corporate sustainability performance (Sihite 2018).

In order to enhance business performance, differentiation simultaneously is important to achieve competitive advantage that leads to organizational performance (Al-Alak and Tarabieh 2011). This study tests previous research, which stated that competitive advantage influences business performance (Meutia 2013), and also supports the resources-based theory (Barney 1991). Previous studies have illustrated that there is a significant relationship between competitive advantage and performance, in other words, competitive advantage is regarded as part of the foundation for high level performance (Ismail et al. 2010). Studies of different researchers support the association between competitive advantage and business performance in a positive way (Majeed 2011).

The concept of quality has been adopted as a major theme in services research (Parasuraman et al. 1988). The philosophy has been the subject of many conceptual and empirical studies, and it is generally accepted that quality has positive implications for an organization's performance and competitive position

(Harrington & Akehurst 1996). The gap between perceived service and expectation has given rise not only to the construct of service quality, but also to satisfaction, which has also been linked to performance (Parasuraman et al. 1988). Some results suggest that the service quality delivered by a business has an effect on performance (Caruana & Pitt 1997) However, until now, this empirical study has become a major concern, especially for those concentrated on tourism services, and it does not include an analysis of service quality in the halal tourism industry. This study is aimed to test the effect of service quality and competitive advantage on business performance, and also explains the relationship between competitive advantage and business performance due to service differentiation as a moderator.

1.1 *Business performance*

Previous studies have used different variables to measure the performance of tourism business with regard to operational and financial performance (Ab-Talib et al. 2017). The financial performance of tourism businesses was evaluated in different industries, then the generic performance measured the confusion of the comparison of performance among those industries. Furthermore, it is highlighted that the application of inappropriate financial performance measures hampers the assessment of firm performance. Apart from objective measures of performance, managers' perceptions of firm performance were used as a subjective measure for overall firm performance compared to competitors (Robinson et al. 1986). Competitive advantage and firm performance are two different constructs with an apparently complex relationship (Ma 2000). Overall, studies have shown a significant relationship between competitive advantage and performance (Ma 2000; Ray et al. 2004). Furthermore, a positive relationship between service quality and business performance was found in former research (Chang & Chen 1998; Ramayah et al. 2011; Sampaio et al. 2018). "Business performance as nonfinancial implication is capturing the range of possible outcome measures that have nonfinancial implications" (Kotler & Keller 2016).

1.2 Competitive advantage

When understanding the concept of competitive advantage, Michael Porter laid a clear foundation that competitive advantage is a function of the way the firm organizes and manages activities. Organizations create value through these activities. Porter explained that a competitive advantage can be gained by offering more value to the customers than competitors. Michael Porter described a category scheme consisting of three general types of strategies that are commonly used by businesses to achieve and maintain competitive advantage (Tanwar 2013). Competitive advantages are formed by such activities, features, and qualities of a hotel organization that are better than their competition. Competitive advantage creation is possible to define only in comparison to rivals/competitors; meaning that the company needs to create more value than its competitors. Two criteria exist for creating and maintaining competitive advantage (Bahtijarević-Šiber 1999 in Wijetunge 2016):

1. Activities that are unique for the company help the company to produce goods or offer services valued by the consumers.
2. Competitors cannot easily copy them.

Service quality is considered an important tool for a firm's struggle to differentiate itself from its competitors. Warraich et al. (2014) conducted a study using 320 companies in the telecom sector in Pakistan and found that service quality can be considered as a source of competitive advantage. Further, it was found that the tangibility and the reliability are the most important dimensions in determining the competitive advantage.

- H1: Service Quality has a positive impact on Competitive Advantage.
- H2: Competitive Advantage has a positive impact on Business Performance.

1.3 Service quality

In the service literature, service quality is interpreted as perceived quality, which means a customer's level of judgment about a service (Barbara 2010). According to Bitner & Hubbert (1994), quality is the overall impression of the relative inferiority/superiority of the organization and its services. "Quality is the totality of features and characteristics in a product or service that bear upon its ability to satisfy needs" (Haider 2001).

The concept of quality gains is important only in the event that the product or services meet the needs and expectations of the guest. Indeed, this is the reason that all strategies are based on quality standards stemming from exceptional knowledge about the guest. Service quality can be defined as "the customer's assessment of the overall excellence or superiority of the service" (Parasuraman et al. 1988). Thus, the quality is "the ability of a product or services to continually fulfil, or even surpass the customer's expectations" (Stevenson 1993). Schroeder (2000) also believes that quality means "to fulfil or surpass customers' requests now and in the future." The SERVQUAL framework developed by Parasuraman et al. (1988) is a method of evaluating service quality for service industries and came about from research that suggests that customer satisfaction is based on multiple factors rather than one factor. The five dimensions identified to measure service quality are tangibles, reliability, responsiveness, assurance, and empathy (Harr 2008). In the SERVQUAL model, the above-mentioned dimensions are defined as follows (Parasuraman et al. 1988):

- Reliability (delivering the promised outputs at the stated level).
- Responsiveness (providing prompt service and help to customers; the reaction speed plays a vital role here).
- Assurance (ability of a service firm to inspire trust and confidence in the firm through the knowledge, politeness, and trustworthiness of the employees).
- Empathy (willingness and capability to give personalized attention to a customer).
- Tangibles (appearance of a service firm's facilities, employees, equipment, and communication materials).

In an attempt to develop a model for service quality, it was found that a variety of factors, namely resource constraints, market conditions, and management indifferences, may result in discrepancy between management perceptions of customer expectations and the actual specifications established for service (Parasuraman et al. 1985).

- H3: Service Quality has a positive impact on Business Performance.

1.4 Service differentiation as moderator

In order to test the influence on customer satisfaction, Cahya-Wulan et al. (2013) used variables of service differentiation, which are understanding customer need, best value proposition, people skills, and a great recovery plan.

A previous study found that service differentiation has a moderating effect in the relationship between the complexity of customer needs, innovation, and business performance. The results of the study show that service differentiation reinforces the positive relationship between looking at the customer's perspective and business performance (Gebauer et al. 2011).

- H4: The positive relationship between competitive advantage and business performance becomes stronger due to service differentiation.

2 METHODS

Based on the hypotheses compiled, the model formed is as illustrated in Figure 1:

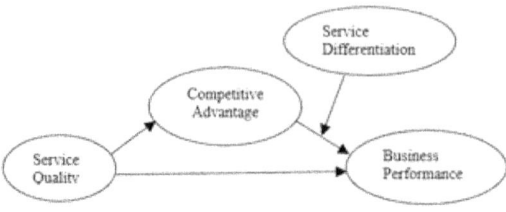

Figure 1. Conceptual framework.

2.1 Research method

This study used quantitative, descriptive, and verification methods. The data were obtained by distributing questionnaires. As many as 100 responses of tourists who have visited halal tourism destinations in Indonesia were collected. The data obtained were processed using the SmartPLS application.

3 RESULTS AND DISCUSSION

3.1 Respondent (descriptive statistic)

Table 1. Respondent identification.

Descriptive Statistics

		N	%
Gender	Male	54	54
	Female	46	46
Age	18–25 years old	4	4
	26–39 years old	52	52
	40–59 years old	44	44
Education	High School	6	6
	Diploma (D1–D4)	10	10
	Undergraduate Degree	50	50
	Master Degree	33	33
	Doctoral Degree	1	1
Occupation	Students/University Students	5	5
	Civil Servants/State Owned enterprise employee/ Teacher/Lecturer	46	46
	Private Employee	26	26
	Entrepreneurs	16	16
	Housewife/Retired	7	7

Continued

Table 1. (*Continued*)

Descriptive Statistics

		N	%
Income	<IDR 2,000,000	6	6
	IDR 2,000.000–2,500,000	12	12
	IDR 2,550.000–3,000,000	5	5
	IDR 3,050.000–5,900,000	26	26
	> IDR 6,000,000	51	51
Visit Frequency	< 3 times a year	51	51
	3–6 times a year	31	31
	> 6 times a year	18	18
Informed by	Friends/Family	39	39
	Workplace recommendation	8	8
	Ads/Newspaper/Print media	2	2
	Internet/Social media	51	51
Destination	Aceh	12	12
	Jakarta	5	5
	West Java	40	40
	Central Java	3	3
	East Java	3	3
	West Nusa Tenggara	8	8
	Riau	1	1
	South Sulawesi	1	1
	West Sumatera	10	10
	Yogyakarta	17	17

3.2 Measurement model

3.2.1 Outer model test

In the initial stage, testing the PLS model aimed to determine the existence of inter-construct collinearity and the predictive ability of the model (Sarstedt et al. 2017). Evaluation of this model could be seen from the following indicators:

- Reliability Indicator
  Reliability indicators aimed to assess whether the measurement indicators for latent variables were reliable or not. It could be seen from the value of the outer loading of each indicator. A loading value above 0.7 indicated that the construct was able to explain more than 50% of the indicator variance (Wong 2013).

  Table 1 shows that the outer loading values for all variables are above 0.7, which means the construct is able to explain more than 50% of the indicator variance.

- Internal Consistency Reliability
  Internal consistency reliability measured how capable indicators are in measuring their latent constructs (Sarstedt et al. 2017). The values used were composite reliability and Cronbach's alpha. For composite reliability values, Sarstedt et al. (2017) reveals that values between 0.6 and 0.7 are considered to have good reliability, and the expected Cronbach's alpha value is above 0.7 (Ghozali & Latan 2015).

Table 2. Final outer loading.

Variable	Outer Loading
SQ 2	0.798
SQ 3	0.819
SQ 4	0.859
SQ 5	0.792
SQ 6	0.866
SQ 7	0.841
SQ 8	0.861
SQ 9	0.870
SQ 10	0.889
SQ 11	0.860
SQ 12	0.830
SQ 13	0.826
SQ 14	0.879
SQ 15	0.889
SQ 16	0.862
SQ 17	0.868
SQ 18	0.822
SQ 19	0.813
SQ 20	0.815
SQ 21	0.839
SQ 22	0.869
CA 4	0.786
CA 5	0.829
CA 6	0.866
CA 7	0.901
CA 8	0.849
CA 9	0.738
CA 10	0.734
CA 11	0.760
BP 1	0.842
BP 2	0.851
BP 3	0.844
BP 4	0.885
BP 5	0.833
BP 6	0.875
BP 7	0.844
SD 1	0.832
SD 2	0.853
SD 3	0.889
SD 4	0.756
SD 5	0.813
SD 6	0.744
SD 7	0.825
SD 8	0.877
SD 9	0.899
SD 10	0.886
SD 11	0.890
SD 12	0.860
SD 13	0.782
SD 14	0.793
SD 15	0.826

Table 3. Cronbach's alpha, rho_a, composite reliability, and average variance extracted value.

	Cronbach's Alpha	Rho_A	Composite Reliability	Average Variance Extracted
SQ	0.980	0.981	0.981	0.717
CA	0.925	0.932	0.938	0.656
BP	0.938	0.939	0.950	0.729
SD	0.969	0.973	0.972	0.699

should be highly correlated (Ghozali & Latan 2015). Judging from the value of average variance extracted (AVE). The expected AVE value was equal to 0.5 or more, which means that the construct can explain 50% or more of the item variance. From Table 1 it can be seen that the AVE value obtained for each variable is above 0.5.

• Discriminant Validity

Discriminant validity aimed to determine if a reflective indicator is a true good measure of the construct based on the principle that each indicator must have a high correlation to the construct. Different construct gauges should not be highly correlated (Ghozali & Latan 2015). In the SmartPLS 3.2.8 application, the discriminant validity test was carried out using the value of cross loadings, and Fornell–Larcker criterion (Henseler et al. 2015).

The expected value of cross loadings was greater than 0.7 and construct correlation with the measurement item was greater than the other constructs. Table 3 shows that the value of cross loadings was greater than 0.7 and construct correlation with the measurement item was greater than the other constructs, which means that each indicator correlates highly with the construct.

Another method for assessing discriminant validity in PLS was to look at the Fornell–Larcker criterion value. If the AVE square root value of each construct was greater than the correlation value between constructs and other constructs in the model, it can be said that the model has good discriminant validity values of Fornell–Larcker (Wong 2013). Table 5 shows this, so it can be concluded that the model has good discriminant validity.

3.2.2 Inner model test

After testing the outer model, the next step was to test the inner model or predictive ability of the model. Model prediction capabilities can be assessed from two criteria, namely the coefficient of determination (R^2) and cross-validated redundancy (Q^2).

• Coefficient of Determination

This value was a way to assess how large an endogenous construct could be explained by exogenous constructs. The expected value was between 0 and 1.

From Table 6, the coefficient of determination (R^2) of this model was 0.794, which categorized the value as moderate (Ghozali & Latan 2015).

Table 3 shows that the Cronbach's alpha values obtained for all variables are greater than 0.7 and for the composite reliability values, the value for all variables is above 0.7, so it can be assumed that the model is considered to have good reliability.

• Internal Consistency Reliability

Convergent validity was determined based on the principle that the measurements of a construct

Table 4. Cross loadings.

	SQ	CA	BP	SD
SQ2	0.798	0.637	0.701	0.678
SQ3	0.819	0.633	0.672	0.671
SQ4	0.859	0.699	0.687	0.722
SQ5	0.792	0.542	0.626	0.624
SQ6	0.866	0.638	0.643	0.732
SQ7	0.841	0.625	0.611	0.696
SQ8	0.861	0.608	0.591	0.661
SQ9	0.870	0.638	0.637	0.636
SQ10	0.889	0.680	0.637	0.680
SQ11	0.860	0.648	0.620	0.644
SQ12	0.830	0.642	0.634	0.600
SQ13	0.826	0.590	0.605	0.669
SQ14	0.879	0.655	0.646	0.720
SQ15	0.889	0.716	0.675	0.754
SQ16	0.862	0.674	0.668	0.739
SQ17	0.868	0.641	0.637	0.724
SQ18	0.822	0.650	0.599	0.720
SQ19	0.813	0.617	0.567	0.634
SQ20	0.815	0.584	0.573	0.627
SQ21	0.839	0.665	0.618	0.706
SQ22	0.869	0.615	0.608	0.745
CA 4	0.593	0.786	0.686	0.643
CA 5	0.581	0.829	0.736	0.673
CA 6	0.661	0.866	0.754	0.684
CA 7	0.722	0.901	0.806	0.741
CA 8	0.674	0.849	0.664	0.706
CA 9	0.515	0.738	0.529	0.531
CA 10	0.520	0.734	0.533	0.500
CA 11	0.594	0.760	0.580	0.652
BP 1	0.666	0.675	0.842	0.700
BP 2	0.654	0.751	0.851	0.771
BP 3	0.594	0.681	0.844	0.720
BP 4	0.651	0.727	0.885	0.755
BP 5	0.570	0.643	0.833	0.657
BP 6	0.676	0.729	0.875	0.706
BP 7	0.648	0.720	0.844	0.773
SD 1	0.690	0.754	0.762	0.832
SD 2	0.701	0.791	0.826	0.853
SD 3	0.748	0.714	0.823	0.889
SD 4	0.489	0.605	0.553	0.756
SD 5	0.561	0.561	0.634	0.813
SD 6	0.545	0.538	0.502	0.744
SD 7	0.613	0.756	0.726	0.825
SD 8	0.763	0.669	0.723	0.877
SD 9	0.756	0.736	0.762	0.899
SD 10	0.771	0.714	0.774	0.886
SD 11	0.762	0.715	0.787	0.890
SD 12	0.722	0.677	0.771	0.860
SD 13	0.614	0.547	0.618	0.782
SD 14	0.638	0.566	0.610	0.793
SD 15	0.699	0.575	0.685	0.826

- Cross-validated Redundancy

This value was used to know predictive relevance. The expected Q^2 value was greater than 0, indicating that the model had an accurate predictive relevance to certain constructs (Sarstedt et al. 2017).

Table 5. Fornell–Larcker criterion.

	SQ	CA	BP	SD
SQ	0.846			
CA	0.755	0.810		
BP	0.747	0.826	0.854	
SD	0.811	0.798	0.853	0.836

Table 6. Coefficient of determination.

Construct	R square	R square adj
Competitive Advantage	0.571	0.566
Business Performance	0.794	0.785

Table 7. Cross-validated redundancy.

	SSO	SSE	Q^2
SQ	2.100.000	2.100.000	
CA	800.000	529.216	0.338
BP	700.000	331.606	0.526
SD	1.500.000	1.500.000	

Table 8. Model fit.

	Saturated Model	Estimated Model
SRMR	0.063	0.075

It can be seen in Table 7 that the model had a Q^2 value that was greater than 0, which means that the model had accurate predictive relevance to the construct.

3.2.3 Model fit

To measure the model fit in SmartPLS, the value of the standardized root mean square residual (SRMR) was used. It is the difference between the observed correlation and the model that states the correlation matrix. Thus, it was possible to assess the magnitude of the average difference between the observed and expected correlations as the absolute size of the (model) match criteria. The expected value was less than 0.1, or 0.08 was a fit criterion.

From Table 8, the SRMR of this model is 0.075, so the model can be categorized as fit.

3.3 Path coefficient

After the model was tested, measurements of path coefficients were carried out between constructs to see the

Table 9. Path coefficient.

	O	M	STDEV	T Stat	P Values
Moderating Effect (SD)->BP	−0.069	−0.064	0.032	2.175	0.030
SQ->CA	0.755	0.760	0.043	17.395	0.000
SQ->BP	0.034	0.040	0.104	0.331	0.741
CA->BP	0.383	0.380	0.081	4.738	0.000
SD->BP	0.494	0.493	0.099	4.983	0.000

Table 10. Total effect.

	O	M	STDEV	T Stat	P Values
Moderating Effect (SD)->BP	−0.069	−0.064	0.032	2.175	0.030
SQ->CA	0.755	0.760	0.043	17.395	0.000
SQ->BP	0.323	0.329	0.104	3.111	0.002
CA->BP	0.383	0.380	0.081	4.738	0.000
SD->BP	0.494	0.493	0.099	4.983	0.000

significance and strength of the relationship, as well as to test the hypothesis.

Hair et al. (2017) state that the path coefficient values range from −1 to +1, where the closer it is to the +1, the relationship between the two constructs is stronger, and the closer it is to −1, it indicates that the relationship is negative. The significance seen from P values whose value is smaller than 0.005.

H1: Service Quality has a positive impact on Competitive Advantage.

For the first hypothesis, from Table 8, it can be seen that the path coefficient obtained for Service Quality on Competitive Advantage is 0.760 with P values of 0.0; which means that H1 can be accepted or Service Quality has a positive impact on Competitive Advantage and the effect is significant.

H2: Competitive Advantage has a positive impact on Business Performance.

For the second hypothesis, also from Table 8, it can be seen that the path coefficient obtained for Competitive Advantage to Business Performance is 0.380 with P values of 0.0; which means that H2 can be accepted or Competitive Advantage has a positive impact on Business Performance and the effect is significant.

H3: Service Quality has a positive impact on Business Performance.

For the third hypothesis, from Table 8, it can be seen that the path coefficient obtained for Service Quality on Business Performance is 0.040 and the P values is 0.741; which means that there is not enough evidence to accept H3 or Service Quality has a positive impact on Business Performance and the effect is not significant.

H4: A positive relationship between Competitive Advantage and Business Performance becomes stronger due to Service Differentiation.

Based on Table 10, it can be seen that the T-statistic (2.175) > 1.96, meaning that service differentiation moderates the effect of competitive advantage on business performance, so that the hypothesis for moderating effects is supported. Thus it can be concluded that service differentiation has an effect on competitive advantage to improve business performance. But because of its negative value, it can be interpreted that increasingly doing service differentiation can weaken the ability of halal tourism, which has competitive advantage in improving business performance.

The discussion of this hypothesis is intended to be able to provide problem solving solutions so that this research can contribute to the development of halal tourism destinations in Indonesia.

1. Relationship between Service Quality and Competitive Advantage.

   H1: Service Quality has a positive effect on Competitive Advantage.

   The acceptance of this hypothesis shows that there is an influence of service quality on competitive advantage. Thus the quality of service that has been carried out by halal tourist destinations can directly influence the creation of a competitive advantage. The competitive advantage achieved by halal tourist destinations will make these attractions survive against increasingly fierce competition. The owners of halal tourist destinations in Indonesia have realized that only halal tourist destinations that provide quality services are able to survive amid competition because the service has a competitive advantage in the minds of customers. The results of this study are in line with the results of research conducted by Wijetunge (2016) that there is a strong positive relationship between service quality and competitive advantage.

2. Relationship of Competitive Advantage with Business Performance.

   H2: Competitive Advantage has a positive effect on Business Performance

   The acceptance of this hypothesis indicates that there is a positive influence between competitive advantage and business performance. Thus the competitive advantage that has been obtained from halal attractions can directly influence the improvement of business performance. As seen from the respondents' answers, it can be concluded that there is a relationship between competitive advantage and business performance. The owner of a halal tourist destination realizes that in order to improve business performance, companies must have a competitive advantage first. The results of this study support the statements contained in the research of Wang et al. (2011). Li et al. (2006) states that there is a positive influence between competitive advantage and performance.

3. Relationship between Service Quality and Business Performance.

H3: Service Quality has a positive effect on Business Performance

The acceptance of this hypothesis indicates that there is a positive influence between service quality and business performance. Thus halal tourist destinations that provide quality services will find it easier to improve business performance. The owners of halal attractions are well aware that service quality is the key to the company's success in achieving optimal business performance. The results of this study support the statement contained in a study conducted by Wijetunge (2016) who found results that there is a strong positive relationship between service quality and various measures of business performance.

4. Differentiation of Services as a Moderator

H4: The positive relationship between Competitive Advantage and Business Performance gets stronger due to Service Differentiation

The acceptance of this hypothesis indicates that it can be concluded that service differentiation influences competitive advantage to improve business performance. However, due to its negative value, it can be interpreted that increasingly differentiating services can weaken the ability of halal tourism, which has a competitive advantage in improving business performance. The results of this study support the statements contained in the research of Cahya-Wulan et al. (2013) who tested the effect on customer satisfaction, using a variety of service differentiation variations, namely understanding customer needs, best value propositions, employee skills, and a great recovery plan. In contrast to previous studies, this study contradicts the results of research showing that service differentiation strengthens the positive relationship between looking at customer perspectives on business performance (Gebauer et al. 2011) in that service differentiation has a moderate effect in the relationship between the complexity of customer needs, innovation, and business performance.

This research has proven that service differentiation as a moderator impacts the impact of halal tourism competitive advantage on business performance. In contrast, Gebauer et al. (2011) found that service differentiation had a moderate effect in the relationship between the complexity of customer needs, innovation, and business performance, which strengthened the positive relationship between looking at the customer's perspective on business performance. The results of this study indicate that service differences have a moderating effect, but are in negative values. A negative sign implies that service differentiation weakens the positive association between competitive advantage and business performance.

This research broadens the existing literature by studying service differentiation; the results of the analysis of the basic model agree with previous research

that wants to improve business performance (Gebauer et al. 2011). First, strong service differentiation by a halal tourism destination reduces the uniqueness of the destination. The premise is that service differentiation is a valuable resource that makes the offer of halal tourist destinations more difficult to emulate, which makes halal travel destinations less sensitive to the needs of more general customers. Halal travel destinations utilize service differentiation to handle changes in customer needs. Service differentiation must be reflected by understanding customer needs that can be emulated by other tourists in general.

Second, strong service differentiation can give employees a better understanding of the process of creating customer value, thus, providing better services, which makes other customers who are not Moslem also want to go to halal travel destinations to get better service.

Third, in agreement with Gebauer et al. (2011), service differentiation not only weakens the association of customer needs and innovation, but also impacts business performance. Investigation in this study shows that customers are doubtful about the lower cost leadership of halal tourism destinations. This is indicated by the question of competitive advantage because low-cost leadership is not reliable.

4 CONCLUSION

From the results of this study, it can be concluded that service quality has a positive effect on customer satisfaction, customer satisfaction has a positive effect on business performance, and customer satisfaction also has an effect on competitive advantage, while competitive advantage has a positive effect on business performance, service quality has a positive effect on business performance, customer satisfaction mediates the relationship of service quality to business performance, and competitive advantage mediates the relationship of service quality to business performance, and all of these relationships are in fact strengthened if service differentiation is carried out.

ACKNOWLEDGMENTS

Special thanks are given to the team of this article Prof. Dr. Hj. Ratih Hurriyati, M. P. and Dr. H. Mokh. Adib Sulthan, S. T., M. T. for their support in this article so that it could be completed and submitted in this conference.

REFERENCES

Ab-Talib, M. S., Chin, T. A., & Fischer, J. 2017. Linking halal food certification and business performance. *British Food Journal* 119(7).

Al-Alak, B. A. & Tarabieh, S. A. 2011. Gaining competitive advantage and organizational performance through customer orientation, innovation, differentiation and market differentiation. *International Journal of Economics and Management Sciences* 1(5): 80–91.

Barbara, C. 2010. Identifying service quality dimensions as antecedents to customer satisfaction in retail banking. *Economic and Business Review* 12(3): 151–16.

Barney, J. B. 1991. Firm resource and sustained competitive advantage. *Journal of Management*: 99–120.

Bitner, M. J. & Hubbert, A. R. 1994. Encounter satisfaction versus overall satisfaction versus quality. *Service quality: New directions in theory and practice* 34(2): 72–94.

Cahya-Wulan, A. N., Suharyono, S., Latief, W. A. 2013. Pengaruh langkah diferensiasi pelayanan terhadap kepuasan pelanggan. *Jurnal Administrasi Bisnis Universitas Brawijaya* 3(1).

Caruana, A. & Pitt, L. 1997. Intqualan internal measure of service quality and the link between service quality and business performance. *European Journal of Marketing* 31(8): 604–616.

Chang, T. Z. & Chen, S. J. 1998. Market orientation, service quality and business profitability: A conceptual model and empirical evidence. *Journal of Service Marketing* 12(4): 246–264.

Gebauer, H., Gustafsson, A., & Witell, L. 2011. Competitive advantage through service differentiation by manufacturing companies. *Journal of Business Research* 64: 1270–1280.

Ghozali, I. & Latan, H. 2015. Partial Least Squares, konsep, teknik dan aplikasi menggunakan program Smartpls 3.0 untuk penelitian empiris. *Semarang: Badan Penerbit UNDIP.*

Haider, S. 2001. ISO 9001:2000 Document Development Compliance Manual. *Florida: St. Lucie Press.*

Hair, J. F., Hult, G. T. M., Ringle C. M., & Sarstedt, M. 2017. A primer on partial least squares structural equation modeling. 2nd Ed. *Thousand Oaks: Sage.*

Harr, K. K. L. 2008. Service dimensions of service quality impacting customer satisfaction of fine dining restaurants in Singapore. *UNLV Theses/Dissertations/Professional Papers/Capstones:* 686.

Harrington, D. & Akehurst, G. 1996. Service quality and business performance in the UK hotel industry. *Int. J. Hospitality Management* 15(3): 283–298.

Henseler J., Ringle C. M., & Sarstedt M. 2015. A new criterion for assessing discriminant validity in variance-based structural equation modeling. *Journal of the Academy of Marketing Science* 43.

Ismail, A. I., et al. 2010. The relationship between organizational competitive advantage and performance moderated by the age and size of firms. *Asian Academy of Management Journal, Vol. 15, No. 2, pp. 157–173.*

Kotler, P. & Keller, K. L. 2016. Marketing management global edition. *Edinburgh: Pearson Education Limited.*

Li, S., Ragu-Nathan, B., Ragu-Nathan, T. S., & Rao, S. S. 2006. The impact of supply chain management practices on competitive advantage and organizational performance. *Omega* 34(2): 107–124.

Ma, H. 2000. Competitive advantage and firm performance. *Competitiveness review* 10(2): 15–32.

Majeed, S. 2011. The impact of competitive advantage on organizational performance. *European Journal of Business and Management*, 3(4): 191–196.

Meutia, M. 2013. Improving competitive advantage and business performance through the development of business network, adaptability of business environment and innovation creativity: An empirical study of batik small and medium enterprises (SME) in Pekalongan, Central Java, Indonesia. *Aceh International Journal of Social Sciences* 2(1): 11–20.

Parasuraman, A., Zeithaml, V. A. & Berry, L. L. 1988. Servqual: A multiple item for measuring consumer perceptions of service quality. *Journal of Retailing* 64:12–40.

Ramayah, T., Samat, N. & Lo, M. C. 2011. Market orientation, service quality and organizational performance in service organizations in Malaysia. *Asia-Pacific Journal of Business Administration* 3(1): 8–27.

Ray, G., Barney, J. B., & Muhanna, W. A. 2004. Capabilities, business processes, and competitive advantage: Choosing the dependent variable in empirical tests of the resource-based view. *Strategic Management Journal* 25: 23–37.

Robinson, R. B., Logan, J. E. & Salem, M. Y. 1986. Strategic versus operational planning in small retail firms. *American Journal of Small Business* 10(3):7–16.

Sampaio, C. A. F., Hernandez-Mogollon, J. M. & Rodrigues, R. G. 2018. Assessing the relationship between market orientation and business performance in the hotel industry – the mediating role of service quality. *Journal of Knowledge Management.*

Sarstedt M., Ringle C. M., & Hair J. F. 2017. Partial Least Squares Structural Equation Modeling.

Schroeder, R. G. 2000. Operations management contemporary concepts and cases. *Boston: Irwin McGraw Hill.*

Sihite, M. 2018. Competitive advantage: Mediator of diversification and performance. *Article of IOP Conference Series: Materials Science and Engineering.*

Stevenson, W. J. 1993. Production/operations management (4th Ed.). *Boston: Irwin Homewood.*

Tanwar, R. 2013. Porter's Generic competitive strategies. *IOSR Journal of Business and Management* 15(1): 11–17.

Warraich, K. M., Warraich, I. A. & Asif, M. 2014. Achieving sustainable competitive advantage through service quality: An analysis of Pakistan telecommunication sector. *Global Journal of Management and Business Research* 13(2).

Wang, W. C., Lin, C. H., & Chu, Y. C. 2011. Types of competitive advantage and analysis. *International Journal of Business and Management* 6(5): 100.

Wijetunge, W. A. D. S. 2016. Service quality, competitive advantage and business performance in service providing SMEs in Sri Langka. *International Journal of Scientific and Research Publications* 6(7): 720–728.

Wong K. K. 2013, Partial Least Squares Structural Equation Modeling (PLS-SEM) Techniques Using SmartPLS. *Marketing Bulletin* 24.

Analysis of commission reduction effects to the travel agent's sales in the domestic market. Case study: Indo Airways

Hendra & T.E. Balqiah
Universitas Indonesia, Jakarta, Indonesia

ABSTRACT: Indo Airways is an Indonesian full-service **airline**. The traditional and online travel agents are contributing 68% of their domestic sales. Those **travel agents** gained commission from the percentage of the sold ticket's basic fare. Indo Airways had reduced the travel agents' commission in the domestic area since April 2017. However, agent contribution to total Indo Airways sales was not significant. This research investigated how the **commission reduction** affected the traditional travel agents' sales in each region of the domestic area. The analysis was based on the theory of **sales promotion**, marketing mix, and the cost leader approach and a hypothesis on approaching the analysis of the effect. The analysis of the secondary data used descriptive analysis based on time series data comparison. The study found that all sales in the different regions decreased, but significant for some region only.

1   INTRODUCTION

IndoAirways or further in this paper will be called IndoAir, is an airline in Indonesia that provides full-services air-transportation. Beside transporting people, IndoAir is providing Cargo services and some ancillaries products. IndoAir's market share in Indonesia in 2017 is 20% (CAPA 2018).

The airline sells the ticket to the customer through many channels. Alamdari (2002) categorized those channels into direct channel and indirect channel. The direct channel consisted of own sales office; call center; own website; and corporates. The indirect channel consisted of traditional travel agents and online travel agent. Based on IndoAir's revenue data, most of the IndoAir tickets were sold through indirect channels which contributed more than 60% of total IndoAir's sales in 2017. According to Harteveld (2016), indirect contribution in 2016 was 54% and will become 48% in 2021.

Companies' important goal in the trade marketing efforts is to have more products display in the retailers in various areas of their stores where related products are sold (Belch & Belch 2018). To support that goal, company can use trade-oriented promotions where the programs are designed to motivate distributors and retailers to carry a product and make an extra effort to push it to their customers. Commission or incentive is one of the promotions that related to it. Airlines as the producers in this industry usually need a broad coverage, which is costly and cannot be provided by their own. To fulfil this need, airlines are supported by agents.

Stephen Shaw (2016) defined the travel agent as a typical relationship in service industries. Products offered to travel agents for sale is more intangible rather than tangible. They sell it with a mutually beneficial relationship. Travel agents will work harder and use their market leverage to sell the products from the company that gives higher benefits.

Agents will get a commission from airlines for each ticket sold. Before April 2017, IndoAir gave traditional travel agent in the domestic area 5% commission for sales of domestic routes and 7% commission for sales of international routes. (JakartaPost 2017). Online travel agent in the domestic area will get 3% commission for sales of domestic routes and 3% commission for sales of international routes.

IndoAir has observed the similar strategy of several major Airlines in the USA and India (O'Malley 2002), and found that airlines offered incentives for travel agents and used commission for tactical purposes. The incentive could encourage the travel agent to sell more tickets. Based on that observation, IndoAir seeks opportunities to decrease the channel's cost in the domestic area and at the same time to effectively use that channel's cost. The Cost Leader (Porter 1980) approach was strategy chosen to make the distribution channel more efficient and productive.

Started from April 2017 IndoAir has decided to change the travel agents' commission policy in the domestic area. A conventional travel agent will get 3% commission for sales of domestic and international routes. Meanwhile, online travel agents will get 2% commission for sales of domestic and international routes. The reduction will be compensated with an incentive program that is given for every achievement to the target (JakartaPost 2017).

According to IndoAir's sales data in 2017, IndoAir's domestic sales were handled by conventional and online travel agents with contribution 68% of total domestic sales. Those travel agents got the percentage of commission from the basic fare of each ticket sold. Before April 2017, the conventional travel agent got about 5%–7% commission from the basic fare on the ticket while online travel agent got 3%–4% commission. However, since April 2017, the commission was reduced for all agents. The conventional travel agent was given 3% commission from the basic fare on the ticket while online travel agent was given 2% commission. This reduction trigger objection and anger from travel agent association around Indonesia.

However, agent contribution to total IndoAir's sales is high and the commission reduction effect has not been anticipated. The main problem in this research was to find out how the commission reduction affected the travel agents' sales. The objective of the research was to analyze the commission reduction influence on each category of travel agent in the domestic market. Expected benefit from this research was a contribution to the airlines business especially about the influence of the new commission policy to airline's sales.

The object of this research was IndoAirways as a leading brand. This research was limited to travel agents based in Indonesia and sold IndoAir's ticket only for flights operated by IndoAir in domestic and international routes, using monthly sales data from January 2016 until August 2018 with sales value after discounted, without taxes, surcharges, and other fees or recognized as Net Sales.

The IndoAir sales data was analyzed based on mean comparison using paired t-test method. The comparison of the travel agent's sales in each region between the period before the implementation of commission reduction and after the commission reduction was provided. This research found that the effect of commission reduction was significant to some regions.

The paper is organized as follows. Section 1 presents the introduction and explain background, problem identification, objectives, research scope and the structure of this paper. Section 2 presents the methodology, which explain the several steps taken in finishing this study. Section 3 presents the result and analysis of the study, which explains about the findings and analysis of the findings. The last section presents the result and further analysis that can be developed from this topic.

2 METHODS

This research was based on theory of sales promotion or trade promotion and distribution channel. Those theories were tested based on explanation by Belch & Belch (2018).

2.1 Hypothesis

The traditional travel agents in Indonesia spread on an extensive location. IndoAir consider regional location in their sales, which can be seen on their organization structure where domestic region is divided into four regions and each lead by Vice President. Based on Wirtz's (2018) explanation in 7p of service marketing, where place and time is one of the essential elements of the marketing mix, the researcher assumed that the commission reduction should give a difference to the sales of the travel agent on each region. The researcher formulated the following hypothesis according to the assumption above.

The hypothesis (H1): there was a difference of agents' average sales for the ticket sold before the commission reduction and after the commission reduction in all region.

2.2 Data

The data analyzed in this research was secondary data, i.e. IndoAir's sales data. Sales data was chosen because it contained the data of factors that were required for this research which may illustrates the effect of the commission reduction. Data were presented in monthly basis with detail of total sales per month, and total number of travel agents for each travel agent type and for each regional area.

IndoAir's sales data used for this research were classified based on the combination of channel type. From the channel type perspectives, travel agents were divided based on Alamdari (2002) segmentation which consisted of indirect channel or traditional travel agents; and online travel agent. Traditional travel agent is the agents that sell the ticket through their outlet. When making purchase, customers must come to this place or through travel agent call center. Online travel is the agents with no physical outlet and sell the ticket only through the internet.

Another data combination was based on the regional area of the traditional travel agents. The regions were divided into Region A (Sumatra island), Region B (Jakarta, Bogor, Depok, Tangerang, and Bekasi), Region C (all Java island except Region 2, and Nusa Tenggara island), and Region D (Kalimantan island, Sulawesi Island, Ambon Island, and Papua island).

There were some criteria that limit the usability of the data for the research. First, the data of domestic travel agents that sold IndoAir's ticket only for flights operates by IndoAir in domestic and international routes. Second, sales data analyzed was net sales on other words, the sales value after discounted, without taxes, surcharges, and other fees. Third, the data period was for three years, started from January 2016 until August 2018. Forth, all sales data currency was in US Dollars, with currency exchange based on the actual currency of transaction dates.

2.3 Research design

Malhotra et al. (2017) classified research design into exploratory research and conclusive research.

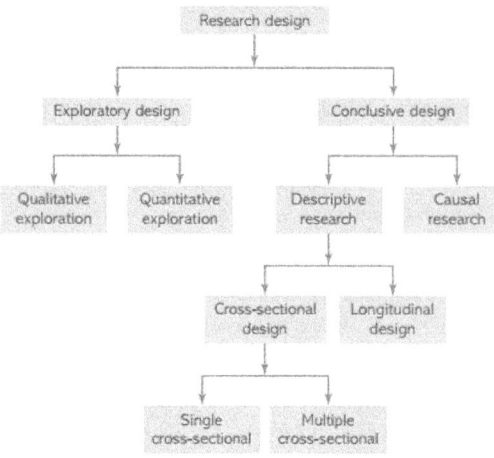

Figure 1. A classification of marketing research design (Malhotra et al. 2017).

Exploratory research is a research design characterized by a flexible and evolving approach to understanding marketing phenomena that are inherently difficult to measure. Conclusive research is a research design characterized by the measurement of clearly defined marketing phenomena. More over the research design classification is illustrated in Figure 1.

This research used the quantitative analysis. Previously related journals, theory, and research were explored to determine the appropriate theories related to the study and models to be applied for this research.

2.4 Data analysis

The following methods were employed to see the effect of commission reduction to the travel agent population and sales before and after the commission reduction were implemented based on time line of the implementation in the sales data.

2.5 Descriptive analysis

This analysis helped to display profile of the data population by measuring properties of the numeric variable in this research which contained the central tendency, variation, and shape of the data. The analysis was applied to both channel type and regional area combination and based on the number of each travel agent criteria.

2.6 Paired T-Test

According to Levine & Krenning (2017), if the assumption is the difference scores are randomly and independently selected from a population that is normally distributed, then the paired t test for the mean difference can be used in related populations to determine whether there is a significant population mean difference. The paired t test statistic followed the t distribution with n − 1 degrees of freedom. The paired

t test assumed that the population was normally distributed, and this test was robust. This mean this test could be used as long as the sample size was not very small, and the population was not highly skewed. To test the null hypothesis that there was no difference in the means of two related populations:

$H0 : \mu D = 0$ (where $\mu D = \mu 1 - \mu 2$)

against the alternative that the means were not the same:

$H1 : \mu D \neq 0$

the t_{STAT} computed test statistic using Equation below:
Paired T Test for The Mean Difference

$$t_{STAT} = \frac{\bar{D} - \mu_D}{\frac{S_D}{\sqrt{n}}} \quad (1)$$

where μD = hypothesized mean difference.

$$\bar{D} = \frac{\sum_{i=1}^{n} D_i}{n} \quad (2)$$

$$S_D = \sqrt{\frac{\sum_{i=1}^{n} \left(D_i - \bar{D}\right)^2}{n - 1}} \quad (3)$$

The t_{STAT} test statistic followed a t distribution with n − 1 degrees of freedom.

For a two-tail test with a given level of significance, a, the null hypothesis was rejected if the computed t_{STAT} test statistic was greater than the upper-tail critical value $t\alpha/2$ from the t distribution, or, if the computed t_{STAT} test statistic was less than the lower-tail critical value − $t\alpha/2$, from the t distribution. The decision rule was

Reject H0 if $t_{STAT} > t\alpha/2$ or if $t_{STAT} < -t\alpha/2$; otherwise, do not reject H0.

3 RESULTS AND DISCUSSION

Table 1 summarizes of the sales data in each region. The average total monthly sales of region B were higher than the others. The average total monthly sales in other region were between 16% and 34% of total region B sales. The maximum and minimum value of region B was also the biggest compared to others. However, region A had a more extensive standard deviation of total monthly sales compared to others with 23% from the mean when others only have 19%–20% deviation. This deviation indicated a high variance in those data.

Based on Table 1, the difference of average sales and total sales in all region meant that geographical segmentation for traditional travel agent was needed. Each region reacted differently to the strategy employed by IndoAir and this should be anticipated in a good manner to avoid lose in sales.

Table 1. Descriptive analysis of each region's sales (part 1).

	Region A	Region B	Region C	Region D
Mean	104,995.50	627,236.33	210,821.06	132,326.19
Standard Error	3,963.93	19,701.23	6,572.11	4,499.82
Median	103,945.00	622,317.00	202,720.50	129,290.00
Mode	n.a	n.a	n.a	n.a
Standard Deviation	23,783,55	118,207.37	39,432.68	26,998.93
SampleVariance	565,657,296.90	13,972,981,231.00	1,554,936,054.00	728,942,197.40
Kurtosis	−0.17	0.07	0.34	−0.48
Skewness	0.29	−0.02	0.47	0.38
Range	105,519.00	540,002.00	172,761.00	105,924.00
Minimum	56,337.00	329,867.00	119,274.00	81,503.00
Maximum	161,856.00	869,889.00	292,035.00	187,427.00
Sum	3,779,838.00	22,580,508.00	7,589,558.00	4,763,743.00
Count	36	36	36	3636
Confdence Level (95%)	8,047.20	39,995.62	13,342.10	9,135.12

Table 2. Descriptive analysis of each region's sales (part 1).

Region		Mean	N	Std. Deviation	Std. Error Mean
Region A	Sales Before	118,411.00	12	16,782.65	4,844.73
	Sales After	89,663.83	12	23,730.84	6,850.50
Region B	Sales Before	652,840.33	12	119,504.58	34,498.00
	Sales After	601,802.08	12	128,143.54	36,991.85
Region C	Sales Before	231,779.08	12	36,314.24	10,483.02
	Sales After	198,011.08	12	42,115.26	12,157.63
Region D	Sales Before	144,568.67	12	20,764.88	5,994.30
	Sales After	119,104.00	12	32,531.93	9,391.16

According to the data on Table 2 it can be seen that sales contribution of each region before the commission reduction was dominated by Region B, followed by Region C, Region D, and Region A. The sales of each region after commission reduction also had the same contribution rank.

The monthly sales data from each region were analyzed using data's mean comparison with the paired t-test method. The results are provided in Tables 2–3. From Table 2, it can be observed that average monthly sales in all region decreased after the commission reduction was implemented. Table 3 shows that the highest decrease was contributed by Region B followed by Region C, Region A, and Region D.

The significances of the commission reduction impacted each region, as seen from data on Table 3. The confidence interval was set to 95% and provided the result of the decrease in Region A was 0.001, Region B was 0.111, Region C was 0.020, and Region D was 0.005.

From the result it can be seen that the impact of commission reduction in Region B (Jakarta, Depok, Bogor, Tangerang, and Bekasi Area) was insignificant, however, for other regions the impact of commission reduction was significant. Based on this result it can be concluded that the hypothesis was accepted.

If product life cycle is used for analysis, based on the characteristic of sales, the data above shows the phase of decline. Some criteria are achieved, for instance the sales are declining, and the cost per customer is lower due to commission reduction. In other hand, the marketing objectives for commission reduction is to reduce expenditure and increase the brand. Strategy for a decline phase in life cycle should be adapted, such as; phase out weak product, cut price, phase out unprofitable channels, advertising must be reduced to level needed to retain hard-core loyalists, and sales promotion should be reduced to minimal level.

In the meantime, IndoAir can implement the overall cost strategy, as explained by Porter (1980) by reducing commission but still focus on differentiation. One of the differentiation requirements is active cooperation from the channel, which works well, especially with the digital channel. It helps IndoAir to keep increasing the sales. Moreover, the importance of the time factor, as explained by Wirtz (2018), where speed and convenience of place and time are necessary for the effective distribution and providing the services become a reason that make digital channel is preferable.

Table 3. Significances analysis of each region's sales (part 2).

Region		Paired Differences				
		95% Convidence Interval of Difference		t	df	Sig. (2-tailed)
		Lower	Upper			
Region A	Sales Before – Sales After	15,058.64	42,435.69	4.622	11	0.001
Region B	Sales Before – Sales After	−13,882.43	115,958.93	1.73	11	0.111
Region C	Sales Before – Sales After	6,462.48	61,073.52	2.722	11	0.02
Region D	Sales Before – Sales After	9,378.50	41,550.84	3.484	11	0.005

4 CONCLUSION

According to place and time concept in marketing mix (Wirtz 2018) it is understood that geographical area is something that need to be considered in the marketing activities. The indirect channels like wholesales and agency, which are also the outsourcing strategy for airlines sales activity, must be supported with the appropriate sales promotion activity as explained by Belch & Belch (2018). An extra value or incentive that offered to that channels will help to create immediate sales.

Based on the result of the research above, travel agent's sales in domestic area were impacted by the commission reduction. But not all domestic region was significantly impacted. This different result should be anticipated with the different strategy. The approach to significantly impacted agent must be focus on the push strategy and the other with the pull strategy.

Further analysis can be provided to answer some questions that related to this commission reduction. Those questions can be (a) the reason why the travel agents in Region B was not significantly impacted, (b) the correlation of the commission reduction with other factors such as capital of the agent, market size, buying power, and (c) how strong the influence of those factors to the decrease of travel agent sales is.

The lack of data is problem that should be anticipate for further research. But the existing data can be compared with external data that possibly become factors impacting the travel agent sales. However, with this limitation we can learn that the treatment for traditional agents in different regions should be specific if an airline willing to do cost efficiency by reducing travel agent commission. Not all region really impacted with that reduction.

REFERENCES

Alamdari, F. 2002. Regional development in airlines and travel agents relationship. *Journal of Air Transport Management*, 8(5), 339–348.

Belch, G.E. & Belch, M.A. 2018. Advertising and promotion: an integrated marketing communications perspective. MCGRAW-HILL.

CAPA. 2018. Indonesia domestic airline market: rapid growth, rivalry intensifies. https://centreforaviation.com/analysis/reports/indonesia-domestic-airline-market-rapid-growth-rivalry-intensifies-410650. CAPA – Center of Aviation

Harteveld, E. 2016. Winning the 'losers' but losing the 'winners'? The electoral consequences of the radical right moving to the economic left. *Electoral Studies*, 44, 225–234.

JakartaPost. 2017.

Levine, R. & Krenning, E.P. 2017. Clinical history of the theranostic radionuclide approach to neuroendocrine tumors and other types of cancer: historical review based on an interview of Eric P. Krenning by Rachel Levine. *Journal of Nuclear Medicine*, 58(Supplement 2), 3S–9S.

Malhotra, N.K., Nunan, D. & Birks, D.F. 2017. Marketing Research: an apllied approach (Fifth edit). Pearson.

O'Malley, C. 2002. Economy devours nest eggs; falling markets, failing companies halt plans to retire. *The Indianapolis Star*, 20.

Porter, M.F. 1980. An algorithm for suffix stripping. *Program*, 14(3), 130–137.

Shaw, S. 2016. Airline Marketing and Management, Seventh Edition. Ashgate.

Wirtz, J. 2018. Cost-effective service excellence. *Journal of the Academy of Marketing Science*, 46(1), 59–80.

Advances in Business, Management and Entrepreneurship – Hurriyati et al. (Eds)
© 2021 Taylor & Francis Group, London, ISBN 978-0-367-67471-7

Impact of customer equity and affective commitment toward loyalty intention

E. Supardi, R. Hurriyati & M.A. Sultan
Universitas Pendidikan Indonesia, Bandung, Indonesia

ABSTRACT: The purpose of this research was to examine the impact of value equity, relationship equity, and affective commitment to loyalty intention on Go-Jek Indonesia. The sample was selected from students at several polytechnics and universities in Bandung who used Go-Jek services through their mobile phone application. Ninety-one questionnaires were analyzed in order to get a meaningful conclusion. The hypotheses were tested using SPSS Smart PLS to discover the relationship between value equity, relationship equity, affective commitment, and loyalty intention, and all of them were accepted. The results indicated that nearly all young people in Bandung used Go-Jek services and they would be loyal in the future as Go-Jek customers. This indicated that value equity, relationship equity, and affective commitment had a positive impact on Go-Jek customer loyalty intention.

1 INTRODUCTION

Researches on loyalty intention are varied. Vogel et al. (2008) found that customer equity (value equity, relationship equity, and brand equity) had an influence on loyalty intentions. Upamannyu & Sankpal (2014) found that brand image and customer satisfaction influenced loyalty intentions. Previous research focused on traditional products and services, which were not impacted by the disruption era.

In this study, an affective commitment variable and the disruption era that forces transportation businesses to reform their services were included to find whether they had an impact on online transportation customer loyalty intention.

2 LITERATURE REVIEW

2.1 Value equity

Value equity is defined as the perceived ratio of what is received to what must be sacrificed (Williamson 1988). Equity theory maintains that perceived equity produces positive affective states that lead to positive attitude such as satisfaction and loyalty (Homans 1961; Walster et al. 1978). Kosarizadeh & Hamdi (2015) defined value equity as "overall customer evaluation on what was already received to what was already paid or sacrificed. Value received by customer based on quality, price, access, easiness, service, comfort, etc."

Value equity works in two ways; first, a company offers a product or service beyond customer expectation, or a company reduces what customers have to give to get what customers want (Torres Telles & Mazhari 2011).

H1: Value Equity has a positive impact on Loyalty Intentions.

2.2 Relationship equity

Relationship equity is defined as the received value from the process to create, keep, and increase strong relations, which involves customers and other stakeholders (Vogel et al. 2008). It is also customer tendencies to stay connected with company brand, beyond objective and subjective judgment (Kosarizadeh & Hamdi 2015).

H2: Relationship Equity has a positive impact on Loyalty Intentions.

2.3 Affective commitment

Affective commitment is defined as the emotional factor related to the degree to which a customer identifies and was personally involved with a company and the resulting degree of trust and commitment (Bendapudi & Berry 1997). Fullerton (2003) described affective commitment as relationships including friendship, rapport, and trust.

H3: Affective Commitment has a positive impact on Loyalty Intentions.

2.4 Loyalty intention

Dick & Basu (1994) defined loyalty intention as "the relative attitude of the customer to the product and the repetitive purchasing behavior," they also mentioned that customer loyalty had cognitive, emotional, and behavioral components. Thus it will be more accurate to define loyalty as a combination of attitudes and behavior.

Zeithaml & Bitner (1996) argued that customer loyalty was "the frequency of being a customer for a product, constantly choosing the same goods and services or a company".

Therefore, behavioral loyalty can be evaluated by using the customer's purchase intentions that result in behavior, in other words, the purchase frequency.

2.5 Conceptual model

Based on the hypotheses, the conceptual model construct is presented in Figure 1:

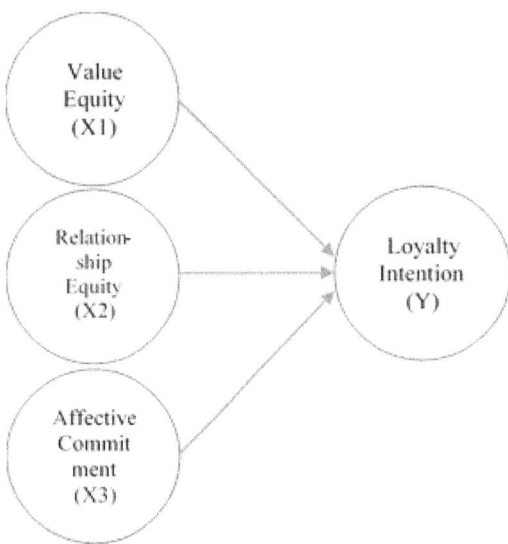

Figure 1. Research model.

3 METHODS

3.1 Research methods

This research was a quantitative research with descriptive and explanatory methods. To obtain data, questionnaires were distributed using Google forms to 122 respondents who used Go-Jek services though their mobile phone. The respondents used Go-Jek not only for transportation purposes, but also for food, cargo, and other delivery services provided on the Go-Jek application.

3.2 Demographic characteristic

There were 122 respondents who filled out the questionnaires and only 91 were selected for this study. The respondents were 44 men (48%) and 47 (52%) woman. Their age ranged between 18 and 23 years old and all of the respondents used Go-Jek services for the different purposes, namely 97% for transportation by bike or motorcycle, 64% for transportation by car (Go-Car), 83% for food delivery services (Go-Food), 27% for

delivering goods (Go-Send), and 9% for other Go-Jek services.

The SmartPLS application was used to process and measure these data.

4 RESULTS AND DISCUSSION

4.1 Measures

4.1.1 Outer model measure

The objective of PLS model measurement was to find construct collinearity and model predictability (Sarstedt et al. 2017). The model evaluation was based on several indicators:

1) Reliability Indicator
   Reliability indicator objective was to examine whether the latent variable indicator measurement was reliable or not. It can be seen from the outer loading value for each indicator. A value loading > 0,7 means the construct was able to explain 50% if its indicator variant (Wong 2013).

   Figure 2 shows that all the outer loading values were above 0,7. It means the constructs were able to explain 50% of the indicator variables.

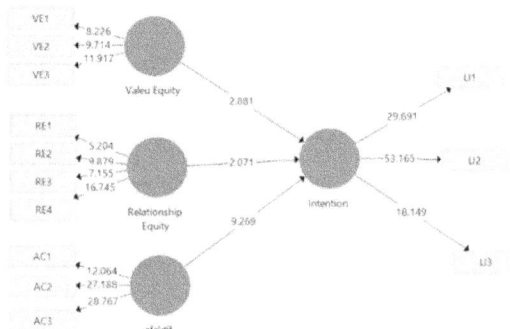

Figure 2. Outer loading value.

2) Internal Consistency Reliability.
   Internal Consistency Reliability examined the indicator ability to measure its latent construct (Sarstedt et al. 2017). Composite reliability and Cronbach's alpha were used to measure internal consistent reliability. For the composite reliability value, Sarstedt et al. (2017) stated that a range value of 0,6 to 0,7 was considered to have a good reliability, while for Cronbach's alpha, the expected value was above 0,7 (Ghozali & Latan 2015).

   Table 1 shows that the Cronbach's alpha values for two variables were above 0.7, while another two variables were nearly 0.7, and the composite reliability values for all of variables were above 0.7.

3) Discriminant Validity
   Discriminant Validity was aimed to determine the correct reflective indicator for measuring its construct, each indicator must have a high correlation with its construct. (Ghozali & Latan 2015).

Table 1. Cronbach's alpha value, composite reliability, and average variance extracted (AVE) value.

Description	Cronbach's Alpha	Rho_A	Composite Reliability	AVE
Intention Loyalty	0.847	0.848	0.907	0.766
Relationship Equity	0.659	0.680	0.795	0.495
Value Equity	0.631	0.634	0.802	0.576
Affective Commitment	0.771	0.792	0.865	0.681

Source: SmartPLS

Table 2. Cross loadings value.

Code	Intention Loyalty	Relationship Equity	Value Equity	Affective Commitment
AC1	0.493	0.308	0.343	0.780
AC2	0.597	0.458	0.533	0.877
AC3	0.751	0.449	0.431	0.815
LI1	0.846	0.479	0.539	0.637
LI2	0.915	0.487	0.503	0.716
LI3	0.845	0.557	0.595	0.655
RE1	0.318	0.596	0.438	0.230
RE2	0.457	0.710	0.411	0.393
RE3	0.331	0.701	0.266	0.339
VE1	0.411	0.412	0.757	0.346
VE2	0.489	0.616	0.708	0.316
VE3	0.508	0.388	0.809	0.537

Source: SmartPLS

At SmartPLS 3.2.8, we used cross loadings of Fornell—Larcker Criterion values and HTMT values were used (Henseler et al. 2015).

The expected cross loading was > 0.7 as shown in Table 2. It means each indicator had a strong correlation with its constructs.

4.1.2 Inner model measure

The next step after measuring the outer model was measuring the inner mode or model prediction ability. The model prediction ability was measure by two criteria, that is, determination coefficient (R^2) and cross-validated redundancy (Q^2).

1) Determination coefficient (R2)

The expected value range for the determinant coefficient was between 0 and 1. Based on Table 3, R2 is 0,665 or categorized as strong, as affirmed in the work of Ghozali & Latan (2015).

2) Cross-validated redundancy (Q^2)

Cross-validated redundancy was measured to get predictive relevance. The results showed that the expected Q^2 value > 0, which meant the model had an accurate predictive relevance to specific constructs. (Sarstedt et al. 2017).

Table 3. R square value.

Description	R Square	R-Square Adjusted
Intention Loyalty	0.665	0.653

Source: SmartPLS

Table 4. Q^2 value.

Description	SSO	SSE	Q^2
Intention Loyalty	273.000	148.844	0.455
Relationship Equity	364.000	364.000	
Value Equity	273.000	273.000	
Affective Commitment	273.000	273.000	

Source: SmartPLS

Table 5. Fit model.

Description	Saturated	Estimated
SRMR	0.108	0.108
D_ULS	0.903	0.903
D_G	0.437	0.437
Chi-Square	217.540	217.540
NFI	0.617	0.617

Source: SmartPLS

Base on Table 4, the Q^2 value is beyond 0, which means the model has an accurate predictive relevance to specific constructs.

4.2 Model fit

To examine a fit model in SmartPLS, Standardized Root Mean Square Residual (SRMR) value was used to distinguish between the correlation and matrix correlation model.

Fit criteria expected value is 0.1 or 0.08. Based on Table 5, the SMRS value for this model was 0,108, which was considered a fit criterion.

4.3 Path coefficient

The next step was to measure path coefficients between constructs in order to obtain significant level, relationship level, and to test whether the hypotheses were accepted or not.

Sarstedt et al. (2017) stated that path coefficient values range from −1 to +1, +1 meant the construct had a strong relationship, while −1 meant the construct had a weak relationship or a negative indication, while P values < 0,005 showed significance.

H1: Value Equity has a positive impact on Loyalty Intention.

Table 6 shows the path coefficient for value equity on loyalty intention is 0.235 and the P value is

Table 6. Path coefficient.

Description	Sample	P Values
Relationship Equity	0.155	0.039
Value Equity	0.285	0.004
Affective Commitment	0.561	0.000

Source: SmartPLS

0.004, which means H1 was accepted. Therefore, value equity had a positive and significant impact on loyalty intention.

H2: Relationship Equity has a positive impact on Loyalty Intention.

The path coefficient for relationship equity on loyalty intention was 0.155 and the P value was 0.039, which means H2 was accepted. It proved that relationship equity had a positive and significant impact on loyalty intention.

H3: Affective commitment has a positive impact on Loyalty Intention.

The path coefficient for affective commitment to loyalty intention was 0.561 and the P values was 0.0; which indicated that H3 was accepted, in other words, affective commitment had a positive and significant impact on loyalty intention.

5 CONCLUSION

This research was conducted to study and analyze the customer loyalty intention, especially of those who use online transportation applications. There was correlation between the independent variables and dependent variable, and all independent variables had an influence on the dependent variable. The current study can be concluded in such manner that value equity, relationship equity, and affective commitment have a positive and significant impact on Go-Jek customer loyalty intention.

ACKNOWLEDGEMENT

The researcher would like to thank Professor Ratih Hurriyati and M. Adib Sultan as lecturers in Management Science at Universitas Pendidikan Indonesia Bandung, who provided guidance in this research.

REFERENCES

Bendapudi, N. & Berry, L. L. 1997. Customers' motivations for maintaining relationships with service providers. *Journal of retailing*, 73(1), 15–38.

Dick, A. S. & Basu, K. 1994. Consumer loyalty: Towards an integrated conceptual approach. *Journal of the Academy of Marketing Science*, 22(2), 99–113.

Fullerton, G. 2003. When does commitment lead to loyalty? *Journal of service research*, 5(4), 333–344.

Ghozali, I. & Latan, H. 2015. Partial least square, konsep, teknik, dan aplikasi menggunakan program SmartPLS 3.0. Edisi 2. *Badan Penerbit Universitas Diponegoro Semarang*.

Henseler, J., Ringle, C. M., & Sarstedt, M. 2015. A new criterion for assessing discriminant validity in variance-based structural equation modeling.

Homans, G. C. 1961. Its elementary forms. *Social Behavior*, 488–531.

Kosarizadeh, M., & Hamdi, K. 2015. Studying the effect of social media on consumer purchase intention. Case study: Leather products. *Journal of Applied environmental and Biological Science*, 5(7), 171–181.

Sarstedt, M., Ringle, C. M., & Hair, J. F. 2017. Partial Least Squares Structural Equation Modeling. *Academy of Marketing Science*. Vol.43

Torres Telles, M. V., & Mazhari, S. 2011. Customer equity drivers and purchase intentions: Examining the customer equity framework in the retail clothing industry in a Swedish context: H&M and Gina Tricot as the case study.

Upamannyu, N. K. & Sankpal, S. 2014. Effect of brand image on customer satisfaction and loyalty intention and the role of customer satisfaction between brand image and loyalty intention. *Journal of Social Science Research*, 3(2), 274–285.

Vogel, V., Heiner, E., & Ramasheshan, B. 2008. Customer Equity drivers and future sales. USA, *American Marketing Association* Vol. 72.

Walster, E., Walster, G. W., & Berscheid, E. 1978. Equity: Theory and research.

Williamson, O. E. 1988. Corporate finance and corporate governance. *The journal of finance*, 43(3), 567–591.

Wong K. K. 2013. Partial Least Squares Structural Equation Modeling (PLS-SEM) Techniques Using SmartPLS. *Marketing Bulletin*. Vol. 24.

Zeithaml, V. A. & Bitner, M. J. 1996. Services marketing. International Editions. The McGraw-Hill Companies.

Advances in Business, Management and Entrepreneurship – Hurriyati et al. (Eds)
© 2021 Taylor & Francis Group, London, ISBN 978-0-367-67471-7

The power of EWOM for online business

C. Savitri, R. Huriyati & H. Hendrayati
Universitas Pendidikan Indonesia, Bandung, Indonesia

ABSTRACT: Promotion is a key to successful companies in the business world. Technology development gives opportunity for business to promote through various links such as social media and the electronic word of mouth (eWOM) and family influence interest in buying goods or services. This research was aimed to empirically examine the impact of social media and family influence on eWOM and how it affects purchase intention in the online shopping platform Shopee. The method of the research was descriptive and explanatory by path analysis. Data were collected from a survey and interviews with 230 respondents. The result of this research proved that family had an influence more than social media and eWOM had a partial influence on purchase intention, while social media and family influence simultaneously.

1 INTRODUCTION

Technological improvement influences culture and behavior and the ability to make choices. Market places become one of the business challenges to reach as many customers as possible. Shopee is one of the market places that has become favorable by customers (Iprice.co.id 2019). However, fluctuating consumer visits have an impact on the market place. This problem is a phenomenon that needs to be investigated.

Word of mouth has an influence toward the purchase decision of consumers (Jalilvand & Samiei 2012). Word of mouth is considered reliable information since it is a noncommercial type of communication. It has higher persuasive, trust, and credibility (Jalilvand et al. 2011). The digital era changes the channel for WOM, where information can be delivered without face-to-face meetings (De Bruyn & Lilien 2008). The levels of trust a person has toward any information is determined by the sources of the information (Mohammed Abubakar 2016).

Accessibility and high coverage are becoming more important than offline word-of-mouth communication (Mohammed Abubakar 2016; Shen 2015). Confirmed by Chevalier & Mayzlin (2003), communication via eWOM has become an important platform where eWOM plays an important role in the purchase decision affecting attitude as well as consumers (Mohammed Abubakar 2016).

Social media is defined as online services that assists users to find different content and make content easily (Prasad et al. 2017). The rapid development of social media provides opportunity for marketer to offer the best service for the consumers. Social media is one of the most attractive media to influence the consumer

behavior in the world (Brown et al. 2007; Jalilvand & Samiei 2012; Prasad et al. 2017; Shen 2015). EWOM had a very strong impact on intention to buy, either directly or indirectly; moreover, the level of trust to a source of information is determined by the source of the information.

Lifestyle and fashion have become an inseparable part from modern society where globalization industry changes lifestyle, especially that of the millennial generation. Family also plays a role in consumer behavioral changes in making choices in product or service purchases. This will be an opportunity for businesses to expand their target. Although research about eWOM and purchase intention is plentiful, it is important to find out how millennials use social media as a source of information in fashion purchase behavior (Bolton et al. 2013) and how family influence their decision making.

1.1 *Electronic word of mouth*

Social media used by millennials has created eWOM, which is one of the ways consumers share their views and steer other consumers to support or oppose certain products (Cheung et al. 2012). WOM communication occurs when consumers give advice or opinion or share their experience to other consumers about a product, services, or a brand. When receiving WOM from family or friends, the message is more persuasive since the informant is considered to not getting any advantage from forwarding the message (Schiffman & Wisenblit 2005). Kotler (2017) stated that Word of Mouth Communication is the communication process to recommended personal or groups toward a product or services and gives personality about the information.

1.2 Social media

Social media is composed of many based with specific characteristics such as blog, content, social networking sites, virtual game, and virtual social world (Andreas 2010), it is supported by the 2.0 web technology that enables users to exchange or manufacture new content (Kaplan & Haenlein 2010). Social media has an important role in the development of marketing, as seen from the number of social networks users like Facebook who made about 56% recommend to a friend in the network and 33% has made a purchase based on recommendations on a social platform. Kaplan & Haenlein (2010) stated that social media is a web-based group, application, and the base of technology.

1.3 Family influence

Family, according to Schiffman & Wisenblit (2005), consists of 2 or more persons connected by blood, marriage, or adoption and who are living together. A nuclear family contains parents and children and it has four functions, namely finance, emotional support, lifestyle, and socialization. Consumer purchase behavior is influenced by individual decision making or family decision making. Studies show that family decision making impacts consumer behavior. Solomon (2018) stated that family is a large department store on the purchase of an organization that plays particular roles in purchase decision.

1.4 Purchase intention

Purchase intention is the phase before consumers plan to buy a product (Kotler 2017). Purchase intention is the stages in which consumers form their choice from among brands combined with the option for alternatives.

2 METHOD

This research was categorized into explanatory research. The population of this research was UBP Karawang colleges who used the Shopee application and the sample size was 230. The sampling method was probability sampling (simple random sampling) in order to give an equal chance to anyone to be chosen as a sample.

3 RESULTS AND DISCUSSION

1. Social media and Family Influence had a correlation coefficient of 0,466 or 46,6% and a strong relationship. The research by Zhang (2015) and Xie et al. (2018) stated that when members of family actively use social media and they get positive experience from a service or products they will share with other members. Kotler (2017) said that family environment and culture in family provides

a change in behavior to choose the information needed.

2. Partial influence of social media toward eWOM had a correlation of 0,269 or 26,9% and the partial influence of Family Influence toward eWOM had a correlation of 0,343 or 34,3%, and, thus, the coefficients are 0,343 > 0,261. It can be stated that family influence had a stronger effect on eWOM than social media. Research by Thoumrungroje (2014) discovered a significant and direct relationship between the intensity of social media and reliance on eWOM and consumption as much as 52,2%. Zhang (2015) mentioned that in sub groups of the Y generation ranging between 21 and 24 years who use social media, mobile technology had a positive influence on eWOM behavior at 0,30 or 30%. The partial influence of eWOM toward Purchase Intention has a coefficient of 0,358 or 35,8% and the variable of eWOM had a strong influence on Purchase Intention. This result is supported by Putri & Prabowo (2015) who found out that dimensions in e-WOM: intensity, positive valence, negative valence, and content simultaneously had an effect toward Purchase Intention of 182,777 with a significant rate of 0,05.

3. Social media variable and Family Influence simultaneously influenced eWOM by 0,269 or 26,9%. It means that social media and Family Influence had a contribution of 26,9% toward eWOM.

4 CONCLUSION

This research supports the research conducted by Chevalier & Mayzlin 2003; Mohammed Abubakar 2016; Goyette et al. 2010; Xie et al. 2018; Prasad et al. 2017; and Jalilvand & Samiei 2012 where social media and Family Influence have a positive influence toward eWOM and eWOM has a strong influence with Purchase Intention.

ACKNOWLEDGEMENT

This research was supported by Shopee (Market place) and students at Buana Perjuangan Karawang University.

REFERENCES

Andreas, D. 2010. Why did the chicken browse the social media? *Elex Media Komputindo.*

Bolton, R. N. et al. 2013. Institutional Repository Understanding Generation Y and their use of social media: A review and research agenda. *Loughborough University Institutional Repository,* 24(3), 245–267. Retrieved from https://dspace.lboro.ac.uk/dspace-jspui/bitstream/2134/13896/3/Understanding Generation Y and Their Use of Social Media_A Review and Research Agenda.pdf

Brown, J., Broderick, A. J., & Lee, N. 2007. Word of mouth communication within online communities. *Journal of Interactive Marketing,* 21(3), 2–21. https://doi.org/10.1002/dir

Cheung, C. M. K., Xiao, B., & Liu, I. L. B. 2012. The impact of observational learning and electronic word of mouth on consumer purchase decisions: The moderating role of consumer expertise and consumer involvement. *Proceedings of the Annual Hawaii International Conference on System Sciences*, 3228–3237. https://doi.org/10.1109/HICSS.2012.570

Chevalier, J. A. & Mayzlin, D. 2003. The effect of word of mouth on sales. *National Bureau of Economic Research*, 40. https://doi.org/10.1509/jmkr.43.3.345.

De Bruyn, A. & Lilien, G. L. 2008. A multi-stage model of word-of-mouth influence through viral marketing. *International Journal of Research in Marketing*, 25(3), 151–163. https://doi.org/10.1016/j.ijresmar.2008.03.004.

Goyette, I., Ricard, L., Bergeron, J., & Marticotte, F. 2010. E-WOM Scale: Word-of-mouth measurement scale for e-services context. *Canadian Journal of Administrative Sciences*/Revue Canadienne des Sciences de l'Administration, 27(1), 5–23.

Iprice.co.id. 2019. Shopee Jadi E-Commerce Paling Top dari Masa ke Masa. Iprice.co.id, 2019. Retrieved from https://databoks.katadata.co.id/datapublish/2019/09/03/shopee-jadi-e- commerce-paling-top-dari-masa-ke-masa

Jalilvand, M. R., Esfahani, S. S., & Samiei, N. 2011. Electronic word-of-mouth: Challenges and opportunities. *Procedia Computer Science*, 3, 42–46. https://doi.org/10.1016/j.procs.2010.12.008

Jalilvand, M. R. & Samiei, N. 2012. The impact of electronic word of mouth on a tourism destination choice: Testing the theory of planned behavior (TPB). *Internet Research*, 22(5), 591–612. https://doi.org/10.1108/10662241211271563.

Kaplan, A. M. & Haenlein, M. 2010. Users of the world, unite! The challenges and opportunities of Social Media. *Business Horizons*, 53(1), 59–68. https://doi.org/10.1016/j.bushor.2009.09.003.

Kotler, P. 2017. [Philip_Kotler] Marketing_4.0_Mo(z-lib.org) (14th ed.). *Prentice Hall.*

Mohammed Abubakar, A. 2016. Does e-WOM influence destination trust and travel intention: A medical tourism perspective. *Economic Research-Ekonomska Istrazivanja*, 29(1), 598–611. https://doi.org/10.1080/1331677X.2016.1189841.

Prasad, S., Gupta, I. C., & Totala, N. K. 2017. Social media usage, electronic word of mouth and purchase-decision involvement. *Asia-Pacific Journal of Business Administration*, 9(2), 134–145. https://doi.org/10.1108/APJBA-06-2016-0063

Putri, L. E. D. & Prabowo, F. S. A. 2015. Pengaruh electronic word of mouth (e-WOM) terhadap purchase intention (studi kasus pada Go-Jek Indonesia). *eProceedings of Management*, 2(3).

Schiffman, L. G. & Wisenblit, J. 2005. Consumer Behavior.

Shen, X. 2015. Herd behavior in consumers' adoption of online reviews. 2009. https://doi.org/10.1002/asi

Solomon, M. R. 2018. Consumer Behavior Buying, Having, and Being. *Pearson.*

Thoumrungroje, A. 2014. The influence of social media intensity and EWOM on conspicuous consumption. *Procedia-Social and Behavioral Sciences*, 148, 7–15.

Xie, X. Z., Tsai, N. C., Xu, S. Q., & Zhang, B. Y. 2018. Does customer co-creation value lead to electronic word-of-mouth? An empirical study on the short-video platform industry. *Social Science Journal*, (2017). https://doi.org/10.1016/j.soscij.2018.08.010

Zhang, Y. 2015. The impact of brand image on consumer behavior: A literature review. *Open Journal of Business and Management*, 03(01), 58–62. https://doi.org/10.4236/ojbm.2015.31006

Advances in Business, Management and Entrepreneurship – Hurriyati et al. (Eds)
© 2021 Taylor & Francis Group, London, ISBN 978-0-367-67471-7

Corporate reputation and its impact on customer citizenship behavior

R.S. Billqis, L.A. Wibowo & H. Hendrayati
Universitas Pendidikan Indonesia, Bandung, Indonesia

ABSTRACT: Customer Behavior Citizenship (CCB) is a voluntary behavior of the customers in providing mental and physical input that can improve the efficiency of a company as well as new ideas that can be the source of business strategy. Lack or absence of CCB can impede companies in evolving in improving services. This research was conducted to identify the influence of corporate reputation and its impact on customer citizenship behavior in one of the four-star hotels in Bandung. The analysis technique used was the path analysis with a sample of 106 respondents. The results showed that the influence of the corporate reputation had a positive and significant influence on CCB. The dimensions quality products and services had the highest score, proving that the performance of corporate reputation has received high appreciation.

1 INTRODUCTION

Customers nowadays are not only playing an active role as buyers, but they can be profitable sources for every company (Bowen & Schneider 1985; Rosenbaum & Massiah 2007). Customer citizenship behavior (CCB) is a voluntary behavior of customers outside their duties such as providing important mental and physical input which can improve the company efficiency. The new ideas they give become valuable sources for developing business strategies which lead to improvement in service delivery (Abbasi et al. 2014; Bailey et al. 2001; Rosenbaum & Massiah 2007). CCB is very important for a company because customers are the main source of profits. The lack or absence of CCB among customers impedes the development of the company since they do not know what their customers want and expect (Groth 2005).

CCB has become a study in manufacturing field before being adopted in the service field. Kotler et al. (2014) argues that CCB is easier to create and build in service companies than manufacturing because customers in the service company are often physically present when the services are provided, unlike the manufacturing companies where customers are rarely present during production (Foote & Tang 2008; Okurame 2012; Sahertian 2010; Zayas-Ortiz et al. 2015). There are a lot of limitations in the previous research on CCB such as the different objects in which the CCB was studied with the discussion, as well as unclear and even biased literatures. This research focused on CCB in hotel industry, since some service industries cannot obtain production only with capital substitution. It requires the survival of the company through customers so that the companies or hotels are able to survive and compete.

The hotel used as the object of research is one of the best four-star hotel in Bandung City (Tripadvisor 2017). This hotel utilizes corporate reputation they have for a long time as a corporate strategy to build and create CCB.

This research was aimed to see how much the influence of corporate reputation is and its impact on customer citizenship behavior. Therefore, strategic development and implementation to understand CCB effectively can be obtained.

1.1 Corporate reputation

Corporate reputation is a part of a market-driven strategy what put customers in the main position as market makers in business strategies (Cravens & Piercy 2013). Grant in Padgett & Leite (2014) states that reputation has been identified as one of the most important intangible resources and strategies that provide a competitive and sustainable advantage for the company.

This means that reputation creates certain expectations of an organization, in this case, the hotel. Positive reputation facilitates effective communication between companies and customers, which causes behaviors and the spread of word of mouth and good customer behavior (Roper & Fill in Manukian 2015). Bartikowski & Walsh (2011) and Abbasi et al. (2014) proposed the dimensions of corporate reputation, i.e. customer trust, customer satisfaction, customer loyalty, customer commitment, quality of employees & management, quality of products & services and attractiveness or emotional appeal of the organizational.

1.2 Customer Citizenship Behavior (CCB)

Customer behavior is defined as the decision making process and activities of individuals involved in it. Customer behavior is essentially an understanding of why customers do what they do and what are they doing. Schiffman & Kanuk (2012) suggest that research on customer behavior is a research about how an individual makes a decision to allocate the available resource, while according to Kotler & Keller (2016) customer behavior is how individuals place goods, services, ideas or experiences to satisfy their needs and desires.

Customers' individual behavior is divided into two parts, namely in-role behavior, and extra-role behavior. In-role behavior refers to behavior from and for customers themselves. Extra-role behavior is the pro-social behavior, actions, and policies shown by customers that can benefit service providers and other customers (Bove et al. 2009; Garma & Bove 2011; Yi et al. 2011). This behavior is considered important for the success and benefits of a company (Yi et al. 2013). Organ (1988) defines customer citizenship behavior (CCB) is an individual's behavior that results from individuals' understanding which as a whole helps the performance of a company effectively. Customer can provide important mental and physical input that can improve company efficiency and can become a valuable source of new ideas for business strategy. This behavior is included into extra-role behavior, because customers voluntarily engage during or after purchasing products and services (Groth 2005; Gruen et al. 2000). This behavior is not directly valued by the company but it can help to promote the effective function of a company so that it improves the quality of the company's services (Groth 2005). Abbasi et al. (2014) defines that the CCB attitude and behavior can play a role as the source of support to develop the company. CCB has highly benefited the company because today, customers are not only playing an active role as buyers of products or services.

Their attitudes and behaviors are increasingly involved in promoting a company and improving service quality (Bowen & Schneider 1985; Zeithaml et al. 2005). CCB in this research referred to the behavior of customers who have purchased products or services more than once or repeat customers. The dimensions of CCB in this research were help other customers, help other company, altruism, courtesy, civic virtue, conscientiousness (Anaza & Zhao 2013; Bartikowski & Walsh 2011; Di et al. 2010; Gruen et al. 2000; Yi et al. 2013).

2 METHODS

The object of this research was one of the four-star hotels in Bandung City. This research consisted of two variables; customer citizenship behavior (dependent variable) and corporate reputation (independent variable). The research was quantitative research with explanatory survey method. Since it was conducted in less than one year, it was also a cross-sectional research. The data analysis technique used was path analysis with 106 respondents as samples, which were hotel repeater guests. Data was collected from interviews, observations, questionnaires, and literature studies.

3 RESULTS AND DISCUSSION

3.1 Demographic subject

Most of the respondents came from Jakarta, Bandung, and Surabaya respectively.

3.2 Reability

Reliability test was conducted to obtain the level of accuracy of data collection tools used. Reliability refers to an understanding that an instrument is reliable enough to be used as a data collection tool because the instrument is good. A reliable and trusted instrument will result in reliable data too.

With reliable instruments, the data produced by such instruments can be trusted too. The reliability testing to instruments was conducted by internal consistency with split half technique, which was conducted by dividing the two scores for each item, which was analyzed using the Brown Spearmen formula. Based on the numbers of tested questionnaires, 20 respondents with significance level of 5% and degree of freedom (dk) of $n - 2$ ($20 - 2 = 18$) and the r table value obtained was 0.468. The results of the instrument reliability testing shows that all variables are reliable because the rcount is bigger than the r table value.

3.3 Path analysis

Direct and indirect influence of the factors forming corporate reputation (X) which consists of trust (X1), satisfaction (X2), loyalty (X3), commitment (X4), quality of employee and management (X5), quality product and service (X6) and attractiveness of the organizational (X7) on CCB (Y) was analyzed using path analysis. The model is presented in Figure 1.

Based on the calculation result above, it can be seen that that H0 is rejected and H1 is accepted. The total influence of corporate reputation is 0.722 or 72.2%, while the other variable path coefficients outside the dimensions of trust (X1), satisfaction (X2), loyalty (X3), commitment (X4), quality of employee and management (X5), quality product and service (X6) and attractiveness of the organizational (X7) are determined using the formula $1 - 0.722 = 0.278$.

This means that the corporate reputation dimension impacts the CCB of 72.2% and the remaining 27.8% is influenced by other factors not included in this research. This condition indicates that corporate reputation has a bigger impact on CCB.

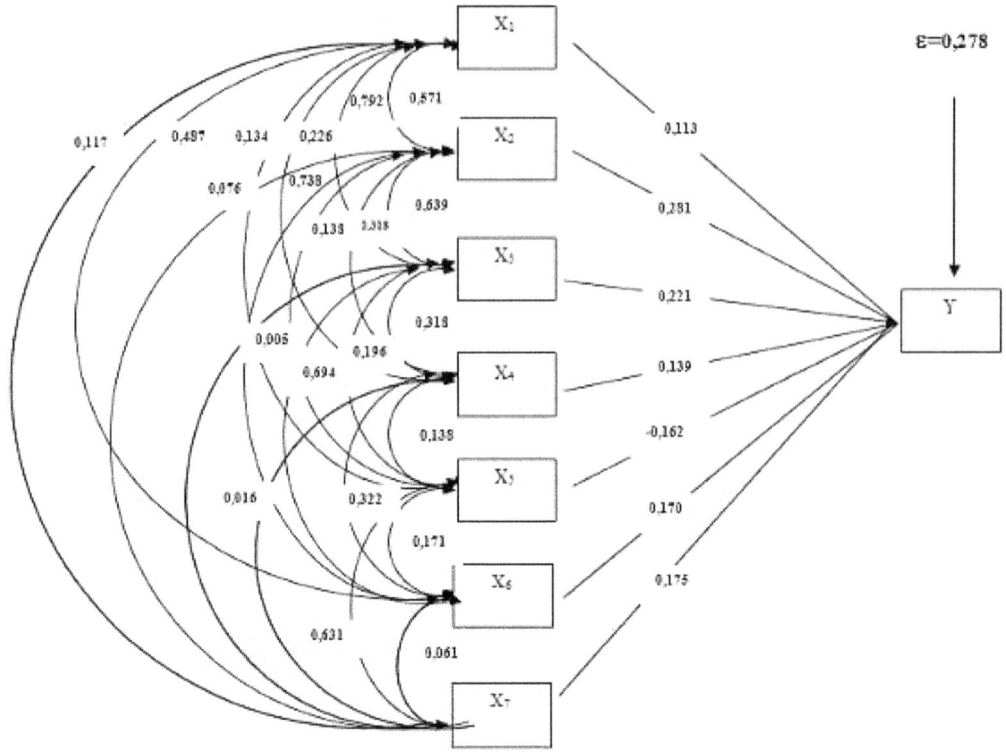

Figure 1. Diagram of analysis of corporate reputation and its impact on customer citizenship behavior.

The results of this research are supported by several previous research. According to Roper & Fill in Manukian (2015), reputation creates certain expectations of the organization, in this case, the hotel. The positive reputation facilitates effective communication between companies and customers, which leads to the spread of word of mouth and good behavior, while according to Grant in Padgett & Leitte (2014), reputation has been identified as one of the most important intangible resources that provide sustainable competitive excellence for the company. Shahsavari & Faryabi (2013) mention that company reputation through customer loyalty and customer commitment affects the dimensions of customer citizenship behavior. Karami et al. (2013) explains that the dimension of reputation influences customer behavior. The statement can strengthen the assumption that corporate reputation has an impact on customer citizenship behavior. Hotels must maintain a strategy of corporate reputation that has been held for a long time in an effort to improve customer citizenship behavior.

4 CONCLUSION

Based on the results of this research, it can be concluded that corporate reputation has a positive and significant effect on customer citizenship behavior.

The dimension of quality of product and service has the highest score, which proves that the performance of the corporate reputation of the hotel has received appreciation or positive response.

Response of repeater guest to corporate reputation can be seen from each dimension – trust, satisfaction, loyalty, commitment, quality employee and management, quality product and service and attractiveness of the organization – which obtained very high score. Quality of product and service have the highest score, which proves that the performance of hotel corporate reputation has gained appreciation or positive response.

Response on customer citizenship behavior which consists of help other customer, help the organization, altruism, courtesy, civic virtue, and conscientiousness, results in a very high value. The highest value comes from help organization dimension because guests are given convenience by the hotel to convey what the guests feel in the form of ideas or comments to the hotel.

Based on the conclusion there are several suggestions proposed. First, the attractiveness of the organization obtained the lowest value, therefore it is recommended that the hotel realize the program that has been approved, and improve the existing program. Programs should be adjusted with holidays so there will be many people participate. Furthermore, there should be addition in in-room facilities to increase service quality. To improve altruism, hotel can invite guests to attend various social or charity events and offer them to participate and to maintain organizational assistance, online guest comment should be provided so that guests can leave comment from their mobile phone. Giving gift should also be considered to attract guest to leave comment.

REFERENCES

Abbasi, A., Zivarmoghbeli, & Ebrahimi, A. 2014. Survey Impact Bank Reputation in Customer Citizenship Behavior; case study of customer agri bank of kerman. Technical Journal of Engineering and Applied Sciences, Vol. 4, pp. 359–399.

Anaza, N.A. & Zhao, J. 2013. Encounter-based antecedents of e-customer citizenship behaviors. Journal of Services Marketing.

Bailey, J.J., Gremler, D.D. & McCollough, M.A. 2001. Service encounter emotional value: The dyadic influence of customer and employee emotions. *Services marketing quarterly,* 23(1), 1–24.

Bartikowski, B. & Walsh, G. 2011. Investigating mediators between corporate reputation and customer citizenship behavior. *Journal of Business Reserach,* Vol. 64, pp. 39–44.

Bove, L., Pervan, S., Beatty, S.E. & Shiu, E. 2009. Service worker role in encouraging customer organizational citizenship behaviors. *Journal of Business Research,* Vol. 62, No. 7, pp. 698–705.

Bowen, D.E. & Schneider, B. 1985. Boundary spanning-role employees and the service encounter: Some guidelines for management and research. *The service encounter,* 127, 148.

Cravens, D.W. & Piercy, N.F. 2013. Strategic Marketing 10th Edition. New York: McGraw-Hill International Edition.

Di, E., Huang, C-J., Chena, I-H. & Yu, T-C. 2010. Organisational justice and customer citizenship behaviour of retail industries. *The Service Industries Journal,* Vol. 30, No. 11.

Foote, D.A. & Tang, T.L.P. 2008. Job satisfaction and organizational citizenship behavior (OCB). *Management Decision.*

Garma, R. & Bove, L.L. 2011. Contributing to well-being: customer citizenship behaviors directed to service personnel. *Journal of Strategic Marketing,* 19(7), 633–649.

Groth, M. 2005. Customers as good soldiers: examining citizenship behaviors in internet service deliveries. *Journal of Management,* Vol. 31, No. 1, pp. 7–27.

Gruen, T.W., Summers, J.O. & Acito, F. 2000. Relationship marketing activities, commitment and membership in profesional assosiations. *Journal of Marketing,* Vol. 64, No. 3, pp. 34–49.

Karami, S., Soltanpanah, H. & Rahmani, M. 2013. The relationship between corporate reputation and organizational citizenship beavior in Private Bank City of Sanandaj in Iran. *Interdiciplinary Journal of Contemporary Research in Business,* Vol. 5, No. 3.

Kotler, Bowen & Makens. 2014. Marketing for Hospitality and Tourism, Sixth Editions. Pearson International.

Kotler, P. & Keller, K.L. 2016. Marketing Management 15th Editions. New Jersey: Prentice Hall.

Manukian, R. 2015. Corporate Reputation Evaluation and Service Quality. University of Applied Sciences.

Okurame, D. 2012. Impact of career growth prospects and formal mentoring on organisational citizenship behaviour. *Leadership & Organization Development Journal.*

Organ, D.W. 1988. Organizational Citizenship Behavior: The Good Solider Syndrome, Lexington Books, Lexington, MA.

Padgett & Leite. 2014. The impact of R&D intensity on corporate reputation: Interaction effect of innovation with high social benefit. *Intangible Capital.* Vol. 216–238.

Rosenbaum, M.S. & Massiah, C.A. 2007. When customers receive support from other customers: Exploring the influence of inter-customer social support on customer voluntary performance. *Journal of service research,* 9(3), 257–270.

Sahertian, P. 2010. Perilaku kepemimpinan berorientasi hubungan dan tugas sebagai anteseden komitmen organisasional, self-efficacy dan Organizational Citizenship Behavior (OCB). *Jurnal Manajemen Dan Kewirausahaan,* 12(2), 156–169.

Schiffman, G.L & Kanuk, L.L. 2012. Consumer Behaviour a European Outlook. Pearson: USA.

Shahsavari, A. & Faryabi, M. 2013. The effect of customer-based corporate reputation on customers' citizenship behaviors in banking industry. *Research journal of applied sciences, engineering and technology,* 6(20), 3746–3755.

Tripadvisor.co.id. 2017 Retrieved from www.tripadvisor.co.id Accessed on February 24, 2017.

Yi, Y., Gong, T. & Lee, H. 2013. The impact of other customers on customer citizenship behavior. *Psychology & Marketing,* 30(4), 341–356.

Yi, Y., Nataraajan, R. & Gong, T. 2011. Customer participation and citizenship behavioral influences on employee performance, satisfaction, commitment, and turnover intention. *Journal of Business Research,* Vol. 64, No. 1, pp. 87–95.

Zayas-Ortiz, M., Rosario, E., Marquez, E. & Gruneiro, P. C. 2015. Relationship between organizational commitments and organizational citizenship behaviour in a sample of private banking employees. *International journal of sociology and social policy.*

Zeithaml, V.A., Bitner, M.J. & Gremler, D.D. 2005. NEW Services Marketing: Integrating Customer Focus Across The Firm.

Price and service quality of online transportation provider

R. Setiawan, N.A. Hamdani, G.A.F. Maulani & I. Permana
Universitas Garut, Garut, Indonesia

ABSTRACT: The purpose of this study was to examine how price and service quality affected consumer purchase decision in online transportation services. To achieve this purpose, this study was conducted using descriptive and verification approach. Data were collected through questionnaires distributed to 96 users of Go-Jek application using accidental sampling. Data analysis was performed using multiple regression. The results revealed that prices and service quality had influence on consumer purchase decision. It is; therefore, necessary for online translation companies to improve their service quality and set appropriate prices to improve their sales performance.

1   INTRODUCTION

Online transportation services have recently gained its popularity in Indonesia. One of the biggest players in this niche is Go-Jek, a company founded by Nadiem Anwar Makarim 2010. Its competitors are Grab, Uber, Blue Jack, and so on. Figure 1 presents the most frequently used online transportation applications in Indonesia.

Figure 1 shows that Go-Jek has more users than other applications. In Garut, particularly, the number Go-Jek drivers continue to increase. On the one hand, Go-Jek emerges as an employment opportunity, but on the other hand it is a threat to other transportation service businesses such as public minivan (in Indonesia known as angkot, abbreviation for angkutan kota, urban transportation), taxi, and other conventional motorcycle taxi (in Indonesia known as ojek). As a form of e-commerce application, Go-Jek offers more economical, comfortable, and personalized services, things that conventional transportation services cannot (Kotler 2002).

Figure 2 shows that online transportation users outnumber the conventional ones in Garut. The result of our preliminary study reveal that people choose Go-Jek because it offers economical and efficient transportation services. The use of m-commerce system will increase if the system is user friendly, meets customer needs and provides customer support (Cass & Carlson 2012). There are five steps in customer purchase decision making process: need identification, information search, alternative comparison, purchase, and post-purchase behavior (Kotler 2010 in Shantanu et al. 2019).

In Indonesia, online transportation services are getting so popular that many international companies start to expand their market to Indonesia. But what is an online transportation service? Sometimes, the term ride-sharing is used to refer to online transportation service (Silalahi et al. 2018). Literatures on the trading system and electronic information also devotes their attention to the reputation system, in the broader field of research on platform ecosystem design and governance (Basili & Rossi 2019). When creating a product line, the retailer must make several decisions simultaneously; choosing the type of product to be included in the product line and the order quantity and price of each type of product (Moon et al. 2017). Price is set accord-ing to product categories (Fecher et al. 2019).

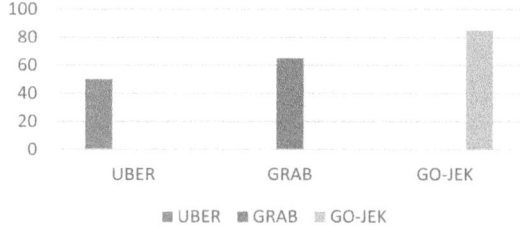

Figure 1.   Popular online transportation applications in Indonesia.

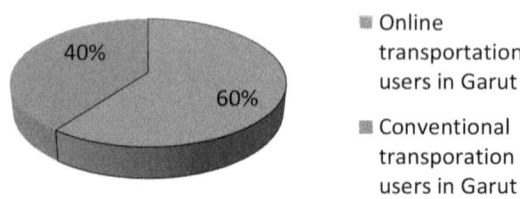

Figure 2.   Online transportation service users in Garut.

Price is also the amount of money used as a means of exchange for getting a product or service (Tjahjono & Rezza 2015). The way to succeed as a content company is to supply exclusive content at a price that is agreeable to the customer (Leland & Bob 2009).

Purchase targets can be related to search characteristics and certain search patterns, which in turn affect customer purchase decision (Weisstein et al. 2017). Price expectations and service epectations should not be confused. Price expectation refers to predictions about the price attribute. In the expected service construct, it is the predicted service that is formed (Toncar et al. 2014).

SERVQUAL is an instrument to measure a service quality and can be used as a contemporary re-search framework (Parasuraman & Zeithaml 1985). Service quality rates are calculated for the industry overall and by individual carrier (Dawna & Blaise 2008). Service quality in particular has repeatedly proven to be a key indicator of customer satisfaction (Brusch et al. 2019).

It is very important to measure service quality both using objective and subjective approaches (Deb & Ahmed 2019), including services provided in m-commerce environment (Huang et al. 2015). Customer purchase of Go-Jek services is facilitated by price and service quality. Customers will always expect the products or services in this matter to be of high quality at a low price (Wu et al. 2017). Hence, this study examines how these two factors, prices and service quality, affect purchase decision.

2 METHODS

This study was conducted on Go-Jek users in Garut. The sample was chosen using a non-probability sampling technique. The respondents had to be 17 years or older to ensure that they understood the nature of questionnaires addressed to them. Since the total number of populations was uncertain, the number of samples was calculated using Lemeshow's (2010) formula as follows.

$$n = 1.962x(0.5(1 - 0.5))/(0, .) = 96.04 \rightarrow 96 \qquad (1)$$

Therefore, the number of respondents in this study were 96 Go-Jek users. The researcher formulated the conceptual model as presented in Figure 3.

The hypothesis were formulated as follow:

H1: Price has influence on Go-Jek customer purchase decision

H2: Service quality has influence on Go-Jek customer purchase decision

3 RESULTS AND DISCUSSION

3.1 *Multiple regression test*

Table 1 presents the results of data processing. The multiple regression equation is as follows:

$$Y = 22.283 + 0.330X1 + 0.159X2 \qquad (2)$$

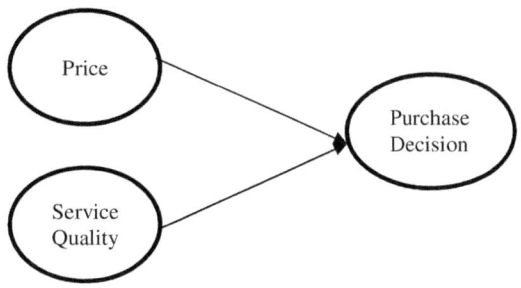

Figure 3. Conceptual model.

Table 1. Coefficients.[a]

Model	Unstandardized coefficients		Standardized coefficients		
	B	Std. error	Beta	t	Sig.
1 (Constant)	22,283	6,876		3,241	,002
price	,330	,125	,260	2,645	,010
Service quality	,159	,072	,217	2,204	,030

a. Dependent Variable: Purchase Decision

The equation shows that the constant is 22.283, meaning that without the variables price and service quality, the purchase decision is 22.83. Therefore, the multiple regression equation can be interpreted as follows:

1. The value of +0.330 indicates that if price increases by 1 point, purchase decision will in-crease by 0.330 points.
2. The value of +0.159 indicates that if service quality increases by 1 point, purchase decision will increase by 0.159 points.

3.2 *Coefficient of determination*

The coefficient of determination tells us how much prices and service quality influence customer purchase decision. This can be calculated using the following formula:

$$D = r^2 X 100\% \qquad (3)$$

Note:
D = coefficient of determination
r2 = 0.807
D = 0.807 × 100%
= 80.7%

The above equation means that prices and service quality contribute to purchase decision as much as 80.7%, the rest is influence by factors other than the studied variables.

Table 2. ANOVA.[a]

Model		Sum of Squares	df	Mean Square	F	Sig.
1	Regression	233,872	2	116,936	5.517	,005b
	Residual	1971,034	93	21,194		
	Total	2204,906	95			

a. Dependent Variable: Purchase Decision
b. Predictors (Constant), Service Quality, Price

3.3 Hypothesis testing

3.3.1 Goodness of fit test (F-test)
The measure of goodness of fit was carried out by means of SPSS.20 by comparing the Sig. value with the probability value of 0.05.

1. If the observed F value > critical F value, the proposed model is good fit for hypothesis testing.
2. If the observed F value ≤ critical F value, the proposed model is not fit for hypothesis testing.

The result of SPSS.20 data processing is presented in Table 2.

Table 2 shows that the observed F value was 5.517, higher than the critical F value of 3.0933. Therefore, the proposed regression model is fit for hypothesis testing. Consumers continue to perceive a positive relationship between product quality and price or product quality, which can serve as a proxy for price (Rezaei 2015).

3.4 Influence of prices on purchase decision

To measure the influence of prices on customer purchase decision, the following hypotheses are stated:

H0: $\beta yx \leq 0$, meaning that price has no influence on purchase decision.

H1: $\beta yx > 0$, meaning that price has influence on purchase decision.

The partial test was performed by comparing the Sig. value with the alpha value of 0.05. The decision is based on the following criteria:

1. H0 is rejected if Sig < alpha (0.05)
2. H0 is accepted if Sig > alpha (0.05)

It was revealed that the Sig. value was 0.01, lower than 0.05. Therefore, H0 was rejected, meaning that price has influence on customer purchase decision.

3.5 Influence of service quality on purchase decision

To measure the influence of service quality on customer purchase decision, the following hypotheses are stated:

H0: $\beta yx \leq 0$, meaning that service quality has no influence on purchase decision.

H1: $\beta yx > 0$, meaning that service quality has influence on purchase decision.

The partial test was performed by comparing the Sig. value with the alpha value of 0.05. The decision is based on the following criteria:

1. H0 is rejected if Sig < alpha (0.05)
2. H0 is accepted if Sig > alpha (0.05)

It was revealed that the Sig. value was 0.03, lower than 0.05. Therefore, H0 was rejected, meaning that prices have influence on customer purchase decision. This finding is in line with that of previous studies (see et al. 2019; Zhang et al. 2013).

4 CONCLUSION

Based on the respondents' responses, it can be concluded that:

1. Go-Jek has set affordable prices to the customers.
2. Respondents believe that Go-Jek offers sufficiently good services.

Customer purchase decision in Go-Jek services depends on the service price and service quality. Therefore, it is necessary for Go-Jek and other companies of its like to take into account service price and service quality dimensions such as tangibles, empathy, responsiveness, reliability and assurance.

ACKNOWLEDGEMENT

This research work is supported by Faculty of Entrepreneur Garut University.

REFERENCES

Basili, M. & Rossi, M.A. 2019. Platform-mediated reputation systems in the sharing economy and incentives to provide service quality: the case of ridesharing services. Electronic Commerce Research and Applications, 100835. https://doi.org/10.1016/j.elerap.2019.100835
Brusch, I., Schwarz, B. & Schmitt, R. 2019. David versus Goliath – Service quality factors for niche providers in online retailing. Journal of Retailing and Consumer Services, 50(April), 266–276. https://doi.org/10.1016/j.jretconser.2019.05.008
Cass, A.O. & Carlson, J. 2012. An e-retailing assessment of perceived website-service innovativeness: Implications for website quality evaluations, trust, loyalty and word of mouth. Australasian Marketing Journal (AMJ), 20(1), 28–36. https://doi.org/10.1016/j.ausmj.2011.10.012
Dawna, R.L. & Blaise, J.W. 2008. Twenty years of service quality performance in the US airline industry. Managing Service Quality: An International Journal, 18, 20–33. https://doi.org/10.1108/09604520810842821
Deb, S. & Ahmed, M.A. 2019. Quality assessment of city bus service based on subjective and objective service quality dimensions. https://doi.org/10.1108/BIJ-11-2017-0309

Fecher, A., Robbert, T. & Roth, S. 2019. Same price, different perception: Measurement-unit effects on price-level perceptions and purchase intentions. *Journal of Retailing and Consumer Services*, 49(March), 129–142. https://doi.org/10.1016/j.jretconser.2019.03.017

Huang, E. Y., Lin, S. & Fan, Y. 2015. Electronic Commerce Research and Applications M-S-Qual: Mobile service quality measurement. *Electronic Commerce Research And Applications*, (January). https://doi.org/10.1016/j.elerap.2015.01.003

Kotler, P. 2002. Marketing in the twenty-first century. Marketing Management, Millenium Edition (10th ed.). New Jersey: Prentice-Hall. https://doi.org/10.1016/j.ssi.2013.04.024

Lee, C., Zhao, X. & Lee, Y. 2019. Service Quality Driven Approach for Innovative Retail Service System Design and Evaluation: A Case Study Corresponding Author: School of Design, South China University of Technology, China. Computers & Industrial Engineering. https://doi.org/10.1016/j.cie.2019.06.001

Leland, H. & Bob, H. 2009. Digital Engagement: internet marketing that captures customers and builds intense brand loyalty (1st ed.). New York: AMACOM, American Management Association.

Moon, I., Park, K.S., Hao, J. & Kim, D. 2017. PT US CR. *European Journal of Operational Research*. https://doi.org/10.1016/j.ejor.2017.03.062

Parasuraman, A. & Zeithaml, V.A. 1985. A Conceptual Model of Service Quality and Its I-mplications for Future Research, 49(1979), 41–50.

Rezaei, S. 2015. Segmenting consumer decision-making styles (CDMS) toward marketing practice: A partial least squares (PLS) path modeling approach. *Journal of Retailing and Consumer Services*, 22, 1–15. https://doi.org/10.1016/j.jretconser.2014.09.001

Shantanu, P., Arushi, G. & Saroj, P. 2019. Purchase decision of generation Y in an online environment. *Marketing Intelligence & Planning*. https://doi.org/10.1108/MIP-02-2018-0070

Silalahi, S.L.B., Handayani, P.W. & Munajat, Q. 2018. Science Direct Service Quality Analysis for Online Transportation Services: Case Study of GO-JEK. Procedia Computer Science, 124, 487–495. https://doi.org/10.1016/j.procs.2017.12.181

Tjahjono, D. & Rezza, P. 2015. Brand Image and Product Price; Its Impact for Samsung Smartphone Purchasing Decision, 219, 221–227. https://doi.org/10.1016/j.sbspro.2016.05.009

Toncar, M. F., Alon, I., Misati, E. & Toncar, M.F. 2014. Pricing strategy & practice The importance of meeting price expectations: linking price to service quality. https://doi.org/10.1108/10610421011059612

Weisstein, F. L., Song, L., Andersen, P. & Zhu, Y. 2017. Examining impacts of negative reviews and purchase goals on consumer purchase decision. *Journal of Retailing and Consumer Services*, 39(July), 201–207. https://doi.org/10.1016/j.jretconser.2017.08.015

Wu, J., Hwang, J., Sharkhuu, O. & Tsogtochir, B. 2017. Asia Pacific Management Review Shopping online and off-line? Complementary service quality and image congruence. *Asia Pacific Management Review*, 1–7. https://doi.org/10.1016/j.apmrv.2017.01.004

Zhang, M., Xie, Y., Huang, L. & He, Z. 2013. Service quality evaluation of car rental industry in China. https://doi.org/10.1108/IJQRM-11-2012-0146

Analyzing customer equity in the telecommunication industry

V. Silviana, H. Hendrayati & P.D. Dirgantari
Universitas Pendidikan Indonesia, Bandung, Indonesia

ABSTRACT: The business tendency in telecommunications industry is influenced by the high level of competition, fare wars, as well as high customer demands for the quality of service provider. The purpose of this research was to explore the implementation of customer equity in companies engaged in telecommunications industry. The dimensions used were value equity, brand equity, and relationship equity. To support this research, 20 students from one of the universities in Bandung were interviewed, then processed using descriptive method. The results showed that the implementation of customer equity fosters loyalty. Customer equity can keep long term customers to remain loyal while also getting new customers.

1 INTRODUCTION

Customer Equity is a long-term asset that is based on relationships that have been built since the beginning and last for a long time with consumers (Kim & Ko 2012). Furthermore, Hossain (2017) argued that Customer Equity is a concept resulting from the effort of maintaining a lifelong relationship with customers that utilizes the sophistication of marketing and direct marketing technologies that are related to creation of value resulting from profits, costs, and cash flows. By considering aspects of consumer behavior, existing research proposes three drivers of customer equity, i.e. value equity, brand equity, and relationship equity. (Hossain 2017). To create a competitive advantage, companies must improve the equity of their customers (Lee et al. 2014). For the continuity of a business, customer equity can contribute to growth and return on investment (Matsuno et al. 2014). A company needs steps that focus on individual customers to achieve the best performance they want (Mizit 2014).

Another concept of customer equity by Rust et al. (2001) integrates a model of customer equity into a model to compute customer lifetime value in a service industry. The model uses individual customer approach and calculates the value of the project in the future. The factors taken into account are the age of the customer relationship and life companies, time period, frequency of customer arrival, expected contribution per subscriber, the arrival, and the discount rate. Furthermore, Rust et al. (2001) stated that customer equity is supported by value relationship equity. customer equity is one of the methods of connecting between marketing programs with the possibility of a customer providing benefits for the company in the future.

Rust et al. (2001) added the definition of customer equity as the total lifetime value of customer base. Total lifetime value calculates the value of customers to a company by taking into account factors such as acquisition, retention, profits, and costs. However, the equivalence of these customers is viewed from financial perspective. On the other hand, considering aspects of the consumer behavior, there is research that suggests three driving factor of customer equity i.e. value equity, brand equity, and relationship equity (Huo et al. 2013). This research adopted the dimensions of customer equity i.e. its implementation and impact, so that the company can maintain customer loyalty while getting new customers.

A company can optimize customer equity by retaining existing customers or attracting new customers. (Kim et al. 2014). Various methods such as telemarketing, door-to-door sales, and direct mail have been used to be more involved with customers and increase their lifetime value (Nguyen et al. 2014). This opinion was proven by a research in the field of banking (Sitorus 2013) which showed that the higher the level of customer equity, the higher the brand equity. The respondents stated that they did not have a plan to switch to another bank as well as agreeing to recommend to others, including the next generation. Other research (Alfifto 2017) in the field of social media saw the influence of customer equity on young entrepreneurs in Medan, the results of these studies indicated that social media marketing consisting of consumption, curation, creation, and collaboration had a positive and significant effect on customer equity in young entrepreneurs in Medan.

This research focused on telecommunications industry, where the aspects examined were forms of activities that created customers loyalty. It aimed to explore the implementation of customer equity drivers – value equity, brand equity, and relationship equity in one of the telecommunications companies in Indonesia.

2 METHOD

This research used qualitative methods and was conducted for one month in November 2018. Data was collected through interviews, the sampling technique used was purposive sampling with a sample of 20 Postgraduate students of Universitas Pendidikan Indonesia who were Telkomsel users. The data was then processed using descriptive analysis techniques.

3 RESULTS AND DISCUSSION

According to Sauter (2010) customer decision making process is short when customers feeling positive emotions and is considerably long when customers experiencing negative emotions. Relevant with the results of the researchs, in implementing customer equities, Telkomsel continues to maintain quality so that old customers remain loyal while recommending Telkomsel to colleagues or family. A similar result was produced from a research by Sitorus in 2013 in banking industry, where customers agreed to recommend Bank Danamon to children or grandchildren and had no plans to move to another bank. Furthermore, on the aspect of brand equity, customers recognize that Telkomsel is a telecommunications company with a good reputation. The presence of more than 400 GraPARI spread throughout Indonesia became a brand reinforcement for Telkomsel to further grow and develop.

Social media allows consumers to be storytellers, whereby they may be critical about bad brand experiences or narrate positive experiences and makes it easier for them to engage with a brand and share brand experience through the social networks (Gökerik et al. 2018). With the sophistication of technology and the ease of accessing the internet, social media is one of the access points between customers and Telkomsel, so customers can get faster responses to complaints submitted through customer service on Twitter.

A research by Alfifto (2017) on the influence of social media marketing on young entrepreneurs in Medan mentioned various benefits of social media for business development such as maximizing network, brand awareness, increasing traffic and search engine rankings, maximizing customer satisfaction, and attracting new customers where the benefits are the same benefits if a company implements customer equity activities well. Entrepreneurs argue that the existence of social media can facilitate effective communication between SME managers, consumers, and suppliers. From several previous studies, it can be concluded that customer equity practices have been successfully implemented in telecommunications, banking and entrepreneurship companies.

4 CONCLUSION

The researcher found that the implementation of Customer Equity at PT Telekomunikasi Selular (Telkomsel) has been running well in every aspect of so as to bring benefits in the form of profits and the number of customers in large numbers. With the implementation of customer equity, Telkomsel has maintained a good relationship between the company and its customers as an asset. Telkomsel has a lot of users who are loyal even though the price offered is higher than other providers but it gives the best quality that makes its customer stay.

ACKNOWLEDGEMENT

This research would not have done without the support of various parties. Therefore the researcher thanked the Postgraduate School of the Indonesian University of Education. In obtaining the data, the authors are very grateful to all respondents who have spent their time and all those who have supported this article from the research that has been made.

REFERENCES

Alfifto. 2017. Pengaruh social media marketing terhadap customer equity pada pengusaha muda di Kota Medan. Universitas Sumatera Utara.

Gökerik, M., Gürbüz, A., Erkan, I., Mogaji, E. & Sap, S. 2018. Surprise me with your ads! The impacts of guerrilla marketing in social media on brand image. *Asia Pacific Journal of Marketing and Logistics*, Vol. 30 No. 5, pp. 1222–1238.

Hossain, T.M. 2017. The impact of integration quality on customer equity in data driven omnichannel services marketing. *Procedia Computer Science*, Volume 121, 784–790.

Huo H., Xu W. & Chen, K. 2013. The influence of customer equity drivers on specific purchasing behavior in retail industry. *Information Technology Journal*, 12: 2236–2240.

Kim, A.J. & Ko, E. 2012. Do social media marketing activities enhance customer equity? An empirical study of luxury fashion brand. *Journal of Business Research*, Volume 24, 1480-1486.

Kim, K., Ko, E., Lee, M., Mattila, P. & Kim, K. 2014. Fashion collaboration effects on consumer response and customer equity in global luxury and SPA brand marketing. *Journal of Global Scholars of Marketing Science*, Volume 24, 350–364.

Lee, C.H., Ko, E., Tikkanen, H., Phan, M.C.T., Aiello, G., Donvito, R. & Raithel, S. 2014. Marketing mix and customer equity of SPA brands: Cross-cultural perspectives. *Journal of Business Research*, Volume 67, 2155–2163.

Matsuno, K., Zhu, Z. & Rice, M.P. 2014. Innovation process and outcomes for large Japanese firms: Roles of entrepreneurial proclivity and customer equity. *Journal of Product Innovation Management*, Volume 31, 1106–1124.

Mizit, N. 2014. Assessing the total financial performance impact of brand equity with limited time-series data. *Journal of Marketing Research*, Volume 51, 691–706.

Nguyen, B., Chang, K. & Simkin, L. 2014. Customer engagement planning emerging from the "individualist-collectivist"-framework: an empirical examination in China and UK. Mark. Intell. Plan. Volume 32, 41–65.

Rust, Zeithaml & Lemon. 2001. Driving customer equity how customer lifetime value is reshaping corporate strategy. New York: The Free Press.

Sauter, D. 2010. More than happy the need for disentangling positive emotions. *Current Directions in Psychological Science*, Vol. 19 No. 1, pp. 36–40.

Sitorus, T. & Ardy. 2013. Analisis efek mediasi customer equity atas pengaruh customer relationship marketing dan customer satisfaction terhadap brand equity. *Jurnal Manajemen*, 104–127.

The impact of brand equity and customer satisfaction on bank customer loyalty

F.A. Fauzan & H. Hendrayati
Universitas Pendidikan Indonesia, Bandung, Jawa Barat, Indonesia

ABSTRACT: Nowadays, customers are getting smarter, price conscious, demanding, less forgiving, and approached by many products, which also happens in the banking industry. The aim of this study is to find out the impact of brand equity and customer satisfaction to customer loyalty of BUMD banks. This research used descriptive and verification method. The population in this study was 4192 by involving 110 respondents as the sample. Meanwhile, data collection was carried out through observation, interviews and questionnaires to obtain primary data. Multiple linear regression analysis was used in this study to analyze data statistically. The results of multiple regression studies indicated that brand equity variables and customer satisfaction had a sig-nificant contribution to loyalty. Therefore, companies must be able to maintain the existing clients/customers seen from various factors to get sustainable value in carrying out its business activities. The hypothesis that brand equity and customer satisfaction had positive and significant influence to the loyalty of BUMD Bank customers was accepted. Further study can be carried out by perceiving other factors that have contributions in customer loyalty.

1 INTRODUCTION

The dynamic economic growth triggers fiercer competition, one of which is in the banking sector. In the midst of banking competition, one of the main factors that can be implemented to increase the growth of the number of customers is offering inno-vative products that meet customer expectations. Customers also now have an increasingly intelligent character, price conscious, demanding, less forgiv-ing, and are approached by many products. Moreo-ver, advances in banking technology have increasingly intensified the competition level, because it opens up wider access to information about various types of products offered to the customers (Mardalis 1998).

When looking at the customer conditions, the views of the company will also change, which can be seen from customer loyalty as a valuable asset of the company that become the goal of many companies. Customer loyalty is considered as an important key to the success of the organization, which contributes profit for a long period of time (Oliver 1999). For companies or organizations competing in the business environment, certainly, customer loyalty is placed in a strategic portion, and it can be managed with a variety of efforts and great innovations so that the company can achieve its goals (Srivastava & Rai 2018). Analyst Meeting 4Q 2013 BJB Bank Presentation in Figure 1:

The savings program is part of the company's strat-egy to encourage the growth of the number of customer accounts. The growth target can be even doubled compared to the previous year. The company was very

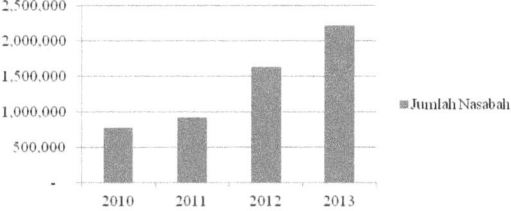

Figure 1. Regional bank customer growth.

Table 1. Number of regional bank customers.

	Year	Savings	Current Account	Deposit	Total
1	2013	4337	101	48	4486
2	2014	4055	99	86	4240
3	2015	4011	105	124	4240
4	2016	3964	102	126	4192

optimistic if the target could be achieved. Moreover, until September 2013, the number of customers of BJB Bank almost reached 2.5 million customers. However, based on data in one of the sub-branch offices, the number of customers has actually decreased from year to year. This can be seen in terms of savings, current accounts, and deposits as in Table 1.

This study aims to look at customer perceptions from the perspective of corporate brand equity and satisfaction that can lead to consumer loyalty from the banking industry.

Brand equity is considered a very effective tool for increasing marketing productivity (Cai et al. 2014). By evaluating brand equity, it means that a company can measure customer satisfaction and brand performance, which is generally conducted by marketing managers (Sung Ho et al. 2015). Brand equity from the consumer's point of view is a suitable initial factor for valuing product equity with many interrelated dimensions such as brand awareness, brand quality, brand association, and brand loyalty (Jing et al. 2015).

Customer satisfaction is the main goal for most companies (Jones & Sasser 1995). Increased customer satisfaction and customer retention will have a positive effect on the improvement of profits, positive news from mouth to mouth, and lower market-ing costs (Cook 1996; Sasser et al. 1997). The value of customer satisfaction is based on how much the company can provide more value to consumers. It is expected that customers will feel something more than they projected. Consumer satisfaction will also affect market share but does not have a direct impact on newcomers (Zahorik & Roland 1993).

Customer loyalty is also one of the essential assets for the company and precisely is a part of the company's goals. Customer loyalty influences the success of the organization and the increase in company profits (Oliver 1997). Selin et al. (1987) state that consumers show the greatest degree of loyalty to products or services, and they tend to repurchase. Therefore, customer loyalty is always placed in a strategic position in almost every organization that faces a business environment every day (Srivastava & Rai 2018). Loyal customers are generally willing to contribute to control the products, offer refer-ences, and indirectly run word-of-mouth (Bowen & Shoemaker 1998). Thus, research attention will in-creasingly focus on identifying the effectiveness of methods to increase loyalty actively conducted Lach (2000). Customer loyalty has the potential to encourage differentiation and maintain competitive advantage (Nastasoiu & Vandenbosch 2019).

2 METHODS

This study aims to see the effect of brand equity and customer satisfaction on the loyalty of Bank customers in West Java. There were two variables used, namely independent and dependent. The first independent variable consisted of brand equity with dimensions of brand awareness, perceived quality, brand association, and brand loyalty. The second one is independent variable that was about satisfaction, with tangibility, empathy, reliability, responsiveness, and assurance dimensions. Furthermore, the dependent variable included loyalty with the dimensions of repurchase, recommending products, and consumer's loyalty. The researcher used mixed methods qualitative and quantitative, descriptive and verification, in which the research results were processed to formulate the

conclusions. Sample collection was completed using purposive sampling method. As many as 110 customer respondents were involved in this study. In addition, the test statistic utilized multiple regression tests.

3 RESULTS AND DISCUSSION

Based on the results of the accumulation of sub-variable average scores, the Brand Equity variable turned out to have a pretty good value because the brand equity of BUMD (Regionally-Owned Enterprises) banks was still not strong enough, considering the very tight competition at this time. In addition, regional bank brand equity was still lower than national or private banks. The result of the accumulation of the average scores of the sub-variable of Customer Satisfaction showed that the value was also quite good. Good points in respondent satisfaction caused this possible to happen.

Responses regarding loyalty variables showed the same good value because bank customers felt safe and comfortable to conduct banking transactions. Hence, it formulated the multiple linear regression equations of:

$$Y = 0.667 + 0.382X1 + 0.382X2 \qquad (1)$$

Based on the regression analysis above, it can be explained that the independent variable of Brand Equity had a positive effect as much as 0.382 (38.2%) on the dependent variable Loyalty. As a result, H0 was rejected and H1 was accepted, which was happened because the local government owned the bank. It can be clearly assumed that regional bank brand equity was still lower than other national or private banks. The independent variable of Customer Satisfaction had a positive effect as much as 0.406 (40.6%) on the dependent variable of Loyalty. Therefore, H0 was rejected and H1 was accepted. The results were quite good because as a company engages in services, customer satisfaction was the main factor that had to be considered since the impact could be felt directly by bank customers.

This study also used the F Test that assessed regression relationships simultaneously. The level of significance taken for this study was 5%. Based on the results obtained from the comparison of Fcount with Ftable, H0 was rejected and H1 was accepted because Fcount 47,969> Ftable 3.08. The results from the significance level comparison showed that Ho was rejected, because of the sig. value was 0,000 <0,05. This means that brand equity and customer satisfaction had a significant influence on customer loyalty.

4 CONCLUSION

The description above shows that the brand equity status and customer satisfaction were considered to be in good level. It was based on the results of the study using descriptive analysis and verification between brand equity and satisfaction with customer loyalty. The results of multiple linear regression test showed both variables had an influence on increasing customer loyalty. From the results of the study, other variables had considerable influence apart from brand equity and satisfaction, which can affect customer loyalty. Thus, the researcher suggests the next researcher to develop another model in analyzing how much influence it has on customer loyalty.

ACKNOWLEDGMENT

The researcher would like to thank the UPI Postgraduate Campus, Bandung, especially the Advisor for the Management Study Program and for all par-ties who helped during the process, so that this research can be completed.

REFERENCES

Bowen, J., & Shoemaker, S. 1998. Antecedents and consequences of customer loyalty. Cornell hotel and restaurant triwulan administration, 39 (1), 12–25

Cai, Y. Y., Zhao, G., He, J. 2014. Influences of two modes of intergenerational communication on brand equity, *Journal of Business Research*, JBR-08154; No of Pages 8

Cook, S. D. 1996. The quest for loyalty: creating value through partnership (p. MA). F. F. Reichheld (Ed.). Boston, MA: *Harvard Business School Press*.

Jing, Z., Yanxin, J., Rizwan, S., Mingfei Du 2015. Building industrial brand equity by leveraging firm capabilities and co-creating value with customers, Industrial Marketing Management, IMM-07207; 12

Jones, T. O., & Sasser, W. E. 1995. Why satisfied customers defect. *Harvard business review*, 73(6), 88.

Sasser, W. E., Schlesinger, L. A., & Heskett, J. L. 1997. Service profit chain. Simon and Schuster.

Lach, J. 2000. Quality of redemption, American demographics, Vol.22 No. 5, page. 36–38.

Mardalis, A. 1998. 1217-2029-1-Sm. Meraih Loyalitas Pelanggan, 111–119. Retrieved from http://journals.ums.ac.id/index.php/benefit/article/viewFile/1217/781

Nastasoiu, A., & Vandenbosch, M. 2019. Competing with loyalty: how to design successful customer loyalty reward programs. *Business horizons*, 62(2), 207–214. https://doi.org/10.1016/j.bushor.2018.11.002

Oliver, RL 1997. Satisfaction: a behavioral perspective on the consumer, Irwin-McGraw, Hill, New York, NY.

Oliver, R. L. (1999). Whence consumer loyalty?. *Journal of marketing*, 63(4_suppl1), 33–44.

Selin, S. W., Howard, D. R., Udd, E., dan Cable, T. T. 1987. Analysis of consumer loyalty for city recreation programs. *Leisure sciences*, Vol.10, hlm. 217–223.

Srivastava, M., & Rai, A. K. 2018. Mechanics of engendering customer loyalty: a conceptual framework. *IIMB Management Review*, 30(3), 207–218. https://doi.org/10.1016/j.iimb.2018.05.002

Sung Ho H., Bang N., Leec, T. J.2015. Consumer-based chain restaurant brand equity, brand reputation, and brand trust. *International Journal of Hospitality Management*, 50 (2015) 84–93.

Zahorik, Anthony J., & Rust, Roland. T. 1993. Customer satisfaction, customer retention, and market share. *Journal of retailing*, 69(2), 193–215.

The creation of a tourism experience to increase tourist satisfaction after visiting Ciletuh-Palabuhanratu geopark

R.T. Hidayah, R. Hurriyati & H. Hendrayati
Universitas Pendidikan Indonesia, Bandung, Indonesia

ABSTRACT: Ciletuh-Palabuhan Ratu Geopark is a leading tourist destination owned by West Java Province at this time. It has a resource of natural panoramas that have received international recognition from UNESCO. However, it has not been able to significantly increase the level of tourist visits in Sukabumi district even though it is said to have strong potential. This research aimed at determining how much the experience of tourism influence tourist satisfaction after visiting Ciletuh-Palabuhanratu Geopark. The research method used in this research was descriptive and verification method. The population was local tourists who have visited Ciletuh-Pelabuhanratu Geopark tourist attraction with a total sample of 100 respondents. The data analysis method used in this study was Path Analysis. The results of the study revealed that the response of domestic tourists regarding the experience of tourism and tourist satisfaction categorized as good, and the experience of tourism influenced the satisfaction of local tourists.

1 INTRODUCTION

Many researchers suggest that tourist satisfaction is one of the key factors in ensuring the level of tourist arrivals (Alegre & Cladera 2009). In relation to economic growth in the destination area, the achievement of satisfaction has a positive impact on local people's income because tourist satisfaction can affect the amount of expenditure that will be spent by tourists in a tourist destination (Smolèiæ Jurdana & Soldiæ Frleta 2017).

Thus, from various perspectives the tourist satisfaction is seen as an important research topic in tourism studies because tourist satisfaction ensures continuity of acceptance of a destination (Xia et al. 2009).

Tourist satisfaction is closely related to expectation and is an emotional reaction as a result of the experience of visiting (Prayag et al. 2017). Tourist satisfaction will be determined by the experience of tourists with the attractions experienced in a particular tourist attraction (Suhartanto et al. 2019). The experience felt will be compared with expectations, resulting in a sense of satisfaction. (Chen & Chen 2010; Pratminingsih et al. 2014).

In tourism industry, the studies on satisfaction have been conducted in various types of attractions including creative tourism (Suhartanto et al. 2019), urban tourism (Chiu et al. 2016), nature tourism (Smolèiæ & Soldiæ 2017), Island tourism (Prayag et al. 2017), heritage tourism/historical sites (Chiu et al. 2016; Wu & Li 2017), and playground tourism (Wu & Li 2017). Research related to geological tourism is limited, so further studies are certainly needed.

Geological tourism is a form of tourism that is gaining attention at various levels of the tourism community such as local, national and international (Farsani et al. 2011; Newsome & Dowling 2010; Ruban 2015). Geo-tourism as a form of sustainable tourism industry will be able to contribute to the sustainability and improvement of the economic prosperity of the local community, which is reflected in the reduced level of unemployment and urbanization as well as the development of sustainable areas in the geo-tourism destination (Cheung 2016; Farsani et al. 2011; Jorgenson & Nickerson 2016).

Geopark Ciletuh-Palabuhanratu located in Sukabumi Regency is a leading destination in the West Java region. However, these destinations are not without problems. Ciletuh-Palabuhanratu Geopark is not able to boost the level of tourist visits in the period of time. As in the long Christmas holiday season 2018 and the New Year 2019, the tourist attraction located on the South Coast of Sukabumi experienced deserted visitors, which affected the achievement of the number of tourist visits (www.pikiran-rakyat.com).

In order to strengthen the above findings, a pre-survey of 30 tourists who had visited the Ciletuh Geopark-Palabuhanratu was conducted. The results of the majority of respondents expressed satisfaction with the quality of travel they experienced on their visit. However, different results appeared when they were asked about their desire to make a visit in the future. Eight tourists said "no" while 9 tourists said "did not know" and only 13 tourists who firmly an-swered would make another visit.

The problem was also triggered by the accessibility problem to the tourist area, which can be seen from the damaged road conditions in several locations and landslides in several points. This road damage affected the length of time taken from the City of Sukabumi to Palampang Beach, Ciemas District that takes 4 to 5 hours by car (www.tribunnews.com).

The above problems arises when all the efforts made by the Government of West Java that succeeded in bringing the Ciletuh-Palabuhanratu Geopark were determined to become UNESCO Global Geopark (UGG) and this predicate brought Ciletuh into an international scale destination (www.travel.kompas.com).

The research aimed at determining the response of tourists who have visited the Ciletuh-Palabuhanratu Geopark of their experiences while traveling and their level of satisfaction. This study also examined the effect of travel experiences on tourist satisfaction after visiting Ciletuh Geopark-Palabuhanratu.

The tourist satisfaction is important to form a successful destination because it can influence tourists in the selection of tourism destinations, as well as tourists' decision to make a return visit (Kozak & Rimmington 2000). Satisfaction is a positive reaction to the results of the assessment of consuming an experience (Babin 1998) caused by two dimensions, namely the expectations of tourists before carrying out tour activities and the experience of tourists after a visit (Pratminingsih et al. 2014). Satisfaction will be achieved when it can fulfill the expectations (Chen & Chen 2010).

The concept of satisfaction is defined as the extent of the tourists' evaluation towards the destination attributes that exceeds expectations (Alegre & Cladera 2009). The comparison involves the expectations of tourists with what tourists see, feel and experience directly when visiting (Meng et al. 2008). Because the essence of the tour activity is experience (Lofman 1991; Mansour & Ariffin 2017), the management of tourist destinations is demanded to be increasingly focused in the creation, management, and delivery of experiences for tourists (Tung & Ritchie 2011).

In the form of experience, the emotional side (Bigné & Andreu 2004; Bigné et al. 2008) and tourist involvement are important conditions for the creation of experience (Moon & Han 2018, 2019). Experience is formed in every part of the service process and develops at all points of contact during the process of interaction with tourists (Mascarenhas et al. 2006).

In an effort to measure the experience of traveling, it is necessary to use the right dimensions, including; 1) hedonism, 2) refreshment, 3) novelty, 4) meaningfulness, and 5) local culture (Kim et al. 2012).

2 METHODS

This research used descriptive – verification analysis to conduct a test on the effect of traveling experience on tourist satisfaction after visiting Ciletuh-Palabuhanratu Geopark, as well as hypothesis testing.

Data collection was done through survey methods through questionnaires. Questionnaires were distributed to tourists who had visited Ciletuh tourism object-Palabuhanratu Geopark with a total sample of 100 respondents. In determining the sample in this research, the authors used non-probability sampling – purposive sampling.

The data analysis method used in this research was Path Analysis (PATH) using the Statistical Package for the Social Science (SPSS) program to test the relationships between variables.

3 RESULTS AND DISCUSSION

The descriptive analysis of the results of respondents' responses was used to enrich the discussion. Through the description of respondents' responses data, it can be seen how tourists respond to the traveling experience and their satisfaction with the tourist destination of the Ciletuh Geopark-Palabuhanratu in Table 1.

Based on the response of local tourists, the traveling experience in the Ciletuh-Palabuhanratu Geopark are in good category in which the refreshment, novelty, and local culture dimensions get the highest value. Based on the data above, it can be said that the Geopark tourist destination Ciletuh Palabuhanratu is able to offer a complete tour package by giving a good impression even though in some points the geopark needs to have strong and distinctive areas compared to other similar destinations.

The interesting additions such as supporting facilities also need to be provided around the tourist area by working with communities and SMEs to decorate the facilities that have been provided by the tour manager. Utilizing local culture through colossal storytelling cultural performances in each destination object to provide a different atmosphere, training local people to talk to foreigners, and also training on how to serve tourists are more than enough in Table 2.

The satisfaction of Ciletuh-Palabuhanratu Geopark tourists can be categorized as very good. This is also supported by good values in the 3 dimensions used. This is in accordance with tourist expectations for information on international awards and tourism potentials owned by the Ciletuh-Palabuhanratu Geopark.

Table 1. The recapitulation of respondents average scores on travel experience variables.

No.	Dimension	Score	Criteria
1	Hedonism	3.8	Good
2	Refreshment	4.2	Good
3	Novelty	4.2	Good
4	Meaningfulness	3.9	Good
5	Local Culture	4.2	Good
	Grand Mean	4.1	Good

Table 2. Recapitulation of respondents response scores on tourist satisfaction variables.

No.	Dimension	Score	Criteria
1	Tourist Satisfaction	4.7	Very Good
2	Destination Recommendations	4.3	Very Good
3	The Wish of Returning	3.4	High enough
	Grand Mean	4.2	Good

However, some efforts are needed to encourage tourists to make recommendations through information channels that they have, such as social media, through the creation of attractive spots for selfie, so that tourists will naturally upload their photos.

Even so, the desire of tourists to visit again is quite high on the criteria. These findings indicate that there needs to be an improvement in the future to increase tourists' desire to revisit.

Through the R-square value, it can be seen that travel experience contributes to 12.6% and the results of the research shows that the significance level of 0.005 is less than 0.05, which means that the experience of traveling influences the satisfaction of local tourists.

The results of this research provide empirical evidence that the higher the tour experience which is designed by the destination manager, executed correctly and consistently, the higher the satisfaction level of these tourists when they are making a visitation.

These results indicate that the tour experience has a direct impact on satisfaction and determines the behavior of these tourists in the future (Otto & Ritchie 1996).

Because mutual satisfaction with emotion and memory are the result arising from the experience gained while traveling, which will later lead to further actions such as repeated visits in the future (Ek et al. 2008). The above statement is supported a research finding by Kim and Ritchie (2013) which showed the positive influence of the traveling experience on behavioral intentions, their relationship with satisfaction, destination image, and loyalty. However, the impression of a traveling experience will depend on the level of tourist involvement on all tourism activities presented in a destination (Kim et al. 2012).

Based on the above findings and their conformity with the references used as a basis in research, it is found that the tourism industry needs to prioritize the formation of experiences that are able to offer a tourism activity that offers a new and different experience. Tourists tend to visit a destination that offers a different culture and lifestyle, thus the tourists get something new (Pearce 1987).

4 CONCLUSION

The recognition from UNESCO to Ciletuh Geopark-Palabuhanratu provides a proof that West Java has a world-class destination, but that it has not been able to boost visit levels even though this destination has international recognition.

The experience of traveling experienced by tourists was categorized in the good category, although it needs to be increased in terms of recreation and entertainment. Thus, tourists can not only enjoy panoramic nature and culture, but also artificial tours that need to be considered as additional offer.

Tourists who have visited Ciletuh Geopark-Palabuhanratu have shown very high satisfaction but these achievements have not been able to encourage tourists to make recommendations and encourage themselves to make a return visit. It should be suspected that accessibility is one of the obstacles.

This study also showed that the experience of traveling affected the satisfaction of tourists. These results are in line with some of the previous research.

ACKNOWLEDGMENTS

The author would like to thank the Head of Management Study Program in Doctoral Study, UPI Postgraduate School as well as the Conference Chair of the 4th Global Conference on Business, Management and Entrepreneurship (GCBME), Prof. Dr. Hj. Ratih Hurriyati, M.P. This paper will not be completed without the process of teaching, fostering and motivation from her.

Thank you also conveyed to the guardian lecturer and supervisor in the completion of this paper, that is Dr. Heny Hendrayati, S.IP., M.M. Thanks to her guidance and direction, the author was able to complete one of the papers which would later be followed by other works.

REFERENCES

Alegre, J., & Cladera, M. 2009. Analysing the effect of satisfaction and previous visits on tourist intentions to return. *European journal of marketing*, 43(5–6), 670–685. https://doi.org/10.1108/03090560910946990

Babin, B. J. 1998. The nature of satisfaction: an updated examination and analysis variance extracted. 2963(97), 1–5.

Bigné, J. E., & Andreu, L. 2004. Emotions in segmentation: an empirical study. *Annals of tourism research*, 31(3), 682–696. https://doi.org/10.1016/j.annals.2003.12.018

Bigné, J. E., Mattila, A. S., & Andreu, L. 2008. The impact of experiential consumption cognitions and emotions on behavioral intentions. *Journal of services marketing*, 22(4), 303–315. https://doi.org/10.1108/088760408 10881704

Chen, C. F., & Chen, F. S. 2010. Experience quality, perceived value, satisfaction and behavioral intentions for heritage tourists. *Tourism management*, 31(1), 29–35. https://doi.org/10.1016/j.tourman.2009.02.008

Cheung, L. T. O. 2016. The effect of geopark visitors' travel motivations on their willingness to pay for accredited geo-guided tours. *Geoheritage*, 8(3), 201–209. https://doi.org/10.1007/s12371-015-0154-z

Chiu, W., Zeng, S., & Cheng, P. S. T. 2016. The influence of destination image and tourist satisfaction on tourist loyalty: a case study of chinese tourists in korea. *International journal of culture, tourism, and hospitality research*, 10(2), 223–234. https://doi.org/10.1108/IJCTHR-07-2015-0080

Ek, R., Larsen, J., Hornskov, S. B., & Mansfeldt, O. K. 2008. A dynamic framework of tourist experiences: space-time and performances in the experience economy. *Scandinavian journal of hospitality and tourism*, 8(2), 122–140. https://doi.org/10.1080/15022250802110091

Farsani, N. T., Coelho, C., & Costa, C. 2011. Geotourism and geoparks as novel strategies for socio-economic development in rural areas. *International journal of tourism research*, 13(1), 68–81. https://doi.org/10.1002/jtr.800

Jorgenson, J., & Nickerson, N. 2016. Geotourism and sustainability as a business mindset. *Journal of hospitality marketing and management*, 25(3), 270–290. https://doi.org/10.1080/19368623.2015.1010764

Kim, J. H., Ritchie, J. R. B., & McCormick, B. 2012. Development of a scale to measure memorable tourism experiences. *Journal of travel research*, 51(1), 12–25. https://doi.org/10.1177/0047287510385467

Kozak, M., & Rimmington, M. 2000. As an off-season holiday destination. https://doi.org/10.1177/004728750003800308

Mansour, J. S. A., & Ariffin, A. A. M. 2017. The effects of local hospitality, commercial hospitality and experience quality on behavioral intention in cultural heritage tourism. *Journal of quality assurance in hospitality and tourism*, 18(2), 149–172. https://doi.org/10.1080/1528008X.2016.1169474.

Meng, F., Tepanon, Y., & Uysal, M. 2008. Measuring tourist satisfaction by attribute and motivation: the case of a nature-based resort. 14(1), 41–56. https://doi.org/10.1177/1356766707084218

Moon, H., & Han, H. 2018. Destination attributes influencing chinese travelers' perceptions of experience quality and intentions for island tourism: a case of jeju island. *Tourism management perspectives*, 28(August), 71–82. https://doi.org/10.1016/j.tmp.2018.08.002

Moon, H., & Han, H. 2019. Tourist experience quality and loyalty to an island destination: the moderating impact of destination image. *Journal of travel and tourism marketing*, 36(1), 43–59. https://doi.org/10.1080/10548408.2018.1494083

Newsome, D., & Dowling, R. 2010. Setting an agenda for geotourism. *Geotourism: the tourism of geology and landscape*, (October 2016), 1–12.

Otto, J. E., & Ritchie, J. R. B. 1996. The service experience in tourism. *Tourism management*, 17(3), 165–174.

Pratminingsih, S. A., Rudatin, C. L., & Rimenta, T. 2014. Roles of motivation and destination image in predicting tourist revisit intention: a case of bandung – indonesia. 5(1). https://doi.org/10.7763/IJIMT.2014.V5.479

Prayag, G., Hosany, S., Muskat, B., & Del Chiappa, G. 2017. Understanding the relationships between tourists' emotional experiences, perceived overall image, satisfaction, and intention to recommend. *Journal of travel research*, 56(1), 41–54. https://doi.org/10.1177/0047287515620567

Ruban, D. A. 2015. Geotourism – a geographical review of the literature. *Tourism management perspectives*, 15, 1–15. https://doi.org/10.1016/j.tmp.2015.03.005

Smolèiæ Jurdana, D., & Soldiæ Frleta, D. 2017. Satisfaction as a determinant of tourist expenditure. *Current issues in tourism*, 20(7), 691–704. https://doi.org/10.1080/13683500.2016.1175420

Suhartanto, D., Brien, A., Primiana, I., Wibisono, N., & Triyuni, N. N. 2019. Tourist loyalty in creative tourism: the role of experience quality, value, satisfaction, and motivation. *Current issues in tourism*, 0(0), 1–13. https://doi.org/10.1080/13683500.2019.1568400

Tung, V. W. S., & Ritchie, J. R. B. 2011. Exploring the essence of memorable tourism experiences. *Annals of tourism research*, 38(4), 1367–1386. https://doi.org/10.1016/j.annals.2011.03.009

Wu, H. C., & Li, T. 2017. A study of experiential quality, perceived value, heritage image, experiential satisfaction, and behavioral intentions for heritage tourists. *In journal of hospitality and tourism research* (Vol. 41). https://doi.org/10.1177/1096348014525638

Xia, W., Jie, Z., Chaolin, G. U., & Feng, Z. 2009. Examining antecedents and consequences of tourist satisfaction: a structural modeling approach. *Tsinghua science and technology*, 14(3), 397–406. https://doi.org/10.1016/S1007-0214(09)70057-4

349

Advances in Business, Management and Entrepreneurship – Hurriyati et al. (Eds)
© *2021 Taylor & Francis Group, London, ISBN 978-0-367-67471-7*

The influence of perceived value to purchase intention: Evidence from Moslem fashion Indonesia

A. Rachman, S. Sarah & N. Maulid
Sekolah Tinggi Ilmu Ekonomi Membangun, Bandung, Indonesia

ABSTRACT: Fierce competition forces companies to pay attention to Consumer Perceived Value (CPV). CPV is a key to compete with other business competitors. This research aims at proposing an integrated model by combining variables based on previous research and testing them with empirical data. This research employs descriptive and verification methods. The empirical data were collected through a questionnaire that was administered to Moslem fashion users. This study used confirmatory factor analysis as an analytical tool in data processing. Moreover, in this study, the CPV determinant variables of previous studies were combined with innovation value on the model and then it is named the "Integrated Model of CPV." This study discovers that the integrated model of CPV consists of price value, social value, emotional value, quality value, perceived risk value, aesthetic value, and innovation value.

1 INTRODUCTION

Currently, the business world has shifted from the product-centric era into the costumer-centric era. Businesses that only focus on product development can no longer be a brilliant way to win consumers and the market, since the best way for business today is to focus on the consumers' needs and wants. A company that strives to fulfil consumers' needs and wants, will attract consumers' perceived value, then will attract the consumers to buy their product. This definitely will make the company survive the competition. One of the important concepts in the consumer-centric era is understanding consumers' perception about the value of the product or services offered by the company, namely consumer perceived value (CPV).

The CPV concept roots from the consumer's point of view related to overall assessment of the product and services based on their perception of "what I get from what I give" (Zeithaml 1988). Previous studies have largely focused on the identification of variables that determine the CPV, stemming from the quality and price (Manroe 1990); they then become more complex and continue to develop until today. Research conducted by Chi & Kilduff (2011), which adopted Sweeney & Soutar's (2001) framework, affirmed that CPV consists of price, social, emotional, and quality variables, and already gives a comprehensive point of view to determine CPV. On the contrary, Chi & Kilduff (2011) and Gallarza & Saura (2006) conducted research in the service sector and found more complex CPV variables that consist of efficiency, service quality, social value, play, aesthetics, perceived

monetary cost, perceived risk, and time and effort spent. Therefore, it is assumed that the sciences of CPV need to be improved in order to make it more applicable for different kinds of business.

However, it becomes disputable whether these variables can determine CPV for different business scopes, as business expands rapidly in line with consumers' needs, like technology products. This trend demands the company to keep offering something new. Hence, it can be denied that innovation is an important toehold as a response to consumers' needs and wants and to determine purchase decisions.

According to Khin et al. (2010), to be a competitive technology company, the company will no longer offer the same products or compete in terms of price and quality, but innovation should be something that the company can offer to the consumer to get competitive advantages compared to competitors. Brillet et al. (2014) focuses on service innovation, and mentions that the novelty of the service innovation determines the perception of value. That concept can be adopted in the mentioned context of technology products.

Product innovation is one of the consumers' considerations in buying a technology product. This study proposes an integrated model by adding "product innovation" as a determinant variable of CPV, which has not been explored by previous studies.

In Indonesia, the fashion market is dominated by the growth of Moslem fashion or the hijab that continues to increase significantly. PT. Alghaniy Faza Utama is one of the companies engaged in Moslem fashion for teenagers to adults. The products currently sold are scarfs, sweaters, and mukena under the Maliqa

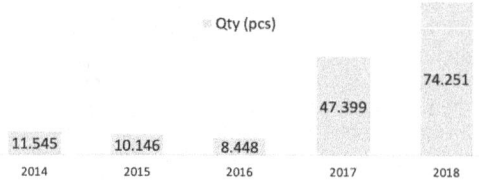

Sales of Maliqa Products

Figure 1. Sales data of PT. Alghaniy Faza Utama.

brand. Following are the sales data of PT. Alghaniy Faza Utama for the past three years:

Based on the data in the Figure 1, it can be seen that the sale of Maliqa products at PT. Alghaniy Faza Utama in 2015 and 2016 experienced a decline from the previous year. This can be caused by the decrease in buying interest. This decrease in buying interest can be influenced by perceived value, which can be seen in the five previous journals. However, none of the five journals have conducted research at PT. Alghaniy Faza Utama.

In this study, the combination of various CPV determinant variables from previous studies and the value innovation variable in the model is named the "Integrated Model of CPV."

2 LITERATURE REVIEW

2.1 Value Concept

The value concept can be described in a social context and marketing context. In the marketing context, the value is regarded as a core part of marketing; what a company sells is its value. Kotler & Keller (2016) define value as a benefit received by consumers compared to the costs incurred.

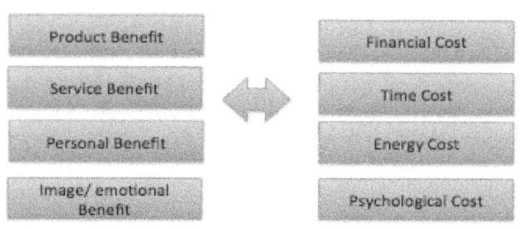

Figure 2. Value concept.

According to Kotler (2016), the benefit gained by the consumer may include product benefits, service benefits, personal benefits, and emotional benefits, which are then compared to the costs incurred by the consumer. The cost that is incurred by the consumer is not only reflected in the cost of the product, but also the time cost, energy spent, and the psychological cost. Talking about value, Kotler & Keller (2016) associate value with consumer value, where the concept of marketing is addressed, namely the identification, creation, communication, achievement, and monitoring

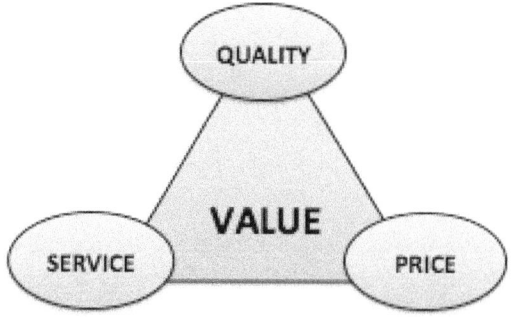

Figure 3. Consumer value triad.

of consumer value. Kotler (2016) gives the concept of value as a combination of three factors, namely quality, service, and price; known as the consumer value triad (See Figure 3).

Kotler & Keller (2016) explain that value will increase in line with the increase of benefit that the consumer gets from the quality and service offered, and will decrease in line with the increasing price. Zeithaml (1988) states that value consists of four aspects, namely (1) value is low price, (2) value is whatever I want in a product, (3) value is the quality I get for the price I pay, and (4) value is what I get for what I give. Gallara & Saura (2006) define the value based on two approaches, namely transactional value and psychological value. Transactional value is defined as the price aspect, while the psychological value is defined as the emotional aspect, which influence the product or service choice of the consumer. Based on those definitions, the researchers underline that value is perceived by price, quality, and emotional value.

2.2 Consumer Perceived Value

CPV is a difference evaluation between all the benefits and rewards as well as other alternatives considered. It is the result of the total consumer value compared to the total cost. Total consumer value covers the total monetary value of the economic, financial, and psychological benefits that are expected by the consumer through a product or service. Whereas the total consumer cost covers the total cost incurred by the consumer to evaluate, acquire, use, and dispose of products and services (monetary cost, time, energy, and psychic). In the concept of CPV, companies need to analyze consumer value so that the CPV would be good in the eyes of consumers; even greater than its competitors. In doing consumer value analysis, companies need to (1) identify the attributes and benefits that are considered valuable for consumers, (2) prioritize the order among the attributes that have been identified, (3) assess the performance of the company and competitors based on attributes and interest level, (4) assess how companies compare to other competitors specifically, and (5) oversee the development of CPV from time to time.

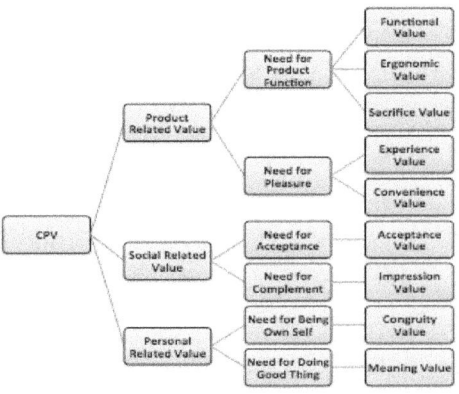

Figure 4.   Consumer perceived value.

Aulia (2006) categorizes CPV into three categories (Product Related Value, Social Related Value, and Personal Related Value) with dimensions attached to them.

1) Product Related Value: The consumers' perspective in assessing a product can be seen as the "need for product function to solve consumers' problems" and "the need for a pleasure when using the product." The first perspective, the perspective that a product with its attributes can be seen as a tool that serves to provide benefit and solve consumer problems, includes functional value, ergonomic value, and sacrifice value. The second perspective, the perspective that consumers' assessment of the product gives pleasure, includes experience value and convenience value.

2) Social Related Value: The consumers' perspective of assessing a product of its social aspects covers the "need for acceptance" and "need for compliment or appreciation." In this condition, a consumer perspective will depend on whether the products used give a high level of confidence, status, prestige, and impression in social environments.

3) Personal Related Value: This consumers' perspective covers the consumption benefit or value held by consumers as individuals. Value and belief characteristics held by each individual give a different perspective to the product; health concern, environment concern, religion concern, life goals concern, and the principle concern. This perspective refers to the personal characteristics of each consumer (what they like and do not like).

Based on the proposed model by Monroe (1990), who states that there is a tradeoff between price and quality, Woodruff (1997) states that to understand CPV, one has to understand things beyond the price and quality dimensions. Meanwhile, Sheth et al. (1991) state social, emotional, functional, epistemic and conditional value dimensions; Kamtameni and Coulson (1996) mention societal value, experimental value, functional value, and market value dimensions; and

Grewal et al. (1998) mention transaction value, in-use value, redemption value, and acquisition value. Then, Sweeney & Soutar (2001) propose PERVAL, including emotional value, social value, quality or performance value, and price value. Chi & Kilduff (2014), in their research using PERVAL dimensions, include price value, social value, emotional value, and quality value. Based on the discussion above, the researchers find price value, social value, emotional value, and quality value.

HI: Price Value as a determinant factor of Consumer Perceived Value
H2: Social Value as a determinant factor of Consumer Perceived Value
H3: Emotional Value as a determinant factor of Consumer Perceived Value
H4: Quality Value as a determinant factor of Consumer Perceived Value

The research model conducted by Gallarza & Saura (2004) on service sector CPV can be applied to the concept of product sector CPV. Gallarza & Saura (2004) mention that the variables that influence the CPV are efficiency, service quality, social value, play, aesthetics, perceived monetary cost, perceived risk, and time and effort spent. Some variables such as efficiency, service quality, social value, perceived monetary cost. and the time and effort spent have been reflected in the variable price value, social value, and quality value in the variables found previously. The researchers found other variables that can be adopted into the Integrated Model of CPV, namely perceived risk and aesthetics. Specifically, in technology products, the perception of risk becomes a strong consideration. Moreover, technology products are closely related to the beauty that can be associated with the aesthetics value. In the context of technology products, especially in Moslem fashion products, it is closely related to the innovation of the product, so the researchers underline another dimension, that is, innovation value as a determinant of CPV.

There are some research works conducted to discover the importance of CPV and the impact of CPV on behavior intention (Khan et al. 2011; Shen et al. 2014; Wahyuningsih 2012), consumer satisfaction (Jabaly & Kharaim 2014; Khan et al 2011; Lin 2003), purchase decisions (Wang & Chen 2016; Yee & San 2011), and loyalty (Chuah et al 2014; Hasan et al. 2014). Furthermore, Aulia (2006) explains that CPV is the key to survive in the tight competition era, since it influences the behavior and attitude of the consumer.

2.3   *Perceived Risk Value*

When consumers want to buy a product service, they will think about the risk that they will received if they buy it. The perceived risk theory was initially proposed by Baurer (1960). It appears based on uncertainty as a result of buying the product or service. Samadi & Nehaji (2009) define perceived risk as a subjective belief of an individual about potentially negative consequences from purchase decisions, and consists of social, psychological, monetary, functional, physical, and convenient. Bhukya & Singh (2015) classify perceived risk dimensions into of functional risk, financial risk, perceived risk, and psychological risk. Some studies proved perceived risk as a variable that determines CPV. Son et al. (2015) mention that manufacturers should be active to communicate the psychological, performance, and financial risk of the product compared to the competitors' product, which will give impact to many aspects such as behavior intention and consumer satisfaction.

Snoj et al. (2004) prove that risk value is a determinant variable of CPV, namely financial, psychological, physical, functional, and social risk to measure overall risk. The definition of each risk is adopted from Murphy & Enis (1986). Financial risk refers to a consumer losing their money when the product does not give satisfaction to them. Psychological risks refer to consumers' ego who think that they chose the wrong product. Physical risk refers to risk that gives a damage to the consumer or others when they use the product. Functional risk refers to risk that occurs when the products do not work properly as the consumer expected. Social risk refers to risk that will be received by the consumer about their status in social environments.

H5: Perceived Risk Value as a determinant factor of Consumer Perceived Value

2.4 Aesthetics Value

Mumcu et al. (2015) describes aesthetics as a factor that influences consumers' perception and purchase decisions. In terms of purchase decisions, aesthetics influence consumers in selecting and evaluating the product. Mumcu et al. (2015) emphasizes that an aesthetic product will add product value and contribute to the uniqueness of the product. Furthermore, designers and manufacturers should focus on visual product aesthetics; visual product aesthetics influence price sensitivity when consumers buy the product. Gazalla & Saura (2004) correlate the aesthetic as a determinant variable of CPV in the tourism sector.

H6: Aesthetics Value as a determinant factor of Consumer Perceived Value

2.5 Product Innovation (Innovation Value)

Khin et al. (2010) summarizes the product innovation definition as the introduction of a new product or service with improvement in product and process, technical and administrative, radical and incremental. The benefit of innovation is for gaining market position, achieving sustainable competitive advantages, reaching growth, business success, and better performance, and also obtaining advantages for product users. According to Khuong & Giang (2014), innovation is defined as a novelty in an idea, practice, and object that should be achieved by a company in order to

create additional value to the consumer, face the fierce competition, and give economic viability to the firm. Rioche & Ackermann (2013) outline product innovation, regarded from a market point of view, as a new product and perceived as such by consumers. They also state the design newness as a construct in product innovation and will provide the perceived newness value. Moreover, Tjiptono et al. (1997) mention that improvement to existing products will affect the higher perception value in the form of adding new characteristics or a new model; transforming into a new process; or transforming into new attributes. Besides, Khalil (2000) mentions innovation as a creative factor in managing technology. It involves the introduction into the market place either by utilization or by commercialization. It is also associated with the creation of value and satisfaction of consumer needs. The study underlined that successful innovations are those perceived by consumers to add value. Hence, to gain marker acceptance, innovation must contribute to the creation of value.

In service innovation, the dimensions to measure the innovation consist of marketing focus innovation, process innovation, pricing innovation, product type innovation, customize the services, and use information technology. In CPV product, Khin et al. (2010) mention that innovation product, where firms compete on grounds of new products with new features, new design, and new function.

H7: Innovation Value as a determinant factor of Consumer Perceived Value

3 METHODS

Samples in this study are Maliqa brand users. Maliqa brand is one of the newcomers in Indonesia, but has

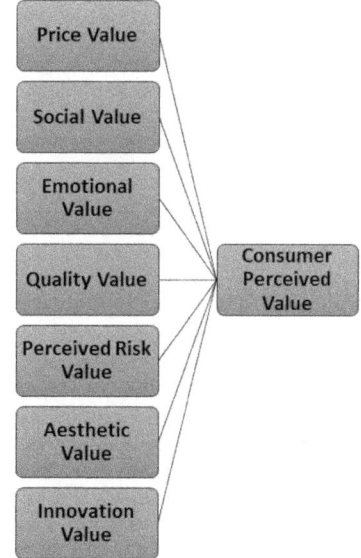

Figure 5. Research framework.

Table 1. Variables and indicators.

Variable	
Price Value (Adapted from Chi & Kilduff 2011)	Is reasonably priced
	Offers value of money
	Is a good product for price
	Would be economical
Social Value (Adapted from Chi & Kilduff 2011)	Would help me to feel acceptable
	Would improve the way I perceived
	Would make a good impression on other people
	Would give its owner social approval
Emotional Value (Adapted from Chi & Kilduff 2011)	Is one that I would enjoy
	Would make me want to wear
	Would make me feel good
	Would give me pleasure
	Is one that I would feel relaxed about using
Quality Value (Adapted from Chi & Kilduff 2011)	Has consistent quality
	Is well made
	Has an acceptable standard of quality
	Would perform consistently
	Has poor workmanship *
	Would not last a long time *
Perceived Risk Value (Modified from Gallarza & Saura 2004 and Mwencha et al. 2014)	Has financial risk after using the product *
	Has a wrong choice feeling after using the product *
	Gives damage impact to yourself and others *
	Has functional risk *
	Has status in social environment risk*
Aesthetic Value (Modified from Gallarza & Saura 2004)	Has a beauty design
	Has a beauty color
	Has a beauty shape
	Has a beauty sound
Innovation Value (Modified from Khin et al. 2010)	Has a new design
	Has a new feature
	Has a new function

Table 2. Price value.

Variable	Indicator	Score Total	%	Average
Price value	Is reasonable priced	128	85.33%	81.83%
	Offers value of money	128	85.33%	
	Is a good product for price	115	76.67%	
	Would be economical	120	80.00%	

Table 3. Social value.

Variable	Indicator	Score Total	%	Average
Social Value	Would help me to feel acceptable	59	39.33%	42.33%
	Would improve the way I perceived	64	42.67%	
	Would make a good impression on other people	67	44.67%	
	Would give its owner social approval	64	42.67%	

a significantly increasing number of sales in Bandung City. The researchers administered a questionnaire with questions that consist of the items listed in Table 1. The scale for answering the questions is 1–5. This study uses confirmatory factor analysis as an analytical tool in data processing.

4 RESULTS AND DISCUSSION

The questionnaire administered to the respondents each had four indicators of price value, social value, perceived risk value, and aesthetics value, five indicators of emotional value, six indicators of quality value, and three indicators of innovation value. Respondents' answers to the price value variable are shown in Table 2.

Based on Table 2, the average percentage of CPV for price value was 81.83%. The CPV for price value was classified in the "good" category, meaning that consumers perceive the product offers a reasonable price and value for money. The product is a product with a good price.

Compared to other players in the Moslem fashion market, Maliqa brand offers cheaper prices, especially compared to incumbent competitors such as Zoya and Shafira. With the same quality and design of Moslem fashion, Maliqa brand offers it at half the price. When, Zoya and Shafira launched the newest product above 1 million, with the same quality and design, Maliqa brand launched the price below 1 million.

Associated with the technology adoption curve, there are several groups in the adoption curve. The first group is innovator. The innovator tends to be not selective in choosing a product, they always buy the product early. They only see the product as a brand and based on its price or its quality. The amount is comparatively lower than other groups. The second group is early adopter, they are very selective in price and quality. It has a bigger percentage compared to the innovator group. As manufacturers, they should choose the strategy to produce the cheap or expensive one related to their objectives to be achieved through the company's capabilities.

Respondents' answers to the social value variable are shown in Table 3. Based on Table 3, the average percentage of CPV for social value is 43,23%. The CPV of social value was classified in the "bad" category.

It means that the product has low ability in helping consumers to feel acceptable in social environments, improving the way they perceived the value, making a good impression on other people, and giving the owners a social approval. According to demographic characteristics, the majority of the respondents come from a metropolis city. A metropolis city has different characteristics related to the income, social economics, culture, and soon. Several of those characteristics influence people in looking at social value. Most of them have higher income than other cities, which means they have more power to buy products that make them acceptable in social environments and also make them have a good impression in front of their friends, family, and also environment. Metropolis city people tend to buy branded products, expensive products, and prestige products. And they will see people from based on what products they have. In this case, Maliqa brand still can't reach a social value in front of the environment. They should compete with other products that have more prestige and give high impression to the consumers.

Respondents' answers to the emotional value variable are shown in Table 4. Based on Table 4, the average percentage of CPV for emotional value is 82.00%. The CPV of emotional value was classified in the "bad" category.

Based on Table 4, it can be seen that consumers think that Maliqa brand is an enjoyable product, Maliqa brand can make consumers want to wear, Maliqa brand can make a consumer feel good and pleasure for the product. Consumers also feel relax when using the product. As newcomers in the Moslem fashion business, Maliqa brand has proved that an Indonesian product can make people want to use it and feel relaxed.

Respondents' answers to the quality value variable are shown in Table 5. Based on Table 5, the average percentage of CPV for quality value is 75.44%. The CPV of quality value was classified in the "good" category.

Based on Table 5, it can be seen that Maliqa brand has consistent quality, is well made, and has consistent performance. Based on inverse indicators, Maliqa brand hasn't a poor workmanship. Consumers think that Maliqa brand would not last a long time related to product quality. As manufacturers, they should prove the long-time quality of the product. Consumers will consider the long-time quality compared to the risk they will get from the product. Since the consumers perceive the product would not last a long time, they will perceive that product as an expensive product because they should pay for the product risk. Manufacturers need continuous quality product control that makes sure that the product gives the best quality related to time for consumption.

Respondents' answers to the perceived risk value variable are shown in Table 6. Based on Table 6, the average percentage of CPV for perceived risk value is 66,40%. The CPV of the perceived risk value was classified in the "sufficient" category. From Table 6, it can be seen that Maliqa brand has high financial, functional, and social environment risk. But consumers have low wrong choice feeling after using the product. Consumers perceive that Maliqa brand is not a product that can give damage impact to themselves and others.

As a manufacturer, Maliqa brand should minimize the financial risk, functional risk, and also social environment risk. For financial risk and functional risk, Maliqa brand should maintain the quality related to the cost. With making qualified product, Maliqa brand can

Table 5. Quality value.

Variable	Indicator	Score Total	%	Average
Quality value	Has consistent quality	127	84.67%	75.44%
	Is well made	127	84.67%	
	Has an acceptable standard of quality	119	79.33%	
	Would perform consistently	121	80.67%	
	Has good workmanship	121	80.67%	
	Would last a long time	64	42.67%	

Table 6. Perceived risk value.

Variable	Indicator	Score Total	%	Average
Perceived risk value	Has a financial risk after using the product	121	80.67%	66.40%
	Has a wrong choice feeling after using the product	64	42.67%	
	Give damage impact to yourself and others	67	44.67%	
	Has functional risk	127	84.67%	
	Has status in social environment risk	119	79.33%	

Table 4. Emotional value.

Variable	Indicator	Score Total	%	Average
Emotional value	Is one that I would enjoy	127	84.67%	82.00%
	Would make me want to wear	127	84.67%	
	Would make me feel good	119	79.33%	
	Would give me pleasure	121	80.67%	
	Is one that I would feel relaxed about using	121	80.67%	

minimize consumers' financial and functional risk. Consumers don't need to spend more money for after sales service and for fixing the product, and it has a low depreciation expense since the product gives a long use time. Maliqa brand should intensively brand the product as a prestigious product so as to minimize the consumer social environment risk.

Respondents' answers to the aesthetic value variable are shown in Table 7. Based on Table 7, the average percentage of CPV for aesthetic value is 91.50%. The CPV of the aesthetic risk value was classified in the

Table 7. Aesthetic value.

Variable	Indicator	Score Total	%	Average
Aesthetic value	Has a beauty design	138	92.00%	91.50%
	Has a beauty color	140	92.33%	
	Has a beauty shape	136	90.67%	
	Has a beauty sound	135	90.00%	

Table 8. Innovation value.

Variable	Indicator	Score Total	%	Average
Innovation value	Has a new design	102	68.00%	84.44%
	Has a new feature	138	92.00%	
	Has a new function	149	93.33%	

Table 9. Kaiser-Meyer-Olkin.

KMO and Bartlett's test			
Kaiser-Meyer-Olkin measure of sampling adequacy			.899
Bartlett's test of sphericity	Approx. Chi-Square		182.709
	Df		21
	Sig.		.000

Table 10. Total variance.

"very good" category. From Table 7, it can be seen that Maliqa brand is perceived by consumers as a product that is very good in aesthetic value. Maliqa brand has a beauty design, shape, and color. Compared to the other variables, aesthetic value gives a very high score. Maliqa brand should maintain this position to reach competitive advantages.

Respondents' answers to the innovation value variable are shown in Table 8. Based on Table 8, the average percentage of CPV for innovation value is 84.4%. Based on Table 8, it can be seen that Maliqa brand has new features and function. But for the new design aspect, the level is still in the sufficient level. Product design as a component of innovation should be paid attention to. Maliqa brand product still has a similar design to other competitors. For improving this, manufacturers should pay attention to developing product design. The more interesting product design will influence consumers' perceived value, increasing the purchase intention and purchase decision of the product. Rapid economic, business, and social change will give impact in faster product life cycle. As consumers, innovation is a need. Innovation as a consideration variable makes consumers want to buy new product, new technology, or shift to another product. Manufacturers that can't respond to this condition with innovative products can't compete with others.

Table 11. Confirmatory factor analysis.

Component Matrix	
	Component
	1
Price value	.893
Social value	.931
Emotional value	.922
Quality value	.873
Perceived risk value	.820
Aesthetic value	.896
Innovation value	.751

Extraction method: Principal component analysis.
a. 1 component extracted.

Total Variance Explained

Component	Initial Eigenvalues			Extraction Sums of Squared Loadings		
	Total	% of Variance	Cumulative %	Total	% of Variance	Cumulative %
1	5.315	75.934	75.934	5.315	75.934	75.934
2	.522	7.453	83.387			
3	.430	6.138	89.525			
4	.279	3.981	93.506			
5	.250	3.574	97.080			
6	.118	1.680	98.759			
7	.087	1.241	100.000			

In first step of data processing, Kaiser-Meyer-Olkin (KMO) should be measured to know whether factor analysis can be used. Based on Table 9, it can be seen that the KMO is 0.899. The KMO value is above the Bartlett's test and Chi Square (0,899 > 182,709) and is significant with 0.000.

Table 10 shows that there are seven factors with eigenvalues more than 1, which is 5,315. Based on Table 10, it can be concluded that those factors can define variation by 79.93%. In the last step, confirmatory factor analysis should be analyzed to conclude variables that are determinants of CPV in this study.

Table 11 shows that all of the variables' values are more than 0,5, which means that all of the variables are factors that determine CPV. This study proves that price value, social value, emotional value, quality value, perceived risk value, aesthetic value, and innovation value are determinant factors of CPV.

5 CONCLUSION

This study found that price value, social value, emotional value, quality value, perceived risk value, aesthetic value, and innovation value are determinants of CPV. This study proposed the Integrated Model of CPV that consists of seven variables, namely price value, social value, emotional value, quality value, perceived risk value, aesthetic value, and innovation value, and 31 indicators to measure the variables. This research gives implication in measuring CPV with a comprehensive framework, especially in technology products that are very related to the innovation value. Future research can add and test other variables that have not been explored in this study such as brand value as determinant variables of CPV. Future research can increase the sample size for generalization.

REFERENCES

Aulia. 2006. A review: Consumer perceived value and its dimension. *Asian Journal of Social Sciences and Management Studies* 2(2): 150–162.

Bhukya, R. & Singh, S. 2015. The effect of perceived risk dimensions on purchase intention: An empirical evidence from Indian private labels market. *American Journal of Business* 30(4): 218–230.

Chi, T. & Kilduff. 2011. Understanding consumer perceived value of casual sportswear: An empirical study. *Journal of Retailing and Consumer Services*.

Chuah, Hui Wen, Marimuthu, Malliga, & Ramayah, T. 2014. The effect of perceived value on the loyalty of generation y mobile internet subscribers: a proposed conceptual framework. *Procedia of Social and Behavioral Science Volume 130 4th International Conference on Marketing and Retailing* 2013.

Gallarza, M. G. & Saura, I. G. 2004. Value dimensions, perceived value, satisfaction, and loyalty: An investigation of university students' travel behavior. *Tourism Management* 27(2006): 437–452.

Hasan, H., Kiong, T. P., & Ainuddin, R. A. 2014. Effect of perceived value and trust on consumer loyalty towards foreign banks in Sabah, Malaysia. *Global Journal of Emerging Trends in e-Business, Marketing and Psychology* (GJETeMCP) 1(2): 137.

Jabaly, S., & Kharaim, A. S. 2014. The effect of perceived value and consumer satisfaction on perceived price fairness of airline travelers in Jordan. *Universal Journal of Management* 2(5): 186–196.

Khalil, T. 2000. Management of technology, the key to competitiveness and wealth creation. *Mc Graw Hill International Editions*.

Khan, N., Latifah, S., Kadir, Sharifah, L., & Syed A. 2011. The impact of perceived value dimension on satisfaction and behavior intention: Young-adult consumers in banking industry. *African Journal of Business Management* 5(16): 4087–4099.

Khin, S., Ahmad, Noor Hazlina., & Ramayah, T. 2010. Product innovation among ICT technopreneurs in Malaysia. *Business Strategy Series* 11(6): 397–406.

Khuong, M. N., & Giang, T. H. 2014. The effect of service innovation on perceived value and guest's return intention—a study of luxury hotels in Ho Chi Minh City, Vietnam. *International Journal of Trade, Economics, and Finance* 5(6).

Kotler, P. & Keller, K. L. 2016. Marketing management (16th Edition). *Edinburgh Pearson Education*.

Lin, C. C. 2003. The role of consumer perceived value in generating consumer satisfaction: An e-business perspective. *Journal of Research in Marketing and Entrepreneurship* 5(1).

Monroe, K. B. 1990. Pricing: Making profitable decisions. *Mc Graw-Hill New York*.

Mumcu, Y., Kimzan, & Halil, S. 2015. The effect of visual product aesthetics on consumers' price sensitivity.

Rioche, L. F. & Ackermann, C. L. 2013. Consumer innovativeness, perceived innovation, and attitude towards "neonetro" product design. *European Journal of Innovation Management* 16(4): 495–516.

Shen, H., Fan, S., Zhan, J., Zhaor, J. 2014. A study of the perceived value and behavioral intentions of Chinese marine cruise tourists. *Tourism, Leisure and Global Change* 1.

Snoj et al. 2004. The relationship among perceived quality, perceived risk and perceived product. *Journal of Product & Brand Management* 13(3): 156–167.

Son, Y. H., Kang, M. S., & Park, S. K. 2015. Investigating temporal effect of value and risk perceptions, and product satisfaction on behavioral intention: Longitudinal study. *International Journal of Software Engineering and its Applications* 9(8): 133–148.

Sweeney, J. C. & Soutar, G. N. 2001. Consumer perceived value: The development of a multiple item scale. *Journal of Retailing* 77: 203–220

Wahyuningsih. 2012. The effect of customer value on behavioral intention in tourism industry. *International Research Journal of Business Studies* 5: 1–12.

Wang, Y. H. & Chen, L. Y. 2016. An empirical study of the effect of perceived price value on purchase intention from low cost carriers. *International Journal of Business and Social Science* 7(4).

Yee, C. J. & San, N.C. 2011. Consumers' perceived quality, perceived value and perceived risk towards purchase decision on automobile. *American Journal of Economics and Business Administration* 3(1): 47–57.

Zeithaml, V. A. 1988. Consumer perceptions of price, quality and value: A means end model and synthesis of evidence. *Journal of Marketing*.

Advances in Business, Management and Entrepreneurship – Hurriyati et al. (Eds)
© 2021 Taylor & Francis Group, London, ISBN 978-0-367-67471-7

The effect of Instagram ad disclosure position on attitude toward ad, attitude toward brand, and intention to share (e-WOM)

D. Anggraeni & T.E. Balqiah
Universitas Indonesia, Jakarta, Indonesia

ABSTRACT: In recent years, marketers have used social media as their marketing communication platform, which is caused by the growth of social media users. Instagram, as one of the fastest growing social media platforms, enables its users to share photos or videos through their platform. Ads on Instagram can be categorized as In-Feed Native Advertising, which has been proved by recent studies as the promising solution for ad avoidance. However, consumers' acceptance of native advertising can be a double-edged sword due to the potential of consumer deception because of consumers' inability to recognize the ad. Thus the Federal Trade Commission enforces the marketer to disclose if the post is an ad. Recent studies about advertising disclosure have focused on position, language, duration, and timing, while the study for implementing disclosure in social media ad was limited. This study examined the effect of Instagram ad disclosure position on attitude toward ad, attitude toward brand, and intention to share via e-WOM. An experimental study was conducted to test the effect of top (Group A) and bottom (Group B) disclosure position among 220 Instagram users. Results indicated that the response for Group A had lower attitude toward ad, attitude toward brand, and intention to share via e-WOM due to the activation of persuasion knowledge model than Group B. This study provides an empirical contribution and strategic implication for marketers' decision-making process in implementing the ad disclosure on their Instagram ad.

1 INTRODUCTION

The increasing number of internet users from year to year has caused business people to switch from conventional media to digital media in delivering marketing communications. At first, marketers used ad pop ups and ad banners that were able to attract the attention of internet users. However, both types of methods have been considered intrusive over time so that users tend to avoid and ignore advertisements. This behavior is reflected by 47% of internet users who have used ad blockers to prevent pop up ads and banner ads from appearing so that the use of these two types of ad is considered ineffective and can even lead to negative perceptions from consumers (Truong & Simmons 2010). This triggers the marketers' creativity to find another way in using the internet as a marketing communication media that doesn't cause negative perceptions of consumers by compiling marketing content that has a similar appearance to the platform used called as Native Advertising.

Native Advertising can be applied in various forms such as in-feed units, paid search units, widget recommendations, promoted listings, in ad with native element units, and custom. Currently in-feed units are the most widely used form as social media usage has increased to 3.4 billion users with an average usage of

2 hours 16 minutes per day (Kemp 2018). One of the fastest growing social media platforms is Instagram, which is a social media platform that allows users to share photos and videos with other users. Since its introduction in 2010, Instagram has grown rapidly due to its superior features. It al-lows users to share through media photos and videos and provide feedback in the form of likes and comments. This makes Instagram users compete to create interesting content that is liked by other users. Various content categories make Instagram users able to choose other user accounts that will be followed according to the suitability of the image of other users with their image, or better known as the concept of self-congruity. For example, following the accounts of celebrities who currently have a number of followers of up to hundreds of millions. This phenomenon is used by brand owners to conduct marketing communication activities through collaboration with celebrities. This is known as endorsement where celebrities will display and give positive reviews about the products of the brand. Based on the theory of self-congruity, someone will tend to follow or imitate the behavior of others who are considered to have similarities with the image of themself. Thus, when a celebrity displays and gives positive reviews about the product, the followers will also have the same tendency to like the product. Positive reviews submitted will also

activate e-WOM (electronic word of mouth). E-WOM can be disseminated by individuals who have the desire to share information with other users on their social media networks (Evans et al. 2017) Basically e-WOM is an organic and unpaid communication carried out by consumers who voluntarily act as supporters of a brand because they previously had a positive affection for the brand. But this is utilized by brand owners by using paid e-WOM to strengthen the positive image of a brand through a process known as influencer marketing.

The use of endorsement on celebrity accounts makes it difficult for customers to distinguish between the original content and the paid content. So that this practice is considered to have a tendency to deceive social media users. Therefore, the FTC (Federal Trade Commission) has stipulated provisions in the use of celebrities as endorsers of brands on social media. In displaying advertisements or giving a review of a product, if it is a collaboration with a brand owner where celebrities will receive a reward, a marker must be indicated that the content is paid content. Various studies were carried out to analyze the way of delivery and the impact of disclosure on the perceptions of internet users. Some studies revealed that when individuals recognize that they are being persuaded, the process of receiving persuasion will become more difficult and there is a tendency to avoid or ignore the message (Tutaj & van Reijmersdal 2012). The lack of clarity regarding this regulation makes marketers need to determine the right way to disclose the content is an advertisement while maintaining the appearance of advertisements so that they are not considered disturbing. Disclosure can be done in various ways such as variations in the language used, position, time, and duration of appearance. Wojdynski & Evans (2015) conducted a study to find out the effective language and positioning of labelling for news platforms on the internet. The results of the study showed that the use of the word advertising and sponsored in the middle and lower parts of the news had the highest ad recognition rate. The research is the basis for re-search conducted by Evans & Phua to find out the use of language that can produce high ad recognition on tagging disclosure on Instagram social media. Based on the research, it was found that the inclusion of tagging using "Paid Ad" writing resulted in high ad recognition compared to "Sponsored" or "SP."

At first there was no Instagram feature to mark paid content displayed by endorsers so that the endorsers only included the brand account without clearly informing them that the content was paid content. However, in 2017, in order to answer the provisions of the FTC, Instagram brought up the paid partnership feature to mark the paid content. Therefore, in order to be able to help marketers in determining the right way of marking, this study would like to compare the two disclosure positions for advertisements using an endorser currently available on the Instagram platform. Instagram's caption, ad recognition, and its impact on brand image and intention to share that information with other Instagram users are examined.

2 LITERATURE REVIEW

2.1 Native advertising and the role of disclosure

Prior studies posit that persuasive intent in nontraditional advertising formats such as social media campaigns is less recognizable than in traditional commercials (Sahni & Nair 2016). Advertising executions in social media that are often shared or recommended by strong ties in the user's network such as family, friends, and peers may increase the appeal of the advertising message, and, thus, subsequently impact advertising related attitudes, brand attitudes, and behavioral intent in a positive manner (Noort et al. 2012). However, conflicting results also warn that social media users are less patient with advertising whenever they perceive the advertisement's persuasive intent (Bang & Lee 2016).

Individuals may feel higher irritation when commercial content appears with no social connection (Ellison et al. 2007). In an attempt to persuade consumers without triggering advertising recognition and the associated coping mechanisms that include resistance and skepticism, advertisers have increasingly incorporated into their strategies the use of "native advertising," which minimizes advertising's interruption of social media usage (Lee et al. 2016). Native advertising is a method of digital advertising that looks very similar to news articles or content that already exists online. It is narrowly defined as a paid form of advertising whose appearance is often in the form of editorial content from the publisher (Brown et al. 2017) or broadly defined as various types of branded content that are similar to the format or design of the platform. Both definitions of native advertising share a fundamental commonality in that the format of native advertising should be similar to its surrounding media content (Lee et al. 2016).

Due to the obfuscation of the advertising and editorial or entertainment content, there are concerns that native advertising's effectiveness is based on viewers' lack of awareness or understanding that it is advertising to begin with (Brown et al. 2017). Thus, the covert nature of native advertising might prevent consumers from recognizing it as advertising and applying subsequent coping mechanisms. The FTC workshop on native advertising in 2013 (FTC 2013b), and subsequent guidelines for recommended best practices in 2015 (FTC 2015), addressed whether consumers could recognize native content as advertising and how best to employ labels and visual cues to differentiate editorial content from commercial or advertising content. These labels or cues, which are referred to as disclosures, are designed to clearly identify the persuasion attempt of an advertisement and protect consumers from being deceived or misled (Hoy &

Andrews 2004). Effective and clear disclosures should make the nature of the persuasive message and the intention behind the message clear to the consumer and in turn, aid consumers in thinking what the message is trying to accomplish (Boerman van Reijmersdal et al. 2018). The role of disclosure effectiveness in the context of native advertising plays a very important role in regards to consumer understanding and recognition of the content as advertising because oftentimes the presence of a disclosure is the only piece of information that delineates the communication as an advertisement.

2.2 Persuasion knowledge model

Individuals learn to approach different types of information through experience. On this basis, Friestad & Wright (1994) proposed that consumers' experience with various persuasive messages helps them to develop an understanding and awareness of persuasive intent in the marketplace. The Persuasion Knowledge Model (PKM) provides a conceptual understanding of how consumers understand and respond to persuasive messages. Persuasion knowledge is defined as the knowledge that "enables them (consumers) to recognize, analyze, interpret, evaluate and remember persuasion attempts and to select and execute coping tactics believed to be effective and appropriate" (Friestad & Wright 1994). Accordingly, individuals learn over time from experience (Friestad & Wright 1994) what constitutes persuasive communication, and how to appropriately carry out coping strategies designed to defend against the persuasive episode (Friestad & Wright 2005).

The consequence of recognition of the content (i.e., a persuasive episode) as advertising entails the use of coping strategies such as heightened skepticism, resistance, and counter-arguing, which in turn, have the potential to negatively affect brand and advertising related attitudes as well as behavioral intent (Nelson et al. 2009).

2.3 The effects of disclosure position on advertising recognition

For disclosures to be effective at conveying information, two sequential processes must occur: consumers must first notice the disclosures and then be able to understand the messages they convey. Research has validated that disclosures in advertising lead to advertising recognition only when consumers view them (Boerman et al. 2018), but the characteristics that increase the likelihood of disclosures in Instagram have not been systematically examined. The model of visual hierarchy suggests that users navigate information through two sequential phases, namely first scanning the page for entry points and then processing the information more deeply around entry points found (Faraday 2000). Where such entry points lie may be dictated by users' preexisting schemata that shape

expectations regarding where on the page particular content is located (Roth et al. 2010) and these patterns may vary based on the content domain of the page being viewed. Studies on online reading behavior confirmed that information near the top left corner of the page was most likely to be seen, followed by information horizontally branching rightward from the top left, and then down the page, in the shape of an F. The guidelines for the positioning of advertising disclosure labels generally suggested that placement near the top of the content was preferable, and a content analysis of online native advertising articles found that disclosures above the story headline were most frequent (Brown et al. 2017).

While this might suggest the supremacy of a top disclosure position, there is also evidence that users begin their general F-shaped viewing pattern further down the page, leaving information above that area unnoticed. Users expect advertising to be toward the right side or top of a web page, and display advertising at the top of the page is also most likely to be ignored. On an Instagram post, when users are scrolling down they will see the Instagram account and then they will focus on the picture and the caption in the post. Overall, research suggests that disclosure timing and location have varying levels of effectiveness for promoting consumer understanding and advertising recognition. One line of research suggests that consumers better locate, and, thus, recognize, advertising when disclosures are placed before or above the content (Boerman et al. 2018). However, competing research suggests that consumers' reading patterns and expectations may lead them to engage with editorial or entertainment content first and disclosure information later, which in turn reduces their odds of recognizing advertising. On the basis of the existing literature, we propose the following competing hypotheses:

H1a: Disclosure at the top of the Instagram post will lead to greater ad perception, ad credibility, informativeness, entertainment, attitude toward ad, attitude toward brand, and intention to share via e-WOM.

H1a: Disclosure at the top of the Instagram post will lead to greater ad perception, ad credibility, informativeness, entertainment, attitude toward ad, attitude toward brand, and intention to share via e-WOM.

2.4 Effects of ad recognition on the attitudes and behavioral intention

Previous research indicates "people need to be aware of a persuasion attempt before they can activate persuasion knowledge" (Boerman et al. 2018). According to PKM, the change of meaning principle (Friestad & Wright 1994) suggests that communications not previously considered the domain of advertising are recognized as advertising when elements of the communication such as the structure, presence of disclosures, or format lead consumers to that conclusion (Nelson et al. 2009). Once viewers conclude that a communication is an advertisement, they then use existing persuasion knowledge to make inferences about the persuasive intent of the communication or the communicator (Boerman et al. 2018). While advertising recognition and persuasion knowledge have often been considered interchange-able concepts, research suggests that advertising recognition is a separate and initial step that leads to the subsequent activation of persuasion knowledge (Evans et al. 2017).

The recognition of advertising elicits protective mechanisms such as increased skepticism and critical processing that influence attitudes toward and perceptions of advertising content in a negative manner (Boerman et al. 2018). While the link between advertising recognition and attitudes toward the ad, brand, or sponsor can be negative, less is known regarding consumers' attitudinal perceptions of story quality and credibility in the context of native advertising. Perceptions of story quality and credibility warrant empirical investigation because PKM predicts the occurrence of a negative relation-ship between advertising recognition and subsequent consumer attitudes.

In this study, ad recognition will be a screening question so that all respondents are aware that the Instagram post was an ad. The analysis conducted is whether the disclosure position will moderate the effect of each independent variable to the dependent variable:

H2: (a) Ad Perception of the top disclosure position will have a significantly more positive effect on Attitude toward ad than the Ad perception of the below disclosure position.
H2: (b) Ad Credibility of the top disclosure position will have a significantly more positive effect on Attitude toward ad than the Ad Credibility of the below disclosure position.
H2: (c) Entertainment of the top disclosure position will have a significantly more positive effect on Attitude toward ad than the Entertainment of the below disclosure position.
H2: (d) Informativeness of the top disclosure position will have a significantly more positive effect on Attitude toward ad than the Informativeness of the below disclosure position.

While the link between advertising recognition and attitudinal measures has been substantiated in previous research (Boerman et al. 2018), it can be posited that consumers' intention to pass along or share a native advertising–style story can be subsequently reduced when their ability to recognize the communication as advertising increases. This pro-posed relationship is theoretically sound based on the change of meaning principle (Friestad & Wright 1994). For example, the recognition of content as advertising when it is not previously anticipated as such may trigger attitudes and responses that are in-formed by increased feelings of skepticism and defensiveness (Nelson et al. 2009).

Recent studies have proved the relationship between the attitude toward ad and attitude toward brand, and attitude toward brand and intention to share. Therefore, we hypothesize the following relationships:

H3: Attitude toward ad of the top disclosure position will have a significantly more positive effect on Attitude toward brand than the Attitude toward ad of the below disclosure position.
H4: Attitude toward brand of the top disclosure position will have a significantly more positive effect on Intention to share than the Attitude toward brand of the below disclosure position.

3 METHODS

This study was designed for all participants in 2 disclosure positions, namely top and bottom, one shot between-subjects experiment was randomly assigned to view one of 2 versions of an Instagram post. Once they had completed seeing the Instagram post, participants completed dependent measures about the post. A total of 300 Instagram users were recruited for participation. Participants ranged in age from 18 to 36 years of age, and 52.5% of participants were male, and had the highest completed level of education (14% graduate degree, 42.6% college graduate, 32.2% some college, 10.7% high school only). Most of the participants used Instagram for 15–30 mins/day and uploaded 0–2 post/week.

The researcher created the stimulus material. Each of the 2 versions of the post consisted of an account name, a picture of food, a caption, and disclosure. Participants were randomly assigned to one of 2 versions of the post, which differed in the position of the "Paid Partnership" in the top and below the picture. Participants accessed the study via a link sent by the researcher. After providing informed consent, participants were taken to a page that contained a dummy Instagram post. After participants had finished seeing the post, they were taken to a page containing dependent measures. Participants completed, in order, measures of ad perception, ad credibility, informativeness, entertainment, attitude toward ad, attitude toward brand, and intention to share via e-WOM. Upon completing those measures, participants completed several demographic questions and were provided additional information about the study and its purpose. The responses from the participants were collected through online questionnaire and it continued with the data analysis for each dependent measure.

Participants were asked "Would you consider the Instagram post was an ad?" and those who checked "yes" were asked to provide detail regarding what on the post made them think portions were advertising. Participants' open-ended responses were coded as 1 (mentioned disclosure) or 0 (did not mention disclosure) based on the procedure used by Tutaj & van Reijmersdal (2012). Ad perception was measured using three 5-point semantic differential-scale items. Participants rated their perception of the ad by selecting one of five numbered points between a word pair (Not Interesting/Interesting, Not Enjoyable/Enjoyable, and Not Informative/Informative). Ad credibility was measured using three 5-point Likert-type items. Participants answered three statements about the post, each starting with "I think the post was..." (credible, trustworthy, and believable) on a scale ranging from strongly disagree to strongly agree. Entertainment was measured using three 5-point Likert-type items.

Participants answered three statements about the post, each starting with "I think the post was..." (entertaining, enjoyable, and pleasing) on a scale ranging from strongly disagree to strongly agree. Informativeness was measured using three 5-point Likert-type items. Participants answered three statements about the post, each starting with "I think the post was..." (a good source, relevant, timely information) on a scale ranging from strongly disagree to strongly agree.

Attitude toward ad was measured using three 5-point semantic differential-scale items. Participants rated their perception of the ad by selecting one of five numbered points between a word pair (not like/like, not favorable/favorable, not good/good). Attitude toward brand was measured using three 5-point semantic differential-scale items. Participants rated their perception of the ad by selecting one of five numbered points between a word pair (not good/good, not attractive/attractive, not likeable/likeable). Intention to share via e-WOM was measured using three 5-point Likert-type items. Participants answered three statements about the post, each starting with "I would like to..." (give positive review, recommend, and share the post) on a scale ranging from strongly disagree to strongly agree.

4 RESULTS AND DISCUSSION

To test hypotheses 1(a) and 1(b), an independent t-test was performed to ascertain the effects of dis-closure position on the ad perception, ad credibility, informativeness, entertainment, attitude toward ad, attitude toward brand, and intention to share (e-WOM). The overall independent t-test result was statistically significant, $p < .05$, except for the informativeness. The different disclosure positions did not affect the ability of the post to give relevant and timely information. The disclosure also existed in both stimulus and all respondents could recognize the disclosure. By comparing the mean of each variable for top and below disclosure position the result show that the top disclosure has higher ad perception, ad credibility, entertainment, attitude toward ad, attitude toward brand, and intention to share via e-WOM. Thus supporting hypothesis 1(a) and failing to support hypothesis 1(b) as shown in Table 1.

Hypotheses 2 to 4 were tested using moderating regression analysis to examine whether the different

Table 1. Result of independent t-test.

Variable	Top Disclosure Mean	Below Disclosure Mean
Ad Perception	3.74	3.46
Ad Credibility	3.29	2.84
Entertainment	3.74	3.37
Informativeness	3.33	3.13
Attitude toward ad	3.64	3.39
Attitude toward brand	3.51	3.24
Intention to share	3.03	2.83

Table 2. Result of regression.

Variable	Sig.	Koef.
Ad Perception	0.000	0.112
Ad Credibility	0.000	0.036
Entertainment	0.000	0.184
Informativeness	0.000	0.001
Attitude toward ad	0.000	0.059
Attitude toward brand	0.069	0.294

disclosure positions will have a moderating effect on the effect of each independent variable to the dependent variable. At first the interaction for each variable to the dummy variable for the top disclosure position and below disclosure position was calculated. Position was recoded into two dummy variables, with the top positioning serving as the referent category; the results of the regression are shown in Table 2.

The results indicate that all independent variables have a significant effect on the independent variable; p value $<.05$. Although the overall analysis for the moderating effect of the disclosure position on each independent variable to dependent variable was not statistically significant, $p > .05$, except the moderating effect of Attitude toward brand to sharing intention. The results from the regression analysis show the negative coefficient.

Based on the results obtained in this study, it is known that the perception of advertising (Ad Perception) has a significant effect on Attitude toward ad. This is indicated by the significance value of the results of the regression test, which is less than 0.05. The Ad Perception variable, which is moderated by the disclosure position, has an effect of 39.9% on Attitude toward ad. Whereas the moderation of the disclosure position on the effect of Ad Perception on Attitude toward ad did not show significant results. So that it can be said that, in this study, the disclosure position variations do not strengthen or weaken the influence of Ad Perception on Attitude toward ad. Based on the regression coefficient values, it is obtained that the disclosure position at the bottom produces a more negative response than the response to the disclosure position at the top; this is indicated by the regression coefficient value at the Ad

Perception moderation with a negative value position (−0.112) toward Attitude toward ad.

Based on the results obtained in this study, it is also known that the assessment of the credibility of advertising (Ad Credibility) has a significant effect on Attitude toward ad. This is indicated by the significance value of the results of the regression test, which are less than 0.05. The variable Ad Credibility, which is moderated by the disclosure position, has an effect of 24.5% on Attitude toward ad. As for the moderation of the disclosure position on the effect of Ad Credibility on Attitude toward ad, it does not show significant results. Hence, it can be said that, in this study, the disclosure position variations do not strengthen or weaken the effect of Ad Credibility on Attitude toward ad. Based on the regression coefficient values, it is obtained that the disclosure position at the bottom produces a more positive response than the response to the disclosure position at the top; this is indicated by the regression coefficient value at the moderation of Ad Credibility with a positive value position (0.036) toward Attitude toward ad.

Based on the results obtained in this study, the assessment of Entertainment has a significant effect on Attitude toward ad. This is indicated by the significance value of the results of the regression test, which are less than 0.05. The Entertainment variable, which is moderated by the disclosure position, has an effect of 48.2% on Attitude toward ad. As for the moderation of the disclosure position on the effect of Entertainment on Attitude toward ad, it does not show significant results. So it can be said that, in this study, the disclosure position variations do not strengthen or weaken the influence of Entertainment on Attitude toward ad.

Based on the value of the regression coefficient, it is obtained that the disclosure position at the bottom produces a more negative response than the response to the disclosure position at the top, which is indicated by the regression coefficient value in the moderation of Entertainment with a negative value position (-0.184) toward Attitude toward ad. Based on the value of the regression coefficient, it is obtained that the disclosure position at the bottom produces a more positive response than the response to the disclosure position at the top; this is indicated by the regression coefficient value on the Informativeness moderation with a positive value position (0.001) to Attitude toward ad. Thus it can be concluded that hypothesis 5 is rejected. Based on the results of regression tests that have been carried out, it is known that the four antecedents, namely Ad Perception, Ad Credibility, Entertainment, and Informativeness, have a significant influence on Attitude toward ad. So as to cause a good attitude toward an advertisement, the company that owns the product or service must not only pay attention to aspects of consumers' perceptions of the advertisement displayed, but also must pay attention to aspects of the credibility of the advertisement that the advert can be trusted to be true (Bracket & Carr 2001). The advertisements that

are displayed must also be able to provide information that is in accordance with the needs of consumers and interesting and fun to look at because advertisements that are considered disturbing can reduce consumers' affection for advertising (Ducoffe 1996). The existence of disclosure at the bottom is considered more disturbing than disclosure at the top so as to produce a more negative perception and entertainment. Based on previous research, it is known that placing disclosure at the bottom will increase ad recognition and activate the PKM, causing consumers to be more skeptical of these advertisements (Wojdynski & Evans 2014). Whereas the effect of Ad Credibility and Informativeness on Attitude toward ad on advertisements with disclosure positions at the top and bottom produces almost the same effect because both stimuli have the same content arrangement so that the variation of disclosure position on stimulus types A and B does not make a significant difference to the ability of advertisements to provide relevant information because the words "Paid Partnership" still exist in both stimuli. It can be concluded that the below disclosure position tends to lower the attitude toward ad, attitude toward brand, and intention to share.

The results also showed that variations in both disclosure positioning and disclosure language influenced the likelihood that participants would perceive the content as advertising. Interestingly, the results showed that although the position of a sponsorship disclosure did affect ad recognition, the traditionally recommended top-of-the-page position was less effective than disclosures in the bottom of the page. This is in contrast with FTC recommendations, which suggested that disclosures "appeared immediately before the ad or at the top left corner of the content" (FTC 2013c, p. 4). One explanation for this may have to do with users' starting point on the page (e.g., F-shaped viewing pattern). A disclosure above the headline may be ignored, and when readers get far enough into the article to view the middle- or bottom-positioned disclosures, they may have been less likely to ignore them due to their proximity to the content of the article itself. Another potential explanation for the effectiveness of a middle-placed disclosure in the context of native advertising may relate to its ability to break up the story content. By interrupting the story content, a middle-placed disclosure attracts more attention to itself, which in turn, may help the viewer recognize the content as advertising. While vertical positioning may have been a key determinant of whether users noticed the disclosure at all, disclosure language also influenced advertising recognition.

REFERENCES

Bang, H. & Lee, W. N. 2016. Consumer response to ads in social network sites: An exploration into the role of ad location and path. *In journal of current issues & research in advertising* (Vol. 37). https://doi.org/10.1080/10641734.2015.1119765

Boerman, S. C., van Reijmersdal, E. A., Rozendaal, E., & Dima, A. L. 2018. Development of the persuasion knowledge scales of sponsored content (pks-sc). *International journal of advertising,* 37(5), 671–697. https://doi.org/10.1080/02650487.2018.1470485

Brown, S. P., Stayman, D. M., Wojdynski, B. W., Golan, G. J., Manoochehr, N., Mirbagheri, A. Y., Hoekstra, J. C. 2017. Insight the efficacy of pop-ups and the resulting effect on brands. *Journal of advertising*, 17(1), 129–140. https://doi.org/10.1016/j.tele.2018.01.006

Ellison, N. B., Steinfield, C., & Lampe, C. 2007. The benefits of Facebook "friends:" Social capital and college students' use of online social network sites. *Journal of computer-mediated communication*, 12(4), 1143–1168. https://doi.org/10.1111/j.1083-6101.2007.00367.x

Evans, N. J., Phua, J., Lim, J., & Jun, H. 2017. Disclosing Instagram influencer advertising: The effects of disclosure language on advertising recognition, attitudes, and behavioral intent. *Journal of interactive advertising*, 17(2), 138–149. https://doi.org/10.1080/15252019.2017.1366885

Faraday, P. 2000. Visually critiquing web pages. https://doi.org/10.1007/978-3-7091-6771-7_17

Friestad, M. & Wright, P. 1994. The persuasion knowledge model: How people cope with persuasion attempts. *Journal of consumer research*, 21(1), 1. https://doi.org/10.1086/209380

Friestad, M. & Wright, P. 2005. The next generation: Research for twenty-first-century public policy on children and advertising. *Journal of public policy & marketing*, 24(2), 183–185. https://doi.org/10.1509/jppm.2005.24.2.183

Hoy, M. G. & Andrews, J. C. 2004. Adherence of prime-time televised advertising disclosures to the "clear and conspicuous" standard: 1990 versus 2002. *Journal of public policy & marketing*, 23(2), 170–182. https://doi.org/10.1509/jppm.23.2.170.51397

Kemp, S. 2018. Digital Report.

Lee, J., Kim, S., & Ham, C. D. 2016. A double-edged sword? Predicting consumers' attitudes toward and sharing intention of native advertising on social media. *American behavioral scientist*, 60(12), 1425–1441. https://doi.org/10.1177/0002764216660137

Nelson, M., Wood, M., & Paek, H. J. 2009. Increased persuasion knowledge of video news releases: Audience beliefs about news and support for source disclosure. *In journal of mass media ethics* (Vol. 24). https://doi.org/10.1080/08900520903332626

Noort, G., Antheunis, M., & Reijmersdal, E. 2012. Social connections and the persuasiveness of viral campaigns in social network sites: Persuasive intent as the underlying mechanism. *In journal of marketing communications* (Vol. 18). https://doi.org/10.1080/13527266.2011.620764

Roth, S. P., Schmutz, P., Pauwels, S. L., Bargas-Avila, J. A., & Opwis, K. 2010. Mental models for web objects: Where do users expect to find the most frequent objects in online shops, news portals, and company web pages? *Interacting with computers*, 22(2), 140–152. https://doi.org/10.1016/j.intcom.2009.10.004

Sahni, N. S. & Nair, H. 2016. Sponsorship disclosure and consumer deception: Experimental evidence from native advertising in mobile search. Ssrn. https://doi.org/10.2139/ssrn.2737035

Truong, Y. & Simmons, G. 2010. Perceived intrusiveness in digital advertising: Strategic marketing implications. *Journal of strategic marketing*, 18(3), 239–256. https://doi.org/10.1080/09652540903511308

Tutaj, K. & van Reijmersdal, E. A. 2012. Effects of online advertising format and persuasion knowledge on audience reactions. *Journal of marketing communications*, 18(1), 5–18. https://doi.org/10.1080/13527266.2011.620765

Advances in Business, Management and Entrepreneurship – Hurriyati et al. (Eds)
© 2021 Taylor & Francis Group, London, ISBN 978-0-367-67471-7

Brand image in the selection of perfume refill

H. Sujono & H. Hendrayati
Universitas Pendidikan Indonesia, Bandung, Indonesia

ABSTRACT: Refill perfume products are still in great demand by various groups, especially the millennial generation. However, in reality there are still refill perfume brands that have not shown their brand image on the market. The purpose of this study is to find out the brand image of foreign brands compared with local brands, and to find out the factors that make up the selection of perfume refill among the millennial generation. The population in this research was perfume refill users in Cimahi City. The sampling technique used was random sampling, where researchers used an age-limited sample, that is, respondents were aged between 17 years and 25 years. The limitation of the age of respondents was made to obtain relatively homogeneous data. The samples taken in this study amounted to 138 respondents. The data collection technique of this study were questionnaires and interviews. Processing data in this study was done by confirmatory factor analysis with the help of SPSS statistical tools. The results of the study showed that the brand image of foreign perfume brands was better known than domestic/local brands. Attributes that were used as factors for selecting perfume, especially among millennials, are quality, safety, and packaging design, with results indicating that the forming factor of refill perfume selection was determined very closely by quality, safety, and packaging design.

1 INTRODUCTION

Indonesia's economic development is expected to bring improvements in various economic sectors. Increased economic growth has an impact on increasing public consumption, including the consumption of primary, secondary, and tertiary products. So that with increasing public consumption, it will become a challenge for product suppliers/providers to be able to capture opportunities from increasingly dynamic market growth.

To attract the attention of consumers of product or service providers, they must be able to provide excellent service to consumers to obtain superior customer value. According to Raharjani (2005), producers of various products must be able to make consumers interested in coming and then making transactions, which is not an easy thing considering that consumers coming to shopping centers have different goals and motives ranging from just looking around, recreation, to specifically for shopping for their needs. Alexander et al. (2015) firmly stated that each customer is unique with differences in needs and desires as well as purchasing choices that are influenced by habits and choices that follow the nature of psychology and social impulses that influence the selection process.

One of the good cosmetic products that is experiencing growth today is perfume. Perfume is a cosmetic product that is widely used by the public, especially the millennial generation. Perfume is a product that has a very broad market both in the local market

and international markets where the opportunities for Indonesia to develop perfume products that have high competitiveness can be supported by the availability of various resources in Indonesia.

Cosmetics are a unique product because, in addition to these products, the ability to meet women's basic needs for beauty at the same time is often a means for consumers to clarify their social identity in the eyes of the people by Fabricant & Gould (Ferrinadewi 2005)

Perfume purchases made by consumers are influenced by various factors ranging from age, in-come level, employment, economic situation to lifestyle as well as personality. Nowadays the need for perfume is very high, so the opportunity is to increase demand from consumers should be managed optimally in order to be able to meet the needs and desires of consumers. This research was conducted aimed to determine the factors of product attributes that influence the purchasing decision of perfume products in the millennial generation.

2 LITERATURE REVIEW

According to Kotler in Habibah & Sumiati (2016), the understanding of products is everything that can be offered to the market to get attention, bought, used, or consumed that can satisfy desires or needs. Conceptually the product is a subjective understanding of the producer for something that can be offered as an effort to achieve organizational goals through meeting the

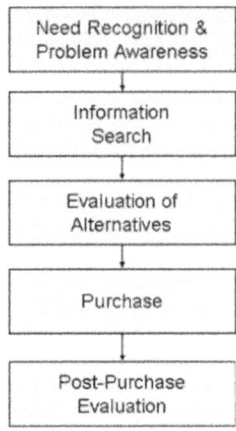

Figure 1. A buyer's decision-making process.

needs and activities of consumers, in accordance with the competence and capacity of the organization and the purchasing power of the market.

Consumer needs for branded products need to be considered. According to Kotler in the work of Sitio (2017), the value of a brand is divided into four levels from the lowest to the highest through (1) brand awareness, (2) brand acceptability, (3) brand preference, and (4) brand loyalty. Of the four levels categorized, the recognition of a brand is the one that must exist to raise the level of awareness of a brand.

Sitio (2017) stated that brand equity, which means the value contained in the name, symbol, and other brand combinations on products and services of a company can increase or reduce the value of customer trust and loyalty, with dimensions of (1) inherent value (attached value), (2) respect, and (3) judgment. In line with the work of Godey et al. (2016), brand equity contributes to increasing brand preference, willingness to pay at premium prices, and consumer loyalty. While the application of consumer purchasing decision behavior refers to marketing strategies, regulator policies, social marketing, and individual information. Purchasing behavior is the whole of consumer attitudes, preferences, intensity, and decisions regarding consumer behavior in the market when purchasing a product or service (Alexander et al. 2015).

An individual's decision is influenced by individual factors such as the age of the buyer and the circumstances of the life, work, and lifestyle of the economic situation, personality, and self-concept. Product purchase decisions are not always resolved by the user. Buyers need to buy products. Marketers must decide to direct the promotion efforts to buyers or users. They must ensure identifying the people who influence decision making the most. If marketers understand consumer behavior well, they are able to predict how customers tend to react to various information and environmental issues and are able to shape their marketing strategies as mentioned by Kotler (Alexander et al. 2015) and shown in Figure 1:

The type of classification of recognition of needs includes (Perreau 2014):

1. Functional needs: These needs are related to functional problems. Consumers buy a washing machine to avoid washing by hand.
2. Social needs: Needs come when consumers want social recognition or from desire to have. Consumers buy luxury items to look good in front of others.
3. Needs to change: Consumers feel the need to change. This can result in the purchase of new clothes or furniture to change the current appearance.

In purchasing decisions, consumers will always see the product as a set of attributes. Product attributes are often regarded as something that is valued by consumers as a factor that determines their relevance to the product. Thus asking the consumer which attributes are considered important is the right way to find out the consideration of the purchase or use of a product by consumers (Suhardi, in Sadeli and Utami 2013). Kotler & Armstrong (2010) states that product attributes have product characteristics, namely brand, packaging, and product quality.

The factors measured in the decision on the selection of perfume products in this study are brand, quality, pride, loyalty, product reliability, design, and perfume needs, and price and place of purchase

3 METHODS

The research method in this study was quantitative descriptive analysis. According to Sugiyono (2010), quantitative methods, as research methods based on the philosophy of positivism, used to examine a population or a particular sample in which data collection using research instruments and data analysis are quantitative or statistical with the aim of testing the hypothesis have been established.

The variable in this study was brand equity. The population in this study were perfume users in Ci-mahi City. The sampling technique used was free random sampling. The researchers limited the age of respondents to between 17 and 25 years of age to obtain relatively homogeneous data. The samples taken in this study amounted to 138 respondents. The method of collecting data in this study was by distributing a questionnaire. Processing data in this study was done by confirmatory factor analysis with the help of SPSS statistical tools

4 RESULTS AND DISCUSSION

Respondents in this study were students in the Management Department of the Faculty of Economics, UNJANI, who represented the millennial generation. The number of respondents was 138 respondents.

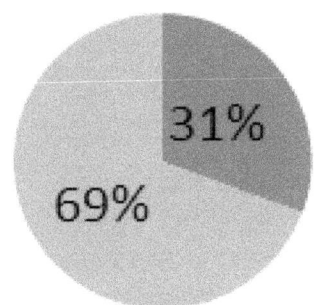

Figure 2. Gender of respondents.

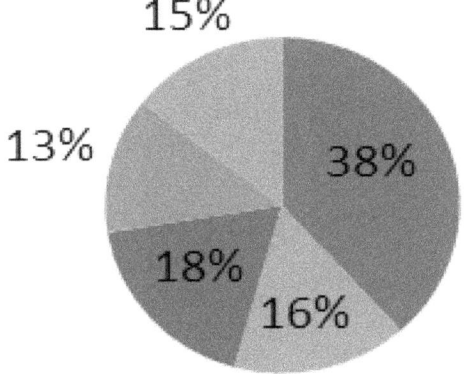

Figure 3. Income (pocket money).

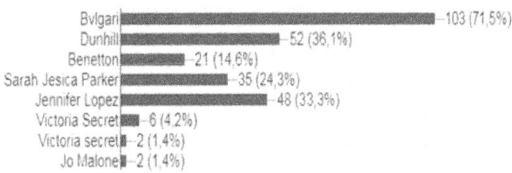

Figure 4. The known perfume brands from overseas brands.

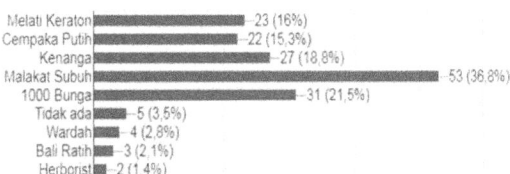

Figure 5. Perfume brands that are recognized from domestic (local) brands.

The profile of respondents used by researchers included gender, pocket money/income, known international perfume brands, known local perfume brands, and perfumes used. The results of the respondents' sex are shown in Figure 2.

In this study 69% of the respondents (95 people) were women and 31% (43 people) were men. The sex of the respondents shows that women use more perfume than men.

The second profile in this study is the allowance received by the respondents. The results are presented in the diagram shown in Figure 3.

Based on Figure 3, it can be seen that the majority of respondents (52 respondents) have pocket money/income of more than 2 million rupiahs, then respondents who have money/income between 500 thousand rupiahs to 1 million rupiahs as much as 18% (25 respondents). There are also quite a number of respondents who have pocket money between 100 thousand and 500 thousand rupiahs as much as 16% (22 respondents) and those who have pocket money/income between 1.5 million rupiahs and 2 million rupiahs of 15% (21 respondents), while the remaining 13% (18 respondents) had pocket money between 500 thousand rupiahs and 1 million rupiahs.

Based on Figure 3, it shows that perfume users come from various circles, thus, showing that per-fume is needed by the millennial generation.

Furthermore, it is explained that the perfume brand most known is from overseas. The results are presented in the bar diagram in Figure 4.

The most-known perfume brand from overseas is the Bvlgari brand. The brand is known by more than 71% of the respondents. The Dunhill brand is known by 36% of the respondents and the Jennifer Lopez brand is known by more than 33.38% of the respondents. The brands Sarah Parker and Benetton are also quite well known by the respondents by respectively 24.3% of the respondents and 14.6% of the respondents.

Furthermore, the known perfume brands from local brands are presented in Figure 5.

Based on Figure 5, the majority of respondents know the Malakat Subuh brand as much as 36.8%. The consecutive brand names known to the respondents were 1000 flowers (31%), Kenanga (18.8%), Melati Keraton (16%), and White Campaka (15.3%).

From the results, the respondents were more familiar with foreign perfume brands compared to domestic perfumes.

Part two discusses the dimensions of perfume product selection. The statements in part two consist of brand, quality, pride, loyalty, product safety, design, and public needs.

The first statement in this study is the satisfaction of the perfume used, in which the majority of respondents agreed and strongly agreed (96%). This showed that respondents were very satisfied with the perfume products they used.

The majority of respondents stated that they agreed and strongly agreed (93%), which showed that the quality of perfume products used was in accordance with the quality expected by the respondents.

Furthermore, from the respondents' statements regarding pride using foreign brand perfume products, the majority of respondents agreed and strongly agreed (57%) with the statement of pride in using perfume from abroad. On the other hand, the number of respondents who expressed doubts, disagreed, and strongly

Table 1. I always use the same brand of perfume.

Information	Frequency	Percentage	Weight	F X B
Strongly Agree	22	16	5	110
Agree	46	33	4	184
Doubtful	17	12	3	51
Disagree	46	33	2	92
Strongly Disagree	7	5	1	7
Total	138	100		444
Average		3.22		
Category		Loyal enough		

Table 2. KMO and bartlett's test.

Kaiser-Meyer-Olkin Measure of Sampling		0.614
Bartlett's Test of Sphericity	Approx. Chi-Square	124.488
	Df	28
	sig	0.000

Table 3. Communalities.

	Initial	Extraction
Quality	1.000	.533
Pride	1.000	.298
Loyalty	1.000	.388
Security	1.000	.513
Packaging Design	1.000	.606
Needs	1.000	.357
Location	1.000	.275

disagreed (43%) was also quite high. So the category for pride in using domestic perfume was in the good category.

Furthermore, from the respondents' statements related to pride using perfume products of domestic and local brands, the majority of respondents agreed and strongly agreed (68%) with the statement of pride using local brand perfumes. Even so, some respondents expressed doubts and disagreements and strongly disagreed (32%) with the statement. Based on these results, the statement of pride in using domestic perfume brands was in the proud category in Table 1.

Based on Table 1, most of the respondents agreed and strongly agreed (49%) with the statement that they always used the same perfume brand. Although many respondents also expressed disagreement, strongly disagreed, and were hesitant (51%). So that the loyalty of local perfume consumers in the category was quite loyal.

Next is the explanation about the safety of perfume products used. The majority of respondents agreed and strongly agreed (70%) with the statement that the product used was safe from side effects. Although there were also quite a number of respondents who expressed doubts and disagreements (30%); so the safety of products from side effects was in the safe category. The majority of respondents said they agreed and strongly agreed (78%) on the design statement of perfume packaging that was used interestingly. There were also respondents who expressed doubts, disagreed, and strongly disagreed (32%) with the statement. This showed that the perfume packaging used was attractive. The majority of respondents (88%) agreed and strongly agreed with the statement that perfume is a need. Only 8% of respondents expressed doubt, disagreement, and strongly disagreed. So based on the results of the statement of re-sponsorship, the statement of perfume is needed is in the category of desperate need.

Furthermore, the majority of respondents (74%) agreed and strongly agreed to the statement of the affordable price of the perfume used. Although there were also respondents who expressed doubts, disagreements, and strongly disagreed (26%) with the

statement. So the affordable perfume price statement was in the affordable category.

Next is about buying perfume offline. The majority of respondents (87%) agreed and strongly agreed to the statement of purchasing products offline. Whereas there were only 13% of respondents who expressed doubt and were not consistent with the statement.

The next one is about the results of the respondents' answers to the statement of purchasing online. The majority of respondents said they did not agree, strongly disagreed, and were hesitant (82%) on the statement of purchasing perfumes online. Only 18% of respondents agreed and strongly agreed to the statement. So that purchasing perfume products online is in the category of not liking. Brand variable factor analysis image perfume refill is presented in Table 2.

Based on the results of the Kaiser-Meyer-Olkin (KMO) and Bartlett's tests shown in Table 2, the results of KMO Measure of Sampling Adequacy (KMO MSA) are greater than 0.5 and the Sig smaller than 0.05. So that factor analysis can be continued because it meets the first requirements shown in Table 3.

Based on the value of communalities, it indicates whether the variables studied will be able to explain the factors. Value extraction must be greater than 0.5 in order to explain the factor. Based on Table 3, there are only three variables that fulfill to be able to explain the factors, namely quality, safety, and packaging design as seen in Table 4.

Based on the Table 4, two factors can be formed from the components that have more than one value, namely components 1 and 2. Factor 1 is able to explain by 27.662% and factor 2 is able to explain by 44.280%. So that the total addition of factors 1 and 2 is 71.942%.

Table 4. Component analysis.

Total Variance Explained

Component	Initial Eigenvalues			Extraction Sums of Squared Loadings		
	Total	% of Variance	Cumulative	Total	% of Variance	Cumulative
1	2.213	27.662	27.662	2.213	27.662	27.662
2	1.329	16.618	44.280	1.329	16.618	44.280
3	.972	12.174	56.428			
4	.935	11.691	68.119			
5	.852	10.652	78.771			
6	.713	8.918	87.689			
7	.513	6.408	94.097			
8	.472	5.903	100.000			

5 CONCLUSION

The results of the study showed that the brand image of foreign perfume brands was better known than domestic/local brands. From the results of the discussion, it revealed that the attributes of perfume selection factors by the millennial generation were quality, safety, and packaging design. The total variation that was able to explain the product selection factor was by 71.942%. So the forming factors of perfume selection were determined very closely by quality, safety, and packaging design.

ACKNOWLEDGEMENT

The study was funded by the Ministry of Research and Universities through the Higher Education Primary Research Scheme in 2019.

REFERENCES

Alexander, R. Khonglah, Oliver. Subramani, A, K. 2015. Customer buying behaviour towards branded casual shoes, ayanavaram, chennai. *Zenith international journal of business economics & management re-search zijbemr*, Vol.5 (6), June (2015)

Ferrinadewi, Erna. 2005. Atribut produk yang dipertimbangkan dalam pembelian

Godey, B., et al., Social media marketing efforts of luxury brands: Influence on brand equity and consumer behavior, *Journal of business research* (2016), http://dx.doi.org/10.1016/j.jbusres.2016.04.181

Habibah, Ummu, Sumiati. 2016. Pengaruh kualitas produk dan harga terhadap keputusan pembelian produk kosmetik wardah di kota bangkalan madura. *Jurnal ekonomi & bisnis*, Hal 31–48 Volume 1, Nomor 1, Maret 2016.

Kosmetik dan pengaruhnya pada kepuasan kon-sumen di surabaya. *Jurnal manajemen & kewirausahaan*, Vol. 7, no. 2, September 2005: 139–151 jurusan ekonomi manajemen, fakultas ekonomi – universitas kristen petra http://puslit.petra.ac.id/~puslit/journals/

Kotler, P. & Amstrong, G. M. 2010. Principles of Marketing. 13 ed. Pennsylvania: Prentice Hall.

Kotler, P., Keller, K., Brady, M., Goodman, M., & Hansen, T. 2016. *Marketing management*.

Perreau, f. N.d. The 4 factors influencing consumer behavior. 2014. http://theconsumerfactor.com/en/4-factors-influencing-consumer-behavior/

Raharjani., Jeni. 2005. Analisis faktor - faktor yang mempengaruiji keputusan pemilihan pasar swalayan sebagai tem-pat berbelanja. *Jurnal studi manajemen dan organisasi*.

Sitio. Arifin. 2017. Analisis pengaruh antara mutu produk, pe-layanan purna jual dan ekuitas merek terhadap cit-ra merek smartphone Samsung. *Journal management and business review*. Vol.14, No.1, January 2017: 120–147

The effect of human-centric brands and online store image on social media engagement in Bandung city small and medium enterprises

R.C. Jaya, A. Yudanegara & S.P.H. Triono
STIE Indonesia Membangun, Bandung, Indonesia

ABSTRACT: Nowadays many small and medium enterprises (SMEs) in Indonesia use social media as a marketing channel. They utilize social media for free promotional media, providing information quickly, and increasing brand awareness. However, the percentage of their engagement is below 1%, which will make the use of social media ineffective. Published research has dealt with social media engagement, but has not yet addressed the human-centric brands toward social media engagement. No research has been developed regarding human-centric brands in SMEs in developing countries. This research was aimed at analyzing the application factors of human-centric brands and online store image on social media engagement using a descriptive verification method. The population of this research was Bandung community. The number of samples in this study were 140 people. The results of this study indicated a positive interaction of human-centric brands on social media engagement. However, the presence of online stores was not positive toward social media involvement. This research is expected to be a guide for SMEs to increase the effectiveness of social media so that it can be used to promote channel, provide effective information, and increase awareness of SME brands.

1 INTRODUCTION

Social media is the most widely used communication and information media today. Social media has complete features for the social needs of its users. Social media is often used by users to disseminate information about themselves, information, and news to introduce a product or service company.

Many social media users in Indonesia use SMEs to create social media accounts for their brands. One important factor in using social media is having a high involvement. Social media involvement is the willingness of consumers to take their time and energy to talk about the business processes carried out by SMEs as shown in Figure 1.

Social media engagement can be done in two directions so that information conveyed about company product and service information, promotions, and after sales services to customer care centers can be received completely and can be understood by both parties. However, the percentage of their engagement is still below 1%, which will make the use of social media ineffective.

The measurement criteria are based on the following data (scrunch.com, June, 2019):

a. Less than 1% = low engagement rate
b. Between 1% and 3.5% = average/good engagement rate
c. Between 3.5% and 6% = high engagement rate
d. Above 6% = very high engagement rate.

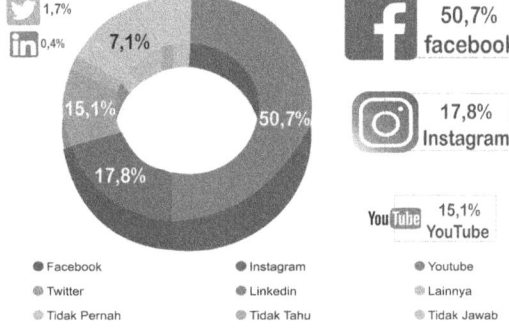

Figure 1. Indonesian social media users.

Based on Table 1, only 3 out of 20 SMEs received good criteria, while the other 17 have low criteria. Social media engagement has an important role in a company. Social media engagement has a positive influence on the customers of an organization or company (Jiang et al. 2015). Social media can be a free channel for sales promotion, product and service information, and customer service. There are many studies on social media engagement (Dhar & Abhishek 2014; Frutos et al. 2014; Jiang et al. 2015), but there is no research examining social media engagement to SMEs. Previous research discusses more on large organizations.

Previous research by Jaya (2015) found that there was a positive effect of online store images on a

Table 1. SME engagement rate.

Number	Instagram Account	Posts	Followers	Average Likes	Percentage	Criteria
1	Mizzleisme	80	320000	90	0,028%	Low
2	Vanillahouse.id	750	170000	1.169	0,688%	Low
3	Taswanita.bandung	2999	150000	80	0,053%	Low
4	Rumahwarna_corp	2890	130000	274	0,211%	Low
5	Levaya.id	549	120000	152	0,127%	Low
6	Gudamgtas_bandung1	1600	55000	86	0,156%	Low
7	Nerobags	2810	22500	431	1,916%	Good/Average
8	Toko_cilina	243	16200	6	0,037%	Low
9	Kumsi_bdg_	311	6921	51	0,737%	Low
10	Bywell.bag	6500	5958	2	0,034%	Low
11	Asteroidfashion_	2798	4399	66	1,500%	Good/Average
12	Bagstore.y	130	3900	8	0,205%	Low
13	Tokosikami	377	3691	8	0,217%	Low
14	Ha.nacollection	354	2510	4	0,159%	Low
15	Taskitabandung	90	1610	12	0,745%	Low
16	Dreameut_bags	50	1500	15	1,000%	Good/Average
17	Tasaku_collection	200	1002	2	0,200%	Low
18	Shoppimgamanah1	800	910	5	0,549%	Low
19	Gufelainc	361	5531	6	0,108%	Low
20	Lapak_tas_bandung	731	3023	2	0,066%	Low

product's customers. On the other hand, in previous research, there was no mention of the relationship between online store images and social media engagement. Human-centric brands have a positive influence on customers (Kotler 2017), but the effect of social media engagement has not yet been seen.

2 LITERATURE REVIEW

2.1 Social media engagement

Social media engagement is a derivative of social web engagement, which means the willingness of customers to spend their time and energy talking to you, about you in a conversation, and through processes that affect your business (Prasetio 2014).

Indicators of social media engagement are the number of brand page content, the number as generated per visitor, the number of users who responded (games, contests, and coupons (participation)), the average number of minutes that visitors remained on the page (duration), like rate per post or other content and applause rate (Laudon & Traver 2017).

Social media engagement is a process where online communication and the content that is posted online helps to build connections with other people in the online community (Sherman & Danielle 2013). Extrovert personalities are more social in nature and more involved in social media activities, whereas introvert personalities are less involved in social media activities. Thus targeting extroverted populations will increase the chance of product purchase trends compared to introverted personality populations (Dhar 2014). Research by Frutos (2014) showed that companies could take advice and opinions from

consumers through interactions via Facebook with consumers.

2.2 Social media engagement

A brand is a name, term, sign, symbol, design, or a combination of all, intended to identify goods or services or sales groups and to differentiate from competing goods or services (Kotler 2009). More and more brands are made not only for the benefit of the company. Brands have also begun to adopt human values to attract customers in the human-centric era. A human-centric brand is the application of human values that are physically attractive, intellectually attractive, socially attractive, and emotionally attractive while at the same time show strong personality and morality (Kotler 2017). Indicators of human-centric brands:

a. Physicality: logo shape, logo color, logo design
b. Intellectuality: product innovation
c. Sociability: communication with customers
d. Emotionality: inspirational messages
e. Personality: show their flaws, take full responsibility
f. Morality: having integrity

2.3 Social media engagement

Store image is a whole picture that is more than just a sum per section, where each part interacts with each other in the minds of consumers (Utami 2010). According to Aghekyan et al. (2012) researchers have revealed that there are many similarities in characteristics between store image and online store image, so that it can be a reference for online store image research. Chen & Teng (2013) define online store image as consumers' perceptions of the nature, functions, and psychology that affect their behavior when they interact with an online store.

On the other hand, Jaya (2015) states that online store image is an overall picture that consumers think about an online store, which arises because of the perceptions and attitudes felt in the sensations of stimulation related to the online store environment. Indicators of online store image are the attractiveness of online shopping layout, the attractiveness of online shopping design, the fairness of product prices, the attractiveness of product displays, the attractiveness of discounts, the ease of remembering online shopping addresses, the convenience of display, completeness of information, and the ease of payment methods

3 METHODS

3.1 Model and concept research

In this research model there are two kinds of arrows, namely one-way arrows that indicate the direct influence of an independent variable on a dependent variable, for example, $X1 \rightarrow Y$, and the second is a two-way arrow that states the correlation between the relationship of independent variables, for example, $X1 \leftrightarrow X2$. A figure model of the diagram path for the relationship between variables can be seen in Figure 2.

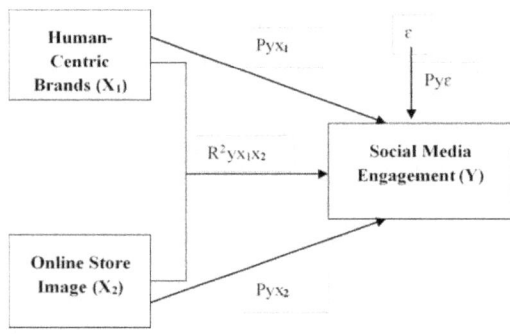

Figure 2. Path diagram model.

ε = Residual Variable (error factor) (1)

Pyx1 = Path coefficient from X_1 to Y (2)

Pyx2 = Path coefficient from X_2 to Y (3)

R^2yx1x2 = Path coefficient from X1 and X2 to Y (4)

3.2 Research methodology

This research used quantitative explanatory design methods to determine the causal relationship between human-centric brands and online store images on social media engagement.

Table 2. Descriptive analysis.

Variable	Percentage	Rate
Social Media Engagement	81.80%	Very Good
Human-Centric Brands	81.30%	Very Good
Online Store Image	80.10%	Good

3.3 Population and sample

This research was conducted in the city of Bandung, Indonesia, starting from September, 2018, until March, 2019. The samples in this study were all female residents of Bandung and its surroundings, amounting to 140 people. With a total female population of 1,245,778 (bps.go.id 2019).

The sampling method was judgment sampling by determining samples that are easy and in accordance with the criteria. The research sample consisted of internet users who had an Instagram account. Population representation by the sample in research is an important requirement for generalization. The object of this research was SMEs in Bandung that had Instagram and 20 SME online stores were used.

From the data, the age of the samples was from 15 to 51 years. With the highest number of respondents being 21 years old, followed by 22 years old. Meanwhile, for the education level of respondents, the highest proportion (71%) was in high school. Furthermore, the most prevalent types of work were private-employees (56%) and students (22%).

4 RESULTS AND DISCUSSION

4.1 Results

The results of the study were presented in two forms, namely descriptive and verification. Descriptive analysis was used to examine respondents' description in the variables of social media engagement, human-centric brand, and online store, and also to determine the effect of each variable, partially and simultaneously. Descriptive analysis in table 2.

Table 2 shows that the description of respondents to the variable social media engagement is very good, this can be seen from its value of 81.8%. Respondents' description of the human-centric brand variable is very good with a percentage of 81.3%. Respondents' description of the online store image variable is good, this is indicated by a percentage of 80.1%. The simultaneous coefficient values are presented in Table 3.

Table 3 shows the testing results of the coefficient of human-centric brands and online store image on social media engagement simultaneously. Based on Table 3, a value of R^2yx1.x2 = 0,193 was obtained. Next, the coefficient values of human-centric brands and online

Table 3. Simultaneous coefficient values.

Change Statistics			
Model	R Square	df1	df2
	Change		
1	,193	2	140

Table 4. Partial coefficient values.

	Standardized Coefficients	
Model	Beta	t
X1	0,440	5,684
X2	–0,007	–,089
Dependent Variable: Y		

store image variables on social media engagement partially are shown in Table 4. Partial coefficient values in Table 4.

A framework for causal relationships between paths (X1 against Y, X2 against Y, X1 and X2 against Y) can be made through the structural equation:

$$Y = pyx1 \ X1 + pyx2 \ X2 + py \ \varepsilon \qquad (5)$$

It can be seen in the data in the Tables that:
Value $R^2yx1.x2 = 0,193$
$\rho \ y \ \varepsilon = 1–R^2yx1.x2 = 1–0,193 = 0,807$
$\rho \ yx1 \ X1 = 0,440$
$\rho \ yx2 \ X2 = –0,007$
So it can be described as:

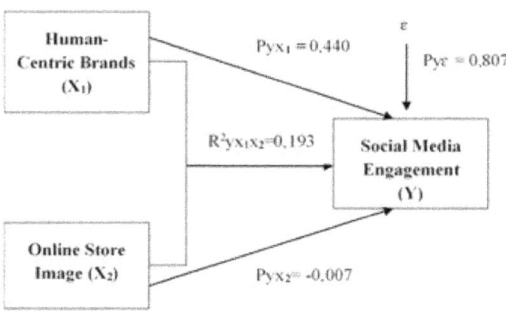

Figure 3. Path analysis diagram of effect of X1 and X2 against Y.

The contribution of human-centric brands that directly affect social media engagement is 0.440. Contributions to the online store image that are influenced by social media engagement are –0.007. The amount of contribution simultaneously is 0.296. The direct influence of other factors outside the variable (?) is 0,807. The direct effect of variables X1 and X2 against Y is shown in Table 5.

Table 5. The direct effect of variables X1 and X2 against Y.

Variable	Coefficient	Direct Influence	Total	Contribution
X1	0,440	0,440	19,3%	–
X2	–0,007	–0,007	0,0049%	–
ε	0,807	0,807	65,12%	–
X1 and X2	–	–	–	0,193 or 19,3%

Based on the test results above, it can be seen that other factors not examined in this research have a large percentage contribution to the influence of social media engagement amounting to 65.12%. While the contribution of the human-centric brand is 19.3%, and the contribution of online store image is only 0.0049%.

4.2 Discussion

Simultaneous and partial tests have been carried out to obtain the results of all variables. In simultaneous testing with the F value of 16.405 with a probability value (Sig) = 0.000 and the value of sig <0.05, then H0 is rejected and Ha is accepted. It means that human-centric brand and online store image have an effect simultaneously and significantly toward social media engagement.

In the first partial testing obtained t_{hit} for the variable X1 is 5.684, while the value for t_{Table} is 1,288, and $t_{value} > t_{Table}$, which means that Ha is accepted. This means that human-centric brands have a positive effect on social media engagement. While in the second calculation, the t_{value} obtained for variable X2 is 0.089, while the value for t_{Table} is 1,288, so $t_{value} > t_{Table}$, which means that H0 is accepted. That is, online store image does not have a positive effect on social media engagement intentions.

Based on the results above, the role of human-centric brand is very important toward social media engagement, amounting to 19.3%. This reinforces the theory of Kotler et al. (2017), which states that human-centric brand needs to be applied to a brand because it will have a positive effect on products, companies, and customers.

The online store image variable test results have an insignificant effect, in contrast with previous research (Aghekyan et al. 2012; Chen 2015; Tang et al. 2019). This can occur because there are other variables not examined such as privacy and security factors of online stores (Hong et al. 2019).

Social media engagement in SMEs is very important and can even be a company's business strategy in creating marketing success opportunities and brand awareness. Social media engagement also has an effect on improving the company's customer service (Li Loic et al. 2017) so that SMEs can continue to maintain engagement with their followers.

5 CONCLUSION

The results showed that the influence of human-centric brands had the strongest contribution to the social media engagement of SME's social media accounts, while the online store image had no effect on social media engagement. SMEs that have taken a physicality approach to the logo shape, logo color, logo design; intellectuality in product innovation; sociability of communication with customers; emotionality to inspirational messages; personality by showing their flaws, taking full responsibility; and a morality approach by having integrity in the SME's brand, are expected to continue. Although, in this study, the online store image has no effect on social media engagement, the criteria for the online store image needs to be maintained. This finding implies that in order to maximize social media engagement, SMEs must pay attention to human-centric brands in all content uploaded on social media.

Thus followers' engagement (duration of visits, level of likes per post, and positive comments) on Instagram accounts increase. To achieve that, all it takes is the participation of the government and academics in providing counseling and training and other parties to provide assistance in supporting the success of SMEs.

REFERENCES

Adhi Prasetio. 2014. *Buku ajar e-commerce & its business*. Mediakita.

Aghekyan, Marine, Forsythe, Sandra, Kwon, Wi-Suk, Chattaraman, & Veena. 2012. The role of product brand image and online store image on perceived risks and online purchase intentions for apparel. *Journal of retailing and consumer services*. 19. 325–331. 10.1016/j.jretconser.2012.03.006.

Chen, Ming-Yi, and Teng Ching-I. 2013. A comprehensive model of the effects of online store image on purchase intention in an e-commerce environment. Electron commerres 2013. Vol.13:1–23. *New York: springer science business media*.

Chen, Ming-Yi. 2015. Do the factors of online store image have a parallel relationship. 10.1007/978-3-319-11806-2_76.

Dhar, Joydip, & Abhishek Kumar. 2014. Analyzing social media engagement and its effect on online product purchase decision behavior. Journal of Human Behavior in the social environment, 24:791–798, 2014. Taylor & Francis Group, LLC. Routledge.

Frutos, Sergio Martín, Ferran Giones, Francesc Miralles. Social media engagement as an e-commerce driver, consumer behavior perspective. Conference: cisti 2014 - 9th Iberian conference on information systems and Technologies, Barcelona: La Salle – Ramon Llull University

Hong, Lu & Che Nawi, Noorshella & Zulkiffli, Wan Farha & Mukhtar, Dzulkifli & Ramlee, Shah. 2019. Perceived risk on online store image towards purchase intention. *Research in World Economy*. 10. 48. 10.5430/rwe.v10n2p48. https://instagram. com/. Instagram Apps 10/02/2019. 08.15 PM. https://instagram. com/

Jaya, Rama Chandra. 2015. Pengaruh diskon harga dan online store image terhadap niat beli

Jiang, Hua, & Luo, Yi & Kulemeka, Owen. 2015. Social media engagement as an evaluation barometer: Insights from communication executives. Public relations review. 10.1016/j.pubrev.2015.12.004.

Kotler, Phillip. 2009. *Manajemen pemasaran*, edisi 13. Jakarta; erlangga.

Kotler, Philip. Hermawan Kertajaya, & Iwan Setiawan. 2017 Marketing 4.0 moving from traditional to digital. New jersey: John Wiley & Sons, inc, hoboken.

Laudon, Kenneth C., Carol Guercio Traver. 2017. E-Commerce 2017, Pearson

Li, Loic, & Brodie, Roderick & Juric, Biljana. 2017. Social media engagement in a service crisis. ANZMAC 2017, Melbourne, Australia

Produk pada online store hiffu bandung. Jurnal indonesia membangun. ISSN: 1412–6907, Vol. September 2915 – December 2015.

Scrunch.com 16/06/2019. 08.5 PM. What is a good engagement rate on Instagram. https://blog. scrunch.com/what-is-a-good-engagement-rate-on-instagram

Sherman, Aliza, & Danielle Elliott Smith. 2013. Social media engagement for dummies. New jersey: John Wiley & Sons, inc., hoboken.

Tang, X. & Zhou, J., & Zhao, L. 2019. Research on the influence of online store image on consumer's willingness of online shopping. *Journal of Silk*. 56. 54–60. 10.3969/j.issn.1001-7003.2019.01.009.

Utami, C. W. 2010. *Manajemen ritel*: Strategi dan Implementasi Ritel Modern. Jakarta: Salemba Empat.

Advances in Business, Management and Entrepreneurship – Hurriyati et al. (Eds)
© 2021 Taylor & Francis Group, London, ISBN 978-0-367-67471-7

Students interest in studying at the Islamic university As-Syafi'iyah dy on students: Motivation, perception, and trust

H. Herawati & H. Hendrayati
Universitas Pendidikan Indonesia, Bandung, Jawa Barat, Indonesia

ABSTRACT: A university is a place for people to gain academic skills and abilities that is chosen as the rapid growth of technology demands people with good academic qualities. The quality of university alumni are among the factors contributing to the motivation, perception, and trust of prospective students before deciding to choose a university. This present study is aimed at exploring students' interest to study at Islamic University As-Syafi'iyah based on the aspect of motivation, perception, and trust. The methodology used in this study was an associative method and a survey which was distributed to 423 students as the participants. Furthermore, the instrument of this study was a Likert scale and the data were analyzed using multiple linear regression analysis. The results revealed that the three aspects previously mentioned significantly contribute to the decision of choosing Islamic University As-Syafi'iyah.

1 INTRODUCTION

Along with the changing times, human resources is increasingly needed to bring a fresh air to edu-cational institutions, especially in higher education. Higher education is a place that can provide knowledge and experience and can also be one of the formation of human resources that are professional, intellectual, and highly competent. Indonesia is a country with a majority Muslim population, but in reality there are more public education institutions than Islamic tertiary institutions in Indonesia. This is proven by the number of State Universities in Indonesia, amounting to 372 State Universities, only 58 of which are State Islamic Universities. In addition, among 3940 Private Universities in Indonesia, only 781 of them are Private Islamic Universities.

Even though in every region in Indonesia there are several Islamic tertiary institutions, they are still very small in number compared to other public universities.

As-Syafi'iyah Islamic University (UIA) is one of the Private Islamic University in the East Jakarta region which is a University under the guidance of Kopertis region III, with the motto "The Integration of Science and Religion" by having 13% of the subjects in it consisting of Islamic subjects . However, in recent years the interest of students to study at UIA has decreased, this is because the perception, motivation and trust of the community began to decrease with the emergence of new universities around the UIA.

Research conducted previously states that motivation, perception, attitude, trust influence in decision making (Dewi 2008; Jamaluddin 2012; Le-nora 2019).

2 METHODS

This descriptive study employed an explanation design. It is intended to explain the influence of a variable on the other variables. This study examined the influence and the causal relationship between the independent variable (i.e., motivation, perception, and trust) and the dependent variable (i.e., decision to study at Islamic University As-Syafi'iyah). A survey was distributed and further examined through a relationship causal study or a study that analyzes the cause and effect between the two variables in this present study.

2.1 Population

Population refers to an area of generalization consisting of the object(s) or subject(s) possessing certain qualities or characteristics that meet the needs of the research. The subject(s) or object(s) are further analyzed before coming into a conclusion (Sugiyono 2011: 90). Based on observations, there were 423 new students in the 2018/2019 academic year in the management program of the Economic Faculty at the Islamic University As-Syafi'iyah

2.2 Sample

To determine the sample size of a population, the Slovin formula was used with the percentage of non-attachment clearances because the desired sampling error is (5%) with the formula: $N = N / 1 + N (e) 2$, the results obtained were 205 people.

2.3 Operational definition

The variables used in this study consisted of three independent variables (X1, X2, X3) and one dependent variable (Y) with a Likert scale as the measurement, which was defined as follows:

1. Motivation (X1) is the drive or driving force of one's work to achieve certain goals. With indicators:

   a. Introduction to the problem
   b. Searching for information
   c. Evaluation
   d. To do

2. Perception (X2) is a process that makes a person to choose, organize and interpret the stimuli received into a meaningful picture with indicators:

   a. Responsible
   b. Achievement
   c. Self-development
   d. Independence in acting
   e. Physiological needs

3. Trust (X3) is the belief of one party regarding the intentions and behavior of the other party. With indicatorsv

   a. Choose
   b. Organize
   c. Interpret
   d. Doing

4. Learning Interest (Y) is Stages – stages in the decision making process to find out something, with indicators

   a. Quality assurance of Education
   b. Environment
   c. Infrastructure
   d. Guarantee to be hired at work

2.4 Data collection technique

Through library research (field research), field research (Field research) by employing interviews and questionnaires (questionnaire) on students of the Islamic University As-Syafi'iyah. MODEL AND CONCEPT OF RESEARCH HYPOTHESIS in figure 1.

Correlation between Variables

1. The aspect of motivation significantly and positively influences the decision of selecting Islamic University As-Syafi'iyah
2. The aspect of perception significantly and positively influences the decision of selecting Islamic University As-Syafi'iyah.
3. The aspect of trust significantly and positively influences the decision of selecting Islamic University As-Syafi'iyah

3 RESULTS AND DISCUSSION

Validity test is a sample test used to test the instruments used in the study, as many as 20 people from a total

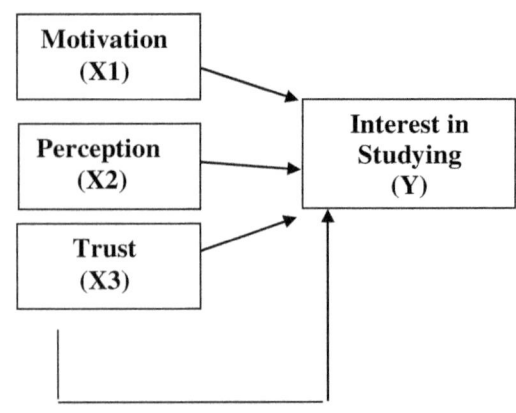

Figure 1. Model and concept of research hypothesis.

of 205 study samples. The results of testing validity and reliability using the Cronbach Alpha Coefficient method (Riduwan 2011: 353). To find out the level of validity (Riduwan 2011: 348) with the provisions of rcount> rtable, then it is said to be valid and If rcount <rtable then it is said to be invalid. The validity test results was valid because it was above the value rtable = 0.444 (from the Rho Spearman table, degree of freedom (df) = n − 2). From the df used is 20−2 = 18 with an alpha of 5%. Cronbach's alpha reliability test results showed that the overall instrument trial was> 0.60 then the research variable was declared reliable, with Motivation (X1) of 0.870, perception variable (X2) of 0.891 and confidence variable (X3) of 0.883. So that data analysis can be continued to predict the relationship between variables in accordance with the proposed hypothesis (Riduwan 2011: 349).

From the Analysis of Multiple Linear Regression to test the hypotheses that have been formulated namely; Motivation (X1), perception (X2), Trust (X3), and Learning Decision (Y) as dependent variables using SPSS 22 computer program packages, the results are obtained:

$$Y = 2,650 + 0,375X1 + 0,111X2 + 0,388X3 \quad (1)$$

From the above equation can be explained as follows:

1. A value = 2,650. This means that if there is no change in the variables Motivation (X1), perception (X2), Trust (X3), the learning decision (Y) is 2,650.
2. Value of b1 = 0.375. This means that each addition of one value to Motivation (X1), then the performance will experience a value increase of 0.375 when X2 and X3 = 0.
3. Value of b2 = 0.111. This means that each addition of one value to perception (X2), then the performance will experience a value increase of 0.388 when X1 and X3 = 0
4. Value of b3 = 0.388. This means that each addition of one value to Trust (X3), then the performance

will experience an additional value of 0.111 when X1 and X2 = 0.

Multiple Coefficient Correlation Analysis (R) and Determination (r) influence between Motivation (X1), Perception (X2) and Trust (X3) can be seen through the correlation coefficient mainly used on the Decision to Register at As-Syafi'iyah Islamic University. The result of the correlation coefficient or R of 0.801 showed that the influence of Motivation, Perceptions and Trust in enrolling at As-Syafi'iyah Islamic University had a positive and very strong influence that was equal to 80.1%. The result of the coefficient of determination or R square (r2) of 0.642 showed that 64.2% of Purchasing Decisions were influenced by Motivation, Perceptions and Trust while the remaining 35.8% was influenced by other factors not examined in this study.

The Testing Hypotheses whether it is simultaneously (Test F) or influential together between variables Motivation, perceptions and Belief in the Decision to Register at As-Syafi'iyah Islamic University was known through the hypothesis test. The test was done by comparing the significant level of calculation results with a significant level of 0.05 (5%) with the following criteria: If Fcount (sig)> 0.05 then Ho is accepted and Ha is rejected and If Fcount (sig) <0, 05 then Ho is rejected and Ha is accepted. From the F test results it was known that the f-count value of 20.920 was greater than the f-table of 2.87 (20.920> 2.87) with a significance value of 0.000 less than 0.05 (0.000 <0.05). This means that the variables Motivation, Perception and Trust simultaneously influenced the decision of learning in the management study program of the Faculty of Economics of the Islamic University of As-Syafi'iyah, so that the hypothesis stating that the alleged variable Motivation, perception and Trust together (simultaneously) influences the Decision studying in the Management study program of the Faculty of Economics of the As-Syafi'iyah Islamic University can be accepted.

Partial Testing (t Test) on the effect of motivation, perception and trust on learning decisions at As-Syafi'iyah Islamic University was known by using the hypothesis test criteria as follows: tcount <ttable (=0.05), then Ho is accepted so that Ha is rejected. - tcount> t table (=0.05), then Ho is rejected so Ha is accepted. Based on the calculation results, it was obtained that:

1. The calculated X1 variable value was 2.979 with a significance value of 0.005 while the value of the table was 2.030 with a significance of 0.05. Thus, it can be concluded that Motivation (X1) partially had a significant positive effect on learning decisions (Y) with a significant value of 0.005 < 0.05.
2. For variable X2, it was obtained that the tcount was 0.695 with a significance value of 0.492 while the ttable value was 2.030 with a significance of 0.05. It can be concluded that Perception (X2) partially had a negative effect on learning decisions (Y) with a significant value of 0.492 > 0.05

3. For variable X3, it was obtained that the tcount was 2.261 with a significance value of 0.030 while the ttable value was 2.030 with a significance of 0.05. It can be concluded that Trust (X3) partially had a significant positive effect on learning decisions (Y) with a significant value of 0.030 < 0.05.

4 CONCLUSION

From the results of the study, it can be concluded that Motivation (X1) had a significant positive effect on learning decisions (Y), Perception (X2) had no partially significant effect on learning decisions (Y), and Trust (X3) had a significant positive effect on learning decisions (Y). While from the F test, Motivation, Perceptions, and Trust variables influenced jointly on the Decision to Register at As-Syafi'iyah Islamic University with an f-calculated value of 20.920 greater than the f-table of 2.87 (20.920 > 2.87) with a significance value of 0,000 less than 0.05 (0,000 < 0.05)

This study resonates to the results seen in other studies (Akbar 2010; Bella 2009; Clow & Donald Baack 2012; Miller 2013; Satya 2012; Suprianto;t al. 2014; Wahyuni 2008; Wang & Lee 2006; Yusnindar et al. 2014; Zhang 2017), where motivation, perception, and trust had a positive influence on buying decision.

ACKNOWLEDGMENT

This study focuses only on the students' interest to choose Islamic University As-Syafi'iyah. It is expected that future research can expand the subject of the study, not only limited to the variable of motivation, perception, and trust, but also other relevant variables, e.g., price, location, and product design (or the available study programs) (Nursento 2014). Exploring more variables is useful to measure the students' interest to pursue their study at Islamic University As-Syafi'iyah. It is also aimed at providing a reference for the development of the admission of new students

REFERENCES

Ahmed, Z. (2014). Effect of brand trust and customer satisfaction on brand loyalty in bawalpur. Journal of sociological research. Vol. 5, No. 1, pp. 306–326

Akbar, Y. J. 2010. Analisis pengaruh motivasi konsumen, persepsi kualitas, dan sikap konsumen terhadap keputusan pembelian sepeda motor honda. Fakultas ekonomi universitas diponegoro semarang.

Assauri, S. 2004. Manajemen pemasaran: dasar, konsep dan strategi. Rajawali Press: Jakarta.

Bella, B. 2009. Analisis pengaruh faktor program promosi, persepsi merek, motivasi, dan sikap konsumen terhadap proses pengambilan keputusan pembelian sepeda motor suzuki di kota semarang. Skripsi tidak dipublikasikan. Fakultas ekonomi universitas diponegoro semarang.

Clow, K. E., & Donald Baack. 2012. Integrated advertising, promotion, and marketing communications. 5th ed. Pearson Education. Prentice Hall, England

Kotler, Philip. (2015). Manajemen pemasaran, analis perencanaan, implemetasi dan kontrol;, Jakarta: PT. Prinhallindo.

Miller, D. (2013). Konsumsi dan konsekuensinya. Hoboken: wiley. Monroe, kb (2003). Harga dan persepsi konsumen terhadap nilai (3rd ed.). New York: mcgraw-hill. Monroe, KB,

Nagadeepa, C., J. Tamil Selvi., & A. Pushpa. 2015. Impact of sale promotion techniques on consumers' impulse buying behaviour towards apparels at bangalore. Journal of management sciences & education. Vol. 4(1)

Peter, J. Paul & Olson, Jeryy C. (1999). Consumer behavior: perilaku konsumen dan strategi pemasaran, Jakarta: Erlangga. [27]

Prasetijo, Ristiyanti, & Ihalauw, John J. O. I. (2004). Perilaku Konsumen, Yogyakarta: CV Andi Offset.

Ramadhani, A. 2011. Pengaruh faktor psikologis terhadap keputusan pembelian sepeda motor merek yamaha. Jurnal ekonomi manajemen, fakultas ekonomi – universitas gunadarma.

Satya, A. P. 2012. Analisis faktor-faktor yang mempengaruhi keputusan pembelian pada toko buku gramedia di kota semarang. (http://www.fisip.undip.ac.id). Diakses maret 17, 2013

Shimp, Terence A. (2009). Periklanan promosi aspek tambahan komunikasi pemasaran terpadu jilid ii, university of south carolina, penerbit erlangga.

Sidharta, I., & Sidh, R. (2013). Analisis faktor-faktor sikap yang membentuk niat maha-siswa menjadi teknopreneur. Jurnal computech & bisnis, 7(2), 117–128.

Suryabrata, Sumadi. (1998). Metodologi penelitian, universitas gadjah mada, jakarta: pt.raja grafindo persada-Suprianto, D.; Susanta, H. and Nurseto, S. (2014). The effect of product design, price and loca-tion toward purchase decision. Jurnal ilmu administrasi bisnis. Vol. 3 No. 4.

Swastha DH, Basu. (1984). Azaz Azaz Marketing, Yogyakarta: Liberty.

Swasta DH, Basu & Irawan (1980). Manajemen pemasaran modern, lembaga akademi manajemen perusahaan, Yogyakarta: YKPN.

Terplan, Kornel. (1992). Communication network management, second edition, Englewood Cliffs New Jersey: Prentice Hall.

Terplan, Kornel. (1997). Telecom operations management solution with net expert, Englewood Cliffs New Jersey: Prentice Hall.

Wahyuni, D. U. 2008. Pengaruh motivasi, persepsi, dan sikap konsumen terhadap kepu-tusan pembelian sepeda motor merek honda di kawasan surabaya barat. (http://www.petra.ac.id / puslit /journals). Di-akses maret 17, 2013

Wang, Jing & Angela Y. Lee. 2006. "the role of regulatory fit on information search and judgement", dissertation, kellog school of management, Northwestern University, Evanston

Weston, J.Fred; E.Copeland, Thomas; Wasana, Jaka, & Kirbrandoko. (1998). Manajemen keuangan jilid 1, edisi kedelapan, erlangga.

Yusnindar et al, 2014 "pengaruh kepercayaan dan persepsi resiko terhadap minat beli dan keputusan pembelian produk fashion secara online di kota pekanbaru", jurnal sosial ekonomi pembangunan universitas riau, volume 12

Zekiri, J. And Hasani, V. V. (2015). The role and impact of the packaging effect on consumer buying behaviour. ECOFORUM. Vol. 4, Special Issue 1, pp. 232–240

Zhafira, N. H.; Andreti, J.; Akmal, S. S. and Kumar, S. (2013). The analysis of product, price, place, promotion and service quality on customer's buying decision of convenience store. A survey of young adult in bekasi, west java, indonesia. International journal of advances in management and economics. Vol. 2, Issue 6, pp. 72–78.

Advances in Business, Management and Entrepreneurship – Hurriyati et al. (Eds)
© 2021 Taylor & Francis Group, London, ISBN 978-0-367-67471-7

The effect of advertising media and message towards purchasing decisions

D.N. Nisrina & H. Hendrayati
Universitas Pendidikan Indonesia, Bandung, Indonesia

ABSTRACT: Advertising media and message are two important factors in attracting consumers' interest in purchasing a product. Sprite is one of the soft drink products from PT. Coca-Cola Amatil, which has long been known in Indonesia. This study aims to determine the effect of advertising media and messages of Sprite product toward customer purchasing decisions in Bandung. There are several measured dimensions from each variable. The variable of media consists of reach, frequency, and time dimensions. The advertising massage variable involves content, format, and source dimensions. Then the purchasing decision variable consists of evaluation of alternatives and purchase decision dimensions. This study used explanatory descriptive with quantitative methods. The sample was 100 respondents selected by using random sampling techniques. The data were gained from an offline questionnaire that was distributed to the Sprite advertisement audience in Bandung. The data were then analyzed by using Kendall and Spearman's correlation analysis. The results indicated that advertising media and messages had an influence on purchasing decisions.

1 INTRODUCTION

To deal with competition, companies carry out promotions to market their products. One of the promotions that companies can do is by advertising their products. Promotion is a very common thing among the public because promotion is a company activity that can attract consumers to buy the goods or services produced by the company (Kotler & Amstrong 2014). Moriarty et al. (2009) stated that "advertising is a type of marketing communication, which is in a broad term refers to all communication techniques marketers used to deliver their message to their customers." As an effort to face competition, many companies carry out activities to be able to market their products, and one of them is by advertising programs. Kotler & Amstrong (2014) argued that advertising is any paid form of non-personal presentation and promotion of ideas, goods, or services by an identified sponsor. Advertising is paid nonpersonal communication from an identified sponsor using mass media to persuade or influence an audience. Wells et al. (2013) stated that in advertising, there are five main decisions: mission, money, message, media, and measurement.

The research of Prof. Deshmukh & Pawar (2015) found that there was a significant difference of television advertisement elements toward purchase decisions and the effort to build a brand. Another research was performed by Woo et al. (2015). Woo conducted a research to explore the factors that determined which communication media influenced consumers to purchase a product. The results showed that it was varied according to the socio-demographic and product category variables.

One of the promotional media used by PT. Coca-Cola Bottling Indonesia is television. The television advertisement by Sprite is "Refreshing Reality." In a research carried out by Frontier Consulting Group, it was stated that every year, Sprite products were included in the top brand list compared to other soda products. It experienced an increase in 2016, and a decrease in 2018 and 2019. Therefore, this research was conducted in order to find out how much influence television advertisements have to cause consumers to buy Sprite products, and how much the influence of the advertising media and messages had on purchasing decisions.

1.1 Advertising media

Advertising media are the channels used by advertisers in mass communication (Lamb et al. 2015). According to (Kotler & Keller 2012), choosing the advertising media is to find the most cost-effective media to convey the desired number and type of exposure to the target audience. This includes deciding on the desired range, frequency, and impact; choosing the main media types; choosing special media means; deciding on the right time for the media; and deciding on media allocation geographically.

Russel & Lane (2006) stated that the success of advertising is measured by good products, the right time, product differentiation, and price competition. The success of product advertising occurs when the products are needed by consumers, have competitive prices, good quality, and no substitute items. Through advertisement, people can have pleasant or unpleasant opinions on the product being advertised. In addition, advertisements are also able to create an attraction that can make the advertised products at-tractive to consumers. The attractiveness is used to create product quality (Wells et al. 2013).

1.2 Advertising messages

The message of the advertisement is what the company plans to convey on its advertisement and how it plans to deliver the message verbally and nonverbally (Sumartono 2002). Whereas, according to Bovee (1996), the advertising message is ideas or news that are communicated or delivered to the audience through advertising media. A good advertisement is usually focused on one core sale, containing interesting things from the advertised brand, stating exclusive and different things that will not be found in other brands, and can convince or is needed by the consumers (Kotler & Keller 2012).

1.3 Purchasing decisions

According to Kotler & Keller (2012), the purchasing decision process is the evaluation stage, in which the consumers form preferences among brands and an intention to buy the most preferred brand. The purchasing decision is the stage in the buyers' decision-making process where consumers actually buy (Kotler & Armstrong 2014). Companies need to understand the consumers' behavior in making the purchasing decision. Schiffman & Kanuk (2010) defined consumer behavior as the behavior shown by consumers in finding, giving, using, evaluating, and spending a product or service that is expected to satisfy their needs

2 METHODS

Based on the theoretical foundation and the framework described above, the hypotheses of this study are:

H1: The advertising media and messages of PT. Coca-Cola Amatil Indonesia have attracted the interest of prospective consumers.
H2: The purchasing decisions on Sprite products tend to be positive.
H3: There is an effect of TV advertisement on the purchase decisions of Sprite consumers in Bandung.

The design used in this research was quantitative design. It used a survey method. Regarding the types, it is categorized as descriptive explanatory since it provided an accurate description of several aspects in the market environment (Aaker et al. 2011).

There were two variables in this research. The independent variables consisted of advertising media and message, while the dependent variable was purchasing decision. There were 100 people chosen as the sample. The sample was gained from a random sampling technique in which the population was studied according to the problem and purpose. In this case, the sample were consumers of Sprite products who have watched the "Refreshing Reality" advertisement.

The data consisted of primary and secondary data. The primary data were gained from the questionnaire distributed to the respondents.

3 RESULTS AND DISCUSSION

The reliability test results show that Cronbach's alpha is 0.939, the value is in the interval between 0.81 and 1.00. Thus it can be concluded that the instrument is very reliable. The results of the average value of the questionnaire items from the media and advertising information variables is 3,952. This indicates that the media and advertising messages carried out by PT. Coca-Cola Amatil Indonesia in Bandung are categorized as good.

The results of the Kendall correlation analysis were carried out using IBM SPSS 24 software. The correlation results of advertising media and purchasing decision is 0.338 (33.80%). This implies that the relationship between advertising media and purchasing decision is weak. On the other hand, the correlation of the advertising message and purchasing decision is equal to 0.395 (39.50%). This means that the relationship between the advertising message and the purchasing decision is fair.

The dimensions of the XI variable (advertising media) are range and frequency and time-based. Meanwhile, the dimensions of the Y variable (purchasing decision) are the evaluation of alternatives and purchase decision dimensions. Based on the results of the Kendall correlation analysis, the dimensions of the X1 variable have a weak positive correlation with the Y variable. It is said to be uncorrelated if the correlation value is 0. Moreover, the correlation of the advertising messages and the purchase decision is fair. The dimensions of the X2 variable (advertising message) are content, format, and source. These dimensions have a fair positive relationship with the dimensions of the Y variable.

To find out whether the independent variables in the study influence the dependent variable, a t-test was conducted. Based on the calculation result, media has a significance value of (Sig 1– tailed) = 0,000. This value is lower than the significance level of 0.05. In other words, t count is greater than t table, and, thus, Ho is rejected and Ha is accepted (3,555 > 1,980). This shows that advertising media has a significant relationship to purchasing decisions. Moreover, t count of advertising message is also greater than t table. Then Ho is rejected and Ha is accepted (4,256 > 1,980). Hence the value of the correlation coefficient is significant. Both are greater than the t table, which is 1,980. This means that there is a consistency between the results of the Kendall correlation analysis and the results of the t-test analysis.

It is known that the result of F count is 24.440. The F table for the data is obtained from the formula (k; n–k), where (k) is the number of independent variables and (n) is the number of respondents. Then 2; 98 is obtained. Referring to F table, the result is 3.09. By comparing F count and F table, which is 24.440 > 3.09, it can be concluded that the advertising media variable and advertising message variable simultaneously influence the purchasing decision.

Based on all the above results, there is a significant influence of media and advertising messages on the decision to purchase Sprite products in Bandung. It strengthens the research conducted by Prof. Deshmukh & Pawar (2015) in a journal entitled "Effects of Television Advertising Elements on Customer's Purchase Decision and Brand Building." In their study, there were six elements of television advertisements, namely celebrities, messages, jingles, characters, punch lines, and logos. One of the elements in television advertising was message, which is one of the variables used in the present study. The research found that six elements of television advertisement had a significant impact on customers' purchase decision.

4 CONCLUSION

Based on the results and discussion, several points can be concluded about the influence of advertising media and messages on the customers' purchasing decision of the Sprite products in Bandung. Advertising media and messages have a strong correlation with the purchase decision. Meanwhile, the correlation of each variable has different strengths. This is based on the result of the F-test, which is 24.440 > 3.09, the results of the Kendall advertising media correlation analysis of the purchase decisions of Sprite customers in Bandung, which is 33.80% (low correlation), and advertising messages with a correlation value of 39.50% (fair correlation). To sum up, the purchase decision of Sprite customers is influenced by media and advertising messages.

For the design, more comfortable implementations are also needed for the exposure and interface design specification process in more detail, so that it becomes an easy reference in the implementation, integration, testing, and deployment of web applications for SME LSP IMC. Analysis and design of IMC in the study is intended that the SMEs LSP has an overview of the activities in an integrated campaign that is contained in the application of the company's website.

ACKNOWLEDGEMENTS

The researcher would like to thank the UPI Postgraduate School, especially the Supervisors of the Management Study Program. Moreover, thank you to the research respondents who have contributed to the success of this research. This research would not have been carried out without the support from various parties.

REFERENCES

Aaker, D. A., Kumar, V., Leone., Robert, P., & Day, G. S. 2011. Marketing Research, 10th Edition. *United States of America: John Wiley & Sons, Inc*

Deshmukh, A.V. & Pawar, V. 2015 Effects of Television Advertising Elements on Customer's Purchase Decision and Brand Building, *International Journal of Research in Finance and Marketing* (5)6:9–35.

Kotler, P. & Keller L. K. 2012. Marketing Management, 13th Edition.

Kotler, P. & Armstrong. 2014. Principle of Marketing, 15th edition. New Jersey. *Pearson Prentice Hall.*

Lamb, C. W., Hair, J. F., & Mc. D. C. 2001, Pemasaran, *Buku Satu, Edisi Pertama, Jakarta, Salemba Empat.*

Moriarty, S., Nanc M., & William, W., 2009. Advertising: Principles and Practice, Upper Sadle River. *Pearson Education*

Russel, T. W. & Lane, R. K. 2006. Advertising Procedure, Prentice Hall, 13th, *New Jersey.*

Schiffman, L. G. & Kanuk, L. L. 2010. Consumer Behavior. New Jersey: *Prentice Hall.*

Sumartono 2002. Terperangkap dalam Iklan: Men-eropong Imbas Pesan Iklan Televisi. Bandung: *Penerbit Alfabeta.*

Top Brand Award. Top Brand Survey Fase 1. Retrieved from http://www.topbrand-award.com, April 4, 2019

Wells, W., Jhon B., & Sandra, M. 2013. Advertising, Principles and Practice, sixth edition. New Jersey: *Pearson Education, Inc.*

Woo, J., Joongha, A., Jongsu, L., & Yoonmo, K. 2015. Media channels and consumer purchasing decisions, *Industrial Management & Data Systems* (115) 8:1510–1528.

The effect of atmosphere on visiting decision

Y. Yuliawati, R. Hurriyati, L.A. Wibowo & V. Gaffar
Universitas Pendidikan Indonesia, Bandung, Indonesia

ABSTRACT: Museum is a permanent and non-profit institution. It serves public needs by collecting, conserving, researching, communicating, and showing real objects for the needs of study, education, and pleasure. One factor that must be considered in museum is atmosphere. The influence of atmosphere on the museum is very important. It affects the attractiveness and the interest of consumers to visit. This research used Atmosphere and Visiting Decision as variable. The method was descriptive analysis with survey methods. The data used were primary data and secondary data. Primary data was collected by accidental sampling. The sample were 100 visitors of MONPERA museum. The result of multiple regression analysis revealed that atmosphere had a positive effect on visiting decisions.

1 INTRODUCTION

In 2010, the Ministry of Culture and Tourism (Ministry of Tourism and Creative Economy) launched Visit Museum program. A Website of Museumku states that the program is supported by various activities in museum. It aims to increase the number of visitors, public appreciation and a concern for the nation's cultural heritage (Source: http: //ejournal.upi.edu74)

Bandung has so many museums, such as Geological museum, the Asian-African Conference, Sri Baduga, Pos Indonesia, Barli, Mandala Wangsit, Train, Wolf Schoemaker, Toys, National Education, Monumen Perjuangan Rakyat (Monpera), Puspa IPTEK, Virajati Seksoad, Bank OCBC NISP museum, and Galeri Soemardja ITB. The two newest museums are Kota Bandung Museum and Gedung Sate (Source: Disbudpar Bandung).

Many people have lack of interest in visiting museum. This is because limited number of collection objects on the museum. Moreover, the atmosphere seems unattractive (product choice) or boring. Turley & Ronald (2007) stated that atmosphere give an impact to the customer when they are in a room, and it influences them to make purchases.

Atmosphere is a combination of physical characteristics (architecture, display, lighting, color, temperature, music), the scents having emotional responses, and customer perceptions which influence customers in buying products (Utami 2008). According to Berman & Evans (2012), store atmosphere is environmental design through visual communication, lighting, color, music, and perfumery to design emotional responses and customer perception which affect customers in buying goods. Store Atmosphere affects the emotional state of a consumer which will cause an increase or decrease in purchases. Emotional state

will make two dominant feelings (feeling happy and arousing desires) in which both of them arise from psychological sets or impulse desires (Fuad 2010).

From the above explanation, it can be sum up that store atmosphere can provide not only pleasant buying environment, but also adding value to the products sold. Store atmosphere will also determine the image of the store itself. As a mediums of communication, it can have a positive and beneficial effect. On the other side, it can also hinder purchasing process.

Store atmosphere that is designed properly and favorable will be able to encourage consumers to buy products. There are several supporting elements of store atmosphere, such as interior, layout, and interior display. All of them will be integrated to form an image or store image. The store image can be a stimulus to enter the store, which continues to the interaction until purchasing process. Therefore, it is important for all relevant parties to be able to understand both the store object and the desired image, so that the produced design is not only attractive but also ideal and sellable.

The study aims to analyze the effect of atmosphere in MONPERA museum to customers' visiting decision. In this case, the atmosphere in MONPERA Museum seems unattractive. It even looks dull and spooky. This might affect the attractiveness and the interest of consumers to visit.

1.1 *Atmosphere*

Atmosphere is a combination of physical characteristics (architecture, display, lighting, color, temperature, music), the scents having emotional responses, and customer perceptions which influence customers in buying products (Utami 2008). According to Berman & Evans (2012), store atmosphere is environmental design through visual communication, lighting, color, music, and perfumery to design emotional responses and customer perception which affect customers in buying goods.

Berman & Evan (2012) divide store atmosphere into four elements:

1. Store exterior. It has a very strong influence on the store's image and must be planned as well as possible. Exterior consists of store front, shop sign (Marquee), shop entrance, building height, store and surrounding area, and parking facilities.
2. General interior. It consists of the type of floor, color and lighting, scents and music, shop furniture, wall texture, air temperature, hallway space, fitting rooms, inter-floor transportation, shop employees, technology, and cleanliness.
3. Store Layout. It is a plan to determine the specific location and arrangement of equipment, merchandise, shop aisles and shop facilities. Store Layout consists of allocation of floor space (selling space, merchandise space, personnel space, and customer space) and shop offer classification.
4. Interior display. It consists of assortment displays, theme-setting displays, ensemble displays, rack displays, and cut case.

Meanwhile, Gilbert in Foster (2008) defined store atmosphere as a combination of planned physical messages. It is described as a change in purchasing environment planning that produces special emotional effects that can cause consumers make a purchasing action.

1.2 Visiting decision

In this study, the concept of a visiting decision is explained by using the theory of purchasing decisions. Consumer purchasing decisions is one part of consumer behavior. The purchasing decision is the attitude of the termination result determined by the buyer after considering the type of product, brand, quantity, time, producer, salesperson, and payment method to meet wants and needs.

According to Kotler & Keller (2012), purchasing decision is a basic psychological process that plays an important role in understanding how consumers really make purchasing decisions. Kotler & Armstrong (2012) defined purchasing decisions as a stage in the process of making purchasing decisions where consumers actually buy. Decision making is an activity of individuals who are directly involved in obtaining and using the offered goods. (Soewito 2013) said that purchasing decisions are decisions taken by consumers to purchase product through the stages that is passed by consumers before making a purchasing. Meanwhile, (Sumarwan 2011) stated that purchasing decision is consumer's selected decision of two or more alternative options.

Purchasing decision has six dimensions: product selection, brand choice, channel selection, number of purchases, timing of visits and payment methods.

Consumers can make a decision to visit a place for various purposes. Every consumer is different in terms of determining suppliers due to location factors, low prices, complete product inventory, comfort, and breadth of space and so on. Consumers can make decisions about how many products/services to visit at a time. Consumers' decision in the selection of visiting times is variant. For example, there are those who visit every day, once a week, once a month, and maybe once a year. When visiting a place, consumers must make a payment in cash or credit.

2 METHODS

This research was classified as non-experimental design or precisely survey research (Creswell 2014). Robson (2002) & Nazir (2005) described explanatory survey as a research design intended to explain the phenomena that occurs by examining the relationships between the variables studied. An explanatory survey design was considered appropriate to be used in this study because it was in accordance with its purpose, which was to obtain an overview of the relationship between the atmosphere and visiting decisions.

There were three dimension of atmosphere variable: 1) Exterior dimensions consisting of museum entrance, parking facilities, and nameplate as a sign used to display the name or logo of a store; 2) General interior including lighting, aroma, and music, the air temperature, and the texture of the walls in the museum; 3) Interior display consisting of display assortment and display racks.

The population was visitors of MONPERA museum. The sample was 100 visitor which was chosen by using accident sampling technique. Primary data, including indicators of each variable, were collected using a closed questionnaire instrument by which the statements were formulated in positive sentences and arranged on a Likerts scale.

3 RESULTS AND DISCUSSION

Table 1. Model summary.

Model	R	Adjusted R R Square	Std. Error of Square	the Estimate
1	,513a	,263	,255	2,849

a. Predictors: (Constant), Atmosphere.

The value of R2 = 0.534 in the change of atmosphere is 26.3 percent. Seventy three point seven percent is determined by other variables which are not discussed in this study.

Based on the the result of ANOVA test, a sig value (0.00) is lower than <(0.05). This indicates that

Table 2. ANOVA[a].

Model		Sum of Squares	df	Mean Square	F	Sig.
1	Regression	283,419	1	283,419	34,930	,000[b]
	Residual	795,171	98	8,114		
	Total	1078,590	99			

a. Dependent Variable: Berkunjung.
b. Predictors: (Constant), Atmosphere.

Table 3. Coefficients[a].

Model		Unstandardized Coefficients		Standardized Coefficients		
		B	Std. Error	Beta	t	Sig.
1	(Constant)	1,540	1,911		,806	,422
	Atmosphere	,294	,050	,513	5,910	,000

Dependent Variable: Berkunjung.

the effect of the atmosphere dimension on visiting decisions to Monpera museum.

The coefficient table depits that the atmosphere has an effect on visiting decisions to Monpera (sig value <0.005). As previously stated that store atmosphere affects the emotional state of a consumer which will cause an increase or decrease in purchases. Emotional state will make two dominant feelings (feeling happy and arousing desires) in which both of them arise from psychological sets or impulse desires (Fuad 2010). Meanwhile, (Utami 2008) stated that the store atmosphere can provide a response and create convenience for consumers while enjoying the atmosphere in the store, and in its continuation can influence consumer purchasing decisions

There are several supporting elements of store atmosphere, such as interior, layout, and interior display. All of them will be integrated to form an image or store image. The store image can be a stimulus to enter the store, which continues to the interaction until purchasing process. Therefore, it is important for all relevant parties in Monpera museum to be able to understand both the store object and the desired image, so that the produced design is not only attractive but also ideal and sellable.

Yang & Lee (2016) said that in-store promotions (besides coupons, sweepstakes, free samples, attractive packaging) including shop atmosphere and interaction with sales people, will attract the consumers to make an impulse buying decision. (Bonn et al. 2007; Regan 2013) showed that atmosphere factor (including color schemes, lighting, and signs) had a signal of impacting visitor intention to revisit and will to recommend attractions.

4 CONCLUSION

Based on the results of the analysis, there are three main points that can be summed up:

1) Based on the results of simple linear regression test, atmosphere has an influence on visiting decisions with a contribution of 0.590 or 59.0%.
2) The result of f-test analysis (simultaneous) is 34,930 with significance value of 0,000 <0,05. This means that atmosphere has a significant influence on visiting decisions.
3) The results of the coefficient of determination test is 0.263 or 26.3%. Hence, it can be concluded that consumers' visiting decisions (dependent variable) is influenced by atmosphere (independent variable). The influence of other variable outside the model is 0.737 or 73.7%.

REFERENCES

Berman, B. & Evans, J.R. 2012. Retail Management: A Strategic Apporach.
Creswell, J.W. 2014. Research Design. Yogyakarta: Pustaka Pelajar.
Fuad, S.M. 2010. Store Atmosphere & Perilaku Konsumen. Malang. Toko Buku Gramedia.
Gilber, C. 2008. Retinopathy of prematurity: a global perspective of the epidemics, population of babies at risk and implications for control. Earlt human development.
Kotler, P. & Armstrong, G 2012. Prinsip-prinsip Pemasaran, Edisi 13. Jilid 1. Jakarta: Erlangga.
Kotler, P & Keller, K.L. 2012. Maeketing Management edisi 14. Global. Editions.pearson Prentice Hall.
Nazir, M. 2005. Metodologi Penelitian. Jakarta. Ghalia Indonesia.
Regan, F 2013. Museum Atmospherics: The Role of the Exhibition Environment in the Visitor Experience (16)2: 201–216
Robson, C. 2002. Real world research: a resource for social scientists and practitioner researchers. Blackwell.
Soewito, Y. 2013. Kualitas produk, merek, dan desain pengaruhnya terhadap keputusan pembelian sepeda Motor Yamaha Mio, Jurnal EMBA (1)3: 2303–1174.
Sumarwan, U. 2011. Perilaku Konsumen: Teori dan Penerapannya Dalam Pemasaran. Ghalia Indonesia.
Turley, L.W. & Ronald, M.E. 2007. Atmospheric Effects on Shopping Behavior: A Review of the Experimental Evidence. Journal of Business Research (49): 139–211.
Utami, C.W. 2008. Manajemen Barang Dagangan dalam Bisnis Riteil. Malang Bayumedia.
Yang, D.J. & Lee, C.W. 2016. In Store Promotional Mix and the Effects on Female Consumer Buying Decision In Relation to Cosmetic Products. International Journal of Management, Economics and Social Science (IJMESS) (5)2: 35–56.

The influence of brand preference on purchase intention of female E-Commerce consumers

B. Prasetiyo, R. Hurriyati & H. Hendrayati
Universitas Pendidikan Indonesia, Bandung, Indonesia

ABSTRACT: Indonesia is the highest share of mobile traffic in Southeast Asia. It has 87% of mobile traffic. On the other hand, based on the number of conversion rates, Indonesia only occupies the third position. Female consumers who do online shopping through E-commerce are relatively fewer than male consumers. The purpose of this study is to examine more deeply about the influence of brand preference on purchase intention of female E-Commerce consumers. This research used simple linear regression. The focus was on the use of E-Commerce. The results showed that brand preferences had influence on purchase intention of female E-commerce consumers. This impacts on the strategy used by E-Commerce companies, in which they have to create websites that provide clear information, contain audiovisual content, give some product choices, maintain product quality and authenticity, give the comfort for female consumers when doing online shopping, and keep transaction secure.

1 INTRODUCTION

The rapid development of information technology has changed the activities of human life in various fields. Technology makes people expect things to be done more practically and efficiently especially when consumers buying the product.

Conventional shopping methods are not efficient because it spends more time. The online shopping method has become commonly method used by consumers at this time. Consumers do not need to visit shops, queue up, and get traffic jams when they come back home from shopping. Online transactions can be done anywhere through a computer, laptop or mobile phone connected to the internet. When you enter the sales site, consumers can easily search, select the items, and make transactions.

Based on the research, Indonesian people have high activity in visiting sales sites (Iprice 2019). There has been a massive increase of mobile visits on E-Commerce in Southeast Asia. Within 12 months, mobile traffic has increased by an average of 19%. Now mobile visits contribute 72% of the total web traffic. Indonesia is the country with the highest share of mobile traffic, which is 87%.

The conversion rate is the percentage of site visits that leads to product purchases. With an average value as a benchmark (1x), Vietnamese E-Commerce has the highest conversion rate, which is 30% (higher than the average). Indonesia is on the third place with a very tight number of conversion rates with Singapore on the second place.

Regarding this, there is an interesting phenomenon (Iprice 2019). Based on mobile traffic, Indonesia ranks is the highest in Southeast Asia. On the other hand, based on the conversion rate, it turns out that Indonesia only occupies the third position under Vietnam and Singapore.

Shopee is E-commerce that is mostly used by female users (dailysocial.id 2019). Related to the product categories, they prefer to buy beauty product while shopping online. On the contrary, male users tend to buy daily necessities.

Purchase intention is the preference of consumers in buying products or services. Many factors influence purchase intention when choosing a product and making final decisions influenced by important external factors (Keller 2001). Based on the explanation above, researchers are interested in conducting research related to brand preference and purchase intention. The title of this study is "The Investigation of Brand Preference on Purchase Intention for Female Consumer through E-Commerce".

Based on the above background, the problem can be formulated as follow: How is the impact of brand preference to purchase intention?

E-Commerce is one of the ways to improve the performance and mechanism of the exchange of goods, services, information, and knowledge by utilizing digital network-based technology. (Indrajit 2002). The internet offers opportunities to sell customers' daily needs directly to consumers. Selling goods or services directly through the internet is called e-commerce (Musimah & Mursid 2019).

Consumer brand preference is the essence of understanding consumer choice behavior; and it always gets great attention from marketers (Ebrahim et al.. 2016). It is defined as bias behavior tendency of consumers of a brand. Brand as a preference for consumers to prefer one brand to another brand. Creating strong bonds between brands and customers is one of the goals of marketing. A strong preference for a brand can help the brand to exist in the long term even though it is in a tight competition. In this competition, there are more brands coming to market selling the same product category so that consumers have more choices. (Ebrahim 2013).

Regarding brand preference, consu mers cannot form their preferences for brands on rational attributes alone. They are looking for brands that create experiences: intrigue through sensory stimulation, emotional and creative ways (Ebrahim 2013). Brand preference has long explained the use of traditional models, with most of the focus on determining decisions that in line with brand attributes on a rational basis. However, it has now changed to experiential marketing.

Purchase intention is the preference of consumers in buying products. In other words, purchase intention has another aspect so that consumers will buy products after being evaluated. Many factors influence purchase intention especially when choosing a product and make a final decision (Keller 2001). The quality of a website has an important role in increasing purchase intention. High-quality websites will increase customer purchase intention (Rezaei et al. 2016). Online purchase intention is the willingness and desire of consumers to participate in online transactions, including the process of evaluating web quality and product information (Wang et al. 2015).

People today tend to appreciate a product from its brand. Products with strong brand equity can have a positive effect on the product. Thus, it affects preference and purchase intention as well. Brand preference will be based on how customers understand their brand equity. However, if the brand is equipped with well-maintained equity, it will significantly affect customer preferences. The intention to buy is based on the brand product in how well brand equity increases and is maintained. The better brand equity is enhanced, the better the customer trust and loyalty to buy a trusted brand (Emor & Pangemanan 2015). Purchase intention can arise due to various factors such as the presence of a stimulus, the results to be obtained, expectations of aspirational values, recommendations, and emotional relationships (Cobb et al. 2013).

2 METHODS

This study examined women who shopped through E-Commerce. Fifty four women chosen as a sample by using purposive random sampling technique. Data was collected through Likert scale questionnaire. It was then analyzed by using simple linear regression.

The Hypothesis (H1) is that there is influence of brand preference on purchase intention.

3 RESULT AND DISCUSSION

In this research, the questionnaire was tested for validity and reliability. The results.

Table 1. Validity and reliability test.

Variable	Question	r	α-Cronbach
Brand Preference	BP 1	0.648	0.786
	BP 2	0.777	
	BP 3	0.781	
	BP 4	0.857	
	BP 5	0.604	
Purchase Intention	PI 1	0.905	0.760
	PI 2	0.891	

In this study, the linear regression method was used to test the research hypothesis. The results of hypothesis testing.

Table 2. Regression Output.

R square	0,237
t-Value	4,014
Sign	0,000

Based on table 1, it can be explained that in the R Square section, the effect of brand preference on purchase intention is 23.7%. The hypothesis (H1) is accepted. There is the effect of brand preference on purchase intention. It is viewed from the coefficient table in the sig section for t equal to 0.000 (less than 0.05)

Brand preference influences purchase intention. It means that if the brand preference level is increased, it will increase purchase intention. This is consistent with previous research from (Emor & Pangemanan 2015; Moradi & Zarei 2011; Pool et al. 2018; Sambath & Jeng 2014; Soenyoto 2015; Vinh & Huy 2016). Although the final conclusions are relatively similar, there is something different. As previously stated, the focus of this study is on the purchase intention of female consumers who use E-Commerce.

Planned behavior proposed (Ajzen 1991) can be used to predict whether someone will do or not do a behavior. This theory uses three constructs namely attitudes toward the behavior, subjective norms, and the ability to control everything that influences if consumers want to carry out the behavior. Related to this theory, then forming a brand preference is one thing that can increase consumers buying interest.

There are three concept to predict women's satisfaction who purchase through e-commerce, namely emotion, trust, and convenience (Rodgers & Harris 2003). Based on this theory, E-commerce companies have to do several things, such as: (1) create websites that provide clear information; (2) create websites that contain audiovisual content; (3) give some product choices; (4) maintain product quality and authenticity;

(5) give the comfort of female consumers when doing online shopping; and (6) keep transaction secure.

4 CONCLUSION

Based on the results of this study, it can be concluded that brand preferences have influence on purchase intention of female E-commerce consumers. Thus, it is suggested for E-commerce companies to apply the following strategies: (1) create websites that provide clear information; (2) create websites that contain audiovisual content; (3) give some product choices; (4) maintain product quality and authenticity; (5) give comfort of female consumers when doing online shopping; and (6) keep transaction secure

REFERENCES

Ajzen, I. 1991. The theory of planned Behavior. Organizational Behavior and Human *Decision Processes*.

Cobb, W.C., Ruble, C.A. & Donthu, N. 2013. Brand Equity, Brand Preference, and Purchase Inten. *Journal of Advertising*: 25–40.

Dailysocial.id. 2019. Retrieved from dailysocial.id: https://dailysocial.id/post/e-commerce-di indonesia 2018

Ebrahim, R.S. 2013. A Study of Brand Preference: An Experiental View. London: *Brunel University* .

Ebrahim, R., Ghoneim, A., Irani, Z. & Fan, Y. 2016. A Brand preference and repurchase intention model: the role of consumer experience. *Journal of Marketing Management*:1230–1259.

Emor, A.M. & Pangemanan, S.S. 2015. Analyzing Brand Equity on Purchase Intention through Brand Preference of Samsung Smartphone User In Manado. *Jurnal Emba*: 124–131.

Indrajit, R.E. 2002. Electronic Commerce. Jakarta:*APTIKOM*.

Iprice 2019 Model for website service quality and technology acceptance factors: A case study of online shoppers in indonesia.[Online]. Retrieved from https://iprice.co.id/insights/stateofecommerce2017/

Keller, K.L. 2001. Building customer-based brand equity: creating brand resonance requires carefully sequenced brand-building efforts. *Marketing Management* 15–19.

Moradi, H. & Zarei, A. 2011. The Impact of Brand Equity on Purchase Intention and Brand Preference-the Moderating Effects of Country of Origin Image. Australian *Journal of Basic and Applied Sciences*: 539–545.

Musimah, N. & Mursid, M.C. 2019. The Effect of Online Consumer Riview on the Intention of Buying Products on Social Commerce. *ATM*: 22–28.

Pool, J.K., Asian, S., & Abareshi, A. 2018. An Examination of the Interplay between country-of-origin, brand equity, brand preference and purchase intention toward global fashion brands. Int.J.Business Forecasting and *Marketing Intelligence*: 43–63.

Rezaei, S., Faizan, A., Amin, M., & Jayashree, S. 2016. Online Impulse buying of tourism products. *Journal of Hospitality and toursm Technology*: 60–83.

Rodgers, S., & Harris, M.A. 2003. Gender and E-Commerce: An Exploratory Study. *Journal of Advertising Research*: 322–329.

Sambath, P., & Jeng, D.J.F. 2014. The Effects of Celebrity Endorsers on Brand Personality, Brand Trust, Brand Preference and Purchase Intention. *The Sustainable Global Marketplace*: 435–439.

Soenyoto, F.L. 2015. The Impact of Brand Equity on Brand Preference and Purchase Intention in Indonesia's Bicyle Industry: A Case Study of Polygon. *Ibuss Management*:99–108.

Vinh, T.T., & Huy, L.V. 2016. Relationships among Brand Equity, Brand Preference, and Purchase Intention: Empirical Evidence from the Motorbike Market in Vietnam. International *Journal of Economics and Finance*: 75–84.

Wang, L., Law, R., Guillet, B.D., Hung, K.., & Fong, D.K. 2015. Impact of hotel website quality on online booking intentions: eTrust as a Mediator. International *Journal of Hospitality Management:* 108–115.

Advances in Business, Management and Entrepreneurship – Hurriyati et al. (Eds)
© 2021 Taylor & Francis Group, London, ISBN 978-0-367-67471-7

The influence of social media on international students higher education choice

O. Intan & T.E. Balqiah
Universitas Indonesia, Depok, Indonesia

ABSTRACT: International student mobility has shown a positive trend in the past decade. The competition among higher education institutions is now more open and global. Social media makes it possible for higher education institutions to market themselves and improve their positioning on the regional level as well as global. This study tries to identify the influence of social media on international students' higher education decision-making process, with an aim to understand how social media can be used as a strategy for higher education institutions for international students' recruitment. This study was carried out through quantitative descriptive approach using nonprobability sampling method. Self-administered questionnaires were distributed to 144 international students from several higher education institutions in Indonesia. The survey results demonstrated a certain level of social media usage during the students' search for higher education, while the primary influences still came from traditional sources. The findings of this study provide practical implications of social media use in higher education institutions' marketing strategy.

1 INTRODUCTION

International student mobility trend is largely driven by the changes in demographic and economics. The population of 18 to 22 year-olds have grown steadily in developing countries and are expected to grow stronger over the next decade. The recession in 2008 and 2009 that resulted in a sharp fall of global trade and GDP have rebounded, and have been gradually normalizing since 2010. The east to west student movement has shifted because the developing countries that used to have limited domestic capacity are now building global partnerships to increase the higher education sector (British 2014). The competition among higher education institutions is more open in the globalization and regionalization era; hence these institutions must know how to continually improve their quality and make themselves visible and accessible to the international market.

Increasing competition and constantly changing social conditions are combinations that push institutions of higher education to be more active and market-oriented. It is crucial for higher education institutions to give attention to their positioning strategy, as the competitive advantage must be maintained. This is somewhat difficult because educational institution's products are considered to be similar (Alexa et al. 2012). With social media, the opportunities are open for institutions of higher education to introduce their program and other competitive advantages to college-bound students in a borderless market. As all other institutions are doing the same, students become

exposed to vast amount of information and communication. The following questions are formulated to identify the research questions of this study:

RQ1. What are the international students' activities in social media?

RQ2. Do international students use both traditional and social media for their higher education selection process?

This study predicts the information that the students receive through social media can influence the international students' higher education choice. Following hypothesis is proposed:

H1. Social media can influence international students' higher education choice. This hypothesis may be verified or disproved by empirical evidence of the findings of this study.

1.1 International students

Scholars have defined higher education as learning that occurs at a university, college, or institute beyond a high school level (Clemmons et al. 2014).

In explaining the student's motivation to study abroad, the "push–pull model" is a dominant framework used in many literatures (Jiani 2017). 'Push' factors work in the country of origin and instigate the choice of student to study overseas (Mazzarol & Soutar 2002). While for the host country, 'pull' factors work to make the country somewhat desirable for international students. Prospective students will take different aspects into account of secondary services in relation to living in the host country, and host city in particular. Among them are university environment, international background, cultural activities, safety, security, and visa requirements (Cubillo et al. 2006).

1.2 Decision making

When a student signs up for a degree, the student actually signs up for a lifetime relationship with the university because the name of that institution will always be associated with their own (Rutter et al. 2016). The university or colleges someone attends may have impact on one's future career, friendships, choice of partner, future residential location, and satisfaction with life; thus makes this as high involvement decision making process.

Marketing scholars have developed a stage model of typical consumer buying process consisting stage of events before the consumer arrives at the buying decisions.

Consumer buying decision process starts with problem recognition or recognizing the unmet need. The second stage is the information search where the consumer will engage in external search for additional information from sources like internet, personal reference or recommendation, commercial sources, public sources, and personal experience (Belch & Belch 2015). The consumer moves to evaluation of alternative stage after collecting information in the previous stage. Purchase decision stage is when the consumer finally stops searching and prefers the beneficial one out of several brands, and develops purchase intention. For highly involved and complex purchases, time delay between purchase intention and the actual purchase is often expected (Belch & Belch 2015). Last is the post purchase behaviour stage where the behaviour is influenced by the degree of satisfaction from the product or service the consumer purchases (Kotler & Keller 2012).

1.3 Traditional higher education marketing

Marketing happens when individuals agree in meeting their wants and needs through an exchange (Kotler & Fox 1995). In higher education field, the institutions offer their academic programs in exchange for which consumers, or students in this case, offer their commitment in terms of efforts, time, and money, in the form of tuition fees (Alexa et al. 2012)

There are many marketing approaches that higher education institutions use to attract and retain students. Activities like advertising in media, outreach endeavours, education fairs or exhibitions, engaging with agents of recruitment, and creating activities for public relations become a choice. The main characteristic of these traditional marketing methods is face-to-face interaction (El & Vrontis 2015).

Mazzarol & Soutar (2002) indicates that personal recommendations provided by close family and friends during the student's purchase decision are still important (Mazzarol & Soutar 2002). Students learn and receive information about overseas study from friends and acquaintances; and they trust that this information is more dependable and accurate if they receive it from personal networks compared to official sources (Jiani 2017). The research also shows that there is a tendency for students to select a country first before they look for a higher education institution within that country (Mazzarol & Soutar 2002).

1.4 Social media

Social media, as defined by (Kaplan& Haenlein 2010), is "a group of Internet-based applications that builds on web 2.0's ideological and technological foundations, and enables user-generated content to be created and exchanged" (Kaplan & Haenlein 2010).

The types of technologies involved in social media are meant to create ways for individuals to make new connections and maintain existing relationships. People can make their own content and share them, and to some extent, make user-specific social networks observable to others (Boyd & Ellison 2007; Kietzmann et al. 2011). News can be spread along with information, photos, videos, and without a lengthy editorial process. Social media is all about conversations and interactions.

1.5 Social media in higher education marketing

The age of students for university or college enrolment are 17 to 18 years old, meaning they are part to the Millenials or Generation Y who have grown up in a digital environment, and surrounded by computers and technologies that are almost always available (Thompson 2007). These generations are socialising online and have intense participation in social media networking. They want everything to be quickly processed with the advance of technology. In other words, they are emotionally open and confident, knowing that information is accessible at all time.

With the use of social media, companies are enabled to engage with the end-consumers with higher speed and greater efficiency at lower cost than what traditional communication tools can achieve (Kaplan & Haenlein 2010). As the behaviour change for people to spend more time on digital devices and most information are available online, the marketers use the digital phenomenon to affect the targeted audience.

1.6 Research framework

A combination of customer segmentation and market basket analysis is proposed to address the goal to analyse the customer profiles and purchasing patterns of online customers in Indonesia. In general, this part is divided into several parts: customers' segmentation, customer profiling, and market basket analysis. Every part will have a different analysis treatment respectively. Research framework is presented in Figure 1.

1.7 Data processing & analysis method

The data used in this study was a real transaction data from one of the e-commerce websites in Indonesia, ranging from January to December 2018.

The average value of F and M variables have an upward arrow ↑ if it exceeds the overall average F and M from all customers, and have a downward arrow ↓ sign otherwise. A contrast rule applied for R parameter. This technique differ online consumers form segments based on their relationship with the website.

Consumer's purchasing patterns was exploited through market basket analysis in every customer cluster and customer profile to provide more precise and personalized items in product recommendation. This whole data processing was conducted using Rstudio software version 3.5.2 on iMac Mojave iOS system and windows 10.

2   METHOD

This research was carried out using descriptive approach. The aim was to identify the use of social media; and whether it influenced international students' higher education decision-making process. As the data collected at one time, it classified into a cross-sectional study. The data were selected using nonprobability sampling procedure. The size of the sample was determined by the optimum number needed to allow valid population inferences to be made. The sample was 144 international students which were selected by using snowball sampling, in which the referral group was then used to find others who possessed similar characteristics and in turn participate as the respondents for the study.

This research used survey questionnaire for primary data collection. A self-completion questionnaire was distributed online using Google form. Data was collected and processed statistically in quantitative methods.

The data analysis was conducted using the statistical tools of SPSS. Descriptive statistics were used to describe the data variables. For predicting dependent variable that had only two possible values (binary), a logistic regression was performed. SPSS was also used to calculate the p-values for the estimated coefficients and predictions for Y. The statistical significance of the independent variable needed to be evaluated so that the hypotheses formulated can be either accepted or rejected to understand the influence of independent variable on the dependent variable.

3   RESULT AND DISCUSSION

This survey collected data from 144 international students in universities in Indonesia between the period of April to May 2019.

3.1 Respondents' profile

In the first section of the questionnaire, the respondents were observed according to the questions describing the basic information of the respondents. The first question is which country they are from. The highest proportion of respondents by country of origin was Malaysia (28), South Korea (26) and Thailand (17). The list is followed by Timor Leste (14), Japan (12), China (10), India (7), France (5), and other neighboring countries as well as more distant countries in Europe and Africa.

From 144 respondents, 54.9% respondents are male; and 45.1% respondents are female. The majority of the respondents are pursuing Bachelor's degree (54.9%), Master's degree or higher (20.8%), and the rest are language program (13.2%) and diploma (11.1%).

3.2 Social media activity

This section discusses the findings in order to answer RQ1: "What are the international students' activities in social media?" The respondents were asked to answer with how they used social media in general, and the kind of activities they engaged while they were using social media. Almost all of the respondents (96.5%) currently have social media. The majority 85.20% of respondents are Facebook users, followed by Instagram (83.1%), Twitter (51.4%), LinkedIn (45.8%) and YouTube (45.1%).

Although Facebook has the highest number of users, the respondents' answer indicates Instagram as the highest level of activity. Respondents are highly engaged in Instagram compared to Facebook which are used more to stay connected instead of sharing daily content.

The kind of activities respondents participate on social media are as follows: creating entry (66%), searching for health and fitness information (65%), sending instant messages (58%), searching for travel and tourism information (56%), searching for information about mobile phone (55%), posting comment (54%), doing online shopping (53%), reading information about technology (52%), and looking for fashion (50%). Creating entry on social media shows the highest response, and this is consistent with Instagram as the most actively used social media platforms. It is also widely used for marketing activities, as the circle is more open and users can follow other people, brands, or other organizations that they do not know directly.

3.3 Traditional and social media for higher education information

This section answers RQ2: Do international students use both traditional and social media for their higher education selection process? When a list of traditional media sources given to the respondents, the result shows that the highest number of students heard about the course they currently attended from internet search (22.2%), university website (16.7%), education exhibition/fairs (13.2%), and friend (11.1%).

The answer for the kind of higher education sought are only slightly different between searching for the university information (72.9%) and course information (70.1%). Student activities information and sample of lectures are 37.5% and 22.9%.

The highest answer for interactions the respondents made when they are searching for higher education course is that they download the university information from the university website (55.6%). This is followed by requesting the information through e-mail (34%), speaking to a university staff at an education exhibition/fair (27.8%), and speaking to staff on the phone (25.7%).

The last question in this section is about the traditional information sources that influences the decision-making process. The highest percentage of respondents are influenced by the university's website (56.6%). The second is influenced by internet search (48.3%), followed by education exhibition/fairs (29.4%); friends and family (28.7%).

The next section discusses the social media information sources. The respondents were asked to think back when they were seeking course information and how they used social media. The choice of social media platform they visited for higher education information are as follows: Facebook (60.5%), LinkedIn (33.3%), YouTube (24%) and Twitter (21.7%).

Students are seeking the information on the university first (76.3%) before they try to find out further information regarding the course (64%). Course recommendation and course opinion come after with 38.8% and 36%, video of university and sample lectures come later with 34.5% and 21.6%.

3.4 The influence of social media on higher education choice

This section discusses the findings in order to prove the hypothesis: "H1. Social media can influence international students' higher education choice".

Respondents were asked to rank the impact of social media in the process of their higher education selection. The dependent variable is whether the respondents contact the higher education institutions' student or staff using social media ("Contact SM"). The independent variables are investigated using cross tabulation with the dependent variable Contact SM. Logistic regression is performed using the three variables:

a. It provides additional information for the university.
b. It has some influence on the university course decision.
c. It helps to make a decision on the university course.

The determinant of whether the use of social media has influence on respondents' higher education selection and whether social media helps them to make

Table 1. Contact students or staff using social media.

| Observed | | | Predicted | | |
| | | | Contact_SM | | Percentage |
			No	Yes	Correct
Step 0	Contact_SM	No	80	0	100.0
		Yes	64	0	.0
Overall Percentage				55.6	

a. Constant is included in the model.b. The cut value is .500.

Table 2. Determinants to contact via social media.

	B	S.E.	Wald	df	Sig.	Exp(B)
AddInfo	0.914	0.278	10.794	1	0.001	2.494
Influence Course	0.601	0.377	2.539	1	0.111	1.824
DecideUniv	0.454	0.326	1.944	1	0.163	1.575
Constant	−3.737	0.665	31.618	1	0	0.024

a. Variable(s) entered on step 1: Impact_AddInfo, Impact_InfluenceCourse, Impact_DecideUniv.

a decision on their higher education choice have p-values higher than 0.05. It is proved to be not significant. However, the first determinant, that social media provides additional information for the respondents' higher education choice, has p-value 0.001. Thus, it is significant. The results indicate that international students use social media sites for assisting them with additional information in their selection process rather than giving them influence in choosing the course they finally attended or the choice of the university or college.

4 CONCLUSION

Traditional marketing strategies work well for local students, but it could only reach up to a certain distance. This can be seen from the majority of international students who study in Indonesia, in which most of them are from the neighboring countries and South Korea. Other than general and trusted information from web search and relatives, education exhibition/fairs are still relevant and influential in higher education decision making.

Although it does not influence the higher education selection process, students visit social media sites for additional information. The fact shows that 96.5% of students are on social media. This is an opportunity for higher education institutions to effectively use this medium for international recruitment, information dissemi-nation and at the same time maintaining the position. The use of social media is more borderless and limitless. It can reach the prospective students at their own convenient time and more personal. The immediate access to heavy loads of information is what the international students need when they have distance and time problem at the same time.

The flow of information on social media are two-ways and overflowing. In one hand, universities can market themselves with tailored information and actually be specified in reaching the students of a certain country, like providing a video advertising with students from that respective country. In this way, the prospective students can feel more related to the image. In the other hands, the information from opinions and electronic word of mouth will be an important influence for Gen Y or Millennials since it comes directly from those who are studying in the university or those who live in the country. In that sense, it is useful for higher education institutions to have their own social media administrator to monitor and direct the information. Moreover, the university can always be informed about what is happening online.

REFERENCES

Alexa, E.L., Alexa, M. & Stoica, C.M. 2012. The use of online marketing and social media in higher education institutions in Romania. *Journal of Marketing Research & Case Studies.*

Belch, G.E. & Belch, M.A. 2015. Advertising and promotion. Singapore: *McGraw-Hill.*

Boyd, D. & Ellison, N.B. 2007. Social network sites: Definition, history, and scholarship. *Journal of Computer-Mediated Communication*, 13(1):210–230.

British, C. 2014. Postgraduate student mobility trends to 2024. *British Council.*

Clemmons, K., Nolen, A. & Hayn, J.A. 2014. Constructing community in higher education regardless of proximity: Re-imagining the teacher education experience within social networking technology. In S. Mukerji, & P. *Tripathi, Handbook of Research on Transnational Higher Education*: 713–729.

Cubillo, J.M., Sánchez, J. & Cerviño, J. 2006. International students' decision-making process. International *Journal of Education Management* 20(2): 101–115.

El, N.S. & Vrontis, D. 2015. Universal recognition for the need of marketing of universities. *In Management innovation and entrepreneurship: A global perspective* : 337–351

Jiani, M. 2017. Why and how international students choose Mainland China as a higher education study abroad destination. *Higher Education* 74(4):563–579.

Kaplan, A.M. & Haenlein, M. 2010. Users of the world, unite! *The challenges and opportunities of social media. Business Horizons* 59–68.

Kietzmann, J. H., Hermkens, K., McCarthy, I. P. & Silvestre, B.S. 2011. Social media? Get serious! *Understanding thefunctional building blocks of social media. Business Horizons* 54(3):241–251.

Kotler, P. & Fox, K.F. 1995. Strategic marketing for educational institutions (second ed.). *New Jersey: Prentice-Hall*

Kotler, P. & Keller, K. L. 2012. Marketing management. Upper Saddle River: *Prentice Hall.*

Mazzarol, T. & Soutar, G.N. 2002. The push-pull factors influencing international student selection of education destination. International *Journal of Educational Management* 16(2): 82–90.

Rutter, R., Lettice, F. & Nadeau, J. 2016. Brand personality in higher education: Anthropomorphized university marketing communications. *Journal of Marketing for Higher Education*: 19–39.

Thompson, J. 2007. Is education 1.0 ready for web 2.0 students? Retrieved from Innovate: *Journal of Online Education* (3)4:5.

How can digital marketing influence the purchasing decisions on lazada.co.id?

R.L. Batu, I. Krisnawati, I.A. Ubaidi & H. Rais
Universitas Singaperbangsa, Karawang, Indonesia

ABSTRACT: Digital business is a business that requires an internet network to operate and market the products. E-Commerce is a fast growing digital business in Indonesia that connects companies with customers to conduct transactions, transport goods, and transport information via the internet. The purpose of this research is to study the influence of Digital Marketing on Online Purchasing Decisions at Lazada.co.id. This research used descriptive and verification research with explanatory survey method. The sampling technique was proportional stratified random sampling which used Slovin formula. There were two types of data: primary and secondary data. The data were processed by using simple linear regression analysis that has been converted into natural log (Ln) with SPSS (Stastistic Product and Service Solution) software version 16. The results of this study proved that Digital Marketing had a positive and significant influence on Online Purchasing Decisions.

1 INTRODUCTION

1.1 Background

Digital technology such as the internet is one of the media that can be used by business people to communicate, and has had the most significant impact on business (Dahiya & Gayatri 2017; Penard & Perrigot 2017). Technological developments trigger the occurrence of digital market systems which is caused by the rapid spread of smartphones and Internet access; where companies must be aware of the importance of creating digital relationships (Busca & Bertrandias 2019; Kannan & Li 2016). Digital breakthroughs continue to challenge market understanding and marketing practices that bring the opportunity (Ruyter et al. 2018). This is proven by the high interest of digital market. Consumers are actively looking for brands that provide interesting content (Geraint & Rowley 2014; Halligan & Shah 2010). They are looking for information and ordering goods via internet (Yang et al. 2015). Digital market, at this point, becomes the most efficient way for businesses to reach potential customers (Kannan & Li 2016; Kumar et al. 2019). It has a greater impact on company profits (Yan & Yeh 2009).

E-Commerce can be defined as a modern business methodology that answers the needs of organizations, traders, and consumers to be more efficient, improve the quality of goods/services, and increase the speed of delivery (Abdullah et al. 2019). The essence of E-Commerce is to enable the process of buying and selling without distance and time limitation. (Billy 2017) The dramatic growth of e-commerce has greatly motivated the development of the retail and logistics industry (Vakulenko 2019).

Table 1. Indonesia's most visited e-commerce.

E-Commerce	2015	2016	2017	2018
Lazada	18,6%	34,7%	36,6%	28,8%
Tokopedia	21,2%	28,1%	26,4%	28,7%
Bukalapak	9,58%	18,2%	20%	22,9%
Shopee	0%	0%	14%	8,4%
Blibli	5,94%	19,1%	3%	11,2%

*Source: Statistik.kominfo.go.id, accessed on October 14th, 2018 at 08:00 am; www.cnnindonesia.com, accessed on October 14th, 2018 at 19:34 pm.

Table 1 shows the most visited e-commerce in Indonesia. In 2015 up to 2018, there were five most popular e-commerce sites; and Lazada became the most visited one. There were the large number of visitors who wanted to do online shopping in Lazada.co.id.

Lazada.co.id is a part of Lazada Group, which operates in Indonesia, Malaysia, the Philippines, Singapore, Thailand and Vietnam. It is a subsidiary of Rocket Internet from Germany, which has successfully created many innovative online products in the world. Lazada is an online shopping destination in Southeast Asia, and becomes a fairly influential E-Commerce network. It grew into a big company after JP Morgan and several other partners joined entrusting investments in Lazada and its parent company, Rocket Internet (Azmi 2015).

Based on Alexa.com ranking, Lazada.co.id has become the largest online store in Indonesia. It has so many visitors as it offers a high concern to the products, such as brands. It provides the clear information

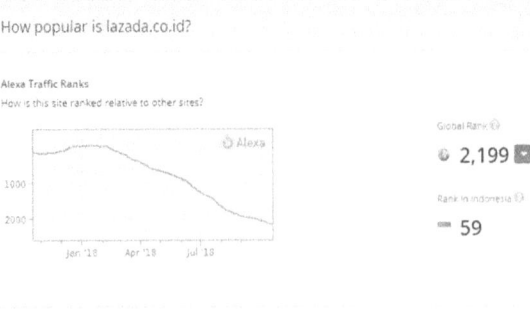

How popular is lazada.co.id?

Alexa Traffic Ranks
How is this site ranked relative to other sites?

Global Rank
2,199

Rank in Indonesia
59

Monthly Unique Visitor Metrics
Past 30 Days — Last Updated October 20, 2018

Figure 1.   Track ranking in Alexa.*Source: www.alexa.com accessed on October 21st, 2018 at 20:11 pm.

Table 2.   Top brand index e-commerce Di Indonesia.

E-Commerce	TBI (2015)	TBI (2016)	TBI (2017)	TBI (2018)
Lazada	–	41,9%	18,0%	31,8%
Tokopedia	1,2%	7,2%	13,4%	18,5%
Blibli	–	–	–	14,7%
Bukalapak	0,7%	6,6%	6,8%	8,7%
Shopee	–	–	–	8%

*Source: www.topbrand-award.com, accessed on 10 Oktober 2018 pukul 14:00 WIB.

Table 3.   Data purchases of the fashion category in Indonesian e-commerce.

E-Commerce	Index
Lazada	19%
Tokopedia	12%
Blibli	17%
Bukalapak	16%
Shopee	24%

*Source: dailysocial.id, accessed on October10th, 2018 at 11:56 am.

of the products quality, and even the information of how to use them

Table 2 shows that in 2016 to 2018, Lazada was included in the top 5 brands, and got the first rank in Indonesia's top brands. The index of Tokopedia and Bukalapak have increased every year. Meanwhile, Blibli and Shopee just entered Indonesia's top brand in 2018.

Purchasing products online is more dominated by young people (Sorce et al. 2005). A variety of products are sold at online stores such as fashion goods, electronics, tickets, machine tools, sports equipment, and other items. The fashion category is the most sought product compared to other type of products (Azmi 2016). The following are purchase data in the fashion category at the Indonesian E-Commerce company in 2018.

Based on Table 3, Shopee became the main choice of consumer decisions in the online purchase of fashion category in 2018. Lazada, as the most popular E-commerce, only stayed in second position. Consumers prefer to purchase fashion product in Shoppe rather than Lazada. This shows that Lazada has several limitation compared to the other E-commerce sites.

1.2   Research problem

Based on the above background, the research problem in this research are:

1. How is digital marketing on Lazada.co.id?

2. How is online purchase decision on Lazada.co.id?
3. To what extent does digital marketing influence online purchase decisions?

1.3   E-Commerce

E-Commerce is the process of trading and exchanging goods, services and information through electronic media, especially the internet (Turban et al. 2008). E-Commerce is an electoric trade media that has certain characteristics. E-Commerce is the buying, selling and marketing goods and services through electronic systems such as radio, television and computer networks or the internet. E-Commerce is a business strategy that allows wider geographical reach of consumers in various markets and ais no restricted to any boundaries (Kremez et al. 2019; Laudon 2013; Wong & Endang 2017).

1.4   Digital marketing

Digital marketing is a promotional mix element that allows interactive or two-way interaction of information where users can participate and modify the form and content of received information (Azmi 2016; Belch & Belch 2015). This is in line with research which found the use of digital marketing in marketing a brand products and services and in building relationship with customer 24-hour through internet services

(Tiffany et al. 2018) Digital marketing has been considered as a new form of marketing that provides new opportunities for companies to do business. The development and the widespread use of internet technology has changed the way people communicate, both in their daily lives and professional life. One of the most important indicators of this transformation is the emergence of new communication tools (Bokde & Seshan 2019). This research uses the concept of Digital Marketing dimension proposed by Aditya M (Salya 2010), namely website, search engine marketing, web banner, social network and affiliate marketing. Social media marketing is part of digital media marketing which can also be defined as marketing goods and services. Based on the results of previous research social media marketing had a significant influence on purchasing decisions (Krisna 2018).

1.5 *Online purchase decision*

Thomson (2013) explained that purchasing decisions are activities or behaviors that arise in response to objects (Baskara 2014). The dimensions of online purchase decision are: Need recognition, information search, alternative evaluation, purchase decision, and post purchase behavior (Fika 2015).

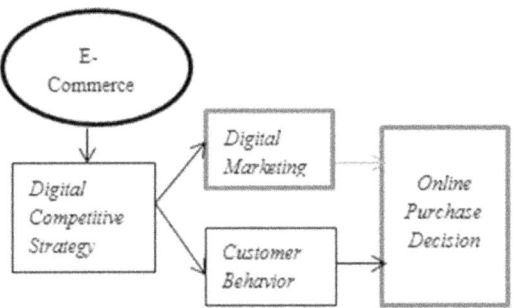

Figure 2.   Research framework.
Source: Research studies, 2019.

2   METHODS

This research used descriptive and verification research with explanatory survey method. The data used are primary data and secondary data. The population were 384 management students from Faculty of Economics and Business of Univesritas Singaperbangsa, Karawang. The sample was 38 students which were selected by using probability sampling technique, especially proportional random sampling. Regarding sampling formula, it used Slovin formula with an error rate of 5%. To collect the data, questionnaire and literature study were conducted. The data was processed by using simple linear regression equation with SPSS (Statistical Product and Service Solution) software version 16

3   RESULT AND DISCUSSION

3.1   *Validity and reliability*

Table 4.   Digital Marketing Validity Test Results.

Digital Marketing			
Dimension	r-count	r table	Desc.
Website	0,542	0,1402	Valid
	0,479	0,1402	Valid
	0,604	0,1402	Valid
SEM	0,65	0,1402	Valid
	0,588	0,1402	Valid
	0,658	0,1402	Valid
Web Banner	0,58	0,1402	Valid
	0,525	0,1402	Valid
	0,532	0,1402	Valid
Social Networking	0,667	0,1402	Valid
	0,593	0,1402	Valid
	0,62	0,1402	Valid
Affiliate Marketing	0,62	0,1402	Valid
	0,54	0,1402	Valid
	0,668	0,1402	Valid

*Source: Data Processing, 2019.

Table 4 shows that all questions of digital marketing are valid. The results reveal that there is one question item with the highest value (0.668). This item is on affiliate marketing dimension. The question here is related to the customers' interest to shop in Lazada.co.id. On the other hand, the lowest value (0.479) is found in website dimension. The question is asked about the customers' awareness of the item they want to buy in Lazada.co.id.

Table 5.   The results of online purchase decision validity test.

Online Purchase Decision			
Dimension	r count	r table	Des.
Need Recognition	0,579	0,1402	Valid
	0,649	0,1402	Valid
	0,643	0,1402	Valid
Information Search	0,615	0,1402	Valid
	0,567	0,1402	Valid
	0,642	0,1402	Valid
Alternative Evaluation	0,697	0,1402	Valid
	0,62	0,1402	Valid
	0,606	0,1402	Valid
Purchase Decision	0,706	0,1402	Valid
	0,588	0,1402	Valid
	0,602	0,1402	Valid
Post purchase Behavior	0,633	0,1402	Valid
	0,608	0,1402	Valid
	0,639	0,1402	Valid

*Source: Data Processing, 2019.

Table 5 depicts that all question items of online purchase decision are valid. One question with the highest

Table 6. Reliability test results.

Variable	r hitung	r table	Ket.
Digital Marketing (X)	0,883	0,6	Reliabel
Online Purchase Decision (Y)	0,909	0,6	Reliabel

*Source : Data Processing, 2018.

Table 7. Summary of digital marketing dimensions.

Dimension	Total Score	Average	Ideal Score	%
Website	2207	735,7	980	20,7%
SEM	2130	710	980	20%
Web Banner	2060	686,7	980	19,3%
Social Networking	2123	707,7	980	20%
Affiliate Marketing	2123	707,7	980	20%
Digital Marketing	10643	3547,8	4900	100,0%

*Source: Data Processing, 2018.

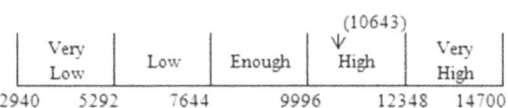

Figure 3. The value of the digital marketing continuum. Source: Data Processing, 2018.

values (0.697) is on alternative evaluation dimension. This questions asked the customers about their willingness to recommend Lazada.co.id to the other people. On the contrary, the lowest values (0.567) is on information search dimension.

Table 6 shows the reliability test of two investigate variables. The result reveals that the variables are reliable.

3.2 Descriptive analysis

3.2.1 Overview of digital marketing
Table 7 present the score recapitalization of digital marketing dimension. The highest score (20.7%) is on website dimension, while the lowest score (19.3%) is on web banner dimension.

Based on Figure 3, the digital marketing continuum value is 10643 or 72.4%. This value is categorized as high. Thus, it can be concluded that all digital marketing dimension used have implemented well

3.2.2 Overview of online purchase decision
Table 8 describes the score recapitalization of online purchase decision dimension. The highest score (20.4%) is on information search dimension, while the lowest score (19.5) is on alternative evaluation score.

Figure 4 shows that the continuum of online purchase decision value is 10411 or 70.8%. This value is considered high. Therefore, it can be summed up

Table 8. Recapitulation of online purchase decision dimension.

Dimension	Total Score	Average	Ideal Score	%
Need Recognition	2067	689	980	19,9%
Information Search	2123	707,7	980	20,4%
Alternative Evaluation	2025	675	980	19,5%
Purchase Decision	2108	702,7	980	20,2%
Post purchase Behavior	2088	696	980	20,0%
Online Purchase Decision	10411	3470,7	4900	100%

*Source: Data Processing, 2018.

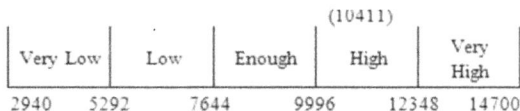

Figure 4. Online purchase decision continuation value. Source: Data Processing, 2018.

Table 9. Linearity test regression model.

		Log_Y	Log_X
Pearson Correlation	Log_Y	1.000	.686
	Log_X	.686	1.000
Sig. (1-tailed)	Log_Y	.	.000
	Log_X	.000	.
N	Log_Y	196	196
	Log_X	196	196

Source : Data Processing menggunakan Software SPSS Versi 16.

that all dimensions of online purchase decision have implemented well

3.3 Verificative analysis

Classical Assumption Test
Table 9 provides the correlation result of explanatory variables and response variables. P-value (third column) 0,000 <0.05. Thus, H0 is rejected. This means that explanatory variable has a real linear relationship (correlation) to the response variable

Table 10 shows the results of normality test regression model. The significance value (Asymp. Sig. (2-tailed) from the Kolmogorov-Smirnov Log X test is 0.178 and Log Y is 0.082. These two values are

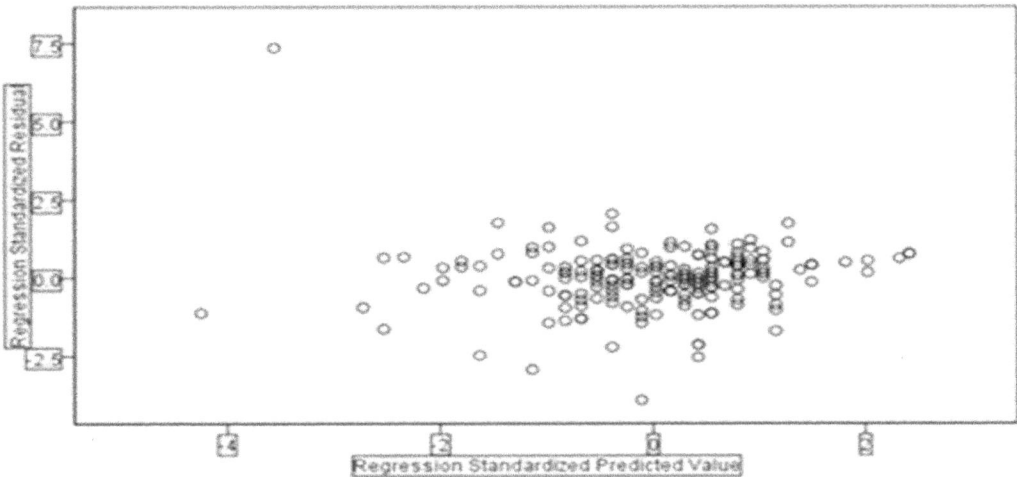

Figure 5. Heteroscedasticity test.Data Processing, 2018.

Table 10. Normality test regression model.

		Log_X	Log_Y
N		196	196
Normal Parameters	Mean	1.7309	1.7217
	Std. Deviation	.05966	.06934
Most Extreme Differences	Absolute	.079	.090
	Positive	.066	.047
	Negative	−.079	−.090
Kolmogorov-Smirnov Z		1.099	1.264
Asymp. Sig. (2-tailed)		.178	.082

*Source: Data Processing, 2018.

Table 11. Non-multicollinearity test.

	Correlations			Collinearity Statistics	
Model	Zero-order	Partial	Part	Tolerance	VIF
1 (Constant)					
Log_X	.686	.686	.686	1.000	1.000

*Source: Data Processing, 2018.

Table 12. Simultaneous coefficient of determination.

Model	R	R Square	Adjusted R Square	Std. Error of the Estimate
1	.686[a]	.471	.468	.05056

*Source: Data Processing, 2018.

greater than 0.05. To sum up, the regression model has fulfilled the normality assumption.

Figure 5 shows no clear patterns and spread points. It can be concluded that there is no heteroscedity in the regression model, so that the regression model is feasible to predict the effect of digital marketing on online purchase decisions.

Table 11 portrays the results of the non-multicollinearity test. The tolerance value of the independent variable, digital marketing (X), is 1.000. This value is less than 10. Hence, it can be concluded that the non-multicollinearity assumption test is fulfilled.

3.3.1 Hypothesis testing

Based on table 11, R2 value is 0.471. This means that 47.1% of online purchase decision (Y) can be explained by digital marketing (X). Meanwhile, 52.9% is determined by other variables which are not examined in this study.

Based on testing the partial determination coefficient (Table 13), the following calculations can be used:

Effect of X on Y = 0.686 × 0.686 = 0.470596 or 47.1%

The calculation results show that digital marketing variable has 471% impact to online purchase decision. The other percentage (52.9%) is determined by the other variables which are not examined in this research.

Table 13. Partial determination coefficient.

	Standardized Coefficients	Correlations
	Beta	Zero-order
(Constant)		
Log_X	.686	.686

*Source: Data Processing, 2018.

Table 14. Simultaneous hypothesis testing (F test).

Model	Sum of Squares	Df	Mean Square	F	Sig.
1 Regression	.442	1	.442	172.768	.000ᵃ
Residual	.496	.194	.003		
Total	.937	195			

*Source: Data Processing, 2018.

Table 15. Partial hypothesis testing (T test).

Model	Unstandardized Coefficients		Standardized Coefficients		
	B	Std. Error	Beta	T	Sig.
1 (Constant)	.341	.105		3.244	.001
Log_X	.798	.061	.686	13.144	.000

*Source: Pengolahan Data, 2018.

Table 14 shows the simultaneous hypothesis testing which uses F statistical test. The significant value of the simultaneous regression model is 0,000. This value is lower than the significance level of 0.05 (5%), 0,000 <0.05. It is in accordance to the comparison between Fcount and Ftable. Fcount is 172.768, while Ftable is 3.89. At this point, Fcount is greater than Ftable. Thus, Ho is rejected. In other words, digital marketing significantly influences online So that the decision to reject H0. This means that digital marketing significantly influences online purchase decision.

Table 15 presents partial hypothesis testing (T test). The table shows that a significant value of partial regression model is 0.001. This value is lower than the significant significance level of 0.05 (0.001 <0.05). Besides, the comparison results between t-count and t-table reveals that the t-value of 13,144 is 1, 97227. In other words, t count is greater than t table (13.144> 1.97227). Thus, it can be concluded that H0 is rejected. This means that there is partial influence of online purchase decisions on Lazada's digital marketing.

4 CONCLUSION

From the results, it can be concluded that there are five dimensions used to measure digital marketing. They are web-site, search engine marketing, web banner, social net-working, and affiliate marketing. Among these five dimensions, affiliate marketing get the highest score. This indicates a good assessment from respondents regarding Lazada.co.id website, in which it provides information and promotion to the customers. Meanwhile, the lowest score is web banner dimension.

There are also five dimension used to measure online purchase dimension. They are need recognition, information search, alternative evaluation, purchase decision, and post purchase decision. Of five dimensions, information search gains the highest score. This also implies good assessment from respondents regarding information search provided by Lazada.co.id. It can make consumers to do an online purchase decision. On the other hand, the lowest score is Alternative Evaluation dimension.

REFERENCES

Abdullah, L.R., Ramli, H.O.B. & Othman, M 2019. Developing a Causal Relationship among Factors of E-Commerce: A Decision Making Approach. *Journal of King Saud University Computer and Information Sciences* 1–8.

Azmi, U 2015. profil perusahaan lazada Indonesia kaskus 2015.

Azmi,U. 2016. Pengaruh Program Komunikasi Pada Digital Marketing Dan Sales Promotion Terhadap Impulse Buying Pada Konsumen Situs Lazada.Co.Id. Manajemen.

Baskara, G.T.H.I.P. 2014 keputusan pembelian melalui situs jejaring sosial (social networking websites) (Studi Pada Mahasiswa Di Kota Semarang). *Jurnal Ekonomi Bisnis* 1–15.

Belch, G.E. & Belch, M.A. 2015. Advertising and promotion in integreted marketing commuinication perspective (10thed) New York. *Mcgraw hill Education.*

Billy, F. 2017. Perkembangan E-Commerce Di Indonesia.

Busca, L. & Bertrandias, L. 2019. Investigating the Four Cultural Eras of Digital Marketing. *Journal of Interactive Marketing* 49: 1–19.

Chumnumpan, P. & Shi, X. 2019. Understanding New Products Market Performance Using Google Trends. *Australasian Marketing Journal (AMJ)* 27(2): 91–103.

Dahiya, R. & Gayatri 2017. A Research Paper on Digital Marketing Communication and Consumer Buying Decision Process: An Empirical Study in the Indian Passenger Car Market A Research Paper on Digital Marketing Communication and Consumer Buying Decision Process: An Empirical Study. *Journal of Global Marketing* 1–23.

Fatta, D.D., Patton D. & Viglia, G. 2017. The Determinants of Conversion Rates in SME E-Commerce Websites. *Journal of Retailing and Consumer Services.* 41: 161–68.

Fika, A. 2015 Pengaruh Keamanan, Kemudahan, Dan Resiko Kinerja Terhadap Keputusan Pembelian Secara Online Di Tokopedia.

Geraint, H. & Rowley, J. 2014. .Business to Business Digital Content Marketing. *Marketers Perceptions of Best Practice.*

Halligan, B. & Shah, H. 2010 Inbound Marketing: *Get Found Using Google, Social Media, and Blogs* 181–87.

Kannan, P.K.. & Li, H.A. 2016. Digital Marketing: A Framework, Review and Research Agenda. *International Journal of Research in Marketing.*

Kremez, Z., Frazer, L. & Thaichon, P. 2019. The Effects of E-Commerce on Franchising: Practical Implications and Models. *Australasian Marketing Journal (AMJ)* 27(3): 158–68.

Krishna, M.K. 2018. Influence of Digital Marketing on Consumer Purchase Behavior. *International Journal of Trend in Scientific Research and Development* 3(1): 836–39.

Kumar, R., Gunasekaran, A., Gupta, S. & Kamboj, J. 2019. Marketing Recommender Engine. *Journal of Retailing and Consumer Services Personalized Digital.*

Penard, T. & Perrigot, R. 2017. An Empirical Analysis of Franchisor Website Functionality. *Journal of Retailing and Consumer Services Online Search Online Purchase in Franchising* 39(7): 164–72.

Ruyter, K.D., Isobel, D. & Viet, L. 2018. When Nothing Is What It Seems: A Digital Marketing Research Agenda." Australasian Marketing *Journal (AMJ)* 26(3): 199–203.

Seshan, U.B.D.S. 2019. To Study the Impact of Digital Marketing on Purchase Decision of Youth in Nagpur City. *International Journal of Advance Research, Ideas and Innovations in Technology* 5(3): 105–12.

Sorce, P., Perotti, P., Widrick, S., Sorce, P., Perotti, V. & Widrick, S. 2005. Attitude and Age Differences in Online Buying.

Tagashira, T. & Minami, C. 2016 The Effects of Online and Offline Information Sources on Multiple Store Patronage. *Australasian Marketing Journal (AMJ.*

Tiffany, R.S., Kamala, S. & Phorkodi, M. 2018 A Study on Impact of Digital Marketing in Customer Purchase Decision in Thoothukudi. *International Journal of Science, Engineering and Management* 3(4): 613–17.

Trinh, G.T., Anesbury, Z.W. & Driesener, C. 2017. Has Behavioural Loyalty to Online Supermarkets Declined? *Australasian Marketing Journal (AMJ).*

Turban, E., Ing, D., Lee, J.K. & Viehland, D. 2008. Electronic Commerce: A Managerial Perspective 2006. *(Fourth Edition).*

Vakulenko, Y., Shams P., Hellström, H. & Hjort, K. 2019. Service Innovation in E-Commerce Last Mile Delivery: Mapping the e-Customer Journey. *Journal of Business Research* 101: 461–68.

Wong, J. & Endang, A.H. 2017 Analisis Bisnis E-Commerce Pada Mahasiswa Universitas Islam Negeri Alauddin Makassar.

Yan, R. & Yeh, R. 2009. Consumer's Online Purchase Cost and Firm Profits in a Dual-Channel Competitive Market.

Yang, Z., Yong, S. & Wang, B. 2015. Search Engine Marketing Financing Ability and Firm Performance in E-Commerce. *Procedia Computer Science* 55: 1106–12.

Advances in Business, Management and Entrepreneurship – Hurriyati et al. (Eds)
© 2021 Taylor & Francis Group, London, ISBN 978-0-367-67471-7

The impact of divergence and relevance as determinants of advertising creativity to consumer responses (study on AirAsias billboard #BikinJadiNyata)

S. Nurfebiaraning & L. Mutia
Telkom University, Bandung, Indonesia

ABSTRACT: Many competitors require companies as advertisers to promote their products or services through advertising. One important step that is carried out in the process of making advertising is planning advertising messages. Creativity is needed to determine how advertising message is delivered so that it attracts attention and is easily remembered by the target audience. There are two determinants of advertising creativity: divergence and relevance. Advertising creativity impacts consumer responses to advertising messages in various stages of response, including cognitive, affective, and behavioral stages. One of advertising media that can maximize the creativity of visual advertising is outdoor media. Outdoor advertising in the form of billboard is utilized by AirAsia to promote its services as Low Cost Airline through #BikinJadiNyata advertising campaigns. This research was conducted using quantitative methods. The technique of collecting data was surveys by distributing questionnaires to 100 respondents who had seen AirAsias billboard #BikinJataNyata on Jalan Asia Afrika Bandung. The results showed that variables of Divergence (X1) and Relevance (X2) simultaneously influence the Consumer Response (Y).

1 INTRODUCTION

The number of competitive competitors requires companies as advertisers to promote their products or services by attracting the attention of consumers through advertisements. Advertisements are messages of commercial and persuasive communication aimed at consumers with the purpose of informing, reminding or persuading advertiser's products or services based on their wants and needs through certain media. The maximum planning for advertising is needed, because basically to produce effective advertising, good advertising management is needed. Not a few advertisers work with advertising agencies in conducting advertising management activities: analyzing products or services based on research, planning advertising message placement, advertising media, monitoring the running of advertising and making decisions at each stage of the advertising campaign. In an advertising agency, there are usually three main divisions: account, media and creative division. These division which work together in accordance with their responsibilities.

One of important step taken in the advertising process is the planning of advertising messages that carried out by creative division. Advertising messages are made to provide solutions to the problems of the target audience according to their needs and desires. In this case, creativity is needed to determine how the advertising message is delivered so that

it attracts attention and is easily remembered by the target audience. There are two determinant factors of creativity: divergence and relevance. In addition, advertising message creativity can determine the success of a brand from an advertiser's product or service. Even so, in developing advertising messages, creative division not only prioritize creativity but also help advertisers to achieve advertising goals that can generate positive consumer responses through cognitive, affective and behavioral aspect.

One advertising media that can maximize the creativity of advertising messages in the form of visuals is outdoor media. Advertising media placed outdoor can be displayed in large sizes with colors, designs, layout, photos/illustrations/graphics and flexible text. Outdoor advertising is utilized by AirAsia to promote its service as Low Cost Airline through the #BikinJadiNyata advertising campaign by creating a billboard placed on Asia Afrika Street Bandung. Bandung is one of the cities that is preferred by AirAsia's city to be visited by travelers. The text of advertising message is "Popotoan Deui Wae, Mending #BikinJadiNyata ka Nu Aslina Kuy!. The message uses Sundanese, which is regional language that directly elevates local culture. The illustrations of advertising message are tourist places, icons and flags of several countries in Asia such as Indonesia with Barong Bali, South Korea with N Seoul Tower, Hanbok and Taekwondo, Japan with Tokyo Tower, Kimono, Mount

Fuji and Kendo, India with Taj Mahal and Sari, Singapore with Merlion and Marina Bay Sands. Based on the background above, the research aims to find out how much the effect of divergence and relevance as determinants of creativity on the billboard of AirAsia #BikinJadiNyata on Asia Afrika Street in Bandung towards consumer responses. The research hypothesis.

H0: There is simultaneously significant effect of divergence (X1) and relevance (X2) towards consumer responses

Ha: There is no simultaneously significant effect of divergence (X1) and relevance (X2) towards consumer responses

1.1 Advertising message

Advertising is a message of commercial and persuasive communication aimed at consumers with the purpose of informing, reminding or persuading advertiser's products or services based on their wants and needs through certain media (Moriarty et al. 2015). One of the key components of advertising is a message based on research and consumer characteristics with an emphasis on creativity and art (Moriarty et al. 2015). The effectiveness of visuals is measured when its messages can get better attention (grab attention), be easily remembered in the mind (stick in memory), increase credibility of messages (seeing is believing), provide attractive visual stories (tell interesting stories), present illustration that tells faster than text (communicate quickly), and connect products with visual associations that represent lifestyle and type of users (anchor association) (Moriarty et al. 2015). Meanwhile, some of the text characteristics are effective if using short words, sentences and paragraphs (succinct); not general (specific), aimed at the target audience (personal), focusing on one idea (single focus), using daily language (conversational), original, news, adding magic phrases, variety, imaginative description, interesting stories and structures that attract attention and build interest (a story with feeling) (Moriarty et al. 2015).

There are visual elements in advertising: illustration and photo, color, typography, design, layout and style (Moriarty et al. 2015). Visual complexity has an important role in advertising messages. Advertisers must consider the visual complexity so that advertising have "stopping power" (Schiffman & Wisenblit 2015). One of the advertising media that can be maximized to develop creative messages is outdoor advertising. The advantages of outdoor advertising are wide coverage of local markets frequency, geographic flexibility and creativity, ability to create awareness, efficiency, effectiveness, production capabilities, and timeliness. Whereas outdoor advertising disadvantages are waste coverage, limited message capabilities, wearout, cost, measurement problems, and image problems (Belch & Belch 2015). Within the outdoor advertising, creativity

is particularly important. It can attract attention and encourage further processing (Baack et al. 2008).

1.2 Divergence and relevance as determinants of creativity

Advertising creativity is the ability to generate fresh, unique, and appropriate or relevant ideas that can be used as solutions for communication problems. Those who study as well as work in advertising generally agree on these two central determinants of creativity, which are often viewed in term of divergence and relevance. Advertising creativity will refer to the divergence or relevance of an advertising or campaign as perceived by the target audience (Smith & Yang 2004). Divergence refers to the extent to which and advertisement contains elements that are novel, different, or unusual. Robert Smith and his colleagues has identified five major factors that could be used to create divergence advertising. These factors are: 1) originality, ads that contain rare elements, move away from the obvious and commonplace; 2) Flexibility, ads that contain different ideas or switch from one perspective to another; 3) Elaboration, ads that contain unexpected details or finish and extend basic ideas so they become more intricate, complicated, or sophisticated; 4) Synthesis, ads that combine, connect, or blend normally unrelated objects or ideas, and 5) Artistic value; ads that contain artistic verbal impressions or attractive shapes and colors.

The second major determinant of creativity is relevance, which reflects the degree to which the various elements of the ad are meaningful, useful, or valuable to the consumer. Smith et al. suggested that relevance can be achieved in two ways. Ad-to-consumer relevance refers to situations where the ad contains execution elements that are meaningful to consumers. Brand-to-consumer relevance refers to situations where the advertised brand of a product or service has personal interest to consumers. Relevance or appropriateness can also be viewed in terms of the degree to which an advertisement provides information or an image that is pertinent to the brand (Belch & Belch 2015).

1.3 Consumer responses

Consumer response consists of cognitive, affective and behavioral aspects. The cognitive stage refers to the knowledge and understanding of consumers about products or services in advertisements. According to (Moriarty et al. 2015), the factors that encourage cognitive are need, cognitive learning, comprehension, differentiation and recall. Furthermore, the affective stage refers to the feelings and emotions. The factors that encourage affective are wants and desires, excitement, feelings, liking and resonance. Behavioral is the last step in the response which is derived from mental rehearsal, trial, buying, contacting, advocating and referrals and prevention (Moriarty et al. 2015).

2 METHODS

The paradigm used in this study was positivist. The research method was quantitative as there were hypothesis. The population of travelers who see the billboard of AirAsia #BikinJadiNyata on Asia Afrika Street on Bandung was basically unknown. Sample was needed as it was impossible to find aevryone in all places that did everything (Punch 1998). The sample 100 respondents who saw the billboard of AirAsia #BikinJadiNyata on Asia Afrika Street on Bandung. The sample was chosen by using non probability sampling-accidental sampling with an error level of 5%. To collect the data needed, questionnaire was used. It was then analyzed by using multiple linear regression test. Divergence as independent variable ($X1$) was measured based on originality, flexibility, elaboration, synthesis, and artistic value factors. Relevance ($X2$) was measured based on ad-to-consumer relevance and brand-to-consumer relevance factors. Meanwhile, consumer responses as dependent variable (Y) was measured based on cognitive, affective and behavioral stages of responses.

3 RESULT AND DISCUSSION

The data shows that most of the respondents were female (66 people). Of 100 respondents, eight five people were 17–22 years old. Forty two people were students. This indicates that the billboard of AirAsia #BikinJadiNyata on Jalan Asia Afrika Bandung more attracted female respondents' attention than men. Visual advertisement illustration in the form of tourist places, and icons of several countries in Asia and its flags was also more grasped by female respondents than men.

Table 1. Respondents data.

No	Respondent	Number
1	Gender	
	Male	34
	Female	66
2	Age	
	17–22 yo	85
	23–28 yo	11
	29–34 yo	4
3	Profession	
	Student	42
	College Student	30
	Entrepreneur	10
	Private Employees	5
	Government Employees	2
	Others	11

*Source: Processed Research Data, 2019.

The Billboard of AirAsia #BikinJadiNyata on Asia Afrika Street on Bandung represents the respondents aged 17–22. Through visual anchor association elements that function to differentiate non-differentiated products with less attention, advertisers often connect products with visual associations that represent lifestyle and type of users. This age represents the millennial generation (18–34 years) who are close to technology and love traveling so they can publish their vacation activities through Instagram social media. In this study, the AirAsia brand is Asia's Best Low Cost Airline that offers a wide selection of cities favorite tourist destinations at affordable prices.

As determinant factors of creativity outdoor advertising billboard of AirAsia, divergence and relevance has impact on consumer response. Consumers are dominated by students. In this case, most of the students travelled with their friend on the weekend and passed Asia Afrika Street, which is one of historical area. They saw the billboard, took its picture, and shared it on their Instagram.

Table 2. Results of multiple regression data.

Variable	Sig.	F_{count}	F_{table}	Coefficient of Determination
X_1 & X_2	0,00	36,3	3,09	0,428

*Source: Processed Research Data, 2019.

The table shows a significance value of 0,00 is smaller than 0.05. This implies that there is a simultaneously significant effect of the variable Divergence ($X1$) and Relevance ($X2$) on the variable Consumer Response (Y). The result of F value (36.3) is greater than F table (3.09). This implies that H0 is rejected and Ha is accepted. Thus the hypothesis which states that there is an influence of the Divergence ($X1$) and Relevance ($X2$) variables on the Consumer Response (Y) variable can be accepted. Based on the significant value of the coefficient of determination, there is a simultaneously influence of the Divergence ($X1$) and Relevance ($X2$) variables on the Consumer Response (Y) variable of with the amount of 42.8%. The remaining percentage (57.2%) is influenced by other variables that are not examined in this study.

Belch & Belch (2015) stated that the shortcomings of outdoor advertising are limited message capabilities. Seeing a long size and wide billboard wall cause consumers who pass it to not receive all of AirAsia's visual messages. This is inlie with (Duncan 2008) who stated that the limitations of outdoor advertising is the "passing" condition. However, it will be different if there is a traffic on the street or the consumers drive slowly. Luckily, there are many consumer who pass the street by walking so that the message of outdoor advertising of AirAsia can be accepted.

The advantages of outdoor advertising are wide coverage of local markets frequency, geographic flexibility and creativity, ability to create awareness, efficiency, effectiveness, production capabilities, and timeliness. (Belch & Belch 2015). Through the media of outdoor advertising in the form of billboard, AirAsia has expanded its reach to the selected city for tourism destinations, historical areas in Bandung, precisely on Asia Afrika Street. Outdoor advertising can increase the frequency of AirAsia #BikinJadiNyata advertising messages to the consumer who passes Asia Afrika Street. AirAsia gives consumers the freedom to take photos or videos in billboard and to publish it on Instagram. AirAsia's billboard #BikinJadiNyata can also create awareness through the size of layouts (layout), typography (letters), and the design of visual messages for AirAsia #BikinJadiNyata. The visual message is very large and flexible. It was put on the roadside of Asia Afrika; moreover colors can build AirAsia brand identity.

According to Belch & Belch (2015), advertising creativity is the ability to produce fresh, unique, and relevant ideas that can be used as solutions for communication problems. There are two main factors determining advertising creativity. First is divergence. It consists of five factors: originality in the form of unusual elements; flexibility ads in the form of ideas with different perspectives; elaboration in the form of complicated or sophisticated ads created from unexpected details or extended basic ideas; synthesis ads in the form of a combination of unrelated objects; and artistic value ads in the form of artistic verbal or interesting shapes and colors. Second factor is relevance which consists of two factors: ad-to-consumer relevance referring to ads that are meaningful to consumers; and brand-to-consumer relevance referring to the extent to which certain brand products or services can meet consumer needs and can provide information related to brands. AirAsia's #BikinJadiNyata billboard can meet the needs of consumers who like traveling, taking photos or videos, and sharing it to Instagram.

Advertising creativity is very important to compete with competitors and to attract consumers' attention. Effective advertising requires not only creativity in the form of ideas, but also communication in different ways. (Moriarty et al. 2015) stated that effective advertising is an art in creativity and science in its strategy. Divergence and relevance factor in outdoor advertising billboard of AirAsia affects consumers' responses. In the cognitive stage, consumers know and understand AirAsia. In affective stage, the consumers have impression on AirAsia based on excitements, feelings, liking, and resonance. Then, behavioral stage refers to the actions of consumers on AirAsia. When consumers already know and believe AirAsia can fulfill their needs and desires through #BikinJadiNyata, they will not hesitate to seek more information about domestic or international flight fares through AirAsia social media and use AirAsia flights. Apart from the influence of divergence and relevance as a determinant of advertising creativity towards consumer response, advertising is successful because the right media sends the right message to the right target audience at the right time (Moriarty et al. 2015) and successful message strategies also make the brand relevant to consumers (Duncan 2002).

4 CONCLUSION

There is a simultaneously influence of the Divergence (X1) and Relevance (X2) variables on the Consumer Response (Y) variable of with the amount of 42.8%. The remaining percentage (57.2%) is influenced by other variables that are not examined in this study.

REFERENCES

Baack, D.W., Rick, T.W., & Brian, D.T. 2008. Creativity and Memory Effects: recall, Recognition, and Exploration of Nontraditional Media. *Journal of Advertising* 37(4).

Belch, G.E.,& Belch, M.A. 2015. Advertising and Promotion. An Integrated Marketing Communications Perspective. Tenth Edition. New York. *McGraw-Hill Education*

Duncan, T. 2008. Principles of Advertising & IMC. Second Edition. International Edition. New York: *McGraw-Hill Education*.

Moriarty, S., Nancy, M., & William, W. 2015. Advertising & IMC. Principles and Practice. Tenth Edition. England: *Pearson Education Limited.*

Punch, K.F. 1998. Introduction to Social Research Quantitative and Qualitative Approach. London: *Sage Publications, Ltd*.

Schiffman, L.G., & Wisenblit, J.L. 2015. Consumer Behavior. Eleventh Edition. Global Edition. England: *Pearson Education Limited*

Smith, R.E., & Yang, X. 2004. Toward a General Theory of Creativity in Advertising: Examining the Role of Divergence *Marketing Theory* 4:31.

Advances in Business, Management and Entrepreneurship – Hurriyati et al. (Eds)
© *2021 Taylor & Francis Group, London, ISBN 978-0-367-67471-7*

The analysis of brand trust in increasing brand loyalty

P.D. Dirgantari, F.A. Pratiwi, Rd.D.H. Utama & Sumiyati
Universitas Pendidikan Indonesia, Bandung, Indonesia

ABSTRACT: The number of local cosmetics in Indonesia has increased the business competition. Companies need to pay attention to customers' loyalty to brand as it is an important aspect in dealing with competition. One of them is through brand trust. The purpose of this study is to obtain a description of brand trust, brand loyalty, and the influence of brand trust on brand loyalty. The method used was an explanatory survey with 107 respondents as sample. The data was analysed by using simple linear regression analysis. The findings of this study indicated that both brand trust and brand loyalty was in quite good category. Brand trust had a positive and significant effect on brand loyalty.

1 INTRODUCTION

Brand loyalty has a function to increase sales volume, get high prices, and retain customers (Rasheed 2015). Marketers have used brand loyalty as a strategy to propose sustainable competitive advantages (Chinomona 2016), even though maintaining loyal customers is very difficult in this competitive era (Schoenbachler et al. 2004).

Brand loyalty was basically conceptualized for the first time in the 1940s (Kuikka & Laukkanen 2012). However, it still becomes an important issues in the business and marketing; and has been used by marketers for company benefit (Bennett 2001; Chinomona 2016; Moller et al. 2006).

In building brand loyalty, company needs to have good relationships with customers. When a company is able to meet customers' desires and expectations, they will be loyal to the brand offered by the company (Kamal & Hashmi 2014; Kuikka & Laukkanen 2012; Wolfling 2013). Brand loyalty tends to trigger customers to repurchase the product in the same period. They often refuse to switch to another brand. This is in line with the theory which stated that brand loyalty triggers customers not only to be loyal, but also to have psychological commitment to the brand (Schoenbachler et al. 2004).

Research on brand loyalty has been carried out in several industries ranging from the telecommunications industry (Daniel 2008; Nawaz 2008) fashion (Esmaeilpour & Ali 2016), retail and technology industries (Schoenbachler et al. 2004), banks (Brink et al. 2006), food and beverages (Rasheed 2015), tourism (Sea et al. 2015), hospitality (Tepeci 1999), mobile phones (Martense 2010) to the cosmetics industry (Yin & Mansori 2016).

Regarding cosmetics industry, the data from Ministry of Industry shows that there is an increase of market in 2010–2018. It can be seen in the following table:

Table 1 shows that cosmetics market in 2017 has increased up to ±4 times compared to 2016. In 2018, it was expected to increase Rp. 80.0 trillion. In the past 9 years, the grown of this market was 22.91% per year. This shows that the attractiveness of the cosmetics industry is quite high, and has the potency to increase.

Table 2 shows that Wardah has the largest cosmetic market share in 2016–2017. It is followed by Mustika

Table 1. Development of cosmetic industrial markets in Indonesia, 2010–2018.

years	Market (Rp. Trillion)
2010	8.9
2011	8.5
2012	9.7
2013	11.2
2014	12.8
2015	13.9
2016	14.8
2017	46.4
2018	80.0

Kenaikan Rata-rata, % tahun 22.91%

Table 2. The share of the cosmetic market in Indonesia in 2015–2017.

Brand Name	Years		
	2015	2016	2017
Wardah	16%	50%	50%
Mustika Ratu	50%	20%	20%
Sariayu	24%	20%	17%
Produk lainnya	10%	10%	13%

Ratu and Sariayu. Unfortunately, Sariayu experienced a decline in 2015–2017. In this case, market share can be used as one of parameter for the brand success (Deniarni & Lisnawati 2016). In other words, market share can be used as a tool to measure the brand loyalty.

Top Brand Award is an award given to the best brands of customer choice. It indicates the level of customers loyalty and attitude towards the brand that can be seen from customers' satisfaction, brand awareness, brand knowledge (customers knowledge of a brand), and future intention. Customers' loyalty can be influenced by behaviour, brand trust, and commitment dimension.

Table 3. Cosmetic top brand index (TBI) in 2014–2017.

Name Brand	Top Brand Index (TBI)			
	2014	2015	2016	2017
Wardah	13,0%	14,9%	22,3%	25,5%
Revlon	12,6%	12,8%	13,3%	12,7%
Pixy	9,0%	11,0%	9,3%	9,6%
Viva	8,2%	–	8,9%	8,8%
Sariayu	9,2%	7,6%	7,7%	7,5%

Table 3 shows that Sariayu stayed at the 3rd position in 2014. The position was dropped to 5 in 2016–2017. In this case, the brand confidence of Sariayu continually decreased; thus, it had bad impact on its brand loyalty. As stated by Zehir et al. (1995) that brand trust can build brand loyalty. Brand trust is the customers' safe feeling with the brand. This feeling is created from the perception that the brand is reliable and responsible for customers' interests and safety (Delgado 2006).

The ignorance of brand loyalty can cause the difficulties in expanding the market share and in maintaining the relationship with the customers (Ahmed & Ahmad 2014). It even can reduce brand awareness and company income (Daniel et al. 2016).

2 METHODS

Based on the above explanation, this study aims to investigate the influence of brand trust to brand loyalty at Sariayu cosmetic company. This study was verification research as it aimed to test the hypothesis and determined the effect of brand trust on brand loyalty. The method used was explanatory survey method. It was used as its capability to explore or examine problems or situations to gain insight and understanding.

To collect the data, a questionnaire was used. The questionnaire consisted of indicators to measure brand trust and brand loyalty of Sariayu cosmetic customers in Indonesia. It was distributed to 107 customers. Documentation analysis was also used to support the primary. The data was then analysed by using simple linear regression analysis.

3 RESULT AND DISCUSSION

In this study, the independent variable is brand trust (X), while dependent variable is brand loyalty (Y). To test the influence of dependent variable, simple regression testing was carried out.

3.1 Simple linear regression equation.

By using the SPSS 24.0, the result of regression coefficient is obtained as follows.

Table 4. Simple linear regression model of brand trust towards brand loyalty.

Coefficients[a]

Model		Unstandardized Coefficients		Standardized Coefficients		
		B	Std. Error	Beta	T	Sig.
1	(Constant)	2.704	3.413		0.792	.430
	Brand Trust	1.060	.073	.818	14.553	.000

a. Dependent Variable: Z

Based on Table 4 column B, there is constant values and simple linear regression coefficients for independent variables. Based on these values, we can determine the simple linear regression model in following formula:

$$Y = a + bX \qquad (1)$$

Source: (Sugiyono, 2014)

$$Y = 2.704 + 1,060X \qquad (2)$$

The result of linear regression equation reveals that constant value is 2.704. This indicates that there is no brand trust. Meanwhile, the value of brand loyalty is 2.704. The regression coefficient on the brand trust variable is 1,060. This implies that the increase of brand trust value is followed by the increase of brand loyalty in the amount of 1,060. Conversely, the decrease in brand trust value is followed by the decrease of brand loyalty for about 1,060. In other words, brand trust will affect the level of brand loyalty. If brand trust is low, so is with brand loyalty.

3.2 Coefficient of determination analysis

Analysis of the determinant coefficient is used to determine the influence percentage of independent variable on the dependent variable. Regarding this, this study used the formula proposed by Riduwan (2013):

$$\boxed{KD = r^2 \times 100\%} \qquad (3)$$

Figure 1. Formula for the coefficient of determination.

Information:

r^2 = correlation coefficient

Table 5. The determinant coefficient of brand trust to brand loyalty.

Model Summar

Mode	R	R Square	Adjusted R Square	Std. Error of the Estimate
	.818	.669	.665	6.33

a. Predictors: (Constant), X
b. Dependent Variable: Y

The following is the calculation result of the determinant coefficient from X to Y:

$$KD = r2 \times 100\%$$
$$= r2 \times 100\%$$
$$= (0,818)2 \times 100\%$$
$$= 0,6691 \times 100\%$$
$$= 66.91\%$$

The calculation results of determinant coefficient for Brand Trust (X) for Brand Loyalty (Y) is 66.91%. In other words, brand loyalty is influenced of 66.91% by brand trust, while 33.09% is influenced by the other factors which is not examined in this study such as cause-related marketing (CRM) (Brink et al. 2006) and country-of-origin image (Esmaeilpour & Ali 2016).

Loyal customer is a customer who believes to use a product in a company. This kind of customer is very important for the company to survive. That is why they should be maintained. Brand loyalty can predict and secure the demand of the company; then it causes the difficulties for other companies to enter the market (Chaudhuri & Holbrook 2001; Delgado & Luis 2001; Sahin et al. 2011).

3.3 The influence of brand trust on brand loyalty

To investigate the influence of brand trust on brand loyalty, T-test was used to compare Tcount and Ttable. The result of t-test formula is explained in Table 6.

Table 6 shows the significance value (t-count) is 14.553. With the significance 0.05, t-table is 1.65922. At this point, the value of t-count is higher than t-table.

Table 6. The significance value of the t-test.

Coefficients

Model	Unstandardized Coefficients B	Std.Error	Standardized Coefficients Beta	t	Sig.
1 (constant)	2.70	3.41		.792	.003
X	1.06	.073	.818	14.553	.000

Dependent Variable: Y

Thus, Ho is rejected. Then, it can be inferred that there is an influence of customers brand trust on customers' brand loyalty at Sariayu Company.

4 CONCLUSION

Brand trust has an influence on brand loyalty. The results of this study shows that the better the customer believes in the brand, the better brand loyalty is. At this point, companies have to attract the customer to the brand. They can make a good testimony to the sold product, provide a good service, or provide a good product quality. Thus, the customers will feel satisfied and repurchase the product.

REFERENCES

Ahmed, Z. & Ahmad, M. 2014. Effect of brand trust and customer satisfaction on brand loyalty in Bahawalpur. *Journal of Sociological Research* 5(1):306–326.
Bennett, S.R.R. 2001. A brand for all seasons? A discussion of brand loyalty approaches and their applicability for different markets. *Journal of Product & Brand Management* 10(1): 25–37.
Brink, D.V.D., Pieter, G.O. & Pauwels, P. 2006. The effect of strategic and tactical cause-related marketing on consumers' brand loyalty. *Journal of Cus* 23(1):15–25.
Chaudhuri, A. & Holbrook, M. B. 2001. The chain of effects from brand trust and brand affect to brand performance: the role of brand loyalty. *Journal of marketing*, 65(2), 81–93.
Chinomona, R.2016. Brand communication, brand image and brand trust as antecedents of brand loyalty in Gauteng Province of South Africa. Africa *Journal of Economic and Management Studies* 7(1): 124–139.
Daniel, E.Y., Simon, M.Q., Nimako, G., Yeboah-asiamah, E. & Quaye, D.M. 2016. The effects of lucky draw sales promotion on brand loyalty in mobile telecommunication industry. Africa *Journal of Economic and Maangement Studies* 7(1):109–123.
Delgado, B.E. 2006. Applicability of a brand trust scale across product categories a multigroup invariance analysis. European *Journal of Marketing* 38(5/6): 573–592.
Delgado, B.E. & Luis M.A.J. 2001. Brand trust in the context of consumer loyalty. European *Journal of marketing*, 35(11/12), 1238–1258.
Deniarni, L. & lisnawati. 2016. Analisis persepsi virtual brand community terhadap kinerja ekuitas merek disposable diaper merek sweety (survei pada anggota fan page facebook bunda cermat). *Jurnal of Business Management and Enterprenuership Eduction* 1(1): 158–173.
Esmaeilpour, F. & Ali, A.M. 2016. The impact of country-of-origin image on brand loyalty: Evidence from Iran. Asia Pacific *Journal of Marketing and Logistic* 28(4): 709–723.
Kamal, F. & Hashmi, H. 2014. An empirical study of brand loyalty on samsung electronics in Pakistan. *Journal of Service Marketing* 5(1): 350–364.
Kuikka,A. & Laukkanen, T. 2012. Brand loyalty and the role of hedonic value. *Journal of Product & Brand Management* 21(7): 529–537.
Moller, J., Torben, J. & Jensen, J.M. 2006. An empirical examination of brand loyalty. *Journal of Product & Brand Management* 15(7): 442–449.

Nawaz, S. 2008. Crossed swords: Pakistan, its army, and the wars within. USA. *Oxford University Press*

Rasheed, K.O. 2015. Product package as determinant of brand loyalty in food and beverages markets of lagos state. American *Journal of Marketing Research* 1(3): 150–157.

Riduwan. 2013. Belajar mudah penelitian. Bandung: *Alfabeta*.

Sahin, A., Zehir, C. & Kitapçı, H. 2011. The effects of brand experiences, trust and satisfaction on building brand loyalty; an empiricsahinal research on global brands. *Procedia-Social and Behavioral Sciences* 24: 1288–1301.

Schoenbachler, D.D., Gordon, G.L. & Aurand, T.W. 2004. Building brand loyalty through individual stock ownership. *Journal of Product and Brand Management* 13(7): 488–497.

Sea, D., Jraisat, L.E., Akroush, M.N., Jaser, R., Laila, A., Dina, T. & Kurdieh, D.J. 2015. Perceived brand salience and destination brand loyalty from international tourists' perspectives: The case of Dead Sea destination Jordan. International *Journal of Culture, Tourism and Hospitality Research* 9(3): 292–315.

Sugiyono, S. 2014. Metode penelitian manajemen. Bandung: *Alfabeta.*

Tepeci, M. 1999. Increasing brand loyalty in the hospitality industry. Internasional *Journal of Contemporary Hospitality Management* 11(5): 223–230.

Wolfling, A. 2013. Impact of quality inconsistency on brand loyalty: 491–506.

Yin, C. & Mansori, S. 2016. Factor that influences consumers' brand loyalty towards cosmetic products. *Journal of Marketing Management and Customer Behavior* 1(1): 12–29.

Zehir, C. Kitapcı, H. & Ozsahin, M. 1995. The effects of brand communication and service quality in building brand loyalty through.

Customer profiling and market basket analysis using k-means algorithm and association rule mining: Evidence from Indonesia e-commerce company

P. Arizona & A. Hananto
Universitas Indonesia, Depok, Indonesia

ABSTRACT: Online customers' segmentation could be valuable research topic of marketing strategy. Previous literature mainly studied the differences between non-purchasers and purchasers, lacking further segmentation of online customers. This research focuses on online customer segmentation based on a large volume of real transaction data in one of Indonesia's e-commerce website. The research proposed a customer clustering technique using the K-Means algorithm and RFM Patterns as an analysis of the customer's profile. Then, the market basket analysis was conducted using the Apriori algorithm for every customer profile and cluster to obtain the association rule as well as product relationships purchased by customers. Later, the result of market basket analysis was utilized as an input for e-commerce companies in designing promotions such as bundling or product recommendation system for segmented customers

1 INTRODUCTION

1.1 Background

Indonesia e-commerce are heading towards big data analytic to understand consumers' behaviour and predicting trends in the market (Frost & Sullivan 2018). Understanding customers' behaviour with similar profile can be performed by analysing transaction data that is recorded on the system.

In this research, customers' transaction data from one of Indonesia's e-commerce is utilized to get a better understanding of Indonesia's online consumer purchasing patterns. There are some usages of customers' information that will be used to help online retailers make a suitable recommendation system for different customers' types.

Initially, clustering is performed in order to obtain a group of customers that have a similar behaviour based on their previous purchases. Then, the association rule will be executed on the segmented customers within the same cluster to capture relationships between products.

This research is aimed to extract knowledge that delivers effective decisions for e-commerce companies and offers a value-added manner to demanding customers (Griva et al. 2018).

1.2 Related works

For e-commerce website, understanding customer behaviour can be identified through how recent customers purchase in the website (recency), how often they purchase in certain time period (frequency), and how much they spend to buy items (monetary).

Products can be recommended based on the highest sales, customers' demography, or analysis of the transaction records of purchased items (Schafer et al. 2001). The e-retailer Amazon (www.amazon.com) successfully applies some types of recommendation systems by (i) providing another similar items to the products that has just being viewed or purchased, and (ii) giving further recommendations system in "customer who bought this item also bought" section that is built upon the transactional information from other customers (Li & Karahanna 2015). In the 2nd fiscal quarter of 2012, the company reported a 29% increase in sale (Mangalindam 2012). McKinsey also estimated that product recommendation contributes around 35% of Amazon's total purchases (MacKenzie et al. 2013).

Thus, by understanding the customers' behaviour, retailers can encourage customers to add more items to their cart by providing a suitable product recommendation list (Rodrigues & Ferreira 2016).

1.3 RFM variables & customer profiles

Recency (R) variable explains how many days since the last time a consumer made a purchase. Frequency (F) variable explains how many times that particular customer make purchases. Monetary (M) variable will reveal total amount of consumers' spending during the transaction period, using the same calculation method with frequency.

Different customer clusters can be strategically positioned using RFM point of view (Ha & Park 1998). These variables are then become the basic of clustering in data mining technique (Birant 2011).

Table 1. Customer profile based on RFM pattern.

Customer Group	RFM Pattern			Customer Profile
	R	F	M	
Group 1	R ↑	F ↑	M ↑	Best
Group 2	R ↑	F ↑	M ↓	Valuable
Group 3	R ↑	F ↓	M ↑	Shopper
Group 4	R ↑	F ↓	M ↓	First time
Group 5	R ↓	F ↑	M ↑	Churn
Group 6	R ↓	F ↑	M ↓	Frequent
Group 7	R ↓	F ↓	M ↑	Spender
Group 8	R ↓	F ↓	M ↓	Uncertain

*Source: Adapted from Birant, 2011

1.4 K-Means algorithm

K-means algorithm is an approach of classifying data based on the centroid data on each cluster. This technique is widely used and desired in marketing research (Chen et al. 2004) because of its capability in similarity grouping, relevant classification, and approximating a general distribution (MacQueen 1967). The value of will be determined in advance as the basis for group distribution clustering, then, looking at its 'elbow

1.5 Market basket analysis & association rule mining

The concept of market basket analysis lies on the association rule mining concept using Apriori algorithm. It is presented as X → Y, where X is a left hand side/lhs (antecedent), and Y is a right hand side/rhs (consequent). This notation means if product X is composed in one basket, it exists a likeliness that product Y appears in the same transaction. X and Y represent two disjoint set of items (Rodrigues & Ferreira 2016).

There are 3 indicators to measure the quality of the rules. Support explains how frequent the set of items X and Y occurs towards the total transaction. Confidence measures the probability of rule X → Y occurs. Lift measures a confidence of co-occurrence between X and Y more than expected.

1.6 Extended view of intercategory relationship

To determine the product relationships, Shocker et al. 2004) differentiate the intercategory relationship into static cases and dynamic cases. Static relationships are prone to be stable and has fewer changes in a relatively long period. Types of static intercategory relationships across buyers are (1) substitutes-in-use; (2) complements-in-use; (3) occasional substitutes; and (4) occasional complements.

Dynamic product relationships show a regular product transition over the times. Some types of dynamic intercategory product relationships are (1) product displacement; (2) enhancing complement; (3) product perseverance; and (4) augmenting complement.

1.7 Recommender system & bundling concept

According to Schafer et al. (2001), recommender system, or in this case a product recommendation, enhances e-commerce's revenue channels in following manners: (1) converting browsers into buyers; (2) increasing cross-sell; and (3) building loyalty. Implementation of the product recommendation reflected in bundling concept offered by retailers. The combination of packages depends on the industry they are in. Therefore, the term 'bundling' here has a wide interpretation. In this research, we refer to five types of bundling by (Stremersch & Tellis 2002): bundling, price bundling, product bundling, pure bundling, and mixed bundling.

2 METHODS

2.1 Research framework

A combination of customer segmentation and market basket analysis is proposed to address the goal to analyse the customer profiles and purchasing patterns of online customers in Indonesia. In general, this part is divided into several parts: customers' segmentation, customer profiling, and market basket analysis. Every part will have a different analysis treatment respectively. Research framework is presented in Figure 1.

2.2 Data processing & analysis method

The data used in this study was a real transaction data from one of the e-commerce websites in Indonesia, ranging from January to December 2018.

The average value of F and M variables have an upward arrow ↑ if it exceeds the overall average F and M from all customers, and have a downward arrow ↓ sign otherwise. A contrast rule applied for R parameter. This technique differ online consumers form segments based on their relationship with the website.

Consumer's purchasing patterns was exploited through market basket analysis in every customer cluster and customer profile to provide more precise and personalized items in product recommendation. This whole data processing was conducted using Rstudio software version 3.5.2 on iMac Mojave iOS system and windows 10.

3 RESULT AND DISCUSSION

3.1 Customer segmentation

Initially, the average value of customers' RFM is calculated by aggregating all data as a comparison to differentiate customer profiles. Table 2 shows the average RFM values from all customers' transactions.

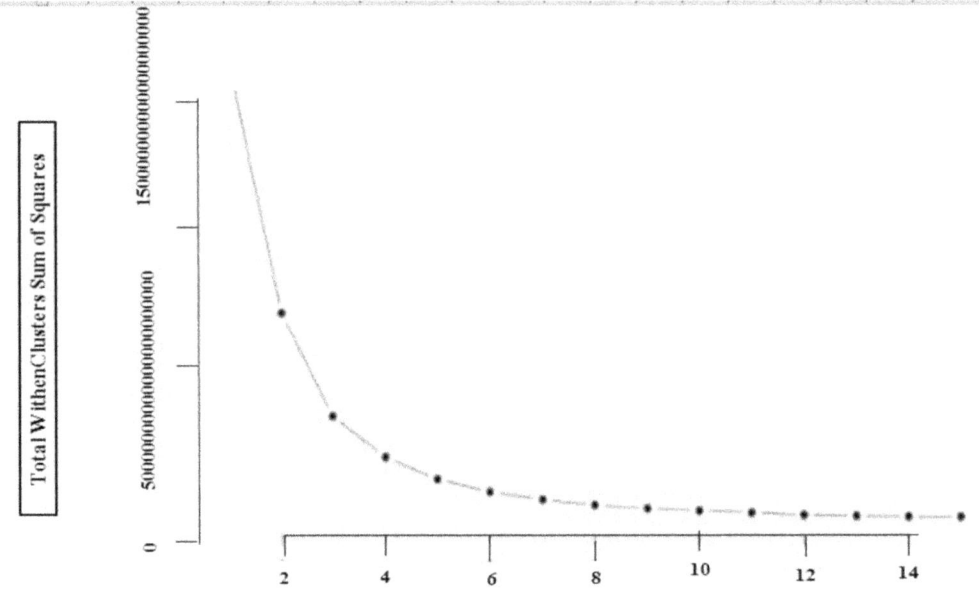

Figure 1.   K-Means algorithm result

Table 2.   Average RFM values of all customers.

Recency	Frequency	Monetary
116.3 days	5.8 times	4.1 M (Rp)

3.1.1   K-Means algorithm

Based on the execution of K-Means algorithm, elbow method with = 5 as the clustering value results in 90.1% within cluster-sum of squares, indicating the probability of error occurred is 9.9%.

Table 3.   Clustering profiling result.

Customer					
#	R	F		M	Pattern	Profile
1	31.95	232.93	↑	3,476	R ↑F ↑	Best
2	30.03	235.62	↑	1,454	R ↑F↑	Best
3	116.8	5.2670	↓	2,3	R ↓F↓	Uncertain
4	40.81	71.283	↑	158,9	R ↑F↑	Best
5	28.73	156.31	↑	552,2	R ↑F↑	Best

3.1.2   Customer profile

The pattern of the average RFM values on Table 3 are then compared to the reference on Table 1.

Cluster 1, 2, 4, 5 fall under loyal or the 'Best' customer profile (0.7%). This customer profile includes customers who did their regular purchases recently, and spent quite a generous amount of money on purchases. On the other hand, Cluster 3 belongs to the potential or 'Uncertain' customer profile (99.3%). It is important to be noticed that there is no specific order regarding the cluster number. The number used on Cluster 1, Cluster 2, etc. is only for nomenclature purposes.

3.2   Market basket analysis

In order to discover the association rules, it is mandatory to set the value of indicators prior to the implementation of the Apriori algorithm. In this study, the support and confidence values are set to 0.0001 and 0.7 respectively to obtain the strong product associations.

3.2.1   The 'Best' customer profile

Transactions from the 'Best' customer profile comes from the customers that are included in Cluster 1, 2, 4, and 5. Figure 2 depicts 17 rules (all) that are generated from those transactions.

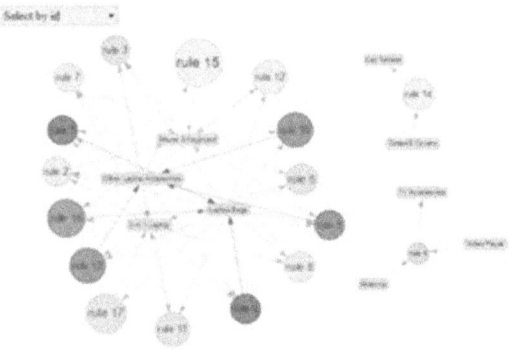

Figure 2.   All rules generated in the 'Best' customer profile

410

To get the clarity of the rules, Table 4 describes the top 3 rules generated from the 'Best' customer profile transaction data set, and sorted based on its confidence level.

Table 4. Description of the top 3 rules in the 'Best' customer profile.

lhs	Rhs	Support	Confidence	Lift	count
{2-in-1 Laptop, Mouse & Keyboard}	{Laptop Bags}	0.0002	1	2,878	149
{Antenna, Video Player}	{TV Accessories}	0.0001	0.989	851.6	95
{Laptop Bags, Mouse & Keyboard}	{2-in-1 Laptop}	0.0002	0.975	621.2	161

3.2.2 The 'Uncertain' customer profile

Transactions of 'Uncertain' customer profile only comes from Cluster 3. Figure 3 depicts the top 15 rules from 104 rules that are generated from those transactions.

Figure 3 Top 15 rules generated in the 'Uncertain' customer profile.

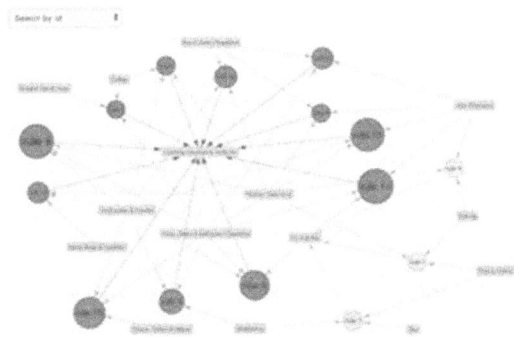

Figure 3. Top 15 rules generated in the 'Uncertain' customer profile

To get the clarity of the rules, Table 5 describes the top 3 rules generated from the 'Uncertain' customer profile transaction data set, and sorted based on its confidence level.

3.2.3 Extended view of the 'Uncertain' customer profile

As mentioned in the previous analysis, the 'Uncertain' customer profile covers around 99.3% of all customers. Therefore, the further analysis of this customer profile is also needed to give a deeper understanding of customer segments by performing the same way as previous steps.

Table 5. Description of the top 3 rules in the 'Uncertain' customer profile.

lhs	rhs	Support	Confidence	Lift	count
{Rice & Grains, Seasoning, Tea}	{Oil & Butter}	0.0001	0.78	6.16	1,008
{Coffee, Kitchen Cleansing, Toothpaste & Powder}	{Clothing Cleanser & Softener}	0.0001	0.764	14.2	680
{Coffee, Floor, Glass & Bathroom cleanse Cleansing & Shower Gel	{Clothing {Softener & Soap}	0.0001	0.762	14.2	663

Table 6. Average RFM values of the 'Uncertain' customer profile.

Recency	Frequency	Monetary
116.8 days	5.2 times	2.3 M (Rp)

Based on the execution of K-Means algorithm, elbow method with = 4 as the clustering value results in 90.5% within cluster-sum of squares, indicating the probability of error occurred is 9.5%.

Table 7. Clustering using = 4 and customer profiling result Profile.

Customer					
# R		F	M	Pattern	Profile
1	60.61 ↑	24.3 ↑	23,4 M ↑	↑R↑F↑M↑	Uncertain A
2	123.2 ↓	3.49 ↓	0.74 M ↓	↓R↓F↓M↓	Uncertain B
3	51.06 ↑	36.9 ↑	54,8 M ↑	↑R↑F↑M↑	Uncertain A
4	77.17 ↑	14.4 ↑	7,82 M ↑	↑R↑F↑M↑	Uncertain A

To be clear in explaining the analysis, customer sub-cluster with R↑F↑M↑ pattern is named 'Uncertain A' (12.83%), and customer sub-cluster with R↓F↓M↓ pattern is named 'Uncertain B' (87.17%). There is no specific order regarding the sub-cluster number. The numbers are only for nomenclature purposes.

Transactions of the 'Uncertain A' comes from sub-cluster 1, 3, and 4, resulting 117 rules. Table 8 describes the top 3 rules that are generated from the 'Uncertain A' customer profile transaction data set, and sorted based on its confidence level.

Transactions of 'Uncertain B' only comes from sub-cluster 2, and resulting 85 rules. Table 9 describes the top 3 rules that are generated from the 'Uncertain B'

Table 8. Description of the top 3 rules in the 'Uncertain A' customer profile.

lhs	rhs	Support	Confidence	Lift	count
{Coffe, Glass & Bathroom Cleansing, Toothpaste & Powder}	{Clothing Cleanser & Softener}	0.0001	0.79	14.3	268
{Coffee, Kitchen Cleansing, Toothpaste & Powder}	{Clothing Cleanser & Softener}	0.0001	0.77	14.0	340
{Floor, Glass & Bathroom Cleansing, Hair Shampo Rat & Insect Repellent}	{Clothing Cleanser & Softener}	0.0001	0.77	13.8	345

Table 9. Description of the top 3 rules in the 'Uncertain B' customer profile.

lhs	rhs	Support	Confidence	Lift	Count
{Rice & Grains, Seasoning, Tea}	Oil & Butter	0.0001	0.83	5.79	516
{Baking, Kitchen Cleansing, Rice & Grain}	{Oil & Butter}	0.0001	0.77	5.40	403
{Baking, Hair Shampoo, Seasoning}	{Oil & Butter}	0.0001	0.77	5.38	379

customer profile transaction data set, and sorted based on its confidence level.

3.3 The use of market basket analysis

The market basket analysis in the 'Best' customer profile shows strong association rules on some electronic products and their accessories. The rule of {Laptop Bags, Mouse & Keyboard} => {2-in-1 Laptop} implies that the products have a complements-in-use relationship, with a 97% confidence value. To extract the benefit from this kind of product relationship, e-commerce players can do a product bundling promotion to enhance the sales on each product category.

On the other hand, there is an 'Uncertain' customer profile which represents the 99.3% of e-commerce customers. In this customer profile, the products are dominated by Oil & Butter, Clothing Cleanser & Softener, and other groceries items. This association rule tells that these product categories also have a complement-in-use intercategory product relationship. Thus, online retailers can get benefit through this matter by implementing more bundling promotions to increase the occurrence of cross-selling.

By examining the market basket analysis in the 'Best' and 'Uncertain' customer profiles, there are some recognized patterns that can be used by online retail players to increase the customers' basket size and gain more profit from their customers. In both customer profiles, most of the product categories which are frequently purchased categories are relatively independent to one another, since only a few of them appear as the top rules. For the 'Best' customer profile, electronic categories dominate the top rules, while in the 'Uncertain' customer profile, groceries items are on the top ranks.

This event implies that those product categories are not necessarily have to be triggered by another category to be purchased by customers, or on the other word, they have a dynamic intercatgeory relationship. To utilize this advantage, online retailers can use these product categories to increase the sales of the other less purchased product categories by doing a cross-selling strategy. For example, {Feature Phones} => {Smart Phones} rule. The purchase of these product categories are considered as enhancing complements relationship because owning a Feature Phones may create a feeling of easiness in terms of communication that will subsequently be enhanced by having a Smart Phones gadget as well.

In addition, the augmenting complement products are also able to be recommended for both customer profile since the product relationship is synergistic, e.g. {Kids Milk & Nutrition} => {Diaper Pants}. Kids Milk & Nutrition would positively affects the purchase of Diaper Pants because the more food consumption would also influence the baby's digestive system. This limitation would legitimized the existence of Diaper Pants in customers' basket.

Equally important, the assessment of the generated top association rules in every customer profile should also being treated in a similar fashion. According to the results, the patterns for the generated top rules are relatively contained by product categories which have a static intercategory relationships.

4 CONCLUSION

Based on the analysis that has been carried out in the previous chapters, it can be concluded that customers' transaction data can be utilized to obtain customer segmentation through RFM variables using K-Means algorithm.

The association rules applied in the 'Best' and 'Uncertain' customer profiles shows the product relationships that could be used by online retailers as the basis in determining personalized product offering, bundling products, and the cross-category product recommendation system to increase customers' average basket size.

For future improvements, it is better to use the supervised product categories to avoid the misinterpretation of product types in generating the association rules. This research uses the 3rd product category names that are provided by the company. However, this makes the nomenclature used in this research is quite ambiguous since some different products are combined as one category, for instance, "Mouse & Keyboard", "Oil & Butter", etc. Another suggestion for future improvement is by combining several prediction models as well as bigger data range in clustering process to get a better result.

REFERENCES

Birant, D. 2011. K. Funatsu (eds). Data mining using *RFM Analysis*: 91–108.

Chen, J.-S., Ching, R. K. & Lin, Y.-S. 2004. An extended study of the K-Means algorithm for data clustering and its application. *Journal of the Operational Research Society* 55(9): 976–987.

Frost & Sullivan. 2018. White Paper. Digital market overview: Indonesia. HM Government, Santa Clara CA. [Online] Retrieved from https://ww2.frost.com/files/3115/2878/4354/Digital_Market_Overview_FCO_Indonesia_25May18.

Griva, A., Bardaki, C., Pramatari, K. & Papapkirakopoulos, D. 2018. Retail business analytics: Customor visit segmentattion using market basket data. *Expert system with applications*.

Ha, J.O. & Park, K.Y. 1998. Comparison of mineral contennts and external structure of various salts. *Journal korean society food science and nutrition* 27.

Li, S.S. & Karahanna, E. 2015, February. Onlune recommendation systems in a B2C e-commerce context: a review and future directons. *Journal of the Associations for Information Systems* 16(2): 72–107.

MacKenzie, I., Meyer, C. & Noble, S. 2013, October. How retailers can keep up with consumers. [Online]. Retrieved from McKinsey & Company: https://www.mckinsey.com/industries/retail/our-insights/how-retailers-can-keep-up-with-consumers.

MacQueen, J. 1967. Some methods for classification and analysis of multivariate observatios. *Proceedings of The Fifth Berkeley Symposium on Mathematical Statistics and Probability* 1: 281–297.

Mangalindam, J.P. 2012. Amazon's recommentation secret. *CNN Money*.

Rodrigues, F. & Ferreira, B. 2016. Product recommendation based on shared customer's behaviour. *Procedia Computer Science* 100: 136–146.

Schafer, J. B., Konstan, J. A. & Riedl, J. 2001. E-commerce recommendation applications. *Data Mining and Knowledge Discovery* 5: 115–153.

Shocker, A.D., Bayus, B. L. & Kim, N. 2004, January. Product complements and substitutes in the real world: The relevance of "other products". *Journal of Marketing* 68(1): 28–40.

Stremersch, S. & Tellis, G. J. 2002, January. Strategic bundling of products and prices: A new synthesis for marketing. *Journal of Marketing* 66: 55–72.

Analyzing the influence of firm-generated content on consumer purchase intention

C.R. Joyosugito & N. Sobari
Universitas Indonesia, Depok, Indonesia

ABSTRACT: Social media is one of the most effective marketing strategies for companies to market their products to wide consumer target in a digital platform. Before deciding on having a purchase intention, most consumers use social media first to get products information as many as possible. This will be a challenge for companies to gain consumer's attention when they want to use digital marketing by using firm generated content on social media, because companies are competing on social media and consumer's information becomes broaden. The objective of this research is to get to know how effective the firm-generated content influences consumer purchase intention directly and with brand awareness, brand loyalty, and electronic word of mouth as mediator variables as well. The methodology of this research was using SPSS Amos system to process 205 people between 25–58 years old as the sample data who answers 41 questionnaires using 5-likert scale. This research has found that, in a case on one of the leading online travel agencies in Indonesia, firm generated content has positive impact on brand awareness, brand loyalty, and electronic word of mouth (e-WOM). Also, there are findings that brand loyalty has positive impact on e-WOM, and e-WOM has positive impact on consumer purchase intention.

1 INTRODUCTION

Along with the development of technology and the digital era, marketing strategies through digital media have become mandatory for companies to communicate with target consumers. Social media is an effective communication for companies in marketing their brands by posting information about their products and as a channel for consumers to get information to confirm their decision to make a purchase (Kim & Ko 2012). Thus, companies must do firm generated content to create, maintain and strengthen their relationships with target consumers. In addition to creating the popularity of corporate brands (Rodriguez et al. 2012), firm generated content is seen as being able to contribute to increased sales and profits (Monica & Balas 2014). Competition in the business world is getting tougher and almost all businesses use social media to carry out marketing strategies. Thus, companies must create firm generated content that able to increase consumer awareness and loyalty towards a brand. Consumer awareness needs to be fostered by the company to ensure that consumers realize the existence of the brand and are able to distinguish it from competitors (Aaker 1996). Consumer loyalty needs to be formed by ensuring that their experience when using products are suit their expectations, so that the company's brand will always be in their minds and the products become the main choice in purchasing (Chi et al. 2009). The level of high and

low consumer awareness and loyalty can affect the impact of consumer behavior in communicating with other consumers (Hutter et al. 2013). A study from Balakrishnan et al. (2014) also states that communication between consumers through social media marketing companies has a positive impact on consumer loyalty. With interaction between consumers who discuss positive experiences about company products, consumers who receive this information will feel confident in making product purchases (Barreda et al. 2015). Thus, the purpose of this study is to find out whether firm generated content has an influence on brand loyalty, brand awareness, and e-WOM on consumer purchase intention.

2 METHODS

2.1 Firm generated content

Marketers take advantage of the opportunity to use social media to approach the target consumers. The function of social media for marketers includes three things: (1) to manage social media user traffic to enter the company's website, (2) to communicate with consumers, and (3) to increase brand exposure (Belch & Belch 2018). Social media is also a marketing platform for conducting corporate brand promotion and communication activities, with the aim of forming consumer decision-making processes (Tuten &

Solomon 2017). The aim of creating firm generated social media content is to bridge the relationship between new consumers and other consumers and companies, invite new customers to be present, and open up topics of conversation between consumers (Ceballos et al. 2016).

2.2 Brand awareness

Brand awareness is an illustration of the level of awareness of consumers in knowing and remembering a product or service of a company's brand (Aaker 1996). Bruhn et al. (2012) state that consumer experience of a certain brand increases because they receive various information about the brand from various types of marketing strategy activities carried out by the company to increase consumer brand awareness. According to Bruhn et al. (2012), marketing communication activities through social media are able to have a positive impact on consumer brand awareness.

H1: FGC has a positive influence on BA.

2.3 Brand loyalty

Brand loyalty is a consumer selection process where consumers decide to choose and buy a company's product brand regardless of the various alternative products available (Yoo & Donthu 2001). Usually this happens to consumers when they are satisfied with the product and have positive experience with the company's brand. Based on research by Yoo et al. (2000), the company's marketing strategy activities have a positive impact on brand loyalty. As stated by Erdogmus & Cicek (2012), corporate social media content is an important thing to encourage relationships between companies and consumers in order to increase brand loyalty of consumers.

H2: FGC has a positive influence on BL.

2.4 Electronic word of mouth

Hutter et al. (2013) inform when consumers feel loyal to a product brand they like, they will inform their experiences and positive things to other consumers. E-WOM itself also is a communication between consumers through social media by sharing opinion on corporate social media. In the research conducted by Barreda et al. (2015) it has been proven that e-WOM has a positive relationship with brand awareness, because through e-WOM consumers can be aware of the existence of the company's brand and know the experiences from other consumers of the company's products. In addition to brand loyalty and brand awareness, it has also been found that the company's social media content has a positive relationship with e-WOM (Barreda et al. 2015).

H3a: BL has a positive influence on e-WOM.
H3b: BA has a positive influence on e-WOM.
H3c: FGC has a positive influence on e-WOM.

2.5 Purchase intention

Balakrishnan et al. (2014) inform that consumers who are social media users feel positive with companies that have social media and information through social media is able to encourage consumers to decide on purchase intention. From the research by Tolba (2011), consumers who loyal to a brand can influence consumers to buy the same product even though there are other alternatives. Companies that have good consumer marketing strategies can increase consumer brand awareness in their brand, so when consumers are aware of the brand's presence they will start making purchases (Sasmita & Mohd Suki 2015). And from the research by Bailey (2004), consumers who do positive e-WOM about a brand also have positive influence on other consumers to have purchase intention.

H4a: FGC has a positive influence on PI.
H4b: BL has a positive influence on PI.
H4c: BA has a positive influence on PI.
H4d: e-WOM has a positive influence on PI.

2.6 Design and collection of data

The research model that is use for this research is based on Poulis et al. (2018) research, which has 9 hypotheses.

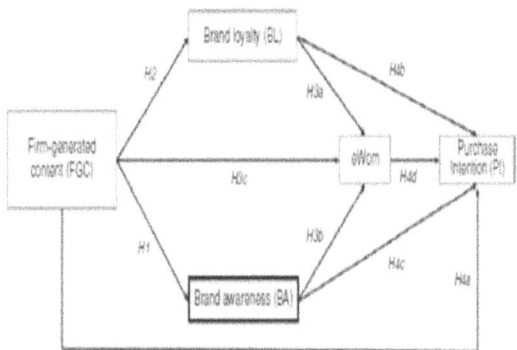

Figure 1. Research model.

In conducting quantitative research, the technique used in sampling is a non-probability sampling with a purposive sampling method. The samples used are respondents with ages of 25–58 years old, have a social media account, have seen or followed the online travel agencies (OTA) social media, and use the OTA application for the past 1 year from January 2018. 51 items questionnaire distributed was a self-administered questionnaire. The available answer choices are 5 likert scales to find out the value of each variable item. The number of samples to be taken and examined for this study are 205 people.

2.7 Reliabilty and validity

From the results of the validity test using Kaiser-Mayer-Okin (KMO) on the pretest, it can be concluded that the Firm Generated Content (FGC), Brand Awareness (BA), Brand Loyalty (BL), E-WOM (EWOM), and Purchase Intention (PI) variable value is valid with a KMO value above 0.5 with range starting from 0.632 to 0.806 with 1 BL item eliminated (BL11) and 9 E-WOM items eliminated (EWOM4, EWOM6, EWOM7, EWOM9, EWOM12, EWOM13, EWOM17, EWOM18, and EWOM20). Then, from the results of the Component Matrix and Measures of Sampling Adequancy (MSA), it is also declared valid for all items because each has a value above 0.5 with range starting from 0.523 to 0.910 for Component Matrix score and 0.530 to 0.913 for MSA score. From the results of the reliability test using Cronbach's Alpha, it can be concluded that all variables and dimensions are reliable because the value of each variable has Cronbach's Alpha coefficient above 0.5 with range starting from 0.708 to 0.913. Based on the results of the validity and reliability test, the number of questionnaire items are eliminated from 51 to 41 questionnaires.

3 RESULTS AND DISCUSSION

Using SPSS AMOS system for the measurement model test, based on the factor loading value in all 41 attributes used in each variable, it is known that there are 33 attributes that have values above 0.60. Thus 8 attributes with factor loading values below 0.60 are eliminated (BL7, BL10, EWOM8, EWOM10, EWOM11, EWOM14, EWOM22, and EWOM23) and 33 attributes have met the validity requirements of the

model. The 33 attributes are then used to measure the reliability of the model by analyzing construct reliability (CR) and average variance extracted (AVE). All CR and AVE scores meet the requirement where CR more than 0.7 and AVE more than 0.5.

Table 1. Construct Reliability and Average Variance Extracted

Variable	CR	AVE
FGC	0.868	0.626
Brand Awareness	0.915	0.643
Brand Loyalty	0.922	0.602
E-WOM	0.924	0.506
Purchase Intention	0.912	0.775

Based on the results of the path diagram of the research model, all values of standardized loading factors on all indicators in latent variables: FGC, BA, BL, and EWOM have values above 0.50. Thus, all indicators of each latent variable are valid and can be used to form structural model.

Structural equation analysis (t-value) was conducted to see how the t-value of the causal relationship between latent variables used in the research model and the results of the research hypothesis testing. Table 2 is a figure of research hypothesis testing.

Based on the results on Table 2, it can be concluded that that firm generated content on social media has a positive influence on brand awareness. As competitors continue to grow in various industries, consumers

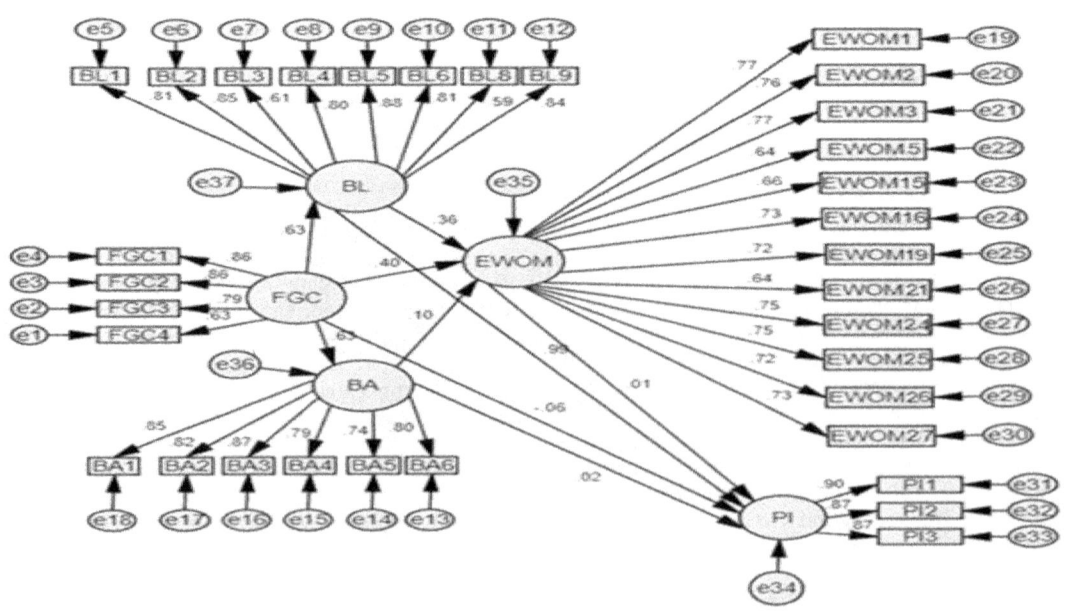

Figure 2. Path diagram.

416

Table 2. t-value analysis.

Path	t-value	Acceptance	Conclusion
FGC → BA	6.814	Significant	H1 accepted
FGC → BL	6.909	Significant	H2 accepted
BL → E-WOM	4.667	Significant	H3a accepted
BA → E-WOM	1.952	Not Significant	H3b unaccepted
FGC → E-WOM	3.978	Significant	H3c accepted
FGC → PI	−1.037	Not Significant	H4a unaccepted
BL → PI	−0.083	Not Significant	H4b accepted
BA → PI	0.228	Not Significant	H4c unaccepted
E-WOM → PI	12.416	Significant	H4d accepted

have a variety of products from various companies and become more complex in making choices. With the presence of social media content, the company can help consumers keep in mind the company's presence among other competitors.

Firm generated content has a positive influence on brand loyalty. Attractive social media content can help consumers stay loyal to the company's brand. For consumers who have good experience, with product information on social media, consumers are reminded by the presence of the product regardless of the competitors. Companies need to improve their social media content to be more interesting and up-to-date to gain customer's attention.

Brand loyalty and firm generated content have a positive influence on e-WOM. As long as consumers feel loyal to the company, they will provide information about their good experiences and provide tips and advice to other consumers through electronic media on social media. And then, as long as the company's social media content attracts, fulfills consumer expectations, and up-to-date consumers feel that the company's social media is effective to be a place to communicate with the admin or company and other consumers to discuss the company's social media content. Brand awareness does not have a positive relationship with e-WOM consumers. Based on the analysis of the research, this can occur because the company is already well known and has become a top of mind consumers.

E-WOM has a positive influence on purchase intention. When consumers want to make a purchase, they are looking for information about products from product providers and other consumers' experiences. Thus, when consumers do and receive e-WOM positively about a brand, they have a positive influence on themselves and other consumers to make purchase intentions.

Firm generated content, brand loyalty, and brand awareness do not have a positive influence on purchase intention. Based on the analysis of the research, this can occur because the company is already well-known and has become a top of mind consumers. So that when consumers know the existence of a company among other competitors, it does not affect consumers to have purchase intentions. And, because many competitors are investing heavily in terms of product promotion and consumers are sensitive to prices, if product promotion information is not as attractive as competitors, this can affect consumer loyalty to the product and move to competitors. In terms of firm generated content, it needs to provide interesting information related to products and promotions that are beneficial to consumers because many companies participate in promotions strategy through social media.

4 CONCLUSION

This study concluded there is a significant positive effect of firm-generated content on brand awareness, brand loyalty, and e-WOM. Then there is a significant positive effect of brand loyalty towards e-WOM, and e-WOM on purchase intention. This research provides an overview for companies to find out how the company's social media content influences consumer buying decisions. There are some managerial implications that can be used as company evaluations for future marketing strategies: Firm generated content is the right marketing communication strategy to maintain consumer loyalty and so consumers can keep in mind the OTA's presence among various online travel agencies in Indonesia. Companies need to develop and update their social media content to meet consumer expectations, so consumers feel informative when looking at corporate social media and feel it's value. Consumers can share their experiences with other consumers and provide advice to other consumers who need information related to OTA. Interactive communication with consumers through social media is needed by responding consumer message more quickly, so consumers believe that the operator able to help them. The company needs to make investments in increasing customer loyalty to OTA so that consumers are more actively communicating on OTA's social media and share their experiences. Overall the level of brand loyalty of consumers in OTA is high but does not have a positive effect with purchase intention. Companies need to be aware there are many OTAs in the market and doing price war to gain sales and profit. Thus, companies need to reevaluate their marketing strategies to increase consumers purchase. Company needs to make an investment to increase e-WOM activities so that it can increase consumers' desire to buy tickets or book hotels in OTA. E-WOM activities that are often used by consumers include communication through OTA's social media on Youtube, Twitter, Instagram, and Facebook. With interactive e-WOM communication on social media, consumers share their experience and can give influence to other consumers in making purchases at OTA.

This study still has limitations. Thus, the researcher propose suggestion that can be used as references for further research. User generated content variables can be involved by further research to measure the effect of consumer purchase intention. Thus, researchers can see a comparison of the effectiveness of communication with social media based on firm generated content and user generated content.

REFERENCES

Aaker, D.A. 1996. Measuring brand equity across products and markets. *California Management Review* 38(3): 102–120.

Bailey, A.A. 2004, Thiscompanysucks.Com: The use of the internet in negative consumer to consumer articulations. *Journal of Marketing Communications* 10(3):169–182.

Balakrishnan, B.K., Dahnil, M.I., & Yi, W.J. 2014. The impact of social media marketing medium toward purchase intention and brand loyalty among generation Y. *Procedia Social and Behavioral Sciences* 148: 177–185.

Barreda, A.A., Bilgihan, A., Nusair, K., & Okumus, F. 2015. Generating brand awareness in online social networks. *Computers in Human Behavior* 50: 600–609.

Belch, G. E. & Belch, M. A. 2018. Advertising and promotion: an Integrated Marketing Communication perspective. New York: McGraw-Hill Education.

Bruhn, M., Schoenmueller, V., & Schäfer, D.B. 2012. Are social media replacing traditional media in terms of brand equity creation?. *Management Research Review* 35(9): 770–790.

Ceballos, M., Crespo, Á.G., & Cousté, N.L. 2016. Impact of firm created content on user generated content: using a new social media monitoring tool to explore Twitter. *Rediscovering the Essentiality of Marketing*: 303–306.

Chi, H.K., Yeh, H.R. & Yang, Y.T. 2009. The impact of brand awareness on consumer purchase intention: the mediating effect of perceived quality and brand loyalty. *The Journal of International Management Studies* 4(1): 135–144.

Erdogmus, I.E. and Cicek, M. 2012. The impact of social media marketing on brand loyalty. *Procedia Social and Behavioral Sciences* 58: 1353–1360.

Hutter, K., Hautz, J., Dennhardt, S., & Füller, J. 2013. The impact of user interactions in social media on brand awareness and purchase intention: the case of mini on facebook. *Journal of Product & Brand Management* 22(5–6): 342–351.

Kim, A.J. & Ko, E. 2012. Do social media marketing activities enhance customer equity? an empirical study of luxury fashion brand. *Journal of Business Research* 65(10): 1480–1486.

Monica, B. & Balas, R. 2014. Social media marketing to increase brand awareness, Journal of Economics and Business Research, Vol. 20 No. 2, pp. 155–164

Poulis, A., Rizomyliotis, I., & Konstantoulaki, K. 2018. Do firms still need to be social? firm generated content in social media. *Information Technology & People* 32(2): 387-4-4

Rodriguez, M., Peterson, R.M., & Krishnan, V. 2012. Social media's influence on business tobusiness sales performance. *Journal of Personal Selling & Sales Management* 32(3): 365–378.

Sasmita, J. & Mohd Suki, N. 2015. Young consumers' insights on brand equity: effects of brand association, brand loyalty, brand awareness, and brand image. *International Journal of Retail & Distribution Management* 43(3): 276–292.

Schivinski, B. & Dabrowski, D. 2016. The effect of social media communication on consumer perceptions of brands. *Journal of Marketing Communications* 22(2): 189–214.

Tolba, A.H. (2011), The impact of distribution intensity on brand preference and brand loyalty. *International Journal of Marketing Studies* 3(3): 56–66.

Tuten, T. L. & Solomon, M. R. 2017. Social media marketing. New Jersey: Sage.

Yoo, B. & Donthu, N. 2001. Developing and validating a multidimensional consumer based brand equity scale. *Journal of Business Research* 52(1): 1–14.

Yoo, B., Donthu, N., & Lee, S. 2000. An examination of selected marketing mix elements and brand equity. *Journal of the Academy of Marketing Science* 28(2): 195–211.

Advances in Business, Management and Entrepreneurship – Hurriyati et al. (Eds)
© 2021 Taylor & Francis Group, London, ISBN 978-0-367-67471-7

Determinant of service value and its implication on public trust: A study on the social security agent members

S. Mufattahah, I. Primina, Sucherly & W.O.Z. Muizu
Universitas Padjadjaran, Bandung, Indonesia

ABSTRACT: The purpose of this study is to find out, analyze, and examine the effect of service operations and service delivery, both simultaneously and partially, on the value of health insurance services in West Java, the effect of service operations and service delivery, both simultaneously and partially, on the trust of service users of Health Insurance Services in West Java, the influence of service value on trust of service users through the value of Health Insurance Services in West Java, and the effect of service operations and service delivery on service users' trust, through the value of Health Insurance services in West Java. The method used in this study is a literature study, which is to find relevant references to cases or problems found. The results of the study show that service operations and service delivery, both simultaneously and partially, influence the value of health insurance services in West Java, service operations and service delivery, both simultaneously and partially, affect the trust worthiness of health insurance service users in West Java, the value of services influences the trust of service users through the value of health insurance services in West Java, and service operations and service delivery affect the trust of service users through the value of health insurance services in West Java.

1 INTRODUCTION

Indonesia is the fourth most populous country in the world, after China, India, and the United States. Due to the large population, Indonesia has become very retarded in terms of health problems (BPS-SUPAS 2015). Therefore, the government seeks to help the Indonesian people with these health issues. In the 1945 Constitution, it is affirmed that "Every person has the right to live a healthy life and has the right to receive health services (Article 28 H number (1))" and "The state is responsible for the provision of adequate health service facilities and public service facilities" (9 34 points (3)). In this constitution, Law No. 40 of 2004 concerning the National Social Security System (SJSN) has been promulgated, in Article 19 paragraph (1) it is affirmed that, "Health insurance is held nationally based on nationally based principles of social insurance and the principle of equity."

West Java is the most populous province in Indonesia, becoming one of the barometers in the implementation of health services by Health Insurance Organizing Agency (BPJS) along with data on the number of health care facilities such as health centers and hospitals, number of hospitals according to managers, number of hospitals, beds, and bed ratio per 1,000 residents, the number of beds in the hospital according to the treatment class, the number of distribution facilities in the field of pharmacy and medical devices, and the number of human health data sources.

There is a tendency of low trust amongst BPJS participants, as seen from the lack of hospital services for BPJS participant patients, the bad BPJS reputation in the eyes of the public and health facilities, and the lack of consistency in providing services to BPJS participants. The low trust of BPJS health participants is allegedly caused by the value of BPJS services, which have an impact on health facilities and their services are relatively low, this can be seen from the long treatment queues, the fact that BPJS coverage is too tenuous, BPJS patient halls are often full, their work systems are slow, and the verification of slow action claims is homework for BPJS Kesehatan.

Kantsperger & Kunz (2010) state that different employee service inconsistencies might erode the trust of service users. This standard helps to ensure that service users are served fairly and of high quality. Furthermore, the perceived intelligence in service performance is important. If service success is caused by external factors (i.e., luck), the company cannot benefit from its performance. Therefore, employees must be trained and prepared for tasks outside of their daily business. This provides an opportunity to build trust based on credibility in what is called the moment of truth. This was also highlighted by Park et al. (2012).

The poor value of BPJS Health services and the low trust of BPJS Health service users tend to be caused by inappropriate service delivery; this can be seen from the system and referral flow of BPJS and the fact that their hospitals are still less effective and simple, service applications BPJS health are still less attractive to be used by BPJS participants, so information about BPJS health is still relatively minimal, and there are several studies regarding the lack of effectiveness of the BPJS health program.

Then in the later stages, namely with regard to service user acceptance of service operations, several studies have been carried out in various industries, including those by Grönroos 2016; Hsu 2013; Johnson 2012; Kumar et al. 2011; Lovelock & Wright 2011; Park et al. 2012; Szu et al. 2015). The latter researcher found the role of demography and healthy social and life construction as determinants of service delivery to the public.

Furthermore, studies of service operations extend to secondary issues, contexts, and subjects. Service value studies from the perspective of operations management are among others developed by Davis & Vollmann (1990) and Chen et al. (1994). They support the findings of Kwan et al. (1988), who criticize consumer waiting periods in fast-food restaurant services and their implications for customer satisfaction. Similar studies on similar subjects were conducted by Verma et al. (1999), but with different companion issues, namely quality, cost, delivery, and flexibility. In addition, studies of operational service delivery were carried out on the aspects of electronic-based service value (e-service operation). These studies studied e-service optimization efforts (Ghosh et al. 2003) and e-service determinants (Ghosh et al. 2004).

Meanwhile, many studies have been carried out on this issue. Kantsperger & Kunz (2010) examine the determinants of confidence of service users, but with different determinant factors (not the delivery and management services) and using different approaches (QUAL). Similar studies that focus on determinants of service provision include the study of Jarvenpaa et al. (1999), which uses cross-cultural variables, and raises the issue of involving public funds in measuring service operations in a comprehensive manner.

Likewise, improper service operations are alleged to be the cause of the low value of services and trust of BPJS Health participants; this can be seen from the health facilities of BPJS service users, where Health is relatively poor, health equipment provided by health facilities for Health BPJS participants are incomplete, so must improve services in a more complete way that is not guaranteed by BPJS Health, as well as the response of BPJS employees themselves, as well as less responsive health facilities.

Based on the above, it is very important and crucial to do research about how the services and service delivery operations are having an effect on the value of the services, as well as the confidence of the participants of BPJS health services in West Java.

2 RESULTS AND DISCUSSION

The increasingly complex needs of affordable health services and those provided by the state require attention, not only on affordability, but also on the quality of health insurance services. Therefore, the level of accessibility and the existence of a health insurance service unit are very important.

The effectiveness of providing health insurance services is determined by the economic conditions and infrastructure in local service units. The better the economic condition of a local area, the better the activities of the health insurance service units in the region. The labor factor will also determine the high competitiveness of the health insurance service unit. Labor costs, the availability of trained medical workers, and medical training facilities will be the main determinants of a service office in determining its service location.

2.1 Relationship between service operation variables with service value variables

Quality in the health insurance service unit industry is very sensitive in influencing trust in reusing certain services. Kantsperger & Kunz (2010) state that different employee service inconsistencies might erode the trust of service users. This standard helps to ensure that service users are served fairly and of high quality. Furthermore, the perceived intelligence in service performance is important. If service success is caused by external factors (i.e., luck), the company cannot benefit from its performance. Therefore, employees must be trained and prepared for tasks outside of their daily business. This provides an opportunity to build trust based on credibility in what is called the moment of truth.

2.2 Relationship between service delivery variables and service value

Good service will basically form an impression on service users due to the excellent process of providing services. From the impression that has been formed, it will enable the company to have different and best service values that are felt by service users, so that companies can use those things to increase the number of service users who want to use company services in the future. The management of the quality value of a service is closely related to service delivery and quality provided, so a strategy that can be taken in winning competition with business competitors is by delivering high-quality services consistently compared to competitors and higher than the expectations of service users. To avoid failure in the delivery of services, it needs to be considered.

According to Brady & Cronin (2001), if the exchange between sacrifice and service quality is the same, then a service can be said to have value or be valuable according to consumer perceptions where value is not always positive, sometimes the relationship between sacrifice and service quality does not meet consumer expectations so that the value created will be negative.

Kettinger & Smith (2009) found that service values and satisfaction can moderate the relationship with service quality. Service value significantly affects the intention to continue to use system services. The study is in line with that of Naeem et al. (2009), who found that a service can be valuable because of the quality of services provided that have an impact on repurchase.

2.3 Relationship between service value variable and variable of user trust in services (customer trust)

The better the quality of services provided by the company and perceived by service users, the better the value of services received by service users, and the higher the level of trust of service users, on the other hand, if service users feel the quality of services provided by the company is not good and is going down, the trust of service users will also decrease.

In carrying out activities, companies must formulate and implement the best strategies in accordance with the company's resources and competencies in facing competition. One alternative strategy that can be developed by a company is a service strategy for consumers.

Service strategies can be implemented at the division level and emphasize service improvements in specific industries or market segments served by the division. Service strategies also emphasize improving competitive quality so that service strategies should also integrate various functional activities to achieve the objectives of the division.

Service in the health insurance service unit is an activity of service due to several activities and functions of the health insurance service unit that support smoothness, security, order traffic or traffic flow, safeguard sailing safety, and intra- or intermodal transfer sites (Gurning et al. 2007) so that these activities do not produce products that are physical.

Confidence of service users is an important factor that should be cultivated and maintained by a company. Whether or not a company develops is highly dependent on the level of trust of service users of the company; the higher the level of trust in service users of the company, the younger it will develop.

The study is in line with that of Naeem et al. (2009), who found that a service can be valued because of the quality of services provided that have an impact on repurchases, which indicate trust in a product or a particular service; besides, Aldlaigan & Buttle (2011) identify that there is a positive and significant correlation between the dimensions of service and overall satisfaction.

To achieve the trust of good corporate service users, a good service process is needed and to maintain the process, it is necessary to study the variables that are determinants in the success of achieving the trust of the company's service users.

Many of these important variables will be examined by researchers, including several internal and external variables that influence customer trust, namely service quality, service value, operational services, and service delivery.

2.4 The relationship between the variable of operation services and the variable of user trust in services

Properly managed services will produce good quality services and potentially satisfy service users. Satisfied users of services will in turn trust the service provider. Even if consistently managed services are properly received by users, long-term trust will arise among service users.

2.5 Relationship between service delivery and user trust

Service delivery must meet the criteria that allow service users to build trust in the reliability of service delivery. Service users need services in the appropriate form, quality, and appropriate momentum (Park et al. 2012). If services are delivered in a form, quality, and momentum that is not right, service users will feel disappointed. In certain proportions, the disappointment of the service user will prevent the use or repurchase of services.

2.6 Theoretical model

Based on existing theories and previous studies as in the literature study, it is proven that service operations, service delivery, service quality, and service value will have an impact and influence on service users' trust; the research paradigm is in accordance with Figure 1. Where each of these variables will be defined in the variable operational definition.

The theoretical studies that have been carried out will be used to compile the Structural Equation Modeling (SEM) of trust of service users in the health insurance service unit in Indonesia, where the variables that influence it have been identified, namely Service Management, Service Delivery, and Operational services. The variables described the causality relationship into an SEM diagram. The researcher then compiled a list of questions that will be used to collect

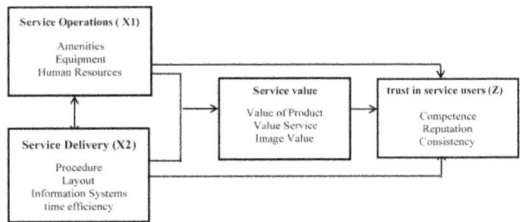

Figure 1. Research paradigm.

data. The collected data will then be tested for reliability and validity by taking a portion of the sample. If the results are valid, then data retrieval is continued until all samples are collected. Data are then used to generate model estimates. Significant variables can then be analyzed and used to create the LCC company's marketing strategy to increase its retention of service users.

3 CONCLUSION

1. Service operations and service delivery, both simultaneously and partially, affect the value of Health Insurance Services in West Java
2. Service operations and service delivery, both simultaneously and partially, affect the trust of health insurance service users in West Java.
3. The value of services affects the trust of service users through the value of health insurance services in West Java.
4. Service operations and service delivery affect the trust of service users through the value of health insurance services in West Java.

REFERENCES

BPS-SUPAS. 2015. Penduduk Indonesia hasil survei penduduk antar sensus 2015. Jakarta: Badan Pusat Statistik.

Brady, M. K. & Cronin Jr, J. J. 2001. Some thoughts on conceptualizing perceived service quality: A hierarchical approach. *Journal of marketing* 65 (3): 34–49.

Chen, I. J., Gupta, A., & Rom, W. 1994. A study of price and quality in service operations. *International journal of service industry management.*

Davis, M. M. & Vollmann, T. E. 1990. A framework for relating waiting time and customer satisfaction in a service operation. *Journal of Services Marketing.*

Ghosh, S., Surjadjaja, H., & Antony, J. 2003. Determining and assessing the determinants of e-service operations. *Managing Service Quality: An International Journal.*

Ghosh, S., Surjadjaja, H., & Antony, J. 2004. Optimisation of the determinants of e-service operations. *Business Process Management Journal.*

Grönroos, C. 2016. Internationalization strategies for services: A retrospective. *Journal of Services Marketing.*

Gurning, G., Saut, R. O., & Budiyanto, E. H. 2007. Manajemen bisnis pelabuhan. Surabaya: PT Andhika Prasetya Ekawahana.

Jarvenpaa, S. L., Noam, T., & Lauri, S. 1999. Consumer trust in an internet store: A cross-cultural validation. *Journal of Computer-Mediated Communication* 5(2).

Kantsperger, R. & Kunz, W. H. 2010. Consumer trust in service companies: A multiple mediating analysis. *Managing Service Quality: An International Journal.*

Kettinger, W. J. & Smith, J. 2009. Understanding the consequences of information systems service quality on IS service reuse. *Information & Management* 46(6): 335–341.

Kumar, V., Batista, L., & Maull, R. 2011. The impact of operations performance on customer loyalty. *Service Science* 3(2): 158–171.

Kwan, S. K., Davis, M. M., & Greenwood, A. G. 1988. A simulation model for determining variable worker requirements in a service operation with time-dependent customer demand. *Queueing Systems* 3(3): 265–275.

Lovelock, C. & Wright, L. 2011. Wirtz services marketing: People, technology, strategy.

Naeem, H., Akram, A., & Saif, M. I. 2009. Service quality and its impact on customer satisfaction: An empirical evidence from the Pakistani banking sector. *International Business & Economics Research Journal (IBER)* 8(12).

Park, J., Lee, J., Lee, H., & Truex, D. 2012. Exploring the impact of communication effectiveness on service quality, trust and relationship commitment in IT services. *International Journal of Information Management* 32(5): 459–468.

Sun, S. Y., Huang, K. L., Scott, S. C., & Lee, C. P. 2015. The influence of service value and service quality on the continuance adoption of SNS. *International Journal of Electronic Business Management* 13.

Verma, R., Thompson, G. M., & Louviere, J. J. 1999. Configuring service operations in accordance with customer needs and preferences. *Journal of Service Research* 1(3): 262–274.

Advances in Business, Management and Entrepreneurship – Hurriyati et al. (Eds)
© 2021 Taylor & Francis Group, London, ISBN 978-0-367-67471-7

Consumer-brand relationship of courier service users: The role of brand experience as an antecedent of CBR

A.M. Rezza, R. Hurriyati, Disman & L.A. Wibowo
Universitas Pendidikan Indonesia, Bandung, Indonesia

ABSTRACT: This study examines how consumer-brand relationship (CBR) can be formed by increasing perceived brand experience of consumer. This research used cross-sectional survey with explanatory method and used structural equation modelling approach to test the hypothesis. A questionnaire was sent to 200 courier service consumers in West Java. The result indicates that brand experience is a good predictor of consumer-brand relationship especially to brand-self distance variable. Experiencing the brand is indeed prerequisite for developing customer-brand relationship over time.

1 INTRODUCTION

Today, the courier service industry in Indonesia is experiencing a positive increase. It is estimated that by 2020, this industry will grow by 15.4% with a value of Rp. 4,396 Trillions (ARF 2018). The factors that influence the growth of this industry are in large part due to the increase in e-commerce activities in Indonesia which in 2013 alone reached a valuation of 94.5 trillion (Depkominfo 2016) and are expected to grow 8-fold by 2020 (CNBC & Kusumajaya 2018).

The growth potential of e-commerce businesses that will continue to increase in the future will also affect the positive growth of the courier service industry in Indonesia because e-commerce activities will not be separated from shipping activities. Shipping activity is one of the activities in the order fulfillment stage, where consumers now see the fast delivery time as one of the important values for e-commerce companies. Consumers comprehend fast delivery time as one of the added values for e-commerce companies (Lapeyrolerie & Salnas 2019).

The increasing shipping activity as a result of the increase in online business activities makes the business of courier services more attractive, and this can be seen from the increasing number of companies engaged in courier activities in Indonesia. According to the data from the Association of Indonesian Express Delivery and Logistics Services Companies (Asperindo), in 2016 there were around 277 companies engaged in express, post and logistics shipping services in Indonesia and this number is likely to be higher because not all courier service companies have entered Asperindo membership (Prahadi 2016).

More and more companies that enter the industry will also increase the business competition in it. Therefore, there is a need for a strategy in order to maintain their business and also create a competitive advantage.

Several studies on consumer-brand relationships (CBR) stated that benefits from creating strong long-term relationships with their customers are repeat purchase, price tolerance, consumer advocacy, and the creation and also sustaining of competitive advantages (Aggarwal & Shi 2018) and the antecedent for this concept could come from brand commitment, attachment, trust, and love (Giovanis & Athanasopoulou 2017).

The consumer-brand relationship is a concept in marketing that is currently being explored by researchers around the world. Fournier first introduced this concept in 1998, and until now, studies of this concept continue to develop. This concept emerges as a consequence of the notion that consumers not only distinguish brands based on what they get from the brand but also from the relationships created between them and the brand. There is also a change in ideology from transactional relations to fostering long-term relationships (Sreejesh & Mohapatra 2013).

Research of consumer-brand relationship continues to grow until now and led to several theoretical paradigms in CBR themes such as Brand Relationship Quality (Fournier 1998), brand commitment (Morgan & Hunt 1994), brand attachment (Whan Park et al. 2010) and brand love (Batra et al. 2012). This diversity in conceptualization and operationalization of the CBR paradigm reflect the vigorousness of theoretical construction, which needs to be identified (Tsai 2011).

Park et al. (2013) introduced a relatively new concept in CBR; attachmentaversion model (AA), which emerges from the notion of relationship, is not only talking about positive feeling (that produces brand attachment) but also negative feeling (that produce brand aversion). Brand attachment and brand aversion represent opposite ends of the relationship spectrum at any point in time, while the transition from one end to the other is also possible over time

(Johnson, et al. 2011). AA model has two components: brand self distance and brand prominence.

Brand self distance is operationally defined as the perceived distance between a brand and the self and Brand prominence is operationally defined as the perceived memory accessibility of a brand to an individual (Park et al. 2013).

Schmitt (2013), stated that brand experience is an important determinant of customer-brand relations. The brand exists in the mind of the consumer as the consequences of consumer experience with the brand. Positive or negative experience inevitably will produce an image of the brand in the consumer mind that will also become a relationship base of a consumer with the brand. Experiencing the brand may be a prerequisite for developing a customer-brand relationship over time (Schmitt 2013).

There are only a few researches in CBR themes that are related to the AA model and also its relevance with brand experience. Thus, it makes this research is very important to further testing of the model to predict consumer's psychological consequences measures of close feeling and emotional valence, and also to test whether this model can fit with brand experience as its antecedent.

This research focused on the AA model as a relatively new theory in CBR themes and also like to explore the construction of CBR from the perspective of brand experience.

2 METHODS

2.1 Data collection

The research was conducted using online survey to collect the data. The target population of this study consists of all consumers who used courier service and was conducted in April to June 2019. In total, there were 250 respondents but only 147 respondents met the research requirements. The sample characteristics indicated that 59.9% were woman. The majority of respondent (44.2%) belong to below 20 age group. Meanwhile, 70% of the respondent have 1–3 years of experience with the courier service brand.

2.2 Measure development

The measurement was adopted from previous studies. Respondents were asked to rate each of the 31 items on a 7-point Likert scale. Brand experience was measured with twenty six items based on Brakus et al. (2009) and attachment-aversion model was measured with five items based on Park et al. (2013).

2.3 Data analysis

Structural equation modelling (SEM) using IBM AMOS 22.0 software was used for data analysis. We estimated our hypothesized measurement model using confirmatory factor analysis (CFA).

3 RESULTS AND DISCUSSION

3.1 Confimatory Factor Analysis (CFA)

We estimated our hypothesized measurement model using confirmatory factor analysis (CFA). Conceptual models with 31 indicators showed that the model acceptable fit with the data (CFI = 0.832; SRMR = 0.067; RMSEA = 0.113). We made some modification to improve fit indices and come up with the 2nd model (CFI = 0.952; SRMR = 0.056; RMSEA = 0.079). Table 1 reports the standardized item loadings, reliability estimates, and average variance extracted (AVE) scores. All item loadings are >0.5, suggesting adequate convergent validity (Hair et al. 2014). They are further supported by AVE scores that are greater than 0.50 (Fornell and Larcker 1981). Cronbach's Alpha and Composite Reliability estimates were above 0.80 for all constructs, indicating acceptable reliability.

Table 1. Reliability and validity estimates of first-order constructs.

1st order constructs & items	loading	α	CR	AVE
Sensory		0.8	0.9	0.6
bexs1	0.8			
bexs2	0.851			
bexs6	0.819			
bexs7	0.851			
Affective		0.9	0.9	0.7
bexa8	0.784			
bexa7	0.873			
bexa5	0.902			
bexa4	0.876			
bexa3	0.838			
Intellectual		0.9	0.9	0.8
bexi3	0.86			
bexi2	0.919			
Behavior		0.9	0.8	0.7
bexb1	0.857			
bexb2	0.825			
Brand Prominence		0.9	0.9	0.8
aabp1	0.904			
aabp2	0.906			
aabp3	0.888			
Self-brand distance		0.9	0.9	0.8
aabs1	0.921			
aabs2	0.832			

3.2 Model fit

The examination of the model fit was conducted using AMOS 24. The Cut off criteria for fit indexes in covariance structure analysis used fit index combination of Comparative Fit Index (CFI) and the Standardized Root Mean Square Residual (SRMR) (Hu & Bentler 1999). The expected value CFI is 0.96 or higher and

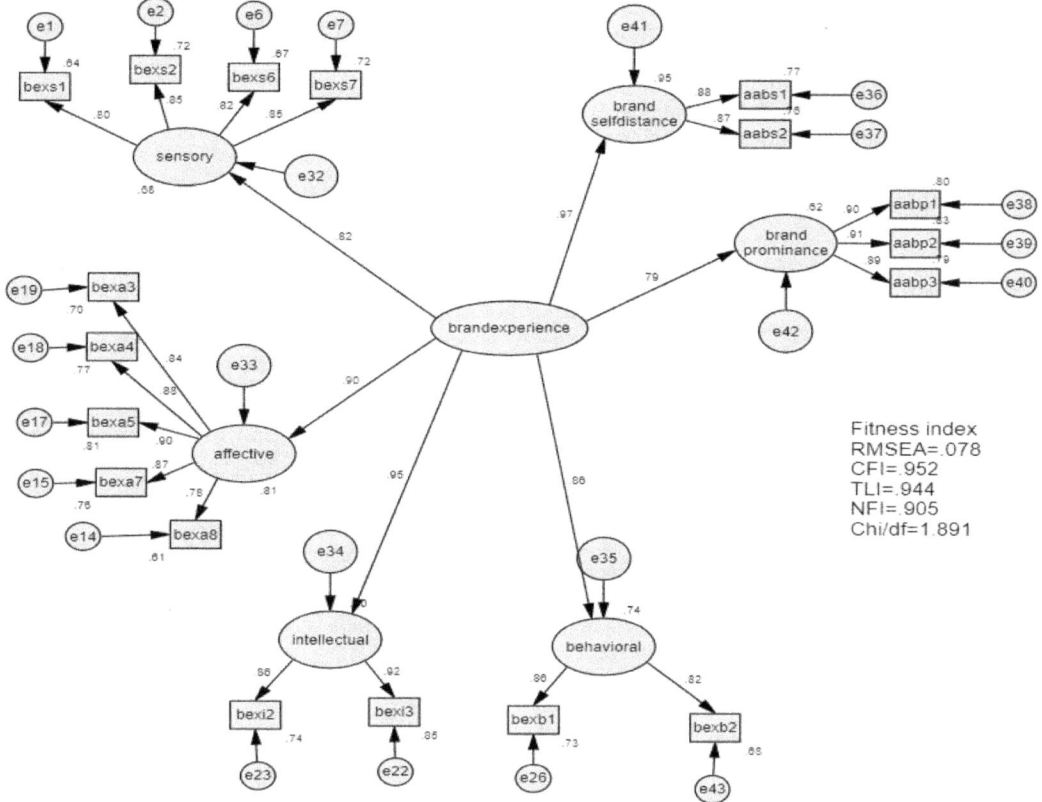

Figure 1. Research diagram

Table 2. Model fit indicators.

Measure	Estimate	Threshold	Interpretation
CFI	0.952	>0.95	Excellent
SRMR	0.056	<0.08	Excellent

the expected value for SRMR is 0.09 or lower. Table 2. reported the result, that the values are satisfactory and suggest an adequate fit of the model.

Figure 1 shows the result of structural model analysis. It also demonstrates significant and positive influences of brand experience on brand-self distance and brand prominence.

3.3 Hypotesis evaluation

Hair et al. (2014) states that the path coefficients values range from −1 to +1, where the closer the values to +1, it means the relationship between the two constructs is stronger, and the closer the values to −1, it indicates that the relationship is negative. For the significance seen from P Values whose value is smaller than 0.005.

The result of the path analysis (Table 3) show that brand experience (BEx) has a positive and significant impact on brand-self distance (b = 1.074; p < 0.05) supporting hypothesis H1. Brand experience (BEx) has a positive impact on brand prominence (b = 0.894, p < 0.005) supporting hypothesis H2.

Table 3. Regression weight.

	Est.	S.E	C.R	P Value
BEx → BSD	1.074	.103	10.423	***
BEx → BPr	.894	.102	8.782	***

This study proposes and test the path between brand experience to the development of customer-brand relationship. It shows that brand experience can enhance the relationship between brand and the customer through brand-self distance and brand prominence. The result of this study shows that the model has a good measurement fit, as all the value of the constructs proved to have good values that confirm their reliability and validity. As for the structural model the results are flawless. All the research hypotheses were confirmed during the analysis. Experiencing the brand is indeed prerequisite for developing customer-brand relationship over time. The result also confirms that experiencing process occurs before relational assessment which consistent with the result of Schmitt (2013) and Park et al. (2013). Paradigm of experience is currently dominant for courier service brand. These significant impact is in line with other research result like Trudeau and Shobeiri (2016) that also showed the significant effect of brand experience on customer-brand relationship. Shamim et al. (2016) shows that brand experience play an important role in increasing the tendency of customers to interact and dialogue with service providers in the contexts of retailing service. In other word, good brand experience makes the customers feel closerto the brand and potentially can make deeper and stronger relationship with it as well.

4 CONCLUSION

This study develops and empirically tests a conceptual model of attachment-aversion as CBR concept. The result showed that brand experience is good predictors of CBR. It positively and significantly affects brand-self distance and brand prominence. The influence of brand experience to the consumer-brand relationship formation is in line with the results of other research in customer-brand relationship and the result of this research can validate it.

REFERENCES

Aggarwal, Pankaj, & Shi, M. 2018. Monogamous versus polygamous brand relationships. *Journal of the Association for Consumer Research* 3(2): 188–201.

ARF. 2018. Industri logistik Indonesia bernilai Rp. 4.396 T di 2020. [Online]. Retrived from www. Motoris.id https://www.motoris.id/industri/industri.

Batra, Rajeev, Ahuvia. A., & Bagozzi R.P. 2012. Brand love. *Journal of Marketing* 76(2): 1–16.

Brakus, Josko J., Schmitt, B.H., & Zarantonello L. 2009. Brand experience: what is it? how is it measures? does it affect loyalty?. *Journal of Marketing* 73: 52–68.

CNBC, & Kusumajaya, E. 2018. Menilik jasa pengiriman. [Online]. Retrieved from https://www.cnbcindonesia.com/news/20181127154856-8-43917/e-commerce-dorong-per tumbuhan-jasa-pengiriman.

Depkominfo. 2016. Jasa logistik melesat di era E-Commerce. [Online]. Retrived from https://kominfo.go.id/index.php/content/detail/6707/Jasa+Logistik+Melesat+di+Era+e-Commerce+/0/sorotan_media.

Fornell, Claes, & Larcker, D.F. 1981. Evaluating structural equation models with unobservable variables and measurement error. *Journal of Marketing Research* 18(1): 39–50.

Fournier, & Susan. 1998. Consumers and their brands: developing relationship theory in consumer research. *The Journal of Consumer Research* 24(4): 343–73.

Giovanis, Apostolos, N., and Athanasopoulou, P. 2017. Consumer brand relationships and brand loyalty in technology mediated services. *Journal of Retailing and Consumer Services*: 126–36.

Hair, J.F., Black W.C., Babin, B.J., & Anderson R.E. 2014. Multivarieate data analysis. Pearson Education Limited.

Hu, L.T., & Bentler P.M. 1999. Cutoff criteria for fit indexes in covariance structure analysis: conventional criteria versus new alternatives. *Structural Equation Modeling: A Multidisciplinary Journal* 6(1): 1–55.

Johnson, Allison, R., Matear, M., & Thomson, M. 2011. A coal in the heart: self-relevance as a post-exit predictor of consumer anti-brand actions. *Journal of Consumer Research* 38(1): 108–25.

Lapeyrolerie, S., & Salnas, E. 2019. Is The Battle For Fastest Shipping worth fighting?. *alixpartners.com*. Accesed June 2019.

Morgan, Robert, M., & Hunt S.D. 1994. The Commitment-trust theory of relationship marketing. *Journal of Marketing* 58(3): 20–38.

Park, Whan, C., Eisingerich, A.B., & Park, J.W. 2013. Attachment – Aversion (AA) model of customer brand relationships. *Journal of Consumer Psychology* 23(2): 229–48.

Prahadi, Y.Y. 2016. Ketum asperindo terpilih, ini janji presdir JNE." *swa.co.id*. Accesed February 2017.

Schmitt, B. 2013. The consumer psychology of customer – brand relationships: extending the AA relationship model. *Journal of Consumer Psychology* 2: 249–52.

Shamim, Amjad, Ghazali, Z., & Albinsson, P.A. 2016. An integrated model of corporate brand experience and customer value co-creation behaviour introduction. *International Journal of Retail & Distribution Management* 44(2).

Tsai, S.P. 2011. Fostering international brand loyalty through committed and attached relationships. *International Business Review* 20(5): 521–34.

Whan Park, C. et al. 2010. Brand attachment and brand attitude strength: conceptual and empirical differentiation of two critical brand equity drivers. *Journal of Marketing* 74(6): 1–17.

Advances in Business, Management and Entrepreneurship – Hurriyati et al. (Eds)
© *2021 Taylor & Francis Group, London, ISBN 978-0-367-67471-7*

The effect of service and brand image on customer value and its implications on visitor trust (a study in the Bandung metropolis area)

R.D. Pertiwi & D.K. Priatna
Universitas Winaya Mukti, Bandung, Indonesia

ABSTRACT: This study examines the trustworthiness of the management of historic buildings in the Bandung metropolis by assessing the performance of service experience, brand image, and customer value. This study uses (1) Service Experience received by visitors, (2) Brand Image according to visitors, (3) Customer Value, (4) Visitor Trust, (5) the amount of influence Brand Experience and Service Image have on Customer Value, either simultaneously or partially, and (6) the amount of influence Customer Value has on Visitor Trust, as variables. This study uses descriptive survey and explanatory survey methods. The nature of the research is descriptive and verification research. Primary data were distributed to a sample of 375 respondents. Respondents had come to visit the heritage building. Descriptive analysis is done by tabulating data for the average value category, for verification analysis using Structural Equation Modeling (SEM) and processing using Lisrel 8.7. The feasibility of the tested model is carried out through the conformity criteria with theoretical logic, accuracy of parameter estimates, explanatory abilities, and predictability. The results show that (1) the Service Experience that has been received by visitors to the Heritage building had an average answer of 3,221, so it was included in the fairly good category, (2) Heritage Brand was stated to be quite good, (3) Customer Value had an average of 3.349,which puts it into the fairly good category, (4) the Heritage Visitor was in the average good category, (5) Brand Service and Image Experience significantly influenced Customer Value on visitors both simultaneously and partially and Service Experience gave a greater influence than the Brand Image of Customer Value, and (6) Customer Value had a significant effect on Visitor Trust.

1 INTRODUCTION

The existence of relics and culture in Bandung Metropolis in the form of buildings, places, or regions is a city heritage that must always be maintained properly in order to maintain its authenticity and can be a very supportive history for the community and the country.

The existence of heritage buildings in Indonesia is regulated based on the legislation concerning Heritage preservation expressed as follows:

1. Law No. 28 of 2002 is concerning Buildings (UUBG) was ratified on December 16, 2002, consisting of 10 Chapters and 49 articles, regulating the provisions of buildings, which includes the requirements of building, Function Building, and Organizing Building Buildings.
2. Law Number 5 of 1992 is concerning Cultural Heritage objects, namely to emphasize the protection and preservation of Historic Development.
3. Law No. 11 of 2010 is concerning Cultural Heritage Objects, Cultural Heritage Buildings, or Cultural Heritage structures.

4. Presidential Regulation No. 78 of 2007 is concerning Heritage.
5. Minister of Public Works and Public Housing Regulation No. 01/PRT/M Year 2015 was effective from 24 February 2015, specifically regulating the preservation of Cultural Heritage Buildings.

The existence of heritage buildings in Bandung is currently not optimal in accordance with the designation of the buildings themselves, as expressed by the Chairperson of Bandung Heritage Aji Bimarsono in Bandung, (Thursday, 07/11/2013), that "the function of heritage buildings is a place of entertainment that accentuates rah-rah, actually has a negative effect on the morality and creative power of the citizens of Bandung City."

The use of cultural heritage buildings for positive arts/cultural activities makes people more inspired and is more meaningful than activities that lead to mere entertainment. Although the heritage building has survived, its human attitude has deteriorated, thus, the cultural ministry is considered incomplete. Its formation values are not only physical buildings, but nonphysical as noble values.

The existence of Heritage in Bandung Metropolitan in the form of buildings, places, or regions, is a wealth of city and county heritage and must always be guarded and maintained properly so that it still maintains its authenticity and becomes history, which is very valuable for the Community and the State.

Data on Heritage buildings in Bandung are grouped into various regions. The data regarding the area of Heritage buildings in the Bandung Metropolis area includes (1) Region I (City Center), (2) Region II (Chinatown/Trade), (3) Region III (Defense and Security/Military), (4) Region V (Villa and non-Villa Housing), (5) Region IV (Ethnic Sundanese), and (6) Region VI (Industry).

The presurvey revealed that visitors showed a lack of trust in management in carrying out management, caring to maintain the sustainability of the Heritage building in Bandung as a valuable historical heritage, and would not return. Thus, this is necessary research to find out what causes visitor to have less Caring for our society's historical heritage.

In the context of the understanding discussed in this study regarding customer trust, this research shows that if management gives its promise to maintain, manage, and preserve the historical heritage in the city of Bandung, consumers will also participate in doing what is done by the management itself.

The presurvey revealed that visitors showed a lack of loyalty to the Heritage building in Bandung as a valuable historical heritage site and would not return. Thus it is necessary to maintain management to find out what causes visitors or even the public to be loyal to the inheritance of this history.

The presurvey revealed that visitors showed a very low benefit for the existence of a Heritage building in Bandung as a valuable historical heritage and will not return. Thus research is needed to find out what causes this or even the community to care less about this historical legacy.

The presurvey revealed that visitors showed a lack of a very high image of the existence of a Heritage building in Bandung as a valuable historical heritage and would not return. Thus this is necessary research to find out what causes visitors or less people to Care for the inheritance of this history.

The presurvey revealed that visitors showed a lack of very high service experience in the presence of the Heritage building in Bandung as a valuable historical heritage and would not return. Thus this is necessary research to find out what causes a visitor or less people to Care for the inheritance of this history.

Communication performed by employees with the consumer must be able to attract and give a positive impression to be able to retain the services provided. Besides that, good communication will make consumers share their experiences while visiting historical tours. This communication is not only done directly by consumers with companies, but can be done through the internet, brochures, and advertisements that are displayed in electronic media or print media.

Based on the background above, the researchers are interested in examining heritage, especially heritage buildings as study loci, then researchers pour into the research title "The effect of service and brand image on Customer Value and Implications on Visitor Trust (A Study in the Bandung Metropolis area)."

2 METHODS

The research method used in this research is a descriptive explanatory survey. Descriptive surveys are conducted to get an overview of the variables under study. While the explanatory survey is performed to obtain a picture of the causal link between the variables studied through hypothesis testing based on the data obtained in the field.

The technique used in this study is a cross section, namely research carried out over a period of time, carried out on various Heritage buildings in the Metropolis Bandung area. The type of research used is descriptive and verification, meaning that the researcher tries to test the answers to problems whose truth is temporary (hypothesis) based on empirical data.

There are four main variables studied in this study, namely (1) Service experience, which is the total functional and emotional value of services consumed. There are unique service experiences for each individual customer and service consumption situation. The value used is an evaluation of cognitive service experience (Sandstrom et al. 2008). (2) Brand Image (Brand Image), which is the level of recognition of a brand by consumers, such as the introduction of logos, taglines, product designs, and other things as the identity of the brand. (3) Customer Value (Customer Value), which is a Comparison between Benefits with Sacrifice (Benefits: Product, Service, HR, and Image) (Sacrifice: money, time, energy, and psychology). (4) Trust (Customer Trust) is trust as a business relationship dimension that determines the rate at which people feel they can depend on the integrity of promises offered by others. This is basically a belief that someone will give what is promised (Morgan & Hunt 1994, Barnes 2001). Position of Service Experience and Brand Image variables are used as Independent variables and Customer Value, Intervening and Trust variables are used as Dependent variables.

3 RESULTS AND DISCUSSION

Based on the results of processing with the LISREL 8.72 program, the measurement model (CFA) for each variable and indicator relationship is shown by the loading factor of each indicator as follows:

3.1 Service experience

Using the Service Experience Variable, which is formed by six dimensions, namely Incentives,

Accessibility, Comfort, Benefits, Environment, and Trust, the following results are obtained:

Table 1. Results of analysis of variable measurement model X1.

Item	Loading Factor Variable Service Experience	T-value	R2	Error
$X_{11,\xi1}$	0.8633	20.6073	0.7452	0.2548
$X_{12,\xi1}$	0.7813	17.6743	0.6104	0.3896
$X_{13,\xi1}$	0.7714	17.3489	0.5950	0.4050
$X_{14,\xi1}$	0.8642	20.6431	0.7468	0.2532
$X_{15,\xi1}$	0.7125	15.5195	0.5077	0.4923
$X_{16,\xi1}$	0.6483	13.7004	0.4203	0.5797

Source: LISREL program data processing results.

Based on the results of the processing data shown in Table 1 by using SEM analysis, the results obtained show that the Benefits have the largest loading factor value compared to other dimensions, which is equal to 0.8642. This shows that the benefits are the biggest forming factor of Service Experience for Respondents; this indicates that the Respondents really hope that by visiting these Heritage buildings they will get great benefits, which include knowledge of the past or past life as a reflection of life in the future, therefore, the ability of HR Management of the Historic Buildings (Heritage) in Bandung Metropolis is required to be better so that the preservation of this history can be maintained.

3.2 Brand image

The Brand Image variable, which is formed by four dimensions, namely Introduction, Reputation, Attraction, and Loyalty, obtained the following results:

Table 2. Results of analysis of variable measurement model X2.

Item	Loading Factor Variable Brand Image	T-value	R2	Error
X21,$\xi2$	0.7697	16.8650	0.5924	0.4076
X22,$\xi2$	0.8147	18.3120	0.6637	0.3363
X23,$\xi2$	0.7811	17.2223	0.6101	0.3899
X24,$\xi2$	0.6828	14.3122	0.4663	0.5337

Source: LISREL program data processing results.

Based on the results of the processing data in Table 2 using SEM analysis, the results obtained show that Reputation has the largest loading factor value compared to other dimensions, which is equal to 0.8147. This shows that Reputation is the biggest factor forming the variable Brand Image. A good reputation possessed by a product of the work of institutional managers shows that managers have good

performance; this indicates that good reputation is the superior result from the Manager of the Historic Building (Heritage) in Bandung Metropolis.

3.3 Customer value

The Customer Value variable is formed by four dimensions, namely Product Benefits, Service Benefits, Personal Benefits, and Citrader's Benefits obtained as follows:

Table 3. Results of analysis of variable measurement models Y.

Item	Loading Factor Variable Customer Value	T-value	R2	Error
Y1,$\eta1$	0.7723		0.5965	0.4035
Y2,$\eta1$	0.6892	13.5694	0.4750	0.5250
Y3,$\eta1$	0.5701	10.9794	0.3251	0.6749
Y4,$\eta1$	0.5027	9.5792	0.2527	0.7373

Source: LISREL program data processing results.

Based on the results of processing data using SEM analysis, the results obtained that the Product Benefits has the largest loading factor value compared to other dimensions, which is equal to 0.7723. This shows that Product Benefits are the biggest factor forming the Customer Value variable; this indicates that the Product Benefits that contribute greatly to building value is a good reputation possessed by a product produced by an institutional manager showing that managers have good performance. This indicates that Great Product Benefits is the superior result from the Management of the Historic Building (Heritage) in Bandung Metropolis and is an advantage of the Management of the Historic Building (Heritage) in Bandung Metropolis.

3.4 Customer trust

The Customer Trust variable is formed by three dimensions, namely Ability, Integrity, and Virtue as follows:

Table 4. Results of the variable Z measurement model analysis.

Item	Loading Factor Customer Trust	T-value	R2	Error
Z1,$\eta2$	0.7478		0.6299	0.4408
Z2,$\eta2$	0.7780	21.7910	0.5902	0.3947
Z3,$\eta2$	0.8129	21.0086	0.5550	0.3392

Source: LISREL program data processing results.

Based on the results of processing data using SEM analysis, the results obtained show that the Virtue dimension has the largest loading factor value compared to other dimensions, which is equal to 0.8129.

This shows that the Benevolence dimension is the biggest factor forming the Customer Trust variable, this indicates that the virtue that contributes greatly to building trust is a good reputation owned by a company showing that managers have good performance; this indicates that the great virtue is the superior result from the Manager of the Historic Building in Bandung Metropolis and is an advantage of the Management of the Historical Building (Heritage) in Bandung Metropolis.

Nevertheless there are still a number of factors that have not been optimal and can be called weaknesses in the customer trust variable with the smallest value of loading factor, namely the ability dimension, with the value of loading factor equal to 0.7478. This indicates that the ability dimension of the Manager of the Historical Building (Heritage) in Bandung Metropolis is not optimal.

Table 5. Direct and indirect variable influences of experience in brand services and image towards customer values and trust.

| | Path coefficient | Direct Influence | Influence Through | | Total |
			Service Experience	Brand Image	
Service Experience	0.6486	42.07%		14.03%	56.10%
Brand Image	0.2665	7.10%	14.03%		21.13%
Total	49.17%	14.03%	14.03% Service Value	77.23% Trust	
	Path coefficient			Direct Influence	
Service Value	0.9132			83.39%	

Source: Data processing results (2018).

Based on Table 5 above, the Value of Customers is influenced by direct and indirect influences. The direct effect of the Service Experience variable is 42.07%, while the indirect effect through the Brand Image in sequence is 14.03%. The direct effect of the Brand Image variable on Customer Value is 7.10%, while the indirect effect through Service Experience sequentially is 14.03%.

Based on the results of the calculation of the partial effect of the largest part is the Service Experience variable with a total influence on the Customer Value of 56.10%. So the conclusion can be drawn that to increase Customer Value, efforts must be supported by the existence of a good Service Experience. Nevertheless the influence of other variables that influence the Customer Value studied in this study is also quite large, namely that of Brand Image, with a total influence partially of 21.13%.

3.4.1 Simultaneous hypothesis testing

Based on the calculation, the value of F amounted to 621.2288, where the criteria for rejection of H_0 is if F is larger than F_{table} or $F_0 > F_{table}$, with degrees of freedom of $v1 = 2$ and $v2 = 375 - 2 - 1$ and the level of trust at 95%, then from the distribution of the Ftable, the F_{table} value for $F_{0.05}$ is $2.375 = 3.020$. With 621.2288 being greater than 3.020, H_0 is rejected, meaning that it can be concluded that there is a linear relationship between Service Experience and Brand Image against Customer Value, or it can be interpreted that there is a joint effect between Service Experience and Brand Image on Customer Value.

3.4.2 Partial testing of hypotheses

3.4.2.1 Partial influence of service experience on customer values:

The partial influence of the Service Experience variable (X1) on Customer Value (Y) needs to be tested statistically, then the statistical hypothesis is as follows:

Ho: $\gamma_1 = 0$. There is not a significant influence of Experienced Service on Customer Value.

Ha: $\gamma_1 \neq 0$. There is an influence of Service Experience on Customer Value.

Criteria for rejecting H_0 are if t_{count} is greater than t_{table} or $t_0 > t_{table}$, with $df = 375 - 2 - 1$.

Table 6. Partial test results of service experience with customer value.

Structural	Path coefficient	t-count	t-table	Conclusion
$\gamma 1$	0.6486	8.0204	1.9663	H_0 is rejected, there is an influence of y 'sg significant of Experience Service on Customer Value

Source: Data processing results (2018).

The path coefficient X_1 against $Y = 0.6486$ obtained the value of $t_{arithmetic}$ of 6.8626 by taking the significance level α of 5%, then the value of t_{table} or $t_{0.05.734} = 1.9632$; so because $t_{count} = 6.8626$ is greater than $t_{table} = 1.9632$, then H_0 is rejected, or in other words, the Service Experience takes effect towards the Customer Value of 0.6486 so that every increase in Service Experience will increase the Customer Value by 0.6486 units.

3.4.3 Influence of customer value process towards customer trust

The structural model 2 illustrates the relationship between Customer Values against Customer Trust, which is stated in the hypothesis as follows: Customer Values Influence Customer Confidence. Based on the results of the LISREL program data processing for structural model 2, according to the hypothesis, the following results are obtained:

$$Z = 0.9132 * Y, \text{Errorvar.} = 0.05288, R^2 = 0.8339$$
$$(0.06812) \quad (0.03444)$$
$$14.2872 \quad 1.5354 \tag{1}$$

Based on the equation above, it can be explained that the Customer Confidence variable positively affected by the variable of Customer Value with a path coefficient of 0.9132, meaning that if the Customer Value increases, customer confidence will be increased by the path coefficient that is equal to 0.9132, or any increase of Customer Value will contribute to an increase in Customer Trust by 0.9132 units.

Thus the proposed conceptual hypothesis has been tested and accepted. The complete structural models for substructure 2 can be described as follows:

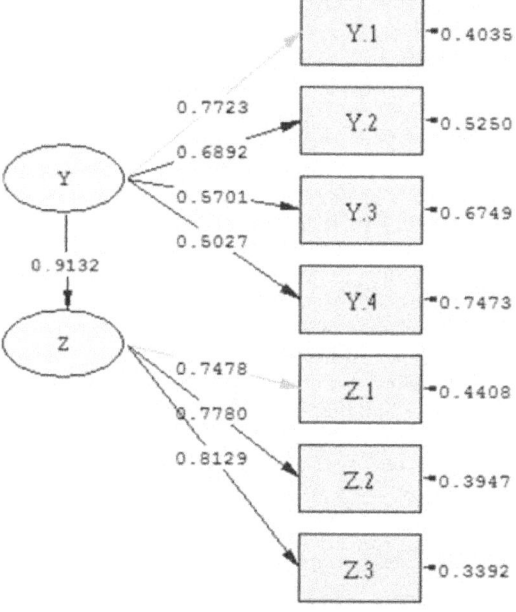

Figure 1. Customer value path coefficient against customer trust.

3.4.3.1 Hypothesis testing
The effect of the variable Customer Value (Y) on Customer Trust (Z) needs to be tested statistically, thus, the statistical hypothesis is as follows:

Ho: $\beta_1 = 0$. There is not a significant influence of Customer Value against Customer Trust.
Ha: $\beta_1 \neq 0$. There is a significant influence of Customer Value against Customer Trust.

Test criteria: Reject H_0 if t count is greater than t table or t Calculate > t table, with $df = 375 - 2 - 1$.

Table 7. Partial test results of customer value against customer trust.

Structural	Path coefficient	t-count	t-table	Conclusion
β_1	0.9132	14.2872	1.9663	H_0 is rejected, there is an influence of y 'sg significant of Customer Value on Customer trust

*Source: Results of data processing (2016).

For the path coefficient Y to $Z = 0.9132$, obtained by value $t_{arithmetic}$, amounted to 14.2872 by taking a significance level α of 5%, then the value of $_{table}$ or $t_{0.05,375} = 1.9663$; so because $t_{count} = 14.2872$, which is greater than $t_{table} = 1.9663$, H_0 is rejected, or in other words, the Customer Value effect on customer confidence is 0.9132, so any increase in Customer Value will increase Customer Trust by 0.9132 units.

3.4.4 Model feasibility testing
The results of the model feasibility test show that the research model meets the criteria of the goodness of an econometric model or characteristics that can be expected and described as follows:

3.4.4.1 Theoretical plausability
This research model shows that the test results are in accordance with their expectations and Marketing management theories that form the basis of the study by studying the influence of Brand Service and Image Experience on Customer Value and their Implications on Customer Trust.

Table 8. Model suitability test results.

Inter-Variable Relationships	Pre-estimation	Post estimation	Suitability
Service Experience with Customer Value	+	+	Corresponding
Effect of Brand Image on Customer Value	+	+	Corresponding
Influence of Customer Value on Customer Trust	+	+	Corresponding

3.4.4.2 Accuracy of the estimates of the parameters
This research model produces an accurate or unbiased and significant path coefficient estimator. The assumption of analysis is fulfilled and the probability of statistical errors from the model is very low (p-value = 0,000) or below the level of significance set at 0.05 for all hypotheses.

3.4.4.3 Explanatory ability
This research model has a high ability in explaining the relationship between the phenomena of management

variables studied. Standard Error (SE) is smaller than 1/2 times the absolute value of the path coefficient (SE $< 1/2 \rho$).

a. Hypothesis Test 1
There is an influence of the Brand Service and Image Experience on Customer Value SE. Service Experience $= 0.08087 < 1/2 (0.6486)$.
SE Brand Image $= 0.07633 < 1/2 (0.2665)$.

b. Hypothesis Test 2
There is an influence of Customer Value on Customer Trust.
SE Customer Value $= 0.06812 < 1/2 (0.9132)$.

3.4.4.4 Forecasting ability
This research model has a high predictive ability for the behavior of the dependent variable as indicated by the high coefficient of determination of the model that approaches or exceeds 50% with the following details:

a. The influence of Service Experience and Brand Image against Customer Value amounted to 77.23%.

b. The influence of Customer Value on Customer Trust amounted to 83.39%.

Thus it can be concluded that the compiled model meets the goodness of econometric criteria based on a strong theoretical perspective, so that it can contribute to the development of science and policy or problem solving.

4 CONCLUSION

In accordance with the results of the research and discussion, the research conclusions are as follows:

1. Visitors' responses about the variable of Service Experience received by visitors had an average answer of 3.221 and a standard deviation of 0.437. This finding is stated to be in a fairly good category, indicating that the Service Experience for visitors is said to not be optimal. There are several indicators that have a value below the average, namely (a) usability of history for visitor Historical Building (Heritage), (b) increased knowledge for Visitor Historical Building (Heritage), (c) Increased experience for visitor Building Historic (Heritage).

2. The visitors' responses to the variable of Brand Image had an average value of 3,201 and a standard deviation of 0.517, thus, it was included in the fairly good category; this indicates that the Brand Image is said to not be optimal. Other findings are that there are some indicators that have a value below the average, namely (a) the characteristic color of Historical Building (Heritage), (b) the brand name of Historical Building (Heritage), (c) the risk and the dangers that can arise from Building Historical (Heritage).

3. The visitors' responses about the Customer Value variable had an average value of 3.349 and the standard deviation of 0.714 categorizes this variable

in the fairly good category; this indicates that the Customer Value can be said to be not optimal. Other findings are that there are some indicators that are still below the average value, namely (a) psychological sacrifice issued in proportion to the service, and (b) money sacrifice spent comparable to service personnel in Building Heritage.

4. The visitors' responses about the Customer Trust variable had an average value of 3.222 and a standard deviation of 0.542, thus, it is categorized as quite well; this indicates that the Customer Trust in the management of the Heritage Building in the Bandung Metropolis Region is not optimal. The findings show that there are several indicators that are still below the average value, namely (a) Honesty in explaining the existence of Heritage Building, and (b) Quality assurance of building maintenance by management of the Heritage Building.

5. Brand Service and Image Experience have a significant effect on Customer Values in Bandung Metropolis Region, both simultaneously and partially, and Service Experience has a greater influence than Brand Image on Customer Value.

6. Customer Value has a significant effect on Visitor Trust in the Bandung Metropolis Area.

7. Conclusion after that with Expert Judgment:

a. The relevance between the findings and the results of empirical research with the opinion of the Expert Judge results are consistent with no crucial differences of opinion. But even so, according to the Expert Judgments, the service experience must be more optimally received by the visitors of the management can explain and tell the history of the existence of the Heritage building and optimize the function of the Heritage building so as to provide satisfaction to visitors.

b. The relevance of the findings of the empirical research with the opinion of the judgment experts is that there is no significant difference. According to them, the heritage brand image needs to be improved. The officers need to improve their knowledge and innovate services and information better, because brand image is a factor supporting the achievement of customer value in order to increase public trust.

c. The relevance between the findings of empirical research and the opinion of the expert judgment results are consistent and there are no very important differences. According to Expert Judgments, to get customer value that contributes greatly to visitor trust, it requires the honesty and openness of managers in conveying a lot of history and knowledge in explaining the existence and function and benefits of heritage building in Bandung Metropolis accompanied by quality care, security guarantees, and maintenance and preservation of cultural heritage and historic buildings (Heritage).

d. The relevance between the findings of empirical research and the opinion of the expert judgment results are consistent and there are no very important differences. However, in the opinion of the experts, to gain the trust of customers or visitors, the managers must convey the value of the historical value and the value of the benefits of the superior Heritage building to the visitors so that a high trustworthiness of visitors can be obtained in a greater manner.

8. Conclusions from understanding the Expert Judgment: The results of the study can reveal the compatibility between the proposed research plan answered in the results of the study and supplemented by discussions supported by theory and previous research and reinforced by the opinions of Expert Judgments.

REFERENCES

Barnes, J. G. 2001. Secrets of customer relationship management: It's all about how you make them feel. New York: McGraw-Hill.
Morgan, R. M. & Hunt, S. D. 1994. The commitment-trust theory of relationship marketing. *Journal of Marketing* 58: 20–38.
Sandstrom, S., Edvardsson, B., Kristensson, P., & Magnusson, P. 2008. Value in use through service experience. *Managing Service Quality* 18(2): 112–126.

The effects of store atmosphere and sales promotion toward impulsive purchase in supermarkets in Karawang District

I. Ratnasari & D. Kusnanto
Universitas Singaperbangsa Karawang, Karawang, Indonesia

ABSTRACT: This research studied the influence of store atmosphere and sales promotion on impulsive purchase in a supermarket in Karawang District. The sample in this study were 258 respondents using the Slovin formula with an error rate of 5%. The analysis employed the path analysis. The results showed that there was a partial and simultaneous influence of store atmosphere and sales promotion on impulsive purchase. The hypothesis test showed positive and significant results. It means store atmosphere and sales promotion has an influence on impulsive purchase.

1 INTRODUCTION

Retailing includes all activities in selling goods or services directly to end consumers for personal and non-business needs. Retailers or retail stores are all business entities whose sales volume mainly comes from retail sales. This retail business is a business that is quite popular in this country because retail business is one of the businesses that is quite profitable for Indonesian businesses. Released from the business webpages, Asosiasi Pengusaha Ritel Indonesia (Aprindo) has released data that project the growth of retail industry for up to 10% until the end of 2019. This figure is higher than last year's realization claims, which were in the range of 8% to 8,5% (September 14, 2019). Therefore, more and more business people who want to try their luck in the retail business. The emergence of many retail businesses, especially in the supermarket sector, becomes evidence of the growing retail business in this country. Almost in every area, both in urban area and in villages, many new retail shops are found. This indicated that the retail business is growing rapidly in Indonesia.

One kind of retail that is currently growing is supermarkets. Supermarkets are increasingly popular in Indonesia. It can be seen from the data released by (www.duniaindustri.com) that modern trade is in the fast moving consumer goods (FMCG) category in Indonesia with an average growth of 6,6% in the period of April 2018 to April 2019, with the highest growth occurring in the minimarket segment by 12,1% and super/hypermarkets by 6,8%. Sales of modern stores per capita in Indonesia are estimated at US$60 with a composition of 56% in minimarkets and 44% in super/hypermarkets. It shows

that the existence of retail businesses, especially in supermarkets in Indonesia, shows positive results for retail entrepreneurs.

Karawang is one of the districts in West Java Province. Karawang is also a popular city for transmigrations because there are many industrial factories in Karawang. So, that is one of the reasons for retail businesses to make a fortune by opening a retail business in Karawang. At present, there are four of the largest supermarkets in Karawang, those are Carrefour, Giant, Transmart, and Superindo with an average of 1000 customer visits per day. The competition to win consumers is quite visible. Consumers are treated to a large selection of products when shopping at supermarkets. Supermarkets provide all the customer needs. A neat arrangement is also one of the factors why consumers choose to shop at supermarkets, thus, consumers do not need to be bothered to move store to store for daily needs whether it's planned needs or unplanned needs. Speaking of unplanned needs, in the economic world, there is such a thing as the impulsive purchase term.

Based on the results of preliminary research of supermarket consumers in Karawang conducted by researchers, it can be concluded that at least 24% of consumers buy unplanned products. Although it is only 24%, impulsive purchases still contribute to the rate of supermarket income. It is indeed a consumptive human nature that causes these unplanned purchases to occur. However, it is not only consumptive behavior that is the sole factor in impulsive purchase. The strategy created by the store can be the cause of impulsive purchase. For instance, the store owner who arranges their shop/store so neat and attractive, so that consumers, when shopping, will subconsciously buy things that they don't really need or are not on their shopping list. Commonly, it is called the Store Atmosphere.

Beside the store atmosphere factor, sales promotions can also be one of the reasons for consumers to make unplanned buying. Sales promotions are designed to stimulate faster or larger purchases of products or services made by consumers. Sales promotions include samples, coupons, cash refund offers, discounts, premiums, gifts, patronage awards, free trials, guarantees, related promotions, cross promotions, point of purchase displays, and demonstrations. It is hoped that sales promotion activities can increase impulsive purchase at supermarkets in Karawang. According to Hidayat (2016), store atmosphere and sales promotion partially and simultaneously influence impulsive purchase. As well as the result conducted by Aini (2016) that store atmosphere and sales promotion have a positive effect on impulsive purchase. It means that if supermarkets in Karawang district want to increase unplanned buying, a comfortable store atmosphere and attractive sales promotion must be created.

2 METHODS

The research method used was descriptive and verification methods. To find a description of each variable, a descriptive method was used with a scale range analysis tool. Meanwhile, to test hypotheses with statistical calculations, the verification method is used with a path analysis tool.

Operationalization of variables used for store atmosphere variable refers to the instruments used by Syafitri (2016), namely the arrangement of light, music, air regulation, layout, display, room color, and aroma. Whereas to operationalize the variable of sales promotion the researcher referred to the instrument used by Hidayat (2016), that was pull strategies that include samples, coupons, contests, rebates and discounts, and bonuses. Then to operationalize the variable of impulsive purchase the researcher referred to the instruments used by Taufiqi (2016), which consisted of spontaneity, strength, compulsion, and intensity, excitement, and stimulation of ignorance of the consequences.

The population in this study were costumers who did shopping at the four biggest supermarkets in Karawang, namely Carrefour, Giant, Transmart, and Superindo. Then a sample of 258 guests were taken. The sampling technique employed incidental accidental sampling, which is a sampling technique based on coincidence, that is, anyone who incidentally met with the researcher could be used as a sample. In addition, the researchers also used cluster sampling only for those who did shopping in the four biggest supermarkets in Karawang, namely Carrefour, Giant, Transmart, and Superindo, who shopped at least once a month, so that the results of filling the sample did not appear to be playing or writing.

The validity of the data was tested using a test of validity, reliability, and normality. In the validity test, the instrument is declared valid or considered eligible if the value of the coefficient rcount $\geq 0,300$, while in the reliability test, the instrument was declared reliable if the coefficient alpha > 0.600. Then, in the normality test, the instrument is declared normal distribution if $Z > 0.05$ and assymp. value 2 tailed > 0.05.

Descriptive analysis consisted of descriptive statistical analysis such as frequency distribution, probability distribution, and normal distribution. Descriptive analysis uses ordinal scale and range of scales to analyze data by describing the atmosphere, sales promotion, and impulsive purchase of supermarket consumers in Karawang. To find out the relationship between independent variables, then Spearman correlation analysis was used. Meanwhile, to analyze partial and simultaneous influence, a path analysis tool is used. The independent (exogenous) variables in this study were store atmosphere and sales promotion, while the dependent variable (endogenous) in this study was impulsive purchase.

For partial hypothesis testing, the t-test was used. T-test is a test used to determine the closeness of the influence between the independent variables (X1, X2) with the dependent variable (Y) partially, or the influence of the independent variables that are smaller or equal to the dependent variable (Sugiyono 2001). The amount of α used in this study is 10%.

To test the effect simultaneously between the independent variables and the dependent variable, the F-test was used. F-test is a test used to determine the closeness of the effect between the independent variable (X) with the dependent variable (Y) together against the dependent variable (Sugiyono 2004).

3 RESULTS AND DISCUSSION

The results of the validity and reliability tests stated that all instruments used were valid and reliable with r count $>$ r critical. The results of the normality test stated that all variables used were normally distributed with a Z score > 0.05 and an assymp. value of 2 tailed > 0.05.

The descriptive analysis in this study showed that the Store Atmosphere (X1) in supermarkets in Karawang obtained a total score of 7557 with the highest indicator being Lighting Arrangement and the lowest indicator located was Product Display (2). Product Display was the lowest indicator because it displays the product on the floor, causing it to be seen untidy and making consumers felt less comfortable in shopping. Sales Promotion (X2) at supermarkets in Karawang obtained a total score of 6920 with the highest indicator being Rebates and Discounts (2), while the lowest indicator was Sample (2). Sample was the lowest indicator because supermarkets in Karawang are limited in providing samples in advance to consumers to provide experience prior to the buying. Impulsive Purchase

(Y) at supermarkets in Karawang obtained a total score of 7451 with the highest indicator being the Buyer Reaction (2) and the lowest indicator being the Impact of Massive Purchases. In this case, supermarket employees in Karawang are less adept at influencing consumers to make large-scale purchases without thinking about the effects that will occur to them if they make large-scale purchases.

Table 1. Simple linear regression model.

Coefficients

Model	Unstandardized coefficient		Standardized coefficient		
	B	Std error	Beta	t	Sig
(Constant)	−1.685	1.466		−1.149	.251
1 Store Atmosphere	.540	.035	.518	15.368	.000
Sales Promotion	.489	.034	.487	14.453	.000

a. Dependent variable: Impulsive purchase.

From Table 1, we knew that path coefficient of the store atmosphere was 0,518 and sales promotion was 0,487. Thus, we can calculate the direct and indirect influence of store atmosphere and sales promotion on impulsive purchase. The equation formula is:

$$Y = 0,518X1 + 0,487X2 + \gamma \qquad (1)$$

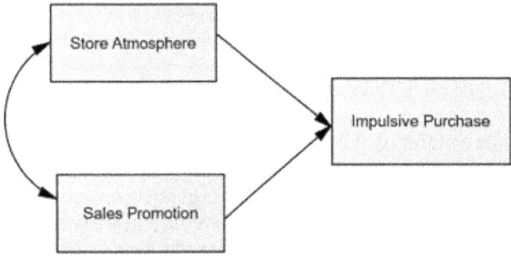

Figure 1. Research paradigm.

The influence of Store Atmosphere on Impulsive Purchase was 42.7%. This showed the positive influence between Store Atmosphere on Impulsive Purchase. This also showed the similarity with previous research conducted by Taufiqi (2016), who argues that Store Atmosphere has a positive influence on Impulsive Purchase, and Syafitri (2016) that store atmosphere has an influence on impulsive purchase.

The Influence of Sales Promotion on Impulsive purchase was 39.6%. This shows the positive influence between Sales Promotion on Impulsive purchases. This also shows the similarity with previous research

Table 2. The influence of sales promotion on impulsive purchase.

Variable	Path Analysis	Calculation	Scale of Influence
Sales Promotion	Direct influence Y	$0,487^2$	0,237
	Indirect influence Y	$0,487 \times 0,518 \times 0,632$	0,159
	Partially influence		0,396

conducted by Aini (2016), who argues that sales promotions have a significant and positive influence on impulsive purchase. It can be stated that the Store Atmosphere had more influence on impulsive purchase compared to Sales Promotions. This was because respondents perceived that Store Atmosphere affects Impulsive purchase.

The simultaneous influence of Store Atmosphere and Sales Promotion on Impulsive purchase with Sig test criteria. $0,000 < 0,05$ and f count (604,639) > f table (3,031) means that Ho is rejected. The total influence of Store Atmosphere and Sales Promotion on Impulsive purchase was 0.796. This showed that Store Atmosphere and Sales Promotion contributed to Impulsive purchase of 79.6%, while the rest are contributed by other variables that were not examined. Thus, proving that the Store Atmosphere and Sales Promotion of Impulsive purchase contribute simultaneously to each other. The results of this study showed similarities with previous research conducted by Syafitri (2016) that Store Atmosphere and Sales Promotion has a positive influence on impulsive purchase.

4 CONCLUSION

Based on the results of the research that has been conducted, the following conclusions are obtained:

a) There is a partial influence between Store Atmosphere and Sales Promotion on Impulsive purchase at supermarkets in Karawang District.
b) There is a simultaneous influence between Store Atmosphere and Sales Promotion on Impulsive purchase at supermarkets in Karawang District.

REFERENCES

Aini, Q. N. 2016. Pengaruh atmosfer toko dan promosi penjualan terhadap shopping emotion dan pembelian tidak terencana (survei terhadap konsumen Giant Mall Olympic Garden). [Online]. Retrieved from studentjournal.ub.ac.id.
Ekonomi bisnis. 2019. Meneropong prospek industri ritel modern. [Online]. Retrived from https://ekonomi.bisnis.com/read/20190904/12/144434/meneropong-prospek-industri-ritel-modern-hingga-2021.

Hidayat, E. 2016. Effect of store atmosphere, sales promotion, service quality and product quality on impulsive purchase. Essay. Jakarta: UIN Syarif Hidayatullaah.

Sugiyono. 2016. Quantitative, qualitative, and R&D research methods. Bandung: Alfabeta.

Syafitri, Rosyida. 2016. The effect of store atmosphere and sales promotion on impulsive purchases with positive emotions as intervening variables. *Journal of Research and Economics and Management* 16(1).

Taufiqi, M. 2016. Effect of store atmosphere on unplanned purchases and women's consumer satisfaction at Matahari Department Store Jember. Essay. Jember: Universitas Jember.

Advances in Business, Management and Entrepreneurship – Hurriyati et al. (Eds)
© 2021 Taylor & Francis Group, London, ISBN 978-0-367-67471-7

The behavioural analysis of #Racheluntukdonggala campaign from Rachel Vennya account on Kitabisa.com

I.I. Wahyuni, D.S. Fithrah & G.F. Adilla
Telkom University, Bandung, Indonesia

ABSTRACT: Kitabisa.com is a first-place crowd-funding platform that has successfully funded social projects. This study aims to identify the stimuli supporting the #Racheluntukdonggala campaign from Rachel Vennya's account on the Kitabisa.com as well as the donators' motives in choosing her account. This account collected a target donation of up to 101% or similar to Rp. 804,357,444 with 7,064 donors. This qualitative study employed the post-positivism paradigm as a research design. To collect data, the interviews were done to the three primary informants and one secondary informant, including a communication psychologist. The findings of this study present the development of customer behavior in supporting crowdfunding projects.

1 INTRODUCTION

On September 28, 2018, the natural disasters in Palu and Donggala, Central Sulawesi lead some crowdfunding platforms to conduct online donation. Kitabisa.com as one of the platforms aims to help people outside of Palu and Donggala to participate in the donation since it concerns on social activities, including donation and fundraising. In reality, campaigns in Palu and Donggala have the highest number of fundraisers compared to the other cities (Lombok, Banten, and Lampung). The initiative has 164 fundraisers while the other added just between 107–119 to fundraisers.

On the other side, an influential factor in platform usage is the support from leaders, public figures, or influencer who create campaign account. The crowdfunding platform connects donors and receivers by pursuing people who are interested to donate. Regarding this, it is very important to give a story effect by telling personally about the issues. The campaign from Rachel Vennya already invited 7,064 people that collected more than the specified target. However, from all campaigns in Kitabisa.com accounts to South Sulawesi, the biggest amount targeted was Rachel Ven. Also, Rachel Vennya received an award from Kitabisa.com because she often opened campaigns.

This study involved five most popular influencer accounts in Indonesia (see Table 1). Among these accounts, Rachel Venya had exceeded the specified target by involving 7,064 participants to collect the fund. By November 18, 2018 at 10:13 AM (West Indonesian Time), the target percentage was 10% to 300% ranging from 1,000–6,000 participants. This percentage showed that Rachel Venya collected the highest amount among other accounts.

Before conducting in-depth research, the researchers conducted pre-research. The pre-research results were the donors that donate to Kitabisa.com who could not help directly to the victims affected by the disaster, therefore they set aside a little of their sustenance to help the victims in Palu and Donggala. In addition, the donors making donations through their Rachel Ven account are following Instagram accounts owned by Rachel Ven. They see that this influencer made an invitation to donate through social media Instagram where

Table 1. Donggala's campaign account.

Campaign Accounts	Target Achieved (IDR)	Collected Funds (IDR)	Number of Donators	Target Percentage
Racheluntukdonggala	800,000,000	804,357,444	7,064	101%
Dwiuntukdonggala	500,000,000	363,948,634	2,414	73%
Ariptipangforsulteng	100,000,000	305,770,121	3,613	306%
Kitapedulidonggala	1,000,000,000	143,351,396	1,228	14%
Pedulipalubyateam	200,000,000	201,195,320	1,154	101%

social activities are carried out on the Kitabisa.com platform. Another reason for donors to donate through their Rachel Ven account is that they admire the personality of a kind-hearted Rachel Ven and also the content that is presented by Rachel Ven to encourage followers to donate. Therefore, the researcher wants to find out what the motives of the donors who want to donate have which will later be conveyed to the City of Palu and its surroundings through its Rachel Ven campaign account on the Kitabisa.com platform.

1.1 *Crowdfunding*

Crowdfunding is a collecting funds method aiming at connecting donors and recipients. Raising funds through this method is believed as an effort to attract people to contribute in the donation program. This is due to the growing numbers of the internet uses in Indonesia which brings an impact to the rising fund programs. There are several crowdfunding available in Indonesia, such as Kitabisa, Hand in Hand, Indiegogo, Kickstarter, Kolase (kumparan.com accessed on May 25, 2019 at 01:02 PM (West Indonesian Time)).

1.2 *Influencer*

Influencers are people who have a lot of followers on social media and have a big influence on their followers, such as artists, celebrities, YouTubers, bloggers, vloggers, and so on. They intend to inspire people or followers to do positive activities since it is believed as an effective promotional method substituting advertisement. In doing their job, they follow three main principles, namely Reach, Resonance, and Relevance (Solis 2012). An influencer posts to social media aiming at engaging their followers through like, share, and comment features. A large number of followers does not necessarily guarantee the success of the campaign. It highly depends on the resonance as the biggest factor of engagement between influencers and followers.

1.3 *The classification of motives*

1. The Curious Motive
The curious motive is the motive to understand, organizes, and predicts used by people to construct the meaning from the world. Such motive provides a reference frame to evaluate new situations and direct appropriate actions. People who are not aggravated in an ambiguous, erratic, or unpredictable atmosphere have the tendencies to give meaning. Meanwhile, due to the limited information, people tend to search for answer to draw conclusions without confirming the complete information.

2. The Competence Motif
Since people want to prove that they can overcome any life problems depending on intellectual, social, and emotional development, competency motives are closely related to the need for security. Being able to

love and be loved is essential for personality growth. The warmth of friendship, sincerity of affection, acceptance of other warm people is needed. Various studies have shown that the unfulfilled love will lead to poor human behavior; people will be aggressive, frustrated, and suicidal.

3. Self-Esteem and Personality identity motives
Self-esteem and personality identity motives show the ability to reveal someone's existence. We want our presence to be considered. Therefore, self-esteem support people to seek their identity. The loss of self-identity will lead to pathological behavior (disease): impulsive, restless, easily affected, and so on.

4. Self-fulfillment Needs
We do not only maintain life but also improve the quality of life. It improves life and potential quality. This is supported by Maslow who claimed that "What a man can be, he can be." The need for self-fulfillment is in the various forms, including:

1. Develop and use our potential creative ways, such as art, music, science, or things that encourage creative self-expression
2. Enriching the quality of life by expanding the range and quality of experience and satisfaction, e.g.traveling
3. Establish a meaningful relationship with other people around us
4. Trying to "humanize".
5. Trust

Trust is a cognitive component of socio-psychological factors. Trust can be realistic or irrational. Trust provides a perspective on humans in understanding reality, providing a basis for decision making and determining attitudes towards attitudinal goals. According to Solomon on Rakhmat 2015, trust is shaped by knowledge, needs, and interests. Knowledge is related to the quality of information. We have a lot of trust in incomplete knowledge.

1.4 *Customer-behavior motive*

Consumer behavior is an action that directly involved in obtaining, consuming, and spending products or services, including the process of decisions that precede and follow this action. In its development experts define consumer behavior as a process that occurs not only the process but also things that affected consumers' behavior, before, after making a transaction. While The American Marketing Association (AMA) defines consumer behavior as a dynamic interaction between affection and cognition, behavior, and environment where humans carry out exchange activities in their lives. In the field of science, consumer behavior learns how individual groups and organizations choose, buy, use, and utilizes goods, services, ideas, or experiences that aim to satisfy their needs and desires by marketing, psychological, socio-cultural, and social media behavior. The study of consumer behavior will be a very important basis in marketing management. The results of the study will help marketers to design marketing mixes, define segmentation, formulate positioning, and differentiate products, formulate an analysis of their business environment, and develop marketing research. Trust is a cognitive component of socio-psychological factors. Trust can be realistic or irrational. Trust provides a perspective on humans in understanding reality, providing a basis for decision making and determining attitudes towards attitudinal goals. According to Solomon on Rakhmat 2015, trust is shaped by knowledge, needs, and interests. Knowledge is related to the quality of information. We have a lot of trust in incomplete knowledge.

2 METHODS

The methodology is a framework that explains how the researchers think about the facts of social life and the science or theory treatment. The research paradigm also explains how the researcher finding the issue to solve the problem. The research needs a clear idea besides the paradigm. Paradigm is the basis of research beliefs that conducted the researchers. (Pambayun 2013). This research uses a post-positivist paradigm. The Post positivism previewed the probabilities issue.

Post positivism characteristics are reductionist, logical, empirical, causally-oriented, and deterministic. This paradigm is seen as a series of logically connected steps and believes rather than a single reality. It supported by precise and thorough data collection analysis (Creswell 2014).

3 RESULTS

Curious motives are motives that understand, organize, and predict. Everyone tries to understand and gain meaning from his world. We need a frame of reference to evaluate new situations and direct actions accordingly. People are impatient in ambiguous, uncertain, or difficult to predict situations. Because of the tendency to give meaning to what is experienced, if the information obtained is limited, people will find their answers. In this motive, people will immediately conclude without waiting for the information to be complete first.

The understanding of the curiosity motive above is based on data from a key informant named Nadia Afifatur Rahman, who said that he understood and knew information about the disasters that occurred in Palu and Donggala. He learned the information through one of the Instagram account owners named Rachel Vennya, but not only that he also knew the information from various media, such as the internet and Instagram accounts from other influencers. She stated that:"...

from Rachel Ven. I know directly from Instagram on Rachel Ven. Then I look for it from other sources, rich from the internet or Instagrams from other influencers too. "

By knowing the information obtained from her Rachel Ven Instagram account and some information obtained from other media, Nadia decided to donate through a campaign account owned by Rachel Ven. Nadia said that she was interested in donating through her Rachel Ven account because she believed in her Rachel Ven compared to other influencers. As Nadia said: "... its interesting, it can be said that there is an interest in raising funds, maybe there might be an interest, but if I look at Rachel Vennya influencers, it might be because I trust her more than other influencers ..." (Results of an interview with Nadia Afifatur Rahman, March 14, 2019)

Based on data from the second informant, Rory stated that he knew the information about the Palu and Donggala disasters began with the television. However, Rory got references or information from other sources, namely those who accidentally when Rory opened the Instagram app, precisely on the Instagram explore screen that spread information about the disaster. As Rory said below: "From that, I got the info from TV. Then, when I looked at the rooftop, I saw there a lot of news that spread about the disaster.

After getting enough information, Rory decided to donate through the Kitabisa.com platform. Rory told his friends that if you want to donate, you should go through the Kitabisa.com platform and a campaign account owned by Rachel Ven. The reason Rory decided to donate through his Rachel Ven campaign account was that he trusted his Rachel Ven more than anyone else. There is a fear that Rory has when donating through other influencer accounts. As Rory said: "... at that time I was gathering with my friends like that, and in discussing the disaster that was the Palu and Donggala disaster, then I had an idea what if the donation was like that. , I was playing Instagram and I saw the instastory, Rachel, and I said 'this way later, if I wanted to donate', then my friend said 'through which', then I answered 'through Kitabisa' then my friends say 'okay, let's donate' there. " Besides that, the third informant who knew information about the disaster that occurred in Palu and Donggala through Twitter. On Twitter, Tara gets a website link that is shared, so she knows and the disasters in Palu and Donggala. Tara also seeks information from Instagram social media: "What is clear from social media, but if I'm not mistaken about Twitter. Actually, it is as amazing as mine, the first time if I am not mistaken it was given a link too, so I opened the link, then started opening and opening Instagram also seeing people posting it "(Adellyn Dwitara Ramadhanty interview result, March 22, 2019).

The curios motive supported the donators to share the information about the disasters that occurred in Palu and Donggala through various media, one of them being Rachel Vennya Instagram account.

Being able to love and be loved is essential for personal growth. People want to be accepted into the group as voluntary members. The warmth of friendship, the sincerity of affection, warm acceptance of others is needed by humans. Various studies have shown that unmet needs for love will lead to unfavorable human behavior; people will become aggressive, frustrated, and kill themselves.

The love motive in this study can be seen from the donors who immediately moved to make donations that occurred in Palu and Donggala by donating in Kitabisa.com, the informants did that because there was an urge within them to do so. Having a great love makes them have an even greater drive to donate in Palu and Donggala: "Yeah yeah, I feel it too, Because my neighbor also lost something when it happened, so you can't just stay quiet or you do not care... maybe because I care about someone else so if there is a disaster or there are people need help, I try it" (Nadia Afifatur Rahman's interview, March 14, 2019). Rory donated with other friends by inviting and gathering what is appropriate. Then the researcher asked for an argument from Kak Ola as the expert informant in this study, Kak Ola said that: "... it makes all people drive to donate can channeling what they want to donate in the form of money, account numbers can be included, both in the form of goods can also be conveyed to be sent, and so on" (Debora Basaria interview results, M .Psi., Psychologist, 14 April 2019).

The loving motive supported the donators that had a high sense of affection to donate to Palu and Donggala. The self-esteem motive supported donators to shared stories with closest people than driven to donate in Rachel Ven account in kitabisa.com.

In facing life, humans need values to demand or giving meaning to their lives. Included in this motif are religious motives. When humans lose value, do not know what the real purpose of life, he has no certainty to act. Thus, he will quickly despair and lose his grip.

In the need for values, longing, and meaning of life in this study, the informants did donation activities because they held the social values that existed in their environment, such as social beings who had to help each other, not only that the informants also gave meaning to every action they took.

A said by the other informant, Nadia, by good action: "I still push myself to good action to other people. I wish someone also does the same thing to me later... hehe.." (Nadia Afifatur Rahman interview results, March 14, 2019).

The other informant said:" as human beings, someone needs help from us and we can help, why not" (Rory Sabina Anindita Semestriono's interview, March 11, 2019).

"Donation is a form of social responsibility." (Adellyn Dwitara Ramadhanty's interview, March 22, 2019).

The researcher asked about this motive with the expert informant named Kak Ola. Kak Ola said that people can make donations either by mentioning his name or not mentioning his name at all, according to Ms. Ola when the person does make donations anonymously, by the understanding of the motive of the need about the values, longing, and meaning of life above which states that there are values that are always used as principles or with the religious values they hold. As said by Ms. Ola below: "... other people may donate but not mentioning a name because he considers that it is part of his worship" (Interview results Debora Basaria, M.Psi., Psychologist, April 14, 2019).

The Value Motive supported the donators to share the principles of social value in their lives.

The self-fulfillment motive supported the donators to help the victims of the disaster in Palu and Donggala through their Rachel Vennya account. Rachel Vennya personality has a good influence on the community.

Basically, we are not only preserving life, but rather we improve its quality and optimize our potential. As stated by Maslow: "what a man can be, he must be". The realization of needs fulfilment can be done in several activities, namely: (1) developing and using our potentials in constructive and creative ways, e.g. art, music, science, or things which encourage creative self-expression; (2) enriching the quality of life by extending the range of experience and satisfaction; (3) creating warm and meaningful relatiship; (4) humanizing specific persons.

In fulfilling their needs, the informants claimed that they have personal satisfaction because they helped the victims of natural disasters in Palu and Donggala. This is further elaborated by one of the informants named Nadia who stated that:

I will also be happy to take part in this program (Nadia Afifatur Rahman's interview result, 14 March, 2019)

Another informant said that he was very happy to contribute in the program. He mentioned that it was his new experience in helping others and wished another opportunity in helping people as stated below:

"Well, this is the first time and I am so happy that I can help our relatives in Palu. I also get new experience and social values from this activity. I am so enthusiast to support donations program".
(Interview with Rory Sabina Anindita Semestriono, March, 11, 2019).

Meanwhile, Tara, the third informant, stated that he was very happy because his money was very helpful for other as seen from the following excerpt:

"Hm..I feel unselfish, I am not spending my money for my personal needs, but it is very useful for others".
(Interview with Adellyn Dwitara Ramadhanty, 22 March 2019)

To support the opinions above, the researcher asked a Communication Psychologist named Kak Ola. She believes that every human being has personal needs that should be fulfilled. These needs derived from the encouragement which may produce happiness for the doers. In detail, her acknowledgment is seen below:

"I can explain psychologically there are needs that should be fulfilled and drives encouragement. Once it has achieved, the doers are happy"
(The interview with Debora Basaria, Communication Psychologist, 14 April 2019)

Such motive, indeed, supports the donators to help the victims of the natural disasters in Palu and Donggala through their Rachel Ven account. A good personality of Rachel Vennya influences the donators to fulfil the needs of the community.

REFERENCES

Creswell, 2014. Penelitian Kualitatif dan Desain Riset. Yogyakarta: Pustaka Pelajar

Pambayun, Ellys Lestari. 2013. One Stop Qualitative Research Methodology In Communication. Jakarta: Lentera Ilmu Cendikia.

Rakhmat, Jalaluddin. 2015. Psikologi Komunikasi Edisi (Revisi). Bandung: PT Remaja Rosdakarya.

Solis, Brian. 2012. The Rise of Digital Influence a "how to" guide for businesses to spark desirable effects and outcomes through social media influence. Sans Fransisco, California: Altimeter.

https://kumparan.com/@millennial/5-situs-crowdfunding-untuk-membantu- mewujudkan-proyek-dan-ide-kamu

Advances in Business, Management and Entrepreneurship – Hurriyati et al. (Eds)
© 2021 Taylor & Francis Group, London, ISBN 978-0-367-67471-7

Managing service quality toward customer satisfaction of Panin Dubai Syariah Bank

D. Mayangsari, R. Hurriyati, D. Disman & L.A. Wibowo
Universitas Pendidikan Indonesia, Bandung, Indonesia

ABSTRACT: This research aimed to analyze the influence of 5 dimensions of service qualities tangibles, reliability, responsiveness, assurance, and empathy toward customer satisfaction. The research subject is a company engaged in sharia banking that has been listed in Bursa Efek Indonesia (BEI), PT Bank Panin Dubai Syariah, Tbk. The research samples were 100 customers of PT Bank Panin Dubai Syariah, Tbk, throughout Indonesia. The technique of collecting research samples used simple random sampling. It was due to all customers of PT Bank Panin Dubai Syariah, Tbk, having the same opportunity to become samples without any conditions. Five hypotheses were found in this research. The research concludes that quality service, which consists of tangible, reliability, responsiveness, assurance, and empathy dimensions, has a significant influence on the level of customer satisfaction of PT Bank Panin Dubai Syariah, Tbk.

1 INTRODUCTION

Previous researches still showed model similarity, especially a review on service model dimensions of service quality (Emel 2014; ;ahnali & Esmaeili 2015; Naik et al. 2010; ;aspor 2010; Rehaman & Husnain 2018).

This uniformity implicated on the model's ability to be applied to all phenomena.

In the research of Emel (2014), globally, all variables (physical environment, people, and process) influenced customer's satisfaction.

The conclusion in the research of Naik et al. (2010) that was conducted at a retail unit in Pantaloons Future Group, was that the state of South India Andhra Pradesh is as follows:

- Pantaloons Future Group has worked very well in the retail segment and been able to provide qualified services at their retail outlets.
- Reliability, responsiveness, empathy, and service assurance has high satisfaction value instead of tangibility (physical facilities).
- Service quality has an impact on customer satisfaction in the retail unit.

The conclusion in the research of Kahnali & Esmaeili (2015) identified five different dimensions for service quality where customer's assessment on service quality was based on those five dimensions (tangibles, reliability, assurance, empathy, and responsiveness).

The conclusion in the research of Rehaman & Husnain (2018) (the research was conducted in a hospital in Sargodha District, Pakistan), showed that most the important factors that impact the service quality are tangible, empathy, and assurance from the other five factors of the SERQUAL model. The research conclusion showed that assurance, empathy, and tangible were significant for patient satisfaction, while reliability and responsiveness were not.

The conclusion in the research of Raspor (2010) the research was conducted in 15 hotels in Opatija Riviera (Kroasia).

The conclusion in the research of Kang & James (2004) states that service quality researchers to date have paid scant attention to the issue of the dimensions of service quality. Much of the earlier work accepted the content measured by the SERVQUAL instrument. Following the argument that SERVQUAL only reflects the service delivery process, the study empirically examines the European perspective (i.e., Grönroos' model), suggesting that service quality consists of three dimensions, namely technical, functional, and image, and that image functions as a filter in service quality perception. The results from a cell phone service sample revealed that Grönroos' model is a more appropriate representation of service quality than the American perspective with its limited concentration on the dimension of functional quality.

The conclusion in the research of Lee et al. (2000) deals with three issues in the area of perceived service quality. First, it compares the gap model with the performance model. Second, it investigates the direction of causality between service quality and satisfaction. Finally, it examines whether the influences of some dimensions of service quality vary across service industry types. Three service firms were selected and respondents were interviewed in each firm.

As hypothesized, the performance model appeared to be superior to the gap model. Besides, the result shows that perceived service quality is an antecedent of satisfaction, rather than vice versa. Finally, tangibles appeared to be a more important factor in the facility/equipment-based industries, whereas responsiveness is a more important factor in the people-based industries. Managerial implications and future research directions are discussed.

The conclusion in the research of Patrick Asubonteng et al. (1996) is that, as competition becomes more intense and environmental factors become more hostile, the concern for service quality grows. If service quality is to become the cornerstone of marketing strategy, the marketer must have the means to measure it. The most popular measure of service quality is SERVQUAL, an instrument developed by Parasuraman et al. (1988). Not only has research on this instrument been widely cited in the marketing literature, but also its use in industry has been quite widespread (Brown et al. 1993).

The research result showed that all variables had an impact on the hotel customer's satisfaction. It also showed that among the five dimensions, "reliability" was the most important factor of the service quality. In the hotel industry, these dimensions refer to problem-solving for the guests, providing services without any mistakes on time and fast, proper, open schedule for hotel facilities.

Based on the previous articles, the previous researchers explained that there were five dimensions of customer service quality. This means, among those five dimensions of service quality, which one has the most impact out of five dimensions at each company in line with their business field.

Based on the phenomena, those two variables are described as variables with influence. Therefore, the problems are formulated as the following:

1. Does the higher quality of Tangibles cause higher customer satisfaction?
2. Does the higher quality of Reliability cause higher customer satisfaction?
3. Does the higher quality of Responsiveness cause higher customer satisfaction?
4. Does the higher quality of Assurance cause higher customer satisfaction?
5. Does the higher quality of Empathy cause higher customer satisfaction?

This research aimed to explain dimensions of service quality with a strong impact on customer satisfaction in Bank Sharia that referred to interrelationship variables pattern. However, specifically, the aims of this research are:

1. To find the impact of the reliability variable toward customer's satisfaction;
2. To find the impact of the responsiveness variable toward customer's satisfaction;
3. To find the impact of the assurance variable towards customer's satisfaction;

4. To find the impact of the empathy variable towards customer's satisfaction.

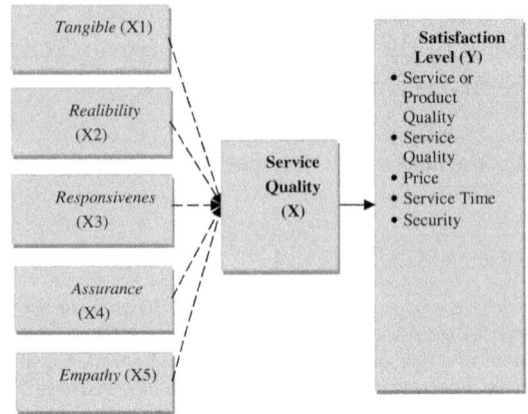

Figure 1. The conceptual framework of this research.

2 METHOD

The population of the analysis unit in this research consists of the customers of PT. Bank Panin Dubai Syariah, Tbk., as mentioned in the beginning, and the sample collecting technique used simple random sampling. It was to give all customers of PT. Bank Panin Dubai Syariah, Tbk., the same opportunity to be the samples unconditionally. The samples consisted of 100 customers of PT. Bank Panin Dubai Syariah, Tbk., throughout Indonesia.

3 RESULTS

After conducting a validity exam on the data, the results show that all identified variables have been valid and reliable and based on the assumption exam, the classic regression of test results show that the data are normally distributed and the heteroscedasticity test shows that the result of scatterplot data is randomly spreading so that it can be said to be homogenous.

Table 1. F-test results.

Model	Sum of Squares	df	Mean Square	F	Sig.
1 Regression	17670.841	1	17670.841	203.182	.000[b]
Residual	9305.838	107	86.970		
Total	26976.679	108			

a. Dependent Variable: Customer's Satisfaction
b. Predictors: (Constant), Service Quality

Based on Table 1, it is indicated that F count = 203 with Sig value $0.00 = 0.05$, which shows that the regression model is said to be proper.

Table 2. Regression results.

Coefficients[a]

| Model | Unstandardized Coefficients | | Standardized Coefficients | | |
	B	Std. Error	Beta	T	Sig
1 (Constant)	−2.777	5.868		−.473	.637
Service Quality	.612	.043	.809	14.254	.000

a. Dependent Variable: Customer's Satisfaction

Table 3. Correlation coefficient test.

Correlations

		Service Quality	Customer's Satisfaction
Service Quality	Pearson Correlation	1	.809**
	Sig. (2-tailed)		.000
	N	109	109
Customer's Satisfaction	Pearson Correlation	.809**	1
	Sig. (2-tailed)	.000	
	N	109	109

**. Correlation is significant at the 0.01 level (2-tailed).

From Table 2, the value of constant a = −2.777 and the value of coefficient b = 0.612. Therefore, the regression model is as the following:

$$Y = -2.777 + 0.612X$$

The above equation can mean:

A = −2.777: Means if the service quality value (X) is zero (0), so the customer's satisfaction value (Y) will be −2.777

B = 0.323: Means if the service quality (X) is increasing as much as one unit, so the customer satisfaction (Y) will increase by 0.612 units.

Based on Table 2, it is indicated that the value of t count = 14.254 with Sig = 0.00 < 0.05. This shows that service quality (X) provides a significant impact on customer's satisfaction (Y).

According to Table 3, it is indicated that the correlation coefficient value is 0.809. It shows a significant positive correlation between the service quality variable and customer satisfaction, in which the higher the service quality value, the higher customer's satisfaction will be and vice versa.

The determination coefficient is applied to determine the percentage of service quality correlation (X) toward customer's satisfaction (Y). Table 3 further shows that the correlation between service quality variable is 0.655 or 65.5%. It suggests that 65.5% of the customer's satisfaction variable can be determined by the service quality variable. On the other hand, the other 34.5% is influenced by other variables apart from the existing model.

4 CONCLUSION

Based on the results of the analysis of service quality toward customer's satisfaction it can be concluded that the service quality, which consists of tangible, reliability, responsiveness, assurance, and empathy dimensions, have a significant influence toward customer's satisfaction level in PT. Bank Panin Dubai Syariah Tbk.

REFERENCES

Asubonteng, P., McCleary, K. J., and Swan, J.E., 1996. SERVQUAL revisited: A critical review of service quality. *Journal of Services marketing.*

Brown, T. J., Churchill Jr, G. A., and Peter, J. P., 1993. Research note: Improving the measurement of service quality. *Journal of retailing*, 69(1), p. 127.

Kahnali, R. A. and Esmaeili, A., 2015. An integration of SERVQUAL dimensions and logistics service quality indicators (A case study). *International Journal of Services and Operations Management*, 21(3), pp. 289–309.

Kang, G. D. and James, J., 2004. Service quality dimensions: An examination of Grönroos's service quality model. *Managing Service Quality: An International Journal.*

Lee, H., Lee, Y., and Yoo, D., 2000. The determinants of perceived service quality and its relationship with satisfaction. *Journal of services marketing*

Naik, C. K., Gantasala, S. B., and Prabhakar, G. V., 2010. SERVQUAL, customer satisfaction and behavioural intentions in retailing. *European Journal of Social Sciences*, 17(2), pp. 200–213.

Parasuraman, A., Zeithaml, V. A. and Berry, L. L., 1988. SERVQUAL: A multiple-item scale for measuring consumer perc. *Journal of retailing*, 64(1), p. 12.

Raspor, S., 2010. Measuring Perceived Service Quality Using SERVQUAL: A Case Study of the Croatian Hotel Industry. *Management (18544223)*, 5(3).

Rehaman, B. and Husnain, M., 2018. The impact of service quality dimensions on patient satisfaction in the private healthcare industry in Pakistan. *J Hosp Med Manage [Internet].*

Advances in Business, Management and Entrepreneurship – Hurriyati et al. (Eds)
© 2021 Taylor & Francis Group, London, ISBN 978-0-367-67471-7

The role of online community in maintaining the brand image

D.S.F. Ali & I.I. Wahyuni
Telkom University, Bandung, Indonesia

ABSTRACT: The new media creates an online community to interact or communicate among its member through online media, especially social media. The advantage of online community is widely used by business people to develop marketing communication strategies in addition to maintaining brand image. The research aims at investigating the role of an online community in maintaining the brand image. This research used descriptive qualitative methods by taking data through in-depth interviews with the management and observation of their online community members. The findings revealed that the role of an online community in maintaining the brand image is as the builder of perception, provider of communication, and developer of experience.

1 INTRODUCTION

The presence of new media assisted by rapid technological developments indeed helps audiences access information anytime and anywhere. The development of new media rises to what is called social media. The definition of social media, according to Nasrullah (2015: 11), is a medium on the internet that allows users to present themselves and interact, work together, associate, communicate with other users, and form social bonds virtually. They describe social media as a collection of software that allows individuals and communities to gather, share, communicate, and in certain cases collaborate or play with each other. In this media era, there is a growing community that usually interacts or communicates among its members through online media using the internet and is usually done online, which can be called an online community (Ali & Wahyuni 2018).

On the other side, Bike to Work (B2W) Indonesia is a group of bicycle enthusiasts who go to work by using a bicycle. These activities are often carried out mainly by people in big cities to reach workplaces usually using environment-friendly and healthy transportation equipment. This Indonesian B2W community was formed on August 27, 2005, along with the Declaration and Joint Statement of the B2W Indonesia Community in DKI Jakarta City Hall. This community aims to campaign for the use of bicycles as alternative transportation, especially to support daily activities that are under the mission of B2W Indonesia, which has declared "Increasing the Number of Bicycle Users for Activities," so that people are encouraged to use their bicycles for daily activities (Source: http://www.b2w-indonesia.or.id; accessed on Dec. 21, 2018, at 21.03 WIB). As time went on, the B2W community spread to every major city in Indonesia. On September 5, 2007, one of the B2W Indonesia branches was established,

namely the B2W community. The purpose of the establishment of the B2W community was to invite the community, especially Bandung city, to use bicycles to be the daily transportation to reduce air pollution and congestion in Bandung (Source: Instagram b2w_bandung; accessed on Dec. 21, 2018, 9:51 p.m.).

The B2W Bandung community is not only active in offline activities, but they are also active in doing online community activities. B2W Bandung members are more active in doing online activities on Facebook social media platform. The Facebook group for B2W Bandung contains 4,897 members who provide information about bicycles, activities, or member activities during daily biking posted to the B2W group in Bandung, as well as information about events or invitations from members to ride together, both individuals and communities from various backgrounds (www.facebook.com/groups/Bike2Workbandung/; accessed on Dec. 21, 2018 at 01:16 WIB).

This research is ultimately focused on giving the role of the B2W Bandung community using Facebook as a place of interaction for members due to the limited time to meet personally and interact with others. In the beginning, they join the community to find people who are interested in the world of bicycles, but later on the company found that social media is the right place to provide information about marketing the products. Many companies tend to market their products by using online media or social media, one of which is a Facebook application that can provide stories or experiences, photos, and images. Through the Facebook application, the online community can share stories of their experiences, take pictures with brands that are used, and exchange information between fellow members of the community (Ali & Wahyuni 2018).

The emergence of social media provides a new direction for redefining media-public relations

(Holmes 2012). Interactive social media characters, open in creating content to wide networks, provide a kind of affirmation that the relationship is like two sides of a coin. On the one hand, the media institutionally provides tools and applications that can be arranged on the internet; on the side, audiences fully create content and utilize media devices according to their own needs. Audiences are no longer passive, not centralized, and isolated, but they are active in producing content and at the same time they are also distributing content to consumers. This continuous turnaround between producers cannot be found in the practice of traditional media, even the presence of social media together with the internet also involves audiences to create new languages that are more universal/global than traditional local media (Palfrey & Gasser 2008).

The development of new communication media, especially the internet, has also transformed interactions between individuals as entities that bring new and different social phenomena from those that have been understood. The communication system on a computer is just a device or tool. Therefore, the community that appears in the cyber world is the next stage of the use of this tool (Nasrullah 2014).

An online community is a place where a group of people gather to share senses of the community as people who do not know each other have similar interests in an internet site that offers several online services, including some access to social environments, community services, official information, and e-commerce services to its residents (Schau et al. 2009).

Polygon, as a famous bicycle brand, has been known for a long time, especially in Indonesia. Polygon marketing does not only focus on making bicycles as a means of transportation, but also wants to make cycling a lifestyle for consumers. All efforts must be made by these well-known bicycle manufacturers to market their products, including starting from making bicycle outlets or shops, distributors, accessories, and so on. During these 28 years, Polygon recorded a production growth of 30% each year; products from Polygon, approximately 60%, have been exported to 60 European countries and Southeast Asia. France, Germany, England, Italy, and Spain are countries that are the main targets of the Polygon brand (http://www.gowes.org; accessed on November 22, 2018 at 02.05 WIB).

PT. Sena Insera was established in 1989 and it is located in the village of Wadungasih, Buduran, Sidoarjo, East Java. At the beginning of the manufacturing, bicycle products were widely exported to approximately 17 countries, including countries in Europe. Then in the following year, Polygon spread its wings in various parts of the world, especially in Southeast Asia and other parts of the world. In Indonesia, Polygon products are marketed in various regions, including Aceh, Medan, Sumatra, and East Kalimantan (http://www.gowes.org; accessed on November 22, 2018 at 02.00 WIB).

The target market characteristics are increasingly widespread. Currently, Polygon is not only doing retail sales, but Polygon also serves the cooperative (company), campus, and community sectors. This is done solely to raise the sales of Polygon. The marketing objectives carried out by Polygon are in line with the mission of B2W, specifically in the city of Bandung, which is to distinguish the people of Bandung and promote bicycles, or it can be said to make cycling a lifestyle for the community, especially in the city of Bandung. With this similarity, a company needs members' community roles to help them in terms of product development or marketing their products. Communities can be a powerful media for marketing strategies. Community marketing is a powerful strategy to engage customers actively and without the impression of being forced, creating a stream of useful conversations for companies, consumers, and markets (http://www.dictio.id/t/mengapa-komunitasmenjadi-sangat-penting; accessed on Nov. 24, 2018 at 11:16 WIB).

Nowadays, the public interest in using bicycle transportation is greatly reduced and replaced by motorbikes, therefore, Polygon Indonesia has a marketing strategy of embracing bicycle communities, one of them is B2W. In Bandung, the bicycle manufacturer has partnered with at least seven bicycle communities, including B2W and Bike to Campus (http://bandung.bisnis.com/read/20110109/3/15330/polygon-gandeng-komunitas-sepeda-bandung; accessed on Nov. 20, 2018 at 11.32 WIB).

A brand is a name or symbol that is distinguishing (such as a logo, stamp, or packaging) to identify goods or services from a seller or a certain group of sellers. Thus differentiating it from goods and services produced by competitors (Aaker 1997: 9).

Brand image is closely related to brand perception in the minds of consumers, which refers to meaning. The consumer experience and information obtained further strengthens the meaning of the brand image. The brand image includes the knowledge and belief in brand attributes known as cognitive aspects, while the consequences of brand use include evaluation, feelings, and emotions into the affective domain. Keller (2013: 93) states that brand image is a perception of a brand that is described by brand associations in consumer memory. Keller suggests the factors of brand image formation include product superiority as one of the factors forming a brand image, where the product excels in competition. Because of the superiority of quality (model and comfort) the characteristic causes a product to have a special attraction for consumers.

Favorability of brand association is where consumers believe that the attributes and benefits provided by the brand will be able to fulfill or satisfy their needs and desires so that they form a positive attitude toward the brand. Then, brand strength is a brand association depending on how information enters consumer memory and how the process survives as part of the brand image.

The strength of this brand association is a function of the amount of information processing received in the coding process. When a consumer actively elaborates on the meaning of information on a product or service, a stronger association will be created in consumer memory. The importance of brand associations in consumer memory depends on how a brand is considered.

Thus the uniqueness of the brand as the association of a brand inevitably must be divided with other brands. Therefore, competitive advantage must be created, which can be used as a prominent reason for consumers to choose a particular brand. Later on, positioning the brand rather leads to experience or self-benefits of the product image. The differences that exist from products, services, personnel, and channels are expected to provide differences from competitors, which can provide benefits for producers and consumers.

Previous researchers focused on brand community influences on the marketing strategy of some products. The innovation in this paper is to explore the role of the members of online community in maintaining the brand. Some companies use the strength of their brand equity through symbols, names, and the quality of the brand to the cycling communities. From the symbol, logo, and brand name, Polygon makes differentiation or differentiation with competing bicycle brands.

2 METHOD

This research used a descriptive qualitative method by taking data through in-depth interviews and observation with the online community members using the consumer-based model. The brand image was adapted from the marketing literature and combined with in-depth interviews as the process of obtaining information for research purposes.

Interviews are part of qualitative methods. In this qualitative method, it is known as in-depth interviews. Understanding of in-depth interviews (In-depth Interview) is the process of obtaining information for research purposes using a question and answer while in a face-to-face meeting between the interviewer and the respondent or the person being interviewed, with or without using the interview guide where the interviewer and informant are involved in social life (Creswell 2014).

The research used an in-depth interview to find out the role of the online community B2W Bandung in maintaining the brand image by taking data three times through interviews with the members and observation of account members.

3 RESULTS

Regardless of diversity, Polygon is a brand that is often used and emerged as a sponsor in the activities of the B2W Bandung community. In line with the aim to increase brand image that is presented through the brand attributes, the name and logo attached to the uniform of the community clothing helped the product build character. Brand character was supported by the management of Polygon itself by producing a new type of pf bicycle. The bicycle was produced in 2011, with the design of a yellow bicycle frame that was in harmony with the color of B2W. Thus it was the B2W logo that was tightened on the bicycle frame. Polygon provides a guarantee of providing comfort when riding on the highway. The presence of the brand in the B2W Bandung community builds benefits as a positive image and involves many people who have a similar impression and feel a sense of comfort when riding the product. This process helps to develop a perception in the social interactions between members, as well as the social interaction between the customer and the brand itself.

The management of Polygon released many channels of communication linked to the B2W community as well. In certain circumstances, most members of the community associate the brand with the part of the lifestyle with value for money products. The members of the community bring the topic to the conversation through online and offline media. The fairly affordable price and the full support from the producer of bicycles become the content that will be provided by the members of the community.

The members of the community associate their experience with the apparel they used during their activities. It develops involvement in social interaction through social media. This activity explains the perception, the benefit becomes a genuine experience they want to share.

The well-established brand association leads to the formation of a brand image that is maintained positively in the eyes of customers through the impression and experience they feel when riding the bicycle. Similar experience when dealing with brand brings more social relationships between members and fosters equality and a similar image. This experience shows the role of the community has a great influence on building brand image, especially at the subjective rating level of each member, and especially in the virtual world. The content is distributed as well as developed by its users. The members actively distribute the experience while developing it. The experience associated with the brand that is generally was excavated from members of the B2W.

4 CONCLUSION

The conclusions that can be drawn from the research about the role of B2W Bandung community in building the image of the brand are as follows:

At the product attributes phase, the role of community members is as a presenter of the attributes itself meanwhile building their character. This activity explains how their riding needs confirmation from the brand through its identity. At the consumer benefits phase, the role of members increased depending on the benefit that they get at the same time they become the accessor or evaluator of the product and service based on their experience, whereas they have value and benefit assessment about the product so that the role becomes adhesive between members and with the brand itself. At the brand personality phase, the members of the community demand the uniqueness and quality of the product. The primary product will lead to brand equity. The result is the combination of perception and benefit and experience itself.

However, there is much yet to be done. Future work will include a much more in-depth study about the role of online brand community members, but also the engagement and influences to society.

REFERENCES

Aaker, David A. 1991. Managing Brand Equity: Capitalizing of the Value of a Brand name. *New York*: The Free Press

Aaker, David A (1996), Measuring Brand Equity Across Product and Market. *California Management Review*, Vol 3, pp. 102–121

Ali, D. S. F., & Wahyuni, I. I. 2018. The role of the hijabers community in building the brand equity of muslim fashion. In *Proceeding of International Seminar & Conference on Learning Organization.*

Holmes, D. 2012. Indonesian Translation of Communication Theory: Media, Technology, Society Yogyakarta, Pustaka Pelajar

Keller, Kevin, L. 2013. Strategic Brand Management; *London, Boston Mass*

Nasrullah, Rulli. 2014. Theory and Cybermedia *Jakarta: Kencana Prenadamedia Group.*

Nasrullah, Rulli. 2015. Social Media Bandung: *PT Remaja Rosdakarya*

Palfrey, J. & Gasser, U. 2008. Born digital: Understanding the first generation of digital natives. *New York: Basic Books*

Schau, Albert Hope Jensen, & M. Muñiz Jr., & Eric J. Arnould. 2009. How Brand Community Practices Create Value, *Journal of Marketing* Vol. 73

Advances in Business, Management and Entrepreneurship – Hurriyati et al. (Eds)
© 2021 Taylor & Francis Group, London, ISBN 978-0-367-67471-7

The relationship user satisfaction with service website: An empirical study of STIE Inaba Bandung

A. Rachman, E. Herlinawati & R.N. Sumawidjaja
Sekolah Tinggi Ilmu Ekonomi Indonesia Membangun (STIE INABA) Bandung, Bandung, Indonesia

ABSTRACT: This study discusses the importance of using the internet to improve competitiveness with other universities. School of Economics (STIE) INABA Bandung has a website that is www.inaba.ac.id that serves as a communication medium that can provide information for students, lecturers, and the general public. This research was made to analyze the effect of website design and the quality of information on user satisfaction over website service by distributing 39 questionnaires to postgraduate students of semester 1 academic year of Management Studies class of 2016 as respondents. This research uses a quantitative method. Using the structural equation method model – partial least square SEM–PLS. Research shows that website design and quality of information affect user satisfaction over website services.

1 INTRODUCTION

The use of the internet today become a medium of communication that is quite important for an organization. Tertiary Institution of Economics (STIE) INABA Bandung has a website that is www.inaba.ac.id which serves as a medium of communication that can provide information for students, faculty and the general public.

Many factors affect user satisfaction (students) on a website such as a web design, quality of information, speed, web navigation, personalization and more. Satisfaction is their suitability student expectations on the service website in terms of both the website design and quality of the information displayed. The website design good quality information that will create satisfaction for users, because that's School of Economics (STIE) INABA Bandung should pay attention to this so that students are not disappointed and feel comfortable with the services provided by the college.

The results of the study by Mohammed (2010) tested the online system users' satisfaction at the University of Petra, Jordan. Dimensions examined are website design, navigation, and personalization. The results showed that all three dimensions affect the satisfaction of the online system at the University of Petra, Jordan. Subsequent research by Al-Manasra et al. (2013) investigates customer satisfaction telecommunications sector in Jordan seen from the dimensions of website quality is the quality of information, interaction and service quality, and ease of use. The results showed that the quality of information, interaction, and service, as well as the convenience of having a positive and significant effect on the satisfaction of users of telecommunications in Jordan. Maditinos et al. (2011)

in his research found that some website attributes contributing to user satisfaction. User satisfaction with regard to how satisfied users associated with the website(website). The results showed that each attribute has an influence on the quality of information, quality of systems, security, privacy, and ultimately affect user satisfaction.

Further research conducted by Ranjbarian et al. (2012) conducted a study on the satisfaction of customers using online booking. There are five models studied to customer satisfaction, namely convenience, merchandising, web design, security, service. The results showed that there are four variables that significantly influence the satisfaction that is the convenience, merchandising, security and service. While web design does not have a significant impact on customer satisfaction. Of course, it is different with the results of research conducted by Mohammed (2010) which states that web design has a significant influence on the satisfaction of Internet users.

Results of research conducted by Sanjaya (2012) on the service ministry's website Kominfo, the results showed that of the 3-dimensional WebQual 4.0, only the dimensions of usability and quality of interaction are considered influential to the satisfaction of the user, while the dimensions of the quality of information considered to be influential for user satisfaction of websites services.

In this study, the authors will discuss the quality of a website that is only two-dimensional web design and quality of information. Web design can be a visual display of the website, the use of so graphic, color, photo, user the type of text that can be felt attractive. While the quality of information is the provision of information on the website that gives a high usefulness for the user.

Based on the above, the panelist would like to do further research with the title "User Satisfaction on Service Website STIE INABA Bandung"

1.1 Theological issues

1. Are there significant effect between the web design of the website user satisfaction with services at the School of Economics (STIE) Inaba Bandung?
2. Is there any significant influence of the quality of information on user satisfaction with the service website at the School of Economics (STIE) Inaba Bandung?

2 METHODS

2.1 Population and sample

According to Sugiyono (2012) population is generalization region consisting of subject/object that has certain qualities and characteristics defined by the researchers to learn and then drawn conclusions. Husein Umar (2008) states that the population is a collection of elements that have certain characteristics in common and have equal opportunity to be selected into the sample.

The population in this study is a graduate student Semester 1 Management Studies academic year 2019 amounted to as many as 39 people. 2. Sample The sample is part of the population. According Sugiyono (2012) sample is part of the number and characteristics possessed by this population. Total side is a sampling technique when all members of the population used as a sample.

Sampling was done by total sampling technique, meaning that all the population sampled in this study.

2.2 Operational definition of variables

2.2.1 Dependent variables

1. Satisfaction with the service Website: satisfaction with web services, satisfactionon web design, satisfaction with the quality of information, on an easy to use web satisfaction.

2.2.2 Independent variables

1. Website quality: visual, color, graphics, text, free.
2. Of information Quality: Accuracy information, information trustworthy, helpful information, relevant to the user information.

2.2.3 Data analysis techniques

Data analysis technique used in this research is the analysis of track/path analysis using software 3.0 Smart PLS (Partial Least Square).

Model evaluation Partial Least Square (PLS) based on measurement predictions that have the nature of not parametric (Ghozali 2010).

1. Measurement model or a model with indicators reflexive outer evaluated by the convergent and discriminant validity of indicators and composite reliability for the block indicator.
2. The inner structural model of the model evaluated with a view percentage variance explained by looking tilapia R2.
3. Dai stability of these estimates was evaluated using the t test statistic obtained through bootstrapping procedure.

"Descriptive analysis is used to describe the respondents in this study are based on a percentage value of a respondent's answers." (Santoso 2010). Descriptive analysis of this frequency is used to present the data that has been obtained for further description. The view that there were served separately which consists of one variable only.

3 RESULTS AND DISCUSSION

3.1 Profile of respondents

The following is a descriptive profile of respondents of this study. Of the total 39 respondents can be seen:

Table 1. Descriptive profile of respondents.

Category	Description	Percentage
age>	<20	0
	21–30	15
	31–40	65
	41–50	8
	50	2
	Male	28
	Female	72

3.2 Validity and reliability test of research variables

3.2.1 Reliability test

Test reliability is used to look at the consistency of the variables used in the study. And variable should have an alpha value Cronbach >0.6. It can be seen that all variables used reliable and qualified to proceed to the next stage.

Table 2. Reliability test.

Variable	Cronbach Alpha	Description
User Experience Web	0.834	Reliable
Design Web	0.812	Reliable
Quality Information	0881	Reliable

3.2.2 Validity test

Validity test used to measure the accuracy of the indicators in the variables used in the study. Each indicator must have a Pearson r values above 0.197. It can be seen all the indicators that are used are valid and eligible to proceed to the next stage.

Table 3. Validity test.

Variable	Indicator	R Pearson	Ket.
User Website	Y1	0.624	Valid
Satisfaction	Y2	0.796	Invalid
	Y3	0.797	Invalid
	Y4	0.798	Valid
Web Design	X1.1	0.799	Valid
	X1.2	0.800	Valid
	X1.3	0.801	Valid
	X1.4	0.802	Valid
	x1.5	0.803	Valid
	X2.1	0.804	Valid
Quality. Information	X2 2	0.805	Valid
	X2.3	0.806	Valid
	X2.4	0.807	Valid
	X2.5	0.721	Valid

3.3 Descriptive variable research

Based on the results of a questionnaire given to respondents with each indicator of the satisfaction of users of the website, design a website, and quality of information. Respondents answered 'as shown in the following table:

Table 4. Website user satisfaction.

Variable	Indicator	Score Total	Total %	Average
Satisfaction	Satisfaction with web services	152	95.5%	91.5%
Userweb	Satisfaction on web design	144	91.5%	
Value %	satisfaction with the quality of information	146	92.5%	
	satisfaction with the ease of use of the web	137	88%	

Based on Table 5, the average satisfaction of users of the website amounted to 91.5% it means that the user satisfaction as a student at the College website Economics STIE INABA Bandung classified in the category of "good".

Respondents answered 'variable web design can be demonstrated in Table 6, the average of 92% of its percentage. The value of web design are classified in the category of "good" means that the design on the website of the School of Economics STIE INABA Bandung has appeal for students and prospective students to obtain more information about the almamater to be chosen as well as help increase the tally of students in the new academic year.

Based on Table 7, the average percentage of 96.5% quality information. This value is classified in the category of "good" this is due to the website High School of Economics STIE INABA Bandung all parties participate in providing the latest information

Table 5. Website design.

Variable	Indicator	Score Total	Total %	average
Web Design Value %	visual presentation on the web	149	94%	92%
	Use of color on the web	151	95%	
	The use of graphics and photos on the web	145	92%	
	The diversity of text on a web	145	92%	
	Free access to the web	140	89.5%	

Table 6. Information quality.

Variable	Indicators	Score Total	Total %	Average
Value% Quality Information	site to provide accurate information	155	97%	96.5%
	Web provides information that can be trusted	154	96.5%	
	Use of graphics and photos on the web	155	97%	
	the diversity of text on a web web access	152	95.5%	
	speed at	149	94%	

updates such as public relations and marketing division, the academic, as well as its student activity unit.

3.4 Partial least square

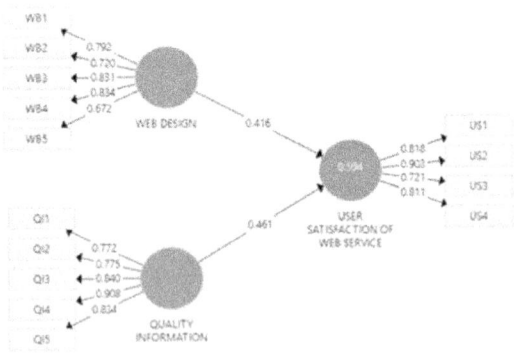

Figure 1. Partial least square.

3.4.1 Indicator reliability

Indicator reliability is used to test the consistency of indicator used in the study. Have a minimum requirement of 0.4. Can be seen all the variables that are used to qualify to proceed to the next stage.

Table 7. Indicator reliability.

Variable	Indicator	Reliability
Satisfaction Website Users	Satisfaction withweb services	0.818
	Satisfaction with web design	0.903
	Satisfaction with the quality of information	0.721
	Satisfaction with the ease of use of the web	.811
design Web	visual presentation on the Web	0.792
	The use of color in web	0.720
	Use of graphics and photos on the web	0.831
	Diversity the text on the web	.834
	access speeds on the web	0.672
Quality information	site to provide accurate information	0.772
	web provides information that can be trusted	0.775
	Usagegraphics and photos on the web	.840
	diversity of text on a web	0.908
	Speed of access to web	0.834

3.4.2 *Internal consistency reliability*

Internal consistency reliability was used to measure the consistency of the variables used in the study and have provided the composite reliability above 0.6. Can be seen all variables are eligible. So research can proceed to the next stage.

Table 8. Internal consistency reliability.

Variable	Composite Reliability	Description
User Experience Web	0.915	Reliable
Design Web	0.888	Reliable
Quality Information	0.88	Reliable

3.4.3 *Convergent validity*

Validity Convergentis used to measure the accuracy of the variables used in the study, have the requisite values above 0.5 ave, can be seen all the variables used are valid and can proceed to the next stage.

Table 9. Convergent validity.

Variable	AVE	Description
Web User Experience	0.685	valid
Web Design	0.666	valid
Quality Information	0597	valid

3.4.4 *Rated R-square and Q-square*

It can be seen that the percentage of the influence of website design and information on user satisfaction with the services the website of 59.4%.

Table 10. Rated R-square and Q-square.

Variable	R-square
Web User Satisfaction	–
Web Design	–
Quality Information	0594

Value Q2

$$= 1 - (1 - 0.594)$$
$$= 1 - 0406$$
$$= 0.594$$

The value of Q-Square used to look at the PLS model used in the study. It can be seen the goodness of fit of the model used was 59.4%. It can be concluded that the model used is quite good.

3.5 *Hypothesis testing*

Hypothesis testing is used to see the value of each hypothesis used in the study. With a minimum value of 5% that can be expressed significant. It can be seen every hypothesis in the study had a path coefficient value above 5%.

Table 11. Hypothesis testing.

Effect of	Path coefficient	T statistics
Web Design -> Web User Satisfaction	0.461	3.355
Quality Information -> Web User Satisfaction	0.416	2819

3.6 *Discussion*

Based on our research and the previous discussion, the authors concluded as follows:

a. Web design has a positive influence and significant to user satisfaction with the service website at the High School of Economics (STIE) Inaba Bandung
b. the quality of information has a positive influence and significant to user satisfaction with the service website on the High School of Economics (STIE) Inaba Bandung
c. the quality of information has the most dominant influence on user satisfaction at the High School of Economics (STIE) Inaba Bandung.

ACKNOWLEDGEMENTS

The results of the research show that the influence of the quality of the website to the user satisfaction according to results of the study by Al-Manasra et al (2013) who studied the impact of website quality on customer satisfaction in the telecommunications company in Jordan. The results showed that there is significantly positive and significant correlation between the quality of information to customer satisfaction telecommunications in Jordan. Reinforced by the results of research by Mohammed (2010) who studied the factors affecting user satisfaction Universitas website at Petra Jordania with a sample of 615 students. The results showed that web design has a positive and significant influence on the satisfaction of Internet users.

REFERENCES

Al-Manasra, E., Khair, M., Zaid, S.A., & TaherQutaishat, F. 2013. Investigating the impact of website quality on consumers' satisfaction in Jordanian telecommunication sector. *Arab Economic and Business Journal* 8(1–2): 31–37.

Ghozali, I. 2010. *Aplikasi analisis multivariate dengan program spss, edisi keempat.* Semarang: Penerbit Universitas Diponegoro.

Maditinos, D., Keisidou, E., & Sarigiannidis, L. 2011. Consumer characteristics and their effect on accepting online shopping, in the context of different product types. *International Journal of Business Science & Applied Management (IJBSAM)* 6(2): 31–51.

Mohammed, M.A. 2010. Factors affecting e-service satisfaction. *Communications of the IBIMA* 2011(2011): 12.

Ranjbarian, B., Sanayei, A., Kaboli, M.R., & Hadadian, A. 2012. An analysis of brand image, perceived quality, customer satisfaction and re-purchase intention in Iranian department stores. *International Journal of Business and Management* 7(6): 40–48.

Sanjaya, I. 2012. Pengukuran kualitas layanan website kementerian kominfo dengan menggunakan metode webqual 4.0. *Jurnal penelitian iptek-kom* 14(1): 1–14.

Santoso, A. 2010. Studi deskriptif effect size penelitian-penelitian di fakultas psikologi universitas sanata dharma. *Jurnal Penelitian* 14(1).

Sugiyono. 2012. *This method business research.* Bandung: Alfabeta.

Umar. (2008). *Research methods for thesis and thesis business.* Jakarta: PT. Raja Grafindo Persada.

Advances in Business, Management and Entrepreneurship – Hurriyati et al. (Eds)
© 2021 Taylor & Francis Group, London, ISBN 978-0-367-67471-7

An analysis of destination image and personality on behavioral intention (survey on domestic tourists in Pariaman city)

Verinita & Refyanto
Andalas University, Padang, Indonesia

ABSTRACT: This study aims to analyze the influence of destination image and destination personality on behavioral intention. This purposive sampling study used a sample of 130 respondents with the criteria for the samples taken being domestic tourists who were visiting Gandoriah beach. The results of the study show that the destination image does not have a positive effect on the behavioral intention and destination personality has a positive effect on behavioral intention. This indicates that the destination image does not have a positive influence on behavioral intention. This is because the coast of Gondariah, located in the city of Pariaman, is a new trainee destination, so it is not popularly known by tourists. Destination personality has a positive effect on behavioral intention. This means that tourists visiting Gondariah beach in Pariaman city feel happy so that it influences their intention to behave again in the future. To increase tourist visits can be recommended to managers of tourist destinations to add tourist attractions so that it can add to the appeal of the coast of Gondariah in the future through the addition of play facilities for children.

1 INTRODUCTION

West Sumatera is one of the Indonesian provinces that has become the main attraction for tourism in Indonesia because it has a relatively complete type of tourist attraction such as the sea, beaches, lakes, and mountains. Currently, the development of the tourism sector in West Sumatra is increasing. This can be seen from the number of visitors who continue to increase.

The West Sumatra Central Statistics Agency (BPS) noted that tourist arrivals to West Sumatra reached 5,209 people, an increase of 27.67% compared to 2016. Archipelago tourist visits rose by 7% in 2016 (BPS Sumatera Barat Website 2017).

Moreover, Pariaman city is one of the cities in West Sumatera Province that has tourist destinations that are visited by many domestic and foreign tourists. Gandoriah Beach is a beach attraction that is about 100 meters from the center of Pariaman city. Gondariah Beach has a beautiful panorama and is the most popular beach attraction in the city of Pariaman. Gandoriah Beach is expected to build a good destination image and be liked by tourists and become a leading tourist attraction in the city of Pariaman. Positive tourist destinations will increase return visits so tourists will recommend it to others.

The destination image is the most important part of a tourist destination. Echtner & Ritchie (2003) define destination image is the response of potential tourists to a destination. According to Oleary & Deegan (2003), destination image makes it easier for tourists when making their vacation travel decisions

and determining whether their travel experience will be very satisfying. Oter & Ozdogan (2005) stated that destination image is a description of the responses of tourists about these tourist destinations. Destination image is an individual's response to destination characteristics that can be influenced by promotional information, mass media, and many other factors (Tasci & Kozak 2006: 304).

According to Ekincy & Hosany (2006) destination personality is individual perceptions about the purpose of the individual visit to a destination. So destination personality is an effective component of the desire of tourists to recommend these destinations to others.

This is also supported by Court & Lupton (1997) who stated that images positively influence the intention to visit again in the future. According to Chen (2006), behavioral intention is an option to visit and evaluate the intention to behave in the future. Evaluation in the form of a quality travel experience because of the perceived value and high satisfaction. By understanding the relationship between intention to behave in the future, destination managers will know more about how to build an attractive image for tourists.

An interesting question to examine is the influence of destination image and destination personality on behavioral intention. And the survey was conducted on tourists who were visiting Gandoriah beach in Pariaman city. This study aims to determine the influence of the destination image on behavioral intention and the influence of Destination Personality on the behavioral intention of Gandoriah beach tourists.

According to Tasci & Kozak (2006), the destination image is an individual's perception of the characteristics of a destination that can be influenced by promotional information, mass media, and many other factors. Hsu et al. (2010) state that the image of a destination is an important factor in a visit to bring more tourists to tourist attractions. According to Echtner & Ritchie (2003), destination image simply refers to the impression of a place or someone's perception of a particular area. Qu et al. (2011), in their research, measure destination image based on three elements, namely cognitive image, unique image, and affective image.

Ekincy & Hosany (2006) define destination personality as a perception of human characteristics regarding individual goals in traveling. Ekincy & Hosany (2006) reveal that destination personality is a metaphor for understanding perceptions of goals from visitors, building brand goals, and creating unique identities for tourist destinations.

According to Ekincy & Hosany (2006), there are three dimensions of destination personality, which are namely sincerity, joy, and friendliness. This shows that tourists rate personality traits differently for each individual. Sincerity consists of sincere, healthy, reliable, and intelligent. The fun consists of authenticity, fun, and courage, while friendliness covers friendly, family-oriented, and charming

Intention to behave is formed by attitudes toward behavior, social norms, and control over perceived behavior. According to Peter & Olson (2002), behavioral intention is a proposition that connects itself with future actions.

Based on the theory of planned behavior model (TPB), the main factors that influence a person's behavior are their intentions or their tendency to take action. TPB, according to Bigne et al. (2001), claims that tourist behavior is a term that includes before a visit, decision making, is visiting a destination, evaluating, and behavior after a visit to a destination. According to Zeithaml & Bitner (2003), behavioral intention can be seen as an indicator that gives a sign that a customer remains a customer or leaves a company that has been serving them.

From the background above, the hypotheses of this study are:

H1: It is assumed that destination image does not have a positive effect on the behavioral intention variable.

H2: It is assumed that destination personality has a positive effect on the behavioral intention variable.

2 METHOD

This study uses a quantitative approach to analyze how one variable influences other variables and the causal relationship between variables is explained by the hypotheses. All data and information were collected using a questionnaire. This study used a sample of 130 respondents. The technique used for sampling was the purposive sampling technique. Samples in this study were visitors to the beach of Gondariah in Pariaman city. Data sources are primary data obtained directly from respondents. Secondary data were obtained from journals and other reading sources.

This study will use Partial Least Square 2.0 (PLS) analysis methods. PLS-SEM analysis usually consists of two sub-models, namely the measurement model, or commonly known as the outer model, and the structural model (structural model), or the inner model (Ghozali 2011).

Testing the hypotheses is seen from the magnitude of the statistical values. Because PLS does not assume normality and data distribution, PLS uses a nonparametric test to determine the significance level of the path coefficient, where the value of t (t-statistics) generated by running the Bootstrapping algorithm on SmartPLS is used to determine whether or not the hypotheses are accepted. The hypotheses will be supported if the T-statistic value exceeds the T-table ranging from -1 to $+1$ because a value close to 0 indicates an explanation of a weak relationship between independent and dependent constructs. Path analysis was tested on the critical value t-statistic 1.65 ($\alpha = 0.10$), 1.96 ($\alpha = 0.05$), and 2.57 (0.01) for the two-tailed cut off (Hair et al. 2010). The results of this hypothesis test use a significance of at least 5% (1.96).

3 RESULTS

The number of respondents was 130 people, consisting of 93 men and 37 women. Based on the above data, it can be analyzed that respondents who are more than 40 years old prefer recreation or a vacation to the beach. The majority of respondents who go on vacation are those who work as civil servants and have a high school education. Most of the respondents' income is between Rp. 3,000,000 and Rp. 5,000,000. And tourists from Pekanbaru are more than tourists from other provinces.

This is because the distance between the provinces of West Sumatra and Pekanbaru is not too far away. So as to enable tourists from Pekanbaru to take a vacation and visit the beach of Gondariah. From the data above, it can be analyzed that the majority of visitors who came to Gandoriah beach had only visited one time. This is because Gandoriah beach is still not widely known by tourists and is still less popular than other excellent tours in the city of Pariaman. Respondents learned information about Gandoriah beach from their family or relatives. Because the family are the closest people who can provide information directly. Respondents who visited Gandoriah beach only wanted to see the beach. This is due to the absence of entertainment venues or games on the Gandoriah beach, so tourists visiting only want to enjoy the beach atmosphere. There are no additional attractions such as bikes and clowns.

The number of respondents is 130 people consisting of 93 men and 37 women. Based on the above data, it can be analyzed that respondents who are more than 40 years old prefer recreation or a vacation to the beach. The majority of respondents who go on vacation are those who work as civil servants and have a high school education. Most of the respondents' income is between Rp. 3,000,000 and Rp. 5,000,000. And tourists from Pekanbaru are more than those of other provinces, this is because the distance between West Sumatra and Pekanbaru is not too far away. So that allows tourists from Pekanbaru to often go on vacation and visit the coast of Gondariah.

From the data above, it can be analyzed that the majority of visitors who came to Gandoriah beach had only visited once. This is because Gandoriah beach is still not widely known by tourists and is still less popular than other excellent tours in the city of Pariaman. Respondents learned information about Gandoriah beach from their family or relatives. Because the family are the closest people who can provide information directly. Respondents who visited Gandoriah beach only wanted to see the beach. This is due to the absence of entertainment venues or games on Gandoriah beach so that tourists who visit can only enjoy the beach atmosphere.

Based on the descriptive analysis of nine indicators from the image of the destination, tourists consider that the city of Pariaman is a city that is visited by many tourists. Because currently, the city of Pariaman is revamping the coastal tourism object, which aims to attract tourists to come to visit the coastal destination. Based on the Descriptive Analysis of Destination Personality, tourists state that Gandoriah beach has attracted visitors personally. This section shows the alternative answers chosen by each respondent in each statement for the behavioral intention variable. This shows that tourists will report if there is a service problem with the officer. This shows the attention of many tourists on the coast of Gondariah.

Based on the results of the study, hypothesis 1, which states that the image of the destination influences the purchase decision is not acceptable. This hypothesis has not been proven because t count $(0.255) >$ t table (1.96), then H1 is not accepted, meaning that the destination image variable does not have a positive effect on behavioral intention. This is different from the results of Farida's research (2013) entitled "Effect of service quality, tourism facilities, promotion of destination imagery and intention to behave in Karimunjawa Tourism Object, Jepara Regency," where the results of their research show that the destination image has a positive effect on the variable intention to behave at 0.277 with a significance of 0.009.

Based on the results of the research, which claims that destination personality influences the behavioral intention, this hypothesis can be accepted. This hypothesis has been proven because t count $(5.048) >$ t table (1.96), then hypothesis 2 is accepted, meaning that destination personality has a significant effect

on behavioral intention. Travelers feel the destination personality effectively influences them before the intention to behave. Evidently, in this study, tourists understand that destination personality will also influence their intention to behave in the future where they will recommend to family and friends to visit the Gandoriah beach.

Based on the results of the study, it can be concluded that the destination image does not have a positive and insignificant effect on behavioral intention of tourists of the Gandoriah beach. This means that the destination image assessed by tourists does not affect tourists to intend to revisit the coast of Gondariah. This is due to the fact that Gondariah beach is a new tourist destination known by tourists so that the image of the Gondariah coastal destination has not yet formed in the minds of tourists. Therefore, the formation of a destination image is very important for the Gondariah coast to increase the intention to behave again. One way that can be done is to increase the beauty and attractiveness of the coast of Gondariah.

Destination personality has a positive and significant effect on behavioral intention of tourists of Gandoriah beach. This is because the destination personality of tourists will make tourists want to behave in the future. Because tourists visiting the coast of Gondariah provide a sense of happiness and happiness and give joy to most tourists.

REFERENCES

Artuger, S., et al. (2013). The effect of destination image on destination loyalty: Application in Alanya. *European Journal of Business and Management, 5* (13): 124–136.

Bigne, J. E., Sanchez, M, I., & Sanchez, J. (2001). Tourism image, variable evaluation and after purchase behavior: Inter-relationship. *Tourism Management, 22* (6), 607–616.

Chon, K. S. (1990). The role of destination image in tourism: A review and discussion. *Tourist Review, 45* (2), 2–9.

Coban, Suzan. (2012). The Effect of Tourist Destination Satisfaction and Loyalty: The Case of Cappadocia. European *Journal of Social Science, 29* (2): 222–232.

Cooper, Donald R., & Pamela, S. Schindler. (2006). *Business Research Methods*, Volume I. PT Media Global Education. Jakarta.

Court, B. & Lupton, R. A. (1997). Customer portfolio development: Modeling destination adopters, inactive and rejecters. *Journal of Travel research, 36* (1), 35–43.

Crockett, S. R. & Wood, L. J. (2002). Brand Western Australia: Holidays of Entirely Different Nature. Morgan, N., Pritchard, A., & Pride, R. (Ed.), Inside Destination Branding: Creating the Uniqe Destination Proposition, Butterworth-Heinemann: Oxford, 124–147.

Ekinci, Y. & Hosany, S. (2006). Destination Personality: An Application of Brand personality to tourism destination. *Journal of Travel Research, 45,* 127–139.

Ekinci, Y., Sirakaya-Turk, E., & Baloglu, S. (2007). Host Image and destination personality. *Tourism Analysis, 12,* 433–446.

Etchner, C. M. & Ritchie, J. R. B. (2003). The meaning and measurement of destination image. *Journal of Tourism Studies, 2,* 2–12.

Farida Naili (2013) Effect of Service Quality, Tourism Facilities, Promotion on Destination Imagery and Intention to Behave in Tourism Objects of Karimunjawa, Jepara Regency

Ghozali, Imam. (2011). *Application of Multivariate Analysis with the SPSS Program*. Five Edition. Semarang: Diponegoro University.

Hair, J. F. Jr., R. E. Anderson, R. L. Tatham, & W. C. Black. (2010). *Multivariate DataAnalysis*. 5th edition. New York: Prentice Hall

Hosany, S., Ekinci, Y., & Uysal, M. (2006). Destination image and destination personality: An application of branding theories to tourism places. *Journal of Business Research, 59*, 638–642.

Hsu, C. H. C., Cai, L. A., and Li, Mimi., (2010). "Experience, Motivation, and Attitude: A Tourist Behavioral Model," *Journal of travel research, 49* (3) 282–296.

Mowen, J, C. & M. Minor. (2010). *Quality Perception*. Erlangga Publisher. Jakarta.

Oleary, S. & Deegan, J. (2003), "People, Pace: Qualitative and Quantitative Image of Ireland's Seize Tourism Destination in France," *Journal of vacation Marketing, 9* (3), pp. 213–226.

Peter, J. P. & J. C. Olson. (2002). Consumer Behavior: Consumer Behavior and Marketing edition fourth edition. Jakarta: Erlangga.

Qu, H., Hyunjung Kim, L., & Hyunjung Im, H. (2011). A model of destination branding: Integration and destination image. *Tourism Management, 32*, 465–476.

Now, Uma. (2006). Reaserch Methods for Business. Volume 2. Fourth Edition. Jakarta: Salemba Empat.

Schiffman, Leon G., and Leslie Lazar Kanuk. (2010). *Tenth Edition Consumer Behavior*. Pearson Education.

Tasci, A. D. A. & Kozak, M. (2006). Destination brands vs. destination image: Do we know what we mean? *Journal of Vacation Marketing, 12* (4), 299–317.

Wasesa, Supreme Silih. (2006). Public Relations Strategy. Jakarta: Pt. Gramedia Main Library.

Zeithaml, Valarie A, & Bitner, MJ. (2003). Marketing Service. Tata McGraw-hill.

The effect of servicescape in Tasik Halal Culinary Festival on the image of Tasikmalaya

T.A. Koeswandi, S. Sulastri & A. Fauziyah
Universitas Pendidikan Indonesia, Bandung, Indonesia

ABSTRACT: This study aims to determine the effect of the servicescape of Tasik Halal Culinary Festival on the image of Tasikmalaya City. This research is a descriptive quantitative study involving the visitors of the Tasik Halal Culinary Festival as the respondents. The data were collected through interviews, observations, questionnaires, and literature studies. The data were analyzed by simple regression where the variable (x) is servicescape and the variable (y) is the image of the City of Tasikmalaya. The data obtained were analyzed using SPSS 20.0 software. The results of this study show that t-count $8.644 >$ t-table 1.665 with Sig. equal to 0,000, which means that the variable of servicescape is positive and partially gives a significant effect on the image of the city of Tasikmalaya. The R-square number of 43.2% shows the influence of the free variable (servicescape) on the dependent variable (the image of Kita Tasikmalaya). This implies that the more representative the concept of halal servicescape in a culinary festival, the better the image of the city will be built.

1 INTRODUCTION

Tasik Halal Culinary Festival is a program of the Tasikmalaya City Government and the Ministry of Tourism in promoting the image of the new Tasikmalaya City as a halal food destination city (Prodjo 2017). The decision that was initiated in early 2017 is based on the high local potential in terms of facilities, natural resources, and human resources. In addition, forecasting until 2025, Indonesia is predicted to have a great opportunity in developing the halal food product industry as shown in Figure 1. According to the Islamic (2019) "Muslim consumer spending on halal food & lifestyles in 2014 was reported worth 1.8 trillion US dollars which is expected to increase to 2.6 trillion US dollars by 2020. These figures show excellent growth potential for halal food products and even industrial service categories."

Figure 1. Forecasting data of halal food in Indonesia according to the market research report.

As a manifestation, the Tasikmalaya City Government is increasingly active in supporting and facilitating the formation of a new city image. One way is to help small and medium industry businesses through free culinary certification assistance (Suryarandika 2018). The aim is to provide a halal label and National Agency of Drug and Food Control (BPOM) number on the product being marketed. As a religious city of Santri, of course halal certified culinary products can add value to the new Tasikmalaya City image and become a superior product because of the long-life product lifecycle on the market.

During the three years the program was designed, the Tasik Halal Culinary Festival involved more than 100 MSMEs. Although the theme is halal food, this program is designed with the physical environment (servicescape) of local wisdom where the design theme actually refers to the creative industries of Tasikmalaya. After this event was held, the number of tourists, both local and foreign, did not experience a significant increase (Department of Youth, Sports 2019).

Visitors as consumers have dynamic natures and behaviors. Shifting the consumer paradigm from product oriented to service and experience oriented challenges entrepreneurs and marketers to provide memorable services and experiences for consumers (Kertajaya 2015). Through sensory, the value of service and experience are created. So an integrated servicescape is needed to be able to create a picture of what consumers want to receive.

Therefore, the author would like to know and study the description of the servicescape formed by the

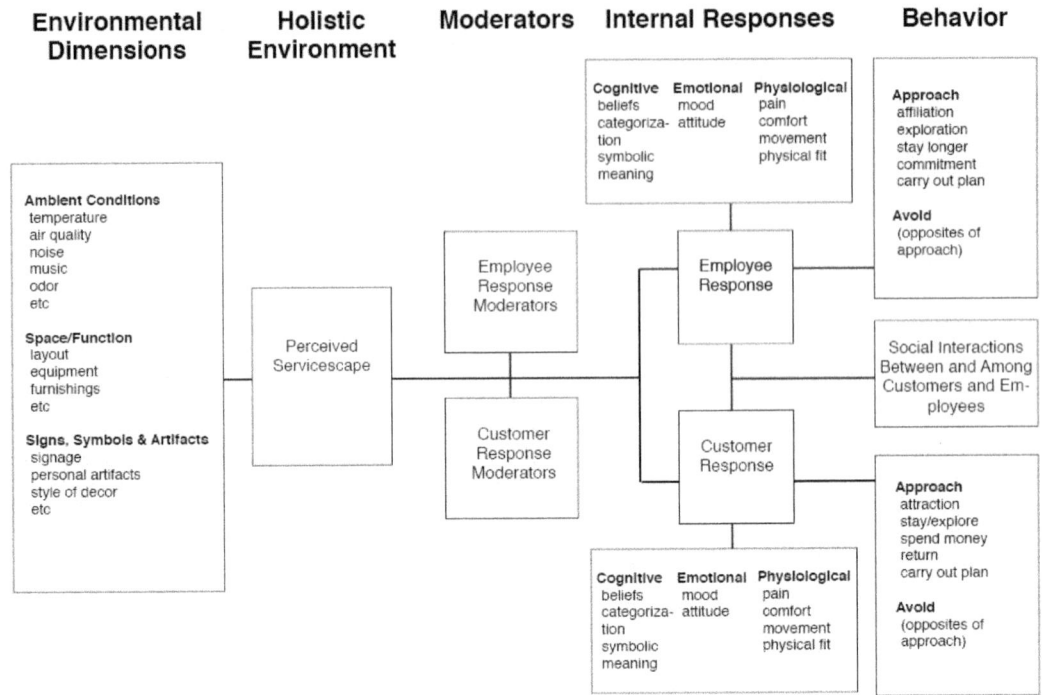

Figure 2. Bitner's (1992) servicescape model.

participants of the Halal Culinary Festival and its influence on the new image of the city of Tasikmalaya as a new image of the halal food destination city.

1.1 Servicescape

Servicescape or physical environment can be conceptualized as intangible environments that are formed through the dimensions of Ambient Condition, Spatial Layout and Functionality, Sign Symbol, and Artifact and perceived by consumers through stimuli and senses to create the atmosphere, impression, or other goals of the company (Bitner 1992). This concept was introduced in 1992, and is applied in the service sector and several industries such as hospitality, hospitals, and culinary. Servicescape is widely used, especially in the era of consumer paradigm shifting where consumer orientation has changed from product oriented to service and experience oriented. Indirectly, servicescape is able to exert influence on perceptions, behavior, and repurchase decisions (Bitner 1992; Bora et al. 2018). So that the management of service servers is required to be able to multiply and build an atmosphere. Through this, a communication of service can be delivered (Mohd et al. 2015). The model shown in Figure 2 is the servicescape model according to Bitner (1992).

This research involved 7 dimensions of servicescape suggested by Rahayu et al. (2017), namely music, cleanliness, temperature, lighting, color, layout/design, and fragrance.

1.2 City image

The image of the city cannot be separated from the role of the community because the image of the city is formed due to the perception built by the surrounding community. The image of the city is formed through the physical environment and is supported by meeting the needs of its people. A good city image usually provides good facilities to the needs of the community. A positive city image can also be defined because the city is able to provide a pleasant experience for the community and visitors (Atik 2015). The dimensions of the city image consist of path, edge, district, nodes, and landmarks (Lynch 1960).

1.3 Tasik halal culinary festival

The Tasikmalaya City Government and the Ministry of Tourism together have a program to redefine the image of the city of Tasikmalaya to become a halal food destination in the east. For this reason, an event program was designed and has been implemented from 2017 until now. The main series of the program is called the Tasikmalaya October Festival (TOF), which is participated in by 75 UMKM halal food businesses. In this festival, the Tasik Halal Culinary Festival, Tasik Investment Expo & Conference (TIEC), Tasikmalaya Culture & Craft Festival, and Tasikmalaya Creative Festival (TCF) were held. More than 200 booths will decorate the exhibition stage, which is centered on HZ Mustafa and Yudhanegara Streets. This activity is a

Figure 3. A stand bazaar in Tasik Halal Culinary Festival, 2019.

Figure 4. Another stand bazaar in Tasik Halal Culinary Festival, 2019 (Source: Author).

manifestation of the vision and mission of the city government to make Tasikmalaya become a trading center by 2025. The ambiance of Tasik Halal Culinary Festival can be seen in Figures 3 and 4:

2 METHODS

This research is a quantitative descriptive research which included 120 visitors of the Tasik Halal Food Festival, 2019, as the sample, both local visitors from Tasikmalaya and non-local visitors. The sample technique chosen was incidental sampling, which is a sampling technique based on coincidence such as anyone who incidentally meets with researchers at the Tasik Halal Culinary Festival, 2019. Interviews using questionnaires, observations, documentation, and also literature studies were conducted to collect data. The data were then analyzed using SPSS 20.

3 RESULTS & DISCUSSION

The result of the influence of the servicescape to the image of Tasikmalaya City as a halal culinary destination can be seen in Table 1.

Table 1 shows the magnitude of the correlation or relationship (R) value, which is equal to 0.314. From the output, a coefficient of determination (R square) of 0.098 was obtained, which implies that the effect of

Table 1. Model summary.

Model	R	R Square	Adjusted R Square	Std. Error of the Estimate
1	.314[a]	.098	.091	3.62107

a. Predictors: (Constant), Servicescape

Table 2. Anova.

Model		Sum of Squares	Df	Mean Square	F	Sig.
1	Regression	168.695	1	168.695	12.866	.000[b]
	Residual	1547.230	118	13.112		
	Total	1715.925	119			

a. Dependent Variable: City Image
b. Predictors: (Constant), Servicescape

Table 3. Coefficients.[a]

Model	Unstandardized Coefficients		Standardized Coefficients		
	B	Std. Error	Beta	t	Sig.
(Constant)	13.509	3.422		3.948	.000
Servicescape	.322	.090	.314	3.587	.000

a. Dependent Variable: City Image

the variable servicescape on the image of the city of Tasikmalaya is 9.8%.

Table 2 shows the calculated F value $= 12,866$ with a significance level of $0,000 < 0.05$. The regression model can be used to predict participation variables, or in other words, servicecape has an effect on the image of the city of Tasikmalaya.

Table 3 shows the constant value (a) is 13.509, while the value of servicescape (b/regression coefficient) is 0.322. So that it can be explained that the constant of 13.509 implies that the consistent value of the Tasikmalaya City's image variable is equal to 13,509. Meanwhile the servicescape coefficient regression that is 0.322 tells that for each addition of 1% servicescape value, the value of the image of the city of Tasikmalaya will increase by 0.322. The regression coefficient is considered positive, so it can be said that the direction of the influence of servicescape on the image of the city of Tasikmlaya is also positive.

From Table 3, it can be found that the significance value is $0,000 < 0.05$. Thus, it can be concluded that servicescape affects the image of the city of Tasikmalaya. From Table 3 it can also be seen that the t-value is $3,587 > $ t-table $2,618$. So, it can be concluded that the servicescape variable affects the image of the city of Tasikmalaya.

So that the model of the influence of servicescape on the image of the city of Tasikmalaya as a Halal Culinary Tourism City can be drawn as follows:

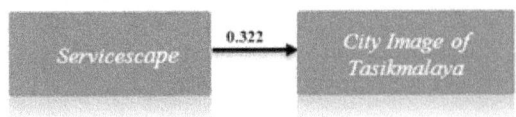

Figure 5. Model of the influence of servicescape to the city image of Tasikmalaya as a halal culinary city.

To sum up, this research is in line with previous studies of Bitner (1992) and Bora et al. (2018) who stated that servicescape is able to exert influence on perceptions, behavior, and repurchase decisions, which leads to a perceived image. This also supports a previous study of Atik (2015) who stated that a positive city image can also be defined because the city is able to provide a pleasant experience for the community and visitors.

4 CONCLUSION

Servicescape or physical environment can be conceptualized as intangible environments that can be divided into several dimensions such as music, cleanliness, temperature, lighting, color, layout/design, and fragrance, which can influence how people perceive an image of a city. In the case of Tasikmalaya, it can be concluded that the city government of Tasikmalaya and the Ministry of Tourism had successfully redefined the image of the city of Tasikmlaya to become a halal food destination in the East Parahyangan through the Tasik Halal Culinary Festival event. The servicescape that had been built could make the visitor feel and perceive the image of " halal." As a further suggestion, it is suggested to find out how Tasikmalaya can maintain the image and what kind of strategy will be the best for marketing the image and also the event.

ACKNOWLEDGEMENTS

This research was supported by Universitas Pendidikan Indonesia with the Research Grant Number: Universitas Pendidikan Indonesia Fiscal Year 2019 With Chancellor's Decree Number: 5493/UN40/KP/2019. Dated May 28, 2019.

REFERENCES

Atik, D. 2015. The importance of city image elements through recognizing and introducing of cities: ed i rne example the importance of city image elements through recognizing and introducing of cities: ed i rne. *Trakia Journal of Science* 7 (August).

Bitner, M. J. 1992. Servicescapes: The impact of physical surroundings on customer and employees. *Journal of Marketing* 56 (April): 57–71.

Dedeoglu, B. B., Bilgihan, A., Ye, B. H., Buonincontri, P., & Okumus, F. 2018. The impact of servicescape on hedonic value and behavioral intentions: The importance of previous experience. *International Journal of Hospitality Management* 72: 10–20.

Department of Youth, Sports. 2019. *Open Data Tasikmalaya.* [Online]. Retrieved from https://data.tasikmalayakota.go.id/dinas-kepemudaan-olahraga-kebudayaan-danpariwisa ta/data-tempat-wisata-kuliner/ (Accessed March 2, 2019).

Islamic, G. E. 2019. Indonesia economic Islamic masterplan 2019–2024. Jakarta: Ministry of Development National Planning.

Kertajaya, H. 2015 Indonesia WOW markplus WOW we are WOW. Jakarta: Gramedia Pustaka Utama.

Lynch, K. 1960. The image of the city (Vol. 11). Cambridge: MIT press.

Mohd, N., Ma, H., & Ariffin, N. 2015. Servicescape: Understanding how physical dimensions influence Exhibitors Satisfaction in Convention Centre. *Elsevier B.V* 211(September): 776–782.

Prodjo, W. 2017. *Tasikmalaya Didorong Menjadi Kota Kulier halal.* [Online]. Retrieved from http://travel.kompas.com/read/2017/10/06/062100827/tasikmalaya-didorong-menjadikota-kuliner-halal (Accessed March 3, 2019).

Rahayu, A., Koeswandi, T., & Wibowo, L. A. 2017. Korean restaurant atmosphere and impression: Impact on repurchase intention. *Advanced Science Letters* 24(12): 9324–9326(3).

Suryarandika, R. 2018. *Pemkot tasikmalaya siap bantu sertifikasi 200 usaha kuliner.* [Online]. Retrieved from https://www.republika.co.id/berita/nasional/daerah/18/02/17/p4aing438pemkot-tasikmalaya-siap-bantu-sertifikasi-200-usaha-kuliner (Accessed March 2, 2019).

Advances in Business, Management and Entrepreneurship – Hurriyati et al. (Eds)
© 2021 Taylor & Francis Group, London, ISBN 978-0-367-67471-7

The effect of e-servicescape on e-trust (online survey on Shopee mobile application's users in Bandung)

T.A. Koeswandi & E.A. Primaskara
Universitas Pendidikan Indonesia, Bandung, Indonesia

ABSTRACT: This study aims to find out the effect of e-servicescape on e-trust on Shopee website. This study is a descriptive quantitative involving 100 Shopee's users who live in Bandung and had accessed the mobile application. The data were collected using online questionnaire and interviews. The data were analysed by simple regression where the variable (x) is the e-servicescape and variable (y) is the e-trust. The result shows that e-servicescape has a positive and partially gives significant effect on the e-trust in Shopee mobile application. This implies that the more representative the e-servicescape in a mobile application, the deeper trust will be built by the users.

1 INTRODUCTION

Indonesia is one of the countries with the highest number of internet users in the world. Until December 2017, 143.3 million out of the country's total population of over 260 million were active internet users. However, almost 30% of the accessed the internet from their mobile phone. This figure is expected to raise to almost 36 percent by 2023 (Muller 2019).

The internet exposure comes along with the rapid advancement in technology. Utami (2012) states that technology plays an important role in the development of online business. One of the form of the technology advancement is the existance of mobile application that is commonly used by e-commerce. GlobalWebIndex reports that Indonesia has the highest rate of ecommerce use of any country in the world, with 90 percent of the country's internet users between the ages of 16 and 64 reporting that they already buy products and services online.

Shopee is one of the e-commerce that has its own mobile application in running its business. It is an online marketplace that comes in the form of a mobile application and offers a variety of products ranging from fashion products to products for everyday needs. It began entering the Indonesian market at the end of May 2015 and only began operating at the end of June 2015 in Indonesia. Shopee is a subsidiary of Singapore-based Garena. Shopee has been present in several countries in Southeast Asia such as Singapore, Malaysia, Vietnam, Thailand, the Philippines and Indonesia. Shopee presents in Indonesia to bring a new shopping experience. Shopee facilitates sellers to sell easily and provides buyers with a secure payment process and integrated logistics arrangements. At present, Shopee download numbers have reached one million downloads on the Google Play Store. However, Bayu (2018) states that online marketplace like shopee is facing some complaints from customer. Bayu (2018) add that during 2017, there are several things that become the concern from customers such as the slow response of complaints (44%), goods have not been received (36%), the system is detrimental (20%), no refunds (17%), suspected fraud (11%), items purchased are not appropriate (9%), suspected cyber crime (8%). There are also complaints about product defects (6%), service (2%), price (1%), information (1%), and late receipt of goods (1%). Thus, Shopee changes the look of the mobile app or e-servicescape to make it easier to use & completes it with the detailed product features.

According to Harris & Godee (2010), e-servisccape is "the online environment factors that exist during service delivery" in which it has 3 dimensions, those are aesthetic appeal, online layout and functionality, and financial security. (Tankovic & Benazic 2018) finds consumers' interpretation of e-servicescape shows a positive influence over perceived e-shopping value and loyalty which indicates trust from consumers. Thus, further research should be done to find out whether e-servicescape affect the e-trust of shopee consumers.

1.1 E-servicescape

There are two terms that the definition looks alike, they are servicescape and e-servicescape. According to Rahayu et al. (2017), servicescape can be considered as physical environment that can help costumer to perceive what company wants to present. In their research, there are 7 dimensions of Servicescape namely music, cleanliness, temperature, lighting, color, layout/design, and fragrance.

Meanwhile, according to Harris & Godee (2010), e-servicescape represents the online environment. Previous studies have proven that it could affect some other areas. Wu et al. (2017) found that e-servicescape dimensions (aesthetic appeal, customization, usability, and financial security) have significant impacts on consumer attitudes and trust toward a website. In the previous research about e-servicescape, Man et al. (2014) stated that e-servicescape has significant and positive effect on customer flow experience. Meanwhile, (Hakim & Deswindi 2015) found e-servicescape dimensions have a positive and strong influence on the perceived quality.

1.2 E-trust

Harris & Goode (2004) stated that e-trust is central to online service dynamics. It represent how consumers feel toward the online product/service. Harris & Goode (2010) found that trust and purchase intentions are affected by consumers' interpretations of online environments. Meanwhile (Kassim & Asiah 2010) found that perceived service quality have a significant impact on customer satisfaction which in turn have a significant effect on trust.

1.3 Shopee

Shopee is an electronic trading platform which is headquartered in Singapore under the SEA Group (formerly known as Garena), which was founded in 2009 by Forrest Li. Shopee was first launched in Singapore in 2015, and has since expanded its reach to Malaysia, Thailand, Taiwan, Indonesia, Vietnam and the Philippines. Because the mobile element is built according to the concept of global electronic commerce, Shopee is one of the "5 most disruptive e-commerce startups" published by Tech in Asia. Shopee itself is led by Chris Feng. Chris Feng is one of the former Rocket Internet activists who once headed Zalora and Lazada. When compared with other sites such as Bukalapak, Tokedia, OLX and others, shopee is the youngest and lacks experience. But with the promotion that was able to stand in line with these previous competitors.

The ambiance of Shopee Mobile Application can be seen in the following figure:

2 METHODS

This study is a descriptive quantitative involving 100 Shopee's users who live in Bandung and had accessed the mobile application. The data were collected using online questionnaire and interviews. The data were analysed by simple regression where the variable (x) is the e-servicescape and variable (y) is the e-trust.

3 RESULTS & DISCUSSION

The result of how influence the servicescape to the image of Tasikmalaya as city of halal culinary destination can be seen in the following table:

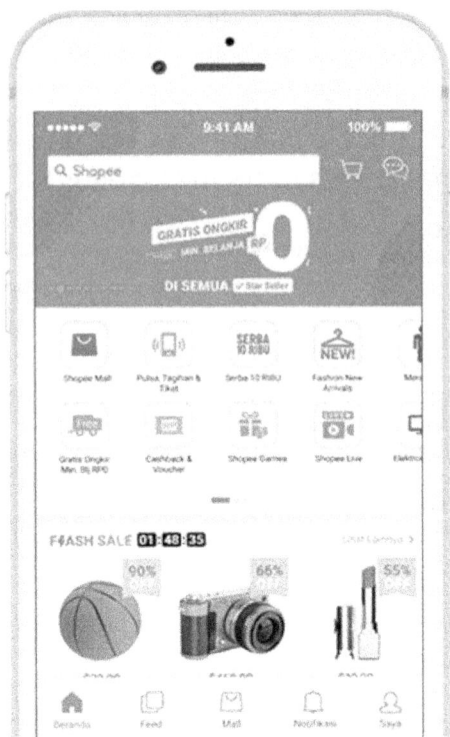

Figure 1. Shopee mobile application's e-servicescape.
* Source: Shopee.co.id.

Table 1. Model summary.

Model	R	R Square	Adjusted R Square	Error of Std. the Estimate
1	.309[a]	.096	.087	3.55640

a. Predictors: (Constant), e-Servicescape

Table 2. Anova.

Model		Sum of Squares	df	Mean Square	F	Sig.
1	Regression	131.247	1	131.247	10.377	.002[b]
	Residual	1239.503	98	12.648		
	Total	1370.750	99			

a. Dependent Variable: e-Trust
b. Predictors: (Constant), e-Servicescape

Table 1 shows the magnitude of the correlation or relationship (R) value that is equal to 0.309. From the output obtained a coefficient of determination (R square) of 0.096 which implies that the effect of the variable e-servicescape on e-trust is 9.6%.

Table 2 shows the calculated F value = 10.377 with a significance level of $0.000 < 0.05$. The regression

Table 3. Coefficients.[a]

Coefficients[a]

Model	Unstandardized Coefficients B	Std. Error	Standardized Coefficients Beta	t	Sig.
1 (Constant)	14.046	3.620		3.880	.000
e-Servicescape	.307	.095	.309	3.221	.002

a. Dependent Variable: e-Trust

model can be used to predict participation variables or in other words, e-servicecape has an effect on the e-trust.

Table 3 shows the constant value (a) of 13.509 while the value of eservicescape (b/regression coefficient) of 0.307. So that it can be explained that the constant of 14.046 implies that the consistent value of e-trust is equal to 14.046. Meanwhile the e-servicescape's coefficient regression that is 0.307 tells for each addition of 1% e-servicescape value, the value of the e-trust will increase by 0.307. The regression coefficient is considered positive, so it can be said that the direction of the influence of e-servicescape on e-trust is also positive.

From Table 3, it can be found that the significance value is $0,000 < 0.05$. Thus, it can be concluded that e-servicescape affects the e-trust. From Table 3 it can also be seen that the T-value is $3.221 > \text{T-table } 2.618$. So, it can be concluded that the e-servicescape variable affects the e-trust.

So that, the model of the influence of e-Servicescape on the image of the e-trust can be drawn as follows:

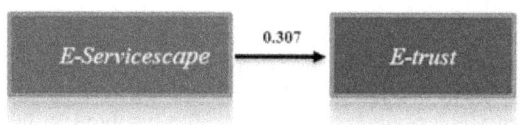

Figure 2. Model of the influence of e-servicescape to e-trust (Survey online on Shoppe mobile application's users in Bandung).

To sum up, this research is in a line with previous study of Wu et al (2017) that e-servicescape dimensions (aesthetic appeal, customization, usability, and financial security) have significant impacts on e-trust in a mobile application of e-commerce.

4 CONCLUSION

This study concludes that e-servicescape has a positive and partially gives significant effect on the e-trust in Shopee mobile application. This implies that the more representative the e-servicescape in a mobile application, the deeper trust will be built by the users.

As a further suggestion, it is suggested to find out how e-trust can influence the customer loyalty in e-commerce mobile application.

REFERENCES

Bayu, D.J. 2018. *YLKI: keluhan terbanyak konsumen selama 2017 soal toko online*. [Online]. Retrieved from katadata.co.id: https://katadata.co.id/berita/2018/01/19/ylki-keluhan-terbanyak-konsumen-selama-2017-soal-toko-online.

Hakim, L. & Deswindi, L. 2015. Assessing the effects of e-servicescape on customer intention: a study on the hospital websites in South Jakarta. *Procedia – Social and Behavioral Sciences* 169(August 2014): 227–239.

Harris, L.C. & Goode, M.M. 2004. The four levels of loyalty and the pivotal role of trust: a study of online service dynamics. *Journal of retailing* 80(2): 139–158.

Kassim, N. & Asiah, A.N. 2010. The effect of perceived service quality dimensions on customer satisfaction, trust, and loyalty in e-commerce settings: A cross cultural analysis. *Asia Pacific Journal of Marketing and Logistics* 22(3): 351–371.

Man, L., Qinhai, M., & Xiaoyu, Z. 2014. A study on the effects of e-servicescape on online experience and behavior intention. *Journal of Management Science* 27(4): 86-96.

Muller, J. 2019. *Internet usage in Indonesia – Statistics & Facts*. [Online]. Retrieved from www.statista.com: https://www.statista.com/topics/2431/internet-usage-in-indonesia/.

Rahayu, A., Koeswandi, T., & Wibowo, L.A. 2017. Korean restaurant atmosphere and impression: impact on repurchase intention. *Advanced Science Letters* 24(2): 9324–9326.

Tankovic, A.C. & Benazic, D. 2018. The perception of e-servicescape and its influence on perceived e-shopping value and customer loyalty. *Online Information Review* 42(7): 1124–1145.

Utami, S.S. 2012. Pengaruh teknologi informasi dalam perkembangan bisnis. *Jurnal Akuntansi dan Sistem Teknologi Informasi* 8(1).

Wu, W., Quyen, P.T.P., & Rivas, A.A.A. 2017. How e-servicescapes affect customer online shopping intention: the moderating effects of gender and online purchasing experience. *Inf Syst E-Bus Manage* 15: 689–715.

Advances in Business, Management and Entrepreneurship – Hurriyati et al. (Eds)
© 2021 Taylor & Francis Group, London, ISBN 978-0-367-67471-7

Exploring halal destination perceived value in tourist destination: Insight from Indonesia

S. Lenggogeni & S. Febrianni
Universitas Andalas, Padang, Indonesia

ABSTRACT: Although the significance of perceived value and perceived risks is well documented in the tourism discipline, the exploration on the religious segment-based together with a specific type of destination and the process through its underlying dimensionalities identification have not yet to be understood. The principal objective of this study was to investigate the underlying dimensions of perceived value and risks for Muslim travelers from the perspective of bottom up approach in the heritage destinations. Using a qualitative approach, semi structured interview were conducted with a 32 Indonesian Muslim domestic market who had a visit experience in Indonesia heritage destinations in the last two years. Seven underlying dimension of Indonesian Muslim travelers perceived values and risks were identified. The study implies that tourists' religiosity and their cultural background have formed the underlying dimensionalities of perceived value for Muslim tourist market. These findings provide a solid evidence based for Global Muslim Perceived Value and Muslim Travel Risks in tourism marketing discipline.

1 INTRODUCTION

Recently, there has been a renewed interest in customer perceived value by hospitality and tourism scholars. Perceived value is a key aspect of tourist destination as well as other hospitality industry. Previous studies have reported that this concept mostly operationalized by single-item scale and single method approach. Yet, the validity of unidimensional measure is frequently attracting the criticism due to its assumption that consumers have a shared meaning of value. On the other hand, multidimensional scale is possible to overcome the validity problem by their operationalization of perceived value variable in various setting of hospitality industry ecosystem (Choe & Kim 2019; El-Adly 2019). Scholars also have documented the role of perceived value with satisfaction and behavioral intentions in tourism industry (Gallarza & Saura 2006; Pham et al. 2020).

In the same way, perceived risk is a primary concern of tourist in behavioural intention (Lenggogeni et al. 2019). Risk perception defined as the uncertainty and consequence of decision outcome (Bauer 1960; Sharifpour et al. 2014). It firstly introduced by Bauer (1960) in marketing discipline. In the history of development of perception of risk in marketing to tourism and leisure studies, scholars have concluded that perceived or travel risk is multidimensional perception (Lenggogeni 2015; Park & Tussyadiah 2017) and contextual based (Dolnicar 2005). It has been discussed from a general travel risk to a crises and disaster context travel risks (Lenggogeni & Saito 2018). Determining

the impact of perceived risk on decision making is important, since it most likely to reduce tourism product purchase (Rittichainuwat, et al. 2009; Sharifpour et al. 2014a; Wang et al. 2019). Since the tourism is very contextual based (Ritchie 2009), travel risk dimensionalities less likely to be generalized accross tourist segmentation, setting and type of travel.

Both of perceived value and perceived risk is simultaneously assessed by tourism scholar in tourism industry (Şen Küpeli & Özer 2020). Literature has shown different type of value and risk dimensionalities across many contexts of travel type, destination and tourist profile. For instance, since Roehl & Fesenmaier first introduced them in the context of leisure in 1992; Sönmez & Graefe (1998) has brought this concept into the context of terrorism. Since perceived value and risk perception is very contextual, the exploration should be based on the traveller's perspective (Simpson & Siguaw 2008). Hence, Lenggogeni et al. (2019) has introduced a bottom up approach to explore tourist perception associated risk in tourism crises and context studies.

One of the movement of new segmentation on religious based is Muslim travellers, a market that requires a specific-needs based on Islamic sharia compliant. According to Battour & Ismail (2016) this market is captured as one of the rapid growing market segments values in tourism industry today's. This market, interestingly, has spent about US $177 billion in 2017 and expected to reach US $274 billion by 2023 (Thomson & Dinar 2018). Therefore, investigating their travel perceived values for a tourism industries, is one of prospective business values that witnessed by destination marketer and hospitality industries.

Despite the concept of Halal, sharia or pilgrim tourism are closely intertwined (Battour & Ismail 2016; Jafari & Scott 2014), but the concept of halal tourism is one of the most often practically adopted in many Muslim and Non Muslim countries. Halal tourism is leisure tourism for Muslim tourists where there is for the availability of tourism products and services in accordance with Islamic norms as well as the convenience of performing worship services for Muslim during their trip (Battour & Ismail 2016). The phenomenon of Halal tourism most likely developed not only due to their belief needs but also their travel risk, due to the emergence of Islamophobia that exist in global world today's. In particular, since there is a significant increase in the number of terrorism act after 9/11 event in United States, followed by the media exposures that highlight the appearance of Muslim as terrorist, Muslim traveller market has raise a demand for safety needs during their Internasional trips. That Islamophobia fears may have an impact not only physically but also psychologically for Muslim tourist. Thus, safety issue could be potentially important for Muslim tourist prior their holiday planning trip particularly when they plan to travel to a countries where Muslim as a minorities.

With the various religious travel needs attached for Muslim tourist in travel industries, there is a limited study that explored travel perceived values associated to destination and tourism industries. While there is a growing numbers of Sharia accomodation around the world that cater Muslim tourist needs, there is still a fuzzy concept for this industries in providing basic elements of Sharia compliant values that need to be fulfilled by this market. In addition, inavalibillity of regulation to certify this industries making the sharia compliant industries difficult to be standardized. For example, while some sharia compliant accomodation provide hotel service that comply with Islamic teaching such as Qibla direction, Halal food, alcohol-free beverages, and prayer room with call of prayers (Battour & Ismail 2016) others may only serve the Qibla direction and prayer rug only. For example, most of Muslim tourist put the Halal food is the most important travel needs that should be provided in a hotel when they visit a destiantion where Muslim is a minority or non Muslim destination (Battour et al. 2011). This is the basic needs that needs to be fulfilled during their trip because according to Islamic law, eating pork or pork by products, animals that were dead prior slaughtered properly or not slaughtered with pronouncing on the name of Allah, blood and blood by products, alcohol, carnivour animals, birds of prey is prohibited. This is also to true to others tourism services industries such as destinations, travel agent, restaurant, or airport (Kamali 2013)

The Muslim market perceived values attached in their travelling ecosystem that based on Sharia compliant such as the needs of halal food and drinks, halal services and entertainment, Islamic dress codes, prayer facilities and other factors (Battour et al. 2011). Furthermore, Muslims travellers typically require the specific values during their trip, for example Female Muslim travellers that demand to wear hijab or specific a dress code for male, they would rather to avoid alcohol, they also require Halal food (Eid & El-Gohary 2015). Undoubtedly, religious beliefs influence and direct Muslim adherents to travel to particular sites and influence their attitudes and behavior, perceptions, and perhaps emotions at those sites (Jafari & Scott 2014). To explore what Muslim tourist perceive about tourist destination, an existing literature that provide single method research in exploring Muslim tourist perceived values is less likely to capture a deeper insight of true values that Muslim tourist really need.

While there is a growing number on halal tourism to serve the muslim traveller market needs, few studies explored unique value proposition on the basis of traveller perspective in tourism literature. Early work by Battour et al. (2011), Eid & El-Gohary (2015), Henderson (2016), Jafari & Scott (2014) have introduced the importance of Muslim traveller market in tourism industries, yet the exploration of their value needs and risks has not been extensively much evaluated in tourism literature. In fact, Muslim market accounted as the largest tourist market in the world (Eid & El-Gohary 2015) and projected to lead as the biggest religion amongst others population in the world (Pew Research Centre 2017) by 2060. A number of Muslim countries as well as non Organization Islamic Coorporation (OIC) has prepared a strategic plan to work with this market in tourism industries. Indonesia for example. The Indonesian Ministry of Tourism set the top ten portofolio of Halal tourist destination including Lombok, Aceh, Jakarta, West Sumatra, West Java.

Accordingly, the need to adopt a bottom up approach study using a qualitative perspective is essential to a gain a deeper insight of tourist perceived value and risk across contextual, segmentation and type of destination. Since religion and tourist decision making is detachable (Lenggogeni & Saito 2018), understanding the perceived value and perceived risk for the perspective of Muslim traveller important. This study aims to identify the Muslim travel perceived value and perceived risk in the heritage destination using the bottom up approach.

2 METHODS

2.1 *Qualitative approach*

To identify the underlying dimensions of Muslim Tourist Perceived Value and their perceived risk, this study adopts a qualitatitve research using a semi structured interview technique. Qualitative approach allows the researcher to obtain a deeper insight of phenomenon as well as to understand the lived experience from informants to enhance the meanings of the study's result (Hillman & Radel 2018). After a pilot test with ten interviewees to check the interview guidelines realibility and validity, a thirty-two interviewees from domestic tourists were obtained by using a semi-structured interview. The target informants were a tourist who experienced who experienced visiting cultural tourist destination in Indonesia. According to Morse (2000) a minimum of 30 and a maximum of 60 respondents for semi-structured interviews is necessary to obtain the richness of data that is required. Using a convinience sampling, the respondents were approached on the spot survey in several heritage sites in West Sumatra. Once the respondents agreed, they were asked to fill the informed consent and permission to have the conservation recorded. The duration of interview was about 30 minutes for each of respondents. This study use the Sweeney & Soutar (2001) and Eid & Gohary (2016) guidelines in thematic data (Clarke & Braun 2013) process. It was validated by member checking of random respondents.

3 RESULTS AND DISCUSSION

The interview were conducted in several heritage tourists attraction in West Sumatra from November to December 2018. The respondent profile associated to demographic and destination preference are presented in Table 1.

As presented in Table 1, in all, a thirthy-two respondents indicated themselves as Muslim, therefore according to this present study's research aim, all informants were eligible to participated in this study. In the response to the question of their favourite cultural and heritage destination choice, almost half of respondents nominated Pagaruyung Great Castle of West Sumatra (45,4%) as the first choice to visit, meanwhile Borobudur and Prambanan temples in Central Java (27.3%) ranked second and third favourite cultural destination respectively and rest of them choosed the Big Ben tower in Bukittinggi, West Sumatra and Siak Palace in Riau.

Over half of respondents were female (72.7%) while male is only accounted for 27.3% and they were dominated by generation Z (78.8%) followed by generation Y (12.1) % and the rest of 9.1 % were from generation X. The majority of respondents were well educated as they attained senior high school (78.8%), followed by respondent with bachelor's degree (12.1%) and the rest hold their master degree (9.1%). The majority of respondents indiated them selves as single single traveller (81.8%), and the rest are family travellers (18.2%). With regard to respondent's occupation, the largest respondents were students (75.8%), while 18.2% of them are white collars (18.2%) and the rest were engaged as entreprenuer (6%).

In response to the question: "As a Muslim traveler, tell me about your opinions about the values should be attached in Cultural Tourist Sites?", seven major themes of Muslim Tourist Perceived Value associated to in cultural and heritage tourist attraction, identified in this study as summarized in Table 2.

Seven major themes of perceived values factors were identified in this study. **The first factor** was Quality Values. This factor with eight indicators describes the values of quality that cultural tourist destination needs to perform for tourist. **The second factor** was Emotional Values. The nine indicators retrieved in one factor reflects respondents' emotional and learning values when visiting the destination. **The third and the fourth factors** was Halal Physical Values and Non-Physical Values. Generally, this factor reflects the values related to destination management and people management in tourist cultural sites related that reflect the need of Islamic sharia compliant in the destination. **The fifth factor** was Price Values. The five indicators reflect values related to financial values related to Islamic sharia compliant. **The sixth factor** was cultural destination values. This factor has one indicator that expressed the need for interpretation and knowledge when visiting cultural destination and three indicators related to the needs of cultural atmosphere in destination. **The seventh factor** was Social Values. This factor has two factors that represent respondents' social values when visiting the destination. Related to the travel risk perceived by Muslim traveller, this study found one major themes only that describe tourist uncertainty of inavalaibilty and hygienity issues in worship place, non halal souvenirs (souvenirs that contain pig leather material or display sexual visualization in the products), other female tourist that wear exposed clothes (that generate a fear for sin by Muslim male travellers if they see them according to Islam teaching); inavalaibilty of halal food and non halal tourism events (that perform dancing or dress that not suit to Islamic teaching).

Table 1. Respondent demographic and destinaton preference profile.

Respondent Profile	Response	Percentage
Cultural Site Preferences in Indonesia	Pagaruyung	45.5%
	Candi Prambanan	12.1 %
	Candi Borobudur	15.2 %
	Jam Gadang	21.2%
	Istana Siak	6.1 %
Marital Status	Single	81.8%
	Married	18.2%
Age	Gen Z (17–23 years old)	78.8%
	Gen Y (24–38 years old)	12.1%
	Gen Y (39–53 years old))	9.1%
Education	Senior High School	78.8%
	Bachelor	12.1%
	Master	9.1%
Occupation	Private	3.0 %
	Civil servant	18.2 %
	Student	75.8 %
	Entrepreneur	3.0 %

Source: Result from this study.

Table 2. Major and minor themes of muslim perceived value associated to cultural and heritage tourist attraction using a thematic analysis.

Major Themes	Minor Themes
Quality Value	1. Tour guide & Education
	2. Knowledge
	3. Crowded
	4. Local wisdom experience
	5. Unique Experience
	6. Cultural Preservation
	7. Hospitality / Service
	8. Halal Destination Zone
Price Value	1. Riba'
	2. Reasonable Price
	3. Pricelist
	4. The suitability of price label and product
	5. The suitability of price and experiences
Emotional Value	1. Safety
	2. Excitement
	3. Relax
	4. Happy
	5. Get reward while travelling
	6. Religious
	7. Curiousity
	8. Learning Experience
	9. Religion Learning/Education
Social Value	1. Islam asIdentity
	2. Friendly local people
	3. Ethical and Responsibility tourist based on Islamic sharia compliant
Halal Physical Value	1. History and Culture
	2. The staff uses Shari'a compliant Islamic Muslim clothing
	3. Worship place management
	4. Halal attributes associated to tourist signage
Non-Physical Value	1. Regulations related to Islamic sharia compliant
	2. The certainty of tourist information
	3. Staff with their honest personality
	4. Free from non halal activities
	5. Islamic service care
	6. Halal Tourism Branding
	7. Destination and tourist information

Source: Result from this study.

This study highlight that general perceived values may not be generalized accross religion segmentation based. The general customer perceived value comprises emotional reactions of customer that appears to vary according to cultural differences and the context (Prebensen et al. 2013; Sweeney & Soutar 2001). It includes several attribute such as perceived quality, price elasticity, and consumer satisfaction (Al-Sabbahy et al. 2004). This concept of perceived value began with the works of Anantharanthan Parasuraman et al. (1998) stated "the consumer's overall assessment of the utility of a product based on perceptions of what is received and what is given". Hence, as perceived value is a subjective and dynamic construct that varies among different tourists and cultures at different times, it is necessary to include subjective or emotional reactions that are generated in the consumer's mind (Bolton & Drew 1991; Havlena & Holbrook 1986; Prebensen et al. 2013; Sweeney & Soutar 2001).

1) Quality value

Sweeney & Soutar (2001) emphasizhed that quality is the utility derived from the perceived quality and expected performance of the products. Within the major theme of Halal perceived Value are eight major themes, which were summarized from the 32 comments of respondents made related to this theme. The first most cited minor theme within Halal Perceived Value is the tour guide & education. Seeing the need for tour guides who have insight and knowledge about the history of the culture itself, the respondents highlight this values as follows:

"The tour guides were educated and attained their licenses." (R2)

"It is unfortunate that not all tour guide that well educated and incapable enough to explain the story. There are licensed tour guide from their formal association like HPI (Indonesian Tour Guide Association) which sound goods. However, not all destination that provide the tour guide service... In fact, tourists is really curious to know the real story in the destination. Thus, this potentially to reduce the quality of service in the tourist place itself." (R22)

"I think either Muslim and Non Muslim Tourist want to learn more about Islam and their history as well as what is halal." (R25)

The second most cited minor theme within the Halal Perceived Value is the knowledge. Unlike other tourist attractions, visitors feel the need of additional knowledge when visiting historical sites.

"Cultural tourism is very important issue when visiting heritage destination. That's not only gain the knowledge, but also learning experience that is very useful for future generations to know Indonesian culture is our identity particularly as a Minangkabau cultural." (R15)

"This cultural heritage is interesting to know, I think this is also true for Non Muslim traveller t get the knowledge and historical value of the culture itself." (R7)

The third most cited minor theme within Halal Perceived Value is the destination management. Several respondents stated that they had experienced a crowded situation and lack of regulation about the destination area.

"I found the car park is very crowded, I am not only feeling unsafe but also uncomfortable. The trips also not well organized." (R24)

"Lack of management of local managers regarding regulation, order in the halal tourist destination area." (R19)

The fourth most cited minor theme within Halal Perceived Value is the unique experience. Several

469

respondents merasakan adanya new unique experience if they visit the heritage tourism

"The reason to visit cultural destination is depends on the destination of the tourists themselves. To me I found a "memorable" experience in my memory." (R13)

"Tourists understand more about ethics and behavior in accordance with Islamic teachings. The things that affect me visiting a halal tourist destination area, I want to experience a new tour with Islamic experiences." (R14)

The fifth most cited minor theme within Halal Perceived Value is the Ethnic Culture. Specifically, tourist that visit West Sumatra a province that very based on Islamic culture, eager to learn about Minangkabau history.

"They can see some Islamic cultures served by cultural tourism in Minangkabau. From that point, I can add to my knowledge and insights about the history of Minangkabau.., with these things I am enjoying the atmosphere of the Minang culture given in the tourist destination." (R18)

The sixth most cited minor theme within Halal Perceived Value is the Cultural Preservation. A similar perception about culture, Minangkabau culture to illustrate of historical preservation that should be preserved.

"Like in Minangkabau, the unique value propotion from cultural tourism is traditional or local houses and its history, so besides visiting, they also get knowledge about it." (R29)

The seventh most cited minor theme within Halal Perceived Value is the Hospitality or service. Respondents also reported that they had good service experiences at travel destinations, but most of destinations provide poor services.

"Too many procedures are complicated, and there is no direction or lack of information about tourist attractions for tourists." (R12)

2) Price value

Sweeney & Soutar (2001) emphasized that the price as the utility derived from the products and services due to the reduction of its perceived short term and long term costs. The variable price is a critical role in influencing customer satisfaction level (Ananthanarayanan Parasuraman & Grewal 2000; Bolton & Lemon 1999; Varki & Colgate 2001).

Price was the second most cited concern by 28 respondents. Price can be explained as "The possibility price to a tourism destination will result in halal tourism". The most frequently cited minor themes from price value is Reasonable price. First response when describing their price in the destination, *"In terms of price perspective, when they offer prices, I thought they already know that visitors had previous information about the price. Therefore, according to Islamic law, we need an honest. Thus price should be reasonable so all tourist can afford to buy it." (R9)*

The second most cited minor theme of price value is the value that cost by the price.

"There is still complex system in destination that making a double price. So, there are still some places that charge additional fees. And what make it worst, if they cost the ticket that is not valuable with their product in the desination" (R3)

The third most cited type of price value is Riba' in tourist destination financial system. Riba' in the the Shari'ah, technically refers to the 'premium' that must be paid by the borrower to the lender along with the principal amount as a condition for the loan or for an extension in its maturity" (Chapra 1985). Talking about this issue an interviewee said,

"There is still a Riba' practice in the cultural tourist destination. For example, using a creditcard." (R5)

"It is necessary to clarify the rates set by the local government. So, I was given the clarity of the money I spent in accordance with what I wanted or not, because,if we want to apply w halal tourism 'It is certain that anything related to halal cannot be Riba'." (R7)

The least cited minor theme of price value is Prices are in accordance with the experience, *"The price is higher while the experience received is not worth the price. There are a number of places that have provided tariff clarity, but there are some places that have not explained the clarity of the prices." (R30*

3) Emotional value

The variable of emotions means the utility derived from the feelings or affective state generated by a product and it is a fundamental factor in increasing tourist's satisfaction; such as relaxation, family, togetherness, excitement, fun, and safety are important (Sweeney & Soutar 2001, Yoon & Uysal 2005). This values attracted 25 comments from respondents.

The most cited minor theme of emotional value is safety. These respondents were concerned about safety in tourism destination.

"I feel comfort and my emotion develop positivelly. I gain the knowledge, and honest and friendly tour guides." (R24)

"I haven't felt the right secure and comfort. There is no feedback that I feel, or the special pleasure I get." (R17)

The second most cited minor theme of emotional value is relax, as one interview said,*"I feel relaxed with the Islamic nuances that are given." (R11)*

The third most cited minor theme is Excited. Some respondents provide comments about the feeling of relaxation they received while visiting tourist attractions.

"Already feeling excited because they can see some Islamic culture served from cultural tourism in Minangkabau." (R31)

The fourth most cited minor theme is get reward while travelling. Their comments were then visited tourist attraction was they felt the reward that was obtained when visiting the destination, *"If there is a halal tour package offered, I will be definitely feels the reward from God, as well as I would be very happy to enjoy the tour. Shortly, while I learn about Islam from this visit, I also gain God reward." (R32)*

The least cited minor theme in emotional value is Religion learning. These respondents preferred to choose halal tourism for the destinations travelled so that they additional about the religion learning self.

"In addition to providing learning, halal tourism can also provide useful religious value for tourists." (R14)

4) Social value

The variable of social refers to the perceived utility derived from the ability of products and services in enhancing their self-image and become a famous products and services provider (Cengiz & Kirkbir 2007; Sweeney & Soutar 2001). This attracted 18 comments from interviewees.

The most cited minor theme in social value is Ethical and responsible tourism based on Islam, *"My concern for visiting cultural tourism destinations is the lack of notice of tourists to use polite clothing. Supposedly, we are copying the regulations that have been implemented by other places of descent outside of West Sumatra. Like, the use of cloth cover aurat (Bali cloth) which is used as a cover for tourists who use shorts."* (R19). Meanwhile other respondents nominated ethics as values that need to be attached in tourist destination, *"Ethics, how to dress, be polite and behave in accordance with Islamic Shari'a." (R21)*

The least cited minor theme in social value is Islam as identity, *"Halal tourism is a Minangkabau cultural identity that always upholds customs. And hope, if we always apply halal tourism, Minangkabau authentic identity can still be felt". (R23)*

And other comments from informant is *"Culture is an identity that underlies us to be able to determine life choices later. If culture is forgotten, it's the same as we have forgotten who we really are." (R9)*

5) Halal physical value

According to the Eid & El-Gohary (2015) variable of physical attributes derived from the value of the products and services offered by tourism Industry, such as all Halal food and drinks, utensils and equipment used by Muslim as described by 20 respondents.

The most cited minor theme of halal physical value is the staff use Shari'a compliant islamic clothing, *"Muslim Friendly here is more accessible in all aspects. Like, maybe Muslim women sellers use hijab, as well as adequate prayer places for tourists." (R25)*

6) Non-physical value

The Non physical attributes are the intangible attributes that compy with the sharia'a law, such as

Table 3. Major and minor themes of Muslim tourist's perceived risk.

Major Themes	Minor Themes
Muslim Perceived Risk	Worship place with hygiene issues
	Inavailability of worship palce
	Non-halal souvenirs
	Tourist with exposed clothes
	Inavailability of halal food
	Non-halal tourism event

Source: Result from this study.

prayer space, separate swimming pool for Muslim, Separate Spa and Hair Saloon for Men and Woman. This value was captured from 18 comments from respondents.

For example, one of the most cited minors of Non physical value is Regulations related to halal as described by one interviewee, *"Provide regulations and directives on halal tourism itself. In terms of price perspective, there are several tourist attractions that have followed the teachings of Islam and some have not." (R18)*

The least cited minor theme is Halal tourism branding. *"In addition to providing learning, halal tourism can also provide useful religious value for tourists." (R19)*

Muslim Halal Travel Risks

Related to perceived risk, this study found one major themes with six indicators of travel risk perceived by Muslim travellers in tourist destination as presented in Table 3 which attracting 20 comments in total from respondents.

The tourism industry is often associated with risky activities, the role of risk perception in the tourism decision-making process is the subject of current tourism marketing research (Lenggogeni et al. 2019; Moutinho 1987). The finding revealed six minor themes of halal perceived risk from Muslim tourist, consisted of the inavailability of praying room, non halal tourist product , hygienity, non halal event that involve the porn issue, and inavalability of halal food.

The most cited minor theme in Halal Perceived Risk is inavailability of praying room or place of worship in destinations. This concern was described by respondents,

"The perception of the risks that might exist is concern the inavalability of mosque that in the destination. It might be there, but it might take a relatively long time to go to the place of worship. And what worst, there is the tour manager did not aware our needs" (R23)

"The concern that I felt was the difficulty in finding a place to worship. Because if I visit a tourist destination that is not a Muslim culture, most places of worship are far from the location of tourist destinations." (R18)

The second most cited minor theme is Souvenirs that have non halal elements. *"In this tourist* destination, there is enough handling of products or souvenirs that contain pornographic elements. To avoid being worried by visitors, here, there has been a handling in terms of Minangkabau culture, which sells or trades merchandise originating from the Minangkabau area, such as miniature jam gadang and so on. So, in this tourist place far from the elements of pornography." (R15)

The third most cited minor theme is higienity in the place of worship. "There are still some things that must be considered to reduce the sense of concern of the tourists, such as the lack of cleanness provided by the managers for tourists. The hygiene is not managed properly and correctly so, that the management of the place of worship is inadequate to make it a place for worship." (R11)

"The concern that I feel is that when visiting cultural tourism destinations, is the lack of cleanliness of places to worship. Poorly managed clean place of worship." (R1)

The fourth most cited minor theme is Events with porno-action elements. "The existence of an event that contains pornographic elements, makes me worry when visiting these tourist destinations. Like, maybe there should be no pornographic elements that include direct involvement between event providers who are not their norm." (R16)

The least cited minor theme is inavailability of halal food. "My concern when visiting cultural tourism destinations is that there are foods that contain foods that are not halal. There is no difference between halal and non-halal food. The cooking method that is equated by the restaurant. As well as not written down the ingredients contained in these foods. So that there is an anxiety of tourists in carrying out tourism activities." (R2)

3.1 Implication

Guided by Sweeney & Soutar (2001) and Eid & El-Gohary (2015), this study identified six underlying dimension of Muslim tourist perceived value and an underlying dimension Muslim travel risks associated to cultural and heritage destination. While most of multi dimensionalities of Muslim Tourist perceived value reflects the study of Eid & El-Gohary (2015) for international Muslim tourist market, our findings extend the perceived value in a heritage and cultural destination for domestic market. However, despite the major dimensionalities is consistent with their studies, the underlying indicators for each dimensionality is interestingly different. For example, while study of Eid & El-Gohary (2015) described the emotional values with the feeling of customer's toward tourism package, our study revealed that an emotional values of Muslim tourist are more to capture their pride to their religion as their self identity. It consisted of Islam as my identity, the warmth feeling of friendly local people and ethical and responsibility as a tourist based on Islamic sharia compliant. Likewise, while Eid & El-Gohary (2015) proposed the price value dimensionality as an

important factor for Muslim tourist market, our study found that Riba' is one of avoidance for tourist to visit the tourist attraction, such as the use of credit cards. Thus, the beliefs of tourist most likely to influence the values needed by tourist market, confirmed the argument proposed by (Jafari & Scott 2014).

In addition, tourist perception is a very contextual based. While Eid & El-Gohary (2015) showed the multidimensionalities for Muslim tourist market associated to tourism package service, our study investigate the Muslim tourist perception for both values and risks with regards to tourist attraction or destination, particularly a heritage and cultural type of tourist site Therefore, this current study support that tourism is contextually influenced by local cultural factors (Devianto et al. 2019; Ritchie et al. 2017) and is multifaceted (Dolnicar 2005).

Using an initial stage of bottom up approach by qualitative study, our result provides a deeper insight of Muslim Perceived Values to server a future direction for Muslim traveler behavior in Halal Tourism. In addition, this study show that perceived values dimensionalities are broad and less likely be subsumed into single dimensions. Perceived values is multidimensional and it needs to be fit into the context and segmentation (Sharifpour et al. 2014a). Furthermore, the finding also highlights that religion factor is undetachable factors that influence the traveller behavior. It enhances what the true travel values mostly needed by Muslim visitor. Thus, this can be seen as one of first perceived values studies using a bottom up approach which contributes into tourism halal marketing studies.

4 CONCLUSION

A key strength of the present study was the religion and cultural is a determinant factor of tourist value and risk perceptions and subsequently to their travel behaviour. Despite this study extend the multidimensionalities of tourist market based of religion by Eid & El-Gohary (2015), our findings show that each of their underlying indicators is significantly different.

Therefore, this study has extended to the halal tourism marketing theory that explains the use of perceived values and risk perception dimensions helps explain the fundamental perception of destination and tourism products by religion based market. First, this study has demonstrated that bottom up approach allows to gain a better understanding of multidimensionalities of tourist perceptions. Second, this study extends our knowledge that the perceived values and perceived risk are segmented and contextual based. Third, this study has provided that religion is an inevitable factor that contribute to the formation of tourist behaviour in tourism and hospitality study. Fourth, this study has offered a thirthy six indicators of domestic Muslim market perceived values in heritage and cultural destination.

Besides, this study also provides several managerial impacts in tourism industry. First, this study suggests the importance of several sharia compliants values attributes related to heritage destination for destination management organizations. Second, this present study highlights the importance of Islamic sharia values such as the muslim staff attitude for being honest, friendly of staff in the destinations. Third these findings provide an important insight of physical and amenities to support the muslim market needs in the destination. Above all, the present study could be used to improve the marketing communication strategies to attract Muslim tourist market.

However, being limited to single method approach, this result may not be applicable to the wider population. Therefore, future study of quantitative and mixed method approach may help to generalize the Muslim tourist perceived value and risk perception. In addition, different setting or context may help to help a better understanding of multidimensionalities of perceived value and risk perception across different segment.

REFERENCES

Al-Sabbahy, H.Z., Ekinci, Y., & Riley, M. 2004. An investigation of perceived value dimensions: Implications for hospitality research. *Journal of Travel Research* 42(3): 226–234.

Battour, M., & Ismail, M.N. 2016. Halal tourism: concepts, practises, challenges and future. *Tourism management perspectives* 19: 150–154.

Battour, M., Ismail, M.N., & Battor, M.J. 2011. The impact of destination attributes on Muslim tourist's choice. 13(6): 527–540.

Bauer, R.A. 1960. Consumer behaviour as a risk taking. In R. S. Hancock (Ed.). *Dynamic marketing for a changing world*: 389–398. Chicago: American Marketing Association.

Bolton, R.N., & Drew, J.H. 1991. A longitudinal analysis of the impact of service changes on customer attitudes. *The Journal of marketing*: 1–9.

Bolton, R.N., & Lemon, K.N. 1999. A dynamic model of customers' usage of services: Usage as an antecedent and consequence of satisfaction. *Journal of marketing research*: 171–186.

Cengiz, E., & Kirkbir, F. 2007. Customer perceived value: the development of a multiple item scale in hospitals. *Problems and Perspectives in Management* 5(3 (continued)): 252.

Chapra, M.U. 1985. *Towards a just monetary system* 8: International Institute of Islamic Thought (IIIT).

Choe, J.Y.J., & Kim, S.S. 2019. Development and validation of a multidimensional tourist's local food consumption value (TLFCV) scale. *International Journal of Hospitality Management*, 77: 245–259.

Clarke, V., & Braun, V. 2013. Teaching thematic analysis: Overcoming challenges and developing strategies for effective learning. *The Psychologist* 26(2): 120–123.

Devianto, D., Ridho, M., Maryati, S., & Lenggogeni, S. 2019. Path analysis of entrepreneurial motivations in tourism based on local resources and creative economy in Nagari Salayo of West Sumatra. *Paper presented at the Journal of Physics: Conference Series*.

Dolnicar, S. 2005. Understanding barriers to leisure travel: Tourist fears as a marketing basis. *Journal of Vacation Marketing* 11(3) : 197–208. [Online]. Retrived from doi:10.1177/1356766705055706.

Eid, R., & El-Gohary, H. 2015. Muslim tourist perceived value in the hospitality and tourism industry. *Journal of Travel Research* 54(6): 774–787.

El-Adly, M.I. 2019. Modelling the relationship between hotel perceived value, customer satisfaction, and customer loyalty. *Journal of Retailing and Consumer Services* 50: 322–332.

Gallarza, M.G., & Saura, I.G.J.T. 2006. Value dimensions, perceived value, satisfaction and loyalty: an investigation of university students' travel behaviour. 27(3): 437–452.

Havlena, W.J., & Holbrook, M.B. 1986. The varieties of consumption experience: comparing two typologies of emotion in consumer behavior. *Journal of consumer research* 13(3): 394–404.

Henderson, J.C. (2016). Muslim travellers, tourism industry responses and the case of Japan. *Journal Tourism Recreation Research* 41(3): 339–347.

Hillman, W., & Radel, K. 2018. Introduction. In *Qualitative Methods in Tourism Research*: xix). Bristol: Channel View Publication.

Jafari, J., & Scott, N.J.A. 2014. Muslim world and its tourisms. 44, 1–19.

Kamali, M.H. 2013. The Parameters of halal and haram in Shari'ah and the Halal industry. 23: IIIT.

Lenggogeni, S. 2015. Exploring travel risks in the natural disaster context: A domestic tourist perspective. *CAUTHE 2015: Rising Tides and Sea Changes: Adaptation and Innovation in Tourism and Hospitality*: 597.

Lenggogeni, S., Ritchie, B.W., & Slaughter, L. 2019. Understanding travel risks in a developing country: a bottom up approach. *Journal of Travel & Tourism Marketing* 36(8): 941–955.

Lenggogeni, S., & Saito, H. 2018. Does a religion matter in travel risk and behavioural intention in earthquake vulnerable cities?: A case for domestic traveller in Indonesia and Japan. *CAUTHE 2018: Get Smart: Paradoxes and Possibilities in Tourism, Hospitality and Events Education and Research*: 836.

Morse, J.M. 2000. Determining Sample Size. *Qualitative Health Research* 10(1): 3–5. [Online]. Retrieved from doi:10.1177/104973200129118183.

Moutinho, L. 1987. Consumer behaviour in tourism. *European Journal of Marketing*.

Parasuraman, A., & Grewal, D. 2000. The impact of technology on the quality-value-loyalty chain: a research agenda. *Journal of the academy of marketing science* 28(1): 168–174.

Parasuraman, A., Zeithaml, V.A., & Berry, L.L. 1998. Alternative scales for measuring service quality: a comparative assessment based on psychometric and diagnostic criteria. In *Handbuch Dienstleistungsmanagement*: 449–482).

Park, S., & Tussyadiah, I.P. 2017. Multidimensional facets of perceived risk in mobile travel booking. *Journal of Travel Research* 56(7): 854–867.

Pew Research Centre. 2017. *The Changing Global Religious Landscape*. Retrieved from Washington, USA.

Pham, L., Williamson, S., Lane, P., Limbu, Y., Nguyen, P.T. H., & Coomer, T. 2020. Technology readiness and purchase intention: role of perceived value and online satisfaction in the context of luxury hotels. *International Journal of Management and Decision Making* 19(1): 91–117.

Prebensen, N.K., Woo, E., Chen, J.S., & Uysal, M. 2013. Motivation and involvement as antecedents of the perceived value of the destination experience. *Journal of Travel Research* 52(2): 253–264.

Ritchie, B.W. 2009. *Crisis and Disaster Management for Tourism* (1 ed.). Bristol: Channel View Publication.

Ritchie, B.W., Chien, P.M., & Sharifpour, M. 2017. Segmentation by travel related risks: an integrated approach. *Journal of Travel Tourism Marketing*, 34(2): 274–289.

Rittichainuwat, B.N., Qu, H., & Mongkhonvanit, C. 2009. A study of the impact of travel inhibitors on the likelihood of travelers' revisiting Thailand. *Journal of Travel & Tourism Marketing* 21(1): 77–87. [Online]. Retrieved from http://www.informaworld.com/10.1300/J073v21n01_06.

Şen Küpeli, T., & Özer, L. 2020. Assessing perceived risk and perceived value in the hotel industry: an integrated approach. *Anatolia*: 1–20.

Sharifpour, M., Walters, G., & Ritchie, B.W. (2014). Risk perception, prior knowledge, and willingness to travel investigating the Australian tourist market's risk perceptions towards the Middle East. *Journal of Vacation Marketing* 20(2): 111–123.

Simpson, P.M., & Siguaw, J.A. 2008. Perceived travel risks: The traveller perspective and manageability. *International Journal of Tourism Research* 10(4): 315–327.

Sönmez, S.F., & Graefe, A.R. 1998. Influence of terrorism risk on foreign tourism decisions. *Annals of tourism research* 25(1): 112–144. [Online]. Retived from doi:10.1016/s0160-7383(97)00072-8.

Sweeney, J.C., & Soutar, G.N. 2001. Consumer perceived value: The development of a multiple item scale. *Journal of retailing* 77(2): 203–220.

Thomson, R., & Dinar, S. 2018. *State of Global Islamic Economy Report 2018*.

Varki, S., & Colgate, M. 2001. The role of price perceptions in an integrated model of behavioral intentions. *Journal of Service research* 3(3): 232–240.

Wang, J., Liu-Lastres, B., Ritchie, B.W., & Mills, D.J. 2019. Travellers' self-protections against health risks: An application of the full Protection Motivation Theory. *Annals of Tourism Research* 78: 102743.

Yoon, Y., & Uysal, M. 2005. An examination of the effects of motivation and satisfaction on destination loyalty: a structural model. *Tourism Management* 26(1): 45–56.

Advances in Business, Management and Entrepreneurship – Hurriyati et al. (Eds)
© 2021 Taylor & Francis Group, London, ISBN 978-0-367-67471-7

Studies of sales promotion, brand image, and product quality to Mitsubishi Truck repurchase intention in Indonesia

A.P. Kusuma & E.Z. Rusfian
Universitas Indonesia, Depok, Indonesia

ABSTRACT: Mitsubishi truck vehicle is a pioneer of commercial vehicles and has been engaged for 49 years in Indonesia and become a market leader for commercial vehicles in Indonesia. Jabodetabek (Jakarta-Bogor-Depok-Tangerang-Bekasi) is one megapolitan area in Indonesia that produces the largest gross domestic product (GDP) for Indonesia. This research was conducted quantitatively by distributing questionnaires to respondents. Data were analyzed using the structural equation modeling (SEM) method using the SmartPLS software application. The results of this study indicate the existence of a positive impact relationship of sales promotion, brand image, and product quality toward repurchase intention. The results of this study can be used as a reference for the company in making strategic decisions to increase repurchase intention.

1 INTRODUCTION

Repurchase intention is a behavior and intention from consumers to make a repeat purchase of a product or service from a company on the next purchase. Some things that can be done to increase intention in repeat purchases are by increasing sales promotion, brand image, and quality of products sold, so that it can make truck customers have a repurchase intention for Mitsubishi trucks in Indonesia, especially in Jabodetabek. For this reason, the researcher wants to examine the factors that influence consumer repurchase intention, especially in terms of sales promotion, brand image, and product quality of ATPM products, especially Mitsubishi trucks in Jabodetabek, Indonesia.

Jabodetabek (Jakarta-Bogor-Depok-Tangerang-Bekasi) is one of the 5 metropolitan areas on the island of Java. On Java, there are 5 metropolitan areas, namely Greater Jakarta (Jabodetabek), Greater Bandung, Semarang, Yogyakarta, and Surabaya. It is known that Java Island is the center of population and economic activity in Indonesia. Where 60% of the population in Indonesia is located on the island of Java. Java Island contributes to almost 60% of the GDP in Indonesia (Andrea et al. 2018).

The object observed by researchers today is the "Truck" commercial vehicle of the Mitsubishi brand. Where previous studies have not discussed similar objects, the variable used in this research is a combination of variables that have a significant positive effect on repurchase intention from some previous studies.

To support this research, the researchers refered to and used previous research that still has relevance to this research. Previous studies functioned as a reference and support for conducting this research. Previous studies used as a reference and support are researches with one or more of the same variables of this study such as repurchase intention variables (interest in buying), sales promotion, brand image, and product quality.

Repurchase intention is a consumer behavior that results in the purchase of the same product or service in subsequent purchases (Eliasaph et al. 2016). Consumers buy the same product repeatedly from the same seller.

Repurchase intention indicates that customers have a great intention in buying continuously from a company. Consumer drivers for driving repeat purchases consist of several variables, namely Customer Satisfaction, Customer Loyalty, Perceived Value, and Trust (Zeeshan et al. 2016).

According to Huang et al. (2019), repurchase intention can be measured through three dimensions/indicators, namely:

Repeated purchases: To measure the customer's intention to repurchase the company's products or services in the future. This is an important indicator of the customer's future behavior intentions.

Willingness to recommend: Refers to customers who have willingness to introduce, recommend, and discuss positively the company. Such behaviors are primary for the company.

Loyal customer: Including time, frequency, amount, and quality of purchase. Although major behaviors are a key factor in measuring actual behavior, they would change with time and other customers might give wrong data.

According to Kotler, sales promotion (sales promotion) is the activity requested to help getting customers to buy a company's product or service. Sales promotion aims to increase buyers to use the product regularly.

Sales promotions produce benefits for consumers. The types of benefits for suppliers are divided into two, namely: Utilitarian benefits (utilization benefits) and hedonic benefits (hedonic benefits). (Chandon et al. 2000).

A brand is a name and symbol. This is a very important tool for creating positive images in customers. A brand has a very important role in creating loyal customers and maintaining the market share of the company (Vahidreza et al. 2015).

According to Huang et al. (2019), brand image has three dimensions, namely as follows:

1) Functional: The actual benefits of a product or service, emphasizing helping consumers in solving problems related to consumption. These types of products are designed to solve the basic needs of consumers. The indicators are reasonable price, good service, and high quality.
2) Symbolic: Added value from products or services, focusing on satisfying consumers who need brand products.
3) Experiential: Perception after using a product or service, emphasizes to satisfy the consumer's need for pleasure, diversity, and cognition provided by the product.

Quality is the ability of the product to meet the expectations of the customer. Quality is the total features and characteristics of a product or service that depend on its ability to meet needs (Sultana 2017).

According to Kotler, product quality is a characteristic of a product or service that depends on its ability to meet needs.

2 METHODS

The research approach chosen by the researchers in this study is to use a quantitative research approach. The sample in this study consisted of 135 respondents from consumers who have purchased Mitsubishi product trucks.

For data analysis in this research, researchers will use structural equation modeling (SEM). SEM is a multivariate analysis technique that combines aspects of factor analysis and multiple regression that allows researchers to simultaneously examine a series of interrelationships of measured variables and latent constructs and between several latent constructs (Hair et al. 2014).

SEM is divided into two models, namely covariance-based SEM (CB-SEM) and partial least square SEM (PLS-SEM) (Hair et al. 2011). In this study, the analysis model used was PLS-SEM. The determination of the use of the analysis model was because the number of samples in this study was below 200 respondents; it only had 135 respondents. Data processing for the PLS-SEM model in this study was by using the SmartPLS 3 software application.

2.1 Conceptual framework

The following research model used sales promotion, brand image, and product quality variables to see how they affect repurchase intentions as follows:

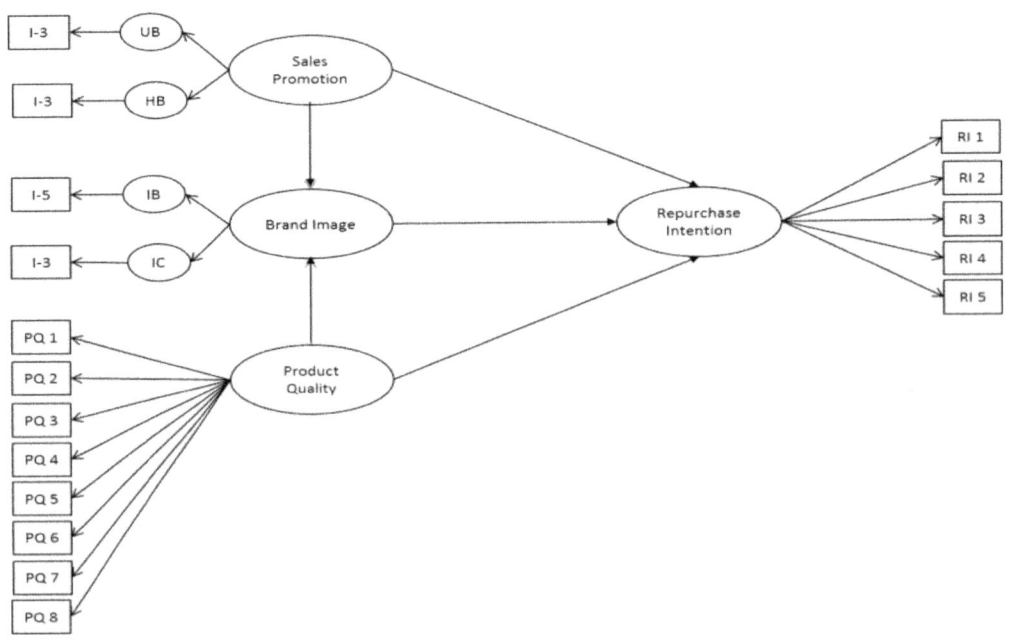

Figure 1. Conceptual framework of research. Source: Research results, 2019.

The study on three variables of repurchase intention consisted of sales promotion, brand image, and product quality.

From the research model and the relationship between variables, the researcher wanted to see the influence of each variable, namely sales promotion, brand image, and product quality, on repurchase intention. For this reason, several hypotheses were obtained to be analyzed in this study, namely as follows:

H1 = Sales Promotion has a positive influence on repurchase intention.
H2 = Brand Image has a positive influence on repurchase intention.
H3 = Product Quality has a positive influence on repurchase intention.
H4 = Product Quality has a positive influence on brand image.
H5 = Sales Promotion has a positive influence on brand image.

3 RESULTS AND DISCUSSION

3.1 *Reliability test*

Reliability testing was conducted by calculating composited reliability (CR), Cronbach's Alpha, and Average Variance Extracted (AVE). Where data was generally considered reliable if Cronbach's Alpha had a value of > 0,700, CR had a value of > 0,700, and AVE had a value of > 0,500. According to Malhotra (2010), the conditions considered reliable are:

1) CR value ≥ 0.7. But 0.6–0.7 was still accepted if the estimated validity of the model was good.
2) AVE value ≥ 0,5.

The following were the results of the measurement of the values of Cronbach's Alpha, CR, and AVE. The results were as follows.

Table 1. Measurement of data reliability.

Dimension	Cronbach's Alpha (>0,700)	Composite Reliability (>0,700)	AVE (0,500)	Conclusion
Brand image benefit	0.928	0.939	0.608	Reliable
Brand image component	0.871	0.904	0.611	Reliable
Product quality	0.894	0.916	0.579	Reliable
Repurchase intention	0.931	0.945	0.710	Reliable
Utilitarian benefit	0.818	0.872	0.579	Reliable
Hedonic benefit	0.832	0.879	0.550	Reliable

Source: Research results using smart PLS, 2019.

Table 2. R^2 (R-Square) test.

	R-Square	R-Square Adjusted
Brand image benefit	0.782	0.777
Brand image component	0.792	0.787
Repurchase intention	0.850	0.844

Source: Research results using smart PLS, 2019.

Based on the data in Table 1 above, it can be observed that all values of Cronbach's Alpha and CR were above the expected limit, which is >0.700. Beside of the AVE value was above the value >0.500. So that these results indicated that all statements and indicators in the latent structure of this research can be said to be reliable.

1) Structural Model (*Inner Model*)
This second step was to describe the test results from R2 from the latent construct of the study.

a) R^2 (R-Square) Test
The following R^2 value was the amount of the determination coefficient of the endogenous variable, which was repurchase intention. The measurement results show how strong endogenous variables can be explained by exogenous variables contained in the study, namely sales promotion, brand image, and product quality. In addition, to see how strong the brand image variable can be explained by the sales promotion and product quality variables. More clearly, the value of R^2 can be seen in Table 2.

Based on Table 2, we could observe that the value of R^2 (R-Square) of the repurchase intention variable was 0.850, while the brand image was 0.782–0.792. This value indicated that repurchase intention can be influenced by sales promotion, brand image, and product quality that was equal to 85%. While the condition, which was 15%, can be influenced by other variables outside of the research model. Likewise, with the brand image, the value indicates that the brand image could be influenced by sales promotion and product quality, which is 78.2%–79.2%.

2) Structural Model
The structural model was used to test the research model hypothesis by looking at whether there was a significant influence between each latent in the research model. The results of the structural model in this study were as follows:

Based on the data in Table 1 above, it can be observed that all values of Cronbach's Alpha and CR were above the expected limit, which is > 0.700. Beside of the AVE value was above the value > 0.500. So that these results indicated that all statements and indicators in the latent structure of this research can be said to be reliable.

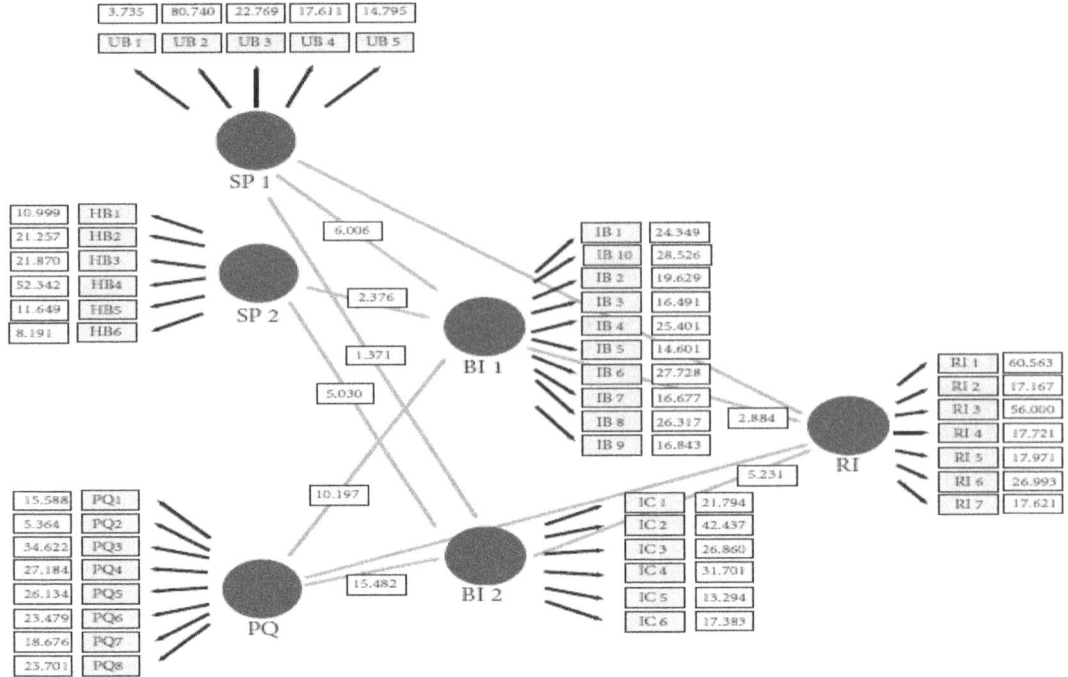

Figure 2.   Structural model. Source: Research results using smart PLS, 2019.

Table 3.   Hypothesis test results.

Hypothesis		Original Sample	T Statistics	P Values	Remark
H1	Sales promotion has a positive influence on repurchase intention				
a	Utilitarian benefit has a positive influence on repurchase intention	0.216	5.03	0	Accepted
b	Hedonic benefit has a positive influence on repurchase intention	−0.129	1.971	0.049	Accepted
H2	Brand image has a positive influence on repurchase intention				
a	Brand image benefit has a positive influence on repurchase intention	0.312	2.894	0.004	Accepted
b	Brand image component benefit has a positive influence on repurchase intention	0.498	5.231	0	Accepted
H3	Product quality has a positive influence on repurchase intention	0.614	11.769	0	Accepted
H4	Product quality has a positive influence on brand image				
a	Product quality has a positive influence on brand image benefit	0.558	10.197	0	Accepted
b	Product quality has a positive influence on brand image component	0.645	15.482	0	Accepted
H5	Sales promotion has a positive influence on brand image				
a	Utilitarian benefit has a positive influence on brand image benefit	0.33	6.006	0	Accepted
b	Hedonic benefit has a positive influence on brand image benefit	0.154	2.376	0.018	Accepted
c	Utilitarian benefit has a positive influence on brand image component	0.067	1.371	0.171	Accepted
d	Hedonic benefit has a positive influence on brand image component	0.281	5.03	0	Accepted

Source: Research results, 2019.

3.2   Discussion of the research hypothesis

In this research, broadly outlined, there were 5 hypotheses that were tested along with 10 derived hypotheses. Hypothesis testing was done by comparing the probability value (p value) obtained from the results of PLS-SEM processing and the level of significance (α) used in the study. The level of significance (α) used in this study was 5% or 0.05. If the p value was < 0.05, then the hypothesis could be accepted and it indicated a significant influence, but if the p value was > 0.05, then the hypothesis was rejected and showed no significant influence. The following is the results of testing the hypothesis that has been done.

From Table 3, it could be concluded that if observed in overall, all of variables (sales promotion, brand image, and product quality) had a significant positive effect on repurchase intention with p values < 0,05. Beside of that, sales promotion had a significant positive effect on brand image, but only utilitarian benefits (monetary saving, quality, and convenience) did not have a significant positive impact on the brand image component (product image, user image, and corporate image). This was because the utilitarian benefits were primarily monetary saving such as discounts, price cuts, etc., and this would lead to the perception of the down spec of the products being sold so that the image of the product will also decrease. In fact, overall, sales promotion had a positive influence on the brand image if used correctly and strategically, including premium brand positioning, prices, etc.

The conclusions about sales promotion described were in line with the research of Danijela (2009) where the results of these studies indicate that the long-term positive impact of sales promotion if used correctly and strategically included the premium brand positioning. A strategic marketing communication plan will clearly explain elements such as objectives, target consumers, and positioning, which will help companies decide which sales promotion method is most suitable for the company. Strengthening planning in the sales promotion process, in line with a closer analysis of all sales promotion methods, would guide a company with premium brand positioning in a more creative form that did not depend on price discounts. Due to price discounts (monetary saving) tending to be short term, for long term, it will have a negative impact on the image of the product.

4 CONCLUSION

Overall, the following conclusions are obtained as follows:

1) Sales promotion, brand image, and product quality have a positive effect on the interest of repurchase intention.
2) Product quality has a positive effect on the brand image of the product.
3) Sales promotion affects the brand image of the product.
4) The dimension of sales promotion (utilitarian benefit) does not have a positive impact on brand image.

In other hand, based on the analysis of this research, there are several academic suggestions that can be submitted by the researchers based on the findings of the field, that is:

a) The next research can examine consumers of Mitsubishi trucks in specific areas such as Kalimantan.

Because one of the other big markets of trucks besides the manufacturing, logistics, and transportation industries is the mining industry and it is known that Kalimantan is one of the areas in Indonesia that will have many coal mines.
b) The next research can examine other exogenous variables that influence repurchase intention such as customer satisfaction, brand loyalty, service quality, and others.

REFERENCES

Andrea, E. P., Ernan, R., Setyardi, P. M., Lutfia, N. F., Nur. E. K., Alfin, M. 2018. Measuring urban and regional sustainability performance in Java: A comparison study between 5 metropolitan areas.

Chandon, P., Brian, W., Gilles L. 2000. A benefit congruency framework of sales promotion effectiveness. *Journal of Marketing* 64.

Danijela, M. 2009. Long-term impact of sales promotion on brand image. Bosnia and Herzegovina.

Eliasaph, I., Farida, B., & Balarabe, J. 2016. Consumer satisfaction and repurchase intentions. Bayero University Kano.

Hair, J. F., Black, W. C., Babin, B. J., & Anderson, R. E. 2014. Multivariate data analysis. London: Pearson Education Limited.

Hair, J. F., Ringle, C. M., & Sarstedt, M. 2011. PLS-SEM: Indeed a silver bullet. *Journal of Marketing Theory and Practice*: 139–151.

Huang, L. C., Ming, G., Ping-Fu, H. 2019. A study on the effect of brand image on perceived value and repurchase intention on ecotourism industry. *Ekologi* 28(107).

Kotler, P., & Waldemar, P. 2006. B2B brand management. Germany: Springer Berlin Heidelberg.

Kotler, P., & Armstrong, G. 2008. Prinsip-prinsip Pemasaran 2008 edisi 12. Jakarta: Penerbit Erlangga.

Malhotra, N. K. 2010. Marketing research: An applied orientation: global edition (6th). New Jersey: Pearson Education.

Sultana, R. C. 2017. Measuring the relationship between product quality dimensions & repurchase intention of smart phone: A case study on Chittagong City. *International Journal of Scientific & Engineering Research* 8. Bangladesh.

Zeeshan, A., Meng, J., Imran, K. M., & Tauqir, A. 2016. Examining mediating role of customer loyalty for Influence of brand related attributes on customer repurchase intention. *Journal of Northeast Agriculture University* 23. Pakistan.

Does halal lifestyle influence the purchase intention of Muslim fashion consumers?

A. Haro
Universitas Negeri Jakarta, Jakarta, Indonesia

A. Suangkupon
Bank Indonesia, Jakarta, Indonesia

ABSTRACT: This study aims to determine the influence between religiosity, lifestyle, and attitude toward the purchase intention of Muslim fashion products in creating a halal fashion value chain. By using a survey in the form of a questionnaire, the total sample of research obtained 200 respondents who consisted of Muslim men and women with the requirement that they had already bought and used Muslim fashion products. Multiple linear regression was used to analyze the data and to test the research hypotheses. The results indicate that attitude and lifestyle have an influence toward purchase intention on Muslim fashion products. Surprisingly, there is no influence between religiosity toward their purchase intention.

1   INTRODUCTION

Indonesia is a country with a majority Muslim population, which is around 209 million or around 87.2% of the total population of Indonesia. This fac-tor led to the rapid development of sharia-based business (halal businesses) such as Islamic financial services, halal food, Muslim fashion, halal media and recreation, as well as halal pharmaceuticals and cosmetics. There is an increasing demand for sharia-based products (halal) one of which occurs in Muslim fashion products (Win & Jan 2016).

The Ministry of Trade said that throughout 2015, the value of Muslim fashion exports reached Rp 58.5 trillion. This figure is only 20% of Muslim clothing sales, which are indeed distributed for export markets. That is, 80% of the products are actually traded to the domestic market. According to Bonne et al. (2007), religion is one of the cultural forces that can influence consumer behavior, including interest in product purchases and lifestyle. The influence of religion depends on the extent to which individuals interpret and follow the teachings of their religion (Haro 2018). Religiosity is the extent to which an individual is committed to the religion they acknowledge and their teachings such as individual attitudes and behavior that reflect this commitment (Sadra 2012). The level of religiosity of each individual has a positive influence on attitudes toward halal products (Mukhtar & Butt 2012). Riptiono and Setyawati (2019) state that Islamic religiosity has no significant effect directly on female Muslim fashion purchase intention, but Islamic religiosity has a signif-icant effect indirectly toward female Muslim fashion purchase intentions through consumer attitudes. The finding obtained by Haque et al. (2018) is actually different, that is, religiosity shows an increase in the intention to purchase a fashion hijab.

Lifestyle is also one of the factors that cause an increase in interest in buying Muslim fashion prod-ucts. Lifestyle, according to Pebriani et al. (2018), is a person's lifestyle in the world that is reflected in activ-ities, interests, opinions, and lifestyle and portrays the interaction of "a person in its entirety" with their envi-ronment. Therefore, consumers, in choosing a product, will choose based on what is most needed and in accor-dance with the buying interest, one of which is lifestyle (Tufail et al. 2018). Nora and Minarti (2016), in their study, found that lifestyle has a significant influence on the purchase of halal products.

In addition, the purchase intention that arises in the minds of consumers, is not only based on the consideration of religiosity and lifestyle alone, but can be driven by other factors, namely attitude. Atti-tudes toward an object can affect related information, assessments, and the resulting behavior (Ajzen 2015). Negative attitudes can have a greater or longer impact than positive or neutral attitudes (Petty & Krosnick 2014). According to Kotler and Armstrong (2014), attitude is the evaluation, feelings, and tendencies of individuals towards an object that are relatively con-sistent. The research conducted by Winahjoe et al. (2018) confirm that attitude has a significant pos-itive impact in predicting the intentions of Islamic women to wear the hijab. This study contributes in the empirical evidence regarding religiosity, lifestyle, and attitude towards purchase intention on Muslim fashion products. The purpose of this research is to analyze the influence of religiosity towards purchase intention in Muslim fashion products, to analyze the influence of lifestyle towards purchase intention in Muslim fashion products, and to analyze the influence of attitude towards purchase intention in Muslim fashion products.

2 METHOD

This research employed a quantitative method by using a questionnaire. The research population consists of men and woman Muslim consumers who have bought and used Muslim fashion products in the area of Jakarta. The total sample of research obtained 200 respondents by using the purposive sampling technique and the range of observation in the questionnaire used the Likert scale with scores of 1–5, where 1 means strongly disagree (STS) and 5 means strongly agree (SS). This research is analyzed by using multiple linear regression analysis with SPSS software.

2.1 Analysis of multiple linear regression

Multiple linear regression analysis is one of the tools to predict the influence of two independent variables or more on one dependent variable. The aim is to prove the existence or absence of functional relationships and causal relationships between two or more independent variables (Ghozali 2012). The following is the multiple linear regression:

$$Y = a + b1X1 + b2X2 + b3X3 + e. \quad (1)$$

2.2 Hypothesis test (t test)

Partial test shows the influence of an independent/independent variable individually in explaining the variation of the dependent variable. The level of trust used is 95% or a significant level of 5% (Sekaran 2003).

2.3 Coefficient of determination

The test for coefficient of determination (R2) is a measurement of the ability of the model in ex-plaining the variation of the dependent variable. The coefficient of determination ranges from zero to one. Each addition of one independent variable in the model will increase the value of R2, although the addition of these variables does not necessarily have a significant effect on the dependent variable. There-fore, many researchers recommend to use the value of adjusted R2 in evaluating the best regression model (Hair et al. 2006).

3 RESULTS AND DISCUSSION

3.1 Demographic statistic

The demographic conditions that could be de-scribed from the 200 respondents are 45 people under the age of 20, 113 people aged 20 to 25 years, 13 people aged 26 to 30 years, 14 people aged 31 to 35 years old, 36 years old up to 40 years as many as 6 people, and aged over 40 years as many as 9 people. For the most prevalent job, there are 165 students. Gender consists of 54 men and 146 women. Income levels obtained 113 people with income of less than 1 million rupiahs, 36 people with income of 1 million rupiahs to 2 million 500 thousand rupiahs, 30 people with income of 2 million 500 thousand to 5 million rupiahs, and 21 people with income above 5 million rupiahs.

3.2 Hypothesis test (t-test)

The survey conducted in this study used a questionnaire that was given personally to a number of respondents where respondents in this study were also the population of this research. The selection of samples is both of Muslim men and women who domiciled in Jakarta who had bought and used Muslim fashion products. Research samples collected were 200 respondents.

Based on Table 1, it could be seen that the partial calculation of the effect of religiosity towards purchase intention in Muslim fashion products obtained a regression coefficient value (b) of −0.005, where at a significance level of 5% obtained a t-count of −0.088 with a significance value of 0.930. Then the result of testing this hypothesis is that H0 is accepted (Ha is rejected), meaning that there is no significant influence between religiosity towards purchase intention in Muslim fashion products. This is because the significance value (p-value) > 0.05. The influence of attitudes towards purchase intention in Muslim fashion products obtained a regression coefficient value (b) of −0.634, where a significance level of 5% obtained a t-count of 14.3323 with a significance value of 0.000. Then the result of testing this hypothesis is that H0 is rejected (Ha is accepted), meaning that there is a significant influence between attitudes towards purchase intention in Muslim fashion products. The influence is positive, where the higher the attitudes, the more interest in buying Muslim fashion products. This is because the significance value (p-value) > 0.05. For the influence of lifestyle towards purchase intention in Muslim fashion products obtained a regression coefficient value (b) of −0.095, where a significance level

Table 1. Result of hypothesis test (t-test).

	Unstandardized Coefficients			
	B	Std. Error	t	Sig.
(Constant)	3.216	0.845	3.808	0.000
Religiousity	−0.005	0.056	−0.088	0.930
Attitude	0.634	0.044	14.323	0.000
Lifestyle	0.095	0.044	2.133	0.034

Source: Proceed by researcher, 2019.

of 5% obtained a t-count of 2.133 with a significance value of 0.034. Then the result of testing this hypothesis is that H0 is rejected (Ha is accepted), meaning that there is a significant influence between lifestyle towards purchase intention in Muslim fashion products. The influence is positive, that is, the higher the level of lifestyle, the more interest is in buying Muslim fashion products. This is because the significance value (p-value) > 0.05.

3.3 Coefficient of determination

The results in this study use the value of Adjusted R Square of 0.530. This shows that the effect of the variables of religiosity, lifestyle, and attitude simultaneously on the variable of purchase intention is 53%, while the portion of 47% is explained by other factors not examined.

Table 2. Result of coefficient of determination.

Model	R	R2	Adj. R2	Std. Error of Estimate
1	0.732	0.535	0.530	1.949

Source: Proceed by researcher, 2019.

Based on the results of the tests that have been carried out, it shows that there is no significant influence between religiosity towards purchase intention in Muslim fashion products. The results of this study contradict some previous studies, which stated that there was a significant influence on the variables of religiosity on purchase interest. Haque et al. (2018) found that the higher the level of religiosity increase, it will increase the consumer purchase intention. Rohmatun and Dewi (2017) found that religiosity has a positive influence towards purchase intention in halal products. This might happen because respondents did not involve religiosity to purchase in Muslim fashion products. Sadra (2012) said that religiosity is a feeling, thought, and motivation that encourages religious behavior. But the opposite thing occurs in the attitude and lifestyle variables. Based on the results of the tests that have been conducted, there is a significant influence between attitudes towards purchase intention in Muslim fashion products. This finding is in accordance with the results of research conducted by Haque et al. (2018).

Kim and Chung (2011), Lada et al. (2009), Bonne et al. (2007), and Winahjoe et al. (2018), state that attitudes have an influence on purchase intention. Attitudes toward an object could affect related information, assessment, and the resulting behavior. Negative attitudes can have a greater or longer impact than positive or neutral attitudes (Petty & Krosnick 2014) so that consumers are expected with better attitudes and intentions to buy will be more likely to receive, buy, and consume products directly. The same thing happened to the influence of lifestyle towards purchase

intention in Muslim fashion products. Based on the results of the tests that have been conducted, there is a significant in-fluence between lifestyle towards purchase intention in Muslim fashion products. Qing et al. (2012) found a similar thing, which stated that lifestyle had a significant effect on purchase intention. Nora & Minarti (2016) in their research also found that lifestyle had a significant influence towards purchase intention in halal products.

4 CONCLUSION

Regarding this result, it can be concluded that lifestyle and attitude have a significant positive in-fluence on the creation of interest in buying Muslim fashion products. This certainly supports the link be-tween consumer behaviors in creating halal value chain creation among Indonesian consumers. These results can be used as an analysis material for Indonesian Muslim fashion producers to pay more attention to the lifestyle factors and attitudes of Indonesian consumers in making an interest in buying Muslim fashion products, which will certainly have an impact on the profit of the company itself. However, different things happen to religiosity. In this result, religiosity has no influence towards purchase intention in Muslim fashion products. This might happen because respondents did not involve religiosity towards purchase intention in Muslim fashion products. This study confirmed consumer behavior in general, so it needs to evaluate more for other factors that influence in consumer purchase intention. For the next research, another research needs to be done on certain brands of Muslim fashion.

REFERENCES

Ajzen, I. 2015. Consumer attitudes and behavior: The theory of planned behavior applied to food consumption decisions. Rivista di Economia Agraria LXX (2): 121–138.
Bonne, K., Vermeir, I., Bergeaudblacker, F., & Verbeke, W. 2007. Determinants of halal meat consumption in France. British Food Journal 109(5): 367–86.
Ghozali, I. 2012. Aplikasi analisis multivariate dengan program IBM SPSS 20. Semarang: Universitas Diponegoro.
Hair, J.F., Black, W.C., Babin, B., Anderson, R.E., & Tatham, R.L. 2006. Multivariate Data Analysis, 6th edition. NJ: Prentice Hall International, Inc.
Haque, A., Anwar, N., Tarofder, A.K., Ahmad, N.S., & Sharif, S.R. 2018. Muslim consumers' purchase behaviour towards halal cosmetic products in Malaysia. Management Science Letters 1: 1305–1318.
Haro, A. 2018. Determinants of halal cosmetics purchase intention on Indonesian Female Muslim Customer. Journal of Entrepreneurship, Business and Economics 6(1): 78–91.
Kim, H. Y. & Chung, J. E. 2011. Consumer purchase intention for organic personal care products. Journal of consumer Marketing 28(1): 40–47.
Kotler, P., & Armstrong, G. 2014. Principle of marketing, 15th edition. New Jersey: Pearson Prentice Hall.

Lada, S., Tanakinjal, H. G., & Amin, H. 2009. Predicting intention to choose halal products using theory of reasoned action. *International Journal Islamic Middle East Finance Management* 2(1): 66–76.

Mukhtar, A. & Butt, M. M. 2012. Intention to choose halal products: The role of religiosity. *Journal of Islamic Marketing* 3(2), 108–120.

Nora, L. & Minarti, N. S. 2016. The role of religiosity, lifestyle, attitude as determinant purchase intention. *International Multidisciplinary Conference*. Jakarta: Indonesia.

Pebriani, W. V., Sumarwan, U., & Simanjuntak, M. 2018. The effect of lifestyle, perception, satisfaction, and preference on the online re-purchase intention. *Independent Journal of Management & Production* (IJM&P) 9(2): 545–561.

Petty, R. E., & Krosnick, J. A. 2014. Attitude strength: Antecedents and consequences. New York: Taylor and Francis.

Qing, P., Lobo, A., & Chongguang, L. 2012. The impact of lifestyle and ethnocentrism on consumers' purchase intentions of fresh fruit in China. *Journal of Consumer Marketing* 29(1): 43–51.

Riptiono, S., & Setyawati, H. A. 2019. Does Islamic religiosity influence female Muslim fashion trend purchase intention? an extended of theory of planned behavior. Iqtishadia 12(1): 12–29.

Rohmatun, K. I. & Dewi, C. K. 2017. Pengaruh pengetahuan dan religiusitas terhadap niat beli pada kosmetik halal melalui sikap. *Jurnal Ekonomi, Manajemen, dan Bisnis 1*.

Sadra, T. 2012. The role of animosity, religiosity, and ethnocentrism on consumer purchase intention: A study in Malaysia toward European brands. *African Journal of Business Management* 6(23): 6890–6902.

Sekaran, U. 2003. Research Methods for Business. USA: John Wiley & Son.

Tufail, H. S., Humayon, A. A., Shahid, J., Murtza, G., Luqman, R., Riaz, H. 2018. Impact of life style and personality on online purchase intentions of internal auditors through attitude towards brands Hafiza Sobia. *European Online Journal of Natural and Social Sciences* 7(3s): 72–83.

Win, H. T. & Jan, M. T. 2016. Muslim consumers' online purchase intention towards Islamic fashion products: A clothing market case. *Amity Journal of Marketing* 1(2): 72–81.

Winahjoe, S., Sutikno, B., & Sudiyanti. 2018. Testing the robustness of theory of planned behavior in predicting women's intention to wear jilbab. *Kawistara* 8(3): 213–309.

*Section 5   Organizational behavior, leadership and
human resources management*

Advances in Business, Management and Entrepreneurship – Hurriyati et al. (Eds)
© 2021 Taylor & Francis Group, London, ISBN 978-0-367-67471-7

Work–life balance and subjective well-being among employees on life science company in Indonesia

G. Gunawan, Y. Nugraha, M. Sulastiana & D. Harding
Universitas Padjadjaran, Bandung, Indonesia

ABSTRACT: Regarding the World Health Organization's (WHO) plan for 2020 to eradicate polio globally, Indonesia is one of the countries that supported this, the program is called polio eradication. A life-science company in Indonesia revealed that the company's profit had decreased significantly because of the program. In order to survive in a competitive environment, the company must be able to follow the pattern of change. Organizational culture naturally depends on the market situation and needs to adapt to it in order to survive or to maintain its competitive position. Organizational culture change in the life science company brought about problems in balancing in work and family life of the employees. This work–life balance concept has not yet received special attention in Indonesia. Employees with high work–life balance tend to have high subjective well-being and have positive outcomes at work. Employees with higher subjective well-being tend to be more productive in the work place and also predicted organizational performance. This study examines employee's work–life balance and the subjective well-being of 146 employees at a life science company in Indonesia. We used the work–life balance questionnaire from Fisher et al. (2009). Subjective well-being instrument has two parts, namely the Satisfaction with Life Scale (SWLS) to see the degree of overall life satisfaction, and the Scale of Positive and Negative Experience (SPANE) to see how often respondents experience positive or negative feelings. Results of the correlation between work–life balance and subjective well-being are low (0.240). Demographic characteristics also did not impact the work–life balance and subjective well-being of respondents. Future research should provide a greater understanding regarding how subjective well-being is related to other dependent variables such as person organization fit because employees will be more interested in organizations that have the same values, beliefs, and so on instead, especially in Indonesia. Research in Indonesia showed that person organization fit has a strong correlation with subjective well-being, the same result as research by Park et al. (2011). The higher alignment of values, goals, needs, and personality of the individual with the organization, the higher the subjective well-being of employees.

1 INTRODUCTION

The polio eradication campaign began in 1988 by the World Health Organization (WHO). This initiative drove WHO to support countries, including Indonesia, to further develop their polio control capacity in any necessary aspects, from laboratory reagents to national-campaigns. In February, 2006, 305 polio cases had been identified in 10 provinces of Sumatra and Java. WHO's South East Asia Region, which includes Indonesia, is expected to be certified polio free in 2020 (World Health Organization 2014).

Such a positive development has also consequences for the manufacturers of vaccines and antisera in Indonesia as they have had to close the polio department forces and to find a new market. Many efforts to improve organizational performance fail because the fundamental culture of the organization—values, ways of thinking, managerial styles, paradigms, approaches to problem solving—remain the same (Cameron & Quinn 2011).

Dynamic changes in the industrial field trigger the competition in the industry. In order to survive in a competitive environment, the company must be able to follow the pattern of change, otherwise the company will not survive. The culture of an organization naturally depends on the market situation and needs to adapt to it in order to survive or to maintain its competitive position.

The Indonesian state-owned company in the health sector in Bandung, Indonesia, was established on August 6, 1890. As the one and only manufacturer of vaccines and antisera Company in Indonesia, it controlled 90% of the domestic market and 3% of the export market, and is now developing into a life science company with a more competitive culture than before. "Life science" refers to activities in the fields that have to do with organisms, like plants, animals, and human beings to produce vaccines and antisera.

Such organizational changes, however, can create a new problem. One of the problems in this company is a significantly higher level needed for counselling. Based on data obtained by interviewing the Human Resources Directorate Head, it became clear that the major problem of employees is about balancing work and family life. Family issues, such as family/life satisfaction, involvement in family, and marital satisfaction, trigger other problems like work attitude problems, such as absenteeism, discipline, involvement, and low job satisfaction.

A key concept here is "work–life balance." Work–life balance is defined as a multidimensional construct regarding issues of time, energy, goal accomplishment, and strain on the two domains of life, namely work and personal life (Fisher 2002). Much of the previous research has been limited by its emphasis on family aspects in the non-work domain. Therefore, the present study was undertaken to extend previous research by developing a broader construct, namely work/personal-life balance. Workers have important roles and responsibilities in the work and non-work (personal life) domains that affect each other. Roles and responsibilities in one domain of life can become enhancement/resources or interference/demands to roles and responsibilities in other domains of life (Fisher et al. 2009).

People who are able to manage their roles and responsibilities enhancement or interference in each domain of life, thus, more effectively allocate their time and energy and also can increase their well-being, job/life satisfaction, and productivity.

A four-dimensional typology of work–life balance has been suggested (Rantanen et al. 2016). According to the proposed four-dimensional typology, individuals can belong to beneficial, harmful, active, or passive work–life balance types. Based on Kaiser et al. (2011) this four-dimensional typology is in line with role conflict theories (Frone 2003; Greenhaus & Beutell 1985; Kahn et al. 1964), role enhancement theories (Barnett & Hyde 2001; Marks 1977; Sieber 1974; Wayne et al. 2007), and the demands–resources approach (Bakker & Geurts 2004; Voydanoff 2005). The Beneficial type, also named the Positive interaction or Balanced type consist of high resources and low demands, the Harmful type, also named the Negative interaction or Imbalanced type, is consisting of high demands and low resources, the Active type, also named the Negative and positive interaction or Blurred type, is consisting of high demands and high resources, and the Passive type, also named the No interaction or Segmented type, is consisting of low demands and low resources. The Beneficial type is expected to facilitate psychological functioning and well-being, whereas the Harmful type threatens it.

Research from Singh et al. (2013) shows the link between the work–life balance and the Subjective Well-Being. Subjective well-being shows life satisfaction and evaluation of important life domains such as work, health, relationships, and leisure. Also includes individual emotions, such as joy and involvement, and negative emotional experiences, such as anger, sadness, and low fear. In other words, happiness is the name given to positive thoughts and feelings towards one's life (Diener 2009). In other words, subjective well-being is explained as a person's subjective evaluation of their life, which includes life satisfaction as a component of cognitive and happiness as an affective component.

According to De Neve et al. (2013), subjective well-being has an important role for employees and organizations. The benefits of subjective well-being are in terms of health, productivity, and organizational behavior, and in relation to individuals with social behavior. In terms of productivity and organizational behavior, what is meant is increasing productivity, increasing income, decreasing absenteeism, creativity and the ability to think flexibly, collaborate, and improve organizational performance.

This work–life balance concept has not yet received special attention in Indonesia. Due to organizational culture change in this company, as described above, which brings about problems in balancing in work and family life, there is an opportunity to investigate the work–life balance and subjective well-being among employees of the life science company in Indonesia. Thus, this research tries to answer the questions stated below:

(1) What is the distribution of the type of employee's work–life balance in a company undergoing organizational change (i.e., a state owned-company in the health sector in Bandung, Indonesia); (2) How is the employee's subjective well-being on state owned-company in the health sector in Bandung, Indonesia; (3) How is the correlation between the type of employee's work–life balance and subjective well-being on state owned-company in the health sector in Bandung, Indonesia.

2 METHOD

2.1 Design and sample

The study is a quantitative cross-sectional study. We selected employees from a life science company in Indonesia who were married and have children. This company was selected because of a current change in the organizational culture of this company. Dynamic changes in the industrial field triggered this company to be more competitive than before, in order to survive in a competitive environment. It can be expected to have an effect on the work–life balance and subjective well-being of the employees. From 6 directorates in the company, we took 1 directorate as a sample. Our sample size consists of 146 employees.

2.2 Procedure

First, we approached the company with a letter from the university to ask for their permission to perform our study there. After the permission was granted, potential participants were asked to complete a survey on work–life balance and subjective well-being issues. Next, the researchers gave participants an informed consent letter that gave brief information about the questionnaire, the time usually needed to fill in the questionnaire, and asked for participants' willingness to fill in the questionnaire by signing it. Participants were also assured that the confidentiality of their answers was guaranteed. After that, the participants received the questionnaire individually at their office.

2.3 Instrument

We used the Work–Life Balance Questionnaire (Fisher et al. 2009). The questionnaire consists of 17 items divided into four dimensions, namely WEPL, PLEW, WIPL, and PLIW to be answered on a 5-point Likert scale. The 5 answering options were never (1), seldom (2), sometimes (3), often (4), and very often (5). It divides the concept of 'work–life balance' into two aspects, that is, 'resources' and 'demands.' Resources contain two sub-aspects, namely 'work enhancement personal life' (WEPL) and 'personal life enhancement work' (PLEW). 'Demands' had two sub-aspects too, that is, 'work interfere personal life' (WIPL) and 'personal life interfere work' (PLIW). The original English questionnaire was translated into Bahasa Indonesia by the first author (GG) and adapted into Indonesian culture, and the questionnaire was also checked by an expert in English literature. All items are valid with an SLF score of more than 0.5. The Cronbach's alpha reliability of WEPL, PLEW, WIPL, and PLIW are 0.714, 0.783, 0.748, and 0.787 respectively, which can be categorized as high reliability (Guilford 1956).

Subjective well-being measures were compiled by Diener (2009). This measuring instrument has two parts, namely the SWLS to see the degree of overall life satisfaction, and SPANE to see how often respondents experience positive or negative feelings. The original English questionnaire was translated into Bahasa Indonesia by the first author (GG) and adapted into Indonesian culture, and the questionnaire was also checked by an expert in English literature. All items are valid with an SLF score of more than 0.5. The Cronbach's alpha reliability of SWLS is 0.947, which can be categorized as very high reliability. The Cronbach's alpha reliability of SPANE positive and negative are 0.915 and 0.889, which can be categorized as very high reliability (Guilford 1956).

2.4 Statistical analysis

The scores of a participant's types of resources and demands in the work–life balance questionnaire were dichotomized in high and low by using group norms represented by the median score. Based on this, we classified the type of work–life balance for each participant in 'beneficial', 'active', 'passive', or 'harmful'. See Table 1.

Table 1. Work–life balance type.

		Demands	
		High	Low
Resources	High	Active Balance	Beneficial Balance
	Low	Harmful Balance	Passive Balance

The first step for the statistical analysis for the subjective well-being questionnaire determined the maximum total score and minimum total score for both measuring instruments by:

1. Maximum score SWLS of 5 items with a scale of 1–7.

   Then the highest value of SWLS is $5 \times 7 = 35$.
   SWLS' Livestock value is $5 \times 1 = 5$.
   SWLS Absolute Norm: (max SWLS score + SWLS min score): $2 = (35 + 5): 2 = 40:2 = 20$
   Low SWLS: Score 5–20
   High SWLS: Score 21–35

2. Maximum score SPANE of 12 items with a scale of 1–5 consisting of 6 items SPANE-P (Positive Affect), 6 SPANE-N items (Negative Affect), and SPANE-B (Affect balance).

   The highest value of SPANE-P is $6 \times 5 = 30$.
   The lowest value of SPANE-P is $6 \times 1 = 6$.
   The absolute norm SPANE-P: (Score Max SPANE-P + min SPANE-P score): $2 = (30 + 6): 2 = 36: 2 = 18$
   The highest value of SPANE-N is $6 \times 5 = 30$.
   The lowest value SPANE-N is $6 \times 1 = 6$.
   The absolute norm SPANE-N: (Score Max SPANE-N + min SPANE-N score) $= (30 + 6): 2 = (36:2) = 18$
   SPANE Category:
   Low SPANE: Score 6–18
   High SPANE: Score 19–30

3. For determining the number of categories used, in this research, scores will only be categorized into high subjective well-being and low subjective well-being. The following guidelines can be used:

   High satisfaction Life + high positive Affect + negative low Affect = High SWB
   High satisfaction Life + low positive Affect + negative high Affect = low SWB
   High satisfaction Life + high positive Affect + negative high Affect = low SWB
   Low satisfaction Life + low positive Affect + negative high Affect = low SWB

For testing the correlation, Chi-square tests were conducted. The SPSS statistical software version 22.0 was used. For testing the influence of demographic characteristics, a cross tabulation technique was used.

3 RESULTS AND DISCUSSION

Data were obtained from 146 employees. The correlation between work–life balance and subjective well-being is shown in Table 2:

Table 2. Correlation between work–life balance and subjective well-being.

		Value	Asymp. Std. Error	Approx. Tb	Approx. Sig.
Nominal by Nominal	Contingency Coefficient	.240			.031
Interval by Interval	Pearson's R	−.102	.090	−1.233	.220c
Ordinal by Ordinal	Spearman Correlation	−.101	.090	−1.217	.226c
N of Valid Cases		146			

Ho: There is no relationship between subjective well-being and work–life balance.

H1: There is a relationship between subjective well-being and work–life balance.

If Sig 0.031 < α (5%) then Ho is rejected. If Ho is rejected it means H1 will be a conclusion. Conclusion: There is a relationship between work–life balance and subjective well-being. The correlation is 0.240, which means low correlation.

Table 3. Employee's work–life balance type.

Work Life Balance Type	Percentage
Beneficial	12 (8.20%)
Harmful	16 (10.95%)
Active	43 (29.45%)
Passive	75 (51.40%)
Total	146 (100%)

From Table 3, it appeared that, most of the employees belong to the passive type, followed by the active, harmful, and beneficial types.

Table 4. Employee's subjective well-being.

Subjective Well-Being	Percentage
High	87 (59.60%)
Low	59 (40.40%)
Total	146 (100%)

On the other hand, most of the employees have low subjective well-being.

Research by Jeffrey et al. (1998) shows that behaviors, attitudes, and subjective well-being are determined by individuals and the environment. High subjective well-being is characterized by a lack of difference between organizational and individual value.

Person-Organization Fit also has a correlation with work–life balance in research by Jeffrey et al. (1999). Work–life balance and subjective well-being are related to another dependent variable, which is person organization fit. Employees will be more interested in organizations that have the same values, beliefs, and so on instead, especially in a collectivistic culture like Indonesia.

4 CONCLUSION

The result correlation between work–life balance and subjective well-being is low (0.240). Demographic characteristics, such as gender, age, marital status, number of children, race, job position, length of work, education, and salary also did not impact the work–life balance and subjective well-being of the respondents.

The organizational culture change that was followed by decreased work–life balance and its impact on the decline of the subjective well-being of employees was not very strongly correlated. The decline of subjective well-being can also be influenced because there is a considerable gap between the value in the new organizational culture that is more competitive than the value that employees have.

Future research should provide a greater understanding regarding how work–life balance and subjective well-being are related to other dependent variables, such as person organization fit, because employees will be more interested in organizations that have the same values, beliefs, and so on instead, especially in Indonesia.

The novelty of this research is that research about work–life balance in Indonesia is still rare. Organizations that are concerned about work–life balance also very limited in Indonesia; very few companies have regulations about work–life balance in Indonesia.

The limitation of this research was the difficulties to meet all employees at the same time, this leads us to limitation of not being able to represent the whole company's opinion and compile thorough results.

REFERENCES

Bakker, A. B. & Geurts, S. A. 2004. Toward a dual-process model of work-home interference. *Work and occupations* 31(3): 345–366.

Barnett, R. C. & Hyde, J. S. 2001. Women, men, work, and family: An expansionist theory. *American psychologist* 56(10): 781.

Cameron, K. S. & Quinn, R. E. 2011.Diagnosing and changing organizational culture based on the competing values framework (third edit). *San Francisco: John Wiley & Sons.*

De Neve, J. E., Diener, E., Tay, L., & Xuereb, C. 2013. The Objective benefits of subjective well-being. *New York: UN Sustainable Development Solutions Network.*

Diener, E. D. 2009. Assessing well-being: The collected works of Diener (social indicators research series 39). *New York: Springer.*

Edwards, J. R., Caplan, R. D., & Harrison R. V. 1998. Person-environment fit theory: Conceptual foundations empirical evidence, and directions for future research. *Oxford: Oxford University Press.*

Edwards, J. R. & Rothbard, N. P. 1999. Work and family stress and well-being: An examination of person–environment fit in the work and family domains. *Journal Organizational Behavior and Human Decision Processes* 77(2): 85–129.

Fisher, G. G. 2002. Work/personal life balance: A construct development study. *Unpublished Doctoral Dissertation.*

Fisher, G. G., Bulger, C. A., & Smith, C. S. 2009. Beyond work and family: A measure of work/nonwork interference and enhancement. *Journal of Occupational Health Psychology* 14(4): 441–456.

Frone, M. R. 2003. Work-family balance.

Greenhaus, J. H. & Beutell, N. J. 1985. Sources of conflict between work and family roles. *Academy of management review* 10(1): 76–88.

Guilford, J. P. 1956. Fundamental statistic in psychology and education 3rd Ed. New York: McGraw-Hill Book Company, Inc.

Kahn, R. L., Wolfe, D. M., Quinn, R. P., Snoek, J. D., & Rosenthal, R. A. 1964. Organizational stress: Studies in role conflict and ambiguity.

Kaiser, S., Ringlstetter, M., Eikhof, D. R., Pina, E. C. M. 2011. Creating balance? International perspectives on work-life integration of professionals. *Springer-Verlag Berlin Heidelberg* 84.

Marks, S. R. 1977. Multiple roles and role strain: Some notes on human energy, time and commitment. *American sociological review*: 921–936.

Park, H. I., Monnot, M. J., Jacob, A. C., Wagner, S. H. 2011. Moderators of the relationship between person-job fit and subjective well-being among Asian employees. *Int J Stress Manag* 18(1):67.

Rantanen, J., Kinnunen, U., & Pulkkinen, L. 2016. The role of personality and role engagement in work-family balance. *Psihološka Obzorja / Horizons of Psychology* 22: 14–26.

Sieber, S. D. 1974. Toward a theory of role accumulation. *American sociological review*: 567–578.

Singh, A. K. & Amanjot, A. 2013. Work life balance and subjective well-being: An empirical analysis using structural equation modeling.

Voydanoff, P. 2005. Toward a conceptualization of perceived work-family fit and balance: A demands and resources approach. *Journal of marriage and family* 67(4): 822–836.

Wayne, J. H., Grzywacz, J. G., Carlson, D. S., & Kacmar, K. M. 2007. A multi-level perspective on the synergies between work and family. *Journal of Occupational and Organizational Psychology* 80(4): 559–574.

World Health Organization. 2014. Programme Budget Matters: Programme Budget 2012–2013. [Online]. Retrieved from http://www.searo.who.int/mediacentre/events/governance/rc/sea-rc69-6rev1_7.2.pdf?ua=1.

Advances in Business, Management and Entrepreneurship – Hurriyati et al. (Eds)
© 2021 Taylor & Francis Group, London, ISBN 978-0-367-67471-7

A behavioral model of unethical behavior in public service for villagers

R. Ambarwati & A.W. Mudjib
Universitas Muhammadiyah Sidoarjo, Sidoarjo, Indonesia

ABSTRACT: Issues of morality and ethics have increasingly become more important in local government and public service settings. Unethical behaviors refer to all activities considered non-ethical and immoral. The main causes of characters that form unethical action in public service are corruption and nepotism. The purpose of this research was to measure the level of unethical behaviors in a village public service. The design of this study was a survey and the data were collected through a cross-section method using a questionnaire. The unit of analysis was all villagers who used village services regularly. The results of this study indicated that corruption was more dominant compared to nepotism in forming unethical actions in public services. The practical implications of this research were useful for the local government, especially for village public service. They needed to improve the effort to reduce unethical behaviors in public service.

1 INTRODUCTION

Unethical behaviors were currently found in many government institutions, and public service sectors were no exception. In the science of state administration, the state civil apparatus or the member of bureaucracy have a function to formulate, implement, and evaluate public policy (Belle & Cantarelli 2017). Public service work is a practical implementation of public policy.

Citizens certainly need good quality public services. It is not an excessive demand because it is also their right as citizens who fulfill their obligations such as paying taxes, complying with legal rules and procedures, and maintaining environmental stability (Lourenço 2016). The question is then whether they are satisfied with the public service provided. Particularly in a village area, the public service provides all forms of government administrative services, especially in the case of essential documents needed by villagers.

Several previous studies analyzed a lot about the implementation of good governance. However, only very few focused on measuring unethical actions. A survey of 600 Australian consumers revealed that both empathy and moral identity were related to detrimental beliefs regarding the passive and the active/legal dimensions of consumer ethics and were related to definite conclusions regarding the "doing good"/recycling dimension. Cynicism was related to positive beliefs regarding the passive aspect of consumer ethics and was referred to as detrimental ideas regarding the "doing good"/recycling dimension. The role of moral disengagement in mediating these relationships was examined. Empathy and moral identity were only indirectly negatively related to the "no harm,

no foul" dimension of consumer ethics through moral disengagement. At the same time, cynicism was indirectly positively related to this dimension through moral disengagement (Chowdhury & Fernando 2014). Even more frequent and pervasive were cases of "ordinary" unethical behavior. Unethical actions committed by people who value and care about morality, but failed to maintain their ethics when faced with an opportunity to cheat (Gino 2015; Kouchaki & Gino 2016). Engaging in unethical behavior produces changes in memory so that memories of unethical actions gradually become less bright and vivid than memories of ethical actions or other types of actions that are either positive or negative in valence. This memory obfuscation of one's unethical acts over time is called "unethical amnesia." Because of unethical amnesia, people are more likely to act dishonestly repeatedly over time (Kouchaki & Desai 2015; Kouchaki & Gino 2016). Their research investigates exposure to in-group members who misbehaved or to others. The cluster members during this analysis have the benefit of unethical actions, greed, self-concern, and pardon. They also expose incremental dishonesty, loss aversion, challenging performance goals, or time pressure to increase unethical behavior.

In contrast, monitoring employees, giving useful reminders, and individuals' willingness to maintain a definite self-view decrease unethical conduct. Findings on the effect of self-control depletion on unethical behavior are mixed (Belle & Cantarelli 2017). Building on the idea that we need to develop a more comprehensive and complete understanding of the value and how it influences actions and decisions as well as the importance and relevance of also adopting a descriptive approach that is grounded in the behavioral sciences (De Cremer & Vandekerckhove 2017). Unethical behaviors, in a sense, are all actions that are not considered ethical and moral (Kouchaki & Gino 2016).

This study aims to measure unethical behavior in the public service sector in rural areas, which includes corruption and nepotism. Corruption and nepotism have a terrible impact on the economy, unfortunately most people in Indonesia view unethical behavior as something normal.

1.1 Corruption

Corruptions have a significant impact on the country's economy because economic growth is hampered by the involvement of employees and government officials in corruption (Purcell 2016). Corruption reduces investments made by the government, both domestically and abroad. Corrupt acts are dishonest behaviors that violate the truth. Corruption is the actions of public officials who abuse their authority, position, or power, resulting in violating some state legal norms (Transparency International 2016). Acts of corruption are usually carried out in secret and for personal gain for wealth or status or because of family, friends, ethnicity, or religious groups. One form of corruption is fundamental bribery and a wrong way that directly impacts it (Liu et al. 2016).

The impact of corruption is so high that it can reduce the quality of people's welfare; the high loss due to corruption will have an effect on the state's obligation to provide welfare rights (Yan & Oum 2014). Therefore, community participation in the prevention of acts of corruption is very much needed and has a critical role as a form of social control. High social power will be able to narrow the space for corruption and widen the scope for anti-corruption (De Cremer & Vandekerckhove 2017).

1.2 Nepotism

Nepotism is an action that refers to giving improper assistance to someone who has a closeness to government officials, such as family members, members of political parties, tribal members, or members of the same religious group (Baumeister & Alghamdi 2015). Although nepotism is not recommended in the public sector, some researchers see its positive aspects, especially in the business context (Chowdhury & Fernando 2014). Nepotism rules when authorities employ each other by providing ways to create relationships and support among many families in a network (Chowdhury & Fernando 2014). Besides that, nepotism is preferred in small-scale family business companies that have smaller systems.

The practice of corruption is rampant, and so is nepotism, however, discussion about nepotism is still rare. Research about nepotism has only been developed after 2010, where several studies showed the impact of nepotism on performing family and corporate companies. The results show that nepotism

produces unbalanced decisions, unfair treatment, and damages the company's performance in the long run (Kouchaki & Desai 2015). Recent research also shows that nepotism causes loss of motivation, self-confidence, alienation, and it also discards highly skilled employees, and limits competition and innovation. The consequences of nepotism undermine the foundation of the organization, which will ultimately impact overall economic development (Stellar & Willer 2018). Nepotism causes a lot of negative impacts on organizational performance, and the lack of interest among researchers in this can have a more significant effect than imagined (Birch & Chiang 2014).

2 METHOD

The stages in this study included (1) literature study, (2) problem formulation and research objectives, (3) data collection, (4) testing research instruments, (5) data processing, (6) interpretation of results, and (7) recommendations at the initial stage described in the introduction to the study of literature and research purposes. The purpose of this study was to measure the level of unethical actions using two dimensions, that is, corruption and nepotism (see Figure 1). The method of data collection was using questionnaires and direct interviews with villagers in East Java, Indonesia, which is the second largest region in Indonesia. This research was conducted over a three-month period, that from January to March, 2019.

The population of this study were all villagers in the Sidoarjo District who used and had been involved in public services in the village administration, aged between 17 and 65 years with a minimum education of high school or equivalent. 185 to 200 questionnaires were distributed to all villagers and interviews were

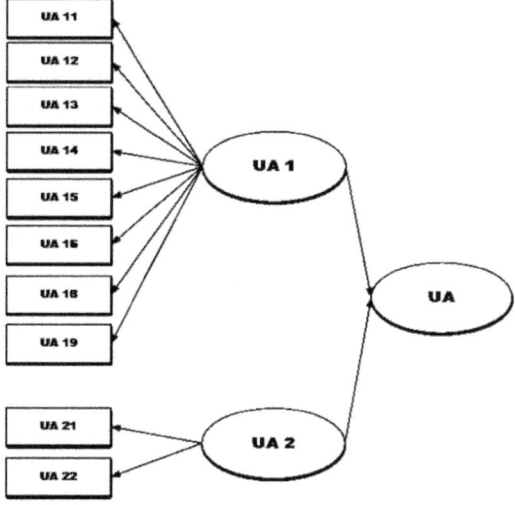

Figure 1. Research framework.

conducted directly by visiting all respondents in each village. Based on these criteria, for obtaining the population in this study, the sampling method of probability sampling was used, and the technique of determining the sample was simple random sampling. A Likert scale was used to measure the attitudes, opinions, and perceptions of the respondents on the object.

Testing the instrument of this study included validation and reliability to verify whether the instrument met the requirements of research method standards. The instrument is said to be good if it meets three main elements, namely being valid, reliable, and practical [10]. At the data processing stage, researchers used SEM (structural equation modelling) analysis. The data analysis was carried out by interpreting the assessment of dimensions that influence the Unethical Actions. The recommendations in this study were the results of the discussion and interpretation of data analysis processed by SEM.

3 RESULTS AND DISCUSSION

The testing instrument in this study was aimed at measuring formative indicators from Unethical behavior. Model evaluation was completed by looking at the outer weight significance value with T-statistics > 1.96, and obtained through a resampling (bootstrapping) procedure, thus, validity and reliability testing was not required.

The results of path coefficient analysis explained that the Corruption indicator was the most dominant form of Unethical Actions with a value of 0.816 with CR = 37.369 (Table 1). The indicators of corruption show that corruption reduces satisfaction with service quality, accountability, and the responsiveness of the government. Corruption also causes a lack of information about government actions and low trust in the government to solve problems. As a disease, corruption endangers not only state finances, but also the nation's condition and state since it causes an imbalance in the share of income received by various groups of society (McLeod & Harun 2014). In such conditions, the most disadvantaged ones are the people at

Table 1. Path coefficients of unethical behavior indicators.

	Mean	Std Dev	T-Stat	P-Val
UA11 <- UA1	0,861	0,026	33,174	0,000
UA12 <- UA1	0,858	0,035	24,398	0,000
UA13 <- UA1	0,872	0,024	35,894	0,000
UA14 <- UA1	0,819	0,035	23,589	0,000
UA15 <- UA1	0,782	0,042	18,674	0,000
UA16 <- UA1	0,700	0,068	10,411	0,000
UA18 <- UA1	0,588	0,064	9,240	0,000
UA19 <- UA1	0,773	0,049	15,709	0,000
UA21 <- UA2	0,939	0,010	90,203	0,000
UA22 <- UA2	0,914	0,026	35,453	0,000
UA1 ->UA	0,816	0,022	37,369	0,000
UA2 ->UA	0,235	0,016	14,286	0,000

the grassroots level, whose welfare is actually under guarantee by the constitution (Liu et al. 2016).

However, in law enforcement, there is community participation, which functions as social control. Corruption uses power basically because of the weakness of social control, or the social environment that shapes it, especially in an environment with lost power and responsibility (Gong 2015). So corruption encompasses standard behavioral deviations, which are violating or contrary to the law. Social control is a normative aspect of social life that inhibits deviant behavior and its consequences, such as prohibitions, demands, punishment, and compensation, and according to abnormal behavior, depends on social control (Goddard et al. 2016). It means that social control determines how behavior is a deviant behavior. The attitude of society's rejection of deviant behavior can be qualified as a crime, where the offense is a shocking thing to the city (Baron et al. 2015). An act is a crime when the action violates a sharp and defined joint consciousness.

The indicator of nepotism with an outer weight value of 0.235 is significant with CR = 14.286 (Table 1). This indicator explains how nepotism reduces our satisfaction with service quality, and nepotism shows a low level of transparency. This indicator is a less dominant factor in shaping Unethical behavior, because it has the smallest value outer weight compared to other indicators. The practice of giving privilege to certain people based on personal preferences, blood ties, and family relations is still rampant today (Gino 2015). Regional leaders exercise their power by privileging their close family within the government. When local leaders with families and officials in the administration are no longer in force, their influence and political heritage will remain active (Kalshoven et al. 2016). Nepotism affects how one determines socio-economic classes based on skin color, appearance, and preferences. The practice of nepotism usually starts very early when parents differentiate their children based on who their parents like most (Kouchaki & Gino 2016). This behavior then enters the sub consciousness of the child, thus, shaping their behavior. The same thing also happens in government bureaucracies when many people choose officials based on personal subjective judgment rather than on quality and qualifications, assuming as long as the selected person is sufficiently qualified, then they can practice lawful nepotism (Baron et al. 2015). The justification of nepotism can affect how a country understands the practice. Therefore, people need to know the impact of nepotism behavior. The government should also make regulations that can prevent the exercise of nepotism from taking place in the government bureaucracy (Yan & Oum 2014).

4 CONCLUSION

The research was aimed to measure unethical behaviors in villages; the results show that corruption is the dominant factor that forms unethical behavior. The perception that corruption reduces the importance of regulations so that there is an increase in unethical actions in the village, especially by public services. The perception of the regional government is more corrupt than the central government. In this case, the government needs to pay attention to the performance of the civilian state apparatus so that the prevention of unethical actions can be done early. This study implies that public perceptions and concerns about unethical actions are increasing. It needs cooperation between citizens and government officials to prevent and control unethical actions in government public services. Implementation of good governance requires the oversight of citizen involvement.

The limitation of this study was only in measuring the concept of unethical behavior in the village government office. As a complement to this research, further research can be carried out at the central government and the national private sector. In addition, future researchers can add several measurements of both variables and indicators related to the implications of unethical actions on good governance.

ACKNOWLEDGMENT

We would like to thank The Directorate General of Higher Education and Universitas Muhammadiyah Sidoarjo for supporting the publication of this research.

REFERENCES

Baron, R. A., Zhao, H., & Miao, Q. 2015. Personal Motives, Moral Disengagement, and Unethical Decisions by Entrepreneurs: Cognitive Mechanisms on the "Slippery Slope." *Journal of Business Ethics*. https://doi.org/10.1007/s10551-014-2078-y

Baumeister, R. F. & Alghamdi, N. G. 2015. Role of self-control failure in immoral and unethical actions. *Current Opinion in Psychology*. https://doi.org/10.1016/j.copsyc.2015.04.001

Belle, N. & Cantarelli, P. 2017. What causes unethical behavior? A meta-analysis to set an agenda for public administration research. *Public Administration Review*. https://doi.org/10.1111/puar.12714

Birtch, T. A. & Chiang, F. F. T. 2014. The influence of business school's ethical climate on students' unethical behavior. *Journal of Business Ethics*. https://doi.org/10.1007/s10551-013-1795-y

Chowdhury, R. M. M. I. & Fernando, M. 2014. The relationships of empathy, moral identity and cynicism with consumers' ethical beliefs: The mediating role of moral disengagement. *Journal of Business Ethics*. https://doi.org/10.1007/s10551-013-1896-7

De Cremer, D. & Vandekerckhove, W. 2017. Managing unethical behavior in organizations: The need for a behavioral business ethics approach. *Journal of Management and Organization*. https://doi.org/10.1017/jmo.2016.4

Gino, F. 2015. Understanding ordinary unethical behavior: Why people who value morality act immorally. *Current Opinion in Behavioral Sciences*. https://doi.org/10.1016/j.cobeha.2015.03.001

Goddard, A., Assad, M., Issa, S., Malagila, J. & Mkasiwa, T. A. 2016. The two publics and institutional theory – A study of public sector accounting in Tanzania. *Critical Perspectives on Accounting*. https://doi.org/10.1016/j.cpa.2015.02.002

Gong, T. 2015. Managing government integrity under hierarchy: Anti-corruption efforts in local China. *Journal of Contemporary China*. https://doi.org/10.1080/10670564.2014.978151

Kalshoven, K., van Dijk, H., & Boon, C. 2016. Why and when does ethical leadership evoke unethical follower behavior? *Journal of Managerial Psychology*. https://doi.org/10.1108/JMP-10-2014-0314

Kouchaki, M. & Desai, S. D. 2015. Anxious, threatened, and also unethical: How anxiety makes individuals feel threatened and commit unethical acts. *Journal of Applied Psychology*. https://doi.org/10.1037/a0037796

Kouchaki, M. & Gino, F. 2016. Memories of unethical actions become obfuscated over time. *Proceedings of the National Academy of Sciences*. https://doi.org/10.1073/pnas.1523586113

Liu, Q., Luo, T., & Tian, G. 2016. Political connections with corrupt government bureaucrats and corporate M&A decisions: A natural experiment from the anti-corruption cases in China. *Pacific Basin Finance Journal*. https://doi.org/10.1016/j.pacfin.2016.03.003

Lourenço, R. P. 2016. Evidence of an Open Government Data Portal Impact on the Public Sphere. *International Journal of Electronic Government Research*. https://doi.org/10.4018/ijegr.2016070102

McLeod, R. H. & Harun, H. 2014. Public Sector Accounting Reform at Local Government Level in Indonesia. *Financial Accountability and Management*. https://doi.org/10.1111/faam.12035

Purcell, A.J. 2016. Australian local government corruption and misconduct. *Journal of Financial Crime*. https://doi.org/10.1108/JFC-10-2013-0060

Stellar, J. E. & Willer, R. 2018. Unethical and inept? The influence of moral information on perceptions of competence. *Journal of Personality and Social Psychology*. https://doi.org/10.1037/pspa0000097

Transparency International. 2016. Transparency International—The Global Anti-Corruption Coalition. *Transparency International*. https://doi.org/10.1177/0115426504019003290

Yan, J. & Oum, T. H. 2014. The effect of government corruption on the efficiency of US commercial airports. *Journal of Urban Economics*. https://doi.org/10.1016/j.jue.2014.01.004

Advances in Business, Management and Entrepreneurship – Hurriyati et al. (Eds)
© *2021 Taylor & Francis Group, London, ISBN 978-0-367-67471-7*

Citizen trust in local government: Explaining the role of village service

R. Ambarwati & F.F. Lestariana
Universitas Muhammadiyah Sidoarjo, Sidoarjo, Indonesia

ABSTRACT: Trust in local governments is needed in crises such as natural disasters, economic emergencies, or political un-rest. The purpose of this study was to measure citizen trust in government public services, particularly in village areas. The data used in this survey was collected through questionnaires distributed to rural communities that use government public services. Explanatory Factor Analysis was applied to analyze the data. The results of this study indicated that the quality of public services was the dominant factor. However, partisan involvement was a factor of concern in increasing citizen trust. The practical research is useful to increase trust in village government services as part of the supporting factors for the success of central government development programs.

1 INTRODUCTION

The current decline in the image of public institutions can cause degradation of public trust. Public trust in public institution is an absolute necessity since the purpose of the institution itself is to serve and meet the needs of the community both directly and indirectly (Bertot et al.. 2016), therefore building relationships and trust in the community is very important for public institutions. A good relationship between the community and the public institution result in confidence on the service process and that the institution is one of the right choices (Kim 2014), so that the desire to use services outside public institutions is smaller. Therefore, public trust is a critical to understand so that the agency can increase community trust and in turn will lead to trust in their organization (Danish, et al. 2013). Service performance is another critical problem that civil servants try to improve by doing excellent service based on what the community expects and desires (Taylor 2014). Improving public services is a long-term effort, to realize a real bureaucratic concept in which the public is primary right holder is not an easy task (Graham et al. 2015). Public service laws are enacted to strengthen, monitor, and provide direction, such as Law no. 25 of 2009 regarding Public Services. It aims to improve the delivery of public services by creating a mechanism to ensure public expectation and demands of the citizens are met.

Implicitly, bureaucratic trust is the primary function of government performance. The bureaucratic reform policies are aimed to increase citizens trust in one country and are equally suitable for other countries (Grimmelikhuijsen & Knies 2017). Trust can be defined as relationships between two or more individuals, between individuals and organizations (such as companies or social service departments), or between several organizations" (Ma & Wang 2014). Research in public administration discussed citizen attitudes on public sector, public trust, citizens' assessment on public institutions officials (Ramesh 2017). Some of them assessed the competence of public officials in carrying out designated assignments and affective evaluation of ethical and caring behavior or having the interests in service users (Griffin & Halpin 2018). The trust in institutions requires employees to be competent, credible, and willing to act for the greater public interest. Public trust concerns the extent to which citizens have confidence in public institutions to operate in the best interests of society and their constituents (Ferry et al. 2018).

Regarding bureaucratic performance, a lack of trust relates to limited administrative authority delegation, excessive oversight, and excessive dependence on formal rules and procedures (Katoch et al. 2017). Besides, the low level of public trust undermines morale, retention, and recruitment of civil servants. Apart from evidence, policymakers concern with the status of trust in the public bureaucracy (Mourtada & Salem 2015) as it often offers poor administrative performance and cause declining public satisfaction.

Public service plays an essential role in general welfare. The main problem for civil government services is the lack of perceptions of citizen trust. It needs the correct steps to increase citizen trust (Agyemang & Ofei 2013). Several previous studies analyzed about the implementation of good governance. However, only very few focused on measuring citizen trust. The studies found that ethnic majority had more trust in public institutions than minority. Identity, language of administration, and lack of equal representation in the bureaucracy were also considerably determined the level of trust citizens had in public institutions.

Further, political patronage in service delivery has subverted the quality of public institutions and trust. Interestingly, citizens who support and being affiliated with the ruling party tend to have more trust in government. Police and village-level officers are least trusted, owing to endemic corruption (Ramesh 2017). The models reveal that the movement of power to select local leaders from the hands of upper-level government to residents erodes the influence of the so-called 'traditional authority orientation' among citizens, and enables them to assess trust level according to the competence of the government. Furthermore, elections cut back the trust-generating result of institutional tendency as voters. The voters are enfranchised within the 'input' method of the presidency, so accentuation 'outputs' less (Ma & Wang 2014). Trust is correlated with both subjective (at the individual-level) and objective (at the national level) indicators of performance. The quality of institutions also matters as countries with lower levels of public sector corruption experience higher levels of trust in the civil service (Houston et al. 2016).

This study measures citizen trust in public services of the government, especially in the village. The measurement used several indicators, including Quality of service, Information and Knowledge, Partisans, and Promise of Politicians.

2 METHODS

This study measures citizens' trust in government public services village areas. The measurement used several indicators i.e. quality of service, information and knowledge, partisans, and promise politicians promise. This study was a survey that collected the data using questionnaires.

The unit of analysis was rural communities that used public services. The data was analyzed using Structural Equation Modeling. The population in this study was all villagers in the Sidoarjo District who used and had been involved in public services in the village administration, aged between 17–65 years. The minimum education level was high school or equivalent. The sampling method was probability sampling and the technique of determining the sample unit was simple random sampling. The sample size in this study was 185–200.

Data was collected by distributing questionnaires and direct interviews with respondents. The measurement of data used a Likert scale to measure attitudes, opinions, and perceptions of the respondents on the object. This research was conducted within three months period from January to March 2019.

In the data processing stage, researchers used SEM (Structural Equation Modelling). The results of data analysis were interpreted to find out dimensions that formed citizen trust. The recommendations in this study were the results of discussion and interpretation

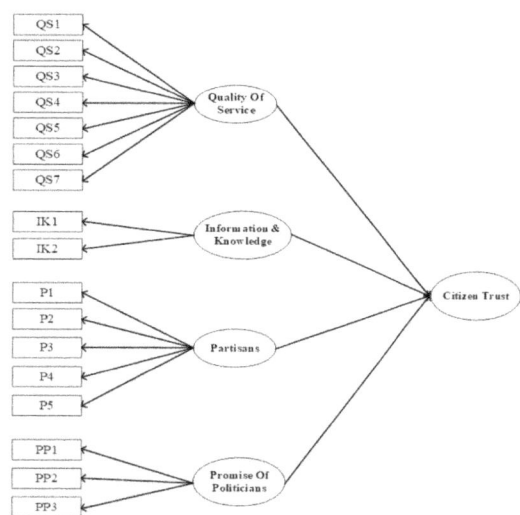

Figure 1. Research framework.

of data analysis processed by SEM. The researcher formulated the research framework as presented in Figure 1.

3 RESULTS AND DISCUSSION

The testing instrument was aimed to measure formative indicators of Citizen Trust. The outer weight significance value with T-statistics > 1.96, it was obtained through a resampling (bootstrapping) procedure which meant construct validity and reliability was not required.

As shown as Table 1, Citizen Trust measurement used four indicators, namely: quality of service, information and knowledge, partisans, and promise of politicians. Quality of service indicator had an outer weight, which was the most dominant in forming Citizen Trust with a value of 0.401 significant with CR = 34.46. Quality of service showed that the government is responsive in terms of policymaking and promotes their ideology. It also indicates political parties continue the previous program when in power, public services are provided on time, the behavior of friendly public service providers, public service providers, can always solve problems, and the quality of public services is excellent. Awareness of the need for unique and satisfying public services has grown from the government before the reform era but has not been followed by implementing public service providers as expected (Houston et al. 2016). The village officials hold responsibilities to provide public services. Good or bad services offered to the community will depend on the quality and quantity, effectiveness, and efficiency (Ma & Wang 2014). The population as the party being served will receive the service in various perceptions.

Table 1. Path coefficients of citizen trust indicators.

Latent variable	Outer weight	Critical ratio (CR)
CT11 <- CT1	0,601	8,820
CT12 <- CT1	0,698	12,232
CT13 <- CT1	0,784	21,394
CT14 <- CT1	0,827	32,371
CT15 <- CT1	0,701	17,129
CT16 <- CT1	0,646	14,739
CT17 <- CT1	0,766	22,675
CT21 <- CT2	0,878	32,214
CT22 <- CT2	0,897	50,948
CT31 <- CT3	0,802	25,907
CT32 <- CT3	0,868	45,523
CT33 <- CT3	0,804	14,574
CT34 <- CT3	0,633	8,057
CT35 <- CT3	0,763	22,293
CT41 <- CT4	0,901	30,989
CT42 <- CT4	0,915	68,754
CT43 <- CT4	0,775	20,726
Quality of Service (CT1) -> CT	0,401	34.46
Information & Knowledge (CT2) -> CT	0,157	21.01
Partisans (CT3) -> CT	0,324	14.70
Promise of Politicians (CT4) -> CT	0,232	16.74

Source: Results of data processing from SmartPLS.

The indicator of Information and Knowledge with outer weight value of 0.157 significant with CR = 21.01. This indicator explained how information about services can increase citizen satisfaction and confidence in good public service performance. This indicator was a less dominant factor in shaping Citizen Trust because it had the smallest value outer weight compared to other indicators. Ease of access to information and knowledge can increase citizen trust in the government (Ma & Wang 2014). Communities need information about the government and legal services. The public will feel justice and security once it is fulfilled.

Indicator of Partisans with a value of 0.324 was significant with CR = 14.70. This indicator explained how partisan affiliation in government increased citizen trust, satisfaction with service quality, trust in government policies. Non-partisans also share the same beliefs with individuals, increased trust in non-partisans through excellent performance. This indicator was also the dominant factor to form Citizen Trust because as it was the second largest value outer weight after the quality service indicator. The importance of community participation in the formulation of public policies in the region's public administration is the result of the collaboration of various actors, both government, society, experts, and social institutions (Graham et al. 2015).

The Politician Promise had an outer weight value at 0.232 significant with CR = 16.74. This indicator explained that honest public service provider, fair public service providers, promises made by politicians are fulfilled, and local politicians are more reliable than the national level. Trust in politics is inseparable from the political actors themselves. Here the existence of a political elite is one of the focus. The political elite is a small group of people who have a significant influence in making and implementing political decisions. The political elite, in this case, has a source of power that includes political power, and the existence of this political elite is little when compared to the population in a country (Hessami 2014).

4 CONCLUSION

This research focused on all government public service in village areas. From the result, it can be concluded that quality of service is a dominant factor that can increase citizen trust. The village officials should also foster good relationships with the partisans and increase the involvement of non-partisans in village development activities.

This research has an implication on public services in villages which function as the extension of central government.

It also expected that this research shed light not only on Citizen Trust to build excellent public service, but also on how to create good governance in local government

ACKNOWLEDGEMENT

The authors would like to thank Indonesia Directorate General of Higher Education and Universitas Muhammadiyah Sidoarjo for supporting the publication of this research.

REFERENCES

Agyemang, C.B. & Ofei, S.B. 2013. Employee work engagement and organizational commitment: a comparative study of public sector organizations in Ghana. *European Journal of Business and Innovation Research.*

Bertot, J., Estevez, E. & Janowski, T. 2016. Universal and contextualized public services: Digital public service innovation framework. *Government Information Quarterly.* https://doi.org/10.1016/j.giq.2016.05.004.

Danish, R.Q., Ramzan, S. & Ahmad, F. 2013. Effect of perceived organizational support and work environment on organizational commitment; mediating role of self-monitoring. *Advances in Economics and Business.* https://doi.org/10.13189/AEB.2013.010402.

Ferry, L., Glennon, R. & Murphy, P. 2018. Local government. *In Public Service Accountability: Rekindling a Debate.* https://doi.org/10.1007/978-3-319-93384-9_3.

Graham, M.W., Avery, E.J. & Park, S. 2015. The role of social media in local government crisis communications. *Public Relations Review.* https://doi.org/10.1016/j.pubrev.2015.02.001

Griffin, D. & Halpin, E. 2018. Local government: A digital intermediary for the information age? *Information Polity.* https://doi.org/10.3233/ip-2002-0019

Grimmelikhuijsen, S. & Knies, E. 2017. Validating a scale for citizen trust in government organizations. *International Review of Administrative Sciences*. https://doi.org/10.1177/0020852315585950

Hessami, Z. 2014. Political corruption, public procurement, and budget composition: Theory and evidence from OECD countries. *European Journal of Political Economy*. https://doi.org/10.1016/j.ejpoleco.2014.02.005

Houston, D.J., Aitalieva, N.R., Morelock, A.L., & Shults, C.A. 2016. Citizen Trust in Civil Servants: A Cross-National Examination. *International Journal of Public Administration*. https://doi.org/10.1080/01900692.2016.1156696.

Katoch, D., Sharma, J.S., Banerjee, S., Biswas, R., Das, B., Goswami, D. & Mukherjee, P. K. 2017. Government policies and initiatives for development of Ayurveda. *Journal of Ethnopharmacology*. https://doi.org/10.1016/j.jep.2016.08.018

Kim, H. 2014. Transformational Leadership, Organizational Clan Culture, Organizational Affective Commitment, and Organizational Citizenship Behavior: A Case of South Korea's Public Sector. *Public Organization Review*. https://doi.org/10.1007/s11115-013-0225-z

Ma, D. & Wang, Z. 2014. Governance Innovations and Citizens' Trust in Local Government: Electoral Impacts in China's Townships. *Japanese Journal of Political Science*. https://doi.org/10.1017/s1468109914000152

Mourtada, R. & Salem, F. 2015. Citizen Engagement and Public Services in the Arab World: The Potential of Social Media. SSRN. https://doi.org/10.2139/ssrn.2578993

Ramesh, R. 2017. Does Trust Matter? An Inquiry on Citizens' Trust in Public Institutions of Sri Lanka. *Millennial Asia*. https://doi.org/10.1177/0976399617715820

Taylor, J. 2014. Public service motivation, relational job design, and job satisfaction in local government. *Public Administration*. https://doi.org/10.1111/j.1467-9299.2012.02108.x

Knowledge sharing as mediator between organizational commitment and employee performance

A. Firdaus & E. Suryadi
Universitas Pendidikan Indonesia, Bandung, Indonesia

ABSTRACT: The aim of this study was to see the effect of organizational commitment on employee performance directly or indirectly with knowledge sharing as a mediating variable. This study was conducted on 61 employees who worked for several multi finance service companies in Jambi City. The results of this study indicated that organizational commitment had a direct and indirect effect on employee performance, knowledge sharing had a direct effect on employee performance, knowledge sharing can be a mediating variable between organizational commitment and employee performance.

1 INTRODUCTION

One of the key factors that increase employee performance is organizational commitment (Gautam & Campus 2017; Ghosh & Swamy 2014; Gorondutse & Hilman 2016; Zangaro 2001). Committed employees increase retention, work more, reduce operating costs and increase efficiency (Liou 2008; Tolentino 2013). Committed employees have a tendency towards making strong efforts for the organization and a strong tendency to continue membership in the organization, since they feel as part or true member of the organization (Kashefi et al. 2013; Meyer & Allen 2004). Individuals who have high commitment will be more motivated, and more focused on the success of the organization (Irefin & Mechanic 2014).

Zheng et al. (2009) shows that committed employees have a sense of responsibility towards the organization, communicate with each other and share new knowledge. Lin (2007) defines the behavior of knowledge sharing as a culture of social interaction that involves transferring knowledge, experience, and skills between members of the organization. Individuals in an organization will be more productive when they share experiences, information and knowledge with each other (Argote & Ingram 2000).

The three main variables of this study were organizational commitment, knowledge sharing and employee performance. The analysis unit was employees who worked in multi-finance service companies in Jambi city. Due to time constraints, only employees working in companies that have been operating for more than ten years examined. The aim of this study was to investigate whether organizational commitment and knowledge sharing have a direct and indirect effect on employee performance.

2 LITERATURE REVIEW

2.1 Knowledge sharing

Knowledge-based economics forces every organization to abandon traditional ways and adopt a new approach in knowledge management (Trivellas et al. 2015). Debowski (2006) explains that the most important part of knowledge management is how each individual can be motivated to share knowledge.

Hansen & Avital (2005) said that knowledge sharing can be understood as a behavior where individuals sincerely provide access to others regarding their knowledge and experience, meanwhile Hoof & Ridder (2004) explain knowledge sharing from the process side, that is the process by which individuals exchange their knowledge (tacit knowledge and explicit knowledge).

In this study the author uses two main dimensions of knowledge sharing, namely knowledge donating, which is an active understanding of employees in communicating their intellectual capital or information sharing in organizations, and knowledge collecting, or an understanding in gathering information for employees using a network of knowledge in organizations (Dysvik et al. 2015; Hoof & Ridder 2004; Lin 2007).

2.2 Organizational commitment

Santrock (2008) explains that organizational commitment is an attitude that reflects the extent to which an individual knows and is bound to his organization. According to Robbins and Timothy (2011), organizational commitment is defined as a situation where an employee sits with the organization and its objectives, and maintains itself as a member of the organization.

The three main approaches that are used as indicators to measure commitment are affective commitments, normative commitments and continuous commitments. These commitment have tremendous impact on company performance (Allen & Meyer 1990; Robbins & Timothy 2011).

2.3 Employee performance

Employee performance is the most important factor that can influence decisions to achieve organizational success (Alromaihi et al. 2017; Sonnentag et al. 2008), it is related to employee behavior that leads to the success of organizational goals (Armstrong & Taylor 2014). According to Griffin (2005), employee performance is a set of individual behaviors that are displayed related to work within the organization.

There are various theories in measuring employee performance, Mathis & Jackson (2010) explain employee performance, as how much they contribute to organizations, that can be measured among other by quantity, quality, duration of output, level of attendance, and cooperation.

2.4 Theoretical framework and hypothesis

Every employee who has a commitment to his organization has strong beliefs and acceptance of the goals and values of his organization, which will increase his sense of responsibility as a member of the organization. Allen & Meyer (1990) explain that committed employees will sincerely provide extra performance for the organization, the additional performance in question also includes that employees try to provide their knowledge to the organization. Hoof & Ridder (2004) state that employees with a strong commitment will contribute their intellectual capital -or knowledge- to fellow colleagues in the organization, Furthermore, Cabrera et al. (2006) proved in their study that there was a strong influence of employee commitment on knowledge sharing behavior in the organization.

H_1: *Organizational commitment has a positive and significant effect on employee performance*

H_2: *Organizational commitment has a positive and significant effect on knowledge sharing*

The study of knowledge sharing was also conducted by Ngah & Ibrahim (2010) they proved that knowledge sharing affect employee performance, as corroborated by Kuzu & Özilhan (2015)

H_3 : *Knowledge sharing has a positive and significant effect on employee performance*

The researcher formulated the conceptual framework as presented in Figure 1.

3 METHODS

This research was a survey with the number of sample was 61 employees who worked for several

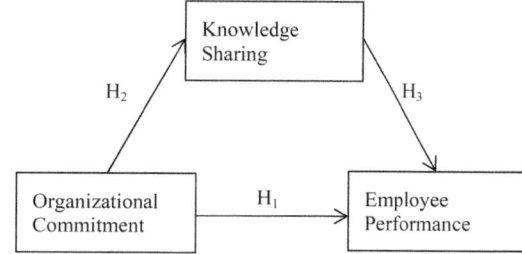

Figure 1. Conceptual framework.

Table 1. AVE, composite reliability, Cronbach Alpha.

Variable	AVE	Composite reliability	Cronbach Alpha
Organizational commitment	0.789	0.801	0.734
Knowledge sharing	0.734	0.823	0.788
Employee performance	0.641	0.789	0.697

multifinance companies in the Jambi city. Data was collected by distributing questionnaires directly to respondents, where each question was rated with a Likert scale. Organizational commitment was measured by 3 dimensions; affective commitment, normative commitment and continuous commitment with 6 item statements. Knowledge sharing was measured by 2 dimensions; knowledge donating and knowledge collecting with 4 item statements, and performance was measured by 5 dimensions: quantity, quality, duration of output, level of attendance, and cooperation. The respondents' answers were processed and analyzed using Component Based SEM (Structural Equation Modeling) is PLS (Partial Least Square).

4 RESULTS AND DISCUSSION

The results of outer loading of all statements have met the requirements with a range of 0.564–0.823, the average variance extracted (AVE), composite reliability and Cronbach Alpha values can be seen in Table 1.

Based on Table 1, it can be seen that the values of AVE, composite reliability and Cronbach Alpha of all variables > 0,5. This indicates that all variables and indicators are valid and reliable.

Based on Table 2, it can be seen that the R2 coefficients determination of employee performance is 0.632 it means that employee performance can be explained by knowledge sharing and organizational commitment variables of 63.2%, while knowledge sharing R2 coefficients are 0.588 which means that knowledge sharing can be explained by organizational commitment of 58.8%.

The goodness of fit test of the structural model of this study uses Q-Square Predictive Relevance (Q^2),

Table 2. Determination coefficients (R2).

Variable	R^2 coefficients
Employee performance	0.632
Knowledge sharing	0.588

Table 3. Direct and indirect effect.

Direct and indirect effect	Total effect	P-value	Information
OC → EP	0.569	0.002	Significant
OC → KS	0.473	0.014	Significant
KS → EP	0.812	0.000	Significant
OC → KS → EP	0.270	0.045	Significant

with the formula:

$$Q^2 = 1 - (1 - R_21)(1 - R_22)$$
$$Q^2 = 1 - (1 - 0.632)(1 - 0.588)$$
$$Q^2 = 1 - (0.1516)$$
$$Q^2 = 0.8484$$

(1)

The results of the Q^2 show that the value of goodness of fit is 0.8484, which means the model is strong and is able to explain employee performance by 84.84%. To see the direct and indirect effects tested by bootstrapping on SmartPLS can be seen in Table 3.

From Table 3 it can be explained that organizational commitment has a positive and significant effect directly on employee performance which means H_1 is accepted; organizational commitment has a positive and significant effect directly on knowledge sharing which means H_2 is accepted; Knowledge sharing has a positive and significant effect directly on knowledge sharing which means that H_3 is accepted. The results of this hypothesis support previous research that commitment and knowledge sharing is one of the factors that can support employee performance, employees who are committed will continue to improve knowledge sharing behavior, which will create and develop new knowledge that can be useful for completing their tasks (Akram & Bokhari 2011; Gold et al. 2015; Kuzu & Özilhan 2015; Ngah & Ibrahim 2010).

5 CONCLUSION

From the results of the study, it can be concluded that organizational commitment directly and indirectly influenced employee performance. Knowledge sharing as mediating variable increases the influence of organizational commitment on employee performance. The results of this study supports the results of previous researchers which asserted that increasing knowledge sharing among employees increased organizational commitment, employee performance and organizational performance.

REFERENCES

Akram, F. & Bokhari, R. 2011. The role of knowledge sharing on individual performance, considering the factor of motivation- The conceptual framework", *International Journal of Multidisciplinary Science and Enggineering*, Vol. 2 No. 9, pp. 44–48.

Allen, N.J. & Meyer, J.P. 1990. The measurement and antecedents of affective, continuance and normative commitment to the organization. *Journal of Occupational Psychology*, Vol. 63, pp. 1–18.

Alromaihi, M.A., Alshomaly, Z.A. & George, S. 2017. Job satisfaction and employee performance: A theoretical review of the relationship between the two variables. *International Journal of Advanced Research in Management and Social Sciences*, Vol. 6 No. 1, pp. 1–20.

Argote, L. & Ingram, P. 2000. Knowledge transfer: A basis for competitive advantage in firms. *Organizational Behavior and Human Decision Processes*, Vol. 82 No. 1, pp. 150–169.

Armstrong, M. & Taylor, S. 2014. Armstrong's Handbook of Human Resource Management Practice, 13th ed., London.

Cabrera, Á., Collins, W.C. & Salgado, J.F. 2006. Determinants of individual engagement in knowledge sharing. *International Journal of Human Resource Management*, Vol. 17 No. 2, pp. 245–264.

Debowski, S. 2006. Knowledge Management. John Wiley & Son Australia, Ltd, Melbourne & Sydney.

Dysvik, A., Buch, R. & Kuvaas, B. 2015. Knowledge donating and knowledge collecting: The moderating roles of social and economic LMX. *Leadership and Organization Development Journal*, Vol. 36 No. 1, pp. 35–53.

Gautam, P.K. & Campus, S. 2017. Issue of Organizational Commitment: Evidence from Nepalese Banking Industry.

Ghosh, S. & Swamy, D.R. 2014. A literature review on organizational commitment – a comprehensive summary. *Journal of Engineering Research and Applications*, Vol. 4 No. 12, pp. 4–14.

Gold, A.H., Malhotra, A. & Segars, A.H. 2015. Knowledge management: An organizational capabilities perspective knowledge management. Vol. 1222, available at:https://doi.org/10.1080/07421222.2001.11045669.

Gorondutse, A.H. & Hilman, H. 2016. The moderating e?ect of organisational Culture on the commitment to corporate.pdf.

Griffin, R. 2005. Management, 8th ed., Houghton Mifflin Company, New York.

Hansen, S. & Avital, M. 2005. Share and share alike: The social and technological influences on knowledge sharing behavior. *Working Papers on Information System*.

Hoof, B.V.D. & Ridder, J.A.D. 2004. Knowledge sharing in context: the influence of organizational commitment, communication climate and CMC use on knowledge sharing. *Journal of Knowledge Management*, Vol. 8 No. 6, pp. 117–130.

Irefin, P. & Mechanic, M.A. 2014. Effect of employee commitment on organizational performance in Coca Cola Nigeria Limited Maiduguri, Borno State Peace Irefin, 2 Mohammed Ali mechanic. *Journal of Humanities and Social Science,* Vol. 19 No. 3, pp. 33–41.

Kashefi, M., Adel, R.M., Rahimi, H. & Abad, G. 2013. Organizational commitment and its effects on organizational performance. *Interdisciplinary Journal of Contemporary Research in Business*, Vol. 4 No. 12, pp. 501–510.

Kuzu, Ö.H. & Özilhan, D. 2015. The effect of employee relationships and knowledge sharing on employees' performance: An empirical research on service industry sciencedirect the effect of employee relationships and knowledge sharing on employees' performance: An empirical research o. *Procedia – Social and Behavioral Sciences*, Vol. 109 No. January 2014, pp. 1370–1374.

Lin, H. 2007. Knowledge sharing and firm innovation capability: an empirical study. *International Journal of Manpower*, Vol. 28 No. 3, pp. 315–332.

Liou, S-R. 2008. An analysis of the concept of organizational commitment. *Nursing Forum*, Vol. 43 No. 3, pp. 116–125.

Mathis, R.L. & Jackson, J.H. 2010. Human Resources Management, South Western Collage Publishing, Ohio.

Meyer, J.P. & Allen, N.J. 2004. TCM employment comitment survey: academic users guide 2004. London, Canada: The University of Western Ontario.

Ngah, R. & Ibrahim, A.R. 2010. The effect of knowledge sharing on organizational performance in small and medium enterprises. *Proceedings of Knowledge Management 5th International Conference 2010*, No. Stewart 2000, pp. 503–508.

Robbins, S.P. & Timothy, J.A. 2011. Perilaku Organisasi, 12th ed., Jakarta: Salemba Empat.

Santrock, J.W. 2008. Educational Psychology, 2nd ed.

Sonnentag, S., Volmer, J. & Spychala, A. 2008. Job Performance, Sage Hand book of Organizational Behavior.

Tolentino, R.C. 2013. Organizational commitment and job performance of the academic and administrative personnel. *International Journal of Information Technology and Business Management*, Vol. 15 No. 1.

Trivellas, P., Akrivouli, Z., Tsifora, E. & Tsoutsa, P. 2015. The impact of knowledge sharing culture on job satisfaction in accounting firms. The mediating effect of general competencies. *Procedia Economics and Finance*, Elsevier B.V., Vol. 19 No. 15, pp. 238–247.

Zangaro, G.A. 2001. Organizational Commitment: A Concept Analysis. *Nursing Forum*, Vol. 36 No. 2, pp. 14–21.

Zheng, M., Bao, G. & Qian, Y. 2009. Employee commitment, knowledge sharing and knowledge integration: An empirical study of professional staffs in chinese firms. PICMET Proceedings, pp. 974–983

Advances in Business, Management and Entrepreneurship – Hurriyati et al. (Eds)
© 2021 Taylor & Francis Group, London, ISBN 978-0-367-67471-7

The power of knowledge donating and knowledge collecting for academic performance: Developing Indonesian students' knowledge sharing models

F.J. Islamy, T. Yuniarsih, Kusnendi & L.A. Wibowo
Universitas Pendidikan Indonesia, Bandung, Indonesia

ABSTRACT: The purpose of this paper is to examine the input of student knowledge sharing on academic performance of universities in Indonesia. The sample of this paper is students at universities of the Indonesian Economy Building School a total of 346 students using probability sampling techniques. Data were collected using a conventional questionnaire and an online questionnaire. Data analysis for this paper included t-tests, one-way ANOVA and multiple regression analysis. The findings of this paper are knowledge sharing input conducted by students, namely through knowledge donating and knowledge collecting activities that have a positive and significant effect on academic performance. The practical implication of the overall findings is that in achieving high academic performance, one must first establish strategies on how to increase knowledge sharing between students in tertiary institutions. This paper contributes to the development of a theoretical model of knowledge sharing in improving academic performance.

1 INTRODUCTION

The role of education in life is very important, as explained in the 1945 Constitution that every citizen is entitled to get education, teaching and the government seeks to organize a system national education whose implementation is regulated by law. UU no. 20 of 2003 on the National Education System, education is a conscious and planned effort to create an atmosphere of learning and learning activities so that the participants increase their potential to have spiritual strength, self-control, personality, intelligence, noble character and skills that need him, society, nation and state. Education can not be separated from the academic performance (academic achievement) produced by each education provider.

Academic performance is an indicator of the Academic performance is an indicator of the (Guney 2009) argue that the academic performance of university students has become an important topic for higher education institutions (HEIs). Indonesian Economy Building School (STIE INABA), is a college in Bandung that has a vision to become the School of Economics which excels in the field of Accounting and Management and Entrepreneurship. Therefore STIE INABA should be more free from the result of academic that is produced. STIE INABA is a private university that still has B accreditation and still can not compete with other private campus in Bandung city.

We are not only in the millennium era, but also in a new era of knowledge era. The competitive advantages that occur in building and utilizing the core. Assessment conducted as a strategic asset with potential becomes very important (Adams & Lamont 2003; Halawi et al. 2005). Knowledge management process according to Probst et al. in Tobing (2011), are: (1) Knowledge Identification; (2) Knowledge Acquisition; (3) Knowledge Development; (4) Sharing Knowledge/Distribution; (5) Knowledge Utilization; and (6) Knowledge Retention.

One of the KM processes is knowledge sharing. Alawi explained that the sharing of knowledge is an important factor in the success of a company, it is better for parts of an organization and it is very profitable for the company (Ismail et al. 2007) without sharing, the learning process and knowledge creation will be hampered. Without sharing, the scale of knowledge utilization will also be very limited, because knowledge is only used by people or units on a limited basis (Tobing 2011). Sharing knowledge is a process that works with others to share knowledge and mutual benefit (Ismail et al. 2007). According Tobing (2011), at this time the organization or tasks undertaken to discover new knowledge of the process of distribution of knowledge is done to remain exist and ready in an increasingly competitive future. Implementation Knowledge sharing in companies can create competitive advantage for companies (Kearns & Lederer 2003).

From various experts discussing knowledge sharing, Van & Ridder (2004), states that knowledge sharing consists of two processes. Two processes in sharing such knowledge are: (1) Knowledge of donations; and (2) Knowledge gathering. Donating knowledge is how employees communicate existing knowledge to others and the collection of knowledge is how one works with them.

One of the factors that influence academic performance is knowledge sharing (Moghavvemi et al. 2018). Linking research on knowledge sharing and academic performance is still very limited, researchers only found two research results that link these variables, the results of the two studies have conflicting results. The results of research on knowledge sharing and academic performance, Moghavvemi et al. (2018) suggest that knowledge sharing has a negative effect on academic performance, while Aslam (2013) found that knowledge sharing has a significant positive effect on academic performance. Looking at the problem and gap research about knowledge sharing and academic performance, the researchers are interested in researching these variables.at STIE INABA Bandung.

1.1 Academic performance

The academic performance of students is highly valued within these advanced economies, as well as the inclusion of substantial implications (Popopat 2009).

Organizational capabilities that lead to effective knowledge creation and transfer are essential components of organizations' competitive advantage. (Haris 2013). Knowledge sharing leads to better team performance, better problem solving, and enhanced creativity Knowledge Sharing (Huang 2009).

1.2 Knowledge sharing

Tobing (2007), the core of KM (Knowledge Management) is knowledge sharing or knowledge transfer, because through knowledge sharing there is an increase in value of knowledge owned by the company.

According to Tobing (2011), the role of knowledge sharing is increasingly important, especially when the traditional KM, which is dominated by knowledge-based IT engineering processes, has shifted to the increasingly soft, social and humanist KM.

According to Devenport & Prusak in Tobing (2011), the term knowledge sharing and knowledge transfer is often used with the assumption of having the same meaning using only the term knowledge transfer to replace the term knowledge sharing. According to them, the word transfer describes the level of effectiveness of the distribution of better knowledge. Because the transfer term consists of two actions: the transmission of knowledge to the recipient and the absorption of knowledge by the recipient.

Knowledge sharing according to Raskov in Tobing (2011), knowledge sharing occurs between individuals within a community, where individuals interact and share knowledge with other individuals through virtual space or face to face.

According to Ismail et al. (2007), "knowledge sharing (knowledge transfer) requires that an individual or a group cooperate with others to share knowledge and achieve mutual benefits". After studying various writings on knowledge sharing Van & Ridder (2004)

states that knowledge sharing consists of two processes, namely:

a. Knowledge donating: communicating to others the knowledge possessed by an individual. (1) Capability. Capability is the extent to which a person has knowledge and willingness to spread his knowledge; (2) Credibility. Credibility reflects the level of trust in support for colleagues; (3) Seriousness. Seriousness reflects the level of seriousness a person is willing to share knowledge possessed.

b. Knowledge collecting: consult with colleagues to gain knowledge from them. (1) The sense of belonging. Smooth the level of willingness and willingness to receive new knowledge; (2) Commitment It is to what extent a person's commitment in applying new knowledge; (3) Satisfaction It is to what extent a person can adapt to new knowledge.

From various opinions on the above knowledge sharing, the authors can conclude that knowledge sharing is the process by which individuals or groups of people interact with each other share knowledge with other individuals or groups.

1.3 Knowledge sharing and academic performance

There have been many studies on knowledge sharing, (Eid & Al-Jabri 2016) argued that knowledge sharing has a positive effect on learning performance. Knowledge sharing is knowledge donating and collecting influences innovation (Camelo-Ordaz et al. 2011; Elrehail et al. 2018; Kamas & Bulutlar 2009; Lin 2007). Knowledge sharing has a positive effect on job performance (Kwahk & Park 2016). Knowledge sharing has a significant positive effect on employee productivity (Aboelmaged 2018). But there are still very few studies linking knowledge sharing and academic performance, the results of research on knowledge sharing and academic performance, Moghavvemi et al. (2018) suggests that knowledge sharing has a negative effect on academic performance, while Aslam (2013) found that knowledge sharing has a positive effect significant to academic performance.

1.4 Research framework

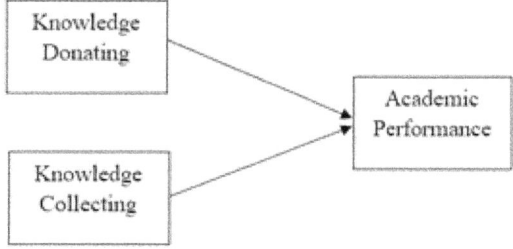

Figure 1. Research framework.

Hypothesis 1: There is a positive influence between knowledge donating and academic performance.

Hypothesis 2: There is a positive influence between knowledge collecting and academic performance.

Hypothesis 3: There is a positive influence between knowledge donating and knowledge collecting on academic performance.

2 METHODS

2.1 Sample and procedure

In this study, the population was all students remain STIE INABA Bandung which in 2018 they were 1848 people. This study used probability samples with proportional stratified random sampling technique based on the study program. The number of samples in this study were 346 respondents.

For the number of elements of each different sub-population, the first fraction (fi) is searched by the number of subpopulations divided by the total of the existing population then searched for the value of the sample taken by the fi t value multiplied by the total number of samples available, the results can be seen in Table 1.

Table 1. Proportional stratified random sampling.

No.	Sub population	Number of student elements	(fi)	Sample
1.	Accounting	1028	0.542	188
2.	Master of	107	0.056	19
3.	management			
4.	Amount	1895	1.000	346

2.2 Testing the hypothesis

The number of questions in the questionnaire to measure the knowledge donating are seven questions, knowledge collecting five questions while the academic performance is eight questions, so that there are 20 items of questions. The validity test shows two invalid question items on the knowledge donating dimension, an invalid question item on the knowledge collecting dimension and one invalid question item on the academic performance dimension so that the researcher discards the four items of the statement so that the number of statement items used is 16 statements because the value of r arithmetic the item < of the value r table that is equal to 0.09. To know the influence between variable of knowledge donating, knowledge collecting, to academic performance can be done by F test. The result of F test calculation can be seen in Table 2.

From the calculation results obtained sig $f = 0.000 < \alpha = 0.05$, then H0 rejected and H1 accepted. Based on the above calculation can be concluded that knowledge donating (X1), knowledge collecting (X2), to academic performance (Y).

To know the influence of knowledge donating (X1) variable, knowledge collecting (X2) together influence

Table 2. ANOVA.

Model	Sum of squares	df	Mean Square	F	Sig.
Regression	5.581	2	2.790	35.852	.000
Residual	26.695	343	.078		
Total	32.275	345			

a. Dependent Variable: AP.
b. Predictors: (Constant), KC, KD.

to academic performance (Y), can be seen based on result of calculation as follows.

Table 3. Model summary.

Model	R	R square	Adjusted R square	Std. error of the estimate
1	.416[a]	.173	.168	.27898

a. Predictors: (Constant), KC, KD.
b. Dependent Variable: AP.

The magnitude of the R of squared (r2) is 0.173. The R quadratic value is used to measure the effect of knowledge donating (X1), knowledge collecting (X2) to academic performance (Y), with the following formula.

$$KD = r2 \times 100\%$$

$$KD = 0.173 \times 100\%$$

$$KD = 17.3\%$$

This means that the influence of knowledge donating (X1), knowledge collecting (X2) to academic performance (Y), is 17.3% combined. While the remaining 82.7% (100% −17.3%) is influenced by other factors. In other words, the variability of academic performance that can be explained by using knowledge discating and knowledge collecting variables is 17.3% while the effect of 82.7% is caused by other variables outside this model.

In accordance with the principle of multiple regression then can be tested partially (t test). The result of t test can be seen in Table 4.

Table 4. Coefficients.

Model	Unstandardized coefficients		Standardized coefficients		
	B	Std. error	Beta	t	Sig
1 (Constant)	1.703	.188		9.057	.000
KD	.268	.055	.254	4.902	.000
KC	.223	.045	.257	4.966	.000

a. Dependent Variable: AP.

The t test results in Table 4 show that the variables X1 and X2 both have a significant and positive influence on the variable Y, because it has a significance value <0.05.

From these findings it can be concluded that Hypothesis 1 which states that knowledge donating has a significant positive effect on academic performance is accepted, Hypothesis 2 which states that knowledge collecting has a significant positive effect on accepted academic performance and the last hypothesis is hypothesis 3 which states that knowledge donating and knowledge collecting Simultaneously significant positive effect on academic performance is accepted.

3 CONCLUSION

Based on research conducted on student respondents of Indonesian High School of Economics Building Bandung on the influence of knowledge donating and knowledge collecting on academic performance at students of Indonesian Higher School of Economics Building Bandung, it can be drawn conclusions that are expected to answer the formulation of problems that have been made before. The conclusions of this study are as follows.

1. Knowledge donating has a significant positive effect on academic performance.
2. Knowledge collecting has a significant positive effect on academic performance.
3. Knowledge donating and Knowledge collecting have a significant positive effect on academic performance.

Seeing the results of the study which states

That knowledge sharing has a significant positive effect on academic performance, this agrees with the results of research conducted by Aslam (2013). Because there are still research results stating that knowledge sharing has a negative effect on academic performance (Moghavvemi et al. 2018) Therefore, to be able to be more convincing about the results of research on knowledge sharing variables and academic performance, further research should be conducted.

REFERENCES

Aboelmaged, M.G. 2018. Knowledge sharing through enterprise social network (ESN) systems: motivational drivers and their impact on employees' productivity. *Journal of Knowledge Management*.

Adams, G.L. & Lamont, B.T. 2003. Knowledge management systems and developing sustainable competitive advantage. *Journal of Knowledge Management* 7(2): 142–154.

Aslam, H. 2013. Social capital and knowledge sharing as determinans of academic performance. *University Of Management and Technology, Proceedings of 3rd International Conference on Business Management*.

Camelo-Ordaz, C., Garcia-Cruz, J., Sousa-Ginel, E. & Valle-Cabrera, R. 2011. The influence of human resource management on knowledge sharing and innovation in Spain: the mediating role of affective commitment. *The International Journal of Human Resource Management* 22(07): 1442–1463.

Eid, M.I.M. & Al-Jabri, I.M. 2016. Computers & education social networking, knowledge sharing, and student learning: the case of university students. *Computers & Education* 99: 14–27.

Elrehail, H., Lawrence, O. & Alsaad, A. 2018. Telematics and informatics the impact of transformational and authentic leadership on innovation in higher education: the contingent role of knowledge sharing. *Telematics and Informatics* 35(1): 55–67.

Guney, Y. 2009. Exogenous and endogenous factors influencing students' performance in undergraduate accounting modules. *Accounting Education: an international journal* 18(1): 51–73.

Halawi, L.A., Aronson, J.E. & McCarthy, R.V. 2005. Resource-based view of knowledge management for competitive advantage. *The Electronic Journal of Knowledge Management* 3(2): 75–86.

Haris, A. 2013. Social Capital and knowledge sharing as determinants of academic performance. *University Of Management and Technology*.

Huang, C.C. 2009. Knowledge sharing and group cohesiveness on performance: An empirical study of technology R&D teams in Taiwan. *Elsevier Ltd. Allrights reserved*.

Ismail, A.A., Yousif, A.N. & Fraidoon, M.Y. 2007. Organizational culture and knowledge sharing: critical success factors. *Journal of Knowledge Management* 11(2): 22–42.

Kamas, R. & Bulutlar, F. 2009. The influence of knowledge sharing on innovation.

Kearns, G.S. & Lederer, A.L. 2003. A resource-based view of strategic IT alignment: How knowledge sharing creates competitive advantage. *Decision Sciences* 34(1): 1–29.

Kwahk, K. & Park, D. 2016. Computers in human behavior the effects of network sharing on knowledge-sharing activities and job performance in enterprise social media environments. *Computers in Human Behavior* 55: 826–839.

Lin, H. 2007. Knowledge sharing and firm innovation capability: an empirical study. 28(3): 315–332.

Moghavvemi, S., Sharabati, M., Klobas, J.E. & Sulaiman, A. 2018. Effect of trust and perceived reciprocal benefit on students' knowledge sharing via Facebook and academic performance. *The Electronic Journal of Knowledge Management* 16(1): 23–35.

Popopat, A.E. 2009. A meta-analysis of the five-factor model of personality and academic performance. *Psychological Bulletin © 2009 American Psychological Association 2009* 135(2): 322–338.

Tobing, P.L. 2007. *Knowledge management*. Yogyakarta: Graha Ilmu.

Tobing, P.L. 2011. *Manajemen knowledge sharing berbasis komunitas*. Bandung: Knowledge Management Society Indonesia

Undang-Undang No. 20. 2003. *Tentang Sistem Pendidikan Nasional*. Jakarta

Van, D.H.B & De Ridder, J.A. 2004. Knowledge sharing in context, the influence of organizational commitment, communication climate and CMC use on knowledge sharing. *Journal of Knowledge Management* 8(6): 117–130.

Advances in Business, Management and Entrepreneurship – Hurriyati et al. (Eds)
© 2021 Taylor & Francis Group, London, ISBN 978-0-367-67471-7

Organizational citizenship behavior analysis based on transformational leadership, commitment and organizational work satisfaction of education staff in Universitas Islam Bandung

M.V. Romi & E. Suryadi
Universitas Pendidikan Indonesia, Bandung, Indonesia

ABSTRACT: The purpose of this study was to analyze organizational citizenship behavior based on transformational study using a survey approach. The sample of this study was 135 educational staff at the Universitas Islam Bandung. The sampling technique used proportional random sampling, and the data analysis technique was Structural Equation Modeling (SEM) using AMOS. The results of this study showed that transformational leadership had a significant and positive effect on organizational citizenship behavior; transformational leadership had a positive but not significant effect on job satisfaction; transformational leadership had a positive and significant effect on organizational commitment; job satisfaction had a positive and significant effect on organizational citizenship behavior; job satisfaction had positive but not significant effect on organizational commitment.leadership, commitment and job satisfaction of education staff at the Universitas Islam Bandung, where employee evaluation results were often ineffective and inefficient. This research was descriptive explanatory.

1 INTRODUCTION

1.1 *Research background*

Efforts to improve workforce performance require management's role to conduct leadership approaches, where the success of an organization is highly dependent on the ability of the leaders. Good leadership can influence employees to work accordingly and to anticipate problems who can see the conditions and needs of employees (Rivai & Mulyadi 2012)

Organizational commitment is also a dominant factor in the formation of employees' organizational citizenship behavior (OCB) (Greenberg & Baron 2000). To discover issues related to OCB a pre survey was conducted on educational staff in Universitas Islam Bandung, the result showed that 29% of work had not been done proportionally, teamwork had not been implemented optimally (31%), and there were still many employees who rely on their partners in the implementation of work (40%).

Related to this problem, the researcher was interested in analyzing organizational citizenship behavior based on leadership and organizational commitment at the Islamic university. To overcome these problems, extra-role behavior (organizational citizenship behavior) and organizational commitment by education personnel were urgently needed. Organ (2006) defines OCB as individual behavior that is free, not directly receiving award from the formal reward system, but as a whole can improve the efficiency and effectiveness of organizational functions.

In the dynamic world of work as it is today, where tasks are increasingly being done in teams and require flexibility, organizations need employees who have OCB behavior, such as helping other individuals in the team, volunteering to do extra work, avoiding conflicts with co-workers, obey the rules, and tolerate the occurrence of losses and disruptions related to work (Robbins & Judge 2008).

According to Luthans (2006) organizational commitment is the attitude of employee loyalty and the ongoing process of organizational members expressing their attention to the success and goodness of the organization. Organizational commitment make employees give their best to the organization. Employees with high commitment tend to have a high work orientation, and to be happy to help and work together. In order for OCB's behavior and organizational commitment to be well demonstrated, the effectiveness of the role of a leader, in this case a direct superior, is needed. To be an effective leader, a leader must be able to influence all the subordinates they lead to achieve the goals of the higher education. Transformational leadership is very suitable to be applied in dynamic universities with 314 educational staff who are professional, educated and have a high level of intellect. Transformational leaders are able to pay attention to the self-development needs of their followers, change the awareness and perspective of towards problems that occur, and be able to please and inspire followers to work hard to achieve common goals (Robbins & Judge 2008). According to Robbins & Judge (2008), job satisfaction is a major determinant of OCB behavior. Hughes et al. (2012) states that job satisfaction is related to one's attitude about work. Satisfied workers are more likely to stay working for the organization. Dissatisfaction is also the main reason someone leaves the organization.

Based on the background described above, then problem statement were formulated as follows: 1) Does transformational leadership have a significant effect on OCB education staff Universitas Islam Bandung? 2) Does transformational leadership have a significant effect on job satisfaction of teaching staff at Universitas Islam Bandung? 3) Does transformational leadership have a significant effect on organizational commitment in the teaching staff at the Universitas Islam Bandung? 4) Does job satisfaction have a significant effect on OCB education staff at Universitas Islam Bandung? 5) Does job satisfaction have a significant effect on organizational commitment of educational staff at Universitas Islam Bandung?

2 LITERATURE REVIEW

Leaders who have a transformational leadership style have vision, rhetoric skills, and manage impressions well and use them to develop strong emotional ties with followers, thereby encouraging them to work towards realizing the leader's vision (Hughes et al. 2012). According to Antonakis et al. (2003) transformational leadership as a behavior that is proactive, raises concern for the common good, and helps followers achieve their goals at the highest level. Transformational leader is a leader who encourages followers to change their motives, beliefs, values, and abilities so that the personal interests and goals of followers can be aligned with the vision and goals of the organization (Goodwin et al. 2001). Khuntia & Suar (2004) assert that leaders who implement transformational leadership influence their followers by involving followers in setting goals, solving problems, making decisions, and providing feedback through training, direction, consultation, guidance, and monitoring of tasks was given. According to Robbins & Judge (2008) and Cavazotte et al. (2012), there are four components of transformational leadership, namely idealized influence, inspirational motivation, intellectual stimulation, and individualized consideration.

Robbins & Judge (2008) define job satisfaction as a positive feeling about one's work that is the result of an evaluation of his characteristics. Job satisfaction is the result of employee perceptions of how well their work provides what is considered important (Luthans 2006). Schleicher et al. (2004); Luthans (2006); Robbins & Judge (2008); Azeem (2010) reveal that there are five components of job satisfaction, namely payment, employment, promotional opportunities, bosses, and colleagues.

Robbins & Judge (2008) define OCB as a behavioral choice that is not part of an employee's formal work obligations, but supports the effective functioning of the organization. Organ (2006) describe OCB as individual behavior that is free, which does not directly and explicitly get rewards from the formal reward system, but overall increases the efficiency and effectiveness of organizational functions. OCB dimensions used in this research are that proposed by Konovsky & Organ (1996); Jahangir et al. (2004); Organ (2006); DiPaola & Neves (2009); Ahmed et al. (2012), Chiang & Hsieh (2012), i.e. altruism, courtesy, sportsmanship, conscientiousness, and civic virtue.

Organizational commitment is defined by Durkin & Bennet (1999) as strong and close feelings of a person towards the goals and values of an organization in relation to their role in the effort to achieve those goals and values. Luthans (2006) states that organizational commitment is an attitude that shows employee loyalty and is an ongoing process of an organization member expressing their attention to the success and goodness of his organization. Curtis & Wright (2001) suggests that commitment is the power of identification of individuals who are in an organization. If a person is committed to the organization, he will have strong identification with the organization, have membership values, agree with the goals and value system, will likely remain in it, and be ready to work hard for the organization. Tett & Meyer (1993); Meyer et al. (2002); Karakus & Aslan (2008); Luthans (2006); Aydogdu & Asikgil (2011) put forward three dimensions of organizational commitment, they are affective commitment, continuous commitment, normative commitment.

Based on the the empirical studies described above, the research model is presented on Figure 1.

Therefore, the research hypotheses are formulated as follows:

H1: *Transformational leadership has positive and significant effect influence on OCB.*

H2: *Transformational leadership has a positive and significant effect on job satisfaction.*

H3: *Transformational leadership has a positive and significant effect on organizational commitment*

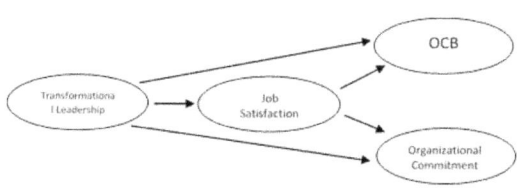

Figure 1. Research model.

H4: *Job satisfaction has a positive and significant effect on OCB.*

H5: *Job satisfaction has a positive and significant effect on organizational commitment*

3 METHODS

3.1 Population and research sample

The population in this study was educational staff at the Universitas Islam Bandung with a total of 314 people. The sampling technique was proportional random sampling. This study used a sample size of 135 respondents who were categorized into 22 sections and represented various levels of education. This number is considered appropriate to represent the population.

3.2 Research variable

The variables in this study consisted of independent variables (transformational leadership) and dependent variable (organizational citizenship behavior and organizational commitment)

3.3 Data analysis and research instruments

Data collection instruments in this study was a questionnaire distributed to respondents who met the criteria. The data analysis technique used Structural Equation Modeling (SEM) with AMOS program.

3.4 Hypothesis test

The hypothesis testing was carried out based on the results of the analysis of causality relationships between research constructs, as presented in Table 1 and Table 2.

Table 1. Measurement of unstandardized and standardized regression weight on structural models.

			Estimate	P	Results
kt	—>	ocb	0.567	0.000	Significant
kt	—>	kk	0.153	0.142	Not significant
kt	—>	ko	0.355	0.000	Significant
kk	—>	ocb	0.164	0.053	Significant
kk	—>	ko	0.080	0.404	Not significant

Source: Data processed

Table 2. Analysis of direct effects, indirect effects, and total effects.

	direct	KT KT	Total	direct	KK KK	Total
KK	0.153	–	0.153	–	–	–
OCB	0.567	0.025	0.592	0.164	–	0.164
KO	0.355	0.012	0.367	0.080	–	0.080

Source: Data processed

4 RESULTS AND DISCUSSION

1. The results of hypothesis testing prove that transformational leadership has a positive and significant effect on OCB. This means that the implementation of transformational leadership at the Universitas Islam Bandung is high. The findings of this study are in line with the results of research conducted by Lian & Tui (2012), Jahangir et al. (2004), Barbuto (2005), Lamidi (2008) and Nguni (2005).

2. Hypothesis testing results indicate that transformational leadership has no significant positive effect on job satisfaction. This means that transformational leadership has no real impact on job satisfaction. The results of this study are in opposition to that of Griffith (2004), Zahari & Shurbagi (2012), Yang & Islam (2012), Al-Swidi et al. (2012), and Yang (2012).

3. The results of hypothesis testing prove that transformational leadership has positive and significant effect on organizational commitment. The results of this study are consistent with research conducted by Tuna et al. (2011), Ismail et al. (2011), Farahani et al. (2011), Yang (2012), and Dunn et al. (2012).

4. Hypothesis testing results prove that job satisfaction has positive and significant effect on OCB. The findings of this study are in line with the results of research conducted by Murphy et al. (2002), Schappe (1998), Krishnan et al. (2009), Williams & Anderson (1991), Mohammad et al. (2011), Foote & Tang (2008), Alotaibi (2001), Huang et al. (2012), and Jahangir et al. (2004)

5. Hypothesis testing results indicate that job satisfaction has no significant positive effect on organizational commitment. The results of this study are opposed to the research of Aydogdu & Asikgil (2011), and Koh & Boo (2004). Based on testing of mediating variables, it is evident that the variable job satisfaction not as a mediating variable.

5 CONCLUSION

Transformational leadership has significant positive effect on OCB by 56.7% with two highest dimensions; idealized influence and inspirational motivation. Transformational leadership has no significant positive effect on job satisfaction by 15.3% with the highest dimensions intellectual stimulation and individualized consideration and it also has significant positive effect on organizational commitment by 35.5%. Job satisfaction has a significant positive effect on OCB by 16.4% but it has no significant positive effect on organizational commitment by 8%. The total direct influence of Transformational leadership to OCB is 59.2%, the total effect of organizational commitment to transformational leadership is 36.7%.

REFERENCES

Ahmed, N., Rasheed, A. & Jehanzeb, K. 2012. An exploration of predictors of organizational citizenship behaviour and its significant link to employee engagement. *International Journal of Business, Humanities and Technology*, Vol 2, No 4, pp. 99–106.

Alotaibi, A.G. 2001. Antacedents of organizational citizenship behavior: a study of public personnel in Kuwait. *Journal Public Personnel Management*, Vol 30, No 3, pp. 363–376.

Al-Swidi, A.K., Nawawi, M.K. & Al-Hosam, A. 2012. Is the relationship between employees' psychological empowerment and employees' job satisfaction contingent on the transformational leadership? A study on the Yemeni Islamic Banks, *Asian Social Science*, Vol 8, No 10, pp. 130–150.

Antonakis, J., Avolio, B.J. & Sivasubramaniam, N. 2003. Context and leadership: an examination of the nine factor full-range leadership theory using the multifactor leadership questionnaire. *The Leadership Quarterly*, Vol 14, No 2, pp. 261–295.

Aydogdu, S. & Asikgil, B. 2011. An empirical study of the relationship among job satisfaction, organizational commitment and turnover intention. *International Review of Management and Marketing*, Vol 1, No 3, pp. 43–53.

Azeem, S.M. 2010. Job satisfaction and organizational commitment among employees in the Sultanate of Oman. *Journal of Psychology*, Vol 1, pp. 295–299.

Barbuto, J.E. 2005. Motivation and transactional, charismatic, and transformational leadership: a test of antecedents, *Journal of Leadership & Organizational Studies*, Vol 11, No 4, pp. 26 40.

Cavazotte, F., Moreno, V. & Hickmann, M. 2012. Effects of leader intelligence, personality and emotional intelligence on transformational leadership and managerial performance. *The Leadership Quarterly*, Vol 23, pp. 443–455.

Chiang, C.F. & Hsieh, T.S. 2012. The impacts of perceived organizational support and psychological empowerment on job performance: The mediating effects of organizational citizenship behavior. *International Journal of Hospitality Management*, Vol 31, pp. 180–190.

Curtis, S. & Wright, D. 2001. Retaining employees - the fast track to commitment. *Management Research News*, Vol 24, No 8, pp. 59–64.

DiPaola, M.F. & Neves, P.M.M.C. 2009. Organizational citizenship behaviors in American and Portuguese public schools: Measuring the construct across cultures. *Journal of Educational Administration*, Vol. 47, No 4, pp. 490–507.

Dunn, M.W., Dastoor, B. & Sims, R.L. 2012. Transformational leadership and organizational commitment: a cross-cultural perspective. *Journal of Multidisciplinary Research*, Vol 4, No 1, pp. 45–59.

Durkin, M. & Bennet, H. 1999. Employee commitment in retail banking: identifying and exploring hidden dangers. *The International Journal of Bank Marketing*, Vol 17, No 3, pp. 124–137.

Farahani, M., Taghadosi, M. & Behboudi, M. 2011. An exploration of the relationship between transformational leadership and organizational commitment: the moderating effect of emotional intelligence: case study in Iran. *International Business Research*, Vol 4, No 4, pp. 211–217.

Foote, D.A. & Tang, T.L. 2008. Job satisfaction and Organizational Citizenship Behavior (OCB): Does team commitment make a difference in self directed teams? *Management Decision*, Vol 46, No 6, pp. 933–947.

Goodwin, V.L., Wofford, J.C. & Whittington, J.L. 2001. A theoretical and empirical extension to the transformational leadership construct. *Journal of Organizational Behavior*, Vol 22, No 7, pp.759–774.

Greenberg, J. & Baron, R. A. 2000. Behavior in Organizations. Prentice Hall. Inc.

Griffith, J. 2004. Relation of principal transformational leadership to school staff job satisfaction, staff turnover, and school performance. *Journal of Educational Administration*, Vol 42, No 3, pp. 333–356.

Huang, C.C., You, C.S. & Tsai, M.T. 2012. A multidimensional analysis of ethical climate, job satisfaction, organizational commitment, and organizational citizenship behaviors. *Nursing Ethics*, Vol 19, No 4, pp. 513–529.

Hughes, R.L., Ginnett, R.C. & Curphy, G.J. 2012. *Leadership: Memperkaya Pelajaran dari Pengalaman*. Edisi Ketujuh, Jakarta: Salemba Humanika.

Ismail, A., Mohamed, H., Sulaiman, A.Z., Mohamad, M.H. & Yusuf, M.H. 2011. An empirical study of the relationship between transformational leadership, empowerment and organizational commitment. *Business and Economics Research Journal*, Vol 2, No 1, pp. 89–107.

Jahangir, N., Akbar, M. & Haq, M. 2004. Organzational citizenship behaviors: its nature and antecedents. *BRAC University Journal*, Vol I, No 2, pp. 7585.

Karakus, M. & Aslan, B. 2008. Teachers' commitment focuses: a three-dimensioned view. *Journal of Management Development*, Vol 28, No 5, pp.425–438.

Khuntia, R. & Suar, D. 2004. A scale to assess ethical leadership of indian private and public sector managers. *Journal of Business Ethics*, Vol 49, No 1, pp. 13–26.

Koh, H.C. & Boo, E.H.Y. 2004. Organizational ethics and employee satisfaction and commitment. *Management Decision*, Vol 42, No 5, pp. 677–693.

Konovsky, M.A. & Organ, D.W. 1996. Dispositional and contextual determinants of organizational citizenship behavior. *Journal of Organizational Behavior*, Vol 17, No 3, pp. 253–266.

Krishnan, R., Arumugam, N., Chandran, V. & Kanchymalay, K. 2009. Examining the relationship between job satisfaction and organizational citizenship behavior: a case study among non academic staffs in a public higher learning institution in Malaysia. *Global Business Summit Conference*, Vol 2, No 43, pp. 221–232.

Lamidi. 2008. Pengaruh kepemimpinan transformasional terhadap organizational citizenship behavior: dengan variabel intervening komitmen organisasional. *Jurnal Ekonomi dan Kewirausahaan*, Vol 8, No 1, pp. 25–37.

Lian, L.K. & Tui, L.G. 2012. Leadership styles and organizational citizenship behavior: the mediating effect of subordinates' competence and downward influence tactics. *Journal of Applied Business and Economics*, Vol 13, No 2, pp. 59–96.

Luthans, F. 2006. *Perilaku Organisasi*, Edisi Sepuluh, Yogyakarta: Penerbit Andi.

Meyer, J.P., Stanley, D.J., Herscovitch, L. & Topolnytsky, L. 2002. affective, continuance, and normative commitment to the organization: a metaanalysis of antecedents, correlates, and consequences. *Journal of Vocational Behavior*, Vol 61, pp. 20–52.

Mohammad, J., Habib, F.Q. & Alias, M.A. 2011. Job satisfaction and organisational citizenship behaviour: An empirical study at higher learning institutions. *Asian Academy of Management Journal*, Vol 16, No 2, pp. 149–165.

Murphy, G., Athanasou, J. & King, N. 2002. Job satisfaction and organizational citizenship behaviour: a study of australian human-service professionals. *Journal of Managerial Psychology*, Vol 17, No 4, pp. 287–297.

Nguni, S.C., 2005. A study of the effects of transformational leadership on teachers' job satisfaction, organizational commitment and organizational citizenship behaviour in Tanzanian Primary and Secondary Schools. *Doctoral thesis*, Universiteit Nijmegen.

Organ, D. W. 2006. Treating employees fairly and OCB: sorting the effect of job satisfaction, organizational commitment and procedural justice. *Citizenship Behavior: Its Nature, Antecedents, and Consequences*, SAGE Publications.

Rivai, V. & Mulyadi, D. 2012. *Kepemimpinan dan Perilaku Organisasi*, Jakarta: Rajawali Pers.

Robbins, S.P. & Judge, T.A. 2008. *Perilaku Organisasi*, Edisi Kedua belas, Jakarta: Salemba Empat.

Schappe, S.P. 1998. The influence of job satisfaction, organizational commitment, and fairness perceptions on organizational citizenship behavior. *The Journal of Psychology*, Vol 132, No 3, pp. 277–290.

Schleicher, J.D., Watt, J.D. & Greguras, G.J. 2004. Reexamining the job satisfaction-performance relationship: The complexity of attitudes. *Journal of Applied Psychology*, Vol 89, No 1, pp. 165–177.

Tett, R.P. & Meyer, J.P. 1993. Job satisfaction, organizational commitment, turnover intention, and turnover: path analyses based on meta-analytical findings. *Personnel Psychology*, Vol 46, No 2, pp. 259–293.

Tuna, M., Ghazzawi, I., Tuna, A.A. & Çatir, O. 2011. Transformational leadership and organizational commitment: the case of Turkey's hospitality industry. *S.A.M. Advanced Management Journal*, Vol 76, No 3, pp. 10–25.

Williams, L.J. & Anderson, S.E., 1991. Job satisfaction and organizational commitment as predictors of organizational citizenship and in-role behavior. *Journal of Management*, Vol 17, No 3, pp. 601–617.

Yang, M-L. 2012. Transformational leadership and Taiwanese public relations practitioner' job satisfaction and organizational commitment. *Social Behavior and Personality*, Vol 40, No 1, pp. 31–46.

Yang, Y-F. & Islam, M. 2012. The influence of iransformational leadership on job satisfaction: The balanced scorecard perspective. *Journal of Accounting & Organizational Change*, Vol 8, No 3, pp. 386–402.

Zahari, I. & Shurbagi, A. 2012. The effect of organizational culture and the relationship between transformational leadership and job satisfaction in petroleum sector of Libya. *International Business Research*, Vol 5, No 9, pp. 89–97.

The effect of compensation and job satisfaction on employees' performance

M.O. Fauzan & Disman
Universitas Pendidikan Indonesia, Bandung, Indonesia

ABSTRACT: Compensation is an important factor in maintaining employee performance, which is needed to achieve the company's vision and mission. To improve and maximize employee performance, a company must also be able to create job satisfaction. Job satisfaction will make employees stay and survive in the company. The population in this study was employees of PT. Inti Indo Sawit Subur, Batanghari Regency, as many as 85 employees. Since the population was less than 100 people, the sampling method applied was the census method. The analysis technique used path analysis to see the direct effect, and was continued by testing the R-square and testing the hypotheses partially and simultaneously with the t-test and F-test. This research concluded that compensation and job satisfaction were considered sufficient by the employees and compensation and job satisfaction had a positive and significant effect on employees' performance.

1 INTRODUCTION

1.1 Research background

An organization is a place for humans to fulfill all their needs in life (Akmal & Tamini 2016; Onsardi et al. 2017). In order to fulfill these needs, people work for one of the organizations that they consider capable of supporting their needs. People who are in the organization will do, obey, and are willing to be given direction by people in higher positions, even though sometimes those people are younger than themselves so that the goals of the company can be achieved by sharing responsibilities. This responsibility in organizations is usually known as performance.

Performance is a tangible result that is seen in the quality and quantity achieved by employees in carrying out their duties in accordance with the roles and responsibilities given to them (Mangkunegara 2013; Rivai & Deddy 2011; Sagala 2011). Performance is the way a manager maximizes results so that the goals of the company can be achieved (Qustolani 2017). There are two factors that affect employee performance, namely internal factors and external factors. Internal factors are factors related to a person's characteristics, including attitudes, personality traits, physical characteristics, desires or motivations, age, sex, education, work experience, cultural back-ground, and other personal variables. External factors are factors that influence employee performance that originated from the environment, leadership, the actions of coworkers, types of training and supervision, the wage system, and the social environment (Riyadi 2011). Organizational performance depends on individual performance, or in other words, individual performance will contribute to organizational performance. This means that the behavior of organizational members both individually and in groups gives strength to organizational performance. Motivation and ability to interact determine performance because both are interrelated to advance the organization (Firmandari 2014).

Among many organizational paradigms in current literature, one of the paradigms describes the organization as a machine that works with a certain order and discipline, which emphasizes the existence of a certain level of productivity, by achieving a certain level of efficiency and that is controlled by a legitimate leadership authority (Thoha 2015).

In Indonesia, there are organizations or companies that operate privately and non-privately. A high interest in the palm oil sector makes many private companies operate in the palm oil products processing field. One of the organizations is PT. Inti Indo Sawit Sub-ur Batang Hari Regency. In improving employee performance, PT. Inti Indo Sawit Subur, Batang Hari Regency, formulate performance improvement strategies; one of which is compensation for the employees. Compensation is a service provided by the company to its employees, both financial and nonfinancial (Kasmir 2016). This is an encouragement for employees to improve their performance (Potale & Uhing 2015).

Another factor to increase performance is job satisfaction. According to Robbins & Hakim, a person with a high level of job satisfaction has positive feelings about the job, while someone who is dissatisfied has negative feelings about the job (Ilahi et al. 2017).

1.2 Research urposes

The purpose of this research is to find out the response on the implementation of compensation, job satisfaction, and employee performance; to investigate the effect of compensation on employee performance; to investigate the effect of job satisfaction on employee performance; and to find out the effect of compensation and job satisfaction on employee performance.

2 LITERATURE REVIEW

2.1 Compensation

According to Panggabean (Sukidi & Wajdi 2016), compensation covers all types of awards in the form of monetary or nonmonetary, given to employees in a proper and fair way for their services in achieving company goals.

According to Panggabean (2004), the types of compensation can be categorized as follows:

a. Salary, which is a financial reward paid to employees on a regular basis, such as yearly, quarterly, monthly, or weekly.
b. Wages are direct financial rewards paid to workers based on working hours, number of items produced, or number of services provided. So, unlike a relatively fixed salary, the amount of wages can change. Basically, a salary or wage is given to attract prospective employees to become employees.
c. Incentive, which is direct benefits paid to employees because their performance exceeds the specified standards. By assuming that money can be used to encourage employees to work harder, those who are productive prefer their salary to be paid based on work.
d. Fringe benefit or additional compensation given based on company policy toward all employees in an effort to improve the welfare of employees.

2.2 Job Satisfaction

According to Luthans (2006), job satisfaction is a result of employees' perceptions of how well their work provides things that are considered important.

Smith, Kendall, & Hulin (Luthans 2006) proposed several dimensions of job satisfaction that can be used to express important characteristics about a job. The dimensions are:

a. The job itself. Every job requires a certain skill in accordance with their respective fields. Difficulty level and expertise needed in doing the job will increase or reduce job satisfaction.
b. Supervisor. A superior needs to respect the work of subordinates. For subordinates, superiors can be regarded as a father/mother/friend or superior.
c. Workers. Entities that relate with each other and their superiors in a company both in the same and different types of work.

d. Promotion. A factor associated with the presence or absence of an opportunity to obtain a career increase during work.
e. Salary/Wages. A factor in fulfilling the life needs of employees who are deemed feasible or not.

2.3 Performance

According to Sedarmayanti (2011), performance is defined as a completion of a task, a management process, or an organization as a whole, where the results of the work must be demonstrated concretely and can be measured (compared to predetermined standards).

The performance dimension according to Sedarmayanti (2011) includes:

a. Work result
b. Workers, processes, or organizations
c. Proof of achievement
d. Measurement
e. Comparison with predetermined standards.

2.4 Conceptual framework

The conceptual framework of this research can be seen in Figure 1.

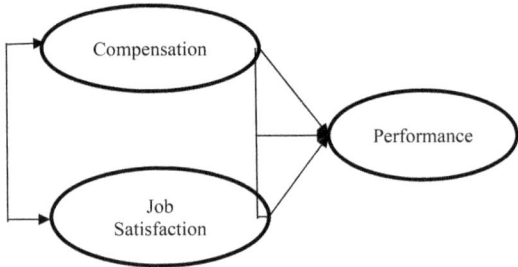

Figure 1. Conceptual framework.

3 METHOD

3.1 Characteristics of respondents

The number of respondents was 95 people; 31 females and 64 males. Their education backgrounds were master degree (7), bachelor degree (25), diploma 3 (8), diploma 2 (15), high school graduate (30), and junior high school graduate (4).

3.2 Description of research variables

The score for the employee compensation variable was 5274, which indicates that the compensation of employees at PT Inti Indo Sawit Subur in Batanghari Regency received is very good.

Table 1. Multiple linear regression analysis results.

| | Unstandardized coefficients | | Standardized coefficients | | |
Model	B	Std. Error	Beta	t	Sig.
(Constant)	19,875	6,344		3,133	,002
Compensation	,327	,103	,314	3,170	,002
Job satisfaction	,130	,105	,123	1,248	,215

The score for the job satisfaction variable was 3532, which shows that employee satisfaction at PT Inti Indo Sawit Subur, Batanghari Regency, is good.

The score for the employee performance variable was 4074, which means that employees at PT Inti Indo Sawit Subur, Batanghari Regency, have very high performance in work.

3.3 Testing of research instruments

3.3.1 Validity test
A validity test is used to find out how far statement items are developed based on each indicator of variables, such as compensation, job satisfaction, and employee performance, to be used in analyzing further data in a study. In other words, this test was conducted to determine the validity of the questionnaire used in this study.

The testing of validity was carried out by means of Product Moment r of 0.202 (n = 95) according to Riduwan & Akdon (2009: 292), with a significant level of 5% ($\alpha = 0.05$).

3.3.2 Reliability test
A reliability test is used to see whether the statement items are reliable or not. In other words, this test is intended to see that the questionnaire distributed to the respondents has a degree of consistency over time. The measurement is only done once, and the results are compared with other statements, or for measuring the correlation between the answers of each statement item that exists. The reliability test resulted in a compensation variable of 0.725, a job satisfaction variable of 0.834, and a performance variable of 0.868.

3.3.3 Multiple linear regression analysis
To analyze the data obtained, qualitative and quantitative analyses were used. The quantitative analysis was used to prove the hypothesis proposed by using multiple linear regression, while the qualitative analysis was used to examine the proof of the quantitative analysis. The multiple linear regression analysis results can be seen in Table 1.

Based on the values obtained, the multiple linear regression equation is as follow:

$$Y = 19,875 + 0,327X_1 + 0,130X_2 + e \qquad (1)$$

Table 2. T-test results (partial influence of each independent variable).

| | Unstandardized coefficients | | Standardized coefficients | | |
Model	B	Std. Error	Beta	T	Sig.
(Constant)	19,875	6,344		3,133	,002
Compensation	,327	,103	,314	3,170	,002
Job satisfaction	,130	,105	,123	1,248	,215

Table 3. F-test results (simultaneous effect of independent variables)

Model	Sum of Squares	df	Mean Square	F	Sig.
Regression	214,472	2	107,236	6,706	,002b
Residual	1471,254	92	15,992		
Total	1685,726	94			

a. Dependent Variable: Performance
b. Predictors: (Constant), Job satisfaction, Compensation

3.4 Hypothesis testing

3.4.1 T-test
The results of the hypothesis testing as shown in Table 2 are:

1. The t_{count} value of the compensation variable is 3,170, which is greater than the t-table (3,170 > 1,661). Therefore, H0 is rejected and H1 is accepted, which means compensation has a partial effect on employee performances.
2. The t_{count} value of the job satisfaction variable is 1.248, which is lower than the t-table (1.661). Therefore, it can be concluded that H0 is accepted and H1 is rejected, which means that job satisfaction does not partially affect employee performance.

3.4.2 F-test
From Table 3, it can be seen that the F-count is 6.706, which is greater than the F-table of 3.09. Therefore, H0 is rejected and H1 is accepted, which means compensation and job satisfaction simultaneously influence employee performance.

4 RESULTS AND DISCUSSION

The first hypothesis in this study was to see the effect of compensation on employee performance. Partial testing proved that compensation is very influential on employee performance with a high level of significance; this indicates that employee performance at

PT Inti Indo Sawit Subur, Batanghari Regency, can be improved by increasing the compensation given to the employees. This is in line with research by Sopiah (2013) and Retnoningsih et al. (2016).

The second hypothesis was to see the effect of job satisfaction on employee performance. Interestingly, partial testing found that job satisfaction had no effect on employee performance. This is a concern for leaders of PT Inti Indo Sawit Subur to provide more job satisfaction to their employees in order to achieve the company's vision and mission.

The last hypothesis was aimed to see the simultaneous influence of compensation and job satisfaction on employee performance. The test results showed that both compensation and job satisfaction simultaneously had a positive influence on employee performance. Thus, if PT Inti Indo Sawit Subur increases compensation and provides job satisfaction for its employees, their performance will also increase. Similar research by Sukidi & Wajdi (2016) also found the same results.

5 CONCLUSION

From the results of the research described previously, conclusions can be drawn as follows:

1. Compensation and job satisfaction simultaneously have a positive effect on employee performance, which means that if compensation and job satisfaction increase, the performance of employees at PT Inti Indo Sawit Subur will also increase.
2. Compensation and job satisfaction partially have a positive influence on employee performance, which means that if compensation increases or job satisfaction increases, the performance of employees at PT Inti Indo Sawit Subur in Batanghari Regency will increase as well.

REFERENCES

Akmal, A. & Tamini, I. 2016. Pengaruh kompensasi terhadap kepuasan kerja karyawan Gaya Makmur Mobil Medan. *Jurnal Bisnis Administrasi*, 35(2), pp. 53–59.

Firmandari, N. 2014. Pengaruh kompensasi terhadap kinerja karyawan dengan motivasi kerja sebagai variabel moderasi (Studi pada Bank Syariah Mandiri Kantor Cabang Yogyakarta). *Ekbisi*, 9(1).

Ilahi, D. K., Mukzam, M. D., & Prasetya, A. 2017. Pengaruh kepuasan kerja terhadap disiplin kerja dan komitmen organisasional (Studi pada karyawan PT. PLN (Persero) Distribusi Jawa Timur Area Malang). *Jurnal Administrasi Bisnis*, 44(1), 31–39.

Kasmir, S. 2016. The mondragon cooperatives: successes and challenges. *Global dialogue: Magazine of the International Sociological Association*, 6(1).

Luthans, F. 2006. Perilaku organisasi.

Mangkunegara, A. A. A. P. 2013. Manajemen sumber daya manusia perusahaan. Bandung: Remaja Rosdakarya.

Onsardi, Asmawi, M., & Abdullah, T. 2017. The effect of compensation, empowerment, and job satisfaction on employee loyalty. *International Journal of Scientific Research and Management*, 05(12), pp. 7590–7599. doi: 10.18535/ijsrm/v5i12.03.

Panggabean, M. S. 2004. Manajemen sumber daya manusia. Bogor: Ghalia Indonesia.

Potale, R. & Uhing, Y. 2015. Pengaruh kompensasi dan stres kerja terhadap kepuasan kerja karyawan Pada PT. Bank Sulut Cabang Utama Manado. *Jurnal EMBA: Jurnal Riset Ekonomi*, Manajemen, Bisnis dan Akuntansi, 3(1).

Qustolani, A. 2017. Pengaruh kepuasan kerja, keadilan prosedural dan kompensasi terhadap kinerja karyawan (Studi kasus pada Industri Rotan Sekecamatan Leuwimunding Majalengka). *MAKSI*, 4(2).

Retnoningsih, T., Sunuharjo, B. S., & Ruhana, I. 2016. Pengaruh kompensasi terhadap kepuasan kerja dan kinerja karyawan (Studi pada karyawan PT. PLN (Persero) Distribusi Jawa Timur Area Malang). *Jurnal Administrasi Bisnis* (JAB), 35(2), pp. 1–7. doi: 10.1016/j.fcl.2007.01.001.

Riyadi, S. 2011. Pengaruh kompensasi finansial, gaya kepemimpinan, dan motivasi kerja terhadap kinerja karyawan pada perusahaan manufaktur di Jawa Timur. Jurnal manajemen dan kewirausahaan, 13(1), 40–45.

Rivai, V. & Deddy, M. 2011. Kepemimpinan dan Perilaku Organisasi. Ketiga. Jakarta: Rajawali Pers.

Sagala, H. S. 2011. Manajemen strategik dalam peningkatan mutu pendidikan, pembuka ruang kreativitas, inovasi, dan pemberdayaan potensi sekolah dalam sistem otonomi daerah. Bandung: Alfabeta.

Sedarmayanti, A. 2011. Manajemen sumber daya manusia, reformasi birokrasi dan manajemen Pegawai Negeri Sipil (cetakan kelima).

Sopiah. 2013. The effect of compensation toward job satisfaction and job performance of outsourcing employees of Syariah Banks in Malang Indonesia. *International Journal of Learning and Development*, 3(2), p. 77. doi: 10.5296/ijld.v3i2.3612.

Sukidi & Wajdi, F. 2016. Pengaruh motivasi, kompensasi, dan kepuasan kerja terhadap kinerja pegawai dengan kepuasan kerja sebagai variabel intervening daya saing. *Jurnal Ekonomi Manajemen Sumber Daya*, 18(2), pp. 79–91. doi: 10.1095/biolreprod.110.087411.

Thoha, M. 2015. Kepemimpinan dalam manajemen cetakan ke 18. Jakarta Raja Grafindo Persada.

The exitence of human capital in increasing organizational performance in Industry 4.0 era

S. Yulianty & H.S. Hadijah
Universitas Pendidikan Indonesia, Bandung, Indonesia

ABSTRACT: In the fourth industrial revolution era, digitalization is increasing rapidly, including among provider companies. In order to improve their performance, some of the companies start to implement human resource efficiency through technology which will threaten the existence of human capital. Therefore, this research was conducted to see the extent of human capital in improving organizational performance, which was measured through competency, product knowledge and employee motivation. The method used in this research was literature review from more than 50 journals about human capital. After the data was collected, the testing of the causality between these variables was carried out using Lisrel program. The results can be concluded theoretically that the quality of human capital had a positive effect on the organizational performance in provider company.

1 INTRODUCTION

Today, we have entered the fourth industrial revolution era where technology, information, and internet are some of the most important needs in life and various human activities.

Some scholars and practitioners have considered four main industry changes throughout the history, while the Industry 4.0 is the last one and an ongoing industry transformation (Qin et al. 2016). While according to (Zawadzki et al. 2016) the fourth industrial revolution, is the recent movement on intelligent automation technology".

Consequently, this condition encourages companies' effectiveness and efficiency which become the main demand in technological advances in the industrial revolution, and as a result, requires every individual to be skilled in utilizing them. It is depicted in human capital, which is inseparable from the effort of every human resource to explore all their capabilities in adjusting and supporting the improvement of the company's organizational performance. (Kiseleva 2013) follows the evolution of human models since the beginning of the twentieth century, from the "Economic human" to the "Socioeconomic human". Economic and social security, labor and intellectual capital, as well as life, innovation and creativity, and entrepreneurial potential stand out among the studied Human characteristics.

Human capital constitutes the basis for innovative development of a company, it is the source of the strategic rebuilding through the usage of – for example – brainstorming in a research lab, re – engineering new processes improving personal skills or developing

new leads. The essence of human capital is the fact of intelligence sharing among the members of the organization (Bontis 1998; Wang & Swanson 2008).

The human capital referred to this paper has a number of important indicators, namely; knowledge that is specified by product knowledge, competence and motivation to measure the existence of human capital and to support the company's organizational performance.

Companies studied in this literature study were companies that are related to the 4.0 industrial era, specifically several provider companies in Indonesia. This study found a positive relationship between human capital and organizational performance in provider companies where similar company rarely conducted research on human capital. Therefore, the aims of this study was to find the existence of human capital in supporting the progress of company's organizations and to find which supporting indicators of human capital are dominant in showing the existence of human capital in company's organizational performance.

2 LITERATURE REVIEW

2.1 *Human capital*

Human capital is the investment in human resources in order to increase their efficiency. The costs of this investment are provided for future use, therefore the organization chooses the investment in individuals, because people are valuable human capital with different qualities (Burund & Tumolo 2004; Guest

1997). In the recent decade, the management of organizations has found that human resources have the greatest importance in gaining sustainable competitive advantage and efficiency. In the world where knowledge and communication with customers have gained increasing importance, human capital, which shows the volume of knowledge, technical skills, creativity, and experience of the organization, gains great importance. Thus labor force is considered as productive assets (Garavan et al. 2001; Hendricks 2002; Pasban & Nojedeh 2016),

Human capital refers to the knowledge, skill and experience of employees (Nerdrum & Erikson 2001; Nordhaug 1993; Pena 2002; Roos 1998). Human capital is categorized into three categories, i.e. capability and potential, motivation and commitment and innovation and learning. Capability and potential refers to the educational level, professional skills, experience, attitudes, personal networks, values, and the ability of current employees to evolve within the organization. Motivation and commitment refers to whether employees align their own interests with those of the firm, while innovation and learning shows the degree to which employees are open to change (Mayo 2001). Human capital refers to knowledge, education, work competence, and psychometric evaluations (Namasivayam & Denizci 2006).

2.2 Knowledge (Product Knowledge)

Knowledge has been conceptualised and characterised in a number of ways in the literature (Becker 2002; Maruping 2002; Nonaka & Reinmoeller 2000; Rastogi 2002). The rapid growth of the global knowledge economy requires firms in various industries and sizes to be embedded with intellectual capital in assuring sustainability and competitiveness. In today's knowledge economy, the influence of human capital, structural capital and relational capital on business performance indicates that the investments in intangible resources gain their importance over tangible resources (Clarke et al. 2011; Cohen & Kaimenakis 2007; Groves 2002; Užienė 2015). There are three classes of knowledge:

Explicit knowledge provides explanations about products and processes (rationale), while understanding of products (strategies) and processes (relationships) is classified as implicit knowledge. Tacit knowledge is defined as intuition, about produces and processes. Knowledge that is extracted from the human's mind can be stored externally as information. Information provides descriptions about products and processes (Wallace et al. 2005).

2.3 Competency

Nowadays, there are a number of discussions and arguments about competency. Many companies and organizations put efforts in implementing competency based systems. In order to ensure the successful future, the process of understanding of real situation, learning from challenges and failures need to be conducted (Duaring 2006; Jamshidia et al. 2012). Great and successful competency allows an organization to become constant in purpose of hiring people, train and develop their skills, measure their performance, and develop leadership potential. It identifies competency gaps in mission – critical occupations (Amrizah & Rashidah 2009; Carmeli & Schaubroeck 2005; Duaring 2006; Jamshidia et al. 2012). In competency – based approaches, the employee is evaluated based on a particular standard but not in comparison with others (Falender et al. 2004).

(OECD 2015; Skorkova 2016) Implementation of competency framework into human resource management of the institutions in public sector would requires a central reform. As any other change it can be difficult for implementation and all employee's commitment and involvement. But benefits – that are already seen in private sector and countries that have already implemented competency management framework – are very attractive, (Daft 1998).

2.4 Motivation

Factors affecting staff motivation at a period where the ?nancial rewards are kept to the least leads to stimulate employee performance (Panagiotakopoulos 2013; Wayne et al. 1999). The job satisfaction works toward making good relationships with staff and colleagues, control of time off, enough resources, and bring autonomy for employee in the organization. It is essential in the stages of employment i.e. early, middle, and late career stage of life because it brings any combination of physiological, psychological, satisfaction that invokes a person truthfully to they are satisfied with their current job and it leads to employee motivation to achieve goals of the organization (Williams et al. 2003).

2.5 Organizational performance

Nowadays, human capital is one of the factors that determine organizational competitiveness, given that competencies, knowledge, creativity, capacity to resolve problems, leadership and personal compromise are some of the assets required to meet the demands of turbulent environments and reach organizational goals (Littlewood 2004; Schultz 1961).

Performance is a broader indicator that can include productivity, quality, consistency, and so forth. On the other hand, performance measures include results, behaviors (criterion – based) and relative (normative) measures, education and training concepts and instruments, including management development and leadership training for building necessary skills and attitudes of performance management (Richard 2002). The organizational performance referred to in this study is the result of human capital itself which is implemented through better service quality and high levels of customer loyalty to reflect improvements in organizational performance.

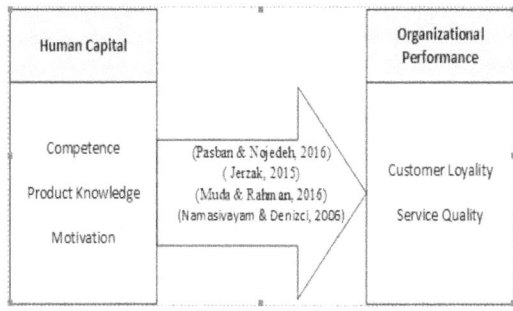

Figure 1. Research Model.

Awareness also has been increasing that services, like products may be guaranteed as tools of implementing a total quality management orientation in the organization (Lawrence & McCollough 2001). Technology offers to develop new services and deliver it better, more efficient services to customers as well as the contrasts and dark side of technology and services (Bitner 2001) and the importance of the concept derives from the benefits associated with retaining existing customers (McMullan 2005).

2.6 Relationship between human capital and organizational performance

A research conducted by (Pasban & Nojedeh 2016), indicated that to achieve the objective of good organization, human skill is depicted on human capital, where one must have good knowledge and competence (Muda & Rahman 2016). Stiles & Kulvisaechana (2003) mentioned in their research that to obtain superior performance, business should rely on labor with attributes of human capital such as education, experience, motivation, talent and skills to achieve superior performance

According to (Namasivayam & Denizci 2006), human capital refers to knowledge, education, competence, and psychometric evaluation. A results of the research conducted by (Jerzak 2015), strengthens that product knowledge as one of the strongest indicator of human capital that can improve the performance of the organization.

Figure 1 shows the research model that illustrates the causality between human capital and organization performance.

3 METHODS

This research method was qualitative using a literature study which consists of several stages; identifying, analyzing and collecting the results of previous research. Literature studies were conducted to find the concepts and theories of causality between human capital and organizational performance, followed by developing a theoretical model based on the results of the literature review.

4 RESULTS AND DISCUSSION

In the literature review, some research stated that human capital can be described by several components mentioned earlier, namely knowledge (product knowledge), competence, and motivation. It is also discovered that human capital is an important element in influencing the progress of organizational performance of a company. Therefore, this literature reviews acts to support the role of human capital in improving the performance of corporate organizations.

As there are not many journals that discuss human capital in provider companies, this journal has limited references, but some of these references would be discussed and designed according to the needs of the journal objectives. The following are previous studies which support the concept of the journal written:

In previous studies, controlling human resources was viewed as more difficult than controlling other resources of the organization, for the rarest and most complicated resources in knowledge – based economics are human resources. Mos managers focus their strategies on real and clear factors of organizations such as technology and using physical and financial resources (Pasban & Nojedeh 2016). Meanwhile, study by (Jerzak 2015; Xiuli & Li 2015) found the essence of human capital roles applied to the Polish construction market, which initially was only considered an executing force, not as a force that built business competitiveness. Simultaneously many businesses tried to change this approach, in order to benefit the market and become more competitive for investors by highlighting the role of human capital. As a result, knowledge, which is one indicator of the human capital system, claims a very important role, especially in developing and stabilizing fairly volatile market conditions.

(Jerzak 2015; Karoglu & Eceral 2015; QiDong et al. 2012) stated that the role of human capital in improving company performance is very suitable to be applied in provider and service companies field. In provider companies, human capital will be very much needed to support the company's organizations performance, where the results of organizational performance are depicted through service quality and customer loyalty. Service quality can be seen from the employees' speed in serving and handling complaints as well as accuracy in providing solutions. Service quality is strongly associated with provider companies, as product knowledge has a very important role in maintaining organizational performance that is assessed from the service quality (Zeithmal et al. 1996). As stated in some of previous references, one of the indicators of human capital, knowledge, is able to improve company performance. With good knowledge, the company's HR are able to improve their insight in developing performance and even solving problems immediately. Aside from knowledge, competency is also one of the indicators of human capital that must be owned by every company, with a high level of competence, the human resource of the company are expected to be able to work effectively and efficiently. Motivation is also an essential indicator of human capital, since the higher the motivation, the stronger the desire of the company's employee to provide the best performance. Motivation is created, one of which, through a training program. These three indicators of human capital are able to increase organizational performance in several business fields depicted through good service quality, which then increase number of customer loyalty.

Another statement from the research conducted by Muda & Rahman (2016), found that the role of human capital is viewed from the elements that influence it and in general, the factor which is important at all levels of management in organizations is human skills. Those who work in the central core of the organization must develop higher skills. They must have sufficient knowledge, information, innovation, and creativity to improve customer satisfaction and create a competitive advantage for the organization. (Namasivayam & Denizci 2006), Human Capital refers to knowledge, education, work competence, and psychometric evaluation.

Furthermore, in research conducted by (Muda & Rahman 2016), the findings state that Small and Medium – sized Enterprises (SMEs) must rely on labor with attributes of human capital such as education, experience, motivation, talents, and skills to achieve superior performance. Overall, human capital plays an important role in improving business performance. In these studies, education is not the only elements that plays a role in the human capital operation, motivation, which is one of the elements of human capital, also plays an important role in the company's organization performance as demonstrated by business performance.

It was explained that organizational performance can be depicted through service quality and customer loyalty, which is supported by the existence of human capital through several indicators, including product knowledge, competence, and motivation. Sufficient knowledge that is owned by the company, can develop insight and make service quality better, competency makes the performance more effective and efficient, while motivation encourages human resource in the company give their best performance. For further research, the researcher suggests indicators of integrity in human capital, which is expected to play a role in maintaining the sustainability of the company.

5 CONCLUSION

In some journals, it is very difficult to find an account which states that human capital cannot improve organization performance. Some of the reference journals state that human capital plays an important role in improving the performance of organizations, which shows the existence of human capital itself. The elements of human capital that were often discussed in several reviews are knowledge, specifically product knowledge, which is a strong indicators of human capital in increasing the company's organization performance.

REFERENCES

Amrizah, K. & Rashidah, A.R. 2009. Enhancing organisation effectiveness through human, relational and structural capital: an empirical analysis. *Malaysian Accounting Review* 8 (1), 1–17.

Becker, G.S. 2002. The age of human capital. Education in the Twenty – First Century, 3–8.

Bitner, M.J. 2001. Service and technology: opportunities and paradoxes. *Managing Service Quality*, Volume: 11 Issue: 6 2001.

Bontis, N.1998. Intellectual capital: an exploratory study that develops measures and models. *Management Decision*, vol. 36, No.2, pp. 6376.

Burund, S. & Tumolo, S. 2004. Leveraging the new human capital: Adaptive strategies, results achieved, and stories of transformation. Boston, USA: Nicolas Brealey America.

Carmeli, A. & Schaubroeck, J. 2005. How leveraging human resource capital with its competitive distinctiveness enhances the performance of commercial and public organizations. *Human Resource Management*, 44(4), 391–412.

Clarke, M., Seng D. & Whiting, R.H. 2011. Intellectual capital and firm performance in Australia. *Journal of Intellectual Capital* 12 (4), 505–530.

Cohen, S. & Kaimenakis, N. 2007. Intellectual capital and corporate performance in knowledge intensive SMEs. *The Learning Organization,* 14 (3), 241–262.

Daft, R.L. 1998. Essentials of organization theory and design. South – Western College Publishing.

Duaring, T. 2006. Implementing a Successful Competency Model. Human Capital Institute.

Falender, C.A., Cornish, J.A.E., Goodyear, R., Hatcher, R., Kaslow, N.J., Leventhal, G. & Grus, C. 2004. Defining competencies in psychology supervision: A consensus statement. *Journal of Clinical Psychology*, 60(7), 771–785.

Garavan, T.N., Morley, M., Gunnigle, P. & Collins, E. 2001. Human capital accumulation: The role of human resource development. *Journal of European Industrial Training*, 25, 48–68.

Groves, S. 2002. Knowledge wins in the new economy. *Information Management*, 36(2).

Guest, D.E. 1997. Human resource management and performance: A review and research agenda. *The International Journal of Human Resource Management*, 8(3), 263–276.

Hendricks, L. 2002. How important is human capital for development? Evidence from immigrant earnings. *American Economic Review*, 92(1), 198–219.

Jamshidia, M.H.M., Raslib, A. & Yusof, M. 2012. Essential Competencies for the Supervisors of Oil and Gas Industrial Companies. Elsevier Ltd, 40 (2012) 368–37.

Jerzak, K. 2015. The essence of human capital in a building company. Elsevier Ltd, 122 (2015) 95–103.

Karoglu, B.A. & Eceral, T.O. 2015. Human capital and innovation capacity of firms in defense and aviation industry in ankara. Procedia – *Social and Behavioral Sciences*, (195), 1583–1592.

Kiseleva, L. 2013. Health as an economic resource in the context of contemporary theories. *Czech Journal of Social Sciences*, Business and Economics, 2(3), 62–71.

Lawrence J.J. & McCollough, M.A. 2001. Quality Assurance in Education Volume: 9 Issue: 3. A conceptual framework for guaranteeing higher education.

Littlewood, H. 2004. Analisis factorial conformatorio y modelamiento de ecuación estructural de variables afectivas y cognitivas asociadas a la rotación de personal. *Revista Interamericana de Psicología Ocupacional*, 23 (1), 27–37.

Maruping, L.M. 2002. Human capital and firm performance: Understanding the impact of employee turnover on competitive advantage. *Proceedings of the Academy of Management Conference*, Denver.

Mayo, A. 2001. The Human Value of the Enterprise: Valuing People as Assets: Monitoring, Measuring, Managing. Nicholas Brealey Publishing, London.

McMullan, R. 2005. A multiple – item scale for measuring customer loyalty development. *Journal of Services Marketing*, 19(7), 470–481.

Muda, S. & Rahman, M.R.C.A. 2016. Human Capital in SMEs Life Cycle Perspective. Procedia Economics and Finance 35. 683–689.

Namasivayam, K. & Denizci, B. 2006. Human capital in service organizations: Identifying value drivers. *Journal of Intellectual Capital*, 7(3), 381–393.

Nerdrum, L. & Erikson, T. 2001. Intellectual capital: A human capital perspective. *Journal of Intellectual Capital*, 2(2), 127–135

Nonaka, I. & Reinmoeller, P. 2000. Dynamic business systems for knowledge creation and utilization. In C. Despres & D. Chauvel (Eds.), Knowledge horizons: The present and the promise of knowledge management. Oxford: Butterworth Heinemann.

Nordhaug, O. 1993. Human capital in organizations: Competence, training and learning. Oslo, Norway: Scandinavian University Press Publication.

OECD, Public Governance Review. 2015. Slovak Republic: Better Co – ordination for better policies, services and results, OECD Publishing Paris.

Panagiotakopoulos, A. 2013. The impact of employee learning on staff motivation in Greek small ?rms: The employees' perspective. *Development and Learning in Organisations*, 27(2), 13–15.

Pasban, M. & Nojedeh, S.H. 2016. A Review of the Role of Human Capital in the Organization. *Social and Behavioral Sciences* 230 (2016) 249–253.

Pena, I. 2002. Intellectual Capital and Business Start – up Success. *Journal of Intellectual Capital* 3 (2), 180–198.

QiDong, J., He, J. & Karhade, P. 2012. Human capital and information technology capital investments for innovation: curvilinear explanations. In: Shaw, M., Zhang, D.,

Yue, W.D.(Eds.), E – Life: Web – Enabled Convergence of Commerce, Workand Social Life, vol. 108. Springer Link, Shanghai, China, pp. 334–346.

Qin, J., Liu, Y. & Grosvenor, R. 2016. A categorical framework of manufacturing for industry 4.0 and beyond. Proced. CIRP 52, 173–178.

Rastogi, P.N. 2002. Knowledge management and intellectual capital as a paradigm of value creation. *Human Systems Management*, 21(4), 229240.

Richard, C. 2002. Experiments with New Teaching Models and Methods.

Roos, J., 1998. Exploring the concept of Intellectual Capital (IC). Long Range Planning. 31(1), 150–153.

Schultz, T.W. 1961. Investment in human capital. *The American Economic Review*, 51(1), 1–17.

Skorkova, Z. 2016. Competency models in public sector. Elsevier Ltd, 230 (2016) 226–234.

Stiles, P. & Kulvisaechana, S. 2003. Human capital and performance: A literature review. University of Cambridge: Cambridge.

Užienë, L. 2015. Open Innovation, Knowledge Flows and Intellectual Capital. Procedia – *Social and Behavioral Sciences*, 213, 1057–1062.

Wallace K.M., Ahmed, S., Bracewell, R., Engineering Knowledge Management. In: Clarkson J, Eckert C, editors, Design Process Improvement: A Review of Current Practice. First ed. London: SpringerVerlag. 2005. p. 326–343

Wang, G.G. & Swanson, R.A. 2008. Economics and human resource development: A rejoinder. *Human Resource Development Review*, (7), 358–362.

Wayne, S.J., Liden, R.C., Kraimer, M.L. & Graf, I.K. 1999.The role of human capital, motivation and supervisor sponsorship in predicting career success. *Journal of Organizational Behavior*, vol.20, no.5, pp. 577–595.

Williams, E.S., Konrad, T.R., Linzer, M., McMurray, J., Pathman, D.E., Gerrity, M., et al. 2003. Re?ning the measurement of physician job satisfaction: Results from the physician work life survey. *Medical Care*, 37(11), 1140–1154.

Xiuli, S.X. & Li, H. 2015. Firm – level Human Capital and Innovation: Evidence from China. *Preliminary Program of the Allied Social Sciences Association*, January 2015, Boston.

Zawadzki, P., Zywicki, K. & Manag. 2016 Prod. Eng. Rev. 7. 105–112.

Zeithmal, V.A., Berry, L.L. & Parasuraman, B.A. 1996. The behavioral consequences of service quality. *Journal of Marketing*, 60, 31–46.

Advances in Business, Management and Entrepreneurship – Hurriyati et al. (Eds)
© 2021 Taylor & Francis Group, London, ISBN 978-0-367-67471-7

Redesigning job analysis to improve the performance of education personnel at Universitas Jenderal Achmad Yani

M.V. Romi, N. Maryani & E. Ahman
Universitas Jenderal Achmad Yani, Cimahi, Indonesia

ABSTRACT: This study aimed to redesign problems related to job analysis that occurs at the University of General Achmad Yani, where the workload was still not optimal. The cause of the many unclear job analyses that exist in the institution is that all sections perform a variety of additional work that is not related to their duties and obligations. This research was intended to redesign the position analysis at the managerial level of the University of General Achmad Yani. Mix methods were employed in this research, the first approach used was a qualitative approach based on case studies. There were three stages of analysis, namely organizing various related data, grouping data based on categories, themes and answer patterns, then testing the assumptions or problems that exist in the data. The quantitative approach used was explanatory research through a quantitative approach, where responses were measured using a Likert scale. The sample of this study was 71 employees at General Achmad Yani University chosen by proportional stratified random sampling. The results of this study showed that all the hypotheses were proven.

1 INTRODUCTION

In general, organizational structure is usually depicted in an organization chart that contains boxes and lines. Daft (2001) mentions that an organization chart is a visible representation that describes all the activities that occur in an organization. Taxonomically, the organization chart according to Daft (2001), describes three main things, namely (1) the level of specialization or complexity of the organization, (2) the level of organizational formalization, and (3) the level of organizational centralization/decentralization. Since the organizational structure is a necessity, it should be designed so that the goals—efficiency, effectiveness, and company value—can be achieved. The organizational structure according to Robbins (2003), is a chain of command, which is graphically depicted using an organization chart. Meanwhile, according to Daft (2010), organizational structure is a formal organizational structure, which includes the design of the system to ensure the effectiveness of communication, coordination, and integration (control) that comes from inter-departmental relations. The process of rationalization and specialization that began during the industrial revolution has prompted the creation of hierarchical, mechanistic, autocratic, and confrontational organizational structures that have become tall and fat (Idrus 2008 in Widjaja 2012).

Universitas Achmad Yani (UNJANI) is one of the higher education institutions in Cimahi. UNJANI as an institution that has employees in all institutional, administrative, academic, staffing, and financial management positions certainly has rules in managing and running the system that has been created. UNJANI has an obligation to prepare qualified human resources. There are two important things in preparing qualified human resources; first, employees must understand their tasks through clarity of the roles as described in the job analysis, and secondly, technological developments require employees to have certain skills and knowledge so that training must be formulated in accordance with their roles and responsibilities.

Job analysis is a systematic way of being able to identify and analyze what is required in a job as well as the personnel required in a task so that human resources are able to complete the task well. From the results of the job analysis, the organization will be able to determine what kind of characteristics the prospective employees must have before occupying a position, the output of which is job specifications and job descriptions. Job description contains the duties, functions, authority, and responsibilities of a person in charge, whereas job specification contains the personnel who will do the work as well as the requirements needed, especially those relating to individual skill issues.

2 LITERATURE REVIEW

2.1 *Job analysis*

According to Mondy (2010), job analysis is a process carried out systematically to determine the skills, tasks, and knowledge needed to complete a task in an organization. By analyzing the answers, the tasks needed to do a job can be identified properly.

The results of this approach are often referred to as job descriptions, in which there are also details of the behavior needed to successfully complete the job (Cascio & Aguinis 2005).

There are two methods of conducting job analysis, namely conventional methods and quantitative methods (Milkovich & Newman 2007). Conventional methods can be done through questionnaires, interviews, and observations. The advantage of this method is that there is a clearer understanding and the disadvantage is the possibility of bias and favoritism. Examples of the conventional method are questionnaires interviews, and observations. The advantages of quantitative methods are that they are more practical and low cost while the disadvantage is the possibility of many aspects that cannot be explored; so the job description which is the output of the job analysis process can become invalid.

2.2 *Job aescription*

Job descriptions describe tasks, responsibilities, terms of work, and main activities. Job descriptions vary in terms of the level of content detail (Mangkuprawira 2003). According to Rivai (2006), a job description generally contains some information such as job name, summary, equipment, environment and activities, description of work assignments, responsibilities, and behavioral appearances at work. It also describes social interactions related to work (for example, the size of work groups and the level of freedom in carrying out work).

2.3 *Job specifications*

Job specifications describe employee qualifications, such as experience, knowledge, expertise, or abilities required to complete the work. Qualifications are required by the employee to perform the tasks and responsibilities described in the job description. Job specifications detail the level of knowledge, expertise, and abilities that are relevant for a job, including education, experience, special training, personal traits, and manual skills.

In addition, a company may also include physical requirements, including the ability to walk long hours, to stand up, and to lift what may be required by the employer. All physical and nonphysical requirements will ideally be related to the type of work that will be performed by employees who meet these requirements (Mangkuprawira 2003).

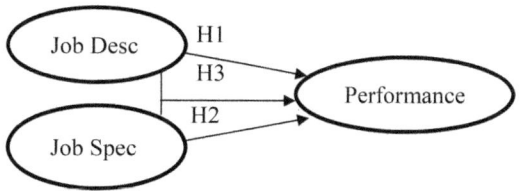

Figure 1. Research paradigm.

2.4 *Research hypothesis*

In this study, job descriptions (X1), and job specifications (X2) were the independent variables and the employee's performance (Y) was the independent variable. Figure 1 shows the research paradigm of this study.

3 METHODS

3.1 *Research approach*

The research approach used in this research was a qualitative approach based on case studies, while the focus was on finding real meaning and new insights (Zikmund et al. 2010). This research used explanatory research with a quantitative approach. Explanatory research is research that explains the causal relationship between variables through hypothesis testing.

3.2 *Research design*

According to Yin (2013), in designing research with the case studies method, as in this research, a sample is drawn from a population using a questionnaire as a data collection tool. The population were all employees at UNJANI. Using the proportionate stratified random sampling technique and the Slovin formula (Umar 2010), the samples used in this study consisted of 71 people.

3.3 *Unit of analysis*

The unit of analysis in this study was an organization and the research subjects were educational staff within UNJANI.

3.4 *Criteria for interpretation of findings*

In this study, several criteria were used to interpret the findings and data collected; these are as follows:

1. Data were obtained from relevant sources, namely institutional documents and key informants that were relevant to the research and not from subjective assumptions.
2. The data obtained were interpreted objectively, in accordance to the actual situation, by eliminating the element of bias.
3. Data interpretation was carried out in line with the research design referred to in this study.

3.5 Criteria for interpretation of findings

One of the features of a case study is the use of several data sources, which is a strategy to improve data credibility (Yin 2003). In this research, the types and sources of data used were archive notes and documentation.

3.6 Data collection procedure

In this study, data collection was carried out in a number of ways, which were:

1. Preliminary survey
    A preliminary survey was conducted with the aim of obtaining a general picture of the object to be examined.
2. In-depth interview
    The key informants used in this study were UNJANI staff members. Interviews were conducted to explore information related to the institution's vision, mission, goals, and strategies and other insights that could support the research.

3.7 Analysis techniques

After the collection stage, the data were analyzed using the data processing methods of Marshall & Rossman (2007), which include several stages:

1. Organizing Data
2. Grouping by Category
3. Testing the Assumptions

4 RESULTS AND DISCUSSION

In an effort to align the roles and responsibilities of each employee with the company's strategic plan, the job analysis must be in line with the company's goals, which include their vision, mission, goals, and strategies. In the previous stage, the organizational structure design was formulated to suit the company's objectives. Then the next stage was to analyze the position in accordance with the organizational structure that refers to the results of the formulation of the company's vision, mission, goals, and strategy. The output of the job analysis used was the job description.

Job analysis was done through documentation and interview processes. The documentation process was carried out by collecting documents related to the job description. Document analysis was carried out to ensure that the interview results were accurate and in accordance with the results of the company's objectives. The results of job analysis process were then analyzed and job description and job specifications were formulated. For now, the job descriptions that exist only summarized the main responsibilities of employees without any additional information. However, not all of the job descriptions referred to company goals. Therefore, to compile the job description, the

Table 1. Interview guidelines.

No	Question	Results
1	State the name of the position and stakeholders of this position.	Identity
2	To whom do these positions have to be held accountable for the work to be done? Who is the staff member that must be held accountable to the stakeholders regarding the tasks performed by the staff?	Superiors and subordinates
3	What are the main responsibilities of this job?	General Description
4	Explain the job descriptions and responsibilities that must be done routinely or non-routinely every day by these position holders	Duties and responsibilities
5	In addition to members of the company, with whom do these positions have to be related in relation to the implementation of duties in accordance with the position, both internally and externally?	Work relationship
6	Mention in detail the form of authority of the office holder associated with the determination of policies in the institution?	Authority
7	1. What physical requirements do you need to carry out in this position? (age, gender, health)	Qualifications and experience
	2. What level of education do you need to carry out this position? (education)	
	3. How long is the experience required to do this position? (year)	
	4. What knowledge do you need to be able to carry out this position? (knowledge)	
	5. What skills/training/courses do you need to have to carry out this position? (skills)	
	6. What nature/personality do you need to have to support someone in order to carry out this position? (ability)	

next step taken was to interview the key informant to get information about the job description and job specifications for each position at UNJANI. This interview was also intended to find out the duties and responsibilities, work relationships, authority, as well as the qualifications and experience needed as seen in Table 1.

The test results show a multiple correlation coefficient (R) of 0.765, which means that the relationship

Table 2. Job analysis.

Position identity	
Part	: Rectorate
Sub Division	: Academic
Position Name	: Head of Academic Bureau
Kedudukan dalam Organisasi	: Position in the Organization
Direct supervisor	: Vice Chancellor I
Direct Subordinates	: Head of Renbangmik, Head of Minmik, Head of Central Library, Head of Labsa, Head of Office. Central Laboratory

Requirements and tenure	Main Duties and Functions of Position
1. Permanent lecturer/ assistant lecturer/DPK or educational staff or retired army.	1. Assist in the formulation of strategic plans in the academic field
2. Maximum age of sixty two years.	2. Formulate programs and activities in the academic field.
3. Have a minimum of S-2/ Sp-1 education and have served as a leader in UNJANI, or S-2/Sp1 with an academic position as an assistant professor, or a class/ space education minimum IV/a with a minimum education of S2/Sp-1, or retired army TNI with a minimum rank of Pamen with a minimum education of S-2/Sp-1.	3. Propose a budget plan in the academic field
	4. Coordinating the implementation of academic planning and development as well as lecturers from each department in an integrated manner and carrying out academic administrative functions.
4. Proposed to and or approved by the Chair of the Foundation Management and appointed by the Chancellor.	5. Optimizing academic collaboration so as to produce outputs that can benefit UNJANI and partners.
5. Having expertise in accordance with the structural position required.	6. Coordinating the Head of Research and Study Programs & Lecturers, the Head of Cooperation and the Head of Academic Administration.
6. The term of the Head of Academic Bureau is four years and can be reappointed no more than two consecutive terms of office are allowed.	7. Coordinate with Karo Sisfo & HR in developing information systems in the academic field.
	8. Prepare and submit reports on the implementation of programs and activities periodically in their fields to Deputy Rector I.

between all independent variables, namely job description variables (X1) and job requirements (X2), to employee performance (Y) is relatively strong.

The test results show a multiple correlation coefficient (R) of 0.765, which means that the relationship between job description (X1) and job requirements (X2) and employee performance (Y) is relatively strong.

The discussion of each determining variable that affects student satisfaction is as follows:

1. Regression coefficient of the job description variable (X1) is 0.912, which shows that every increase in one unit of job description (X1) will increase employee performance (Y) by 0.912 units, assuming the other independent variables are constant.
2. Regression coefficient of the job requirements variable (X2) is 0.764, which indicates that every increase in one unit of job requirements (X2) will increase employee performance (Y) by 0.764 units, assuming the independent variables are constant.

Table 2 shows the example of the output of the job analysis.

5 CONCLUSION

The conclusions of this study are summarized from the results and analysis of the discussion. This research resulted in the formulation of vision, mission, goals, and strategies for UNAJNI, a new organizational structure design that was adjusted to the needs of the institution as well as job analysis and job specifications from all positions in UNJANI.

From the f test it can be concluded that the job description (X1) and the job specification (X2) simultaneously have a significant effect on employee performance (Y) since the f value is greater than the f table (47.844 > 3.13). The influence of the job description variable (X1) on employee performance (Y) showed that t count is greater than t table, that is, 4.733 > 1.99 and the Significance value is $0.000 < \alpha$ 0.05. It can be concluded that H0 is rejected, which means that the job description variable (X1) has a significant effect on the employee performance variable (Y), if the other independent variable is constant. The influence of the job specification variable (X2) on employee performance (Y) has a t count greater than the t table, that is, 3.649 > 1.99 and a Significance value of $0.001 < \alpha$ 0.05. It can be concluded that H0 is rejected, which means job specification (X2) has a significant effect on employee performance (Y), if the other independent variable is constant.

The support or contribution of job descriptions (X1) and job specifications (X2) to UNJANI's teaching staff performance is 58%, while the remaining 42% is influenced by other variables that are not included in this research model.

REFERENCES

Cascio, W. F. & Aguinis, H. 2005. Applied psychology in human resource management.

Daft, R. L. 2001. Manajemen: Edisi Kelima. Jakarta: Erlangga.

Daft, R. L. 2010. Era Baru Manajemen, Edisi 9, Penerbit: Salemba Empat, Jakarta.

Mangkuprawira, S. 2003. Manajemen Sumber Daya Manusia Strategik. PT Ghalia Indonesia, Jakarta.

Marshall & Rossman. 2007. *Designing Qualitative Research*. London Sage Publication.

Milkovich, George, & Newman, J. 2007. Compensation (9International Edition: McGraw-Hill, Inc.

Mondy, R. W. 2010. Human Resource Management. Edisi Kesebelas. New Jersey:Pearson, Inc.

Rivai, H. V. 2006. Manajemen Sumber Daya Manusia untuk Perusahaan, DariTeori ke Praktik. PT Rajagrafindo Persada, Jakarta.

Robbins, S. P. 2003. Essentials of Organization Behavior (7 ed.). International Edition: McGraw-Hill, Inc.

Umar, R. 2010 Compressive Sampling meets Information Theory.

Widjaja, T. 2012. Measuring the Benefits of E-Procurement System in Asia Pulp And Paper (Doctoral Dissertation, Binus).

Yin, R. K. 2003. Case Study Research: Design and Methods (3rd ed.). London and New Delhi: Sage Publications.

Yin, R. K. 2013. Studi Kasus: Desain & Metode. (Mudzakir, Trans). Boston:Massachusetts Institute of Technology. (Original work diterbitkan tahun1987).

Zikmund, W. G., et al. 2010. Business Research Methods (eight edition). South Western, USA: Cengage Learning

Advances in Business, Management and Entrepreneurship – Hurriyati et al. (Eds)
© 2021 Taylor & Francis Group, London, ISBN 978-0-367-67471-7

Do creativity and innovation affect employee performance

J. Alisa & E. Ahman
Universitas Pendidikan Indonesia, Bandung, Jawa Barat, Indonesia

ABSTRACT: In industry 4.0 era, creativity and innovation are important to reach a company's purpose. This research was driven by the non-optimal employee performance which might be affected by creativity and innovation. This research investigates how creativity and innovation can affect the employees' performance at Tridaya Course Bandung. The type of the research is descriptive involving 100 employees as its population with the sample of 50 respondents by using multiple regression analysis technique. It was found from the F test that the creativity and innovation simultaneously influenced employee performance. However, the T-test showed that creativity had a partial effect on employees' performance, and innovation had no partial effect on the employees' performance. Hence, it can be concluded that creativity can affect employee performance and innovation doesn't affect employee performance.

1 INTRODUCTION

The employee's low performance in this competitive era is still a major issue in management human resources. An effective performance of the employee is essential to be improved for the organization to achieve its goals (Noe 1996).

The employees' performance problems is faced by the companies in various sectors, such as banking, entertainment, hotel, health institutions, until small companies including education (Inuwa 2016, Jha & Mishra 2015; Khalaf et al. 2016; Salman & Hassan 2016)

It has been realized that employees must develop unique dynamic characteristics that empower their competitive advantage to survive in a changing market environment. Employees should focus on exploiting their human resources, especially on Employees Performance (EP), as a source of strategic excellence (Wright et al. 2001). Employee performance is still a concern because performance is at the core of the problems and challenges in human resource management of all organizations (Khan et al. 2011; Schaefer et al. 2015). This has always been a problem as it was stated by Jablonsk et al. (1972) that reinforcement techniques can be used to modify behavior and improve employee performance in the organization.

According to the Partner-Lawyer model by Donnelly et al. (1991) individual performance is influenced essentially by many factors, such as (1) expectations for rewards; (2) motivation; (3) capability needs and characteristics; (4) perception of a given task; (5) internal and external rewards; (6) perception level of remuneration and job satisfaction (7) internal and external rewards; and (8) perception level of rewards and job satisfaction.

Innovation and creativity are intrinsically linked (Robinson & Beesley 2010) where creativity is the emergence of new ideas (Beesley & Cooper 2008) innovation requires the implementation of those creative ideas (Robinson & Beesley 2010). Thus, to develop a sound business framework, organizations must promote creative behavior among their employees (Nieves et al. 2014).

This paper aims to integrate creativity and innovation to employee performance in the education institution at Tridaya Course Bandung .

It is expected that this research can contribute to the development of knowledge about human resource management and theories of creativity, innovation and employee performance and to give a solution for company about employee perfomance through creativity and innovation.

2 LITERATURE REVIEW

2.1 Creativity

Creativity is defined as producing ideas that meet the criteria of novelty and usefulness (Amabile 1988). An idea is considered to be novel if it differs from the traditional practices in an organization or industry, and useful if its outcome is adaptable to reality (Godart et al. 2015). On the other hand, innovation comprises two processes – the generation of ideas and the selection and implementation of a useful idea (Leung et al. 2008). In other words, creativity (i.e., generating ideas) can be seen as a part of the innovation process (Erez & Nouri 2010). In particular, at the organization and country levels, innovation is the final performance outcome of the creative activities of employees (Xie & Paik 2018).

Some researches argued that creativity has special characteristics distinguishing it from other mental behaviors, specifically, relying more heavily on intrinsic motivation (Eisenberg 1999).

According to Munandar (2004), there are several dimensions of creativity: 1) Process; creativity is a process or ability that reflects fluency, flexibility, and originality in thinking, and the ability to elaborate (develop, enrich, analyze) an idea. This definition emphasizes aspects of the change process (innovation and variation). 2) Person; creative action arises from the uniqueness of the overall personality in interaction with the environment. Personality or motivation dimensions such as flexibility that can produce new works, encouragement to achieve and gain recognition so as to be able to provide benefits, tenacity in facing obstacles, and moderate risk-taking so that problems can be solved.

2.2 Innovation

Kanter (2000) defined innovation as the creation and exploitation of new ideas. Innovation also encompasses the adaptation of products or processes from outside an organization. Finally, researchers exploring innovation have explicitly recognized that idea generation is only one stage of a multistage process on which many social factors impinge by Kanter (1988). Thus, innovation is viewed here as a multistage process, with different activities and different individual behaviors necessary at each stage. This is because innovation is actually characterized by discontinuous activities rather than discrete, sequential stages (Schroeder et al. 1989). Innovation is a process that begins with a new idea and concludes with market introduction (Freeman & Engel 2007).

According to Inkeles et al. (1974), innovation has important aspects, including opportunity exploration, generativity, formative investigation, championing, and application.

2.3 Employee performance

Performance is a multi-component concept and on the fundamental level one can distinguish the process aspect of performance, that is, behavioral engagements from an expected outcome (Borman & Brush 1993; Campbell et al. 1993; Roe 1999).

Performance in the form of task performance comprises job explicit behaviors which include fundamental job responsibilities assigned as a part of job description. Task performance requires more cognitive ability and is primarily facilitated through task knowledge (requisite technical knowledge or principles to ensure job performance and having an ability to handle multiple assignments), task skill (application of technical knowledge to accomplish a task successfully without much supervision), and task habits (an innate ability to respond to assigned jobs that either facilitate or impede the performance) (Conway 1999).

According to Schuler & Jackson (2003), there are three types of performance criteria; criteria based on the personal characteristics of an employee, criteria based on behavior which focused on the work carried out, and criteria based on the result.

3 METHODS

The research methodology used was the cross-sectional method by employing a descriptive study. The method of this research is an explanatory survey which analyzed the influence of creativity and innovation of the employee performance involving Tridaya Course Bandung as the subject. This research used analysis technique of multiple linear regression. A literature study and survey by distributing questionnaires using Likert Scale (1 for disagree and 5 for agree) were employed to collect the data. The data were analyzed using statistical data processing software that is SPSS version 25.

In this research, creativity was measured by dimension's process and person. The innovation was measured by using the dimensions of opportunity exploration, generativity, formative investigation, championing, and application. Meanwhile, the employee performance was measured based on personal characteristics of an employee, based on behavior which focused on the work carried out, and based on result dimensions.

4 RESULTS AND DISCUSSION

This research was conducted in Tridaya Course Bandung with a population of 100 employees and a sample of 50 respondents with 39 women and 11 men. According to the age groups, most of them are 20–25 years old amounted to 27 respondents (54%). Based on the length of employment, most respondents have been employed for 1–2 years with a number of 17 respondents (34%).

Based on the equation model of multiple regression analyses used in this research, it can be formulated as shown:

$$Y = 3.884 + 0.362X_1 + 0.306X_2 \qquad (1)$$

Hypothesis testing was done by coefficient of correlation test, coefficient of determination, T-Test, and F test. The coefficient of correlation (R) is 0.585 which means that the creativity and innovation to employee performance is considered moderately correlated. In addition, the coefficient of determination (R^2) is 0.343, which shows that the linear relationship in this model is able to explain the employee performance for 34.3%, while 65.7% is influenced by other factors outside creativity and innovation.

The result of Fcount is 12.249 with the level of significance is 0.000. By using the level of significant of 0.05 ($\alpha = 5\%$) and the value of Ftable = 3.19, it can be concluded that Fcount = 12.249 > F-table = 3.19 and the significant value is 0.000 < 0.05. It means that there is a significant influence of creativity and innovation on employee performance.

Meanwhile, the result of tcount for creativity is 2.856, while the value on ttable = 2.021. Value on ttable used for the comparison is found at the level of significance of 0.05, which is at the confidence level of 95%. The result for creativity is tcount = 2.856 > ttable = 2.021 and the significant value is 0.006 < 0.05. It means that H0 is accepted and Ha is rejected. The result of this test can be used to declare that creativity has significant partial influence on employee performance. According to Lakoy (2015), creativity simultaneously had a significant effect on employee performance.

The result for innovation is opposite from creativity, for the tcount = 1.622 < ttable = 2.021 and the significant value is 0.112 > 0.05. It means that H0 is rejected and Ha is accepted. The result of this test can be used to declare that innovation has insignificant partial influence on the employee performance. This supports previous research from Allifiana et al. (2011) that innovation has insignificant effect to employee performance at PT. Telekominikasi Indonesia, Tbk Kandatel Joglo.

These findings reveals that creativity has a positive and significant impact on employee performance. Meanwhile, innovation has a positive and insignificant impact on employee performance.

5 CONCLUSION

Based on the results of this research using multiple linear regression, creativity has a positive and significant impact on employee performance. Meanwhile, innovation has a positive and insignificant impact on employee performance.

It is expected that this research can help other researchers to conduct a research on the effect of creativity and innovation on employee performance by applying different indicator from other theories to get a better result.

REFERENCES

Allifiana, Anna., Dafyenti, Fenny., & Utomo, Idi Setyo. 2011. Impact of creativity and innovation on employee performance at PT. Telekomunikasi Indonesia, Tbk Kandatel Joglo. Binus University.

Amabile, T.M. 1988. *A model of creativity and innovation in organizations*. Research in organizational behavior, 10, 123–167.

Beesley, L.G. A., & Cooper, C. 2008. Defining Knowledge Management (KM) activities: towards consensus. *Journal of knowledge management,* 12(3), 48–62.

Borman, W.C., Brush, D.H. 1993. *More progress toward a taxonomy of managerial performance requirements.* Human performance, 6(1), 1–21.

Campbell, J.P., McCloy, R.A., Oppler, S.H., & Sager, C.E. 1993. A theory of performance. In C.W., Schmitt, W.C.A., Borman (Eds), *personnel selection in organizations.* 35–70. *San Francisco. CA*: Jossey Bass.

Conway, J.M. 1999. Distinguishing contextual performance from task performance for managerial jobs. *Journal of applied psychology*, 84(3), 3–13.

Donnelly, JH., Gibson, JL., & Ivancevich, JM. .1991. *Fundamentals of management*. New York. U.S.A: McGraw-Hill Education Group.

Eisenberg, J. .1999. *How individualism-collectivism moderates the effects of rewards on creativity and innovation: a comparative review of practices in japan and the US*. Creativity & innovation management. Vol. 8 No. 4, pp. 251–261.

Erez, M., & R. Nouri. 2010. *Creativity: the influence of cultural, social, and work contexts*. Management and organization review 6 (3):351–370.

Freeman, J. & Engel, J. S. 2007. *Models of innovation: startups and mature corporations*. California management review, 50 (1), 94–119.

Godart, F.C., W.W. Maddux, A.V. Shipilov, & A.D Galinsky. 2015. Fashion with a Foreign Flair: Professional Experiences Abroad Facilitate the Creative Innovations of Organizations. *Academy of Management Journal,* 58 (1): 195–220.

Inkeles, Alex & Smith, David. 1974. Becoming modern: individual change in six developing countries. *London: heinemann educational.*

Inuwa, Mohammed. 2016. Job satisfaction and employee performance: an empirical approach. *The millennium university journal.*

Jablonsk, S. F., Devries, D. L., Fennessey, J., Graen, G., Peterson, S., Rowland, K., & Devries, C. 1972. *Operant conditioning principles extrapolated to the theory of management* 1. 358, 340–358.

Jha, R. & Mishra, M.K. 2015. A study of HRM and employees performance in banking sector in India. IJARIIE.

Kanter, R. M. 1988. *When a thousand flowers bloom: structural, collective, and social conditions for innovation in organization*. Research in organizational behavior, 10, 169–211.

Kanter, R. M. 2000. *A culture of innovation.* executive excellence, 17 (8).

Khalaf, S.N., Morsy, M.A., Ahmed, G.S., Ali, N.A. 2016. Impact of effective training on employee performance in hotel establishments. *Journal of faculty of tourism and hotels, Fayoum University.*

Lakoy, A.C. 2015. The effect of communication, teamwork, and creativity on the employees performance in hotel Aryaduta Manado. *Jurnal EMBA*. Vol. 3 No. 3. ISSN 2303-11.

Leung, A. K. Y., W.W. Maddux, A D. Galinsky, and C.Y. Chiu. 2008. Multicultural experience enhances creativity: the when and how. *American psychologist* 63 (3): 169–181.

Munandar, Utami. 2004. *Development of creativity of talented children,* Jakarta: Rineka Cipta.

Nieves, J., Quintana, A., & Osorio, J., 2014. *Knowledge-based resources and innovation in the hotel industry*. Int. j. hosp. manag. 38, 65–73.

Noe, R. A. 1996. Is career management related to employee development and performance? *Journal of organizational behavior,* 17(2), 119–133.

R A Khan, R. A. G., F A Khan, F. A., & M A Khan, M. A. 2011. Impact of training and development on organizational performance. *Global journal of management and business research,* 11(7), 63–69.

Robinson, R.S., & Beesley, L.G. 2010. *Linkages between creativity and intention to quit: an occupational study of chefs*. Tourism management, 31(6), 765–776.

Roe, R.A. 1999. Work performance: *A multiple regulation perspective*. In C.L., Cooper, I.T., Robertson, International review of industrial and organizational psychology (pp. 231–335). Chichester: Wiley Publishers.

Salman, W.A, & Hassan, Z. 2016. Impact of effective teamwork on employee performance. *International journal of accounting & business management* .

Schaefer, C. P., Cappelleri, J. C., Cheng, R., Cole, J. C., Guenthner, S., Fowler, J., & Mamolo, C. 2015. Health care resource use, productivity, and costs among patients with moderate to severe plaque psoriasis in the United States. *Journal of the american academy of dermatology*, 73(4), 585–593.

Schroeder, R. G., Van de Ven, A. H., Scudder, G. D., & Polley, D. 1989. The development of innovation ideas. I: Van de Ven, AH; Angle, HL; Poole, MS (red.), *Research on the management of innovation*: the minnesota studies.

Schuler, R.S dan Jackson, S.E. 2003. *Human resource management: facing the 21st century*, 6th edition. Jakarta: erlangga.

Wright, P. M., Dunford, B. B., & Snell, S. A. 2001. Human resources and the resource based view of the firm. *Journal of management*, 27(6), 701–721.

Xie, G., & Paik, Y. 2018. Cultural differences in creativity and innovation: are Asian employees truly less creative than western employees?. Asia pacific business review.

Comparison of ISO 9001:2015 and ISO 21001:2018 for implementation in educational institution

R. Aurachman
Telkom University, Bandung, Indonesia

L. Studiyanti
Universitas Trisakti, Jakarta, Indonesia

A. Febriani
IT Telkom Purwokerto, Purwokerto, Indonesia

ABSTRACT: In 2018, a new standard for educational institution management was issued, namely ISO 21001:2018. This standard has similarities with ISO 9001:2015, which will help educational institutions to implement both ISO 21001:2018 and ISO 9001:2015. An institution doesn't need large and drastic change requirements and standard certification. Organizations only need to modify the existing management system. This paper displays the comparison between ISO 9001:2015 and ISO 21001:2018. The method of semantic analysis of standard documents was employed in this research. Following the analysis, it was found that 47 sub-standards require adjustment and 14 sub-standards do not require adjustment in its implementation.

1 INTRODUCTION

ISO 9001 is widely used by various organizations with various forms and business models. ISO 9001 itself has standards that are in accordance with the pattern of operation in the production of certain goods in factories or in manufacturing. However, other organizations engaged in manufacturing also implement ISO 9001, such as in the service sector. One form of the service sector that implements ISO 9001 is the education sector. Education can include colleges, secondary schools, elementary schools, and even kindergartens. Because the ISO 9001 standard fully guides the manufacturing industry and can guide the service industry, some adjustments are needed in implementing ISO 9001 in educational organizations for the performance of management system designers.

The international standard of quality management system ISO 9001 has a potential for developing a robust management/governance system. Research results showed that an ISO 9001 certified senior high school has better students' perceived service quality than a non-ISO 9001 certified school (Sumaedi & Bakti 2011). The perceived compatibility, relative advantage, and adaptability of the ISO 9001 standard had a positive effect, meanwhile its cost and duration have a negative effect on the adoption of the ISO 9001 standard within HEIs in Lithuania, and the perceived complexity and observability of the ISO 9001 standard had no effect on its adoption (Kasperavičiūtė-Černiauskienė & Serafinas 2018). However, above all the benefits mentioned, there is a research that found that the members of the management team of the schools perceived a higher impact of the implementation of ISO 9001:2008 standards than teachers in the different dimensions evaluated (Rodríguez-Mantilla et al. 2019). Even other research examined the phenomenon of the decertification of ISO 9001 in an SME in Italy. The reason of the desertification should be a lesson learned so other organizations can prevent the same failure for other implementations. Costs related to consultancy and the certification body were no longer a difficulty for maintaining ISO 9001 and neither were misinterpretations with the external auditors or paperwork. Some of the reasons for the desertification are:

1. Internal audits were considered a problem when they were not managed with the aim of measuring performances.
2. Staff could represent a cost when dedicated only to administrative and bureaucratic activities.
3. The more relevant difficulties were top management commitment.
4. The measurability of performances.
5. The customers' current lack of interest in ISO 9001.

In obtaining a system change guidance from ISO 9001 to 21000, the process carried out was to examine standard documents. The process of the study includes:

1. Reading and observing the standard
2. Comparing the standard
3. Observing differences between the standard
4. Observing points existing in standard 21000, which do not exist in standard 9001
5. Formulating system modification form existing due to clause addition to ISO 21000

In the process of system formulation, it is assumed that there is no system component that needs to be eliminated even though there is a clause existing in ISO 9001, but not in ISO 21000. It is assumed that what is already in ISO 9001 is useful even if it does not become an additional requirement in ISO 21000. Another reason is that an organization may not want to do certification migration from ISO 9001 to 21000, but wants to add certification with ISO 21000 without having to abolish ISO 9001 certification. Certification is a form of communication between the organization, its customers, and the certification registrar. ISO 9001 is much more widely known than the 21000. There are several customers and interested parties who are still more familiar with ISO 9001 than with ISO 21000.

The comparing process is done by viewing the two standard documents side by side. Because ISO 21000 and 9001 are relatively similar in structure, coupled with an HLS (High Level Structure), which positions each clause in a row, this process easily finds differences and additions between the two standards. These additions are relatively clearly portrayed through standard sentences that are expressed differently. The standard sentence that contains the same content is not significantly changed compared to ISO 9001:2015. This causes the process to be effective in identifying changes in standards and clauses.

The FGD process was carried out with the aim of agreeing on the indicators and target values of each of these indicators. The negotiation and bargaining process was carried out so that a mutual agreement could be obtained. The agreement was determined for the wellness of the university, and at the same time was in accordance with the ability of the recipient of the workload. The indicators of performances were determined based on a strategic plan that has been approved by the leadership of the university. The university has a vision and long-term plan that is outlined in the form of annual plans and annual targets.

The results of the comparison are shown in Table 1 (in below). The first column describes the chapters and sub-chapters of ISO 21000. The second column explains what needs to be followed up by organizations that want to apply ISO 21000 if they have applied ISO 9001 before.

In carrying out the implementation of ISO 21001, several changes are needed for the organization. The change is needed so that the organization is able to meet existing requirements. However, if the organization has implemented ISO 9001:2015, no effort is needed for organizations that have not implemented ISO 9001:2015 to apply ISO 21001. There are several similarities in the clause between ISO 9001:2015 and ISO 21001. This study seeks to propose what changes need to be made by ISO 9001:2015 certified organizations in order to implement ISO 21001:2018 as well. These changes are designed as simple as possible so as to reduce the difficulty that the organization has in making changes. In addition, it mainly reduces the level of hassles from implementing these changes, which certainly changes many ways of working. It was proven after being mapped that most clauses can be fulfilled through the modification and enrichment of existing systems.

Other research can be carried out to improve the awareness about the implementation of ISO 21000 as has been done for ISO 9001. For example, calculating the cost of adopting the standard (Kartikasari 2018) measuring the contribution of increasing institutional quality (Celik & Olcer 2018), compiling a balanced score card based on the standard (Ju et al. 2014), finding the key attribute for the successful implementation of the standard (Sukwadi & Ching-Chow 2012), examining the influence and role of the organizational committee in applying the standard (Koloor & Rahimi 2018), discussing the effect of a standard on the teaching methods used (Ishihara & Kazuyoshi 2014), investigating the role of standards in strengthening government participation in organizations (Chaudhry 2011), and many other developments that could be done.

4 CONCLUSION

Several articles and clauses of ISO 21001 strengthen the weak points existing in ISO 9001:2015. Some clauses mandate new procedures. The organization needs to add no less than 5 new procedures if it wants to meet ISO 21001. Some of the new procedures include security audits procedures, managing the learning environment procedures, admission procedures, and complaint handling procedures. Then, it

Table 1. Additional change from ISO 9001 to ISO 21000.

Section	Change aspect
4.1 Understanding the Organization and Its Context	–
4.2 Understanding the needs and expectations of stakeholders	Interested parties should include Learners, other beneficiaries, dan staff dari organisasi
4.3 Determining the scope of the EOMS	There is a requirement that EOMS include all service received by student
4.4 Management System	–
5.1.1 General	Additional poin of job description for leader which relevant to EOMS pada poin k, l, m, dan s
5.1.2 Customer Focus	5.1.2 Focus on Learners and other beneficiaries. eliminate point c at ISO 9001 about enhancing customer focus 5.1.3 Additional Requirement for Special Needs Education
5.2.1 Developing Policy	There is an additional point about a)support the educational organization mission and vision, f)takes into account relevant educational, scientific and technical development, g) includes a commitment to satisfy the organization's social responsibility, h)describes and includes a commitment towards managing intellectual property, i) considers the needs and expectations of relevant interested parties
5.2.2 Communicating the quality policy	Add information about Annex D
5.3 Organizational roles, responsibilities and authorities	Add point b), g, h, i, dan j
6.1 Action to Address Risk and Opportunities	–
6.2 Quality objectives and planning to achieve them	
6.3 Planning of changes	Add Point e)
7.1.1 General	there is an explanation about the purpose of resource management and need to explain what resources are needed
7.1.2 People	It is necessary to explain in detail the human resource requirements needed and communicate these needs to third parties
7.1.3 Infrastructure	7.1.3.2 Organization needs to review of facility security and dimensions of facilities, according to its users 7.1.3.3. There needs to be a more detailed explanation of the facilities and it's designation in Educational activities
7.1.4 Environment for the operation of educational process	There is a more detailed explanation of what environmental factors are meant
7.1.5 Monitoring and measuring resources	There is a detailed explanation about education delivery
7.1.6 Organizational Knowledge	There is an encouragement for academics to share knowledge 7.1.6.2 Learning Resources are new requirements. There are special requirements regarding the regulation of learning resources
7.2 Competence	7.2.1 There are additional points c) and e) about evaluating staff competency and periodically increasing the competency 7.2.2 Additional Requirement for Special Needs Education is a completely a new requirement. Organizations need to implement development of staff competencies that interact directly with special needs education
7.3 Awareness	Policy is made more broadly, not only related to quality but also related to education policy
7.4 Communication	There is a new clause number 7.4.2 regarding communication purposes and 7.4.3 regarding communication arrangement
7.5.1. General (Documented Information)	Explained what are the forms of the documented information
7.5.2. Creating and Updating (Documented Information)	Must be accessible to people with special needs
7.5.3. Control of Documented Information	There are additional requirements in points b), f), and g) NOTE 2 requires that legality must be ensured when media changes are made

(Continued)

Table 1. (Continued)

Section	Change aspect
8.1 Operational Planning and Control	There is clause 8.1.2 which explains in detail the processes that need to be planned and controlled related to learning process there is clause 8.1.3 which regulates the process modification for participants with special needs
8.2 Requirement for the Educational Products and Services	8.2.1. There are needs for future needs analysis as the basis for determining educational service requirements, Need to consider international demand and development, labor market, research, and HSE Detailed in 8.2.2 that the organization ensures that the customer understands all information related to the education services provided
8.3 Design and Development of the Educational Products and Services	On 8.3.2 there are additional arrangements for controlling the development of educational services, especially in point l), m), n)
8.3.4 Design and Development Controls	8.3.4.2 Educational Service Design and Developmental Controls. There are several things that need to be explained in the development of educational services including prerequisite, student characteristics, and others 8.3.4.3 Curriculum Design and Development Controls. Mengatur perihal Learning Outcomes, Learning Activities, dan lain-lain 8.3.4.4 Summative Assessment Design and Development Controls. It regulates the design control method of assessment for students
8.3.5.Design and Development Output	Organizations need to regulate that the results of designing educational services are retained as documented information
8.3.6.Design and Development Changes	–
8.4 Control of Externally Provided processes, products and services	–
8.5.1. Control of The Educational Producst and Services	8.5.1.2 There are requirement about the Admission Process 8.5.1.3 Requirement about what processes must be present in learning process 8.5.1.4 Requirement about Assessment Process 8.5.1.5 Organization should regulating about recognizing assessments that have been carried out outside the institution 8.5.1.6 Requirement about arrangements for learning for students with special needs
8.5.2 Identification and Traceability	Arrange more details about what needs to be traceable related to the education process
8.5.3 Property belonging to Interested Parties	–
8.5.4 Preservation	–
8.5.5 Protection and transparency of learner's data	Requirement for managing data student's data
8.5.6.Control of changes in the educational products and services	–
8.6 Release of the educational products and services	–
8.7 Control of the educational nonconforming outputs	–
9.1 Monitoring, measurement, analysis, and evaluation	9.1.1. Additional requirements about the need to determine acceptance criteria from measurements 9.1.1. There is an opportunity for each individual to see the measurement of his work results as a form of reflection 9.1.2. must consider positive input and negative input 9.1.2.2. There are guidelines in managing handling and appeals 9.1.3. A new guide to Other monitoring and measuring needs 9.1.4. A new guide about methods for monitoring, measurement, analysis and evaluation
9.2 Internal Audit	–
9.3 Management Review	–
10. Improvement	–

is proven that in implementing ISO 21001, organizations can make efficient efforts, provided that they have applied ISO 9001:2015 beforehand.

REFERENCES

Celik, Bunyamin, & Olcer., O. H. 2018. What is the contribution of ISO 9001 quality management system to educational institutions?. *International journal of academic research in business and social sciences* 8.6, 445–462.

Chaudhry, S. 2011. ISO 9001: A standard to develop a robust governance system in higher education institutions. *International journal of excellence in public sector management* 81.183, 1–15.

Ishihara, Masahiko, Makoto, N., & Kazuyoshi, I. 2014. Development of educational program for production managers based on a symbiotic competition with abc-g network. *Industrial engineering and management systems* 13.3, 258–266.

Ju, Y., Sohn, S., Ahn, J., & Choi, J. 2014. Balanced scorecard based performance analysis of accreditation for engineering education. *Industrial engineering and management systems*, 13(1), 67–86.

Kartikasari, D. 2018. Cost-benefit analysis on ISO 9001 certification and higher education accreditation. *Cakrawala Pendidikan*.

Kasperavičiūtė-Černiauskienė, Ramunė, & Serafinas, D. 2018. The adoption of ISO 9001 standard within higher education institutions in Lithuania: Innovation diffusion approach. *Total quality management & business excellence* 29.1–2, 74–93.

Koloor & Rahimi, H. 2018. Studying the role of organizational climate and organizational commitment in predicting service recovery (Case study: Free zone border market, maku, west azerbaijan). *Industrial engineering & management systems* 17.4, 642–652.

Rodríguez-Mantilla, Miguel, J., Fernández-Cruz, F. J., & Fernández-Díaz, M. J. 2019. Comparative analysis between management team and teachers on the impact of ISO 9001 standards in educational centres. *International journal of quality and service sciences*.

Sukwadi, Ronald, & Ching-Chow Yang. 2012. Determining critical service attributes and appropriate improvement actions in indonesian heis. *Industrial engineering and management systems* 11.3, 241–254.

Sumaedi, Sik, & Bakti, G. M. 2011. The students' perceived quality comparison of ISO 9001 and non-ISO 9001 certified school: An empirical evaluation. *International journal of engineering & technology IJETIJENS* 11.1, 104–108.

Advances in Business, Management and Entrepreneurship – Hurriyati et al. (Eds)
© 2021 Taylor & Francis Group, London, ISBN 978-0-367-67471-7

Problem and challenge in university goal setting for performance management system in Indonesia

R. Aurachman
Telkom University, Bandung, Indonesia

D.B. Baskara
Institut Teknologi Telkom, Surabaya, Indonesia

A. Febriani
IT Telkom Purwokerto, Purwokerto, Indoneisa

ABSTRACT: One of the processes in the performance management system is the goal setting and target setting. There are several studies that examined the development of performance management systems and goal settings. The results of the research converged on the conclusions about the importance of employee participation in target setting. On the other hand, organizations need competitive targets to keep growing. Through 23 FGDs with 69 respondents as well as simulations to determine the target of 400 performance indicators, the proposed method for determining the target size between several work units was obtained. The proposal carries the principle of fairness and at the same time supports the progress of the organization.

1 INTRODUCTION

One important stage in the performance management system process is determining achievement targets. Goal setting is the key to individual and organizational effectiveness (Hughes & Hughes 1965). Goal setting is a problem of defining relationship between an organization and its environment. Goal setting behavior is sometime purposive but not necessarily rational, sometime it can be determined by accident (Thompson & McEwen 1958). There is a research proving that setting clear standards and opening participation in goal setting increased the level of satisfaction of the relevant employees (Arvey et al. 1976). Other studies proved that clear targets, evaluations, and giving positive feedback increased productivity (White et al. 1977). Through goal setting, self-regulation can be formed (Latham & Locke, "Self-regulation through goal setting." 1991), and motivation can be managed (Locke 2000) (Latham & Locke 2007). However, in other studies goal setting and feedback did not directly increase the commitment of human resources to the organization, but rather affected commitment indirectly through their effects on employee perceptions of support from the organization (Hutchison et al. 1996). Another paper discussed the importance of goal acceptance in moderating goal setting effects and showed how workers' acceptance of goals could be influenced at various stages of the progression from goal setting to goal attainment (Erez et al. 1983). On the other hand, some negative effects

of goal setting also need to be taken into consideration. Side effects associated with goal setting include a narrow focus that neglects non-goal areas, distorted risk preferences, a rise in unethical behavior, inhibited learning, corrosion of organizational culture, and reduced intrinsic motivation. Rather than dispensing goal setting as a benign, over-the-counter treatment for motivation, managers and scholars need to conceptualize goal setting as a prescription-strength medication that requires careful dosing, consideration of harmful side effects, and close supervision. The practice of setting goals should be done under careful observation (Ordóñez et al. 2009).

In the process of discussion and negotiation of target decisions, specifically at the university, there are several lessons that can be taken. The situation and conditions in discussions and negotiations are strongly influenced by the type of organizational culture, the culture of Indonesian society, and the designed system. Certainly, the impressions, messages, and images that arise from this discussion process can be the insights for the members of the organization on the proposed system. Therefore, this study is expected to provide insight and knowledge for the performance of management system designers.

2 METHODS

In obtaining general guidelines for the process of determining performance indicators, especially at

Table 1. Existing target.

Variable	Multiplicator	FTE	FIF	FRI	FEB	FKB	FIT	FIK	Total	Publication-lecturer ratio
Lecturer		20	15	20	10	5	25	20	115	1
Publication target	2	40	30	40	20	10	50	40	230	2
Achievement		80	30	40	25	15	55	45	290	2.521
Next year publication target	2.522	51	38	51	26	13	64	51	294	2.56

universities in Indonesia, a Focus Group Discussion (FGD) was conducted. The FGD was conducted with 23 groups where each group consisted of 2 to 3 people. Therefore, there were approximately 69 respondents in that intensive discussion.

Each FGD represented a work unit at the universities. One work unit covered one directorate or faculty One on the Table 1 directorate consisted of two until five sections. Each work unit had 20 to 30 performance indicators.

The FGD process was carried out with the aim of agreeing on the indicators and target values of each indicator. The negotiation and bargaining process was carried out so that a mutual agreement could be obtained. The agreement was determined for the wellness of the university, and at the same time was in accordance with the ability of the recipient of the workload. The indicators of performances were determined based on a strategic plan that has been approved by the leaders of the university. The university has a vision and long-term plan which is outlined in the form of annual plans and annual targets.

3 RESULTS AND DISCUSSION

The division of targets between faculties needs to be selected whether it is based on ability or is evenly divided based on the number of lecturers and students. If it turns out that the target distribution is conducted based on ability, then the performance target owner must accept when there is an increase in the number of performance target due to good achievement last year. However, this seems unfair. The party that achieved the target seemed to be convicted with a heavier target in the next period. In addition, those whose achievements were poor in the previous period get a gift in a form of a small target in the next period. This also causes the organization to be unmotivated to achieve high targets and would probably choose to play safe (low achievement) so that achievement in the next period is not too heavy. This is not healthy for the organization. Therefore, compared to these two things, it is considered more appropriate to take things proportionally based on the number of lecturers and students as multipliers. This means that no matter what achievements they had in the previous period, the number of students and lecturers will determine the size of the target in each period. The multiplier may change or

increase. For example, if the multiplier is twice the number of lecturers, regardless of the achievement in the past period, in the future the target achievement must be multiplied twice by the number of lecturers. These multipliers can increase in the context of organizational growth, for example, increasing to 3 times by the number of lecturers. For instance, the target of one faculty publication is twice by the number of lecturers this year. Then, in the coming year it is targeted to be three times the number of lecturers.

The increase of multipliers can be based on several things, such as the accumulation of achievements from all faculties in the previous period. For example, if targeted at the same level last year, the achievement of each faculty was two times by the number of lecturers. However, it turned out that there were faculties that excelled in an achievement accumulatively. In the following illustration table, the achievement of FTE reached 80. This was due to the achievement of all universities, which is 290, that has a multiplier ratio of 2.5. This ratio was greater than the previous ratio of 2.

In the following year, the ratio was increased by 2.5 from the number of lecturers for all faculties and not just for the FTE. It did not mean that the high achievers were burdened with bigger targets even though it showed that a faculty was guilty by increasing achievement, but at the end all faculties were burdened by increasing targets. This needs to be anticipated with the multiplier adjustment described in the next paragraph. The amount of multiplier is not necessarily fully adopted. It could be that even though the achievement in the previous period was 2.5, what was adopted was only 2.3 on the basis of a mutual agreement. The risk is that the unit will reduce its achievement so that large targets in the next period will be minimized. This is because of the impact of increasing achievement that the increase in targets is not felt by only the related faculty units, but also is divided into other units. It is also possible to adjust targets on the basis of mutual agreement, eliminate fear from increasing achievement.

A greater achievement in one faculty causes an increase in targets in other faculties. This can be the basis for rejection of new targets. This can be refuted by an argument that each faculty is assumed to share knowledge with each other so that the target increase is still acceptable. The existence of faculties that refuse to increase the target also shows that the faculty is a burden for other universities in pursuing the common goal adjust target in Table 2.

Table 2. Adjusted target.

Variable	Multiplicator	FTE	FIF	FRI	FEB	FKB	FIT	FIK	Total	Publication-lecturer ratio
lecturer		20	15	20	10	5	25	20	115	1
publication target	2	40	30	40	20	10	50	40	230	2
achievement		80	30	40	25	15	55	45	290	2.52
next year publication target	2.52	51	38	51	26	13	64	51	294	2.55
adjustment		3	2	2	2	2	−13	2	0	
adjusted target		54	40	53	28	15	51	53	294	
adjusted ratio		1.06	1.05	1.04	1.08	1.154	0.797	1.039	7.22	

Table 3. Adjusted and harmonized target.

Variable	Multiplicator	FTE	FIF	FRI	FEB	FKB	FIT	FIK	Total	Publication-Lecturer Ratio
Lecturer		20	15	20	10	5	25	20	115	1
Publication Target										
Publication Target	2	40	30	40	20	10	50	40	230	2
Achievement		80	30	40	25	15	55	45	290	2.522
Next year Publication Target	2.52	51	38	51	26	13	64	51	294	2.557
Adjustment		3	2	2	2	2	−13	2	0	
Adjusted Target		54	40	53	28	15	51	53	294	
Adjusted Ratio		1.06	1.05	1.04	1.08	1.154	0.797	1.039	7.22	
Intellectual Property Target										
Next year Publication Target	1.3	26	20	26	13	7	33	26	151	0.514
Adjustment		−1	−1	−1	−1	−2	7	−1	0	
Adjusted Target		25	19	25	12	5	40	25	151	
Adjusted Ratio		0.96	0.95	0.96	0.92	0.714	1.212	0.962	6.68	
Average of Adjusted Ratio										
Average of Adjusted Ratio		1.01	1	1	1	0.9	1.01	1	6.95	

Adjusting multipliers can also be done in order to see the weaknesses and potential of each faculty. Using the previous example, it can be seen that each faculty was required to reach the publication target of 2.5 times by the number of lecturers in the faculty. However, it is possible for a faculty not to have the ability that is in accordance with that target. Then, adjustments can be made so that the target is not a hundred percent in accordance to the standard.

As an example, in the table it can be seen that both the FTE and FIK target was 51. However, adjustments were reduced by 13 targets. It was therefore could be covered by other faculties where other faculties increased their target by 2. Then, there was a kind of cooperation between the faculties which bore one another's burdens. Furthermore, the calculation of the ratio between the initial target and the target of the adjustment was done, which showed how much the target increased compared to the amount of resources owned. In the illustration that can be seen in the table, it shows that the biggest burden is actually borne by FKB which is equal to 1,154, even though the FTE gets an additional target of 3. This is because of the small number of FKB lecturers. The increase in the target of just two people gives a burden for each FKB lecturer which will be slightly greater than the lecturers in other faculties.

Possible trading of the number of multipliers between the comparison units. There are faculties that excel in aspects of scientific publications. On one hand, there are other faculties that excel in producing intellectual property. It can be that the target of publications of a faculty is reduced and the reduction is borne by other faculties. However, the faculty that has experienced a reduction in the target, as compensation, accepts the target more on other factors. The illustration can be seen in the table. FIT has reduced in Table 3.

The number of scientific publications by 13 or by a ratio of 0.79. Therefore, as compensation, the intellectual property target of +7 is charged to FIT, which means the ratio is 1.21. The excess target of intellectual property is then used as a deduction for other faculties, which is −1. Furthermore, the ratio of the target and target of the adjustment results can be calculated. Each faculty will have two adjustment ratios; adjustment ratio of publication target and intellectual property targets. The average of the two is calculated to see whether the adjustment load is evenly distributed.

In the average line of adjustment ratio, it can be seen that the number is close to 1, which means that equity of target has begun. However, there are still two faculties whose average ratio reaches more than 1, that is 1.01. On the other hand, FIT has an implied average adjustment ratio of 0.9 which shows that the accumulative FIT was alleviated in achieving its targets. This is difficult to avoid but can be minimized to the extent that can be agreed upon by all faculties. The ideal thing is achieved when all the average adjustment ratios are 1.

A more appropriate recommendation is to accumulate all achievements in the following period.

Each unit should not be given a target that seems to be only a collector. This risks cause a perception of injustice between work units. As an illustration, if a faculty is given a target for scientific publications, the directorate of research and community service should not only have the target of accumulated publications from each faculty. If the total target of publications from all faculties is 294, the target of the PPM directorate then is not just to reach 294 publications. They can reach the target by only waiting.

Targets that support the process of scientific publications can be charged to PPM so that faculty can reach the targeted number. The equated faculties and PPM's targets can cause PPM to simply take the chance; however, the faculty cannot do the same thing. On the other hand, the equalization of targets is acceptable if indeed the PPM has a role in increasing publications, for example, allocating incentive funds for scientific publications. Targeting can also be done by burdening the faculty with the target of implementing the process while the supporting directorate is charged with achieving the target. Another example is that the faculty focuses on strengthening human resources while the achievement targets are charged to the directorate. The directorate can also focus on providing opportunities and information for researchers to reach.

The solution proposed through this research is based on a problem that occurs in the organization. Each organization has characteristics that can be different one another. Subsequent research can examine how the effects of applying the solutions produced by this research. Another form of development of this research is the application of performance indicators for individuals and not work units. Survey and formulation can be done by considering more types of organizations and adequate samples.

4 CONCLUSION

Through simulation and 23 sessions of FGD with 69 respondents to determine the target of 400 performance indicators, a proposed method for determining the target size between several work units was obtained. The proposal carries the principle of fairness and at the same time supports the progress of the organization.

This method requires several weaknesses that need to be anticipated. One disadvantage of this method is the complexity created when the number of work units increases. This method also still provides disincentives for achieving high performance.

REFERENCES

Arvey, R., Dewhirst, H., & Bowling, J. (1976). "Relationships between goal clarity, participation in goal setting, and personality characteristics on job satisfaction in a scientific organization.". Journal of Applied Psychology 61.1, 103.

Erez, Miriam, & Kanfer, F. (1983). "The role of goal acceptance in goal setting and task performance.". Academy of management review 8.3, 454–463.

Hughes, C., & Hughes, C. (1965). Goal setting: Key to individual and organizational effectiveness. New York, NY: American Management Association.

Hutchison, Steven, & Garstka, M. (1996). "Sources of Perceived Organizational Support: Goal Setting and Feedback 1." Journal of Applied Social Psychology 26.15, 1351–1366.

Latham, G., & Locke, E. (1991). "Self-regulation through goal setting." Organizational behavior and human decision processes 50.2 , 212–247.

Latham, G., & Locke, E. (2007). "New developments in and directions for goal-setting research." European Psychologist 12.4, 290–300.

Locke, E. (2000). Motivation by goal setting. Handbook of organizational behavior, 43–56.

Ordóñez, L., Schweitzer, M., Galinsky, A., & Bazerman, M. (2009). Goals gone wild: The systematic side effects of overprescribing goal setting. Academy of Management Perspectives, 23(1), 6–16.

Thompson, J., & McEwen, W. (1958). "Organizational goals and environment: Goal-setting as an interaction process.". American Sociological Review 23.1, 23–31.

White, S., Mitchell, T., & Bell, C. (1977). "Goal setting, evaluation apprehension, and social cues as determinants of job performance and job satisfaction in a simulated organization." Journal of applied psychology 62.6, 665.

Advances in Business, Management and Entrepreneurship – Hurriyati et al. (Eds)
© 2021 Taylor & Francis Group, London, ISBN 978-0-367-67471-7

Organizational culture and small and medium-sized enterprises performance: An empirical investigation in the Jakarta context

C. Hakim & Disman
Universitas Pendidikan Indonesia, Bandung, Indonesia

ABSTRACT: This study investigated the postulated relationship between organizational culture and SME's performance. The necessity of the assessment of this relationship is driven on the basis of the past literature. For the present inquiry, the survey method was employed with a total number of 265 questionnaires. The population was drawn using stratified random sampling technique. The Smart-PLS 3.0 was used for data analysis due to its increasing popularity in presenting authentic calculations. The reporting of the results was based on Smart-PLS standards followed by a two-step approach: first, the assessment of reliability and validity was conducted using measurement model, and secondly, assessment of hypothesized relationship was done using structural model. This study underlined that organizational culture had significant relationship with SME's performance

1 INTRODUCTION

Small and Medium Enterprises (SME) sector in Indonesia has been developing rapidly. This sector is a part of the national economy which has a strong position in the national economy development. SME sector has the ability to widen job opportunities, provide wide economic services to public, take role in the process of equalizing and improving people income, drive economic growth and take roles in realizing national stability in general and especially economic stability (Ardiana et al. 2010). In addition to taking role in the economic growth and absorbing employment, SME also takes role in distributing the development results. When economic crisis happened in Indonesia, SME sector had proven tough in overcoming the crisis by Kristiyanti (2012).

The performance of SMEs becomes a tool to assess and evaluate the success of SME's goals. Performance is defined as a description of level or achievement of the execution process on an activity, program, or policy in order to realize what have been formulated in the organization's strategic scheme namely targets, goals, vision and missions to build a good organization by Bastian (2001). The level of SME's performance can be seen from how far SME can achieve its predetermined target, goals, vision, and missions by Mahsun (2006).

The strong performance of SME must be supported by strong organizational culture. Strong culture means all personnel have one common perception in achieving the organizational goals. Common perception is based on the same values which are believed, the highly upheld norms, and the obeyed behavior patterns (Darsono & Ashari 2010). Organizational culture is

the "soul" of an organization because the philosophy, vision, and missions of organization which reside there will become an important strength for the company to compete. Organization culture is believed as an important variable in the field of organizational behavior (Kilman et al. 1985; Ouchi & Wilkins 1985; Schein 1990). Attention paid to organizational behavior is due to its great impact on performance, in this case the performance of SME. Theorists also consider it suitable for developing organizational procedures (Deal & Kennedy 1982; Jarnagin & Slocum 2007), but in Indonesia with its eastern culture, it must be tested again whether organizational culture influences the performance of SMEs.

2 LITERATURE REVIEW

2.1 *Organizational culture*

Organizational culture is a set of behavior, feeling, and psychological frame which is very deeply internalized and commonly shared by the members of organization (Osborne & Plastrik 2000). Organizational culture as the basic assumption is found and developed by a group of people while they are learning to solve problems, to adapt themselves to external environment, and to integrate with internal environment by Edgar (1995). Organizational culture is a pattern of trust and expectation embraced by organization's members. Trust and expectation result in the values which strongly build the behavior of individuals and groups of organization's members (Schwartz & Davis 1981). Organizational culture indicates a unique configuration of norms, values, trust, and behavior that characterize the group and individuals when they cooperate to finish their jobs (Eldride & Crombie 1974).

2.2 SME's performance

SME's performance is the totality of performance by an SME for the achievement of its goals. SME's performance can be seen from how far SME can achieve its goals based on the previous plan by Surjadi 2009). SME's performance is the description of the achievement level of job implementation in an SME to realize the organization's target, goals, missions and vision by Bastian (2001).

2.3 Relationship between organizational culture and SME's performance

Some literatures proved the correlation between organizational culture and organization's performance (Hansen & Wernerfelt 1989; Lewis 2002; Lee & Yu 2004). On the other side, organizational culture is found to be the predictor of short-term performance (Denison 1984; Gordon & DiTomaso 1992). Furthermore, Lee & Yu (2004) reported a stronger correlation between organizational culture and organization's performance in manufacturing companies than in service companies.

Scholz (1987) stated that organizational culture had a significant role as the source of competitive for an organization. Likewise, Van der Post et al. (1998) in their research found that organizational culture had a positive correlation with the financial performance of organization. This correlation is also supported by organization's strategy (Choe 1993; Schwartz & Davis 1981; Scholtz 1987). Pool (2000) stated that there was a positive correlation between organizational culture and job stressor. This proved that organizational culture had a vital impact on performance.

From those literatures and research, it can be concluded that organizational culture is one of the key factors of organization's performance. However, empirical research done to investigate the relationship of SME's performance is still limited, especially SME's performance in Jakarta. This study aims to examine the possible relationship between organizational culture and SME's performance in Jakarta. Therefore, the following hypothesis is proposed:

H1: There is a positive correlation between organizational culture and SME's performance.

3 METHODS

3.1 Participants and procedure

This research used survey method to collect data, where the respondents were from the management or owners of SMEs who have had business legality in Jakarta, Indonesia. Since the total population was big and spread in five municipalities and one regency,

as many as 300 questionnaires were taken as sample (Krejcie & Morgan 1970). However, to increase the probability, 500 questionnaires were sent to SME managers in a random way from the available list. As a result, 265 questionnaires were received in total.

3.2 Measures

In order to measure the organizational culture, an 18-item scale taken from Denison (2000) was used. Whereas to measure the SME's performance, a 4-item scale was adopted from Deshpandé et al. (1993) and a 3-item scale was adopted from (Jaworski & Kohli 1993).

4 RESULT AND DISCUSSION

4.1 Measurement model results

Before doing analysis, some tests of linearity, normality, and multicollinearity were examined. After this assumption was fulfilled, a path modeling of partial least square (PLS) was used by implementing Smart PLS 3. To assess the psychometric nature of the adopted scales that have been adopted in this research, the reliability of individual item, the reliability of internal consistence, and the validity of discriminant were ensured.

First, the reliability of individual item was examined using outer loading from the measurement of each construction (Hair et al. 2014). As a practical rule, the item with content more than 0.50 was retained (Barclay et al. 1995; Chin 1998).

Second, to ensure the reliability of internal consistence of coefficient size, composite reliability was used. The interpretation on the reliability of internal consistence was based on the practical rule that the coefficient of composite reliability must be at least 0.70 or above 0.70 (Bagozzi & Yi 1988). The coefficients of composite reliability of latent construct reported in Table 1 are 0.870 and 0.836 exceeding the minimum level that can be accepted, that is 0.70 (Bagozzi & Yi 1988). Thus, the reliability of

Table 1. Results of measurement model.

Latent Variable	Items	Loadings	AVE	CR	ALPHA
SME's Performance	UP1	0.728	0.526	0.870	0.821
	UP3	0.705			
	UP5	0.764			
	UP6	0.750			
	UP7	0.728			
	UP8	0.652			
Organizational culture	OC1	0.711	0.506	0.836	0.757
	OC2	0.665			
	OC3	0.764			
	OC4	0.780			
	OC5	0.650			

internal consistence of the measures for this research is considered as sufficient.

Third, based on Fornell & Bookstein (1982) recommendation, AVE was used to ensure the validity of discriminant. It was done by comparing the correlation between latent construct and the root-square of extracted average variances where he thinks AVE root-square must be bigger than the correlation of latent construct in Table 1.

Table 2 shows the comparison of correlation between latent construct and AVE root-square (the value given in bold), indicating that all AVE root-squares are bigger than the correlation of latent construct, which indicates a sufficient discriminant validity in Table 2.

4.2 Structural model results

To assess the significance of path coefficient, bootstrap standard procedures were implemented using 5.000 bootstrap samples and 265 cases (Hair et al. 2012; Henseler et al. 2009) to result in standard error and t-statistics. As a technique, bootstraps results in estimated standard error (Tenenhaus et al. 2005). Table 3 describes the relationship between organizational culture and SME's performance. The path coefficient of organizational culture to the performance is 0.541 (value-t = 10.723, p < 0.000). Thus, H1 is accepted in Table 3.

Table 4 describes the value of R2 from the SME performance endogenous construct that is 0.27. This indicates that organizational culture can explain 27% of the variances of SME's performance in Table 4:

The aim of this study was at examining the relationship between organizational culture and SME's performance. The result of this research confirmed the correlation between organizational culture and SME's performance. This is in line with the findings of previous researches (Deshpande et al. 1993; Marcoulides & Heck 1993). The result also indicated

Table 2. Correlation and discriminant validity.

Latent variables	1	2
SME's Performance	0.727	
Organizational culture	0.539	0.713

Table 3. Path coefficients and hypothesis testing.

Hypothesis	Relations	Beta	SE	t-statistics	p-value
H1	OC>UP	0.541	0.050	10.723	0.000

Table 4. Variance explained in the endogenous variable.

Latent variable	Variance explained
SME's Performance	0.27

that organizational culture was an important factor that influenced the performance of SME (Deshpande et al. 1993). From this research data, it can be seen that organizational culture influenced performance.

4.3 Implications for theory and practice

This study has given a theoretical implication by providing additional empirical evidences in the domain of resources-based view theory stating that organizational culture is a unique and inimitable ability of an organization as well as an important factor in explaining how an organization works and how they can lead themselves so as to achieve the company's goals in effective ways (Barney 1986; Hall 1993; Peteraf 1993). The result indicated that organizational culture was a potential predictor of SME's performance, so the company must consider to develop a better organizational culture to improve the performance in order to get competitive advantages.

4.4 Limitations and future research direction

Based on the previous theory, it is said that with a good organizational culture, an SME will result in a better organization's performance. The result of this research indicated a positive correlation between organizational culture and business performance. In this research, there is a wider terminological gap of culture which has been operationalized and segmented to various features (Denison 1990; Wilson 2001). Therefore, further investigation on the dimension of organizational culture is necessary.

Although this research has given insights on the role of organizational culture to improve SME's performance, there are some points that can be underlined. First, the cross-sectional design adopted in the research that makes causal conclusion cannot be made. Therefore, it is recommended to study longitudinal design to find the changes that happen overtime. Second, the reported measures possibly have social defect. Although this research tried to alleviate such a problem by ensuring the anonymity and improve the scale item, but there is a possibility for the problem to happen. Therefore, it is suggested for the next research to use other strategies to assess the relationship between organizational culture and SME's performance.

As mentioned before, there are 73% variables which cannot be explained by organizational culture, which is potential to influence the performance of SME. So, further researchers can investigate other variables such as the effectiveness of leadership and knowledge management so as to get a complete research.

5 CONCLUSION

Regardless of the limitation of this research, it was found that there was a positive correlation between organizational culture and SME's performance. Thus, it is important for SMEs to implement the culture-performance relation. The SMEs which implement the culture of focusing on result will make their employees motivated to result in a better performance of SME.

REFERENCES

Bagozzi, R. P., & Yi, Y. 1988. On the evaluation of structural equation models. *Journal of the academy of marketing science*, 16(1), 74–94.

Barclay, D., Higgins, C., & Thompson, R. 1995. The partial least squares (pls) approach to causal modeling: personal computer adoption and use as an illustration. *Technology studies*, 2(2), 285–309.

Barney, J. B. 1986. Organizational culture: can it be a source of sustained competitive advantage?. *Academy of management review*, 11(3), 656–665.

Bastian, Indra. 2001. Akuntansi Sektor Publik di Indonesia, Yogyakarta: BPFE UGM.

Chin, W. W. 1998. The partial least squares approach to structural equation modeling. *Modern methods for business research*, 295(2), 295–336.

Choe, M. K. 1993. An empirical study of corporate strategy and culture in korea. *Quarterly review of economics and business*, 21(2), 73–92.

Darsono & Ashari. 2010. Pedoman praktis memahami laporan keuangan (tips bagi investor, direksi, dan pemegang saham). Yogyakarta: Penerbit Andi.

Deal, T. E., & Kennedy, A. A. 1982. Corporate cultures reading. MA: Addison-Wesley.

Denison, D. R. 1984. Bringing corporate culture to the bottom line.organizational dynamics, 13(2), 5–22.

Denison, D. R. 1990. Corporate culture and organizational effectiveness. John Wiley & Sons.

Denison, D. R. 2000. Organizational culture: can it be a key lever for driving organizational change. The international handbook of organizational culture and climate, 347–372.

Deshpande, R., Farley, J. U., & Webster Jr, F. E. 1993. Corporate culture customer orientation, and innovativeness in japanese firms: a quadrad analysis. *Journal of marketing*, 57(1).

Edgar H. Schein. 1995. The role of the founder in creating organizational culture. V8-Issue 3. JWS Inc

Fornell, C., & Bookstein, F. L. 1982. Two structural equation models: lisrel and pls applied to consumer exit-voice theory. *Journal of marketing research*, 440–452.

Gordon, G. G., & DiTomaso, N. 1992. Predicting corporate performance from organizational culture. *Journal of management studies*, 29(6), 783–798.

Hair, J. F., Hult, G. T. M., Ringle, C. M., & Sarstedt, M. 2014. A primer on partial least squares structural equation modeling (pls-sem). Thousand oaks, CA: SAGE.

Hair, J. F., Sarstedt, M., Ringle, C. M., & Mena, J. A. 2012. An assessment of the use of partial least squares structural equation modeling in marketing research. *Journal of the academy of marketing science*, 40(3), 414–433.

Hall, R. 1993. A framework linking intangible resources and capabiliites to sustainable competitive advantage. *Strategic management journal*, 14(8), 607–618.

Hansen, G. S., & Wernerfelt, B. 1989. Determinants of firm performance: the relative importance of economic and organizational factors. *Strategic management journal*, 10(5), 399–411.

Henseler, J., Ringle, C. M., & Sinkovics, R. R. 2009. The use of partial least squares path modeling in international marketing. *Advances in international marketing (AIM)*, 20, 277–320.

Jarnagin, C., & Slocum, J. 2007. Creating corporate cultures through mythopoetic leadership. *Smu cox school of business research paper series*, (07-004).

Jaworski, B. J., & Kohli, A. K. 1993. Market orientation: antecedents and consequences. *The journal of marketing*, 53–70.

Krejcie, R. V., & Morgan, D. W. 1970. Determining sample size for research activities. *Educational and psychological measurement*, 30(3), 607–610.

Kristiyanti, M. 2012. Peran strategis usaha kecil menengah (UKM) dalam pembangunan nasional. *Majalah ilmiah informatika*, 3(1), 63–89.

Lee, S. K. J., & Yu, K. 2004. Corporate culture and organizational performance. *Journal of managerial psychology*, 19(4), 340–359.

Lewis, D. 2002. Five years on–the organizational culture saga revisited. *Leadership & organization development journal*, 23(5), 280–287.

Mahsun, Mohamad. 2006. Pengukuran Kinerja Sektor Publik,. Penerbit BPFE,Yogyakarta.

Marcoulides, G. A., & Heck, R. H. 1993. Organizational culture and performance: proposing and testing a model. *Organization science,* 4(2), 209–225.

Osborne, David & Peter Plastrik. 2000. Memangkas birokrasi: lima strategi menuju pemerintahan wirausaha (*terjemahan ramelan abdul rosyid*), Jakarta: PPM.

Ouchi, W. G., & Wilkins, A. L. 1985. Organizational culture. Annual review of sociology, 11(1), 457–483.

Peteraf, M. A. 1993. The cornerstones of competitive advantage: a resource-based view. *Strategic management journal*, 14(3), 179–191.

Pool, S. W. 2000. Organizational culture and its relationship between job tension in measuring outcomes among business executives. *Journal of management development*, 19(1), 32–49.

Schein, E. H. 1990. Organisational culture, *american psychologist*, 45 (2), 109–119.

Scholz, C. 1987. Corporate culture and strategy—the problem of strategic fit.long range planning, 20(4), 78–87.

Schwartz, H. & Davis, S.M. 1981. Matching corporate culture and business strategy. *Organizational dynamics*, 10, 30–48.

Surjadi, H. 2009. Pengembangan Kinerja Pelayanan Publik. Malang: Refika Aditama

Tenenhaus, M., Vinzi, V. E., Chatelin, Y. M., & Lauro, C. 2005. PLS path modeling. Computational statistics & data analysis, 48(1), 159–205.

Van der Post, W. Z., De Coning, T. J., & Smit, E. V. 1998. The relationship between organizational culture and financial performance: some south african evidence. *South african journal of business management*, 29(1), 30–41.

Wilson, A. M. 2001. Understanding organisational culture and the implications for corporate marketing. *European journal of marketing*, 35(3–4), 353–367.

Advances in Business, Management and Entrepreneurship – Hurriyati et al. (Eds)
© *2021 Taylor & Francis Group, London, ISBN 978-0-367-67471-7*

The influence of organizational climate and working ability on employee performance

I. Gumelar & E. Ahman
Universitas Pendidikan Indonesia, Bandung, Indonesia

ABSTRACT: Employee performance still becomes a problem that is interesting to investigated. This is a core of human resource management. This research attempts to analyze employee performance by using the Miles theory. The research design was a time series with descriptive and verification methods. In analyzing the data, multiple regression was used. The results indicated that organizational climate and employee working ability affected employee performance. In other words, the better organizational climate and employee working ability, the better employee performance will be.

1 INTRODUCTION

Employee performance has always been a serious problem (Sinex & Chapman 2015). In the past, the problem of employee performance dealt with working results. To date, it focuses on the comfort and safety condition in a company (Lipman 2016). Most employees prefer to not be supervised even though their performance is considered low (DeWeese et al. 2019). To solve this problem, a company's management should act based on a concept (Schaefer et al. 2015).

In the concept of performance, companies must provide feedback to the employees by providing the same objectivity and perception through new knowledge-based programs (Chen 2015). Managerial systems that are not concept-based will create a gap (Shalhoub et al. 2013). This is proven by a study in China, which found the disruption of employee psychology due to imbalance between employee dedication and justice.

The company basically has several basis problems, such as lack of equipment and expertise, high work load, unrepresentative work environment, lack of administrative staff, a small number of expert employees, low and high pressure, and a lack of contribution (Asim 2013). In a manufacturing company, employee performance is indicated by the ability in mastering the technology used. Here, the employee is demanded to be able to operate technology (García et al. 2015). They need to increase their competence (Taormina 2011). Employees with low technology competence will influence the production time (García et al. 2015). This kind of employee is perceived as a company burden (Ulku & Pamukcu 2015), so that the company applies the termination of employment (Ulku & Pamukcu 2015). This concept is relevant. However,

the problem occurs when this concept is applied to the employee who works on time (Rostami et al. 2015) or does not reach the production target (Gabrielsson et al. 2014).

Another problem is that a company provides tight supervision to control the quality of its product. Most of the employees have no space to work (DeWeese et al. 2019). This high demand causes low mental employees to resign (DeWeese et al. 2019). There is also a gap between employee knowledge and expertise. This makes production delays, and has different quality standards (Chen et al. 2014). The other problem takes place in the maintenance of equipment to support employees in providing the output. Without maintenance, employees will work unproductively (Kahya 2015) as they are not facilitated with appropriate equipment. This potentially creates bad behavior (Yun & Takeuchi 2015).

This study aims to analyze the above problems based on the Miles theory. There are three variables in the Miles theory, such as achievement, affiliation, and power.

Organizational climate is the perception and description of internal members of the organization (Stringer 2002). This perception creates consensus between individuals in the form of global influences of organizational members that are gained through interaction with organizational policies, structures, and processes (Wrench 2012). Organizational climate occurs in every organization, and affects the behavior of the members of the organization.

In the Miles theory, organizational climate is formed through management style that creates job satisfaction for employees. It is interrelated with the other variables, such as achievement, affiliation, and power. On the other hand, the Steers theory conveys

4 determinant factors in organizational climate. Those are management policies and practices, organizational structure, technology, and environment outside the company.

Human Resource Management (HRM) has a function to accommodate workability variables (Armstrong 2016). In the concept of HRM, working ability becomes a factor that supports the achievement of the organization's vision and mission (Kumar 2016). HRM, however, has a role in developing working ability, expertise, experience, and training (Mehmood et al. 2017).

Ability is the capacity of an individual to perform various tasks. An individual with high levels of ability tends to complete work assignments well, quickly, and precisely (Taormina 2011). Working ability is occupational competence that an individual has to perform the task according to the provisions (Zainullah et al. 2016).

Organizational culture is included in HRM, particularly in directional function. This function provides an overview about employee performance, tasks, and timing (Armstrong 2016). At this point, an organization will be successful if employees are oriented in time and performance in completing the charged tasks (Jyoti & Dev 2016).

There are two important elements in employee performance (Bigliardi et al. 2012). The first is functional tasks, which is related to how well employees complete work. Then the second is behavioral tasks. It is related to how well the employee handles personal and organizational activities, including dealing with conflict, managing time, managing the environment, empowering other individuals, working in a group, and working individually.

According to Armstrong (2016), performance is often regarded as simply the outcomes achieved, that is, a record of a person's accomplishments. This means that performance is considered as an achieved result, that is, a record of one's achievements. Employees should have performance that is in accordance with their job description (Shalhoub et al. 2013). This is in line with the work of Robbins & Judge (2013), who stated that performance is the way organizations, teams, and individuals do some works. It is the end result of an activity. Employee performance is how much the employee contributes to the company in output quantity, output quality, time period, attendance, and attitude (Unud 2016).

2 METHOD

This research was conducted to determine the effect of organizational climate and working ability on employee performance. The independent variables were organizational climate and working ability, while the dependent variable was employee performance.

This research was conducted at one company in Bandung. The research design was a time series with

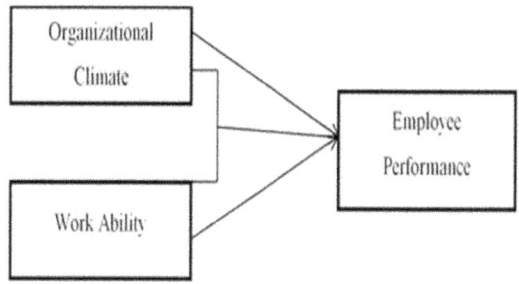

Figure 1. Research methodology.

descriptive and verification methods. The data were gained from library research and documentation.

3 RESULTS AND DISCUSSION

The result of multiple linear regression between the influences of organizational climate on employee performance is 0.984. Meanwhile, the result of working ability on employee performance is 0.84.

The result of the F-test in the hypothesis shows that the value of the t-count of organizational climate is 8.031 and that of the working ability is 2,569. With a significance level of $\alpha = 0.05$, the t-table value is 2.026. At this point, the value of the f-count is higher than that of the f-table for organizational climate and working ability. Thus, it can be concluded that Ho is rejected and Ha is accepted.

This means that there is a significant effect of the organizational climate on employee performance. This is in line with the research conducted by Chang & Lee 2010; Gokhan 2016; Mcinerney & Alexander 2016:11. They found that organizational climate influenced employee performance. The result of this research is also in accordance with the research performed by Balkin 2017; Sakamoto 2017; Zachary 2017, which found the significant impact of working ability on employee performance.

4 CONCLUSION

Based on the variable of organizational climate in the Miles theory (achievement, affiliation, and power) and the result of linear regression analysis, it can be concluded that:

1. The organizational climate is quite good. This can be seen from the representative dimensions of the affiliation variable.
2. The working ability is good enough. It can be seen from the credible dimension of the affiliation variable.
3. The employee performance is quite good. It is proven by the working ability dimension of achievement.

4. Organizational climate affects employee performance.
5. Working ability has an impact on employee performance.
6. Organizational climate and working ability have a simultaneous impact on employee performance.

REFERENCES

Armstrong. 2016. Human Resource Management Practice. (K. Page, Ed.) (10th Edition). Philadelphia: *British Library*.

Asim, M. 2013. Impact of Motivation on Employee Performance with Effect of Training: Specific to Education Sector of. International *Journal of Scientific and Research Publications* 3(9):1–9

Balkin, D. 2017. Managing Human Resource. (Pearson, Ed.) (7th Edition). *Texas: Pearson Inc*.

Bigliardi, B., Dormio, A. I., Galati, F., & Schiuma, G. 2012. The impact of organizational culture on the job satisfaction of knowledge workers. *Vine* 42(1):36–51.

Chang, S. C. & Lee, M. S. 2010. A study on relationship paternalistic leadership, organizational culture, the operation of learning organization and employees' job satisfaction. *The Learning Organization* (14).

Chen, A. 2015. Operationalizing physical literacy for learners: Embodying the motivation to move. *Journal of Sport and Health Science* 4(2):125–131.

Chen, S., Zhu, X., Welk, G. J., Kim, Y., Lee, J., & Meier, N. F. 2014. ScienceDirect Using Sensewear armband and diet journal to promote adolescents' energy balance knowledge and motivation. *Journal of Sport and Health Science* 3(4):326–332.

DeWeese, B., Hornsby, G., Stone, M., & Stone, M. H. 2019. The training process: Planning for strength-power training in track and field. Part 2: Practical and applied aspects. *Journal of Sport and Health Science*

Gabrielsson, J., Politis, D., Dahlstrand, Å. L., & Patents, A. 2014. Productivity in Indian manufacturing: Evidence from the textile industry," *Journal of Economic and Administrative Sciences JEAS* (3).

García, P. F., Párraga, M. J. A., Soto, H. V. M., & Latorre, R. P. A. 2015. Changes in balance ability, power output, and stretch-shortening cycle utilisation after two high-intensity intermittent training protocols in endurance runners. *Journal of Sport and Health Science*.

Gokhan, T. 2016. An Integrative Model of Job Characteristics, Job Satisfaction, Organizational Commitment, Organizational Citizenship Behavior Ability and Motivation. *IJJHS* 13(3): 205.

Jyoti, J. & Dev, M. 2016. Perceived High-performance Work System and Employee Performance: Role of Self-efficacy and Learning Orientation. SAGE, 15 *Employee Performance* 19.

Kahya, E. 2015. The effects of job characteristics and working conditions on job performance. *International journal of industrial ergonomics* 37(12):515–523.

Kumar, S. 2016. Innovative Motivational Techniques and Their Impact on Performance and Job Satisfaction. *ORP* 393–397.

Lars, P. 2011. Organizational Climate and Performance. *TUDELF* 66.

Lipman, V. 2016. Research on the Mechanism that Paternalistic Leadership Impact on Employee Performance. *Organizational Justice as an Intermediary* 3(4).

Mcinerney, M. L. & Alexander, R. P. 2016. Employee Performance a Case for a Manufacturing Firm. *IJHSS* 1(1), 63–71.

Mehmood, W., Corresponding, K., Muhammad, H., Tariq, A., Ghaffar, A., Anjum, M.Z., & Bajwa, E.U. 2017. Empirical study of Employee job Satisfaction. *Journal of business management* 6(2):29–35.

Robbins, S. P. & Judge, T. A. 2013. Organizational behavior. *Person education limited*.

Rostami, A., Sommerville, J., Wong, L. I., & Lee, C. 2015. Risk management implementation in small and medium enterprises in the UK construction industry. *Engineering, Construction and Architectural Management* 22(1): 91–107.

Sakamoto, S. 2017. Beyond World Class Productivity. (Spinger, Ed.) (First Edit). London: *Springer London*.

Schaefer, C. P., Cappelleri, J. C., Cheng, R., Cole, J. C., Guenthner, S., Fowler, J., & Mamolo, C. 2015. Health care resource use, productivity, and costs among patients with moderate to severe plaque psoriasis in the United States. *Journal of the American Academy of Dermatology* 73(4):585–593.

Shalhoub, J., Giddings, C. E. B., Ferguson, H. J. M., Hornby, S. T., Khera, G., & Fitzgerald, J. E. F. 2013. Developing future surgical workforce structures: A review of post-training non-Consultant grade specialist roles and the results of a national trainee survey from the Association of Surgeons in Training. International *Journal of Surgery (London, England)* 11(8):578–583.

Sinex, J. A & Chapman, R. F. 2015. Hypoxic training methods for improving endurance exercise performance. *Journal of Sport and Health Science* 1–8.

Stringer, C. 2002. Modern human origins: Progress and prospect. *Philosophical Transaction of yhe royal society of london* 563–579.

Taormina, R. J. 2011. Organizational socialization: The missing link between employee needs and organizational culture. *Journal of Managerial Psychology* 24(7).

Ulku, H. & Pamukcu, M. T. 2015. The impact of R&D and knowledge diffusion on the productivity of manufacturing firms in Turkey. *Journal of Productivity Analysis* 44(1):79–95.

Wrench, J. 2012. An Introduction to Organizational Communication. (Narissa Punyanunt, Ed.) (1st Edition). Virginia: *Virginia University*.

Yun, S. & Takeuchi, R. 2015. Employee Self-Enhancement Motives and Job Performance Behaviors: Investigating the Moderating Effects of Employee Role Ambiguity and Managerial Perceptions of Employee Commitment. *JOAP* 92(3):745–756.

Zachary, L. 2017. Creating a Mentoring Culture, The Organization Guide. (John Willey, Ed.) (First Edit). San Francisco: *John Wiley & Sons Inc*.

Zainullah, A., Suharyanto, A., & Budio, S.P. 2016. Pengaruh upah, kemampuan dan pengalaman kerja terhadap kinerja pekerja pelaksanaan bekisting pada pekerjaan beton. *Rekayasa sipil* 6(2)

Advances in Business, Management and Entrepreneurship – Hurriyati et al. (Eds)
© 2021 Taylor & Francis Group, London, ISBN 978-0-367-67471-7

The effect of organizational structure on the implementation of knowledge sharing

F.J. Islamy & D.A.A. Mubarok
Sekolah Tinggi Ilmu Ekonomi Indonesia Membangun, Jakarta, Indonesia

ABSTRACT: This research aims to understand the effect of organizational structure on knowledge sharing. There are three criteria for organizational structures that support the implementation of knowledge sharing: participative decision-making, ease of information flow, and cross-functional teams. This research is intended to contribute to academic institutions in improving knowledge sharing. The sample in this study was 119 lecturers at state universities in Bandung. The sample was taken by using probability sampling techniques. The analysis technique used was multiple linear regression. The variables studied were organizational structure (participative decision-making, ease of information flow and cross-functional teams) and knowledge sharing. The research findings indicated that organizational structure was positively significant related to knowledge sharing.

1 INTRODUCTION

Knowledge does not only exists in documents and repositories, but it becomes embedded in people's minds over time and it is demonstrated through their actions and behaviors (Al-Alawi et al. 2007). Given the importance of knowledge to organizations, knowledge management has become an integral and important facet of corporate strategy (Masa'deh et al. 2016). In the last decade, knowledge management (KM) has become a research attracting much interest. The increasing spread of theoretical works on knowledge management is due to its importance for the firm as well as the development of the competence-based view (CBV) (Marqués & Garrigós 2006). Nowadays, knowledge management strategies have become an important topic for any organization (Darroch 2005; Marqués & Garrigós 2006; Masa'deh et al. 2017; Yousif et al. 2013). Knowledge management is management of organizational knowledge for creating business value and generating a competitive advantage (Tiwana 2002). It relates to knowledge acquisition, knowledge sharing, and organizational memory. Knowledge acquisition refers to the acquisition of new knowledge internally and externally; while knowledge sharing refers to transfer / shared knowledge; and organizational memory relates to storing knowledge for future use either in the form of designing organizational system or in the form of rules, procedures, etc (Jyoti & Rani 2015). Regarding the process, there are three processes of knowledge management: (1) knowledge acquisition, (2) knowledge sharing, and (3) knowledge utilization (Tiwana 2002).

The process of knowledge management involves several activities. The most commonly discussed activity in the process of knowledge management now-adays is knowledge transfer (knowledge sharing) (Al-Alawi et al. 2007). Research (Dang et al. 2018) shows that knowledge sharing is a factor that has the first ranking as knowledge enabling factors (KEFs). It is critical to a firm's success as it leads to faster knowledge deployment (Al-Alawi et al. 2007), (Islamy 2013). Knowledge sharing is considered the first generation of knowledge management. It should be made visible in organizations by making it as a part of business strategy, initiating it obliquely on to another key business, routinizing, matching the organization's style, and aligning reward (Vorakulpipat & Rezgui 2008). Knowledge sharing is a systematic process in sending, distributing, and disseminating multidimensional knowledge and contexts from one person or organization to other people or organizations through varied methods and media. (Tobing 2011)

(Tobing 2011) explained that the form of the organizational structure affects knowledge sharing.

Traditional structures with complex hierarchies are not the structures that support the flow of knowledge within the organization. Al-Alawi (2007) explained that the organizational structure that supports the implementation of knowledge sharing is: (1) participative decision-making. In the implementation of knowledge sharing, the organizational structure must be able to support decisions that come from various functions or parts of the company. (2) The ease of information flow: organizational structure must be able to support the ease of information flow in the company. (3) Cross-functional teams. Organizational structures must be able to support teams or groups that come from various functions. Research conducted by (Al-Alawi et al. 2007) and Islamy (2013) found that there was a positive correlation between certain aspects of the organizational structure and knowledge sharing within an organization.

The importance role of education is highlighted in the 1945 Constitution, in which each citizen has the right to education, and the government strives to organize a national education system whose implementation is regulated in law. Higher education is one of the dimensions of national education that is expected to be the center of the implementation and development of higher education, which can improve the quality of life in a community, nation, and state (Absah 2009). Every college has and requires human resources to realize every set goal. Universities have a lot of knowledge in the minds of their HR, especially lecturers. Lecturers are required to do a research and store the results of the research in the available knowledge system, participate in uploading the results of the research, or conduct scientific publications to share their knowledge. In fact, there are still many lecturers who have not done it. Therefore, the authors want to find out whether the results of this study support the results of the study conducted by (Al-Alawi et al. 2007) regarding organizational structure factors that can affect the implementation of knowledge sharing. There are several differences between previous research and this present research, especially in the organizational structure variables. Previous research was conducted in private and public companies. This present research, however, is conducted in tertiary institutions. This research also adds the number of universities and samples.

Based on the above background, the problem of this research is the effect of organizational structure on knowledge sharing. More specifically, it attempts to find out the following research problems:

1. What is the organizational structure of Participative decision making according to lecturers' perception?
2. What is the ease of information flow's organizational structure according to the lecturers' perception?
3. What is the organizational structure of Cross-functional teams according to lecturers' perceptions?
4. What is the organizational structure according to the lecturers' perception?

1.1 Knowledge management

According to Davidson & Voss in knowledge management is how people from different places start talking to each other. Meanwhile, Horwitch and Armacost (Sopiah & Sangadji 2018) stated that knowledge management is the implementation of creation, capture, transfer, and access to the right knowledge and information to make better decisions, act appropriately and provide results to support business strategies. There are three pocess of knowledge management: (1) Knowledge acquisition, (2) Knowledge sharing, and (3) Knowledge utilization. (Tiwana 2002)

1.2 Knowledge sharing

Knowledge sharing is a systematic process in sending, distributing, and disseminating multidimensional knowledge and contexts from one person or organization to other people or organizations through variate methods and media. (Tobing 2011) and (Al-Alawi 2007) stated that knowledge sharing (knowledge transfer) requires an individual or a group cooperate with others to share knowledge and achieve mutual benefit. (Hooff & Ridder 2004) argued that knowledge sharing consists of two processes: (1) Knowledge donating, communicating the knowledge to other people or other individuals. (2) Knowledge collecting, consulting and discussing with other colleagues to get knowledge. After studying various writings on knowledge sharing, (Hooff & Ridder 2004) revised the process of knowledge sharing. The revision is as follows:

a. Knowledge donating: communicating the knowledge possessed by an individual with others
   1) Capability
      Capabolity is the extent to which a person has knowledge and willingness to spread his knowledge
   2) Credibility
      Credibility reflects the level of trust in support of colleagues
   3) Seriousness
      Seriousness reflects the level of seriousness a person is willing to share knowledge possessed
b. Knowledge collecting: consulting with colleagues to gain knowledge from them.
   1) The sense of belonging
      Smooth the level of willingness and willingness to receive new knowledge
   2) Commitment
      It is to what extent a person's commitment in applying new knowledge
   3) Satisfaction
      It is to what extent a person can adapt to new knowledge

1.3 Organizational structure

An organizational structure defines how job tasks are formally divided, grouped, and coordinated (Robbins & Judge 2013). (Al-Alawi 2007) explained that the organizational structure that supports the implementation of knowledge sharing is:

a. Participative decision-making: The organizational structure must be able to support the opinions of every employee in the company's decisions
b. The ease of information flow: The organizational structure must be able to support the ease of any information flow that exists within the company in reaching an agreement.
c. Cross-functional teams: Organizational structures must be able to support teams or groups that come from various functions to work together and fight information to achieve joint decisions.

2 METHODS

In line with the above explanation, the framework of this research is presented:

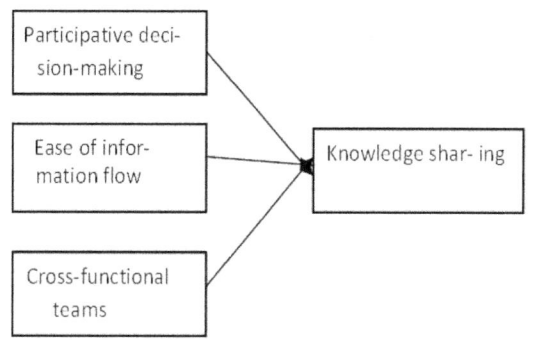

Figure 1. Research framework.

Based on the framework, the formulated hypothesis are:

Hypothesis 1: There is a positive influence between participative decision-making and knowledge sharing

Hypothesis 2: There is a positive influence between ease of information flow and knowledge sharing

Hypothesis 3: There is a positive influence between Cross-functional teams and knowledge sharing

Hypothesis 4: There is a positive influence between organizational structure and knowledge sharing

This study aims to determine the effect of organizational structure (participative decision-making, the ease of information flow, and cross-functional teams) on the implementation of lecturer knowledge sharing at state universities in Bandung. In carrying out this research, quantitative method was used. Regarding the study, it used causal study. The population was 5,927 permanent lecturers in state universities in Bandung. A probability sample with a proportional stratified random sampling technique was used. At this point, the population was firstly grouped based on their college. The sample was then selected randomly. As a result, there were 119 lectures chosen as the sample. The classification of the sample is described in Table 1.

The data was collected from the questionnaire. It was then analyzed by using correlation analysis with aimed to know the relationship between variables. The calculation of correlation coefficients is done using the SPSS program. The results can be seen in Table 2.

The results of multiple regression analysis can be seen in Table 4.

From the results of the significance test in Table 4, It can be concluded that Ho is rejected and Ha is accepted. This indicates that there is simultaneously and significantly influence of organizational structure

Table 1. Proportional stratified random sampling.

Sub population	Total element	Fraction value	Sample
Universitas Pendidikan Indonesia	1235	0,25	25
Universitas Padjajaran	1921	0,32	39
Institut Teknologi andung	1427	0,24	29
Universitas Negeri Islam Bandung	461	0,07	9
Politeknik Manufaktur Bandung	98	0,01	2
Politeknik Negeri Bandung	498	0,08	10
Sekolah Tinggi Pariwisata Bandung	125	0,02	2
Institut Seni Budaya Indonesia Bandung	125	0,02	3
Jumlah	5927	1,000	119

*Source: 2019 data processing results

Table 2. Correlations.

			SO1	SO2	SO3	K5
1	SO	Person Correlation	1	637**	578**	498**
		Sig. (1-tailed)		000	000	000
		N	119	119	119	119
2	SO	Person Correlation	637**	1	536**	396**
		Sig. (1-tailed)	000		000	000
		N	119	119	119	119
3	SO	Person Correlation	578**	536**	1	468**
		Sig. (1-tailed)	000	000		000
		N	119	119	119	119
	K5	Person Correlation	498**	396**	468**	1
		Sig. (1-tailed)	000	000	000	
		N	119	119	119	119

**Coreelations is signitificant at the 0.001 (1-tailed)
*Source: 2019 data processing results

Table 3. Conclusion correlation relations.

Variabel	korelasi	Sig.	Conclusion
SOI (Participative decision-making) to KS (Knowladge Sharing)	0,498	0,000	Strong enough, in the same direction, singnificant
SOI (Ease of information Flow) to KS (Knowladge Sharing)	0,396	0,000	Low, in the same direction, singnificant
SOI (Cross-funcional teams) KS (Knowladge Sharing)	0,468	0,000	Strong enough, in the same direction, singnificant

*Source: 2019 data processing results

Table 4. ANOVA.

Model		Sum of square	df	Mean Square	F	Sig.
1	Regresion	7.542	3	2.514	16.344	.000
	Residual	17.689	115	154		
	Total	17.689	118			

a. Dependent Variable: KS
b. Predictors: (Constant), SO3, SO2, SO1
*Source: 2019 data processing results

Table 5. Model summary.

Model	R	R Square	Adjusted R Square	Std. Error of the Estimate
1	547*	.299	.281	.39220

a. Predictors: (Constant), SO3, SO2, SO1
b. Dependent Variable. KS
Source: 2019 data processing results

Table 6. Coefficients.

Model (Constant)	Unstandarized Coefficients B	Standardized Coefficients Std.Error	Beta	t	Sig.
	2.457	.160		15.33	.000
SO1	.189	.066	.312	2.871	.005
SO2	.030	.053	.061	.576	.566
SO3	.116	.045	.255	2.570	.011

a. Dependent Variable. KS
Source: 2019 data processing results

variables on knowledge sharing variable. The magnitude of the influence of organizational structure variables on knowledge sharing can be known by the R square value of $0.299 = 29.9\%$.

T table at 0.05 significance level is 1.66. Based on Table 6, it can be concluded that t count O1 variable and knowledge sharing $= 2.871 > 1.66$ with a significance value 0,000. Hence, Ho is rejected and Ha is accepted. This means that SO1 variable has a significant positive effect on knowledge sharing.

T count SO2 variable and knowledge sharing is $0.576 < 1.66$ with a significance value 0.566. This implies that Ho is accepted and Ha is rejected. In other words, SO2 variable does not significantly influence knowledge sharing.

T count SO3 variable and knowledge sharing is $2.570 > 1.66$ with a significance value 0.011. Therefore, Ho is rejected and Ha is accepted. This indicates that SO3 variable has a significant positive effect on knowledge sharing.

3 CONCLUSION

The conclusions of this study are as follows.

1. Participative decision-making has a significant positive effect on the implementation of knowledge sharing
2. Ease of information flow does not have a significant effect on the implementation of knowledge sharing
3. Cross-functional teams have a significant positive effect on the implementation of knowledge sharing
4. Organizational structure has a significant positive effect on the implementation of knowledge sharing

Based on the conclusions and limitations of this research, the further research can examine the role of other organizational cultural factors by combining qualitative research methods and longitudinal studies to get a deeper understanding. It can also conduct in-depth research on organizational culture and knowledge sharing to fully comprehension the role of organizational culture in the success of knowledge sharing.

REFERENCES

Absah, Y. 2009. Pengaruh Pembelajaran Organisasi terhadap Kompetensi, *Tingkat Diversifikasi dan Kinerja Perguruan Tinggi Swasta di Sumatera Utara* 6(3)

Al-Alawi, A.I., Al-Marzooqi, N.Y. & Mohammed, Y.F. 2007. Organizational culture and knowledge sharing: Critical success factors. *Journal of Knowledge Management*. 11(2):22–42.

Dang, C.N., Le-Hoai, L. & Kim, S.Y. 2018. Impact of knowledge enabling factors on organizational effectiveness in construction companies. *Journal of Knowledge Management*. 22(4):759–780.

Darroch, J. 2005. Knowledge management, innovation and firm performance. *Journal of Knowledge Management*. 9(3):101–115.

Hooff, V.D., & Ridder, D.J.A. (2004). Knowledge Sharing in Context: The Influence of Organizational Commitment, Communication Climate and CMC use on Knowledge Sharing *Journal of Knowledge Management* 8(6): 117–130

Islamy, F.J. 2013. Pengaruh Budaya Organisasi Terhadap Implementasi Knowledge Sharing Dosen Tetap Universitas Pendidikan Indonesia Bandung Tahun 2013: 1–13.

Jyoti, J. & Rani, A. 2015. High performance work system and organisational performance: role of knowledge management.

Marqués, D.P. & Garrigós, S.F.J. 2006. The effect of knowledge management practices on firm performance. *Journal of Knowledge Management*.10(3):143–156.

Masa'deh, R., Obeidat, B.Y.& Tarhini, A. 2016. A Jordanian empirical study of the associations among transformational leadership, transactional leadership, knowledge sharing, job performance, and firm performance: A structural equation modelling approach. *Journal of Management Development*.35(5): 681–705.

Masa'deh, R., Shannak, R., Maqableh, M. & Tarhini, A. 2017. The impact of knowledge management on job performance in higher education: The case of the University of

Jordan. *Journal of Enterprise Information Management.* 30(2): 244–262.

Robbins, S.P. & Judge, T.A. 2013. Organizational Behavior (Fifteenth). Pearson.

Tiwana, A. 2002. Knowledge Management Toolkit The Amrit Tiwana Knowledge Management Toolkit, *The. In Knowledge Management Toolkit.*

Tobing, P.L. 2011. Manajemen Knowledge Sharing Berbasis Komunitas. Bandung: Knowledge *Management Society Indonesia*

Undang-Undang No. 20 Tahun 2003. Tentang Sistem Pendidikan Nasional. JakartaVorakulpipat, C. & Rezgui, Y. 2008. An evolutionary and interpretive perspective to knowledge management. *Journal of Knowledge Management* 12(3): 17–34.

Yousif, A., L. Hyphen, Hakim, L.A. & Hassan, S. 2013. Knowledge management strategies, innovation, and organisational performance: An empirical study of the Iraqi MTS. *Journal of Advances in Management Research.* 10(1): 58–71.

The role of women leadership in improving employee performance in educational organizations

M. Fadil & H.S. Hadijah
Universitas Pendidikan Indonesia, Bandung, Indonesia

ABSTRACT: Several researches show that women leaders in educational organizations had limitations. They could not give better influence than men leaders. This study aims to show that women leaders can motivate their staff members to have a good performance. This was qualitative research, which used the findings from previous research as the primary data. The results showed that women leaders in educational organizations had a democratic leadership style. They applied participative, interactive, and collaborative methods. They focused on interpersonal relationship in organizations. They also had a parental "approach" with "maternal" characteristics. As a result, their staff members had strong commitment, motivation, and creativity. In other words, the women leaders had the ability to lead the people, and their leadership affected employee performance.

1   INTRODUCTION

In this development era, an organization uses some strategies to achieve their expected goals. One of the strategies is by selecting a leader. A leader is someone who has a certain position in an organization and functions in designing a plan, organizing, supervising, and making effective decisions. A leader should have a skill in involving and affecting the members of an organization to achieve the goal together. This skill is known as leadership.

According to Fitriani (2015), leadership is often divided into two perceptions, namely leadership as a position and leadership as a social process. In the first perception, leadership is viewed as a complex thing of rights and obligations possessed by someone or an organization. In the second perception, leadership covers all actions done by someone or an organization that results in a community movement.

Regarding the function, Duryat (2016) stated that the leader of an educational organization functions as a manager and a head of educational supervision. As a manager, a leader should have an accurate strategy to empower the teachers and the education system through a cooperative work system, give a chance and an opportunity to improve their professionals, motivate them to be actively involved in organization, and have a sense of belonging. Meanwhile, as a head of educational supervision, a leader should be able to give planned training to help the teachers and the staff members to work effectively. In performing this function, it is not quite important whether it is done by men or women leaders. In other words, leadership does not distinguish the actor based on the gender.

As stated by Gary et al (2008), a leader that shows similar behavior and achievement will ideally receive the same evaluation, regardless of the differences in gender.

Nevertheless, Mitroussi & Kyriaki (2009) argued that diversity in the educational system is formed by a different cultural context and other parameters such as gender. The ability of women leaders, for example, is still doubtful. There are hesitations toward women leaders' performances, including in an educational organization.

Stavroula et al. (2017) stated that most of the women leaders in educational organizations experience barriers such as lack of opportunity for career advancement, family support, gender bias (discrimination), isolation, lack of social support, a culture dominated by men, and negative environment in an organization. Vassiliki (2012) argued that the staff members from a woman leader tend to consider them as an emotional, sensitive, and hesitant person when facing a difficult situation. Jouharah (2017) mentioned some terms for the barriers of women leaders, namely the glass-ceiling, labyrinth, and inequal regime. All of those terms lead to the view that obstacles faced by women leaders are caused by some practices occurring in various levels of organizational structure. Those terms also show how women are excluded and isolated from a top management level where they should be responsible for managing and supervising all organizations. This is support by the work of Jouharah (2017) who found that women in a workplace frequently received discrimination related to a job promotion. In conclusion, gender equality does exist in educational organization policy. However, in practice, men are more preferred than women in all aspects. They are perceived as the head of the household, and they deserve to get a higher salary for their position as a leader.

Based on the above fact, this study attempts to analyze the literature study to show that women leaders can motivate their staff members in an educational organization to produce good performance.

1.1 Women leadership

Bagilhole & White (2011) said that women leadership has become an interesting subject in the past few years. Some women leaders have stabilized themselves in various fields of public life, such as education, politics, economics, and business.

Idris (2017) suggested that the spiritual intelligence owned by women leaders has contributed to staff members growing and becoming excellent during their career. They have a willingness to share assistance with their staff members to know what to do. Moreover, the studies showed that spiritual intelligence can change the workplace into more a significant and directed situation by building a balance and harmonious relationship between the staff members.

A research done by Yanez & Moreno (2007) revealed that the competence of women leaders to read the organization culture and to stimulate changes can maintain the climate of social harmony in the workplace. Here, they adopt a flexible leadership style. Arar (2017) conveyed that women leaders succeed to make their staff members to have the purpose in working, build a fair system, create a harmonious environment, and have some time to do some activities outside the school as a part of the community. Cubillo & Brown (2003), in their study, mentioned that women leaders are a special group since they have higher self-confidence, a remarkable quality of tenacity, and higher courage and independence.

Even though women leaders face various obstacles in the job promotion process and gender stereotypes, their leadership competency is better than men (Chesterman 2006). The research revealed that the success level of women leaders is not extremely different from men; even, at the professor level, their success level can exceed men leaders' success. Lund (2008) mentioned that women tend to adopt a strict ethics rule. They can develop transformational leadership (Gardiner & Grogan 2000). They spend more time to train and develop the staff members in an organization than the men do (Krishnan & Park 005). The preference of women leaders for a relational leadership approach can improve the focus of the staff members on the relationship between the stakeholders, which are proven to be able to give benefits for organizations.

1.2 Employee performance

Borman & Motowidlo (1997) defined performance as the effectiveness of the staff members in performing the assigned tasks that can fulfil the organization vision by always appreciating the individuals' proportionally. Tripathy (2014) stated that performance is divided into two segments, namely administrative technical duties and leadership duties. Administrative technical duties include planning, organizing, and administration of daily tasks through technical ability, business evaluation, etc. Leadership duties establish strategic goals and uphold the required work standard. It also includes directing the staff members to accomplish their job through motivation, recognition, and constructive criticism. Meanwhile, the indicators of good employee performance are creativity and reactivity in facing difficulties and the ability for interpersonal adaptation (Audrey & Patrice 2012).

1.3 The influence of women leadership on employee performance

Dezso and Ross (2012) proposed that a women leader can create an inclusive culture that can promote diversity in an organization based on the democratic and participative management style. A woman leader tends to offer a critical insight about the way an organization can make a strategic decision by estimating the stakeholders including the staff members. Here, the women leaders' orientation is to build a relationship that leads to the staffs' awareness and commitment.

Gupta (2018) stated that women leaders tend to have a stronger impact on the awareness and the attitude for appreciating diversity in an organization. They had a stronger commitment toward equality and justice than the men leaders.

Moreover, women leaders tend to adopt a more democratic and collaborative style than men leaders (Eagly & Johnson 1990). This style focuses more on the people and transformational style than on the tasks and transactional style. Metcalfe (2010) stated that women leaders use this transformational style to motivate other people by changing individual interest into group interests. This is different with men leaders who focus on the tasks and tend to be directive people (Dobbins & Platz, 1986). Women leaders' strengths are participation, sharing authorities, and inclusivity.

Women also use an interactive leadership style (Manjulika 1998) by stimulating participation, sharing the power and information, and increasing environmental motivation. This is in line with the work of Le Breton and Miller (2006), who argued that women prefer to lead the staff members by empowering, stimulating participation, and teamwork, rather than prioritizing individual competence. However, a woman as a leader believes that the staff members will do their best when they feel comfortable with themselves and their job. Besides, they will try to create a situation that can support those things. In other words, women leaders focus more on a relationship, such as interpersonal and cognitive skills (Kairys 2018)

Agezo (2010) showed a unique leadership practice done by a woman in managing a school. This woman has an emotional quotation in inspiring and

Women Leadership		Employees Performance
Democratic (Participatory, Interactive and Collaborative)	Derso and Ross (2012), Gupta (2018), Eagly and Johnson (1990), Dobbins and Platz (1986), Metcalfe (2010), Le Breton and Miller (2006), Dita and Susanti (2015), Manjulika (1998)	Strong Work Commitment Motivated Inspired Creative
Focusing on Interpersonal Relationships	Derso and Ross (2012), Dita and Susanti (2015), Kairys (2018), Gupta (2018), Agezo (2010)	
Sensitivity to Feelings	Dita and Susanti (2015)	
"Parent / "Motherly" Approach	Dita and Susanti (2015)	

Figure 1. The relationship model between women leadership and employee performance.

building a relationship with staff. This woman was proven to be able to create a work environment that stimulated creative thoughts as well as designing and implementing new sophisticated programs. Besides educational organizations, there is also a study investigating women leadership practices in non-educational organizations. Dita & Susanti (2015) found that a woman leader could enhance employee performance as she led with heart. This is a feminine leadership style that emphasizes more on teamwork, participation, information sharing, interpersonal skills, sensitivity to feelings, and staff perceptions. The research also showed that the success of a woman leader is due to the adoption of a "maternal" style and "parental" approach in reprimanding the staff members.

Based on the analysis toward several previous studies, the relationship model between women leaders and employee performance can be described as seen in Figure 1.

Figure 1 shows that there are so many strategies or techniques used by women leaders. Those are using a democratic style, focusing on interpersonal relationships in an organization, having sensitivity toward individual's and other organization members' feelings, and using a "parental" approach with "maternal" characteristics.

2 METHOD

As previously stated, this study was qualitative research, which used the findings of previous studies as the primary data. There were several stages done in analyzing the data. The first stage was conducting a comprehensive search to gain reliable sources (book or indexed journal) written by people who were credible in the field. The second stage was selecting references to produce more specific discussion of the women leadership in an educational organization. Nevertheless, irrelevant references were still used for

supporting the analysis. The third stage was composing a summary based on the conducted analysis.

3 RESULTS AND DISCUSSION

Based on the analysis of the previous studies, it was found that women leaders were able to enhance employee performance in an organization, especially in educational organizations. There are several strategies used. First is using a democratic style. This style prioritizes the group goals rather than the individual goals. It stimulates participative, interactive, and collaborative mechanisms between all members in the organization. This makes the staff members feel respected. As a result, they have a strong commitment in working and achieving the organization's goals. Democratic style, however, definitely impacts to motivation to work better and the freedom of thinking. This is because all staff members are allowed to be actively involved in running and achieving the organization's goals.

Second is using interpersonal relationships. This kind of relationship creates good communication between a leader and the staff members. The staff members will recognize their leader both personally and professionally. They also have a role model of a good leader and organization executor.

And last, women leaders use their sensitivity to recognize the staff's feelings and perceptions. They use a parental" approach with "maternal" characteristics. This approach increases the staff's motivation to work better. They will also be inspired by the mature behavior shown by their leader.

4 CONCLUSION

Based on the results of the analysis, it can be concluded that the women leadership style in leading an educational organization is democratic, in which they use participative, interactive, and collaborative methods. They also use interpersonal relationships, and sensitivity toward individual's and other organization members' feelings. Moreover, they apply a "parental" approach with "maternal" characteristics to increase staff's motivation, creativity, and commitment.

REFERENCES

Agezo, K. W. 2010. Women leadership and school effectiveness in junior high schools in Ghana. *Journal of Educational Administration* (48) 6:689–703.
Arar, K. 2017. Arab women educational leadership and the implementation of social justice in Schools. *Journal of Educational Administration*.
Audrey, C. V. & Patrice, R. 2012. Adaptive performance: A new scale to measure individual performance in organizations. *Canadian Journal of Administrative Sciences* (29) 2: 280–293.

Bagilhole, B. & White, K. 2011. Gender, Power and Management: *A Cross-Culture Analysis of Higher Education*.

Borman, W. C. & Motowidlo, S.J. 1997. Task performance and contextual performance: The meaning for personnel selection research. *Human Performance* (10) 2:99–109.

Breton, L. M. I. & Miller, D. 2006. Why do some family businesses outcompete? Governance, long-term orientations, and sustainable capability. *Entrepreneurship Theory and Practice* 30(6): 731–746.

Chesterman, H. W. 2006. Academic women promotions in Australian universities. *Employee Relations*. (28)6: 505–522.

Cubillo, L. & Brown, M. 2003. Women into educational leadership and management: international differences? *Journal of Educational Administration*. (41) 3:278–291.

Dezso, C. L. & Ross, G. D. 2012. Does women representation in top management affect firm performance? A panel data investigation. *Strategic Management Journal*. (33) 9:1072–1089.

Dita S. N. & Susanti, D. D. 2015. A Women Leader in Executive Service: The Case of Mayor Risma of Surabaya. In *Asian Leadership in Policy and Governance. Asian Leadership in Policy and Governance Public Policy and Governance*. (24):287–304.

Dobbins, G. H. & Platz, S. J. 1986. Sex differences in leadership: How real are they? *Academy of Management* Review (11) 1:118–127.

Duryat, M. 2016. Kepemimpinan Pendidikan: 139. Bandung: CV Alfabeta.

Eagly, A. H. & Johnson, B.T. 1990. Gender and leadership style: A meta-analysis. *Psychological Bulletin*. (108) 2:233–256.

Fitriani, A. 2015. Gaya Kepemimpinan Perempuan. *Teropong Aspirasi Islam Journal* 11 (2).

Gardiner, M. E. & Grogan, M. 2000. Coloring Outside the Lines: Mentoring Women into School Leadership. New York: State University of New York Press.

Gary, N., Powell, D., Kathryn, A. B., & Bartol, M. 2008. Leader evaluations: A new women advantage? Gender in Management: *An International Journal* 23 (3): 156–174.

Gupta, A. 2018. Women leaders and organizational diversity: Their critical role in promoting diversity in organizations. Development and Learning in Organizations: *An International Journal*.

Idris, S. D. 2017. Effectiveness of the use of Spiritual Intelligence in Women Academic Leadership Practice. *International Journal of Educational Management*. (31) 2.

Jouharah, M. A. 2017. Women and leadership: Challenges and opportunities in Saudi higher education. *Career Development International*.

Kairys, M. R. 2018. The influence of gender on leadership in education management. International *Journal of Educational Management*.

Krishnan, H. A. & Park, D. 2005. A few good women on top management teams. *Journal of Business Research* (58)12: 1712–1720.

Lund, D. B. 2008. Gender differences in ethics judgment of marketing professionals in the United States. *Journal of Business Ethics* (77) 4:501–515.

Manjulika, K. A. 1998. Women in management: A Malaysian perspective. *Women in Management Review* (13) 1:11–18.

Metcalfe, B. 2010, An investigation of women and male constructs of leadership and empowerment. *An international Journal*.

Mitroussi, A. & Kyriaki, M. 2009. Women educational leadership in the UK and Greece. Gender in Management: *An International Journal* (24) 7:505–522.

Stavroula, K., Czabanowska, K., Davis, S. F., & Brand, H. 2017. Women leadership barriers in healthcare, academia and business. Equality, Diversity and Inclusion: *An International Journal*.

Tripathy, S. P. 2014. Impact of motivation on job performance of contractual staff in Devi Ahilya University Indore (M.P.). *Paripex-Indian Journal of Research*. (3) 5:1–5.

Vassiliki, B. 2012. Men vs women; educational leadership in primary schools in Greece: An empirical Study. *International Journal of Educational Management*. (26) 2:175–191.

Yanez, J. L. & Moreno, M. S. 2007. Women leaders as agents of change in higher education organizations. *Gender in Management: An International Journal*. (23) 2:86–102.

The effectiveness of gamification on students' attitude toward teamwork

S.M.N. Rahmah & N.A.R. Putranto
Institut Teknologi Bandung, Bandung, Indonesia

N.L. Krisna
Universitas Persada Indonesia YAI, Jakarta, Indonesia

ABSTRACT: Gamification, especially board games, is one of methods that is popularly used in the education world. However, the studies that measure the effectiveness of board games as apparatus for learning and changing attitudes and behaviors are still limited. Therefore, the researchers conducted this research to prove whether a board game can achieve a certain level of attitude toward teamwork and the components of attitude toward teamwork, which are affection, behavior, and cognition. This research was designed to measure the effectiveness of board games by comparing the pretest and posttest results in experimental research methods by involving 49 participants. Then, the results of comparison showed that gamification, especially board games effectively give a positive significant improvement in attitude toward teamwork and attitude components.

1 INTRODUCTION

1.1 *Research background*

Today, sophisticated technology enables companies to look for ways to achieve their vision and can compete in a very volatile and rapidly changing era (Salleh et al. 2010). To solve this problem, besides having to pay attention to financial and technology aspects, the role of human capital is also very important because human capital is the main subject that regulates everything in the company (Nassazi 2013). The result of the Chief Executive Officer Survey by Pew Research Center (2018) shows that talent is a priority for a company to growth. This situation generates an increase in demand for human capital that meets the criteria that match the needs of the organization (Salleh et al. 2010). One of the main criteria that is very important in qualified human capital is soft skills because, not like technical skills, soft skills are a side of human abilities that cannot be replaced by advanced technologies such as automation and artificial intelligence (Hes 2017).

To answer this challenge, higher education as one of the main providers of talent needed by the labor market continues to focus on developing soft skills as early as possible in accordance with Indonesian education goals to provide professional talent with relevant skills to contribute to the society (Ministry of Research, Technology, and Higher Education 2017). Unfortunately, the goal of higher education has not been fully achieved because technical skills are still preferred over soft skills, thus, higher education must evaluate this condition to be able to produce more

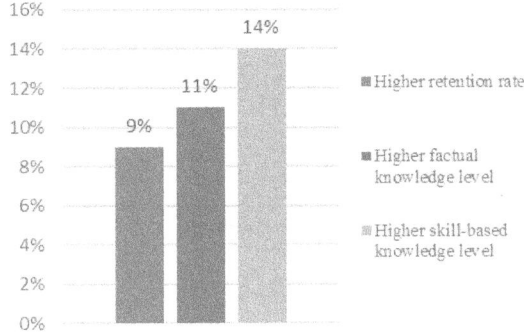

Figure 1. Training outcome percentages with using simulation games by Sitzman.

qualified talent and meet labor market demand criteria (Handayani 2015).

Among the various soft skills that must be mastered by talent, one of the skills that is very important for talent is the ability to collaborate and to work in a team (Workplace Learning Report 2018). Therefore, in the world of education, various methods have emerged to develop teamwork. One of the most effective learning methods to obtain better results from learning, developing a positive attitude toward the course, and achieving the desired behavior is to use gamification (Forbes 2018). Gamification is applying game elements and game-filled experience to learning (Deterding 2012). Figure 1 shows the higher training outcome percentages obtained using gamification.

Research problem, objective and limitation

Kasurinen & Knutas (2018) stated that only 3.6% of research papers about gamification were related to social behavior and education (non-computer science related). Meanwhile, research about the effect of gamifications on attitudes is very important to support higher education and organizations or companies to improve the attitude and validate the method. This research has a purpose to give contribution about the effectiveness of gamification on changing students' attitude toward teamwork. Therefore, this research focuses on the use of board games as a gamification technique on first-year students in BT University, Bandung (not the real name).

2 METHOD

This study was designed to measure the effective-ness of the use of gamification, especially board games. The researchers collected the primary data by conducting experimental research with a comparison of the results of a before–after experiment. The tool that was used for treatment was the Forbidden Island game. Then the questionnaire of the pretest and posttest that was given as treatment referred to research conducted by Mohammed (2010) for question guideline components of attitude (affect, behavior, and cognition) and the work of Kiffin-Petersen & Cordery (2003) for question guidelines of attitude toward teamwork. Here is Figure 2 that shows the treatment design:

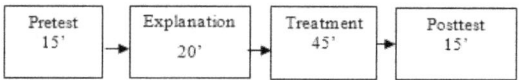

Figure 2. Treatment design.

The first stage for treatment design is the pretest with a questionnaire developed by Mohammed (2010) and Kiffin-Petersen & Cordery (2003). Then, the researchers gave an explanation about the game, gameplay, and rules. After that, respondents were given the treatment, which is playing the board game; one game can be played by maximum of six respondents. Last, after treatment, the researchers gave a posttest with the same questionnaire that was used in the pretest stage.

3 RESULTS AND DISCUSSION

3.1 *Distribution of data*

Based on the data collected, there were 53 respondents that followed this experiment. However, not all the respondents followed all of experiment activity. From all those respondents, only 49 respondents fulfilled all the requirements. This number is more than the minimum number of samples for experimental research according to Hill (1998), who suggests 30 subjects per group. Based on age distribution, the mean age of

the respondents is 19 years old. Then, the gender distribution of respondents consisted of 26 respondents who were female and 23 respondents who were male.

3.2 *Descriptive statistics*

Table 1. Descriptive statistics comparison between pretest and posttest.

	Mean	N	Std. Deviation	Std. Error Mean
Pretest	151.41	49	18.193	2.599
Posttest	164.24	49	18.049	2.578

The researcher collected the data about the component of attitude (affect, behavior, and cognition), and the attitude toward teamwork of the respondents before and after treatment. The measurement conducted using a Likert scale ranging from 1 to 5. The "5" refers to strongly agree and the "1" refers to strongly disagree. According to the data with 49 respondents, there was a positive difference between the results of the pretest (M = 151.41) and the posttest (M = 164.24). Meanwhile, the normality test result using a Kolmogorov–Smirnov test ($p_{pretest}$ = .053, $p_{posttest}$ = .200) had a significant value of more than .05 that indicated the posttest result was normally distributed.

3.3 *Significance of findings*

According to Table 2 that provides information about the paired sample test, the results showed that attitude toward teamwork has a Sig. (2-tailed) with a value of .000, which means it is less than p = .05 that indicated that there were positive significant differences on the attitude toward teamwork results between the pretest and posttest after treatment.

Table 2. Paired t-test results for attitude components.

	Paired Differences				
	Mean	Std. Deviation	Std. Error Mean	t	Sig. (2-tailed)
Attitude toward teamwork	−3.224	3.749	.536	−6.021	.000
Affect	−3.122	3.683	.526	−5.934	.000
Behavior	−3.367	4.885	.698	−4.825	.000
Cognition	−3.122	4.161	.594	−5.252	.000

This result might happen because the gameplay of the game promotes teamwork and collaboration. It requires respondents to discuss about strategy for mutual decision in every turn. Therefore, H1 is supported, meaning there was a positive significant difference in students' attitude toward teamwork before and after treatment using gamification

According to the results of the paired sample test, the components of attitude such as affect, behavior, and cognition had the Sig. (2-tailed) with a value of .000, which means it is less than $p = .05$ that indicated that there were positive significant differences on all of the components of attitude toward teamwork results between the pretest and posttest after treatment.

For affection result, it might have happened because respondents felt satisfied and happy when they could work together and accomplish the challenge in the game with satisfactory results. Therefore, H2 is supported. The subjects exposed to gamification will give positive significant differences in students' level of affection.

Therefore, for the behavior results, this condition probably happened because the respondents practiced to respect each other, were responsible as a team, and gave opinions and contributions at every turn of the game. Then, H3 is supported; the subjects exposed to gamification will give significant differences on students' level of behavior.

Lastly, for the cognition results, it might have happened because during playing the games, there was an exchange of thought and knowledge sharing. That information can be formulated as a new insight for respondents. Then, H4 is supported; the subjects exposed to gamification will give significant differences on students' level of cognition.

4 CONCLUSION

Gamification, especially board games, is one of the methods that is popularly used in the education world. However, the studies that measure the effectiveness of board games as apparatus for learning and achieving desirable attitudes and behaviors are still limited. Therefore, the researchers conducted this research to prove whether a board game can achieve a certain level of attitude toward teamwork and the components of attitude toward teamwork, which are affect, behavior, and cognition. In order to answer the research objective, the results showed there was a significant difference between the pretest and posttest results after treatment for attitude toward teamwork and the components of attitude toward teamwork.

Although it can be said that gamification, especially board games, can effectively improve the positive attitude toward teamwork, certainly, each method has

advantages and disadvantages. The use of gamification to improve positive attitude is not something that can be obtained instantly, it takes time to build a positive attitude permanently. So, assistance during playing is needed. Giving explanations and emphasis about the importance of teamwork during playing can provide more insight and understanding for participants. Other than that, playing more than one game with different themes and with different people can also train participants about adaptations.

REFERENCES

Deterding, S. 2012. Gamification designing for motivation.

Forbes. 2018. Seven learning and development trends to adopt in 2019. [Online]. Retrieved from https://www.forbes.com/sites/forbeshumanresourcescouncil/2018/09/24/seven-learning-and-development-trends-to-adopt-in-2019/#48d7205e104b.

Handayani, T. 2015. The relevance of graduates of higher education in Indonesia. *Jurnal Kependudukan Indonesia*.

Hes, D. 2017. 5 reason why soft skills are more important than ever. [Online]. Retrieved from Oxbridge Academy: https://www.oxbridgeacademy.edu.za/blog/5-reasons-soft-skills-important-ever/.

Hill, R. 1998. What sample size is "enough" in internet survey research? Interpersonal computing and technology: An electronic journal for the 21st century.

Kasurinen, J. & Knutas, A. 2018. Publication trends in gamification: A systematic mapping study. *Computer Science Review*.

Kementerian Riset, Teknologi dan Pendidikan Tinggi. 2017. Laporan kerja Kemenristekdikti.

Kiffin-Petersen, S. & Cordery, J. 2003. Trust, individualism and job characteristics as predictors of employee preference for teamwork. *International Journal of Human Resource Management* 14(1): 93–116.

Mohammed, A. F. 2010. Professional attitudes toward teamwork in the social welfare organization. Fayoum University.

Nassazi, A. 2013. Effects of training on employee performance. Evidence from Uganda.

Pew Research Center. 2018. PwC's 21st CEO Survey. [Online]. Retrieved from https://www.pwc.com/gx/en/ceo-agenda/ceosurvey/2018/gx.html.

Salleh, K. M., Sulaiman, N., & Talib, K. N. 2010. Globalization's impact on soft skills demand in the Malaysian workforce and organizations: What makes graduates employable? *1st UPI International Conference on Technical and Vocational Education and Training 7*.

Sitzman, T. 2010. A meta analytic examination of the instructional effectiveness of computer-based simulation games.

Workplace Learning Report. 2018. [Online]. Retrieved from https://learning.linkedin.com/resources/workplace-learning-report-2018.

An analysis of learning organization application at PT. Setiajaya Mobilindo – a Toyota service center

Sarwania & E. Nurzaman
Universitas Pamulang, Tangerang Selatan, Indonesia

N.L. Krisna
Universitas Persada Indonesia YAI, Jakarta, Indonesia

M.S. Adhy
PT. Setiajaya Mobilindo, Depok, Indonesia

ABSTRACT: This research was conducted based on the current rapid technological advancements and fierce business competition, which force companies to quickly adapt and to compete by implementing learning organization. This research employed a qualitative approach using Senge's Five Disciplines of Learning Organizations and SWOT analysis to observe the company condition. In this research, 10 employees were assigned as the key informants with predetermined criteria. It was found that the company had implemented Senge's five pillars (shared vision, system thinking, mental model, team learning, and personal mastery). On the other hand, SWOT analysis revealed that (1) certified and experienced labors and superior SOP were required in learning organization application, (2) the quality and competitiveness were generally good, but it was necessary to provide advantages as a form of corporate responsibility, (3) learning organization was established by improving communication between management, employees, and consumers using Android for customers the service availability. It can be concluded that the learning organization application increased the company competitiveness and customer satisfaction level.

1 INTRODUCTION

It is essential for organizations to acquire knowledge and improve their technology in order to meet various challenges and problems in this globalization era. External and internal organizational changes have encouraged them to adapt. These changes are based on the paradigm that organizations or institutions must adapt and transform into learning organizations.

Learning organization was popular after Senge (1990) promoted his ideas in Fifth Discipline. Since then the term "learning organization" has been discussed on various occasions. In his book, Senge mentions that the core of a learning organization is the five disciplines, which are personal mastery, mental models, shared vision, team learning, and system thinking.

In order to survive in this era, PT. Setiajaya Mobilindo, as an organization aims at providing the best service for the customer and to be the best in the region, cannot be separated from learning organization. As one of Toyota dealers and service companies, after-sales service is one of the most important business aspects in which reliable service will satisfy customers and making them long-term customers that

could support the company survivability. In addition, excellent service also has an impact on increasing car sales where consumers will not have to worry upon knowing that there are reliable car maintenance and repair service.

The company was conscious of the importance of learning organization application. Accordingly, they had already implemented some of the concepts on their personnel in the form of the 5s briefing, internal and external training, and also Gemba Kaizen. However, the result was found not satisfactory.

Based on the above background, it was deemed necessary to conduct a research about the learning organization application in the company as a strategy in dealing with the complexity of chance in the Toyota Car Service.

2 METHODS

This research employed a qualitative case study method. PT. Setiajaya Mobilindo, a company engaged in the automotive sector, was employed as the subject. Using qualitative case studies, the existed phenomena were explored through various data sources. In this case, the data were collected through interview, observation, and related document. The interview was conducted using an interview guideline. The observation data was written informatively and analyzed based on the manual codes. On the other hand, document study was also conducted to support the existing data. These techniques were integrated forming an in-depth picture of the case.

Thematic and SWOT analyses were employed. SWOT analysis consists of strength, weakness, opportunity, and threat. It is also a systematic identification of various factors to formulate a company strategy. Also, a theory-driven approach was also adopted in analyzing the data based on Senge's learning organization theory.

3 RESULTS AND DISCUSSION

3.1 *Analysis results*

Based on the observation, it was discovered that learning organization has been applied in the company. It was proven by the existence of three programs, namely:

1) Kodawari, a dealer standardization program referring to Toyota standardization regulating building, stall, clothing, etc.
2) Kaizen, a continuous improvement program to solve a small to a large problem. If an employee slips because the floor is slippery due to oil, the employee should immediately clean the floor (small problem).
3) 5S, a part of Kaizen program introducing the method to make work environment clean, tidy, and comfortable.

After questionnaire data were obtained from the employees, the following results were obtained:

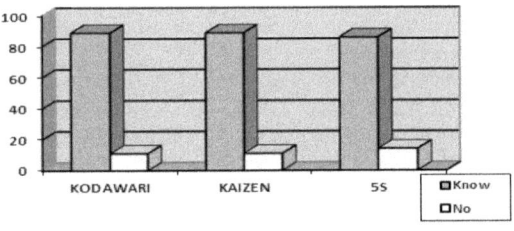

Figure 1.   Employees' awareness diagram of the Kodawari, Kaizen, 5s learning organization at PT. Setiajaya Mobilindo.

Figure 1 depicts that 89% of the employees were aware of the Kodawari program and 11% of them did not. Similarly, 89% of them were aware of the Kaizen program and 11% of them did not. On the other hand, 86% of them were aware of the 5s program and 14% did not.

From the result, it can be seen that not all of the employees were aware of the learning process in their workplace. It shows that the socialization process had not been conducted thoroughly for all employees.

Observation, data collection, and interview results showed that learning organization application in the company can be observed as follow:

1. Shared Vision
The behavioral aspect was deemed important in the learning process because it was about employees' behavior toward learning itself. It was found that not all employees had a similar perception of learning. Therefore, the company must be able to find the right approach so that each employee can be encouraged to learn. Knowledge sharing process in the company can be seen as follow:

a) Discussion forum
   PT. Setiajaya Mobilindo regularly conducted morning briefing or an internal focus discussion. This is a discussion forum conducted in the field or in the meeting room before or after work to discuss the problem and to find for a solution. It was conducted every Wednesday morning before work and Friday afternoon at 16.15 to finish. This discussion forum was a method to accommodate knowledge sharing, which is a group of people who share common interests and collaborate to form communities in learning, solving the problem, and sharing ideas. With increasingly advanced technology and knowledge development, employees' knowledge needs to be updated.
b) Self-learning
   PT. Setiajaya Mobilindo also provides a wall magazine to post the latest information regarding a problem-solving solution for the mechanics. It was also similar to mechanics who had attended training where local web access containing the material and knowledge will be given for them to access through their gadget where they can learn independently.

2. System Thinking
System thinking is one of the most important aspects of learning. PT. Setiajaya Mobilindo has implemented system thinking through:

a) Skill training
   Every employee in Cimanggis branch was required to attend a skill training held by PT. Setiajaya Mobilindo. The training provided employees with the material given by mentors in the classroom and at the end of the training, they would usually be given practical simulations in the field.
b) Product knowledge training
   Product knowledge training was conducted by the company to invent new products. Usually, PT. Setiajaya Mobilindo would send instructors or technical leaders to take part in the head office training. Then, they would distribute the information to PT. Setiajaya Mobilindo employees in the form of training (interview with technical leaders).

3. Mental Model
Training, development, and motivation are the focus of this strategy aiming at increasing the knowledge

quality of the employees in accordance with the three main aspects of capital, which are knowledge, skill, and attitude.

Based on the observation, it was found that learning organization programs, such as Kaizen, were conducted successfully. It was proven by the boarding program that could equate employees' perceptions of Toyota culture. This program was enacted to solve the new mechanics problem, such as:

a) Educational background of new employees graduating from different schools would cause them to have different skill. This perception must be equated or standardized before they do their job.
b) Lack of concern about work Safety and 5s, such as loose bolts, chamois leather left on a car hood, and dirty mechanics' clothing.
c) Lack of understanding and ability regarding basic automotive and periodic maintenance procedure.

With the Onboarding program, it is expected that the company has ideal mechanics as follows:

a) Having good knowledge about Toyota culture
b) Possessing good basic service skills
c) Capable to do periodic maintenance (SBI, SBE, and EM) according to proper SOP

4. Team Learning

In addition to individual learning, it can also be done in a group. Learning in a group allows employees to share their knowledge and experience. It can be done through various scheduled programs in the company such as skill training, product knowledge training, focus group discussion, and cross dealer sharing session. These activities were designed to facilitate employees in acquiring knowledge in order to improve their abilities and competencies. It was proved by Kaizen program, a continuous improvement program to analyze the existing problem and to be solved together.

Problem-solving employing QCC (quality control circle) system was conducted to find the flow of problem. The solution would be adapted as a continuous improvement that could generate large profits and the employees would be rewarded by the management.

5. Personal Mastery

PT. Setiajaya Mobilindo provided opportunities for employees to learn and improve their skills in order to achieve organizational goals. It was according to the company policy requiring employees to have certification according to Toyota Astra Motor (TAM) standard, such as standard training that has to be undergone by all of PT. Setiajaya Mobilindo employees managed by the Training Center Department of PT. Setiajaya Mobilindo.

After completing the on the job training, employees were required to participate in an exam certification with a minimum of 80 graduation score in which they had to retake it next year if they fail. Employees also would be awarded by training incentives according to their level of training if they pass.

To make it easier, after passing the exam, each employee would also be given access to TEAM 21 local web portal (internal learning), which stands for 21st century Toyota Education for Automotive Mastery. This is a learning media for employees in the form of web portals. This web portal contains training material according to their level that can be accessed by all employees who had participated in the training.

It was found that the company had quite a decent training culture where employees had their own personal awareness to expand their knowledge continuously by learning in improving their skills (interview with the result with TL).

Individual learning could also be done independently by employees. They could acquire knowledge from various sources by utilizing various facilities provided by the company such as modules, repair manual CDs, and website (interview with mechanics).

PT. Setiajaya Mobilindo appreciated its employees with learning enthusiasm and they would be provided with a small incentive (interview with the head of the shop). It would certainly encourage them to continue learning.

PT. Setiajaya Mobilindo realized the importance of enhancing and improving employees' capability. Each of them would receive training based on their respective positions. Those who had passed will be rewarded, especially for them who really want to learn. The reward can be in the form of vacation, incentive, or promotion.

3.2 Obstacles in applying learning organization

1. Time

Rapid change often takes place in the automotive industry. In this case, PT. Setiajaya Mobilindo Toyota employees were required to work quickly to be able to keep up with the changes. This demand of work encouraged employees to utilize their work time as optimized as possible in order to achieve their work goal. This type of business caused many employees not having spare time to study or access the portal independently. Their productivity was calculated from the "lifetime", which is the effectiveness of work time utilization. It was used because they were required to complete their work as quickly as possible so that when employees were using their break time, they were required to use it as effectively as possible.

2. Employees Internal Factors

Each employee has their own character and nature. Some liked to and others did not like to learn. It was one of the managerial obstacles in spreading learning awareness for employees. Basically, employee's ability and skill mastery are determined by their comprehension of the given material during training, knowledge application, and experience. There was a relatively new employee but they had better abilities compared to others. They had good skills because they usually helped employees with a higher level of skill (interview with the head of the center).

3. System/Technology

Damaged learning equipment was also an obstacle because they needed to follow the determined procedure in order to buy the equipment, which may interfere with the learning process.

4. Cost

The rapid development of the automotive industry forced PT. Setiajaya Mobilindo to quickly adjust their employees' knowledge in order to adapt to the latest trends. It required training in order to improve

Table 1. EFAS table learning organization application at PT. Setiajaya Mobilindo.

No	Strategies	Grade	Rate	Grade × Rate
OPPORTUNITY				
1	Mechanics are young and interested in technological mastery	0.14	4.76	0.68
2	Technological mastery in finding information is not optimal (gadget), for example utilizing Google for seeking information	0.11	5.27	0.56
3	Mobile phone or other technologies optimization for information gathering	0.11	4.32	0.46
4	Housing complex offers a growing sales opportunity	0.07	5.14	0.37
5	School offers growing sales opportunity	0.07	3.97	0.28
6	Vehicle pickup service	0.14	4.92	0.70
7	After-sales service such as vehicle checking service (home service)	0.14	5.46	0.78
THREAT				
31	Information transfer will be late if you have to be trained first in which can be a threat	0.11	5.38	0.58
32	Emergence of new competitors	0.11	5.11	0.55
TOTAL		1		

Table 2. IFAS table learning organization application at PT. Setiajaya Mobilindo.

No	Strategy Factors	Grade	Rate	Grade × Rate
STRENGTH				
1	Employees are standardized by Toyota certification	0.05	5.76	0.30
2	Experienced employees	0.05	5.78	0.30
3	Complete service center facilities	0.05	5.68	0.29
4	Superior standard (SOP, *Kaizen*, *Kodawari*)	0.05	5.84	0.30
5	Express maintenance service (maximum one hour)	0.04	5.49	0.21
6	Strategic location	0.04	4.95	0.19
7	Community purchasing power	0.04	5.41	0.21
8	Structured training to facilitate good knowledge sharing activity	0.05	5.68	0.29
9	Loyal employees with result orientation	0.04	5.62	0.22
10	Young employees are hungry for knowledge	0.04	5.46	0.21
11	No jealousy among employees if someone is prioritized to have a training	0.03	4.14	0.11
12	Good working cooperation	0.05	5.54	0.28
13	Extra service (the center opens until 8:00 p.m.)	0.04	5.16	0.20
14	Ideal spread of mechanics level	0.04	5.54	0.21
WEAKNESS				
1	Limited equipment forcing customers to wait	0.05	5.35	0.27
2	Traffic jam in the middle of the city	0.04	5.11	0.20
3	SOP is not operated entirely	0.04	5.41	0.21
4	Promotion media is not optimized (no marketing team)	0.04	4.89	0.19
5	A new mechanic has to be accompanied by experienced mechanics, which may interfere with their work	0.04	5.22	0.20
6	New mechanics have different knowledge standard because they were graduated from different vocational schools	0.04	4.27	0.16
7	Lack of cleanliness (dirty clothes, chamois cloth on customer's car, oil marks on the floor, and others	0.04	5.49	0.21
8	Lack of checking on some work such as loose wheel bolts and others	0.04	5.57	0.21
9	You cannot attend training because you have to wait for information from TAM (Toyota Astra Motor)	0.04	5.30	0.20
10	Not every employee can participate in the training because an arrangement has to be made	0.04	4.76	0.18
TOTAL		1		

their knowledge. As it was mandatory and prioritized, the company prepared to cover the cost. However, to send the participant, they had to adapt to the budget.

3.3 SWOT analysis of learning organization at PT. Setiajaya Mobilindo

From the questionnaire, which was distributed to 37 of PT. Setiajaya Mobilindo employees, the following results were obtained.

Table 3. SWOT matrix learning organization application at PT. Setiajaya Mobilindo.

Internal External	Strength (S) – Toyota certified employees – Experienced employees – Complete facilities – Superior SOP – Maximum one hour of service – Strategic location – Available community purchasing power – Structured training – Loyal employees – Young employees – No jealousy among employees – Great teamwork – Available extra service until 8:00 p.m. – Adequate employees' skill	Weakness (W) – Limited equipment – Traffic jam location – Poor SOP – Poor promotional media – New mechanics have to learn from the experienced – Different knowledge of new mechanics – Lack of cleanliness – Lack of control in several works – Employees have to wait for information from TAM for training – Training cannot be conducted simultaneously
Opportunity (O) – Mechanics are thirsty for knowledge – Technology usage is still not optimal – Gadget utilization for obtaining information – Houses are growing – Schools are growing – Vehicle pickup service – After-sales service	SO – Optimize technology as a learning tool – Perform data collection on every house and school with Toyota cars and offer them service discount or promo continuously – Home service availability for who do not have time to go to the center – Shuttle service availability	WO – Android application for customers to check service availability – Separate the learning session so that new mechanics will not interfere with the experienced mechanics work – Improve SOP and control gadget-usage time – Prioritize the given training – Give reward and punishment for implementing or neglecting SOP
Threat (T) – Information transfer will be late if it has to wait for training – Emergence of new competitors	ST – Utilize discussion and knowledge sharing session to solve the problem – More careful customer service preventing them from moving to competitor – Faster customer's complaint handling – Educate customer – Door prize program to attract customer	WT – Give a maximum discount – Create an economic package for customers – Create a door prize for customer – Shuttle service – Conduct self-learning activities

Table 4. Strategy analysis learning organization application at PT. Setiajaya Mobilindo.

Strategy S-O	Strategy S-T	Strategy W-O	Strategy W-T
Utilize technology (Android) as a learning tool	Optimize discussion and knowledge sharing to solve problems	Android application for customers to check service availability	Offer maximum discount
Collect data from houses and schools that possess Toyota and offer them service discounts or promos continuously	More careful customer service preventing them from moving to a competitor	Separate the learning session so that new mechanics will not interfere experienced mechanics work	Create an economic package for customers
Vehicle pickup service for customers	Faster customer's complaint handling	Improve SOP and control gadget-usage time	Create a door prize for customers
Shuttle service for customer's car that broke down or stuck in traffic	Offer door prize for customers	Prioritize the given training. Give reward and punishment for implementing or neglecting SOP	Shuttle service Conduct self-learning activities

4 CONCLUSION

Based on the results, the following conclusions were drawn:

1) PT. Setiajaya Mobilindo Toyota had implemented learning organization concept of Senge's five pillars, namely:
   a) Shared vision: conducting a group discussion and forum to facilitate the application of personal vision into a shared vision.
   b) System thinking: skill and product knowledge training routine and also a comprehensive organizational system.
   c) Mental model: individual performance improvement programs such as Kaizen, 5s, and On-Boarding
   d) Team learning: available collective learning media such as group discussion, forum, training, wall magazine, and, website.
   e) Personal mastery: each individual was given the opportunity to develop themselves in a structured training and technology utilization for self-learning.
2) Learning organization application at PT. Setiajaya Mobilindo that prioritized service had attracted more customers and increased their satisfaction.

REFERENCE

Senge, P.M. 1990. The fifth discipline: The Art & the practice of learning organization. New York: Currency doubleday.

Advances in Business, Management and Entrepreneurship – Hurriyati et al. (Eds)
© 2021 Taylor & Francis Group, London, ISBN 978-0-367-67471-7

Organizational commitment at manufacturing industries in West Java and its implication on employee performance

Z.Z. Noor
Universitas Jayabaya, Jakarta, Indonesia

N. Limakrisna
Universitas Persada Indonesia YAI, Jakarta, Indonesia

ABSTRACT: Manufacturing industries have done bureaucratic reform; one of the elements set back is resource issues of improving mankind. All restructuring management is done to improve the employee's performance. The purpose of this research is to build the model of commitment organization and employee performance. This research took samples from 171 employees. This study is descriptive and examines association causal relationships. The data were processed by using SPSS. The research instruments consisted of data analysis and the hypothesis. The results of the study state that the contribution of competence, leadership, and commitment to explain employee performance is about 81.6%. It means that the competence, leadership, and commitment awarded to an employee together or simultaneously had links and a real impact on the performance of employees.

1 INTRODUCTION

The existence of resources can enhance humanity in a company and also plays a very important role in the company. The potential of every human resource in the company must be good, so that it can provide maximum results. Companies and employees are two things that need each other. If employees succeed in bringing progress to the company, the benefits can be reaped by both parties. For employees, success is self-potential and actual opportunities to fulfill life. For companies, success is a vehicle for company growth and development.

Companies often neglect human resource management; although they have often heard of the importance of human resource management, planned and focused handling, both by companies and individuals as employees themselves, is still rare. In fact, good human resource management will improve employee performance and will increase company productivity.

The business environment is currently facing two main challenges. First, rapid changes in the business environment are accompanied by improvements in the quality and needs of consumers. On the other hand, it turns out that the wants and needs of consumers are not static and continue to develop dynamically. Consumers expect their wants and needs to be met in good quality and satisfaction. Second, increased competition between companies requires each company to run operations in a more efficient, effective and productive manner. To face this challenge, companies need to demand higher employee performance,

which is influenced by the competencies that must be possessed by every employee. If competence is not optimal then employees who perform poorly will be given greater authority and duties. In carrying out their activities, all PT. Yamaha Music Manufacturing Asia factories throughout the world have a strong commitment to creating harmonious relationships as stated in the following commitments:

1. Commitment to customers: As satisfaction for customers, Yamaha will always offer superior quality products and services through sophisticated and traditional technology, creativity and sensitivity to color choices, and they will always be a brand that is recognized for its existence, is trusted, and is full of impressions.

2. Commitment to the collaborating parties: The thing that really makes the Yamaha brand shine is Mintra's working relationship with Yamaha. Yamaha builds trusting relationships within a fair regulatory framework based on the social norms of society, while demonstrating the best performance and self-realization through work, and for the creation of a cheerful corporate climate of pride and confidence.

3. Commitment to society: Prioritizing the safety and preservation of the Earth's environment, adhering to very high moral rules, and as a company that inhabits the community it plays an active role in developing social, cultural, and economic aspects that are good for local communities and the global community.

So, the importance of human resource factors make company management to improve human resources. The main focus of improving human resources, among others, is the factor of employee competency and commitment.

The six characteristics of competency are knowledge, skills, values, self-concept, nature, and motives. These characteristics can be grouped into two large group, namely hard competencies and soft competencies. Hard competence consists of educational background and the training of employees of PT. Yamaha Music Manufacturing Asia, which in general, is quite good, but it must also be upgraded so that employees can carry out production activities well. Soft competence is functional competency, which is leadership, where this competency is an invisible competency, making it more difficult to upgrade.

Based on the above analysis, the important role of competence and commitment to improve employee performance is seen. From the initial observations made on employees at PT Yamaha Music Manufacturing Asia, it can be seen that employee skills, especially competency, leadership, and employee commitment are still low. This can be seen from the fact that there are still employees who are incompetent in carrying out tasks, unable to cooperate in teamwork, and lack a sense of responsibility.

Furthermore, viewed from the performance of employees, it appears that there are limitations to employee performance. This can be seen from the frequent employee procrastination so that the completion of work time is not done on time, is not oriented to the success of the job, and the employment is only given to certain people. Work becomes ineffective and employees work like robots without any creativity and ideas/suggestions. This has caused the performance of employees to be low.

2 RESULTS AND DISCUSSION

In testing, there is a positive influence between competence (x1), leadership (x2), and commitment (x3) on an employee's performance (y).

A. The double correlation test results of the multiple correlation analysis between the competency, leadership, and commitment variables to the performance variable are shown in Table 1:

Table 1. Double correlation between x1; competence, x2; leadership; and commitment, x3, to the y variable of employee performance.

Model Summary[b]

Model	R	R Square	Adjusted R Square	Std. Error of the Estimate
1	.909[a]	.827	.816	.66692

* Source: output data spss.

From the explanation above, a double correlation between competence, leadership, and commitment to employee performance can be seen, indicated by the value of r = 0.909. This means there is a strong and consistent positive correlation between competence, leadership, and shared commitment to employee performance.

B. The correlation significance of the formulation hypothesis is as follows:
$H0 = \beta1 = \beta2 = \beta3 = 0$, meaning that there is no correlation between competence, leadership, and joint commitment to employee performance. $H1 \neq \beta1 \neq \beta2 \neq \beta3 \neq 0$ (is the competence, leadership, and commitment together on performance employees). The significance correlation double at $\alpha = 0.05$ is as follows:

Table 2. Test of significance of multiple correlations ANOVAb.

Model	Sum of Squares	df	Mean Square	F	Sig.
1 Regression	103.548	3	34.516	77.812	.000a
Residual	21.735	49	.444		
Total	125.283	52			

* Source: Output data spss.

The significance value is shown by the amount of statistics produced, namely f-test = 77.812 > f-table = f 0.05; 3; 49 = 2.79 with a p-value of 0,000 < α = 0.05; so, H0 is rejected.

This shows that there is a positive influence between competence (x1), leadership (x2), and commitment (x3) on an employee's performance (y).

C. Multiple regression test results of the multiple regression test analysis on leadership and commitment to employee performance are presented in Table 3:

Table 3. The regression coefficient competence, leadership, and a commitment to employee performance.

Coefficients[a]

Model	Unstandardized Coefficients		Standardized Coefficients		
	B	Std. Error	Beta	t	Sig.
1 (Constant)	32.711	2.791		11.720	.000
Competence	.015	.071	.015	.209	.835
Leadership	.922	.085	.849	10.860	.000
Commitment	.041	.034	.089	1.217	.230

a. Dependent Variable: Performance
* Source: output data spss.

Table 3 above shows the estimation of the model, with a β_0 (intercept) of 32.711; β_1 of

0.015; β_2 of 0.922, and of β_3 of 0.041. The regression equation is as follows:

$$Y = a + b_1 X_1 + b_2 X_2 + b_3 X_{3\backslash} \qquad (1)$$

$$Y = 32.711 + 0.015 X_1 + 0.922 X_2 + 0.041 X_3 \qquad (2)$$

This equation can be defined as the following: (a) A constant of 32.711 means that if the values of competence (x1), leadership (x2), and commitment (x3) are 0, so an employee performance (y) value of 32.711 is positive; (b) the competence regression variable (x1) of 0.015 which means that if competence increased by a unit, then employee performance (y) will increased by 0.015 units assuming the variables of leadership and commitment worth are constant; (c) the regression coefficient variable of leadership (x2) as much as 0.922 means that if leadership is increased by a unit, so employee performance (y) will increase by 0.922 units assuming the variable of commitment and the variable of competence worth are constant; and (d) the regression coefficient variable commitment (x3) as much as 0.041 means that if commitment increased by a unit, so employee performance (y) will increase by 0.041 a unit assuming the variables of competency and leadership worth are constant.

2.1 The significance double regression

The formulation hypothesis is as follows:

$H_0 = \beta_1 = \beta_2 = \beta_3 = 0$ (there is no positive influence of competence, leadership, and commitment together on the performance of employees).
$H_1 \neq \beta_1 \neq \beta_2 \neq \beta_3 \neq 0$ (it is competence, leadership, and commitment together on the performance of employee). The significant regression double is presented in Table 4 below this:

Table 4. Anova.

ANOVA$_b$

Model	Sum of Squares	df	Mean Square	F	Sig.
1 Regression	103.548	3	34.516	77.812	.000a
Residual	21.735	49	.444		
Total	125.283	52			

* Source: output data spss.

Table 4 shows the statistics f-test = 77.812 > f-table = f0.05; 3; 49 = 2.79 with p-score (0.000) < α = 0.05 so H0 is rejected. It means there is an influence of competence, leadership, and commitment together on the performance of employees. Thus, the hypothesis of the research is proven received or supported.

2.2 Coefficient determination

Measures of competence, leadership, and commitment and employee performance variables are measured by the coefficient of determination (R2). The size of the coefficient of determination is presented in Table 5 below:

Table 5. The coefficients of determination of competence, leadership, and commitment to employee performance.

Model Summary^b

Model	R	R Square	Adjusted R Square	Std. Error of the Estimate
1	.909^a	.827	.816	.66602

* Source: output data spss.

Table 5 shows the R value of 0.909 and a value of R2 of 0.816, which means the contribution of the variables of competence, leadership, and committed explain the performance variable of employees by 81.6% and the remaining 18.4% is described by other variables that are not included in the model.

2.3 Hypothesis

The influence of competence, leadership, and commitment of employee performance is positive and significant indicated by a value of correlation coefficient of 0.909 with probabilities significance 0.000 < 0.05. The influence of significance is also shown by the value of double regression, namely an f-test of 77.812 is greater than f-table = f0.05; 3; 49 = 2.79. This means that competence, leadership, and commitment awarded to an employee together or simultaneously had links and a real impact on the performance of employees.

3 CONCLUSION

This study can be concluded that:

1. There is a positive and significant influence of competence on the performance of employees, so that employee performance can be increased by improving competence.
2. There is a positive and significant influence of leadership on the performance of employees, so that employee performance can be increased by improving leadership.
3. There is a positive and significant influence of commitment on employee performance, so that employee performance can be increased by improving commitment.
4. There is a positive and significant influence from competence, leadership, and commitment together on the performance of employees.

5. The results of this research said that the variable of leadership had a dominant influence on employee performance, this can be seen from the value of the coefficients of determination more leadership that is greater than the variables of competence and commitment.

REFERENCES

Achua, Christopher F. and Robert N. Lussier, (2010). Leadership. Ohio: South Western Cengage Learning.

Amirullah, 2002, Perilaku Konsumen, edisi pertama, cetakan pertama, Penerbit: Graha Ilmu, Yogyakarta

Colquitt, Jason A, Jeffery A. Lepine and Michael J. Wesson. 2013. Organization Behavior. Singapore: McGraw-Hill.

Davis, Keith dan John W. Newstrom, (2007), "Organization Behavior"; Human Behavior at Work. Singapore: McGraw-Hill.

Dessler, Gary. "Human Resource Management". Pearson Prentice Hall. 2005.

Febriani, I. S. 2004. Hubungan antara Kepuasan Kerja dan Keterlibatan Kerja dengan Komitmen Organisasi pada Karyawan PT Astra International Tbk-Isuzu Cabang Semarang. Skripsi. Semarang: Fakultas Psikologi Universitas Katolik Soegijapranata.

Gibson et al., (2009) Organization: Behavior, Structure, Processes, thirteenth edition Singapore, McGraw-hill internasional edition.

Gibson, James L., John M. Ivancevich, James H. Donelly, Robert Konopaske, (2012), Organizations, Behavior, Structure Process, Fourth Edition, Boston, USA Mc.Graw Hill Inc.

Gitosudarmo, Indriyo dan I Nyoman Sudita, 2008, Perilaku Organisasi, edisi pertama. BPFE, Yogyakarta

Handoko, Hani, Manajemen Personalia dan Sumber Daya Manusia. Yogyakarta: BPFE UGM, 2000.

Hani Handoko, T. 1995. Manajemen, BPFE, Yogyakarta

Hasibuan, Malayu S. P. 1994. Manajemen Sumber Daya Manusia dan Kunci Keberhasilan, Haji Masyarakat Agung, Jakarta.

Hersey, Paul, Kenneth H. Blanchard and Dewey E. Johnson, 2012. Management of Organization Behavior 10th Edition, Prentice Hall.

Hodge, B. J. & Anthony, W. P. 1988. Organization Theory. America: Allyn & Bacon (third ed.)

Hutapea, Parulian dan Nurianna Thoha, 2008, Kompetensi Plus: Teori, Desain, Kasus dan Penerapan untuk HR dan Organisasi yang Dinamis, Penerbit: Gramedia Pustaka Utama, Jakarta

Ivancevich, John M. (2008). "Human Resources Management 10th". Singapore: McGraw-Hill.

Jonathan, Sarwono. 2006. Metode Penelitian Kuantitatif dan Kualitatif. Yogyakarta: Graha Ilmu

Jones, Gareth R., (2013), Organizational Theory, Design And Change, Seventh Ed., Boston, USA, Pearson.

Kaihatu, Thomas Stefanus dan Wahju Astjarjo Rini, (2007). "Kepemimpinan Transformasional Dan Pengaruhnya Terhadap Kepuasan Atas Kualitas Kehidupan Kerja, Komitmen Organisasi Dan Perilaku Ekstra Peran: Studi Pada Guru-Guru SMU Di Kota Surabaya". Jurnal Manajemen dan Kewirausahaan, Vol.98 No.1, pp: 49–61.

Kerlinger, Fred.N. 2003. Azas-azas Penelitian Behavior. Yogyakarta: Gajah Mada Universitas Press

Knoers, A.M dan Hadinoto, S.R, 2001, Psikologi Perkembangan: Pengantar dalam Berbagai Bagiannya, Penerbit: Gajah Mada University Press, Yogyakarta

Kreitner, Robert dan Angelo Kinicki. " Perilaku Organisasi", Jilid 2, Jakarta:Salemba Empat, 2005.

Kreitner, R. dan Kinicki, A. (2008). "Organizational Behavior: Key Concepts, Skill & Practices". Boston: McGraw-Hill/Irwin.

Kriyantono, Rachmat. 2007. Teknik Praktis Riset Komunikasi. Jakarta: Kencana Prenada Group

Kuntjoro, ZainuddinSri. 2002. Komitmen Organisasi. Http://www.epsikologi.com/masalah/250702.htm.

Lumley, E.J., M. Coetzee, R. Tladinyane, dan N. Ferreira., (2011). "Exploring The Job Satisfaction And Organizational Commitment Of Employees In The Information Technology Environment". Southern African Business Review. Vol. 15, No. 1. Hh. 100–118.

Luthans, Fred, (2008). "Organizational Behavior". Singapore: McGraw-Hill.

Malthis, Robert L. dan John H. Jackson, 2006, Human Resource Management (Manajemen Sumber Daya Manusia), Edisi Sepuluh, Terjemahan: Diana Angelica, Penerbit: Salemba Empat, Jakarta

Mangkunegara , Anwar Prabu. Manajemen Sumber Daya Manusia, Bandung: Badan penerbit. PT. Rosdakarya, 2005.

Manullang, 2002, Manajemen Personalia, Penerbit: Ghalia Indonesia, Jakarta

Mathis, L. Robert dan Jackson, John H. "Manajemen Sumber Daya Manusia". Jilid 1. Jakarta: Salemba Empat. 2009.

Moh As'ad. 2003. Psikologi Industri.Yogyakarta: Libery

Muhammad Agus Tulus. 1995. Manajemen Sumber Daya Manusia Panduan Mahasiswa, P. T. Gramedia Pustaka Umum, Jakarta.

Oktorita, Dkk. 2001. Hubungan Antara Sikap Terhadap Penerapan Program K3 dengan Komitmen Karyawan Pada Perusahaan. Jurnal Psikologi. No 2 (116–132).

Paula, Crouse, Doyle Wendy and Young Jeffrey D. 2011. Trends, Roles, and Competencies in Human Resource Management Practice: A Perspective from Practitioners in Halifax, Canada. Proceedings of ASBBS, Vol. 18 No. 1.

Prasetyo, Lis. " Pengaruh Gaya Kepemimpinan Terhadap Kinerja",Jurnal NeoBisnis,Vol 2,No.2, Desember 2008

Prihadi, S, 2004, Kinerja, Aspek Pengukuran, Penerbit: PT. Gramedia Pustaka, Jakarta

Priyatno, Duwi. 2008. Mandiri Belajar SPSS, Cetakan Ketiga. Yogyakarta: Media Kom

Quick, James Campbell and Debra L. Nelson, (2009). "Organizational Behaviour: Foundations, Realisties and Challenges". The United States: Thomson South-Western.

Reza, Aditya."Pengaruh Gaya Kepemimpinan, Motivasi dan Disiplin kerja Terhadap Kinerja Karyawan". Skripsi Mahasiswa Universitas Diponogoro, 2010.

Robbins, Stephen P. and Timothy A. Judge (2009) Organizational behavior. Singapore: Person international.

Robbins, Stephen P. and Judge Timothy A. (2013) Organizational behavior, fifteenth Edition, Pearson International Edition, USA.

Robbins, Stephen P, dan Timothy A. Judge, (2009). "Organizational Behavior". New Jersey: Prentice Hall.

Robbins, Stephen P, dan Timothy A. Judge, (2013). "Organizational Behavior". New Jersey: Prentice Hall.

Santosa dan Ashari. 2005. Analisa Statistik dengan Microsoft Excel & SPSS. Yogyakarta: Andi

Schermerhorn, Hunt, Richards, Osbron, (2011), Organizational Behavior, Emerging Knowledge And Practice For The Real World, Fifth Edition, USA, Mc.Graw Hill, Inc.

Sedarmayanti."Manajemen Sumber Daya Manusia". Cet kelima. Bandung: PT Refika Aditama, 2011.

Sekaran. 2000. Research Methods for Business, Askill-Building Approach, Thirt Edition. America: John Woley & Sons,Inc.

Siregar, Sofyan. 2010. Statistik Deskriptif Untuk Penelitian. Jakarta: Rajawali Pers

Siswandoko, Tjatjuk, dan Darsono. " Sumber Daya Manusia Abad 21" . Jakarta: Nusantara Consulting.2011.

Sparrow, Paul R, Anil Chandrakumara, and Nelson Perera, (2010). " Impact of work values and ethics on citizenship and task performance in local and foreign invested firms a test in a developing country context". Centre for health development – CHSD Sydeny Business School, pp.1–8.

Sugiyono, 2005, Statistika untuk penelitian, Bandung: Alfabeta.

Sugiyono, 2009, Metode Penelitian Bisnis. Cetakan ke -14. Bandung. Alfabeta

Sukmalana, Soelaiman, (2010), Perencanaan SDM (Konsep, Proses, Strategi dan Implementasi), Jakarta PT. IPU Publishing.

J. Supranto & Nandan Limakrisna. 2009. Statistika Untuk Penelitian Pemasaran dan Sumber Daya Manusia. Jakarta. Mitra Wacana Media.

Susanty, Aries dan Rizqi Miradipta, (2013). "Analysis of the effect of attitude towards works, organizational commitment, and job satisfaction, on employee's job performance". European Journal of Business and Social Sciences, Vol. 1, No. 10, pp 15–24.

Sutrisno, Edy. "Manajemen Sumber Daya Manusia". Ed.1. Jakarta: Kencana Prenada Media Group. 2009.

Suyadi Prawirosentono. 1999. Manajemen Sumber Daya Manusia: Kebijakan Kinerja Karyawan, BPFE Yogyakarta.

Tampubolon, Biatna Dulbert. " Analisis Faktor Gaya Kepemimpinan dan Faktor Etos Kerja terhadap Kinerja Pegawai pada Organisasi yang telah Menetapkan SNI 19–9001-2001", Jurnal Standarisasi Vol. 9 No 3, 2007

Thoha, Miftah. Kepemimpinan Dalam Manajemen, Jakarta, PT. Rajagrafindo Persada, 2010

Umam, Khaerul. "Perilaku Organisasi", Bandung: CV Pustaka Setia, 2010.

Veithzal Rivai, 2006. Kepemimpinan dan Perilaku Organisasi. Jakarta: PT. Raja Grafindo Persada.

Veithzal Rivai dan Ella Jauvani Sagala. "Manajemen Sumber Daya Manusia Untuk Perusahaan". Cet ke 3. Jakarta: Rajawali Pers. 2010

Wibowo, Manajemen Kinerja, Jakarta: Rajawali Pers, 2012

Yukl, Gary A. (2010). "Leadership in Organizations". New Jersey: Prentice Hall.

Yunarsih, Tjutju dan Suwatno, 2008, Manajemen Sumber Daya Manusia, Teori, Aplikasi dan Isu Penelitian, Cetakan Kesatu, Penerbit: Alfabeta. Bandung.

The influence of motivation and climate of work on officer performance at PT Arsel Medical Technology

A. Lasminingrat & N. Djunaedii
Universitas Winaya Mukti, Bandung, Indonesia

ABSTRACT: This report aims to understand the influence of motivation on Officer performance at PT. Arsel Medical Technology, to know the effects of the climate at work on Officer performance at PT. Arsel Medical Technology, and to know the influence of motivation and work climate together on Officer performance at PT. Arsel Medical Technology. The research methods used in this research are the descriptive method and causality method, the investigation type is causality, and the time horizontal is cross-sectional. Based on the research, it was found that the motivation influences on officer performance at PT. Arsel Medical Technology. There are a few things to be considered by the company, that is, a system of compensation such as a transportation allowance, holiday, allowance, overtime rewards, and incentives in value, does not satisfy the desires of employees. One of the compensation systems is to motivate employees to realize the purpose of a company that has been set. In general compensation is given in return for the behavior at work or work performance of a person or group. Compensation system connecting of the incentives and performance is not based on seniority or the number of working hours. Climate work will have an influence on officer performance. Climate work was a sufficient consideration in order to increase the motivation to work. Climate work is associated with performance and compensation; it is accepted that climate work and performance will interact. When working climate will step up the commitment of the company is a good performance, while the working poor will to behavior to the company. And the influence of motivation and performance influence climate work, but if viewed as partial, it turns out that the more dominant influence is the employee performance, so the company is expected to do more repair work, and stimulate motivation in order to keep the employee performance high.

1 INTRODUCTION

Globalization is a phenomenon where the G8 industrialized nations have done massive investments all over the world. Most investments are implanted in developing countries. And Indonesia is a country that is not in the state of being free from the credit of the foreign investment.

The presence of foreign companies in Indonesia has a significant influence to the economy. At the macro level, the entry of foreign direct investment influenced the income levels of countries. At the micro level, local companies have got to defend themselves when dealing with those new competitors, especially against companies that are in the same type of industry.

An increase in the performance of employees cannot be separated from the company capacity in the fulfilment of various demands on the employees, which in this case, employees demand for the provision of compensation in conformity with expectation to the employees' motivation. It is important because motivation can be the cause, channeling, and supporters of a person's behavior, that is, the person is willing to work

hard and be enthusiastic to achieve optimum results. According to Susanty & Baskoro (2012), "motivation is one of the factors either directly or indirectly affecting the low level of employment and high satisfaction of employee performance that will had a substantial impact on the development of the company and, therefore, there should be the provision of motivation toward an employee's need to receive special attention from the company management, so that the employees can be maintained next year and it is expected to keep increasing employee performance." So that it will have a positive impact on employees' work satisfaction.

To anticipate that PT. Arsel Medical Technology should improve the quality of human resources. PT. Arsel Medical Technology is trying to fix its compensation policies to be able to improve the job satisfaction of employees by increasing job satisfaction, and workers tend to be motivated to improve their performance for the company. At the same time, employees are expected to create and work in a climate that is conducive and optimized and sustainable, so that employees can provide the results of the optimal.

Table 1. The annual performance.

No	Year	Performance	
		Target	Achievement
1	2016	>75%	74.50%
2	2017	>75%	74.30%
3	2018	>75%	74.75%

*sources: HRD Div. PT. Arsel medical technology 2018.

The problem is relating to the employee performance at PT. Arsel Medical Technology that is not maximum and there should be a pretty serious handling. It can be seen from the low support from the management and marketing division. This can be seen from Table 1 the achievement of annual performance achieved by PT. Arsel Medical Technology.

Table 1 shows that employee performance experienced a fall in 2017, and in the year 2018, has been on an increase compared to the previous year. However, in overall view of these achievements, it hasn't met the standards of reaching the 75% target. It is suspected that this is because they are not optimal in terms of the level of encouragement to an employee and they also have not fully responded to employees. So that, it had an impact on motivation and the work climate. Even though there has been increases from year to year, but they have not been able to increase the motivation to employees. Employees are still less motivated in working.

According to Robbins (2006), "factors influencing the dissatisfaction of employees include the working conditions and policy organization." The provision of motivation will exactly encourage employees to change their behavior to grow and flourish to find success in working.

From the description above, a research needs to be conducted on motivational influence, working towards the performance of the climate and employees at PT. Arsel Medical Technology.

1.1 Motivation

Motivation, deriving from the Latin word *movemore,* means impulse or to move. Motivation is only shown in human resources in general and subordinates in particular. "Motivation is to steer the potential subordinate and cooperate productivity achieved and created by the set of objectives" (Hasibuan 2009).

According to Kartika & Kaihatu (2010), "a definition of motivation is a whole process of giving motive worked to the subordinate in such a way that they want to work with loan by the achievement of the aims of the organization with efficient and economical."

Sopiah (2008) said that "motivation is a situation where business and volition is given to someone handed to results or specific objectives. Results in this context can be productivity, attendance, or other behavior creative work."

In Maslow's theory, as mentioned by Hasibuan (2006), the theory of a hierarchy is to follow the theory of plural behavior or works because of the push to fulfill the needs.

Herzberg in Thoha (2007) developed the "theory of motivation." Two factors, according to this theory, affect any condition of a job motivation, it is also called satisfier or intrinsic motivation while the factor of health (hygiene) is called a dissatisfier.

1.2 Climate work

The term "climate" here is figurative (metaphor). Figuratively, it is a form of the word that in it is a term or phrase that clears. It means that it can be applied to a different situation depending on the purpose, for example "this place is like a petting zoo." Although comparison is figurative, the comparison gives information about the contents, the structure, and the new the meaning of the situation.

According to Davis (2002), "the climate work can be interpreted as a human environment in which workers or organization work and its existence can be touched or seen."

According to Robbins (2006), "employment climate is a term used to load the variables referring to value, behavior trust, and the basic principles that act as a foundation for a system management organization."

1.3 Employee performance

According to Mangkunegara & Anwar (2003), "performance is the result of a work as the quality and quantity of reached by someone on the carried out in accordance with the responsibility of who is given obtained from a function and a particular occupation or a particular activity for a certain time." Meanwhile, the performance, according to Lazer and Wikstrom in Rivai (2005), is the technical ability, conceptual ability, and the ability of interpersonal relations.

1.4 The framework of ideas

In the activities of the organization, human resources are a very important aspect, so that an attention must be given to the type of activities that were adequate. Failure in the management of human resources in an organization or company will threaten the company's survival.

Motivation and employment climate are important aspects that should get attention from an organization in improving the performance of its employees. Motivation and employment climate are conducive to help employees feel compelled or excited in the completion of a job they brought, as for extrinsic motivators and from the superior and support in the form of support from a situation or work environment conducive to make employees more productive in a company and improve its performance.

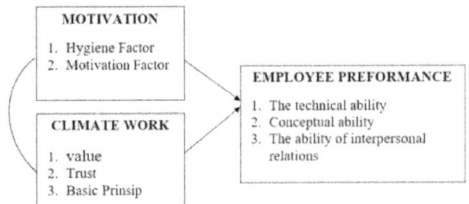

Picture 1. The influence of motivation and climate work on employee performance.

2 METHODS

2.1 Sample and population

The population are objects of research for the source of a research. According to Sugiyono (2007), "populations are objects or subjects that are the quantity and specific characteristics set by researchers to study and then drawn conclusions." The population of this research is 225 employees in PT Arsel Medical Technology located at Soekarno Hatta street no.725, Bandung, and the samples were 83 employees.

This research is designed and descriptive verification. According to Sugiyono (2007), "descriptive research is that research conducted to determine values, or independent variables, either one variable or over, without strike or read with other variables," while according to Arikunto & Suhardjono (2006), "research verification wants to test the hypothesis of a truth research conducted through data collection in the field. Research was intended to know several variables, namely the motivation, climate work, and employee performance."

Methods used in this research include the methods of description survey and explanatory survey implemented through data collection in the field. The type of investigation on this research is causality, because it tested the relationship for a result of the variables. This study of cross-sectional category includes a bunch of data used to examine a phenomenon in one time. The type of investigation in the research is causality. It is the type of research that relates between independent variables and the effect, in this case the motivation and work climate, while the dependent variable is the employee performance.

2.2 Technique data analysis

The technique used in this research was the use of literature study and studying in the field, in which the two techniques were used for research using collected data before the spread instrument. Firstly, the data were tested, in which the reliability and the validity testing of the instrument were performed by the use of correlation product moment (Sugiyono 2007).

$$r_b = \frac{\Sigma XY - \frac{(\Sigma X(\Sigma Y))}{n}}{\sqrt{\left\{\Sigma X^{2-\frac{(\Sigma X)^2}{n}}\right\}\left\{\Sigma Y^{2-\frac{(\Sigma Y)^2}{n}}\right\}}} \quad (1)$$

Information: rb = Pearson item with a correlation coefficient between variables that are concerned; N = the number of respondents; x = the number scores as a whole for item questions; Y = the number scores as a whole for item questions.

According to Sugiyono (2007), "Instrument expressed as valid if a correlation coefficient > 0.3." Meanwhile, this study used Cronbach's alpha using SPSS to test the reliability of the instrument.

Before the data were analyzed, firstly the data were processed. The data processing was collected from the interviews and the questionnaire that was categorized into 3 steps of preparation. This included activities such as research preparation, collecting and checking the questionnaire sheets, and putting a value on scoring/the judgment that has been set.

To know the quality of the motivation, and the instrument of climate work, it was analyzed the data processed attitude respondents of grains of the questionnaire in order to see the result whether it is a positive/negative.

The analysis technique is to take interval measurement data. Therefore, through the interval sequence method, data transformation is carried out in the following stages:

1. Note each item question.
2. Count frequency answer for each item (f), how respondents scored 1, 2, 3, 4, or 5.
3. (p) for the proportion by means of dividing the frequency with the number of respondents.
4. Count the proportion cumulative (p).
5. Count z value for each of the proportions cumulatively obtained by using normal table.

Scale Value
$$= \frac{\text{(Density at lower limit)} - \text{(Density at upper limit)}}{\text{(Area below upper limit)} - \text{(Area below lower limit)}} \quad (2)$$

6. Counting score (value the transformation) for each choice between the score:

$$\text{Score} = \text{scale Value} + \text{Scale Value}_{minimum} + 1 \quad (3)$$

To test the effect of variables, research on employee performance results of the tabulation of applied research and analysis on the path analysis approach. According to Wirasasmita (2004), "contribution influence analysis and comparison of the effect of contributions (overall or partial), the path of influence analysis between variables can be determined and expressed as a percentage of incentives to see the causal relationship (x1), climate variables employment (x2), and employee performance (intermediate variables) y."

The hypothesis proposed is as follows: There is an influence of motivation and work climate on employee performance.

If the research hypothesis is stated in hypothesis statistics, the hypothesis formulation is as follows:

Reject Ho if $F_{\text{hitung}} \geq F_{\text{tabel}(0.05)(n-k-1)} \rightarrow$ Then there is an influence of motivation and job climate against employee performance.

Receive Ho if $F_{\text{hitung}} \leq F_{\text{tabel}(0.05)(n-k-1)} \rightarrow$ Then there is no influence of motivation and climate at work on the performance of employees.

The statistics used are as follows:

$$F = \frac{(n - k - 1) \sum_{i=1}^{k} P_{YX} r_{YX_i}}{k \left(1 - \sum_{i=1}^{k} P_{YX} r_{YX}\right)} \quad (4)$$

The criteria, refuse H_0 if $F > F_\alpha$; $_{(k,n-k-1)}$ are obtained from table with distribution where $f\,\alpha = 5\%$ and $db_1 = k$ and $db_2 = n-k-1$.

After the general hypothesis (simultaneous in significant research), then it can be done testing for sub-hypotheses, these comprise:

1. To depend on the motivation of employee performance at PT. Arsel Medical Technology.
2. Work to depend on the climate of employee performance at PT. Arsel Medical Technology.

The calculation analysis is as follows:

1. Determine a correlation coefficient double Rx1x2, Rx2x3, Rx1x3 using formula:

$$R_{x_1x_2y} = \sqrt{\frac{R_{yx_1}^2 + R_{yx_2}^2 - 2R_{yx_1}R_{yx_2}R_{x_1x_2}}{1 - R_{x_1x_2}^2}} \quad (5)$$

$$R_{x_2x_3y} = \sqrt{\frac{R_{yx_2}^2 + R_{yx_3}^2 - 2R_{yx_2}R_{yx_3}R_{x_2x_3}}{1 - R_{x_2x_3}^2}} \quad (6)$$

$$R_{x_1x_2y} = \sqrt{\frac{R_{yx_1}^2 + R_{yx_2}^2 - 2R_{yx_1}R_{yx_2}R_{x_1x_2}}{1 - R_{x_1x_2}^2}} \quad (7)$$

2. Determine the causality of px1x2, px2x3, px1x3, pyx1, pyx2, pyx3.
3. Other counting the influence:

$$P_{\varepsilon_y} = \sqrt{1 - R^2 Y (X_1 X_2 X_3)} \quad (8)$$

4. Decision H0 acceptance or rejection
5. Decision criteria

Reject Ho if $F_{\text{count}} \geq F_{\text{tabel}(0.05)(n-k-1)}$ $\quad (9)$

Received Ho if $F_{\text{count}} \leq F_{\text{tabel}(0.05)(n-k-1)})$ $\quad (10)$

When, $t = \dfrac{P_{yx_i}}{\sqrt{\dfrac{\left(1 - R_{y(x_1 x_2 x_3)}^2\right) C_{ii}}{(n-k-1)}}}$ $\quad (11)$

Reject H_0, $t_0 i \geq t_{(\alpha;n-k-1)}$ $\quad (12)$

3 RESULTS AND DISCUSSION

3.1 *The research results*

Table 2. Characteristics of respondents on the basis of sex.

Basis of sex	Amount	%
Male	54	65.06
Female	29	34.94
Amount	83	100

* sources: Data processed, 2018.

Table 3. Characteristics of respondents based on age.

Based on Age	Amount	%
<25 tahun	4	4.82
26–30 tahun	7	8.43
31–35 tahun	15	18.07
36–40 tahun	30	36.14
41–45 tahun	13	15.66
46–50 tahun	9	10.84
>51 tahun	5	6.02
Amount	83	100

* sources: Data processed, 2018.

Table 4. Characteristics of respondents based on education.

Based on Education	Amount	%
D3	11	13.25
S1	46	55.42
S2	26	31.33
Amount	83	100

* sources: Data processed, 2018.

Table 5. Characteristics of respondents' time of work.

Time of work	Amount	%
<5 tahun	7	8.43
6–10 tahun	11	13.25
11–15 tahun	47	56.63
16–20 tahun	10	12.05
21–25 tahun	6	7.23
26–30 tahun	2	2.41
Jumlah	83	100

* sources: Data processed, 2018.

Table 6. Test validity of instrument for variable of motivation.

The Statement	r hitung	r kritis	Description
P1	0.823	0.30	Valid
P2	0.319	0.30	Valid
P3	0.407	0.30	Valid
P4	0.695	0.30	Valid
P5	0.642	0.30	Valid
P6	0.505	0.30	Valid
P7	0.873	0.30	Valid
P8	0.590	0.30	Valid
P9	0.634	0.30	Valid
P10	0.385	0.30	Valid
P11	0.560	0.30	Valid
P12	0.519	0.30	Valid
P13	0.558	0.30	Valid
P14	0.320	0.30	Valid
P15	0.722	0.30	Valid

* sources: Data processed, 2018.

3.2 Data analysis techniques

3.2.1 The validity

Table 7. The validity of instrument for the climate work variable.

The Statement	r hitung	r kritis	Description
P1	0.406	0.30	Valid
P2	0.739	0.30	Valid
P3	0.855	0.30	Valid
P4	0.692	0.30	Valid
P5	0.418	0.30	Valid
P6	0.700	0.30	Valid
P7	0.813	0.30	Valid
P8	0.855	0.30	Valid
P9	0.587	0.30	Valid
P10	0.798	0.30	Valid
P11	0.697	0.30	Valid
P12	0.739	0.30	Valid
P13	0.831	0.30	Valid
P14	0.742	0.30	Valid
P15	0.805	0.30	Valid

* sources: Data processed, 2018.

Table 8. The instrument for the validity of employee performance variable.

The Statement	r hitung	r kritis	Description
P1	0.695	0.30	Valid
P2	0.765	0.30	Valid
P3	0.941	0.30	Valid
P4	0.876	0.30	Valid
P5	0.738	0.30	Valid
P6	0.453	0.30	Valid
P7	0.784	0.30	Valid
P8	0.729	0.30	Valid
P9	0.797	0.30	Valid
P10	0.696	0.30	Valid
P11	0.688	0.30	Valid
P12	0.834	0.30	Valid
P13	0.373	0.30	Valid
P14	0.455	0.30	Valid
P15	0.397	0.30	Valid

* sources: Data processed, 2018.

3.2.2 The reliability testing

3.2.2.1 Reliability test for variable of motivation

After calculating the instrument of work motivation, we got $\propto$ of 0.829, and this meets the test criteria. Therefore, all of that, the statement of the instrument of the work motivation variable is declared reliable. So that it can be used as a statement in research. With the reliability of these instruments, it is expected to reduce errors in this study.

3.2.2.2 Reliability test for the variable of climate work

After the completion of the calculation for the instrument of climate at work, it obtained $\propto$ as much as 0.772 in the test and it meets the criteria. That is why the work climate variable is considered reliable.

3.2.2.3 The employee performance reliability for variables

After this calculation on employee performance, it obtained $\propto$ of 0,729 and it meets the criteria. So that, all instruments in the research are considered reliable.

3.2.2.4 Hypothesis

Based on the calculation on statistics, where $1/2 \propto = 0.05$, it shows that the $F_{count} = 89.476$ and $F_{table} = 000$. It means $F_{count} \geq F_{table}$. Thus, it refuses H_0 and accepts that the climate at work and the motivation have an influence on employee performance at PT. Arsel Medical Technology.

H_0: $Px1 = Px2 = 0$, there is no influence of motivation and work climate against the employee performance in PT. Arsel Medical Technology simultaneously.

H_1: $Px1 \neq Px2 \neq 0$, there is an influence of the motivation and climate at work on employee performance in PT. Arsel Medical Technology simultaneously.

The direct effect of the motivation on employee performance is $(PYX1)^2 = (0.55)^2 = 0.3025$ or 30.25%, and the indirect effect on employee performance is $(pyx1) \times (px2\ x1) \times (pyx2) = 0.55\ 0.81 \times 0.33 = 0.14705$ or 14.70%, and the influence of the motivation that can be donated to employee performance is 44.95%.

The direct effect of climate at work on the performance of an employee is $(PYX2)^2 = (0.33)^2 = 0.1089$ or equal to 10.89%, and climate has an indirect effect on the performance of employees work of $(pyx1) \times (px2\ x1) \times (pyx2) = 0.55 \times 0.81 \times 0.33 = 0.14705$ or 14.70%, and the effects of climate at work that can be donated on the performance of employees are as much as 25.59%.

The influence of motivation and climate at work on employee performance in PT. Arsel Medical Technology = 44.95% + 25.59% = 70.54%.

From these results, it proves that motivation and climate at work influence employee performance with the level of influence that could be donated by motivation and climate at work on employee performance by around 70.54%, meanwhile the remaining fund of 29.46% is influenced by other factors outside motivation and climate at work.

4 CONCLUSION

1. There is an influence of the motivation to employee performance in PT Arsel Medical Technology of as much as 44.95% because it is supported by dimensional motivation.

2. There are effects of climate at work on the performance of employees in PT Arsel Medical Technology of as much as 25.59% because it is supported by work climate dimensions at work.
3. There are effects of motivation and work climate of employee performance in PT Arsel Medical Technology of as much as 70.54% because it is supported by dimensions of motivation and climate at work.

REFERENCES

Arikunto, S. & Suhardjono, S. 2006. *Penelitian tindakan kelas*. Jakarta: Bumi Aksara.

Davis, K. 2002. *Human behavior at work, organizational behaviour, seventh edition*. New York: McGraw Hill, Inc.

Hasibuan, M. S. P 2006. *Manajemen: dasar, pengertian, dan masalah edisi revisi*. Jakarta: Bumi Aksara.

Hasibuan, M. S. P. 2009. *Manajemen sumber daya manusia*. Jakarta: Bumi Aksara

Kartika, E. W. & Kaihatu, T. S. 2010. Analisis pengaruh motivasi kerja terhadap kepuasan kerja (studi kasus pada karyawan restoran di Pakuwon Food Festival Surabaya). *Jurnal manajemen dan kewirausahaan* 12(1): 100.

Mangkunegara, A. A. & Anwar, P. 2003. *Evaluasi kinerja SDM*. Bandung: Refika aditama.

Rivai, V. 2005, *Manajemen sumber daya manusia untuk perusahaan dari teori ke praktik*. Jakarta: Rajawali Pers, Raja Grafindo Perkasa

Robbins, S. P. 2006. *Prinsip-prinsip perilaku organisasi*. Bandung: Erlangga.

Sopiah, S. 2008, *Perilaku organisasional*. Yogyakarta: ANDI

Sugiyono, S. 2007. *Metodelogi penelitian bisnis*, Bandung:CV Alfabeta.

Susanty, A. & Baskoro, S. W. 2012. Pengaruh motivasi kerja dan gaya kepemimpinan terhadap disiplin kerja serta dampaknya pada kinerja karyawan (studi kasus pada pt. Pln (persero) apd semarang). *J@ Ti Undip: Jurnal Teknik Industri* 7(2): 77–84.

Thoha, M. 2007, *Perilaku organisasi konsep dasar dan aplikasinya*. Jakarta: Raja Grafindo Persada.

Wirasasmita, Y. 2004. Penggunaan analisis jalur dalam penulisan tesis dan desertasi. Fakutas Ekonomi Universitas Padjadjaran, Bandung.

Advances in Business, Management and Entrepreneurship – Hurriyati et al. (Eds)
© 2021 Taylor & Francis Group, London, ISBN 978-0-367-67471-7

The analysis of economic literacy and social environment on lifestyle and its impact on students' consumption behavior

E. Mulyana, Tetep, Jamilah, A. Maulana & O. Hermanto
Institut Pendidikan Indonesia, Garut, Indonesia

ABSTRACT: The purpose of this research is to analyze the influence of economic literacy and the social environment on lifestyle and its impact on students' consumption behavior. This research uses quantitative approaches with survey methods. The data collection employed a questionnaire as the instrument. The population in this study consists of students in the Department of Office administration of 13 vocational high schools, which are accredited by the Indonesian Board of National Accreditation for Schools with an A (very satisfying) score in Garut City. The determination of the number of samples used the formula described by Taro Yamane. A proportionate random sampling technique was used in pulling samples. The results show that the social environment has a positive impact on lifestyle, while economic literacy does not have any impact on lifestyle. The data analysis techniques employed structural equation modelling (SEM). The social environment affects student consumption behavior both directly and indirectly, while economic and lifestyle literacy directly impact student consumption behavior. The conclusion is that the high and low behavior of student consumption is influenced by economic literacy, social environment, and lifestyle.

1 INTRODUCTION

The lifestyle of hedonism is very appealing to teenagers, as it only wants pleasure. Such behavior will become accustomed to and become a culture. The motivation of hedonists according to Arnold & Reynolds (2003), is a behavioral-driven buying activity with five senses, delusions, and emotions that make the pleasure and enjoyment of the material become the main goal of life. According to Gardner & Steinberg (2005), teenagers are more likely to do risky behavior than adults and the influence of peers plays an important role in explaining their behavior during adolescence. Meanwhile, Harari & Hornik (2010) suggested that peers influence the involvement of adolescent consumer products.

Laursen (2005) suggests that their group activities are also not only through physical meetings, but also through social media networks, such as Facebook, Twitter, Instagram, WhatsApp, line, etc. They spend their free time-sharing information and experience with either a schoolmate or a friend in a club or an organization. According to the research of Ioanas & Stoica (2014), social networks have a role in influencing consumer behavior in a virtualized environment, especially when a level of exposure of messages and relationships is made between the various pieces of information provided and consumers who will make purchasing decisions. When one member has had these

items, they often become the source of information for other members.

The formation of consumptive behavior in adolescents is influenced by several factors. One of the most influential factors is the reference group. Reference groups are groups of people who strongly influence individual behavior. A person will see a group of references in determining the products they consume. This opinion is reinforced by Howkins & Bert (1980), who suggests that a reference group is a group that has the values and views used by an individual that are included as a cornerstone for their behavior.

A consumptive student has no awareness of the importance of the priority scale. Students in their cognitive development and emotions still see that the attributes of a tertiary are equally important (even more important) to their primary needs as a student. This condition especially occurs in those who are gaining a considerable budget from their parents, because the budget will affect the consumption behavior. There are several ethical theories relating to the conduct of consumers according to Porter (1988), namely (a) hedonism, where pleasure and happiness are the purposes of life lived, (b) utilitarianism, an understanding of the moral foundation to seek usability or greatest happiness, (c) self-realization, a theory that looks at the purpose of life. It has full awareness of the personal ability of its self.

The high and low level of understanding of economic fundamentals means it also demonstrates the low and high fundamentals of economic literacy. A partial mastery of the economic concept affects student consumption behavior patterns. The concept of economics belongs to the social sciences, which must be understood and mastered by the students. It is in accordance with the statements of Chapin and Messick (Mulyana 2015), which suggest that the concept of social sciences depicts material derived from seven disciplines, namely (a) History, (b) Geography, (c) Economics, (d) Political science, (e) Sociology, (f) Anthropology, and (g) Psychology.

The mastery of economic concepts in students often causes misconceptions. Klammer (Mulyana 2017) suggests that the existence of this misconception will greatly impede the process of acceptance and assimilation of new knowledge, thus, preventing the success of students in learning the concepts based on experience. According to Mathews (1999), the benefit of studying economic literacy is, among others, becoming savers. Even according to the opinion of Sina (2012), the result of an understanding of economic literacy that is not adequate can be seen from how the person is experiencing errors when making spending decisions; another phenomena is the low level of willingness of the people of Indonesia to save.

Mathews (1999) adds that economic literacy acts as an individual's ability to recognize and use economic concepts and ways of economic thinking to improve and gain well-being. It is revealed in the Salemi theory (2005) that students could achieve economic literacy if they can apply the basic economic concepts in later years, in situations relevant to their lives and different from those taught in class. There is a difference between the age of adolescents and adults in psychosocial ability. In the adulthood phase, people are more mature in making purchasing decisions. Meanwhile in the adolescent phase, they tend to seek sensation because of the influence of their fellow group or peers, so their actions are often full of risks (Cauffman & Steinberg 2000). Loudon & Bitta (1993) argue that the notion saying teenagers are a consumptive-oriented group because teenagers like to try new things, is not realistic and tends to be extravagant. Based on the phenomenon, the researchers are trying to analyze economic literacy, social environment, lifestyle, and their impact on student consumption behavior.

2 METHOD

This research uses a quantitative approach with statistical analysis. The research method employed an explanatory survey, aimed to explain the causal relationship and hypotheses submission. Cresswell (2008) suggests that surveys are used to collect data or information about large populations using relatively small samples.

The population in this study consists of students in the Department of Office administration of 13 vocational high schools, which are accredited by the Indonesian Board of National Accreditation for Schools with an A (very satisfying) score in Garut City both public and private schools. Samples are testers of the population and some of them must be carefully selected to represent the population (Cooper & Schindler 2008). The sampling technique used to determine the analysis unit was a proportionate random sampling technique. The inclusion of sample students is done through calculations using the formula of Taro Yamane. According to the formula, if the sample error is 5%, then the amount in the study is 386 students. The hypothesis testing in this study used structural model analysis or is commonly referred to as SEM. SEM is a statistic technique used to test a series of relationships.

3 RESULTS AND DISCUSSION

3.1 Analysis of the impact of economic literacy on lifestyle

The results of the hypothesis testing showed that economic literacy did not positively impact lifestyle. This means that the high level of student understanding of economics does not affect the lifestyle of the students. A partial mastery of the economic concept affects student consumption behavior patterns. The concept of economics belongs to the social sciences, which must be understood and mastered by the students.

Klammer (Mulyana 2017) suggests that the existence of this misconception will greatly impede the process of acceptance and assimilation of new knowledge in the students, thus, preventing the success of students in learning the concept from daily experience. In fact, it is Sina's opinion (2012) that the result of inadequate economic literacy will be seen from how someone encountered an error when making a spending decision.

3.2 Analysis of the impact of social environment on lifestyle

The results of the hypothesis testing show that the social environment has a positive impact on lifestyle. This suggests that the family environment, residential environment, and existence in the group contribute meaningfully to the students' lifestyle. The indicator of existence in the group contributes the highest.

Students lifestyles are influenced by their social environment. This is in accordance with the statement from Saragih et al. (2013) who suggest that the social environment is all persons/people who affect individuals either directly or indirectly. The student's lifestyle is influenced by socio-cultural factors and social environmental factors. The social or group classes that students follow have a strong impact on the student's lifestyle. Families will also have an impact on the student's lifestyle. It is in accordance with the statement of Mangkunegara (2005).

3.3 Analysis of the impact of economic literacy on consumption behavior

Hypothesis testing results show that economic literacy has a positive impact on consumption behavior. Meaning the higher the understanding of economic knowledge then the behavior of students' consumption is increasingly rational. Contributions to the highest-rated economic literacy indicators are the demand and supply, while the contribution of the lowest-rated economic literacy indicators is industrial development. It is in accordance with the work of Mathews (1999), which suggests that economic literacy acts as an individual's ability to recognize and use economic concepts and an economic way of thinking to improve and gain welfare.

The North Central Regional Educational Laboratory (abbreviated NCREL) (Mercan, et al. 2014) defines economic literacy as the ability to interpret economic problems and examine different options and relate to finding solutions, defining costs and profits, analyzing the results of changes in economic conditions and public policies, collecting and organizing data, and weighing costs and profits. Thus, the education of economic literacy is a process that enhances the understanding of the concept and risk of consumers or investors through informing, instruction, and/or objective recommendation, and developing the skills and trustworthiness necessary to realize the financial risks and opportunities, make conscious choices, know where to consult for assistance, and perform other measures with the intention of repairing financial conditions (OECD in Mercan et al. 2014).

3.4 Analysis of the impact of social literacy on consumption behavior

The results of the hypothesis testing showed that the social environment positively impacts consumption behavior. This implies that the high and low impact of the family's environment, the environment of residence, and existence in the group will have an impact on consumption behavior patterns. The highest contribution of social environment indicators is the existence of the group, while the contribution of the lowest social environmental indicator is the environment of residence.

The factors of hedonism may arise due to social environmental impacts. Students will be easily trapped with the fun and happiness gained in their social environment.

3.5 Analysis of the impact of lifestyle on consumer behavior

The results of the hypothesis testing show that lifestyle has a positive impact on consumption behavior. This suggests that the high and low behavioral consumption of students is influenced by their lifestyles. The highest contribution to lifestyle indicators is the shopping

pattern, while the contribution of the lowest lifestyle indicator is activity.

This is in accordance with the statements of Kotler & Amstrong (2004) that there are two basic factors affecting consumer behavior such as external factors (family, social class, culture, and reference group) and internal factors (motivation, perception, attitude, lifestyle, personality, and learning). Personality is closely related to the understanding of a person's lifestyle, which can be defined as the pattern by which people live and use money and time (Engel et al. 1995).

4 CONCLUSION

Based on the analysis, it can be concluded that:

The impact of each variable's contribution on student consumption behavior can be seen as follows:

a) In the economic literacy variables, the demand and offer indicators contributed the highest, while the industry development indicators contributed the lowest. Meanwhile, in lifestyle variables, the shopping pattern indicator contributes the highest, while the activity indicator contributes the lowest. For social environment variables, the indicator of existence in the group contributes the highest, while the environmental indicators of the residence give the lowest contribution.

b) Economic literacy, social environment, and lifestyle positively impact consumption behavior. This suggests that economic literacy, social environment, and lifestyle can be high predictors of consumption behavior. The highest consumption behavior indicator is conformity to the needs, while the lowest consumption behavior indicator is the budget of how to use.

ACKNOWLEDGEMENT

Thanks to the Department of Office administration of 13 vocational high schools, which are accredited by the Indonesian Board of National Accreditation for Schools with an A (very satisfying) score in Garut city and all team. All researchers at Institut Pendidikan Indonesia (IPI) Garut for the collaboration in supporting this research.

REFERENCES

Arnold, M. J. & Reynolds, K. E. 2003. Hedonic shopping motivation. *Journal of Retailing* 79(2): 77–95.

Cauffman, E., & Steinberg, L. 2000. Researching adolescents' judgment and culpability. In T. Grisso & R. G. Schwartz (Eds.), *Youth on trial: A Development Perspective on Juvenile Justice*: 325–343). Chicago: The University of Chicago.

Cooper, D. R., & Schindler, P. S. 2008. Business research methods. 10th ed, New York: McGraw-Hill/Irwin,

Cresswell, J. W. 2008. Educational research: Planning, conducting, and evaluating quantitative and qualitative research (3rd ed.). Upper Saddle River. NJ: Merril.

Engel, J. F., R. D. Blackwell and P. W. Miniard. 1995. Consumer behaviour. Eight Edition. 449–455 The Dryden Press.

Gardner, M. & Steinberg, L. 2005. Peer influence on risk taking, risk preference and risky decision making in adolescence and adulthood: An Experimental Study. *Temple University Developmental Psychology* 41(4): 625–635.

Harari, T. T. & Hornik, J. 2010. Factors influencing product involvement among young consumers. *Journal of Consumer Marketing* 27(6): 499–506.

Howkins, C., & Bert. 1980. Consumer behavior (implications for marketing strategy). Texas: Business Publication. Inc.

Ioanas, E. & Stoica, I. 2014. Social media and its impact on consumers behavior. *International Journal of Economic Practices and Theories* 4(2).

Kotler, P., & Amstrong, G. 2004. Principles of marketing. 10th ed. New Jersey: Pearson Prentice Hall.

Laursen, E. 2005. Rather than fixing kids-build positive peer cultures reclaiming children and youth. *Proquest Education Journal* 14(3): 137–142.

Loudon, D. L., & Bitta, A. J. D. 1993. Consumer behavior: Concepts and application. Singapore: Mc Graw-Hill Book Company.

Mathews, L. G. 1999. Promoting economic literacy: Ideas for your classroom. *Paper prepared for the 1999 AAEA annual meeting.* Nashville: Tennessee.

Mercan, N., Kahya, V., & Alamur, B. 2014. A research regarding to relationship between economic literacy and consumer preferences in knowledge economics. *European Journal of Research on Education,* 2(6) 1–13.

Mulyana, E. 2015. Comparing the effectiveness of learning of cooperative integrated reading and composition, Group Investigation, and Team Games Tournament types in students understanding of social studies concepts (A quasi experiment to the Eight Grade Students of SMPN 38 Bandung). Thesis. Universitas Pendidikan Indonesia.

Mulyana, E. 2017. Increased understanding of the concept of social studies through the study of controversial issues and Group Investigation. *Scientific Journal of Graduate Students in Educational Administration* 5(2): 131–142.

Porter, B. F. 1988. Reason of living–A basic ethics. Russell Sage College. ISBN-10; 0023960507 ISBN-13; 9780023960505. 1988. Pearson. Paper.

Salemi, M. K. 2005. Teaching economic literacy: why, what and how. *International Review of Economics Education.* 4(2): 46–57.

Sina, P. G. 2012. Analysis of economic literacy. *Journal Economia* 8(2).

Advances in Business, Management and Entrepreneurship – Hurriyati et al. (Eds)
© 2021 Taylor & Francis Group, London, ISBN 978-0-367-67471-7

Financial literacy, rationality, and social control in consumptive students

Tetep, N.A. Hamdani, T. Widyanti & A. Darojat
Institut Pendidikan Indonesia, Garut, Indonesia

ABSTRACT: Consumptive behaviors among teenage students have become common. This study was conducted using a quantitative approach. The purpose was to examine financial literacy, rationality and social control in 75 consumptive students. It was revealed that consumptive behaviors were partially influenced by financial literacy, rationality and social control. As an intervening variable, financial literacy had influence on student consumptive behaviors via rationality and social control. The higher the rationality and the social control are, the lower the consumptive behaviors will become. This implies that students with a good rationality and social control are likely to know how to prioritize their needs.

1 INTRODUCTION

Consumption behavior should essentially be in line with the basic needs not with the desire for excessive satisfaction. Contemporarily, people in society tend to have existence crisis, especially for teenagers. They are tempted to be fully accepted by the society in many ways. Acceptance and equality have caused them to keep in touch with the trends in their surroundings.

Consumers who have financial knowledge will be able to manage their consumption behavior. Knowledge of the information in the future will encourage consumers to spend on food and services according to the information they get. Rational consumers have patterns and ways to be able to meet effective and efficient needs by choosing two or more alternative commodities as a means of satisfying their needs. The difference between primary, secondary and tertiary needs on consumptive behavior tends to be borderless and obscured.

Consumptive behavior usually occurs in adolescents. In this case, students are related to the psychological characteristics of the final phase of adolescence. There are two main factors that influence consumer behavior, namely socio-cultural factors which consist of culture, special culture, social class, social group and family reference. Based on these explanations, the authors are interested in conducting research entitled Analysis of Student Consumptive Behavior Based on Financial Literacy, Rationality and Social Control.

2 LITERATURE REVIEW

According to Nitisusastro (2013), to fulfill the most basic needs and the need to actualize themselves, consumers must buy, use, and consume various needs goods and services. Consumers who have financial knowledge will be able to manage their consumer behavior. Knowledge of information in the future encourages consumers to consume goods and services based on the information they get.

Rational consumers have patterns and ways to be able to meet effective and efficient needs by selecting two or more alternative commodities as a means of satisfying their needs. The difference between primary, secondary and tertiary needs in consumptive behavior tends to be boundless and obscured. Harli et al. (2015) explained that consumptive behavior is consumption activities that are carried out excessively and not according to needs. Consumptive behavior usually occurs in adolescents, this is related to psychological characteristics possessed by adolescents. There are two main factors that influence consumer behavior, namely socio-cultural factors consisting of culture, special culture, social class, social groups and family references. Other factors are psychological factors consisting of motivation, perception, learning process, trust and attitude

This consumer behavior will further influence the decision-making process in consuming a particular product or service, starting from identifying problems to fulfill needs, finding information for satisfying needs, evaluating and selecting several rational alternative choices, and consuming decisions on a product or service as the initial stage of fulfilling the level of satisfaction. One of the internal factors that influence consumption behavior that comes from psychological aspects is the learning process factor, which is an individual process to understand a knowledge. Knowledge about finance is commonly referred to as financial literacy. Financial literacy is knowledge that must be understood by every consumer.

According to Imawati (2013) good financial literacy allows consumers to choose goods, manage finances well and plan for the future. Consumers who have financial literacy will be smarter in choosing and providing complaints about the goods or services they consume.

Houston (2010) states that financial literacy is an additional dimension to financial knowledge that can complement the ability to make financial decisions. Financial literacy is a basic need for everyone to avoid financial problems. Financial problems are limitted to low income, financial problems can also arise from financial management (miss-management) such as misuse of credit, and the absence of financial planning. The low level of financial literacy will have an impact on the low desire to save for planning in the future and excessive spending habits will make the community become consumptive, making it difficult to become a smart consumer. Other factors that influence consumer behavior are emotional factors and rational factors. Consumers who pay attention to rational factors tend to take into account the benefits of a commodity as a means of satisfying their needs. Consumers who pay attention to rational factors tend to take into account the benefits of a commodity as a means of satisfying their needs.

Samuelson & William (1999) explain that economics shows how a person in a rational consumer will maximize marginal utility, maximum satisfaction and happiness of the items he has bought. Consumers with financial literacy at a certain level will be able to increase their rationality in consumption. On the other hand, Torell et al. (2005) argues that high income earning potential indicates an increase in income which can make a high influence of satisfaction. High income potential causes consumptive behavior and tends to ignore economic rationality.

According to the Program for International Student Assessment (PISA) in 2012 (in Imawati et al. 2013) financial literacy is knowledge and understanding of financial concepts used to make effective financial choices, improve financial well being from individuals and groups to participate in economic life. The definition of financial literacy, according to Bhushan & Medury (2013) is that "Financial literacy is the ability to make judgments and decisions that take decisions regarding the use and management of money" which means financial literacy is the ability to consider information and make effective decisions about the financial spending and management. .

Chen & Volpe (1998) argue that students who have low knowledge will make wrong decisions in their finances. Students who have low financial literacy will make wrong decisions in consumption, because they do not take into account the priority needs.

According to Sumartono (2002) the indicators of consumptive behavior are: 1. Buying products because of the prize, 2. Buying a product because of the attractive packaging, 3. Buying products to maintain self-appearance and prestige, 4. Buying products

based on pricing (not on the basis of their benefits or uses), 5. Buying products for status symbol. 6. Using the product because of conformity to the models that advertise, 8. Trying more than two similar products (different brands). Students who have good financial literacy will be selective in consuming. They will prioritize what is needed, and putting aside what they want because they know that they have to face the possibility that might occur if they set the priorities aside.

Consumptive behavior can be interpreted as an act of unfinished use of a product. It means that before the product is used up, someone has used the same product from another brand. Or worse, someone buys goods because of the gift offered or buys a product because many people use it (Sumartono 2002). "Consumptive behavior is the desire to consume goods that are actually not needed excessively to achieve maximum satisfaction" (Tambunan & Tulus 2001). This consumptive behavior does not seem to have good benefits for the consumer because it spends up income. "Consumptive behavior is the tendency of humans to do unlimited consumption, buy something excessively or randomly" (Chita et al. 2015). It is unplanned that the purchase of goods or services is due to the absence of a priority-based budget.

Ritzer & Douglas (2007) explain Coleman's (1990) rational choice theory where the basic idea of this theory is that "basically individual actions lead to a purpose and the purpose (and also action) was determined by value or choice (preference) ". This is also in line with the results of previous research from Harnum (2012) which states that there is a significant negative relationship between self-control techniques and the tendency of consumptive behavior. Students who have low self-control will have consumptive behavior. For the sake of social recognition, students can behave consumptively; buying an item or service not because of their need, but based on the desire or fulfilling satisfaction. This is reinforced by Balakrishnan et al. (2000) which states that individual consumers in choosing a product are more considering the rational aspect than the efficiency aspect, that teenagers are easily persuaded by advertisement, love to follow the friend, not realistic, and tend to be wasteful in using their money. Students who have good self-control can avoid consumptive information, this is because students can control their behavior, cognitive and decisions. Gailliot et al. (2007) say that "self control refers to one's control of override abilities of one's thoughts, emotions, urges, and behavior. Self-control refers to a person's ability to control one's thoughts, emotions, pressures and behavior. Louden & Della Bitta (1979) claims that a person's consumptive behavior will be strongly influenced by his intellectual considerations.

According to the Program for International Student Assessment (PISA) (2012) the aspects that can be obtained from financial literacy are: 1) Money and transactions. 2) Planning and financial management. 3) Risks and benefits. 4) Financial landscape. Those four aspects becomes the assessment to determine one's financial literacy abilities. The ability of the four financial literacy aspects is certainly influenced by many things, as stated by Lusardi & Mitchell (2007) claiming that there are "three things that give influence on financial literacy skills, namely: 1) Sociodemography; there are differences in understanding between men and women. Men are considered to have higher financial literacy abilities than women as well as their cognitive abilities. 2) Family background; the education of a mother in a family has a strong influence on financial literacy, especially for mothers who are college graduates. They are superior to 19 percent higher than those of secondary school graduates. 3) Friendship group (peer group); someone's group or community will influence someone's financial literacy, influence their consumption patterns and use of existing money".

Self-Control According to Mahoney & Thoresen (in Ghufron & Risnawat 2010) is the individual complete construction to his environment. Individuals with high self-control are very concerned about the right ways to behave in various situations. Individuals tend to change their behavior according to the demands of social situations which can then adjust the impression that their behavior is more responsive to situational guidance, more flexible, cooperative in social interaction, warm and open. In brief, good self control can make individuals easily accepted in their environment.

3 METHODS

This study used quantitative methods. The subject of this research is Institut Pendidikan Indonesia students. The sampling technique used proportional technique with the "Slovin" formula to determine the sample of 75 respondents from 2 programs. The independent variable, intervening variable and dependent variables are financial literacy, rationality and consumptive behavior respectively. The data were collected using questionnaire method and the data analysis was carried out by path analysis using SPSS. to test the relationship of intervening variables, Path analysis was also carried out. Social control was used as a supporting variable from this study.

4 RESULTS AND DISCUSSION

Path Analysis is used to analyze student consumptive behavior based on financial literacy, rationality and social control. The results of path analysis (path analysis) using SPSS are: 1. Regression of financial literacy and rationality towards consumptive behavior.

$$Y1 = -b1X1 - b2X2 + e1. \tag{1}$$

The results of the analysis using SPSS resulted in multiple linear regression analysis with consumer behavior as the dependent variable as presented in Table 1.

Table 1. Results of multiple linear regression analysis sub-structural I with consumer behavior as dependent variables.

Variabel	Stand. Coef Beta	t hit	Sig	R^2	Adj R^2	F Hit
Constant	33.361	23.787	.000			
Likeu	−.485	−	.000	.325	.321	78.155
Ras	−.200	−10.259 −4.232	.000			

Source : Research data, 2018.

Based on the linear regression analysis, the following equation was generated:

$$Y1 = -0.485X1 - -0.200X2 \tag{2}$$

0.325 determination coefficient was used to compute residual value (e1) of the linear regression phase I as presented in the following calculation:

$$
\begin{aligned}
e1 &= \sqrt{1 - R^2} \\
&= \sqrt{1 - 0.325} \\
&= 0.821 \\
&= 0.82
\end{aligned}
\tag{3}
$$

The results of multiple regression analysis indicate adjusted R2 value of 0.321 which indicates that the consumptive behavior variables are influenced by financial literacy and rationality by 32.1% while the remaining 67.9% is influenced by other variables. The regression coefficient (X1) of -0.485 states that the financial accounting variable affects consumptive behavior negatively by 48.5%, which means that the higher the financial literacy is, the lower the student consumptive behavior will get. The regression coefficient (X2) is -0.200 which states that the rationality variable influences consumer behavior negatively by 20%, which means that the higher the rationality is, the lower the student consumptive behavior will be.

The results showed that rationality can be used as an intervening variable between financial literacy and consumer behavior. The contribution of financial literacy to consumptive behavior decreases through rationality. This means that the higher financial literacy will increase rationality, and the increase in rationality will reduce student consumptive behavior. The results of hypothesis testing indicate that financial literacy has a positive and significant effect on rationality. High rationality will make consumptive behavior even lower. So that it can be said that student rationality can be used as an intervening variable for financial literacy and student consumptive behavior. This is because financial literacy has a positive and significant influence on rationality and the negative and significant influence of rationality on consumptive behavior will further reduce the level of student consumptive behavior towards his allowance. Someone who has high rationality is usually a person who is psychologically mature and will tend to be rational in making choices

The Effect of Rationality on Consumptive Behavior Rationality has a direct negative effect on consumptive behavior. Consideration of rationality used by students in consuming a particular commodity is indicated by a high total score in the indicator of rationality. The first indicator is basic needs or priority scale, which describes students' understanding of the priority scale itself. The higher understanding of the priority scale they have indicates the more rational students in need fulfillment. The higher the level of rationality the students have, the lower the consumptive behavior will get because with a high level of rationality they are able to arrange the most important and urgent needs that must be met compared to tertiary needs.

5 CONCLUSION

a) Based on the results of the analysis, it was revealed that self-control and social control negatively affect student consumptive behavior.
b) The aspect of rationality is theoretically capable of suppressing excessive consumption behavior. Consumers who have high rationality will be able to choose several alternative choices for commodities that can satisfy their needs. Rationality relates to priority scale aspects of choosing a particular commodity to be able to maximize the benefits of a number of rational choices. Practically, students' financial problems can arise from mismanagement of monthly allowance received from their parents, such as buying unnecessary goods because they have no financial planning so that the allowance supposed to be enough for one month is used up too early.
c) Financial literacy has a direct and negative effect on student consumptive behavior. This means that the higher the student's financial literacy is, the lower the effect on his consumptive behavior will become. Rationality has a negative and significant direct effect on student consumptive behavior; indicating that the higher the student's rationality is, the lower the influence on his consumptive behavior will get. Financial literacy influences consumer behavior through student rationality and social control. This means that the higher the student's financial literacy is, the higher the influence on rationality and social control will get. Furthermore, high rationality and social control will reduce consumptive behavior.

ACKNOWLEDGEMENT

Thanks the Respondents for supporting and collaborating. All researcher at Institut Pendidikan Indonesia (IPI) Garut for the collaboration to support this research.

REFERENCES

Balakrishnan, P.V., Nataraajan, R., & Desai, A. 2000. Consumer rationality and economic efficiency: is the assumed link justified. *The Marketing Manajement Journal* 10(1): 1–11

Bhushan, P., & Medury, Y. 2013. Financial literacy and its determinants. *International Journal of Engineering, Business and Enterprise Applications* (IJEBEA) 4(2): 155–160.

Chen. H., & Volpe, R.P. 1998. An analysis of personal financial literacy among college student. *Financial Services Review* 7(2): 107–128.

Chita, M.C.R., David, L., & Pali, C. 2015. Hubungan antara self control dengan perilaku konsumtif online shopping produk fashion pada mahasiswa Fakultas Kedokteran Universitas Sam Ratulangi angkatan 2011. *Jurnal e-Biomedik* 3(1): 297–302.

Conference on Teacher Education. Join Conference UPI & UPSI: 552560.

Gailliot, M.T., Baumeister, R.F., Dewall, C.N., Maner, J.K., Plant, E.A., Tice, D.M., Brewer, L.E., & Schmeichel, B.J. 2007. Self Control relies on glucose as limited energy source: willpower is niore then a metaphor. *Journal of personality and social Psychology* 92(2): 325–336

Ghufron, M., & Risnawat, R. 2010. Teori-teori psikologi. Yogjakarta: Ar-ruz Media.

Harli, Claresta, F., Linawati, N., & Memarista, G. 2015. Pengaruh financial literacy dan faktor sosiodemografi terhadap perilaku konsumtif. *Jurnal Finesta* 03(1): 58–62.

Harnum, D. 2012. Hubungan antara teknik kontrol diri dengan kecenderungan perilaku konsumtif mahasiswa di Ma'had Sunan Ampel Al-Aly Universitas Islam Negeri (UIN) Maulana Malik Ibrahim Malang.

Imawati, Indah, Susilaningsih, dan Ivada, E. 2013. Pengaruh financial literacy terhadap perilaku konsumtif remaja pada Program IPS SMA Negeri 1 Surakarta tahun ajaran 2012/2013. *Jurnal Jupe UNS* Vol 02(1): 4858.

Lusardi, A., & Mitchell, O.S. (2007). Baby boomer retirement security: the roles of planning, financial literacy, and housing wealth. *Journal of Monetary Economics* 54(2007): 205–224.

Mahoney, M.J., & Thoresen, M.J. 1974. Behavioral self-control. New York: Holt, Rienehart and Winston

Nitisusastro, Mulyadi. 2013. Perilaku konsumen dalam perspektif Kewirausahaan. Bandung: Alfabeta.

Program for International Student Assessment (PISA). 2012. Financial Literacy Assesment Freamwork. Amerika: International Network on Financial Education OECD.

Ritzer, G., dan Goodman, D.J. 2007. Teori sosiologi modern edisi keenam. Jakarta: Kencana Prenada Media Group.

Samuelson, P.A., & William, D.N. 1999. Mikro ekonomi. Jakarta: Erlangga.

Sumartono. 2002. Terperangkap dalam iklan meneropong imbas pesan iklan televisi. Bandung: Alfabeta

Torell, L., Allen, N.R.R, Octavio A.R., & Daniel, W. M. 2005. Income earning potential versus consumptive amenities in determining ranchland values. *Journal of Argicultural and Resource Economics* 30(3): 537–560.

Social capital and entrepreneurial motives among students in East Preanger Indonesia

Tetep, A. Suherman & L. Dianah
Institut Pendidikan Indonesia, Garut, Indonesia

Y. Susanti & G.A.F. Maolani
Universitas Garut, Garut, Indonesia

ABSTRACT: Social capital is a social networks concept that can develop one's life capacities. Framed under a case study research, this study analyzed social capital factors in relation to entrepreneurial motivation for undergraduate students in East Preanger, West Java, Indonesia. This study used a quantitative approach with multiple regression analysis. The results showed that the norms and values in communication and human network systems had a significant effect on the undergraduate students' entrepreneurial motivation in East Preanger, West Java, Indonesia. Better social capital provides the impetus to increase and develop entrepreneurial motivation. The capacity to support today's entrepreneurial motivation is the development online social capital.

1 INTRODUCTION

Universities play an important role to instigate their students' entrepreneurial motivation which will eventually lead them to become job creators. For that reason, it is necessary for universities to provide guidance for their students to be able to carry out entrepreneurship. Additionally, each program in the universities is required to include courses related to Entrepreneurship Courses (KWU), entrepreneurship apprentice (MKU), Business internships (KKU), and Entrepreneurial Student Programs (PMW) to provide the students with more added values. Entrepreneurs are those who create a business with risks and uncertainties to gain profits and develop the business by opening more opportunities. Utilizing every available resources for entrepreneurship needs to be supported by every college in supporting their graduates' entrepreneurial motivation. Universities need to provide more opportunities for their students to become reliable individuals with integrity to have the ability to try, communicate, work together, and have personality. Entrepreneurship has been prepared as a major contributor to the country's economy by promoting innovation, competition and employment (Nenzhelele 2014).

Entrepreneurship for undergraduate students refers to as a process of instilling the spirit of entrepreneurship as a form of attitude. Motivation and interest are needed for entrepreneurial students to identify business opportunities, then utilize these opportunities to create new employment opportunities. Students' interest and knowledge of entrepreneurship are expected to shape their tendency to open new businesses in the future. Research conducted by Mulyaningsih (2012) found that the factors influencing entrepreneurial interest indicate that entrepreneurial interest variables are influenced by 60.4% by capital, skills, place, and entrepreneurial spirit. In the present day, there are many entrepreneurial opportunities for anyone who is keen on seeing opportunities. A big business opportunity can be done at home with low risk such as online business through the internet or social media; Facebook, Instagram, Twitter and blog. An entrepreneurial career can support people's welfare by generating tangible financial rewards. According to Nahaphiet & Ghoshal (1998) in Muniady (2015) there are 3 dimensions of social capital i.e. structural dimensions, relational dimensions, and cognitive dimensions. The structural dimensions include the number of relations, the diversity of relations, and the position of relations. Relational dimensions include trust, quality of relationships, reciprocity. Cognitive dimensions include shared vision, ideas and norms.

Entrepreneurial opportunities can be achieved through the interest in reading opportunities and accelerating economic development with new ideas and turning these ideas into business profits (Turker & Selcuk 2008). BPPS data (June 2019) shows that: The number of workforce in February 2019 was 136.18 million, showing an increase of 2.24 million compared to February 2018. In line with the increase in the number of workforce, the Labor Force Participation Rate (TPAK) also increased by 0.12 percent. In the past year, unemployment decreased by 50,000 people, in line with the TPT (Open Unemployment Rate) which fell to 5.01 percent in February 2019. Based on the level of education, TPT for Vocational High Schools (SMK) was still the highest among other education levels by 8.63 percent.

The decline of unemployment rate in Indonesia can be caused by entrepreneurship. Entrepreneurial activity is one of country development determinants because the country's economic growth can be achieved by having growing entrepreneurs (Wedayanti & Giantari 2016). The decline occurs at the tertiary level of education, because the mindset has changed. In general, students choose to become workers at reputable companies and government agencies (to become civil servants) to guarantee their future. The concept from "looking for work" after graduating from college needs to be changed to "creating jobs" (Nursito & Nugroho 2013). Most students want an established job by getting honorable status and generating maximum profit after graduating from college (Oktarilis 2012). The current trend in university students is that students who will graduate from universities will usually start a business by just continuing the business of their parents (Hongdiyanto 2014). These things inspire the researcher to find out the students' Social Capital factors and entrepreneurial motives in East Preanger, Indonesia.

2 METHODS

The current study used descriptive quantitative study in which the social capital factors and entrepreneurial motivation of undergraduate students in East Preanger was closely observed. The population of this study was active social science education program students in East Preanger Region. The sample was chosen based on nonprobability sampling technique by using purposive sampling. As many as sixty under graduate students was chosen as the main respondents of this study. The obtained data was processed in SPSS 25 for Windows.

3 RESULTS AND DISCUSSION

After the data was processed statistically using SPSS 25.0 for windows, the following results were obtained:

Table 1. Coefficient of determination.

Model summary

Model	R	R Square	Adjusted R Square	Std. Error of the Estimate
1	.957[a]	.915	.912	2.08536

a. Predictors: (constant), human network systems (X2), norms and communication value factors (X1).

The computation results of the Determination Coefficient using SPSS 25 indicate that R Square (Determination Coefficient) is 0.915. This means that 91.5% of the Norm and Value Factors in Communication and Human Network Systems influence the entrepreneurial motivation, the remaining 8.5% are influenced by other factors.

Table 2. Hypothesis testing.

ANOVA[b]

Model	Sum of Squares	df	Mean Square	F	Sig.
1 Regression	2679.106	2	1339.553	308.034	.000[a]
Residual	247.877	57	4.349		
Total	2926.983	59			

a. Predictors: (constant), human network systems (X2), norms and communication value factors (X1).
b. Dependent variable: entrepreneurial motivation (Y).

From the Anova test or F test, it was obtained that Fobserved = 308.034 with a significance value of 0.000. Since the probability is <0.05, the regression equation of $Y = 5.086 + 0.661\ X1 + 0.247\ X2$ is significant. It implies that Self Efficacy, Subjective Norms, Attitude, and Entrepreneurship Education can be used to predict entrepreneurial intentions. Thus the hypothesis which states that the Norms and Values in Communication and Human Network Systems Factors Significantly Influence the Motives of Entrepreneurship is accepted.

Table 3. Analysis of multiple linear regression.

Coefficients[a]

Model	Unstandardized Coefficients B	Unstandardized Coefficients Std. Error	Standardized Coefficients Beta	t	Sig.
1 (Constant)	5.086	2.048		2.483	.016
Norm and Comvalue (X1)	.661	.115	.711	5.760	.000
Human Network System (X2)	.247	.120	.255	2.062	.044

a. Dependent Variable: Entrepreneurial Motivation (Y).

Multiple linear regression analysis was used to determine social capital factors and students' entrepreneurial motivation in East Preanger, Indonesia. According to Sugiono (2011) the regression equation is:

$$Y = a + b1X1 + b2X2 + e \qquad (1)$$

Y = Motivation for Entrepreneurship, a = Constant, b1 = Regression coefficient Norma Factor and Communication Value, b2 = Regression coefficient of Human Network System, X1 = Norm Factor and Communication Value, X2 = human network system.

The calculation results of SPSS 25.0 has resulted in the following regression equation:

$$Y = 5.086 + 0.661X1 + 0.247X2 \qquad (2)$$

It means that:

1) If the value is constant, then the value of Y will increase by 5.086. In other words, the Norm Factors and Communication Values and Human Network Systems are constants, so the entrepreneurial motivation is 5.086.
2) The above equation shows that if the Norms and Communication Values and Human Network Systems increase by one unit while the other variables are constant, the entrepreneurial motivation will increase by 0.661 and 0.247.

This study used a sample of 60 undergraduate students studying entrepreneurship education in the East Priangan Region of Indonesia. The results showed that the Norms and Communication Values affect the entrepreneurial motivation. Effective communication can be seen from openness, empathy, support, positivity, and similarity. A communicator (entrepreneur) in conducting effective communication is driven by two important factors. Trust in entrepreneurship reflects that the message or information received by the customer is considered correct according to empirical reality. In line with the research of Teng-Li Yu & Jiun-Hao Wang (2018) the most prominent factors influencing social entrepreneurship beliefs, followed by stakeholder perspectives and communication efficacy. In addition, the effective management is the strongest factor that influences social entrepreneurship preparation, followed by a stakeholder perspective and affective empathy. In particular, cognitive empathy is revealed to be negatively related to social entrepreneurship preparation. Social capital and support were found to have no relationship with the intention of social entrepreneurship. Costa et al. (2009) found that: (1) students, in general, have a tendency to be involved in entrepreneurial activities; (2) entrepreneurial interest is influenced by perceived entrepreneurial calls from the region, social support, and mastery of perceived strategic entrepreneurial skills.

Human Network System influences entrepreneurial motivation. Social capital is a relational resource that is inherent in cross-sectoral personal relations, which is very useful for individual development in the social community of the organization. Social capital will be of economic value when it can have a positive impact on individuals and groups, such as to access information, find work, start a business, and minimize transaction costs. Research results related to social capital and entrepreneurial motivation; Soogwan Doh & Edmund J. Zolnik (2011) revealed that

"The results of empirical models that simultaneously influence the factors that influence entrepreneurship at both the individual and state level show that there is a positive relationship between social capital and entrepreneurship. Madriz et al. (2018) claims that human and social capital, factors related to knowledge, have a statistically positive relationship with a tendency to become entrepreneurs. There are only a few differences between factors related to knowledge in various countries especially those related to the cultural context, which influences the tendency to become entrepreneurs. Estrin et al. (2016) specific entrepreneurial human capital is relatively more important in commercial entrepreneurship, and general human capital in social entrepreneurship, and that the effects of human capital depend on the rules. Based on the results of previous research, it is clear that the Norms and Values Factors in Communication and Human Network Systems have a strong significant effect on undergraduate students' entrepreneurial motivation in East Preanger, West Java, Indonesia. The better the students' social capital is, the higher the entrepreneurial motivation increase and develop.

4 CONCLUSION

Based on the results of analysis, some conclusions can be drawn as follows:

Norms and Communication Value Factors influence the entrepreneurial motive. The better the norm and value factor of communication gets, the higher entrepreneurial motive will increase.

a) Human Networking System influences the entrepreneurial motive. The better the human network system gets, the higher the entrepreneurial motive will increase.
b) Norms and values in communication and human network systems have a significant influence on the students' entrepreneurial motive in East Preanger, West Java, Indonesia. Better social capital provides the impetus to increase and develop entrepreneurial motivation. The current supporting capacity for entrepreneurial motives is the development of online social capital support capacity.

ACKNOWLEDGEMENT

Thanks to the Respondents for supporting and collaborating. All researcher at Institut Pendidikan Indonesia (IPI) Garut for the collaboration to support this research.

REFERENCES

BPPS data. 2019. The number of workforce. [Online]. Retrieved from https://www.bps.go.id/pressrelease/2019/05/06/1564

Costa, F.J, Soares, A.A.C., & Bonfim, D.G. 2009. Factors of influence on the entrepreneurial interest: an analysis with students of information technology related courses. *JISTEM – Journal of Information Systems and Technology Management.*

Estrin, S., Mickiewicz, T., & Stephan, U. 2016. Human capital in social and commercial entrepreneurship. *Journal of Business Venturing* Volume 31(4): 449–467. [Online]. Retrived http://doi.org/10.1016/j.jbusvent.2016.05.003.

Hongdiyanto, C. 2014. Identifikasi kepemilikan entrepreneurial spirit mahasiswa. *Jurnal Entrepreneurial dan Entrepreneurship* 3(1,2).

Li Yu, T., & Wang, J.H. 2018. Factors affecting social entrepreneurship intentions among agricultural university students in Taiwan. *International Food and Agribusiness Management Review* 22 (1). [Online]. Retrived from https://www.wageningenacademic.com/doi/pdf/10.22434/IFAMR2018.0032.

Madriz, C., Leiva, J.C., & Hen, R. 2018. Human and social capital as drivers of entrepreneurship capital humane social como motores del emprendimiento. *Journal of Small Business and Enterprise Development.*

Mulyaningsih, 2012. Faktor-faktor yang mempengaruhi minat wirausaha pengelolaan pangan organik. Malang: Jurnal Wacana.

Muniady, R. 2015. The effect of cognitive and relational social capital on structural social capital and micro-enterprise performance. Sage Open, 5(4), 2158244015611187.

Nursito, S., & Nugroho, A.J.S. 2013. Analisis pengaruh interaksi pengetahuan kewirausahaan dan efikasi diri terhadap intensi kewirausahaan. *Kiat Bisnis* 5(2).

Oktarilis, S.N. 2012. Pengaruh faktor – faktor yang dapat memotivasi mahasiswa berkeinginan wirausaha. *Jurnal Ekonomi Manajemen Universitas Gunadarma.*

Turker, D., & Selcuk, S.S. 2008. Which factors affect entrepreneurial intention of university students?. *Journal of European Industrial Training* 33(2).

Wedayanti, & Giantari. 2016. Peran pendidikan kewirausahaan dalam memediasi pengaruh norma subyektif terhadap niat berwirausaha. *E-Jurnal Manajemen Unud* 5(1): 533–560 ISSN: 2302-8912.

Advances in Business, Management and Entrepreneurship – Hurriyati et al. (Eds)
© 2021 Taylor & Francis Group, London, ISBN 978-0-367-67471-7

The correlation between transformational leadership and commitment to change in the rapid change organization environment

K. Astari & D.M. Martina
Universitas Indonesia, Depok, Indonesia

ABSTRACT: Human resources in an organization have the important role for supporting the company in doing some significant changes to maintain the productivity. The commitment of change can be built by the leader who has powerful authority through the exact leadership style. The leader can apply transformational leadership as one of the leadership styles that bring a positive impact to the relationship between the leader and the team members. This study aims to examine the correlation between transformational leadership and the commitment to change of employees in the rapid change environment in a multinational company. The level of transformational leadership and commitment to change are measured by valid psychological instrument, supported by secondary data and interview results, and analyzed with inferential statistics. The results of this study indicate that the correlation between the variables is significantly proven. The leader who applied a transformational leadership style correctly has team members who are highly committed to change.

1 INTRODUCTION

Nowadays, we live in society 4.0 where the rapid transformations of digital technologies are bringing drastic changes in society and industry. The wave of digital technologies is increasing the diversity and complexity of people's values (Fukuyama 2018). The previous three society revolutions have been passed. The first society is called the hunter gatherer society where humans lived through hunting animals or collecting plants, and the second era was the Agrarian society (Society 2.0) where humans lived by building and maintaining farmland, the third is the Industrial society (Society 3.0) where humans used advanced technology to drive strong manufacturing industry to support the huge population. In society 4.0, technology and human resources are well integrated to build innovative products through digital transformation, which affects the value creating process in organizations. In society 4.0, knowledge and intelligence are completed by humans with the support of technologies. Now, we are entering society 5.0., where knowledge and intelligence are done by machine through artificial intelligence, big data, etc. (Onday 2019). Society 5.0 is focusing on providing sustainability in economics, social, politics, and people through advanced technologies and efficient ways (Fukuyama 2018). The concept of society 5.0 was introduced by the Japanese government with the aim to create a super intelligent society, which will support us to overcome globalization issues effectively such as climate change, population aging, rural depopulation, etc. (Ferreira & Serpa 2018).

Commitment to change captures the notions of positivity and proactive intent, which involves a lack of resistance to change and the absence of negative attitudes (Herscovitch & Meyer 2002; Kotter & Schlesinger 1979; Piderit 2000). Commitment to change is defined as a positive support toward change, which is illustrated by an attitude that will bolster the smooth running of the changes for achieving organizational goals. Commitment to change consists of affective commitment, normative commitment, and continual commitment to change (Herscovitch & Meyer 2002). The successful implementation of organizational changes is significantly determined by the commitment to change of employees (Gao Urhahn et al. 2016; Herold et al. 2008; Shin et al. 2012). In previous studies, it was identified that positive attitudes to change were found to be vital in achieving organizational goals (Eby & Shope 2000). The commitment to change of employees can be built through developing a sense of urgency that can be conducted by the leader who has been authorized in the organization. Leaders are people who create the work environment, which stimulates the employee to have a commitment toward changes (Brown & Eisenhardt 1997). Leadership is one of the most important elements that is critical in organizational change (Yousef 2000). Leadership can affect the attitudinal dimension in the organization (Jaskyte 2003). Leaders are required to communicate and to provide a good quality of leadership to foster acceptance to the proposed change (Santhidran et al. 2013). Leadership is divided into three types, namely transactional leadership, transformational leadership, and laissez-faire (Sadeghi & Pihie 2012). Transformational leadership is the most effective type of leadership that can be implemented during organizational change (Chou 2015). Transformational leaders can facilitate the creation of shaping certain behavior of employees that will lead them to accept the changes and have advanced commitment to it (Manz & Sims 2001). Based on this explanation, the researchers were interested to examine the relationship between transformational leadership and the commitment to change of the employees who were facing rapid changes in their organization during the new era of digital transformation.

2 METHOD

2.1 Data collection

Data were collected through an ODQ questionnaire using Weisbord's model (Preziosi 1980), the Transformational leadership tools used by Podaskoff et al. (1996), and the Commitment to change scale used by Herscovitch & Meyer (2002) with the details shown in Table 1.

Table 1. Measurement tools.

Instruments	Total Item	Reliability
ODQ by Weisbord's model (Preziosi 1980),	35	$\alpha = .89$
Transformational Leadership tools by Podalskoff et al. (1996) adapted by Rubin et al. (2005)	22	$\alpha = .94$
Commitment to change instrument by Herscovitch & Meyer (2002)	18	$\alpha = .71$

2.2 Measurements

The instruments used were written in English as the original scale because the language was the same between the language in the scale/instrument and the language that participants used in the organization.

ODQ was used to identify the organizational problem based on the perspective of organization members (Preziosi 1980), which consisted of 35 items. The measurement of this variable used a Likert scale of 1–7 (1 = strongly agree, 7 = strongly disagree) with a Cronbach's Alpha coefficient of 0.89. An example of the statement was "The goals of this organization are clearly stated."

Commitment to change was measured using the instrument used by Herscovitch & Meyer (2002) consisting of 18 items. The measurement of this variable used a Likert scale of 1–6 (1 = strongly disagree, 6=strongly agree) with a Cronbach's Alpha coefficient of 0.71. An example of the statement was "This change is a good strategy for the organization."

Transformational leadership was measured using the instrument used by Herscovitch & Meyer (2002) that was written in English as the original instrument consisted of 22 items. The measurement of this variable used a Likert scale of 1–6 (1 = strongly disagree, 6=strongly agree) with a Cronbach's Alpha coefficient of 0.94. An example of the statement was "Providing a good model to follow."

2.3 Participants and procedures

The participants of the study were employees who work in subsidiary of a multinational company in the oil and gas industry. Questionnaires were distributed through an online system and paper pencil (manual), which was participated in by 401 employees, but only 221 participants gave back completed questionnaires.

Table 2. Demographic profile.

Categories	Group	Frequency	Percentage (%)
Employment status	Tetap	121	54.75
	Kontrak	100	45.25
	Staff	101	45.70
	Supervisor/ coordinator/ team leader	41	18.55
Position level	Superintendent/ chief	36	16.29
	Manager	30	13.57
	Senior manager	13	5.88
	ICT	12	5.43
	GAC	11	4.98
	HR	22	9.95
	Security	10	4.52
	General services	12	5.43
	SEQ	19	8.60
	Production & maintenance	20	9.05
	Logistic	10	4.52
	Exploration	13	5.88
	Drilling	14	6.33
Department	Reservoir	16	7.24
	Commercial	5	2.26
	Legal	6	2.71
	Finance	15	6.79
	Procurement	9	4.07
	Project development	11	4.98
	LNG	6	2.71
	Energy solution	2	0.90
	Health	4	1.81
	Compliance	4	1.81

2.4 Data analyzed

Data were analyzed using descriptive analysis for examining the level category of each variable, Pearson product moments for identifying the correlation, and Cronbach's Alpha for measuring reliability.

Table 3. The level of each variable.

	Transformational Leadership		
Categories	Score	Frequency	Percentage (%)
Low	<79.19	40	18.10%
Average	79.19–112.56	165	74.66%
High	>112.56	16	7.24%
Commitment to change			
Low	<41.01	42	19.00%
Average	41.01–55.69	144	65.16%
High	>55.69	35	15.84%

Table 4. Correlation measurement.

	Correlation	
		CC
TL	Pearson product moment	.580
	Sig. (2-tailed)	.000
	N	221

*Correlation is significant at the 0.05 level (2-tailed).
**Correlation is significant at the 0.01 level (2-tailed).

3 RESULTS AND DISCUSSION

In Table 3, the level of transformational leadership and the commitment to change of participants were captured as the average score. It showed that only 7.24% of participants have a good quality of transformational leadership. 18.10% were categorized as a group that has a low level. While in the level of commitment to change, only 15.84% of participants are committed to changes.

Based on Table 4, the result of the Pearson product moments measurement is that there was a significant positive correlation between transformational leadership and commitment to change ($r = .58$, $\rho < 0.05$). It showed that if the transformational leadership is getting higher, the level of commitment to change will be increased too.

4 CONCLUSION

This study examined the correlation between transformational leadership and commitment to change. There was a positive connection between both variables. This finding supported the previous research, which explained that a transformational leader can facilitate

the creation of shaping certain behaviors of employees that will lead them to accept the changes and have advanced commitment to them (Manz & Sims 2001). In addition, this finding showed that leadership has an important role to affect and to shape the attitude of organization members regarding change, especially a leader who has transformational leadership. If the organization members are already committed to the change, the organization can more easily improve and modify their business strategy and system to adapt in this new era where digital transformations are developed rapidly. Organizations need to adapt in the current situation to have competitive advantage.

ACKNOWLEDGEMENT

First of all, I am grateful to the God Almighty for establishing me to complete the research. I wish to express my sincerest thanks to my lecturer who gave full support and valuable guidance in completing this research. I would also like to extend my special thanks of gratitude to all management of the company for providing me with all data and facilities that were needed and required. I also would like to send my gratitude to all who directly or indirectly, have lent their helping hand in this venture

Consistency of style is very important. Note the spacing, punctuation, and caps in all the examples below.

REFERENCES

Brown, S. L., & Eisenhardt, K. 1997. The art of continuous change: Linking complexity theory and time paced evolution in relentlessly shifting organizations. *Administrative qcience Quarterly* 42: 1–34.

Chou, P. 2015. Transformational leadership and employee's behavioral support for organizational change. *Taiwan: European Journal of Business and Management* 7.

Eby, D. L. M., & Shope, J. 2000. Driving decisions work book. University of Michigan Transportation Institute.

Ferreira, C. M., & Serpa, S. 2018. Society 5.0 and social development: Contributions to a discussion. Sciedu Press: 5(4).

Fukuyama, M. 2018. Society 5.0: Aiming for a new human centered society. Japan: Japan Spotlight.

Gao Urhahn, X., Biemann, T., & Jaros, S. J. 2016. How affective commitment to the organization changes over time: A longitudinal analysis of the reciprocal relationships between affective organizational commitment and income. *Journal of Organizational Behavior* 37: 515–536.

Herold, D. M., Fedor, D. B., Caldwell, S., & Liu, Y. 2008. The effects of transformational and change leadership on employees' commitment to a change: A multilevel study. *Journal of Applied Psychology* 93: 346–357.

Herscovitch, L., & Meyer, J. P., 2002. Commitment to organizational change: Extension of a three-component model. *Journal of Applied Psychology* 87(3): 474–487.

Jaskyte, K. 2003. Assessing changes in employees' perceptions of leadership, behavior, job design and organizational arrangements and their job satisfaction and commitment. *Administration in Social Work* 27(4).

Manz, C. C., & Sims, H. P. 2001. The new superleadership: Leading others to lead themselves. Berrett - Koehler: San Francisco, CA.

Onday, O. 2019. Japan's society 5.0: Going beyond industry 4.0. Turkey: Yedetepe University.

Preziosi, R. 1980. Organizational diagnosis questionnaire. *The 1980 Annual Handbook for Group Facilitators*. New Jersey: University Associates.

Sadeghi, A., & Pihie, Z. A. L. 2012. Transformational leadership and its predictive effects on leadership effectiveness. *International Journal of Business & Social Science*.

Santhidran, S., Chandran, V. G. R., & Borromeo, J. 2013. Enabling organization change – leadership, commitment to change and the mediating role of change readiness. *Journal of Business Economics and Management* 14(2): 348–363.

Shin, J., Taylor, M. S., & Seo, M. G. 2012. Resources for change: The relationships of organizational inducements and psychological resilience to employees' attitudes and behaviors toward organizational change. *Academy of Management Journal* 55: 727–748.

Yousef, D. 2000. Organizational commitment: A mediator of the relationships of leadership behavior with job satisfaction and performance in a non-western country. *Journal of Managerial Psychology* 15(1).

Advances in Business, Management and Entrepreneurship – Hurriyati et al. (Eds)
© 2021 Taylor & Francis Group, London, ISBN 978-0-367-67471-7

The role of human capital management in improving the performance of autonomous public university/Perguruan Tinggi Negeri Berbadan Hukum (PTN-BH)

U. Zuraida & S.H. Senen
Universitas Pendidikan Indonesia, Bandung, Indonesia

ABSTRACT: To improve the quality of the university in order to become a World Class University (WCU), Indonesia has changed several State Universities to become autonomous public university (PTN-BH). Using the concept of human capital management that treats employees as assets rather than cost, it will be a revolutionary way of managing employees. This is because human capital is a concept related to the value added of people to organizations. This paper is a literature review that later will be followed up in empirical research using what is expected to be the role of HCM in increasing PTN-BH performance.

1 INTRODUCTION

Since the mid 1990s the global policy trend in developed and developing countries in the field of education has been to develop world class universities (WCU) (Byun 2013). Then followed by the establishment of an international institution that conducted an assessment of the quality of universities in the world in 2003 (QS World University Ranking). According to Altbach and Salmi (2011) universities that are considered to be qualified and have become WCU are research-oriented, and become the "key" to get into the knowledge economy (Altbach 2009). Similarly, according to Orozco et al. (2015), Seta Postiglione & Arimoto (2015), universities that appears in the top ranking in WCU ranking are universities that make research as its main activity. Therefore Mok (2007) states that many Asian countries such as Japan, Malaysia and Singapore have changed the legal status of their national universities, with the hope that legal entities can become more competitive and entrepreneurial. For Indonesia, the change in the university's legal status was carried out in 2012 by establishing an autonomous public university (PTN-BH) in accordance with Law No. 12 on Education. It is done with the hope of improving the quality and competitiveness of Indonesian universities at the global level to become WCU.

The vital and dynamic strategy that is needed for progress, especially in the education sector, according to Armstrong (2007) is through Human Capital Management (HCM). By using the HCM concept that treats employees as assets rather than costs, it will become a revolutionary way of managing employees.

2 LITERATURE REVIEW

2.1 WCU & PTN-BH

According to Zhihong (2008), there are three benchmarks for first-class universities. That is, there must be well-known international professors who conduct important research; universities must make achievements that have had a profound impact on human civilization and social development; and universities must have graduates who have contributed to human civilization. While Salmi (2009) provides the WCU criteria are graduates who are highly sought after, leading research, and do technology transfer. While Altbach (2004) emphasizes excellence in research, high-quality faculties, independent internal governance, academic freedom and adequate facilities and funding.

Meanwhile, PTN-BH status is given to a college that is considered to be well established in the management of academic and non-academic including the areas of organization management, finance, student affairs, energy (personnel) and infrastructure. With PTN-BH status, universities is given full authority by the government to manage its own organization, including the form and mechanism of funding. Thus, it is expected that PTN-BH will be more productive because of the independence that includes autonomy, justice and adaptability (Wahab 2004). In addition, with PTN-BH status, it is expected that universities will be able to compete in the global arena.

One of the autonomy needed in organizing quality universities is autonomy in the management of Human Resources (HR). This is because HR is a 'source of life for an organization (Mahesar et al. 2016) and the contribution of employees becomes organizational competitiveness (Ulrich 1997). In addition, employees will also be the most influential to achieve organizational excellence (Azar et al. 2004). This is in accordance with the opinion of Shattock (2002) which states that with the management of qualified resources, state universities will be able to realize management that will produce qualified and competitive graduates.

2.2 Human capital, human capital management and performance

Weatherly (2003) states that the development of human resources in today's competitive environment has been replaced by a new concept known as HCM. This is supported by Marimuthu et al. (2009) that HCM is one of the fundamental solutions for entering the international world. The HC theory originates from macroeconomic development proposed by Schultz (1971) and Becker (1964). This theory by Rastogi (2002) and Mayo (2001) is applied in the field of corporate value creation. In addition, it is also applied to be a differentiator and to achieve competitive advantage (Chatzkel 2004; Gratton 2000; Pfeffer 1994). Furthermore, according to Tomer (2003) and Chuang (1999), this theory is applied to achieve organizational growth in the long run. The HC concept treats employees as assets, not costs. And this concept relates to the value added of people to organizations and becomes a way of managing revolutionary people.

HC is the sum of the knowledge, skills, experience and other workforce attributes that are relevant in terms of organizational workforce that drives productivity, performance, and achievement of strategic objectives (Baron and Armstrong 2002). While according to Bontis and Fitz-Enz (2002) and McGregor et al. (2004), HC is a combination of knowledge, skills, innovation and individual abilities of company employees to fulfill the tasks at hand. Rastogi (2002) conceptualizes HC as knowledge, competence, attitudes and behaviors embedded in a person. Another definition of HC is the collaborative composition of the knowledge, skills and abilities of employees who work for common goals (Saad 2014) and 'knowledge, exuniversitiesse and skills acquired by people through media education and training' (Abdul & Aziah 2012). Meanwhile, the Organization for Economic Co-Operation and Development/OECD (2001) states that HC is "the knowledge, skills, competencies, and attributes contained in an individual that facilitates the creation of personal, social and economic welfare. According to Wright & McMahan (2011), at the individual level, HC consists of characteristics possessed by individuals who can produce positive results for these individuals, whereas at the unit level, HC can refer to aggregate accumulation of individual HCs that can be combined in ways that create value for the unit.

Meanwhile, HCM deals with the process of obtaining, regulating, examining and reporting data that enlightens the direction of management of value-added people, strategic investment and operational decisions at the company level and at the frontline level of management (Oyinlola & Adeyemi 2014). HCM also deals with obtaining, analyzing, and reporting data that informs the direction of strategic decision making, investment, and human resource management decisions at the company level and at the front line management level (Baron & Armstrong 2007). According to Kannan & Akhilesh (2002), Khandekar & Sharma (2003), and Molina & Ortega (2002), HCM is a systematic way to identify and improve the competence and ability of an employee to achieve company goals, especially in the field of academic management. HCM postulates humans as capital and considers costs for education, health and training as investments (Schultz et al. 1980). The use of strategies to guide the approach to managing people with interests as assets is a key characteristic of HCM (Salau et al. 2016).

On the other hand, the success of each organization depends on the effective application of its human capital management practices (HCM practice) (Bassi & McMurrer 2007). Human capital management practices that implemented in an organization indicate the level of maturity/maturity level of an organization's human capital management. By knowing the level of maturity of its human capital management, the organization/company can find out which factors are the weakest and need special attention to be the basis for developing strategies to enhance the role of human capital management. Increasing the role of human capital management is very important because this division will manage and develop human capital owned by the organization so that it can carry out its role in the organization as stated by Ulrich (2007). The factors driving the practice of human capital management include: 1. Leadership practices; 2. Employee engagement; 3. Accessibility to Knowledge; 4. Employee Optimization and 5. Learning Capacity. The five drivers/dimensions of human capital management can be observed from 23 performance of human capital management practices that show organizational strength in human capital management. The human capital management practices consist of: 1. Leadership practices, consisting of: 1). managerial communication, 2). inclusivity, 3). supervision skills, 4). executive skills, 5). leadership development, and 6). succession of planning systems; 2. Employee Engagement consists of: 7). job design, 8). commitment, 9). time and evaluation of the Employee Engagement system; 3. Accessibility to Knowledge, consisting of: 10). Availability, 11). collaboration, 12). information sharing, 13). information collection system; 4. Employee Optimization, consisting of: 14). process, 15). working conditions, 16). accountability, 17). employee performance system; 5. Learning Capacity, consisting of: 18. innovation, 19. training, 20. development, 22. assessment and 23. leadership support and learning management systems.

Related to company performance, Menon's (2010) research shows that there is a relationship between HC and company performance where education, related competencies and individual skills affect performance, especially productivity. Invesment HC in the company has a significant impact on learning and performance of the company. Meanwhile according to Heckman (2000), the development of human resources directly contributes to innovation which in turn causes positive implications for the company's performance. This statement is reinforced by the results of research by Romijn and Albaladejo (2002) which explain that work experience increases the absorptive and innovative power of companies, which in turn affects their performance. In addition, Hassen (1995), Ashour (1997), Bontis & Fitz-Enz (2002) state that training contributes to building HC and improving organizational performance. Horowitz & Sherman (1980) found that experience, school, and training at work became an important influence on performance. According Hiltrop (1996) a lot of evidence available that represent a strong positive relationship between HC organization and performance. Wright et al. (1995) also state that HC plays an important role in the performance of a company. In addition, the results of the Hitt et al. (2001) found that HC showed a positive effect on performance. Similarly, Alike, Chaudhry & Roomi (2010) found a relationship between the contribution of human resource development and company performance.

3 METHODS

This research is a quantitative research that will be conducted at PTN-BH in Bandung: ITB, UPI, UNPAD. The study population is the head of the study program (PRODI) or his representative at the PTN-BH as well as the lecturers involved in tertiary accreditation both at the National Accreditation and international accreditation. The research sample was selected by simple random sampling technique with the calculation of the minimum sample size based on the Bernaulli formula. Data collection was carried out with the help of questionnaires and interviews. The questionnaire used is a questionnaire about human capital-Management (Bassi & McMurrer 2007). This questionnaire is consisted of 58 items divided into 5 dimensions of Human Capital Management namely: Leadership practices (14 items), Employee Engagement (13 items), Accessibility to knowledge (7 items), Employee optimization (14 items), and Learning capacity (10 items). Each of these dimensions also consists of 4-6 indicators. The second questionnaire about PTN-BH performance was measured by university performance indicators adapted from Haim et al. (2017).

4 RESULTS AND DISCUSSION

This research is only limited to the literature review that produced the research model. Based on the

theories that have been put forward, the following research model is proposed:

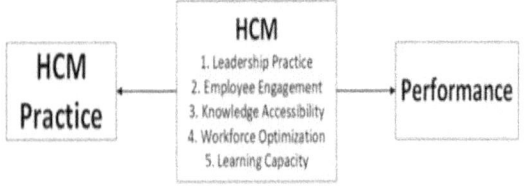

Figure 1.   Research model.

5 CONCLUSION

This paper is only limited to the literature that produced a research model to find out the role of HCM in improving the performance of PTN BH. This research will proceed with searching empirical data to test the research model and hypothesis. As for the results of the empirical research, it is expected to be able to provide input for improving the performance of PTN BH in the city of Bandung.

REFERENCES

Abdul, G.A., & Aziah, I. 2012. An expository analysis on the implementation challenges of human capital development skills through academic programme: a case of public research University in Malaysia. *International Journal of Academic Research in Economics and Management Sciences* 1(6): 144–151.

Altbach, P.G., 2009. Peripheries and centers: Research universities in developing countries. *Asia Pacific Education Review*: 10(1), 15–27.

Altbach, P.G., & Salmi, J. 2011. The road to academic excellence: the making of worldclass research universities. World Bank Publications.

Baron, A., & Armstrong, M. 2007. Human capital management: achieving added value through people. Kogan Page Publishers.

Bassi, L., & McMurrer, D. 2007. Maximizing your return on people. Harvard Business Review.

Becker, G.S. 1964. Human capital: a theoretical and empirical analysis with special reference to education. Chicago: University of Chicago Press.

Bontis, N., & Fitz-Enz, J. 2002. Intellectual capital ROI: a causal map of human capital antecedents and consequents. *Journal of Intellectual capital* 3(3) 223–247.

Byun, K., Jae-Eun Jon; Dongbin Kim, 2013. Quest for building world-class universities in South Korea: outcomes and consequences, High Educ (2013) 65:645–659. DOI 10.1007/s10734-012-9568-6

Chatzkel, J.L. 2004. Human capital: the rules of engagement are changing. Lifelong learning in Europe 9(3): pp.139–145.

Chaudhry, N.I., & Roomi, M.A. 2010. Accounting for the development of human capital in manufacturing organizations. *Journal of Human Resource Costing & Accounting.*

Chuang, Y.C. 1999. The role of human capital in economic development: evidence from Taiwan. *Asian Economic Journal* 13(2): 117–44.

Gratton, L. 2000. Living strategy: Putting people at the heart of corporate purpose. London: Prentice Hall.

Heckman, J.J. 2000. Policies to foster human capital. *Research in economics* 54(1): 3–56.

Hitt, M.A., Bierman, L., Shimizu, K., & Kochhar, R. 2001. Direct and moderating effects of human capital on strategy and performance in professional service firms: A resource-based perspective. *Academy of Management journal* 44(1): 13–28.

Kannan, G., & Akhilesh, B.K. 2002. Human capital knowledge value added: A case study in infotech. *Journal of intellectual capital* 3(2): 167–179.

Khan, E.A., & Quaddus, M. 2018. Dimensions of human capital and firm performance: Microfirm context. *IIMB management review* 30(3): 229–241.

Khandekar, A., & Sharma, A. 2003. Managing human resource capabilities for sustainable competitive advantage: an empirical analysis from Indian global organization. *Education and Training* 47(8/9): 628–639.

Kim, M.S., & Thapa, B. 2018. Relationship of ethical leadership, corporate social responsibility and organizational performance. *Sustainability* 10(2): 447.

Mahesar, H.A., Chaudhry, N.I., Ansari, M.A., & Nisar, Q.A. 2016. Do Islamic HRM practices influence employee outcomes: mediating role of employee engagement. *International Research Journal of Arts & Humanities* (IRJAH), 44(44).

Marimuthu, M., Arokiasamy, L., & Ismail, M. 2009. Human capital development and its impact on firm performance: Evidence from developmental economics. *Journal of international social research* 2(8).

McGregor, J., Tweed, D., & Pech, R. 2004. Human capital in the new economy: devil's bargain?. *Journal of Intellectual Capital*.

Mok, K.H. 2007. Questing for internationalization of universities in Asia: critical reflections. *Journal of studies in international education* 11(3–4): 433–454.

Molina, A.J. & Ortega, R. 2002. Can effective human capital management lead to increased firm performance?. IE Working Paper WP 15/02. [Online]. Retrived from http://ssrn.com/abstract=1024549

Orozco, J.E.F., Becerra J.I.V., & Arellano, C.I.M. 2015. Perspectivas actuales sobre los rankings mundiales de universidades. *Revista de la educacioÁn superior* 44(175): 41 67. [Online]. Retrived from doi:https://doi.org/10.1016/j.resu.2015.09.001

Oyinlola, O.M., & Adeyemi, A.Z. 2014. An empirical analysis of human capital development and organizational performance in banking sector: A Nigerian Experience. *International Journal of Economics, Commerce and Management*, 2 (7).

Pfeffer, J. 1994. Competitive advantage through people. *California Management Review* 36(2): 9–28.

Postiglione, G.A., & Arimoto, A. 2015. Building research universities in East Asia. *Higher Education* 70(2): 151–153.

Rastogi, P.N. 2002. Knowledge management and intellectual capital as a paradigm of value creation. *Human Systems Management* 21: 229–240.

Romijn, H., & Albaladejo, M. 2002. Determinants of innovation capability in small electronics and software firms in southeast England. *Research Policy* 31(7): 1053–1067. [Online]. Retrived from doi:10.1016/s0048-7333(01)00176-7.

Salau, O.P., Adeniji, A.A., & Oyewunmi, A.E. 2014. Relationship between elements of job enrichment and organizational performance among the non academic staff in Nigerian public universities. *Marketing and Management Journal* 12(2): 173–189.

Salau, O.P., Falola, H.O., Ibidunni, A.S., & Ig-binoba, E.E. 2016. Exploring the role of human capital management on organizational success: Evidence from public universities. Management Dynamics in the Knowledge Economy 4(4): 493–513.

Salmi J. 2009. The challenge of establishing worldclass universities. World Bank Publications.

Schultz, T.W. 1971. Investment in human capital. The Role of Education and of Research.

Tomer, J.F. 2003. Personal capital and emotional intelligence: an increasingly important intangible source of economic growth. *Eastern Economic Journal*, 29(3): 453–70.

Ulrich, D. (1997). A new mandate for human resources. *Harvard business review* 76: 124–135.

Weatherly, L.A. 2003. Human capital the elusive asset; measuring and managing human capital: a strategic imperative for HR. *Research Quarterly* 13(1): 82–86.

Wright, P.M., & McMahan, G.C. 2011. Exploring human capital: putting 'human' back into strategic human resource management. *Human resource management journal* 21(2): 93–104.

Advances in Business, Management and Entrepreneurship – Hurriyati et al. (Eds)
© 2021 Taylor & Francis Group, London, ISBN 978-0-367-67471-7

Stakeholder engagement and business sustainability

A. Sulaeman
Universitas Wiralodra, Indramayu, Jawa Barat, Indonesia

E. Tisnawatisule, Hilmiana & M.F. Cahyandito
Universitas Padjadjaran, Bandung, Jawa Barat, Indonesia

ABSTRACT: Soybean-based MSMEs in West Java are businesses that are proven to be able to survive in the long term with very large and ever-increasing consumer needs, pressure from customers, and the public relating to product quality and safety. This often threatens business sustainability. It is necessary to have mutually beneficial cooperation between soybean-based MSMEs and stakeholders. It is expected to be able to overcome these problems. Thus, this study aims to determine the level of engagement with stakeholders, the achievement of business sustainability, and to know the influence of stakeholder engagement on business sustainability. This research was conducted with a survey of 156 soybean-based MSMEs in West Java by distributing valid and reliable questionnaires. The results of this study state that engagement with stakeholders with suppliers, customers, communities and the government has taken place and contribute to improving the sustainability business of MSMEs.

1 INTRODUCTION

1.1 *Soybean-based MSMEs strategicposition*

Compared to other non-oil-and-gas MSMEs, soybean-based MSMEs have advantages that open the opportunities for economic development. Some of these advantages are (1) non-oil and gas export commodity, (2) business that is passed down from generation to generation involving different groups of community members, (3) absorbance of a significant number of the labor force, (4) Indonesian unique commodity (5) a source of inexpensive, convenient and nutritious source of food. Despite their advantages, soybean-based MSMEs also face new and expanding challenges. For example, they are often constrained by problems relating to the negative impacts of their business activities on economic, social, environmental and cultural dimensions, which reduce their business sustainability, which is marked by the reduction of the number of business units due to bankruptcy or business closure caused by violations of those dimensions.

Business sustainability is shown by the ability to survive over a long period as a result of business health, taking into account the stakeholders' welfare and reducing the negative impacts of the business on the environment and community (Edward & Lundrum 2009). Previous studies found that factors influencing business sustainability consist of internal and internal factors. Internal factors include (1) Performance, (2) Owner, and (3) Employee. Whereas external factors include (1) Customer, (2) Supplier,

and (3) Government (Parish 2013). External factors can influence the internal condition of the company. In this research, the external factors of the customer, supplier, and government act as stakeholders (Freeman 1984). Therefore, stakeholders play an important role in determining business sustainability. To achieve success in stakeholder engagement, several conditions are needed, namely: (1) Trust building, (2) Resource adequacy, (3) Information openness, (4) Effective communication, (5) Inclusivity, and (6) Strategic thinking.

1.2 *Previous research of business sustainability*

Previous research revealed that the factors influence business sustainability consist of internal factors. The importance of stakeholder engagement in bolstering business sustainability has been stated by several studies. Bal et al. (2013) and Hoai & Anh (2017) said that stakeholder engagement is a booster of sustainability for the strategic construction business that involves a lot of parties. Stakeholder engagement activities are marked by mutual support and cooperation in providing suggestions and business solution plans and development. Stakeholders who have a wide range of outreach are needed to support the sustainability of finance and *mimicrofinanceusinesses* (Hoai & Anh 2017; Kimando et al. 2012). Stakeholder engagement can create value for parties of interest (Donalson & Preston 1995; Freeman 1984; Perrini & Tencati 2006; Post et al. 2002).

Previous studies on business sustainability were mostly conducted on big corporations, particularly those in industrial and financial sectors, with the qualitative research approach (grounded research) and based on strategic management studies. In this study, the research is conducted on a large number of small scale MSMEs that have large impacts on the social and environmental dimensions. A quantitative approach is used for this research as it has the aim of making predictions by developing the concept of human resources sustainability (Ehnert 2006) through hypothesis testing with an inferential statistics approach.

This research also underlines Previous studies on business sustainability that were mostly conducted on big corporations, particularly those in industrial and financial sectors, with a qualitative research approach (grounded research) and based on strategic management studies. In this study, the research is conducted on a large number of small scale MSMEs that have large impacts on the social and environmental dimensions.

A quantitative approach is used for this research as it has the aim of predicting by developing the concept of human resources sustainability (Ehnert 2006) through hypothesis testing with an inferential statistics approach. This research also underlines human resources management studies as the basis for problem-solving because the unit of analysis being studied is SMMEs that prioritize the role of human resources in their business operations.

Furthermore, human resources management is part of multidimensional studies where humans play a leading role. Therefore this research is aimed at identifying the effect of stakeholder engagement on business sustainability by testing the hypothesis that states that stakeholder engagement affects business sustainability. At the practical level, this research is expected to be able to provide suggestions to soybean-based MSMEs about the need for engagement with stakeholders to develop business sustainability for MSMEs. Academically, this study is expected to be able to provide an additional source of knowledge about MSMEs and policies for their development.

2 METHODS

This research is a descriptive verification study. A descriptive study aims at explaining the condition of the studied variables and present it in a narrative with the support of the central tendency statistical analysis, whereas the verification study aims at discovering the relationship of the variables through hypothesis testing. The unit of analysis of this research is the owners of Micro Small and Medium Enterprises (MSMEs) who operate a soybean-based industry, such as tempeh, tofu, or *kecap* (soybean sauce) manufacturing business, and other products made from soybeans, such as *tauco* (soybean paste), soybean milk and soybean flour.

The research was conducted by surveying 156 soybean-based product entrepreneurs in West Java. The data were collected by means of interviews and questionnaires, the combined instruments of this research. The economic dimension of business sustainability was measured by the level of profit earned, the level of labor force absorption, the duration of the business, and the ability to deal with risks.

The social dimension was measured by the level of wage fairness, the level of use of the dangerous substance, the level of work safety, and the level of work health implementation. The environmental dimension was measured by the level of energy use per output and the level of waste filter system application. The data collected were subsequently analyzed descriptively and inferentially. The descriptive analysis was conducted with the central tendency statistics, whereas the inferential analysis was conducted with structural equation modeling partial least squares (SEM-PLS).

3 RESULT AND DISCUSSION

3.1 *Result*

Business sustainability is shown in 3 dimensions called the triple bottom line (Elkingthon 1997), which consist of economic dimension, social dimension, and environmental dimension. In the economic dimension, 48% of the entrepreneurs could increase their business profits by 5% in 6 months, 78% of the entrepreneurs operated a labor-intensive business that employed 10 to 15 people, 23% of the entrepreneurs could deal with business risks and 68% have operated the business without being interrupted by other businesses for 5 to 15 years. In the social dimension, 25% stated they paid wages based on the principle of reasonability and fairness, 55% stated they did not use dangerous addictive substances, 78% applied work safety regulations and 37% applied work health regulations. In the environmental dimension, 78% of the entrepreneurs stated they still could not afford efficiency in energy use and 84% did not yet have systemic waste filtering facilities.

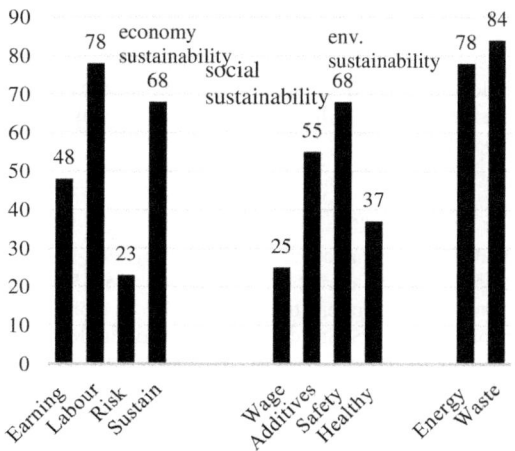

Figure 1. Level of business sustainability.
Sources: primary research

The research also revealed the relationship between the business owners and the stakeholders, shown by information openness, trust-building, and fulfillment of needs. Figure 2 presents the result of stakeholder research:

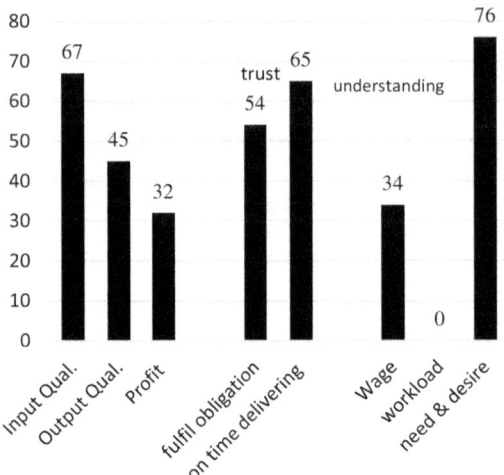

Figure 2.   The percentage of stakeholder engagement level.
Sources : Primary research

Concerning trust-building, the researchers discovered that 54% of the business people built trust by keeping their obligation fulfillment on time and 65% by delivering their products on time. The research also revealed that 34% of the business owners understood their labor's wage needs. 75.6% of the business owners understood the workload of their workers and understood the customers' needs and expectations.

Hypothesis testing was conducted on sample bootstrapping for the overall model. The value used to test the hypothesis was the coefficient path and R2 value (determination coefficient). The endogen construct was used to evaluate the model with value p significance. The results show that the determinant coefficient of the endogen construct (R2) of stakeholder engagement was 0.57, which means 57% of business sustainability was influenced by stakeholder engagement.

3.2   Discussion

Stakeholder engagement in the form of trust-building efforts will create trust from the stakeholders. The customers' and suppliers' trust will result in customer' and suppliers' loyalty. Steady customers and suppliers' loyalty will increase profit (Hallowell 199; Keisidou et al. 2013), reduce the level of business risks as a result of supply dependability (Ahmadi & Sorpong 201; Mahndretos 2014). In addition to increasing revenue, trust from the community stakeholders can also reduce the level of risks from customers and the surrounding communit's claims. Stakeholder engagement is shown

by information openness about business growth and obstacles. A business owne's openness is expected to bring in a lot of input that can improve business performance and anticipate business risks. By understanding the stockholders' needs, such as reward and punishment for employees, good quality products for the consumer, and the return of their stock value, a business can develop employee loyalty, trust, and satisfaction so that a favorable long term relation can be established

A business owner's confidence in the social dimension of sustainability means that he will continue to put the efforts into doing ethical things such as paying reasonable and fair wages, providing safe and healthy products, and paying attention to workplace safety and health. Stakeholder engagement shown by information openness about the fluctuation of business conditions to the employee stakeholder makes it possible for the employees to have a sense of belonging to the business and improve their responsibility for the business' fate. A business owner's understanding of the stakeholder's needs will develop the stakeholder's loyalty and long term mutual relationship.

The stakeholder's confidence in the environmental dimension will encourage the business owner to improve the environmental friendliness of his business operations. Stakeholder engagement through information openness influences how far business ethics is implemented by a company. The understanding of stakeholder's needs relevant to environmentally friendly performance is expected to improve environmental sustainability as MSMEs will continually improve their processing techniques that can reduce waste and create more valuable value from the waste.

4   CONCLUSION

Business sustainability has three dimensions, namely economic, social, and environmental dimensions. In the economic dimension, the development of soybean-based MSMEs is marked by its significant role in absorbing the labor force; however, they are relatively weak in their ability to deal with business risks such as supply stability, a price increase of the input material, and competition. In the social dimension, it is marked by a safe working environment in the technical sense and the low rate of conflicts; however, soybean MSMEs in West Java have not been able to pay reasonable and fair wages to their employees. In the environmental dimension, most MSMEs have not had special facilities and installation for waste management.

In stakeholder relation-building, MSMEs emphasized trust-building with the stakeholders by timely delivery and ensuring the quality of their products. Based on the research, stakeholder engagement has a medium impact on business sustainability. Therefore, to develop business sustainability, it is necessary to build a good relationship with the stakeholders so that it can sustain business sustainability, maintain a good reputation, and avoid arbitrariness.

REFERENCES

Ahmadi, B.D., Sorpong, S. K. 2017. Assesing The Social Sustainability of Supply Chains Using Best Worst Method. *Resources Conservation and Recycling* 126 July 2017 DOI.10.10.101/resconrec.2017.07.020

Bal, M. D. Bryde, Donald, F.2013. Stakeholder Engagement Achieving Sustainability in The ConstructionSector.www. mdpi.com/*journal/sustainability*. ISSN : 2071-1050. DOI:10.3390/s45020695.

Delloite. (2011). Stakeholder Engagement. *AA1000 Stakeholder Engagement Standard 2011 Accomodates* 2008.

Donalson, T. Preston, LE.1995. The Stakeholder Theory of Corporation:Concept, Evidence & Implication. In : *Academic of Management Review* pp 65–1.

Dyllick and Hockert.2002.Beyond The Business Case for Corporate Sustainabiliy Business. *Strategy and The Environmental Quality Management* Vol.II No.2 pp.130–141.

Edward Freeman, Jefrey S Harny. Andre C. Wich. (2010). *Stakeholder Theory : the State Of The Art.* Cambridge University Press.

Ehnert I.2006. Sustainability Issue in Human Resources Management : Linkage Theoretical Approach and Outlined for an Emerging Field. *Paper for 21st EIASM – SHRM, Workshop Aston Birmingham*, March 28th–29th 2006.

Elkingthon, John.1997. Partnership From Canibals with Forks : The Triple Bottom Line of 21st- Century Busines. *Environmental Quality Management* Vol. 8 No.1 pp 37–51.

Freeman.1984. *Strategy Management : A Stakeholder Approach* : Boston : Pitman

Kimando, Lawrence.2012. Factor Influencing The Sustainability of Micro Finance Institution in Murang a Municapality. *International Journal of Business and Commerce* Vol.1. No.10 Juni2012.www.ijbenet.com.

Keisidou, Elissavet, Lazaros, S., Dimitrios, I.M.2013. Customer Satisfaction, Loyalty and Financial Performance : A Holistic Approach of The Greek Banking Sector. International *Journal of Bank Marketing* Vol. 1 No. 4 2013 pp 259–288 ©Emerald Group Publishing Limited 0265-2323-DOI 10.1108/IJBM-11-2012-0114.

Landrum, Nancy R. Edwards, S.2009. Sustainable Business An Executive's Primer. Business Expert. *Press. LLC USA*

Larson, Sand W, LJ.2009. Monitoring The Success of Stakeholder Engagement : Literature Review, In Measam TG Brake L (Ed5). *People Communities and Economies of The Lake EyreBasin.DKCR Research Report 45 Desert Knowledge Cooperative Research Centre Alic Springs* pp 251–298.

Lawrence and Weber.2011. Business and Society Stakeholdes, *Ethics Public Policy* 13th Ed. Mc Graw Hills.

Parish, et al. 2013. Sustainability in small and medium sized Entreprise in Regional Australia : A Framework of Analysis. *Small Entreprise Association of Australia and New Zealand 26th Annual SEAANZ Proceedings.*

Advances in Business, Management and Entrepreneurship – Hurriyati et al. (Eds)
© *2021 Taylor & Francis Group, London, ISBN 978-0-367-67471-7*

The influence of organizational commitment on job satisfaction: An analysis of employees corporate and commercial division in Regional Bank X West Java Indonesia

M. Masharyono, Sumiyati, S.H. Senen & A. Lestari
Universitas Pendidikan Indonesia, Bandung, Indonesia

ABSTRACT: Employees satisfaction is considered as one of the most important elements of a successful orga-nization. The organization must find out how to make their employees satisfied with their job. Problems regarding the achievement of job satisfaction that has not been optimal are experienced by employees in a corporate and commercial division at Regional Bank X. The lower job satisfaction of Regional Bank X employees is charac-terized by the increased employee's turnover. One of the efforts to improve the job satisfaction of employees in a corporate and commercial division at Regional Bank X is to increase their organizational commitment. The study uses descriptive and verification analysis while the method of the study is an explanatory survey. The number of samples taken was 40 respondents, and the study uses a simple linear regression analysis study. The findings of the study show that the description of the organizational commitment is in a high category, the description of the job satisfaction is in a high category, especially in worker relationships dimension, and then, organizational commitment has an influence on job satisfaction by 57.4%. If an increase in organizational commitment to employees is carried out, then job satisfaction will increase.

1 INTRODUCTION

The challenge to improve the quality of human resources with special skills to compete with the business world had to be faced by every company (Mangkunegara & Waris 2016). Human factors are considered as a part of the development of society so that the peak of excellence and sustainable develop-ment will not be possible unless human resources are improved (Arjmandi et al. 2016). Human resources or employees of each organization are the most important part so they must be inclined and affected by fulfilling their duties (Kurniawan 2017; Rahmawati 2017). The more aspects of the work that befits the individual, the higher the level of feeling satisfied. Getting satisfac-tion from work is the hope of every employee (Silen 2016).

Satisfaction in the workplace is a complex subject and difficult to define, because it is too subjective and diverse where everyone has different needs and goals (Lizote et al. 2017). The main determinant of organiza-tional effectiveness is employee satisfaction with work and their commitment to the organization (Kurniawan 2017).

Employee satisfaction holds an important key in every organizational arrangement (Ahmed & Jamil 2015). Research on job satisfaction occurs in vari-ous industries both in the manufacturing sector and in the service sector (Patulak et al. 2013). In the

manufacturing industry several studies have been con-ducted in Turkey, China (Ölçer 2015) and Bali for Indonesia (Bagus et al. 2016). Problems with job sat-isfaction also occur in the banking services industry from Vietnam, Kazakhstan, Greece (Antonaki & Triv-ellas 2014; Erarslan et al. 2018; Nguyen et al. 2014). In Indonesia, research on banking job satisfaction occurs in Bandung, Bali, Kudus, North Sulawesi, North Bar-ito, Jakarta, Kudus and Malang (Ningrum et al. 2018; Prasetio et al. 2017; Purwanti & Rasmini 2015).

Employees with low job satisfaction are character-ized by increased turnover (Nikolajevaite 2016). Low job satisfaction is associated with low performance, limited service quality, and decreased customer sat-isfaction, higher staff turnover and can cause health complaints (Hennekam 2016). A 'good job' satis-faction can be seen from how employees' organiza-tional commitment is characterized by low turnover (Rehman et al. 2013). Based on the background of the study above, the purpose of this study is as follow: (1) organizational commitment, (2) job satisfaction, and (3) the influence of organizational commitment on job satisfaction.

1.1 Literature review

Mathis & Jackson (2011) state that Human Resources Management is designing management systems to ensure that human talent is used effectively and efficiently to accomplish organizational goals. This understanding can be understood that human resource management is the activity of designing a management system to ensure that human talent is used effectively and efficiently to achieve organizational goals.

The level of employees who believe and accept organizational goals and the desire to stay with the organization can also be defined as organizational commitment (Mathis & Jackson 2011). Employee satisfaction with work and commitment to their organization has been seen as the main determinant of organizational effectiveness (Kurniawan 2017). Increasing performance and productivity, improving quality and innovation, higher levels of job satisfaction, lower absenteeism and turnover intention are positive results of organizational commitment to each employee (Jia et al. 2017; Mccunn et al. 2018).

Organizational commitment is as an attitude based on the level of identification or attachment to the organization where someone works and has a strong correlation with job satisfaction (Mccunn et al. 2018). Researchers believe that managing organizational commitment correctly can deliver effective results and include organizational effectiveness, improved performance, fewer employee turnover, and reduced absenteeism (Sheikhy 2015). The dimension of organizational commitment according to Allen and Meyer in Mccunn et al. (2018) presents three types of commitments: (1) Affective Commitment (2) Continuous commitment, (3) Normative commitment.

Human resource management, through internal marketing, can be a solution to create employees who are capable, efficient, and committed, so that the quality of services can be delivered with love (Suprihanto Wrangkani et al. 2018). Employees who are competent and satisfied with their superiors, and who are know what to expect are assets for the organization (Dinc 2017).

Human resource management according to Mathis & Jackson (2011) has 7 functions, one of which is a strategic HR management. In strategic HR management there is a discussion about HR retention which is HR planning where managers anticipate future supply and demand for employees and the nature of labor problems, including employee retention. One of the scopes of HR retention is individual workers and organizational relationships where relationships between individuals and their leaders can be broadly changed from profitable to un-favorable. Job satisfaction is the output of individual workers and organizational relationships.

Job satisfaction is a positive emotional state resulting from the evaluation of a person's work experience (Mathis & Jackson 2011). Locke in Sony & Mekoth (2016) defines job satisfaction as a pleasant or positive emotional state that results from a person's work or work experience. Dimensions of job satisfaction according to Mathis & Jackson (2011) is worker relation-ships, pay and benefits, performance recognition, and communications with managers and executives.

Based on the explanation of the influence of organizational commitment on job satisfaction, the paradigm is illustrated in the frame of mind in Figure 1 as follows:

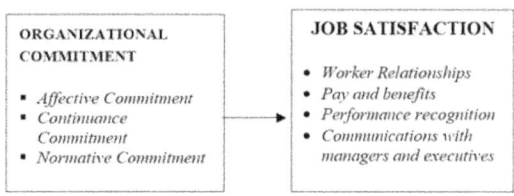

Figure 1.   Research paradigm.

2   METHODS

The research analysis unit in this study was the employees at Regional Bank X. They are employees of the corporate and commercial di-vision of the Regional Bank X in less than one year so that the data was collected by a cross-sectional method. The number of samples used was 40 employees from the corporate and commercial divisions in Regional Bank X.

Data collection techniques was literature studies and field research with questionnaires while data analysis used descriptive technique and verification analysis. The analysis of a verification data used a simple linear regression with SPSS 22.0 for Windows software.

3   RESULTS AND DISCUSSION

3.1   *Descriptive analysis*

3.1.1   *Organizational Commitment*
Based on the results of the questionnaire, a description of organizational commitment to employees in the corporate and commercial divisions of Regional Banks X is shown as in Table 1 below:

Table 1.   Recapitulation of responses to organizational commitments.

No	Dimension	Ideal Score	Total Score	%
1	Affective Commitment	1000	843	84.30
2	Continuance Commitment	800	613	76.63
3	Normative Commitment	800	628	78.50
TOTAL		2600	2084	79,81

*Source: Data processing results 2019 (SPSS 22.0 for windows software).

Table 2. Recapitulation of job satisfaction responses.

No	Dimensions	Ideal Score	Total Score	%
1	Worker relationships	600	514	85.67
2	Pay and benefits	600	433	72.17
3	Performance recognition	600	457	76.17
4	Communication with managers and executives	600	484	80.67
	TOTAL	2400	1888	78.67

*Source: Data processing results 2019 (SPSS 22.0 for windows software).

Based on Table 1, the highest score on organizational commitment is on the dimension of affective commitment which scores 843 or 84.30%, while the lowest score is on the dimen-sion of continuance commitment, which is 613 or 76.63%. Thus it can be said that almost all respondents stated organizational commitment to employees of the corporate and commercial divisions of Regional Banks X included in a high category. A person's affective commitment will be stronger if the experience in the organi-zation is consistent with some expectations and shows a desire to continue working because of agreeing with the organization and indeed wants to do it (Wayan & Cipta 2016).

3.1.2 Job satisfaction

Based on the results of 40 questionnaires, there is an overview of the job satisfaction of the employees in the Corporate and Commercial Division of Bank Daerah X. It can be seen in Table 2.

Based on Table 2, the highest score of job satisfaction based on the results of respondents' answers is in the dimensions of worker relation-ships with a score 514 or equal to 85.67%, while the lowest score found in the pay and benefits dimension which scores 433 or 72.17%, it can be said that almost all respondents stated that job satisfaction for employees in corporate and commercial divisions of the Regional Bank X was in a high category. Employees job satisfaction leads to higher employee organizational commitment and high commitment leads to the overall successful organization (Olawale & Olarewaju 2016).

3.2 Verification analysis

This study consists of independent variables namely organizational commitment (X) while for the dependent variable is job satisfaction (Y).

Based on the results of data processing by SPSS 22.0 for Windows, it can be obtained a simple linear regression coefficient as follows.

Table 3. Regression coefficient.

Coefficients[a]					
	Unstandardized Coefficients		Standardized Coefficients		
Model	B	Std. Error	Beta		
1 (Constant)	5.639	6.207		.909	.369
Organizational Commitment	.276	.118	.295	2.335	.025

a. Dependent Variable: Work Satisfaction
*Source: Data processing results 2019 (SPSS 22.0 for windows software).

Based on Table 3 in column B, there are con-stant values and multiple linear regression coef-ficients for independent variables. Based on these values, it can be determined that multiple linear regression models are expressed in the form of equations as follows:

$$Y = a + bX$$
$$Y = 5.639 + 0.276\,X \tag{1}$$

The above equation can be interpreted as follows: $a = 5.639$ it means that if the variable X is zero (0), then the variable Y will be worth 5.639; $b_2 = 0.276$ it means that that if organizational commitment (X) increases by one unit and the other variable are constant, then Y variable will increase by 0.276 units.

To find out the percentage effect of X on Y, the coefficient of determination can be known by the formula stated by Riduwan (2013), the formula as follows:

$$KD = r2 \times 100\% \tag{2}$$

Information: KD = Determination coefficients; r = Correlation coefficient; 100% = Constant.

The Effect of Organizational Commitment on Job Satisfaction can be seen from the results of the following Table 4:

Table 4. Partial determination coefficient.

Model Summary				
Model	R	R Square	Adjusted R Square	Std. Error of the Estimate
1	.758[a]	.574	550	3.122

a. Predictors: (Constant), Organizational Commitment
b. Dependent Variable: Work Satisfaction
* Source: Data processing results 2019 (SPSS 22.0 for windows software).

$$KD = r2x100\%$$
$$= (0.758)2x100\%$$
$$= 57.4\% \tag{3}$$

The number of the correlation coefficient (R) is 0.758. This means that the relationship between organizational commitment and job satisfaction is about 0.758. Based on the calculation of the determination coefficient for organiza-tional commitment with job satisfaction is about 57.4%. In other words, satisfaction is influenced by organizational commitment which is about 57.4%, while the remaining 42.6% is influenced by other factors that have been ignored by the writer that can be employee engagement. (Albdour & Altarawneh 2014), organizational culture (Troena & Setiawan 2012), motivation (Zainuddin et al. 2012), transformational leadership (Atmojo 2012).

3.3 Partial hypothesis test (T test)

The t-test basically shows how far the influence of one explanatory variable / independent individually in explaining the dependent variable. To find out the percentage of the influence of organizational commitment on job satisfaction, this study uses SPSS 22.0 for Windows program, and is obtained the following output:

Table 5. Significance value of T test.

Coefficients[a]

Model	Unstandardized Coefficients		Standardized Coefficients		
	B	Std. Error	Beta	t	Sig.
1 (Constant)	5,639	6.207		.909	.369
Organizational Commitment	.276	.118	.295	2.335	.025

a. Dependent Variable: Work Satisfaction
* Source: Data processing results 2019 (SPSS 22.0 for windows software).

Based on Table 5, t-count is about 2.335 in organizational commitment. Significant level (α) is about 5%, and degrees of freedom $df = n - k = 40 - 3 = 37$ obtained t-table which is 2.026. Due to t count> t table or 2.335> 2.026 then Ha is accepted, this means that the organizational commitment has an effect on job satisfaction, the influence of organizational commitment on job satisfaction is about 57.4 %, it included in a high category (Sugiyono 2017), while the remaining 42.6% is influenced by other factors that the writer ignores that can be an employee engagement (Albdour & Altarawneh 2014), organizational culture (Troena & Setiawan 2012), motivation (Zainuddin et al. 2012), and transformational leadership (Atmojo 2012).

Organizational commitment has an influence on job satisfaction of corporate and commercial division employees at the Bank's X headquarters. This shows that the higher organizational commitment the

higher job satisfaction felt by the employee. In line with Chandra's research results in Wiswari & Sudibya (2016) states that organizational commitment affects job satisfaction, where employee job satisfaction will increase with organizational commitment.

Organizational commitment is important to make individuals work and encourage to be responsible for their work, so they feel useful, grow feeling satisfied at work (Silen 2016). Lizote et al. (2017) in their research stated that employees feel satisfied when they are affective commitments and are not satisfied if their commitment is only to obey the norm.

4 CONCLUSION

Based on the discussion above, it can be concluded as follows: (1) The picture of organizational commitment employees in the corporate and commercial divisions of Regional Banks X is in a high category; (2) The picture of job satisfaction employees in the corporate and commercial divisions of the Regional Bank X is in a high category; (3) There is an influence between organizational commitment and job satisfaction. This shows that the stronger the organizational commitment, the higher the job satisfaction of employees in the division of corporate and commercial of Regional Bank X.

Based on the results of the study, the writer suggests increasing the organizational commitment by increasing family values in the corporate climate, feeling of belonging to each employee, instilling employee confidence in company values, moral attitudes, and responsibilities. Increasing performance and productivity, improving quality and innovation, higher levels of job satisfaction, lower absenteeism and turn-over intention are positive results of organizational commitment to each employee (Jia et al. 2017; Mccunn et al. 2018).

REFERENCES

Ahmed, K. & Jamil, S. 2015. Impact of HR competencies on employee's job satisfaction. Journal of Resources Development and Management 5: 15–28.
Albdour, A.A. & Altarawneh, I.I. 2014. Employee engagement and organizational commitment: evidence from Jordan. International Journal of Business 19(2).
Antonaki, X.E. & Trivellas, P. 2014. Psychological contract breach and organizational commitment in the Greek banking sector: The mediation effect of job satisfaction. Procedia-Social and Behavioral Sciences 148: 354–361.
Arjmandi, A., Yaghoubi, N., & Doaei, H. 2016. Exploring the dimensions and components of Islamic values influencing the productivity of human resources from the perspective of Mashhad Municipality employees. 230(May): 379–386.
Atmojo, M. 2012. The influence of transformational leadership on job satisfaction, organizational commitment, and employee performance. International Business Study 5(2).

Bagus, I., Dharmanegara, A., Sitiari, N.W., Gde, I.D., & Wirayudha, N. 2016. Job Competency and work environment: the effect on job satisfaction and job performance among SMEs worker. 18(1): 19–26.

Dinc, M.S. 2017. Organizational commitment components and job performance: mediating role of job satisfaction. *Pakistan Journal of Commerce and Social Sciences* 11(3): 773–789.

Erarslan, S., Ya, Ç.K.A., & Altindağ, E. 2018. Effect of oganizational cynicism and job satisfaction on organizational commitment: an empirical study on banking sector. *The Journal of Faculty of Economics and Administrative Sciences* 23: 905–922.

Hennekam, S. 2016. Competencies of older workers and its influence on career success and job satisfaction. *Employee Relations*.

Jia, A., Lim, P., Teck, J., Loo, K., & Lee, P.H. 2017. The impact of leadership on turnover intention: the mediating role of organizational commitment and job satisfaction. *Journal of Applied Structural Equation Modeling* 1(June): 27–41.

Kurniawan, Z. 2017. Influence Of competence, cultural organization, and job satisfaction of career development and implications on the performance of employees (survey on state-owned enterprises (soes) in the Region of Cirebon). *International Journal of Scientific & Technology Research* 6(12).

Lizote, S.A., Verdinelli, M.A., & Nascimento, S.d. 2017. Organizational commitment and job satisfaction: a study with municipal Civil Servants. *Brazilian Journal of Public Administration* 51(6): 947–967.

Mangkunegara, A.P. & Waris, A. 2016. Effect of training, competence and dicipline on employee performance in company (case study in PT. Asuransi Bangun Askrida). *2nd Global Conference on Business and Social Science*.

Mathis, R.L. & Jackson, J.H. 2011. *Human resource management (13th Ed.)*. United States: South WEstern – CENGAGE Learning.

Mccunn, L.J., Kim, A., & Feracor, J. 2018. Reflections on a Retro fit: organizational commitment, perceived productivity and controllability in a building lighting project in the United States. *Energy Research & Social Science* 38(October 2017): 154–164.

Nguyen, T.N., Mai, K.N., & Nguyen, P.V. 2014. Factors affecting employees' organizational commitment–a study of banking staff in Ho Chi Minh City, Vietnam. *Journal of Advanced Management Science* 2(1): 7–11.

Nikolajevaite, M. 2016. Relationship between employees' competencies and job satisfaction: british and lithuanian employees. *Psychology Research* 6(11): 684–692.

Ningrum, F.D.C., Haryono, A.T., & Fathoni, A. 2018. Effect of competence, organizational commitment and career development on employee performance Bank Mandiri in Sub Branch Office Kudus. *Journal of Management* 4(4).

Olawale, R. & Olarewaju, A. 2016. Job satisfaction, turnover intention and organizational commitment. *BVIMSR's Journal of Management Research* 8(2).

Ölçer, F. 2015. Mediating effect of job satisfaction in the relationship between psychological empowerment and job performance. *Theoretical and Applied Economics* XXII(3), 111–136.

Patulak, M.E., Thoyib, A., & Setiawan, M. 2013. The role of organizational commitment as mediator of organizational culture and employees' competencies on employees' performances (a study on irrigation area management in Southeast Sulawesi). 4(5): 166–175.

Prasetio, A.P., Yuniarsih, T., & Ahman, E. 2017. Job satisfaction, organizational commitment, and organizational citizenship behaviour in state-owned banking. *Universal Journal of Management* 5(1): 32–38.

Purwanti, N.W.D. & Rasmini, N.K. 2015. Pengaruh kompetensi, motivasi, komitmen organisasi pada kinerja dewan komisaris BPR sekabupaten Gianyar. *E-Jurnal Akuntansi*: 686–704.

Rahmawati, A. 2017. Effect of competence on organizational citizenship behavior and performance management: the impact on organizational effectiveness. *The International Journal of Engineering and Science (IJES)* 6(11): 74–85.

Rehman, K., Rehman, Z.U., Khan, A.S., Nawaz, A., & Rehman, S. 2013. Impact of job satisfaction on organizational commitment: a theoretical model for academicians in HEI of developing countries like Pakistan. *Internasiona Journal of Academic Research in Accounting, finance and Management Sciences* 3(1): 80–89.

Riduwan. 2013. *Cara menggunakan dan memakai analisis jalur (path analysis)*. Bandung: Alfabeta.

Sheikhy, A. 2015. An investigation into the effect of human resource competencies and organizational commitment on employees' job satisfaction, discipline of work, and job performance in Khuzestan Telecommunication Company. 3(3).

Silen, A.P. (2016). Pengaruh kompetensi dan pengembangan karir terhadap kepuasan kerja dengan komitmen organisasional sebagai variabel mediasi (studi pegawai Politeknik Ilmu Pelayaran (PIP) Semarang). Jurnal Bisnis Dan Ekonomi (JBE) 23(2): 174–187.

Sony, M. & Mekoth, N. 2016. The relationship between emotional intelligence, frontline employee adaptability, job satisfaction and job performance. *Journal of Retailing and Consumer Services* 30: 20–32.

Sugiyono. 2017. *Metode penelitian kuantitatif, kualitatif, dan R&D*. Bandung: Alfabeta.

Suprihanto, J., Wrangkani, T.D., & Meliala, A. 2018. The relationship between internal marketing and the organizational commitment of doctors and nurses at Mardi Waluyo Hospital, Metro Lampung Indonesia. *International Journal of Healthcare Management* 11(2): 79–87.

Troena, E.A. & Setiawan, M. 2012. The influence of organizational culture, organizational commitment to job satisfaction and employee performance (study at municipal waterworks of Jayapura, Papua Indonesia). *International Journal of Business and Management Invention* 1(1): 69–76.

Wayan, B.I. & Cipta, W. 2016. Pengaruh kompetensi dan budaya kerja terhadap kinerja karyawan. *Journal Bisma Universitas Pendidikan Ganesha Jurusan Manajemen* 4(1): 1–10.

Wiswari, N.K.A. & Sudibya, I.G.A. 2016. Pengaruh kepemimpinan transformasional dan komitmen organisasional terhadap kepuasan kerja dan kinerja pegawai. *E-Jurnal Manajemen* 5(12).

Zainuddin, P., Riama, L.V., & Oktarida, A. 2012. Pengaruh kompetensi dan motivasi kerja terhadap kepuasan kerja serta implikasinya pada kinerja dosen (survei pada perguruan tinggi negeri di Kota Palembang). *Prosiding* 229–238.

Advances in Business, Management and Entrepreneurship – Hurriyati et al. (Eds)
© *2021 Taylor & Francis Group, London, ISBN 978-0-367-67471-7*

The effects of organizational development, and work attitudes on internal control systems and their implications for organizational performance

Suhono, Nursito & E. Mahpudin
Universitas Singaperbangsa, Karawang, Indonesia

ABSTRACT: This study aims to examine the effect of organizational development, organizational commitment, and work attitudes, on the systems of internal control and its implications for organizational performance. This study uses a survey method with a causal approach to analyze the influence between variables using structural equation modeling (SEM) as an analytical tool. The findings generated in this study reveal that both simultaneously and partially the development of organizations, organizational commitment work attitude, has a positive and significant effect on the systems of internal control in the village unit in the area, Karawang Regency, West Java. Organizational development, organizational commitment, and work attitude, on the systems of internal control and its implications for organizational performance simultaneously have a positive and significant effect on organizational performance, but also partially, only organizational development and organizational commitment positive and significant effect on organizational performance, while work attitudes do not have a direct effect on organizational performance. An important finding in this study is that from several variants, both directly and indirectly affect and have implications for the performance of the organization.

1 INTRODUCTION

The Law on Villages will be more comprehensive in regulating villages and is expected to be able to provide great expectations for the progress and welfare of the community and village government. So that it can answer various problems in the village, which include various aspects as well as restore the livelihood base of the village community and strengthen the village as a strong and independent community entity. It is also expected to be able to carry out the mandate and assignment of some functions given by the government, and especially the regency/city government that is above it, as well as spearheading in any implementation of development and society. Thus, its arrangement also intended to prepare the village in response to the process of modernization, globalization, and democratization that continues to develop without losing its identity.

This is in line with the Indonesian government's reform of state financial management with the enactment of a package of laws in the field of state finance. Law No. 17 of 2003 concerning State Finance, Law No. 1 of 2004 concerning the State Treasury and Law No. 15 of 2004 states that the accountability of the implementation of the state budget and regional budgets, this strengthens the demands of performance on

public institutions. Performance can be interpreted as a form of obligation to account for the success or failure of the implementation of the organization's mission in achieving the goals and objectives that have been set previously.

Concerning the problem that the village government's internal control system is not yet optimal, the employee's attributes include two aspects, those that can be changed and those that cannot be changed, so it needs to be focused on what can be changed, such as values, attitudes, abilities, and perceptions. According to Winardi (2006) values are a constellation of likes, dislikes, points of view, attitudes, tendencies within oneself, rational and irrational judgments, prejudices, and patterns of association, which affect one's opinions about the world. With the existence of good employee work attitudes that need to be supported by organizational commitment so that it will cause a positive effect that will participate in overseeing the development of the organization or even vice versa.

Based on the description above, a study was carried out on the effect of the influence of organizational development, organizational commitment and work attitude towards the internal control system and its implications for the performance of village government organizations in Karawang Regency.

The formulation of the problem that can be arranged is as follows:

1. Do organizational development, organizational commitment and work attitude influence the internal control system both partially, and simultaneously?
2. Do organizational development, organizational commitment to work attitude, and internal control systems influence organizational performance both partially, and simultaneously?

The research objectives based on the formulation of the problem above are identifying:

1. The effect of organizational development, organizational commitment, and work attitude affect the internal control system both partially and simultaneously;
2. The effect of organizational development, organizational commitment, and work attitude affect the internal control system both partially and simultaneously.

2 LITERATURE REVIEW

Organizational development is a long-term effort, not a short-term effort, in the sense that organizational development is a continuous or ongoing effort and a willingness to make changes sustainably (Soehardi 2003).

Organizational development or often also known as Organization Development, according to Neil (2016) is said to have ambiguity in meaning because its definition depends on the scope and from which side one sees. Meanwhile, according to Hani (2003) states that there are two ways of handling changes in organizational development.

Commitment is a feeling of identification, loyalty, and involvement shown by workers towards the organization or unit of the Wibowo organization (2007), according to Colquitt et al. (2009) organizational commitment; "As the desire on the part of an employee to a member of the organization."

Work attitude is the creation of a new (added) value in a unit of resources. The quality of human resources was determined by two things, namely aptitude and attitude. Aptitude is talent, intelligence, or dexterity. While attitude as a position (Echols & Shadily 2000). Regarding the quality of human resources as forming the organizational culture, there are several elements, including beliefs and values. Value is something that is believed by members of the organization to know what is right and what is wrong. While beliefs are attitudes about how they should work in organizations (Supartha 2008). The success of an organization according to Liker and Hoseus (2008) is determined by two value streams, namely product value and value/assumptions that shape employee attitudes.

Organizational system or structure and all the means and tools used are coordinated within the organization to maintain the security of property belonging to the organization, giving accuracy and correctness of accounting data, promoting efficiency in the operation and helping to maintain compliance with the policy that has been in charge of management have first. (Galbraith & Kazanjid 2016). The internal control structure must include the basis of policies and procedures designed and used to provide adequate confidence. Moeller (2009).

Organizational performance is a form of achieving organizational goals as expected. Sukmalana (2009). According to Mardiasmo (2009) measurement of organizational performance includes financial and non-financial performance. This is related to organizational goals. Government performance indicators include input indicators, process indicators, output indicators, outcome indicators, benefit indicators, and impact indicators. (BPKP 2010).

3 METHODS

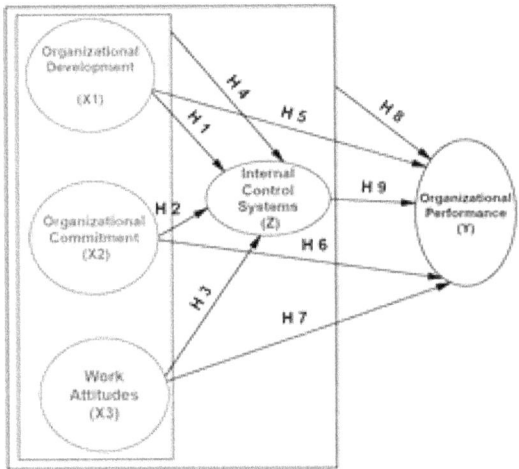

Figure 1. Theoretical research model.

Based on the above framework, the hypothesis can be determined:

1. There is a significant influence on organizational development, organizational commitment and work attitude influence the internal control system both partially and simultaneously;
2. There is a significant influence on organizational development, organizational commitment, and work attitude both partially and jointly through an internal control system on organizational performance.

The research method used in this study is a survey method with a causal approach to analyze the influence between variables using the Structural Equation Model (SEM), which is run through AMOS as a tool for analysis. Respondents used were 221 people. This research was conducted at the village unit in the area, Karawang Regency.

4 RESULTS AND DISCUSSION

4.1 Structural Equation Model (SEM) test

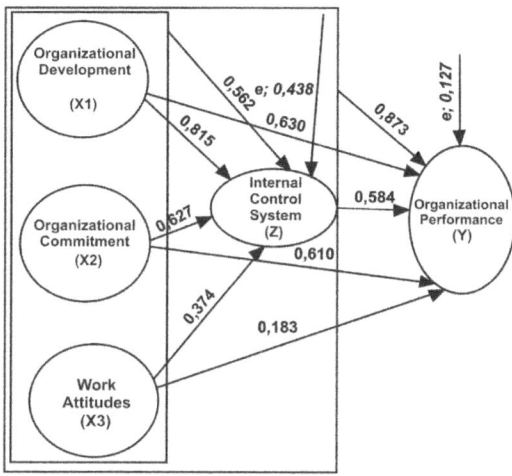

Figure 2. Structural Equation Model (Standardized Model) result.

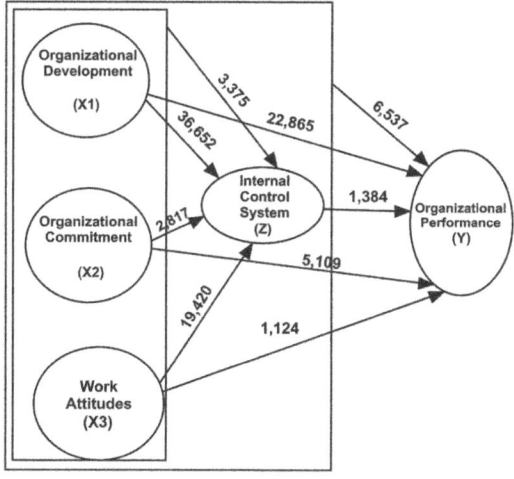

Figure 3. Structural Equation Model result (t-value model).

4.2 Analysis results

Based on the t value of the influence of organizational development on the internal control system is 36,652> 1, 318 so it can be said to be significant, so partially organizational development has a significant effect on the quality of financial reporting, meaning that the better the organizational development planning, the higher the internal control system carried out village government.

The t value for the influence of organizational commitment partially to the internal control system is 2,817> 1,318. That organizational commitment has a significant effect on the internal control system, meaning that the organizational commitment is less useful in building commitment with togetherness then the internal control system will be less good on the run.

The value of t for work attitude towards the internal control system partially is 19,420> 1,318. The working attitude has a significant effect on the internal control system, meaning that the better the work attitude is proven by the employees, the better the internal control system will be in achieving organizational goals.

The influence of organizational development variables, organizational commitment, and work attitude variables together on the internal control system variables are: 0.562 x 100% = 56.2%, the rest is 43.8. % is influenced by other factors. The calculated F value is greater than the F table value that is equal to 3.375> 1.318, so the calculated F value obtained from the sig value of 0.000 <0.05. Thus, the internal control system of the Village Government in Karawang Regency is positively influenced by organizational development, organizational commitment, and work attitude variables.

The t value of the direct influence of the internal control system variables on organizational performance variables is 1,384> 1.318 so that it can be said to be significant, meaning that the more functioning the management and the integration of a good internal control system, the better the performance of the organization.

5 CONCLUSION

a. Organizational development partially has a positive and significant effect on internal control systems. This indicates that the development of village organizations that demand a better system of internal control will make it easier to achieve the expected organizational development together.
b. Organizational commitment partially has a positive and significant effect on the internal control system. This shows that if the commitment of employees is done well, the better the internal control system.
c. Work attitude partially has a positive and significant effect on the internal control system. This means that if the work attitude is getting better, it will further improve the internal control system.
d. The influence of organizational development, organizational commitment, and work attitude simultaneously have a positive and significant effect on the internal control system. Of the three variables, when viewed partially, it turns out that organizational development has the most significant influence on internal control systems.

e. Organizational development partially has a positive and significant effect on organizational performance. This indicates that the better the organization's development, the more it will improve organizational performance.

f. Organizational commitment has a significant effect on organizational performance. This indicates that the better the commitment was given by employees to the organization, the more it will advance better organizational performance.

g. Work attitude has no significant effect on organizational performance. This means, that if the work attitude decreases or is not good it will further worsen the performance of the organization that has been built with difficulty.

h. Organizational development, organizational commitment, and work attitude together through internal control systems affect the performance of the organization Of the four variables if seen partially, it turns out that organizational development has the most significant effect on organizational performance. This means that the better organizational development with the stimulus from the State in the form of Village funds, the better the performance of the organization.

i. Internal control system (Z) has a significant effect on organizational performance (Y). This means, that if the internal control system is increasingly functioning it will further improve organizational performance better this internal control system contributes very well because it is supported by the development of the main organization, the second organizational commitment, and work attitude, although indirectly. h Organizational development, organizational commitment, and work attitude together through internal control systems affect the performance of the organization Of the four variables if seen partially, it turns out that organizational development has the most significant effect on organizational performance. This means that the better organizational development with the stimulus from the State in the form of Village funds, the better the performance of the organization.

j. The internal control system (Z) has a significant effect on organizational performance (Y). This means, that if the internal control system is increasingly functioning it will further improve organizational performance better this internal control system contributes very well because it is supported by the development of the main organization, the second organizational commitment, and work attitude, although indirectly.

REFERENCES

BPKP. 2010. Pedoman Penyusunan Anggaran Berbasis Kinerja (Revisi). [Online]. Retrieved from http://www.bpkpp.go.id.

Colquitt, J.A., Jeffry A.L., & Michael J.W. 2009. Organizational behavior. Mc Graw – Hill Irvin.

Echols, J.M., & Shadily, H. 2000. Kamus Inggris Indonesia. Jakarta: PT. Gramedia.

Galbraith, J.R, & Kazanjid, R.K. 2016. Strategy implementation: The role of structure and process. St paul, MN: West Publishing.

Liker, J.K., & Hoseus, M. 2008. Toyota culture the heart and soul of the Toyota way. Mcgraw-hill.

Mardiasmo. 2009. Otonomi dan manajemen keuangan daerah Ed. II. Yogyakarta: Andi Yogyakarta.

Moeller, R. 2009. Brink's modern internal auditing. John Wiley and Sons. Inc. Hoboken, New Jersey Published Canada.

Soehardi, S. 2003. Esensi perilaku organisasional. Yogyakarta: BPFE UST.

Sukmalana, S. 2009. Langkah dan kebijakan evaluasi kinerja. Palembang: Universitas Tridinanti.

Supartha, W.G. 2008. Budaya organisasi. Denpasar: Udayana University Press.

Undang-Undang No. 17 Tahun 2003 tentang Keuangan Negara, Undang-Undang No. 1 Tahun 2004 tentang Perbendaharaan Negara dan UU No. 15 tahun 2004

Wibowo. 2007. Manajemen kinerja. Jakarta: PT. Raja Grafindo Parsada.

Winardi. 2006. Asas-asas manajemen. Bandung: PT. Alumni.

Advances in Business, Management and Entrepreneurship – Hurriyati et al. (Eds)
© 2021 Taylor & Francis Group, London, ISBN 978-0-367-67471-7

Analysis of attracting and retaining talent based psychological contract at the IT Company in Jakarta

A. Nura & T. Yuniarsih
Universitas Pendidikan Indonesia, Bandung, Indonesia

ABSTRACT: The Company is currently facing the era of industrial revolution 4.0 where one of the influences is the labor market that opens to other countries in the world. The hijack talent phenomenon that results highly talented employees in a company is currently captured and given more attractive offers and facilities by competing companies. Psychological Contract is one way to create attractiveness and retain talent so that the employees don't have to move to another company, or even to other countries in the world. This research uses a quantitative survey approach where the research subjects are in the form of talents at an IT company. The object of this study relates to the characteristics of the psychological contract variables and attracting and retaining talent variables. The population of this study was 221 talented employees. The data were then analysed by using Structural Equation Modeling (SEM) with AMOS program. The results show that there is a positive relationship between the psychological contract and attracting and retaining talent of 0.565 with a t value of 6.251, which means that there is a significant relationship between the psychological contract and attracting and retaining talent.

1 INTRODUCTION

Human resources that are unique and cannot be imitated are recommended as a corporate strategy to achieve sustainable competitive advantage (Abu & Amran 2016). Talented labor is very important, in order to differentiate the organization from its competitors. Today's "Talent War" is not only between companies, but also between governments. This concerns foreign talent, which most countries and companies try hard to attract and retain (Harvey 2014). This results a frequent hijack talent. The company will use various methods to find and retain talent. A successful company will invest time, money and commitment to retention (McLeod 2016).

Therefore, attracting and retaining talent becomes an important issue for the company. Attracting and retaining talent is influenced by the new psychological contract. (Baker 2014). The relational component of psychological contract plays an important role in developing strong organizational- employee relations and reducing employee intention to leave the organization (Abdallah & Kukunuru 2016). HRM practices and psychological contracts in- fluence employee attitudes and behavior and it also influences the organizational effectiveness (Aggarwal & Bhargava 2009). Practitioners must note that the nature of the psychological contract will have an impact on commitment and retention (Behery et al. 2012). However, employees with higher levels of affective commitment and psychological contract violations will think about quitting their jobs. (Addae et al. 2006). Most studies have reported negative relationships between Psychological Contract Violations (PCBs) and attitudes such as Job Satisfaction (JS) and Organizational Commitment (OC) (Antonaki & Trivellas 2014). Psychological contracts have changed significantly in the past three decades. Today's employees expect more than salary. They expect such promotions, en-hancements and professional development, while in the past they expected loyalty, commitment, effort and high results, without a commensurable value proposition for employees (Davis 2015). Being part of a talent pool has a positive impact on psychological contracts and organizational commitment, but it does not have to be translated into trust and intention to stay with the company (Seopa et al. 2015). Employees in the talent pool with those not in the talent pool have the same intention to leave the company, if their expectations are not fulfilled (Seopa et al. 2015). Fulfillment of psychological contracts by increasing relations activities such as good communication and honoring unwritten promises. Psychological contracts offer a useful framework for enhancing overall global mobility and the effectiveness of Global Talent Management (GTM) (Mcnulty & Cieri 2016). Where, attracting and retaining talent is part of the Talent Management, so it can be concluded that the fulfillment of psychological contracts can help in increasing the attracting and retaining talent. Leaders who do not implement concrete plans to utilize technology in the "talent war" will quickly fall behind (Keller & Meaney 2017). A study by MyHiringClub.com obtained information from executives in Asia and the Middle East, about one-third of Technology Information (IT) workers leave the company each year (McLeod 2016). This talent problem also occurs in IT companies in Indonesia.

Table 1. Number of talent turnovers in IT companies.

Year of	Total Annual Turnover
2012	27.4%
2013	18.5%
2014	8.5%
2015	9.2%
2016	7.2%

The talented workforce that resigned reached 27.4% in 2012, this continued until 2016. Despite the decline in the number of resigned talents from year to year, it still remains a big problem for the company. What if the resigned talent is the key person for the company. Low talent changecan be as damaging as high talent change (Matuson 2014). Attracting and retaining talent is concerned with how to become a genuine employer of choice (Baker 2014). This means developing a work culture that reflects the needs and interests to change, both individuals and companies. The essence of attracting and retaining talent is the working relationship of the new model. The study identified eight new psychological contract value models or work relationships that would affect the attractiveness and efforts of organizations to retain talented employees (Baker 2014).

1.1 Psychological contract

Psychological contracts are closely related to human resource management practices (Festing & Schäfer 2014). The relationship between employees and companies is basically a contractual exchange of resources between two parties, namely employees and companies (Aggarwal & Bhargava 2009). These employees and companies are bound by agreement as a result of perceptions regarding the obligations of fulfilling contracts by institutions, or also called psychological contracts. Psychological contracts are beliefs about the terms and conditions of mutual exchange of agreements between individuals and organizations (Rousseau 1998). Psychological contracts are what employees and employers want and expect from each other (Hiltrop 1995). Psychological contracts are employee beliefs, which are based on promises expressed, based on the exchange of agreements between employees and companies (Kotter in Aggarwal & Bhargava 2009). In magnitude, psychological contracts are reciprocal expectations be- tween two parties Guest in (Lee & Liu 2009) Psychological contract is formed from employee's belief that the organization will respect its contribution and fulfill its expectations. The realization of psychological contract can bring about balance, harmony, smoothness

of employees in work, organizational progress, good relations between employees and companies, as well as minimizing the emergence of conflict and social inequality (Sulistiobudi & Kadiyono 2017). So, it can be concluded that the psychological contract is a work relationship between employees and companies that have hopes that they can fulfill their obligations and what will be given in return for one another.

1.2 Attracting and retaining talent

Talent management was first introduced by McKisey's Steven Hankin in 1997 and became a popular book entitled "the war for talent" in 2001 (Beth & Helen 2001). A highly attracting and retaining talented employees and consolidating competitive advantage are important factors for companies all over the world. Understanding what can attract talented employees to work in a company can provide important insights for human resource managers. As stated by Baker: Attracting and Retaining Talent is concerned with how to become a genuine employer of choice. This means developing a workplace culture that reflects the changing needs and interests of both individuals and organizations is needed (Baker 2014). Referring to the opinion above, it can be concluded that now the company must strive to become a company chosen by talented employees and can maintain them within the company. This means companies must develop a work culture that reflects changing needs and interests for both individuals and companies.

1.3 Relationship of psychological contracts with employee attracting and retaining talent

Attracting and retaining talent is influenced by new psychological contracts (Baker 2014). The relation component of psychological contracts plays an important role in developing strong organizational-employee relationships and reducing employee intentions to leave the organization (Abdallah & Ku-kunuru 2016). Human Resource Management Practices and psychological contracts influence employee attitudes and behavior and have an influence on organizational effectiveness (Aggarwal & Bhargava 2009). Perception of organizational support causes affective commitment which can reduce the desire to leave the organization (Addae et al. 2006). Practitioners and academics must note that the nature of the use of psychological contract will have an impact on commitment and retention (Behery et al. 2012). There is a strong positive relationship between relational contracts and all forms of commitment (Cohen 2011). However, employees with higher levels of affective commitment, psychological contract violations think about quitting their jobs. (Addae et al. 2006). Most studies have reported negative relationships between Psychological Contract Violations (PCBs) and attitudes such as Job Satisfaction (JS) and Organizational

Commitment (OC) (Antonaki & Trivellas 2014). Permanent workers will have more relational contracts, involve commitment to the organization, and interest in satisfying work (McDonald & Makin 2000). Psychological pressure is related to NA and procedural justice. Job satisfaction is related to interactions between psychological contract violations and information fairness (Ellershaw et al. 2014). The perception of psychological fulfillment of the contract is positively related to organizational identification and job satisfaction, while the violation of the psychological contract is negatively related to this result (Rodwell et al. 2015). High psychological contract violations, weakening the positive relationship between good leadership and psychological well-being of employees (Erkutlu & Chafra 2016). Psychological contracts have changed significantly in the last three decades. Today's employees expect more than salary. They expect such promotion, improvement and professional development, while in the past they expected loyalty, commitment, effort and high results, without a commensurable value proposition for employees (Davis 2015).

2 METHODS

2.1 Hypothesis

2.1.1 Psychological contracts have a positive effect on attracting and retaining talent

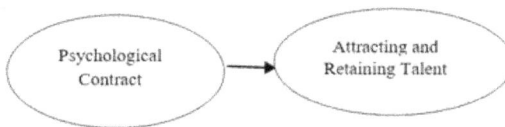

Figure 1. Research conceptual model.

2.2 Research method

2.2.1 Research design
The research method was explanatory survey. Exogenous variables consisting of attracting and retaining talent variables with endogene variables, namely psychological contracts.

2.2.2 Research participants
This study uses a human resource management approach related to talent management theory. The subject of this research is talent in IT companies. IT companies have eight companies, which include one holding company and seven subsidiaries. To deepen the study in a qualitative manner, interviews were also conducted with the company's talent manager.

2.2.3 Population and research samples
The population of this research is all the workforce who have talent in IT companies that is equal to 221 people. Through the cluster random sampling method, a sample of 70% is drawn, which means the number of

Table 2. Variable definition and operations.

Variable	Concept Definition	Indicator
Psychological contract (PC)	Contract Psychological shared values between the employees and management. What employees want to the needs of the organization. (Baker 2014).	1. Flexible deployment 2. Customer focus 3. Performance focus 4. Project- based work 5. Human spirit and work 6. Commitment 7. Learning and development 8. Open information
Work Life Balance (WLB)	A broad term thus emerged in literature to refer to work/non-work conflict and it is "Work Life Balance" (Fisher 2001; Hobson et al. 2001)	1. Work interfere private life (WPLE) 2. Private life interfece work (PLIW) 3. Work private life enhancement 4. Work Enhancement Of Personal Life (WEPL)
Work Satisfaction (JS)	Job satisfaction is a positive feeling from the employee towards a job. (Robbin & Judge 2013)	Work 2. Payment 3. Promotion 4. Supervision 5. Co-workers
Attracting and Retaining Talent (AT&R)	Attracting and Retaining Tal- ent is concerned with how to become a genuine employer of choice. This means developing a workplace culture that reflects the changing needs and interests of both individual and organization (Baker 2014).	1. Employability 2. Communication capacity 3. Roles 4. Cross functional work 5. Meaningful work 6. Short-term commitment 7. Lifelong learning 8. Enterprise

samples is 239 people. The basis for 70% sampling is that each company has different talents, therefore 70% sampling is taken to represent the population.

2.2.4 Types and data sources
The type of data used in this study is primary data and secondary data.

2.2.5 Research instruments
The research instrument was a questioner. The data obtained in this study are sourced from each list of questions regarding the effect of psychological contracts, on attracting and retaining. talent.

Instrument Validity Test Results: To test the validity and reliability of the instrument, a trial of 30 respondents was first performed. The re- sults of the validity test show that all indicators on psychological contract variables and attracting and retaining talent have a p-value smaller than 0.05, meaning that all indicators forming psychological contract variables and at-tracting and retaining talent can be used as valid measurement tools as research instrument.

Instrument Reliability Test Results: All variables have a Cronbach alpha value greater than 0.7. This shows that the level of internal con- sistency is strong and is an indication that the meas- urement scale for all constructs is reliable (Hair et al. 2014).

2.2.6 *Data analysis techniques*

The data processing is carried out with the help of the SPSS program package for windows version 22, and Pro-gram Amos version 20

3 RESULTS AND DISCUSSION

3.1 *Demographic analysis*

Table 3. Characteristics of respondents by age.

Age	number of employees	Percentage
25-30	24	11
31-35	44	20
36-40	44	20
41-45	42	19.5
46-50	40	17.9
51-55	25	11.9
56-60	2	0.8
Total	221	

Table 4. Characteristics of respondents based on gender.

gender	Amount	Percentage
Male	181	82
Female	40	18
Total	221	100

Table 5. Characteristics of respondents by level education.

Level Education	Amount	Percentage
D3	22	10.1
S1	156	70.6
S2	43	19.3
S3	–	–
Total	221	100

3.2 *Description of research results*

a. There is a significant positive effect between psychological contracts with attracting and retaining talent. This means that the better the psychological contract the company, the better the attracting and retaining of employee talent.
b. The value of the influence of psychological contracts (PS) on attracting and retaining talent (ART)

Table 6. Characteristics of respondents by status marriage.

Status Marriage	Amount	Percentage
Married	200	90.4
Single/divorce	21	9.6
Total	221	100

is high. This means that PS has a large role and is one of the important variables that companies need to consider in attracting and retaining talent (ART). The company must realize the activities and promises that have been agreed upon in a psychological contract. If psychological contract indicators have been fulfilled well, employees will continue to work in the company faithfully. This is in line with previous research, including: Attracting and re-taining talent is influenced by the new psychological contract. (Baker 2014). The relational component of psychological contracts plays an important role in developing strong organizational employee relations and reducing employee intentions to leave the organization (Abdallah & Kukunuru 2016). Employees in the talent pool with

c. those not in the talent pool have the same intention to leave the company, if their expectations are not fulfilled (Seopa et al. 2015). Fulfillment of psychologi- cal contracts by increasing relations activities such as good communication and honoring unwritten prom- ises. Psychological contracts offer a useful framework for enhancing overall global mobility and the effectiveness of Global Talent Management (GTM) (Mcnulty & Cieri 2016). Where, attracting and retaining talent is part of the Talent Management, so it can be concluded that the fulfillment of psychological contracts can help in increasing the attracting and retaining talent. To explore the views of HR managers and employees regarding the factors that influence employee retention using the perspective of psychological contracts. Fulfillment of psychological contracts, which reflects the subjective nature of work relations, is more important than working hours and working conditions in predicting employee intentions to quit (Hartwell 2010). Globally, attracting and retaining talented employees is a lasting challenge (Barrick & Zimmerman 2005). One suggested approach to overcoming this challenge is for employers to make it attractive, for employees to continue to meet psychological contract preferences (Bravo 2007). Employees' psychological contracts represent their beliefs about attractive offers from companies, so that they remain (Rousseau 1998).

3.3 *Findings*

The highest loading factor value of a psychological contract (PC) is human spirit and work. The highest loading factor value of attracting and retaining talent (ART) is roles.

4 CONCLUSION

4.1 Conclusions

1. Respondents' perceptions of psychological contract variables in IT companies are high, especially in the human spirit and work variables. How to link employee work with personal meaning is to find meaning in employee work engages and contribute to greater productivity, and lower staff turnover.
2. Respondents' perceptions of attracting and retaining talent variables include high categories, especially in the role variable. Employees understand that the role of their work is broader than the limits of the statement of duties. Therefore, the company must in- volve talented employees in various fields of work, so that talented employees play an important role and feel needed and recognized its existence.
3. There is a significant positive influence between psychological contracts with attracting and retaining talent in IT companies. This means that, the better the psychological contract the company, the better the attracting and retaining talent. Companies must have a psychological contract in accordance with the desires of talent and corporate goals, unite the two, and realize the contract in order to become a selection company.

4.2 Suggestion

1. Along with the changing times and technology, the needs and desires of talented employees and organizations, are changing. For this reason, it is necessary to make adjustments continuously to the psychological contract (new psychological contract).
2. In order to increase attracting and retaining talent, companies must pay attention to psychological contracts.
3. If you want to increase psychological contracts, then increase human spirit and work.
4. If you want to increase attracting and retaining talent, then increase recognition and appreciation for roles.
5. Although psychological contracts are not written contracts that it has legal force, companies should still fulfill implied promises that have been agreed upon together, because psychological contracts have the most influence in carrying out attracting and retaining talent.

REFERENCES

Abdallah, M.B.S. & Kukunuru, M.P.S. 2016. Psychological contracts and intention to leave with mediation effect of organizational commitment and employee satisfaction at times of recession. *Review of International Business an Strategy* 26(2).

Abu, M. & Amran, A. 2016. Corporate sustainable business practices and talent attraction. *Sustainability Accounting, Management and Policy Journal.*

Addae, H.M., Parboteeah, K.P., & Davis, E.E. 2006. Organizational commitment and intentions to quit: An examination of the moderating effects of psychological contract breach in Trinidad and Tobago. *International Journal of Organizational Analysis* 14(3): 225–238.

Aggarwal, U. & Bhargava, S. 2009. Reviewing the relationship between human resource practices and psychological contract and their impact on employee attitude and behaviours: A conceptual model. *Journal of European Industrial Training* (33).

Antonaki, X.E. & Trivellas, P. 2014. Psychological contract breach and organizational commitment in the greek banking sector: the mediation effect of job satisfaction. *Procedia – Social and Behavioral Sciences* 148: 354–361.

Baker, T. 2014. Attracting and retaining talent. London: S. House, Ed.

Barrick, M.R. & Zimmerman, R.D. 2005. Reducing voluntary, avoidable turnover through selection. *Journal of Applied Psychology* 90(1): 159.

Behery, M., Paton, R.A., & Hussain, R. 2012. Psychological contract and organizational commitment. *Competitiveness Review* 22(4): 299–319.

Beth, A., Helen, H. 2001. *The war for talent.* Boston: Harvard Business School.

Cohen, A. 2011. Values and psychological contracts in their relationship to commitment in the workplace. *Career Development International* 16(7): 646–667.

Davis, P.J. 2015. Implementing an employee career-development strategy. *Human Resource Management International Digest* 23: 28–32.

Ellershaw, J., Steane, P., Mcwilliams, J., Dufour, Y., Ellershaw, J., Steane, P., & Dufour, Y. 2014. Promises in psychological contract drive commitment for clinicians.

Erkutlu, H. & Chafra, J. 2016. Benevolent leadership and psychological well-being. *Leadership & Organization Development Journal* 37(3): 369–386.

Festing, M. & Schäfer, L. 2014. Generational challenges to talent management: A framework for talent retention based on the psychological- contract perspective. *Journal of World Business* 49(2): 262–271.

Hartwell, J.K. 2010. Psychological contracts: a new strategy for retaining reduced-hour physicians. *J Med Pract Manage* 25: 285–297.

Harvey, W. 2014. Victory can be yours in the global war for talent. *Human Resource Management International Digest.*

Hiltrop, J.M. 1995. The changing psychological contract: The human resource challenge of the 1990s. *European management journal* 13(3): 286–294.

Keller, S. & Meaney, M. 2017. Attracting and retaining the right talent. *McKinsey Global Institute study.*

Lee, H.W. & Liu, C.H. 2009. The relationship among achievement motivation, psychological contract and work attitudes. *Social Behavior and Personality: an international journal* 37(3): 321–328.

Matuson, R.C. 2014. Increasing the magnetism of your organization. *Human Resource Management International Digest* 22(1): 41–42.

McDonald, D. & Makin, P. 2000. The psychological contract, organisational commitment and job satisfaction of temporary staff. *Leadership & Organizational Development Journal* 21(2): 84–91.

McLeod, S. (2016). *Bandura – social learning theory.*

Mcnulty, Y. & Cieri, H. De. 2016. Linking global mobility and global talent management: the role of ROI. *Employee Relations* 38(1): 8–30.

Rodwell, J., Ellershaw, J., & Flower, R. 2015. Fulfill psychological contract promises to manage in-demand employees John. *Journal of Managerial Psychology* 44(5): 689–701.

Rousseau, D.M. 1998. The'problem'of the psychological contract considered. *Journal of organizational behavior*: 665–671.

Seopa, N., Wöcke, A., & Leeds, C. 2015. The impact on the psychological contract of differentiating employees into talent pools. *Career Development International* 20(7): 717–732.

Sulistiobudi, R.A. & Kadiyono, A.L. 2017. Menemukan kesejahteraan psikologis di balik profesi dosen: Psychological contract sebagai salah satu prediktor tercapainya psychological well being pada dosen. *Humanitas* 14(agustus): 120–138.

Effect of professionalism, work attitude, and accounting systems on accountability and their implications on quality of financial reporting

N. Nursito, S. Suhono & E. Mahpudin
Universitas Singaperbangsa Karawang, Karawang, Indonesia

ABSTRACT: This study aims to examine the effect of Professionalism, Work Attitude, and Accounting System on Accountability and the Implications on the Quality of Financial Report of the Ministry of Health, in West Java Region. This study used descriptive and inferential statistical methods to conclude the influence of exogenous variables on endogenous both simultaneously and partially. The analysis model used the Structural Equation Model (SEM). The main structure consists of Professionalism, Work Attitude, Accounting System and Accountability as exogenous variables and Quality of Financial Reporting as endogenous variables. The results of the research showed that simultaneously Professionalism, Work Attitude and Accounting System had a significant effect on Accountability. Professionalism partially had a positive and significant effect on Accountability. Work attitude partially had a positive and significant effect on Accountability while Accounting System partially did not affect Accountability. Accountability, Professionalism, Work Attitudes and Accounting Systems together had a positive and significant effect on the Quality of Financial Reporting, and Accountability, Professionalism, Work Attitudes and Accounting Systems partially influenced the Quality of Financial Reporting.

1 INTRODUCTION

Public demands on accountability and good financial management state continues to raise. Accountability embodies in form quality financial report. To improve the quality of financial report a compatible accounting system is required to produce an integrated financial report. Government regulations are continuously changing to support the improvement of the reports so that people get transparent report about results of the government performance.

On the contrary, the phenomenon existed is the low quality of financial reporting and government accounting system that is less than satisfactory. In addition, human resource performance has become an issue, particularly regarding with work attitude of government apparatus providing service to the people. As a result, public believe that government officers are lacking of professionalism.

2 LITERATURE REVIEW

2.1 *Professionalism*

A professional one convictions in accordance with the knowledge, experience and values owned. A profession is a work: (a) based on science specific, (b) having morality and the ethical principles professional, (c) meet a public need, and (d) has the authority and liberty to give "recommendations" to solve problems in accordance with their area of expertise (Schein 2014). Professionals are having the characteristics of: (1) a set of knowledge and experience, (2) training, (3) a responsibility to serve public or part of him, (4) the need to have the trust of the public, (5) as a means to make a good living, the implementation of the (6) the career development, thus professional conduct expressed in competence and integrity (Holmes & Burns 2011).

2.2 *Work attitude*

The work was the general feeling when communicate with others, more details that if he is part of a personality of the most important for the various feelings for what he had learned, someone might dislike towards others or someone might did not want to do a thing (Wallace & Masters 2006). According to the organization it was determined by two factors - products and value/assumption. Employees with positive values or attitudes form cultural organization (Liker & Hoseus 2008).

2.3 *Accounting systems*

An accounting system is a set of procedures and computerized manual starting from collecting data, recording, summarizing the financial position and operation and reporting government financial (Zeyn 2011). To the government, accounting is a series of systematic procedure, another equipment or elements to realize the function of analysis transactions to financial reporting in the government organization (Peraturan Pemerintah RI Nomor 71 Tahun 2010).

2.4 Accountability

Accountability and the functioning of all components of the firms based on each duties and authority (Schein 2014). Thus, it is the shareholders accountability mandate to an agent to give an account, presents, report all activities which are the responsibility of the mandate to the contributors who have the right and authority to those asked. Therefore, accountability can be defined as an obligation to serve and report any follow-up to a person or institution especially in terms of financial administration (Mardiasmo 2004).

2.5 The Quality of finance reports

The quality financial report is degrees of a financial statement and financial reporting itself (Cornelli & Yosha 2003). And the quality financial reporting of the government is normative measurements as a source of inside information (Peraturan Pemerintah RI Nomor 71 Tahun 2010).

3 METHODS

The methodology used in this study was quantitative method to analyze causal influence between variables using structural equation model (SEM) using Amos with a sample size of 495. This study was conducted in units of accounting for Health Ministry in West Java (Ferdinand 2010; Hair et al. 1995).

4 RESULTS AND DISCUSSION

Testing structural equation model (SEM) test, professional variable influence the work and accounting system together over the accountability to: 0.702 x 100% = 70.2%, the rest of 29,8 percent is influenced by other factors. The value of f –count is greater than the f table which is 2.155 and 1.318, so that the f count obtained from the sig 0.000; 0.05. Thus accountability is influenced positively by professional, the work attitude and accounting system.

As for the simultaneous influence, professionalism, work attitude, accounting system and accountability on the quality of financial reporting valued at 57.9 % and the remaining percent of 42,1 was influenced by other factors. Thus, the quality of financial report in accounting unit of the ministry of health in west java is influenced positively by professionalism, work attitude, accounting system and accountability.

The test scores and direct influence of accountability on the quality of financial report was significant with a value of 6.331 & gt; 1,318, while the ratio

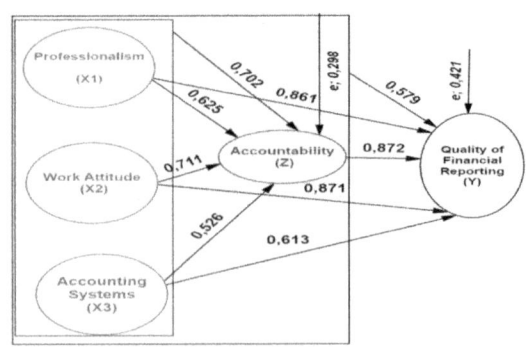

Figure 1. The results of structural equation model (standardized model).

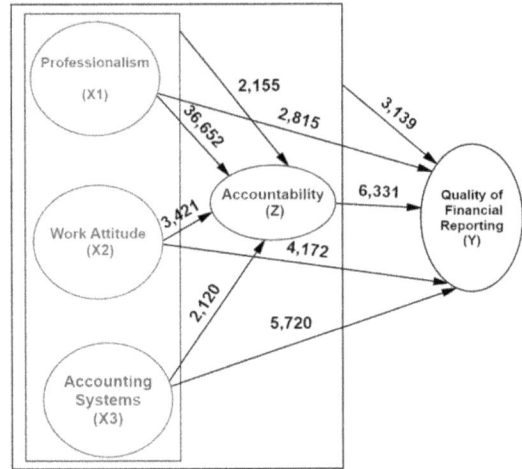

Figure 2. The results of structural equation model (T-value model).

of critical 0,872 means an increased of accountability contributes 87,2 percent on the quality of financial report. The results of structural equation model can be seen in Figure 1 (standardized model) and Figure 2 (t-value model).

5 CONCLUSION

Partially, professionalism has a positive and significant impact on accountability. This means that the professionalism is demanded to enable better accountability in reaching the quality of financial reporting expected.

Work attitude partially has a positive and significant impact on accountability this means that when work attitude is gets better the more one is to promote accountability resulting in the better financial reporting.

Accounting system in full has a positive and significant impact on accountability. It shows that when equipped with a good accounting system, better accountability will be resulted which will help boost the quality of reliable financial reporting.

Accountability in full have had a positive impact on the quality of financial reporting and significant. This indicates that the better accountability the more one is improve the quality of financial reporting.

The quality of financial reporting to his unit accounting health ministry areas in West Java can be achieved this proved accountability contribute quite well on the quality of financial reporting. The contribution of accountability is caused by profesionalisme, the work of civil servants and accounting system that run in well integrated.

REFERENCES

Cornelli, F. & Yosha, O. 2003. Stage financing and the role of convertible debt. Review of Economic Studies 70(1), 1–32.

Ferdinand, A. 2010. Struktur equation modelling dalam penelitian manajemen: aplikasi model-model rumit dalam penelitian untuk Tesis Master & Disertasi Doktor, BP Undip.

Hair, J.F., Anderson, R.E., Tantham, R.L., Black, W.C. 1995. Multivariate Data Analysis with Readnings, Fourth Editions, Practice Hall International.

Holmes, A.W & Burns, D.C. 2011. Auiditing Standards and Procedural. Home wood Illinois: Richard D. Irwin, Inc.

Liker, J.K. & Hoseus, M. 2008. Toyota Culture The Heart and Soul of the Toyota Way, Mcgraw-hill.

Mardiasmo. 2004. Perwujudan transparansi dan akuntabilitas publik melalui akuntansi sektor publik: suatu sarana govermance.

Peraturan Pemerintah RI Nomor 71 Tahun 2010; Standar Akuntansi Pemerintahan Bandung Fokusmedia.

Schein, E.H. 2014. Organization culture and leadership. San Francisco: Jossey Bass Publisher.

Wallace, H.R. & Masters, A.A. 2006. Personal Development for Life and Work. Thompson South Western.

Zeyn, E. 2011. Pengaruh good governance dan standar akuntansi pemerintahan terhadap akuntabilitas keuangan dengan komitmen organisasi sebagai pemoderasi. *Jurnal Ilmiah Universitas Pasundan Bandung*, Jawa Barat.

Advances in Business, Management and Entrepreneurship – Hurriyati et al. (Eds)
© 2021 Taylor & Francis Group, London, ISBN 978-0-367-67471-7

Upgrading global value chain performance through human resources capacity development: A case study at SME's Geopark Belitung

R. Rofaida, A.K. Yuliawati & B.P. Gautama
Universitas Pendidikan Indonesia, Bandung, Indonesia

ABSTRACT: The purpose of this study is to identify problems related to the capacity of human resources in SMEs in the geopark region of Belitung and recommend strategies to increase the capacity of human resources. Geopark tourism development is a development priority in the Belitung district. The declaration of the Belitung Geopark as a national geopark is a very important momentum to improve the value chain performance from upstream to downstream from the process of providing tourism in the geopark. Increasing human resource capacity becomes very important because it is an unlimited asset of the organization and can be a competitive advantage. This re-search was qualitative. The sample size was determined by the snowball method and then the sampling technique used was purposive sampling. The sample size was 50 SMEs in Geopark Belitung. The data collection methods were observation, questionnaires, in-depth interviews, literature study, and Focus Group Discussion. The data were analyzed using resources-based view and descriptive analysis. The result of this study shows that SMEs at Geopark Belitung need human resources capacity development. The priority areas are change in mindset from business actors from mining businesses to service businesses and increasing expertise in the production process.

1 INTRODUCTION

Indonesia has advantages in the diversity of natural and environmental characteristics that can be optimized through various types of economic value activities. There are three types of economic value activities, namely ecotourism (geotourism), ge-otourism (geo-tourism), and the earth park (ge-opark). The most recent concept of utilization and the focus of this article is the geopark. Geopark is considered to be the best concept because it inte-grates the management of geodiversity, biodiversi-ty, and cultural diversity to improve the people's economy while still paying attention to the protec-tion/conservation of the three diversity. One of the destinations that become the priority of geopark development in Indonesia is Aspiring geopark on Belitung Island. The use of geodiversity, biodiversity, and cultural diversity has become an economic activity carried out by various business units, most of which are still micro, small, and medium enter-prises (MSMEs). The product produced is called a geoproduct. The development of the concept of geopark tourism in Belitung needs to be balanced with the development of geoproducts. Geo products are innovative, new, or recreated traditional products related to biodiversity. These products are strategies to promote local identity that can be a way to bring Geopark Home (Rodrigues 2017). Geoproduct is a type of product related to geopark or geotourism in general that includes local products that contribute

actively to local economic growth and also to increase awareness of the values of geodiversity, biodiversity, and culture. Geoproduct not only enhances the local economy but also educates tourists and popularizes geology. (Farsani 2010). In the concept of production, economic activity is carried out through a series of pro-cesses called value chains. Where every value chain activity produces value-added that is of economic value for the geoproduct produced. The development of geopark MSMEs in Belitung Island is still faced with various inhibiting factors, including limitations in obtaining funding sources through banking and non-banking institutions, limited access to scientific resources through training and counseling, limited access to technology, and low competence/quality of human resources. Improving the quality of human resources is the focus of this study because the compet-itive advantage of com-panies in the global business environment lies in human resources through their competencies.

2 METHODS

This research was a qualitative study. The sample size was determined by the snowball method and then the sampling technique used was purposive sampling. The sample size was 50 SMEs in Geopark Belitung. The data collection methods were observations, ques-tionnaires, indepth interviews, literature study, and

Focus Group Discussion. The data were analyzed using resources-based view and de-scriptive analysis.

3 RESULTS AND DISCUSSION

3.1 *Global value chain at SMEs Geopark Belitung*

The SMEs in Belitung Geopark have dominated by-products, namely coffee, processed seafood snacks, batik, and pepper. The coffee product value chain has two channels, namely the cooperative and non-cooperative value chains. In the cooperative value chain, coffee farmers distribute coffee to cooperative units which then through the cooperative will be dis-tributed to exporters or coffee shops/cafes. At this stage, for a wider market reach. The cooperative coop-erates with large exporters who will then dis-tribute coffee to consumers on a national or international scale. In the non cooperative distribution scheme, cof-fee farmers distribute direct coffee to users, namely cafes/coffee shops or end consumers. The value chain in snack food geoproducts tends to channel geo-products directly to the marketing network owned without going through cooperatives. Suppliers produc-ing snack products (such as cassava, tubers) distribute raw materials to traditional markets. In this case the traditional market as the main distribution of snack food raw materials to entrepreneurs. Furthermore, the businessman processes the raw materials into various kinds of snacks which are ultimately conveyed to end users, both local people and tourists visiting Belitung. erica is one of the leading geoproducts in the Belitung region, so the local government has formed a pepper com-modity cooperative. In the process, the produc-tion of pepper in Belitung is under the supervision of the Plantation Office. In the pepper value chain pro-cess, there is a village unit cooperative that facilitates the provision of inputs and out-put / pepper trade. The value chain in snack food geoproducts tends to channel geoproducts directly to the market-ing network owned without going through cooperatives. Suppliers produc-ing snack products (such as cassava, tubers) distribute raw materials to traditional markets. In this case the traditional market as the main distribution of snack food raw materials to entrepreneurs. Furthermore, the businessman pro-cesses the raw materials into various kinds of snacks which are ultimately conveyed to end users, both local people and tourists visiting Belitung.

3.2 *Human resources capacity development at SMEs Geopark Belitung*

Most of the Belitung geopark industries fall into the SMEs category. The scale of micro and small businesses is influenced by the competence of en-trepreneurs in running their businesses, where there are still many deficiencies in entrepreneurial com-petencies that must be improved. The recommenda-tions for developing human resource capacity in the

Geopark Belitung SMEs are based on problems that occur in the competency aspects (evaluation approach) and what competencies are needed by the SMEs in the Belitung Geopark) to be able to compete in the global market (development approach). The identification of the two aspects above will be divided based on the stages of the business activities carried out namely production, finance, and marketing.

In the aspect of production, the results of research through the distribution of questionnaires, indepth interviews show that the positive indicators are: (1). competence to search for product design ideas for example via the internet and mass media, (2). compe-tence in making product designs refers to local culture, for example, typical Belitung fruit motifs such as durian and sempor leaves for batik motifs, (3). com-petency is modeled after product design already on the market, (4). competence to predict the number of requests and pro-duce according to these predic-tions. But indicators that still need to be improved are: (1). competence to make products with good quality, not many defective products occur, (2). competence to use technology in the production process, (3). Prod-ucts produced lack the characteristics (for example in product and packaging variations) that distinguish them from competitors. Plans for developing human resource capacity in production aspects: (1). Train-ing to improve competency in managing raw materials as a first step to improve product quality, (2). Train-ing to design clear operational procedures/stages of the production process and deter-mine product quality standards, (3). Training to improve the quality control process, (4). Technical training in the use of produc-tion technology. The results of research conducted by Hsu et al. (2010) claimed that the ability of sup-ply chain management in small and medium-sized enterprises can improve business performance. This research measures supply chain management compe-tence in terms of five first order constructs: innovation ori-entation, proactiveness orientation, risk-taking characteristics, relational capital, and coordination capability.

In the financial aspect, most of them use their cap-ital without any assistance from the government or the private sector. The capital is capital that is is-sued personally and is family capital, this is done to facili-tate the process of financial management that does not involve too many people. The profit aspect is also con-sidered to be good enough with in-dicators that they can meet the needs of family life. The lack of capi-tal is caused by the lack of willing-ness and ability to find other sources of funds to develop businesses such as finding funding to the banking sector. An under-standing of financial strengths and weaknesses faced by SMEs will help this company to create an appro-priate financial management program (Salikin et al. 2014)

Planning competence still needs to be improved because finance is still very simple (in the form of irregular bookkeeping), cash flows are not clear, financial monitoring is not done, there is only one budget for all organizational activities, meaning there is no clear financial division for each activity organization, so the nominal profit every month is difficult to know. Recommendations for developing human resource capacity in the financial aspect are (1). increasing financial literacy competencies, through participating in training related to financial reports. and (2). training on preparation of business feasibility proposals from financial aspects to in-crease the trust of banks or other financial institu-tions so that access to capital resources increases. Financial management in SMEs has to refer to the Pecking Order Theory, namely SMEs have a pref-erence to choose internal financing before external financing. This theory has been proven in the re-sults of Abanis et al. (2013) research conducted at SMEs in several regions in western Uganda (Abanis et al. 2013).

In the marketing aspect, the results of the survey further identified the marketing aspects that were the problem, namely the lack of promotion media such as not doing online promotions, this happened because of the limited competency of business actors in follow-ing technological advances, lack of information about the market (tastes and marketing channels), not all industries have the competence to do marketing out-side the province and nationally. Recommendations for capacity building for human resources that can be done are training to increase competence in expanding marketing networks, and training in technology adop-tion for online promotion (eg marketing via Instagram, marketing via Facebook, and creating websites.). The results of research conducted by Gill-more (2011) and Franco, M. et al. (2014), produce findings that entrepreneurs are adapting to apply standards of mar-keting practice standards within companies, how they use networks and marketing capabilities to enhance business activities.

4 CONCLUSION

In industries where the level of business com-petition is very high and dynamic, increasing the performance of global value chains in the Belitung geopark SMEs through developing human resource capacity is a very strategic step. Recommendations for develop-ing human resource capacity are carried out in three business activities, namely production, finance, and marketing.

REFERENCES

Abanis et al. 2013. Financial management practices in small and medium enterprises in selected districts in Western Uganda. *Research Journal of Finance and Accounting* 4 (2).

Farsani, N.T. 2010. Geoparks as art museums for geotourists. *Revista Tourisimo And Desenvolvimento*.

Franco M. et al. 2014. An exploratory study of entrepreneurial marketing in SMEs: The role of the founder-entrepreneur. *Journal of Small Business and Enterprise Development* 21 (2). [Online]. Retrived from https://doi.org/10.1108/JSBED-10-2012-0112.

Gillmore, A. 2011. Entrepreneurial and SME marketing. *Journal of Research in Marketing and Entrepreneurship* 13(2). [Online]. Retrived from https://doi.org/10.1108/14715201111176426.

Hsu et al. 2010. Entrepreneurial SCM competence and per-formance of manufacturing SMEs. *International Journal of Production Research* 49 (22). [Online]. Retrived from https://doi.org/10.1080/00207543.2010.537384.

Rodrigues. 2017. The concept of the geoproduct: Successful examples from Naturtejo UNESCO Global Geopark. *14th Europan Geoparks Conference*.

Salikin, Norasikin, Wahab, Norailis, &Muhammad, I. 2014. Strengths and weaknesses among Malaysian SMEs: Financial management perspectives. *Procedia-Social and Behavioral Sciences* 129. [Online]. Retrived from https://doi.org/10.1016/j.sbspro.2014.03.685.

Advances in Business, Management and Entrepreneurship – Hurriyati et al. (Eds)
© 2021 Taylor & Francis Group, London, ISBN 978-0-367-67471-7

The effect of individual characteristics of millennial generation on entrepreneurship behaviors: Empirical evidence on SMEs

Basuki & R. Widyanti
Universitas Islam Kalimantan MAB, Banjarmasin, Indonesia

ABSTRACT: This study aims to examine empirically the effect of individual characteristic of millennial generation towards entrepreneurial behavior in the context of the Millennial Generation Small Business Industry (SMEs). This study used a survey method. The research was classified as explanatory research that explained the relationship between variables through hypothesis testing. There were 150 respondents. The data analysis tool was SPSS for Windows version 20.0. The results showed that the Individual Characteristics of Millennial Generation had an effect on Entrepreneurship Behavior. Regression test results indicated that the regression coefficient (B) was 0.272 with the level of Sig.t (probability) of 0.001, (Sig.t = 0.001 <, 0.05). The t count value was 3.396, while t table was 1.655 (t count = 3.396> t table = 1.655). This meant that the Individual Characteristics of Millennial Generation had a significant effect on Entrepreneurship Behavior. This implies that the stronger the individual character of the millennial generation, the stronger it is to shape entrepreneurial behavior.

1 INTRODUCTION

The success of a business, including millennial generation in carrying out its business activities, is inseparable from its internal characteristics. Donahue and Kentle (in American Psychological Association 2013) put some types of individual characteristics that influence a person's work and company. These characteristics are: openness to ex-perience, conscientiousness, extraversion, agreea-bleness, and neo critics. One of the main characteris-tics of the millennial generation is characterized by the use and the familiarity with communication, me-dia and digital technology. Furthermore, Yoris Se-bastian (in the Profile of the Indonesian Millennium Generation 2018) mentioned several advantages of the millennial generation, namely wanting to be fast, easy to change jobs in a short time, creative, dynam-ic, technology literate, close to social media, etc.

The distinctive characteristics of the millennial generation culture are shown by making technology as a lifestyle (sheltered) because they are born from edu-cated parents. They are multi-talented, multi languages, more expressive and explorative. In life, they are always convinced, optimistic, confident, and sim-ple. They like instant things. In work, looking at achievement is something that must be achieved. They work and learning more interactively through team-work, collaboration and groups. They are independent and well structured. They have digital literacy to the use of technology, gadgets, and in-ternet. They also prefer visual or image instructions. In human relations or communication, they prefer instant communication in real time environments. They develop networks that enable this generation to connect and collaborate with each other.

The purpose of this study is to examine and ana-lyze the influence of individual characteristics of millen-nial generations on entrepreneurial behavior of mil-lennial Small and Medium Enterprises in Banjar-masin.

1.1 *Individuals characteristics of millennium generation (Gen Y)*

Lyons (2004) revealed the characteristics of genera-tion Y. He stated that characteristics of each individ-ual is different, depending on where he grew up, eco-nomic strata, and social family. The communica-tion patterns are very open compared to previous genera-tions. They are fanatical social media users and their lives greatly affected by technological de-velopments. They more open to political and eco-nomic views, so that they look very reactive to envi-ronmental changes that occur around them. They also have more attention to wealth.

Donahue and Kentle (in American Psy-chological Association 2013) put some types of in-dividual char-acteristics that influence a person's work and company, namely: 1) Openness to Experi-ence, 2) Conscien-tiousness, 3) Extraversion, 4) Agreeableness, 5) Neo critics.

1.2 Entrepreneurial behavior

An understanding of behavior is the main concern of psychology (American Psychological Association 2013); therefore, discipline has developed a model of how to predict behavior. A very useful model for understanding behavior is a certain extent under will control is Theory of Planned Behavior (Ajzen 1988). Theory assumes that behavior is best explained by intention to try to do behavior (Ajzen 1988). Intention, in turn, is shaped by attitudes, subjective norms, and perceived behavioral control. Attitudes are determined by the belief that certain behaviors will lead to favorable outcomes. Subjective norms are determined by the beliefs of other important people (friends, family) about certain behaviors and the extent to which a person tends to obey this belief. This independent variable measures the value that people place on the opinions of people who are close to them.

1.3 Conceptual model and hypotheses

The success of an individual in carrying out his business activities is inseparable from the internal characteristics of the individual. Donahue and Kentle (in American Psychological Association 2013) put several types of individual characteristics that influence a person's work and company, such as: openness to experience, conscientiousness, ex-traversion, agreeableness, and neo critics. Mean-while.(In the Indonesian Generation Profile 2018), it has been drawn that the millennial generation has unique characteristics based on region and socio-economic conditions. One of the main characteristics of the millennial generation is characterized by the use and familiarity with communication, media and digital technology. Millennial also have the charac-teristics of being creative, informative, passionate and productive. They involve technology in all as-pects of life. They have characteristics of open communication. They are fanatical users of social media, their lives are greatly affected by technologi-cal developments. They are also more open to politi-cal and economic views. Yoris Sebastian (in the Pro-file of the Indonesian Millennium Generation 2018) mentioned several advantages of the millennial gen-eration, namely wanting to be fast, easy to move jobs in a short time, creative, dynamic, technology literate, close to social media, etc.

Based on theoretical and empirical studies described previously, the framework of conceptual models of in this study can be seen in the following figure:

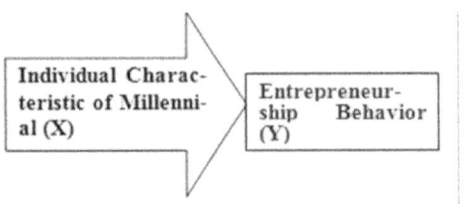

Figure 1. Conceptual research models.

2 METHODS

Two types of data were used: primary data and sec-ondary data. Primary data was obtained by distrib-uting questionnaires to 150 respondents. Secondary data was obtained from various related sources such as the Central Bureau of Statistics, Cooperatives and SMEs, the Provincial Government of Banjarmasin, as well as other agencies.

Descriptive statistical analysis and inferential statis-tical analysis were used in analyzing the data. De-scriptive statistical analysis was used to describe the characteristics of each variable such as respondent characteristics and frequency distribution of items from each studied variable. The collected data was then tabulated into a table, and discussion was car-ried out descriptively. Meanwhile, inferential statis-tical anal-ysis is a data analysis technique using a causality or causal approach between the independ-ent variable and the dependent variable. Inference analysis was used to examine the influence of several variables in this study by using multiple linear re-gression. Data processing was done by using SPSS Version 20.0.

3 RESULT AND DISCUSSION

3.1 Characteristics of respondents

Respondents in this study were millennial generation leaders or owners of small and medium enterprises (SMEs) included in Banjarmasin, South Kalimantan. The number of representative samples were 150 re-spondents.

Based on primary data collected through question-naires, the owner of small and medium enterprises (SMEs) in South Kalimantan were dominated by women (57.70%). The men were only 41.30%. The age was ranged between 18–24 years old (28.47%), 25–2 years old (38.00%), and 33–40 years old (3.33%). Most of the respondents were high school graduates (82.00%), diploma graduates (10.70%), and bache-lor graduates (7.30%). This educational background is very influential on a person's ability to carry out their work. The stronger the level of ed-ucation a person will be, the more capable the person is in car-rying out the work. Based on business capital, most of the spent Rp.1,000,000 $\leq$ Rp.5,000,000 (34.67%); Rp 6,000,000 $\leq$ Rp.10,000,000 (31.33%); Rp.11,000,000 $\leq$ Rp.15,000,000 (14.00%); Rp.16,000,000 $\leq$ Rp.20,000,000 (10.00%); and Rp.20,000,000 (10.00 %). Meanwhile, their income form the busi-ness were varied. Most of get Rp.1,000,000–Rp.3,000,000 (49.33%); Rp.4,000,000–Rp.6,000,000 (32.67%); Rp.7,000,000–Rp.10,000,000 (13,33%) and Rp. 11,000,000–above (4.67%).

3.2 Results of hypothesis analysis

The analysis of hypothesis testing 1 in the Charac-teristic Millennial Generation and its influence on

Table 1. Hypothesis test.

Variable	Regression Coeff (B)	T$_{account}$	Sig.t	r^2
Variable (X)	0,272	3,396	0,001	0,0734
Multi R	= 0.793			
R Square (R^2)	= 0.639			
F account	= 26,455			
Sig. F	= 0.000			
α	= 0.05			

Entrepreneurship Behavior shows that the regression coefficient (B) is 0.272 with a Sig.t (probability) level of 0.001, (Sig.t = 0.001 < 0.05). The value of T count is 3.396, while t table is 1.655 (t count = 3.396 > t table = 1.655). This means that the Charac-teristics of Millennial Generation Individuals have a significant effect on Entrepreneurship Behavior. This indicates that the stronger the individual character of the millennial generation, the stronger it is to shape entrepreneurial behavior. Meanwhile, the partial determination coefficient (r2) is 0.0734. This means that the contribution of the Characteristics of Millennial Individuals to the formation of Entrepre-neurship Behavior is 7.43%. The individual charac-teristics will basically influence the entrepreneurial behavior of millennial generation SMEs. As Donahue and Kentle (in American Psychological Association 2013) stated that the characteristics of in-dividual influence his/her work and company.

4 CONCLUSIONS

4.1 Conclusion

Based on the results, it can be concluded that millennial generation of SMEs in Banjarmasin have good individual characteristics. It is measured from six indicators that are reflected in the form of multi-tasking, independent, and close to technology. Meanwhile, ambition indicator implies that they are ambitious in running their business. However, they will happily share that success with others.

Furthermore, millennial generation of SMEs also have good entrepreneurial behavior. It is reflected in two indicators, namely: trying to start a business and being involved in every step towards venture creation. Finally, hypothesis testing shows that the characteristics of millennial generation have a significant effect on entrepreneurship behavior.

4.2 Recommendations

Entrepreneurial behavior is reflected in four forms: 1) starting a new business today, 2) trying to start a new business with others, 3) spending a lot of time to start a business, and 4) analyzing the market op-portunities that determine a business and money stored in business. Most of the millennial SME en-trepreneurs in Banjarmasin have less interest in trying to do business activities. They are less involved in any steps to create businesses in the past and present. Coaching or training should be done both through educational institutions, so that their entre-preneurship interest can be continually increased.

REFERENCES

Ajzen, I. 1988. Attitudes, personality, and behavior. Chicago, IL: The Dorsey Press.
Donahue, J. & Kentle. 2013. American Psychological Association. [Online]. Retrieved from http://www.apa.org/support/about/apa/psychology.aspx#answer. Accessed 08/27/2013.
Lyons. 2004. An exploration of generational Valui in life and at work. ProQuest Dissertations and Theses 441-4441 [Online]. Retreved from htt:// ezproxy.um.my/docview/305203456?accountid=28930
Patrik, M., Kreiser, L.D., Marino. & Weaver, M.K. 2003. Culture Influences on Entrepreurial Orientation: The Impact of national Culture on Risk Taking Proactiveness in SMEs, Entrepreneurship Theory and practive.

Section 6   Strategic management, entrepreneurship
and contemporary issues

Advances in Business, Management and Entrepreneurship – Hurriyati et al. (Eds)
© 2021 Taylor & Francis Group, London, ISBN 978-0-367-67471-7

Ethical leadership and performance appraisal satisfaction: The mediating role of trust

I. Nazaruddin, H. Sofyani, C.M. Putri, E.S. Fatmaningrum & F. Wahyuni
Universitas Muhammadiyah Yogyakarta, Yogyakarta, Indonesia

ABSTRACT: his study aimed to examine the effect of ethical leadership on performance appraisal satisfaction. The study also examined the role of trust as a mediator in the relationship between ethical leadership and performance appraisal satisfaction. The data were collected through questionnaires and completed by 205 teaching academics in the faculty of economics and business at university in Yogyakarta. Validity and reliability testing were carried out before testing the hypotheses, and the data were analyzed using PROCESS. The results of the study show that ethical leadership and trust have a positive effect on performance appraisal satisfaction. The results also provide empirical evidence of an indirect influence of ethical leadership on performance appraisal satisfaction through the mediation of trust. For higher education management, the results of this study demonstrate the role of ethical leadership in increasing trust as a way to increase lecturer satisfaction with performance appraisal.

1 INTRODUCTION

Increasing global competition means universities are facing major challenges. Global competition acts as a motivator for countries to increase investment in education, including in the form of grants and the certification of educator staff as an effort to increase their competitiveness and compete internationally. Universities are also seeking ways to improve the quality of education (Chen et al. 2009; Lawrence & Mccollough 2001). Lecturers, as a higher education resource, play an important role in improving both the quality of education and the quality of graduates. Improvements must be supported by improvements in the performance of lecturers. The quality of lecturers has an influence on the profitability and sustainable competitive advantage of institutions in today's competitive world (Ahmed et al. 2014; Chang & Hahn 2006; Fakhimi & Raisy 2013) as such, institutions are viewed as being successful if they have skilled and they are highly motivated human resources.

Performance management provides a means for improving the quality of lecturers, with one form of performance management being performance appraisal. The process of performance appraisal is seen as an important part of the strategic management approach and it is a tool that connects competency, behavior and institutional strategic objectives. Performance appraisal enables college leaders and lecturers to define, communicate and review expectations, goals and processes in the context of achieving strategic objectives. The results of the performance appraisal process will improve performance by increasing employee contribution to institutional goals and performance (Kuvaas 2006). The purpose of performance appraisal is to maintain, improve and assist employees in developing and overcoming performance barriers. In addition, employees will focus more on their performance in accordance with the priorities of the university (Lawler 1994). The performance appraisal process is carried out to harmonize employee behavior with organizational strategic goals. Unfortunately, it sometimes fails to change the work behavior of employees and they experience dissatisfaction (Elicker et al. 2006). These conditions can then result in a decrease in commitment and performance. Performance appraisal satisfaction can be used to inform the leadership of higher education institutions with regard to the acceptance of lecturers when faced with the application of a performance appraisal system. Based on this, performance appraisal satisfaction is crucial for higher education because it can promote the positive behavior of employees and improve university performance.

In addition, performance appraisal satisfaction can be improved through the type of leadership (Alonderiene & Majauskaite 2016; Aydin et al. 2013; Chang & Lee 2007). Some researchers argue that crises of leadership and identity are common occurrences for a large number of universities (Bradley et al. 2017; Bryman 2007; Kligyte & Barrie 2014). The leadership model in tertiary institutions is currently limited because it is more administrative and bureaucratic, with managers who lack the personal and strategic analytical skills to effectively lead academic colleagues (Ball 2007; Bradley et al. 2017; Lumby 2012). Relatively little attention continues to be paid to the leadership model at universities, which makes it an interesting area to study. The concept of leadership that will be examined in this study is the model of ethical leadership due to the importance of ethics in education.

Ethical leadership is a leadership concept that focuses primarily on taking the correct action, being fair, having integrity and guiding others by communicating ethics and ethical rules and respecting ethical behavior among subordinates. Ethical leaders can influence the accountability of their followers or subordinates (Brown et al. 2005; Den Hartog 2015; Dionne et al. 2014; Hassan et al. 2013). Ethical leadership will also affect trust. Trust in leaders is influenced by the views of employees on the quality of the relation- ships between leaders and employees. Employees are able to develop a higher level of trust in leaders who keep promises and behave consistently. Leaders who implement ethical leadership will respect ethical behavior and discipline unethical behavior. They clearly communicate to employees what is expected of them and how they can contribute positively to the institution (Kalshoven et al. 2011; Simons 2003). To add, the employees tend to trust leaders when they feel they are being supported and treated fairly. Some of the results of previous studies also show that leadership models will influence the level of trust and have an impact on employee satisfaction (Liu et al. 2003; Mo & Shi 2017).

Based on the previous explanation, this study examines the direct and indirect influence of ethical leadership on trust and performance appraisal satis-faction from the viewpoint of lecturers.

1.1 Literature review

Social learning theory is used to explain the impact of ethical leadership (Feng et al. 2019; Neubert et al. 2009). Leadership behavior patterns will serve as a model for their subordinates; thus, they will observe, imitate and replicate the behavior of the leader. If a leader implements ethical leadership, employees are motivated to strive in order to achieve institutional goals. Leadership will also influence employee trust, which in turn will affect their commitment to achieving institutional goals.

Performance appraisal is an element of a performance management strategy that integrates employee activities with company policy (Fletcher 2001). Performance appraisal procedures are beneficial for employees and institutions because they can be used as references by employees and institutions to evaluate and improve performance in accordance with institutional goals. The effectiveness of the performance appraisal system is demonstrated by the level of satisfaction with the system.

Increased performance appraisal satisfaction can increase the motivation and performance of lecturers.

In other words, lecturer satisfaction with their performance appraisal and its components (such as index valuation, transparency, fairness) can increase the effectiveness of performance appraisal (Toppo 2012).

The phenomenon being experienced is one of continuing low satisfaction with the performance appraisal system (Elicker et al. 2006), which has led institutions to identify the determinants of employee satisfaction in performance appraisal. A low level of satisfaction with performance appraisals can have the impact of reducing organizational performance.

Leaders play a role in shaping the behavior of followers who contribute to the achievement of institutional goals (Van 2003). The results of previous studies indicate that the type of leadership can improve lecturer satisfaction, which in turn motivates employees (Trottier et al. 2008). One type of leadership is ethical leadership.

Ethical leadership is a type of leadership that provides a role model as an example, through the ethical personality of the leader. In addition, ethical leadership promotes ethical behavior to employees through two-way communication, reinforcement and decision-making (Brown et al. 2005). Ethical leadership includes behavior that is inclusive and fair, along with behaviors such as communicating the importance of ethics and sanctions for unethical behavior among employees (Brown & Mitchell 2010).

Based on the above description, the following hypotheses are defined:

H1: Ethical leadership has a positive effect on performance appraisal satisfaction.
H2: Ethical leadership has a positive effect on trust.

Trust refers to the willingness of a party to de- pend on another party to take certain actions, where the first party may not have the ability to monitor the actions of the other party (Saraih et al. 2018). Trust is an important component in the institution that will have an impact on the efficiency of appraisal performance (Naji et al. 2015). Trust in leaders is likely to increase both employee compliance and employee contributions to performance (Bello 2012; Ponnu & Tennakoon 2009). Trust is a positive expectation of those who are trusted, that they will not be disappointed and are able to rely on the other parties. Trust is important as it can affect employee commitment and impact on them improving their performance. From the perspective of satisfaction with performance assessment, trust in leadership will be related to lecturers' satisfaction in the performance appraisal process.

Based on the above description, the following hypotheses are defined:

H3: Trust has a positive effect on performance appraisal satisfaction.
H4: Trust mediates the relationship between ethical behavior and performance appraisal satisfaction.

Table 1. Validity and reliability test results.

Variable	Factor loading	Cronbach's alpha
PAS	.456 - .796	.771
EL	.493 - .757	.830
T	.660 - .859	.722

Table 2. Descriptive statistics.

Variable	Min	Max	Mean	Std. deviation
PAS	18	35	25.46	3.2
EL	28	50	38.24	4,71
T	10	19	15.13	2.07

2 METHODS

This research is a cross-sectional study using a survey method. The research respondents were lecturers from A- and B-accredited colleges in Yogyakarta, Indonesia. A total of 300 questionnaires were distributed, 237 of which were returned and 205 were valid responses that were used.

Performance appraisal satisfaction was measured using an instrument developed by Kuvass consisting of 7 questions (Kuvaas 2006). Ethical leadership was measured using a 10-question instrument developed by (Brown et al. 2005), while the trust was measured by 4 questions (Hennig-thurau et al. 2001). The responses for all of the variables were measured using a 5-point Likert scale, with a range from 1 = Strong Disagree to 5 = Strongly Agree.

The instrument indicators for each variable were all valid (loading factors above 0.4) and reliable (Cronbach's alpha above 0.6), thus satisfying the required criteria. The validity and reliability test results are shown in Table 1. Analysis of the hypothesis testing data was conducted using PROCESS v3.1.

3 RESULTS AND DISCUSSION

The results of testing the descriptive statistics derived from the 205 lecturer respondents in 12 universities in the province of Yogyakarta Special Region, showed that the actual average was relatively higher than the theoretical mean (Table 2).

Based on the test results using PROCESS (Table 3), H1 is supported, as seen from the regression coefficient of .393 and P-value <.01. This shows that a higher level of ethical leadership will improve performance appraisal satisfaction. Leaders who demonstrate integrity and fairness and who provide examples to their subordinates will improve the performance of the lecturer satisfaction appraisal. In this scenario, lecturers will perceive that the feedback they receive for

Table 3. Hypotheses 1, 2, 3 and 4 test results.

Ethical leadership predicting trust (path a)

R^2	F	Coeff	P value
.265	73.177	.226	.000

Trust predicting performance appraisal satisfaction (path b)

R^2	F	coeff	P value
.358	56.220	.323	.002

Ethical leadership predicting performance appraisal satisfaction (path c)

R^2	F	coeff	P value
.326	98.316	.393	.000

Ethical leadership lessened predicting performance appraisal satisfaction (path c)

R^2	F	coeff	P value
.358	56.220	.320	.000

their performance assessment is relevant. The results of the study are in line with the research of Trottier et al (2008), which showed that leadership types can improve lecturer satisfaction, which in turn motivates employees.

The test results for H2 show that ethical leadership has a positive and significant effect on trust (path a), with a regression coefficient value of .226 and P-value <.01. This study supports H2, thus indicating that ethical leadership among higher education leaders will increase the trust placed in leaders by lecturers.

The research results also support H3, thus indicating that trust has a positive effect on performance appraisal satisfaction. The regression coefficient between the two variables is .323, with a P-value <.01. Trust in institutions and leaders will impact lecturer acceptance and lecturer satisfaction with performance appraisal as they believe that the action taken by leaders and institutions is the best thing.

Mediation effects can be seen from the indirect effect of ethical leadership on performance appraisal satisfaction (Table 4), which has a coefficient of .073. The confidence interval results from the bootstrap lower level (BootLLCI) of 0.018 and bootstrap upper level (BootULCI) of 0.143 indicate that trust has a mediating effect on the relationship between ethical leadership and performance appraisal satisfaction. The size of the effect can be seen from the direct effect coefficient of ethical leadership on performance appraisal satisfaction of 0.106. The results of this study are in line with those of several previous studies that have also shown that the leadership model will influence the level of trust and have an impact on employee satisfaction (Liu et al. 2003; Mo & Shi 2017).

Table 4. Total, direct and indirect effects of ethical leadership on performance appraisal satisfaction.

Total effect of ethical leadership on performance appraisal satisfaction	
Effect	P
.392	.000

Direct Effect of ethical leadership on performance appraisal satisfaction	
Effect	P
.3199	.000

Indirect Effect of ethical leadership on performance appraisal satisfaction

	Effect	BootSE	Boot LLCI	Boot ULCI
Trust	.073	.032	.018	.143

Completely standardized indirect effect of ethical leadership on performance appraisal satisfaction

	Effect	BootSE	Boot LLCI	Boot ULCI
Trust	.106	.046	.025	.207

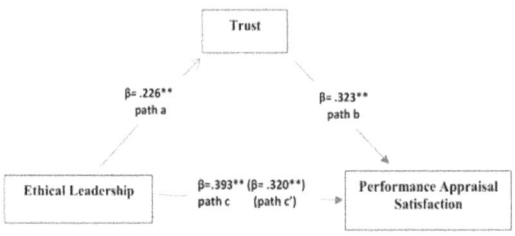

Figure 1. ** statistically significant at the 5% level. Trust mediates the relationship between EL and PAS.

4 CONCLUSION

The results of this study indicate that trust mediates the relationship between ethical leadership and the performance appraisal satisfaction of lecturers in universities. Ethical leadership can increase the lecturers' level of trust in their leaders, which is accompanied by satisfaction with the performance appraisal system applied at universities (Figure 1).

In addition, ethical leadership plays an important role in improving the performance of lecturer appraisal. This is because such leaders model exemplary standards to lecturers, to do what they say, motivate their lecturers and attempt to remain fair and balanced. The research results are in line with social learning theory.

This finding has several implications. Ethical leadership plays an important role as it increases both the trustworthiness of lecturers and their acceptance of the system in place for assessing their work. Higher education institutions need to work on improving the style of leaders to become oriented to ethical leadership models. Leaders with high ethical leadership can foster a climate of trust among the lecturers working at their tertiary institutions, which in turn engenders a tendency for lecturers to be satis fied with the performance assessment system in place, ultimately leading to better performance among lecturers.

ACKNOWLEDGMENTS

The researcher would like to extend his gratitude to the Ministry of Research, Technology and Higher Ed- ucation (Kemenristekdikti) as the main sponsor of this research. This research is funded by the Kemen- ristekdikti under Penelitian Dasar Unggulan Perguruan Tinggi (PDUPT) 2019 scheme entitled "pemodelan dimensi pengukuran kinerja excellent: anteseden dan konsekuensi dalam mewujudkan kinerja dosen berdaya saing global" (modeling excellent performance measurement dimensions: antecedent and consequence in realizing the performance of globally competitive lecturers).

REFERENCES

Ahmed, A., Hussain, I., Ahmed, S., & Akbar, M.F. 2014. Performance appraisals impact on attitudinal outcomes and organisational performance. *International Journal of Business and Management* 5(10): 62–68.

Alonderiene, R. & Majauskaite, M. 2016. Leadership style and job satisfaction in higher education institutions. *International Journal of Educational Management* 30(1): 140–164.

Aydin, A., Yilmaz, S., & Uysal, S. 2013. The effect of school principals' leadership styles on. *Educational Sciences: Theory & Practice* 13(June): 806–811.

Ball, S. 2007. Leadership of academics in research. *Educational Management Administration and Leadership* 35(4): 449–477.

Bello, S.M. 2012. Impact of ethical leadership on employee job performance. International *Journal of Business and Social Science* 3(11): 228–236.

Bradley, A.P., Grice, T., & Paulsen, N. 2017. Promoting Leadership in Australian Universities. *Australian Universities' Review* 59(1): 97–105.

Brown, M.E. & Mitchell, M.S. 2010. Ethical and unethical leadership: exploring new avenues for euture research. *Business Ethics Quarterl* 20(4): 583–616.

Brown, M.E., Treviño, L.K., & Harrison, D.A. 2005. Ethical leadership: A social learning perspective for construct development and testing. Organizational Behavior And Human Decision Processes 97(2): 117–134.

Bryman, A. 2007. Effective leadership in higher education. *Literature review Studies in Higher Education* 32(6): 693–710.

Chang, E. & Hahn, J. 2006. Does pay-for- performance enhance perceived distributive justice for collectivistic employees. *Personnel Review* 35(4): 397–412.

Chang, S.C. & Lee, M.S. 2007. A study on relationship among leadership, organizational culture, the operation of learning organization and employees' job satisfaction. *Learning Organization* 14(2): 155–185.

Chen, S.H., Wang, H.H., & Yang, K.J. 2009. Establishment and application of performance measure indicators for universities. *The TQM Journal* 21(3): 220–235

Den Hartog, D.N. 2015. *Ethical leadership.*

Dionne, S.D., Gupta, A., Sotak, K.L., Shirreffs, K.A., Serban, A., Hao, C., & Yammarino, F.J. 2014. A 25-year perspective on levels of analysis in leadership research. *The Leadership Quarterly* 25(1): 6-35.

Elicker, J.D., Levy, P.E., & Hall, R.J. 2006. The role of leader-member exchange in the performance appraisal process. *Journal of Management* 32(4): 531–551.

Fakhimi, F. & Raisy, A. 2013. Satisfaction with performance appraisal from the employees' perspective and its behavioral outcomes (case study of headquarters offices of Bank Refah). *European Online Journal of Natural and Social Sciences* 22(3): 296–305.

Feng, T., Wang, D., Lawton, A., & Luo, B. 2019. Customer orientation and firm performance the joint moderating effects of ethical leadership and competitive intensity. *Journal of Business Research Journal* 100: 111–121.

Fletcher, C. 2001. Performance appraisal and management: The developing research agenda. *Journal of Occupational and Organizational Psychology* 74(4): 473–487.

Hassan, S., Mahsud, R., Yukl, G., & Prussia, G.E. 2013. Ethical and empowering leadership and leader effectiveness. *Journal of Managerial Psychology.*

Hennig-thurau, T., Langer, M.F., & Hansen, U. 2001. Modeling and managing student loyalty an approach based on the concept of relationship quality. *Journal of Service Research* 3(4): 331–344.

Kalshoven, K., Den Hartog, D.N., & De Hoogh, A.H. 2011. Ethical leadership at work questionnaire (ELW): Development and validation of a multidimensional measure. *The Leadership Quarterly* 22(1): 51-69.

Kligyte, G. & Barrie, S. 2014. Collegiality: Leading us into fantasy – the paradoxical resilience of collegiality in academic leadership. *Higher Education Research and Development* 33(1): 157–169.

Kuvaas, B. 2006. Performance appraisal satisfaction and employee outcomes: Mediating and moderating roles of work motivation. *International Journal of Human Resource Management* 17(3): 504–522.

Lawler, E.E. 1994. Performance management: The next generation. *Compensation & Benefits Review* 26(3): 16–19.

Lawrence, J.J. & Mccollough, M.A. 2001. A conceptual framework for guaranteeing higher education. *Quality Assurance in Education* 9(3): 139–152

Liu, A., Fellows, R., & Fang, Z. 2003. The power paradigm of project leadership. *Construction Management and Economics* 21(8): 819–829.

Lumby, J. 2012. What do we know about leadership in higher education? The leadership foundation for higher education's research. *Review Paper Leadership Foundation for Higher Education*: 1–28.

Mo, S. & Shi, J. 2017. Linking ethical leadership to employee burnout, workplace deviance and performance: Testing the mediating roles of trust in leader and surface acting. Journal of business ethics 144(2): 293–303.

Naji, A., Ben Mansour, J., & Leclerc, A. 2015. Performance appraisal system and employee satisfaction: the role of trust towards supervisors. *Journal of Human Resources Management And Labor Studies* 3(1): 40–53.

Neubert, M.J., Carlson, D.S., Kacmar, K.M., Roberts, J.A., & Chonko, L.B. 2009. The virtuous influence of ethical leadership behavior: Evidence from the field. *Journal of Business Ethics* 90(2): 157–170.

Ponnu, C.H. & Tennakoon, G. 2009. The association between ethical leadership and employee outcomes – the Malaysian case. *Electronic Journal of Business Ethics and Organization Studies* 14(1).

Saraih, U., Mohd Karim, K., Irza, H.A., Amlus, M., & Aida, N. 2018. Relationships between trust, organizational justice and performance appraisal satisfaction: Evidence from Public Higher Educational Institution in Malaysia. *International Journal of Engineering & Technology* 7(2.29): 602.

Simons, T. 2003. Behavioral integrity: The perceived alignment between managers' words and deeds as a research focus. *Organization Science* 13(1): 18–35.

Toppo, M.L. 2012. From performance appraisal to performance management. *IOSR Journal of Business and Management* 3(5): 1–6.

Trottier, T., Van Wart, M., & Wang, X. 2008. Examining the nature and significance of leadership in government organizations. *Public Administration Review* 68(2): 319–333.

Van, W.M. 2003. Public-sector leadership theory: an assessment. *Public Administration Review* 63(2): 214–228.

Advances in Business, Management and Entrepreneurship – Hurriyati et al. (Eds)
© 2021 Taylor & Francis Group, London, ISBN 978-0-367-67471-7

The effect of performance measurement system for lecturers and role clarity on lecturers' performance: Role clarity as an intervening variable

H. Sofyani, I. Nazaruddin, E.S. Fatmaningrum & F. Wahyuni
Universitas Muhammadiyah Yogyakarta, Yogyakarta, Indonesia

ABSTRACT: The low performance of many lecturers in Indonesian universities has led to the development of performance measurement system for lecturers (PMSL) by several universities. Until now, research examining the impact of the implementation of PMSL on lecturers' performance has been very difficult to find. Therefore, this study aimed to empirically examine the effect of PMSL implementation and role clarity on lecturers' performance. The testing of role clarity as an intervening variable was also attempted. Data from 203 questionnaires were obtained from lecturers at leading private universities in Indonesia (accredited "excellent" and having internationalization programs), and the hypotheses were tested using the Partial Least Squares (PLS) approach. The results revealed that the implementation of PMSL, and role clarity significantly, had a positive effect on lecturers' performance. Also, this study concluded that role clarity served as an intervening variable.

Keywords: performance measurement system for lecturers (PMSL), role clarity, performance, lecturers, universities.

1 INTRODUCTION

The development of the world and the influence of globalization require universities to boost their competitive advantage in order to compete with other universities, both nationally and globally. To achieve a high rank and portray a good reputation, the performance of lecturers is a major factor that contribute toward universities' competitive advantage (Rasheed et al. 2016). Therefore, efforts to maintain competitive advantage is visible through the implementation of a performance measurement system (PMS) for lecturers (PMSL). The implementation of PMS support the formation of positive lecturer behavior that is in accordance with the university's objectives, it also encourages performance excellence and strengthens the scientific ethos that upholds the essence of higher education institution as an innovative learning center (Molefe 2010). Several universities in Indonesia – especially those with "A" accreditation – have begun to developed PMSL as an effort to manage the performance quality of their lecturers.

Research on PMS in the public sector had been carried out since the early 1990s (Atkinson & McCrindell 1997). However, the performance review mainly focused on organizational performance and tested the performance of individuals within the organization. In addition, the study of individual's performance in public sector is dominated by government context, with very few studies examining the performance of lecturers at universities, despite the fact that universities as higher education organizations play a vital role in the development of a country.

Several studies have provided empirical evidence of a positive relationship between PMS and performance. Spekle & Verbeeten (2014) found that an exploration of the use of PMS tends to improve the performance of both organizations and employees. Similar results were also indicated by Marchand & Raymond (2008), Lee & Yang (2011), and Phusavat et al. (2011). However, Tahar & Sofyani (2019) presented a conflicting finding that PMSL was not always able to encourage the performance of employees (lecturers). They believed this was due to PMSL policy being perceived by lecturers as more of an administrative policy that must be adhered to, as opposed to a motivational tool fostering a commitment to performance.

From the various research findings outlined above, there remains a gap concerning the relationship between PMS and performance, especially related to the findings from Tahar & Sofyani (2019), who investigated PMS in colleges. Using this research gap as the departure point, this study examined the relationship between PMSL and lecturer performance, with role clarity as a mediating variable. The variable role clarity derives from the empirical claims of Tahar & Sofyani (2019), who argued that the lack of a relationship between PMSL and lecturer performance was due to lecturers regarding PMSL as an administrative goal rather than as a means of harmonizing strategic plans, visions and the university mission. In this regard, it was assumed that in Tahar & Sofyani (2019) research, PMSL did not encourage performance due to the fact that the duties and roles of lecturers were not described in detail, as well as individual performance targets they must meet (Hall 2008).

2 LITERATURE REVIEW

From the standpoint of goal setting theory, the implementation of PMS is expected to lead to individual organizations achieving better work performance based on the acceptance and awareness of performance targets (Basri 2013; Locke 1975). Kaplan et al. (2010) stated that PMS had a close relationship with strategies to create a comprehensive shared understanding of the vision, mission, goals and objectives of the organization. Furthermore, the leader can then align individual actions with mutual understanding. Ittner et al. (2003) stated that PMS could improve the communication needed to achieve the chosen strategy, motivate performance to achieve goals and provide feedback.

H_1: The implementation of a performance measurement system has a positive effect on the performance of lecturers.

The performance improvement sought by the implementation of the PMS basically aims to make the role of the lecturer clear. Individuals need enough information to perform tasks effectively (Hall 2008), while a lack of information about job objectives and guidance on the most effective work behavior being targeted can result in inefficient efforts, misdirection and the completion of tasks that are not on target, thereby reducing job performance (Tubre & Collins 2000). The opposite of role clarity is role ambiguity (Rasit & Isa 2014), which is a situation where individuals do not have sufficient information to choose the most effective work behavior, or when tasks, authorities and responsibilities are unclear (Burney & Widener 2007).

Rainey & Jung (2010) stated that the absence of certain goals and objectives can be a source of ambiguity, confusion and lack of direction for subordinates. Therefore, the adoption of PMSL – a comprehensively formulated performance measures – especially for lecturers, enables the development of more specific goals and objectives and allows clear direction and motivation to subordinates (Lunenburg 2011). This also means that PMSL is able to improve performance providing that it can minimize role ambiguity, in other words, produce role clarity for lecturers (Hall 2008).

H_2: Role clarity has a positive effect on the lecturer performance.

H_3: Role clarity mediates the relationship between the implementation of a performance measurement system and lecturer performance.

Based on the above hypotheses, the researcher formulated the research model as presented in Figure 1.

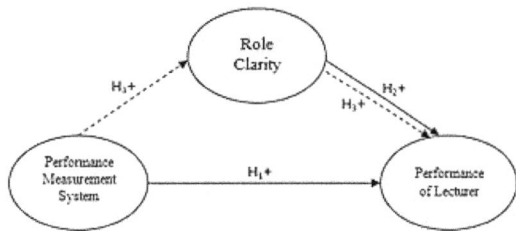

Figure 1. Research model.

3 METHODS

This research was conducted on private campuses in Indonesia based on the development of PMSL, which had just started to be initiated. The campuses chosen held A accreditation, had internationalization programs, and developed a PSML. A purposive sampling was used to obtain respondents. The respondents in this study were permanent lecturers who were directly involved in the implementation of a comprehensive performance measurement system. The data were collected through the distribution of 5-interval Likert questionnaires, both online and via direct distribution. Prior to distribution, the questionnaire was tested three times with a pilot and validation from the researchers and experts. A total of 1000 questionnaires were distributed, 203 of which were used in research.

Hall (2008) explained that PMSL is a system that provides information about the goals and performance targets for each lecturer comprehensively and is one of the tools used in the implementation of the organizational strategy (Malina & Selto 2001). In this study, PMSL was measured using a questionnaire re-fined by Hall (2008) that had previously been developed by Chenhall (2005). Because this research was conducted in a university context and reviewed the management of lecturer performance, the PMS instrument was adjusted to the nature of the lecturer performance indicators that were generally accepted in Indonesia. In addition, performance indicators were developed with reference to international accreditations such as AUN-QA and the QS university rankings. The measurement indicators were formulated by testing their validation through three rounds of discussion between experts and the researchers.

Role clarity refers to an individual's beliefs about the expectations and behaviors related to their work role (Kahn et al. 1964). In this study, role clarity was divided into two aspects: clarity of purpose (the degree to which the results and objectives of the work are clearly stated and well defined), and the clarity of the process (the extent to which individuals are convinced of how to perform their work (Sawyer 1992). The variable measurement indicator were adapted from Hall (2008), including clarity of purpose and clarity of the process.

Performance is the work of individuals in organizations when carrying out their duties, which in this case relates to lecturers (Sukirno & Siengthai 2011). The performance variables were measured using instruments developed by the researchers with reference to instruments developed by Sukirno & Siengthai (2011). This development related to performance indicators that focus on quality and outcomes that are regulated by campus accreditation assessments in Indonesia, AUN-QA and the QS university world rankings.

Data analysis was performed using the structural equation model with partial least squares (PLS) regression. PLS is highly suited to this study because it assumes minimal data and requires a relatively small sample size and a strong theoretical foundation (Chin et al. 2003).

4 RESULTS AND DISCUSSION

Table 1 contains details of the sample and research location, while Table 2 presents the characteristics of the respondents. The respondents in this study were lecturers at universities holding A accreditation in Java and had internationalization programs. Of a total of 1000 questionnaires distributed to respondents, 219 were returned. The removal of incomplete responses left a total of 203 complete questionnaires for further analysis.

The researcher conducted a non-response bias test to determine whether the difference in the timing of the questionnaires (first week versus second week) affected the homogeneity of the data. The test results found no non-response bias, as indicated by the Sig. (2-tailed) greater than 0.05.

The validity and reliability of the measurements were tested before testing the hypotheses. Two types of validity test were conducted, namely convergent validity and discrimination. The results of the convergent validity test and measurement reliability are presented in Table 3, while the results of the convergent validity test are presented in Table 4. From the results of testing, the value of the item, loading was greater than 0.4, and the composite reliability value of all items was greater than 0.5. Furthermore, the AVE, composite

Table 2. Characteristics of respondents.

		Frequency (n = 203)	%
Gender	Female	108	53.2
	Male	95	46.8
Age	25–35	101	49.8
	36–45	32	15.8
	46–55	50	24.6
	>55	20	9.9
Experience	<5years	68	33.5
	5 to <10 years	44	21.7
	10 to <15 years	19	9.4
	>15 years	72	35.5
Functional Tenure	Assoc. Prof	40	19.7
	Senior Lecturer	33	16.3
	Assistant Professor	94	46.3
	Lecturer	36	17.7

reliability and Cronbach's alpha values also exceeded the required rule of thumb, which is 0.5 for composite reliability and AVE, and 0.7 for Cronbach's alpha (Hair Jr. et al. 2014).

The result of cross loading item test for validity are presented in Table 4. The test results showed that the correlation value for the indicator to the item itself was greater than for other items. Thus, it can be concluded that this study passed the results of the measurement test and the hypothesis test was carried out (Sholihin & Ratmono 2013).

The results of the hypothesis testing for this research are presented in Tables 5 and 6. The H1 test results demonstrates that PMSL had a positive effect on the performance of the lecturers, which supported the research of Ittner et al. (2003) and Spekle & Verbeeten (2014). This is in line with the theory of goal setting, in that a performance management mechanism in the form of PMSL is capable of leading lecturers to improve their performance. These results also corroborate the findings of many other studies that emphasize the importance of PMS for performance improvement purposes (Franceschini & Turina 2013; Janudin & Maelah 2016; Molefe 2010). In particular, this is a relatively new finding in the context of PMS in public institutions, particulary in universities.

Furthermore, this study also supports H2, which states that role clarity has a positive effect on lecturer performance. This finding is consistent with the ideas set out in the goal setting theory that the clarity of the tasks and performance targets to be achieved by lecturers, as developed in the form of key performance indicators (KPIs), will lead to the attainment of higher performance by lecturers (Basri 2013; Locke 1975). Thus, the presence of a clear goal setting concept in the formulation of the PMSL will lead to improved performance (Locke & Latham 2013).

Table 1. Demographics data.

University	Location of Sample	Number of Respondent
Universitas Muhammadiyah Yogyakarta	DI Yogyakarta	67
Universitas Ahmad Dahlan	DI Yogyakarta	46
Universitas Islam Indonesia	DI Yogyakarta	15
Universitas Muhammadiyah Surakarta	Central Java	40
Universitas Telkom	West Java	10
Universitas Parahyangan	West Java	25
Total unit sampling (n)		203

Table 3. Convergent validity test results and measurement reliability.

Variable	Item	Loading	AVE	Composite Reliability	Cronbach's Alpha
Performance	P1	0,742163	0,539982	0,889806	0,853940
	P2	0,764336			
	P3	0,797466			
	P4	0,796673			
	P5	0,729310			
	P6	0,499050			
	P7	0,770244			
Performance Measurement System	PMS1	0,790352	0,668813	0,941481	0,928877
	PMS2	0,780790			
	PMS3	0,864962			
	PMS4	0,858227			
	PMS5	0,760299			
	PMS6	0,893199			
	PMS7	0,736863			
	PMS8	0,844200			
Role Clarity	RC1	0,709060	0,602554	0,898649	0,863308
	RC2	0,847727			
	RC3	0,515216			
	RC4	0,826907			
	RC5	0,869047			

Table 4. Cross loading.

	P	PMS	RC
P1	0,770244	0,245312	0,252246
P2	0,742163	0,147724	0,252222
P3	0,764336	0,265236	0,329936
P4	0,797466	0,375769	0,227038
P5	0,796673	0,252689	0,240482
P6	0,729310	0,248171	0,235230
P7	0,499050	0,171446	0,220160
PMS1	0,293939	0,790352	0,265147
PMS2	0,239390	0,780790	0,262520
PMS3	0,214557	0,864962	0,240037
PMS4	0,309810	0,858227	0,329784
PMS5	0,266086	0,760299	0,208520
PMS6	0,310899	0,893199	0,324540
PMS7	0,316827	0,736863	0,162808
PMS8	0,286872	0,844200	0,282341
RC1	0,178481	0,238621	0,709060
RC2	0,326418	0,380384	0,847727
RC3	0,053226	0,225868	0,515216
RC4	0,296757	0,195545	0,826907
RC5	0,275001	0,207080	0,869047
RC6	0,355674	0,227882	0,830333

Table 5. Results of testing hypotheses 1 and 2.

Hypotheses	Original Sample	T Statistics	P-Value	Conclusion
PMS -> P	0,352711	3,762627	0,001	Supported
PMS -> RC	0,336751	3,490886	0,001	Supported

performance of lecturers if the PMSL itself is not able to clearly present the roles and duties of lecturers. This result is in line with the findings of Hall (2008), once again reinforcing the premise of goal setting theory which explains the essential nature of role and task clarity in achieving organizational goals and objectives. Clarity of roles and tasks can also be used by university heads as a tool for managing their lecturers with regard to the extent to which they have adhered to and carried out the roles and tasks assigned to them. The existence of controls will further support efforts to improve the performance of the lecturers themselves. The results of this study responded to the research by Tahar & Sofyani (2019), who found that PMSL did not always have a positive effect on performance. In the context of their findings, there was an assumption that the PMSL developed did not provide role clarity for lecturers, thus causing the implementation of PMSL to fail and the optimal performance not being achieved.

Finally, the results of this study provide evidence that role clarity wholly mediates the relationship between PMSL and performance. In other words, the implementation of PMS will not improve the

Table 6. Results testing of the mediating effects of role clarity (hypothesis 3).

Hypotheses	Original Sample	T Statitics	P-Value	Conclusion	
PMS -> P	0,247460	1,925885		Not supported	Full
PMS -> RC	0,323274	2,951867	0,001	Supported	Mediating
RC->P	0,257253	2,696545	0,001	Supported	

5 CONCLUSION

The results of this study indicate that the implementation of PMS and role clarity can work as a determinant of the performance of lecturers at university. In addition, this study concluded that role clarity became a full mediating variable in relation to the implementation of PMS and lecturer performance. This finding therefore responds to the inconsistency shown in the results of various previous studies which found that PMS implementation did not always affect performance. The substance of the implementation of PMS from the point of view of goal setting theory is that it must be able to generate clarity of both the objectives and roles of lecturers, thus promoting role clarity.

ACKNOWLEDGEMENT

The researcher would like to extend his gratitude to the Ministry of Research, Technology and Higher Education (Kemenristekdikti) as the main sponsor of this research. This research is funded by the Kemenristekdikti under Penelitian Dasar Unggulan Perguruan Tinggi (PDUPT) 2019 scheme entitled "pemodelan dimensi pengukuran kinerja excellent: anteseden dan konsekuensi dalam mewujudkan kinerja dosen berdaya saing global" (modeling excel-lent performance measurement dimensions: anteced-ent and consequence in realizing the performance of globally competitive lecturers).

REFERENCES

Atkinson, A.A. & McCrindell, J. Q. 1997. Strategic performance measurement in government. *CMA magazine*, 71(3), 20–22.

Basri, Y.M. 2013. Mediasi Konflik Peran dan Keadilan Prosedural dalam Hubungan Penguku-ran Kinerja Dengan Kinerja Manajerial. *Jurnal Akuntansi dan Keuangan Indonesia*, 10(2), 225–242.

Burney, L. & Widener, S.K. 2007. Strategic performance measurement systems, job-relevant information, and managerial behavioral responses—Role stress and performance. *Behavioral research in accounting*, 19(1), 43–69.

Chenhall, R.H. 2005. Integrative strategic performance measurement systems, strategic alignment of manufacturing, learning and strategic outcomes: an exploratory study. *Accounting, organizations and society*, 30(5), 395–422.

Chin, W.W., Marcolin, B.L. & Newsted, P.R. 2003. A partial least squares latent variable modeling approach for measuring interaction effects: Results from a Monte Carlo simulation study and an electronic-mail emotion/adoption study. *Information systems research*, 14(2), 189–217.

Franceschini, F. & Turina, E. 2013. Quality improvement and redesign of performance measurement systems: an application to the academic field. *Quality & Quantity*, 47(1), 465–483.

Hair Jr, J.F., Sarstedt, M., Hopkins, L. & Kuppelwieser, V.G. 2014. Partial least squares structural equation modeling (PLS-SEM) An emerging tool in business research. *European Business Review*, 26(2), 106–121.

Hall, M. 2008. The effect of comprehensive performance measurement systems on role clarity, psychological empowerment and managerial performance. *Accounting, organizations and society*, 33(2-3), 141–163.

Ittner, C.D., Larcker, D.F. & Meyer, M.W. 2003. Subjectivity and the weighting of performance measures: Evidence from a balanced scorecard. *The Accounting Review*, 78(3), 725–758.

Ittner, C.D., Larcker, D.F. & Randall, T. 2003. Performance implications of strategic performance measurement in financial services firms. *Accounting, organizations and society*, 28(7), 715–741.

Janudin, S. E. & Maelah, R. 2016. Performance measurement system in Malaysian public research universities: is it contemporary? *International Journal of Management in Education*, 10(3), 219–233.

Kahn, R.L., Wolfe, D.M., Quinn, R.P., Snoek, J.D. & Rosenthal, R.A. 1964. Organizational stress: Studies in role conflict and ambiguity.

Kaplan, R.S., Norton, D.P. & Rugelsjoen, B. 2010. Managing alliances with the balanced scorecard. *Harvard Business Review*, 88(1), 114–120.

Lee, C.L. & Yang, H.J. 2011. Organization structure, competition and performance measurement systems and their joint effects on performance. *Management accounting research*, 22(2), 84–104.

Locke, E.A. 1975. Personnel attitudes and motivation. *Annual review of psychology*, 26(1), 457–480.

Locke, E.A. & Latham, G.P. 2013. New developments in goal setting and task performance: Routledge.

Lunenburg, F.C. 2011. Goal-setting theory of motivation. International journal of management, business, and administration, 15(1), 1–6.

Malina, M.A., & Selto, F. H. 2001. Communicating and controlling strategy: an empirical study of the effectiveness of the balanced scorecard. *Journal of Management Accounting Research*, 13(1), 47–90.

Marchand, M. & Raymond, L. 2008. Researching performance measurement systems: An information systems perspective. *International journal of operations & production management*, 28(7), 663–686.

Molefe, G.N. 2010. Performance measurement dimensions for lecturers at selected universities: An international perspective. *SA Journal of Human Resource Management*, 8(1), 13.

Phusavat, K., Ketsarapong, S., Ranjan, J. & Lin, B. 2011. Developing a university classification model from performance indicators. Performance Measurement and Metrics, 12(3), 183–213.

Rainey, H.G. & Jung, C.S. 2010. Extending goal ambiguity research in government: From organizational goal ambiguity to programme goal ambiguity. *Public management and performance: Research directions*, 34–59.

Rasheed, M.I., Humayon, A.A., Awan, U. & Ahmed, A.D. 2016. Factors affecting teachers' motivation: An HRM challenge for public sector higher educational institutions of Pakistan (HEIs). *International Journal of Educational Management*, 30(1), 101–114.

Rasit, Z.A. & Isa, C.R. 2014. The influence of comprehensive performance measurement system (CPMS) towards managers' role ambiguity. *Procedia Social and Behavioral Sciences*, 164, 548–561.

Sawyer, J.E. 1992. Goal and process clarity: Specification of multiple constructs of role ambiguity and a structural equation model of their antecedents and consequences. *Journal of applied psychology*, 77(2), 130.

Sholihin, M. & Ratmono, D. 2013. Analisis SEM-PLS dengan WarpPLS 3.0 untuk Hubungan Nonlinier dalam Penelitian Sosial dan Bisnis. Yogyakarta: Penerbit ANDI.

Spekle, R.F. & Verbeeten, F.H. 2014. The use of performance measurement systems in the public sector: Effects on performance. *Management accounting research*, 25(2), 131–146.

Sukirno, D. & Siengthai, S. 2011. Does participative decision making affect lecturer performance in higher education? *International Journal of Educational Management*, 25(5), 494–508.

Tahar, A. & Sofyani, H. 2019. Performance Measurement System and Lecturer's Performance: Measurements Using Output-Based Instruments. *Riset Akuntansi dan Keuangan Indonesia*, 4(1), in press.

Tubre, T.C. & Collins, J. M. 2000. Jackson and Schuler (1985) revisited: A meta analysis of the relationships between role ambiguity, role conflict, and job performance. *Journal of Management*, 26(1), 155–169.

Advances in Business, Management and Entrepreneurship – Hurriyati et al. (Eds)
© 2021 Taylor & Francis Group, London, ISBN 978-0-367-67471-7

Strategy paradox and personality perspective to create business transformation

H.S. Hadiyanto
Bina Nusantara University, Jakarta, Indonesia

ABSTRACT: Scientific publication in strategic management and new strategy models are emerging in the last few decades. But in its application, the variations of the strategy model are often contradictory, paradoxical, and debatable. As a result, decision makers rely more on his personality traits to choose the right strategy. The purpose of this research was to create new propositions from past literature and develop new conceptual models about the strategy orientation from personality perspective. Strategy orientation was categorized into two paradoxical strategies, i.e. strategic consistency and strategic flexibility. The result of this research showed that there was conceptual relation between the two paradoxical strategies with personality traits to create business transformation.

1 INTRODUCTION

Research publication on strategic management has grown rapidly in the last few decades, which resulted in various newest version of strategy models. However, some of the models of these strategies are often conflicting, paradoxical, or contradictory to each other that often creates a dilemma and make it difficult for managers to make choices. In addition, turbulence and information update occur faster with a degree of dubious accuracy. Under this circumstance, decision makers often become more adept in their personality to make a decision (Gallén 2010; Hambrick & Fredrickson 2001; Parnell 2005). In this context, personality can be principles, values, thoughts, or beliefs about what strategy to be applied (Beaver 2003).

These perspective are in line with several studies that discuss the influence of personality on a strategy orientation. One mentions that the type of manager with a certain personality will have a tendency in applying a specific strategy that suits his personality (Gallén 2010). Likewise, other researchers reveal that personality will influence the decision-making process (Robbins & Judge 2008). But the relationship between personality and the choice of strategy has not been explored in many studies, especially those that discuss, analyze, and argue a particular strategy approach. Some researchers focus more on the relationship between a particular strategy approach and performance (Cingoz & Akdogan 2013; Moss et al. 2013; Yu 2012), some others focus more on the differences between two contradictory strategies, known as the strategy paradox (De Wit & Meyer 2010; Raynor 2007), and several other researchers emphasize on comparing and analyzing about which one of

the two strategy paradox more effective to improve performance (Hamsal & Agung 2006; Parnell 1994).

Some researchers are aware that there is no conceptual model affirms which strategy is retained by certain individuals when facing with two strategy paradox. Based on this notion, this paper focused on fundamental research, which is defined as a research intended for the development of a science and directed at the development of existing theories (Nasution & Usman 2008). Qualitative method with literature review is used to obtain information and support theory, particularly to find conceptual link between two strategy paradox that is actually believed byan individual and personality traits to create business transformation.

Strategy paradox in this research is devided into two orientations, i.e. strategic flexibility and strategic consistency, while personality traits consisted of four dimension; source of energy, way of gathering information, decision making, and relation with external world. As the final output, the purpose of research was to create propositions and develop new conceptual model between strategy paradox, personality traits, and business transformation. The definitions about each variable were explained in theoritical review below.

2 LITERATURE REVIEW

2.1 *Strategy paradox: Flexibility-consistency*

There are two strategy paradoxes which contradict each other, namely strategic consistency and strategic flexibility (Hamsal & Agung 2006; Lamberg et al. 2009; Parnell & Lester 2003; Moss et al. 2013; Yu

2012). Some other literature use different terminology, but generally still contains relatively the same notions.

2.1.1 Strategic Flexibility

Many researchers have proposed definition of strategic flexibility. One of them defines startegic flexibility as the ability to change tactics in a short time (with low cost) to adapt and take advantage of the uncertainty to succeed in turbulent periods (Ghemawat & del Sol 1998). Some other researchers define it as the ability to change direction and reconfigure the strategy quickly (Combe et al. 2012; Johnson et al. 2003). There are also other studies that interpret strategic flexibility as the ability to adapt to substantial, uncertain, fast, and significant environmental changes that have a significant impact on performance (Aaker & Mascharenhas 1984; Ogunmokun & Li 2012). Based on several definitions from the researchers above, strategic flexibility is synonymous with adaptation, change quickly, and business transformation.

Based on this view, those who embrace strategic flexibility here will prioritize attention to the external conditions of the organization rather than to the internal conditions ("outward looking"). They believe that the process of adjusting to external changes such as changes in demand, consumer behavior, competitive change becomes the crucial factor that must be done immediately if the organization wants to achieve superior performance (Brauer & Schmidt 2006; Combe et al. 2012). They also believe that it is very crucial for the organization to "create the future" by becoming the first-mover in change (Fisscher & de Weerd-Nederhof 2001; Raynor & Leroux 2004). Thus, the adherents to strategic flexibility emphasize on the importance of first-mover advantage (Combe et al. 2012; Eppink 1978; Grewel & Tasuhaj 2001; Parnell 2005; Petterson & Welch 1992).

2.1.2 Strategic Consistency

Strategic consistency is defined as a commitment to implement a common strategy over a period of time to reduce uncertainty (Parnell 2005; Parnell & Lester 2003), it is also a form of action to adjust the organizational structure to fit the established strategy (Sriram & Anikeeff 1995).

People who are more oriented in strategic consistency emphasize on internal conditions rather than external conditions. Therefore, if there is a change in the external environment they are more likely to choose to "wait and see" until conditions become more certain (Lieberman & Montgomery 1988; Parnell 2005). In accordance with the principle of "second but better", it does not matter if they do not become initiators, as long as they can produce something better. Because they are more oriented to the internal organization (inward looking), therefore the alignment process of the organizational structure becomes very crucial. This is necessary in order to create good synchronization and coordination so that the existing bureaucracy does not impede the strategy

implementation process (Saffold 1998). In this case, synchronization and coordination may take a long time to produce optimal result since they cannot be created overnight. Thus, this type of managers are not convinced that performance can produce optimal results if strategy and structure are forced to change in a fast and short time.

2.2 Personality traits

Personality is defined as a characteristic that settled in a person so that it can describe individual behavior when interacting with the environment (Feist & Feist 2006; McCrae 2011). One approach most commonly used is the Myers Briggs Type Indicator (MBTI) based on Jung's personality type (Pittenger 1993). As a derivative of Jung's personality theory, MBTI is widely used in research on personality types, and is relatively easy to run through individuals who show their preferences on the questionnaire (Borg & Shapiro 1996). The Myers Briggs Type Indicator (MBTI) can identify individual personality traits by using eight measurement parameters that grouped into four characteristics: source of energy, way of gathering information, decision making, and relation with external world (Keirsey & Bates 1998).

2.2.1 Source of Energy: extrovert-introvert

This dimension reflects where the origin of a individual's energy source comes from. Individuals with extroverted characteristics are visible from activities oriented to the outside world or the external environment, so they will look more enthusiastic and active. This extrovert characteristic can also be seen from a friendly, sociable, and assertive personality. While individuals with introverted characteristics are more calm and peaceful, and have a tendency to observe first before participating in a process or activity, and generally seen as quiet and shy (Akhavan et al. 2016; Bradley & Hebert 1997; Prakash et al. 2016).

2.2.2 Way of gathering information: sensitive-intuitive

This dimension reflects how an individual processes data to received better information. Individuals with sensitive characteristics are described as practical individuals who prefer routine and systematic steps (Robbins & Judge 2008). They focus on the present, on what things can be improved now by using experience guidance, concrete data, and prefer proven ways so that they are generally good at technical planning and applicative details (Buaton & Astuti 2013). While individuals with intuitive characteristics rely more on the unconscious process and rather see a holistic picture (Robbins & Judge 2008) so they are more focused on how current conditions will affect the future (Wandrial 2014). They process data by looking at patterns and relationships, abstract, conceptual thinkers and see possible possibilities. They are guided by imagination, choosing unique ways, and focusing on what might be

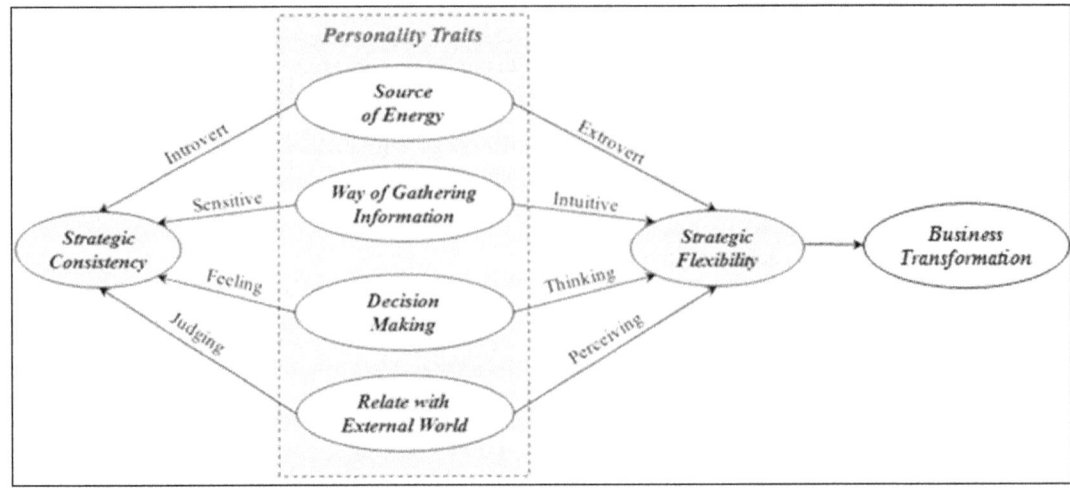

Figure 1. Conceptual research model.

2.2.3 Decision making: feeling-thinking

This dimension reflects how people make decisions. Individuals with the thinking preference use reason and logic to deal with problems and tend to analyze the pros and cons of a situation and inconsistencies that occur (Robbins & Judge 2008; Wandrial 2014). They tend to be task-oriented and objective, apply principles consistently, so they seem stubborn, and are good at analyzing and maintaining standard operating procedure (Buaton & Astuti 2013). While people with feeling preference rely more on their personal values and emotions so they often consider the feelings of others in making decisions, with the aim to maintain harmony among individuals (Robbins & Judge 2008; Wandrial 2014).

2.2.4 Relation with external world: judging-perceiving

This dimension reflect the degree to which a person's flexibility in relation to the outside world. Individuals with judging preference are described as individuals who desire control and prefer their organized and structured world (Robbins & Judge 2008). "Judging" here does not mean judgmental, but is defined as the type of person who always rely on a systematic plan, and always think and act regularly, so they do not like things appear suddenly and out of planning (Buaton & Astuti 2013). Individuals with perceiving are more flexible and spontaneous, therefore, sudden changes are not a problem and situations filled with uncertainty just make them more enthusiastic (Buaton & Astuti 2013; Robbins & Judge 2008).

3 PROPOSITIONS AND CONCEPTUAL MODEL

Based on theoretical review between two paradoxical strategy and personality traits from Myers Briggs Type Indicator (MBTI), conceptual model (Figure 1) was be constructed followed by several propositions.

3.1 Strategy paradox and source of energy

Individuals with extrovert character are seen from activities oriented with the outside world or the external environment, so that they will look more enthusiastic and active. Meanwhile, individuals with introvert character are more calm and peaceful, and have a tendency to observe first before participating in a process or activity (Akhavan et al. 2016; Bradley & Hebert 1997; Prakash et al. 2016). When these two personality characteristics are linked to the concept of strategy paradox, it is noticeable that the perspective of strategic flexibility is more likely to be oriented to external factors first and emphasizes the importance of being first-mover (Combe et al. 2012; Eppink 1978; Grewel & Tasuhaj 2001; Parnell 2005; Petterson & Welch 1992).

While the perspective of strategic consistency emphasizes more on the internal factors of the organization first, and the view that there is no guarantee to achieve first-mover advantage (Parnell 2005). Thus, those who are more oriented toward strategic consistency will have a tendency to observe what happens first. Based on the synthesis between these concepts two proposition were drawn:

Proposition 1: *Person who tends to extroversion will be more oriented towards strategic flexibility.*

Proposition 2: *Person who tend to introversion will be more oriented towards strategic consistency.*

3.2 Strategy paradox and way of gatheringiInformation

Individuals with sensitive character are described as practical individuals who prefer routine and systematic steps (Robbins & Judge 2008). In contrast, individuals with intuitive character rely more on the unconscious process, see the overall picture, and more focused on how the current conditions will affect the future (Wandrial 2014). If these two personality characteristics are related to the concept of strategy paradox, where strategic consistency is defined as the tendency to apply the same strategy and has proven its success in the past (Parnell 2005; Parnell & Lester 2003; and the strategic flexibility is oriented to achieve first-mover advantage in the future (Parnell 2005), a proposition can be formulated as follows:

Proposition 3: *Person with intuitive orientation will have a tendency to emphasize on strategic flexibility.*

Proposition 4: *Person with sensitive orientation will have a tendency to emphasize on strategic consistency.*

3.3 Strategy paradox and decision making

Individuals with thinking orientation are those who use reason and logic to deal with problems, and tend to analyze the pros and cons of a situation and inconsistencies that occur (Robbins & Judge 2008; Wandrial 2014). In contrast, individuals with the feeling orientation rely more on their personal values and emotions. As a result, they often consider the feelings of others in making decisions with a view to maintaining harmony among individuals or groups (Robbins & Judge 2008; Wandrial 2014).

Given perceptions of the impact of strategic flexibility such as restructuring or change in working process that may be perceived to threaten job security (Davis & Fisher 2002; Floyd & Wooldridge 2000), a person with characteristic of feeling tend to be difficult to make changes. Based on the interrelationship between these concepts propositions were formulated as follows:

Proposition 6: *Person with thinking orientation will have a tendency to emphasize on strategic flexibility.*

Proposition 7: *Person with feeling orientation will have a tendency to emphasize on strategic consistency.*

3.4 Strategy paradox and relation with external world

Individuals with judging orientation are described as individuals who desire control and prefer their organized and structured world (Robbins & Judge 2008). This personality is coherent with the the strategic consistency that emphasizes more on certainty and continuity with what has been done before (Parnell 2005; Parnell & Lester 2003). In contrast, more perceiving-oriented individuals have a tendency to be more flexible and spontaneous (Robbins & Judge 2008).

These characteristics are in-line with strategic flexibility that emphasizes the importance of adapting to environmental change. Relationship between concepts here can be synthesized into propositions as follows:

Proposition 7: *Person with perceiving orientation will have a tendency to emphasize on strategic flexibility.*

Proposition 8: *Person with judging orientation will have a tendency to emphasize strategic consistency.*

4 CONCLUSION

Based on the overall theoretical reviews, it can be concluded that there is a relation between the strategy paradox and personality traits as stated in eight propositions above. In this regard, strategic flexibility is more suited to support business transformation rather than strategic consistency. People with extroversion, intuition, thinking, and perceiving preference tend to choose strategic flexibility rather than strategic consistency.

Further research can be conducted to test whether the relationships between variable are valid in the form of examination at individual level. The results of this research implies in business transformation, it is important to find a person who has the tendency to be an agent of change.

ACKNOWLEDGEMENT

I would like to express my gratitude to God and appreciation to my family for helping me through all difficulties. I would also like to thank all researcher from various fields of study who have helped me understand basic concept clearly, link each other, propose new proposition, and develop new conceptual model. At the end, hopefully this paper can be used as a starting point to for further research.

REFERENCES

Aaker, D.A. & Mascharenhas, B. 1984. The Need for strategic Flexibility. *Journal of Business Strategy.*

Akhavan, P., Dehghani, M., Rajabpour, A. & Pezeshkan, A. 2016. An investigation of the effect of extroverted and introverted personalities on knowledge acquisition techniques. *Journal of Information and Knowledge Management Systems.*

Beaver, G. 2003. Beliefs and principles: the compass in guiding strategy. *Strategic Change*, Vol. 12, pp. 1–5.

Borg, M.O., Shapiro, S.L. 1996. Personality type and student performance in principles of economics. *The Journal of Economic Education*, 27(1 -Winter), 3–25.

Bradley, J.H. & Hebert, F.J. 1997. The effect of personality type on teamperformance. *Journal of Management Development*, Vol. 16.

Brauer, M., Schmidt, S.L. 2006. Exploring strategy implementation consistency over time: the moderating effects of industry velocity and firm performance. *Journal of ManagementGovernance* 10:205–226.

Buaton, R., Astuti, S. 2013. Perancangan sistem pakar tes kepribadian dengan menggunakan metode bayes. Binjai: STMIK Kaputama Binjai Sumatra Utara.

Cingoz, A., Akdogan, A.A. 2013. Strategic flexibility, environmental dynamism, and innovation performance: An empirical study. *Social and Behavioral Sciences* 99, 582–589.

Combe, I., Rudd, J.M., Leeflang, P.S.H. & Greenley, G. 2012. Antecedents to Strategic Flexibility. *European Journal of Marketing*.

Davis, D. & Fisher, T. 2002. Attitudes of middle managers to quality-based organizational change. *Managing Service Quality*, Vol. 12, pp. 405–413.

De Wit, B. & Meyer, R. 2010. Strategy synthesis: resolving strategy paradoxes to create competitive advantage. Hampshire: Cengage Learning EMEA.

Eppink, D.J. 1978. Planning for Strategic Flexibility. *Long Range Planning*, 11, 9–15.

Feist, J. & Feist, G. 2006. Theories of Personality, Seventh Edition. New York: McGraw-Hill

Fisscher, O. & de Weerd-Nederhof, P. 2001. Strategic Flexibility. *Creativity and Innovation Management*, 10(4), 223–224

Floyd, S.W. & Wooldridge, B. 2000. Building strategy from the middle: reconceptualizing strategy process. Sage Publications, London.

Gallén, T. 2010. Managers and strategic decisions: does the cognitive style matter? *Journal of Management Development* 25:2, 118–133.

Ghemawat, P. & del Sol, P. 1998. Commitment versus Flexibility? *California Management Review*, 40(4).

Grewel, R. & Tansuhaj, P. 2001. Building organizational capabilities for managing economic crisis: the role of market orientation and strategic flexibility. *Journal of Marketing*, Vol. 65 No. 2, pp. 47–80.

Hambrick, D.C. & Fredrickson, J.W. 2001. Are you sure you have a strategy? *Academy of Management Executive*, Vol. 15 No. 4, pp. 48–59.

Hamsal, M. & Agung, I.G.N. 2006. Paradoxical strategies and firm performance: the case of Indonesian banking industry. *The South East Asian Journal of Management* Vol. I, No. 1.

Johnson, J.L., Lee, R.P., Saini, A. & Grohmann, B. 2003. Market-focused strategic flexibility: conceptual advances and an integrative model. *Journal of the Academy of Marketing Science*, Vol. 31 No. 1, pp. 74–89.

Keirsey, D. & Bates, M. 1998. Please Understand Me II. Prometheus Nemisis Book Company.

Lamberg, J.A., Tikkanen, H., Nokelainen, T. & Suur-Inkeroinen, H. 2009. Competitive dynamics, strategic consistency, and organizational survival. *Strategic Management Journal*, 30(1), 45–60.

Lieberman, M. & Montgomery, D. 1988. First Mover Advantages. *Strategic Management Journal*, 9 (5).

McCrae, R. 2011. Personality Theories for the 21st Century. *Teaching of Psychology* 38(3).

Moss, T.W., Payne, G.T. & Moore, C.B. 2013. Strategic consistency of exploration and exploitation in family businesses. *Family Business Review*, October 1, 2013.

Nasution, M.E., Usman, H. 2008. Proses Penelitian Kuantitatif. Jakarta: Lembaga Penerbit Universitas Indonesia.

Ogunmokun, G.O. & Li, L-Y. 2012. The effect of manufacturing flexibility on export performance in China. *International Journal of Business and Social Science* Vol. 3 No. 6; Special Issue -March 2012.

Parnell, J. 1994. Strategic consistency versus flexibility: does really strategic change enhance performance? *American Business Review.*

Parnell, J. & Lester, D.L. 2003. Toward a philosophy of strategy: reassessing five critical dilemmas in strategy formulation and change. *Strategic Change*, 12, 291–303.

Parnell, John A. 2005. Strategic Philosophy and Management Level. Management Decision, 43(2).

Parnell, John A. 2005. Managing Paradoxes in Strategic Decision-Making, International Journal of Management and Decision Making, Fourthcoming

Petterson, B. & Welch, D.E. 1992. Creating meaningful switching options in international operations. *Long Range Planning*, Vol. 33, pp. 688–705.

Pittenger, D.J. 1993. The Utility of the Myers-Briggs Type Indicator. *Review of Educational Research*, 63(4 – Winter), 467–488.

Prakash, S., Singh, A. & Yadav, S.K. 2016. Personality (introvert, and extrovert) and professional commitment effect among B.Ed teacher educator students. *The International Journal of Indian Psychology.*

Raynor, M.E. 2007. The strategy paradox: why committing to success leads to failure (and what to do about it). New York: Random House Inc.

Raynor, M.E. & Leroux, X. 2004. Strategic Flexibility in RnD. *Research Technology Management.*

Robbins, S.P. & Judge, T.A. 2008. Perilaku Organisasi (12th ed.). Jakarta: Penerbit Salemba Empat.

Saffold, G.S.III. 1998. Culture, traits, strength, and organizational performance: moving beyond strong culture. *Academy of Management Review*, Vol. 13.

Sriram, V. & Anikeeff, M.A. 1995. Strategic consistency amd performance: an analysis of real estate developers. *Journal of Managerial Issues.*

Wandrial, Son. 2014. Tipe Kepribadian Pada Mahasiswa Kelas Manajemen Universitas Bina Nusantara Dengan Menggunakan Myers-Briggs Type Indicator (MBTI). Binus Business Review Vol. 5 No. 1, 344–354.

Yu, F. 2012. Strategic flexibility, entrepreneurial orientation and firm performance: Evidence from small and medium-sized business (SMB) in China. *African Journal of Business Management* Vol. 6(4).

Advances in Business, Management and Entrepreneurship – Hurriyati et al. (Eds)
© 2021 Taylor & Francis Group, London, ISBN 978-0-367-67471-7

Factors affecting micro business perceptions on using electronic payment (e-payment) in Bandung City

N.A. Arafa, C. Wijayangka & B.R. Kartawinata
Telkom University, Bandung, Indonesia

ABSTRACT: This research was based on development of digital technology in the financial sector, i.e., digital systems or electronic payments. Currently, e-payments has started to be adopted by MSMEs. The purpose of this study was to find out the factors that influenced micro business owner perceptions in using e-payment in Bandung City. The variables used in this study were benefits, ease of use, security, trust, self-efficacy and consumer perception. The sampling technique was purposive sampling and the sample size was 185. The data obtained was analyzed using multiple linear regression. The result of partial test showed that benefits, ease of use, and trust did not have significant influence on perceptions to use e-payment as opposed to security and self-efficacy. The result of simultaneous test showed that benefit, ease of use, security, trust and self-efficacy significantly influenced consumer perception on using electronic payments.

1 INTRODUCTION

The development of digital technology continues to grow rapidly reaching various economic sectors particularly financial or banking sector. The latest innovation in the financial sector is called Financial Technology or Fintech. Financial Technology is defined as an industry consisting of companies that use technology to make financial services more efficient. Now, financial technology has been developed and applied on financial sector both all over the world.

Digital-based financial services that have been developed in Indonesia today includes payment channel systems, digital banking, online digital insurance, peer to peer (P2P) lending, and crowd funding. Among these services, the most dominant sector in Indonesia is the payment sector. Accord-ing to Financial Services Authority of Indonesia (OJK.go.id. 2018), Fintech business in Indonesia is dominated by payment sector (43%), loans (17%), and the rest are aggregators, crowd funding, and personal or financial planning. The fundamental reason that payment sector is more dominant because Fintech players comes from various groups, ranging from startup business owners to conglomerates who need payment transactions to meet their financial need.

In 2019, one of fintech start-up dominating In-donesian market is e-payment. Electronic payment (Pei et al. 2015) is a transfer of value from the payer to the recipient of payment via electronic networks that allows customers to access and manage their bank accounts and transactions remotely. In e-payment service system, customers send all data related to

payments to traders, with no further external interaction between traders and customers. E-payment also has many benefits such as increasing payment efficiency, increasing effectiveness and time efficiency and it can even increase customer loyalty.

The application of Fintech especially payment services in MSMEs enables them to serve customers from all over the world in a real time, receive payments instantly, pay employees and manage inventory more easily than traditional methods. According to Fan (2018) MSMEs with high levels of technological involvement are more profitable, last longer and grow bigger. Currently, 3.79 million MSMEs in Indonesia has gone Online and utilized financial technology services to help access payment of their products. This amount is around 8 percent of the total MSMEs owners in Indonesia.

The purpose of e-payment adoption for MSMEs is to benefit consumers in terms of convenience and lower transaction cost. Users can access and manage their transactions via smart phone or web-based user interface as it is supported by the high setup of broadband services and penetration rate.

According to a recent study by Polling Indonesia conducted in cooperation with the Indonesian Internet Providers Association (APJII.or.id. 2019), the number of internet users in Indonesia increased by 10 percent last year. The study showed that 171 million people or 64.8 % of the total population of 264 million Indonesians, were already connected to the Internet in 2018. The figure represented an increase from 54.86% recorded in 2017. MasterCard Indonesia also revealed the level of flexibility and comfort offered by online-shopping via a smartphone which reached 49.9%. About 43.5% of respondents agreed that mobile apps simplify their online shopping activities.

This studies revealed important factors that influence the perceptions consumers on e-payment i.e. benefits, ease of use, security, trust and self-efficacy (Alyabes & Alsalloum 2018; Teoh et al. 2013). The purpose of this study was to identify factors that influenced consumers in using e-payments particularly among micro business owners, so that appropriate strategies can be developed to promote e-payment use among MSMEs. In addition, the results may contribute to online transaction facility providers in understanding consumers' problem when using e-payment systems.

2 LITERATURE REVIEW

2.1 *Micro business*

Micro business in Law of the Republic of Indonesia Number 20 of 2008 are individual business entities that meet the criteria of micro business regulated in Law of the Republic of Indonesia Number 20 0f 2008 regarding micro, small and medium enterprises. The criteria for micro business is having net assets of Rp 50.000.000,00 (fifty million rupiah), excluded of land and building of their place business or having maximum annual sales exceeding Rp 300.000.000,00 (three hundred million rupiah).

2.2 *Electronic payment (e-payment)*

The e-payment system is a financial commitment that includes both sellers and buyers facilitated by the use of electronic communication (Alyabes & Alsalloum 2018). E-payment according to Tan (2004) is one in which monetary value is transferred electronically or digitally between two entities as compensation or consideration for receipt of goods or services. Some other studies (Junadi 2015) stated that e-payments are payment processes carried out without using paper instruments. E-payment systems can be divided into four categories, i.e. online credit card payment systems, electronic online cash, electronic check systems and electronic smart card payment systems.

2.3 *Perceptions of e-payment*

Perception is defined as a process that begins with consumers' exposure and attention to marketing stimuli and ends with consumer interpretation (Hawkins & Mothersbaugh 2010). Abrazhevich (2001) explains that the design of an electronic payment system that is easily understood from the user's point of view is important to attract user acceptance of electronic payments. Further, Arvidsson (2014) confirms that if consumers experience positive experience from a service that they never experienced with any services

before, it is most likely that they will adopt the new service (Arvidsson 2014). In this study factors affecting micro business owners to use e-payment were presented in the following subsections.

2.4 *Benefits*

Barber et al. (1989) defines benefits as the level of a person who believes that using a particular system can improve his performance in work, it means that the benefits of an e-payment facility will be able to improve performance productivity of people using it. In this study, indicators for measuring benefits were: time and costs saving, usage comfort, accuracy, speed, and convenience.

H1: *Benefits significantly influence consumers' perceptions on using e-payment*

2.5 *Ease of use*

Ease of use is the extent to which someone believes that using an e-payment system will be free of effort. In his study, Barber et al. (1989) defines ease of use as the extent to which someone believes that using technology will be free from mental and physical effort. One may find that the system is difficult to use even though they believe this system is useful.

H2: *Ease of Use significantly influences consumers' perceptions on using e-payment*

2.6 *Security*

Tsiakis and Stephanides in Teoh et al. (2013) defines security as a set of procedures and programs to verify information sources and ensure information integrity and privacy. In addition, security is also related to regulations and legal protections felt by consumers, as well as truth, privacy and authenticity.

H3: *Security significantly influences consumer's perceptions on using e-payment*

2.7 *Trust*

Trust is defined as someone's desire to get treatment from others with expectation that they will take important actions to fulfill those expectations, regardless of their ability to monitor or control others. (Mayer et al in Pei et al. 2015)

H4: *Trust significantly influences consumers' perceptions on using e-payment*

2.8 *Self-efficacy*

According to Farrell et al. (2016) in behavioral psychology, the general concept of self-efficacy refers to the feeling of the individual self, related with the belief that the individual can complete the task given and more broadly overcome the challenges of life.

H5: *Self-efficacy significantly influences consumers' perceptions on using e-payment*

3 METHODS

This study uses quantitative methods with descriptive analysis. Descriptive research is defined as a study to determine the value of an independent variable, either one or more variables without making comparisons or connection with other variables (Sarwono 2010).

In this research questionnaire was used as data collection instrument which comprised of three sections; section one contained screening questions to target appropriate respondents, section two was aimed to collect demographic information and section the last section was aimed to investigate factors affecting micro business owners perceptions on using e-payment. There were 19 statements in the last section developed from six dimensions adopted from previous researches by Alyabes & Alsalloum (2018).

The population in this research was MSMEs business owners in Bandung City. Sampling technique used was purposive sampling with the total number of total respondent 185. Once collected, data was analyzed using multiple linear regression analysis with SPSS 22.0. Table 1 shows the indicators used to measure the variables.

Table 1. Models based on factors affecting consumers perception of e-payment.

Models	Indicators	
Benefiits	B1	It saves me time and cost using an e-payment system
	B2	An e-payment system is convenient for me
	B3	The billing and transaction process are accurately handled
	B4	The transaction process with e-payment system is faster than the traditional payment system
	B5	I feel making financial transactions by e-payment system is easier
Ease of Use	E1	I felt concerned about my security when using an e-payment system
	E2	Learning to use an e-payment is easy
Security	S1	Security issues have an important influence on me when using e-payment system
	S2	Matters of security have significant influence on me in using an e- payment system
Trust	T1	I trust the ability of an e-paymentsystem to protect my privacy
	T2	Confidential information is delivered-safely to consumers
	T3	I trust the e-payment system will not lead to transaction fraud
	T4	I feel the risk associated with e-payment system is low
Self-efficacy	SE1	Comments of other people will influence my intention to use an e-payment system
	SE2	I will use an e-payment system when my friends introduce it to me

Table 1. Cont.

Models	Indicators	
Consumers Perception	C1	An e-payment system is better than traditional payment channels
	C2	An e-payment system is much more efficient than traditional payment channels
	C3	I will choose the trusted e-payment system to make transactions
	C4	I feel that a user-friendly e-payment system will influence me to adopt the system

4 RESULTS AND DISCUSSION

4.1 Multiple linear regression

Based on Table 2, regression equation can be formulated as follows:

$$Y = 8,251 + 0,038X1 + 0,002X2 + 0,423X3 + 0,090X4 + 0,268X5 \qquad (1)$$

Table 2. Multiple linear regression test result.

Model	Unstandardized Coefficients		Standardized Coefficients		
	B	Std. Error	Beta	T	Sig
(Constant)	8.251	1.511		5.461	.000
Benefits	.038	.054	.054	.705	.482
Ease of Use	.002	.106	.001	.018	.986
Security	.423	.118	.257	3.577	.000**
Trust	.090	.073	.089	1.242	.216
Self-efficacy	.268	.081	.0234	3.325	.000**

**correlation significant at 0.05, respectively (two-tailed)

a. The value of the constant (a) is positive at 8.251 (82.51%) which means when benefits, ease of use, security, trust, self-efficacy has zero value, consumer perception on using e-payment is 8.251 (82.51%).

b. Regression coefficients on Benefits has a positive value 0.038. It means, if each factor in the Benefits increases by 1 percent, consumer perception on using e-payment will increase by 0.038 (3.8%) thus when benefits increase perception will also increase.

c. Regression coefficients on Ease of Use had a positive value 0.002, which means if each factor in the Ease of Use variable increases by 1 percent, consumer perception on using e-payment will also increase by 0.002 (2%). thus when benefits increase perception will also increase

d. Regression coefficients on Security had a positive value 0,423. It means if each factor in Security increases by 1 percent, consumer perception on using e-payment will increase by 0,423 (42,3%). In other words, when security increases, perception on e-payment will also increase.

e. Regression coefficients on Trust has a positive value 0,090, thus, if each factor in the Trust increases by 1 percent, consumer perception on using e-payment will increase by 0,090 (9,0%). Therefore, increase in perception is influenced by increase in trust.

f. Regression coefficients on Self-efficacy has a positive value 0,268, which means if each factor in the Self-efficacy variable increases by 1 percent, consumer perception on using e-payment by 0,268 (26,8%). Thus, increase in self-efficacy leads to increase in perception

4.2 Classic assumption test

4.2.1 Normality test

Based on Table 3, the results of non-parametric normality test using the Kolmogorov-Smirnov indicates that the value of Asymp.Sig (2-tailed) is 0.200. So the data are normally distributed because the significance results are greater than 0.05.

Table 3. Normality test result. One-Sample Kolmogorov-Smirnov Test.

		Unstandardized Residual
N		185
Normal Parameters[a,b]		.0000000
	Mean	2406.862
	Std.Deviation	18676
Most Extreme	Absolute	.041
Differences	Positive	.041
	Negative	−.036
Test Statistic		.041
Asymp. Sig. (2-tailed)		.200[c,d]

a. Test distribution is Normal.
b. Calculated from data.
c. Lilliefors Significance Correction.
d. This is a lower bound of the true significance.

4.2.2 Heterokedasticity test

Based on Table 4, heterokedasticity test shows that the number of significance in each independent variable >0.05 which means the data is normally distributed.

4.2.3 Multicolinearity test

The VIF value from every dimensions (benefits, ease of use, security, trust and self-efficacy) is 1.00. It can be concluded that there is no multicollinearity problem.

4.2.4 Partial test (T) and simultaneous test (F)

Based on Table 5, the $R2$ value shows 14,3% of variances. Security and Self-efficacy are significantly associated with perception on using e-payment, thus H3 and H5 are accepted. However, benefit, ease of use and trust are not significantly associated e-payment perception. Therefore, H1, H2, H4 are not accepted.

The multiple regression and t test result show that benefit does not significantly effect on consumers perceptions. This result is supported by a previous research conducted by Wibowo & Wijaksana (2016) who found out that some micro business owners were not convinced of the benefits when using electronic payment since they prioritized cash payment for a quick fund turnover. In addition, based on a survey conducted by JakPat (2018), digital payments such as e-money, Go-pay, OVO, T-cash, were still considered less popular among people in consumptive age in Indonesia since they did not know how to use them.

Likewise, ease of use does not significantly affect consumers' perceptions, which is similar to a research by Chanchai et al. (2016). This is because the business owners have relatively low education background that some features and functions in e-payment service are not well understood.

Security is the most crucial feature among all factors and it has a significant effect on consumers perceptions (t = 3.325, p ≥ 1,973), thereby H3 is accepted. This result is supported by Alyabes & Alsalloum (2018) who found that security is an important factor that influences consumer behavior in using e-payment which makes them confident in using an e-payment service.

Table 4. Heterokedasticity test result.

Model	Unstandardized Coefficients		Standardized Coefficients		
	B	Std. Error	Beta	T	Sig.
(Constant)	2.674	.907		2.949	.004
Benefits	−.038	.033	−.095	−1.169	.244
Ease of Use	.086	.064	.110	1.354	.178
Security	−.144	.071	−.154	−2.024	.064
Trust	.014	.044	.025	0.323	.747
Self-efficacy	−.022	.048	−.034	−.454	.651

a. Dependent Variable: RES2

Table 5. Partial and simultaneous test result.

Model	Unstandardized Coefficients		Standardized Coefficients		
	B	Std. Error	Beta	T	Sig.
(Constant)	8.251	1.511		5.461	.000
Benefits	.038	.054	.054	.705	.482
Ease of Use	.002	.106	.001	.018	.986
Security	.423	.118	.257	3.577	.000**
Trust	.090	.073	.089	1.242	.216
Self-efficacy	.268	.081	.0234	3.325	.000**

**correlation significant at 0.05, respectively (two-tailed)

However, trust does not have significance for users of e-payment systems, in other words trus does not affect consumer behavior. This result is similar to previous study by Teoh et al. (2013). Fraud Management Insight (2017) surveyed countries with the lowest level of public trust in digital transactions in telecommunications, financial services and retail sectors. Indonesia ranked 10 out of 10 countries surveyed with an average score 1.8. This is due to the long-term challenges caused by the vulnerability of online transaction systems that becomes major obstacle to build higher trust.

Self-efficacy has a significant effect on consumers' perceptions (t = 3,325, p ≥ 1,973). This outcome is supported by some studies by Teoh et al. (2013) and Alyabes & Alsalloum (2018). Consumers who have good experience with e-payment and feel its usefulness with be motivated to continue using it in the future. In addition, recommendation or positive comments from family, friends and others who have already used e-payment influence them to try using e-payments.

5 CONCLUSION

The purpose of this research was to determine the factors that influence the perception on using e-payment. Based on the results of descriptive test, classic assumption test, partial test (T) and simultaneously test (F), it can be concluded that:

1. The results descriptive analysis showed that benefits, ease of use, security, Trust, Self-efficacy, and Consumer perception on e-payment respectively obtained total score of 75.35%, 62.59%, 75.24%, 73.62%, 72.43%, and 76.05%, which are considered good.
2. Security and self-efficacy partially had significant influence on perceptions of e-payment, while benefits, ease of use, and trust did not.
3. Benefit, ease of use, security, trust and self-efficacy simultaneously influenced perceptions of micro business owners in using e-payment.

ACKNOWLEDGEMENT

My deepest gratitude goes to Mr. Chandra Wijayangka S.T., M.M and Mr. Budi Rustandi Kartawinata S.E., M.M as my advisor for the unending support and guidance in making this research possible and also for all parties who indirectly participated in the process of completing this research.

REFERENCES

Abrazhevich, D. 2001. Electronic Payment Systems: Issues of User Acceptance. *Research Gate Article Database*, 1–7.
Alyabes, A.F. & Alsalloum, O. 2018. Factors Affecting Consumers' Perception of Electronic Payment in Saudi Arabia. European Journal of Business and Management, 36–45.
APJII.or.id. 2019. Retrieved from https://apjii.or.id/content/read/104/398/BULETIN-APJII-EDISI-33—Januari-2019
Arvidsson, N. 2014. Consumer attitudes on mobile payment services–results from a proof of concept test. *International Journal of Bank Marketing*.
Barber, W.B., Davis, W.H. & Rautenkranz, K. 1989. U.S. Patent No. 4,858,121. Washington, DC: U.S. Patent and Trademark Office.
Chanchai, et al. 2016. An Investigation of Mobile Payment (m-payment) Services in Thailand. *Asia. Pacific Journal of Business Administration*, Volume 8.
Farrell, L., Fry, T.R. & Risse, L. 2016. The significance of financial self-efficacy in explaining women's personal finance behaviour. *Journal of Economic Psychology*, 54, 85–99.
Fraud Management Insight. 2017. Retrieved from https://www.experian.com.sg/insights/fraud-management-insights-2017
Hawkins, D.I. & Mothersbaugh, D.L. 2010. Consumer behavior: Building marketing strategy. Boston: McGraw-Hill Irwin.
JakPat.net. 2018. Retrieved from https://blog.jakpat.net/cashless-payment-extended-usage-of-go-pay-and-ovo-survey-report/
Junadi, S. 2015. A model of factors influencing consumer's intention to use e-payment system in Indonesia. *Procedia Computer Science*, 59, 214–220.
OJK.go.id. 2018. Retrieved from https://www.ojk.go.id/id/data-dan-statistik/laporan-kinerja/Pages/Laporan-Kinerja-OJK-2018.aspx
Pei, Y., Wang, S., Fan, J. & Zhang, M. 2015. An empirical study on the impact of perceived benefit, risk and trust on

e-payment adoption: comparing quick pay and union pay in China. *In 2015 7th international conference on intelligent human-machine systems and cybernetics* (Vol. 2, pp. 198–202). IEEE.

Sarwono, J. 2010. Pintar menulis karangan ilmiah-kunci sukses dalam menulis ilmiah. Penerbit Andi.

Tan, M. 2004. E-payment: The digital exchange. NUS Press.

Teoh, W.M.Y., Chong, S.C., Lin, B. & Chua, J.W. 2013. Factors affecting consumers' perception of electronic payment: an empirical analysis. *Internet Research.*

Wibowo, U.N. & Wijaksana, T.I. 2016. Pengaruh pemberian kredit terhadap pengembangan usaha mikro dan kecil di kota Bandung (study kasus Kredit Cinta Rakyat pada Bank BJB KCP Mochamad Toha). *eProceedings of Management,* 3(3).

The impact of adaptive and innovative capabilities on financial performance of small and medium enterprises in Medan

Y. Absah & R.H. Harahap
Universitas Sumatera Utara, Medan, Indonesia

ABSTRACT: Technological developments have caused Small and Medium Enterprises (SMEs) to have the ability to face increasingly fierce business competition. The industrial revolution 4.0 brings changes in the lifestyle of consumers so that SMEs must develop a number of core competencies to adapt to the changes that occur. The optimal use of strategic resources as core competencies by SMEs will improve their financial performance. The adaptive and innovation capabilities are believed to influence an enterprise's financial performance. Provided with the population of SMEs of culinary in Medan, 100 of them were used as samples. The research result simultaneously and partially showed a positive and significant impact of the adaptive and innovation capabilities on the financial performance of culinary SMEs in Medan. Adaptive capabilities have a dominant influence on the financial performance of culinary SMEs in Medan

1 INTRODUCTION

In Indonesia, SMEs is the foundation of the economic growth. SMEs also play an important role in developing the national economy and industry as well as providing job opportunities. In the global economy development, SMEs contribute 80% (Jutla et al. 2002). They are crucial for the economic performance and development of any country and are an important source of flexibility and innovation (OECD 2002).

In a dynamic business environment, renewing dynamic capabilities is inevitable. This situation requires a company not only to adapt to changes in the environment but also to change its capabilities. The situation of a business environment that is full of changes makes its capabilities no longer relevant. In that case, the company must adjust its capability base (Regenerative dynamic capabilities).

Being the 3rd biggest city in Indonesia, Medan is also famous for its culinary. Medan is the melting pot of diverse signature dishes resulting from assimilation of various cultures such as Malay, Batak, Minang, Indian, Chinese, and more. Chinese culinary has been predominant in North Sumatran food tradition. Chinese influence has set Medan culinary different from, for instance, that of Padang. Many tourists have come to Medan for the unique culinary based on social media references. Technology, more specifically the internet, has changed how business goes.

The fast-changing business environment has indirectly demanded that SMEs adapt. The small dimension nature of the company enables SMEs to swiftly respond customers' demand. Thus, SMEs are supposedly to adopt any changes in business environment effectively and relevantly. On the other hand success of the company's adaptation is determined by the attitude of employee ownership of the company (Thompson et al. 2013).

However, few details SMEs owners need to pay more attention is to create more awareness to innovate rather than imitate the existing products in the market. The fact is that customers tend to desire distinct and unique products. An SME who can develop its innovated product is defined superior than its competitors. The development of product innovations must be planned and done carefully suited to the customer's demand.

The success of an organization in any industries depends on its innovative capability (Saunila & Ukko 2013). A research by Flinders et al. (2010) found it is important to focus on the management innovation to develop new products. A company's ability to compete can be built from its various competitive superiority sources. SMEs in China have proven that their key to competitive power lies in its superiority of cost and the strategies of differentiation and innovation (Yan 2010).

This research was aimed to analyze the effect of adaptive and innovative capabilities on the financial performance of culinary SMEs in Medan. It is also expected that this research can be an input for SMEs to upgrade their financial performance through an improved adaptive and innovation capabilities, as well as for academics to develop marketing science, strategic management, and SMEs.

2 LITERATURE REVIEW

Adaptive capability is a skill to identify and capitalize opportunities coming from the market and is measurable from its ability to respond the opportunity, monitor the market, customers and competitors, and allocate resources for marketing purposes (Hofer et al. 2015). Adaptive capacity depends on changing market or product expectations (McKee et al.1989), and the company's ability to meet these expectations with available resources and capabilities (Penrose 1959). Lee (2004) in his research concluded that focusing on internal changes was required to upgrade the competitive power of small manufacturing enterprises in China. Adaptive capability can be utilized as a medium to interact with their external environments such as market scanning, customers, competitors, and technology (Evaristo & Zaheer 2014).

H1: adaptive capability has positive and significant influence on financial performance

Innovative capability is the skill to develop new products or markets through adjustments of the orientation of innovation strategies towards the innovation and process attitudes (Wang & Ahmed 2007). The innovation process is driven by the need to understand how things work, grow revenue, reduce costs, or increase productivity; to solve customer problems; or to keep people living healthy and safe (Estrin 2009). Damanpour (1991) suggests that innovation is an introduction of a new tool, system, product or service, production process technology, structure or administration, or planning program to be adopted by an organization. Damanpour (1991) further classify innovation into administrative innovation, technical innovation, product/service innovation, process innovation, radical innovation, and incremental innovation.

The development of innovative capability is crucial to the survival and development of the company (Francis & Bessant (2005) in Saunila (2014); Larsen & Lewis (2007) in Hadiyati (2011) state that innovation is a highly significant character that an entrepreneur should possess. A research by Hariati & Tjahjadi (2015) reveals that sustainable innovation strategy is influential to the financial performance.

H2: innovative capability has positive and significant influence on financial performance

3 METHODS

This research used quantitative approach the relationship between variables were explained by multiple linier regression analysis. The sampling technique was area clustering and the size was 100 comprising of small and medium culinary businesses in Medan.

Data collection methods were in-depth interviews and questionnaire distribution.

Research variables were measured by several indicators, i.e. adaptive capability was measured by the ability to handle emergency or critical situations, the ability to work under uncertain condition, and the ability to learn new tasks, technology, or procedures; innovative capability was measured by development of new ideas, the ability to accelerate new offers to customers, the ability to manage processes to reduce costs, and the ability to provide a total solution to overcome customer problem; and financial performance was measured by sales growth, debt reduction, increase capital, increase operating profit, increase business assets, and increase profit margins.

4 RESULTS AND DISCUSSION

Table 1 shows that simultaneously adaptive and innovation capabilities influence financial performance of culinary SMEs in Medan, represented by the significant value 0,000 < 0,05, F count (15,571) > F table (2,75).

Table 1. ANOVA.[a]

Model		Sum of Squares	df	Mean Square	F	Sig.
1	Regression	877.570	2	438.785	15.571	.000[b]
	Residual	2733.430	97	28.180		
	Total	3611.000	99			

a. Dependnt Variable: Financial Performance
b. Predictoers: (Constant), Innovation Capabilities, Adaptive Capabilities

Table 2 shows that the value R^2 for adaptive capabilities and innovation capabilities describes a change in the financial performance value of 0,227, which means that those two independent variables simultaneously influence financial performance as much as 22.7%.

Table 2. Determinan.

Model	R	R Square	Adjusted R Square	Std. Error of the Estimate
1	493[a]	.243	.227	5.30846

a. Predictors: (Constant), Innovation Capabilities, Adaptive Capabilities

From Table 3 the regression equation is formulated as follow:

$$Y = -7,666 + 0,386 \text{ X1} + 0,398 \text{ X2} \quad (1)$$

Table 3 also shows that:

1. Adaptive capabilities positively and significantly influences financial performance, with significant value of 0,005 and tcount $(-2,851) < $ t table (1,980). This suggests a stagnant financial performance on an increased SMEs culinaryadaptive capability.

The above result is relevant to several previous researches. Stoica & Schindehutte (1999) finds that the adaptive capability of SMEs is relevant to performance. Such adaptability may empower culinary SMEs to cope with various dynamics of fast changing business environments, most importantly concerning the more predominant use of the internet as the medium of promotion, as a source of inspirations of packaging design, product uniqueness, distinctive taste, and many more. The internet has been increasingly becoming the main source of the hunting spot of trending culinary. Medan's culinary SMEs should be able to seize the current trend and utilize the internet to further build their business for example by adopting e- business applications such as GrabFood, GoFood, or OVO. Wachira (2014) suggested that there is a certain need to create greater awareness among the SMEs on the importance of e-business adoption.

Table 3. Coefficients.[a]

| Model | Unstandardized Coefficients | | Standardized | | |
	B	Std. Error	Beta	t	Sig.
1 (Constant)	−7.666	5.440		−1.409	.162
Adaptive Capabilities	.386	.135	.298	2.851	.005
Innovation Capabilities	.398	.158	.264	2.522	.013

a Dependent Variable: Financial Performance

2. Innovation capabilities positively and significantly influences financial performance, with significant value of 0,013 and t count $(2,522) > $ t table (1,980). This suggests an increased financial performance on an increased innovation capability.

This result is supported by other previous researches such as that of Hariati & Tjahjadi's (2015). Today's customers always want to keep updated on new trends in culinary. The ever-developing culinary business in Medan provides many alternatives for customers. Such development requires that culinary business owners have more innovative capability to produce more varieties for their product design, packaging, and service.

5 CONCLUSION

Adaptive and innovation capabilities simultaneously and partially showed a positive and significant impact of on financial performance of SMEs of culinary in Medan.

ACKNOWLEDGEMENT

The researcher would like to thank Kemenristek Dikti Republik Indonesia for providing lecturer research grants, so that this research can be carried out.

REFERENCES

Damanpour, F. 1991. Organizational Innovation: A Meta-Analysis of Effects of Determinants and Moderators. *The Academy of Management Journal*, 34(3), September, 555–590.

Estrin, J. 2009. Closing the Innovation Gap: Reigniting The Spark of Creativity in a Global Economy. New York: McGraw-Hill.

Evaristo, R. & Zaheer, S. 2014. Making the most of your firm's capabilities. *Bus. Horiz.* Vol.57, 329–335.

Flinders, et al. 2010. Overcoming the barriers to managing innovation in the early stages of new product development in SME's. *IAM Converence* 1st–3rd September, 2010.

Hadiyati, E. 2011. Kreativitas dan Inovasi berpengaruh terhadap Kewirausahaan Usaha Kecil. *Jurnal Manajemen dan Kewirausahaan* , 13(1), 8–16.

Hariati & Tjahjadi, B. 2015. Hubungan antara strategi inovasi dengan kinerja keuangan yang dimediasi oleh modal intelektual dan kinerja pelanggan. *Konferensi Regional Akuntansi*, Vol. II, April 2015.

Hofer, Maria, K., Niehoff, L.M., & Wuehrer, G.A. 2015. The effects of dynamic capabilities on value-based pricing and export performance, in (ed.) *Entrepreneurship in International Marketing (Advances in International Marketing*, 25, 109–127.

Jutla, D., Bodorik, P. & Dhaliqal, J. 2002. Supporting the e-business readiness of small and medium-sized enterprises: approaches and metrics. *Internet Research: Electronic Networking Applications and Policy*, 12(2), 139–164.

Lee, C.Y. 2004. TQM in small manufacturers: an exploratory study in China. *International journal of quality & reliability management*.

McKee, D.O., Varadarajan, P.R. & Pride, W.M. 1989. Strategy cadaptability and firm performance.*A Market-Contingent Perspective*, J. Mark, 53, 21–35.

OECD, 2002. OECD small and medium size enterprise outlook, available at: www.oecd.org.

Penrose, E. 1959. The Theory of the Growth of the Firm. Billing and Sons Ltd.: Guild for d/London/Worcester, UK.

Saunila, M. 2014. The relationship between innovation capability and performance: The moderating effect of measurement. *International Journal of Activity and Performance Management*, 63(2).

Saunila, M. & Ukko, J. 2013. A conceptual framework for the measurement of innovation capability and its effects. *Baltic Journal of Management*, 7(4), 355–375.

Stoica, M. & Schindehutte, M. 1999. Understanding adaptation in small firms: links to culture and performance. *Journal of Developmental Entrepreneurship*, 4(1), Spring 1999.

Thompson, Peter, B., Shanley, M. & McWilliams, A. 2013. *Journal of Business Strategies*. Vol. 30 Issue 2, p.145–179.

Wachira, K. 2014. Adoption of E-Business by Small and Medium Enterprises in Kenya: Barriers and Facilitators. *International Journal of Academic Research in Business and Social Sciences*, 4(11), November.

Wang, C. L. & Ahmed, P. K., 2007. Dynamic Capabilities: A Review and Research Agenda. *The International Journal of Management Reviews*, 9(1), 31–51.

Yan, Shigang, 2010. competitive strategy and businee environment: the case of small enterprises in. *Asian Social Science*, October 2010, 6(11).

Advances in Business, Management and Entrepreneurship – Hurriyati et al. (Eds)
© *2021 Taylor & Francis Group, London, ISBN 978-0-367-67471-7*

Entrepreneurship intention and its affecting factors of private Islamic university students in Jakarta

B.E. Samiono
Universitas Al Azhar Indonesia, Jakarta, Indonesia

M. Akbar & Hamidah
Universitas Negeri Jakarta, Jakarta, Indonesia

ABSTRACT: This study explains entrepreneurship intention of private Islamic University students in Jakarta and its affecting factors. The object of the study was 400 respondents from the Faculty of Economics and Business from Al Azhar University of Indonesia and University of Muhammadiyah Prof. Dr. Hamka. They were considered to represent Private Islamic Universities in Jakarta with accreditation A and B in entrepreneurial concentration, Management Study Program. Compared to previous studies, this study provides an in-depth exploration of dimensions and indicators of entrepreneurship intention through multivariate analysis using Structural Equation Modeling (SEM) method and specifically individual perception as mediation. The results obtained revealed that the entrepreneurial intention of private Islamic University students in Jakarta was formed merely in having serious thoughts of business operations, yet their desire to have their own business was not too strong. Entrepreneurship intention was directly influenced by entrepreneurship attitude and indirectly influenced by entrepreneurship education and entrepreneurial self-efficacy through entrepreneurship attitude.

1 INTRODUCTION

The Indonesian government, through the Ministry of National Education, attempts to encourage the development of high-quality entrepreneurial intention in the younger generation, particularly at the tertiary level. This attempt is significantly appropriate since universities are considered capable of playing a greater role in improving the economic aspects of the nation. Moreover, this attempt aims to increase the low number of entrepreneurs in Indonesia, reaching merely 1% of the total population. This figure indicates that Indonesia is lagging behind Malaysia and Singapore whose number of entrepreneurs reaches 3% and 7.2% respectively. Indonesia is also lagging far behind in terms of the quality and competence of regional and global entrepreneurship. According to the 2018 Global Entrepreneur Index, Indonesia ranks 90th among 137 countries surveyed.

Currently, several universities in Indonesia have a similar issue related to the development of entrepreneurship education. The culture of education in Indonesia appears unable to support the development of entrepreneurial spirit of their students, resulting in the stagnancy of the development of the creativity of young generation (Febriyanto 2013). It is expressed by Wibowo & Pramudana (2016) that entrepreneurship education in Indonesia is inadequate. The Ministry of National Education acknowledges

that the available educators apparently focus on the preparation of the available workforce rather than the development of entrepreneurial behavior of students.

Approximately 52 private universities are under the supervision of Higher Education Service Institutions (LL Dikti) Region III, seven of which are private Islamic universities and five of which are private Islamic universities registered with accreditation A and B. There are merely two private Islamic universities providing special attention to entrepreneurship by opening a special concentration of entrepreneurship, namely Al-Azhar University of Indonesia and University of Muhammadiyah Prof. Dr. Hamka. Both universities encountered a similar issue, namely the small ratio of graduates becoming entrepreneurs.

Observing the position of private Islamic universities under the supervision of Higher Education Service Institutions (LL Dikti) Region III and the demands for quality improvement of the existing entrepreneurship, several issues of entrepreneurship development in private Islamic universities associated with Private Islamic Higher Education were revealed. Entrepreneurship development in private Islamic tertiary institutions is an interesting object to study, considering the large number of universities and students in Indonesia as well as the fact that Indonesia has not been able to grow muslimpreneur despite being the largest Muslim country in the world.

Studies of entrepreneurial intention and the influencing factors have been carried out worldwide. A study by Rengiah & Sentosa (2014) analyzed entrepreneurial intention of final year students having obtained entrepreneurship courses from four universities in Malaysia. Similar studies examined entrepreneurial intention of students from 17 countries in Europe (Küttim et al. 2014) and entrepreneurial intention of students studying entrepreneurship in the third year from three universities in Uganda by Oyugi (2015). These studies have similarities in applying Theory of Planned Behavior (TPB), using a cross-sectional study and the variable of entrepreneurship education to measure entrepreneurial intention. Compared to other studies, this study carried out an in-depth examination of the dimensions and indicators of entrepreneurial intention using Structural Equation Modeling (SEM) method and specifically individual perception as mediation

2 LITERATURE REVIEW

Entrepreneurial intention is defined as a willingness of an individual to engage in entrepreneurial behavior and action, to be self-employed, or to establish a new business (Lee et al. 2012). Meanwhile, Linan et al. (2011) state that entrepreneurship intention is the effort taken by an individual to carry out the entrepreneurial behavior. It is also explained that entrepreneurship intention is a process of information-seeking, in which the information shall be used to achieve the goal of establishing a business (Linan et al. 2011). Ogundipe et al. (2012) argues that entrepreneurship intention is a tendency of an individual to start various entrepreneurial activities in the future. This statement is based on a study on business and counseling students at the Lagos State University Sandwich Program (Ogundipe et al. 2012). Entrepreneurship intention is measurable using the entrepreneurship intention scale with the following indicators; 1) becoming an entrepreneur, 2) preferring to be an entrepreneur rather than an employe, 3) having a really serious thought, and 4) taking every effort to start and establish a firm eventually (Lee et al. 2012).

Turker & Selcuk (2009) state that entrepreneurship education utilizing educational and training programs is helpful for the intention of an individual, such as entrepreneurial knowledge that fosters entrepreneurial desire to carry out entrepreneurial activities. Entrepreneurship education is defined as a professional application process of knowledge, attitudes, skills and competencies. It is more than merely teaching students how to be independent business owners. It is regarding creating and maintaining environmental learning promoting the nature and behavior of entrepreneurs, such as thinking creatively and independently, willing to take risk, assuming responsibility, and respecting diversity by Gautam (2015). Meanwhile, Martinez (2011) states that building knowledge

and skills regarding or for the purpose of entrepreneurship is generally part of recognized education for primary, secondary, or tertiary level educational institutions. An entrepreneurship study on university students in Malaysia by Rengiah & Sentosa (2014) measured entrepreneurship education through curriculum, teaching methods, and the role of universities in supporting entrepreneurship.

In the context of entrepreneurship, behavior (attitude) is defined as the degree to which an individual holds a positive or negative personal valuation regarding being an entrepreneur (Linan et al. 2011). Entrepreneurial attitude is measurable using the Entrepreneurial Attitude Orientation (EAO) survey instrument model developed by Robinson et al. (2018). This model utilizes 4 attitude subscales, consisting of 4 constructs, namely: achievement in business, innovation in business, perceived personal control, and confidence (Robinson et al. 2018). Referring to the theory, entrepreneurial attitude includes three aspects, namely, affection (feelings and emotions), cognition (thoughts and beliefs), and knowledge (actions and behavior). The combination of the three dimensions of entrepreneurial attitude in terms of affection, cognition, and knowledge motivates an individual to be an entrepreneur (Pihie & Bagheri 2011).

Self-efficacy refers to the conviction that an individual is able to succeed in performing the desired behavior required to produce an outcome (e.g., successfully launching a business) by Bandura (2012). In this regard, selfefficacy clearly influences one's belief in success (achieving) or failure (not achieving) the predetermined goals. Ajzen (2001) suggests that self-efficacy is a condition where an individual believes that an action is easy or difficult by understanding various existing risks or barriers should they are willing to perform the action. Related to self-efficacy, an individual shall attempt to carry out what they think they are able to and shall not attempt to carry out what they think they are not able to Maddux (2013). According to McGee et al. (2009), Entrepreneurship Self-Efficacy (ESE) is a construction measuring one's confidence in their ability to successfully launch entrepreneurial ventures. The indicators of Entrepreneurial Self-Efficacy (ESE) include: risktaking, innovation, management, financial, and marketing.

3 METHODS

This study applied a self-administered survey and associative causal quantitative approach to examine the direct and indirect effects of the independent variable on the dependent variables as well as Structural Equation Modeling (SEM) method to analyze the influence between variables. The endogenous variable in this study is Entrepreneurial Intention (Z), the exogenous variable is Entrepreneurship Education (X1), while the intermediate variables are Entrepreneurship Attitude (Y1) and Entrepreneurship

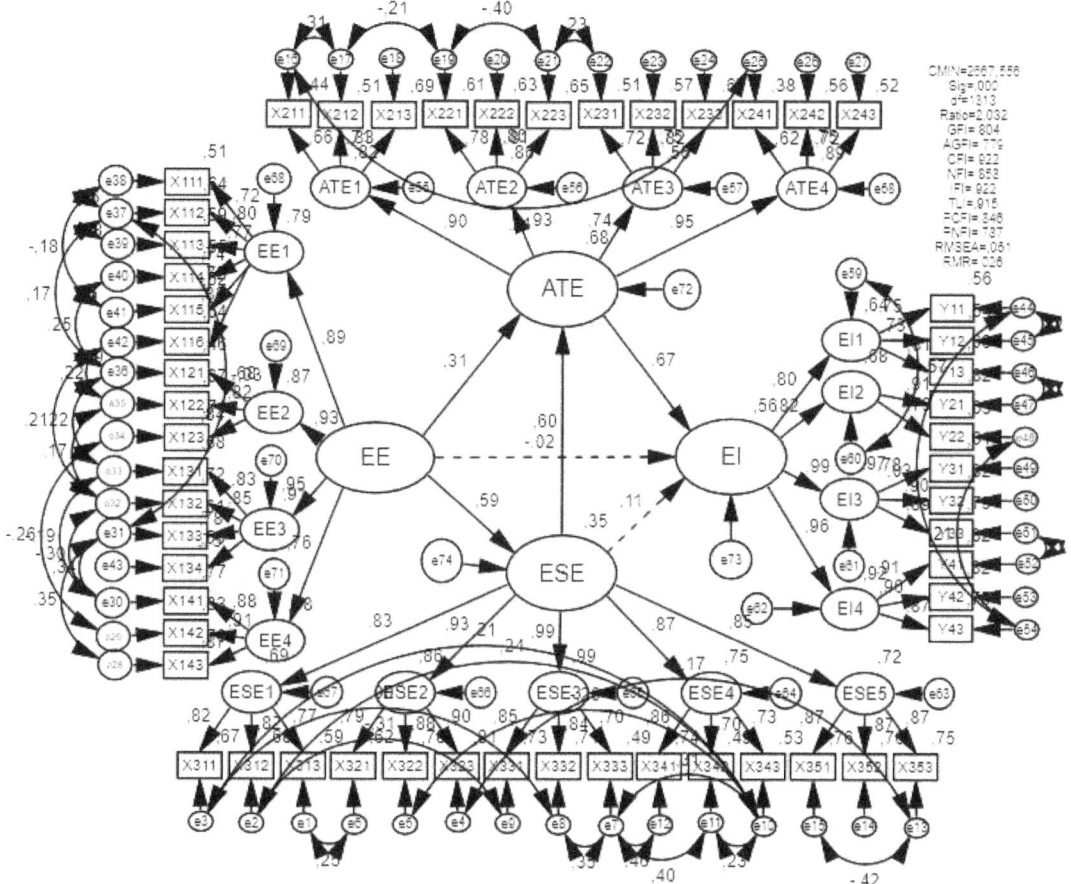

Figure 1. Overall structural model fitreference.

Self-Efficacy (Y2). The object of the study was private Islamic universities students in Jakarta under the supervision of Higher Education Service Institutions. The respondents were the students of Al Azhar University of Indonesia (UAI) and University of Muhammadiyah Prof. Dr. Hamka (Uhamka) since they are private Islamic universities in Jakarta with accreditation A and B in entrepreneurial curriculum and concentration in their study program. The sample of this study was 400 active students having taken entrepreneurship lectures as both faculty and university subjects taken used simple random sampling technique.

4 RESULTS AND DISCUSSION

The analysis of the Overall Structural Model on the combination of all latent variables using SEM analysis is shown in Figure 1

The data of the study obtained a modification of the index to meet the goodness of fit, the results of which are presented in Table 1. This model obtains

the "minimum was achieved" result, evident from Chi-square = 2,667.556, Degrees of freedom = 1,313, and Probability level = 0.000. Despite zero probability (0.000), this result indicates that the overall model is in accordance with the sample data.

The results of the model fit test in Table 1 indicate the measurement of the absolute fit model, in which RMSEA = 0.051 and GFI = 0.804. Thus, it indicates that the model is in good category. The calculation of increm5ental compatibility model reveals that CFI = 0.922, NFI = 0.858 and IFI = 0.922, meaning that the model has good criteria. Meanwhile, the measurement of the conventional compatibility model shows AGFI of 0.779, implying that the model is also in good category. Overall, it was found that the analysis of the goodness of fit full model has been met according to the Table 1.

4.1 The analysis of entrepreneurship intention

The analysis of Entrepreneurship Intention of students of Private Islamic University in Jakarta is presented in Tables 2 and 3.

655

Table 1. GOF full model.

	Result	Term	Note
Absolute Fit Model			
(RMSEA)	0.051	≤ 0.08	Good of Fit
(GFI)	0.804	≥ 0.90	Marginal Good
Incremental Fit Model			
(CFI)	0.922	≥ 0.90	Good of Fit
(NFI)	0.858	≥ 0.90	Marginal Good
(IFI)	0.922	≥ 0.90	Good of Fit
Parsimonious Fit Model			
(AGFI)	0.779	≥ 0.90	Not Good

Table 2. The effect of variables on the dimensions of entrepreneurship intention.

Variable			Dimension	Estimated
EI	→	EI.1	Becoming an entrepreneur	0.799
IE	→	EI.2	Preferring to be an entrepreneur rather than an employee	0.957
EI	→	EI.3	Having a really serious thought	0.986
EI	→	EI.4	Taking every effort to start and establish a firm eventually	0.825

Table 3. The effect of dimensions on the indicators of entrepreneurship intention.

Dimension			Indicator	Estimated
EI.1	→	Z1.1	Interested in becoming an entrepreneur	0.750
	→	Z1.2	Desiring to have a business	0.732
	→	Z1.3	Planning to have a business	0.813
EI.2	→	Z2.1	Deciding to be an entrepreneur	0.906
	→	Z2.2	Desiring to have business as the main job	0.793
EI.3	→	Z3.1	Seriousness in thinking of business concepts	0.784
	→	Z3.2	Seriousness in thinking of business operations	0.903
	→	Z3.3	Seriousness in thinking of capital	0.886
EI.4	→	Z4.1	Seriousness in planning a business concept	0.907
	→	Z4.2	Seriousness in creating a business	0.904
	→	Z4.3	Seriousness in planning business development	0.866

Table 4. Direct and indirect effects of the entire mode.

Variable			Est	Direct	Indirect	Total
EE	→	ESE	0.589	0.589		0.589
EE	→	ATE	0.312	0.312		0.312
ESE	→	ATE	0.601	0.601		0.601
ATE	→	EI	0.670	0.670		0.670
ESE	→	EI	0.113	0.113		0.113
EE (ATE)	→	EI	−0.024	−0.024	0.067	0.043
EE (ESE)	→	EI	−0.024	−0.024	0.209	0.185

There are several dimensions of Entrepreneurship Intention: becoming an entrepreneur (EI.1), preferring to be an entrepreneur rather than an employee (EI.2), having a really serious thought (EI.3), and taking every effort to start and establish a firm eventually (EI.4), which have a strong influence. The biggest influence is provided by the dimension of having a really serious thought (EI.3) with a value of 0.986 with the indicator of having a serious thought of business operations (Z3.2) (0.903). Meanwhile, the smallest influence is provided by the dimension of becoming an entrepreneur (EI.1) with a value of (0.799) with the indicator of desiring to have a business (Z1.2) (0.732).

According to Tables 2 and 3, it is obvious that Entrepreneurship Intention of private Islamic universities students in Jakarta is formed merely in having serious thought of business operations, particularly in business operations they shall create. However, their desire to have their own business is not too strong. This study discovered an interesting finding where efficacy has a considerable influence on entrepreneurship attitude rather than entrepreneurial intention. Therefore, it is fair to assume that positive entrepreneurial attitude is the key for Islamic universities students in Jakarta to be able to increase their entrepreneurial intention. Even so, it is not a necessary increase since the essence is to increase efficacy, yet entrepreneurial attitude is required as an intermediary.

4.2 The factors affecting entrepreneurship intention

The variables affecting entrepreneurship intention of students of Private Islamic University students in Jakarta are presented in Table 4.

Observed from the results obtained in Table 4, it can be concluded that entrepreneurial intention is merely affected by Entrepreneurship Attitude (ATE) with a value of 0.670. It is in line with the research finding by Rahmawaty (2014) regarding the entrepreneurial intensity model with the basic theory of Planned Behavior that there was a positive relationship between entrepreneurial attitude and entrepreneurial intention. The similar finding was also discovered by Andika & Madjid (2012), Rio-Rama et al. (2016) & z (2014). However, Entrepreneurship Intention (EI) in this study is not influenced by Entrepreneurship Education (EE) and Entrepreneurship Self-Efficacy (ESE).

The absence of the influence of Entrepreneurship Education (EE) on Entrepreneurship Intention (EI) is not in line with several previous studies, including a study by Lee et al. (2012)

discovering the influence of entrepreneurship education on Entrepreneurship Intention of young people to become entrepreneurs. Several similar studies (Bae et al. 2014; Dabale & Masese 2014; Hamidi et al. 2008; Karali 2013; Küttim et al. 2014; Malebana & Swanepoel 2017; Muofhe & Du Toit 2011; Oyugi 2015; Tung 2011) discovered similar finding with Bae et al. (2014), that there was a significant correlation yet insignificant effect of entrepreneurship education on entrepreneurial intention. Even by controlling the variables of pre-education entrepreneurial intention, the relationship of entrepreneurship education and post-education entrepreneurial intention was not significant. It is similar to the finding in the study conducted (Patricia & Silangen 2017).

Table 4 also reveals that Entrepreneurial Intention (EI) does not affect Entrepreneurial Self-Efficacy (ESE). This study has different findings from previous dominant studies, concluding that self-efficacy is an important factor in determining entrepreneurial intention in one's career (Andika & Madjid 2012; Malebana & Swanepoel 2014; McGee et al. 2009; Rahmawaty 2014). This phenomenon is supported by a study carried out by Bae et al. (2014) where the risk factors for entrepreneur determined the influence of entrepreneurship self-efficacy on entrepreneurial intention.

5 CONCLUSION

The findings showed that entrepreneurial intention of private Islamic universities students in Jakarta was formed merely in having serious thoughts of business operations, yet their desire to have their own business was not too strong. Entrepreneurship Intention is directly influenced by Entrepreneurship Attitude and indirectly influenced by Entrepreneurship Education and Entrepreneurial Self-Efficacy through Entrepreneurship Attitude.

ACKNOWLEDGEMENT

The authors acknowledge and thank LP2M Universitas Al Azhar Indonesia for supporting this research.

REFERENCES

Ajzen, I. 2001. Consumer attitude and behavior. *Annals of tourism research*.vol. 28. https://doi.org/10.1016/S0160-7383(00)00072-4

Andika, M., & Madjid, I. 2012. Analisis pengaruh sikap, norma subyektif dan efikasi diri terhadap inteansi berwirausaha pada mahasiswa fakultas ekonomi universitas syiah kuala. Eco-entrepreneurship, 1(1), 190–197. https://doi.org/10.1007/978-1-4419-0143-0

Bae, T. J., Qian, S., Miao, C., & Fiet, J. O. 2014. The relationship between entrepreneurship education and entrepreneurial intentions: a meta-analytic review. *Entrepreneurship: theory and practice*, 38(2), 217–254. https://doi.org/10.1111/etap.12095

Bandura, A. 2012. On the functional properties of perceived self-efficacy revisited. *Journal of management*, 38(1), 9–44. https://doi.org/10.1177/0149206311410606

Dabale & Masese 2014, W. P., & Masese, T. 2014. The influence of entrepreneurship education on beliefs, attitudes and intentions: a cross-sectoral study of africa university graduates. *European journal of business and social sciences*, 3(9), 1–13. Retrieved from http://www.ejbss.com/recent.aspx-/

Febriyanto. 2013. Peran mata kuliah kewirausahaan terhadap minat mahasiswa berwirausaha. *Derivatif jurnal manajemen*, 7(2), 43–48. ISSN: 1978-6573

Gautam, M. 2015. Entrepreneurship education: concept, characteristics and implications for entrepreneurship eduction concept, charaterictic. *An international journal of education,* 05(january), *2231–2404.*

Hamidi, D. Y., Wennberg, K., & Berglund, H. 2008. Creativity in entrepreneurship education. *Journal of small business and enterprise development*, 15(2), 304–320. https://doi.org/10.1108/14626000810871691

Karali, S. 2013. Erasmus university of Rotterdam erasmus centre for entrepreneurship the impact of entrepreneurship education programs on entrepreneurial intentions: an application of the theory of planned behavior (master thesis). Co-reader: hendrik halbe msc erasmus cent. Erasmus school of economics Rotterdam.

Küttim, M., Kallaste, M., Venesaar, U., & Kiis, A. 2014. Entrepreneurship education at university level and students' entrepreneurial intentions. *Procedia – social and behavioral sciences*, 110, 658–668. https://doi.org/10.1016/j.sbspro.2013.12.910

Lee, W. N., Lim, B. P., Lim, L. Y., Ng, H. S., & Wong, J. L. 2012. Entrepreneurial intention: a study among students of higher learning institution. Entrepreneurial intention: a study among students of higher learning institution, (August), 15.

Linan, F., Rodríguez-Cohard, J. C., & Rueda-Cantuche, J. M. 2011. Factors affecting entrepreneurial intention levels: a role for education. *International entrepreneurship and management journal*, 7(2), 195–218. https://doi.org/10.1007/s11365-010-0154-z

Maddux, J. E. 2013. Self-efficacy theory. In self-efficacy, adaptation, and adjustment: theory, research, and application *(the springer series in social clinical psychology)* (pp. 3–33). https://doi.org/10.1007/978-1-4419-6868-5_1

Malebana, M. J., & Swanepoel, E. 2014. The relationship between exposure to entrepreneurship education and entrepreneurial. *Southern african business review*, 18(1), 1–26.

Martinez Campo, J. L. 2011. Analysis of the influence of self-efficacy on entrepreneurial intentions. *Prospect*, 9(2), 14–21. Retrieved from http://dialnet.unirioja.es/descarga/articulo/4208261.pdf

McGee, J. E., Peterson, M., Mueller, S. L., & Sequeira, J. M. 2009. Entrepreneurial self-efficacy: refining the measure. *Entrepreneurship: theory and practice*, 33(4), 965–988. https://doi.org/10.1111/j.1540-6520.2009.00304.x

Muofhe, N. J., & Du Toit, W. F. 2011. Entrepreneurial education's and entrepreneurial role models' influence on career choice. *Sa journal of human resource management,* 9(1), 1–15. https://doi.org/10.4102/sajhrm.v9i1.345

Ogundipe, S. E., Kosile, B. A., Olaleye, V. I., & Ogundipe, L. O. 2012. Entrepreneurial intention among business and counselling students in lagos state university sandwich programme. *Journal of education and practice*, 3(14), 64–72.

Oyugi , J. L. 2015. The mediating effect of self-efficacy on the relationship between entrepreneurship education and entrepreneurial intentions of university students. *Journal of entrepreneurship management and innovation,* 11(2), 31–56. https://doi.org/10.7341/20151122

Patricia, & Silangen, C. 2017. The effect of entrepreneurship education on entrepreneurial intention: an experimental study on undergraduate business students. *Journal of management research,* 9(3), 72. https://doi.org/10.5296/jmr.v9i3.11282

Pihie, Z. A. L., & Bagheri, A. 2011. Malay students' entrepreneurial attitude and entrepreneurial efficacy in vocational and technical secondary schools of malaysia. *Pertanika journal of social science and humanities*, 19(2), 433–447.

Rahmawaty, A. 2014. Model intensi kewirausahaan: peran personality traits (upaya mewujudkan kesejahteraan masyarakat) pendahuluan, (06), 103–127. Retrieved from http://repository.iainpekalongan.ac.id/160/1/7-PDF_1_Isi_Proceeding %285%29.pdf

Rengiah, D. P., & Sentosa, P. D. I. 2014. A conceptual development of entrepreneurship education and entrepreneurial intentions among malaysian university students. *IOSR journal of business and management,* 16(11), 68–74. https://doi.org/10.9790/487X-161126874

Robinson, P. B., Stimpson, D. V., Huefner, J. C., & Hunt, H. K. 2018. An attitude approach to the prediction of entrepreneurship. *Entrepreneurship theory and practice,* 15(4), 13–32. https://doi.org/10.1177/104225879101500405

Tung Lo Choi. 2011. The Impact of Entrepreneur-ship Education on Entrepreneurial Intention of Engineering Students. https://doi.org/10.1111/jsbm.12065

Turker, D., & Selcuk, S. S. 2009. Which factors affect entrepreneurial intention of university students? *Journal of european industrial training,* 33(2), 142–159. https://doi.org/10.1108/03090590910939049

Wibowo, S., & Satria Pramudana, K. A. 2016. *Pengaruh pendidikan kewirausahaan terhadap intensi berwirausaha yang dimediasi oleh sikap berwirausaha.* Manajemen, 5(12), 8167–8198.

Advances in Business, Management and Entrepreneurship – Hurriyati et al. (Eds)
© 2021 Taylor & Francis Group, London, ISBN 978-0-367-67471-7

The effect of innovation on increasing business performance of SMEs in Indonesia: Study of SMEs manufacturing industry sector in West Java, Indonesia

E. Herlinawati
Sekolah Tinggi Ilmu Ekonomi Membangun, Bandung, Indonesia

Suryana, E. Ahman & A. Machmud
Universitas Pendidikan Indonesia, Bandung, Indonesia

ABSTRACT: This study aims to analyze the influence of innovation on business performance. This research is motivated by the phenomenon of the important role of small and medium enterprises (SMEs) in national economic growth that should be accompanied by increasing business performance. However, SMEs are having the challenge of losing competition in the global market because of low innovation. The method used in this study is an explanatory survey, with data collected through questionnaires and documentation. The population of the study is SMEs in the manufacturing industry sector in West Java, Indonesia, with a sample size of 346 respondents. The data collection method used a 5-point Likert scale questionnaire, meanwhile the data analysis applied SEM AMOS. The results of the analysis show that the research model is acceptable, which means that innovation has a positive effect on business performance. The implication of this research is that business performance can be improved through product innovation, process innovation, and distribution innovation.

1   INTRODUCTION

SMEs have always been the focus of attention for economic development, economic growth, and job creation in the world (Wanambisi & Bwisa 2013). In an increasingly real and complex era of economic globalization, SMEs must be brave and ready to face the global market not only concentrated in the local market (Suryana & Bayu 2012).

The potential of SMEs currently is not matched with the ability to improve performance and the competitiveness of the global market due to various increasingly complex small business problems where the dominant problems are generally caused by the use of traditional technology, a lack of capital, and managerial aspects, including weak decision making, low quality of human resources, small business scale, lack of experience, and limited access to finance as well as SME managers/owners lacking in creativity and innovation, thus, losing to competition in both local and global markets (Zimmerer et al. 2008).

Likewise, the majority of SMEs in the manufacturing industry sector in Indonesia are still concentrated in the local market, and are not ready to face the competition in the global market. The growth of SMEs in the manufacturing industry sector is still constrained by various problems that hinder the success of SMEs in Indonesia. SME products with minimal innovation with less developed production are feared to threaten business continuity.

The difficulty of product marketing, including the lack of market information, mastery of technology, and networks, has caused SMEs to fail. This condition is strongly suspected to be triggered by the character of a weak entrepreneur, a managerial role that is not yet firm in managing the business as well as low innovation, while the business environment continues to change (Herlinawati et al. 2017, Machmud 2009). These weaknesses can have an impact on the unsuccessful implementation of entrepreneurship, while entrepreneurship is the result of discipline and the systematic process of applying creativity and innovation in meeting market needs and opportunities. The essence of entrepreneurship is the ability to create something new and different through creative thinking and innovative actions (Suryana 2014; Umar & Ngah 2014).

Changes in the business environment are very fast, so innovation becomes important for the sustainability of the company. Innovation is an indicator of the success in winning against the competition. Innovation will bring the organization into a new dimension of performance and become important for all aspects of operations and work systems and processes since innovation is part of the culture of learning (Suryana 2014). The ability to innovate is one of the most important characters of entrepreneurs (Larsen & Lewis 2007). Similarly, Craven and Piercy (2009) stated that creativity and innovation have an important role for the growth of organizational performance in the global market (Hadiyati 2011).

Schumpeter (1934) stated that innovation activities carried out continuously are the main source of long-term success of the company. Artz et al. (2010) states that a company's ability to produce innovation may be more important than ever in order to improve performance and maintain competitive advantage caused by the high level of competition and shorter product life cycles. At present, innovation has become the goal of all companies (Atalay et al. 2013).

Several previous studies have shown that innovation has a positive effect on business performance (Atalay & Sarvanc 2013; Hadiyati 2011; Price et al. 2013; Rosenbusch et al. 2011; Rosli & Sidek 2013; Welsch et al. 2013). But there are also some findings that indicate product innovation does not affect business performance (Ardyan & Putri 2015), and not all indicators of innovation affect performance, where product innovation has no effect, while process innovation, marketing innovation, and organizational innovation affect performance (Hamali & Hidayat 2017). The difference in findings of innovation and performance is a gap for researchers to conduct further research.

Based on the above phenomenon, it is necessary to do further research on improving SMEs business performance through innovation. The purpose of this study is to analyze the effect of innovation on the business performance of SMEs in manufacturing industries in West Java, Indonesia.

2 METHOD

The method used in this study was an explanatory survey to test the conceptual model that describes the relationship between the constructs of innovation and business performance. Business performance is measured by four indicators (Kaplan & Norton 1992, 1996; Neely 2004), namely financial perspective (Y1), customer perspective (Y2), internal business process perspective (Y3), and learning and growth perspective (Y4), while innovation is measured using three indicators (Fontana 2009; OECD 2005, 2010; Wang et al. 2011), namely product innovation, process innovation, and distribution innovation. The data collection used a Likert scale questionnaire ranging from 1 to 5. The population in this study were 203.181 SMEs in the manufacturing sector in West Java, Indonesia, obtained using proportional random sampling techniques, with a sample size of 346 SMEs obtained. The research questionnaire was tested first on 40 SME business actors using Pearson Correlation (r > 0.50 and sign < 0.05) and Cronbach's Alpha (0.971 and 0.726) values; all question items are valid and reliable. To test the effect of innovation on business performance, structural equation modeling (SEM) AMOS was used.

3 RESULTS AND DISCUSSION

SMEs in the manufacturing sector in West Java, Indonesia, which were the research samples, run businesses in the textile and textile products industry (Convection) 54.62% including Batik centers 6.94% and 13.87% Embroidery centers, Leather Industries and Leather goods in the form of Bags, footwear, and leather jackets 35.84%, Mendong Center 0.30%, Ceramic Center 7.51%, and Hat Craft 1.73%. The major marketing area (37%) was at the national level. Regarding business ownership status, the majority of respondents (71%) were owners and managers. The highest number of employees (78%) were between 20 and 99 people. Based on gender, the businesses were dominated by males (92%). The majority of respondents (58%) were between 46 and 55 years old, the majority of respondents (61%) possessed an undergraduate education, and the majority duration of running the business (35%) was for 16–20 years.

The innovation in SMEs in the manufacturing sector in West Java, Indonesia, tends to be low. The dimensions of product innovation were at a low level with a percentage gain of 75.15%. The low level of product innovation was due to the lack of unique product design carried out by 81.50% of respondents as well as the lack of renewal of products produced by 68.79% of respondents. The dimensions of process innovation were at a low level with a percentage of 80.68%. The low process innovation was due to the inefficiency of 88.15% of respondents in controlling inventory with business partners, also due to the reluctance of 67.24% of respondents to run production with business partners. The dimensions of distribution innovation were at a low level with a percentage of 67.48%. The low distribution innovation was due to the low frequency of online-based marketing of respondents, there were also 65.03% of respondents who have not used digital marketing as a media for the promotion and sale of their products.

Business performance in SMEs in the manufacturing sector in West Java, Indonesia, tends to be low. The financial perspective was at a low level with a percentage of 73.56%. This low dimension was due to the low sales growth of 73.70% of respondents and operating profit growth of 73.41% of respondents. The customer perspective had a growth that tends to be low with a percentage of 61.71%. This low dimension was due to the low ability of 67.92% of respondents to get new customers and the low ability of 55.49% of respondents to retain customers. The internal business process perspective tends to be low with a percentage of 68.80%. This low dimension was due to the inefficiency of 73.70% of respondents in running the company's operations. It was likewise with product development, where 63.9% of respondents did not make product changes in the past 3 years. The learning and growth perspective were at a low level with a percentage of 66.61%. This low dimension was caused by the low 76.87% of respondents who made changes in employee specific skills that has an impact on the low performance of 56.35% of respondents. The test results of the measurement model of innovation and business performance presented in Table 1 show the value of loading factor (λ) > 0.5, the value of composite reliability (CR0) above 0.7, and variance extracted (VE) above 0.5; so it can be concluded that innovation and business performance have good validity and construct reliability.

Table 1. Model of measurement of innovation and business performance.

Variable	Indicator	Λ	λ^2	E	CR	VE
Innovation	X1	1,003	1,006	−0,006	0,950	0,864
	X2	0,923	0,852	0,148		
	X3	0,856	0,733	0,267		
Business Performance	Y1	0,882	0,778	0,222	0,944	0,811
	Y2	0,754	0,569	0,431		
	Y3	0,882	0,778	0,222		
	Y4	1,059	1,121	−0,121		

λ = Loading Factor, e = error, CR = composite reliability, VE = variance extracted.
Source: SEM AMOS output, 2019.

The multivariate data normality test provides a CR (9,465) > 2.58, which means that the distribution of data is multivariate with abnormal distribution. The Mahalanobis distance test (d2) shows the value of d2 (40,262) < X2 (40,87), meaning that there are no cases of outliers. Multicollinearity test gives a condition number = 98,199 < 1000 and determinant of sample covariance matrix = 3651,975 > 0; so it can be concluded that there is no multicollinearity problem so that the data are feasible to use.

The test of the goodness of fit model presented in Table 5 shows that not all measures of the research model fit the data, but overall, the research model was fit, because GFI and AGFI $\geq$ 0.90, RMSEA $\leq$ 0.08, and NFI, CFI, and TLI $\geq$ 0.90 (Maholtra 2010).

The research findings presented in Table 6, showed the high and low business performance positively

Table 2. Assessment of normality.

Variable	Min	Max	Skew	CR	Kurtosis	CR
Y4	2,000	10,000	,203	1,542	−1,216	−4,618
Y3	2,000	10,000	,335	2,543	−1,078	−4,092
Y2	2,000	10,000	−,166	−1,263	−,857	−3,256
Y1	2,000	10,000	,523	3,974	−,976	−3,704
X1	2,000	10,000	,160	1,214	−,787	−2,989
X2	4,000	20,000	,401	3,048	−,773	−2,935
X3	4,000	20,000	−,070	−,533	−,907	−3,445
Multivariate					11,423	9,465

Table 3. Outliers data.

Mahalanobis distance (d2)		
Max	Min	X2
40,262	8,010	40,87

Table 4. Multicollinearity.

Determinant of sample covariance matrix	Condition number
3651,975	98,199

Table 5. Goodness of fit.

No.	Goodness of Fit Index	Cut-off Value	Result	Evaluation
1	Significant Prob	$\geq$ 0,05	0,000	Bad Fit
2	RMSEA	$\leq$ 0,08	0,081	Good Fit
3	GFI	$\geq$ 0,90	0,968	Good Fit
4	AGFI	$\geq$ 0,90	0,930	Good Fit
5	RFI	$\geq$ 0,90	0,969	Good Fit
6	IFI	$\geq$ 0,90	0,987	Good Fit
7	TLI	$\geq$ 0,90	0.978	Good Fit
8	CFI	$\geq$ 0,90	0,987	Good Fit
9	NFI	$\geq$ 0,90	0,981	Good Fit

Source: SEM AMOS Output.

influenced by innovation, this could be seen from the value of the path coefficient (SRW) > 0. The SRW value of 0.462 showed that innovation has an effect of $0.462^2 = 0.2134$ on performance business, which means 21.34% of the high and low variations that occur in business performance could be explained by innovation. The remaining 78.64% was the influence of other variables not explained in the model. The highest contribution of each dimension of innovation came from product innovation (X1) with 99.7% and the lowest is contributed by distribution innovation (X3) with 86.1%. While the highest achievement of business performance comes from the learning perspective and growth with 91.3% and the lowest contributed by the customer's perspective with 75.7%. The test results showed that innovation has a positive and significant effect on business performance.

The research findings showed that innovation had a positive effect on business performance. The coefficient was positive, meaning that the higher the ideal innovation will be followed by increasing business performance. This finding was in accordance with the findings Rosenbusch et al. (2011), who highlight that the relationship of innovation and performance

Table 6. Regression weights and standardized regression weight.

			RW	SRW	SE	CR	P
Business_ Performance	<—	Innovation	,267	,462	,031	8,555	***
X3	<—	Innovation	1,000	,861			
X2	<—	Innovation	1,071	,928	,041	25,905	***
X1	<—	Innovation	,612	,997	,021	29,269	***
Y1	<—	Business_ Performance	1,000	,884			
Y2	<—	Business_ Performance	,776	,757	,045	17,307	***
Y3	<—	Business_ Performance	,965	,865	,043	22,210	***
Y4	<—	Business_ Performance	1,048	,913	,044	23,868	***

in SMEs is depended on the context, such as company age, type of innovation, and culture, which influence the impact of innovation on company performance. Atalay et al. (2013), stated that innovation is an important factor because it leads to improvements in products, processes, making continuous progress that helps companies survive, allowing companies to grow faster, more efficiently, and more profitably than non-innovators. While Hamali & Hidayat (2017) found that process innovation, marketing innovation, and organizational innovation affect performance. While product innovation does not affect performance. The findings of Rosli & Sidek (2013) stated that product innovation and process innovation influence company performance significantly, where the stronger influence comes from product innovation. Welsch et al. (2013), added that past performance is a strong indicator of the results of innovation, so that future performance can be more predictable, and innovation is an important factor that has an impact on improving performance.

These findings further strengthen the concept of innovation defined by Schumpeter (1934), that to create economic growth, innovators or entrepreneurs are needed, namely people who are involved in the business world who have the enthusiasm and courage to apply new ideas to reality. Drucker (1991) highlights the need for an innovation process that is consistent with the search for change and systematic analysis of potential innovators as a source of social and economic transformation. Innovation is an important factor and is related to performance; as stated by Suryana (2014), innovation is the implementation of renewal and is important for all aspects of operations, work systems, and processes that will bring the organization into a new dimension of performance.

Referring to the results of the research and discussion described above, it can be explained that improving business performance can be done through increased innovation. Thus, the model of improving business performance (financial perspective, customer perspective, internal business process perspective, and learning and growth perspective) can be determined through innovation (product innovation, process innovation, and distribution innovation).

4 CONCLUSION

Innovation in SMEs in the manufacturing sector in West Java, Indonesia, tends to be low, and the achievement of business performance tends to be low. Innovation has a positive influence on business performance. Innovation can explain variations that occur in business performance according to the research model. The low level of innovation and low business performance, if left unchecked, will hinder the development of SMEs; chances are that SMEs will grow faster and have a smaller competitive advantage. To prevent adverse effects due to low innovation and business performance, SMEs should continually improve indicators that are perceived as low by respondents by increasing their competitive advantage in product renewal, product uniqueness, and technological renewal. The use of resources and inventory control with business partners become important in the effort toward effectiveness and efficiency. Likewise, in product distribution by utilizing offline and online media.

REFERENCES

Ardyan, E. & Putri, O.T. 2015. The positive impact of an entrepreneur who has entrepreneurial competence on the success of product innovation and business performance. *Journal of Entrepreneurship and Small and Medium Enterprises* 1(1): 11–19. ISSN 2477-2836.

Atalay, M., Anafarta, N., & Sarvan, F. 2013. The relationship between innovation and firm performance: An empirical evidence from Turkish automotive supplier industry. *Procedia Social and Behavioral Sciences* 75: 226-235.

Fontana, A. 2009. Innovate we can. Manajemen inovasi dan penciptaan nilai. Jakarta: Gramedia Widiasarana Indonesia.

Hadiyati, E. 2011. Creativity and innovation affect small business entrepreneurship. *Journal of Management and Entrepreneurship* 13(1): 8–16.

Hamali, S., & Hidayat, C. 2017. The influence of innovation dimensions on marketing and financial performance in small industries of Binong Jati Knitwear in West Java. *Journal of Banking & Management.*

Herlinawati, E., Sumawidjaja, R. N., & Machmud, A. 2017. The role of sharia microfinance in SMEs business development. *International Conference on Economic Education and Entrepreneurship. Proceeding* 3(1).

Kaplan, R.S., & Norton, D. P. 1992. The balanced scorecard-measures that drive performance. *Harvard Business Review*: 71–79.

Kaplan, R. S., & Norton, D. P. 1996. Balanced scorecard. Jakarta: Erlangga.

Kaplan, R. S., & Norton, D. P. 2006. Alignment: Using the balanced scorecard to create corporate synergies. Boston: Harvard Business School Press.

Machmud, A. 2009. Partnership model of Bandung small and medium enterprises. *Journal: Buletin Ekuitas* 2(2): 1466–1778.

Maholtra, N. K. 2010. Marketing research an applied orientation. 6th edition. New Jersey: Pearson education Inc, Prentice Hall.

Neely, A. (2004). Business performance measurement: Theory & practice. Cambridge University Press.

Price et al. 2013. The relationship between innovation, knowledge and performance, in family & non family firms: An analysis of SMEs. *Journal of Innovation and Entrepreneurship*.

Rosenbusch, N. et al. 2011. Is innovation always beneficial: A meta-analysis of the relationship between innovation and performance SMEs. *Journal of Business Venturing* 26: 441–457.

Rosli, M. M., & Sidek, S. 2013. The impact of innovation on the performance of small and medium manufacturing enterprises: Evidence from Malaysia. *Journal of Innovation Management in Small & Medium Enterprises* 1(1).

Suryana. 2014. Kewirausahaan: kiat dan proses menuju sukses, edisi keempat. Jakarta: Salemba Empat.

Suryana, Y., & Bayu, K. 2012. Kewirausahaan: Pendekatan Karakteristik Wirausahawan Sukses Ed. 2. Kencana.

Umar, A., & Ngah, R. 2014. The relationship of entrepreneurial competencies and business success of Malaysian SMEs: The mediating role of innovation and brand equity.

Welsch, H., Price, D. P., & Stoica, M. 2013. Innovation, performance and growth intentions in SMEs. *International Journal of Economics and Management Engineering* 3(5): 176.

Zimmerer, T. W., Scarborough, N. M., & Wilson, D. 2008. Essentials of entrepreneurship and small business management. Salemba Empat.

The influence of co-creation on creative industry performance in Indonesia

R.N. Sumawidjaja, S. Suryana, E. Ahman & A. Machmud
Universitas Pendidikan Indonesia, Bandung, Indonesia

ABSTRACT: This research is aimed to analyze the influence of co-creation on creative industry performance in Indonesia. This study is based on the contribution of the creative industry in the national economy. The method of the study was causal explanatory survey research. Co-creation was measured by Across Interaction, Product Options, Access and Price Experience, while Firm Performance (Financial, Customer, Internal Business Process, Learning and Growth Perspective). The population were leather industries, leather goods and foot-wear in West Java Indonesia that involved 252 respondents. All research variables were measured by using scale 1–5 with data collection techniques through questionnaires. The data was analyzed by SEM. The results show that co-creation has a significant influence on firm performance. This study implies to improve the performance of creative industry, it is necessary to improve the co-creation process such as increasing across-interactions, increasing the diversity of product options, and facilitating access to information on products.

1 INTRODUCTION

Performance is the work achieved by a person or group of persons within an enterprise in accordance with their respective powers and responsibilities in achieving the objectives of the enterprise legally, not violating the law and not contrary to morals or ethics (Veithzal & Ahmad 2005). Performance is a description of the level of achievement of task implementation in an organization, in an effort to realize the goals, mission and vision of the organization (Bastian 2001). Performance is defined as the record of outcomes produced on a specified job function or activity during a specified time period (Bernardin & Russel 1998).

The firm's performance is as successful a new product in market development, where firm performance can be measured through sales growth and market share (Pelham & Wilson 1996). Firm performance is a result made by the management on a continuous basis (Helfert 1996). To measure the performance of the firm, one of them using the concept of Balance Scorecard which consider the balance between financial performance and non-financial performance with four perspectives: financial perspective, customer perspective, internal business process perspective and learning and growth perspective (Kaplan & Norton 1996).

Co-creation is an active, creative and social process through collaboration between producers and consumers initiated by companies to create value for customers (Coates 2012). Co-creation is a medium for improvising innovation and the ability to create value in a company along with fostering customer relations (Prandelli et al. 2006; Sawhney 2006).

Co-creation is a concept expressed by Prahalad & Ramaswamy (2004) which describes a new approach to innovation that consists of creating products and experiences through collaboration with consumers, suppliers, other companies and network-connected channel partners that are profitable for innovation. According to Prahalad & Ramaswamy (2004) Co-creating is engaging the active participant in the consumption experience, with interaction with locus of co-creation value. In this study four dimensions of co-creation were expressed by Prahalad & Ramaswamy (2004), namely: across interaction, product options, access, price experience.

The relationship between co-creation and firm performance has not been done enough in research. Roser et al. (2013) suggested that all co-creation approaches have two general qualities, namely: (1) widening organizational boundaries; (2) co-creator involvement. They conclude that firm performance usually uses a collection of ideas, strategies and has its own unique approach in co-creation that is specific in its purpose to increase productivity from the performance of the firm.

Research results of Fatemeh & Naser (2017) through structural equation modeling shows that at the significance level of 0.95 T values are obtained which result in greater co-creation relationships with firm performance. So that the relationship between co-creation and firm performance is accepted. So that it can be said that the co-creation affects the performance of the firm. Similar to the results of re-search conducted by Silva et al. (2013), in a study conducted involving consumers in co-creation, the results of the research conducted showed the results that co-creation had an effect on firm performance, with the greatest influence compared to other variables in the study. Similar to the research conducted by Payne et al. (2008) with the development of innovation models, the results of the relationship show that co-creation affects the performance of the firm; Mortel (2010) found that co-creation had an effect on financial performance in companies that P500.

While Tijmes (2010), Verma et al. (2013) found from his research that there was a positive co-creation relationship with the firm's performance. Research from Kim et al. (2015) shows the results that a group of firms that fully implement co-creation address the significance of significant financial and non-financial performance. Kathryn & Bharat (2015) research shows that there is a direct and positive impact on co-creation on firm performance.

The aim of this study is analyze the relationship between co-creation and firm's performance of creative industry in Indonesia. The study of creative industry in Indonesia is focusing on contribution of creative industry to the national economy of Indonesia. This study showed a significant growth but not followed by the contribution. The contribution of the creative industry for the national economy is only 7.38 percent. Nevertheless, the creative economy is able to greet workforce of 15.9 million people (Bekraf 2016). The development of creative industry in Indonesia is still dominated by small and medium scale industry. During 2011–2015 the creative industry sub-sector leather industry, foot-wear and leather goods sector, which only average grew 0.27%. This condition shows the firm performance of this sector is stagnant. This should be a concern of the stakeholders because the growth and sustainability of the industry will involves many entrepreneurs and absorbs a lot of labor.

2 METHODS

The research method used quantitative method through causal explanatory survey research to test the correlation of co-creation with the performance of creative industry in Indonesia. The measurement of co-creation refers to dimensions due to the opinions of Prahalad & Ramaswamy (2004) namely: across interaction, product options, access, and price experience. To measured firm/company performance refers to measurement of balanced scorecard with four perspectives: financial perspective, customer perspective, internal business process perspective and learning and growth perspective (Kaplan & Norton 1996).

The research was conducted on SMEs industry players of leather, leather goods and footwear in West Java Indonesia with population size amounted to 1.571 units and with sample size referring to Isaac and Michael the sample size are 252 respondents. Characteristics of respondents based on male gender 77%, and female 23%; The majority of respondent ages ranged

from 36–45 years old 49%, 25–35 years old 5%, 46–55 years old 26%, 56–65 years old 9% and more than 65 years old 12%; Based on level of education are High Scholl 67%, Elementary School 14%, Diploma 19% and postgraduate 1%; Based on duration of business between are 16–20 years by 33%, 5–10 years 13%, 11–15 years 18%, 21–25 years 18%, and more than 25 years 17%; Based on Business turnover per year in Rupiah >350 Million–2.5 Billion by 56%, maks. 350 Million 43%, and more than >2,500 Billion 1%.

All research variables are measured using Likert scale 5–1, where the number 5 is strongly agreed and the number 1 is strongly disagree. Data collection is done through questionnaire on the perpetrators of SMEs, and data analysis model using Structural Equation Modeling (SEM) AMOS. SEM analysis is used to test the model in causal form between co-creation and Performance of creative industry in Indonesia. To answer that problem the first stage done normality test, test outliers, and multicollinearity , and then to check the level of compatibility between data and model, validity and reliability, the measurement model and the coefficient significance of the structural model were performed fit tests. The suitability of the measurement model is carried out against each construct by looking at the relationship between latent variables and some indicators through validity and reliability test of the measurement model.

3 RESULTS AND DISCUSSION

Normality tests were performed using critical ratios skewness value and kurtosis at a 0.01 level of significance. The data used are said to be normally distributed when c.r skewness and kurtosis all manifest variables ≤ 2.58 Multivariate test results $8.9497 > 2.58$. This indicates that all indicators are not normally distributed. Furthermore, Mahalanobis distance (d2) test is used to test the possibility of multivariate outliers at 0.001 and df = number of observed variables. It says there is no case of outliers if the value of d2 < X2. Test results show d2 (31.2484) < X2 (50.7955) meaning there is no case of outliers. Last is Multicolinearity can be seen through the determinant of covariance matrix. The determinant value must be greater than 0. The very small determinant value indicates that there is a multicollinearity problem so the data cannot be used for research. The AMOS SEM output results in a value determinant = 2.0554 > 0. So it can be concluded that there is no multicollinearity problem. Based on the test results can be seen that the data is normally distributed, there are no cases of outliers and the da-ta set of the sample empirically meet the main statistical assumption that there is no problem multicollinearity. Thus it can be concluded that the sample data sets deserve to be used in further analysis.

Test results of validity cshows significant and reliability indicating that CR co-creation 0.913 > 0.70 and AVE value 0.725 > 0.50. CR for Firm performance

665

Table 1. Estimation and parameter test of structural model.

			RW	SRW	S.E.	C.R.	P
Firm_Performance	<—	Cocreation	.9549	.5962	.0958	9.9625	***
X4	<—	Cocreation	1.0000	.8808			
X3	<—	Cocreation	.9669	.8400	.0561	17.2321	***
X2	<—	Cocreation	1.1240	.8284	.0660	17.0224	***
X1	<—	Cocreation	1.1872	.8558	.0680	17.4464	***
Y1	<—	Firm Perform.	1.0000	.9520			
Y2	<—	Firm Perform	.7896	.8529	.0366	21.5551	***
Y3	<—	Firm Perform	.8910	.8188	.0462	19.2823	***
Y4	<—	Firm Perform	1.0886	.9203	.0389	27.9731	***

0.936 > 0.70 and AVE value 0.787 > 0.50. It can be concluded that the measurement model has adequate validity and reliability to measure co-creation and firm performance.

The overall model fit test is performed to evaluate generally the Goodness of Fit between the data and the model. Examination result of Model Fit: RMSEA (0.07) < 0.08; CFI (0.9698) > 0.90; TLI (0.9555) > 0.90 and GFI (0.9343) > 0.90, overall the model is fit as stated by Maholtra (2010), that: (1) Use at least one size that is either absolute (eg GFI, AGFI). In this model GFI above 0.90, thus can be interpreted model on Fit condition. (2) Use at least one size that is absolute bad (eg Chi-square, RMSR, SRMR, RMSEA). In this model RMSEA < 0.08 means the model under Fit condition. (3) Use at least one comparative size (eg NFI, NNFI, CFI, TLI, RNI). In this model CFI and TLI & gt; 0.90, so it can be interpreted model on Fit condition.

Based on Table 1. it can be concluded that the magnitude of the influence of co-creation on firm performance is equal to (R2) 0.59622 = 0.3503 which means that the firm's performance is deter-mined at 35.03% by co-creation, the remaining 64.97% is determined by other factors.

The low co-creation has an impact on the low performance of firms in the creative industries SMEs sub-sector of leather, leather goods and footwear in West Java. The positive influence of co-creation on the firm's performance in the creative industry shows that co-creation can explain the variations that occur in the firm's performance in accordance with the re-search model. The low ability of co-creation and firm performance must be anticipated through across in-teraction with increasing community involvement in producing products, product option with the diversi-ty of product choices, access with ease of product formation, so that creative industry SMEs are able to retain customers and reach new customers, who will affect the increase in financial and non-financial per-formance.

REFERENCES

Bastian, I. 2001. *Akuntansi sektor publik di Indonesia*. Yogyakarta: BPFE.

Bekraf, B. 2016. *Database peta kegiatan dalam negeri bekraf 2016–2018*.

Bernardin, J.H. & Russel, J.A. 1998. *Human resource management: An experiental approach*. Mc Graw-Hill.

Coates, N. 2012. *The meaning of co-creation*. London UK.

Fatemeh, H. & Naser G. 2017. Impact of Co-creation on innovation capability and firm performance: A Structural Equation Modeling. *AD-minister* 30: 73–90.

Helfert, E.A. 1996. *Teknik analisis keuangan: petunjuk praktis untuk mengelola dan mengukur kinerja perusahaan, edisi kedelapan*. Jakarta: Erlangga.

Kaplan, R.S. & Norton, D.P. 1996. *The balanced scorecard: translating strategy into action*. London: Harvard Business Press.

Kathryn, B. & Bharat, N. 2015. Co-creation of value in digital ecosystems: a conceptual framework. *AIS Elibrary (AISeL)-AMCIS 2015*.

Kim, D.W., Lee, S.M., Hong, S.G. & Kim, J.W. 2015. The impact of co-creation implementation on the performance of small and medium manufacturers: an empirical study. *The Journal of Information Systems* 24(4): 1–19.

Maholtra, N.K. 2010. *Marketing research an applied orientation. 6th edition*. New Jersey: Pearson education Inc, Prentice Hall.

Mortel, V.D.J.L. 2010. The value of co-creation. *Master thesis. School of Business and Economics, Maastricht University*.

Payne, A.F., Storbacka, K. & Frow, P. 2008. Managing the co-creation of value. *Journal of the Academy of Marketing Science* 36(1): 83–96.

Pelham, A.M. & Wilson, D.T.1996. A longitudinal study of the impact of market structure, firm structure, strategy and market orientation culture on dimensions of small-firm performance. *Journal of the Academy of Marketing Science* 24(1): 27–43.

Prahalad, C.K. & Ramaswamy, V. 2004. Co-creating unique value with customers. *Strategy & Leadership* 32(3): 4–9.

Prahalad, C.K. & Ramaswamy, V. 2004. *The future of competition: co-creating unique value with customers*. Boston, MA: Harvard Business School Press.

Prandelli E.G., Verona, V. & and Raccagni, D. 2006. Diffusion of web-based product innovation. *California Management Review* 48(4): 109–135.

Roser, T., DeFillippi, R. & Samson, A. 2013. Managing your co-creation mix: co-creation ventures in distinctive contexts. *European Business Review* 25(1): 20–41.

Sawhney, M. 2006. Defining, designing, and delivering customer solutions. *The service-dominant logic of marketing: Dialog, debate, and directions*: 365.

Silva, F.J., Camacho, M.A. & Vázquez, M. 2013. Heterogeneity of customers of personal image services: a segmentation based on value co-creation. *International Entrepreneurship Management Journal* 9: 619–630.

Tijmes, A.H. 2010. Co-creation and firm performance: innovation success enhancing effects of anf motives costumer involvement. *University of Twente, the Netherlands.*

Veithzal, R, & Ahmad, F.M.B. 2005. *Performance appraisal; sistem yang tepat untuk menilai kinerja karyawan dan meningkatkan daya saing perusahaan edisi 1.* Jakarta: Raja Grafindo Persada.

Verma, R., Rajagopal, R. & Mercado, P.R. 2013. Impact of service co-creation on performance of firms: the mediating role of market oriented strategies. *International Journal of Services and Operations Management* 15(4): 449–466.

The role of socio-economic background, family economic education, and financial literacy on student decision making

Tetep, E. Mulyana, E. Dimyati & Maskur
Institut Pendidikan Indonesia, Garut, Indonesia

ABSTRACT: This research aims to determine the role of socio-economic background, family economic education, and financial literacy on student decision making. The research method employed a quantitative approach. The theory was analyzed by using deductive logic. The results of the deductive logic were then transformed into the research hypothesis with the measurement and operational concepts. The population in this study were students of the Faculty of Social Sciences, Language, and Literature Education of Institut Pendidikan Indonesia, Garut. Proportional random sampling was employed as the sampling technique. For the research instruments, polls and tests were used. In addition, documentation and polls were used as the data collection techniques. The results showed that (1) the socioeconomic background of parents is very important and has a significant effect on students' decision-making process. (2) The family economic education is very important and has a significant effect on student decision making. (3) Financial literacy is not considered to be notable and has no effect on student decision making. (4) There are simultaneous roles and influences between socio-economic background, family economic education, and financial literacy on student decision making.

1 INTRODUCTION

Students who have good financial literacy will be cautious and selective in spending their money, so that they are more rational and do not behave in a consumptive way. Conversely, if they are lacking financial literacy, they will tend to be extravagant and it will lead to consumptive behavior. PISA (2012) states that financial literacy can encourage a change in the behavior of people in a more positive direction to spend their money. As such, a person with good financial management will limit self-indebtedness to consumptive interest and save their money for better welfare.

Chen and Volpe (1998) argue that students with a low level of knowledge will make wrong decisions in their financial activities. Students who have a low level of literacy skills will make the wrong decision to consume because, in this case, they won't take into account which goods or services they need first (priority needs). The cause of difficulty in taking decisions for students, besides the low level of literacy, is that they have learning difficulties. According to Shuang Li (2016), difficulties in learning means the condition faced by students where the competency achieved is not in accordance with the predefined criteria.

In fact, most children are unable to be accountable for their money. Consequently, the induplication of them becomes uncontrolled and becomes wasteful. According to Bamforth & Geursen (2014), family is the most dominant external factor. Families, especially parents, have an important role in guiding and

influencing a child's financial skills, knowledge, and financial behavior. Based on the descriptions stated above, the researchers are interested in conducting research on the role of socio-economic background, family economic education, and financial literacy toward student decision making.

2 METHOD

This research employed a quantitative approach with deductive logic, then, derived research hypotheses with measurement and operational concepts. Cresswell (2008) states that a quantitative study, consistent with the quantitative paradigm, is an inquiry into a social or human problem, based on testing a theory composed of variables, measured with numbers, and analyzed with statistical procedures, in order to determine whether the predictive generalizations of the theory hold true.

The statistics and types of research used were exploitation research. In this research, the whole student population of the Faculty of Social Sciences, Language, and Literature Education of Institut Pendidikan Indonesia, Garut, were treated as the population. The technique undertaken to take samples in this study was proportional random sampling, which is a random way of taking members of the population without regard to the tiers in members of the population. The research instruments employed polls and tests. Meanwhile, the data collection techniques employed documentation and polls.

3 RESULTS AND DISCUSSION

3.1 *Description of socio-economic background variable.*

The results were obtained from the questionnaire. In order to facilitate the determination of the classification of conditions, the length of the interval class through the highest score of 5 (the highest question value) is multiplied by 7 (number of questions), resulting in the highest score of 35. Likewise, the lowest score of 1 (lowest value) is multiplied by 7 (number of questions), resulting in the lowest score of 7. A more detailed overview of the socio-economic background can be seen in Table 1.

Table 1. Description of socio-economic background variable data.

No	Interval	Criteria	Frequency	Percentage
1	25–34	Medium to Top	33	26%
2	16–24	Medium	68	54%
3	7–15	Medium to Low	26	20%
Total			127	100%

According to Table 1 above, it is known that the socio-economic background of the students, namely (a) the medium to top category is composed of 33 students or 26%, (b) the medium category of 68 students or 54%, and (c) the medium category of 26 students or 20%. Thus, it can be concluded that the socio-economic background of the students is mostly in the medium level category.

3.2 *Description of family economic education variable*

The results were obtained from the questionnaire. In order to facilitate the determination of the classification of conditions, the length of the interval class through the highest score of 5 (the highest question value) is multiplied by 10 (number of questions), resulting in the highest score of 35. Likewise, the lowest score of 1 (lowest value) is multiplied by 10 (number of questions), resulting in the lowest score of 10. A more detailed overview of the family economic education can be seen in Table 2.

According to Table 2, it is known that the family economic education of students, namely (a) the category of "always" is composed of around 42 students or 33%, (b) the category of "often" of around 53 students or 42%, (c) the category of "sometimes" of around 25 students or 20%, (d) the category of "rarely" of around 7 students or 6%, and (e) the category of "never" of

Table 2. Description of family economics education variable data.

No	Interval	Criteria	Frequency	Percentage
1	43–50	Always	42	33%
2	35–42	Often	53	42%
3	27–34	Sometimes	25	20%
4	19–26	Rarely	7	6%
5	10–18	Never	0	0%
Total			127	100%

0 students or 0%. Thus, it can be concluded that the family economic education variable of the students is mostly in the often category.

3.3 *Description of financial literacy variables*

The results were obtained from the questionnaire. In order to facilitate the determination of the classification of conditions, the length of the interval class through the highest score of 2 (the highest question value) is multiplied by 13 (number of questions), resulting in the highest score of 26. Likewise, the lowest score of 1 (lowest value) is multiplied by 13 (number of questions), resulting in the lowest score of 13. A more detailed overview of the family economic education can be seen in Table 3:

Table 3. Description of financial literacy variables data.

No	Interval	Criteria	Frequency	Percentage
1	22–26	High	32	25%
2	18–23	Medium	59	46%
3	13–17	Low	36	28%
Total			127	100%

According to Table 3 above, it is known that the family economic education, namely (a) the high category is composed of around 32 students or 25%, (b) the medium category is composed of around 59 students or 46%, (c) the low category is composed of around 36 students or 28%. Thus, it can be concluded that the family economic education variable of the students is mostly in the medium category.Description of students' decision-making process in consumption variable

The results were obtained from the questionnaire. In order to facilitate the determination of the classification of conditions, the length of the interval class through the highest score of 5 (the highest question value) is multiplied by 20 (number of questions), resulting in the highest score of 100. Likewise, the lowest score of 1 (lowest value) is multiplied by 20 (number of questions), resulting in the lowest score of

20. A more detailed overview of the family economic education can be seen in Table 4:

Table 4. Description of students' decision-making process in consumption variable data.

No	Interval	Criteria	Frequency	Percentage
1	85–100	Very Agree	13	10%
2	69–84	Agree	89	70%
3	53–68	Hesitant	24	19%
4	37–52	Less Agree	1	1%
5	20–36	Disagree	0	0%
Total			127	100%

According to Table 4 above, it is known that the students' decision making in consumption variable data, namely (a) the category of "very agree" is composed of around 13 students or 10%. (b) The category of "agree" is composed of around 89 students or 70%, (c) the category of "hesitant" is composed of around 24 students or 19%, (d) the category of "less agree" is composed of around 1 student or 1%, and (e) the category of disagree is composed 0 students or 0%. Thus, it can be concluded from the result that the student is said to agree to behave rationally in decision-making consumption.

3.4 The role of socio-economic background on students' decision making in consumption

The role of parents' socio-economic background on the decision making in consumption of students of the Faculty of Social Sciences, Language, and Literature Education of Institut Pendidikan Indonesia, Garut, is examined. Regarding consumption, it can be seen from the regression coefficient in the X1 variable, which obtained the number 0.221. Therefore, the role of the socio-economic background of parents on students' decision making in consumption is 0,221. As for the continuity test, the multiple linear regression coefficient for the socio-economic background variable that is obtained is 2,022 > 1.979 with a significance level 0,045 < 0.05. It can be concluded that H0 is rejected and Ha is accepted, which interpreted a significant positive influence from the role of the socio-economic background of parents on the decision making of students in consumption.

Based on the poll, it indicates that the relationship between the social background of the students' parents is mostly found in the middle level with a number of 68 students or 54%, this number is greater than the upper intermediate level, which amounted to 33 students or 26%. Thus, students with a background in the middle-level socio-economic tend to have rational consumption.

3.5 The role of family economic education on students' decision making in consumption

The role of family economic education on the students' decision-making in consumption can be seen from the regression coefficient in the variable X2, which obtained the score of 0.668. Thus, the role of the family economic education on students decision-making process in consumption amounted to 0.668. As for the continuity test, the multiple linear regression coefficient for family economic education variables that is obtained is 7,290 > 1.979 with a significance level of 0,000 < 0.05. It can be concluded that H0 is rejected and Ha is accepted, which interpreted that there was a significant positive influence of the role of family economic education on the decision making in the consumption of students of the Faculty of Social Sciences, Language, and Literature Education of Institut Pendidikan Indonesia, Garut.

Based on the poll, it indicates that the family economic education in the students is mostly in the middle level with the number of 42 students or 33%, this number is greater than the upper middle level, which amounted to 25 students or 20%, and also the category rarely is composed of only 7 students or 6%. While the category of never is composed of 0 students.

3.6 The role of financial literacy on students' decision making in consumption

The role of financial literacy on the decision-making process of students in the Faculty of Social Sciences and Literature, Institut Pendidikan Indonesia, Garut, can be seen from the regression coefficients in X3 variables, which obtained the score of 0.160. Thus, the role of financial literacy on the decision-making process with student consumption is 0.160. Meanwhile, tests of linear regression coefficients for the Multiple Financial literacy variables show that t-count < t-table, which is 0.726 < 1,979, with a significance level of 0.469 > 0.05. It can be concluded that Ho is accepted and Ha is rejected, which can be interpreted that there is no significant positive influence of financial literacy on the students' decision-making process in consumption.

Based on the poll, it indicates the financial literacy rate of the 127 students. The sum of the literacy rate that belongs to the high category is only 25%, this number is less than the student financial literacy rate, which is categorized at low with 28%.

4 CONCLUSION

Based on the results, it can be concluded that:

a) Parental background plays a significant role and effect on student decision making in consumption; this is demonstrated from the analysis of the results that the probability (Sig) magnitude of 0.045 is smaller than 0.05 and t-count (2,022) > (1.979), which means that Ho is rejected. From these results, it can be concluded that the social socio-economic background can affect students in the decision-making process.

b) Family economic education plays a significant role and effect on the decision making of student consumption; this is demonstrated from the analysis result that the probability (Sig) magnitude of 0.000 is greater than 0.05. And t-count (7.290) > t-table (1,979), so Ho is rejected. From these results, it can be concluded that a high family economic education can be influential in the students' decision-making process.

c) Financial literacy plays a significant role and effect on the decision making of student consumption. This is shown from the analysis results of the t-test that the probability of Sig. of 0.469 is greater than 0.05. And t-count (0.726) < t-table (1,979), so Ho is accepted. From these results, it can be concluded that high financial literacy cannot be influential on the decision making of student consumption.

d) Simultaneously, the socio-economic background variable (X1), family economic education (X2), and financial literacy (X3) play roles in students' decision making in consumption. It is shown from the results of the analysis, namely Sig. F (0.000) < 0.05, then Ha was received with the equation $Y = 40.741 + 0.221\ X1 + 0.668\ X2 + 0.160\ X3 + \mu$; then it can be inferred that socio-economic background, family economic education, and financial literacy increased student decision making.

ACKNOWLEDGEMENT

Thanks to the students of the Faculty of Social and Language at Institut Pendidikan Indonesia, Garut, for the collaboration and all the researchers at Institut Pendidikan Indonesia, Garut, for supporting this research.

REFERENCES

Bamforth, J., & Geursen, G. M. 2014. Categorising the money management behaviour of young consumers. *Young Consumers*. 10.1108-00658.

Chen, H & Volpe, R. P. 1998.An analysis of personal financial literacy among college students. *JAI Press Inc*. 7(2): 107–128.

Cresswell, J. W. 2008. Educational research: Planning, conducting, and evaluating quantitative and qualitative research (3rd ed.). Upper Saddle River,NJ: Merril..

Program for International Student Assessment (PISA). 2012. Financial literacy assessment framework. Amerika: international network on financial education OECD.

Shuang Li. 2016. A case study on learning difficult and corresponding supports for learning in cMOOCs. *Journal of Learning and Technology*, 42(2): 12.

Advances in Business, Management and Entrepreneurship – Hurriyati et al. (Eds)
© 2021 Taylor & Francis Group, London, ISBN 978-0-367-67471-7

Homecoming phenomenon in West Java society: Socio-cultural system and entrepreneurial motivation

Tetep, A. Suherman & L. Dianah
Institut Pendidikan Indonesia, Garut, Indonesia

ABSTRACT: The homecoming phenomenon is an annual tradition carried out by Indonesians, usually prior to Eid al-Fitr, Eid al-Adha, Christmas, and New Year. This phenomenon is an attraction for the development of socio-cultural and economic systems of society from time to time. This study revealed the phenomenon of homecoming to the development of socio-cultural systems, entrepreneurial motivation, and the influence of these two variables. The research sample amounted to 55 people who were selected randomly from hawkers found along the Eid (Eid al-Fitr) homecoming flow. By using a quantitative approach to test partially and simultaneously between variables, the data were processed using SPSS analysis. The research findings indicated that the Eid al-Fitr homecoming phenomenon was an annual phenomenon that became a tradition of people particularly in West Java. Partially the homecoming phenomenon affected the socio-cultural system in society, especially in strengthening relationships and kinship visitation. Another factor affecting the encouragement and desire for entrepreneurship of the community was the number of hawkers found along the congestion was increasing. Thus, the homecoming phenomenon had significant impact on the development of the socio-cultural system and motivation for community entrepreneurship.

1 INTRODUCTION

Homecoming is a socio-cultural and religious tradition for Indonesian people, which is carried out annually, especially preceding the Eid al-Fitr. This homecoming phenomenon is an exciting event for the Indonesian people whose birth lands or ancestral lands are located in the rural areas and these people leave for big cities to make a living and do economic activities in the city. In Indonesia, big cities are mostly inhabited by immigrants who are "newcomers" (Abeyasekere 1989) or another term is immigrants from rural areas (see Evers & Korff 2000; Jellinek 1991; Somantri 2007). The homecoming period is an opportunity for people to visit their hometown as well as to express their longing for their relatives, friends, and former neighbors in their village. According to Somantri (2007), currently homecoming has switched its orientation, where residents of big cities return to their hometown for reasons, such as (1) family recreation in a kinship atmosphere, (2) practical and efficient extensive family meetings held at the right time socio-culturally, (3) social and economic lobbying within the framework of strengthening and expanding social capital.

Homecoming has become a socio-cultural order even now turning into socio-economic phenomena in the society. On the one side, the homecoming becomes a tradition that enriches the socio-cultural order of the Indonesian people; on the other side, homecoming

has also increased the society's economic activities. According to Mulder, homecoming is part of internal migration. In the socio-cultural context, the homecoming tradition is part of nonpermanent migration. The migration is the movement of people to settle from one place to another and surpass the area, political boundaries, or the borders of other countries (Mantra 1984; Ravenstein 1885; Shaw 1991; Wirawan 2006). Migration has the potential to create a new socio-cultural system in society. Migration also changes the paradigm of development and economic growth in society. Bacotang et al. (2016) state that migration is actually a global change in the economic, social, cultural, and political situation. According to Purwakanata et al. (2011), homecoming has triggered economic excitement of almost Rp. 84.9 trillion for Dhu'afa wallet (an Islamic philanthropy organization that is devoted to empowering the poor through compassionate social technopreneurship) money turnover, of which 56% is used for accommodation, tourism, and alms, while 44% is used for transportation, food, and souvenirs. The homecoming tradition increased the distribution of money for villagers and it moves the wheels of the Indonesian economy on a global scale (Irianto 2012).

Referring to this homecoming phenomenon, the researchers are interested in examining the tradition of the homecoming phenomenon in relation to the socio-cultural system and the entrepreneurial motivation of society, particularly in West Java. The socio-cultural environment of society, which consists of attitudes, interactions, perceptions, life goals, and the desire to survive, can promote the growth of entrepreneurial motivation. Business instincts arise due to the interaction of the social environment among family, the community, and society (see Adeleke et al. 2003; Robaro & Mamuzo 2012; Sairin 2002). Therefore, homecoming activities as a socio-cultural tradition can contribute to the developing growth of the urge for becoming an entrepreneur among the society.

2 METHOD

In studying the homecoming phenomenon, especially related to the socio-cultural order and entrepreneurial motivation of society, this study used a survey method with quantitative data (Creswell 1994). The survey was conducted on a community of hawkers who took advantage of the congestion during Eid al-Fitr homecoming activities. Data were collected through direct interviews with the formatting of questions in a closed-ended interview on a scale of 1 and 2 due to very limited time since the interviews were conducted when the hawkers were resting and the researchers asked for their time for interviews. The collected quantitative data were then processed by statistical analysis using SPSS analysis to see the partial influence of homecoming phenomenon variables on the socio-cultural order of society, homecoming variables toward entrepreneurial motivation, and the joint influence of homecoming variables on the socio-cultural order and entrepreneurial motivation.

3 RESULTS AND DISCUSSION

Based on data analysis, the results of this study can be described based on the focus of the problems studied in this study as follows:

3.1 Homecoming Phenomenon in West Java

According to the data collected through interviews with hawkers on the streets between Purbaleunyi Toll Road to Garut, West Java, an explanation of the homecoming phenomenon was obtained. All respondents (100% or 55 respondents) provided the perception that homecoming was an annual tradition of the society, especially during Eid al-Fitr. Homecoming became the culture of the people in West Java and Indonesia, since the people who did homecoming activities generally came from rural areas or their family members came from rural areas. Therefore, the Eid holidays, which tended to be long, were often used for homecoming activities and gathering with their families in the rural areas (99% of respondents stated).

Another reason was that the homecoming activity was closely related to religious rituals to forgive each other after one month of fasting (stated by 95%

of respondents). The Eid al-Fitr homecoming was a moment of victory so people wanted to welcome this moment with euphoria and by gathering with distant family members. It could be interpreted that homecoming was a tradition that became a socio-cultural system of society in Indonesia, especially for Muslims, socially as a media to strengthen ties between family and either close or distant relatives. In terms of culture, homecoming was an annual custom that raised a new cultural system, where, after returning to their hometown, many people also took new family members back to the city where they lived or worked. This presented and produced new cultures and customs as well as new social systems for both new members and the environment they visited.

Related to business activities, the phenomenon of homecoming increased and encouraged the entrepreneurial motivation of the respondents. All respondents or 100% (55 respondents) stated that this homecoming phenomenon encouraged them to be hawkers who sold things or food on the street. The number of hawkers who were accustomed to selling things or food increased by around 45%. Out of the 55 respondents, 27 of them were just selling things or food during the congestion. Thus, homecoming activities could produce a new socio-cultural system in people's lives and stimulated the desire for entrepreneurship, such as selling things or food during the homecoming congestion.

3.2 Correlation between Homecoming and Socio-Cultural System in Society

Based on the results of quantitative data, which were processed and analyzed using SPSS analysis, the following data were obtained:

Table 1. Correlation between variables of homecoming phenomenon and the society's socio-cultural system in West Java.

Correlations

		Hometown	Soscul
Hometown	Pearson Correlation	1	.461**
	Sig. (2-tailed)		.000
	N	55	55
Soscul	Pearson Correlation	.461**	1
	Sig. (2-tailed)	.000	
	N	55	55

**. Correlation is significant at the 0.01 level (2-tailed).

Table 1 indicated that there was a very significant correlation between the homecoming phenomenon as variable X to the socio-cultural system of society with a correlation of 461 with a significance value of 000 from the number of respondents (as many as 55 people). Thus, it could be stated that the correlation of the homecoming phenomenon to the socio-cultural system correlated significantly (high) with a correlation of 46% at the 0.01 level of significance (2-tailed). Hence, it could be stated that the homecoming phenomenon contributed to the development and even changes in the socio-cultural system of society in West Java. However, the essence of the homecoming socio-cultural system was maintained, and changes only occurred in the lifestyle brought by people from the city they had migrated to or the workplace environment they had worked at.

Table 2. Correlation between variables of homecoming phenomenon and entrepreneurial motivation of society in West Java.

Correlations

		Hometown	Entrepreneur
Hometown	Pearson Correlation	1	.301*
	Sig. (2-tailed)		.025
	N	55	55
Entrepreneur	Pearson Correlation	.301*	1
	Sig. (2-tailed)	.025	
	N	55	55

*. Correlation is significant at the 0.05 level (2-tailed).

Referring to Table 2, it could be explained that the homecoming phenomenon correlated significantly with the entrepreneurial motivation of society who took advantage of the moment of congestion during Eid al-Fitr. The significance of the correlation was indicated by the Pearson correlation of 301 at the significance level of 0.25. Based on this significance, it could be stated that the variable correlation value of the homecoming phenomenon and the entrepreneurial motivation was at an average position of 30.1%. The homecoming phenomenon increased the desire and encouragement of the society to become entrepreneurs, even though some respondents only became temporary hawkers during the homecoming periods. However, most of respondents became entrepreneurs as daily business habits, and the desire for business got stronger in the homecoming periods. The types of commodities sold by respondents were relatively varied, but especially included drinking water and snacks, which served as a substitute for food during congestions. The food being sold were such as tofu, crackers, mineral water, coffee, fried food, fruits, and other food. For travelers, the situation of congestion was considered a problem, but for hawkers this situation became a blessing as a profit-making business.

Based on the results of the ANOVA calculations shown in Table 3 on the simultaneous influence of factors of homecoming phenomenon on the socio-cultural system and entrepreneurial motivation, the

Table 3. The simultaneous influence of the homecoming phenomenon on the socio-cultural system and entrepreneurial motivation of society in West Java.

ANOVA^b					
Model	Sum of Squares	Df	Mean Square	F	Sig.
1 Regression	32.388	2	16.194	8.252	.001^a
Residual	102.048	52	1.962		
Total	134.436	54			

a. Predictors: (Constant), Entrepreneur, Soscul.
b. Independent Variable: Homecoming.

obtained data were that the F-count value was 8.252 at a 0.01 significance level; while the F-table value for N = 55 was 4.02. Thus, F-count > F-table indicated that the hypothesis was accepted or F-count 8.252 > F-table 4.02 indicated that Ho was rejected and Ha was accepted. This could be interpreted as the factors of the homecoming phenomenon simultaneously affected the socio-cultural system factors and entrepreneurial motivation of the hawker community in West Java, Indonesia. Hence, the tradition of the annual homecoming phenomenon at Eid al-Fitr contributed to the socio-cultural system and also had a significant influence on increasing the entrepreneurial motivation in the hawker community in West Java.

4 CONCLUSION

The phenomenon of homecoming for Eid or Eid al-Fitr is an annual tradition that has become a socio-cultural system, which is inherent particularly among West Java societies, and, in general, this phenomenon also occurs in all regions of Indonesia. As the research findings revealed that aside from being a challenge to changes in the social and cultural systems of society, the homecoming phenomenon also provided a positive contribution to the development of socio-cultural systems and increased motivation for entrepreneurship. This was due to the more congested traffic jams that occurred on the street, more and more homecoming communities who rested in the rest area chatting with other travelers, and more and more people becoming hawkers and moving along congested streets to sell things just to take advantage of the homecoming moments. Therefore, the homecoming phenomenon will always continue to be an interesting topic to be studied in terms of sociological, anthropological, geographical, and other developments.

5 ACKNOWLEDGEMENT

Thanks to the respondents for supporting and collaborating and all researchers at Institut Pendidikan

Indonesia (IPI), Garut, for the collaboration to support this research.

REFERENCES

Abeyasekere, S. 1989. Jakarta a history. OVP, Singapore.

Adeleke, A., Oyenuga, O. O. & Ogundele, O J. K. 2003. Business policy and strategy. Mushin, Lagos: Concept Publications Limited.

Bacotang, Abustam, Hijjang, & Manda, D. 2016. Migration and economic changes: Sociological analysis on the contributions of Bugis Ethnic for the economy of Kupang. *Mediterranean Journal of Social Sciences* 7(2). ISSN 2039-2117 (online) ISSN 2039-9340 (print).

Creswell, J. W. 1994. Research design qualitative & quantitative approaches USA: Sage Publication.

Evers, H. D., & Korff, R. 2000. Southeast Asian urbanism: The meaning and power social space. New York: ST. Martin's Press, Inc.

Irianto, A. M. 2012. Mudik dan keretakan budaya. *Jurnal Humanika* 15(9)

Jellinek, L. 1991. The wheel of fortune: The history of poor community in Jakarta. Sydney: Allen & Unwin.

Mantra, I. B. 1984. Migrasi penduduk Indonesia Yogyakarta: PPK – UGM.

Ravenstein, E. G. 1885. The laws of migration. *Journal of Statistical Society of London* 48(2): 167–235.

Robaro, A., & Mamuzo, M. O. 2012. The impact of socio-cultural environment on entrepreneurial emergence: A theoretical analysis of Nigerian Society. *European Journal of Business and Management* 4(16). ISSN 2222-2839 (Online) ISSN 2222-1905 (Paper)

Sairin, S. 2002. Perubahan sosial masyarakat Indonesia. Yogyakarta: Pustaka Pelajar Offset.

Shaw, R. P. 1991. Migration theory and fact Pennsylvania: Regional Science Research Institute/RSRI.

Somantri, G. R. 2007. Kajian sosiologis fenomena mudik. Universitas Indonesia.

Advances in Business, Management and Entrepreneurship – Hurriyati et al. (Eds)
© *2021 Taylor & Francis Group, London, ISBN 978-0-367-67471-7*

Entrepreneurial environment in entrepreneurial motivations: An exploration study

B.L. Nuryanti, R.D.H. Utama & R.A. Tammie
Universitas Pendidikan Indonesia, Bandung, Indonesia

ABSTRACT: Researches on entrepreneurial motivation level have been done by many researchers. Research on entrepreneurial motivation is important for research to find the motivation and behavior, thus, helping to understand how they carry out their daily activities. This study aims to determine the effect of entrepreneurial environment on entrepreneurial motivation. This research uses a descriptive approach with an explanatory survey method. The respondents are 167 people. The data collection was done with the use of a questionnaire. The analysis technique was a verification technique by using frequency distribution; the results showed that the entrepreneurial environment effected entrepreneurial motivation enough. The differences in this study are located in the object of research, research time, measuring instrument, the literature used, the theory that was used, and the results of the study.

1 INTRODUCTION

Entrepreneurship education has long been used as a pillar of the economy in the face of economic and social changes made by the Indonesian government. The high motivation of persons in entrepreneurship is pushing up the number of entrepreneurs in Indonesia. The determinant key of business success includes the motivation of entrepreneurship (Eijdenberg 2016), risk-taking, and their interest to achieve business success as the focus of entrepreneurial motivation (Ismail et al. 2016). Research in various fields has shown how people are motivated to start entrepreneurship (Knight 2015). Research on entrepreneurial motivation is important for research to find the motivation and behavior, thus, helping to understand how they carry out their daily activities in accordance with the environment, and to understand how they predict the progress of small businesses and medium-sized ones (Eijdenberg 2016).

Researches on entrepreneurial motivation level have been done by many researchers. The problems of entrepreneurial motivation level become good research ideas in science education as well as practical (Boluk & Mottiar 2014) along with understanding the attitude factor that supports entrepreneurship and facilitates students' motivation in entrepreneurship to encourage students toward entrepreneurship (Dehkordi et al. 2012).

Some of the studies said that students are willing to attend training in which there is a tendency toward entrepreneurship curriculum and motivation for entrepreneurship (Sarmento 2016). Training and

education influence the behavior and attitude of students in the entrepreneurial educational environment (Mulyadi 2010).

Academic and non-academic activities can build entrepreneurial motivation in students (Kingdom et al. 2014). Entrepreneurial learning can encourage students to be more motivated toward entrepreneurship. It is an example of academic activities. Schools will be able to create entrepreneurial motivation, born of self-esteem through the development of an entrepreneurial environment to form a more powerful motivation than the motivation that is formed because of environmental or external factors.

Innovative and creative learning models are expected to manage and develop the learning components in a design that is planned by taking into account the actual conditions of the contributing factors in the implementation of learning to be done. To enhance student creativity and interest in learning, the necessary facilities to help to develop the students' cognitive abilities are learning media (Nuryanti 2004).

Increasing a conducive entrepreneurial environment is an effective solution to drive the entrepreneurial motivation (Shetzer et al. 2010). Research conducted by Olvecka (2013) stated that the entrepreneurial environment has a significant positive effect on the motivation of entrepreneurship, while research conducted by Fereidouni et al. (2010) states that the WED environment has a less significant effect on students' entrepreneurial motivation.

The theory by Nafziger (2012) stated that entrepreneurial motivation is influenced by the business environment. Entrepreneurial environment affects the behavior and characteristics of entrepreneurial persons based on the theory that environmental factors influence the entrepreneurial student's entrepreneurial motivation. Vocational students who have entrepreneurial environments tend to be more motivated and conducive to entrepreneurship.

Based on the background of the problems mentioned above, the purpose of this study was to obtain findings regarding: (1) an overview of entrepreneurial environment and (2) entrepreneurial motivation.

Entrepreneurial environment is a combination of several factors that play a role in developing entrepreneurship (Fereidouni et al. 2010). Entrepreneurial environment is a combination of several factors that lead to the entrepreneurial process that can form self-employment (Koranti 2013).

The desire for wealth creation is a significant driver. This factor is assumed to depend on the given environment. For example, entrepreneurial income tax, capital gains, and dividends vary in different environments, states, and countries. Therefore, the financial expectations could also be associated with environment attributes. Opportunities and recognition have been discovered as central phenomena in the field of entrepreneurship. Environment can be attributed to the emergence of opportunities and the perception of their opportunities (Gnyawali & Fogel 2010) and it is assumed to be different from the environment to other environments.

Entrepreneurial environment is divided into three dimensions of social environments, methods, and environment policies (Mitra 2013). Programs are the environments that can affect someone through interpersonal and social networking culture. Programs include parents and school and community friends.

Methods refer to the development situation and the existing market competition, and environmental policy means the government policies and regulations such as laws and legislation that support someone to do entrepreneurship. In this study, the policy is the policy of the school and parents in supporting students' entrepreneurial processes. Students who got support from their school and parents will feel safe and comfortable for entrepreneurship.

Motivation is one of the success factors of entrepreneurs in completing tasks. The greater the motivation, the greater the success achieved. Called push factors and also factors that cause satisfaction. Their satisfaction will add enthusiasm to carry out activities (Herzberg in Rusdiana 2014). Basrowi (2011) stated some motivations for becoming entrepreneurs, those are:

1) Profit. An entrepreneur can determine how much profit they want and the benefits to be obtained and how that will be paid to the other party and their employees.
2) Freedom. Freedom to set their time, freedom from oppressive rule and culture, and freedom from the rules of the organization.

3) Personal dream. Freedom from the boring routine of work. Rewards to determine the mission, vision, and dreams on their own.
4) Self-reliance. Having a sense of pride because they can be self-sufficient in all things by their own efforts.

Meanwhile, Jay Mitra (2012) stated that entrepreneurial motivation is divided into several dimensions, including:

1) Independence. An entrepreneur wants to make the decision by themself. An entrepreneur does not want to depend on others.
2) Self-Confidence. Confidence is divided into two types of general confidence and special confidence. Confidence in general could be in a person who believes that they can face the world, face the challenge, overcome problems, and believe themselves to be capable of reaching their dreams. While the special confidence is confidence in oneself in overcoming specific problems.
3) Achievement Motivation. Achievement motivation is a desire to achieve an excellent standard, for example, to increase the income earnings, business performance, etc.
4) Drive to Action. Directions to act, a person who has the motivation would have proactivity, ambition, and energy.
5) Egoistic Passion. Being selfish in entrepreneurship is the desire to win, be powerful, and be an advanced private entrepreneur.
6) Tenacity. A motivated self-employed person will tend to be resilient in their work and in entrepreneurship.

However, in this study, the researchers took the dimensions of entrepreneurial motivation used by Fayolle et al. (2015), which are Independence, Striving for Achievement, Self-realization, and Ambition for Freedom. Based on a description of the picture of the entrepreneurial environment and entrepreneurial motivation, an analysis paradigm of entrepreneurial environment and entrepreneurial motivation is clearly illustrated in Figure 1.

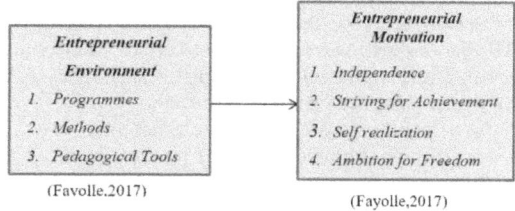

Figure 1. Research paradigm.

2 METHOD

The method used in this research is a verification method. A verification study is conducted to know and be able to explain the characteristics of the variables examined in a situation (have now, 2014). Research verification, according to Suharsimi Arikunto (2010), is basically intended to test the validity of a hypothesis, which is carried out through data collection in the field. Research verification predicts and explains the relationship of variables with other variables. This study aimed to see the effect of entrepreneurial environments on level XII students' entrepreneurial motivation in SMKN 1 Cimahi.

The type of research used in this research was an explanatory survey, which aimed to determine the relationship between variables by hypothesis testing. An explanatory survey method is done to know the whole of the area or object of research (Nasahudin 2012). Conclusions from these studies generally accept all information collected directly from the majority of the population empirically in order to know the opinion of the majority of the population regarding the object under study.

The types and sources of data used in this study can be seen more clearly in Table 1 as follows:

Table 1. Types and sources of data.

No.	Data	Data types types	Data source
1	Entrepreneurship Students of SMKN 1 Cimahi	Primary	Pre-study to the school
2	Entrepreneurial Motivation Overview of level XII Students of SMKN 1 Cimahi	Primary	Pre-study to the school
3	Entrepreneurial Environment Overview Data Students of SMKN 1 Cimahi	Primary	Pre-study for school

Source: Research pre, 2018.

This study uses a random sampling technique simply because the population is more than 600 people, and the samples taken are as many as 167 students. The population in this study consists of level XII students of SMK Negeri 1 Cimahi, amounting to 676 students. The data collection techniques used are literature studies and field studies with questionnaires. The data were analyzed using frequency distribution.

Verification research is research that aims to test a theory or the results of previous studies, in order to obtain results that reinforce the theory or the research.

The formula of coefficient of determination is as follows:

$$D = r2100\% \tag{1}$$

Information:

D = Large or number of the coefficient of determination

$r2$ = Correlation coefficient

The criteria in the analysis of the coefficient of determination are as follows:

a) If Kd is close to zero (0), it means that the influence of the independent variable on the dependent variable is weak.

b) If Kd is approaching one (1), it means the influence of the independent variables on the dependent variable is strong.

To interpret the extent of the influence of entrepreneurial environment on entrepreneurial motivation, certain guidelines are used. Determinant coefficient value should be between 0 and 100%. If the coefficient is getting closer to 100% it means that there is stronger influence of exogenous variables on endogenous variables. Meanwhile the closer it is to 0%, the weaker the influence of exogenous variables on endogenous variables. The interpretation guidelines are described in Table 2.

Table 2. Guidelines for providing interpretation of coefficient of determination.

Interval coefficient	The degree of influence
0 to 19.99%	Very weak
20%–39.99%	Weak
40%–59.99%	Moderate
60%–79.99%	Strong
80%–100%	Very strong

Source: Sekaran, 2014.

3 RESULTS AND DISCUSSION

Toutain et al. (2017) suggest that conscious effort of the school is needed to create and establish activities that make students active in entrepreneurial activism. Facilities supporting entrepreneurship, entrepreneurship training, school support, role models, as well as business licensing schools create an entrepreneurial environment that encourages students to be entrepreneurs (Fayolle 2008; Gnyawali & Fogel 1994; Yao 2016).

The ability of a teacher contributes to the success of the learning process inside and outside the classroom. Teachers should be role models and drive entrepreneurial students in the learning process. Increased capacity through various pedagogical teachers training is needed for the success of the learning process (Purwanto 2017; Shah 2010).

Classical Assumption Test On Simple Linear Regression Model.

Before evaluating the measurement model in a simple linear regression on the entrepreneurial environment

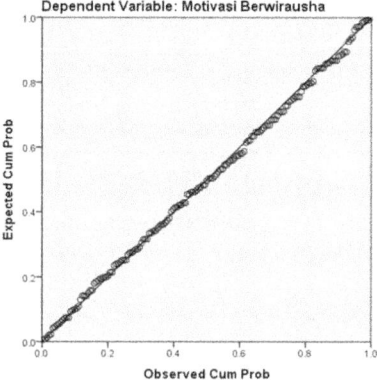

Figure 2. Output normality test of entrepreneurial environment on entrepreneurial motivation.

effects on entrepreneurial motivation, it is necessary to test classic assumptions made in this study.

1) Normality test

Normality testing is done to see if the research data were normally distributed or not. Normality test were performed using the SPSS (Statistical Product and Service) 22.0 application for Windows. In the Kolmogorov–Smirnov test, if the significance is below 0.05, it means that there is a significant difference. Meanwhile if the significance is above 0.05, then there is no significant difference.

Table 3. Kolmogorov–Smirnov test.

One-Sample Kolmogorov–Smirnov Test

		Unstandardized Residual
N		167
Normal Parameters[a,b]	Mean	0.0000000
	Std. Deviation	7.87783235
Most Extreme Differences	Absolute	0.034
	Positive	0.034
	Negative	−0.022
Test Statistic		0.034
Asymp. Sig. (2-tailed)		0.200[c,d]

a. Test distribution is normal.
b. Calculated from data.
c. Lilliefors Significance Correction.
d. This is a lower bound of the true significance.
Source: Appendix through SPSS 24.0 for Windows, 2018.

Based on Table 2, the significance level is found to be 0.200 or greater than 0.05, so it can be said that the data were normally distributed. In addition, normality testing also generated a graphic image of the Kolmogorov–Smirnov test. The output of the normality test can be seen in Figure 2.

Figure 2 shows that the data spread around the diagonal line and followed the direction of the diagonal. Thus it can be concluded that the regression meets the assumption of a normally distributed population. The data will be distributed normally if the expected probability value is equal to the probability of observation and the data criteria can be considered normal. In the graph plot, the similarity between the porbabilitas expectation and observation probability is shown by a diagonal line, which is the intersection of the line of expectation and probability diagonal (Sanusi 2013).

The testing for normality using the Kolmogorov–Smirnov test showed that the data are spread around the diagonal line and followed the direction of the diagonal line. Then, we can conclude that the population meets the normal distribution assumption. In addition, to further strengthen the evidence that the research data were normally distributed, the researchers tested for normality using the Kolmogorov–Smirnov test.

1) Linearity test

It is necessary to know whether these two variables really have a linear relationship, thus, it is necessary to test for the linearity regression of the X variable on the Y variable. The linearity test is intended to determine the possibility of a linear relationship between entrepreneurial environment and entrepreneurship motivation. The results of the data analysis with ANOVA output can be seen in Table 4.

Based on the test results, it is found that the Sig. deviation from linearity is 0.064 or greater than 0.05, which means that the overall entrepreneurial environment variable (X) in this model is fit and there is a relationship between the variables of entrepreneurial environment and entrepreneurship motivation. Based on the linearity test results, it can be said that the data generated meet the assumptions of linearity.

2) Scatter diagrams

A husky scatter diagram or chart (scatter plot) is used to determine and indicate where there is a relationship between the variables of X and Y through the depiction of the value of these variables. The results of data processing in SPSS 22.0 for Windows are presented in Figure 3.

Figure 3 illustrates that the points on the scatter diagram shaped a spread pattern from the lower left to the right. It means that if X changes then the Y changed even more, this shows that there is a relationship between the variables of X and Y. Based on the classical assumption that has been done, then this research can be processed using simple linear regression analysis.

3) Isolated point test

The next step is to notice the points of remote location on the scatter diagram. The test image output isolated point can be seen in Figure 4.

Furthermore, to determine whether the true black point is an isolated point, then statistical calculations using the formula test are performed. Where the criteria used in this test are:

t: Reject H0, meaning that a suspicious point is regarded as remote and must be removed: > -2.

Table 4. Linearity test of entrepreneurial environment on entrepreneurial motivation.

			Sum of Squares	df	Mean Square	F	Sig.
			ANOVA Table				
Entreperneurial motivation Entrepreneurial environment	Between Groups	(Combined)	17972.620	60	299.544	7.889	.000
		Linearity	11695.473	1	11695.473	308.016	.000
		Deviation from Linearity	6277.147	59	106.392	2.802	.064
	Within Groups		4024.853	106	37.970		
	Total		21997.473	166			

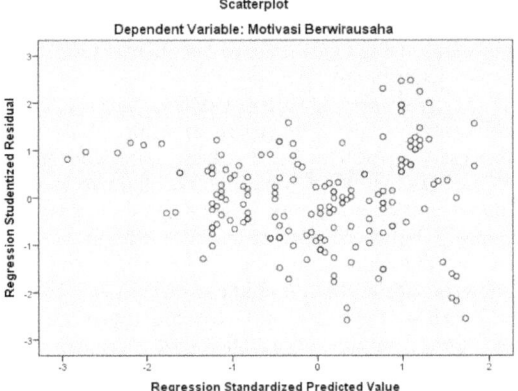

Figure 3. Scatter diagrams of entrepreneurial environment on entrepreneurship motivation.

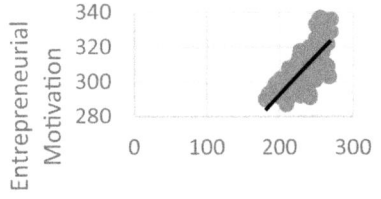

Figure 4. Output of remote test point of entrepreneurial environment on entrepreneurial motivation.
Source: Data processing, 2018.

t: Accept H0, meaning that the suspicious point is not regarded as an isolated point and does not need to be excluded from the analysis: ≤ -2.

The statistical testing used to determine whether a black dot is an isolated point or not is presented in Table 5.

Based on the results, the point is not an isolated point because the results of t indicate a smaller value than t-table. Thus, the criteria used for the results of these calculations received H0, the meaning suspicious point is not regarded as an isolated point and should not be issued: ≤ -2.

Table 5. Isolated point test.

		$Y = 0,4851+$	$S_{Y-\hat{Y}}$	t_{value}	t_{table}	
$\hat{Y}$	X	16,791	$(0,05 - \hat{Y})$	$\left(t = \dfrac{Y - \hat{Y}}{S_{Y-\hat{Y}}}\right)$	(167-2)	Decision
270	336	$Y = 203,651+$ 0,445X $Y = 353,171$	$S_{Y-\hat{Y}} =$ 0,05 − 336 $S_{Y-\hat{Y}} =$ −335,95	$t_{value} =$ $\dfrac{17,171}{-335,95}$ $t_{value} =$ −0,051	1,65141	No isolated point

Source: Data processing, 2018.

Simple Linear Regression Equations
The simple regression model that was established in this study was as follows:

$$Y = a + Bx \qquad (2)$$

Information:
Y = The dependent variable (entrepreneurial motivation)
X = The independent variable (entrepreneurial environment)
a = Price of Y when X = 0 (constant)
b = Score that affects an increase or decrease of Y grounded by X (entrepreneurial environment)

By using SPSS 22.0 for Windows, the results of the regression coefficients were obtained as follows:

Table 6. Simple linear regression model.

Coefficient

Model	Unstandardized Coefficients		Standardized Coefficients		
	B	Std. Error	Beta	t	Sig.
(Constant) 1	203.651	7.677		26.529	0.000
Environment Entrepreneurship	0.445	0.033	0.729	13.686	0.064

a. Dependent variable: Entrepreneurial motivation.

Based on Table 6, column B, it showed a constant value and the value of simple linear regression coefficients for the independent variable. Based on these values, it can then be determined by simple linear regression model that is expressed in the following equation:

$$Y = a + bX$$

$$Y = 203\,651 + 0.445 \qquad (3)$$

Based on the simple linear regression equation above, a constant value of 203.651 states that if there is no entrepreneurial environment, then there is an amount of 203.651 for entrepreneurial motivation. The regression coefficient on the entrepreneurial environment variable is 0.445, which means each of the additions of the value of the entrepreneurial environment increased entrepreneurial motivation by 0.445. Likewise, if there is a decline in entrepreneurial motivation, entrepreneurial environment will reduce the motivation of entrepreneurship by a 0445 unit value.

It can be said that entrepreneurial environment would affect the level of entrepreneurial motivation. If the relationship of entrepreneurial environment is less conducive to the students, it will reduce the motivation of students in schools for entrepreneurship.

Determinant coefficient analysis is used to determine the percentage of the impact that an independent variable has on the dependent variable. Thus, in this study, the determinant coefficient is used to determine the percentage effect of X on Y, so the formula used is as according to Riduwan (2013) as follows:

$$KD = r^2 \times 100\% \qquad (4)$$

Information:
r^2 = coefficient of correlation

Table 7. Entrepreneurial environment coefficient determination against entrepreneurial motivation.

		Model Summary		
Model	R	R Square	Adjusted R Square	Std. Error of the Estimate
1	0.729a	0.532	0.529	7.902

Source: Data processing, 2018.

Here are the results of the calculation of the coefficient of determination from X to Y:

$$
\begin{aligned}
KD &= r\,2 \times 100\% \\
&= R\,2 \times 100\% \\
&= (0.532)\,2 \times 100\% \\
&= 0.532 \times 100\% \\
&= 53.2\% \qquad (5)
\end{aligned}
$$

The results of the calculation of the coefficient of determination for the entrepreneurial environment (X) on entrepreneurial motivation (Y) was 53.2%; in other words, entrepreneurial motivation is 53.2% influenced by the environment of entrepreneurship. Table 7 shows the coefficients using a t-test, where t > t-table. Since the t-table has 167 respondents, at $\alpha = 0.05$ is 1.65141. Thus obtaining 13.686 > 1.65141, then Ho is rejected. Thus, we can conclude that Ho is refused and Ha is received, so that there is an influence between entrepreneurial environment and the motivation of entrepreneurship in level XII students of SMK Negeri 1 Cimahi.

Statistically, the hypothesis being tested was in the framework of decision-making acceptance, or rejection can be formulated as follows:

H0: $\rho \leq 0$, meaning that it cannot influence the entrepreneurial environment to motivate entrepreneurship.

Ha: $\rho > 0$, meaning that there is a positive influence on the motivation of the entrepreneurial environment on entrepreneurship.

4 CONCLUSION

The study showed that the entrepreneurial environment affected the entrepreneurship motivation of students by 53.2% and the remaining 46.8% was affected by factors that were not examined. Some other experts said that it was entrepreneurial support factors. Culture is another factor that can affect the motivation of entrepreneurship. The level of understanding of entrepreneurial risk and ambiguity encourages a person to understand the opportunities and challenges of entrepreneurship to grow entrepreneurial motivation (Ismail et al. 2016).

REFERENCES

Basrowi. 2011. Entrepreneurship for higher education. Bogor: Ghalia Indonesia.
Boluk, A. K., & Mottiar, Z. 2014. Motivations of social entrepreneurs blurring the social contribution and profits dichotomy. 10(1): 53–68. [Online]. Retrieved from http://doi.org/10.1108/SEJ-01-2013-0001.
Dehkordi, A. M., Sasani, A., Candidate, M. A., & Management, E. 2012. Investigating the effect of emotional intelligence and personality traits on entrepreneurial intention using the Fuzzy Dematel Method University of Tehran. 3(13); 286–296.
Eijdenberg, E. L. 2016. Does one size fit all? A look at the entrepreneurial motivation and entrepreneurial orientation in the informal economy of Tanzania. 22(6): 804–934. [Online]. Retrieved from http://doi.org/10.1108/IJEBR-12-2015-0295.
Fayolle, A. 2008. Linking entrepreneurial orientation and dynamic capabilities: Research issues and alternative models. *The Dynamics between Entrepreneurship, Environment and Education*: 308. [Online]. Retrieved from http://doi.org/10.1017/CBO9781107415324.00 4.
Fayolle, A., Kyro, P., & Linan, F. 2015. Developing, shaping and growing entrepreneurship. [Online]. Retrieved from http://doi.org/10.4337/9781784713584.

Fereidouni, H. G., Masron, T. A., & Nikbin, D. 2010. Consequences of external environment on. 15(2): 175–196.

Gnyawali, D., & Fogel, D. 1994. Environments for entrepreneurship development: Key dimensions and research implications. [Online]. Retrieved from http://doi.org/1042258794184.

Ismail, I., Husin, N., Abdul, N. M, Hanum, M., & Che, R. 2016). Entrepreneurial success among single mothers: The role of motivation and passion. *Procedia Economics and Finance* 37(16): 121–128. [Online]. Retrieved from http://doi.org/10.1016/S2212-5671(16)30102-2.

Kingdom, U., Orziemgbe, G., Chukwujioke, K., Aondoaver, T., & Polytechnic, B. S. 2014. Relationship between emotional intelligence and entrepreneurial performance: The mediating. (10): 1–16.

Knight, J. (2015). The evolving motivations of ethnic entrepreneurs. *Enterprising Communities* 9(2): 114–131. [Online]. Retrieved from http://doi.org/10.1108/JEC-10-2013-0031.

Koranti, K. 2013. Analysis of external and internal factors. 5: 8–9.

Mitra, J. 2013. Entrepreneurship, innovation and regional development (1st ed.). New York: Routledge.

Mulyadi, H. 2010. Education and training against influence entrepreneurial attitudes and implications on kewiraushaaan Student Conduct.

Mulyadi, H., & Irawan, A. 2016. Influence business success against entrepreneurial skills. 1(1): 213–223.

Nafziger, E. W. 2012. Economic development (Fourth Ed.). New York: Cambridge University Press.

Nuryanti, B. L. 2004. Learning model e-learning through learning media thus homepage as expected to increase student interests and creativity. 1(1): 7.

Olvecka, V. 2013. Development of environment in Slovakia Entrepreneurial. 7(2): 216–229. [Online] Retrieved from http://doi.org/10.13165/IE-13-7-2-06.

Purwanto, N. 2017. Psychology of education (28th Ed.). Bandung: Youth Rosdakarya.

Riduwan. 2013. Easy learning research. Bandung: Alfabeta.

Rusdiana. 2014. Entrepreneurship theory and practice (1st Ed.). Jakarta: Pustaka Setia. [Online] Retrieved from http://search.proquest.com/docview/1651837258?accountid=14548.

Sanusi, A. 2013. Business research methodology (mold to). Jakarta: Salemba four.

Sarmento. 2016. Predicting entrepreneurial motivation among university students:The role of entrepreneurship education. [Online]. Retrieved from http://doi.org/ http://dx.doi.org/10.1108/ET-01-2016-0019.

Shah, M. 2010. Educational psychology with an entrepreneurial environment. New approach. 8(1). Bandung: Youth Rosdakarya. [Online]. Retrieved from http://doi.org/ http://dx.doi.org/10.1108/JEEE-

Shetzer, L., Stackman, R. W., & Moore, L. F. 2010. Business-environment attitudes and the new environmental paradigm. 37–41. [Online]. Retrieved from http://doi.org/10.1080/00958964.1991.9943057.

Toutain, O., Fayolle, A., Pittaway, L., & Politis, D. 03-2015-0021 (2017). Role and impact of the environment on entrepreneurial learning. *Entrepreneurship Regional Development* 29(9–10): 869–888. [Online]. Retrieved from http://doi.org/10.1080/08985626.2017.1376517

Yao, X. 2016. Effect of students' perceived entrepreneurial environment. [Online]. Retrieved from http://doi.org/ http://dx.doi.org/10.1108/JEEE-03-2015-0021

Advances in Business, Management and Entrepreneurship – Hurriyati et al. (Eds)
© 2021 Taylor & Francis Group, London, ISBN 978-0-367-67471-7

The strategy of digital service differentiation to increase customers' satisfaction of retail third party fund in sharia bank

A.G. Ali, A. Rahayu, L.A. Wibowo & M.A. Sultan
Universitas Pendidikan Indonesia, Bandung, Indonesia

ABSTRACT: Nowadays, digital technology for banking cannot be ignored and has an important role in achieving the business purpose. Digital technology brings up a big potential for product innovation and services which are difficult to control and predict. Therefore, the company needs dynamic tools to support their business by managing new kind of digital innovation process, so that strategy of innovation is necessary for providing services through giving fast response and simple solution for customers or prospective ones. Regarding that situation, sharia banking needs a business innovation strategy through digital services differentiation. The main purpose is to increase customers' satisfaction and the no ones as well, especially the customers who become retail third party funds. Meanwhile, this research aims to assess, theoretically and empirically, the quality of the relationship between the customers/prospective ones and the bank to analyze their satisfaction level at one of the sharia banks in Indonesia towards its digital services.

1 INTRODUCTION

Indonesia is one of the countries with a penetration level of low banking services. Being compared to other developing countries, Indonesia is considered to be left behind in terms of financial inclusion. According to the survey of World Bank (2014), only 36 percent of the adult population has accounts at a formal financial institution. This number is lower than that of in East and Pacific Asia (69 percent) on average, countries with middle to lower-income on average (42 percent). In East Asia, Indonesia is even left behind by Thailand with its 78 percent population has bank accounts (Research of DBS Indonesian Multi-Finance Companies, Bridging Gaps with the Underbanked).

Those numbers are very lame if compared to cellular phone penetration. Survey We Are Social stated that 91 percent of the Indonesia population had cellular phones, while cellular phone owners were 47 percent. Cellular phone card users were even more than the population, 371.4 million, or 142 percent of the population.

The Survey of Asosiasi Penyelenggara Jasa Internet Indonesia (APJII) (Association of Indonesia Internet Service Provider) in 2016 explained that several internet users in Indonesia were 132.7 million people or 51.8 percent of Indonesia total population.

The Digital era has given simplicity to customers to access financial services only from their palms. The trend of using digital transactions keeps on increasing along with internet penetration raise. Otoritas Jasa Keuangan (Financial Service Authorities) stated that e-banking users soared up to 270 percent, from 13.6 million customers in 2012 to 50.4 million customers in 2016. The frequency of e-banking transactions was also increasing by 169 percent, from 150.8 million transactions in 2012 became 406.6 million transactions in 2016.

The more competitive in financial institutions nowadays, the more focus is needed on technology and product service innovation. The future winner will be determined by organizations which able to take advantage of digital technology to provide the customers' experiences beyond common. Recently, technology has developed rapidly and people work hard to create innovations and expect to bring changes for life on earth. By simply access to the internet from people's gadgets and other simplicities becomes a positive impact of the industry evolution.

As a relatively newly developing industry, the development slot for sharia banking is still open. Sharia banking also has better prospects compared to other banking with a similar scale. Sharia banking, with its complete characteristic, enables them to be more creative in innovating various products and services for business support. Other than that, its unique characteristic of products and services enables them to improvise to work on other markets that cannot be done by conventional banking.

Sharia banking needs a strategy to influence customers by seeking opportunities to attract their interest to use sharia banking services either in the product or services so that it is necessary to create service differentiation.

As an effort to broaden financial services access, Bank BJB Sharia innovates a service with a branchless banking concept. Branchless banking means a distribution network to provide financial services outside bank branches by using technology with effective and efficient cost as well as safe and comfortable.

E-banking is an implementation of SSTs (self-service technologies) becomes a new world for the banking sector which will keep innovating so that needs an adjustment in increasing the service quality which then called customer empowerment (Davies and Elliot 2006).

Hunger & Wheelen (2003) suggested two "generic" competitive strategies to surpass other companies in certain industries, they were low cost and differentiation. Low cost means the ability of a company or a business unit to design, to create, and to market, a product more efficiently compared to its competitors. Meanwhile, differentiation means the ability to provide customers unique and superior value in quality, characteristic, or after-sale service. This strategy is called generic because any kinds and sizes of companies – even a non-profit organization – can apply it.

2 METHOD

Malhotra & Dash (2016:108) suggested that an accurate research method can avoid speculative problem solving and can increase objectivity in finding the depth of knowledge. Based on the previous research to achieve clear coverage and description of the object of the study and its variables, the methodology applied in this research was descriptive and verification type, in which the research was conducted by using causalities method, whereas the researcher evaluated the correlation or influence between the dependent and independent variables. Based on a descriptive and verification analysis, the data was collected from the field, so that method of this research was a descriptive and explanatory survey. The research was conducted through causalities analysis which explained the influence of one variable towards another variable. Meanwhile, the time horizon was cross-sectional because this research was conducted at a certain time. The analysis unit of this research was a sharia bank. The observation was done through time horizon and cross-section or one-shot. It means the information or data collection was gained directly and empirically in one place (Sekaran & Bougie 2010).

In this research, the data was quantitative and it provided the data of the 3 in 1 Maslahah service in Bank BJB Sharia. The data was gained through questionnaire distribution. The data analysis used SPSS 21. It is an application program that has a high ability to analyze statistical data and also data management in a graphic by using the descriptive menu and simple dialogue columns.

3 RESULTS

To overcome tighter competition, reliable management that capable to anticipate each competition and run the company effectively and efficiently is very necessary. Various efforts to attract customers' interest is being done, one of which is creating excellent products and services done by Bank BJB Sharia. The services are called 3 in 1 Maslahah Services. These services have collaborated products and services, they are Jemput Maslahah (open an account outside the office), Maslahah Card (instant ATM card & active on the spot), and Maslahah Mobile (mobile banking which is instantly active). It is expected that those products and services can increase service and to reach some retail Third Party Fund as well as an effort to control bank operational cost in accelerating its business growth.

Jemput Maslahah is a service provided by Bank BJB Sharia to fulfill prospective customers' needs. They just call Salam Maslahah on 1500727, the bank officer will visit the prospective customers directly to their places to open accounts and instantly active to do transactions, to provide Kartu.

From the calculation of r count, it can be iv. seen through the output above (Corrected Item – Total Correlation) that if compared to the r table value 0.195, r count > r table. It indicates that all items of questions in the questionnaire are valid or can be applied in the research.

Maslahah an instant ATM card that is instantly active and able to be used for transactions. Mobile banking is also instantly active. Those services are banking services that customers can use anytime and anywhere for 24 hours in real-time online directly for those who own the ATM card of Bank BJB Sharia through cellular phone by using the wifi network.

To find how far the 3 in 1 Maslahah Service supports customers and prospective customers in fulfilling their banking need in Bank BJB Sharia and how far the customers' satisfaction towards the 3 in 1 Service, it is necessary to do a statistic testing. In this case, the writer used validity and reliability testing to find appropriateness from the questions of the questionnaire result. The writer made a questionnaire for customers of retail third party fund, especially savings customers of Bank BJB Sharia.

Table 1. Reliability testing.

Cronbach's alpha	N of Items
0.946	10

Reliability testing was using Cronbach's Alpha with condition Cronbach's Alpha $\geq$ 0.6. From the result of SPSS above, it can be seen that Cronbach's Alpha is

Validity Testing Data

r table: significant level: 5% n $-$ 2 $=$ 106 $-$ 2 $=$ 104

Criteria

r count>r table

Table 2. Item – total statistics.

	Scale Mean if Item Deleted	Scale Variance if Item Deleted	Corrected Item-Total Correlation	Cronbach's Alpha if Item Deleted
P1	75.70	115.203	.628	.948
P2	75.82	115.558	.757	.941
P3	76.24	111.630	.859	.936
P4	75.65	114.153	.864	.938
P5	76.33	114.985	.808	.938
P6	75.91	115.934	.858	.937
P7	76.03	116.085	.788	.939
P8	76.03	113.590	.830	.937
P9	75.77	118.139	.742	.941
P10	76.01	119.848	.651	.945

$0.946 > 0.6$ so that it can be conclude that the result of the questionnaire data is reliable.

From the result of validity and reliability testing above, it is found that the 10 questions given to the customers are valid and reliable, so that it can be applied in this research. See Table 3 for the description of respondents.

In Table 3, it can be seen that from 10 questions distributed to the 106 respondents (customers), each question has average value 8 so that it can be concluded that:

Customers have recognized the 3 in 1 Maslahah service of Bank BJB Sharia (Jemput Maslahah, Maslahah Card, and Maslahah Mobile),

3 in 1 Maslahah Service has accommodated and supported customers' banking needs. It is due to innovation that the customers needn't come to the bank for opening accounts and doing banking transactions.

Rapid response and service of 3 in 1 Jemput Maslahah service has satisfied customers' needs due to providing a solution when they are not able to go out of offices or houses; Maslahah Mobile Service can facilitate banking transactions for 24 hours only through cellular phone; Maslahah Card is used as transaction tool by customers easily.

Customers are satisfied with the 3 in 1 Maslahah Service in Bank BJB Sharia.

From the testing result statistically on 3 in 1 Maslahah Service in Bank BJB Sharia using validity and reliability testing, it is found that the service is an innovation with rapid response and provides a solution

so that the customers' satisfaction of retail third party fund is different from other competitors.

Porter (2011) discussed that "generic strategic" could be applied to products and services in all industries and organizations on any scales. Porter stated that generic strategy: "Leadership Cost", "Differentiation" and "Focus". Focus strategy had been divided into two parts: "Cost Focus" and "Differentiation Focus".

Theoretical review about Service Differentiation, according to Kottler (2012:147) that differentiation is a process to add a serial of valuable differences to distinguish the company's offer from competitors. Next, according to Kutcher (2010:14), that differentiation is integrated with the company's success in a competitive business environment. Differentiation strategy is one of business strategy which main focus on doing an effort to create and to promote special products to various customers so that they can be loyal to the products. According to Ferdinand (2013) that successful differentiation must be a strategy which able: (a) to generate customer value; (b) to emerge characteristic and good perception, and (c) to perform as a different form which cannot be imitated. Effective placement needs a demonstration to target customers to use the company's technology for taking important advantage and experiencing emotional result (Solomon et al. 1985; Young and Feigin 1975). Based on the previous relevant research result, a research was conducted by Subawa & Widhiasthini (2018) found some causal factors of the transformation of customer behavior in the era of industry revolution 4.0 consisted of follower culture which meant as a form of most people behave in their environment that was done simultaneously. Another researcher conducted by Rose et al. (2012) with the topic "Online Customer Experience in e-Retailing: An Empirical Model of Antecedents and Outcomes" discussed online shopping behavior which had drawn attention. It was focused on an interaction between the online shoppers and the e-retailers through scaled-online shopping websites, which then generated experience to online shopping customers (OCE). Other research was conducted by Helm et al. (2015) published in the International Journal of Consumer Studies, "Consumer Cynicism: Developing Scale to Measure Underlying Attitudes Influencing Marketplace Shaping and Withdrawal Behaviours". A research conducted by Nylén & Holmström (2015) explained that digital technology became more

Table 3. Descriptive statistics.

	P1	P2	P3	P4	P5	P6	P7	P8	P9	P10
N Valid	106	106	106	106	106	106	106	106	106	106
Missing	0	0	0	0	0	0	0	0	0	0
Mean	8.69	8.57	8.15	8.74	8.06	8.48	8.36	8.36	8.61	8.38
Median	9.00	9.00	8.00	9.00	8.00	9.00	8.50	8.00	9.00	9.00

important to achieve a business purpose. Other research conducted by Setia et al. (2013), Leveraging Digital Technologies: How information quality leads to localized capabilities and customer service performance, stated that the more increasing role of the customer in creating and sending services, the more increasing encouragement to build an organization focused on customers. Digital technology played a key role in the organization. Next, Teperi & Leppänen (2011) in their article titled From Crisis to Development – Analysis of Air Traffic Control Work Processes, explained that an industry with high technology generated the huge number of information from the system and in the era of big data, there were a lot of opportunities to obtain the huge advantage of the technology. In their recent research, it had been described how digital technology had emerged huge potential for product and service innovation which was difficult to control and to predict. Therefore, the company needed a dynamic tool to support their business.

Ardolino et al. (2018) found that Internet of Things (IoT) was a basic for any transformation service, although most of it was necessary to become supply provider and predictive analytics (PA) was very important to move to performance provider profile. Besides providing scalability for all profiles, cloud computing (CC) specifically was used to apply industrialist strategy, so that it led to the standardized, repeated, and product offer.

Research done by Cortet et al. (2016), the provision of PSD 2 about "Access to Account" for Initiative of Account Payment and Information Service was to insist the bank accelerating customer payment account opening for a licensed and innovative service provider (bank and fintech non-bank). Important element in the payment account opening concept was Application Programming Interfaces (APIs) by the bank. The Fintech players attempted to benefit the emerged landscape of API and to catch customers and mindshare developers as well as payment and non-payment income (data-rich service) which had been long accepted by the financial institution. Meanwhile, a research conducted by Kane et al. (2015), in their article Strategy, not Technology, Drives Digital Transformation Becoming a Digitally Mature Enterprise, found that mature digital business focused on digital technology integration, such as social, cellular, analytic, and cloud, in servicing to change its procedure. Immature digital business focused on solving the problem of discrete business with individual digital technology. The capability to re-arrange digitally business was mostly determined by a digital strategy that was supported by leaders who grew a culture to change and to create a new one. While this insight was consistent with previous technology evolution, the uniqueness of digital transformation was that risk-taking became a cultural norm since more advanced companies sought for new competitive excellence digitally. Schrauf and Berttram (2016) in an article Industry 4.0: How digitization makes

the supply chain more efficient, agile, and customer-focused, stated that if the vision of Industry 4.0 wished to be manifested, most of the company process had to become more digital. The critical element became an evolution of the traditional supply chain headed for a connected, smart, and efficient supply chain ecosystem. Today's supply chain was a serial of steps that most separated and taken through promotion, product development, manufacture and distribution, and finally to the customers.

4 CONCLUSION

The 3 in 1 Maslahah Service is a product differentiation of Bank BJB Sharia due to differentiation of 3 in 1 Maslahah service distinguishes its main service based on the simplicity of ordering, delivery, setting, customer training, consultation, maintaining and improvement. The simplicity of ordering refers to how simple the customers can order to the company. Based on Bank BJB Sharia 3 in 1 Maslahah service that is given consists of (a) Jemput Maslahah is a service that is given to fulfill customers' need by going to their places directly; (b) Maslahah Card is an ATM facility that is given to the customers who want to open a new account in Maslahah iB saving and instantly active; (c) while Mobile Maslahah is a banking service to be used anytime and anywhere for 24 hours real-time online by customers and can be accessed directly via cell phone using data communication/wifi.

Generally, customers of Retail Third Party Fund have recognized the 3 in 1 Maslahah service in Bank BJB Sharia. However, it must be more promoted, for example in social, mass, electronic media to be known by broader people. The customers of Retail Third Party Fund in Bank BJB Sharia need e-service can be seen from the respondents who feel being supported by this service, especially those who don't have much time to come to the bank for opening an account Jemput Maslahah service, the transaction for 24 hours real-time online using Maslahah Mobile and cash withdrawal for 24 hours using Maslahah Card.

The customers just call Salam Maslahah and the call center officer will give clear information about 3 in 1 Maslahah service. Next, after it is clear, the nearest bank officer will come to the customers' places directly with Jemput Maslahah service based on the customers' requests. Jemput Maslahah service will be on time and fast in providing service kindly and emphatically, listening to and giving solutions to the customers' banking needs. The bank officer will help to install Maslahah Mobile service after opening an account by the service of Jemput Maslahah and Maslahah Card being processed.

The available facilities in the 3 in 1 Maslahah service have helped users to promote Bank BJB Sharia products. The customers feel satisfied with the 3 in 1 Maslahah service. It is a prime innovation of Bank BJB Sharia in the banking industry and to keep competing with other competitors and product development must be done continuously.

REFERENCES

Ardolino, M., Rapaccini, M., Saccani, N., Gaiardelli, P., Crespi, G., & Ruggeri, C. (2018). The role of digital technologies for the service transformation of industrial companies. *International Journal of Production Research, 56*(6), 2116–2132.

Cortet, M., Rijks, T., & Nijland, S. (2016). PSD2: The digital transformation accelerator for banks. *Journal of Payments Strategy & Systems, 10*(1), 13–27.

Ferdinand, A. (2013). Structural Equation Modeling Journal Title Manajemen, (3) Universitas Diponegoro

Helm, A. E., Moulard, J. G., & Richins, M. (2015). Consumer cynicism: Developing a scale to measure underlying attitudes influencing marketplace shaping and withdrawal behaviours. *International Journal of Consumer Studies, 39*(5), 515–524.

Hunger, J. D., & Wheelen, T. L. (2003). *Essentials of strategic management.* NJ: Prentice Hall.

Kane, G. C., Palmer, D., Phillips, A. N., Kiron, D., & Buckley, N. (2015). Strategy, not technology, drives digital transformation. *MIT Sloan Management Review and Deloitte University Press, 14*(1–25).

Kutcher, Kevin. 2010. Differentiation. *Rural telecommunications 19* (1) Pp.14

Malhotra, N. K., & Dash, S. (2016). *Marketing research: An applied orientation.* Pearson,.

Nylén, D., & Holmström, J. (2015). Digital innovation strategy: A framework for diagnosing and improving digital product and service innovation. *Business Horizons,* 58(1), 57–67.

Porter, M. E. (2011). *Competitive advantage of nations: creating and sustaining superior performance.* simon and schuster.

Rose, S., Clark, M., Samouel, P., & Hair, N. (2012). Online customer experience in e-retailing: an empirical model of antecedents and outcomes. *Journal of retailing, 88*(2), 308–322.

Schrauf S dan Berttram P. 2016. Industry 4.0: "How digitization makes the supply chain more efficient, agile, and customer-focused". *Strategy &Technology* 1–32

Sekaran, U., & Bougie, R. (2010). *Research Methods For Business.* Chiches: John Wiley & Sons.

Setia, P., Setia, P., Venkatesh, V., & Joglekar, S. (2013). Leveraging digital technologies: How information quality leads to localized capabilities and customer service performance. *Mis Quarterly,* 565–590.

Solomon, M. R., Surprenant, C., Czepiel, J. A., & Gutman, E. G. (1985). A role theory perspective on dyadic interactions: the service encounter. *Journal of marketing, 49*(1), 99–111.

Subawa, N. S., & Widhiasthini, N. W. 2018. Transformasi Perilaku Konsumen Era Revolusi Industri 4.0. In *Conference on Management and Behavioural Studies* (131–139).

Teperi, A. M., & Leppänen, A. (2011). From crisis to development–analysis of air traffic control work processes. *Applied ergonomics, 42*(3), 426–436.

Young, S., & Feigin, B. (1975). Using the benefit chain for improved strategy formulation. *Journal of Marketing,* 39(3), 72–74.

Advances in Business, Management and Entrepreneurship – Hurriyati et al. (Eds)
© 2021 Taylor & Francis Group, London, ISBN 978-0-367-67471-7

Competitive advantages strategy of rural banks in West Java

U. Supriatna, A. Rahayu, L.A. Wibowo & M.D. Sugiharto
Universitas Pendidikan Indonesia, Bandung, Indonesia

ABSTRACT: This study aims to find what factors affecting the rural banks in encountering the bank business competition, reasons why the rural banks in West Java gain difficulties in to possess the competitive advantages and finding ways of the rural banks can surpass the competition. This study involves 87 respondents consisting of the customers of the public banks and the rural ones. The data are analyzed using multiple linear regression. The findings of the research show that the total of the human resources affects the service, the interest rate does not affect the service, the information technology does not affect the service partially and the human resources total, the interest rate and, the information technology affect the service simultaneously. In conclusion, the prominent factor affects the service in rural banks is human resource management.

1 INTRODUCTION

The intense competition in the banking industry, technological developments, and changes in customer tastes can cause customers to move from one bank to another. Banks must have the ability to develop strategic choices so they can adapt to a dynamic environment. Therefore, company resources must be managed systematically to produce a superior value that can truly be valued by the customers (Ferdinand 2000, p. 4–5). The creation of superior value for customers is a stepping stone for companies to obtain competitive advantage (Menon et al. 1997, p.187). In service companies, especially in the banking sectors, competitive advantage is often sought in the form of superior service. According to Wahlers (1994, p. 230), states that an appropriate and accurate strategy in service quality is an important factor that influences a company's competitive advantage if planned and implemented properly.

Accordingly, In the field of banking services, product quality measured is service quality. Parasuraman et al. (1990, p. 42) define service quality as a difference in consumers' perceptions of the quality of service received with consumer expectations of quality. The main things that underlie service quality are its Indicators which can be used to evaluate service quality, regardless of the type of service. Companies and competitors jointly compete in producing and providing value to their customers, as far as possible following the value expected by the customer. The goal of developing a strategy is to produce superior value or better customer service than what competitors can do (Ferdinand 2000, p. 5). Bank customers who have large amounts of funds, of course, want good service, and in this case, the service officers are considered less

than optimal so there are still many complaints about bank services. The teller staff is not proportional to the number of customers, so long queues often occur that cause customer to have to wait long enough to get service. For customers who transact through ATMs, they are sometimes disappointed because they often appear off-line interruptions, money runs out so cash withdrawals cannot be made.

Since the mid-1980s, there has been a paradigm shift in financial policy from subsidized credit to financial system development (Adams and Von Pischke 1984). The old credit paradigm directed at the sector, led by supply, and subsidized is based on false assumptions about the willingness and ability of poor farmers and other entrepreneurs to pay for financial services, which leads to the incorrect design and implementation of policies. The new paradigm departs not from necessity, but from demand (IE willingness and ability to pay market prices) for savings, credit, and insurance services by farmers and other entrepreneurs. The new paradigm recognizes that high transaction costs and risks that partly result from information asymmetry and moral hazard problems (Stiglitz & Weiss 1981) for financial intermediaries and clients are some of the root causes of the gap between supply and demand. Therefore, the new paradigm emphasizes the search for technological and institutional innovations (including appropriate governance and incentive structures) to reduce the costs and risks of financial intermediation. The new paradigm recognizes the possibility of markets and government failures (ie general institutional failures) and negates the thesis put forward by market liberalization advocates that "an undressed financial system will automatically function optimally" (Krahnen & Schmidt 1994, p. 24). The new paradigm, on the other hand, sees financial market liberalization (for example about the formation of interest rates) as a necessary but not sufficient condition to deepen the financial system. Besides, because the technological and institutional innovations needed to deepen the financial system and to serve the poorer segments of the population can be easily copied by non-profit financial institutions.

An important concept that highlights the role of information technology in the competition is the "value chain." This concept divides a company's activities into the technologically and economically distinct activities it performs to do business. We call these "value activities." The value a company creates is measured by the amount that buyers are willing to pay for a product or service. A business is profitable if the value it creates exceeds the cost of performing the value activities. To gain a competitive advantage over its rivals, a company must either perform these activities at a lower cost or perform them in a way that leads to differentiation and a premium price (more value). (Porter & Millar 1985).

A competitive advantage is sustained depends upon the possibility of competitive duplication. Following Lippman & Rumelt (1982) and Rumelt (1984), a competitive advantage is sustained only if it continues to exist after efforts to duplicate that advantage have ceased. In this sense, this definition of sustained competitive advantage is an equilibrium definition (Hirshleifer 1982).

From the above background several problem formulations can be made, namely:

a. What factors do affect the rural banks in facing banking business competition?
b. Why are the rural banks in West Java difficult to have a competitive advantage?
c. What is the strategy taken by the rural banks in encountering the business competition?

2 METHODS

The research method used is based on descriptive and verification research carried out through data collection in the field, the method used in this study is the explanatory survey method. The explanatory survey is carried out to explore the problem situation, which is to get ideas and insights into the problems faced by the management or researchers (Maholtra 2010: 96). The survey method is used to get data from a particular natural place, but the research conducts treatment in data collection.

3 RESULTS AND DISCUSSION

According to Kotler (2003), competitive marketing strategies, strategies positioning a firm strongly against its competitors, and give the company the advantage of the strongest strategic position.

According to Michael Porter, there are several basic competitive strategies (Kotler 2003), namely:

1. Overall cost leadership

It means that the company works hard to get the lowest production and distribution costs. Low cost means making the price of goods and services lower than competitors and winning a large market share.

2. Differentiation

This means that the company concentrates on creating a completely different type of product and marketing program to become a leader in its industrial class. Most consumers prefer to own this brand if the price is not too high.

3. Focus

It means that the company focuses on its efforts to serve a small market segment rather than serving a larger market.

4. Operational excellence

It means that the company provides better value by leading its industry in price and comfort. This needs the target consumer.

5. Product leadership

Every company that competes in an industrial environment has a desire to be superior to its competitors. Companies generally implement this competitive strategy explicitly through the activities of various existing functional departments of the company. The basic thinking of creating a competitive strategy starts with developing a general formula on how the business will be developed, what exactly is the goal, and what policies will be needed to achieve that goal. The definition of competitive advantage itself has two different but interrelated meanings. The tight competition causes the company to try to surpass the competition by implementing the right competitive strategy so that it can implement and realize the objectives following what is expected

Accordingly, the market success is obtained by companies that best match current environmental requirements, namely goods and services that people are ready to buy. Individuals, businesses, and even entire countries must find out how they produce marketable value. Therefore companies must understand what consumers want now and for the future. Thus, the success and failure of a company are very dependent on the competitive advantage possessed by the company.

To realize the realization of the Bank's Business Plan, the rural banks in West Java, in general, will carry out matters relating to raising funds, channeling funds and developing bank employees, setting plans for management policy as follows:

a. Improve financial conditions by maintaining and improving business market share, by improving the composition of funds and credit products, to become a healthy bank and be able to provide profits in the form of dividends to shareholders
b. Improving the ability of employees, expertise, and skills of employees according to their fields of work through intensive and oriented training programs,

which have a direct impact on business activities, to increase the productivity of human resources so that they can become the driving force for BPR growth to be better

c. Arranging and perfecting operational SOPs and developing business networks/office operations, to improve services to customers and the public

d. Maintaining the growth and quality of credit at a healthy rating, to make productive credit facilities by positioning NPLs at a ratio below 5%

e. The amount of interest for collecting funds from third parties, especially time deposits, should not exceed the requirements of the deposit insurance institution (LPS, unless the depositor signs a statement, not in the LPS guarantee)

f. Maximizing the collection of low-cost funds from the public in the form of savings

g. The addition of marketing and collector staff

h. To realize the distribution of funds (lending) must first pay attention to the basis of the Cash Ratio, Maximum Lending Limit (BMPK) and Loan Debt Ratio (LDR) and the size of the ratio must be good according to the provisions of the Financial Services Authority (OJK)

i. The granting of credit must be following the policies and procedures set by the bank, both starting from the application, credit analysis, credit decisions, credit administration, monitoring, and resolution of problem loans.

j. To implement the collection of funds and distribution of funds, stipulated as follows:

- Market segmentation = especially the lower middle class, targeting the location of shops, agriculture, factories in the VPR work area in West Java
- Price (credits rate) by applying a price (rate) that competes with competitors, while funding rates are adjusted to the availability of bank liquidity
- Place of service provided to receive credit deposits/deposits and savings deposits can be done at the customer's place of business or home using the IBS collection facility, whereas deposit deposits must go through the BPR counter directly.
- Promotion is preferred by personal selling media (direct selling / picking up the ball) and mass selling media.

k. The compiled Bank Business Plan (RBB) must be used as a control tool at the end of every month or operational actions by the Directors

l. The compiled Bank Business Plan (RBB) must be evaluated for the level of deviation between the plan and the realization by the Board of Directors, especially by internal audit or the Board of Commissioners.

m. The Board of Directors must follow up and improve the findings of the Financial Services Authority (OJK) audit findings under the agreed commitments.

Based on the data below, the rural banks' market share in West Java in the credit sector, fund mobilization, and total assets are above the average BPR market share in Indonesia, which is 14.13%.

Table 1.   Rural banks market share.

No	Main indicators	Number of conventional rural banks		Market Share National
		West Java	National	
1	Average of Rural Banks Allocation of the fund	290	1.619	17.91%
	Third party fund	12.327	84.861	14.53%
	Saving	3.796	26.723	14.20%
	Deposit	8.531	58.137	14.67%
2	Credit	11.537	89.073	12.95%
	Total of Asset	18.010	125.945	14.30%

Testing hypotheses of H1, H2, and H3 by t test and F test, Coefficient of Determination using the following results.

- Hypothesis Testing of H1

The significance value of the effect of X1 on Y is known to be 0,000 < 0.05 and the value of t observation as much as 11,321 > 1,992, so it can be concluded that H1 is accepted, meaning that there is an effect of X1 on Y.

- Hypothesis Testing of H2

The significance value of the effect of X2 on Y as much as 0.052 > 0.05 and the t observation is as much as 1.969 < 1.992, so it is concluded that H2 is rejected, meaning that there is no effect of X2 on Y

- Hypothesis Testing of H3

The significance value of the effect of X3 on Y as much as 0.057 > 0.05 and t observation as much as

Table 2.   Coefficients.[a]

Model	Unstandardized Coefficients		Standardized Coefficients	t	Sig.
	B	Std. Error	Beta		
(Constant)	−5.902	5.705		−1.035	.304
TOTAL_HRM	2.209	.195	.721	11.321	.000
TOTAL_INTEREST_RATE	.519	.264	.134	1.969	.052
TOTAL_IT	.437	.227	.136	1.927	.057

a. Dependent Variable: TOTAL_SERVICE.

Table 3. ANOVA.[a]

Model	Sum of Squares	df	Mean Square	F	Sig.
1 Regression	12814.401	3	4271.467	90.262	.000[b]
Residual	3880.495	82	47.323		
Total	16694.895	85			

a. Dependent Variable: TOTAL_SERVICE.
b. Predictors: (Constant), TOTAL_IT, TOTAL_HRM, TOTAL_INTEREST_RATE.

Table 4. Model summary.[b]

Model	R	R Square	Adjusted R Square	Std. Error of the Estimate
1	.876[a]	.768	.759	6.87918

a. Predictors: (Constant), TOTAL_IT, TOTAL_HRM, TOTAL_INTEREST_RATE.
b. Dependent Variable: TOTAL_SERVICE.

1.927 <1.992, so that it is concluded that H3 is rejected, meaning that there is no effect of X3 on Y

• Testing simultaneous hypotheses of H4

Based on the results of Table 3, it is known that the significance value of the simultaneous effect of X1, X2, and X3 on Y is 0.00 <0.05 and the calculated F value is as much as 90.262 > 2.71. The F value of this result table is concluded that there is an influence of X1, X2, and X3 simultaneously on Y

In Table 4 it is known that the R Square value of 0.768, this implies that the simultaneous effect of variables of X1, X2, and X3 on Y is 76.8%.

4 CONCLUSIONS

Based on the explanation above, the following conclusions can be drawn:

a. Factors that influence the rural banks' superiority in encountering banking business competition are the quality of human resources, services, interest rates, and by looking at the competitive situation both the attitudes and behavior of consumers, competitors, and change factors. Furthermore determining the marketing position in the competition through the determination of marketing strategies per the competition faced. That competitive advantage is a strategy carried out by a company to master the market to find the maximum profit. Continued competitive advantage will create an advantage that is not easily imitated, which allows a company to seize and maintain its position as a market leader. Because it is not easily imitated, sustainable competitive advantage can support a company's success for a long time. Thus, the company can also achieve the expected benefits.

b. There are three types of strategies used in competitive advantage, namely cost leadership strategy,

differentiation strategy, and focus strategy. Wherefrom the three strategies can make the rural banks in West Java to achieve a competitive advantage, each strategy has advantages and disadvantages of each - then a company if they want to continue to exist mastered competitive advantage, they must also be able to see the state of the market that continues to grow by technological progress by innovation.

c. Based on the results of the research, the strategy that must be carried out by the rural banks in facing business competition is to have a competitive advantage both with similar banks and with commercial banks or other financial institutions by increasing the quality of its Human Resources, due to the quality level of Resources Humans who are high will improve services and with services that increase will improve company performance.

REFERENCES

Adams, D W, and J. Von Pischke (Eds). 1984. *Undermining rural development with cheap credit*. Boulder, Colo., U.S.A.: Westview Press.

Antonioni, David, 1996. Designing an Effective 360 – Degree Appraisal Feedback Process. *Organizational Dynamic*, Autumn: 24–38.

Baker, W.E., and J.M. Sinkula, 1999. "The Effect of Market Orientation and Learning Orientation on Organizational Performance". *Journal of The Academy of Marketing Science. Vol 27*

Barney, Jay. Firm Resources and Sustained Competitive Advantage, *Journal of Management*, (1991)

Beal, Reginald M., 2000. Competing Effectively, Environment Scanning, Competitive Strategy and Organizational Performance in Small Manufacturing Firms. *Journal of Small Business Management*, January: 27–47.

Buttery, Alan, and Rick Tamaschke, 1996. The Use and Development of Marketing Information Systems in Queensland, Australia. *Marketing Intelligence and Planning*, Vol 14 No. 3: 29–35.

Choo, Chun Wei, 1999. The Art of Scanning the Environment. *Bulletin of the American Society for Information Science*, March: 21–47.

Cooper, Donald R., Dan C. William Emory, 1998. *Metode Penelitian Bisnis*. Erlangga, Jakarta

Coyne, Kevin P. 1997. Sustainable Competitive Advantage – What It Isn't. *Strategic Management Journal*, Vol. 10, 75–87 (1989).

Krahnen, J.P., and R.H. Schmidt. 1994. *Development finance as institution-building: A new approach to poverty-oriented banking*. Boulder, Colorado: Westview Press.

Lippman, S., & Rumelt, R. 1982. Uncertain imitability : An analysis of interfirm differences in efficiency under competition. *Bell Journal of Economics, 13*: 418–438.

Manfred Zeller (Professor and Director, Institute of Rural Development, Georg-August-University Gottingen, Germany), *Models of Rural Financial Institutions*: 5.

Michael E. Porter and Victor E. Millar, How Information Gives You Competitive Advantage, FROM THE JULY 1985 ISSUE.

Stiglitz, J.E. And A. Weiss. 1981. "Credit Rationing in Markets with Imperfect Information," *American Economic Review, Volume 71*, No. 3, 1981, pp. 393–410.

Advances in Business, Management and Entrepreneurship – Hurriyati et al. (Eds)
© 2021 Taylor & Francis Group, London, ISBN 978-0-367-67471-7

Determining factors of entrepreneur students

E. Ustha, A. Rahayu & L.A. Wibowo
Universitas Pendidikan Indonesia, Bandung, Indonesia

ABSTRACT: The purpose of this study was to analyze the effect of success itself, risk tolerance, freedom of work, need for achievement, and the readiness of the instrumentation in Bandung on students who wish to become entrepreneurial. Types of data used are primary data. The method of the research was a survey with an inductive analysis approach to looking at ways that are generally used in tested factors influencing students' desire to become entrepreneurs. The method of data collection was using a questionnaire with the number of respondents being as many as 100 people. The sampling techniques used were simple random techniques (Random Sampling Method). The data were processed and analyzed by using SPSS application assistance. Analytical tools used in this study included testing the validity, reliability, and multiple linear regression. The results of this study indicate that the variables of self-success, freedom of work, and the need for achievement are dominant variables affect the desire of the students in Bandung to become entrepreneurs.

1 INTRODUCTION

Unemployment in Indonesia is increasing in number over time. Job seekers who have a bachelor's degree or not have to compete to get work in limited jobs. The cause of the problem of educated unemployment is that many scholars aim only to find work, not create jobs. Whereas being an entrepreneur is one of the supporters that determines the economic slowdown because the field of entrepreneurship gives the freedom to work and be independent. It is this entrepreneur who can create new jobs to be able to absorb labor.

The tendency that occurs in students who are in college today is that most of them prefer an established job by getting honorable status and making a lot of income after completing their education. The tendency is that most students, including final year students, as well as graduates who have just graduated, do not have entrepreneurship plans. Generally, they prefer to become a worker in large companies and government agencies (become civil servants) to guarantee their future. Therefore, university graduates need to be directed and supported to not only be oriented as job seekers, but can and are ready to become job creators as well. The main thing that causes someone to do entrepreneurial activities is the desire for entrepreneurship.

Adi Susanto (2000) suggests that some of the motivations that can encourage a person to become an entrepreneur are the desire to feel free at work, self-success achieved, and the tolerance of risk.

Tolerance of risk is the ability and creativity of a person in completing the size of a risk taken to get the expected income.

Also, entrepreneurs must be able to determine the amount of capital needed to start a business; an entrepreneur must first determine the minimum amount of each resource needed. Some resources are needed in a higher quantity and quality than compared to others (Susanto 2009: 11).

The availability of business information is also an important factor that drives one's desire to open a new business and a critical factor for business growth and sustainability (Indarti 2008). The intervention of others can determine a person's success or failure in the business world. Business relations have the principle of being directly proportional, meaning that the more the number of business relationships, the faster someone reaches success in business, and vice versa (Sudjatmoko 2009).

The availability of capital, the availability of information, and the availability of business relations is called the instrumentation readiness of an entrepreneur (Indarti 2008).

The need for achievement can be interpreted as a unity of character that motivates a person to face challenges to achieve success and excellence. The need for achievement can also encourage the ability to make decisions and the tendency to take risks of an entrepreneur (in Indarti 2008). Achievement needs affect one's entrepreneurial interest in wanting to achieve the desired career path according to the hard work done.

So students are motivated and have an entrepreneurial desire it is felt necessary to analyze the factors that can influence students' desires to become entrepreneurs in the hope that it will later become a consideration for universities in developing courses, especially in the field of entrepreneurship. These factors are first, self-success, second, risk tolerance, third, freedom of work, fourth, achievement needs, and fifth, instrumentation readiness to be an entrepreneur.

2 METHOD

2.1 Sample

The population of this study consisted of students of four tertiary institutions in Bandung, namely the Indonesian University of Education (UPI), Padjadjaran University (UNPAD), Bandung Islamic University (UNISBA), and the Equity College of Economics (STIE Ekuitas).

The number of samples (size of samples) is determined based on calculations from the Slovin formula with a tolerable error rate of 10%, which obtained a sample size of 100 respondents.

The sampling technique used was cluster random sampling. Cluster random sampling is used to maintain the representation of each existing university with the same proportions.

2.2 Variables

This research is explanatory research, which will prove the causal relationship between the independent variables, namely the self-success factor variable, the risk tolerance variable, and the freedom of work variable; and the dependent variable, which is the desire factor to be an entrepreneur. And it is correlational research, namely research that seeks to see whether two or more variables have a relationship or not, how big the relationship is, and to determine the direction of the relationship.

The variables in this study consisted of independent variables (X), namely the Self-Success Factor (X1), Risk Tolerance Factor (X2), Freedom of Work Factor (X3), Need for Achievement Factor (X4), and Instrumentation Readiness Factor (X5). And it includes a dependent variable (Y), which is the Desire to Become an Entrepreneur.

2.3 Model

The model that will be used in this study is illustrated in Figure 1.

The data analysis method used in this research is the quantitative analysis method. It intends to achieve the first goal, which is to analyze the effect of self-success, risk tolerance, and freedom in working toward the desires of entrepreneurial students, by using multiple regression analysis.

The relationship model of these variables can be arranged in functions or equations as follows:

$$Y = a + b_1 X_1 + b_2 X_2 + b_3 X_3 + b_4 X_4 + b_5 X_5 + e \qquad (1)$$

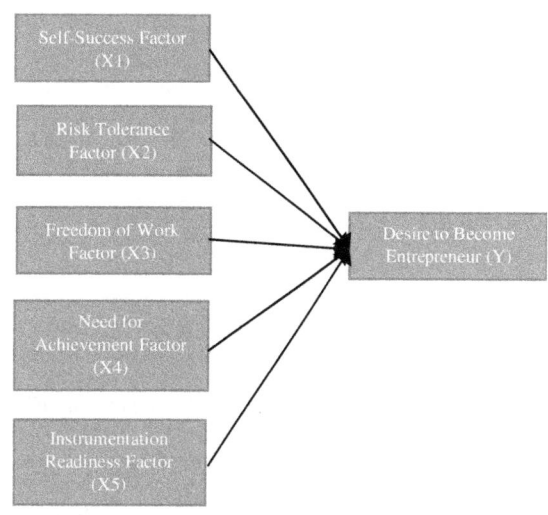

Figure 1. Schematic framework for research model.

Where:
Y = Desire of entrepreneurial students
a = Constant
b = Coefficient of independent variable regression
X1 = Self-success
X2 = Tolerance for risk
X3 = Freedom of work
X4 = Need for achievement
X5 = Instrumentation readiness
e = Error

In this study, the primary data were obtained through direct interviews in the field using a structured research questionnaire, which is divided into four parts, namely self-success factor, risk tolerance factor, freedom of work factor, and the desire to become an entrepreneur.

Overall, for the self-success factor, risk tolerance factor, freedom of work, and desire to be an entrepreneur, statements are measured using a 5-point Likert scale, where respondents are asked to answer with a choice of numbers between 1 and 5 (1 = very not agree and 5 = strongly agree).

3 RESULTS AND DISCUSSION

3.1 Multiple regression analysis

Regression equation models that can be obtained from data processing are as follows:

$$Y = 1.488 + 0.602X_1 + 0.211X_2 + 0.306X_3 + 0.399X_4 + 0.053X_5 + e \qquad (2)$$

This means that with an increase in self-success, the higher the freedom to work, and the higher the need for achievement will increase the soul of an entrepreneur in students.

693

3.2 Discussion of hypothesis test results

The self-efficacy variable has a significant positive effect on a student's desire to become an entrepreneur. The results of this hypothesis are following previous studies conducted by Segal et al. (2005), Adi Tama & Djastuti (2010), Widhari & Suarta (2012), and Adeline (2011), based on a standardized regression analysis, the first hypothesis, which states that there is a significant influence between the variables of self-success on the desire to become entrepreneurs, is proven and the hypothesis is accepted. This shows students have a high enthusiasm for working in running their business and have an optimistic spirit for the goals expected for their future.

The risk tolerance variable does not significantly influence the desire of students to become an entrepreneur. The results of this hypothesis are not proved under previous studies conducted by Segal et al. (2005), Adi Tama & Djastuti (2010), Widhari & Suarta (2012), and Adeline (2011). These state that the risk tolerance variable positively and significantly influences the desires of entrepreneurial students, this is because not all students like the challenges and like to take the opportunities that exist.

The freedom of work variable has a significant positive effect on the desire of students to become an entrepreneur. The results of this hypothesis are under previous research conducted by Widhari & Suarta (2012), who proved a hypothesis that states that there is a significant influence between the variables of freedom in working on the desire to become entrepreneurs. With entrepreneurship, students after graduation will be able to create jobs and have the freedom to choose the desired job.

The variable of the need for achievement has a significant positive effect on a student's desire to become an entrepreneur. The results of this hypothesis are consistent with previous research conducted by Indira (2010), which states that there is a significant influence of the variable of the need for achievement on the desire to become entrepreneurs. Because someone who needs achievement will pursue more entrepreneurial work than other types of work and tend to do a good job at their job.

The instrumentation readiness variable does not significantly influence the desire of students to become an entrepreneur. The results of this hypothesis are not proved under previous research conducted by Agustina (2011), which states that the readiness of instrumentation has a positive and significant effect on the desire of students to become entrepreneurs. This is because not all students feel they have a good instrument readiness, they have a lack of available capital and information, and the social networks owned by each needs to be prepared if they want to become entrepreneurs.

4 CONCLUSION

From the discussion that has been described, the following conclusions can be drawn:

The independent variables (self-success, risk tolerance, freedom of work, need for achievement, and instrumentation readiness) together have a positive and significant influence on students' desire to become entrepreneurs.

The independent variables of self-success, freedom of work, and the need for achievement partially or individually have a positive and significant influence on the desire of students to become entrepreneurs. While the independent variables of risk tolerance and instrumentation readiness have no significant effect; this is because there are still many students who have not carried out entrepreneurial activities and tend to avoid risks in decision making, as well as their limited access to capital, known information, and social networks every student has.

The success factor has the most influence on the students' desire to become entrepreneurs. This can be seen from the coefficient value of 0.602, which is greater than the tolerance of risk (0.211), freedom of work (0.306), the need for achievement (0.399), and instrument readiness (0.053) values.

REFERENCES

Adeline. 2011. Faktor Faktor yang Mempengaruhi Minat Berwirausaha Budidaya Lele Sangkuriang., *Jurnal Ekonomi Manajemen,* Universitas Gunadarma

Adi Tama, A. & Djastuti, I. 2010. Analisis faktor-faktor yang memotivasi mahasiswa berkeinginan menjadi wirausaha. (Doctoral dissertation, Universitas Diponegoro).

Agustina, Cynthia. 2011. Intensi Kewirausahaan Mahasiswa: Studi Perbandingan Antara Fakultas Ekonomi dan Fakultas Ilmu Komputer. Skripsi. Bekasi: Universitas Gunadarma

Indarti, N, 2008. "Intensi Kewirausahaan Mahasiswa: Studi Perbandingan antara Indonesia, Jepang, dan Norwegia", *Jurnal Ekonomi dan Bisnis Indonesia,* (23), 4.

Indira, Christera Kuswahyu. 2010. Student Entrepreneurship Intention: Study of Comparison Between Java and Non Java, *Jurnal Manajemen.* Fakultas Ekonomi. Universitas Gunadarma

Segal, G., Borgia, D., & Schoenfeld, J. (2005). The motivation to become an entrepreneur. *International journal of Entrepreneurial Behavior & research.*

Sudjatmoko, Agung. 2009. Cara Cerdas Menjadi Pengusaha Hebat. Jakarta: VisiMedia.

Susanto, A. B. 2009. *Leadpreneurship.* Jakarta: Esensi

Widhari, C. I. S., & Suarta, I. K. 2012. Analisis faktor-faktor yang memotivasi mahasiswa berkeinginan menjadi wirausaha. *Jurnal Bisnis dan Kewirausahaan, 8*(1), 54–63.

Advances in Business, Management and Entrepreneurship – Hurriyati et al. (Eds)
© 2021 Taylor & Francis Group, London, ISBN 978-0-367-67471-7

Conceptual framework of innovation strategy in SMEs

N.A. Hamdani, D. Disman, A. Rahayu & R. Hurriyati
Universitas Pendidikan Indonesia, Bandung, Indonesia

ABSTRACT: The use of new innovations as tools to face competition is indeed considered to be effective enough to win the market. The absence of innovation will make consumers feel bored, leave the product, and we can be sure the business will sink amid the hustle and bustle of competition. This study analyzes the application of innovation strategy concepts in companies and small and medium enterprises (SMEs). The approach in this study is a scientific analysis of the literature and previous research on innovation strategies in SMEs. Based on the results of the study, the conceptual framework of the application of SME strategies was obtained, including both dimensions of the indicators and keys to success and obstacles. The implication of this research is that there is an appropriate concept that can be used to develop research in the field of innovation strategy.

1 INTRODUCTION

SMEs play a considerable role in the economy in the Asia Pacific region. Data from the Asian Development Bank (ADB) shows that SMEs contribute to up to 62% of employment and constitute 96% of the total number of companies in 20 countries in the Asia Pacific region. In addition, the contribution of SMEs to exports in various countries in the Asia Pacific region is fairly high, for example, it is 40% in China and India, 26% in Thailand, 19% in South Korea, and 16% in Indonesia. In developed countries like Japan, SMEs contribute to 70% of employment, 50% to gross domestic product (GDP), and constitute 99% of the total number of companies in the country (Josephus & Primus 2018).

Previous studies have examined SME performance in terms of different aspects such as ICT use (Ashrafi & Murtaza 2008), intellectual capital approach (Astuti 2005), absorptive capability (Tzokas et al. 2015), distinctive capability (Rahim et al. 2009), marketing strategy (Jaakkola et al. 2007; Morgan 2012), electronic marketing (Hamdani & Maulani 2018; López & Sicilia 2014; Wan et al. 2015), entrepreneurship (Effendi & Hadiwidjojo 2013), and entrepreneurial orientation (Mahmood 2013).

SME business performance is largely determined by innovation because innovation enables SMEs to develop their businesses (Ismail et al. 2010). Innovation can also improve the competitiveness of SMEs (Moghavvemi 2012; Zeebaree & Siron 2017). Financial performance, operating performance, and company performance are also highly determined by and related to innovation (Kim-soon et al. 2017).

Innovation strategy is an important issue to study because the innovation strategy is the key to successful SME performance in various developing countries (Susanto & Wasito 2017). Innovation strategy is also worth studying due to the application of high technology in SMEs to improve business performance (Sebiane 2016). This study examines conceptual innovation strategies that can be applied to SMEs to improve their performance and competitiveness.

1.1 Literature review

Innovation as an economic success introduces new ways or new combinations of old ways of transforming inputs into outputs (technology) to produce a drastic change in the ratio between the perceived use value by consumers of the benefits of a product (goods and/or services) and the price set by the producer (Fontana 2011). Innovation is the process of turning opportunities into new ideas and putting them into practice. There are four types of innovation, namely (1) product innovation, which refers to changes in goods or services, (2) process innovation, which refers to changes in how a product is created or delivered, (3) position innovation, which refers to changes in how a product is introduced, and (4) paradigm innovation, which refers to changes in mental models that underlie organizational activities (Tidd & Bessant 2009).

Innovation as creation (discovery) focuses on using resources (manpower, time, and funds) to create or develop new products, new services, new ways of doing things, and new ways of thinking (Ahmed et al. 2010). Innovation as a product, process, and organizational change does not always originate from new scientific discoveries, but it can be a result of combining existing technologies and then applying them to new contexts (Bozkurt & Kalkan 2014). Innovation is a deliberate change in the framework in which practical innovations extend to the point when demand is fully met. Today, the concept of innovation is quite open, it transcends organizational boundaries and exploits not only internal changes, but also external changes. It is also suggested that innovation is the fundamental basis of competitiveness and a source of progress and

Table 1. Barriers to innovation in SMEs.

Authors	Some Barriers to Innovation in SMEs
Piatier (1984)	(1) Lack of government support as an important barrier to innovation in European countries
Economist Intelligence Unit (2007)	(1) Necessities related to the frequency, timing, and speed of innovation; (2) Organizational culture mutation and reducing time to market as a permanent challenge in the assumption of innovation objectives; (3) Chief Executive Officers (CEOs) of full age have a greater departure from the view against the goals of innovation and innovative capacity of the organization
Baranano et al. (2005)	(1) Lack of qualified human resources; (2) Huge absence of external communication between the knowledge generators
Sierra (2009)	(1) Organizational structure, as well as the climate; (2) Culture and strategy resistance to change; (3) Tradition and cemented rules; (4) Market leadership and absence of rethinking on it; (5) Additional work brought by change; (6) Weak repay on risk assumption
Silva et al. (2007)	(1) High economic cost and risk associated with innovation; (2) Lack of funding; (3) Organizational rigidity; (4) Lack of skilled human resources; (5) Lack of market information and technology; (6) Government regulation; (7) Weak capacity to approach clients, as well as lack of cooperation with centers of learning
Madrid-Guijarro et al. (2009)	(1) External environment; (2) Human resources; (3) Risk; (4) Financial position

*Source: Bozkurt & Kalkan (2014).

development. Companies and nations that continue to innovate will succeed in sustaining their economic sustainability (Dibrov 2015).

The basic thinking of the general strategy of innovation is to create a new product life cycle and, thus, make old similar products obsolete (Al-Battaineh 2018). Innovation is different from creativity. Creativity is new thoughts; on the contrary, innovation is doing something new or transforming new ideas

into a business success (Kabukcu 2015). Innovation is the process of realizing these new ideas in the form of product innovation, service innovation, process innovation, and management innovation (Al-Battaineh 2018). Innovation can be distinguished from the capacity for new ideas, which is the ability of organizations to develop or use new products and processes (Drucker & Peter 1985). Innovation combined with several factors of competition culture can create a large capacity for new ideas, resulting in excellent organizational performance (Dibrell et al. 2008).

2 METHODS

The present study was conducted through a literature review. The literature included relevant books and scientific articles on innovation and innovation strategy. Facts and data were then analyzed using a descriptive approach.

3 RESULTS AND DISCUSSION

Innovation strategies affect company performance. The balance scorecard approach shows that marketing, financial, and business process performance are determined by an innovation strategy (Karabulut 2015). It is also suggested that innovation strategies are linked to formal structures, technological capabilities, and innovation performance (Kamasak 2015).

An innovation strategy plays an important role in determining the innovation capability of a company (Tushman & O'Reilly 1997). It has also a positive impact on the company's financial performance (Nybakk & Jenssen 2012). Innovation is the key to competitive advantage that determines the economic success of every organization (Atalay et al. 2013). It has been proven that there is a positive relationship between competitive advantage and technological innovation. Innovation in one area always affects other areas (Dibrell et al. 2008).

Innovation that systematically exploits changes is very effective. Innovation is a goal-directed search for changes that might offer economic or social innovation (Drucker & Peter 1985). When applied in SMEs, innovation strategies may be restricted by several factors as outlined in Table 1.

Barriers to innovation in SMEs are generally linked to government policies, human resources, risks, weakness in predicting external environment, and limited financial capability.

Innovation strategies play an important part in improving SMEs (Sebiane 2016). Studies suggest that innovation strategy in SMEs is closely related to their capability to adopt technology (Akmal et al. 2017). Dynamic conditions may enable SMEs make innovation to achieve the expected business performance (Wook & Wook 2016).

Innovation strategy is related to a company's strategy response in adopting innovation. Previous studies

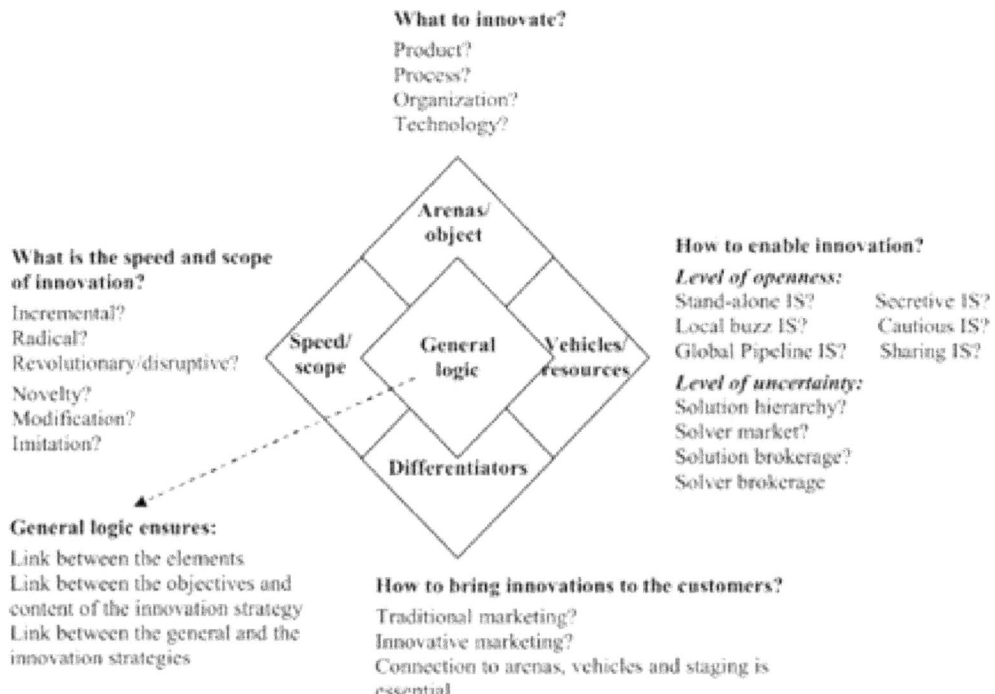

What to innovate?
Product?
Process?
Organization?
Technology?

Arenas/object

What is the speed and scope of innovation?
Incremental?
Radical?
Revolutionary/disruptive?
Novelty?
Modification?
Imitation?

Speed/scope

General logic

Vehicles/resources

How to enable innovation?
Level of openness:
Stand-alone IS?   Secretive IS?
Local buzz IS?   Cautious IS?
Global Pipeline IS?   Sharing IS?

Level of uncertainty:
Solution hierarchy?
Solver market?
Solution brokerage?
Solver brokerage

Differentiators

General logic ensures:
Link between the elements
Link between the objectives and
content of the innovation strategy
Link between the general and the
innovation strategies

How to bring innovations to the customers?
Traditional marketing?
Innovative marketing?
Connection to arenas, vehicles and staging is
essential

Figure 1. Model of Fredickson's innovation strategies.
*Source: Stankevice & Jucevicius (2016).

have shown that there are many types of innovation strategy such as offensive innovation strategy, defensive innovation strategy, imitative innovation strategy, dependent innovation strategy, traditional innovation strategy, and opportunist innovation strategy (Freeman 1978 in Hadjimanolis & Dickson 2000). This typology is based on the speed and time of entry of a company into the new technology area. Urban & Hauser (1980) in Hadjimonalis & Dickson (2000) distinguishes innovation strategy into proactive strategy and reactive strategy. The first is where a company projects and anticipates environmental changes. This strategy is usually employed by first mover companies. This strategy allows companies to build market share and reputation, but it requires high development costs and has the risk of investment in technology or faulty design. Reactive strategy is a strategy in which companies only react to consumer demand and competitors' activities.

Based on Fredickson's model, the appropriate innovation strategies to be applied in SMEs are offensive, defensive, imitation (Jakubavicius 2008), technological process, organizational process, and process innovation (Edquist et al. 2001).

4 CONCLUSION

The appropriate innovation strategies in SMEs are offensive, defensive, and imitation considering their internal and external factors, including technology, process, and product. It is necessary for SMEs to be able to project and anticipate environmental changes.

REFERENCES

Ahmed, A., Pervaiz, K., & Shepherd, C. D. 2010. Innovation management (1 Th Edition). *New Jersey, USA: Pearson Education Inc.*

Akmal, N., Wahab, A., & Jabar, J. 2017. Organizational innovation strategy towards small medium enterprise performance in Malaysia. *International Journal of Arts Humanities and Social Sciences* 2(7): 1–9.

Al-Battaineh, M. 2018. Effect of innovation strategies on the functional performance of SME organizations in (Hassan Industrial City). *International Journal of Business and Management Invention (IJBMI)* 7(5): 12–18.

Ashrafi, R. & Murtaza, M. 2008. Use and impact of ICT on SMEs in Oman. *Electronic Journal of Information Systems Evaluation* 11(3): 125–138.

Astuti, P. D. 2005. Hubungan intellectual capital and business performance. *Jurnal Maksi* 5: 34–37.

Atalay, M., Anafarta, N., & Sarvan, F. 2013. The relationship between innovation and firm performance: An empirical evidence from Turkish automotive supplier industry. *In 2nd International Conference on Leadership, Technology and Innovation Management* 75: 226–235.

Baranano, A. M., Bommer, M., & Jalajas, D. S. 2005. Sources of innovation for high-tech SMEs: A comparison of USA, Canada, and Portugal. *International Journal of Technology Management* 30(1–2): 205–219.

Bozkurt, Ö. Ç. & Kalkan, A. 2014. Business strategies of SMEs, innovation types and factors influencing their innovation: Burdur model. *Akademic Review* 14(2): 189–198.

Dibrell, C., Davis, P., & Craig, J. 2008. Fuelling innovation through information technology in SMEs. *Journal of Small Business Management* 46(2): 203–218.

Dibrov, A. 2015. Innovation resistance: The main factors and ways to overcome them. *Procedia – Social and Behavioral Sciences* 166: 92–96.

Drucker, D. & Peter, F. 1985. Innovation and entrepreneurship, practice and principles. New York: Harper & Row Publisher.

Economist Intelligence Unit. 2007. No Title. [online]. Retrieved from https://www.eiu.com/.

Edquist, C., Hommen, L., & McKelvey, M. D. 2001. Innovation and employment: Process versus product innovation. Edward Elgar Publishing.

Effendi, S. & Hadiwidjojo, D. 2013. The effect of entrepreneurship orientation on the small business performance with government role as the moderator variable and managerial competence as the mediating variable on the small business of apparel industry in Cipulir Market, South Jakarta. IOSR Journal of Business and Management 8(1): 49–55.

Fontana, A. 2011. Innovate we can! Innovation management and value creation revision edition.

Hadjimanolis, A. & Dickson, K. 2000. Innovation strategies of SMEs in Cyprus, a small developing country. *International Small Business Journal* 18(4): 62–79.

Hamdani, N. A. & Maulani, G. 2018. The influence of E-WOM on purchase intentions in local culinary business sector. *International Journal of Engineering & Technology* 7: 246–250.

Ismail, K., Zaidi, W., Omar, W., Soehod, K., Senin, A. A., & Akhtar, C. S. 2010. Role of innovation in SMEs performance: A case of Malaysian SMEs. *Mathematical Methods in Engineering and Economics Role* 1(2): 145–149.

Jaakkola, M., Parvinen, P., & Möller, K. 2007. Strategic marketing and its effect on business performance in Three European Engineering Countries. *In of the 36th Annual Conference of the Helsinki: Reserach Gate.* [Online]. Retrieved from http://www.stratmark.fi/wp-con-tent/uploads/2008/03/Jaakkola_Parvinen_Moller.pdf.

Josephus, J. & Primus, P. 2018. *Makin besar, peran UKM di Kawasan Asia Pasifik.* [Online]. Retrieved from https://ekonomi.kompas.com/read/2018/05/09/14090 4226/makin-besar-peran-ukm-di-kawasan-asia-pasifik.

Kabukcu, E. 2015. Creativity process in innovation-oriented entrepreneurship: The case of Vakko. *Procedia – Social and Behavioral Sciences* 195: 1321–1329.

Kamasak, R. 2015. Determinants of innovation performance: A resource-based study. *In World Conference on Technology, Innovation and Entrepreneurship* 195: 1330–1337.

Karabulut, A. T. 2015. Effects of innovation types on performance of manufacturing firms in Turkey. *In World Conference on Technology, Innovation and Entrepreneurship* 195: 1355–1364.

Kim-soon, N., Ahmad, A. R., Kiat, C. W., & Sapry, H. R. 2017. SMEs are embracing innovation for business performance. *Journal of Innovation Management in Small & Medium Enterprises* 1: 1–17.

López, M. & Sicilia, M. 2014. Determinants of E-WOM influence: The role of consumers' internet experience. *Journal of Theoretical and Applied Electronic Commerce Research* 9(1): 28–43.

Madrid-Guijarro, A., Garcia, D., & Van Auken, H. 2009. Barriers to innovation among Spanish manufacturing SMEs. Journal of small business management 47(4): 465–488.

Mahmood, R. 2013. Entrepreneurial orientation and business performance of women-owned small and medium enterprises in Malaysia: Competitive advantage as a mediator. *International Journal of Business and Social Science* 4(1): 82–90.

Moghavvemi, S. 2012. Competitive advantages through IT innovation. *Adoption by SMEs* 7564(1): 24–39.

Morgan, N. A. 2012. Marketing and business performance. *Journal of the Academy of Marketing Science* 40(1): 102–119.

Nybakk, E. & Jenssen, J. I. 2012. Innovation strategy, working climate, and financial performance in traditional manufacturing firms: An empirical analysis. *International Journal of In-novation Management* 16(April): 1–30.

Piatier, A. 1984. *Barriers to innovation.* London; Dover, NH: F. Pinter.

Rahim, A., Bakar, A., Hashim, F., & Ahmad, H. 2009. Distinctive capabilities and strategic thrusts of Malaysia's Institutions of Higher Learning. *International Journal Marketing Studies* 1(2): 158–164.

Sebiane, F. 2016. The contribution of innovative strategy in the growth of high technology SMEs and their business in Dubai. *International Journal of Research in Management & Business Studies (IJRMBS 2016)* 3(4): 2014–2017.

Sierra González, J. H. 2009. Assessing exporting culture in Colombian SMEs: A look at the Export Promotion Program (EPP). *Cuadernos de administración* 22(39): 99–134.

Silva, M. J., Leitão, J., & Raposo, M. L. B. 2007. Barriers to Innovation faced by Manufacturing Firms in Portugal: How to overcome it?

Stankevice, I. & Jucevicius, G. 2016. Innovation strategy: An integrated theoretical framework innovation strategy: an integrated theoretical framework. *Socialiniai Mokslai* 69(3): 23–31.

Susanto, A. B. & Wasito, W. 2017. Improve the performance of SMEs through innovation strategies in developing countries. *International Journal of Scientific & Technology Research* 6(10): 282–285.

Tidd, J., & Bessant, J. (2009). Managing innovation: Integrating technological, market and organizational change (4th Edition). Great Britain: Wiley.

Tushman, M. & O'Reilly, C. A. 1997. Winning through innovation: A practical guide to leading organizational change and renewal (6th Edition). Boston, USA: Harvard Business School Press.

Tzokas, N., Kim, Y. A., Akbar, H., & Al-Dajani, H. 2015. Absorptive capacity and performance: The role of customer relationship and technological capabilities in high-tech SMEs. *Industrial Marketing Management* 47: 134–142.

Wan, N., Jadhav, V., Khanna, M., Pettersson, J., Kersmark, M., Staflund, L., & Broekhuizen, T. 2015. The conceptualization of Electronic Word-of-Mouth (EWOM) and company practices to monitor, encourage, and commit to EWOM – A Service Industry Perspective. *Exjobb* 14(1): 256.

Wook, Y. & Wook, S. 2016. Market dynamics and innovation management on Performance in SMEs: Multi-agent simulation approach. *Procedia – Procedia Computer Science*, 91(Itqm): 707–714.

Zeebaree, M. R. Y. & Siron, R. 2017. The impact of entrepreneurial orientation on competitive advantage moderated by financing support in SMEs. *International Review of Management and Marketing* 7(1): 43–52.

Advances in Business, Management and Entrepreneurship – Hurriyati et al. (Eds)
© *2021 Taylor & Francis Group, London, ISBN 978-0-367-67471-7*

Lecturer management at As-Syafi'iyah Islamic University (case study on human capital)

S. Lestari, M. Akbar & B. Santoso
Universitas Negeri Jakarta, Jakarta, Indonesia

ABSTRACT: The concept of human capital addresses the added value that employees make to an organization. Human capital is one of the most important elements of intangible assets in universities. Therefore, universities must improve the quality of human resources in facing challenges in order to achieve the university's vision and mission. In this study, using qualitative methods, there are three data validity tests that have been carried out, namely data triangulation, method triangulation, and theory triangulation. To get quality human resources, the process can be done by planning, organizing, directing, and developing. Based on the results of interviews with core informants and supporting informants, it was found that the use of technology in the operational process was still not optimal, there's a lack of support, supervision, and dissemination of the implementation and objectives of the As-Syafi'iyah Islamic University's (UIA) vision and mission, and the training budgeting has not been made at the beginning of the year and was only submitted if there was training. In the case of lecturer development, it was found that there was a lack of control and approach from the leader. However, there are many lecturers who have a long period of work and remain in service in the UIA because it is based on sincerity to worship the God Almighty even though the welfare of lecturers is still lacking.

1   INTRODUCTION

The concept of human capital addresses the added value that can be done by employees (humans) to the organization where they work. The definition of human capital is the knowledge, expertise, ability, and skills that make humans (employees) as capital (assets) of the company. If the company treats employees as capital, the company will get greater profits than only treating employees as human resources (Gaol 2014). The human resource (HR) factor plays an important role, including in tertiary institutions known as the teaching profession. HR management becomes the artery of the institution, because the human factor that is managed becomes the determinant of the path of institutional activities according to Siagian (Arwildayanto 2008). Human capital is one of the most important elements of an organization's intangible assets. Human capital management (HCM) is a way for organizations to organize, recruit, retain, train, and develop their employees and people that are valuable assets to the organization, not only as expenses (Kimgsmil in Sulistiyanto 2019). Quality HR indirectly illustrates the quality of the organization. HR has many advantages compared to technology according to Schermerhorn (Purba 2008).

2   LITERATURE REVIEW

According to Siagian (2008), the HR factor plays an important role, including in tertiary institutions known

as the teaching profession. HR management is the pulse of the institution, because of the human factor that is managed. In management, as generally known, there are three levels that can be described in a pyramid as follows:

Figure 1.   Three levels of management.

1) Top Manager. This includes the board of directors or directors whose job it is to make important decisions, which affect almost the entire course of the organization.
2) Middle Manager. The task of the middle manager is to develop an operational plan and create continuity between what is demanded by their superiors and the capabilities of their subordinates.
3) Executive Manager (Lower Manager). The task of the executive manager is to carry out an operational plan that has been developed by middle management (Gaol 2014).

According to Nalbantian et al., the definition of human capital is as follows:

"The stock of accumulated knowledge, skills, experience, creativity and other relevant workforce attributes and suggest that HCM involves putting into place the metrics to measure the value of these attributes and using that knowledge to effectively manage the organization."

Qualitative data analysis is a series of processes of looking for and compiling qualitative data obtained from interviews, observation notes, and other material systematically. This process aims to make data easily understood and obtain findings that can be shared with others. The qualitative data analysis technique used in this study follows the concept of Miles and Huberman. There are three stages of qualitative data analysis according to Miles and Huberman, namely data reduction, display, and conclusion/verification.

Management is planning, organizing, leading, and controlling human and other resources to achieve organizational goals efficiently and effectively; organizational resources include assets such as people and their skills and knowledge, machines, raw materials, computer and information technology, and financial capital, according to Jones and George (Gaol 2014).

3 METHOD

In this research at the UIA, the method used was a qualitative approach and research sources were obtained from observations and interviews with lecturers, study programs, deans, HRDs, and leaders at UIA. This study discussed how to plan, organize, direct, and supervise lecturers such as for the flow of research (see Figure 2).

Qualitative evaluation uses qualitative data, and to capture the data, qualitative instruments are used. According to Michail Quin Patton (2012), qualitative understanding is as follows: "Qualitative data consists of detailed descriptions of the situations, events, people, interactions, and behaviors observed directly from people about their experiences, attitudes, and beliefs, and thoughts; and experts or all sections of documents, correspondence, notes, and case histories. Detailed, direct negotiations, and documentation of qualitative meaning cases are raw data from the empirical world. Data are collected as an open narrative without trying to fit program activities or people into predetermined categories, standards such as response choices consisting of queries or special tests."

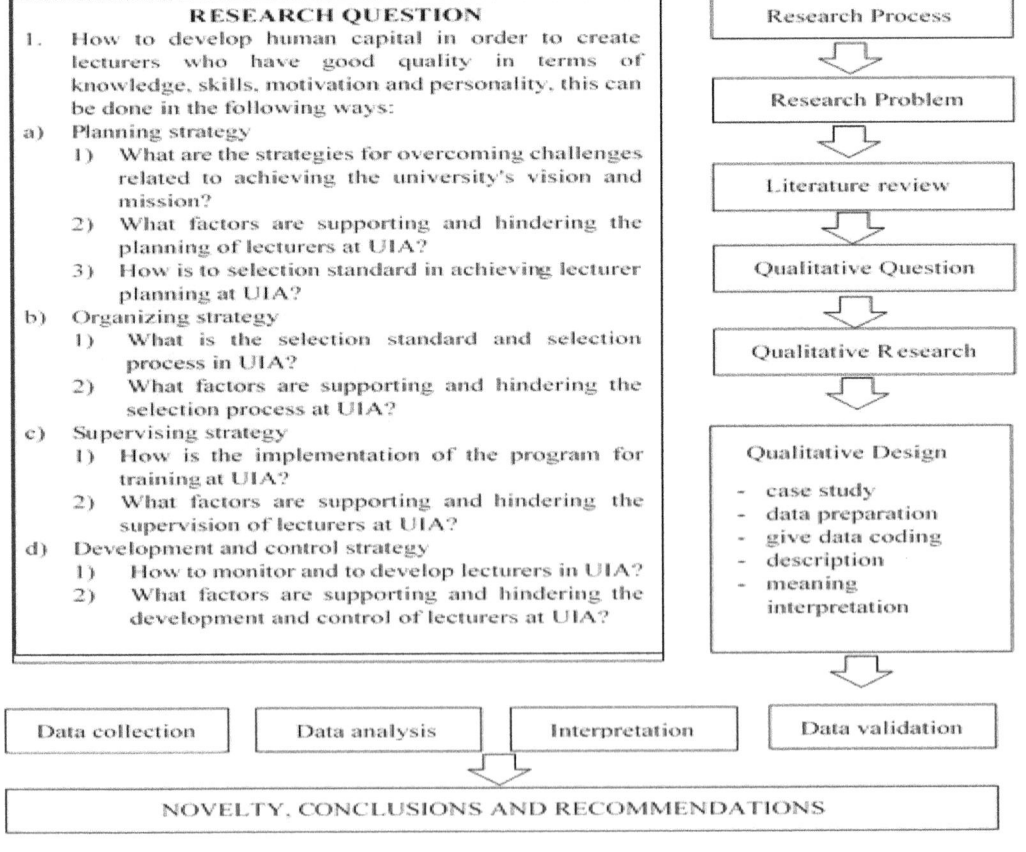

Figure 2. Research flow.

Testing data validity in qualitative research
1) Test credibility.

In research, the data credibility from qualitative research results, among others, can be done with extended observations, increased perseverance in research, triangulation, discussion with colleagues, negative case analysis, and member checks (Sugiyono 2016).

2) Test transparency.

This transferability is external validation in quantitative research. External validation shows the level of accuracy and can be applied to the results of the study into the population where the sample was taken. This transfer value is related to the question, where the results of the study can be applied or used in other situations. Therefore, the researchers hope that people can understand the results of the qualitative research and apply the results of this study. The researchers will provide detailed, clear, systematic descriptions and hope that this research can be applied elsewhere (Sugiyono 2016).

3) Dependability testing.

In qualitative research, dependency testing is done by conducting an audit of the entire research process. This method is conducted by an independent auditor or supervisor to audit the overall activities of researchers in conducting research. According to Sanafiah Faisal, how researchers begin to determine the problem/focus is to go into the field to determine the source of data, to conduct data analysis, to test the validity of the data, and to make conclusions, which must be pointed out by the researcher. If the researcher does not have and cannot show "traces of their field activities," then the reliability of the researcher must be doubted.

4) Confirmability testing.

Confirmability testing in qualitative research is similar to the dependability test, so that the test can be done simultaneously.

5) Test confirmability

It means testing the results of research whether a function of the research process is carried out, then the research can meet the confirmability standards.

Employee development aims to enhance careers, to improve employee skills in the field, and to update new procedures and systems. It can also be used as problem solving by providing various skills, thus, retaining and motivating employees in developing careers. According to Mathis and Jockson (Widodo 2015), employee training and development is designed to help organizations achieve their goals. Therefore, determinants of organizational training need to reflect the diagnostic stage of training determinants. This assessment looks at employee and organizational performance issues to determine whether training will help. With the training held, it is intended to improve the quality of employees and reduce spending to be more efficient and more productive.

4 RESULTS AND DISCUSSION

To achieve the goals of the university's vision and mission, qualified employees who have high loyalty to the university are needed, such as what is done at UIA in the following ways:

1) Planning

The basic reasons for planning are (1) protective benefits as a result from reducing the possibility of mistakes in decision making and (2) positive benefits in the form of increasing success in achieving organizational goals (Rusniati & Ahsanul 2014). The researchers found facts based on the results of interviews with informants in the planning stage indicating that there were still some factors that were lacking, including:

a) Lack of socialization regarding lecturer planning in each faculty.
b) The use of technological systems that have not been optimized so that it can hamper the implementation process.
c) Lack of control from leaders in each faculty, so that needs are not quickly resolved.

2) Organizing

The procurement function includes (1) announcing and receiving application letters either sent via the internet, post, or coming in person, (2) conducting HR selection in accordance with organizational needs, (3) conducting orientation and training before plunging into the field, (4) carrying out human resource appointments if deemed to have good performance and met the requirements set, (5) placement of human resources in accordance to their expertise (Randall et al in Kurnia & Santoso 2018). The researcher found facts based on the results of interviews with informants in the organizing stage indicating that there were still some factors that were lacking, including:

a) Planning for a budget for training has not been made at the beginning of the year, spending is done when needed.
b) Not yet maximizing the use of technology systems.

3) Direction and development

It is a function of giving encouragement to workers to be able to work effectively and efficiently based on the planned goals. The researcher found facts based on the results of interviews with informants indicating that there were still some factors that were lacking, including:

a) Lack of willingness from lecturers to attend training coupled with a lack of support from relevant leaders.
b) Lack of control and approach from the leader.
c) The welfare of lecturers is still lacking.

4) Supervision

The objectives of supervision/control were varied, but the most common were (a) total work output (in terms of quantity) (b) quality of work (in terms of quality) (c). employees (sincerity, craft, and work skills) (d) money (legal and efficient use) (e) supplies (purchase, use, and maintain them correctly) (f) workspace (good arrangement and use) (g) time (use for the benefit of the organization concerned) (h) the method of work (Gie in Iin 2018). The researcher found facts based on the results of interviews with informants in the organization indicating that there were some factors still lacking, including:

a) Lack of socialization regarding the timing of the training to lecturers in each faculty.
b) Lack of monitoring regarding the implementation of teaching and training from relevant superiors.

5 CONCLUSION

In order to achieve the university's vision and mission, quality human resources and loyalty to the university are needed. In order to achieve this goal, the steps that need to be taken include planning, organizing, directing and controlling. In addition, the university need to regulate human resources so that lecturer performance is better. Good performance greatly increases the value of higher education in the eyes of the community. It is also hoped that leaders in higher education understand that human capacity must be well managed because it is an asset for the university (Baron & Armstrong 2013).

REFERENCES

Arwildayanto, Siagian dalam. 2008. Pendekatan budaya kerja dosen profesi. *Jurnal Manajemen Sumber Daya Manusia PerguruanTinggi.*

Bontis et al. 2013. Human Capital Manajemen. 9. Jakarta: PPM.

Davenport dalam Baron A., & Armstrong, M. 2013. Human Capital dalam Organisasi. 9–10. Jakarta: PPM.

Dressley, G. dalam Chr. Gaol, J. L. 2014. Human Capital Manajemen SDM. 82. Jakarta: PT Gramedia Widiasarana Indonesia.

Elliot dalam Baron, A., & Armstrong, A. 2013. Human Capital Manajemen. 8–9. Jakarta: PPM Jakarta.

Gaol, J. L. Siagian dalam Chr 2014. Humana Capital Manajemen SDM. 83. Jakarta: PT Gramedia Widiasarana Indonesia.

Milkovich G. T., & Boudreau, J dalam Chr Gaol, J. L. 2014. Human Capital Manajemen SDM. 83. Jakarta: PT Gramedia Widiasarana Indonesia.

Terry, G. R. dalam Chr Gaol, J. L. 2014. Human Capital Manajemen Sumber Daya Manusia. 39. Jakarta: PT Gramedia Widiasarana Indoensia.

Widodo, S.E. 2015. Manajemen Pengembangan Sumber Daya Manusia. 7–10. Yogyakarta: Pustaka Pelajar.

Widodo, S. E. 2015. Indkator Mengukur SDM, buku Manajemen Pengembangan Sumber Daya Manusia. 368–372. Yogyakarta: Pustaka Pelajar.

Clustering of Micro, Small and Medium enterprises (MSMEs) as a strategy for increasing regional economic competitiveness in Tasikmalaya city

S.S. Maesaroh, A. Hermawan & B.M. Purwaamija
Universitas Pendidikan Indonesia, Bandung, Indonesia

ABSTRACT: The existence of Micro, Small and Medium Enterprises (MSMEs) plays an important and strategic role in national economic development. However, various problems in the development of MSMEs have become obstacles in increasing competitiveness. The aim of the study is to analyze the clustering of MSMEs in Tasikmalaya City as an effort to develop the regional economy. This research used quantitative descriptive method. The data were collected through an interview process using a questionnaire and processed using Geographic Information System (GIS). The results showed that human resources were the most important factor in increasing the competitiveness of MSMEs. The competitiveness level of MSMEs in Tasikmalaya City was still considered low, especially because of the lack of optimal institutional and promotional roles. Improvement of institutional roles is needed to improve the level of competitiveness of MSMEs in Tasikmalaya City.

1 INTRODUCTION

The existence of MSMEs is very important for economic growth in Indonesia (Hayashi 2002). However, the development and improvement of MSMEs often experience several major problems (Hamzani & Achmad 2016). These factors have become obstacles in increasing competitiveness of MSMEs. Identifying the factors that cause the low competitiveness are important in the recent condition (Rifai et al. 2013).

The determinants of competiveness are divided into twolane relation. On the one hand, it is the result of variable environment. On another, it depends on the decisions taken inside the company (Piatkowski 2012). Problems that affect the competitiveness of MSMEs are such as limited capital, low labor competency (Hamzani & Achmad 2016; LPPI & Bank Indonesia 2015), lack of productivity and innovation, ease of doing business, market access, infrastructure, and general macroeconomic conditions (LPPI & Bank Indonesia 2015). To solve these problems, both central and regional government have carried out several efforts such as formulating supporting policy. Unfortunately, these policies have not been carried out optimally (Hamzani & Achmad 2017).

The problems of every MSME cannot be equated because of various characteristics and wide limitations. Identification of characteristics of MSMEs is required in order to make sure the optimal policy. One of the characters of MSMEs identified is the competitiveness level (Sipa et al. 2015). The strategy to identify the competitiveness of MSMEs is by developing clusters (clustering) on existing MSMEs. Clustering is used to connect MSMEs to each other, so that it can optimize MSMEs clusters in certain sectors. Developing clusters is expected to encourage the creation of innovations and synergies among related actors (Bappenas 2015).

The aim of this study is to determine the condition of the MSMEs cluster, analyze the competitiveness of each cluster of MSMEs, and provide strategic recommendations for the development of MSMEs clusters in Tasikmalaya City. Geographic Information System (GIS) in this study used to increase effectiveness and efficiency. GIS technology can provide information on location, attributes, and creative industries based on maps. The map produced used as an appropriate description to make recommendations for the development of MSMEs based on local wisdom.

2 METHODS

This research used quantitative descriptive method consisting of two stages. The first stage was determination of the priority weight of the six variables of MSMEs competitiveness (human resources, technological, managerial, institutional, promotional and capital resources). The weighting of these variables was obtained through in depth interview to the expert that has been determined. The interview method used was a pairwise comparison questionnaire with the Saaty scale 1–9. Relative values were processed to determine the relative rank of all observed variables in Table 1.

Determination of weights was done by comparing each variable with a paired matrix. The matrix Pairwise comparisons results represented in vertical and

Table 1. The fundamental scale of absolute numbers (Saaty 2006).

Intensity	Definition
1	Equal importance
3	Moderate importance
5	Strong importance
7	Very strong demonstrated importance
9	Extreme importance
2,4,6,8	Value between two adjacent elements

horizontal forms in the form of stochastic matrices, called supermatrix.

In calculating these weights, the value of consistency was considered. Consistency Ratio (CR) is a parameter used to examine consistency of pairwise comparisons that have been done. Measurement of CR used the following formula:

$$CI = (Average - n)/(n - 1) \qquad (1)$$
$$CR = CI/RI \qquad (2)$$

Table 2. List of random consistency indexes.

n	1	2	3	4	5
RI	0	0	0.58	0.9	1.12
n	6	7	8	9	10
RI	1.24	1.32	1.41	1.45	1.49

Where CI = Consistency Index, Λ = Consistency Vector, n = Number of alternatives, CR = Consistency Ratio, and RI = Random of Consistency Index.

The second stage of the research was conducting observations and interviews using a questionnaire to the specified MSMEs. Determination of MSMEs point (marking area) was carried out by using Global Positioning System (GPS). Data from the observations and interviews at this stage were processed using the Arc GIS 10.5 software. Exploration of the characteristics of competitiveness data used the overlay technique of each competitiveness variable.

3 RESULTS AND DISCUSSION

Tasikmalaya City consists of ten sub-districts consisting Tawang, Bungursari, Tamansari, Cibeureum, Cihideung, Indihiang, Cipedes, Kawalu, Mangkubumi, and Purbaratu. Every sub-district in Tasikmalaya City has its own characteristics in producing handicraft products.

Handicrafts are very popular products produced from Tasikmalaya City. Tamansari sub-district produces kelom geulis, Mangkubumi sub-district produces footwear, Cipedes sub-district produces batik, Indihiang and Cihideung sub-districts produces payung geulis, Cibeureum sub-district produces jackets, Kawalu sub-district produces bordir, Purbaratu

sub-district produces mendong, Tawang sub-district produces wood products, and Bungursari sub-district produces leather crafts.

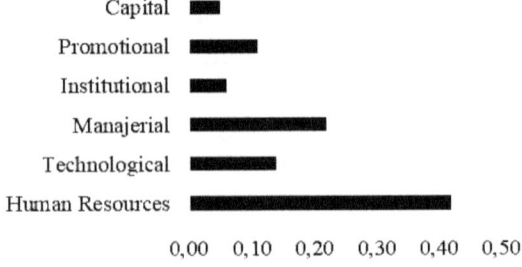

Figure 1. Weight of priority determinants of the competitiveness of msmes in tasikmalaya city.

Products in each sub-district were analyzed using six variables of MSME competitiveness analysis. The results of pairwise comparison analysis indicates that according to experts, the factor that most influenced the competitiveness of MSMEs in Tasikmalaya City is human resources (42%). The priority differences of the six factors are shown in Figure 1.

The results of pairwise comparison show that the value of Consistency Ratio is by 1.8% (<10%). It shows that expert opinion is consistent, so that it can be accepted and used as a basis for consideration in analyzing the competitiveness of MSMEs in Tasikmalaya City. The competitiveness of MSMEs in Tasikmalaya City is generally shown in Figure 2.

Figure 2 shows the level of competitiveness of MSMEs in each sub-district in Tasikmalaya City. In general, the level of competitiveness of MSMEs in

Figure 2. Competitiveness map of msmes in tasikmalaya city.

Tasikmalaya City is relatively low. The development of handicraft products is important in supporting the potential of Indonesia's creative economy. However, the development of handicrafts in Tasikmalaya City was constrained by several factors. Low competitiveness of MSMEs caused products were not able to compete well in the market. Various problems related to human resources, managerial, technological, promotional, institutional, and capital resources are issues that need to be considered.

Fundamental solution to improve the competitiveness is innovation. Studies on the behavior of SMEs indicated that innovation helped such enterprises effectively in creating their competitive advantage (Sipa et al. 2015).

3.1 *Hyman resources*

The main priority for increasing the competitiveness of MSMEs in Tasikmalaya City is human resources (42%). MSMEs are business activities that are able to absorb labor and provide economic services to the community (Sipa et al. 2015). This important role of the MSMEs is not accordance with the conditions of its human resources management. Both culture and informal work structure are in the absence of a clear career path, which cause MSMEs difficulties to improve the quality of human resources (Bank Indonesia 2015).

Based on the interviews results with MSMEs, the level of education of most workers only senior high school graduates. The skills of the average MSME workers in Tasikmalaya City classified as intermediate, where workers already had the talent to make handicrafts obtained from generation to generation in their family.

Unskilled workers were not supported by adequate training (Hamzani & Achmad 2017). The survey results showed that training in MSMEs development was still rarely carried out by MSMEs actors in Tasikmalaya City. The lack of training and assistance led to low innovation, so that almost all MSMEs in Tasikmalaya City were unable to compete with other products including imported products.

The low quality of human resources in MSMEs was seen from several characteristics (Rifai et al. 2016). First, lack of knowledge on the latest technology, so that the products produced were less innovative. Second, lack of owner's ability to read market needs, so that MSMEs have not met the market needs. Third, there was no long-term strategic plan because business owners were often involved in technical issues. In general, MSMEs did not have long-term goals that guide their business development. Fourth, the low level of individual motivation affected the ability of MSMEs actors to capture globalization challenges.

Improving the quality of MSMEs human resources needs to be followed by evaluation and continuity of the program. This program does not only involve MSMEs, but also involves other stakeholders such as the government and the private sector. Entrepreneurship training, recording of financial reports, increasing communication and marketing, and digital training are program priorities (Fouad 2013). Through these programs, entrepreneurs not only learn how to manage their business but also how to be innovative entrepreneurs (Olughor 2015). Innovation drives MSMEs to be sustainable and able to survive both now and future (Koc & Ceylan 2007).

3.2 *Managerial*

The second priority in increasing the competitiveness of MSMEs in Tasikmalaya City is managerial (22%). In general MSMEs in Tasikmalaya City have not yet professionally implemented organizational and management systems. MSMEs in Tasikmalaya City didn't have an adequate financial management system yet. Workers were unable to provide the qualified financial statements based on accounting standards (Hamzani & Achmad 2017). This condition mainly was indicated on MSMEs in Indihiang, Bungusari, Mangkubumi, Tamansari, Tawang and Cibeureum Sub-districts. On the other hand, Kawalu and Cipedes Sub-districts showed a fairly good management organization system. Bordir products produced in Kawalu and Batik products produced in Cipedes showed a higher level of managerial competitiveness. The planning process to supervision in both districts carried out professionally. Both of products produced had premium quality and were supplied to modern supermarkets in big cities, such as Jakarta.

A good organization and management system in MSMEs is an important factor for creating innovation. MSMEs business managed by families usually depends on one particular person. The sustainability of a business that depends on one figure make MSMEs unable to survive. This condition happened in Payung Geulis MSMEs in Indihiang Sub-district. Business activities declined when certain people did not exist.

These issues can be resolved by preparing a trained workforce, adequate capital, appropriate technology, and building a network of cooperation with other parties. For this reason, the organization and management in MSMEs need to improved, from traditional to professional. The planning process until supervision needs to be done professionally to ensure optimal business activities.

3.3 *Technological*

The third priority in increasing the competitiveness of MSMEs in Tasikmalaya City is technological (14%). The problems of MSMEs in Tasikmalaya City were the characteristics of MSMEs, which in general did not have adequate technological capabilities yet. The existing internet technology was not used optimally because MSMEs owners were generally elderly. The willingness and ability to use technology in MSMEs were very limited. This caused MSMEs were not able to compensate for consumers rapid changing tastes. Observed MSMEs have not used information technology such as web creation and social media. Tasikmalaya MSMEs could not compete with large companies that have large capital in accessing technology. Changes of consumer tastes due to globalization make MSMEs owners should be able to adjust the business cycle according to change of the era.

The assistance scheme that focused on giving credit must be changed to technical or technological assistance. Providing technology access to MSMEs is needed to support products produced, so it can compete and be accepted by consumers. The focus of technology upgrades is directed at the production, management, and marketing processes. The assistance in these three aspects is expected to be a driving force for the creation of innovation (Tambunan 2015).

3.4 Promotional

The fourth priority in increasing the competitiveness of MSMEs in Tasikmalaya City is promotional (11%). The promotion of MSMEs in Tasikmalaya City was still considered low. The average MSMEs observed was less concerned about marketing their products. They have not had marketing strategies such as competitive prices, discounts for consumers, and after sales services. In fact, internet progress has not been utilized optimally in promoting the products produced. The awareness to expand the reach of e-commerce marketing was still very low. Product marketing still relies on simple ways, mouth to mouth marketing. Even though there are many promotional media that can be used such as Facebook, Twitter, Instagram, You tube, and other marketing media. Some MSMEs just only depended on orders from existing customers and sold them to markets in Tasikmalaya City.

The low promotion activities carried out have an impact on limited market share only in the region. The marketing range is only limited to the domestic scope which is limited by the area and friendship or family environment. Some MSMEs, such as Payung Geulis, were reluctant to accept orders online because the product was not ready to send. There was no stored product stock due to lack of the capital. Products were only made if there was an order. In this case, products ordered online could not be served because the MSMEs took a long time to complete orders.

The socialization of the benefits of using e-commerce is needed with the assistance by e-commerce and government owners. Trainings especially in digital basic are now expected to increase the marketing reach of MSMEs.

3.5 Institutional

The fifth priority in increasing the competitiveness of MSMEs in Tasikmalaya City is institutional (6%). In general, MSMEs had limitations in building industrial and social relations, resource mobility, information access, and resource development (Broughton 2011). For this reason, an organization in the form of MSMEs institution is needed. Institution is a place expected to be beneficial for the smooth process of production and marketing. The form of institutions in the form of cooperatives or joint MSMEs makes MSMEs business activities stronger and organized better.

The characteristics of the Tasikmalaya City MSMEs were nonstandard of product quality, limited product design, limited product type, difficulty in pricing, nonstandard raw materials, and product continuity that has not been guaranteed. These characteristics make the role of cooperatives or joint MSMEs important. MSMEs joining the group can obtain various benefits that not obtained by independent MSMEs. By forming a group, MSMEs can work together in the process of providing raw materials, pricing, product standardization, product quantity, marketing, etc.

In fact, the role of cooperatives or a combination of MSMEs in Tasikmalaya City has not been run optimally. Cooperation between MSMEs has not been established well, but if cooperation is carried out it will increase the efficiency and effectiveness in running a business. The average MSMEs owners observed stated that they did not have a relationship with cooperatives or MSMEs groups. Every MSME runs its own business. This causes the MSMEs in Tasikmalaya City were not well integrated. MSMEs did not have the power of position that could threat the business sustainability.

In developing a network of cooperation between MSMEs, there needs to be an assistance from out-side parties, in this case is the government. The existence of technical guidance and the provision of comprehension about MSMEs is important so that MSMEs understand the benefits of building cooperative networks. Many among the MSMEs do not understand the benefits if they work together.

3.6 Capital resources

The sixth priority in increasing the competitiveness of MSME in Tasikmalaya City is capital resources (5%). Approximately 60–70% of MSMEs in Indonesia have not had access to bank financing (Bank Indonesia 2015). In terms of limited capital, MSMEs have a poor quality of financial statement (Kinyua 2014). One of the MSMEs capital indicators figure from the value of investment or the average growth rate per year (Tambunan 2015). The investment process is closely related to the addition of production capacity either from the addition of labor or the addition of new production machines.

MSMEs generally had a very limited capital and income from personal funds. Some products such as kelom geulis in Tamansari Sub-district, payung geulis in Indihiang District, Rubber Products in Bungursari

Subdistrict had a low level of competitiveness. The difficulty of funding for investments constrains business sustainability.

From the owner perspective, capital problems can be solved by doing good financial management. In general, MSMEs in Tasikmalaya City did not have a good financial administration system. Financial management were still mixed between household ownership and business. This caused difficulties for the banks that would facilitate credit to find complete information about the MSMEs finances.

The mindset of consumptive oriented MSMEs reduced the confidence of banks in providing credit. Changes in mindset from consumption to investment are needed so that the MSMEs implemented can develop. Changes in the mindset of MSME business actors deal with providing training, guidance, and education held by the government.

4 CONCLUSION

Tasikmalaya City is one of the cities that produces unique handicraft products. Each of sub-district has the characteristics of their respective products. The results of MSME clustering based on six competitiveness variables indicated that MSMEs in Tasikmalaya City had a low level of competitiveness. A strategy to increase the competitiveness of MSMEs in each sub-district in Tasikmalaya City needs to be adjusted by mapping its competitiveness. Development activities carried out can be in the form of developing human resources, increasing access to technology, developing networks of cooperation both internally and externally, increasing access to promotion, and facilitating access to capital.

ACKNOWLEDGMENT

This research supported by Indonesian Education University Grant (No: 5493/UN40/KP/2019). We thank Dr. Ridwan Syaf for providing expert data and MSMEs of Tasikmalaya City as respondents in this study.

REFERENCES

Bank Indonesia. 2015. Pemetaan dan strategi peningkatan daya saing umkm dalam menghadapi MEA 2015 dan pasca MEA 2025. Jakarta: Bank Indonesia.
Bappenas. 2015. Kajian strategi pengembangan kawasan dalam rangka mendukung akselerasi peningkatan daya saing daerah. Jakarta: Bappenas.
Broughton, A. 2011. Smes in the crisis: employment, industrial relations and local partnerships. Dublin: european foundation for the improvement of living and working conditions.
Fouad, M. A. A. 2013. Factors affecting the performance of small and medium enterprises (SMES) in the manufacturing sector of cairo, egypt. *International journal of business and management studies* 5 (2).
Hamzani, U. & Achmad, D. 2016. The performance of Micro, Small and Medium Enterprises (MSMEs): indigenous ethnic versus non-indigenous ethnic. *Procedia-social and behavioral sciences* 219: 265–271.
Hamzani, U. & Achmad, D. 2017. Micro, Small and Medium Enterprises (MSMEs) coaching program. *Journal of Business and Economics Review* 2(3): 20–25.
Hayashi, M. 2002. The role of subcontracting in SME development in Indonesia: Micro-level evidence from the metal-working and machinery industry. *Journal of Asian Economics* 13(1): 1–26.
Kinyua, A. N. 2014. Factors affecting the performance of small and medium enterprises in the jua kali sector in nakuru town, kenya. *Journal of business and management* 6(1):5–10.
Koc T., & Ceylan C. 2007. Factors impacting the innovative capacity in large-scale companies. *Technovation* 27: 105–114.
LPPI & Bank Indonesia. 2015. Profil Bisnis Usaha Mikro, Kecil dan Menengah (UMKM). *Jakarta: LPPI dan Bank Indonesia*.
Olughor, R. J. 2015. Effect of innovation on the performance of smes organizations in nigeria. *Journal of Management* 5(3): 90–95.
Piatkowski, M. 2012. Factors strengthening the competitive position of sme sector enterprises. An example for poland. *Procedia – social and behavioral sciences* 58:269–278.
Rifai, M., Indrihastuti, P., Sayekti, N.C., Gunawan, C.I. 2016. Strategy in enhancing the competitiveness of small and medium enterprises in asean free trade era. *International journal of academic research in business and social sciences* 6(12):76–88.
Rifai, M., Prihatminingtyas, B., Susanto, R. Y., Wani, H. U., Susanto; Wani H. U. 2013. Improvement of competitiveness to encourage the development of nationality based food and beverage industries to the good local industry. *Research report* MP3EI, DP2M Dikti 2013.
Saaty, T.L. 2006. Decision making with the analithic network process. *Pittsburgh: Springer*.
Sipa, M.,Mitka, I.G., Skibinski, A. 2015. Determinants of competitiveness of small enterprises: polish perspective. *Procedia economics and finance* 27 (2015): 445–453.
Tambunan. 2015. Ukuran Daya Saing Koperasi dan UKM. Jakarta: Pusat Studi Industri dan UKM Universitas-Trisakti.

The effect and application of digital marketing in improving a startup company branding in Tasikmalaya

I. Yusuf & G. Ghaida
Universitas Pendidikan Indonesia, Bandung, Indonesia

M. Fahreza
Institut Koperasi Indonesia, Sumedang, Indonesia

ABSTRACT: Digital marketing is an alternative for new startups because of limited promotional budgets. The use of social media and the right website are alternatives that can be used to create an image, attachment, and even sales for new startup businesses. There is a change in marketing style that was originally conventional (offline) to digital (online). The concept of digital marketing for business people is that they can market their products from anywhere and anytime through the internet. However, to be able to optimize digital marketing, an effective marketing communication strategy is needed by new startup businesses in marketing their products or services. The purpose of this study is to find out how the digital marketing communication strategy is carried out by startup businesses in marketing their products or services. In addition, the researchers also aim to know the obstacles and benefits of implementing digital marketing for new startup businesses in the city of Tasikmalaya. This study uses a descriptive quantitative method and the data are collected through an interview process using a closed questionnaire to startup businesses in the city of Tasikmalaya.

1   INTRODUCTION

The industrial world is entering a new era called the Industrial Revolution 4.0, this topic has been widely discussed throughout the world, including in Indonesia. Industry 4.0 sector can contribute to the creation of more jobs and new investments based on technology. The implementation of the fourth-generation industry must be followed by the formation of a healthy and sustainable ecosystem to be effective and able to drive the entire economic sector. To achieve business success in the digital age, it takes ecosystems and communication that are well formed by business people, so that a strong and mutually beneficial ecosystem is achieved.

One example of how digital business ecosystems are well formed is Silicon Valley, which is an area covering the San Francisco Bay Area and California regions in the United States. Silicon Valley is known as the largest information technology industry in the world. However, the name Silicon Valley itself is not legally registered as the name of the area. This is just a designation that refers to the center of the technology industry in the United States. From that region emerged startup companies that are now worldwide, such as Apple and Google. Apple Inc. is a multinational company made by Steve Jobs and is engaged in the design, development, and sale of electronic goods around computers, smartphones, tablets, and the like. Since it was founded on April 1st, 1976, by Steve Jobs,

the company has continued to create innovative electronic products. Then what about the development of startups in Indonesia? Currently the development is quite good and developing. Every year, there are many new startups launching in Indonesia. According to www.dailysocial.net, there are currently at least more than 1500 local startups in Indonesia. The potential of Indonesian internet users, which is increasing from year to year, is certainly an "opportunity" to establish a startup company.

From Figure 1 it can be seen that the number of smartphone and internet users in Indonesia continues to increase. Smartphone CAGR reaches 26% and internet CAGR reaches 13%. The internet and smartphones have penetrated "the bottom of the pyramids." Indonesia entered the era of "the borderless world" on a massive scale (Gathering 2019). However, the growth of internet users is not yet a guarantee that startup companies will succeed. It is not a few local startup companies that fail at the beginning of the business. Gojek Indonesia's CEO, Nadiem Makarim, revealed a research result that 90% of startups will fail and only 10% will survive. One factor that makes startups in Indonesia fail is the weak use of marketing strategy, so it does not reach the right market or consumer.

The purpose of this research is to find out how digital marketing communication strategies are carried out by startup businesses in marketing their products or services. What factors are constraints and what

The Growth Rate of Internet Users (million)

CAGR smartphone: 26%, CAGR Internet: 13%

- Seluler   - SmartPhone   - Mobile Internet

Figure 1. The growth rate of internet users in Indonesia. (Gathering 2019).

Table 1. Characteristics of a startup company.

No.	Characteristics of a Startup Company
1	Company is less than 3 years old
2	Employs less than 20 people
3	Income of less than $100,000/year
4	Generally operates in the field of technology
5	Products are applications in digital form
6	Usually operates through a website
7	Has a very innovative and disruptive idea
8	Undergoing accelerator or incubation program
9	Very fast business growth
10	Entering investor radar coverage

Table 2. Smartphone operating system in Indonesia.

No.	Operating System	Percentage
1	Android	92.31%
2	iOS	5.85%
3	Windows	0.17%
4	Kai OS	1.49%

benefits are able to contribute maximally to the welfare of the society by applying digital marketing to build a startup business image in the city of Tasikmalaya.

2 METHOD

This research uses a quantitative descriptive approach; the method used is a survey through the distribution of questionnaires and the involvement of business actors in a discussion group (FGD). The purpose of this study is so that researchers can find out the influence and application of digital marketing communication strategies carried out by entrepreneurs in marketing products or services in the city of Tasikmalaya. Interviews using a closed questionnaire, observation, and documentation are quantitative approaches obtained by the researchers.

The instrument in this research was a questionnaire consisting of indicators to measure the internal environment, external environment, and digital marketing methods that affect the image of startups in the community.

3 RESULTS AND DISCUSSION

Marketing, as an activity in the field of marketing to expand information, influences or persuades products produced by a company where they can be accepted by the community as potential consumers (Tjiptono 2011). As internet technology develops, digital marketing is better known. Digital marketing can be defined as all marketing efforts using electronic devices or the internet with a variety of marketing tactics and digital media where businesses can communicate directly with potential customers who spend time online. Meanwhile, the term startup is an absorption of English, which means the action or process of starting a new organization or business venture.

A startup refers to a company that has not been operational for long. These companies are mostly newly established companies and are in the development and research phase to find the right market. Starting to develop in the late 1990s to 2000s, the term startup is

associated with everything related to technology, the web, and the internet. Table 1 shows the characteristics of a startup company.

From Table 1, we can see that the characteristics of startup companies are more identical to companies engaged in technology and the web. At present the facts in the field are that, the development of companies that are commonly labeled as a "startup" are companies that are concerned with the technology and online fields. The growth of startup companies in Indonesia is also influenced by the high number of smartphone users.

There are more than 5 million applications running on smartphones, 40% of them run on android. Table 2 shows that Android and iOS still dominate the market in Indonesia (Simpson 2018). The number of applications in the Android operating system also helps startup companies to develop. One of them is in building the company's image through digital marketing.

However, not all digital marketing channels chosen can help build the image of a startup company. Here are some marketing strategies commonly used by startup companies to improve their company image.

3.1 Email marketing

Email is one of the oldest marketing channels. The first email appeared in 1971, and the first spam was detected even present in 1978. Then email continued to grow until now we know what is called email marketing. This tool is the best investment that can be used to achieve growth. Starting from the welcome email series, special offers, newsletters, to all activities for growth and retention. According to the DMA, email marketing has an ROI (return of investment) of 122%. The best thing that can be obtained from email marketing is that it is low cost, even using a local email marketing service provider.

3.2 *Social media marketing*

According to Statista, 90% of companies agree that social media is important for their business, and 89% of marketers say that social media helps boost their company's popularity. The advantages we get from using social media include creating loyal customers, increasing traffic and exposure for websites, influencing search engine optimization (SEO) and search results on Google, and increasing sales. But not all social media can be used to improve the company's image. If a company succeeds in using Twitter, this does not necessarily apply to all companies.

3.3 *Google AdWords*

Google AdWords is often referred to as PPC or Pay-per-Click marketing. It is a method through Google's search engine to bring many visitors to a company website. PPC marketing is useful for generating hot leads or prospective customers who are ready to buy the goods or services that the company makes. The advantages gained with PPC marketing are:

- Reach the right customers at the right time with the right ad.
- Manage marketing expenses.
- Analytics obtained to sharpen the next strategy.
- Target only specific people who are indeed willing to buy company goods or services.

The thing to remember is that PPC can cost in an instant, that's why specific knowledge and goals are needed in each use of PPC marketing.

3.4 *Search engine optimization*

Many startup companies think that by creating a good website, visitors will come by themselves, but this is a wrong idea. There are so many similar websites with company websites in the same area and they also create websites that are no less good. The challenge is how to make the company's website appear at the top when visitors or prospective customers are looking. Here the role of SEO is important for startup companies' digital marketing strategies in building their corporate image. SEO not only increases traffic to a company's website, but also builds user trust. SEO itself is a process and relies heavily on the content that is on the company's website.

3.5 *Content marketing*

Content marketing becomes a necessity for startup companies; in addition to its low costs, content marketing also creates good content, just like how a startup company talks about their startup to everyone that they meet. In addition, content marketing can last a long time. The data prove that 90% of B2C marketing uses content marketing very intensively. Whereas 88% of B2B marketers agree that marketing content is very important in their company. The advantages of using content marketing are:

- Increase brand awareness and engagement with users.
- Build company credibility.
- Long-lasting.
- SEO-friendly.

4 CONCLUSION

There are several obstacles for startup companies in building their company image through digital marketing. The development of startup companies in Tasikmalaya itself still needs to be improved because at this time, there aren't many startup companies from Tasikmalaya. This is happening because of several factors, among others:

4.1 *Internet network*

Technology infrastructure in the city of Tasikmalaya has not been evenly distributed, such as the 4G internet network from several providers that have not reached all the regions, especially in the Tasikmalaya District. This has made it difficult to build the image of a startup company in Tasikmalaya.

4.2 *Startup ecosystem*

To help build the image of a startup company, an ecosystem that continues to accommodate the activities of a startup company is needed.

4.3 *Business incubator*

Business incubators can help prevent many businesses from helping to develop startup companies.

4.4 *Human resources*

To build the image of a startup company, it requires human resources who understand how to utilize digital marketing. In addition, startup companies in Tasikmalaya have not yet developed because of the limitations of human resources in the field of information and technology such as programmers, user interfaces, and user experiences.

4.5 *Government support*

The programs and support from the government, which are still minimal in supporting the development of startups in many cities, including in the city of Tasikmalaya, also influence the image of startup companies in the city of Tasikmalaya.

4.6 *Coworking space*

The trend of coworking spaces in big cities in Indonesia helps the growth of startup companies in building corporate image by building networks and developing business ecosystems.

REFERENCES

Gathering, IT. J. 2019. Sharing vision outlook 2019. *Bandung: IT Journalists Gathering.*

Simpson, R. 2018. Microsoft breakthrough as Windows 10 overtakes 7 globally for first time. [Online]. Retrieved from https://gs.statcounter.com/press/microsoft-breakthrough-as-windows-10-overtakes-7-globally-for-first-time.

Tjiptono, F. 2011. Manajemen & strategi merek. Yogyakarta: Andi.

How important is strategic knowledge creation for SMEs?

N.A. Hamdani & A.O. Herlianti
Universitas Garut, Garut, Indonesia

ABSTRACT: The era of Society 5.0 encourages SMEs to manage, create, communicate and apply knowledge to all kinds of business activities to achieve their business goals. The purpose of this study is to analyze the effectiveness of strategic knowledge creation in SMEs. To this end, an explanatory survey was conducted on small and medium coffee enterprises. The data analysis technique used is SEM-PLS with SmartPLS 3 computer software. The results revealed that strategic knowledge creation was sufficiently effective to be applied in SMEs. Strategic knowledge creation could determine how to manage knowledge, how and when to create knowledge, and how to apply it in order to boost SMEs' productivity.

1 INTRODUCTION

The success of Micro, Small and Medium Industries in the era of society 5.0 can be determined by the Strategic Knowledge Creation owned by entrepreneurs. Strategic Knowledge Creation can be the ability of the company as a whole to create new knowledge, spread it throughout the organization and realize it in products, services, and systems (Giuri et al. 2019). The creation of new knowledge as intangible assets can be a new paradigm shift about resources that will have the potential to move companies to be more intelligent and innovative (Dzenopoljac et al. 2018). Intangible assets are more important than company resources that have been understood as financial resources (Bao & Zhou 2010), buildings, land, technology, market position and other tangible assets (Costa & Rezende 2018). In connection with this the perception formed that the essence of the company is a knowledge organization (Ramírez et al. 2011), supported by the opinion of the existence of uncertainty in the economy (Meriläinen & Halinen 2009), then one of the definite competitive resources is knowledge (Bandera et al. 2017). Knowledge management is able to create business value (Alyoubi 2015), communicate, apply knowledge to all kinds of business activities to achieve business goals and build competitiveness (Marsina et al. 2015). The underlying assumption is that knowledge management theory basically arises to answer the question of how to manage knowledge and how to manage it (Yang et al. 2002), how and when knowledge creation should be done and supported (Nonaka et al. 2005) and how to use accumulated knowledge that has been created so as to increase organizational productivity (Shu et al. 2013). This has become the basis for much research to date.

Business performance is still a major topic in strategic management research, there are opportunities to conduct further research on the business performance of Coffee SMEs in Indonesia by reconstructing innovation strategies and knowledge creation strategies by considering entrepreneurial orientation and dynamic capabilities (Souto 2013), Performance of Coffee SMEs in Indonesia, encourages research into business performance in Coffee SMEs in the era of society 5.0. Indonesia is indeed the third largest coffee producer country in the world, but only a small proportion is produced based on sustainability standards. In fact, the EU is the largest global market for sustainability-based coffee. Indonesian coffee products which have a 1.2 million hectare plantation area are still inferior to Vietnam which only has a coffee bean area of 630,000 hectares. Indonesian coffee production is 500 kilograms of coffee per hectare while in Vietnam 2.7 million tons of coffee per hectare. Coffee SMEs do not have enough knowledge to process coffee of the same quality at every harvest. In the aspect of raw materials, current production tends to be stagnant, no new innovations have been developed, and in the production process, processing and packaging technology in small and medium scale industries is very simple (Hamdani & Maulani 2018).

The application of Knowledge Management as the basis of knowledge creation can not only be applied to large companies, but knowledge management can be applied to SMEs (Desouza & Awazu 2006). Although the results of the study show that knowledge creation and human resources are still low, the application of SKM can improve the competitiveness and performance of SMEs (Cerchione et al. 2015). Knowledge Creation and technology mastery in SMEs are different from theories about Kowledge creations that are often found in large companies. In theory, experts generally state that the role of Knowledge creation and technology is only applied to large companies, but empirically the application of knowledge creation is often found

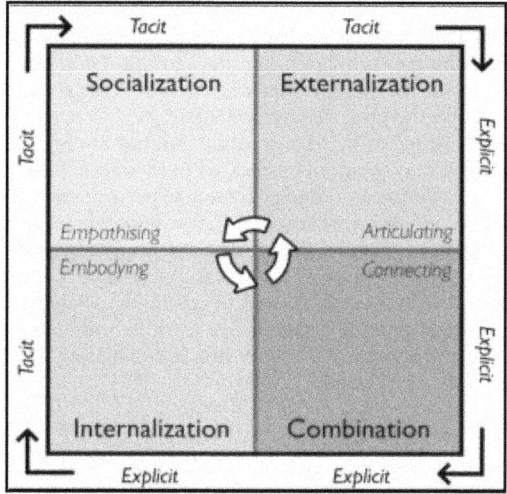

Figure 1. SECI model.
Source: (Zaim et al. 2015).

in SMEs (Rrustemi 2011). Based on this, it was also stated that technology has a major role in the development of SMEs (Carr 2005). In this case there are two knowledge namely tacit knowledge and explicit knowledge, both of these knowledge are in the individual, if there is social interaction between individuals then automatically new knowledge will be formed, this is called the ontology of knowledge creation (Zaim et al. 2015).

Figure 1 shows the SECI Model.

Adoption of technology is needed to support business performance and SMES competitiveness, the use of information technology has a very large role supported by government policies that support the technology sector (Gusaptono et al. 2012; Raf 2000). The process of knowledge creation as a strategy to achieve innovation strategies and business performance (Hamdani 2018a). The dimensions used in this study includes Exploration, Institutional Entrepreneurship, Exploitation and Combination (Yang et al. 2002), this is supported by similar studies such as (Sołek-Borowska 2017), and then (Durst et al. 2013; Hamdani 2018b). The focus of this research is on business performance resulting from knowledge creation strategies in the era of society 5.0.

2 METHODS

The method used was explanatory survey with a sample of 50 respondents. The data analysis technique used was PLS with the help of SmartPLS 3 computer software. The data collection method in this study was using a questionnaire distribution. In this study, the technique used was Probability Sampling, which is a sampling technique to provide equal opportunities for each member of the population to be selected as sample members with Accidental Simple Random Sampling,

a sampling method by selecting who coincidentally exists or is encountered.

3 RESULTS AND DISCUSSION

The measurement model analysis in this research uses SmartPLS3 software with the second order construct (SOC) or higher order construct (HOC) method, which is a modeling method where the construct is reflected or shaped by latent dimensional constructs. The HOC modeled in this study uses HOC type 3 where the measurements of the lower order and the higher order are both reflective. Execution of the modeling with Smart-PLS 3 software using the configuration parameters described in Table 1.

The results of running the initial measurement model of the SKC variable produce the validity and reliability parameter values which all meet the rule of thumb. All outer loading values in the manifest variable against the dimension construct, as well as the outer loading value of the dimension construct for the SKC variable are above 0.7. Next, the AVE value generated in each dimension and variable construct is also above the value of 0.5, so it can be said that the convergent variable in the SKC variable and its festivities are fulfilled. Based on the test results tabulated in table 4.37, all correlation squared values between latent constructs <AVE of each construct are related, so it can be said that the SKC latent variable satisfies discriminant capacity. The measurement of discriminant validity using the cross loading method also shows the results that the outer loading value in each construct is above the cross loading value between constructs measured with other constructs so that the discriminant validity of the SKC variable is met. the alpha cronbach's value for each dimension and variable construct is above 0.7, and the composite reliability value for each dimension and variable construct is also above 0.708 so it can be said that the SKC variable and its manifestations have good relativity. The measurement results of the outer model on the SKC latent variable prove that all dimensional constructs and SKC variables are valid and reliable.

Figure 2 show the measurement model of variable Strategic Knowledge Creation (SKC).

Table 1. Configuring SmartPLS running software 3.

Configuration	Value
Weighting Scheme	Path
Maximum Iterations	5000
Stop Criterion (10-^)	7
Sub Samples	4000
Sign Changes	No Sign Changes
Amount of Results	Complete Bootstraping
Confidence Interval	Bias Corrected and
Method	ASKCelerated Bootstrap
Test Type	Two Tailed
Significance Level	0.05

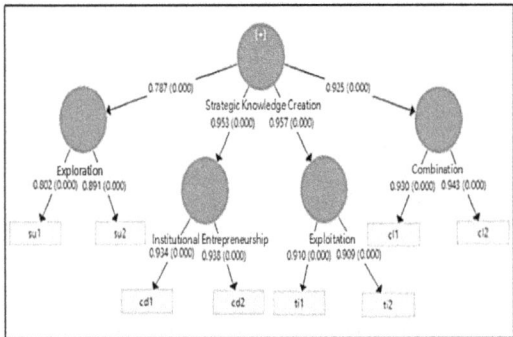

Figure 2. SKC variable initial measurement model outer: outer loadings & P-value; inner: path coefficients & P-value.

It is proven that there is no significant positive effect between strategic knowledge creation on firm performance because statistically, the t-table value is smaller than t-table at 5% significance level, that is 1.682 <1.96. It was also tested that the influence of SKC → BP based on the value of the path coefficients of 0.258 proved insignificant because it was at a p-value of 0.093 which was greater than 0.05.

The concept of strategic knowledge creation in the process of product creation and development carried out by SMEs Kopi together with customers and partners such as suppliers, is measured in the "medium" category. Based on the evaluation results of the inner model on the joint creation variable, exploration dimension is the highest dimension of path coefficient compared to others. This means that the exploration activities carried out by the coffee SMEs together with customers, or also with partners were able to provide the greatest influence in the strategic knowledge creation activities on coffee SMEs. 80% of Indonesia's coffee production volume enters the export market, so quality is one of the guidelines to increase its added value, although domestic consumption is still low and dominated by imported coffee.

The first thing that must be done in an effort to create and manage knowledge is to understand the dimensions of the ontology and epistemology of the process of creating knowledge. Based on the ontology side, knowledge creation basically comes from individuals. Knowledge that comes from an organization is the result of the creation of the people in the organization (Sołek-Borowska 2017). Several studies emphasize the importance of knowledge creation in improving business performance (Bao & Zhou 2010; Durst et al. 2013)

4 CONCLUSION

The overall results of the study indicate that strategic knowledge creation in the category is quite effective to be applied to SMEs. Based on the results of this research, strategic knowledge creation can answer the question of how to manage knowledge and how to manage it, how and when knowledge creation should be done and supported and how to use the accumulation of knowledge that has been created so as to increase the productivity of SMEs.

Strategic knowledge creation that has always been carried out by coffee businesses in Indonesia directly is not able to contribute positively to performance, but the paradigm involving consumers as the center of orientation in creating this product, is positively able to give effect to the coffee business unit in innovating as an effort to mediate so that strategic knowledge creation can provide a significant positive influence on the performance of coffee SMEs in Indonesia.

REFERENCES

Alyoubi, B.A. 2015. Decision Support System and Knowledge-based Strategic Management. *Procedia Computer Science*, 65(Iccmit), 278–284. https://doi.org/10.1016/j.procs.2015.09.079.

Bandera, C., Keshtkar, F., Bartolacci, M.R., Neerudu, S. & Passerini, K. 2017. Knowledge management and the entrepreneur: Insights from Ikujiro Nonaka's Dynamic Knowledge Creation model (SECI). *International Journal of Innovation Studies*, 1(3), 163–174. https://doi.org/10.1016/j.ijis.2017.10.005.

Bao, Z. & Zhou, T. 2010. The strategy of knowledge management and knowledge creation. *Proceedings – 3rd International Conference on Information Management, Innovation Management and Industrial Engineering, ICIII 2010*, 1, 262–265. https://doi.org/10.1109/ICIII.2010.70.

Carr, J. 2005. The Implementation of Technology-Based SME Management Development Programmes Diffusion of management learning technology from HE to SMEs. *Educational Technology & Society*, 8(3), 206–215.

Cerchione, R., Esposito, E. & Spadaro, M. R. 2015. The spread of knowledge management in SMEs: A scenario in evolution. *Sustainability (Switzerland)*, 7(8), 10210–10232. https://doi.org/10.3390/su70810210.

Costa, R.G.G. & Rezende, J.F.C. 2018. Strategic alignment of knowledge management and value creation: implications on to an oil and gas corporation. *RAUSP Management Journal*, 53(2), 241–252. https://doi.org/10.1016/j.rauspm.2017.11.001.

Desouza, K. & Awazu, Y. 2006. Knowledge management at SMEs: five peculiarities. *Journal of Knowledge Management*, 10(1), 32–43. https://doi.org/10.1108/13673270610650085.

Durst, S., Edvardsson, I.R. & Bruns, G. 2013. Knowledge creation in small construction firms. *Journal of Innovation Management*, 1(1), 125–142. Retrieved from http://feupedicoes.fe.up.pt/journals/index.php/IJMAI/article/view/7.

Dzenopoljac, V., Alasadi, R., Zaim, H. & Bontis, N. 2018. Impact of knowledge management processes on business performance: Evidence from Kuwait. *Knowledge and Process Management*, 25(2), 77–87. https://doi.org/10.1002/kpm.1562.

Giuri, P., Munari, F., Scandaro, A. & Toschi, L. 2019. The strategic orientation of universities in knowledge transfer activities. *Technological Forecasting and Social Change*, 138 (April 2017), 261–278. https://doi.org/10.1016/j.techfore.2018.09.030.

Gusaptono, R.H., Effendi, M.I. & Charibaldi, N. 2012. The Information Technology (IT) Adoption Process and E-Readiness to Use within Yogyakarta Indonesian Small Medium Enterprises (SME). *JICT-International Journal of Information and Communication Technology Research*, *2*(1), 29–37.

Hamdani, N.A. 2018a. Building knowledge-creation for making business competition atmosphere in SMEs of Batik. *Management Science Letters*, *8*, 667–676. https://doi.org/10.5267/j.msl.2018.4.024.

Hamdani, N.A. 2018b. Building Knowledge Creation For Making Business Competition Atmosphere in SME of Batik. *Management Science Letters*, *8*, 667–676. https://doi.org/10.5267/j.msl.2018.4.024.

Hamdani, N.A. & Maulani, G. 2018. The influence of E-WOM on purchase intentions in local culinary business sector. *International Journal of Engineering & Technology*, *7*, 246–250.

Marsina, S., Hamranova, A., Okruhlica, F. & Bolek, V. 2015. Knowledge Creation and Learning within the Building Project Orientation of Organizations. *Procedia Manufacturing*, *3*(Ahfe), 723–730. https://doi.org/10.1016/j.promfg.2015.07.315.

Meriläinen, K. & Halinen, A. 2009. Customer knowledge creation in strategic business networks Towards an analytical framework Full competitive paper to be submitted to the 25. *Knowledge Creation Diffusion Utilization*, 1–14.

Nonaka, I., Peltokorpi, V. & Tomae, H. 2005. Strategic knowledge creation: the case of Hamamatsu Photonics. *International Journal of Technology Management*, *30*(3/4), 248. https://doi.org/10.1504/ijtm.2005.006709.

Raf, M. 2000. Studi pada sentra industri kecil batik di Kota Jambi. *Manajemen Dan Keirausahaan*, *14*(1999), 91–101.

Ramírez, A.M., Morales, V.J.G. & Rojas, R.M. 2011. Knowledge Creation, Organizational Learning and Their Effects on Organizational Performance. *Engineering Economics*, *22*(3), 309–318. https://doi.org/10.5755/j01.ee.22.3.521

Rrustemi, V. 2011. Organizational learning and knowledge creation processes in SMEs. *Journal of Knowledge Management, Economics and Information Technology*, (6), 1–21.

Shu, L., Liu, S. & Li, L. 2013. Study on business process knowledge creation and optimization in modern manufacturing enterprises. *Procedia Computer Science*, *17*, 1202–1208. https://doi.org/10.1016/j.procs.2013.05.153

Sołek-Borowska, C. 2017. Knowledge creation processes in small and medium enterprises: A Polish perspective. *Journal, Online Management, Applied Knowledge*, *5*(2), 61–75.

Souto, P.C.D.N. 2013. Beyond Knowledge, Towards Knowing: the Practice-Based Approach To Support Knowledge Creation, Communication, and Use for Innovation. *Review of Administration and Innovation – RAI*, *10*(1), 51–79. https://doi.org/10.5773/rai.v1i1.948.

Yang, C., Fang, S. & Lin, J.L. 2002. Organisational knowledge creation strategies: A conceptual framework. *International Journal of Information Management*, *30*(3), 231–238. https://doi.org/10.1016/j.ijinfomgt.2009.08.005.

Zaim, H., Gürcan, Ö.F., Tarım, M., Zaim, S. & Alpkan, L. 2015. Determining the Critical Factors of Tacit Knowledge in Service Industry in Turkey. *Procedia – Social and Behavioral Sciences*, *207*, 759–767. https://doi.org/10.1016/j.sbspro.2015.10.156.

Entrepreneurial orientation of small and medium coffee enterprises in the era society 5.0

N.A. Hamdani & A.O. Herlianti
Universitas Garut, Indonesia

ABSTRACT: In the era society 5.0, it is necessary for a business to be equipped with an innovative entrepreneurial orientation. The present study describes the entrepreneurial orientation of small and medium coffee enterprises. The study was conducted using an explanatory survey. The samples were small and medium coffee enterprises in Garut, Indonesia. The data analysis technique used was SEM-PLS. The results showed that the small and medium coffee enterprises had good entrepreneurial orientation in a such a way that it could improve their market share and business performance by conducting environmental analysis, making innovations, taking risks, and developing new ideas, products, and services.

1 INTRODUCTION

Entrepreneurial orientation has been an established concept in the entrepreneurship literature. It is based on an organization's creative and innovative ability (Shan et al. 2016) and resources to improve performance (Cui et al. 2018). Entre-preneurial orientation plays an important role in building sustainable competitive advantage (Solano et al 2018) and in enhancing marketing capabilities and innovations (Lee & Chu 2017). Studies in Spain (Alayo et al. 2019; Arzubiaga et al. 2018; Núñez-Pomar et al. 2016) and in Taiwan (Lee & Chu 2017) showed that entrepreneurial orientation played an important role in explaining the level of internalization in SMEs. Further studies on SMEs (Brettel & Rottenberger 2013; Emõke–Szidónia 2015; Mason et al. 2015) indicated that international performance of SMEs was facilitated by network ability and international entrepreneurial orientation, not international market orientation (Solano et al. 2018). It is very likely for companies that adopt an entrepreneurial orientation to perform better than companies that adopt a conservative orientation (Miao et al. 2017). It is also suggested that environmental turbulence can have a positive impact on company performance by encouraging companies to be more effective and achieve greater performance (Jiang et al. 2018) and a negative impact on companies with an excellent entrepreneurial orientation (Pratono & Mahmood 2016).

A dynamic ability is relational and cognitive social capital to develop a higher entrepreneurial orientation (Rodrigo-Alarcón et al. 2018). This study extends the domain of entrepreneurial research into strategic alliances by hypothesizing that there is a positive relationship between entrepreneurial orientation (EO) and the success of company-level alliances (Li, L et al.

2017). A previous study showed that entrepreneurial orientation could significantly improve business performance (Arshad et al. 2014).

The performance of SMEs in Indonesia is an interesting research subject. Most of these SMEs are in the entrepreneurship development zone. Entrepreneurship is one of important factors in a nation's development (Hamdani & Maulani 2018). To develop, a nation requires at least 2% of its population to become entrepreneurs (Hamdani 2018). Indonesia has not even reached that number. The total number of entrepreneurs in Indonesia in 2014 is only 1.5% of its total population (Kemenperin 2015). It is a very small number especially if compared to its neighboring countries. In Singapore, for example, 7% of its population are entrepreneurs (see Figure 1).

One of the most potential sectors is plantation. In 2016, its contribution to GDP reached approximately 3.46%, ranked the first among agriculture, animal husbandry, hunting, and agricultural services sectors. One of the biggest commodities in Indonesian plantation sector is coffee. Coffee is a quite important Indonesia's export commodities besides oil and gas. In addition, coffee is also in great demands in the domestic market. This study specifically examines the entrepreneurial orientation of small and medium coffee enterprises in Garut, Indonesia, in the era of society 5.0.

2 METHODS

This study was conducted using an explanatory survey. Questionnaires were addressed to 50 respondents. The data were collected through surveys selected using accidental random sampling technique. The data was

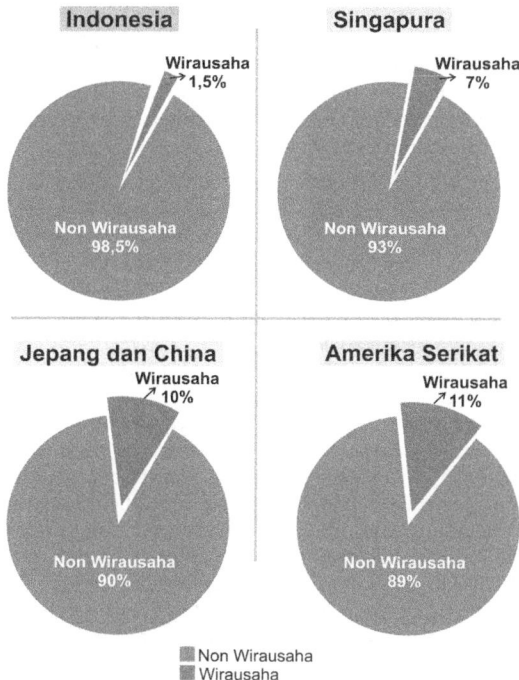

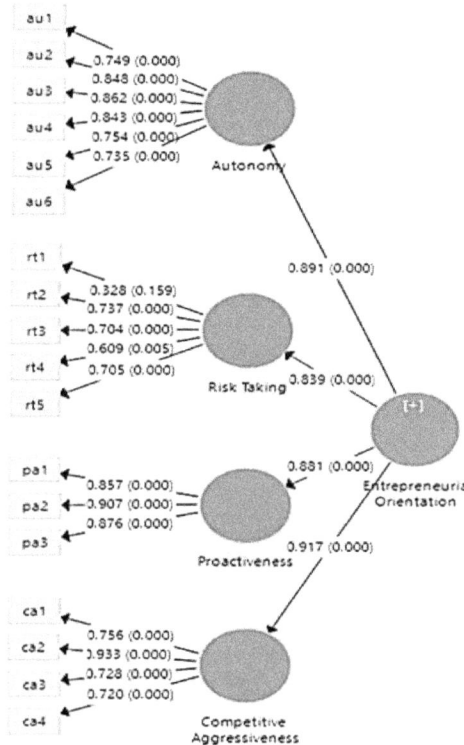

Figure 1.   Number of entrepreneurs in Indonesia compared to those in developed countries.

Table 1. Respondents' responses towards the entrepreneurial orientation of small and medium coffee enterprises in Garut.

Dimension	Total Score	Max. Score	Percentage	Category
Autonomy	1,578	2,100	75.1%	High
Risk Taking	1,111	1,750	63.5%	Moderate
Proactiveness	747	1,050	71.1%	Moderate
Competitive Aggressiveness	923	1,400	65.9%	Moderate
TOTAL	4,359	6,300	69.2%	Moderate

distributed to the respondents through the Google form link. Data were analyzed using SmartPLS.

3   RESULTS AND DISCUSSION

The results of descriptive analysis show that the entrepreneurial orientation of small and medium coffee enterprises in Garut is in "moderate" category. The autonomy dimension scored the highest by 75.1%, and risk taking scored the lowest by 63.5% as illustrated in Table 1.

Figure 2 shows that the outer loadings of indicators rt1 and rt4 in the risk taking dimension are lower than 0.7, and hence are stated as not valid because the AVE value is 0.461 (<0.5). This is due to the fact that the outer loadings of some manifest variables to EO

Figure 2.   Initial measurement model of EO outer: Outer loadings & p-value; inner: path coefficients & p-value.

are lower than 0.7. Therefore, the measurement model needs revising by removing invalid indicators, rt1, rt4 au1, rt2, rt3, ca3, and ca4 in Figure 2.

Figure 3 shows that in the revised model all outer loading values are higher than 0.7 and all AVE values are also above 0.5, meaning that EO and its manifest variables have convergent validity. The discriminant validity was measured using Fornell and Larcker's (1981) criterion, where a latent variable is stated to have discriminant validity if the squared value of correlation between latent constructs is higher than the AVE value of each construct. The result shows that all latent variables have discriminant validity.

It is also revealed that the Cronbach's alpha value of dimension and variable constructs are above 0.7 and that the composite reliability value is 0.708, so it can be concluded that EO and its manifest variables are valid and reliable in Figure 3.

Based on Figure 4, structural model equation is as follows:

$$Business\ performance = 0.811 * Entrepreneurial\ Orientation \quad (1)$$

$$T\text{-}start\ value = 13.966;\ p\text{-}value = 0.000;$$

$$f2 = 1.924 \quad (2)$$

$$R2 = 0.658;\ Q = 0.383 \quad (3)$$

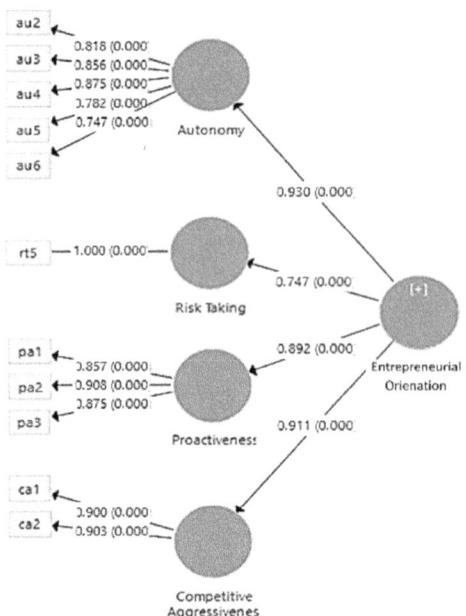

Figure 3. Revised measurement model of EO outer: Outer loadings & p-value; inner: path coefficients & p-value.

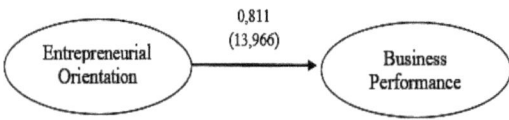

Figure 4. Substructure I, score path coef. (t-stat).

The above model suggests that entrepreneurial orientation has positive influence on the performance of small and medium coffee enterprises in Garut as much as 81.1%. Using the alpha level of 5%, statistically, the influence is significant since the observed t value was 13.966, higher than 1.96 and the p-value was 0.000, lower than 0.05. It can then be said that the higher the entrepreneurial orientation in creative businesses, the more creative the people in the businesses in question.

How much the model of substructure I could make predictions can be seen in the structural model (inner model) evaluation values as follow:

1. Referring to rule of thumb, the R2 value of 0.67, 0.33, and 0.19 means strong, moderate, and weak. The R2 value of 0.658 in the substructure I then shows that using the sample the model can make strong predictions. In other words, entrepreneurial orientation contributes to business performance as much as 65.8%.
2. The obtained f 2 value in the substructure I is 1.924, meaning that entrepreneurial orientation can greatly improve creativity. This is based on the effect size criteria in the rule of thumb of inner model suggesting the effect is small if the f 2 value is 0.02; medium if it is 0.15; and large if it is 0.35.

3. Based on the rule of thumb of Q2 predictive relevance model is deemed relevant in predicting endogen latent variable if Q2 is higher than 0. In substructure I, the Q2 value is 0.383. Therefore, entrepreneurial orientation can be said relevant to be used to make predictions about performance.

Based on the results of structural model analysis and mediating effect analysis in the substructure I, it can be concluded that entrepreneurial orientation statistically has significant positive influence on business performance as, at the level significance of 5%, the observed t of 13.966 is higher than the critical t of 1.96. In addition, the path coefficient of 0.811 and the p-value of 0.000, lower than 0.05, confirm that the influence is significant.

There are four variables studied in the research objects: entrepreneurial orientation, creative people, co-creation, and soft innovation. The variable creative people received high score, and the others received moderate scores. Entrepreneurial orientation involves entrepreneurial intentions and actions. According to Lumpkin & Dess (1996), entrepreneurial orientation (EO) refers to processes, practices, and decision-making activities in entrepreneurship. The term is also used to refer to strategy-making practices and firm behaviors in business activities (Sciascia et al. 2014).

In the present study, four construct dimensions of entrepreneurial orientation of small and medium coffee enterprises including autonomy, risk taking, proactiveness, and competitive aggressiveness were measured. The results of descriptive analysis showed that all dimensions of entrepreneurial orientation of small and medium coffee enterprises in Garut received moderate scores, except for autonomy that received high score. The results of inner model evaluation showed that autonomy construct had the biggest path coefficient, meaning that the research objects were autonomous in establishing and implementing vision, missions, and ideas, in strategic decision making, and in seizing business opportunities.

The risk taking dimension scored the lowest. This is due to the fact that the studied enterprises did not take any risks in making investments in state-of-the-art technology and in human resources. They also avoided business credits despite the fact that loans may bring about more profits (Núñez-Pomar et al. 2016).

This study found that entrepreneurial orientation managed to boost creativity among people in the studied enterprises. This finding is in line with that of a previous study that entrepreneurial orientation, like taking risks in making investments in quality human resources and adaptability to future changes and challenges, turned out to be positively able to exploit elements of corporate asymmetry such as skills and expertise of the workforce (Hamdani 2019). Another study also suggested that entrepreneurial orientation encouraged knowledge creation in companies, and knowledge creation in its turn helped developing strategic and creative resources (Shan et al. 2016).

Empirically, this is clearly reflected in the characteristics of coffee enterprise owners in Garut who are very creative. Their creativity in combination with pop culture among the youths who are fond to flock to socialize and interact in communities becomes a business potential in coffee sectors.

Their creativity is not only proven by their ability to provide coffee products, but also to make coffee as contemporary life style. Coffee shops that have been mushrooming recently further strengthen the identity of Garut as a culinary city.

This study revealed that entrepreneurial orientation had significant influence on the business performance of small and medium coffee enterprises in Garut. This confirmed that entrepreneurial orientation required an alternative contingency models to explore the underlying process, related to entrepreneurial activity in order to be able to improve company performance (Jiang et al. 2018). These models include entrepreneurial orientation model that is able to moderate manufacturing capabilities to performance (Cui et al. 2018), entrepreneurial orientation model that interacts with HRM-based capabilities (Saha et al. 2017), and mediation models that are able to encourage knowledge creation which then has implications for innovation activities (Miao et al. 2017).

The finding of this study also confirmed that EO could directly mediate the influence of the entrepreneurial climate to improve company performance (Martin & Javalgi 2016). EO is a strategy-making process in entrepreneurship (Sirén et al. 2017) that requires other activities. Justifying a previous study (DiVito & Bohnsack 2017), which suggested that entrepreneurial orientation could not be considered as having a direct effect on the improvement of company performance. It required some contingency models such as mediation models, moderation models, interaction models, etc. that are in accordance with the company environments (Zehir et al. 2015).

4 CONCLUSION

The entrepreneurial orientation of small and medium coffee enterprises in Garut was found to be at moderate level. Of its four dimensions, autonomy scored the highest. This comes as no surprise because most of the studied enterprises are medium to small scale, owned by individuals. The other three dimensions received moderate score and needed improvement to accommodate dynamic coffee business environments. It was also revealed that entrepreneurial orientation had proven to significantly influence the business performance of small and medium coffee enterprises in Garut.

REFERENCES

Alayo, M., Maseda, A., Iturralde, T., & Arzubiaga, U. 2019. Internationalization and entrepreneurial orientation of family smes: the influence of the family character. *International business review*, 28(1), 48–59. https://doi.org/10.1016/j.ibusrev.2018.06.003

Arshad, A. S., Rasli, A., Arshad, A. A., & Zain, Z. M. 2014. The impact of entrepreneurial orientation on business performance: a study of technology-based smes in malaysia. *Procedia – social and behavioral sciences*, 130(1996), 46–53. https://doi.org/10.1016/j.sbspro.2014.04.006

Arzubiaga, U., Kotlar, J., De Massis, A., Maseda, A., & Iturralde, T. 2018. Entrepreneurial orientation and innovation in family smes: unveiling the (actual) impact of the board of directors. *Journal of business venturing*, 33(4), 455–469. https://doi.org/10.1016/j.jbusvent.2018.03.002

Brettel, M., & Rottenberger, J. D. 2013. Examining the link between entrepreneurial orientation and learning processes in small and mediumsized enterprises. *Journal of small business management*, 51(4), 471–490. https://doi.org/10.1111/jsbm.12002

Cui, L., Fan, D., Guo, F., & Fan, Y. 2018. Explicating the relationship of entrepreneurial orientation and firm performance: underlying mechanisms in the context of an emerging market. *Industrial marketing management*, 71(November), 27–40. https://doi.org/10.1016/j.indmarman.2017.11.003

DiVito, L., & Bohnsack, R. 2017. Entrepreneurial orientation and its effect on sustainability decision tradeoffs: the case of sustainable fashion firms. *Journal of business venturing*, 32(5), 569–587. https://doi.org/10.1016/j.jbusvent.2017.05.002

Emőke–Szidónia, F. 2015. International entrepreneurial orientation and performance of romanian small and mediumsized firms: empirical assessment of direct and environment moderated relations. *Procedia economics and finance*, 32(15), 186–193. https://doi.org/10.1016/s2212-5671(15)01381-7

Hamdani, N. A. 2018. Building knowledge-creation for making business competition atmosphere in smes of batik. *Management science letters*, 8, 667–676. https://doi.org/10.5267/j.msl.2018.4.024

Hamdani, N. A. 2019. Contributing factors of good corporate governance and employee performance to bank performance. *The journal of social sciences research*, (SPI4), 235–237. https://doi.org/10.32861/jssr.spi4.235.237

Hamdani, N. A., & Maulani, G. A. F. 2018. The influence of e-wom on purchase intentions in local culinary business sector. *International journal of engineering & technology*, 7(2.29), 246. https://doi.org/10.14419/ijet.v7i2.29.13325

Jiang, X., Liu, H., Fey, C., & Jiang, F. 2018. Entrepreneurial orientation, network resource acquisition, and firm performance: a network approach. *Journal of business research*, 87(June 2017), 46–57. https://doi.org/10.1016/j.jbusres.2018.02.021

Lee, T., & Chu, W. (2017). The relationship between entrepreneurial orientation and firm performance: influence of family governance. *Journal of family business strategy*, 8(4), 213–223. https://doi.org/10.1016/j.jfbs.2017.09.002

Li, L., Jiang, F., Pei, Y., & Jiang, N. 2017. Entrepreneurial orientation and strategic alliance success: the contingency role of relational factors. *Journal of business research*, 72, 46–56. https://doi.org/10.1016/j.jbusres.2016.11.011

Martin, S. L., & Javalgi, R. R. G. 2016. Entrepreneurial orientation, marketing capabilities and performance: the moderating role of competitive intensity on latin american international new ventures. *Journal of business research*, 69(6), 2040–2051. https://doi.org/10.1016/j.jbusres.2015.10.149

Mason, M. C., Floreani, J., Miani, S., Beltrame, F., & Cappelletto, R. 2015. Understanding the impact of entrepreneurial orientation on smes' performance. The role of the financing structure. *Procedia economics and finance*, 23(October 2014), 1649–1661. https://doi.org/10.1016/S2212-5671(15)00470-0

Miao, C., Coombs, J. E., Qian, S., & Sirmon, D. G. 2017. The mediating role of entrepreneurial orientation: a meta-analysis of resource orchestration and cultural contingencies. *Journal of business research*, 77, 68–80. https://doi.org/10.1016/j.jbusres.2017.03.016

Núñez-Pomar, J., Prado-Gascó, V., Añó Sanz, V., Crespo Hervás, J., & Calabuig Moreno, F. 2016. Does size matter? Entrepreneurial orientation and performance in spanish sports firms. *Journal of business research*, 69(11), 5336–5341. https://doi.org/10.1016/j.jbusres.2016.04.134

Pratono, A. H., & Mahmood, R. 2016. Entrepreneurial orientation and firm performance: how can micro, small and mediumsized enterprises survive environmental turbulence? *Pacific science review b: humanities and social sciences*, 1(2), 85–91. https://doi.org/10.1016/j.psrb.2016.05.003

Rodrigo-Alarcón, J., García-Villaverde, P. M., Ruiz-Ortega, M. J., & Parra-Requena, G. 2018. From social capital to entrepreneurial orientation: the mediating role of dynamic capabilities. *European management journal*, 36(2), 195–209. https://doi.org/10.1016/j.emj.2017.02.006

Saha, K., Kumar, R., Dutta, S. K., & Dutta, T. 2017. A content adequate five-dimensional entrepreneurial orientation scale. *Journal of business venturing insights*, 8(March), 41–49. https://doi.org/10.1016/j.jbvi.2017.05.006

Sciascia, S., Oria, L. D., Bruni, M., & Larrañeta, B. 2014. Entrepreneurial orientation in low- and medium-tech industries: the need for absorptive capacity to increase performance. *European management journal*. https://doi.org/10.1016/j.emj.2013.12.007

Shan, P., Song, M., & Ju, X. 2016. Entrepreneurial orientation and performance: is innovation speed a missing link? Journal of business research, 69(2), 683–690. https://doi.org/10.1016/j.jbusres.2015.08.032

Sirén, C., Hakala, H., Wincent, J., & Grichnik, D. 2017. Breaking the routines: entrepreneurial orientation, strategic learning, firm size, and age. Long range planning, 50(2), 145–167. https://doi.org/10.1016/j.lrp.2016.09.005

Solano Acosta, A., Herrero Crespo, Á., & Collado Agudo, J. 2018. Effect of market orientation, network capability and entrepreneurial orientation on international performance of small and medium enterprises (smes). *International business review*, 27(6), 1128–1140. https://doi.org/10.1016/j.ibusrev.2018.04.004

Zehir, C., Can, E., & Karaboga, T. 2015. Linking entrepreneurial orientation to firm performance: the role of differentiation strategy and innovation performance. *Procedia – social and behavioral sciences*, 210, 358–367. https://doi.org/10.1016/j.sbspro.2015.11.381

Social media exposure: Effects on barbershop performance in Indonesia

N.A. Hamdani & G.A.F. Maulani
Universitas Garut, Indonesia

ABSTRACT: The industrial revolution era 4.0 requires business entities to use information and communication technology, particularly for marketing purposes, as well as in barbershop business. Businesses now need to involve information and communication technology in driving their business performance so that business processes are optimized. This study examines the extent of the influence of the use of social media among barbershop businesses on their business performance. The respondents are 43 owners of barbershops in Garut, Indonesia. Data processing was carried out using partial least squares structural equation modeling (PLS-SEM) analysis. It was revealed that the use of social media has a significant positive effect on barbershop business performance. Barbershop exposure in social media is one of the factors lined with increased sales.

1 INTRODUCTION

The industrial revolution 4.0 encourages business people in Indonesia to always make innovation (Hamdani et al. 2019). Innovation can be done in various ways, one of which is through social media marketing (Alam Hamdani & Abdul Fatah 2018) as social media can be an effective channel for marketing activities (Lund et al. 2018).

Indonesians are among the most prolific Internet users. It was recorded in 2018 that Internet users in Indonesia reached 171,71 million or 64.8% of its population (APJII 2018), a significant increase from 143,26 million in the previous year (APJII 2017). By 18.9% or 32.4 million of the total Indonesia's Internet users are social media users (APJII 2018). This numbers means marketing huge opportunities for various business entities, including barbershop business.

In Indonesia, barbershop business is closely linked with Garut. Most of finest barbermen in the country comes from this city, particularly from Banyuresmi Subdistrict (Supriadin 2017). For these Garutians, barbershop business has been passed down from generation to generation. Therefore, massive marketing efforts are required to maintain its sustainability, one of which through social media.

Garutian barbers are mostly men with an average net income ranging from 5 to 6 million rupiahs. The present study seeks to examine the effect of social media marketing on their barbershop business performance.

Social media refer to a set of software applications connected to the Internet to facilitate interaction and information exchange between their users (Kaplan & Haenlein 2010). Social media are also considered as one of the important information channels in building an organizational performance perspective (Ahmed et al. 2019). In addition, social media can also serve as a search tool to find information useful for organizational development (Bulearca & Bulearca 2010).

Social media are deemed as a new paradigm of communication. This can be seen in the way people communicate using internet platform applications today (Kietzmann et al. 2011). Figure 1 illustrate this communication paradigm. Social media enable business entities to build a broad network with their customers in Figure 1.

Previous studies showed that social media had seven functional resources: identity, conversation, sharing, presence, relationship, reputation and groups (Kietzmann et al. 2011; Paniagua & Sapena 2014). In addition, business performance could be measured using the following indicators: operational performance, financial performance (Hamdani et al. 2019; Lopez et al. 2005; López-Nicolás & Meroño-Cerdán 2011) and corporate social performance (Paniagua & Sapena 2014). Another study suggested that it could be measured using sales growth, return on investments, and profit (Arief et al. 2013).

Based on data and facts, this study seeks to examine the use of social media by barbers and explain their impact on their business performance.

2 METHODS

This study was carried out using a quantitative approach. The quantitative explanatory method is a research method that aims to determine the characteristics of variables by examining a number of

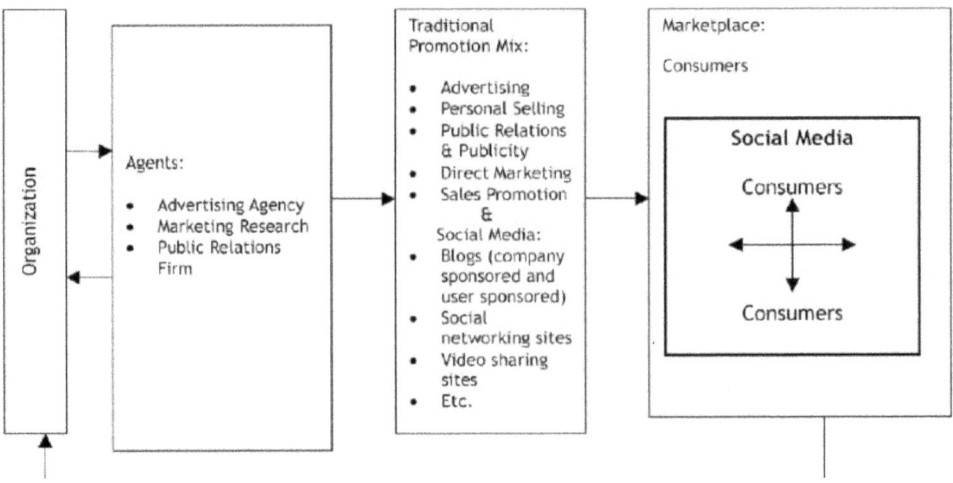

Figure 1. New paradigm of communication (Mangold & Faulds 2009).

samples. Data were collected through a survey. Questionnaires were addressed to 43 barbershop owners in Garut as the respondents to examine the effect of social media exposure on barbershop business performance. Data analysis was carried out using partial least squares structural equation modeling (SEM-PLS). Social media exposure was measured using the following indicators: identity, conversation, sharing, presence, relationship, reputation and groups. Moreover, barbershop business performance was measured using operational performance and financial performance.

3   RESULTS AND DISCUSSION

PLS-SEM modeling was done to measure to what extent social media exposure affected barbershop business performance. This modeling is illustrated in Figure 2:

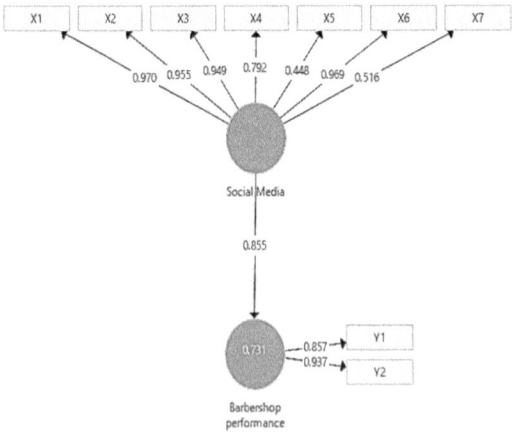

Figure 2.   PLS-SEM modeling and algorithm.

Table 1.   Construct reliability and validity.

	Cronbach's alpha	Rho_a	Composite reliability	Average variance extracted (ave)
Barbershop performance	0.767	0.848	0.892	0.806
Social media	0.909	0.955	0.934	0.683

Figure 2 shows that the path coefficient between social media exposure (X) and barbershop business performance (Y) is 0.855, meaning that social media exposure has influence on barbershop business performance as much as 0.855. In addition, the factor loading value of each social media exposure indicator is as follows: the factor loading of identity (X1) is 0.970, of conversation (X2) is 0.955, of sharing (X3) is 0.949, of presence (X4) is 0.792, of relationship (X5) is 0.448, of reputation (X6) is 0.969, and of groups (X7) is 0.516.

As for barbershop business performance indicators, the factor loading of operational performance (Y1) is 0.857 and of financial performance (Y2) is 0.937.

Indicators of each variable are said to be reliable if their factor loading values are higher than 0.50. Therefore, relationship (X5) with a factor loading value of 0.448 is not reliable and should therefore be eliminated from the modeling.

The average variance Extracted (AVE) value in Table 1 shows that a variable is valid. A variable is said valid if its AVE value is higher than 0.50. Looking at Table 1, it can be said that both research variables barbershop business performance and social media exposure are valid since their AVE values were higher than 0.50.

The reliability of each variable can be determined the composite reliability (CR) and Cronbach's alpha

Table 2. R square.

	R square	R square adjusted
Barbershop performance	0.714	0.705

Table 3. Mean, STDEV, T-values, P-values.

	Original sample (o)	Sample mean (m)	Standard deviation (stdev)	T statistics (|o/stdev|)	P values
Social media -> barbershop performance	0.845	0.846	0.048	17.740	0.000

(CA) value. A variable is said reliable if their CR and CA values are higher than 0.7. Table 1 shows that both variables barbershop business performance and social media exposure are reliable since their CR and CA values are higher than 0.7.

In Table 2, the R-Square value of 0.714 means that the variable barbershop business performance has a variability of 71%.

In Table 3, the original sample value of 0.845 means that social media exposure has positive influence on barbershop business performance. This indicates that the higher the social media exposure, the better the barbershop business performance. The T Statistics value of 17.740, higher than the critical t of 1.96, means that the influence that social media exposure has on barbershop business performance is significant.

Basically, with the presence of social media, the existence of barbershop in Garut Regency was known by the community (Hamdani & Nugraha 2020). This shows that social media did have a good impact on the marketing performance of barbershop, where barbershop could show the characteristics of each shop through social media (Hamdani et al. 2019; Nizar Alam Hamdani et al. 2019). Furthermore, several barbershops showed an increase in customers who had a good influence on the financial performance of their business (Hashi & Krasniqi 2011). This is in line with some previous research findings that the application of information and communication technology could improve business performance and create a competitive advantage for the business. (Hamdani & Maulani 2019; Mark 2009; Yokakul & Booth 2011).

4 CONCLUSION

Social media play a very important role in barbershop business in Garut. They can be utilized for marketing activities and therefore help sustain their business existence in an increasingly competitive environment.

With their functional resources, which include identity, conversation, sharing, reputation and presence, social media had a significant effect on the financial and operational performance of barbershop business in Garut. Therefore, social media can help maximize the potentials of barbershop business, especially in penetrating both national and international market.

ACKNOWLEDGEMENT

We thank all those who helped with this research, especially to the entire barbershop community in Garut Regency. Besides, we appreciate the Fakultas Kewirausahaan Universitas Garut for funding this research.

REFERENCES

Ahmed, Y. A. et al. 2019. Social media for knowledge-sharing: a systematic literature review', telematics and informatics. *Elsevier ltd*, 37, pp. 72–112. doi: 10.1016/j.tele.2018.01.015.

Alam Hamdani, N. and Abdul Fatah Maulani, G. 2018. The influence of e-wom on purchase intentions in local culinary business sector. *International journal of engineering & technology*, 7(2.29), p. 246. doi: 10.14419/ijet.v7i2.29.13325.

APJII. 2017. Penetrasi & perilaku pengguna internet Indonesia, APJII.

APJII 2018. *Penetrasi & profil perilaku pengguna internet indonesia, APJII.* Jakarta. Available at: www.apjii.or.id.

Arief, M. et al. 2013. The effect of entrepreneurial orientation on the firm performance through strategic flexibility: a study on the smes cluster in malang. *Journal of management research*, 5(3), pp. 44–62. doi: 10.5296/jmr.v5i3.3339.

Bulearca, M. and Bulearca, S. 2010. Twitter: a viable marketing tool for smes?. *Global business & management research*, 2(4), pp. 296–309. Available at: http://ezlibproxy.unisa.edu.au/login?url=http://search.ebscohost. com/login.aspx?direct=true&db=bth&AN=57622388&site=ehost-live.

Hamdani, N. A. and Maulani, G. A. F. 2019. The influence information technology capabilities and differentiation on the competitiveness of online culinary smes. *International journal of recent technology and engineering* (IJRTE), 8(1S), pp. 146–150.

Hamdani, N. A. and Nugraha, S. 2020. The influence of information technology and entrepreneurial orientation on competitiveness and business performance, in advances in business, management and entrepreneurship. *London: taylor & Francis group, llc*, pp. 565–569.

Hamdani, N. A., Solihat, A. and Maulani, G. A. F. 2019. The influence of information technology and co-creation on handicraft sme business performance. *International journal of recent technology and engineering*, 8(1S), pp. 151–154. Available at: https://www.ijrte.org/download/volume-8-issue-1s/.

Hashi, I. and Krasniqi, B. a. 2011. Entrepreneurship and sme growth: evidence from advanced and laggard transition economies. *International journal of entrepreneurial behavior & research*, 17(5), pp. 456–487. doi: 10.1108/13552551111158817.

Kaplan, A. M. and Haenlein, M. 2010. Users of the world, unite! The challenges and opportunities of social media. *Business horizons*, 53(1), pp. 59–68. doi: 10.1016/j.bushor.2009.09.003.

Kietzmann, J. H. et al. 2011. Social media? Get serious! Understanding the functional building blocks of social media. *Business horizons*, 54(3), pp. 241–251. doi: 10.1016/j.bushor.2011.01.005.

Lopez, S., Peon, J. M. and Ordas, C. J. 2005. Organizational learning as a determining factor in business performance. *The learning organization*, 12(3), pp. 227–245. doi: 10.1108/09696470510592494.

López-Nicolás, C. and Meroño-Cerdán, Á. L. 2011. Strategic knowledge management, innovation and performance. *International journal of information management*, 31(6), pp. 502–509. doi: 10.1016/j.ijinfomgt.2011.02.003.

Lund, N. F., Cohen, S. A. and Scarles, C. 2018. The power of social media storytelling in destination branding, journal of destination marketing and management. *Elsevier ltd*, 8(January), pp. 271–280. doi: 10.1016/j.jdmm.2017.05.003.

Mangold, W. G. and Faulds, D. J. 2009. Social media: the new hybrid element of the promotion mix. *Business horizons*, 52(4), pp. 357–365. doi: 10.1016/j.bushor.2009.03.002.

Mark, C. 2009 *A handbook of information technology*. First edit. New delhi: global media.

Nizar Alam Hamdani, Galih Abdul Fatah Maulani and Arif Abdullah Muharam. 2019. Entrepreneurial culture in the village of the barbers, garut, Indonesia. *International journal of engineering and advanced technology*. Blue eyes intelligence engineering and sciences engineering and sciences publication – BEIESP, 8(5C), pp. 685–687. doi: 10.35940/ijeat.e1096.0585c19.

Paniagua, J. and Sapena, J. 2014. Business performance and social media: love or hate?. *Business horizons. 'kelley school of business, indiana university*, 57(6), pp. 719–728. doi: 10.1016/j.bushor.2014.07.005.

Supriadin, J. 2017. Banyuresmi, kampung sekolah tukang cukur andal indonesia, www.liputan6.com. Available at: https://www.liputan6.com/regional/read/2968462/banyuresmi-kampung-sekolah-tukang-cukur-andal-indonesia.

Yokakul, N. and Booth, P. 2011. The role social capital, knowledge exchange and the growth of indigenous knowledge-based industry in the triple helix system: the case of smes in thailand. *9th international triple helix conference*, pp. 1–13.

Advances in Business, Management and Entrepreneurship – Hurriyati et al. (Eds)
© 2021 Taylor & Francis Group, London, ISBN 978-0-367-67471-7

Prediction and mitigation of ship accidents mortality rate

A.B. Arnanto & A. Subroto
Universitas Indonesia, Depok, Indonesia

ABSTRACT: The shipping industry is known to have high potential of risk associated with safety. From ship accident data that happened in Indonesia from 2005 until 2018 taken from National Transportation Safety Comittee (NTSC), this study develops a model for decision making methods with a decision tree to predict the probability of fatal shipping accidents and corresponding mortalities. The model results show that both the fatal acci-dents and mortalities are mainly caused by technical and human factors, and only a few accidents occur in the waters caused by weathers. In addition, the number of ship accidents has increased in the past 7 years. The biggest portion of ship accidents is occupied by passenger ships. Mostly, the ship accident occurred in Java Sea and Bangka Strait. From the description of the data and the decision tree model made, it is expected to measure the level of risk of ship accidents. The results of this study are beneficial for policy-makers in proposing efficient strategies to prevent fatal shipping accidents, also for ship owner, operator and insurance institution.

1 INTRODUCTION

As the biggest archipelagic country that has approx-imately 17.000 islands and 95.000 kms coastline length, sea transportation becomes main mode of transportation in Indonesia to transport goods among the islands. Accidents occurring in rivers, lakes, and sea based on the data taken from National Transpor-tation Safety Comittee (NTSC) are mainly caused by technical and human factors, and only a few acci-dents in the sea caused by weathers. In addition, the num-ber of ship accidents has increased in the past 7 years. The biggest portion of ship accidents is occu-pied by passenger ships. The mortality impact from accidents has experienced an increasing trend, especially in the last 2 years. Given the fatal impact, especially result-ing in significant numbers of mortalities, a method or model is needed to predict ship accidents that might occur based on the ship accident database so mitiga-tion can be done to reduce the impact of death and the frequency of ship accidents. From the prediction model that is built, it is expected that it can also identify the factors or variables that determine ship accidents resulting in death. Thus, based on infor-mation on fac-tors or variable that affect ship acci-dents, mitigation can be done as needed.

The purposes of this research are:

1. To develop severity model to predict ship acci-dent mortality rate;
2. To mitigate ship accident severity based on the model.

This study takes the following references.

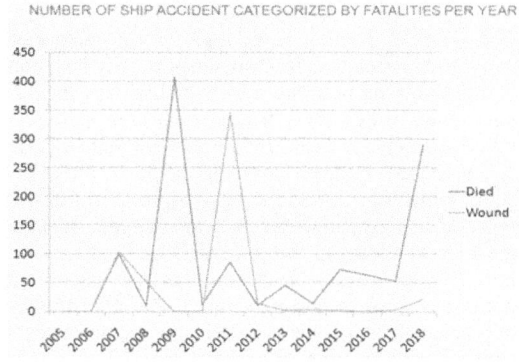

NUMBER OF SHIP ACCIDENT CATEGORIZED BY FATALITIES PER YEAR

Figure 1. Number of ship accident categorized by fatalities year to year.

1.1 Accident analysis and prevention

There have been many studies on the analysis and prevention of ship accidents, one of them in the Else-vier journal entitled Investigation of shipping accident injury severity and mortality (Weng & Yang 2014). The journal is developing a prediction model of fatal probability of an accident using worldwide shipping accident data. The con-tributory factors of an accidents are also discussed in this study.

Table 1 explains the variables used in the study. Vari-able accident time or xday is divided into 2, namely day and night time period. In contrast, the quadrant division of time in this study is divided into 4 where each quadrant has 6 hours starting at 12 at night.

Table 1. Variable description.

Variables	Notation	Description	Proportions
Accident injury severity	y	1 for fatal accident, 0 otherwise	3.77% for fatal accidents
Ship type	x_{st}	1 for cruise ship, 0 otherwise	3.87% for cruise ship
Collision	x_{col}	1 if a collision is occurred, 0 otherwise	19.17% involved in collision accidents
Fire/explosion	x_{fire}	1 if a fire/explosion is occurred, 0 otherwise	8.37% involved in fire/explosion accidents
Machinery/hull damage/failure	x_{hull}	1 if a machinery failure or hull damage occurs, 0 otherwise	37.67% involved in machinery failure or a hule damage
Contact	x_{cont}	1 if a contact occurs, 0 otherwise	19.17% involved in contact accidents
Grounding	x_{grou}	1 if a grounding occurs, 0 otherwise	18.72% involved in grounding accidents
Sinking	x_{sin}	1 if a sinking occurs, 0 otherwise	7.70% involved in sinking accidents
Miscellaneous	x_{mis}	1 if the accident is caused by miscellaneous non-clasified causes, 0 otherwise	12.79% involved in miscellaneous accidents
Weather conditions	x_{we}	1 if the accident is occurred under adverse weather conditions, 0 otherwise	4.33% accident occurred under adverse weather conditions
Accident location	x_{loc}	1 if the accident is occured far away from the coastal area/harbor/ports, 0 otherwise	6.17% accident occurred far away from the coastal area/harbor/ports
Accident time	x_{day}	0 if the accident occurs during the daytime period (daylight conditions) 1 for the night-time period (darkness conditions)	50.43% of accidents occurred during the daylight conditions
Number of people	x_{pas}	The number of people on board the ship	–

The variable location is divided into 2 categories, namely ship accidents far from the port, which is more than 25 km and near the port, which is less than 25 km. While in this study the accident area is divided into 3, namely western, middle and east-ern Indonesia, not based on the proximity to the port. Unlike this study which uses all the data con-tained in the ship type field or column where there are 12 types of ships, the type of vessel in the journal is only categorized into 2 types, namely cruise ships or passengers and not cruise ships or non-passengers. But after the test of p value, the ship type variable is issued because the coefficient value exceeds alpha.

1.2 *Risk matrix*

The level of risk will be measured based on fre-quency or probability of an event risk occurring and based on the severity or impact of risk. From the lev-el of risk measured, treatment can be done in antici-pation of the risks that occur.

From the risk matrix it can be seen that the level of risk that can be accepted is from the level of very low to medium for impact and frequency. As for the risk with treatment by transfer, that is for those who have a high or very high impact level and a very low or low frequency. Treatment with mitigation methods to reduce risk can be done for low or medium impact levels and high or very high frequencies. Risks that must be avoided are for high or very high impact levels and high or very high frequencies.

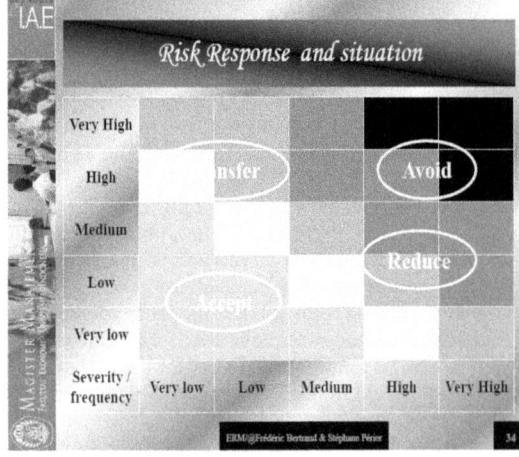

Figure 2. Risk response and situation.
Source: Enterprise Risk Management (ERM), Frederic Ber-trand, Stephane Perier, IAE Grenoble.

Table 2. The risk matrix

16 critical	16	32	64	128
9 major	9	18	36	72
4 significant	4	8	16	32
1 minor	1	2	4	8
Impact/	1	2 not	4	8 highly
Frequency	unlikely	very probable	probable	probable

1.3 Decision tree

Decision trees are prediction models using tree structures or hierarchical structures. The concept of a decision tree is to convert data into decision trees and decision rules. The main benefit of using decision trees is their ability to break down complex decision-making processes to be simpler so that decision makers will better interpret the solution to the problem. The Decision Tree is also useful for exploring data, finding hidden relationships between a number of prospective input variables with a target variable. With a decision tree, people can easily see identifying and seeing the relationship between the factors that influence a problem and can find the best solution by taking into account these factors. This decision tree can also analyze the value of risk and the value of information contained in an alternative problem solving. The role of this decision tree as a decision support tool has been developed by humans since the development of tree theory based on graph theory. This very large use of decision trees has made it useful for humans in various decision-making systems

1.4 Risk matrix

To build a decision tree model, the R programming language will be used, using Graphical User Interface R, called R Studio. R is an open source programming language that can be downloaded for free at https://repo.bppt.go.id/cran/bin/windows/base/, the R version used is version 3.5.2.

2 METHODS

The process for building the decision tree model in this research is as follows:

2.1 Cleaning the data

At this stage, data cleaning and selection of variables will be used in making the model. Variables that do not affect the model will not be used. In addition, classification data will be formed. Data on the inci-dence of ship accidents are classified into 4 time quadrants, with per quadrant having a period of 6 hours. The location of the ship accident is catego-rized as western, middle and eastern Indonesia. The text in the data is likened to the writing so there is no typo data. Data on victims is categorized as mor-tality for ship accident which results in death, and is categorized as zero mortality if there are no fatalities. This mortality data will be predicted in this study.

2.2 Create train and test datates

From 141 records of accident data, it will be divided into data train and test for modeling. Data train is data that is used to form a decision tree classifier model. After the model is formed, then the accuracy level will be tested using test data. The test data used to test must be data that is completely separate from the data train.

2.3 The decision tree model

To make a decision tree model, the Rpart function will be used from the available package R. Whereas to visualize the decision tree model that has been built, it uses the Rpart.plot function which is also one of the packages in R.

2.4 Make prediction

To make predictions, the predict function is used, where predictions will be made using the data set test that has been created.

2.5 Performance measurement

After making a prediction, it will be tested how ac-curate the decision tree model is by using confusion matrix. From the calculation of the right and wrong predictions, the accuracy of the model will be ob-tained in percentage.

2.6 Tune the model

The model that has been built can be increased again according to the level of accuracy with the function rpart.control. In order to get a higher level of accura-cy, it can be done by setting the maximum amount of depth, node and leaf.

2.7 Variable data

Variables that will be used to make decision tree model are as follows:

1. Time Quadrant
2. Area Quadrant
3. Ship Type
4. Category
5. Cause Of Accident
6. Classification Category
7. Classification
8. Mortality

2.8 Decision tree model

Using Rpart from package R with 7 variables, shuf-fled data and data train size 0.6, the decision tree model is obtained as follows:

2.9 Performance measurement

An accuracy measure can be computed for classifi-cation task with the confusion matrix. The confusion matrix is a better choice to evaluate the classification performance.

Each row in a confusion matrix represents an actual target, while each column represents a predicted tar-get. The first row of this matrix considers mortality (the False class): 4 were correctly classified as mor-tality (True negative), while the remaining one was wrongly classified as zero mortality (False positive). The accuracy test the decision tree model for predic-tion in this research is 79%.

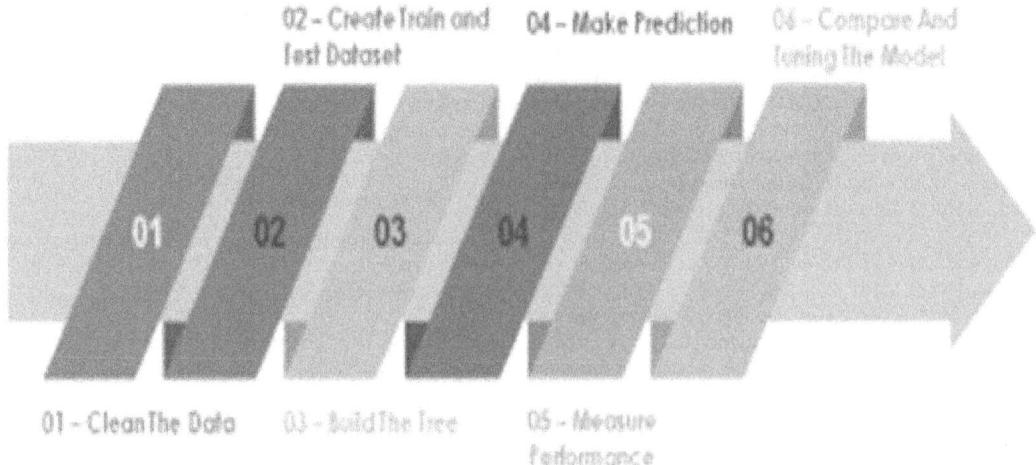

Figure 3.   Building decision tree model.

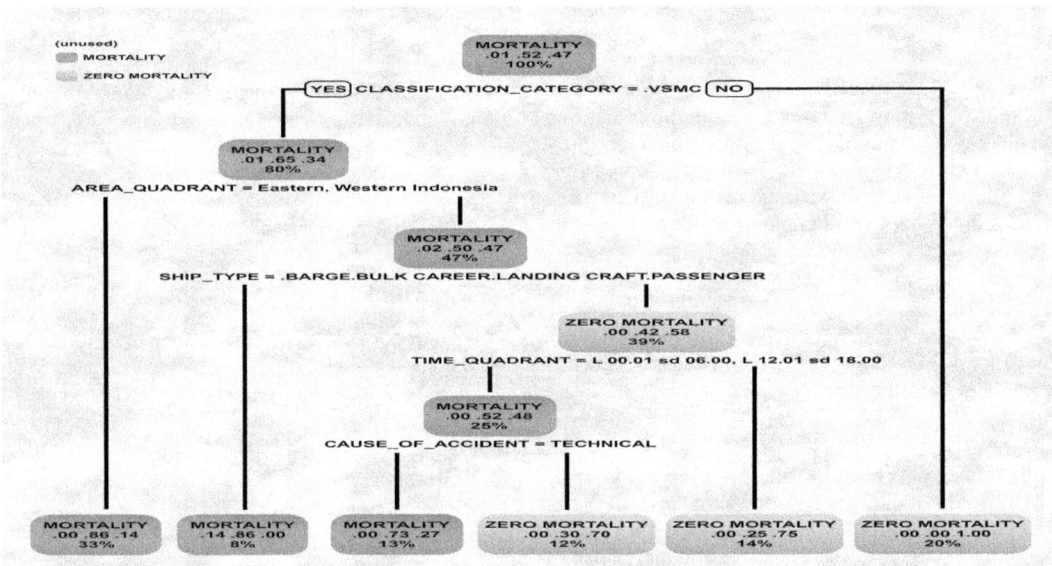

Figure 4.   Decision tree model.

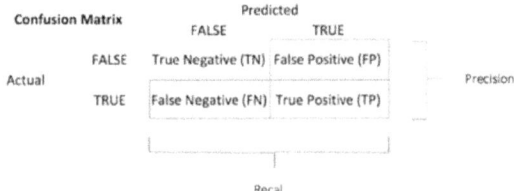

Figure 5.   Confusion matrix.

2.10   *Risk rating and mitigation*

From the description of the data and the decision tree model that has been made, it will be measured the level of risk of ship accidents that occur in Indo-nesian sea in the period from 2005 to 2018 for burn-ing, sinking, collision and aground. The risk rating for those acci-dents is 32. The risk level after mitiga-tion is 9, which is an acceptable risk (risk appetite).

3   RESULTS AND DISCUSSION

When compared to using the Classification And Regression Tree (CART) that is based on testing data, the Cart is slightly more accurate in predicting fatal accidents, which is 82% compared to 81.5%, whereas to predict non fatal accidents, both models have almost

Table 3. Decision tree model comparation.

| | | | Size | | Prop Table | | | | |
| | Number of | Data shuffled/ | | | Train | | Test | | |
No	variable used	not shuffled	Train	Test	M	Z	M	Z	Accuracy
1	7	Shuffled	0.6	0.4	46%	53%	52%	48%	59%
2	7	Not Suffled	0.4	0.4	52%	47%	43%	59%	79%
3	7	Shuffled	0.7	0.3	48%	51%	49%	51%	67%
4	7	Not Suffled	0.7	0.3	48%	51%	49%	51%	77%
5	6	Shuffled	0.8	0.2	48%	51%	48%	52%	55%
6	7	Shuffled	0.8	0.2	49%	51%	45%	52%	66%
7	7	Not Suffled	0.8	0.2	49%	50%	45%	55%	76%
8	7	Shuffled	0.9	0.1	48%	51%	47%	53%	73%
9	7	Not Suffled	0.9	0.1	48%	51%	47%	53%	67%

the same level of accuracy, 87% for CART and 86.9% for logistic regression. Then, it is be compared using adjustments to the variables used, shuffled / no shuffled and adjustments to the percentage data train and test. The result is as follows.

4 CONCLUSION

The most accurate model is the decision tree model that uses data train size 0.6 and shuffled with pre-dictive accuracy of 79%. Whereas for mortality pre-diction, it will be achieved if:

Category of collision, explosion, sink or oth-ers and ship type are bulk carrier, fishing ves-sel, speed boat, tanker or tug boat;

Category of collision, explosion, sink and others with type of ship are not bulk carrier, fishing vessel, speed boat, tanker and tug boat and occur at 1,2 or 3 time quadrants;

Category of collision, explosion, sink and others with type of ship are not bulk carrier, fishing vessel, speed boat, tanker and tug boat and occur in time quadrant 4 and occur in the eastern or western Indonesian quadrant area

ACKNOWLEDGEMENT

I am enormously grateful to Mr. Athor Subroto, Phd for his continuous encouragement, topic, material and kindly advice throughout my study.

REFERENCES

Analytics Vidhya. A guide to machine learning in R for beginners: decision trees. [Online]. Retrived from www.analythicsvidhya.com.
Achmatim Net. Mengukur kinerja algoritma klasifikasi dengan confusion matrix. [Online]. Retrived from achma-tim.net
Bertrand, F., Perier, S., IAE Grenoble. Enterprise Risk Management (ERM).
Chartio. What is a scatter plot and when to use it decision tree in R with example, Guru99. [Online]. Retrived from chartio.com.
Weng, J., & Yang, D. 2014. Investigation of shipping accident injury severity and mortality.

Advances in Business, Management and Entrepreneurship – Hurriyati et al. (Eds)
© 2021 Taylor & Francis Group, London, ISBN 978-0-367-67471-7

Digital entrepreneurship: Platform strategy from the perspective of leather entrepreneurs in Garut

I. Permana, S. Nugraha, N.A. Hamdani & A.O. Herlianti
Universitas Garut, Garut, Indonesia

ABSTRACT: Digital entrepreneurship is an effort to use information technology for businesses, in which entrepreneurs utilize digital platforms in commercial activities to connect with customers, develop products, and renew services. This study describes leather entrepreneurs' perception about platform as a strategy in digital entrepreneurship. The results of data analysis showed that the respondents had good perceptions about platform. They could select, manage, store, and interpret platform. The adoption of digital entrepreneurship by leather entrepreneurs was facilitated by the use of internet, ICT knowledge, prior knowledge and education, family environment, personal needs, easy internet access, media, and programming experts. Leather entrepreneurs used platform as a strategy to adopt digital entrepreneurship.

1 INTRODUCTION

Digital entrepreneurs with their new ways of doing business had an enormous effect on the whole world, especially in the last decade. They are not only completely changed the business world, but also shaped the way we communicate with each other in everyday life. In this study, entrepreneurship refers to a process of designing, launching and running a new business with its distinct characteristic of new value creation (Hsieh & Wu 2018). Open innovation by adopting the platform strategy appears to gain popularity in research over the past few years (Boudreau 2010; Parker & Alstyne 2017). An interesting aspect to study is how platform strategy on digital entrepreneurships of leather as local business in Garut. The other aspect is due to the emergence of many platforms in Indonesia today.

1.1 Digital entrepreneurship

Digital entrepreneurship is a subcategory of entrepreneurship in which some or all of the physical evidence is changed by organizations from traditional to digital (Kraus et al. 2018). The concept of entrepreneurship usually only considers one type of entrepreneurship at a time, rather than focuses on entrepreneurial business, entrepreneurial knowledge, entrepreneurial institutions, sociomaterial practices and digital business (Davidson & Vaast 2010). Thus, it can be seen as a reconciliation of doing business in entrepreneurship in the digital age (Le et al. 2018). Some researchers have begun to explore various practices of interaction and entrepreneurship. The environment as an institution

is able to form ideas about entrepreneurship and how business ventures from time to time restructure existing institutions (Hwang & Powell 2000). A company changes from an entrepreneurial venture to an institutional entrepreneur (Dieleman & Sachs 2008). Digital entrepreneurship is a phenomenon which occurs through technological assets like internet and information and communications technology (Le et al. 2018). Development of a digital business model in three main and broad stages including several substages, respectively, starting with (Le et al. 2018):

1. Idea generation
2. Initial phase
3. Entrepreneurial business management

1.2 Platform strategy

Entrepreneurs utilize digital platforms to commercialize their business. The power of digital networks offers the potential for fast-growing and advanced technology, although this potential can be seen as a major threat as well. On the negative side, rapid growth can mean considerable risk, because competitors who launch more innovative technological advances can destroy all business models (Giones et al. 2017). There are two classification of platform strategies (Hsieh & Wu 2018):

1. Commercialisation capabilities
   Commercialization capabilities criteria can be understood as a set of six main elements: technology, marketing, financial information, intellectual property, resource, impact of utilities (Karaveg et al. 2016).

Table 1. Predictors of new product and/or service (Henard & Szymanski 2001).

Characteristics	Performance
Product	Product advantage
	Product meets customer need
	Product price
	Product technological sophistication
	Product innovativeness
Firm Strategy	Marketing synergy
	Technological synergy
	Order of entry
	Dedicated human resources
	Dedicated R & D resource
Firm Process	Structured approach
	Predevelopment task proficiency
	Marketing task proficiency
	Technology proficiency
	Launch proficiency
	Reduce cycle time
	Market orientation
	Customer input
	Cross-functional integration
	Cross-functional communication
	Senior management communication
Marketplace	Likelihood of competitive response
	Competitive response intensity
	Market potensial

2. Tendency towards new product and/or service development.

There are determinants factors of enterprise resource planning platform selection, such as summarized cost, credibility, functionality, usability, flexibility, implementability, and reputation (Keil & Tiwana 2006). The success of digital entrepreneurs in carrying out business activities depends on the possible interactions offered by the platform (Nambisan 2016). Therefore, the hypothesis of this study are:

H0: Platform strategy performance has a significant positive effect on the entrepreneurship digital adoption.
H1: Platform strategy performance has not a significant positive effect on the entrepreneurship digital adoption.

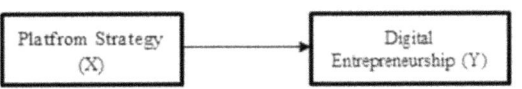

Figure 1. Research framework.

2 METHODS

2.1 Procedure

The population was leather entrepreneurs in Garut. This study used surveys to collect data needed for analysis. The questionnaire consisted of two parts. The first part was about the demographic information of respondents. The second part contained a total of 33 questions about digital entrepreneurship and platform strategy, which were asked using a 1–5 Likert scale. Due to the limited access, the sample was only 30 respondents. However, this number was still in the appropriate sample size range from 30 to 500 (Roscoe 1975). Next, SPSS 24 was used to analyse the collected data. Validity and reliability tests were conducted on the collected data. Finally, descriptive analysis was performed to obtain results.

2.2 Respondent

The survey was conducted in January-April 2019. Thirty respondents were leather entrepreneurs in Garut.

Table 2. Respondent.

Demography		Total
n	Female	30
Sex	Male	15
		15
Age	27 35 year	10
	36 49 year	10
	5 60 year	10
Income (months IDR)	< 10 million	2
	11 20 million	8
	21 30 million	15
	31 40 million	2
	≥41 million	3
Be an entrepreneur	< 1 year	3
	1 5 year	7
	6 10 year	8
	11 15 year	7
	≥16 year	5

2.3 Validity and reliability test result

A validity test was performed using the Pearson Bivariate Correlation formula in SPSS 24 statistical program. The test aimed to compare the value of rxy (Pearson Correlation) or Corrected Item-Total Correlation with the rtable value. For the total of 30 respondents, the rtable value is 0.306. Then, the validity test was conducted for all indicators of brand awareness, purchasing decision, and brand advocacy. If the indicator has a proxy value (Pearson Correlation) that is greater than the rtable value (0.306), then the indicator is valid.

The reliability test was carried out by comparing the Cronbach Alpha value. The values are greater than 0.361, which means that all variables are reliable.

Based on Table 3, it is known that the platform strategy as independent variable has a high reliability value. Meanwhile, digital entrepreneurship as dependent variable has a middle reliability value.

Table 3. Validity and reliability test result.

Variable	Indicators	Pearson Correlation	Result
Platform strategy (Reliability Alpha 0.8)			Reliable
	Product	0.870	
	Firm strategy	0.860	
	Firm process	0.764	
	Marketplace	0.750	
Digital entreprenurship (Reliability Alpha 0.6)			Reliable
	Initial phase	0.647	Valid
	Entrepreneurial	0.883	Valid
	Business management	0.994	Valid

3 RESULT AND DISCUSSION

3.1 Data analysis

The results can be arranged in a linear regression equation, Y = 4.549+1.048 X. Constant (α) of 1.251 means that if the platform strategy variable has a value of 0, then the digital entrepreneurs value is 1.251. The value of the regression coefficient for the strategy platform variable (X) is 1.048. This means that if the platform strategy variable has increased 1%, then the digital entrepreneurship (Y) will increase by 23%. The negative coefficient means that there is no direct relationship between brand awareness and the decision-making process.

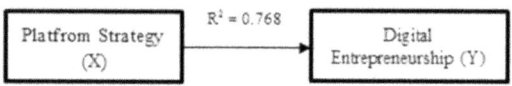

Figure 2. Hypothesis result.

The significance value is 0.00 < 0.05 and the tvalue is 4.549 > ttable 2.048. It means that Ho is accepted. In other words, platform strategy performance has a significant positive effect on the entrepreneurship digital adoption.

3.2 Finding

This research indicates that digitalisation bring a major shift into how entrepreneurs conduct business today. It is not only dealt with business models and their various possible forms or characteristics. It creates new opportunities for entrepreneurs as well as challenges for digital entrepreneurial activities. However, as already indicated, it is assumed that research on digital entrepreneurship is still in its infancy. There are only 35% leather entrepreneurs that consciously use platform strategy for commercialisation capabilities. Most of them (65%) use platform strategy for developing new product and/or service.

Challenges of digital entrepreneurship are considerably diverse. Technological infrastructures are constantly advancing and digital technology continuously offers new developments to society. From one side, challenges, such as low diffusion rates of specific technologies for payments, might be over-come soon. On the other side, new challenges will form by advanced technological opportunities. Entrepreneurship research should continuously investigate the challenges and opportunities emerging in the near future. The aims is to give a feedback for entrepreneurs, as it is related to their interest.

The results of data analysis show that the respondents have good perceptions about platform. They can select, manage, store, and interpret platform. The adoption of digital entrepreneurship by leather entrepreneurs is facilitated by the use of internet, ICT knowledge, prior knowledge and education, family environment, personal needs, easy internet access, media, and programming experts. Leather entrepreneurs use platform as a strategy to adopt digital entrepreneurship.

4 CONCLUSION

This article investigated the effect of platform strategy on the digital entrepreneurship. It is discovered that platform strategy of leather entrepreneurs have significant positive effect on the digital entrepreneurship. The limitation of this research is very small number of respondents, which is only 30 respondents. Further studies is expected to use a large number of sample so that the problem can be investigated in depth.

ACKNOWLEDGMENT

This work was funded by the Fakultas Kewirausahaan Universitas Garut of Indonesia.

REFERENCES

Hsieh, Y. & Wu, Y.J. 2018. Entrepreneurship through the platform strategy in the digital era: Insights and research opportunities. Comput Human Behavior. 95: 315–325.

Parker, G. & Alstyne, M.V. 2017. Innovation, Openness, and Platform Control. Management Sci ence. 64(7): 3015–3032.

Boudreau, K. 2010. Open Platform Strategies and Innovation: Granting Access vs Devolving Control Open Platform Strategies and Innovation: Granting Access vs Devolving. Control. Management Science. 50(10): 1849–1872.

Kraus, S., Palmer, C., Kailer, N., Kallinger, F.L., Spitzer, J. & Kraus, S. 2018. Digital entrepreneurship A research agenda on new business models for the twenty-first century. Internasional Journal Enterperneurship Behaviour research.

Davidson, E. & Vaast, E. 2010. Digital Entrepreneurship and its Sociomaterial Enactment. In: Digital Entrepreneurship and its Sociomaterial Enactment 1–10.

Le, D.T., Vu, M.C. & Ayayi, A. 2018. Towards a Living Lab for Promoting the Digital Entrepreneurship Process. International Journal of Entrepreneurship.

Hwang, H. & Powell, W.W. 2000. Institutions and Entrepreneurship'. International journal: Institutions and Entrepreneurship. 201–302.

Dieleman, M. & Sachs, W.M. 2008. Coevolution of Institutions and Corporations in Emerging Economies: How the Salim Group Morphed into an Institution of Suharto s Crony Regime. Journal Managemen Studies. 45(7): 1274–1300.

Giones, F., Brem, A. & Clark, J.H. 2017. Digital Technology Entrepreneurship: A Definition and Research Agenda. Technol Innovation Management Review.7(5):4 4–51.

Karaveg, C., Thawesaengskulthai, N. & Chandrachai, A. 2016. R&D commercialization capability criteria: Implications for project selection Journal Management Development.35(3).

Henard, D.H. & Szymanski, D.M. 2001. Why Some New Products Are More Successful Than Others. Journa of Marketing Research. 362–375.

Keil, M. & Tiwana, A. 2006. Relative importance of evaluation criteria for enterprise systems: a conjoint study. Infomation System Journal. 16: 237–262.

Nambisan S. 2016. Entrepreneurship: Toward a Digital Technology Perspective of Entrepreneurship. Entrepreneurship Theory and Practice. 41(6).

Roscoe, J.T. 1975. Fundamental Research Statistics for The Behavioral Sciences. 2nd ed. Holt Rinehart & Winston.

733

Rapid changes: Strategic flexibility in higher education

G.A.F. Maulani, D. Disman, A. Rahayu & R. Hurriyati
Universitas Pendidikan Indonesia, Bandung, Indonesia

ABSTRACT: Private universities are business entities in Indonesia. Due to the large number of private universities in Indonesia, the competition between them becomes intense. Therefore, they are required to be able to adapt to any changes in order to sustain. The purpose of this study is to examine the effect of strategic flexibility of private university competitiveness. The samples were 27 private universities in West Java, Indonesia. Data analysis was performed using PLS-SEM. The results showed that strategic flexibility, indicated by attention, assessment, and action had influence on private university competitive advantage. This implies that it is necessary for private universities to take into account their adaptability to changes in order for them to be able to manage risks in their environments.

1 INTRODUCTION

The Industrial Revolution 4.0 and Society 5.0 encourage every organization and business entity, including higher education institutions (HEIs), to strive for their existence (Maulani & Hamdani 2019). As a business entity, a HEI is expected to be competitive to sustain their existence (Rabah 2015; Supriyatna et al. 2019). Therefore, it is necessary for HEIs to seek out relevant strategies to address the globalization (Wangenge-Ouma & Langa 2010).

Table 1 shows that in 2018 there are as many as 3,171 private HEIs in Indonesia (Kementerian Riset 2018), an increase of 17 from the previous year and 47 from 2016 (Kementerian Pendidikan Tinggi 2017; Tinggi 2016). In other words, in terms of quantity, the number of Indonesian private HEIs continues to grow from year to year, leading to more competitive environment in higher education. However, despite this quantitative increase, 130 HEIs are reported to close throughout 2015 – 2019 due to their inability to compete (Mediani 2019). In addition, to justify their low quality, only 30 out of 3,171 private HEIs are accredited A by Indonesian accreditation board (Kementerian Riset 2018). To put it another way, quantity does not mean quality, indicating that Indonesian private HEIs in general do not have competitive advantage.

As a professionally managed organization, a HEI is required to apply several strategies to adapt to the business environment (Wheelen et al. 2018). In dealing with dynamic environments, it is necessary for HEIs to be highly flexible about their resources (Aisjah 2017). Strategic flexibility, if employed, may help them with their business processes (Eryesil et al. 2015) because strategic flexibility emphasizes the optimization of strategic organizational resources that must be

Table 1. Number of private HEIs in Indonesia.

Year	2016	2017	2018
Number	3.124	3.154	3.171

implemented in achieving good performance (Cingöz & Akdoğa 2013). Furthermore, this strategy can make efficient the use of resources in order to maintain the competitive advantage of higher education (Maulani & Hamdani 2018). To create a competitive advantage, it is also necessary for HEIs to ensure that they provide good service quality and offer differentiation (Lewangka et al. 2015).

Strategic management is a set of managerial decisions and actions that determine the long-term performance of a company. This includes external and internal environmental scanning, strategy formulation (strategic or long-term planning), strategy implementation, evaluation and control (Wheelen & Hunger 2012; Wheelen et al. 2018).

Simply put, a strategy describes what managers and organizations do to achieve short and long-term goals. Organizations have broad directions and goals, which must always be clearly articulated and understood and which will sometimes be summarized in the form of mission statements. More specific milestones and targets (goals) can help guide specific actions and measure the progress (Thompson & Martin 2005).

In the classical strategy model, competitive advantage is obtained from a combination of external and internal factors (opportunities and strengths against threats and weaknesses); however, it is important to consider that competition is a complex phenomenon

Table 2. Dimensions of competitive advantage variables.

(Lewangka et al. 2015)	(Nurchayati & Gozali 2010)	(Aydin 2013)
Competitive advantage is measured by parameters: Quality of Service Differentiation Customer satisfaction	Competitive advantage is measured by parameters: Quality of education program Differentiation Cost	Competitive advantage is measured by parameters: Reputation Educational standards and curriculum Cost Location Student activities

Table 3. Dimensions of strategic flexibility variables.

(Supeno et al. 2015)	(Aisjah 2017)	(MacKinnon et al. 2008)
Strategic flexibility is measured by parameters: Attention Assessment Action	Strategic flexibility is measured by parameters: Attention Assessment Organization's ability to demonstrate flexible action	Strategic flexibility is measured by parameters: Operational flexibility Human capital flexibility

Table 4. Research variables.

Variable	Indicator
Strategic Flexibility (X)	Operational Flexibility (X1) Human Capital Flexibility (X2)
Competitive advantage in HEIs (Y)	1. Differentiation (Y1) 2. Quality of service (Y2) 3. Customer satisfaction (Y3)

(Mohebi & Farzollahzade 2014). After all, managers' strategic choices directly affects organizational performance, and organizational strategic adaptation to their external environment is the principle of competitiveness (Mainardes et al. 2011). At present HEIs must have a more entrepreneurial focus and contribute in ways that are relevant to the community, thus completing its mission.

Based on some literatures, an overview of the dimensions of the construct of competitive advantage and its elements can be summarized as in Table 2.

Strategic flexibility is a company's ability to respond and successively adapt to environmental changes. This term is also applicable to strategic decision making because it is the extent to which new and alternative options in strategic decision making are generated and considered (Combe & Greenley 1986).

Strategic flexibility emerges as a response to a concern about the ability to adapt to changes that require competitive advantage. Companies that do not anticipate these changes face difficulties in changing their position in the market, changing their game plans, or dismantling their current strategic posture while younger competitors move on to overcome exogenous changes in their markets by changing the rules of their mutually competitive games (Harrigan 2017).

Based on the review of some literatures, strategic flexibility and its elements can be summarized in Table 3.

This paper examines the effectiveness of the implementation of the flexibility strategy in Indonesian private HEIs and its implications for their competitiveness.

2 METHODS

The present study was conducted using a verification approach. Data were obtained from 27 private HEIs under the Ministry of Research, Technology and Higher Education of the Republic of Indonesia. These universities are located in Garut, Tasikmalaya, and other areas in West Java, Indonesia. The samples were determined using a non-probability sampling technique. Data analysis was performed using PLS-SEM. Table 4 presents the research variables.

3 RESULTS AND DISCUSSION

Based on the results of PLS-SEM data processing, there are two models that can describe the relationship between strategic flexibility, operational flexibility, and human capital flexibility. The variable competitive advantage in HEIs has the following indicators: differentiation, quality of service, customer satisfaction. Figure 2 illustrates the PLS algorithm.

Figure 2 shows that the coefficient value between strategic flexibility and competitive advantage is 0.990, meaning that strategic flexibility has significant positive influence on the competitive advantage in HEIs. In addition, there are some values that can be interpreted as follows:

a. Both indicators of strategic flexibility, operational flexibility and human capital flexibility, have loading factor values of 0.979.
b. As for the variable competitive advantage, the indicator differentiation has a loading factor value of 0.427, service quality has a loading factor value of 0.972, and customer satisfaction has a loading factor value of 0.967.

The loading factor value of the indicator differentiation is lower than 0.50. Therefore, this indicator should be eliminated from the model. Table 5 sums up the construct reliability and validity.

Table 5 shows that the Average Variance Extracted (AVE) values of strategic flexibility and competitive

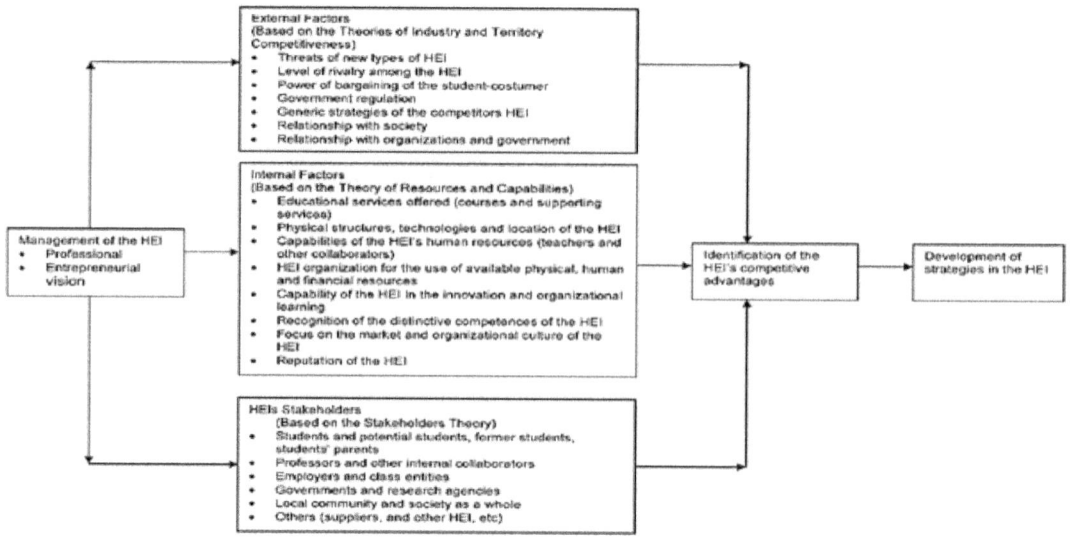

Figure 1. Identification model for competitive advantage in HEIs (Mainardes et al. 2011).

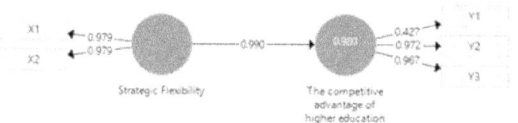

Figure 2. PLS algorithm modelling.

Table 5. Construct reliability and validity.

	Cronbach's Alpha	rho A	Composite Reliability	Average Variance Extracted (AVE)
Strategic Flexibility	0.956	0.956	0.979	0.958
The Competitive Advantage of Higher Education	0.749	0.944	0.856	0.687

advantage are 0.958 and 0.687, higher than the suggested value of 0.5. It means that both variables are valid. These variables are also reliable because their Composite Reliability (CR) and Cronbach's Alpha (CA) values are higher than the suggested value of 0.7 as:

a. The variable strategic flexibility has CA value of 0.956 and CR value of 0.958
b. The variable competitive advantage has CA value of 0.749 and CR value of 0.856.

The results of goodness of fit test show that the value of R-Square was 0.988, meaning that strategic flexibility in private HEIs can contribute to their competitive advantage as much as 98%, and the other 2% is influenced by factor other than the studied variable.

Table 6. Mean, STDEV, t-values, p-values.

	Original Sample (O)	Sample Mean (M)	Standard Deviation (STDEV)	T Statistics (|O/STDEV|)	P Values
Strategic Flexibility -> The Competitive Advantage of Higher Education	0.990	0.990	0.007	151.094	0.000

To determine the significance of the influence of strategic flexibility on the competitive advantage, PLS-SEM bootstrapping was performed. The results are summarized in Table 6.

The positive value of the original sample in Table 6 indicates that strategic flexibility has positive influence on competitive advantage in HEIs. In other words, the more a HEI apply a strategic flexibility in their organization, the more competitive advantage they have. The T Statistics value of 151.094 is higher than the critical t of 1.96, meaning that strategic flexibility has significant influence on competitive advantage. Put another way, operational flexibility and human capital flexibility have significant influence on the competitiveness of private HEIs in Indonesia in terms of quality of service and customer satisfaction. The customer in this case include students, lecturers, and end users.

Based on the results of the calculation show that strategic flexibility is a good strategy and has a significant relationship to the competitive advantage of organizations that move and change in a dynamic environment, this is in line with some previous research findings (Chen et al. 2017; Cingöz

& Akdoğan 2013). Also, operational flexibility and Human Capital Flexibility have a significant role in the creation of excellent strategic flexibility in Higher Education Institutions. (Supeno et al. 2015).

4 CONCLUSION

Private HEIs have always to deal with dynamic environments. Therefore, as an organization, they are expected to quickly adapt to whatever changes occurring in their environments so as to contribute in relevant way to the community. The implementation of strategic flexibility is required to anticipate uncertainty in their dynamic environmental conditions. This can be done by ensuring flexible business and operational processes and human capital in their organization. Statistically, the implementation of strategic flexibility can significantly influence their competitive advantage as seen from such indicators as the quality of service, both academic and non-academic, and student satisfaction.

ACKNOWLEDGEMENTS

We would like to thank all respondents, especially universities in Garut Regency who have supported this research.

REFERENCES

Aisjah, S. 2017. Intellectual capital and strategic flexibility effect the performance companies in small and medium enterprises in Malang-Indonesia. *Australian Academy of Accounting and Finance Review* 3(3): 98–110.

Aydin, O.T. 2013. Location as a competitive advantage to attract students. *International Review of Management and Marketing* 3(4): 204–211.

Chen, Y., Wang, Y., Nevo, S., Benitez, J., & Kou, G. 2017. Improving strategic flexibility with information technologies: insights for firm performance in an emerging economy. *Journal of Information Technology* 32(1): 10-25.

Cingöz, A. & Akdoğan, A.A. 2013. Strategic flexibility, environmental dynamism, and innovation performance: an empirical study. *Procedia - Social and Behavioral Sciences* 99: 582–589. doi: 10.1016/j.sbspro.2013.10.528.

Combe, I.A. & Greenley, G.E. 1986. Capabilities for strategic flexibility: a cognitive content framework. *European Journal of Marketing*: 81–111.

Eryesil, K., Esmen, O. & Beduk, A. 2015. The role of strategic flexibility for achieving sustainable competition advantage and its effect on business performance. *International Journal of Business and Economics Engineering* 9(10): 3456–3462.

Harrigan, K.R. 2017. Strategic flexibility and competitive advantage. *Oxford Research Encyclopedia of Business and Management* 1(c): 1–30.

Kementerian Pendidikan Tinggi. 2017. *Statistik pendidikan tinggi 2017*. Jakarta: Pusat Data dan Informasi Iptek Dikti.

Kementerian Riset. 2018 *Statistik pendidikan tinggi 2018, kementerian riset, teknologi dan pendidikan tinggi*. Jakarta: Pusat Data dan Informasi Ilmu Pengetahuan, Teknologi, dan Pendidikan Tinggi.

Lewangka, O., Kurniaty, K., Sumardi, S., & Jusni, J. 2015. Analysis of competitive advantage through private. *International Journal of Research in Social Sciences* 5(5). pp. 65–70.

MacKinnon, W., Grant, G., & Cray, D. 2008. Enterprise information systems and strategic flexibility. *Proceedings of the 41st Annual Hawaii International Conference on System Sciences (HICSS 2008)*: 402–402.

Mainardes, E.W., Ferreira, J.M., & Tontini, G. 2011. Creating a competitive advantage in Higher Education Institutions: proposal and test of a conceptual model. *International Journal of Management in Education* 5(2/3): 145.

Maulani, G. A. F. and Hamdani, N. A. (2018) 'Perencanaan Strategis Sistem Informasi pada Perguruan Tinggi Swasta di Indonesia (Studi Kasus pada Institut Pendidikan Indonesia Garut)', Jurnal PETIK, 4(September), pp. 162–166. doi: https://doi.org/10.31980/jpetik.v4i2.367.

Maulani, G.A.F. & Hamdani, N.A. 2019. The influence of information technology and organizational climate on the competitiveness of private universities in Indonesia. *International Journal of Recent Technology and Engineering* 8(1S): 142–145.

Mediani, M. 2019. 130 *perguruan tinggi swasta ditutup sepanjang 2015-2019*. [Onlne]. Retrieved from https://www.cnnindonesia.com/nasional/20190802172238-20-417874/130-perguruan-tinggi-swasta-ditutup-sepanjang-2015-2019.

Mohebi, M.M. & Farzollahzade, S. 2014. Improving competitive advantage and business performance of SMEs by creating entrepreneurial social competence. *Management Reserach* 2(spesial Issue): 20–26.

Nurchayati, A. & Gozali, I. 2010. Penerapan model strategi keunggulan bersaing berorentasi lingkungan pada Perguruan Tinggi Swasta (PTS) Di Kota Semarang. *Jurnal Ilmiah Serat Acitya*: 33–45.

Rabah, K. 2015. Effects of competitive advantage on organizational. *Kenyatta University*.

Supeno, H., Made, S., Siti, A., & Arsono, L. 2015. The effects of intellectual capital, strategic flexibility, and corporate culture on company performance: A study on small and micro-scaled enterprises (SMEs) in Gerbangkertosusila Region East Java. *International Business and Management* 11(1): 1-12.

Supriyatna, A., Yulianto, E., Hamdani, N. A., & Maulani, G. A. F. (2019). Budaya Perusahaan: Penerapan Good Corporare Governance Serta Implikasinya Terhadap Keberlanjutan Kinerja Bank. Business Innovation and Entrepreneurship Journal, 1(1), 11–20.

Thompson, J. & Martin, F. 2005. *Strategic management: awareness and change*. New York: Thomson.

Tinggi, K.R.T. 2016. *Statistik Pendidikan Tinggi 2014/2015*. [Online]. Retrieved from http://www.ristekdikti.go.id/wp-content/uploads/2016/11/E-Book-Statistik-Pendidikan-Tinggi-2014-2015-revisi.pdf.

Wangenge-Ouma, G. & Langa, P.V. 2010. Universities and the mobilization of claims of excellence for competitive advantage. *Higher Education* 59(6): 749–764.

Wheelen, T.L. & Hunger, J.D. 2012. *Strategic management and business policy*.

Wheelen, T.L., Hunger, J.D., Hoffman, A.N., & Bamford, C.E. 2018. *Strategic management and business policy: globalization, innovation and sustainability*. Upper Saddle River, NJ: Prentice Hall.

Innovation strategy of coffee industry

N.A. Hamdani, D. Disman, A. Rahayu & R. Hurriyati
Universitas Pendidikan Indonesia, Bandung, Indonesia

ABSTRACT: Coffee shops have been developing quite a lot in Indonesia. This indicates that coffee drinking is now part of lifestyle. This study aims to figure out the appropriate innovation strategies to boost the business performance of coffee industry. The samples were 67 coffee industries in West Java, Indonesia. The data were analyzed using PLS-SEM. The results revealed that an innovation strategy as indicated by aggressiveness, analysis, defensiveness, futurity, proactiveness, and riskiness had influence on the business performance, which is indicated by financial and nonfinancial performance of small and medium coffee industries. It is necessary for small and medium coffee industries to make an innovation strategy as a response to the changes in culture and consumer behaviors and the industrial revolution. That way, they can improve their business performance and sustain.

1 INTRODUCTION

Generally, there are two types of coffee, namely arabica and robusta. Arabica has been dominating world's coffee production by 62%, the rest is robusta (Kufa et al. 2011). In 2013, robusta coffee production reached 7,204.298 tons from a total land area of 15,475.045 hectares. In 2014, total production increased to 7,853.349 tons from a total land area of 15,654.646 hectares. In 2015, there was an increase in production to 8,060.017 tons from a total land area of 15,688.460 hectares. However, in 2016, the total production decreased to 7,247.556 tons from a total land area of 14,384.530 hectares (Perkebunan 2017).

Innovation is of the greatest importance for every business entity to improve their competitiveness and business performance (Reguia 2014). Innovation can be influenced by several factors such as research and development, government, leaders, organizations (Dotun 2015; Fazlzadeh 2010), technology, and managerial capability (Abereijo et al. 2007); (Ghobakhloo et al. 2012). Innovation in SMEs can be influenced by internal factors such as management, human resources, and information system application and by external factors such as business environment and government policy (Fong 2011). The purpose of this study is to examine the implementation of an innovation strategy that can improve the business performance of small and medium coffee enterprises.

Some previous studies have shown that the business performance of SMEs is closely linked with such factors as innovation, technology, knowledge creation, and marketing (Hamdani & Susilawati 2018; Kamasak 2015; Swaminathan 2014). Business performance is

a concept used to measure the achievement of business activities (Wheelen & Hunger 2018). Periodic performance measurement is required by every organization, both profit-oriented and nonprofit-oriented organization. The results of such measurements will be useful for them as the basis for planning. In addition, organizations may become more aware of their internal and external environment, so they can identify their strengths, weak-nesses, opportunities, and threats. The results of periodic performance measurements can also be useful to deal with critical factors that require immediate countermeasures so as not to cause more severe con-sequences (Muangkhot 2015). According to Hubbard & Beamish (2011), business performance can be measured by using marketing and financial performance. Business performance measurement through marketing performance can be done by looking at sales, market growth, and market share. In terms of financial performance, business performance can be measured using return on investment (ROI), revenue mix, asset turn over, and significant cost reduction. Business performance is a multidimensional concept that defines the success and the achievement of a business (Civelek et al. 2015). According to Wheelen & Hunger (2018), business performance can be said as the end result of various business activities. In the strategic management process, the predetermined objectives should be used to measure company performance after the implementation of a strategy (Wheelen & Hunger 2018).

Innovations can be classified into several types: administrative innovation, technical innovation, product/service innovation, process innovation, radical innovation, incremental innovation (Damanpour

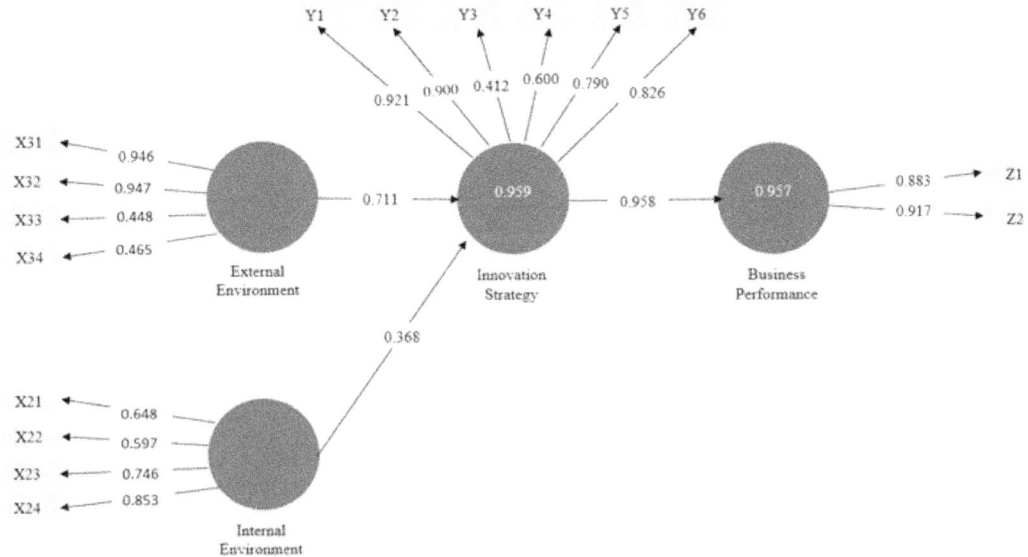

Figure 1. Results of analysis in PLS algorithm.

1991). An innovation strategy is related to a company's strategy to adopt innovation. Innovation can be classified into offensive strategy, defensive, imita-tive, dependent, traditional, and opportunist strategy. This classification is based on the entry speed and time of a company into the new technology area (Hadjimonalis & Dickson 2000). Innovation strategy can be measured by such indicators as aggressiveness, analysis, defensiveness, futurity, and proactiveness (Bozkurt & Kalkan 2014).

2 METHODS

This study involved a sample of 65 small and medium coffee enterprises in West Java, Indonesia. These coffee enterprises were chosen using a simple random sampling technique. Data collection was conducted through surveys of small and medium coffee industry that met certain criteria set by the government.

Data analysis was performed using PLS-SEM. The measurement involved internal variables including production process, marketing, technology, and human resources; external variables including government, economy and business, industrial revolution and consumer behavior; and innovation strategy variables including aggressiveness, analysis, defensiveness, futurity, proactiveness. The business performance was measured using financial and nonfinancial performance.

3 RESULTS AND DISCUSSION

Data on the effects of internal and external environments on innovation strategy and their impact on the

business performance of small and medium coffee enterprises were processed using SmartPLS. Figure 1 presents the modeling.

This model can be interpreted as follows:

1. The path coefficient from external environment to the latent variable innovation strategy is 0.711, it means that external environment had influence on innovation strategy as much as 0.711.

   a. The factor loading of the indicator government is 0.946, it means that external environment contributed to government as much as 0.946.
   b. The factor loading of the indicator economy and business) is 0.947, it means that external environment contributed to economy and business as much as 0.947.
   c. The factor loading of the indicator indus-trial revolution is 0.448, it means that external environment contributed to industrial revolution as much as 0.448.
   d. The factor loading of the indicator consumer behavior is 0.465, it means that external environment contributed to consumer behavior as much as 0.465.

2. The path coefficient from internal environment to the latent variable innovation strategy is 0.368, it means that internal environment had influence on innovation strategy as much as 0.368.

   a. The factor loading of the indicator production process is 0.648, it means that internal environment contributed to production process as much as 0.648.
   b. The factor loading of the indicator technology is 0.597, it means that internal environment contributed to technology as much as 0.597.

Table 1. Construct reliability and validity.

	Cronbach's Alpha	rho_A	Composite Reliability	Average Variance Extracted (AVE)
Business Performance	0.767	0.781	0.895	0.810
External Environment	0.708	0.886	0.815	0.552
Innovation Strategy	0.856	0.905	0.899	0.612
Internal Environment	0.708	0.763	0.807	0.516

Table 2. Path coefficients.

	Original Sample (O)	Sample Mean (M)	Standard Deviation (STDEV)	T Statistics (|O/STDEV|)	P Values
External Environmet → Innovation Strategy	0.819	0.820	0.031	26.336	0.000
Innovation Strategy → Business performance	0.957	0.957	0.016	59.345	0.000
Internal Environment → Innovation Strategy	0.258	0.257	0.038	6.817	0.000

c. The factor loading of the indicator human resources is 0.746, meaning that internal environment contributed to human re-sources as much as 0.746.
d. The factor loading of the indicator marketing is 0.853, it means that internal environment contributed to marketing as much as 0.853.

3. The path coefficient from innovation strategy to business performance is 0.958, it means that innovation strategy had influence on business performance as much as 0.958.

a. The factor loading of the indicator aggressiveness is 0.921, meaning that innovation strategy contributed to aggressiveness as much as 0.92.
b. The factor loading of the indicator analysis is 0.900, it means that innovation strategy contributed to analysis as much as 0.900.
c. The factor loading of the indicator defensiveness is 0.412, it means that innovation strategy contributed to defensiveness as much as 0.412.
d. The factor loading of the indicator futurity is 0.600, it means that innovation strategy contributed to futurity as much as 0.600.
e. The factor loading of the indicator proactiveness is 0.790, it means that innovation strategy contributed to proactiveness as much as 0.790.
f. The factor loading of the indicator riskiness is 0.926, it means that innovation strategy contributed to riskiness as much as 0.926.

The PLS Algorithm shows that the factor loading values of indicators industrial revolution, indicator consumer behavior, technology and defensiveness were below 0.6 and therefore should be removed from the model.

Table 1 shows that the factor loading value of each construct is above 0.6. Therefore, all indicators used in this study were valid. The average variance extracted (AVE) values of all variables are also above the suggested value of 0.5. Therefore, the four variables could be said to have met the requirements. All variables were reliable because their composite reliability values were above the suggested value of 0.7. This is also justified by their Cronbach's alpha values, which were

the suggested value of 0.6. The SmartPLS modeling also resulted in the R-square value of 0.917 for the variable business performance and 0.959 for the variable innovation strategy.

Table 2 shows that external environment had significant influence on innovation strategy because its T Statistics value is 26.336 (<1.66). The original sample estimate shows a positive value of 0.819, meaning that the relationship between these variables was positive. Internal environment also had signifi-cance influence on innovation strategy because its T Statistics value was 6.874 (>1.66). The relationship between these variables was also positive since the its original sample estimate displays a positive value of 0.258. In addition, innovation strategy had signif-icant positive influence on business performance be-cause its T Statistics of 59.345 is higher than the crit-ical T of 1.66 and its original sample estimate had a positive value of 0.957. The formulation of an inno-vation strategy in small enterprises is usually restrict-ed by limited resources and limited technological ca-pabilities (Hadjimonalis & Dickson 2000). The strengths of small enterprises do not lie in their physical resources, but in the characteristics of their be-havior, such as flexibility and management. The choice of innovation strategy itself varies, greatly depending on their environments and their response to environmental changes (Ndesaulwa & Kikula 2016).

Some research supports the results presented. The government has an important role in driving product innovation (Ghobakhloo et al. 2012). The company's external factors and business factors also have an impact on the application of innovation. (Wu & Wang 2009). E-WOM and marketing greatly affect business performance (Hamdani & Maulani 2018). Some research also explains the impact of technology strategies on business (Darbanhosseiniamirkhiz & Wan Ismail 2012; Kumar 2014; Manyati & Mutsau 2019).

4 CONCLUSION

The finding of this study has led to a conclusion that an innovation strategy had influence on the business performance of small and medium coffee enterprises. Their innovation strategy was very much influenced by their internal and external factors. In other words, it could be said that innovation strategies are largely determined by a country's economic conditions and government policies. Stable economic conditions and supports from the government will encourage innovation strategies in SMEs. In addition, innovation strategies in SMEs is also dependent on their human resources and marketing techniques. Aggressiveness and riskiness are among factors should be taken into account when SMEs employ an innovation strategy, meaning that that they need to consider their resources and opportunities and calculate the risks when employing an innovation strategy.

REFERENCES

Abereijo, I.O., Ilori, M.O., Taiwo, K.A., & Adegbite, S.A. 2007. Assessment of the capabilities for innovation by small and medium industry in Nigeria. *African Journal of Business Management* 1(November): 209–217.

Bozkurt, Ö.Ç. & Kalkan, A. 2014. Business strategies of SME's, innovation types and factors influencing their innovation: burdur model. *EGE Academic Review* 14(2): 189–198.

Civelek, M.E., Çemberci, M., Artar, O.K., & Uca, N. 2015. Key factors of sustainable firm performance: a strategic approach.

Damanpour, D. 1991. Organizational innovation: A metaanalysis of effects of determinants and moderators. *Academy of Management Journal* 34(3): 555–590.

Darbanhosseiniamirkhiz, M. & Wan Ismail, W.K. 2012. Advanced manufacturing technology adoption in smes: an integrative model. *Journal of Technology Management & Innovation* 7(4): 112–120.

Dotun, F.O. 2015. The key determinants of innovation in small and medium scale enterprises in Southwestern Nigeria. *European Scientific Journal* 11(13): 465–480.

Fazlzadeh, A. 2010. An investigation of innovation in small scale industries located in Science Parks of Iran. *International Journal of Business and Management* 5(10): 148–155.

Fong, M.W.L. 2011. Chinese SMEs and information technology adoption. *Issues in Informing Science and Information Technology* 8(5): 313–322.

Ghobakhloo, M., Hong, T.S., Sabouri, M.S., & Zulkifli, N. 2012. Strategies for successful information technology adoption in small and medium-sized enterprises. *Information* 3(4): 36–67.

Hadjimonalis, H. & Dickson, D. 2000. Innovation strategies of SMES in Cyprus, a small developing country. *International Small Business Journal* 18(4).

Hamdani, N.A. & Maulani, G. 2018. The influence of E-WOM on purchase intentions in local culinary business sector. *International Journal of Engineering & Technology* 7: 246–250.

Hamdani, N.A. & Susilawati, W. 2018. Application of information system technology and learning organization to product innovation capability and its impact on business performance of leather tanning industry. *International Journal of Engineering and Technology* 7: 393–397.

Hubbard, G. & Beamish, P. 2011. *Strategic management: thinking, analysis, action, frechsfores*. Newcastle: Pearson Australia.

Kamasak, R. 2015. Determinants of innovation performance: a resource-based study. *In World Conference on Technology, Innovation and Entrepreneurship* 195: 1330–1337.

Kufa, T., Ayano, A., Yilma, A., Kumela, T., & Tefera, W. 2011. The contribution of coffee research for coffee seed development in Ethiopia. *Journal of Agricultural Research and Developmen* 1(1): 9–16.

Kumar, P. 2014. Information technology: roles, advantages and disadvantages. *International Journal of Advanced Research in Computer Science and Software Engineering Research* 4(6): 1020–1024.

Manyati, T.K. & Mutsau, M. 2019. Exploring technological adaptation in the informal economy: A case study of innovations in small and medium enterprises (SMEs) in Zimbabwe. *African Journal of Science, Technology, Innovation and Development* 0(0): 1–7.

Muangkhot, S. 2015. Strategic marketing innovation and marketing performance: an empirical investigation of furniture exporting businesses in Thailand. *The Business and Management Review* 7(1): 9–10.

Ndesaulwa, A.P. & Kikula, J. 2016. The impact of innovation on performance of Small and Medium Enterprises (SMEs) in Tanzania: a review of empirical evidence. *Journal of Business and Management Sciences* 4(1): 1–6.

Perkebunan, D. 2017. Produksi kopi jawa barat.

Reguia, C. 2014. Product innovation and the competitive advantage. *European Scientific Journal* 1(June): 140–157.

Swaminathan, A. 2014. *Marketing capabilities, innovation and firm performance*. [Online]. Retrieved from http://lib.dr.iastate.edu/etd.

Wheelen, T.L. & Hunger, J.D. 2018. Strategic management and business policy globalozation, innovation and sustainability.

Wu, Y. & Wang, C. 2009. The impacts of internal and external environments on subsidiary's capabilities. *In International Conference on Business management and Information Technology Application BMITA*.

Advances in Business, Management and Entrepreneurship – Hurriyati et al. (Eds)
© 2021 Taylor & Francis Group, London, ISBN 978-0-367-67471-7

Information technology resources in barbershop MSME business in Indonesia

G.A.F. Maulani, S. Nugraha, N.A. Hamdani & T.M.S. Mubarok
Universitas Garut, Indonesia

ABSTRACT: In the current 4.0 industrial revolution era, the business world has become the most demanding entity to acclimatize the use of information technology including MSMEs, which run medium or micro businesses. The application of information and communication technology is a decent strategy to maintain business continuity. Nonetheless, information and communication technology are still not well utilized in the barbershop industry in Garut Regency. Hence, this study aims to determine the effect of information technology resources on the Barbershop MSME Business Performance in Garut Regency, West Java, Indonesia. This study involved 43 barbershop owners or managers in Garut Regency as the respondents. Based on the PLS-SEM analysis, it was discovered that IT human resources and IT supporting elements had a very large and significant influence on Barbershop. Therefore, it can be concluded that information technology resources had a very significant influence on the Barbershop MSME business performance in Garut Regency.

1 INTRODUCTION

Information technology is one of the key words in the 4.0 industrial revolution civilization (Hamdani & Maulani 2019). It has a very strategic role in the current global economic transformation, including in the business world (Hamdani et al. 2019). It also creates a new atmosphere that can change business processes in all organizations at present (Maulani & Hamdani 2019). Today's business world can survive and withstand if information technology is involved in every process (Ashrafi & Murtaza 2008) in Table 1.

In Indonesia, Micro, Small and Medium Enterprises (MSME) are business entities that have a strategic role in the development of the National Economy. It is renowned that the number of MSMEs in Indonesia is increasing from year to year.

Table 1 indicates that Micro, Small and Medium Enterprises are businesses that are in great demand by the community and businesses contributing around 60% to GDP (Gross Domestic Product), which also provides employment for the community (Sarwono 2015). This also includes barbershop MSMEs in Indonesia, which have improved business growth every year. In 2017, barbershop business in Indonesia reached 5,000 barbershop brands. This shows that public began to pay attention to barbershop business (Rafikasari 2017).

Barbershop in Indonesia has a unique history in every regions, including Garut. Garut is one of the regencies in West Java, Indonesia, known as one of the places of "legend" for the barber men in Indonesia (Teguh 2019). There is an area or village known as a

Table 1. The number of Indonesia MSME in the last 5 years (in million).

Years	2012	2013	2014	2015	2016	2017
MSME Number	55.2	56.5	57.1	58.5	60.8	62.9

high-quality barber "producer". As a result, with this potential, Garut has a large number of barber-shops (Hamdani et al. 2019).

Strategic management is a set of managerial decisions and actions that regulates the long-term performance of a company. This includes environmental analysis (both external and internal), strategy formulation (strategic or long-term planning), strategy implementation, and evaluation and control (Wheelen & Hunger 2012; Wheelen et al. 2018) Information technology is part of the information system science (De et al. 2017). It is one component combining human and material resources to support operational information in business (Abrego Almazán et al. 2017). It is believed to have a successful impact on organizational performance (Su 2014) especially on MSMEs' performance (Levy & Powell 2000; Sarosa & Zowghi 2003). In addition, it also creates knowledge that pushes business entities to be more competitive (Hamdani 2018).

Several studies affirmed that information technology resources can be described in several indicators, including: IT Infrastructure Resources (Windrum &

De Berranger 2010), IT Human Resource (Chen & Tsou 2012) and IT Relationship Resource (Mao et al. 2016b). Some previous studies explain that IT formulations in business include IT resources such as infrastructure and IT knowledge capital for business people and business ability in managing the interaction of business processes with Information Technology processes. These indicators had a significant influence on MSMEs competitive advantage (Mao et al. 2016b). To find out barbershop business performance, financial performance and strategic performance were used as the focal indicators (Hamdani et al. 2019; Tarutë & Gatautis 2014).

IT Resources are the technological foundation of the organization to ensure information and communication are carried out accurately, in real-time, and comprehensively. Whereas IT Human Resources are defined as technical and managerial IT skills of an organization's employees. In addition, there are also IT Resources Relationships that refers to the relationship between IT and business units, reflecting the level of trust and willingness to share risks and responsibilities (Mao et al. 2016a; Maulani & Hamdani 2019)

However, barbershop MSMEs have not used their full potential in the use of information technology as one of their business strategies. Based on the preliminary observations, about 90% of barbershops in Garut have not yet optimized information technology to support their business development. This phenomenon reinforces the purpose of this study to determine the extent to which information technology resources have an impact on the barbershop business performance in Garut Regency

2 METHODS

The quantitative explanatory method is a research method that aims to determine the characteristics of variables by examining a number of samples. This verification study involved data taken from 43 owners or managers from Barbershop in Garut Regency, West Java, Indonesia. Verification study aims to determine the relationship between variables through a hypothesis test based on data in the field. The sample determination in this study was selected by using a simple random sampling technique. The research data was processed using PLS-SEM. This research involved some observed variables including Table 2:

Table 2. Research variables.

Variables	Indicator
IT Resource (X)	IT Infrastructure Resource (X1)
	IT Human Resource (X2)
	IT Relationship Resource (X3)
Business Performance (Y)	1. Financial Performance (Y1)
	2. Strategic Performance (Y2)

3 RESULTS AND DISCUSSIONS

What follow is the result of PLS-SEM analysis and calculation that illustrates the relationship between IT Resource variables on barbershop MSME business performance in Garut Regency in Figure 1:

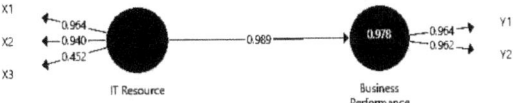

Figure 1. PLS Algorithm modelling.

Based on calculations and analysis using PLS and modeling as in Figure 1, there are several indicators that have interpretable values. The value of the coefficient from the variable IT Resource (X) to the Barbershop MSME Business Performance Variable is 0.989. This indicates that information technology resources has a positive and strong influence on Barbershop MSME business performance. In addition, the following interpretations are also drawn from the indicators:

a. The IT Resource variable has a loading factor value on each indicator, among others IT Infrastructure Resource Indicator (X1) has a loading factor value of 0.964; the Human Resource IT indicator (X2) has a loading factor value of 0.940 and the IT Relationship Resource (X3) indicator has a value of 0.452.

b. Business Performance Variable has a factor loading value on each indicator, among others: Financial Performance Indicator (Y1) has a loading factor value of 0.964; The Strategic Performance indicator has a loading factor value of 0.962

To determine the reliability of an indicator of variables, SmartPLS is used to see if the indicators have a loading factor value below 0.5, the indicators must be discarded from the model.

Based on Figure 1, there is one indicator that has a loading factor value below 0.5. The indicator is IT Relationship Resource (X3). Therefore, IT Relationship Resource is deemed not reliable in relation to the influence of information technology resources on barbershop business performance. The indicator is then removed from the PLS model so that the model performed convergent validity test because all loading factors are greater than 0.5. There are several results displayed on SmartPLS, presented in the matrix in Table 3.

Based on Table 3, there are several values that can be interpreted. To find out the validity of the variable or construct is by looking at its Average Variance Extracted (AVE) value. The required accepted AVE value is supposed to be greater than 0.5. The calculation results show that the Business Performance variable has a value of 0.928 while for the IT Resource variable is 0.673. Therefore, it can be concluded that the two variables are considered valid.

Table 3. Construct reliability and validity.

	Cronbach's Alpha	rho_A	Composite Reliability	Average Variance Extracted (AVE)
Business Performance	0.922	0.922	0.962	0.928
IT Resource	0.739	0.914	0.850	0.673

In reliability test, an analysis was carried out by looking at the appropriate Composite Reliability (CR) and Croncbach Alpha (CA) values in Table 3. The minimum value for the Composite Reliability and Croncbach Alpa values must be greater than 0.7. The following are the values:

a. The IT Resource variable has a CA and CR values of 0.739 and 0.850 respectively.
b. The Business Performance variable has a CA and CR values of 0.922 and 0.962 in that order.

The values confirmed that the variables have decent reliability values. Further testing was done for the structural model (inner model). The test was performed by looking at the R-Square value which belongs to the Goodness-fit model. Based on the calculation results, the R-Square value is 0.978. It explains that the IT Resource Variable can explain the variability of barbershop MSME performance by 97% while the 3% is explained by other variables.

Furthermore, in finding the significance of the influence of all the variables, a test is needed. Then it can be analyzed by using a bootstrapping technique on PLS-SEM described in the parameter coefficient and its statistical significance value in Table 4:

Table 4. Mean, STDEV, T-values, P-values.

	Original Sample (O)	Sample Mean (M)	Standard Deviation (STDEV)	T Statistics (|O/STDEV|)	P Values
IT Resource Business Performance	0.989	0.989	0.007	147.917	0.000

Based on the calculation results in Table 4, there is a value in the original sample scored 0.989. This means that IT Resources has a positive influence on MSME business performance. This also means that the more information technology resources are optimally implemented, the better MSME business performance will get, specifically for barbershop in Garut Regency. In addition, the t observed value (147,917) outscored the t critical value (1.96). Henceforth, it can be concluded that the IT Resource which includes IT Human Resource and IT Infrastructure,

has a very significant rapport on Barbershop MSME Business Performance which emphasizes on financial performance and strategic performance factors.

In this research, IT Resources is described as the relationship between business and information technology. IT Resources at barbershop show a very strong relationship, which shows that managers, staff, customers, and partners always communicate, coordinate, negotiate through the use of useful information and communication technology. The IT Resources dimension shows the technology-based relationship between organizations and business partners, as well as partnerships between IT groups and business units. This is in line with the findings of several previous studies which show that IT Resources had a vital role in increasing the competitive advantage of a business. (Chen et al. 2017; Ladokun et al. 2013; Mao et al. 2016a).

4 CONCLUSION

Barbershop as one of the developing business entities is expected to look more at Information Technology as a strategic effort to improve their business performance. Information technology resources related to infrastructure, such as hardware and software, need to be better optimized. In addition, it also needs to be supported by technology-literate human resources who understand the use and operation of Information Technology. This needs to be done in order to achieve good business performance which includes Barbershop MSMEs financial and strategic performances such as Customer Satisfaction, Employee Satisfaction and Social Performance.

ACKNOWLEDGEMENT

We thank all those who helped with this research, especially to the entire barbershop community in Garut Regency. Besides, we appreciate the Faculty of Entrepreneurship at the University of Garut for funding this research.

REFERENCES

Abrego Almazán, D., Sánchez Tovar, Y. & Medina Quintero, J. M. 2017. Influence of information systems on organizational results', contaduría y administración. Universidad nacional autónoma de méxico, facultad de contaduría y administración, 62(2), pp. 321–338. doi: 10.1016/j.cya.2017.03.001.

Ashrafi, R. & Murtaza, M. 2008. Use and impact of ict on smes in oman. *Electronic journal of information systems evaluation*, 11(3), pp. 125–138. ISSN 1566-6379.

Chen, J. & Tsou, H. 2012. Journal of engineering and performance effects of it capability, service process innovation, and the mediating role of customer service. *Journal of engineering and technology management*. Elsevier b.v., 29(1), pp. 71–94. doi: 10.1016/j.jengtecman.2011.09.007.

Chen, Y. et al. 2017. Improving strategic flexibility with information technologies: insights for firm performance in an emerging economy. *Journal of information technology*. Palgrave macmillan UK, 32(1), pp. 10–25. doi: 10.1057/jit.2015.26.

De, L. et al. 2017. Using information systems to strategic decision: an analysis of the values added under executive's perspective. *Brazilian journal of information studies*: research trends, 11, pp. 54–71.

Hamdani, N. A. 2018. Building knowledge creation for making business competition atmosphere in sme of batik. *Management science letters*, 8, pp. 667–676. doi: 10.5267/j.msl.2018.4.024.

Hamdani, N. A. & Maulani, G. A. F. 2019. The influence information technology capabilities and differentiation on the competitiveness of online culinary smes. *International journal of recent technology and engineering* (IJRTE), 8(1S), pp. 146–150.

Hamdani, N. A., Maulani, G. A. F. & Muharam, A. A. 2019. Entrepreneurial Culture in the Village of the Barbers, Garut, Indonesia. *International Journal of Engineering and Advanced Technology* (IJEAT), 8(5C), pp. 685–687. doi: 10.35940/ijeat.E1096.0585C19.

Hamdani, N. A., Solihat, A. & Maulani, G. A. F. 2019. The influence of information technology and co-creation on handicraft sme business performance. *International journal of recent technology and engineering*, 8(1S), pp. 151–154. Available at: https://www.ijrte.org/download/volume-8-issue-1s/.

Koperasi dan Usaha Kecil dan Menengah, K. and Statistik, B. P. 2017. *Sandingan data umkm 2012–2017*. Available at: http://www.depkop.go.id/uploads/tx_rtgfiles/sandingan_data_umkm_2012-2017_.pdf.

Ladokun IO, Osunwole OO & Olaoye BO. 2013. Information and communication technology in small and medium enterprises: factors affecting the adoption and use of ict in nigeria. *International journal of academic research in economics and management sciences*, 2(6), pp. 2226–3624. doi: 10.6007/IJAREMS.

Levy, M. & Powell, P. 2000. Information systems strategy for small and medium sized enterprises: an organisational perspective. *The journal of strategic information systems*, 9(1), pp. 63–84. doi: 10.1016/S0963-8687(00)00028-7.

Mao, H. et al. 2016a. Information technology resource, knowledge management capability, and competitive advantage: the moderating role of resource commitment. *International journal of information management*. Elsevier Ltd, 36(6), pp. 1062–1074. doi: 10.1016/j.ijinfomgt.2016.07.001.

Mao, H. et al. 2016b. International journal of information management information technology resource, knowledge management capability, and competitive advantage: the moderating role of resource commitment. *International journal of information management*. Elsevier ltd, 36(6), pp. 1062–1074. doi: 10.1016/j.ijinfomgt.2016.07.001.

Maulani, G. A. F. & Hamdani, N. A. 2019. The influence of information technology and organizational climate on the competitiveness of private universities in indonesia. *International journal of recent technology and engineering*, 8(1S), pp. 142–145. Available at: https://www.ijrte.org/download/volume-8-issue-1s/.

Rafikasari, D. 2017. *Pria sadar penampilan, jumlah barbershop di indonesia meningkat*, sindonews.com. Available at: https://lifestyle.sindonews.com/read/1220517/186/pria-sadar-penampilan-jumlah-barbershop-di-indonesia-meningkat-1499966337.

Sarosa, S. & Zowghi, D. 2003. Strategy for adopting information technology for smes: experience in adopting email within an indonesian furniture company. *Electronic journal of information systems evaluation*, 6(2), pp. 165–176.

Sarwono, H. A. 2015. Profil Bisnis Usaha Mikro, Kecil Dan Menengah (UMKM), Bank Indonesia dan LPPI.

Su, H. Y. 2014. Business Ethics and the Development of Intellectual Capital. *Journal of Business Ethics*, 119(1), pp. 87–98. doi: 10.1007/s10551-013-1623-4.

Tarutë, A. & Gatautis, R. 2014. ICT Impact on SMEs Performance. *Procedia-Social and Behavioral Sciences*. Elsevier B.V., 110, pp. 1218–1225. doi: 10.1016/j.sbspro.2013.12.968.

Teguh, I. 2019. *Ada apa di balik banyaknya juru pangkas rambut asli garut*, tirto.id. Available at: https://tirto.id/ada-apa-di-balik-banyaknya-juru-pangkas-rambut-asli-garut-deL7.

Wheelen, T. L. & Hunger, J. D. 2012. Strategic management and business policy, policy. doi: 10.1017/CBO9781107415324.004.

Wheelen, T. L. et al. 2018. Strategic management and business policy: globalization, innovation and sustainability.

Windrum, P. & De Berranger, P. 2010. The adoption of e-business technology by smes. *University business*, pp. 177–201. Available at: http://hdl.handle.net/2173/93160.

Author index

Lightning Source UK Ltd.
Milton Keynes UK
UKHW050835010822
406672UK00007B/835